Glycoconjugates

Glycoconjugates

Composition, Structure, and Function

edited by

Howard J. Allen

Roswell Park Cancer Institute
Buffalo, New York

Edward C. Kisailus

Canisius College
Buffalo, New York

Marcel Dekker, Inc. New York • Basel • Hong Kong

Library of Congress Cataloging-in-Publication Data

Glycoconjugates: composition, structure, and function / edited by
Howard J. Allen, Edward C. Kisailus.
p. cm.
Includes bibliographical references and index.
ISBN 0-8247-8431-6 (alk. paper)
1. Glycoproteins. 2. Glycolipids. 3. Lectins. I. Allen, Howard J. II. Kisailus, Edward C.
[DNLM: 1. Glycoconjugates. QU 75 G5673]
QP552.G59G595 1992
574.19'2483--dc20
DNLM/DLC
for Library of Congress 92-4641
CIP

This book is printed on acid-free paper.

MARCEL DEKKER, INC.
270 Madison Avenue, New York, New York 10016

Current printing (last digit):
10 9 8 7 6 5 4 3 2 1

PRINTED IN THE UNITED STATES OF AMERICA

Preface

Integration of the structural and biosynthetic data with the functional role of glycoconjugates is central to the understanding of glycobiology. A discussion of glycoconjugate function must include those macromolecules that interact with the oligosaccharide moieties of glycoconjugates. In this volume we have brought together chapters that discuss current approaches to elucidating oligosaccharide structures; that give the mechanisms of biosynthesis and turnover of those structures; that discuss the characteristics of macromolecules that interact with glycoconjugates; that present the emerging role of molecular biology in the study of glycoconjugates; and that present plausible concepts concerning glycoconjugate function. Our objective is to present these topics in such a way as to minimize dry data enumeration and to highlight and even speculate on design-order and structure-function correlates of glycoconjugates.

This book presents a sequence of chapters leading from the classical methodology of structural and biosynthetic analysis through to today's new methods in ongoing research on the molecular biology of glycoconjugates and carbohydrate-binding proteins. The authors were encouraged to focus on the importance of their topic in formulating a structure-function relationship that includes the known, the hypothesized, and the speculative.

Several concepts are emerging in the field of glycobiology regarding the presence of carbohydrate-binding proteins in plants, animals, and microorganisms. These proteins, called lectins, appear to mediate activities such as nodule formation in leguminous plants, organization of embryonic tissue, metastatic behavior of cancer cells, and bacterial colonization of mucous membranes. Our objective in including discussions of lectins in this book is to illustrate the specificity of recognition that cells have developed while using only a relatively small array of monosaccharides.

Since the ability to comprehend any scientific presentation is dependent upon an understanding of unique terminology, we have presented a summary of the rules of carbohydrate nomenclature in the opening chapter. We hope that this will remove the mystery, especially for students, as to how terms are derived.

This volume summarizes significant contemporary information in the field of glycobiology in understandable language for students with a background in biochemistry and cell biology, for scientists who lack a detailed background in glycobiology, and for glycobiologists who wish to refresh their knowledge of areas of glycobiology that are different from their own specific area of expertise. The scope of this book provides an overview for the glycoconjugate biochemist while bringing the nonspecialist and student up to date on current concepts and approaches in glycobiology. Researchers in related areas will find stimulating ideas, and both researchers and students will be able to arrive at an understanding of the biological significance of glycoconjugates, lectins, and other glycoconjugate receptors.

We wish to thank Sosimo Fabian, Sean McGhee, and Brian Riegel for their assistance in the editing of this book.

Howard J. Allen
Edward C. Kisailus

Contents

Contributors

Howard J. Allen, Ph.D. Department of Surgical Oncology, Roswell Park Cancer Institute, Buffalo, New York

O. P. Bahl, Ph.D. Department of Biological Sciences, State University of New York at Buffalo, Buffalo, New York

V. P. Bhavanandan, Ph.D., D.Sc. Department of Biological Chemistry, The Milton S. Hershey Medical Center, The Pennsylvania State University, Hershey, Pennsylvania

Inka Brockhausen, Ph.D. Department of Biochemistry, University of Toronto, and The Hospital for Sick Children, Toronto, Ontario, Canada

Richard D. Cummings, Ph.D. School of Chemical Sciences, Department of Biochemistry, The University of Georgia, Athens, Georgia

E. A. Davidson, Ph.D. Department of Biochemistry and Molecular Biology, Georgetown University School of Medicine, Washington, D.C.

Marilynn E. Etzler, Ph.D. Department of Biochemistry and Biophysics, University of California, Davis, California

Minoru Fukuda, Ph.D. Glycobiology/Chemistry Program, La Jolla Cancer Research Foundation, La Jolla, California

Kiyoshi Furukawa, Ph.D. Department of Biochemistry, The Institute of Medical Science, The University of Tokyo, Tokyo, Japan

Nachman Garber, M.Sc., Ph.D. Department of Life Sciences, Bar-Ilan University, Ramat-Gan, Israel

Nechama Gilboa-Garber, Ph.D. Department of Life Sciences, Bar-Ilan University, Ramat-Gan, Israel

Shaheer H. Khan, Ph.D. Department of Gynecologic Oncology, Roswell Park Cancer Institute, Buffalo, New York

Edward C. Kisailus, Ph.D. Department of Biology, Canisius College, Buffalo, New York

Akira Kobata, Ph.D. Department of Biochemistry, The Institute of Medical Science, The University of Tokyo, Tokyo, Japan

Samar K. Kundu, Ph.D., D.Sc. Abbott Laboratories, Abbott Park, Illinois

Roger A. Laine, Ph.D. Departments of Chemistry and Biochemistry, Louisiana State University and A&M College, and The LSU Agricultural Center, Baton Rouge, Louisiana

Joseph T. Y. Lau, Ph.D. Department of Molecular and Cellular Biology, Roswell Park Cancer Institute, Buffalo, New York

Reiko T. Lee, Ph.D. Department of Biology, Johns Hopkins University, Baltimore, Maryland

Y. C. Lee, Ph.D. Department of Biology, Johns Hopkins University, Baltimore, Maryland

Reuben Lotan, Ph.D. Department of Tumor Biology, The University of Texas M.D. Anderson Cancer Center, Houston, Texas

Khushi L. Matta, Ph.D. Department of Gynecologic Oncology, Roswell Park Cancer Institute, Buffalo, New York

Terrance P. O'Hanlon Department of Molecular and Cellular Biology, Roswell Park Cancer Institute, Buffalo, New York

Kenneth Olden, Ph.D. Department of Oncology, Howard University Cancer Center, Howard University Medical School, Washington, D.C., and National Institute of Environmental Health Sciences, Research Triangle Park, North Carolina

Harry Schachter, M.D., Ph.D. Department of Biochemistry, University of Toronto and The Hospital for Sick Children, Toronto, Ontario, Canada

Anthony S. Serianni, Ph.D. Department of Chemistry, University of Notre Dame, Notre Dame, Indiana

Gerardo R. Vasta, Ph.D. Center for Marine Biotechnology, University of Maryland, Baltimore, Maryland

Paul H. Weigel, Ph.D. Department of Human Biological Chemistry and Genetics, University of Texas Medical Branch, Galveston, Texas

Kiang-Teck Yeo, Ph.D. Howard University Cancer Center, Howard University Medical School, Washington, D.C., and Beth Israel Hospital and Harvard Medical School, Boston, Massachusetts

Tet-Kin Yeo, Ph.D. Howard University Cancer Center, Howard University Medical School, Washington, D.C., and Beth Israel Hospital and Harvard Medical School, Boston, Massachusetts

1

An Introduction to Glycoproteins

O. P. Bahl *State University of New York at Buffalo, Buffalo, New York*

In introducing the subject of glycoproteins, attempt has been made to describe briefly the historical developments in the areas of their structure, biosynthesis, and function. In this rather limited coverage of the subject, the focus has been on the major concepts, rather than the individuals responsible for their development. Only a few general references are cited as a source of additional references, pertinent to the material covered.

Glycoproteins are a diverse group of complex macromolecules that are ubiquitous in nature and are present in virtually all forms of life: animals, plants, and microorganisms. Bacteria, in general, do not contain glycoproteins, except the obligately halophilic bacteria of the genus *Halobacterium* and *Bacillus stearothermophilus* NRS 2004/3a, in which their presence has been demonstrated with some degree of certainty. Glycoproteins comprise several important classes of macromolecules such as enzymes, hormones, immunoglobulins, transport proteins, cell adhesion molecules, toxins, lectins, structural proteins, and the recently discovered cytoplasmic and nucleoplasmic proteins. The diverse biological functions that these macromolecules perform, include among others, enzymatic catalysis, hormonal control, immunological protection, ion transport, blood clotting, lubrication, surface protection, structural support, cell adhesion, intercellular interaction and, most important, in recognition in general. They are present in extra- and intracellular fluids, connective tissue, and cell membranes. *Glycoproteins* are best defined as conjugated proteins containing as a prosthetic group of one or more heterosaccharides covalently bound to the polypeptide chain [1]. Although the history of glycoproteins dates back to almost over a century and a half, their emergence to the forefront of scientific investigation has occurred only in the last two decades. The lack of interest in glycoproteins for a protracted period was probably because carbohydrates, as a distinct class, were not deemed as important and crucial to life processes as proteins. They were identified as compounds, completely lacking biological specificity, that merely served as a source of energy or the structural or protective materials for the cells [2]. As a result, even those carbohydrates that were in covalent association with proteins did not attract much attention until the 1960s. The single most important discovery that has aroused an enormous interest in glycoproteins, has been the role of protein-bound carbohydrate in biological recognition, such as that in host–pathogen, cell–cell, and cell–molecule interactions. Other contributing factors to this renewed interest are the findings that indicate alteration in the glycan structures in the various developmental and pathological states. The recent developments in recombinant DNA methodology, which have made it possible to produce glycoproteins in heterologous eukaryotic cells, have further stimulated investigations in this field. Such recombinant glycoproteins generally have carbohydrates different from those in the native glycoproteins. Therefore, it becomes imperative that the effect of such an alteration in glycosylation on their physicochemical properties be known.

I. HISTORICAL DEVELOPMENTS IN PROTEINS AND CARBOHYDRATES

The past 20–30 years have seen a great many strides made in our understanding of the structure, biosynthesis, and function of glycoproteins. Glycoproteins being conjugated molecules, their progress had to await the development of the basic concepts of both carbohydrate and protein structures. It took almost the first 50 years of this century to achieve this. At about the turn of the century, sugar structures and their interrelationship were established by Emil Fischer. The controversy of the five- versus six-membered ring structure for sugars—Fischer being strong proponent of the former proposal—held back the development of the carbohydrates. The field of the carbohydrates in much of the 1920s was in a state of disarray and, as a result, no real progress in the chemistry of carbohydrates was made [3,4]. It was not until 1929 that Haworth put this controversy to rest by unequivocal chemical synthesis of sugars. Nevertheless, during the first 30 years, procedures for the methylation of sugars were developed by Purdie and Haworth. Periodic acid reagent for oxidation of sugars was introduced by Malaparade in 1929. Other landmarks in the first half of the century were the determination of the structures of D-glucosamine, in 1937, and D-galactosamine, in 1945. Sialic acid was isolated from submaxillary mucin in 1936, and later from the nerve tissue in 1941, and by the end of the 1950s, its structure was finally confirmed. Also, the colorimetric methods for the analyses of sugars and hexosamines were developed. This substantial growth of basic structural information led to the study of more complex molecules.

At about the beginning of the second half of the 20th century, semimicroanalytical techniques began to emerge. New separation and analytical methods, based on partition chromatography, such as paper chromatography and column chromatography, including ion exchange chromatography, were developed. Various electrophoretic methods, involving paper, starch, and polyacrylamide and gel filtration using chemically modified dextrans, were also developed. The development of gas chromatography, first introduced for fatty acids in 1952, and later refined and extended to carbohydrate derivatives such as trimethyl silylethers and alditol acetates, had a profound impact on the growth of our knowledge of carbohydrates. These developments made it possible to analyze picomolar quantities of sugars. Chemical methods, such as periodate oxidation, were refined and extensively employed in structural determination. Micromethods for the methylation of carbohydrates and the introduction of combined gas chromatography and mass spectrometry for the identification and estimation of methylated and acetylated sugars accelerated the pace of structural studies. The use of nuclear magnetic resonance (NMR) spectroscopy for structural studies was introduced in the later part of 1950s. Subsequent extension of NMR to ^{19}F, ^{13}C, and ^{31}P provided one of the most powerful tools yet known for conformational studies. Radioactive labeling with ^{14}C and ^{3}H also provided new and powerful methods for the study of carbohydrate metabolism [3,4].

Concerning protein chemistry, at the turn of the century, it was known that proteins were made up of amino acids. However, only 11 amino acids were isolated and identified from 1800 to 1890. During the next decade, three additional amino acids were isolated. Most of the other commonly occurring amino acids in proteins were isolated in the first 20–25 years of this century except L-threonine, which was discovered in 1935 by Rose. The peptide theory for protein structure was proposed by Hofmeister and Fischer at the turn of the century. It was not until 50 years later that the hypothesis was proved correct by the determination of the primary structure of insulin by Sanger, a monumental achievement, considering the limited number of techniques available for the determination of protein structure at that time. Since the elucidation of insulin structure, a large number of new reagents for structural studies, including phenylisothiocyanate by Edman and a host of highly specific exo- and endoproteases, have been discovered. In addition, revolutionary changes have taken place in the development of a number of sensitive and automatic techniques, including those for amino acid analysis, amino acid sequencing in proteins by gas-phase sequence, and peptide and protein purification by high-performance liquid chromatography (HPLC). Automatic peptide synthetic methods by solid-phase synthesis, x-ray crystallography for three-dimensional structure of peptides and proteins, and NMR for their conformational analysis are some of the more recent developments that will continue to have a profound effect on the future advancements in protein chemistry.

II. DEVELOPMENTS ON GLYCOPROTEIN STRUCTURE

Studies on glycoproteins began around mid-19th century. Some of the earliest glycoproteins studied were those from mucous secretions, known as mucins. These are highly viscous proteins secreted by a number of tissues, such as salivary and sweat glands and intestinal and respiratory tracts. Until the end of the last century the physicochemical studies of mucins and other glycoproteins were confined merely to their solubility properties, their behavior toward acid and alkali, and their viscosity. Much effort was also devoted toward demonstrating the presence of carbohydrates in proteins by the furfural reaction after treatment with concentrated sulfuric acid (Molisch, 1886; Schiff, 1880), by testing for reducing substances after acid hydrolysis, and by elemental analysis. By the end of the last century, the ideas concerning the glycoprotein nature of proteins began to take hold. In a communication to the Royal Society of London, an English physician, F. W. Pavy (1893), proposed for the first time the glycosidic nature of proteins, implying that all proteins were glycoproteins. However, even until 1920s, the existence of glycoproteins was not accepted by the leading carbohydrate chemists of the time. The presence of carbohydrates in proteins was dismissed as a contaminant. The term *glycoprotein* appeared for the first time in a textbook by Hammersten in 1895. In the first quarter of this century, several other proteins, such as blood group substances as well as several globular proteins, including ovalbumin, ovomucoid, and orsomucoid (α-acid glycoprotein), were found to be glycoproteins. The confusion still existed, however, over whether or not thc carbohydrates were contaminants or integral components of proteins. One of the first proteins on which detailed structural analysis was carried out, was ovalbumin. Ovalbumin was selected because it had been previously crystallized by Hofmeister (1890) and thus could be obtained in a homogeneous form. With the preparation of a highly pure glycopeptide by controlled acid hydrolysis of ovalbumin by Neuberger in the late 1930s, the question of the carbohydrate as an integral component of proteins was settled once and for all [5]. It took another two decades before the carbohydrate peptide linkage, *N*-acetylglucosaminyl-asparagine (GlcNAc-Asn), in the glycopeptide was established and later confirmed by chemical synthesis.

The introduction of a micromethylation procedure, involving the use of a strong nucleophile dimethylsulfinylcarbanion and methyl iodide, combined with gas chromatography and mass spectrometry; the availability of exoglycosidases in the 1960s; and the other then available methods, such as periodate oxidation and a number of separation and analytical techniques as outlined in the foregoing, led to an explosive growth of glycoprotein research. Numerous glycoproteins were analyzed in the ensuing 15 years. As a result, the following picture of glycoprotein structure emerged. Primarily, there are two major classes of glycoproteins, those containing asparagine, or *N*-linked, and those containing *O*-linked carbohydrates. The *N*-linked carbohydrates fall into three main categories, all of which share a common pentasaccharide core structure consisting of mannose (Man) residues linked to GlcNAC-Asn, Man1,3(Man1,6)Manβ1,4GlcNAcβ1,4GlcNAc-Asn—and differ only in the peripheral sugar residues. In fact, the extensive diversity of *N*-linked structures that exist is due to the differences in the structure of peripheral carbohydrates. The high mannose-type structures contain two to six additional mannose residues linked to the pentasaccharide core. The complex type structures have instead two to four lactosamine or sialyllactosamine units to the α-linked mannose residues of the core forming bi-, tri-, and tetraantennary structures. Sulfated lactosamine has also been found to substitute sialyllactosamine in some glycoproteins, such as the glycoprotein hormones, lutropin, follitropin, and thyrotropin. Other complex-type structures contain polylactosamine in their outer chains, as in erythroglycan, initially isolated from erythrocyte membranes, and later a number of similar glycoproteins were discovered from other tissues. In the developing brain tissue, polysialosyllactosamine units constitute the outer chains such as those in the neural and liver cell adhesion glycoproteins (N-CAM, L-CAM), respectively, which contain a large number of α2,8-linked sialic acid (NeuAc) residues distributed in three or four outer chains. In addition, some complex type structures have 1,4β-linked "bissecting" *N*-acetylglucosamine or 1,6α-linked focuse (Fuc) residues attached to the β-linked mannose and the innermost *N*-acetylglucosamine, respectively, in the peptasaccharide core. Finally, the hybrid-type structures combine the structural features of both the high mannose and the complex types.

Commonly occurring *O*-linked animal glycoproteins contain α-*N*-acetylgalactosaminyl-serine (αGal*N*Ac-Ser) or threonine (Thr) linkage known as mucin-type linkage, whereas the linkage in plant glycoproteins is α-L-arabinosylhydroxyproline. Animal glycoproteins have either linear chains formed by an extension of an oligosaccharide chain through β1,3-linked galactose, or branched structures by additional extension through β1,6-*N*-acetylglucosamine linked to *N*-acetylgalactosamine.

The *N*- and *O*-linked carbohydrate structures elucidated before 1975 mostly represented the composite structures of multiple carbohydrate variants at a single glycosylation site, a phenomenon termed *microheterogeneity* or simply *heterogeneity*. Initially, microheterogeneity was ascribed to minor variations of sugars at the nonreducing termini of carbohydrate chains. Those changes at the nonreducing peripheral monosaccharides were considered to be the result of incomplete glycan synthesis or of artifacts caused by degradation during the isolation of glycoproteins. Therefore, the existence of microheterogeneity was ignored during the structural characterization of the carbohydrates. More recently, however, the heterogeneity has been found to be due to major variations in the degree and location of branching and in the carbohydrate structures. First observation of such microheterogeneity was made in the mid-1960s in ovalbumin, in which at a single glycosylation site was found with more than eight carbohydrate variants with high mannose- and hybrid-type structures. These variants or *glycoforms*, a recently introduced term, were not due to animal variation, since albumin obtained from a single egg laid by a hen of an inbred strain yielded the same results. The average structures of glycoproteins determined in the 1960s or early 1970s in many instances were not completely an accurate representation of the carbohydrates. In the last 15 years, it has become clear that the individual glycoforms might have functional relevance [6]. The recent findings on the dramatic changes in the structures of the carbohydrate moieties of glycoproteins with oncogenic and ontogenic events and a large body of other experimental evidence are suggestive of the functional significance of the individual carbohydrate variants. Therefore, a need was felt to reexamine and revise a large number of earlier structures determined before 1975 after the separation of the individual carbohydrate variants. The isolation of the individual glycoforms was achieved by various chromatographic methods, including the use of lectins. In recent years, a battery of lectins with well-defined carbohydrate specificities have become available. The glycoforms are generally available in limited quantities. However, the development of sensitive and rapid methods of carbohydrate compositional analysis, based on the application of HPLC for the separation of monosaccharides and amperometric titration for their detection and estimation, has contributed substantially to the characterization of glycoforms.

III. DEVELOPMENTS IN THE BIOSYNTHESIS OF CARBOHYDRATE MOIETY

Elucidation of the biosynthesis of the carbohydrates in glycoproteins essentially began with the discovery, in 1950 by Leloir and his group, of the involvement of sugar nucleotides as donors of monosaccharides. The pace of developments in this area increased with the increasing knowledge of structural information. The availability of radioactive sugars and their derivatives further contributed to the rapid growth during the 1960s. It was then understood that peripheral sugars in the *N*-linked carbohydrates such as the incorporation of sialic acid, D-galactose, and D-*N*-acetylglucosamine, proceeded by the stepwise addition of these sugars from their respective sugar nucleotides. The assembly of the core comprising mannose and *N*-acetylglucosamine was not elucidated until the 1970s, although it was known that the innermost *N*-acetylglucosamine was incorporated cotranslationally in the polypeptide chain while it was still attached to the polysomes. It was then known that as the protein moved through the channels of endoplasmic reticulum (ER), membrane-bound transferases attached monosaccharides one at a time. The glycoprotein finally was understood to reach the Golgi for packaging into secretory granules for secretion. It was also considered that the enzymes involved in carbohydrate attachment were located along the membranes, including those of the endoplasmic reticulum, rough (RER) and smooth (SER), as well as those of the Golgi apparatus. In short, in the 1960s, the carbohydrate attachment was believed to be a postribosomal event that was not under primary genetic control As far as the assembly of *O*-linked carbohydrates, such as those present in mucins and blood group substances, it was known that it involved the stepwise additions of individual carbohydrates by specific glycosyltransferases and sugar nucleotides in a specific order. For example, the sialyl-*N*-acetylgalactosamine unit in bovine submaxillary mucin was shown to be formed by two enzymes found in the sheep submaxillary gland: one transferred *N*-acetylgalactosamine from uridine diphosphate (UDP)-*N*-acetylgalactosamine to serine or threonine, whereas the other transferred sialic acid from cytidine monophosphate (CMP)-sialic acid to *N*-acetylgalactosamine. Similarly, the enzymes involved in the synthesis of oligosaccharides with A-, B-, H-, and Le-type blood group activities were found in gastric mucosa or milk and included two fucosyl-, two galactosyl-, one *N*-acetylgalactosaminyl-, and one *N*-acetylglucosaminyltransferase [7]. In brief, this was the status of our knowledge until 1970. The understanding of the biosynthesis, however, made rapid progress in the 1970s with two breakthroughs by the discoveries of dolicholphosphorylmonosaccharide as a sugar donor and a lipid-linked dolicholpyrophosphoryl oligosaccharide, $Glc_3Man_9GlcNAc_2$-P-P-Dol, as a donor of the oligosaccharide involved in the biosynthesis of glycans in glycoproteins. Dolichol (Doc) is a polyisoprenol containing 16–20 isoprene units, the first being saturated. Dolicholphosphorylmonosaccharides such as Glc-P-Dol and Man-P-Dol are formed from dolichol phosphate (Doc-P) as an acceptor and UDP-Glc and GDP-Man as sugar donors, respectively, in the presence of appropriate enzymes. The lipid-linked oligosaccharide, $Glc_3Man_9GlcNAc_2$P-P-Dol is assembled on Dol-P by step wise addition of monosaccharides, with the exception of the first sugar, which is inserted as GlcNAc-1-P from the respective sugar nucleotides and dolicholphosphoryl sugars. The first five mannoses including the one β-linked are derived from sugar nucleotide, GDP-Man, whereas Man-P-Dol serves as the sugar donor for the remaining four mannoses. The involvement of Dol-P-Man as a donor of mannose residues is based on studies with mutant animal cell lines that lack the ability to synthesize Dol-P-Man. Such cell lines are capable of synthesizing only $Glc_3Man_5GlcNAc_2$-P-P-Dol. In the assembly of oligosaccharide lipid, the various sugars are added in a certain order, determined by the specificity of the enzymes. Each of the steps in the synthesis of the oligosaccharide lipid, $Glc_3Man_9GlcNAc_2$-P-P-Dol, is catalyzed by a separate enzyme. The present picture of the pathway of glycan synthesis has emerged from extensive studies, carried out during the 1970s in a number of cell types, such as the Chinese hamster ovary (CHO) cell line, NIL-8 hamster fibroblasts, and chicken embryo fibroblasts. In brief, the glycan biosynthesis involves two main steps: first, the assembly of the oligosaccharide on a lipid intermediate, $Glc_3Man_9GlcNAc_2$-P-P-Dol, and its transfer en bloc to the growing polypeptide chain; and, second, the processing of the oligosaccharide. The enzyme, oligosaccharyltransferase, responsible for the transfer of the oligosaccharide, recognizes the tripeptide sequence, -Asn-X-Ser/Thr, where X is any amino acid other than proline. Subsequent to the transfer of the oligosaccharide, $Glc_3Man_9GlcNAc_2$, to the protein, the oligosaccharide undergoes a series of processing reactions causing first the trimming by exoglycosidases and then elongation of the sugar chains by specific glycosyltransferences. Three glucoses and one mannose are rapidly cleaved off in the RER. The glycoprotein with eight remaining mannoses enters the Golgi membranes, more specifically the *Cis*-Golgi. If a protein is destined for lysosomes, the mannosyl residues are phosphorylated at the C-6 hydroxyl groups. The recognition signal for phosphorylation resides in the protein. The phosphorylation is carried out by two enzymes, the *N*-acetylglucosaminylphosphotransferase and *N*-acetylglucosame-1-phosphodiester-*N*-acetylglucosaminidase. The resulting phosphorylated glycoprotein binds to a specific receptor in the Golgi that recognizes Man-6-P. The lysosomal proteins thus become segregated from the other proteins, and exit either from the *Cis*-Golgi or are further processed by traversing through the medial part and exit from the *trans*-Golgi to the prelysosomal particles, endosomes or CURL, and then to the lysosomes.

Further processing of the oligosaccharide, $Man_8GlcNAc_2$, in glycoproteins to be secreted involves the removal of additional three mannoses to yield $Man_5GlcNAc_2$. Subsequently, *N*-acetylglucosaminyltransferase I inserts *N*acetylglucosamine to 1,3α-linked mannose, followed by the cleavage of two mannoses and further addition of another *N*-acetylglucosamine to the 1,6α-linked mannose. A fucosyl residue is also incorporated at this stage in the medial Golgi. The termination of the oligosaccharide chain occurs in the *trans*-Golgi by the incorporation of galactose and sialic acid. Other modifications, such as the incorporation of *N*-acetylgalactosamine, sulfation, and *O*-acetylation of sialic acids, also occur in the *trans*-region of the Golgi.

Evidence to support the validity of the foregoing pathway of glycan assembly stems from studies on various mutant cell lines lacking in certain enzymes involved in the synthesis of the oligosaccharide–lipid or in the processing of the oligosaccharides. Support also comes from the studies involving the use of inhibitors of glycosidases and glycosyltransferases in which certain carbohydrate intermediates are found to accumulate. Some of the common inhibitors that have been employed in these studies are tunicamycin, deoxynojiromycin, deoxymannojiromycin, and castanospermine.

After the elucidation of the basic biochemistry of glycan biosynthesis, other complex problems involving subcellular organelles, such as transmembrane translocation of sugars from the cytoplasmic face to the lumen face of RER and Golgi apparatus, subcellular sites of biosynthesis, trafficking from organelle to organelle and within an organelle and, most importantly, regulation of processing, became the focus of researchers in the 1980s. The enzymes involved in the synthesis of oligosaccharide–lipid, except a few, and the oligosaccharide-processing enzymes face the lumen of RER and Golgi, whereas the sugars and sugar nucleotides occur on the cytoplasmic face of the membranes. Consequently, for the complete synthesis and processing to occur, transmembrane movement of the sugars must take place. The transport of sugar nucleotides to the lumen face in the Golgi is facilitated by transporter proteins present within the membranes. These permeases or antiports also export the corresponding nucleotides from the Golgi vesicles on a one-to-one exchange basis, such as one molecule of UDP as UMP from the vesicle is exchanged for one molecule of UDP-Gal to the vesicle. The mechanisms of the sugar movement across RER membranes is not as well understood as that in the Golgi. The problem of transmembrane movement of sugar residues in the RER arises because the sugar nucleotides and Man-P-Dol or Glc-P-Dol as donors of sugar residues during the synthesis and processing of the oligosaccharide–lipid are not able to enter the lumen. A number of enzymes involved in the synthesis of $GlcNAc_2$-P-Dol, Man-P-Dol, or Glc-P-Dol are present on the cytoplasmic face of RER, whereas the assembly of the oligosaccharide–lipid $Glc_3Man_9GlcNAc_2$-P-P-Dol, takes place on the lumen face of RER. The sugar is either transported to the lumen of RER by a specific glycosyltransferase or as a part of the lipid-linked intermediate. At present, no decision can be made between these two possibilities [8].

To determine the subcellular sites of glycan biosynthesis and the molecular mechanism of their structural variation, the availability of the purified enzymes, glycosidases and glycosyltransferases, involved in the synthesis of both the *N* and *O*-linked glycans is essential. Therefore, another focus of attention in the 1980s was the purification and characterization of these enzymes and the utilization of the antienzyme antibodies for their subcellular localization. The number of such enzymes is estimated to be over 100 [9]. A detailed knowledge of their properties including their specificity will certainly lead to a better understanding of the regulation of glycan synthesis. Recently, several glycosyltransferases for terminal sugars have been studied at the genetic level. Their amino acid sequences, derived from the nucleotide sequences, have revealed the presence of some common structural domains in these enzymes; however, no significant amino acid sequence homology has been observed [9].

Despite that the *N*-linked glycans are derived from the same oligosaccharide–lipid precursor, there exists an enormous diversity of oligosaccharide structures. Since this structural variation is confined to the outer part of the oligosaccharide chains, the oligosaccharide synthesis must be highly regulated at the processing level. This was another active area of research in the 1980s. The differences in the levels of the processing enzymes, glycosyltransferases in the Golgi, their specificity and regulatory properties in different animal species and tissues, and the polypeptide matrix, are some of the factors that may influence the structural variation of glycans in general. The level of expression of the various enzymes is different in different animal species and tissues and, thus, the same proteins from different animal species and tissues have different carbohydrate structures. The evidence in support of this concept comes from the differences in the oligosaccharides elaborated on viral proteins and recombinant proteins in various cultured cell lines. The differences in the carbohydrates in a protein during development, in the various pathologic states, or in oncogenic growth may also be caused by the changes in the levels of expression of the processing enzymes. Furthermore, the glycosylation machinery can be varied at will by insertion of a glycosyltransferase by transfection with DNA fragments or expression vectors containing cDNA coding for the glycosyltransferases involved in the incorporation of terminal glycosyl residues. Another regulatory factor may reside in the protein matrix itself. The amino acid sequence or the overall conformation of the polypeptide chain may play an important role in the regulation of oligosaccharide processing. This could explain the differences in the glycosylation patterns in different proteins from the same cell type. Furthermore, the amino acid sequences and conformational differences near the individual glycosylation sites may be another important factor that may account for the variation in the oligosaccharides at different sites in the same protein from the same cell type. However, the factors that are responsible for the heterogeneity of glycan structures at a single glycosylation site are not now well understood. Some of the other regulatory factors in the oligosaccharide processing may include the rate of subcellular movement of proteins and the degree of accessibility of the oligosaccharide chains to the enzymes.

IV. FUNCTION OF GLYCANS IN GLYCOSYLATED PROTEINS

As knowledge of the widespread occurrence of glycoproteins in important biological phenomena and their structural diversity grew, so did the interest in defining the functional significance of glycans. The effect of carbohydrates on the physiochemical properties of proteins, such as viscosity, isoelectric pH, the degree of hydration, solubility, thermal stability, and resistance to proteolysis, has been known for some time. Similarly, it has also been known that the carbohydrate is altered during various pathological states, including malignancy. The first clear-cut demonstration of the functional significance of the carbohydrate was in the blood group substances, in which the immunological specificity was found to be dependent on monosaccharides or short oligosaccharide chains. Glycoproteins occur both in the soluble state and as components of insoluble complex intracellular and extracellular structures. A unifying hypothesis must explain the role of carbohydrates in such diverse structures. However, no such hypothesis was proposed until mid-1960s, when the first serious attempt to define the functional significance of carbohydrates was made. The glycosylation was proposed as a means of labeling proteins for intracellular recognition and export from the cells synthesizing these proteins [10]. The hypothesis implied that all extracellular proteins were glycosylated, whereas the intracellular proteins were devoid of carbohydrates. In support of the proposal were cited carbohydrate analyses of a large number of extra- and intracellular proteins. The hypothesis was not challenged and rejected until several years later in the early 1970s. The major criticism of the hypothesis was that a host of extracellular proteins, including albumin, were lacking in carbohydrates and, conversely, a large number of intracellular proteins in the various subcellular organelles were glycosylated [11]. Furthermore, mutant cell lines or cells in the presence of inhibitors of glycosylation, such as tunicamycin, or cells containing modified recombinant DNA were found to secrete deglycosylated proteins in the medium, indicating that glycosylation was not always necessary for secretion. A major breakthrough in the search for functional significance of carbohydrates occurred when it was fortuitously discovered that the removal of sialic acid resulted in the rapid clearance of glycoproteins from circulation. Initially, this observation was made during studies on ceruloplasmin, but later, it was found to be a general phenomena. The removal of sialic acid exposes penultimate galactose residues at the nonreducing termini of the carbohydrate chains. The asialoglycoprotein thus formed interacts with a Gal/GalNAc-binding receptor on hepatocytes and is internalized by receptor-mediated endocytosis. Other specific carbohydrate-binding receptors or lectins for GlcNAc/Man and L-fucose were subsequently found to be located in the reticuloendothelial system. Thus, this was the first demonstration of the biological recognition in the function of glycans. At about the same time, another hypothesis on the functional significance of carbohydrates in intercellular adhesion was advanced. From studies on the neural retinal cells, it was proposed that the adhesion of these cells was caused by the interaction between oligosaccharides in the membranes of one cell with the glycosyltransferases present in the membranes of other adjacent cells. The binding or adhesion occurred as a result of the complex formation between the oligosaccharide substrates and the enzyme transferases.

Further developments in the 1980s have revealed a much broader role of carbohydrates in recognition involving carbohydrate-binding proteins or lectins in diverse biological phenomena. The binding of viruses to host cells involves the viral hemagglutinins and the host cell surface glycoproteins. Conversely, viral surface oligosaccharides may be ligands for host cell surface glycoproteins and, thereby, contribute to the host cell range of the virus [6]. Similarly, the cell surface carbohydrates also act as receptors for other pathogens or toxins, such as the adhesion of human pathogen *Mycoplasma pneumoniae* to erythrocytes [12]. The carbohydrate–lectin interactions are also involved in lymphocyte migration, since treatment of lymphocytes with exoglysosidases or inhibitors of oligosaccharide-processing enzymes has been shown to alter their migratory or the so-called homing properties. Mammalian sperm and egg interaction has been shown to involve the binding of a sperm protein with a zona pellucida protein, ZP-3. The *O*-linked and not the *N*-linked carbohydrates in ZP-3 are involved in the binding, since only the *O*-linked oligosaccharides inhibit this binding [9]. The lectin–carbohydrate recognition has also been implicated in the early development and differentiation of cells. Different carbohydrates appear during different stages of development. In the embryogenesis of the mouse, several stage-specific antigens (SSEA) appear at different stages of development, such as [Galβ1,4(Fuc-α-1,3)GlcNAc-R], which appears at the 8- to 12-cell stage and causes the cells to compact. The participation of the carbohydrate in this process is shown by the disruption of embryogenesis by a carbohydrate analogue. Neural cell adhesion molecules (N-CAM) with polysialyosyl residues, as described earlier, also appear during the early development and are believed to be involved in cellular adhesion.

In the mid-1970s, the role of carbohydrates in recognition was further revealed by the startling discovery of a mannose-6-phosphate receptor. The existence of the receptor was initially discovered as a result of extensive studies on genetic disorders of mucopolysaccharide catabolism. Cells cultured from patients with such disorders were found to accumulate abnormal amounts of mucopolysaccharides, implying a deficiency of hydrolytic enzymes. However, when the diseased-state cells were incubated with the culture medium of cells derived from normal individuals, the levels of the accumulation were considerably reduced, presumably by the uptake of hydrolases. The uptake was inhibited by 6-phosphomannose, indicating the presence of mannose-6-P–recognizing receptor. The primary role of the receptor is to target proteins to lysosomes by interaction with a common recognition marker, mannose-6-P. The phosphorylation of the high mannose-type glycoproteins takes place in the Golgi apparatus, presumably in the *cis*-region by two enzymes, GlcNAc-1-P transferase and GlcNAc phosphodiesterase and a sugar donor, UDP-GlcNAc. The enzymes targeted for lysosomes have a protein determinant that is recognized by the transferase. After the protein is phosphorylated, it becomes segregated from the other secretory proteins and exits from the *cis*- or *trans*-regions of the Golgi. Two mannose-6-P receptors, one with a molecular weight of 215,000 and the other 46,000 have been purified, cloned, and sequenced. The major difference between the two is in their ion dependence. The former is independent of the divalent cation requirement, whereas the latter requires divalent cations for ligand binding.

Although a variety of biological phenomena, such as alteration of carbohydrates in the cellular growth, development, and differentiation, and in various disease states, are suggestive of the biological role of glycans in glycoproteins, it is only in a few instances that their direct involvement in the biological function has been demonstrated. The most well-studied molecules in which the direct participation of the glycan moiety in their function has been shown, are glycoprotein hormones, human chorionic gonadotropin (hCG), lutropin (LH), and thyrotropin (TSH). These heterodimers are made up to two subunits, α and β, the α-subunit being common to all, whereas the β-subunits are hormone-specific. The integrity of the dimeric form of the hormone is essential for receptor-binding activity. The individual subunits do not bind to the receptor and, therefore, are devoid of biological activity. The function of these hormones is mediated by the stimulation of cyclic-adenosine monophosphate (cAMP) accumulation and steroidogenesis. The role of carbohydrate in the biological function of hCG, LH, TSH has been studied by their deglycosylation, followed by the evaluation of the deglycosylated analogues for their biological function. The deglycosylation has been achieved by enzymatic hydrolysis with specific exo- and endoglycosidases, by the use of inhibitors of glycosylation, by chemical methods, and finally, by site-directed mutagenesis. One of the advantages of the latter method is that the individual glycosylation sites can be modified to delete specific carbohydrate chains. The deglycosylation does not affect the receptor-binding activity of the hormones, whereas the ability of the deglycosylated hormones to stimulate cAMP and steroidogenesis is drastically reduced. In fact, the deglycosylated hormones act as antagonists of the native hormones. Since the action of all these hormones is mediated by cAMP, it is conceivable that the common α-subunit may have a critical role in their function. The removal of the carbohydrate from the α-subunit, particularly at Asn-52 resulted in a relatively reduced biological response, compared with the native hCG-α after reconstitution with the native or deglycosylated hCG-β, respectively, implying that the carbohydrate at Asn-52 played a greater role in signal transduction than the carbohydrate in the β-subunit or the carbohydrate at Asn-78 in the α-subunit.

The molecular mechanism of the carbohydrate action in hCG appears to be in the activation of the stimulatory guanosine triphosphate (GTP)-binding protein (Gs). This is shown by the fact that, although the deglycosylated hCG retains its ability to bind to the purified receptor when reconstituted in liposomes along with Gs and GTP, it loses its ability to activate Gs, as determined by the loss of binding of GTP to Gs and GTPase activity. The receptor has two binding sites: one that binds with the hormone and the other that interacts with Gs. It appears that the glycosylated hCG after binding with the receptor, brings the Gs-binding site of the latter in a proper orientation for interaction with Gs and, thereby, cause its activation. The deglycosylation, on the other hand, destroys the ability of the hormones to perform this function. Thus, it appears that the glycan moiety of the hormone plays an important role in maintaining the integrity of the signal transduction chain.

V. FUTURE PERSPECTIVES AND CONCLUDING REMARKS

Glycoproteins are currently an active area of research in which further new and exciting developments are expected. Undoubtedly, several challenging questions such as the functional significance of individual glycoforms or carbohydrate variants of a glycoprotein, the three-dimensional structures of glycoproteins and the enzymes involved in the synthesis of glycans, genetic regulation of glycosidases and glycosyltransferases, and last, but not least, functional significance of changes in glycosylation during development, growth, differentiation, and disease state, would receive further attention in the near future. To probe the role of carbohydrate variants or glycoforms (microheterogeneity) in a glycoprotein, their availability in the purified form would be desirable. Therefore, development of more sensitive methods for the isolation and characterization of carbohydrate variants would facilitate such investigations. Our knowledge of the three-dimensional structure of glycoproteins is still in a state of infancy. The lack of progress in this area has primarily been due to the difficulty in obtaining crystals suitable for x-ray crystallography. However, recent successful crystallization of hCG (1989) and its chemically deglycosylated analogue would certainly stimulate further interest in the crystallization of other glycoproteins by a similar approach. Conformational studies by two-dimensional (2D)-NMR are now limited to only small molecules because of the limitation of the technique. Furthermore, developments in these areas would certainly result in the availability of extensive information on the three-dimensional structures of glycoproteins, and this, in turn, should lead to new insights into the role of carbohydrates. To understand the factors that control the alteration in the glycosylation that occurs during different normal and abnormal physiological states, it would be necessary to study the regulation of expression of glycosidases and glycosyltransferases involved in the assembly of glycans. Furthermore, the functional significance of these changes in glycosylation in various disease states at a molecular level is an important problem and still remains unresolved. Clearly, one hopes that these and other problems would trigger the imagination of researchers in the field of glycoproteins.

REFERENCES

1. Gottschalk, A. (1966). Historical introduction. In *Glycoproteins* (A. Gottschalk, ed.). Elsevier Publishing, New York, pp. 1–19.

2. Sharon, N., and Lis, H. (1982). Glycoproteins. In *The Proteins*, Vol. 5 (H. Neurath and R. L. Hill, eds.). Academic Press, New york, pp. 1–144.
3. Kent, P. W. (1988). Some landmarks in carbohydrate biochemistry: 1921–1986. *Bull. Biochem. Soc. 10*:4–10.
4. Kent, P. W. (1989). Some landmarks in carbohydrate biochemistry: 1921–1986. Part II: Some biological functions of carbohydrates and their biological transformation. *Bull. Biochem. Soc. 11*:4–13.
5. Neuberger, A. (1971). Past and present concepts of glycoproteins. In *Glycoproteins of Blood Cells and Plasma* (G. A. Jamieson and T. J. Greenwalt, eds.). J. B. Lippincott, Philadelphia, pp. 1–15.
6. Rademacher, T. W., Parekh, R. B., and Dwek, R. A. (1988). Glycobiology. *Annu. Rev. Biochem. 57*:785–838.
7. Spiro, R. J. (1970). Glycoproteins. *Annu. Rev. Biochem. 39*:599–638.
8. Kornfeld, R., and Kornfeld, S. (1985). Assembly of asparagine-linked oligosaccharides. *Annu. Rev. Biochem. 54*:631–664.
9. Paulson, J. D., and Colley, K. J. (1989). Glycosyl transferases. *J. Biol. Chem. 264*:17615–17618.
10. Eylar, E. H. (1965). On the biological role of glycoproteins. *J. Theor. Bio.. 10*:89–113.
11. Winterburn, P. J., and Phelps, C. F. (1972). The significance of glycosylated proteins. *Nature 236*:147–151.
12. Feizi, T. (1985). Demonstration by monoclonal antibodies that carbohydrate structures of glycoprotein and glycolipids are onco-developmental antigens. *Nature 314*:53–57.

2

Nomenclature of Monosaccharides, Oligosaccharides, and Glycoconjugates

Edward C. Kisailus *Canisius College, Buffalo, New York*
Howard J. Allen *Roswell Park Cancer Institute, Buffalo, New York*

A prerequisite for being able to communicate in any discipline is a familiarity with the appropriate terminology. The field of glycoconjugate biochemistry is characterized by the application of a large set of rules of nomenclature. These rules were drafted by commissions assembled by the International Union of Pure and Applied Chemistry (IUPAC) and the International Union for Biochemistry (IUB). These rules were published in various scientific journals to make them accessible to working scientists. The editorial office of most scientific journals can direct scientists to a specific published set of rules upon request.

The rules presented here were adapted from the following sources: I. Monosaccharides (*Eur. J. Biochem.* 1971; 21:455–477), II. Oligosaccharides (*Arch. Biochem. Biophys.* 1983; 220:325–329); and III. Glycoproteins, glycopeptides, peptidoglycans (*Glycoconjugate J.* 1986; 3:123–134).

Additional rules with application unique to glycolipids may be found in *Eur. J. Biochem.* 1977; 79:11–21.

The rules of nomenclature presented here were selected to represent those that are likely to be most informative and useful for the beginning glycoconjugate biochemist.

The reader should be aware that the rules of nomenclature are subject to change as new synthetic and natural products are characterized. Efforts at developing rules commensurate with simplicity, while retaining all essential elements of structure, also leads to updates of the rules. A proposed update for *Rules of Carbohydrate (Mono-, Oligo-, and Polysaccharides) Nomenclature* was recently presented by Dr. Derek Horton at the American Chemical Society—Carbohydrate Division Meeting, April 22-27, 1990, Boston, Massachusetts.

It is hoped that a familiarity with the rules of nomenclature reproduced here will assist the reader in understanding the numerous glycoconjugate and carbohydrate structures to be encountered in the scientific literature.

I. MONOSACCHARIDES

A. The Structure Name

These rules are designed to name first a parent monosaccharide represented in the Fischer projection of the acyclic form and then its cyclic forms and derivatives.

The numbering system used in monosaccharides is based on the location of the (potential) carbonyl group. Modification of that group or introduction of further similar groups, therefore, can often destroy the uniqueness of the numbering system and permit a derivative to be named from more than one parent. To determine the unique systematic name of a derivative, it is important to follow the procedure of establishing the Fischer projection of the appropriate parent monosaccharide, naming that according to the rules and, thereafter, deriving the name of any derivative.

B. Conventional Representations

1. The Fischer Projection

In this representation of a monosaccharide, the carbon chain is written vertically with carbon atom number 1 at the top. The groups projecting to left and right of the carbon chain are considered as being in front of the plane of the paper. The optical antipode with the hydroxyl group at the highest-numbered asymmetric carbon atom on the right is then regarded as belonging to the D-series. It is now known that this convention represents the absolute configuration.

2. The Haworth Representation

The Haworth representation of the cyclic form of monosaccharides can be derived from the Fischer projection, as follows: The monosaccharide is depicted with the carbon chain horizontal and in the plane of the paper, the potential carbonyl group being to the right. The oxygen bridge is then depicted as being formed behind the plane of the paper.

The heterocyclic ring is, therefore, located in a plane approximately perpendicular to the plane of the paper, and the groups attached to the carbon atoms are above and below the ring. The carbon atoms of the ring are not shown.

Groups that appear to the right of the vertical chain in the Fischer projection then appear below the plane of the ring in the Haworth representation. However, at the asymmetric carbon atom involved through oxygen in ring formation with the carbon atom of the carbonyl group, a formal double inversion must be envisaged to obtain the correct Haworth representation.

In the pyranose forms of D-aldohexoses, C-6 will always be above the plane. In the furanose forms of D-aldohexoses, the position of C-6 will depend on the configuration at C-4; it will, for example, be above the plane in D-glucofuranoses, but below the plane in D-galactofuranoses.

C. The Reference Carbon Atom and the Anomeric Prefix

1. The Reference Carbon Atom

The *reference carbon atom* is defined as the highest-numbered asymmetric carbon atom in the monosaccharide chain.

2. *The Anomeric Prefix*

In a definitive name, the *anomeric prefix* (α or β) relates the configuration at the anomeric (or glycosidic) center to that of the reference carbon atom.

The anomer having the same orientation, in the Fischer projection, at the anomeric carbon atom and at the reference carbon atom is designated α; the anomer having opposite orientations, in the Fischer projection, is designated β.

The anomeric prefix (α or β) can be used only in conjunction with, and having the aforementioned defined relation to the configurational prefix (D or L) denoting the configuration at the reference carbon atom. Furthermore, it may be used only when the locant of the anomeric center is smaller than that of the reference carbon atom.

D. Numbering Monosaccharides

The basic principle for the numbering of monosaccharides gives the (potential) carbonyl group the lower of the possible numbers (cf. Rule Carb-4). This numbering system is usually retained even when a modification introduces a group which, on the basis of general organic chemical nomenclature, would have priority over the (potential) carbonyl group [e.g., in uronic acids the (potential) carbonyl group retains locant 1, despite the normal priority of the carboxyl group].

In ketoaldonic acids, the carboxyl group that replaces the original (formal) aldehyde retains the locant 1.

E. New Asymmetric Centers

Not infrequently, derivatives of monosaccharides contain asymmetric carbon atoms not present in the parent monosaccharide. Examples include benzylidene derivatives, certain other acetals, *ortho* ester structures, and such. When the stereochemistry at such a carbon atom is known, it will be indicated in the name by use of the appropriate sequence rule symbol, *R* or *S*.

F. Basis of Nomenclature

Rule Carb-1. The basis for the naming and numbering of a monosaccharide or monosaccharide derivative is the structure of the parent monosaccharide ($C_nH_{2n}O_n$), represented in the Fischer projection of the acyclic form.

1. *Choice of Parent Structure*

Rule Carb-2. If, in naming a derivative, a choice of parent monosaccharide is possible, the selection of parent is made according to the following order of preference, treated in the order given until a decision is reached:

1. The monosaccharide, of which the first letter of the trivial name (see Rule Carb-5), or of the configurational prefix, or of the first cited configurational prefix of a systematic name (see Rule Carb-8), occurs earliest in the alphabet. If two possible parents have the same initial letter, then the choice will be made according to the letter at the first point of difference in the trivial name, the configurational prefix of the systematic name, and so on. *Examples*: allose before glucose, glucose before gulose, allo- before gluco-, gluco- before gulo-.
2. The configurational symbol D- before L-.

3. The monosaccharide that gives the point(s) of modification of the >CH(OH) chain the lowest locant(s) (see Note 1).
4. The monosaccharide that gives the lowest locants (see Note 1) to the substituents, present in the derivative.
5. The monosaccharide that, when the substituents have been placed in alphabetical order, results in the first-cited substituent having the lowest locant (see Note 1).

2. *Trivial and Systematic Names*

Rule Carb-3. In naming monosaccharides or monosaccharide derivatives, either trivial or systematic names can be used for the parent monosaccharide.

Trivial names are defined in Rule Carb-5.

Systematic names are formed by adding one or more configurational prefixes (see Rule Carb-8) to the appropriate stem name (see Rule Carb-6).

Rule Carb-4. The names *aldose* and *ketose* are used in a generic sense to denote monosaccharides in which the (potential) carbonyl group is terminal (aldehydic) or non-terminal (ketonic), respectively. In an aldose, the carbon atom of the (potential) carbonyl group is atom number 1; in a ketose, it has the lowest number possible.

Rule Carb-5. The trivial names of the acyclic aldoses with 3, 4, 5, or 6 carbon atoms are retained, and are used in preference to their systematic names for the aldoses and for the formation of names of their derivatives. The trivial names of these aldoses are

Triose: glyceraldehyde (glycerose is not recommended)
Tetroses: erythrose, threose
Pentoses: arabinose, lyxose, ribose, xylose
Hexoses: allose, altrose, galactose, glucose, gulose, idose, mannose, talose

Rule Carb-6. The *stem names* of the acyclic aldoses having 3, 4, 5, 6, 7, 8, 9, 10 . . . , carbon atoms in the chain are triose, tetrose, pentose, hexose, heptose, octose, nonose, decose. . . .

Lowest locants are defined as follows: When a series of locants containing the same number of terms are compared term by term, that series is lowest that contains the lowest number on the occasion of the first difference.

The stem names of the acyclic ketoses having 4, 5, 6, 7, 8, 9, 10 . . . , carbon atoms in the chain are tetrulose, pentulose, hexulose, heptulose, octulose, nonulose, deculose. . . .

3. *Configurational Symbols and Prefixes*

Rule Carb-7. Configurational relationships are denoted by the symbols D and L which in print will be small capital roman letters and which are not abbreviations for *dextro* and *levo*. Racemic forms may be indicated by DL. Such symbols are affixed by a hyphen immediately before the monosaccharide trivial name (see Rule Carb-5) or before each configurational prefix (see Rule Carb-8) of a systematic name, and they are employed only with compounds that have been related definitely to the reference standard glyceraldehyde (see Rule Carb-8). The configurational symbol should not be omitted, if known.

If the sign of the optical rotation under specified conditions is to be indicated, this is done by adding (+) or (–). Racemic forms may be indicated by (±). With compounds optically compensated intramolecularly the prefix *meso* is used where appropriate.

Example: D-glucose or D(+)-glucose, D-fructose or D(–)-fructose, DL-glucose or (±)-glucose.

Rule Carb-8. The configuration of a >CHOH group or a set of 2, 3, or 4 contiguous >CHOH groups (or wholly or partly derivative groups, such as >$CHOCH_3$, >$CHOCOCH_3$ or >$CHNH_2$) is designated by the appropriate one of the following configurational prefixes, *glycero, erythro, threo,* which (except for glycero-) are derived from the trivial names of the aldoses mentions in Rule Carb-5. When used in systematic names these prefixes are to be uncapitalized and are italicized in print. They are affixed by a hyphen to the stem name defined in Rule Carb-6. There may be more than one configurational prefix in a name.

Each prefix is D or L, according to whether the configuration at the reference carbon atom in the Fischer projection is the same as, or the opposite of, that in D(+)-glyceraldehyde.

The systematic name for a monosaccharide is then formed by using the configurational symbols and prefixes with the appropriate stem name (see Rule Carb-6).

For sets of more than four contiguous >CHOH groups see Rule Carb-9.

The trivial names, which are preferred, of the acyclic aldoses with four to six carbon atoms (see Rule Carb-5) thus correspond to the following systematic names:

Trivial	Systematic
D-Erythrose	D-*erythro*-tetrose
D-Threose	D-*threo*-tetrose
D-Arabinose	D-*arabino*-pentose
D-Lyxose	D-*lyxo*-pentose
D-Ribose	D-*ribo*-pentose
D-Xylose	D-*xylo*-pentose
D-Allose	D-*allo*-hexose
D-Altrose	D-*altro*-hexose
D-Galactose	D-*galacto*-hexose
D-Glucose	D-*gluco*-hexose
D-Gulose	D-*gulo*-hexose
D-Idose	D-*ido*-hexose
D-Mannose	D-*manno*-hexose
D-Talose	D-*talo*-hexose

4. *Multiple Configurational Prefixes*

Rule Carb-9. An acyclic monosaccharide containing more than four contiguous asymmetric carbon atoms is named by adding two or more prefixes, together indicating the configurations at all asymmetric carbon atoms, to the stem name defined in Rule Carb-6.

The configurational prefixes employed are given in Rule Carb-8. The sequence of ($4n + m$; where n is 1 or more and m is 0, 1, 2, or 3) asymmetric carbon atoms is divided, beginning at the asymmetric carbon atom next to the functional group, into (n) sets of four asymmetric carbon atoms, and a final set (m) of fewer than four. The order of citation of these prefixes commences at the end farthest from carbon number 1.

The locants corresponding to the configurational prefixes may be inserted in the name, if desired. In such cases, all locants are given and immediately precede the appropriate configurational prefixes.

Compounds that require multiple configurational prefixes, but that do not necessarily contain more than four contiguous asymmetric carbon atoms are covered by Rules Carb-10 (ketoses) and Carb-14 (deoxysaccharides).

5. *Ketoses*

Rule Carb-10. Ketoses are classified as 2-ketoses, 3-ketoses, and so on, according to the position of the (potential) carbonyl group (see Rule Carb-4).

The systematic name of an individual ketose is obtained by affixing, before the stem name (see Rule Carb-6) and by means of a hyphen, the locant of the (potential) carbonyl group. The locant is preceded by the configurational prefix or prefixes (see Rule Carb-8) for the groups of asymmetric centers present.

If more than one configurational prefix is needed, the order of their citation commences at the end farthest from carbon atom number 1.

The locant 2 may be omitted from the name of a 2-ketose when no ambiguity can arise.

When the carbonyl group is at the middle carbon atom of a ketose containing an uneven number of carbon atoms in the chain, two names are possible. That name will be selected as accords with the order of precedence given in Rule Carb-2.

The following are examples of nonsystematic names of ketoses that are established by usage and may be retained:

D-Ribulose	for	D-*erythro*-2-pentulose
D-Xylulose	for	D-*threo*-2-pentulose
Sedoheptulose	for	D-*altro*-2-heptulose
D-Fructose	for	D-*arabino*-2-hexulose
D-Psicose	for	D-*ribo*-2-hexulose
D-Sorbose	for	D-*xylo*-2-hexulose
D-Tagatose	for	D-*lyxo*-2-hexulose

6. *Deoxy-Monosaccharides and Amino-Monosaccharides*

Rule Carb-14.

1. The replacement of an alcoholic hydroxyl group of a monosaccharide, or monosaccharide derivative, by a hydrogen atom is expressed by using the prefix *deoxy* preceded by the appropriate locant and followed by a hyphen together with a systematic or trivial name (see Rule Carb-3). The systematic name consists of a stem name with such configurational prefixes as express the configuration at the asymmetric centers present in the deoxy-compound. The order of citation of the configurational prefixes commences at the end farthest from carbon atom number 1.

   A trivial name may be used only if the transformation of the saccharide into the deoxy-compound does not alter the configuration at any asymmetric center.

   Trivial names of 6-deoxy-hexoses established by usage may be retained and used for the formation of names of derivatives.

   The following is established as a trivial name for biochemical use: deoxyribose for 2-deoxy-D-*erythro*-pentose.
2. The replacement of an alcoholic hydroxyl group of a monosaccharide, or monosaccharide derivative, by an amino group is envisaged as substitution of the appropriate hydrogen atom of the corresponding deoxy-monosaccharide by the amino group.

   Substitution in the amino group is indicated by use of prefix *N* (see Rule Carb-15), unless the substituted amino group has a trivial name (for example: CH_3CONH-, acetamido).

The stereochemistry at the carbon atom carrying the amino group is expressed according to Rule Carb-8.

Trivial names accepted for biochemical usage:

D-Galactosamine for 2-amino-2-deoxy-D-galactopyranose (the name chondrosamine is not recommended)
D-Glucosamine for 2-amino-2-deoxy-D-glucopyranose
D-Mannosamine for 2-amino-2-deoxy-D-mannopyranose
Neuraminic acid for 5-amino-3,5-dideoxy-α-D-*glycero*-D-*galacto*-2-nonulopyranosonic acid
Muramic acid for 2-amino-3-*O*-[1-(*S*)-carboxyethyl]-2-deoxy-aldehydo-D-glucose

When the complete name of the derivative includes other prefixes, "deoxy" takes its place in the alphabetical order of detachable prefixes; in citation the alphabetical order is preferred to numerical order.

Examples:
4-Amino-4-deoxy-3-*O*-methyl-D-*erythro*-2-pentulose
4-Deoxy-4-(ethylamino)-D-*erythro*-2-pentulose

7. *O-Substitution*

Rule Carb-15. Replacement of the hydrogen atom of an alcoholic hydroxyl group of a saccharide or saccharide derivative by another atom or group is denoted by placing the name of this atom or group before the name of the parent compound. The name of the atom or group is preceded by an italic capital letter *O* (for oxygen), followed by a hyphen, to make clear that substitution is on oxygen. The *O* prefix need not be repeated for multiple replacements by the same atom or group.

Replacement of hydrogen attached to nitrogen or sulfur by another item or group is indicated in a similar way with the use of italic capital letters *N* or *S* (for examples see Rules Carb-24 and Carb-36).

The italic capital letter *C* may be used to indicate replacement of hydrogen attached to carbon to avoid possible ambiguity.

Rule Carb-16 (alternative to Rule Carb-15). *O*-Substitution products of saccharides or saccharide derivatives may be named as esters, ethers, and so forth, following the procedures prescribed for that purpose in IUPAC Nomenclature of Organic Chemistry, Section C, 1965.

8. *Acyclic Forms*

Rule Carb-17. The acyclic nature of a monosaccharide or monosaccharide derivative containing an uncyclized carbonyl group may be stressed by inserting the italicized prefix *aldehydo* or *keto*, respectively, immediately before the configurational prefix(es) or before the trivial name. These prefixes may be abbreviated to *al* and *ke*.

9. *Ring Size in Cyclic Forms*

Rule Carb-18. The size of the ring in the cyclic form of a monosaccharide (aldose or ketose) or monosaccharide derivative may be indicated by replacing the terminal letters *se* of the name of the acyclic form by *furanose* for the five-atom ring, *pyranose* for the six-atom ring, and *septanose* for the seven-atom ring.

Rule Carb-19 (alternative to Rule Carb-18). The size of the ring in the cyclic form of a monosaccharide (aldose or ketose) or monosaccharide derivative may be indicated by two numerals, placed in parentheses and joined to the end of the name of the acyclic compound by a hyphen. These numerals denote the two carbon atoms to which the ring oxygen atom is attached, the carbon atom of the potential carbonyl group being cited first.

10. Anomers

Rule Carb-21. The free hydroxyl group belonging to the internal hemiacetal grouping of the cyclic form of a monosaccharide or monosaccharide derivative is termed the *anomeric* or *glycosidic* hydroxyl group.

The two cyclic forms of an aldose or ketose, or aldose or ketose derivative (termed *anomers*), are distinguished with the aid of the anomeric prefixes, α and β, relating the configuration at the anomeric carbon atom to that at the reference asymmetric carbon atom of the compound; the anomer having the same configuration, in the Fischer projection, at the anomeric and the reference carbon atom is designated α.

The anomeric prefix, α or β, followed by a hyphen, is placed immediately in front of the configurational symbol, D or L, of the trivial name or of the configurational prefix denoting the group of asymmetric carbon atoms that includes the reference carbon atom.

11. Glycosides

Rule Carb-23. Mixed acetals, resulting from the replacement of the hydrogen atom of the anomeric or glycosidic hydroxyl group by a group X, derived from an alcohol or phenol (XOH), are named *glycosides.* The term *glycoside* is used in a generic sense only, and may not be applied to specific compounds. Glycosides are named by replacing the terminal *e* of the name of the corresponding cyclic form of the saccharide or saccharide derivative by *ide* and placing before the word thus obtained, as a separate word, the name of the group X.

12. Glycosyl Radicals and Glycosylamines

Rule Carb-24.

1. The radical formed by detaching the anomeric or glycosidic hydroxyl group from the cyclic form of a monosaccharide or monosaccharide derivative is named by replacing the terminal *e* of the name of the monosaccharide or monosaccharide derivative by *yl.* The general name of these radicals is *glycosyl* (glycofuranosyl, glycopyranosyl, glycoseptanosyl) radical.
2. The replacement of the glycosidic hydroxyl group of a cyclic form of a monosaccharide derivative by an amino group is indicated by adding the suffix *amine* to the name of the glycosyl radical.

13. Glycosyloxy Radicals

Rule Carb-25. The radical formed by removal of the hydrogen atom from the anomeric or glycosidic hydroxyl group of the cyclic form of a monosaccharide or monosaccharide derivative is named by replacing the terminal *e* of the name of the saccharide by *yloxy.*

14. Alditols

Rule Carb-26.

1. Names for the polyhydric alcohol (alditols) of the saccharide series are derived from the names of the corresponding acyclic aldoses by changing the suffix *ose* to *itol.*

   If the same alditol can be derived from two different aldoses, preference is given to that name that accords with the order of precedence given in Rule Carb-2.

2. To the trivial names of alditols, optically compensated intramolecularly, that have no D- or L-prefix, the prefix *meso*- may be added for the sake of clarity.

Examples: *meso*-erythritol, *meso*-ribitol, *meso*-xylitol, *meso*-allitol, *meso*-galactitol

The prefixes D or L must, however, be used (1) in the names of *meso*-alditols containing more than four contiguous asymmetric carbon atoms, to define the steric relation of the configurational prefixes cited; and (2) in naming derivatives of *meso*-alditols that have become asymmetric by substitution.

15. *Aldonic Acids*

Rule Carb-28. Monocarboxylic acids formally derived from aldoses, having three or more carbon atoms in the chain, by oxidation of the aldehydic group, are named aldonic acids, and are divided into aldotrionic acid, aldotetronic acids, aldopentonic acids, aldohexonic acids, and so on, according to the number of carbon atoms in the chain. The names of individual compounds of this type are formed by replacing the ending *ose* of the systematic or trivial name of the aldose by *onic acid.*

Derivatives of these acids formed by change in the carboxyl group (salts, esters, lactones, acyl halides, amides, nitriles, or others), are named according to the IUIPAC Nomenclature of Organic Chemistry, Section C, 1965, Rules C-4.

16. *Uronic Acids*

Rule Carb-30. The monocarboxylic acids formally derived by oxidation of the terminal CH_2OH group of aldoses having four or more carbon atoms in the chain, or of glycosides derived from these aldoses, to a carboxyl group are named *uronic acids.* The names of the individual compounds of this type are formed by replacing (1) the *ose* of the systematic or trivial name of the aldose by *uronic acid*, or (2) the *oside* of the name of the glycoside by *osiduronic acid.*

The carbon atom of the (potential) aldehydic carbonyl group (not that of the carboxyl group) is numbered 1.

Derivatives of these acids formed by change in the carboxyl group (salts, esters, lactones, acyl halides, amides, nitriles, or others) are named according to the IUPAC Nomenclature of Organic Chemistry, Section C, 1965, Rules C-4.

17. *Cyclic Acetals*

Rule Carb-32. Cyclic acetals formed by the reaction of saccharides or saccharide derivatives with aldehydes or ketones are named in accordance with Rule Carb-15, bivalent radical names being used as prefixes, the names of such radicals following the rules of general organic chemical nomenclature. In indicating more than one cyclic acetal grouping of the same kind, the appropriate pairs of locants are separated typographically when the exact placement of the acetal groups is known.

18. *Ortho Esters*

Rule Carb-33. The "glycosides of the *ortho* ester structure" are named (1) as cyclic acetals, according to Rule Carb-32, or (2) as *ortho* esters, according to Rule C-464.1 (see Note 2), with the name of the saccharide as the first term of the name.

19. *Acetals and Thioacetals*

Rule Carb-34. The compounds obtained by transforming the carbonyl group of a saccharide or saccharide derivative into the grouping

$$>C<^{OR^1}_{OR^2},\ >C<^{OR^1}_{SR^2},\ \text{or}\ >C<^{SR^1}_{SR^2}$$

are named by placing after the name of the saccharide or saccharide derivative the term *acetal*, *monothioacetal*, or *dithioacetal* as appropriate, preceded by the names of the radicals R^1 and R^2. With monothioacetals, the mode of bonding of two different radicals R^1 and R^2 is indicated by the use of the prefixes *O* and *S*.

20. *Hemiacetals*

Rule Carb-35. The compounds obtained by transforming the carbonyl group of the acyclic form of a saccharide, or saccharide derivative, into the grouping

$$>C<^{OR}_{OH},\ >C<^{OR}_{SH},\ >C<^{SR}_{OH},\ \text{or}\ >C<^{SR}_{SH}$$

are named as indicated in Rule Carb-34, by using the endings *hemiacetal*, *monothiohemiacetal*, or *dithiohemiacetal*, as appropriate. The two isomers of a monothiohemiacetal, or dithiohemiacetal, as appropriate. The two isomers of a monothiohemiacetal are differentiated by use of *O* and *S* prefixes.

21. *Other Sulfur Monosaccharides*

Rule Carb-36. Replacement of a hydroxyl oxygen atom of an aldose or ketose, or of the oxygen atom of the carbonyl group of the acyclic form of an aldose or ketose, by sulfur is indicated by placing the (nondetachable) affix *thio* (see Note 3), preceded by the appropriate locant, before the systematic or trivial name of the aldose or ketose.

Replacement of the ring oxygen atom of the cyclic form of an aldose or ketose by sulfur is indicated in the same way, the number of the highest-numbered carbon atom of the ring being used as locant.

Glycosides in which the glycosidic oxygen atom is replaced by sulfur are designated generically as *thioglycosides*.

Selenium compounds are named analogously, by use of the affix *seleno*.

22. *Intramolecular Anhydrides*

Rule Carb-37. An intramolecular ether (commonly called an intramolecular anhydride), formed by elimination of water from two alcoholic hydroxyl groups of a single molecule of a monosaccharide (aldose or ketose) or monosaccharide derivative, is named by attaching the (detachable) prefix *anhydro* by a hyphen before the monosaccharide name; this prefix, in turn, is preceded by a pair of locants identifying the two hydroxyl groups involved.

II. OLIGOSACCHARIDES

An oligosaccharide is a molecule containing a small number (two to about ten) of monosaccharide residues, connected by glycosidic linkages. A carbohydrate containing two such residues is a disaccharide, a carbohydrate containing three such residues is a trisaccharide, and so on. The following recommendations apply.

A. Trivial Names

1. Certain trivial names firmly established in the literature are specific for particular structures, and their continued use is allowed in instances for which the full name may be unwieldy.

Examples: Disaccharides: cellobiose, chitobiose, gentiobiose kojibiose, lactose, melibiose, sophorose, sucrose, α,α-trehalose, turanose
Tri- and oligosaccharides: melezitose, panose, raffinose, stachyose.

2. Such disaccharide names as xylobiose and mannobiose, which are ambiguous, should be used only when there is no risk of confusion. The systematic name should be given together with the trivial names, at the first mention.
3. The accepted trivial names for disaccharides should be extended to higher oligosaccharides only when the latter contain a single type of sugar residue and linkage (e.g., cellotetraose, maltotriose). When, however, the trivial name is derived from the name of a carbohydrate that contains two or more different sugars, or types of linkage, or both, extension of the names to higher oligosaccharides is not recommended.

Example: Not recommended are such names as agarotetraose and nigerotriose.

B. Systematic Names

Systematic names should be assigned as in Rules Carb-39 and Carb-40.

Rule Carb-39 (Disaccharides).

1. *Nonreducing*: A nonreducing disaccharide is named, from its component monosaccharide parts, as a glycosyl glycoside.
2. *Reducing*: A reducing disaccharide is named, from its component monosaccharide parts, as a glycosylglycose.
3. *Derivatives*: Derivatives of disaccharides are named according to the rules for monosaccharides.

Rule Carb-40 (Trisaccharides and Higher Oligosaccharides).

1. *Nonreducing*: A nonreducing trisaccharide is named as a glycosylglycoslglycoside, from its component monosaccharide parts. Higher nonreducing oligosaccharides are named similarly.
   Between the name of one glycosyl radical and the next are placed two locants that indicate the respective positions involved in this glycosidic union; these locants are separated by an arrow (pointing from the locant corresponding to the glycosyl carbon atom to the locant corresponding to the hydroxylic carbon atom involved) and are enclosed in parentheses.
2. *Reducing*: A reducing trisaccharide is named as a glycosylglycosylglycose, from its component monosaccharide parts. Higher reducing oligosaccharides are named similarly.
3. *Reducing branched*: By inserting one glycosyl substituent in brackets, it is distinguished from the second glycosyl substituent.

C. Abbreviated Names for Use for Oligosaccharides

1. The symbols chosen are derived from the trivial names of the constituent sugars. For the sake of clarity, brevity, and listing in tables, the symbols have, wherever possible, been restricted to three letters, usually the first three letters of the trivial name.

Allose = All	Arabinose = Ara	Rhamnose = Rha
Altrose = Alt	Lyxose = Lyx	Fucose = Fuc
Galactose = Gal	Ribose = Rib	Fructose = Fru

Glucose = Glc
Gulose = Gul
Idose = Ido
Mannose = Man
Talose = Tal
Xylose = Xyl
Neuraminic acid = Neu
Muramic acid = Mur

Symbols derived from less common trivial names may be used, but the systematic names should be given together with the trivial name and the abbreviation at the first mention.

Examples
3,6-Dideoxy-D-*xylo*-hexose (abequose; Abe)
6-Deoxy-D-glucose (quinovose; Qui)
3-*C*-(Hydroxymethyl)-D-*glycero*-aldotetrose (D-apiose; Api)

2. The symbols represent the structural formulas of the compounds and also their names.
3. The symbols represent, the individual sugars or their residues. The use of the symbol to represent the free sugar is not recommended in textual material, but such use may occasionally be desirable in tables, diagrams, and figures.
4. The ring form is indicated, where necessary, by using the first letter of furanose, pyranose, or septanose, italicized and uncapitalized, added after the abbreviated name of the monosaccharide.

Examples
Arabinofuranose; Ara*f*
Glucopyranose; Glc*p*

5. Uronic acid: The suffix A is added to the symbol for the parent monosaccharide.

Examples
Glucaronic acid; GlcA
Galactopyranuronic acid; Gal*p*A

6. Deoxy sugars: Rational names, but no abbreviations are recommended. One exception is 2-deoxy-D-*erythro*-pentose (deoxyribose), which is abbreviated dRib.
7. Aminodeoxy sugars: For 2-amino-2-deoxy sugars, the suffix *N* is added to the symbol of the parent monosaccharide. If the latter is *N*-acetylated, the suffix becomes NAc.

Examples
2-Amino-2-deoxy-D-glucopyranose; D-Glc*p*N
2-Amino-2,6-dideoxy-L-galactose; L-FucN
2-Acetamido-2-deoxy-D-mannopyranose; D-Man*p*NAc

For other, less common, aminodeoxy sugars, the appropriate locants are added before *N*.

Example: 3,6-Bis(acetamido)-3,6-dideoxymannose; Man3,6$(NAc)_2$

8. For anhydro sugars, the appropriate locants and the prefix *An* are added before the symbol of the parent monosaccharide.

Example: 3,6-Anhydrogalactose; 3,6AnGal

9. Configuration: The configurational symbol (D or L) is included before the symbol for the monosaccharide, and is separated therefrom by a hyphen.

Examples
D-Glucopyranose; D-Glc*p*
L-Arabinofuranose; L-Ara*f*
3,6-Anhydro-D-galactose; 3,6An-D-Gal

10. Anomeric configuration: The anomeric symbol (α or β) is included before the configurational symbol and separated therefrom by a hyphen.

    Examples
    α-D-Glucopyranose; α-D-Glc*p*
    β-L-Arabinofuranose; β-L-Ara*f*

11. Structural formulas may be used for complicated features together with the abbreviated notation whenever necessary for clarity.

D. Unbranched Oligosaccharides

Between the symbol (abbreviated name) of one monosaccharide group or residue and the next are placed two locants that indicate the respective positions involved in this glycosidic union. These locants are separated by an arrow (directed from the locant corresponding to the glycosyl carbon atom to the locant corresponding to the carbon atom carrying the hydroxyl group involved) and are enclosed in parentheses (see foregoing, Rule Carb-40). For nonreducing oligosaccharides, double-headed arrows are used between the locants of the appropriate glycosyl carbon atoms when the symbols are used, but not when the names are spelled out.

The hyphens, except that separating the configurational symbol and the symbol for the monosaccharide, may be omitted, e.g.,

αD-Gal*p*(1→4)αD-Glc*p*A(1→4)αD-Xyl*p*(1→4)αD-Glc*p*N(1→3)D-Man

E. Cyclic Oligosaccharides

A cyclic oligosaccharide is symbolized as illustrated in the following:

Example: Cyclomaltohexaose (the older names cyclohexaamylose, Schardinger α-dextrin, or α-cyclodextrin, are not recommended).

→4)-[α-D-Glc]$_6$-(1→

F. Branched and Substituted Oligosaccharides

1. Substituents may be symbolized as follows:

Acetyl	Ac	Methyl	Me
Benzoyl	Bz	Phenyl	Ph
Benzyl	Bzl, $PhCH_2$(Bn)	*p*-toluenesulfonyl	Tos (Ts), tosyl
Ethyl	Et	trimethylsilyl	Me_3Si
Glycolyl	Gl	trityl	Trt,Ph_3C (Tr)
methanesulfonyl	$MeSO_2$(Ms), mesyl		

Most of these symbols are in the lists of those recommended for biochemical use or for symbolizing derivatives of amino acids. Some symbols that are commonly used in the carbohydrate literature are given in parentheses. Substituents will not carry prefixes if attached to oxygen or nitrogen, but will be preceded by an italicized *C* if attached to carbon. The position of the substituent will be shown by the appropriate numeral. Substituents directly follow the abbreviation for the monosaccharide residue.

Examples

Ethyl D-glucopyranuronate	D-Glc*p*A6Et
β-D-Galactopyranose 4-sulfate	β-D-Gal*p*4SO_3^-
2-*C*-Methyl-D-xylose	D-Xyl2*C*Me
3,4 Di-*O*-methyl-L-rhamnose	L-Rha3,4Me_2
N-Acetylneuraminic acid	Neu5Ac
N-Acetyl-2-deoxyneur-2-enaminic acid	Neu2en5Ac

2. For branched oligosaccharides, the main chain and oligosaccharide side chains will be depicted as outlined for unbranched oligosaccharides. The position of a branch is indicated above, or below, the main chain, with the numerals and an arrow indicating the glycosidic linkage.

Examples

α-D-Glc*p*(1→4)-α-D-Glc*p*(1→4)-D-Man*p*
2
↑
1
β-D-Xyl*p*

and

β-D-Glc*p*(1→4)-β-D-Gal*p*A(1→6)-β-D-Glcp(1→5)-L-Ara*f*
2
↑
1
β-D-Xylp(1→3)-β-D-Galp

3. In drawing structural formulas, Haworth perspective formulas, conformational formulas, or Mills formulas may be used. It should be borne in mind that the detailed conformation indicated by a conformational formula has not always been established. Mills formulas are particularly useful in describing synthetic work. Use of a Fischer projection is sometimes advisable.

G. The Condensed System of Symbolism of Sugar Residues in Oligosaccharides and Oligosaccharide Chains

In the condensed system the common configuration and ring size are implied in the symbol. Thus, Glc means D-glucopyranose; Fru, D-fructofuranose; and Fuc, L-fucopyranose. Whenever the configuration or ring size is found to differ from the common one, or is to be emphasized, this may be indicated by using the appropriate symbols from the extended system.

The anomeric descriptor indicates the configuration of the glycoside linkage, and, therefore, is placed before the locant if the direction of the bond is to the right, or after the locant if the direction of the bond is to the left. The two locants are separated by a hyphen. No hyphens are used between the symbol for the sugar and the parentheses indicating the glycosidic bond.

Example: Raffinose = Gal(α1-6)Glc(α1-2β)Fru

The parentheses may be omitted in representing branched oligosaccharides, when parentheses are used to indicate the branches. In this way, it is possible to write branched sequences on one line, as shown in the examples.

The main difference between the extended form and the condensed form is the place of the anomeric descriptor, α or β. In the extended form, the anomeric descriptor is considered to be part of the name (symbol) of the sugar unit, this system is preferred by carbohydrate chemists. In the condensed system, the anomeric descriptor specifies the type of glycosidic linkage. This usage, first codified in *Abbreviations and Symbols for Chemical Names of Special Interest in Biological Chemistry*, is preferred by many biochemists. The JCBN has been unable to reach a consensus on which system should be recommended for general use, so both systems here are given as optional.

As in the extended system, placing a hyphen or parenthesis to the right of the symbol for a monosaccharide residue signifies removal of OH from the reducing carbon. Thus amygdalin may be represented.

Extended: β-D-Glc*p*(1→6)-β-D-Glc*p*-O-CH(C*N*)-Ph
Condensed: Glc(β1-6)Glc(β)-O-CH(C*N*)Ph

This also applies in representing glycolipids.

The following examples illustrate the use of the two different systems. They refer to the numbered structural formulas given in the foregoing. For long or multiple branches, it may be advisable to use the two-line notation even in the condensed system.

Structure 1

Extended
α-D-Gal*p*(1→4)-α-D-Glc*p*A(1→4)-α-D-Xyl*p*(1→4)-α-D-Glc*p*N(1→3)-D-Man*p*

Condensed
Gal(α1-4)GlcA(α1-4)Xyl(α1-4)GlcN(α1-3)Man

Structure II

Extended: α-D-Gal*p*(1→4)-α-D-Gal*p*(1→6)-α-D-Glc*p*(1↔2)-β-D-Fru*f*
Condensed: Gal(α1-4)Gal(α1-6)Glc(α1-2β)Fru

Structure III

Extended: β-D-Xyl*p*(1→4)-β-D-Man*p*(1→4)-β-D-Xyl*p*(1→4)-D-Man*p*
Condensed: Xyl*p*(β1-4)Man(β1-4)Xyl*p*(β1-4)Man

Note: Here, Xyl*p* is used in the condensed system to stress the pyranose form.

Structure IV

Extended: α-D-Glc*p*(1→4)-α-D-Glc*p*(1→4)-D-Man*p*

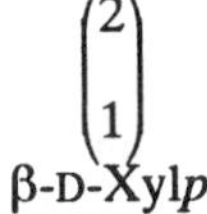

Condensed: Glcα1-4(Xylβ1-2)Glcα1-4Man

Structure V

Extended: β-D-Glc*p*-(1→4)-β-D-Gal*p*A-(1→6)-β-D-Glc*p*-(1→5)-L-Ara*f*

```
                                   2
                                   ↑
                                   1
β-D-Xylp-(1→3)-β-D-Galp
```

Condensed: Glcβ1-4GalAβ1-6(Xylβ1-3Galβ1-2)Glcβ1-5Ara
or
Glcβ1-4GalAβ1-6Glcβ1-5Ara

```
                2
                ↑
Xylβ1-3Gal 1
```

III. GLYCOPROTEINS, GLYCOPEPTIDES, PEPTIDOGLYCANS

Various types of compounds consisting of carbohydrates covalently linked with other types of chemical constituents are classified under the general name of glycoconjugates. The major groups of glycoconjugates are the glycoproteins, glycopeptides, peptidoglycans, glycolipids, and lipopolysaccharides. The first three of these are considered in the present document.

A. Glycoproteins, Proteoglycans, and Glycosaminoglycans

A *glycoprotein* is a compound containing carbohydrate (or glycan) covalently linked to protein. The carbohydrate may be in the form of monosaccharide(s), disaccharide(s), oligosaccharide(s), or their derivatives (e.g., sulfo- or phospho-substituted). One, a few, or many carbohydrate units may be present. *Proteoglycans* are a subclass of glycoproteins in which the carbohydrate units are polysaccharides that contain amino sugars. Such polysaccharides are also known as *glycosaminoglycans.*

B. Glycopeptides, Glyco Amino Acids, and Glycosyl Amino Acids

A *glycopeptide* is a compound consisting of carbohydrate linked to an oligopeptide composed of L- or D-amino acids, or both. A *glyco amino acid* is a saccharide attached to a single amino acid by any kind of covalent bond. A *glycosyl amino acid* is a compound consisting of a saccharide linked through a glycosyl linkage (*O*-, *N*-, or *S*-) to an amino acid. (The hyphens are needed to avoid implying that the carbohydrate is necessarily linked to the amino group.)

C. Peptidoglycans

A *peptidoglycan* consists of glycosaminoglycan formed by alternating residues of D-glucosamine and either muramic acid (2-amino-3-*O*-[(*S*)-1-carboxyethyl]-2-deoxy-D-glucose) or L-talosaminuronic acid (2-amino-2-deoxy-L-taluronic acid) that are usually *N*-acetylated or *N*-glycoloylated. The carboxyl group of the muramic acid is commonly substituted by a peptide that contains residues of both L- and D-amino acids, whereas that of L-talosaminuronic acid is substituted by a peptide consisting of L-amino acids only.

D. Form of Carbohydrate in Glycoproteins

In many glycoproteins (e.g., plasma glycoproteins such as human α_1-acid glycoprotein or fetuin), the carbohydrate is in the form of oligosaccharides, linear or branched, the latter containing up to about 20 monosaccharide residues; glycoproteins containing mono- or disaccharide units are also known [e.g., collagens, fish antifreeze glycoproteins, sheep submaxillary (or submandibular) glycoproteins], as well as those that contain oligosaccharides that consist of repeating units of *N*-acetyllactosamine (e.g., band 3 of the human erythrocyte membrane).

E. Proteoglycans

A *proteoglycan* is a protein glycosylated by one or more (up to about 100) glycosaminoglycans. The glycosaminoglycans of the proteoglycans are linear polymers of up to about 200 repeating disaccharide units that consist of a hexosamine (D-glucosamine or D-galactosamine) alternating with a uronic acid (D-glucuronic or L-iduronic) or a neutral sugar (D-galactose). The hexosamines are usually *N*-acetylated, and in some of the glycosaminoglycans the D-glucosamine is *N*-sulfated. Varying degrees of sulfation occur in other positions of the hexosamines, as well as on the L-iduronic acid. The chain of repeating units is linked to the protein by an oligosaccharide of a structure different from the repeating units. This linkage region is identical in most of the proteoglycans (chondroitin sulfates, dermatan sulfate, heparin, and heparan sulfate), but is different in keratan sulfate. This last glycosaminoglycan contains repeating units of *N*-acetyllactosamine that are *O*-sulfated. Because of their content of uronic acid and/or ester sulfate, the glycosaminoglycans of the proteoglycans are anionic polyelectrolytes and have been referred to as *acidic glycosaminoglycans* (equivalent to the older term *acid mucopolysaccharides*). Proteoglycans may also contain one or more oligosaccharides of structure similar to those found in other glycoproteins.

Several diseases of proteoglycan metabolism are recognized, including the Hunter's and Hurler's syndromes; they are referred to as mucopolysaccharidoses, but strictly speaking, should go by the term *glycoproteinoses*, which encompasses other diseases of glycoprotein metabolism, such as mannosidosis and fucosidosis.

F. Microheterogeneity

A particular glycoprotein may occur in forms that differ in the structure of one or more of its carbohydrate units, a phenomenon known as microheterogeneity. Such differences may affect both the size and charge of individual glycoproteins; occasionally, the differences may be solely due to changes in linkage position in a carbohydrate unit. For example, chicken ovalbumin contains a single glycosylated amino acid residue (Asn-293), but more than a dozen different oligosaccharides have been identified at that site. In proteoglycans, individual glycosaminoglycan chains may differ in structure (e.g., the degree of sulfation, the ratio of D-glucuronic acid to L-iduronic acid, and chain length). Proteoglycans are, therefore, highly polydisperse, in contrast with typical glycoproteins, which, in spite of their microheterogeneity, are not markedly polydisperse in their molecular size.

G. Glycopeptides and Related Glycoconjugates

Glycopeptides, glyco amino acids, glycosyl amino acids, and glycosylpeptides are obtained by enzymic or chemical cleavage of glycoproteins, or by chemical or enzymic synthesis. Examples of glycosyl amino acids that constitute the common linking compounds between the carbohydrate and protein in glycoproteins are as follows: 2-acetamido-*N*-(L-aspart-4-yl)-2-deoxy-β-D-glucopyranosylamine (i.e., N^4-(*N*-acetyl-β-D-glucosaminyl) asparagine, which is abbreviated to (GlcNAc-)Asn (parentheses here around the carbohydrates placed next to the symbol for an amino acid residue indicate substitution on its side chain: O^3-(*N*-acetyl-α-D-galactosaminyl)serine or threonine, (GalNAc-)Ser, or (GalNAc-)Thr; *O*-β-D-xylosylserine, (Xyl-)Ser; O^5-α-D-galactosylhydroxylysine (see Note 4), (Gal-)Hyl; and β-L-arabinosylhydroxyproline (see Note 5), (L-Ara-)Hyp. Another example of a glyco amino acid [Man_9GlcNAc(β1-4)GlcNAc-]Asn, isolated from a proteolytic digest of soybean agglutinin. Such a compound may be referred to as an oligosaccharylasparagine.

H. *N, O*-Glycoproteins

If desirable, the linkage between carbohydrate and protein may be indicated by the locants *N*- or *O*-. The locant *N*- is used for the *N*-glycosyl linkage to asparagine. *N*-linked oligosaccharides are divided into two major classes: the *N*-acetyllactosamine type, containing *N*-acetyl-D-glucosamine, D-mannose, D-galactose, L-fucose, and sialic acid; and the oligomannose type, containing *N*-acetyl-D-glucosamine and a variable number of D-mannose residues. Structures containing both oligomannose- and *N*-acetyllactosamine-type oligosaccharides are designated as hybrid type. Examples of *N*-glycoproteins (or *N*-glycosylproteins) are chicken ovalbumin, pig ribonuclease, human α_1-acid glycoprotein, and soybean agglutinin.

The locant *O*- is used for *O*-glycosyl linkage to serine, threonine, hydroxylysine, or hydroxyproline. Sheep submaxillary glycoprotein, collagen, fish antifreeze glycoproteins, and potato lectin are *O*-glycoproteins (or *O*-glycosylproteins).

Two types of carbohydrate–peptide linkage in the same protein or peptide chain may be indicated by a combination of the locants. Thus, calf fetuin, procollagen, human erythrocyte membrane glycophorin, and human chorionic gonadotropin are *N*-,*O*-glycoproteins (or *N*-,*O*-glycosylproteins).

I. Asialoglycoproteins and Asialo-agalactoglycoproteins

Glycoproteins from which the sialic acid has been removed (by treatment with enzymes or mild acid) are designated by the prefix *asialo* (e.g., asialo-α_1-acid glycoprotein and asialofetuin). Removal of both sialic acid and galactose results in asialo-agalactoglycoproteins.

J. Condensed Representation of Sugar Chains

For writing the structure of sugar chains, the nonreducing terminus of the carbohydrate chain should always be on the left-hand end. The following simplification might be suitable. Current practice allows the use of either an extended form or a condensed form, which allows structures to be shown in one line as well as in two or more, and in which the longest chain should always be the main chain.

The condensed form is still unnecessarily long, however, and there should be no serious loss if it is shortened further by (1) omitting locants of anomeric carbon atoms,

(2) omitting the parentheses around the specification of linkage, and (3) omitting hyphens if desired. We, therefore, suggest a more condensed or short form of writing:

A glycopeptide sequence, represented in the condensed form in two lines as:

```
Gal(β1-3)GalNAc(α1-)┐
                    -Ala-Thr-Ala-
```

or in the condensed form in one line as:

-Ala-[Gal(β1-3)GalNAc(α1-)]Thr-Ala-

may be written in the short form in two lines as:

```
Galβ3GalNAcα┐
            -Ala-Thr-Ala-
```

or in the short form in one line as:

-Ala-(Galβ3GalNAcα)Thr-Ala-

K. Representation of *N*-Linked Oligosaccharides

As a rule, *N*-linked oligosaccharides contain a common pentasaccharide core as follows:

```
Manα6
     \
      Manβ4GlcNAcβ4GlcNAc
     /
Manα3
```

For the sake of uniformity, the location of substitution should be written as in the foregoing in accordance with Haworth's representation of the pyranose structure of monosaccharides, in analogy with the glycogen molecule.

L. Peptidoglycans

In peptidoglycans, peptide units of adjacent polysaccharides (glycosaminoglycans) may be cross-linked by a peptide bond between the COOH-terminal alanine residue of one peptide subunit and the ϖ-amino group of the diamino acid residue of the other (e.g., L-lysine or *meso*-diaminopimelic acid), thereby giving rise to a giant macromolecule forming the rigid cell wall ("sacculus"). This macromolecule is known to occur as a monomolecular layer between the inner and outer membrane in gram-negative bacteria and as a multimolecular layer, often associated covalently or noncovalently with various additional compounds (teichoic acid, neutral polysaccharides, and such) in gram-positive bacteria.

Extensive investigations of peptidoglycans in thousands of bacterial strains demonstrated the existence of more than 100 chemotypes in the eubacteria. The peptidoglycans of eubacteria have been classified into two major groups (A and B) and several subgroups, according to the mode of cross-linkage. Within group A, two main subgroups are recognized: one in which the COOH-terminal alanine residue is directly bound to the ϖ-amino group of the diamino acid at position 3 of the peptide subunit of an adjacent polysaccharide (glycosaminoglycan), and one in which the cross-linkage is mediated by an interpeptide bridge consisting of either one or up to five amino acid residues (e.g., Lys-Gly_5-type). Depending on the kind of diamino acid at position 3 and the amino acids serving as interpeptide bridges, many variations of these subgroups of murein are known. In the peptidoglycan types of group B, the cross-linkage does not occur at position 3 of the peptide subunit, but at position 2, utilizing the α-carboxyl group of the D-glutamic residue. The interpeptide bridge must contain a diamino acid, which may be lysine, ornithine, or diaminobutyric acid, in either the L- or D- configuration. As a second special characteristic of the group B peptidoglycan types, the L-alanine residue at position 1 of the peptide subunit is replaced by either glycine or serine.

In the archaebacteria, several organisms contain a peptidoglycan that differs in certain respects from those described in the foregoing, typical for the eubacteria. It consists of a polysaccharide formed by alternating (β1-3)-linked *N*-acetylated residues of D-glucosamine or D-galactosamine and (β1-3)-linked *N*-acetylated residues of L-talosaminuronic acid, and a peptide containing exclusively L-amino acids attached to the carboxyl group of L-talosaminuronic acid. The peptide units of adjacent polysaccharides may be cross-linked by a peptide bond between the γ-carboxyl group of the glutamic acid of one peptide subunit and the α-amino group of the lysine residue of the other, thus giving rise to a huge macromolecule that forms the rigid cell wall (or sacculus) of the methanogenic bacteria. This multimolecular layer is often associated covalently, but also noncovalently, with neutral polysaccharides. Only a few chemotypes of such peptidoglycans have so far been described.

M. Implied Configurations and Ring Sizes

In the condensed system of symbols for sugar residues, the common configuration and ring size (usually pyranose) are implied in the symbol. Thus, Gal denotes D-galactopyranose; Man, D-mannopyranose; Fuc, L-fucopyranose; GlcNAc, 2-acetamido-2-deoxy-D-glucopyranose or D-glucosamine; Neu5Ac (which may be abbreviated to NeuAc), *N*-acetylneuraminic acid. The symbol Sia stands for sialic acid, a general term that can also be used when the exact structure is known.

Whenever the configuration of ring size is found to differ from the common one, it must be indicated by using the appropriate symbols for the extended system. The configuration of amino acids is L, unless otherwise noted. Although symbols such as Gal and Man are useful in representing oligosaccharide structures they should not be used in the text to represent monosaccharides.

NOTES

1. *Lowest locants* are defined as follows: when a series of locants containing the same number of terms are compared term by term, that series is lowest that contains the lowest number on the occasion of the first difference.
2. IUPAC Nomenclature of Organic Chemistry, Section C.
3. It should be noted that the appropriate affix is *thio*, not *thia*; the latter is used in systematic organic chemical nomenclature to indicate replacement of CH_2 by S.
4. This compound could be more strictly described as 5(α-D-galactopyranosyloxy)-L-lysine, but hydroxylysine is regarded as a trivial name, so names are based on it.
5. Hydroxyproline is the trivial name for *trans*-4-hydroxy-L-proline.

3

Methods of Carbohydrate Analysis and Structural Determination: Chemical and Enzymatic Methods

Akira Kobata and Kiyoshi Furukawa *The Institute of Medical Science, The University of Tokyo, Tokyo, Japan*

Studies on glycoproteins were evoked by the advancement of the chemical knowledge of human blood group antigens in the 1960s. Later, development of cell biology revealed that the carbohydrate moieties of glycoconjugates play key roles in cellular recognition. This finding focused the attention of many biologists on the field of glycoprotein research. In the past 10 years, genetic engineering has reached its "golden age." The target of this new technique is now expanding from simple proteins to glycoproteins, because many bioactive proteins contain carbohydrates. Development of a variety of vector–host systems for the transfection of cloned human DNAs has enabled us to handle even rare biological glycoproteins on a large scale. However, the proteins produced by using bacteria as hosts lack the sugar moieties. Proteins produced by yeast and animal cells contain sugars, but there is no guarantee that the glycoproteins have the same set of carbohydrate chains as their natural counterparts of human origin. Some of the recombinant DNA products have already been found to have modified or reduced biological activities, compared with natural glycoproteins. Hence, the importance of carbohydrate chains has been increasingly recognized year by year, as is evident from the recent issues of *Annual Review of Biochemistry* [1–9].

In response to the new direction of glycoprotein research, more sensitive analytical methods have been required, because the conventional methods used in the past required large amounts of samples.

In this chapter, a series of sensitive methods that meet the current demands of biological and gene-cloning studies will be reviewed, to assist readers in the application of such to their own research projects.

I. ANALYSIS OF CARBOHYDRATE COMPOSITION

Analytical data on the composition of monosaccharides are necessary to develop a strategy for the structural study of glycoproteins' carbohydrate chains. So far, the major carbohydrate chains of glycoproteins can be classified into two groups according to their sugar–peptide linkage regions (Fig. 1). Those linked to an asparagine (Asn) residue of the polypeptide portion are called Asn-linked sugar chains. Others that are linked to either serine or threonine residue, are called mucin-type sugar chains. Since the mucin-type sugar chains do not contain a mannose (Man) residue, which is always found in the Asn-linked sugar chains, the distribution of the two sugar chain groups in a glycoprotein can be presumed by monosaccharide analysis. This information is important because the structural studies of the two groups of polysaccharide chains are performed with different procedures, as will be described later. The data on monosaccharide composition are also essential as the basis for the structural analysis of an oligosaccharide.

Because of the importance, development of a sensitive method to determine the monosaccharide composition of polysaccharide chains has been a target of eager study. Especially, the introduction of gas chromatography [10] and liquid chromatography [11] has substantially decreased the amount of samples needed for the analysis of monosaccharide composition.

The susceptibility of each glycosidic linkage to acid hydrolysis is quite different for various carbohydrate residues. Sialic acid linkages are the most labile, and almost all sialic acid residues can be released by heating the sugar chains in either 0.01 N HCl, 0.01 N H_2SO_4, or 1 N acetic acid at 100°C for 10 min. Fucosyl residues are released as fucose by heating in 0.01 N HCl or 1 N acetic acid at 100°C for 40 min. Most hexoses are quantitatively released as monosaccharides by heating a polysaccharide chain in 1 or 2 N HCl at 100°C for 2 hr. In contrast, *N*-acetylhexosaminyl residues can be released quantitatively as monosaccharides by heating a sugar chain in 2–4 N HCl for 3–4 hr. Therefore, it is impossible to obtain data on monosaccharide composition by one-point acid hydrolysis. Reliable data can be obtained only by performing a time-course study of monosaccharide release by acid hydrolysis and then extrapolating the data of hydrolysis to zero hours to compensate for the degradation of each sugar in acidic solution. It is also recommended that

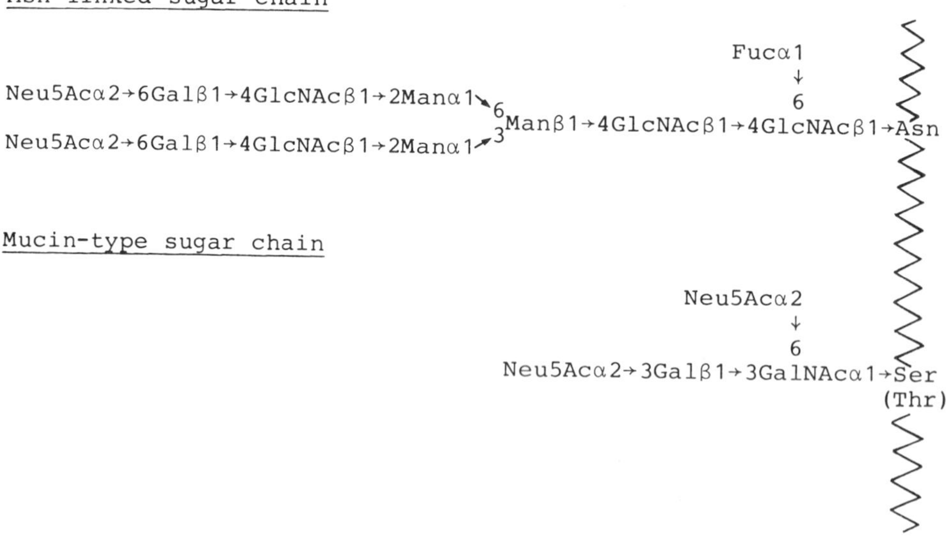

Figure 1 Two major groups of sugar chains found in glycoproteins.

the number of sialic acids and fucosyl residues be determined independently, because they are extensively degraded under conditions that hydrolyze hexosyl and *N*-acetylhexosaminyl linkages.

All monosaccharide residues, obtained by acid hydrolysis of sugar chains, contain an aldehyde group. This aldehyde group can be quantitatively converted to a tritium-labeled carbinol residue by the following reaction.

$$-\mathrm{CHO} \xrightarrow{\mathrm{NaB^3H_4}} -\underset{\substack{|\\ ^3\mathrm{H}}}{\mathrm{CHOH}}$$

Since the radioactivity incorporated into each monosaccharide is proportional to its molar amount, data on monosaccharide composition can be obtained by measuring the radioactivity, if the monosaccharide alcohols are successfully separated. We found that high-voltage paper electrophoresis, using a borate buffer, pH 9.0, is suitable for separating the reduction products of the six monosaccharides: glucose, galactose, mannose, fucose, glucosamine, and galactosamine, which are detected in glycoproteins. With use of this technique, a very sensitive and simple method for analysis of the monosaccharide composition of sugar chains was developed [12]. A time-course study of sugar chain hydrolysis can be easily performed by spotting, side by side, the samples hydrolyzed for different lengths of time and subjecting them to paper electrophoresis.

Although inexpensive, the method is inconvenient for the analysis of numerous glycoprotein samples because it contains many processes, including the separation of labeled sugar alcohols from radioactive contaminants after reduction of the acid hydrolysates of sugar chains with NaB^3H_4, the detection of their radioactivities by scanning the papers after electrophoresis, and the extraction and precise determination of their radioactivity. To overcome this problem, a high-performance liquid chromatographic (HPLC) method for fractionating all the sugar alcohols derived from monosaccharides found in glycoproteins and glycolipids was developed [13]. As summarized in Table 1, many additional sugar alcohols can be separated by the Shodex Sugar SP-1010 column, equipped with a SP-1010 precolumn (Showa Denko Co., Tokyo). Therefore, the method can be expanded to analyze the sugar compositions of glycoproteins in bacteria and plants as well as animals. Not only does this save time, but it reduces the chances for introduction of experimental errors in this novel method because the procedures are simpler than those of paper electrophoresis. The method has already been applied successfully for the analysis of carbohydrate composition of several glycoproteins obtained by recombinant DNA methods [14].

II. RELEASE OF OLIGOSACCHARIDES FROM POLYPEPTIDE BACKBONE

Many glycoproteins contain more than one oligosaccharide chain in one molecule. In many cases, these multiple oligosaccharide chains have different structures. Even in the case of a glycoprotein with only one sugar chain. The microheterogeneity of carbohydrate chain structure has to be considered. Therefore, each oligosaccharide chain must be separated for its structural elucidation.

When heated in an alkaline solution, the mucin-type carbohydrate chains are released as oligosaccharides by β-elimination [15]. The released oligosaccharides are then cleaved stepwise from their reducing termini by a peeling reaction [16]. Since the peeling reaction

Table 1 Retention Times of Alditols and *N*-Acetylamino Alditols on a Shodex Sugar SP-1010 Column

Group	Sugar alcohols	Retention times
Hexitols	Allitol	37.40
	Mannitol	46.38
	Talitol	50.22
	Galactitol	61.56
	Sorbitol	67.62
	Iditol	83.94
Pentitols	Ribitol	35.10
	Arabinitol	44.40
	Xylitol	55.40
6-Deoxyhexitols	Rhamnitol	46.00
	Fucitol	53.94
	Quinobitol	63.90
N-Acetylaminohexitols	*N*-Acetylmannosaminitol	33.66
	N-Acetylgalactosaminitol	33.90
	N-Acetylglucosaminitol	39.00

proceeds rapidly, it is impossible to release mucin-type carbohydrate chains quantitatively as intact oligosaccharides. Addition of $NaBH_4$ in the reaction mixture will convert the oligosaccharides to the corresponding sugar alcohols as soon as they are released from the polypeptide. Because the oligosaccharide alcohols are alkaline-stable, the mucin-type carbohydrate chains can be quantitatively converted to oligosaccharide alcohols by warming glycoproteins in alkaline borohydride solution [17]. This method has been widely used for the structural study of the mucin-type sugar chains of many glycoproteins.

Because of the lack of suitable method to release the Asn-linked sugar chains from the polypeptide backbone, a tedious and time-consuming procedure was used as a general method for the separation of each of them. First, a sample glycoprotein was exhaustively digested with pronase, and the resulting Asn-oligosaccharides were fractionated by an appropriate method, such as ion-exchange column chromatography. By using each glycopeptide as a probe, the structure of the sugar moiety was investigated by ordinary analytical techniques. With this method, Sheares and his colleagues have recently determined the structure of the carbohydrate chain of ovalbumin, the gene of which was transfected and expressed in cells other than ovary [18]. Through this experiment, they could estimate the factors that control the carbohydrate structures of glycoproteins. For comparative studies of the carbohydrate structures of a well-known glycoprotein, this method might still be of use. However, it must be emphasized that ovalbumin is rather a rare case from which most of the sugar chains can be obtained as Asn-oligosaccharides [19]. In many other cases, sugar chains were still bound to peptides of various sizes [20]. Therefore, structural heterogeneities in both the peptide portion and sugar moiety make it hard to fractionate each oligosaccharide component. Even in the case of an Asn-oligosaccharide mixture, many fractions, which were considered to be a single component, were found to be mixtures of several Asn-oligosaccharides with the same number of monosaccharide units [21]. Because of the problems inherent in the exhaustive pronase digestion method, development of enzymatic and chemical methods to release Asn-linked sugar chains of glycoprotein as intact oligosaccharides has been a target of keen investigation.

A. Endo-β-*N*-acetylglucosaminidase Digestion

Introduction of endo-β-*N*-acetylglucosaminidases has opened a new era in the history of glycoprotein research. These enzymes hydrolytically cleave the *N,N'*-diacetylchitobiose moiety of Asn-linked sugar chains and release oligosaccharides by the following reaction:

R→GlcNAcβ1→4GlcNAc→Asn-peptide ➡ R→GlcNAc + GlcNAc→Asn-peptide
(R = sugar chain)

Endo-β-*N*-acetylglucosaminidases were found in animal organs, plants, as well as microorganisms. Among them, endo D, H, C_{II}, and F were satisfactorily purified and have been used for the structural study of the Asn-linked sugar chains of glycoproteins (Table 2).

Table 2 Structural Requirements of Various Endo-β-*N*-acetylglucosaminidases (Endo) for Their Substrates

Enzymes	Structures required for substrates[a]	Source
Endo D	R ↓ 6 R→4Manα1→3Manβ1→4GlcNAcβ1¦→4R_1	*Streptococcus pneumoniae*
Endo H	R R ↓ ↓ 2 6 Manα1→3Manα1↘ 6 R→4Manβ1→4GlcNAcβ1¦→4R_1 3 R↗	*Streptomyces griseus* and *S. plicatus*
Endo C_{II}	R→2Manα1→3Manα1↘ 6 R→4Manβ1→4GlcNAcβ1¦→4R_2 3 R→2Manα1↗	*Clostridium perfringens*
Endo F	The structural requirement is not clear, but it seems to cleave almost all Asn-linked sugar chains at their *N,N'*-diacetylchitobiose moieties.	*Flavobacterium meningosepticum*

R, sugar or hydrogen; R_1, (±Fucα1→6)GlcNAc→Asn(peptides), (±Fucα1→6)GlcNAc, or (±Fucα1→6) $GlcNAc_{OH}$; R_2, GlcNAc→Asn, GlcNAc, or $GlcNAc_{OH}$.
[a]Dashed lines indicate where sugar chains are cleaved by the enzyme.

1. Endo D

The enzyme, secreted into the culture medium of *Streptococcus pneumoniae*, was purified by ammonium sulfate precipitation, followed by two-step column chromatography on Sephadex G-50 and DEAE-Sephadex A-25 [22] or DEAE-cellulose [23]. The properties of this enzyme were almost completely characterized. The basic glycon structure required by the enzyme is the trisaccharide, Manα1→3Manβ1→4GlcNAcβ1→ [24, 25] (see Table 2). The most important part is the α-mannosyl residue of the trisaccharide. Substitution at the C-4 position of this mannose residue with other sugars is not harmful [26], but at the C-2 position, substitution will inhibit the action of endo D [24, 25]. The aglycon specificity of this enzyme is rather loose, such that R_1 in the table can be either the (±Fucα1→6)GlcNAc groups, their peptide derivatives or their reduced forms. Therefore, the enzyme releases Manα1→6(Manα1→3)Manβ1→4GlcNAc from the complex-type Asn-linked sugar chains after their outer chains are removed by the concerned action of sialidase, β-galactosidase, and β-*N*-acetylhexosaminidase or Manα1→6(Manα1→3)Manα1→6(Manα1→3) Manβ1→4GlcNAc from high mannose-type sugar chains after removal of Manα1→2 residues by *Aspergillus saitoi* α-mannosidase I [24]. The enzyme can also be used to discriminate the isomeric monosialyl biantennary sugar chains shown in Figure 2. After incubation with a mixture of jack bean β-galactosidase and β-*N*-acetylhexosaminidase, radioactive *N*-acetylglucosaminitol is released from decasaccharide I but not from II by endo D treatment.

2. Endo H

Endo H is secreted by *Streptomyces griseus* [27] and *Streptomyces plicatus* [28]. The enzyme from the latter was purified up to 3700-fold by precipitation of filtered culture medium with zinc acetate and then with ammonium sulfate, followed by sequential chromatography on columns of DE-52-cellulose, SP-Sephadex C-25, and Sephadex G-75 [28]. The basic glycon structure required by the enzyme is the tetrasaccharide: Manα1→3Manα1→6Manβ1→4GlcNAcβ1→ [29] (see Table 2). The nonreducing terminal α-mannose residue can be substituted with other sugars at its C-2 position. Therefore, the enzyme can hydrolyze all high mannose-type sugar chains and most of hybrid-type sugar chains. Because the complex-type sugar chains do not contain the Manα1→3 residue linked to the Manα1→6 arm, they cannot be the substrate of endo H, even after their outer chain moieties are removed.

The aglycon specificity of this enzyme is similar to that of endo D. It hydrolyzes the sugars in which R_1 in Table 2 is either GlcNAc, its peptide derivative, or its reduced form [30]. Hughes et al. reported that an unusual hybrid-type sugar chain, in which the proximal

```
      Neu5Acα2→6Galβ1→4GlcNAcβ1→2Manα1↘6
I.                                       Manβ1→4GlcNAcβ1→4GlcNAcOT
                Galβ1→4GlcNAcβ1→2Manα1↗3

                Galβ1→4GlcNAcβ1→2Manα1↘6
II.                                      Manβ1→4GlcNAcβ1→4GlcNAcOT
      Neu5Acα2→6Galβ1→4GlcNAcβ1→2Manα1↗3
```

Figure 2 Two isomeric monosialyl complex-type sugar chains.

N-acetylglucosamine residue is fucosylated at its C-6 position, can also be cleaved by endo H [31]. Therefore, fucosylation will not inhibit the action of endo H as estimated before.

3. *Endo C_{II}*

Two endo-β-*N*-acetylglucosaminidases (C_I and C_{II}) were found in the culture medium of *Clostridium perfringens* [32]. The purified endo C_I was found to have the same property as endo D. In contrast, endo C_{II} has a unique specificity that is different from endo D and endo H [29–33]. The minimum requirement of this enzyme is similar to endo H, but is more complicated. It requires the pentasaccharide Manα1→3Manα1→6(Manα1→3)Manβ1→4GlcNAcβ1→ as glycon (see Table 2). The two α-mannosyl residues located at the nonreducing termini can be substituted with other sugars at their C-2 positions. However, substitution at the C-4 position of the α-mannosyl residue of the Manα1→3Manβ1→ group will inhibit the action of endo C_{II} [29]. The subtle difference in the substrate specificities of endo H and endo C_{II} were ingeniously used by Kornfeld and his colleagues to elucidate the processing pathway of biosynthesis of the Asn-linked sugar chains [34].

4. *Endo F*

Endo F was initially found as a contaminant of the commercial carboxypeptidase A. Further study revealed that the enzyme was produced by the contaminating bacteria, *Flavobacterium meningosepticum* [35]. A partially purified enzyme preparation released the complex-type sugar chains from virus coat glycoproteins and the human major histocompatibility antigens HLA-A and HLA-B [35]. Further studies revealed that sugar chains can also be released from α_1-acid glycoprotein and ovalbumin, which were pretreated by reductive alkylation or with mercaptoethanol. Therefore, the enzyme was considered to have a broad substrate specificity, acting not only on complex-type, but also on hybrid-type and high mannose-type sugar chains [35]. However, it has been more recently found that the enzyme preparation is also contaminated with another endoglycosidase called peptide:*N*-glycosidase F (glycopeptidase F) [36], which will be described later.

Since the pH optima of these two enzymes are quite different, it is possible to preferentially use the hydrolyzing activity of endo F at pH 4.0, and of peptide: *N*-glycosidase F at pH 9.3. Although the possibility of removing all Asn-linked sugar chains from glycoproteins by incubation with the mixture of these enzymes has been considered, it is necessary to separate them and to determine their substrate specificities for their safer use in the structural study of the sugar chains of glycoproteins.

B. Endo-α-*N*-acetylgalactosaminidase Digestion

Enzymes that catalyzes the following reaction were found in the culture media of *Clostridium perfringens* [37] and *Streptococcus pneumoniae* [38].

Galβ1→3GalNAcα1→Ser (or Thr)-peptide ➡
Galβ1→3GalNAc + Ser (or Thr)-peptide

The streptococcal enzyme was purified and characterized [38–39]. Its glycon specificity is very strict and releases only the Galβ1→3GalNAc from various mucin-type sugar chains (Table 3). Both serine- and threonine-linked disaccharide can be released by the action of this enzyme. Therefore, the enzyme in concerted action with sialidase can be used to remove the sialylated Galβ1→3GalNAc, which occurs widely in glycoproteins.

Table 3 Action of Streptococcal Endo-α-*N*-acetylgalactosaminidase on Various Mucin-Type Sugar Chains

Mucin-type glycopeptide[a]	Enzyme action[b]
GalNAcα1→R	–
Galβ1→3GalNAcα1→R	+
Fucα1→2Galβ1→3GalNAcα1→R	–
Neu5Acα2 ↓ 6 Galβ1→3GalNAcα1→R	–
Fucα1→2Galβ1→3GalNAcα1→R 6 ↑ Neu5Acα2	–

[a]R, Ser- or Thr-peptides.
[b]+, hydrolyzable; –, not hydrolyzable.

C. Endo-β-galactosidase Digestion

Enzymes that cleave the β-galactosyl linkages within sugar chains have been found in culture media of various bacteria. The enzyme purified from *S. pneumoniae* cleaves the β-galactosidic linkage shown in Figure 3 and releases oligosaccharides [40]. Therefore, the enzyme releases GalNAcα1→3(Fucα1→2)Gal and Galα1→3 (Fucα1→2)Gal from type II but not type I A and B blood group determinants, respectively. The enzyme also acts on A and B blood group substances.

The enzyme purified from *Escherichia freundii* was initially found to act on keratan sulfate [41]. Later, this enzyme was found to cleave the β-galactosyl linkage within poly-*N*-acetyllactosamine residues included in glycoproteins and glycolipids as follows [42]:

R-GlcNAcβ1→3Galβ1→4R′ ➡ R-GlcNAcβ1→3Gal + R′

The glycon (R) and aglycon (R′) specificities of the enzyme have been well studied. In Table 4, structures of oligosaccharides released from human granulocyte and erythrocyte membrane glycoproteins by digestion with the *E. freundii* enzyme are summarized [43, 44]. Judging from these data, the enzyme is considered to act on site I, but not on site II, of the following fucose-containing structure:

```
               II                        I
               ↓                         ↓
–GlcNAcβ1→3Galβ1→4GlcNAcβ1→3Galβ1→4GlcNAc–
                         3
                         ↑
                       Fucα1
```

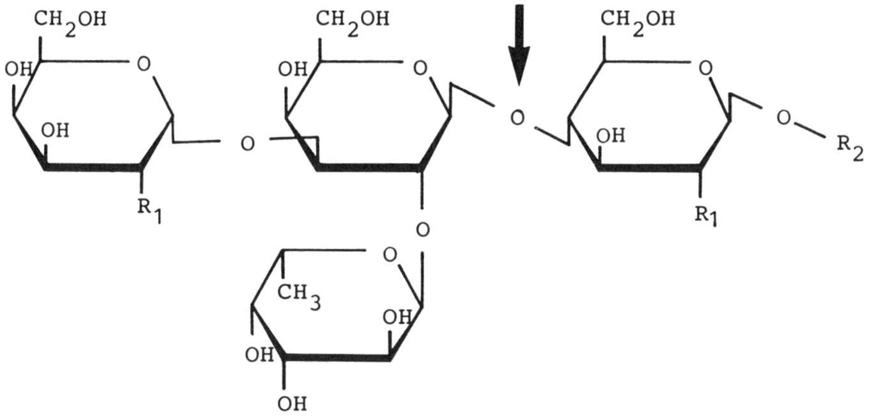

Figure 3 General structure susceptible to the action of streptococcal endo-β-galactosidase. Arrow indicates the linkage hydrolytically cleaved by the enzyme. R_1 represents either OH or $NHCOCH_3$ group, and R_2 represents either hydrogen or sugar chains.

Furthermore, the enzyme can cleave the β-galactosidic linkage, the C-3 and C-6 positions of which are substituted with *N*-acetylglucosamine residues. Interestingly, a similarly branched structure in glycolipid could not be cleaved by the enzyme [45]. Therefore, one must be careful in using the enzyme for the structural study of the sugar chains of complex carbohydrates. A similar enzyme has also been purified from *Flavobacterium keratolyticus* [46] and *Bacteroides fragilis* [47].

Quite recently, another endo-β-galactosidase was found to occur in the culture fluid of *Clostridium perfringens* [48]. This enzyme specifically catalyzes the following reaction:

$$\text{Gal}\alpha 1\rightarrow 3\text{Gal}\beta 1\rightarrow 4\text{GlcNAc}\rightarrow \text{R} \Rightarrow \text{Gal}\alpha 1\rightarrow 3\text{Gal} + \text{GlcNAc}\rightarrow \text{R}$$

The enzyme might be effectively used to detect the Galα1→3Gal residue in the outer chain moieties of Asn-linked sugar chains as well as in mucin-type sugar chains.

D. Peptide: *N*-glycosidase (Glycopeptidase) Digestion

An enzyme that belongs to this group was found, for the first time, in almond emulsion by Takahashi and her co-workers and was named glycopeptidase [49]. The enzyme was unique because it acts on the β-aspartylglycosylamine linkage and releases oligosaccharide-glycosylamine and peptide. The glycosylamine was then spontaneously cleaved to an oligosaccharide and ammonia (Fig. 4). Plummer and Tarentino confirmed the presence of this enzyme [50], and also found another enzyme with similar specificity in the preparation of endo F from *Flavobacterium meningosepticum*. They named this enzyme peptide:*N*-glycosidases [36].

1. *Glycopeptidase*

The enzyme was purified from commercial preparation of β-glucosidase from almond emulsin (Sigma Chemical Co., St. Louis, Missouri) by sequential chromatography on columns of Sephadex G-200 and DE-52-cellulose, followed by isoelectric focusing [51], or

Table 4 Structures of Oligosaccharides Released by Endo-β-galactosidase (*E. freundii*)

Structures
GlcNAcβ1→3Gal
Galβ1→4GlcNAcβ1→3Gal
Fucα1→2Galβ1→4GlcNAcβ1→3Gal
Galβ1→4GlcNAcβ1→3Gal 3 ↑ Fucα1
GlcNAcβ1→3Galβ1→4GlcNAcβ1→3Gal 3 ↑ Fucα1
Galβ1→4GlcNAcβ1→3Galβ1→4GlcNAcβ1→3Gal 3 ↑ Fucα1
Galβ1→4GlcNAcβ1→3Galβ1→4GlcNAcβ1→3Gal 3 3 ↑ ↑ Fucα1 Fucα1
Neu5Acα2→6Galβ1→4GlcNAcβ1→3Gal
Neu5Acα2→3Galβ1→4GlcNAcβ1→3Gal 3 ↑ Fucα1
Neu5Acα2→3Galβ1→4GlcNAcβ1→3Galβ1→4GlcNAcβ1→3Gal 3 ↑ Fucα1
Galβ1→4GlcNAcβ1↘ 6 Gal 3 GlcNAcβ1↗
Galβ1→4GlcNAcβ1↘ 6 Gal 3 Galβ1→4GlcNAcβ1↗

Table 4 (Continued)

```
                    Galβ1→4GlcNAcβ1↘
                                    6
                                     Gal
                                    3
          Fucα1→2Galβ1→4GlcNAcβ1↗

                    Galβ1→4GlcNAcβ1↘
                                    6
                                     Gal
                                    3
      GlcNAcβ1→3Galβ1→4GlcNAcβ1↗

                    Galβ1→4GlcNAcβ1↘
                                    6
                                     Gal
                                    3
Galβ1→4GlcNAcβ1→3Galβ1→4GlcNAcβ1↗

Galβ1→4GlcNAcβ1↘
                6
                 Galβ1→4GlcNAcβ1→3Gal
                3
Galβ1→4GlcNAcβ1↗

                         Galβ1→4GlcNAcβ1↘
                                         6
                                          Gal
                                         3
Fucα1→2Galβ1→4GlcNAcβ1→3Galβ1→4GlcNAcβ1↗

                 Fucα1→2Galβ1→4GlcNAcβ1↘
                                         6
                                          Gal
                                         3
Fucα1→2Galβ1→4GlcNAcβ1→3Galβ1→4GlcNAcβ1↗
```

Source: Refs. 43, 44.

on chromatography columns of DE-52-cellulose, CM-52-cellulose, and Sephacryl S-200 [52]. By the former procedure, the enzyme has been separated into three isozymes [51]. The enzymes have broad substrate specificity and can release all types of Asn-linked sugar chains after removal of sialic acid residues. However, the degree of hydrolysis to release the sugar chains from glycoproteins depends on the size and conformation of the peptide portion and also on how crowded the sugar chains are on a polypeptide chain. Takahashi and Nishibe [53] showed that the enzymes act on glycopeptides containing more than three amino acids and do not act on those that contain one or two amino acids (Table 5). It was also reported that release of oligosaccharides from ovalbumin, stem bromelain, desialylated transferrin, and fibrinogen was incomplete [51, 54]. To avoid such inconvenience, Tarentino and Plummer [52] pretreated glycoproteins with chaotropic reagents, such as

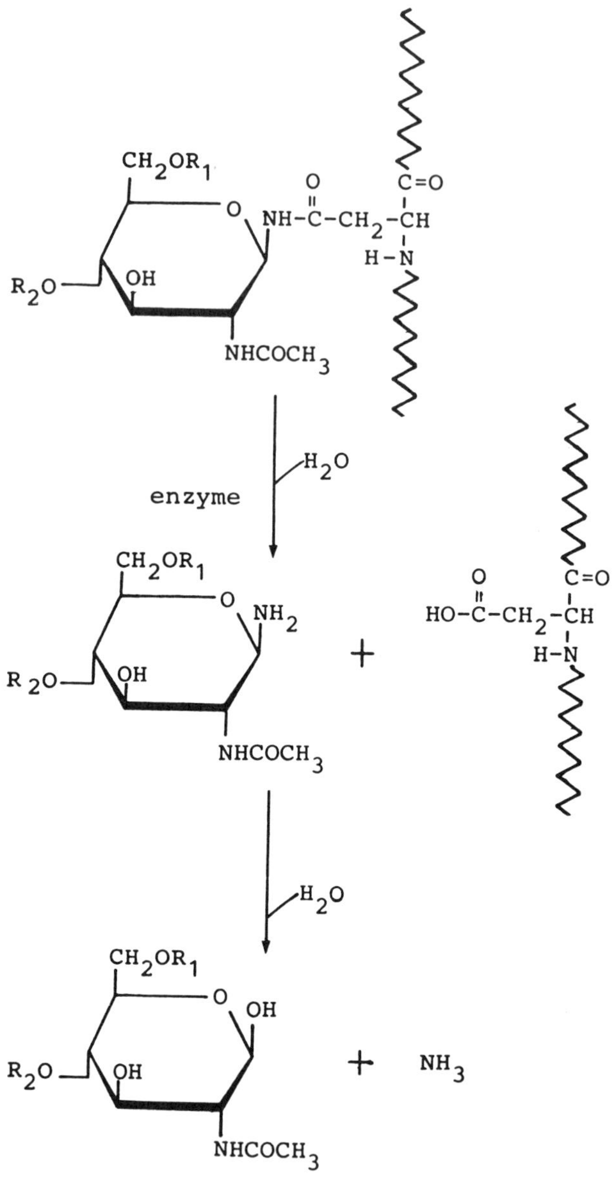

Figure 4 Action mechanism of peptide-N^4-(N-acetyl-β-glucosaminyl)asparagineamidase. R_1 represents either hydrogen or fucose residue and R_2 represents sugar chains.

NaSCN and $NaClO_4$ and treated them with the enzyme in the presence of β-mercapto-ethanol. Under such condition, sugar chains can be quantitatively released as oligosaccharides from ribonuclease B and the μ-chain Fab fragment. It is noteworthy that the purity of the enzyme varies from lot to lot of the commercial β-glucosidase preparation, which is the starting material for purifying the enzyme. Especially, contamination with endo-β-N-acetylglucosaminidase is critical.

Table 5 Action of Glycopeptidase on Various Glycopeptides

Glycopeptides	Rate of hydrolysis[a]
$Man_{5\sim6}GlcNAc_2$-Asn	0
Man_3·$HexNAc_{5\sim6}$-Asn-Arg	0
Man_3·Xyl·Fuc·$GlcNAc_2$-Asn-Glu-Ser	100
Man_3·Xyl·Fuc·$GlcNAc_2$-Asn-Asn-Glu-Ser-Ser	95
Man_3·Xyl·Fuc·$GlcNAc_2$-Asn-Asn-Glu-Ser-Ser-Met	84
Man_3·Xyl·Fuc·$GlcNAc_2$-Ala-Arg-Val-Pro-Arg-Asn-Asn-Glu-Ser-Ser-Met	72

[a]The numbers indicate the relative hydrolysis rate by taking the value of Man_3·Xyl·Fuc·$GlcNAc_2$-Asn-Glu-Ser as 100.
Source: Ref. 53.

2. *Peptide-N^4-(N-acetyl-β-glucosaminyl)asparagine amidase*

As described before, this enzyme, originally called peptide:*N*-glycosidase and now officially named peptide-N^4-(*N*-acetyl-β-glucosaminyl)asparagine amidase (EC 3.5.1.52), was found in the preparation of endo-β-*N*-acetylglucosaminidase F that was synthesized and secreted by *F. meningosepticum*. Although the enzyme was purified by ammonium sulfate fractionation and HPLC with use of a Mono S column [36], it was still contaminated with endo F. Recently, an enzyme preparation freed from endo F has been sold by Genzyme Co. (Boston, Massachusetts). Since the pH optima of the two enzymes are quite different, it is possible to inhibit the action of endo F by performing the enzyme digestion at pH 9.3. Effective release of oligosaccharides with *N,N′*-diacetylchitobiose at their reducing termini was achieved from human IgM glycopeptides, Rauscher virus GP70, and Sindbis virus glycoprotein, at pH 9.3 [36]. Other exoglycosidase activities were not found in the enzyme preparation, but it was found to be contaminated with protease. Addition of EDTA (50 mM) to the reaction mixture prohibited the protease activity [36]. With the characteristics so far described, the enzyme preparation can be effectively used to study the functional role of the Asn-linked sugar chains of glycoproteins as well as their structures. The enzyme has been used to estimate the numbers of Asn-linked sugar chains in glycoproteins obtained by recombinant DNA technique by chasing the change in molecular weight after various incubation times [35].

E. Chemical Methods to Release Sugar Chains from Glycoproteins

The enzymatic method to release sugar chains of glycoproteins as oligosaccharides is mild and sugar-free proteins can occasionally be recovered. However, the most critical point of these methods is the accessibility of the enzymes to the sugar chains of glycoproteins. Most chemical methods cannot save the protein part intact. Therefore, they cannot be used for studies of the functional role of sugar chains in glycoproteins. However, sugar chains of glycoproteins can be quantitatively recovered as oligosaccharides by these methods.

Therefore, for the structural study of the sugar chains of glycoproteins, these methods are considered more suitable.

1. *Hydrazinolysis of Asparagine-Linked Sugar Chains*

Hydrazinolysis was initially used for determining the COOH-terminal amino acid of proteins. Introduction of this method for the study of carbohydrate portions of glycoproteins was made for the first time by Matsushima and Fujii in 1950 [55]. Later, this method was further developed by Bayard and Montreuil to cleave the GlcNAc-Asn linkage and *N*-acetylglucosaminyl linkages [56]. However, the developed method was insufficient for the quantitative release of Asn-linked sugar chains as oligosaccharides. Mizuochi et al. [57] succeeded in applying the method for the structural study of the sugar chains of C1q, a subunit of the first component of complement. They improved the method by applying it to studies of the sugar chains of many glycoproteins and established it as a reliable quantitative method to release the Asn-linked sugar chains of glycoproteins as oligosaccharides [58].

The principle of hydrazinolysis is that the glycosylamine linkage is cleaved by heating in anhydrous hydrazine at 100°C for 10 hr, whereas the glycosidic linkages remain untouched. Therefore, all Asn-linked sugar chains are released from polypeptide portion as intact oligosaccharides. The released oligosaccharides remain linked to hydrazine, even after addition of water to the reaction mixture, because the glycosylamine linkage is stabilized in water by the electronegative effect of the neighboring amino group. Furthermore, all *N*-acetylaminosugar residues including *N*-acetylneuraminic acid residues are de-*N*-acetylated by hydrazinolysis. These modified oligosaccharide derivatives can be recovered as the original oligosaccharides by exhaustive *N*-acetylation, because the *N*-acetylated hydrazine derivatives can be easily cleaved in water as follow.

$$\text{–CH=N–NH}_2 \xrightarrow{N\text{-acetylation}} \text{–CH=N–NHAc} \xrightarrow{\text{H}_2\text{O}} \text{–CH=O} + \text{H}_2\text{NNHAc}$$

The glycolyl group of *N*-glycolylneuraminic acid is more resistant to hydrazinolysis than the acetyl group of *N*-acetylneuraminic acid. However, it is impossible to avoid de-*N*-glycolylation. Therefore, identification of sialic acid residues must be performed on intact glycoproteins.

2. *Trifluoroacetolysis*

The method of trifluoroacetolysis was developed by Nilsson and Svensson [59]. When glycoproteins are heated with a mixture of trifluoroacetate (TFA) and anhydrous trifluoroacetic acid (TFAA) (1:100 v/v), all hydroxyl groups are trifluoroacetylated. Since trifluoroacetyl groups strongly withdraw electrons, they stabilize the glycosidic linkages. Therefore, only the glycosylamine linkage is cleaved by heating in TFA-TFAA, and oligosaccharides are released as trifluoroacetate derivatives. To recover the original oligosaccharides, the trifluoroacetyl group must be removed by appropriate chemical methods. For this purpose, Nilsson and Svensson devised the following procedure [60]: The oligosaccharide derivative is de-*O*-trifluoroacetylated by heating in acetic acid. After reducing the oligosaccharide with $NaBH_4$, the *N*-trifluoroacetyl residue was removed by treating the sample in 0.1 N NaOH at room temperature for 24 hr. The oligosaccharide was then *N*-acetylated to recover the original oligosaccharide. Because of the complexity of the procedure, the method has not been as widely used as hydrazinolysis for the structural study of Asn-linked sugar chains.

3. *Alkaline-Borohydride Treatment*

As described already, mucin-type sugar chains are released from polypeptide portion by mild heating in alkaline solution. To protect oligosaccharides from the peeling reaction, a high concentration of $NaBH_4$ is added to the reaction mixture. Various conditions have been reported for releasing the mucin-type sugar chains as oligosaccharides. One of the most widely used conditions is to incubate glycoproteins in 0.5 N NaOH solution containing 1M $NaBH_4$ at 60°C for 48 hr [17]. Approximately 70–80% of the mucin-type sugar chains of most glycoproteins are recovered as oligosaccharides.

III. STRUCTURAL ANALYSIS OF OLIGOSACCHARIDES

A. Sequential Exoglycosidase Digestion

In contrast with polypeptides and polynucleotides, which are linear chains of the constructing units, sugar chains can be branched. This situation will make the sequencing of sugar chains more complicated than that of proteins and nucleic acids. The most widely used method for determining the monosaccharide sequence of sugar chains is the sequential exoglycosidase digestion. Exoglycosidases are hydrolases that remove monosaccharide residues from the nonreducing terminal of sugar chains. All of them have two kinds of specificities—the glycon specificity and aglycon specificity—toward their substrates. The glycon specificity is directed to the monosaccharide to be removed, including its anomeric configuration. Therefore, the enzymes that hydrolyze β-galactosidic linkages are called β-galactosidases. The glycon specificity directed to anomeric configuration is very high, and none of the β-galactosidases, yet reported, can hydrolyze both α- and β-glycosides is not known. The glycon specificity toward the monosaccharide moiety is usually also strict. For example, α-fucosidases and α-mannosidases can hydrolyze only α-L-fucopyranosides and α-D-mannopyranosides, respectively. However, some enzymes show wider specificities. For example, most of the enzymes that cleave β-*N*-acetylglucosaminyl linkages also hydrolyze β-*N*-acetylgalactosaminyl linkages. Because of this, they are called β-*N*-acetylhexosaminidases. The hydrolytic ratio of β-*N*-acetylglucosaminyl and β-*N*-acetylgalactosaminyl linkages varies by enzyme, indicating that such enzymes discriminate the epimeric configuration of the C-4 hydroxyl group of the two amino sugars to different extents. A similar situation is observed for β-galactosidase purified from almond emulsin. This enzyme cleaves β-glucopyranosyl linkages as well as β-galactopyranosyl linkages [61]. These enzymes, however, should be considered as extraordinary cases, and most exoglycosidases hydrolyze only a single glycosyl linkage, thereby permitting position assignment of the monosaccharide residue hydrolyzed.

The principle of sequencing monosaccharides in a sugar chain depends on the glycon specificities of exoglycosidases. For example, the Neu5Acα2→6Galβ1→4 GlcNAcβ1→ . . . group, which is widely found at the nonreducing termini of glycoprotein sugar chains releases an *N*-acetylneuraminic acid when incubated with sialidase. No other exoglycosidases can act on this sugar chain. However, after removal of the *N*-acetylneuraminic acid residue, the newly exposed β-galactosyl residue can be removed by β-galactosidase digestion. Accordingly, the trisaccharide group can be removed only by sequential digestion with sialidase, β-galactosidase, and β-*N*-acetylhexosaminidase. Many times, a sugar chain may contain two nonreducing termini because of branching, as follows:

```
Galβ1→4GlcNAcβ1↘
                6
                Galβ1→ - - - - -
                3
Galβ1→4GlcNAcβ1↗
```

There might also be a sequence, in which the same monosaccharide residues are linked sequentially, Galβ1→3Galβ1→4GlcNAcβ1→ . . . . Therefore, the amount of each monosaccharide must be determined after each exoglycosidase digestion by an appropriate method.

An oligosaccharide with a reducing terminal can be converted to tritium-labeled oligosaccharide by reduction with NaB^3H_4. Since the tritium-label is located at the reducing end of the oligosaccharide, and exoglycosidase releases monosaccharide from the other end of the sugar chain, the amount of monosaccharide released by each exoglycosidase digestion can be estimated by analyzing the change in effective size of the radiolabeled oligosaccharide by gel permeation chromatography before and after the enzymatic digestion. For this purpose, column chromatography with use of Bio-Gel P-4 (under 400 mesh), which can separate most of the sugar chains of glycoproteins, has been developed [62]. The radiolabeled degradation product can then be used for the next exoglycosidase digestion. For example, a complex-type sugar chain (A-4N) released from fetuin by hydrazinolysis was sequenced as follows [63]: After NaB^3H_4 reduction, the oligosaccharide fraction was eluted at the position equivalent to 15.8 glucose units by Bio-Gel P-4 column chromatography (Fig. 5A). When A-4N was digested with jack bean β-galactosidase, a radioactive product eluting at 12.8 glucose units (see Fig. 5B) was obtained. Since the effective size of most hexoses are approximately 1.0 glucose unit, the result indicated that three galactose residues were removed by the β-galactosidase digestion. Oligosaccharide A-4N was totally resistant to jack bean β-*N*-acetylhexosaminidase. However, the radiolabeled product in Fig. 5B was hydrolyzed by the enzyme and converted to a radiolabeled product with the mobility of 7.2 glucose units (see Fig. 4C). Since one *N*-acetylglucosamine residue behaves in many cases as approximately 2.0 glucose units, the result indicated that three *N*-acetylglucosamine residues were removed by the β-*N*-acetylhexosaminidase digestion. The radiolabeled oligosaccharide was then sequentially degraded by incubating with α-mannosidase (see Fig. 5D), β-mannosidase (see fig. 5E), and β-*N*-acetylhexosaminidase (see Fig. 5F), releasing two mannoses, one mannose, and one *N*-acetylglucosamine, respectively, and was finally converted to *N*-acetyl-[^{3}H]glucosaminitol. These results indicated that the structure of the oligosaccharide A-4N can be written as $(Gal\beta1\rightarrow GlcNAc\beta1\rightarrow)_3(Man\alpha1\rightarrow)_2Man\beta1\rightarrow GlcNAc\beta1\rightarrow GlcNAc_{OT}$.

Because the principle of sequential exoglycosidase digestion relies on the substrate specificity of each enzyme, special care must be taken to avoid contamination of the enzyme with other exoglycosidases. Most commercially available enzymes are purified by using artificial substrate such as *p*-nitrophenyl glycosides. Some of these enzymes are not useful for monosaccharide sequencing because many exoglycosidases, which cleave sugar chains but do not act on *p*-nitrophenyl glycosides, have been reported to occur. Therefore, the purity of each enzyme should be confirmed by using an appropriate standard oligosaccharide, such as a biantennary oligosaccharide obtained from transferrin or milk oligosaccharides.

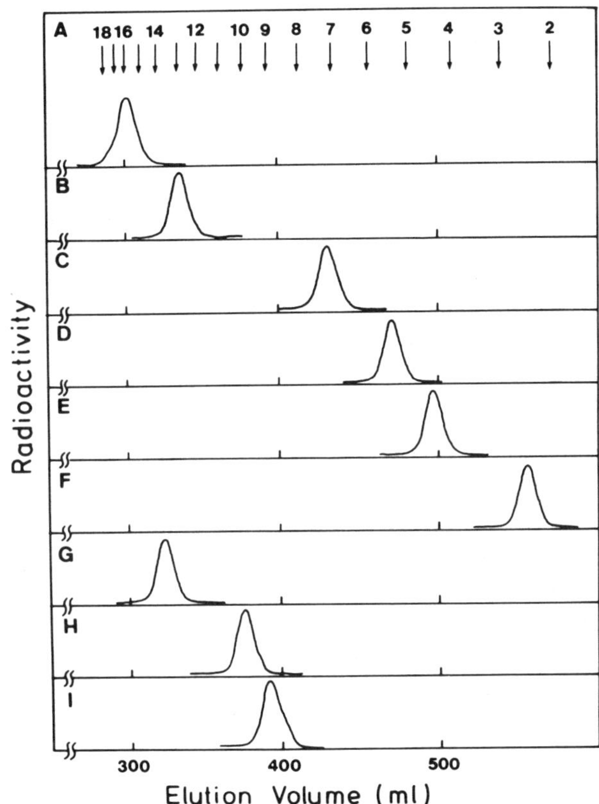

Figure 5 Sequential exoglycosidase digestion of fetuin oligosaccharide A-4N. Exoglycosidase digests were analyzed by Bio-Gel P-4 column chromatography: (A) A-4N; (B) A-4N digested with jack bean β-galactosidase; (C) the peak in *B* digested with jack bean β-*N*-acetylhexosaminidase; (D) the peak in *C* digested with jack bean α-mannosidase; (E) the peak in *D* digested with snail β-mannosidase; (F) the peak in *E* digested with jack bean β-*N*-acetylhexosaminidase; (G) A-4N digested with streptococcal β-galactosidase; (H) the peak in *G* digested with streptococcal β-*N*-acetylhexosamindase; (I) the peak in *H* digested with jack bean β-galactosidase. (From Ref. 63.)

B. Aglycon Specificities of Exoglycosidases

In actually performing the sequential exoglycosidase digestion, special care must be taken with the aglycon specificity of each enzyme used. The specificity is directed to the sugar chain structures in the vicinity of the position at which monosaccharide to be hydrolyzed is linked. This specificity is useful for the structural study of sugar chains because the enzymatic digestion will afford information about the structure of the sugar chains to which the target monosaccharide is linked. In the following part of this section, current knowledge of aglycon specificities of exoglycosidases, which are useful for the structural study of the sugar chains, will be described.

1. α-Mannosidases

α-Mannosidases have been purified from various sources including fungi, bacteria, and mammalian organs. Four types of α-mannosyl linkages: the Manα1→2Man, the Manα1→3Man, the Manα1→4Man, and the Manα1→6Man are currently found in the sugar chains of glycoproteins.

Aspergillus saitoi contains two *a*-mannosidases with different aglycon specificities [64]. α-Mannosidase I cleaves only the Manα1→2Man linkage [65]. Therefore, it has been used as an effective tool to identify high mannose-type sugar chains, because all of them will be converted to a heptasaccharide, Manα1→6(Manα1→3)Manα1→6(Manα1→3)Manβ1→4GlcNAcβ1→4GlcNAc by incubation with this enzyme. Because the heptasaccharide can be separated from most of the complex-type and hybrid-type oligosaccharides, the total amount of high mannose-type oligosaccharides in a very complex mixture of oligosaccharides released from glycoproteins can be easily determined with use of this α-mannosidase. Another enzyme, α-mannosidase II, from the same source, releases one mannose residue from the R→Manα1→6(Manα1→3)Manβ1→4GlcNAcβ1→$GlcNAc_{OT}$ group, but not from the isomeric Manα1→6(R→Manα1→3)Manβ1→4GlcNAβ1→$4GlcNAc_{OT}$ group, in which R represents any sugars [64]. Therefore, this enzyme is useful for assigning the position of monoantennary complex-type sugar chains.

Although jack bean α-mannosidase cleaves the Manα1→6Man and the Manα1→2Man linkages much faster than the Manα1→3Man linkage [24], it cleaves all α-mannosyl linkages at high enzyme concentrations. Therefore, this enzyme can be used to determine the number of all α-mannosyl groups in a sugar chain.

2. β-Galactosidases

Enzymes of this group have been purified from many sources. However, two of them have been widely used for structural study of the sugar chains of glycoproteins. The enzyme purified from jack bean meal cleaves the Galβ1→4GlcNAc linkages 50 times faster than the Galβ1→3GlcNAc linkage, when the enzyme concentration is limited to 0.6 unit/ml. At higher enzyme concentrations, it cleaves both linkages readily [66].

The aglycon specificity of the β-galactosidase purified from *S. pneumoniae* is narrower. It readily cleaves the Galβ1→4GlcNAc linkage, but cannot cleave the Galβ1→3GlcNAc linkage, even at a high enzyme concentration [67]. Therefore, the jack bean enzyme is suitable for determining the total β-galactosyl residues at the nonreducing terminal of sugar chains, and the streptococcal enzyme is used to determine the number of the Galβ1→4GlcNAc groups.

3. β-N-*Acetylhexosaminidases*

β-*N*-Acetylhexosaminidase, purified from jack bean meal, has a wide aglycon specificity and cleaves all β-*N*-acetylglucosaminyl linkages. In contrast, the enzyme purified from the culture medium of *S. pneumoniae* has a very narrow aglycon specificity. Among the five β-*N*-acetylglucosaminyl linkages shown in the following structure, which are found in the complex-type oligosaccharides, the enzyme can cleave only the GlcNAcβ1→2Man linkage [68].

```
GlcNAcβ1↘
            6
             Manα1↘
            2
GlcNAcβ1↗               6
           GlcNAcβ1→4Manβ1→4GlcNAcβ1→4GlcNAc
GlcNAcβ1↘               3
            4
             Manα1↗
            2
GlcNAcβ1↗
```

Furthermore, the enzyme cannot cleave the GlcNAcβ1→2Man linkage of the GlcNAcβ1→6(GlcNAcβ1→2)Man group and of the GlcNAcβ1→2Manα1→6 (GlcNAcβ1→4)Man group, although it cleaves the GlcNAcβ1→2Man linkage of the GlcNAcβ1→4(GlcNAcβ1→2)Man group (Table 6). Therefore, these two β-*N*-acetylhexosaminidases are very useful tools for determining the structures of complex-type oligosaccharides.

For example, more detailed structural information of oligosaccharide A-4N, which was described previously, was obtained by using the β-galactosidase and β-*N*-acetylhexosaminidase purified from *S. pneumoniae* [63]. When radiolabeled A-4N was incubated with streptococcal β-galactosidase, only two galactose residues were removed, indicating that the remaining one galactose residue occurs as the Galβ1→3GlcNAc group (see Fig. 5G). When the radiolabeled product in Figure 5G was incubated with streptococcal β-*N*-acetylhexosaminidase, two *N*-acetylglucosamine residues were removed (Fig. 5H). Therefore, these two *N*-acetylglucosamine residues should occur as the GlcNAcβ1→2Man group. Incubation of the radiolabeled product in Figure 5H with jack bean β-galactosidase released the remaining galactose residue (see Fig. 5I). This radioactive oligosaccharide was resistant to streptococcal β-*N*-acetylhexosaminidase digestion, but was converted to (Manα1→)$_2$Manβ1→GlcNAcβ1→GlcNAc$_{OT}$ by incubation with jack bean β-*N*-acetylhexosaminidase. These results indicated that the structure of the outer chain moiety of A-4N can be written as follows:

```
Galβ1→4GlcNAcβ1→2Manα1↘
                                  6
Galβ1→3GlcNAcβ1↘                 Manβ1→GlcNAcβ1→GlcNAcOT
                          4       3
                           Manα1↗
                          2
Galβ1→4GlcNAcβ1↗
```

4. α-Fucosidases

Since most fucose residues found in nature have an L-configuration, opposite that of other monosaccharides, the enzymatic properties of α-L-fucosidases will be described here. Fucosidases purified from *Aspergillus niger* [69], *Clostridium perfringens* [70], and *Bacillus fulminans* [71] do not hydrolyze synthetic substrates. They cleave the Fucα1→2Gal linkage, but not the Fucα1→3GlcNAc, the Fucα1→4GlcNAc, or the Fucα1→6GlcNAc linkage. An interesting piece of evidence is that the enzymes readily release fucose residues from blood type H hog submaxillary mucin, but not from blood type A mucin [72].

Table 6 Action of Streptococcal β-*N*-Acetylhexosaminidase on Various Oligosaccharides

Structures of oligosaccharides
GlcNAcβ1→2Manα1→3Manβ1→4GlcNAc$_{OT}$
GlcNAcβ1→2Manα1→6Manβ1→4GlcNAc$_{OT}$
GlcNAcβ1→4Manα1→3Manβ1→4GlcNAc$_{OT}$
GlcNAcβ1→6Manα1→6Manβ1→4GlcNAc$_{OT}$
GlcNAcβ1→2Manα1↘ 6 Manβ1→4GlcNAc$_{OT}$ 3 GlcNAcβ1→2Manα1↗
GlcNAcβ1↘ 6 Manα1↘ 2 6 GlcNAcβ1↗ Manβ1→4GlcNAc$_{OT}$ 3 GlcNAcβ1→2Manα1↗
GlcNAcβ1↘ 6 Manα1↘ 2 6 GlcNAcβ1↗ Manβ1→4GlcNAc$_{OT}$ GlcNAcβ1↘ 3 4 Manα1↗ 2 GlcNAcβ1↗
GlcNAcβ1↘ 6 Manα1→6Manβ1→4GlcNAc$_{OT}$ 2 GlcNAcβ1↗
Manα1↘ 6 Manα1↘ 3 Manα1↗ 6 GlcNAcβ1→4Manβ1→4GlcNAc$_{OT}$ 3 GlcNAcβ1→2Manα1↗
Manα1→3Manα1↘ 6 GlcNAcβ1→4Manβ1→4GlcNAc$_{OT}$ GlcNAβ1↘ 3 4 Manα1↗ 2 GlcNAcβ1↗
GlcNAcβ1→2Manα1↘ 6 GlcNAcβ1→4Manβ1→4GlcNAc$_{OT}$ 3 GlcNAcβ1→2Manα1↗

The underlined *N*-acetylglucosamine residues are cleaved by the enzyme.

blood type H

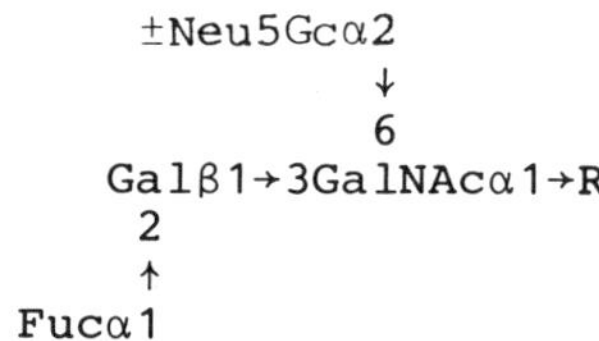

blood type A

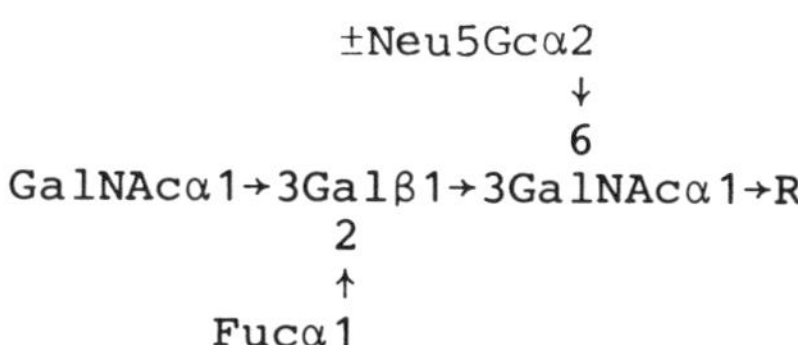

Figure 6 The major sugar chains of submaxillary mucin obtained from hogs with blood type H and blood type A. R represents Ser- or Thr-peptides.

Submaxillary mucins purified from hogs with blood type H and blood type A contain the sugar chains shown in Figure 6, respectively. Therefore, addition of an α-*N*-acetylgalactosamine residue at the neighboring C-3 position of the galactose residue will sterically inhibit the action of the enzymes on the Fucα1→2Gal linkage.

Two α-fucosidases with different aglycon specificities were purified from almond emulsin [73]. α-Fucosidase I cleaves the Fucα1→3GlcNAc and the Fucα1→4GlcNAc linkages, but not the Fucα1→2Gal and the Fucα1→6GlcNAc linkages and synthetic substrates. α-Fucosidase II does not act on synthetic substrates, the Fucα1→3GlcNAc, the Fucα1→4GlcNAc, or the Fucα1→6GlcNAc linkage, but cleaves the fucosyl linkage of Fucα1→2Galβ1→4Glc. Interestingly, the enzyme cannot release fucose from Fucα1→2Galβ1→3GlcNAcβ1→3Galβ1→4Glc. This indicates that the aglycon specificity of α-fucosidase II is similar, but not the same as those of bacterial α-fucosidases described earlier.

α-Fucosidases purified from the hepatopancreas of mollusca, such as *Turbo cornutus, Charonia lampas*, and *Chamelea gallina*, cleave all α-fucosyl linkages yet known [74,75]. Enzymes with broad aglycon specificities have also been obtained from various mammalian organs. Among them, the enzymes from bovine epididymis and kidney are commercially available. By proper use of these α-fucosidases, with broad and narrow aglycon specificities, the α-fucosyl linkages in sugar chains can be predicted.

5. *Sialidases*

Sialidases from *C. perfringens* [76], *Vibrio cholerae* [77], *Arthrobacter ureafaciens* [78], and influenza virus [79] are commercially available. Among them, the enzyme from *A. ureafaciens* is the most useful for sequential exoglycosidase digestion because it cleaves all sialyl linkages as yet reported. The enzyme is also useful for removal of sialic acid residues from glycoproteins, because it is totally free from protease activity.

Sialidase from Newcastle disease virus (NDV) was reported to cleave the Siaα2→3Gal linkage, but not the Siaα2→6Gal/GalNAc linkage, of glycoproteins [80]. Later, a peculiar finding about the aglycon specificity of NDV sialidase was made [81]. Although the enzyme could not cleave the Siaα2→6GalNAc$_{OT}$ linkage, it slowly releases sialic acid from Siaα2→6(Galβ1→3)GalNAc$_{OT}$ and GalNAcα1→3(Fucα1→2) Galβ1→3(Siaα2→6)GalNAc$_{OT}$. Therefore, the Siaα2→6GalNAc$_{OT}$ linkage becomes susceptible to sialidase digestion when the C-3 position of the GalNAc$_{OT}$ residue is substituted by sugars. In support of this rule, one sialic acid residue of the Neu5Acα2→3Galβ1→3(Neu5Acα2→6)GalNAc$_{OT}$ was readily cleaved by NDV sialidase and the other residue was cleaved at a much slower rate. With this limitation in mind, NDV sialidase can be used effectively to assign the position of the Siaα2→3Gal group in a sugar chain.

C. Methylation Analysis

Methylation analysis is an essential procedure for the structural study of the sugar chains of glycoproteins. As described in the previous section, some of the glycosidic linkages in an oligosaccharide can be predicted by using the aglycon specificities of exoglycosidases. Yet, methylation analysis is the most reliable method for the elucidation of the whole structure of an oligosaccharide. Furthermore, methylation data is necessary for predicting the structure of oligosaccharides by a series of lectin column chromatographies, as will be described in the following section. If an unusual sugar linkage is detected, one must be careful in interpreting the result of lectin column affinity chromatography.

The principle of methylation analysis of sugar chains is to methylate completely the free hydroxyl groups of an oligosaccharide by an appropriate method. The permethylated oligosaccharide will then be hydrolyzed, and the partially *O*-methylated monosaccharides are quantitatively determined after being separated by an appropriate method. Therefore, effective and complete methylation is essential for obtaining reliable data. The methodology of effective methylation of oligosaccharides has been improved by many researchers. Among them, the most brilliant improvement was achieved by Hakomori [82]. This procedure uses a mixture of methylsulfinylcarbanion and methyl iodide as the methylation reagent and can convert any oligosaccharides and glycopeptides by a single-step reaction. Therefore, the method greatly contributed to decreasing the amount of sample needed to obtain the permethylated form. It is also convenient to label oligosaccharides by reduction with NaB^3H_4 because the radioactivity can be used as an effective marker during the purification of permethylated samples. When the oligosaccharide is small, the permethylated oligosaccharide can be purified by thin-layer chromatography using benzene/methanol (9:1) as a solvent [83]. The permethylated sugar will be located by a radiochromatoscanner and recovered almost quantitatively by elution with $CHCl_3/CH_3OH/H_2O$ (10:10:1). When the oligosaccharide is large, the permethylated product can be purified by passing through a small column of Sephadex LH-20 or of silica gel [83].

The permethylated oligosaccharide is then hydrolyzed in 93% acetic acid, containing 0.5 N H_2SO_4 at 76°C for 4 hr [84]. In some cases, hydrolysis time must be extended up to 16 hr because of the relative resistance of the highly branched complex-type oligosaccharides [85]. The hydrolysis products are then converted to partially *O*-methylated alditol acetates and partially *O*-methylated 2-*N*-methylacetamido-2-deoxyalditol acetates by reduction with $NaBH_4$ followed by acetylation with acetic anhydride [86].

Separation and quantitation of various partially *O*-methylated alditol acetates can be performed by gas chromatography using a Gaschrom Q column coated with ECNSS-M, OV-225, or OS-138 [87]. Partially *O*-methylated 2-*N*-methylacetamido-2-deoxyalditol acetates can be quantified by a 2–3% OV-17-Gaschrom Q column, which does not irreversibly absorb the amino sugars [88]. An OV-225-Gaschrom Q column can also be used for this purpose.

Recent introduction of capillary columns in the field of gas chromatography affords higher resolution and sensitivity. For the quantitative analysis of partially *O*-methylated sugars, bonded-phase fused silica capillary columns, such as DB-1 and DB-5 (J & W Scientific, Rancho Cordova, California), Ultra 1 and Ultra 2 (Hewlett Packard), and SPB-1 and SPB-5 (Supelco, Bellefonte, Pennsylvania) can be effectively used. In Figure 7, the data of methylation analysis of an oligosaccharide A-3 obtained from fetuin [63] is shown. It is noteworthy that the fractionation pattern was obtained by using only 1–0.2 nmol of sample.

However, as the sensitivity of methylation analysis has increased, a new problem has arisen. A series of alkyl esters of phthalic acid, which are widely included in the plastic materials, often contaminate the samples to be analyzed. Since these esters have T values in the same range as those of partially *O*-methylated alditols and amino alditols, they often cause trouble in the interpretation of chromatograms and in the quantitative analysis of each sugar component. Introduction of gas chromatography–mass spectrometry has solved this problem, as identification of sugar peaks can rely on mass patterns, and even quantitative data can be obtained by using mass fragmentography in the presence of contaminating phthalic acid esters.

Other powerful tools introduced in recent years for the structural study of sugar chains are high-resolution NMR spectroscopy [89,90] and fast-atom bombardment mass spectrometry [91]. Although they still require significantly large amounts of

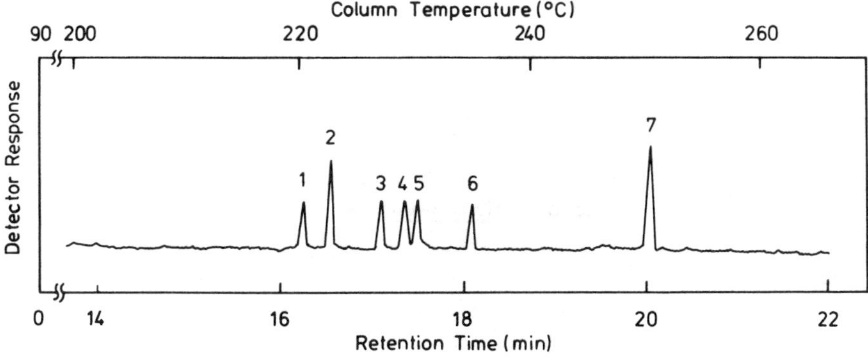

Figure 7 Total ion chromatograms of partially *O*-methylated alditol acetates and aminoalditol acetates obtained by methylation analysis of oligosaccharide A-3 obtained from fetuin; (1) 3,4,6-tri-*O*-methylmannitol; (2) 2,4,6-tri-*O*-methylgalactitol; (3) 2,3,4-tri-*O*-methylgalactitol; (4) 1,3,5,6-tetra-*O*-methyl-2-*N*-methylacetamido-2-deoxyglucitol; (5) 3,6-di-*O*-methylmannitol; (6) 2,4-di-*O*-methylmannitol; and (7) 3,6-di-*O*-methyl-2-*N*-methylacetamido-2-deoxyglucitol, on a 30 m DB-5 capillary column programmed from 90 to 260°C at 8°C/min.

samples, they will, no doubt, replace a large portion of the structural analysis methods for oligosaccharides in the future. Since NMR can yield data under nondestructive condition, it will be used for the first step in the structural study of oligosaccharides to obtain preliminary structural information of the sample.

D. Analysis of Glycosidic Linkages of Oligosaccharides by Smith Degradation

Periodic acid cleaves the C–C bonds by preferential oxidation of adjacent hydroxyl groups (vicinal diols), as shown in Figure 8. Since oligosaccharides are rich in vicinal diol groups, they can be easily oxidized with periodic acid. After the resulting aldehyde groups are converted to carbinol groups by reduction with $NaBH_4$, the cleaved monosaccharides can be selectively hydrolyzed by mild acid treatment [92]. This reaction, which was named Smith degradation after Dr. Smith who developed the method, has been widely used for determining the structures of sugar chains. With use of tritium-labeled oligosaccharides, important structural information can be obtained from only 100 pmol of oligosaccharides.

During the structural study of oligosaccharides in the urine of patients with mannosidosis, a genetic defect of lysosomal mannosidase, a pentasaccharide, composed of four mannose and one *N*-acetylglucosamine residues, was isolated [93]. Analysis of the tritium-labeled monosaccharide, released from the tritium-labeled oligosaccharide alcohol by acid hydrolysis, revealed that *N*-acetylglucosamine was located at the reducing terminal. When the tritium-labeled pentaitol was digested with jack bean α-mannosidase, Manβ1→4GlcNAc$_{OT}$ was obtained as a sole radiolabeled product. Methylation analysis of the pentaitol revealed that Man1→, →3Man1→, and $^{\rightarrow 6}_{\rightarrow 3}$Man1→ occur in 2:1:1 molar ratio. From these results, the following three structures could be considered for the pentaitol:

```
A.   Manα1→3Manα1↘
                   6
                    Manβ1→4GlcNAcOT
                   3
          Manα1↗

B.        Manα1↘
                   6
                    Manβ1→4GlcNAcOT
                   3
     Manα1→3Manα1↗

C.   Manα1↘
              6
               Manα1→3Manβ1→4GlcNAcOT
              3
     Manα1↗
```

When the tritium-labeled pentaitol was subjected to Smith degradation, all radioactivity was recovered as Manα1→6Manβ1→4XylNAc$_{OT}$. Since hexopyranoside residues

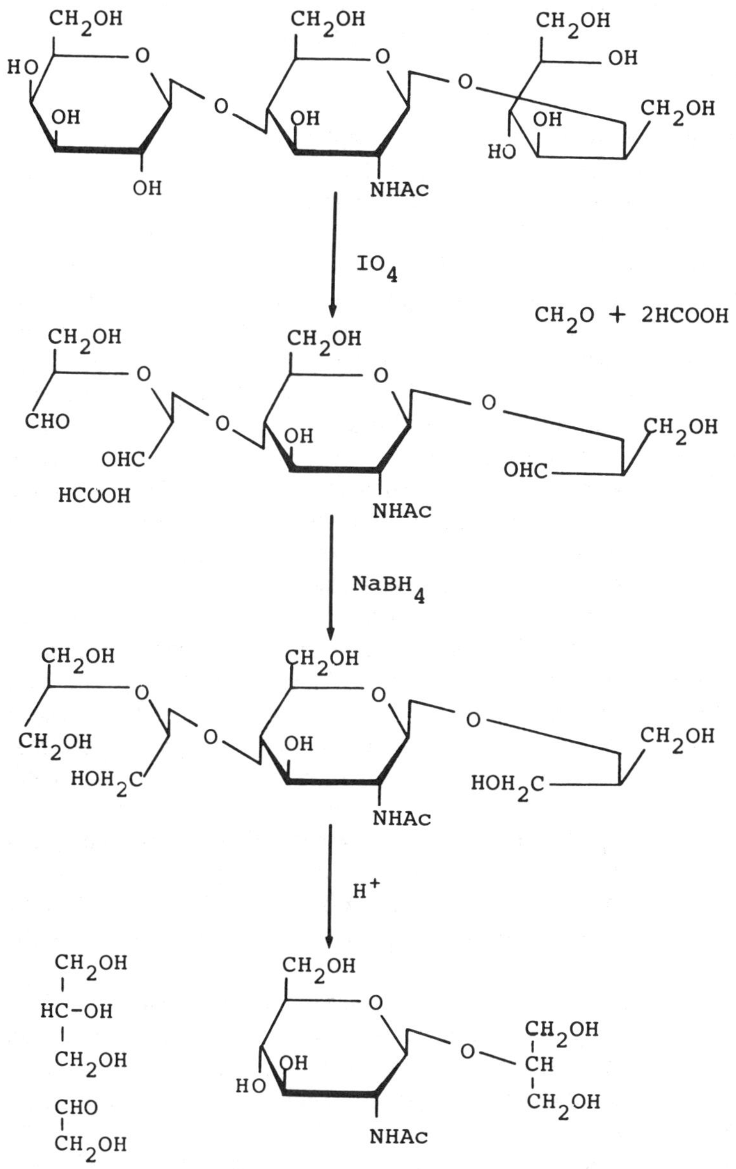

Figure 8 Smith degradation of Galβ1→4GlcNAcβ1→2mannitol.

that are substituted at the C-3 position by other sugars are resistant to periodate oxidation, the underlined monosaccharide should be oxidized and removed by Smith degradation. Therefore, Manα1→6Manβ1→4XylNAc$_{OT}$ will be produced from oligosaccharide A and Manα1→3Manβ1→4XylNAc$_{OT}$ from oligosaccharides *B* and *C*. Given this evidence, structure A was assigned for the pentasaccharide [93].

IV. FRACTIONATION AND STRUCTURAL ESTIMATION OF OLIGOSACCHARIDES BY IMMOBILIZED LECTIN COLUMN CHROMATOGRAPHY

Although plant lectins will be described in detail in Chapter 10 of this book, we would like to discuss here the use of immobilized lectin columns, because they are becoming indispensable tools for the structural study of the sugar chains of glycoproteins.

The first trial of a lectin column for the structural study of the sugar chains of glycoproteins was reported by Ogata et al. [94]. Study of the behavior of various radiolabeled Asn-linked oligosaccharides and oligomannosides in a concanavalin A (Con-A)-Sepharose column revealed that at least two nonsubstituted or 2-*O*-substituted α-mannosyl residues must be present in a sugar chain to be retained on the immobilized lectin column. This finding was particularly useful for the structural analysis of complex-type Asn-linked sugar chains. All complex-type oligosaccharides contain a pentasaccharide, Manα1→6(Manα1→3)Manβ1→4GlcNAcβ1→4GlcNAc as a common core. Since the core has two α-mannosyl residues, this core binds to a Con A–Sepharose column. Because the first outer chain is linked at the C-2 position, the mannose residue can still interact with Con A. However, substitution with two outer chains totally abolishes the interaction of the α-mannosyl residue. Therefore, a biantennary complex-type sugar chain binds to a Con A–Sepharose column, but those with more than three outer chains pass through the column without interaction. High mannose-type oligosaccharides, which have many Manα1→ residues and →2Manα1→ residues, will bind more tightly to a Con A–Sepharose column. With use of this character of a Con A–Sepharose column, Ogata et al. [20] found that the glycopeptides, obtained from the plasma membrane of polyoma-transformed baby hamster kidney cells by pronase digestion, are not only enriched in sialic acid residues, but contain more branched structures than those from baby hamster kidney cells. Later, Narasimhan et al. [95] added the evidence that bisected biantennary complex-type sugar chains do not bind to a Con A–Sepharose column. They also reported that biantennary complex-type oligosaccharides can be eluted from the column with 10 mM α-methylglucopyranoside, and high mannose-type oligosaccharides can be eluted with 250 mM α-methylmannopyranoside.

Complex-type sugar chains can be classified by the presence or absence of an α-fucosyl residue linked at the C-6 position of the proximal *N*-acetylglucosamine residue of the trimannosyl core. It was found that some of the hybrid-type sugar chains are also fucosylated at their core [96]. The fucosylated complex-type and hybrid-type sugar chains can easily be separated from their nonfucosylated counterparts by passing through an immobilized *Aleuria aurantia* lectin (AAL) column [97]. Sugar chains with a nonfucosylated core pass through the column without interaction. In contrast, sugar chains with a fucosylated core bind to the column and are eluted with a buffer containing 0.5 mM L-fucose. Because the binding is affected neither by the status of antennary nor by the presence of bisecting *N*-acetylglucosamine residue, the fractionation can be used as a more reliable method to determine the fucosylated sugar chains than immobilized lentil and pea lectin columns [98]. In these cases, not only the fucosyl residue linked to the core portion, but the structure of the outer chains are required for the binding. The Fucα1→2Galβ1→4GlcNAc, the Galβ1→4(Fucα1→3)GlcNAc, and the Galβ1→3(Fucα1→4)GlcNAc groups weakly interact with an AAL column, whereas the Fucα1→2Galβ1→3GlcNAc, and the Galβ1→4GlcNAcβ1→3Galβ1→4(Fucα1→3)GlcNAc groups show almost no interaction with the matrix [97]. However, oligosaccharides with more than two of these groups bind more strongly than those with only one of the groups. Therefore, care must be taken in interpreting the results of affinity chromatography for the structural determination of oligosaccharides.

Because most of the Asn-linked sugar chains of glycoproteins are included in complex-type sugar chain subclasses, lectins that interact with the β-galactosyl residues, which are widely found in the outer chain moieties of complex-type sugar chains, are of particular use. Many galactose-binding lectins have been reported [99]. Among them, the binding specificity of *Ricinus communis* agglutinin I (RCA I) has been investigated most extensively [100].

An immobilized RCA I column retains oligosaccharides with nonreducing terminal β-galactose residues. The affinity of oligosaccharides to the column is proportional to the number of unsubstituted β-galactosyl residues. Therefore, the following biantennary complex-type oligosaccharides can be readily fractionated by passing through an RCA I–agarose column:

```
A.           GlcNAcβ1→2Manα1↘
                              6
                               Manβ1→4GlcNAcβ1→4GlcNAc
                              3
             GlcNAcβ1→2Manα1↗

B.  Galβ1→4GlcNAcβ1→2Manα1↘
                              6
                               Manβ1→4GlcNAcβ1→4GlcNAc
                              3
             GlcNAcβ1→2Manα1↗

C.  Galβ1→4GlcNAcβ1→2Manα1↘
                              6
                               Manβ1→4GlcNAcβ1→4GlcNAc
                              3
    Galβ1→4GlcNAcβ1→2Manα1↗
```

Oligosaccharide A passes through the column without any interaction, and oligosaccharide B is retarded in the column, but is eluted with buffer only. In contrast, oligosaccharide C is firmly bound and eluted with a buffer containing 10 mM lactose [100]. A triantennary complex-type oligosaccharide with three Galβ1→4GlcNAcβ1→ residues can also be separated from oligosaccharide C by passing through an RCA I–agarose column [100]. Study of the behavior of human milk oligosaccharides revealed that the Galβ1→3GlcNAc and the Galβ1→4(Fucα1→3)GlcNAc groups interact, but more weakly than the Galβ1→4GlcNAc group.

Several complex-type oligosaccharide contain a bisecting *N*-acetylglucosamine residue. Irimura et al. [101] reported that bisected biantennary complex-type oligosaccharides with the Galβ1→4GlcNAcβ1→ outer chains bind to an erythroagglutinin (E-PHA)–agarose column. Cummings and Kornfeld [102] confirmed this evidence and indicated that none of the nonbisected sugar chains will bind to the column. Yamashita et al. [103] investigated the behavior of 12 bisected complex-type and hybrid-type oligosaccharides in an E-PHA–agarose column and concluded that the minimal structural unit required for binding to the column is as follows:

```
Galβ1→4GlcNAcβ1→2Manα1↘
                              6
                    GlcNAcβ1→4Manβ1→4GlcNAcβ1→4R3
                 R2↘          3
                     4
                      Manα1↗
                     2
  R1→4GlcNAcβ1↗
```

In this structure, R_1 and R_2 represent either hydrogen or sugars, and R_3 represents either (±Fucα1→6)GlcNAc→Asn or (±Fucαa→6)GlcNAc$_{OT}$. With this basic information, the immobilized lectin column can be used as an effective tool to fractionate and determine the structures of bisected complex-type oligosaccharides. However, we have recently found that some of the commercial E-PHA–agarose is contaminated with leukoagglutinin (L-PHA)–agarose, which will cause a misinterpretation of oligosaccharide structures (unpublished data). Therefore, each researcher should be cautious in using the column for studies of glycoprotein sugar chains.

Another variation in the complex-type sugar chains is antennary. Although mono- and biantennary sugar chains can be separated from other highly branched complex-type sugar chains by a Con A–Sepharose column, tri- and tetraantennary oligosaccharides cannot be separated by this column. An immobilized *Datura stramonium* agglutinin (DSA) column can be effectively used to fractionate these highly branched oligosaccharides. A study of the behavior of 34 complex-type oligosaccharides in a DSA–Sepharose column [104] showed that oligosaccharides that contain the Galβ1→4GlcNAcβ1→4(Galβ1→4GlcNAcβ1→2)Man group are retarded by the column, as long as the pentasaccharide group is not substituted by other sugars. Oligosaccharides that contain unsubstituted Galβ1→4GlcNAcβ1→6(Galβ1→4GlcNAcβ1→2)Man group bind to the column and are eluted with buffer containing *N*-acetylglucosamine oligomers. This binding characteristic was not affected by either the structure of the inner core portion or by the presence of bisecting *N*-acetylglucosamine residue. Therefore, the column can fractionate a mixture of complex-type oligosaccharides into three fractions: the "pass-through fraction" contains biantennary oligosaccharides; the "retarded fraction" contains triantennary oligosaccharides, with a C-2, 4 branch; and the "bound fraction" contains triantennary oligosaccharides, with a C-2, 6 branch and tetraantennary oligosaccharides. Oligosaccharides with the Galβ1→4(Fucα1→3)GlcNAc groups will be recovered in the pass-through fraction. However, they will be recovered either in the retarded or in the bound fraction after almond emulsion α-fucosidase I digestion. Complex-type oligosaccharides with at least one Galβ1→4GlcNAcβ1→3Galβ1→4GlcNAc group in their outer chain moieties bind to the column and elute with buffer containing *N*-acetylglucosamine oligomers. However, these oligosaccharides can be discriminated from C-2, 6 branched triantennary and tetraantennary oligosaccharides containing only the Galβ1→4GlcNAc outer chains by their change in column behavior after digestion with *E. freundii* or *F. keratolyticus* endo-β-galactosidase.

As an example, the usefulness of serial lectin column chromatography for the separation of the Asn-linked sugar chains of human immunoglobulin G will be documented. Human IgG is a glycoprotein composed of two types of polypeptide chains: heavy (H) and light (L), with the stoichiometry H_2L_2. Each heavy chain contains an Asn-linked sugar chain at asparagine-297 in the C_H2 domain of the Fc region [105]. Structural study of the oligosaccharides released from whole serum IgG by hydrazinolysis revealed that extremely high structural multiplicity is present in the sugar chains. The basic structure of the oligosaccharides is of biantennary complex type. However, more than 30 different structures were found [106]. Even after desialylation, there were still 16 different oligosaccharides, as summarized in Table 7, in the sugar mixture liberated from human IgG by hydrazinolysis. These oligosaccharides can be satisfactorily separated by passing them through three immobilized lectin columns (Fig. 9) [107]. As the first step, the neutral oligosaccharide mixture was separated into two fractions: a pass-through fraction and bound by the affinity chromatography on an AAL–Sepharose column. The pass-through fraction (AAL-) contains all nonfucosylated oligosaccharides, whereas the bound fraction (AAL+) contains all fucosylated oligosaccharides. Each fraction was separated into eight oligosaccharides by sequentially passing through an RCA–agarose and an E_4-PHA–agarose column or a Con A–Sepharose column as shown in Figure 9. These oligosaccharides were assigned the 16 different oligosaccharide structures shown in Table 7.

The data presented here indicate that serial lectin column chromatography can be effectively used for fractionating and for structural studies of the glycoprotein sugar chains. Development of lectin columns suitable for HPLC will save the time required for the whole fractionation. We have recently found that enzymes located at the apical membrane of epithelial cells have Asn-linked sugar chains with extraordinarily complicated structures [108]. Serial lectin column chromatography will become an essential tool for the study of such complicated sugar chains.

V. STRUCTURAL ANALYSIS OF MUCIN-TYPE SUGAR CHAINS

As already described in Section II, mucin-type sugar chains have been studied after they were released from the polypeptide backbone by β-elimination. Because the oligosaccharides obtained have already been reduced, they could no longer be labeled as could Asn-linked sugar chains released by hydrazinolysis. To overcome this bottleneck, addition of tritiated $NaBH_4$ to the β-elimination reaction mixture was tried to obtain the labeled oligosaccharide [109]. Since the method was simple, it has been applied for the studies of mucin-type sugar chains of glycoproteins obtained in small amounts. A drawback of this method is that an extremely high concentration of tritiated $NaBH_4$ is required to protect the released oligosaccharides from a peeling reaction. Production of a large amount of $^3H_2$ gas has also hampered the frequent use of the method for study of mucin-type sugar chains. Kabat and his colleagues used acetone to destroy excess tritiated $NaBH_4$ in the reaction mixture [110]. By this method, excess tritium is converted to [^{3}H]isopropanol. However, from the standpoint of environmental pollution, $^3H_2$ gas is much better than [^{3}H]isopropanol because it is inactive and lighter than the atmosphere.

When an oligosaccharide alcohol contains sialic acid, the $NaIO_4$–NaB^3H_4 method [111] can be used to label the sialic acid moiety. The galactose oxidase–NaB^3H_4 method [112] can also be used to label the oligosaccharides that have galactose residues at their nonreducing termini. However, the label can not be used effectively for the sequencing of the sugar chain, because it is removed from oligosaccharides by the first exoglycosidase digestion.

Amano et al. [81] devised a new method to label the oligosaccharides released from glycoproteins by alkaline-$NaBH_4$ treatment. The oligosaccharide alcohol mixture was de-*N*-acetylated by heating in anhydrous hydrazine at 106°C for 24 hr. Hydrazine was removed by repeated evaporation with toluene, and the residue was then *N*-acetylated with

Table 7 Structures of Desialylated Asparagine-Linked Sugar Chains Found in Human IgG

	Structures
I	GlcNAcβ1→2Manα1↘ 6 Manβ1→4GlcNAcβ1→4R 3 GlcNAcβ1→2Manα1↗
II	GlcNAcβ1 ↓ GlcNAcβ1→2Manα1↘ 4 6 Manβ1→4GlcNAcβ1→4R 3 GlcNAcβ1→2Manα1↗
III	GlcNAcβ1→2Manα1↘ 6 Manβ1→4GlcNAcβ1→4R 3 Galβ1→4GlcNAcβ1→2Manα1↗
IV	GlcNAcβ1 ↓ GlcNAcβ1→2Manα1↘ 4 6 Manβ1→4GlcNAcβ1→4R 3 Galβ1→4GlcNAβ1→2Manα1↗
V	Galβ1→4GlcNAcβ1→2Manα1↘ 6 Manβ1→4GlcNAcβ1→4R 3 GlcNAcβ1→2Manα1↗
VI	GlcNAcβ1 ↓ Galβ1→4GlcNAcβ1→2Manα1↘ 4 6 Manβ1→4GlcNAcβ1→4R 3 GlcNAcβ1→2Manα1↗
VII	Galβ1→4GlcNAcβ1→2Manα1↘ 6 Manβ1→4GlcNAcβ1→4R 3 Galβ1→4GlcNAcβ1→2Manα1↗
VIII	GlcNAcβ1 ↓ Galβ1→4GlcNAcβ1→2Manα1↘ 4 6 Manβ1→4GlcNAcβ1→4R 3 Galβ1→4GlcNAcβ1→2Manα1↗

R, (Fucα1→6)GlcNAc or GlcNAc.
Source: Ref. 107.

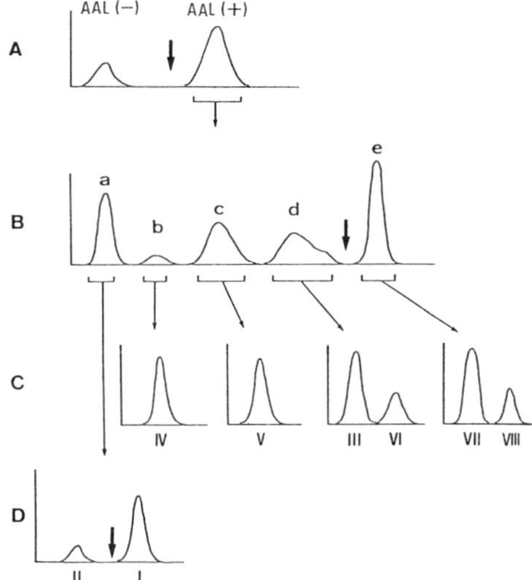

Figure 9 Fractionation scheme of asialo-oligosaccharides of human IgG by serisal lectin column chromatography. Asialo-oligosaccharides of human IgG were applied to an AAL–Sepharose column (A). The fraction bound to an AAL–Sepharose column was further fractionated by chromatography on an RCA 120-WG003 column (B). The four fractions (b, c, d, and e) in *B* were separately subjected to E_4-PHA–agarose column chromatography (C) and the fraction a in *B* was subjected to Con A–Sepharose column chromatography (D). Bold arrows indicate the positions at which the elution buffers were changed to buffer containing 1 mM focuse (A), 10 mM lactose (B), or 100 mM α-methyl-D-mannopyranoside (D). The ordinate of each figure represents radioactivity. The detailed condition of each chromatography run and the carbohydrate structures of fractions III, IV, V, VI, VII and VIII are described in reference 107. Structures of peaks I–VIII are shown in Table 7.

either [^{3}H]acetic anhydride or [^{14}C]acetic anhydride in 0.1 M $Na_2CO_3/NaHCO_3$ buffer, pH 10.7. Since all mucin-type sugar chains contain at least one *N*-acetylhexosamine residue and the total amount of amino sugars in an oligosaccharide sample can be estimated by analysis of its sugar composition, the re-*N*-acetylation method can be used for the radiolabeling of the mucin-type sugar chains, without producing too much radioactive waste.

When a glycoprotein with mucin-type sugar chains is warmed at 37°C for 36 hr in 0.1 N NaOH containing 0.5 M Na_2SO_4, the serine and threonine residues linked to sugar chains are converted to cysteinic acid and aminosulfonyl butyric acid [113]. Although this reaction does not contribute to the structural analysis of sugar chains, the method can be used for calculating the number and distribution of the two hydroxyamino acids of the mucin-type sugar chains in one molecule of glycoprotein. With use of $Na_2{}^{35}SO_4$, this analysis could be performed with a small amount of sample [38].

VI. CONCLUDING REMARKS

The use of methods described in this chapter in various combination will enable us to elucidate the structures of almost all of the sugar chains of glycoproteins. To avoid the use of radioactive material, several new methods for labeling the reducing terminal of an oligosaccharide by converting it to a fluorescent derivative have been developed [114]. Detection of the fluorescent derivatives is as sensitive as tritium-labeled oligosaccharides. However, we would like to stress here that the sensitivity for detection of oligosaccharide by currently used tritiated $NaBH_4$ reduction can be increased by approximately 100 times, because NaB^3H_4 with higher specific activity is commercially available. Therefore, NaB^3H_4 reduction will afford the most sensitive detection of sugars.

As described in this chapter, the discovery of new exoglycosidases with strict aglycon specificities will in the future surely advance the structural study of sugar chains. Study of details in the binding specificity of immobilized lectin columns as well as finding new lectins with different binding specificities will also afford useful methods to elucidate the structures of sugar chains.

Quite recently, a new endoglycosidase, which cleaves various mucin-type sugar chains as oligosaccharides, has been reported to occur in the culture fluid of *Streptomyces*, sp. OH-11242 [115]. Purification of this enzyme to a level that is free from other exoglycosidases will afford another convenient method for the study of structures of mucin-type sugar chains, because the released oligosaccharides can easily be labeled by NaB^3H_4 reduction, and analyzed with use of the various methods described in this chapter.

REFERENCES

1. Edelman, G. M. (1985). Cell adhesion and the molecular processes of morphogenesis. *Annu. Rev. Biochem. 54*:135–169.
2. Kornfeld, R., and Kornfeld, S. (1985). Assembly of asparagine-linked oligosaccharides. *Annu. Rev. Biochem. 54*:631–664.
3. Lis, H., and Sharon, N. (1986). Lectins as molecules and as tools. *Annu. Rev. Biochem. 55*:35–67.
4. Quiocho, F. A. (1986). Carbohydrate-binding proteins: Tertiary structures and protein–sugar interactions. *Annu. Rev. Biochem. 55*:287–315.
5. Osawa, T., and Tsuji, T. (1987). Fractionation and structural assessment of oligosaccharides and glycopeptides by use of immobilized lectins. *Annu. Rev. Biochem. 56*:21–42.
6. Hirschberg, C. B., and Snider, M. D. (1987). Topography of glycosylation in the rough endoplasmic reticulum and Golgi apparatus. *Annu. Rev. Biochem. 56*:63–87.
7. Wiley, D. C ., and Skehel, J. J. (1987). The structure and function of the hemagglutinin membrane glycoproteins of influenza virus. *Annu. Rev. Biochem. 56*:365–394.
8. Elbein, A. D. (1987). Inhibitors of the biosynthesis and processing of *N*-linked oligosaccharide chains. *Annu. Rev. Biochem. 56*:497–534.
9. Kikuruzinska, M. A., Bergh, M. L. E., and Jackson, B. J. (1987). Protein glycosylation in yeast. *Annu. Rev. Biochem. 56*:915–944.
10. Laine, L. A., Esselman, W. J., and Sweeley, C. C. (1972). Gas-liquid chromatography of carbohydrates. *Methods Enzymol. 28*:159–167.
11. Lee, Y. C., Johnson, G. S., White, B., and Scocca, J. (1971). An accelerated system for analysis of neutral sugars in complex carbohydrates. *Anal. Biochem. 43*:640–643.
12. Takasaki, S., and Kobata, A. (1974). Microdetermination of individual neutral and amino sugars and *N*-acetylneuraminic acid in complex sugars. *J. Biochem. (Tokyo) 76*:783–789.
13. Takeuchi, M., Takasaki, S., Inoue, N., and Kobata, A. (1987). Sensitive method for carbohydrate composition analysis of glycoproteins by high-performance liquid chromatography. *J. Chromatogr. 400*:207–213.
14. Takeuchi, M., Takasaki, S., Miyazaki, H., Katoh, T., Hoshi, S., Kochibe, N., and Kobata, A. (1988). Comparative study of the asparagine-linked sugar chains of human erythropoietin purified from urine and the culture medium of recombinant Chinese hamster ovary cells. *J. Biol. Chem. 268*:3657–3663.

15. Carubelli, R., Bhavanandan, V. P., and Gottschalk, A. (1965). Studies on glycoproteins. XI. The *O*-glycosidic linkage of *N*-acetylgalactosamine to seryl and threonyl residues in ovine submaxillary gland glycoprotein. *Biochim. Biophys. Acta 101*:67–82.
16. Whistler, R. L., and BeMiller, J. N. (1958). Alkaline degradation of polysaccharides. *Adv. Carbohydr. Chem. 13*:289–329.
17. Carlson, D. M. (1968). Structures and immunochemical properties of oligosaccharides isolated from pig submaxillary mucins. *J. Biol. Chem. 243*:616–626.
18. Sheares, B., and Robbins, P. W. (1986). Glycosylation of ovalbumin in a heterologous cell: Analysis of oligosaccharide chains of the cloned glycoproteins in mouse L cells. *Proc. Natl. Acad. Sci. USA 83*:1993–1997.
19. Huang, C. C., Mayer, H. E., Jr., and Montgomery, R. (1970). Microheterogeneity and paucidispersity of glycoproteins. Part I. The carbohydrate of chicken ovalbumin. *Carbohydr. Res. 13*:127–137.
20. Ogata, S., Muramatsu, T., and Kobata, A. (1976). New structural characteristics of the large glycopeptides from transformed cells. *Nature 259*:580–582.
21. Endo, Y., Yamashita, K., Tachibana, Y., Tojo, S., and Kobata, A. (1979). Structures of the asparagine-linked sugar chains of human chorionic gonadotropin. *J. Biochem. (Tokyo) 85*: 669–679.
22. Koide, N., and Muramatsu, T. (1974). Endo-β-*N*-acetylglucosaminidase acting on carbohydrate moieties of glycoproteins. *J. Biol. Chem. 249*:4897–4904.
23. Muramatsu, T., Koide, N., and Maeyama, K. (1978). Further studies on endo-β-*N*-acetylglucosaminidase D. *J. Biochem. (Tokyo) 83*:363–370.
24. Tai, T., Yamashita, K., Ogata, A. M., Koide, N., Muramatsu, T., Iwashita, S., Inoue, Y., and Kobata, A. (1975). Structural studies of two ovalbumin glycopeptides in relation to the endo-β-*N*-acetylglucosaminidase D. *J. Biol. Chem. 250*:8569–8575.
25. Ito, S., Muramatsu, T., and Kobata, A. (1975). Release of galactosyl oligosaccharides by endo-β-*N*-acetylglucosaminidase D. *Biochem. Biophys. Res. Commun. 63*:938–944.
26. Mizuochi, T., Amano, J., and Kobata, A. (1984). A new evidence about the substrate specificity of endo-β-*N*-acetylglucosaminidase D. *J. Biochem. (Tokyo) 95*:1209–1213.
27. Arakawa, M., and Muramatsu, T. (1974). Endo-β-*N*-acetylglucosaminidases acting on the carbohydrate moieties of glycoproteins: The differential specificities of the enzymes from *Streptomyces griseus* and *Diplococcus pneumoniae*. *J. Biochem. (Tokyo) 76*:307–317.
28. Tarentino, A. L., and Maley, F. (1974). Purification and properties of an endo-β-*N*-acetylglucosaminidase from *Streptomyces griseus*. *J. Biol. Chem. 249*:811–817.
29. Tai, T., Yamashita, K., and Kobata, A. (1977). The substrate specificities of endo-β-*N*-acetylglucosaminidases C_{II} and H. *Biochem. Biophys. Res. Commun. 78*:434–441.
30. Tarentino, A. L., and Maley, F. (1975). A comparison of the substrate specificities of endo-β-*N*-acetylglucosaminidases from *Streptomyces griseus* and *Diplococcus pneumoniae*. *Biochem. Biophys. Res. Commun. 67*:455–462.
31. Foddy, L., Feeney, J., and Hughes, R. C. (1986). Properties of baby-hamster kidney (BHK) cells treated with swainsonine, an inhibitor of glycoprotein processing. *Biochem. J. 233*:697–706.
32. Ito, S., Muramatsu, T., and Kobata, A. (1975). Endo-β-*N*-acetylglucosaminidases acting on carbohydrate moieties of glycoproteins: Purification and properties of the two enzymes with different specificities from *Clostridium perfringens*. *Arch. Biochem. Biophys. 171*:78–86.
33. Tai, T., Yamashita, K., Ito, S., and Kobata, A. (1977). Structures of the carbohydrate moiety of ovalbumin glycopeptide III and the difference in specificity of endo-β-*N*-acetylglucosaminidase C_{II} and H. *J. Biol. Chem. 252*:6687–6694.
34. Kornfeld, S., Li, E., and Tabas, I. (1978). The synthesis of complex-type oligosaccharides. *J. Biol. Chem. 253*:7771–7778.
35. Elder, T. H., and Alexander, S. (1982). Endo-β-*N*-acetylglucosaminidase F: Endoglycosidase from *Flavobacterium meningosepticum* that cleaves both high-mannose and complex glycoproteins. *Proc. Natl. Acad. Sci. USA 79*:4540–4544.

36. Plummer, T. H., Jr., Elder, J. H., Alexander, S., Phelan, A. W., and Tarentino, A. L. (1984). Demonstration of peptide:*N*-glycosidase F activity in endo-β-*N*-acetylglucosaminidase F preparations. *J. Biol. Chem. 259*:10700–10704.
37. Huang, C. C., and Aminoff, D. (1972). Enzymes that destroy blood group specificity. *J. Biol. Chem. 247*:6737–6742.
38. Endo, Y., and Kobata, A. (1976). Partial purification and characterization of an endo-α-*N*-acetylgalactosaminidase from the culture medium of *Diplococcus pneumoniae. J. Biochem. (Tokyo) 80*:1–8.
39. Umemoto, J., Bhavanandan, V. P., and Davidson, E. A. (1977). Purification and properties of an endo-α-*N*-acetyl-D-galactosaminidase from *Diplococcus pneumoniae. J. Biol. Chem. 252*: 8609–8614.
40. Takasaki, S., and Kobata, A. (1976). Purification and characterization of an endo-β-galactosidase produced by *Diplococcus pneumoniae. J. Biol. Chem. 251*:3603–3609.
41. Kitamikado, M., Ueno, R., and Nakamura, T. (1970). Enzymatic degradation of whale cartilage keratan sulfate. I. Isolation of keratan sulfate degrading bacterium. *Bull. Jpn. Soc. Sci. Fish. 36*:592–596.
42. Fukuda, M. N., and Matsumura, G. (1976). Endo-β-galactosidase of *Escherichia freundii*. Purification and endoglycosidic action on keratan sulfates, oligosaccharides, and blood group active glycoprotein. *J. Biol. Chem. 251*:6218–6225.
43. Fukuda, M., Spooncer, E., Oates, J. E., Dell, A., and Klock, J. C. (1984). Structure of sialylated fucosyl lactosaminoglycan isolated from human granulocytes. *J. Biol. Chem. 259*:10925–10935.
44. Fukuda, M., Dell, A., Oates, J. E., and Fukuda, M. N. (1984). Structure of branched lactosaminoglycan, the carbohydrate moiety of band 3 isolated from adult human erythrocytes. *J. Biol. Chem. 259*:8260–8273.
45. Scudder, P., Hanfland, P., Uemura, K., and Feizi, T. (1984). Endo-β-D-galactosidases of *Bacteroides fragilis* and *Escherichia freundii* hydrolyze linear but not branched oligosaccharide domains of glycolipids of the neolacto series. *J. Biol. Chem. 259*:6586–6592.
46. Kitamikado, M., Ito, M., and Li, Y.-T. (1981). Isolation and characterization of a keratan sulfate-degrading endo-β-galactosidase from *Flavobacterium keratolyticus. J. Biol. Chem. 256*: 3906–3909.
47. Scudder, P., Uemura, K., Dolby, J., Fukuda, M. N., and Feizi, T. (1983). Isolation and characterization of an endo-β-galactosidase from *Bacteriodes fragilis. Biochem. J. 213*:485–494.
48. Fushiku, N., Muramatsu, H., Uezono, M. M., and Muramatsu, T. (1987). A new endo-β-galactosidase releasing Galα1→3Gal from carbohydrate moieties of glycoproteins and from a glycolipid. *J. Biol. Chem. 262*:10086–10092.
49. Takahashi, N. (1977). Demonstration of a new amidase acting on glycopeptides. *Biochem. Biophys. Res. Commun. 76*:1194–1201.
50. Plummer, T. H., Jr., and Tarentino, A. L. (1981). Facile cleavage of complex oligosaccharide from glycopeptides by almond emulsin peptide:*N*-glycosidase. *J. Biol. Chem. 256*:10243–10246.
51. Takahashi, N., and Nishibe, H. (1981). Almond glycopeptidase acting on aspartylglycosylamine linkages. *Biochim. Biophys. Acta 657*:457–467.
52. Tarentino, A. L., and Plummer, T. H., Jr. (1982). Oligosaccharides accessibility to peptide: *N*-glycosidase as promoted by protein-unfolding reagents. *J. Biol. Chem. 257*:10776–10780.
53. Takahashi, N., and Nishibe, H. (1978). Some characteristics of a new glycopeptidase acting on aspartylglycosylamine linkages. *J. Biochem. (Tokyo) 84*:1467–1473.
54. Nishibe, H., and Takahashi, N. (1981). The release of carbohydrate moieties from human fibrinogen by almond glycopeptidase without alteration in fibrinogen clottability. *Biochim. Biophys. Acta 661*:274–279.
55. Matsushima, Y., and Fujii, N. (1957). Studies on amino-hexose. IV. *N*-Deacylation with hydrazine and deamination with nitrous acid, a clue to the structure of aminopolysaccharides. *Bull. Chem. Soc. Jpn. 30*:48–50.

56. Bayard, B., and Montreuil, J. (1974). Méthodologie de la structure et du Métabolisme des Glycoconjugués. CNRS, Paris, pp. 208–218.
57. Mizuochi, T., Yonemasu, K., Yamashita, K., and Kobata, A. (1978). The asparagine-linked sugar chains of subcomponent Clq of the first component of human complement. *J. Biol. Chem. 253*:7404–7409.
58. Takasaki, S., Mizuochi, T., and Kobata, A. (1982). Hydrazinolysis of asparagine-linked sugar chains to produce free oligosaccharides. *Methods Enzymol. 83*:263–268.
59. Nilsson, B., and Svensson, S. (1978). A new method for *N*-deacetylation of 2-acetamido-2-deoxy sugars. *Carbohydr. Res. 62*:377–380.
60. Nilsson, B., and Svensson, S. (1979). A new method for degradation of the protein part of glycoproteins: Isolation of the carbohydrate chains of asialofetuin. *Carbohydr. Res. 72*:183–190.
61. Von Helferich, B., and Kleinschmidt, T. (1967). Zur Kenntnis des Süssmandel-Emulsins. Kristallisation der Komponente B. *Hoppe-Seyler Z. Physiol. Chem. 348*:753–758.
62. Yamashita, K., Mizuochi, T., and Kobata, A. (1982). Analysis of oligosaccharides by gel filtration. *Methods Enzymol. 83*:105–126.
63. Takasaki, S., and Kobata, A. (1986). Asparagine-linked sugar chains of fetuin: Occurrence of tetrasialyl triantennary sugar chains containing the Galβ1→3GlcNAc sequence. *Biochemistry 25*:5709–5715.
64. Amano, J., and Kobata, A. (1986). Purification and characterization of a novel α-mannosidase from *Aspergillus saitoi. J. Biochem. (Tokyo) 99*:1645–1654.
65. Yamashita, K., Ichishima, E., Arai, M., and Kobata, A. (1980). An α-mannosidase purified from *Aspergillus saitoi* is specific for α1,2 linkages. *Biochem. Biophys. Res. Commun. 96*:1335–1342.
66. Arakawa, M., Ogata, S., Muramatsu, T., and Kobata, A. (1974). β-Galactosidases from jack bean meal and almond emulsin: Application in the enzymatic distinction of Galβ1→4GlcNAc and Galβ1→3GlcNAc linkages. *J. Biochem. (Tokyo) 75*:703–714.
67. Paulson, J. C., Prieels, J.-P., Glasgow, L. R., and Hill, R. L. (1978). Sialyl- and fucosyltransferases in the biosynthesis of asparaginyl-linked oligosaccharides in glycoproteins. *J. Biol. Chem. 253*:5617–5624.
68. Yamashita, K., Ohkura, T., Yoshima, H., and Kobata, A. (1981). Substrate specificity of diplococcal β-*N*-acetylhexosaminidase, a useful enzyme for the structural studies of the complex type asparagine-linked sugar chains. *Biochem. Biophys. Res. Commun. 100*:226–232.
69. Bahl, O. P. (1970). Glycosidases of *Aspergillus niger* II. Purification and general properties of 1,2-α-L-fucosidase. *J. Biol. Chem. 245*:299–304.
70. Aminoff, D., and Furukawa, K. (1970). Enzymes that destroy blood group specificity. I. Purification and properties of α-L-fucosidase from *Clostridium perfringens. J. Biol. Chem. 245*:1659–1669.
71. Kochibe, N. (1973). Purification and properties of α-L-fucosidase from *Bacillus fulminans. J. Biochem. (Tokyo) 74*:1141–1149.
72. Kochibe, N., and Furukawa, K. (1980). Purification and properties of a novel fucose-specific hemagglutinin of *Aleuria aurantia. Biochemistry 19*:2841–2846.
73. Ogata, A.-M., Muramatsu, T., and Kobata, A. (1977). α-L-Fucosidase from almond emulsin: Characterization of the two enzymes with different specificities. *Arch. Biochem. Biophys. 181*: 353–358.
74. Nishigaki, M., Muramatsu, T., Kobata, A., and Maeyama, K. (1974). The broad aglycon specificity of α-L-fucosidases from marine gastropods. *J. Biochem. (Tokyo) 75*:509–517.
75. Reglero, A., and Cabegas, J. A. (1976). Glycosidases of mollusca: Purification and properties of α-L-fucosidase from *Chamelea gallina* L. *Eur. J. Biochem. 66*:379–387.
76. Cassidy, T. T., Jourdian, G. W., and Roseman, S. (1965). The sialic acids. VI. Purification and properties of sialidase from *Clostridium perfringens. J. Biol. Chem. 240*:3501–3506.
77. Schramm, G., and Mohr, E. (1959). Purification of neuraminidase from *Vibrio cholerae. Nature 183*:1677–1678.

78. Uchida, Y., Tsukada, Y., and Sugimori, T. (1974). Production of microbial neuraminidases induced by colominic acid. *Biochim. Biophys. Acta 350*:425–431.
79. Gottschalk, A. (1957). Neuraminidase: The specific enzyme of influenza virus and *Vibrio cholerae. Biochim. Biophys. Acta 23*:645–646.
80. Paulson, J. C., Weinstein, J., Dorland, L., van Halbeek, H., and Vliegenthart, J. F. G. (1982). Newcastle disease virus contains a linkage-specific glycoprotein sialidase. *J. Biol. Chem. 257*: 12734–12738.
81. Amano, J., Nishimura, R., Mochizuki, M., and Kobata, A. (1988). Comparative study of the mucin-type sugar chains of human chorionic gonadotropin present in the urine of patients with trophoblastic diseases and healthy pregnant women. *J. Biol. Chem. 263*:1157–1165.
82. Hakomori, S. (1964). A rapid permethylation of glycolipid, and polysaccharide catalyzed by methylsulfinyl carbanion in dimethyl sulfoxide. *J. Biochem. (Tokyo) 55*:205–208.
83. Stellner, K., Saito, H., and Hakomori, S. (1973). Determination of aminosugar linkages in glycolipids by methylation. *Arch. Biochem. Biophys. 155*:464–472.
84. Li, E., Tabas, I., and Kornfeld, S. (1978). The synthesis of complex-type oligosaccharides. *J. Biol. Chem. 253*:7762–7770.
85. Yamashita, K., Kamerling, J. P., and Kobata, A. (1982). Structural study of the carbohydrate moiety of hen ovomucoid. *J. Biol. Chem. 257*:12809–12814.
86. Sawardeker, J. S., Sloneker, J. H., and Jeanes, A. R. (1965). Quantitative determination of monosaccharides as their acetates by gas-liquid chromatography. *Anal. Chem. 37*: 1602–1604.
87. Lindberg, B. (1972). Methylation analysis of polysaccharides. *Methods Enzymol. 28*:178–195.
88. Tai, T., Yamashita, K., and Kobata, A. (1975). Synthesis and mass fragmentographic analysis of partially *O*-methylated 2-*N*-methylglucosamines. *J. Biochem. (Tokyo) 78*:679–686.
89. Montreuil, J., and Vliegenthart, J. F. G. (1979). Primary structure and conformation of glycans *N*-glycosidically linked to peptide chains. In *Glycoconjugate Research*, Vol. 1 (J. D. Gregory and R. W. Jeanloz, eds.), Academic Press, New York, pp. 35–78.
90. Vliegenthart, J. F. G., van Halbeek, H., and Dorland, L. (1981). The application of 500-MHz high-resolution ^{1}H-NMR spectroscopy for the structural determination of carbohydrates derived from glycoproteins. *Pure Appl. Chem. 53*:45–77.
91. Dell, A., Morris, H. R., Egge, M., von Nicolai, H., and Strecker, G. (1983). Fast-atom-bombardment mass-spectrometry for carbohydrate-structure determination. *Carbohydr. Res. 115*: 41–52.
92. Smith, F., and Van Cleve, J. W. (1955). Reduction of the products of periodate oxidation of carbohydrates. I. Hydrogenation with Raney nickel of the dialdehydes from the methol glycopyranosides. *J. Am. Chem. Soc. 77*:3091–3098.
93. Yamashita, K., Tachibana, Y., Mihara, K., Okada, S., Yabuuchi, H., and Kobata, A. (1980). Urinary oligosaccharides of mannosidosis. *J. Biol. Chem. 255*:5126–5133.
94. Ogata, S., Muramatsu, T., and Kobata, A. (1975). Fractionation of glycopeptides by affinity column chromatography on concanavalin A–Sepharose. *J. Biochem. (Tokyo) 78*:687–696.
95. Narasimhan, S., Freed, J. C., and Schachter, H. (1986). The effect of a "bisecting" *N*-acetylglucosaminyl group on the binding of biantennary, complex oligosaccharides to concanavalin A, *Phaseolus vulgaris* erythroagglutinin (E-PHA), and *Ricinus communis* agglutinin (RCA-120) immobilized on agarose. *Carbodydr. Res. 149*:65–83.
96. Taniguchi, T., Adler, A. J., Mizuochi, T., Kochibe, N., and Kobata, A. (1986). Structures of the asparagine-linked sugar chains of bovine interphotoreceptor retinol-binding protein—occurrence of fucosylated hybrid type oligosaccharides. *J. Biol. Chem. 261*:1730–1736.
97. Yamashita, K., Kochibe, N., Ohkura, T., Ueda, I., and Kobata, A. (1985). Fractionation of L-fucose-containing oligosaccharides on immobilized *Aleuria aurantia* lectin. *J. Biol. Chem. 260*:4688–4693.
98. Kornfeld, S., Reitman, M. L., and Kornfeld, R. (1981). The carbohydrate-binding specificity of pea and lentil lectins. *J. Biol. Chem. 256*:6633–6640.

99. Goldstein, I. J., and Poretz, R. D. (1986). Isolation, physicochemical characterization, and carbohydrate-binding specificity of lectins. In *The Lectins: Properties, Functions and Applications in Biology and Medicine*, Sect. 2. (I. E. Liener, N. Sharon, and I. Goldstein, eds.), Academic Press, New York, pp. 33–247.
100. Baenziger, J. U., and Fiete, D. (1979). Structural determinants of *Ricinus communis* agglutinin and toxin specificity for oligosaccharides. *J. Biol. Chem. 254*:9795–9799.
101. Irimura, T., Tsuji, T., Tagami, S., Yamamoto, K., and Osawa, T. (1981). Structure of a complex-type sugar chain of human glycophorin A. *Biochemistry 20*:560–566.
102. Cummings, R. D., and Kornfeld, S. (1982). Characterization of the structural determinants required for the high affinity interaction of asparagine-linked oligosaccharides with immobilized *Phaseolus vulgaris* leukoagglutinating and erythroagglutinating lectins. *J. Biol. Chem. 257*:11230–11234.
103. Yamashita, K., Hitoi, A., and Kobata, A. (1983). Structural determinants of *Phaseolus vulgaris* erythroagglutinating lectin for oligosaccharides. *J. Biol. Chem. 258*:14753–14756.
104. Yamashita, K., Totani, K., Ohkura, T., Takasaki, S., Goldstein, I. J., and Kobata, A. (1987). Carbohydrate binding properties of complex-type oligosaccharides on immobilized *Datura stramonium* lectin. *J. Biol. Chem. 262*:1602–1607.
105. Sutton, B. J., and Phillips, D. C. (1983). The three-dimensional structure of the carbohydrate within the Fc fragment of immunoglobulin G. *Biochem. Soc. Trans. 11*:130–132.
106. Parekh, R. B., Dwek, R. A., Sutton, B. J., Fernandes, D. L., Leung, A., Stanworth, D., Rademacher, T. W., Mizuochi, T., Taniguchi, T., Matsuta, K., Takeuchi, F., Nagano, Y., Miyamoto, T., and Kobata, A. (1985). Rheumatoid and primary osteoarthritis are associated with change in the glycosylation pattern of total serum immunoglobulin G. *Nature 316*: 452–457.
107. Harada, H., Kamei, M., Tokumoto, Y., Yui, S., Koyama, F., Kochibe, N., Endo, T., and Kobata, A. (1987). Systematic fractionation of oligosaccharides of human immunoglobulin G by serial affinity chromatography on immobilized lectin columns. *Anal. Biochem. 164*:374–381.
108. Yamashita, K., Tachibana, Y., Matsuda, Y., Katsunuma, N., Kochibe, N., and Kobata, A. (1988). Comparative studies of the sugar chains of aminopeptidase N and dipeptidylpeptidase IV purified from rat kidney brush-border membrane. *Biochemistry, 27*:5565–5573.
109. Aminoff, D., Gathmann, W. D., McLean, C. M., and Yodomae, T. (1980). Quantitation of oligosaccharides released by the β-elimination reaction. *Anal. Biochem. 101*:44–53.
110. Wu, A. M., Kobat, E. A., Pereira, M. E. A., Gruezo, F. G., and Liao, T. (1982). Immunochemical studies on blood groups: The internal structure and immunological properties of water-soluble human blood group A substance studied by Smith degradation, liberation, and fractionation of oligosaccharides and reaction with lectins. *Arch. Biochem. Biophys. 215*:390–404.
111. Lenten, L. V., and Ashwell, G. (1972). Tritium-labeling of glycoproteins that contain sialic acid. *Methods Enzymol, 28*:209–211.
112. Morell, A. G., and Ashwell, G. (1972). Tritium-labeling of glycoproteins that contain terminal galactose residues. *Methods Enzymol. 28*:205–208.
113. Harbon, S., Harbon, G., and Clauser, H. (1968). Quantitative evaluation of *O*-glycosidic linkages between sugars and aminoacids in ovine submaxillary gland mucoprotein. *Eur. J. Biochem. 4*:265–272.
114. Hase, S., Ibuki, T., and Ikenaka, T. (1984). Reexamination of the pyridylamination used for fluorescence labeling of oligosaccharides and its application to glycoproteins. *J. Biochem. (Tokyo) 95*:197–203.
115. Iwase, H., Ishii, I., Ishihara, K., Tanaka, Y., Omura, S., and Hotta, K. (1988). Release of oligosaccharides possessing reducing-endo *N*-acetylgalactosamine from mucous glycoprotein in *Streptomyces* sp. OH-11242 culture medium through action of endo-type glycosidase. *Biochem. Biophys. Res. Commun. 151*:422–428.

4

Nuclear Magnetic Resonance Approaches to Oligosaccharide Structure Elucidation

Anthony S. Serianni *University of Notre Dame, Notre Dame, Indiana*

Despite the many technological advances in biochemical instrumentation and techniques during the past two decades, the geography of the cell surface remains essentially uncharted territory. The desire to map these surfaces at the molecular level is driven by the fact that many biological processes are mediated by specific recognition between biochemical signals in the cell environment and target molecules on the cell surface. Carbohydrates play a key role in these recognition events, commonly occurring in oligomeric forms covalently attached to proteins (e.g., glycoproteins) or lipids (e.g., glycosphingolipids, glycosyl phosphatidylinositol membrane anchors) that constitute, in part, cell membrane structure. In addition to determining blood group specificity, cell surface oligosaccharides serve as receptors for the binding of toxins, viruses, and hormones. Some diseases are accompanied by structural changes in cell surface carbohydrates; for example, the autoimmune response elicited by type I diabetes mellitus in humans is caused by an altered ganglioside structure on the cell surfaces of pancreatic islet cells. Cell surface oligosaccharides may alter drug pharmacokinetics and efficacy, control key events in fertilization and early embryonic development, and target aging cells for destruction. Thus, understanding these diverse biological processes at the molecular level requires a knowledge of the structure and biochemistry of cell surface oligosaccharides.

Glycoproteins contain an oligosaccharide component covalently linked to the protein backbone by *N*-glycosyl (*N*-linked oligosaccharides) or *O*-glycosyl (*O*-linked oligosaccharides) linkages, with the former being more common; *S*-linked oligosaccharides have been found in a few instances [1] in which a D-glucosyl (D-Glc) or D-galactosyl (*D*-Gal) residue is linked to the sulfur atom of L-cysteine. In mammalian cells, *N*-linked glycoproteins have a 2-acetamido-2-deoxy-β-D-glucopyranosyl (*N*-acetyl-β-D-glucosamine; β-D-GlcNAc) ring linked to the 4-position of one or more L-asparagine (Asn) residues in protein **1** (see Scheme 1). *O*-Linked oligosaccharides are covalently attached to the protein by joining a monosaccharide, commonly a 2-acetamido-2-deoxy-α-*D*-galactopyranosyl moiety (α-D-GalNAc), to the 3-position of one or more L-serine or L-threonine residues in protein **2** (see Scheme 1), although other alcohol-containing amino acids (e.g., hydroxylysine or hydroxyproline) can be involved.

N-Linked oligosaccharides have a common core structure consisting of two *N*-acetyl-D-glucosamine (GlcNAc) units linked β-(1→4), followed by a D-mannose (D-Man) residue, from which branch two D-Man units **3** (see Scheme 2). The three major classes of *N*-linked oligosaccharides contain this core structure, but differ in the sugar residues extending from this core. These three classes are (1) the *high mannose*-type **4**, containing 4–11 more mannose units attached in various ways to the core; (2) the complex-type **5**, containing additional D-GlcNAc, D-Gal, sialic acid (*N*-acetyl-D-neuraminic acid; SA), and L-fucose (L-Fuc; 6-deoxy-L-galactose) units; and (3) the *hybrid*-type **6**, containing both high mannose and complex units (see Scheme 2).

In animal glycoproteins, the following simple monosaccharides are commonly found as constituents of *N*-linked oligosaccharides: β-D-GlcNAc, α-D-Man, β-D-Man, α-D-Gal, β-D-Gal, α-SA, and α-L-Fuc. Monosaccharides found in animal *O*-linked oligosaccharides include α-D-GalNAc, α-D-GlcNAc, β-D-GlcNAc, α-D-Gal, β-D-Gal, α-L-Fuc, α-SA, α-D-Xyl (α-D-xylose) and β-D-Glc. Other sugars may also be found as constituents of these oligosaccharides; for example, sulfated monosaccharides are sometimes found in *N*-linked oligosaccharides in animals [2].

Adding to this complexity is that glycoproteins are generally not chemically homogeneous compounds. A "purified" glycoprotein, for example, may be a heterogeneous mixture of molecules having the same polypeptide primary sequence, but different oligosaccharides appended to the same site or to differing numbers of sites on the protein. This *microheterogeneity* gives rise to protein subsets called *glycoforms* that have

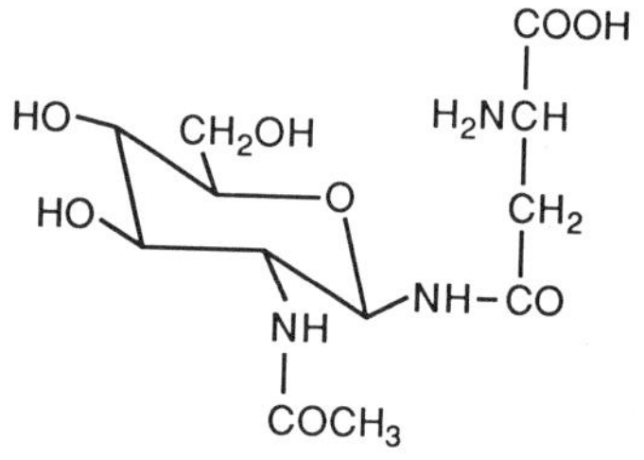

COOH
H_2NCH
CH_2
HO CH_2OH O HO NH–CO
NH
$COCH_3$

1

HO CH_2OH COOH
HO O H_2NCH
O–CH_2
NH
$COCH_3$

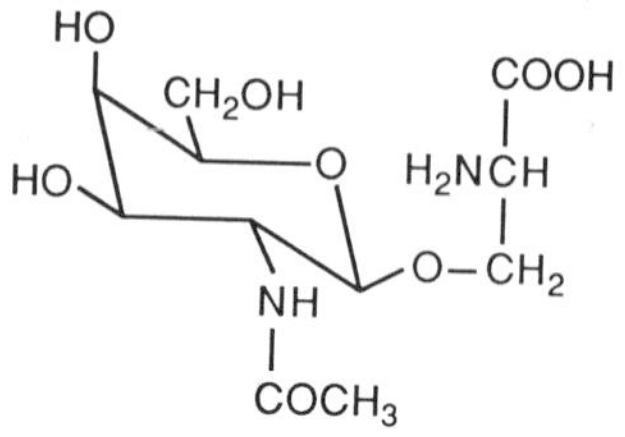

2

Scheme 1

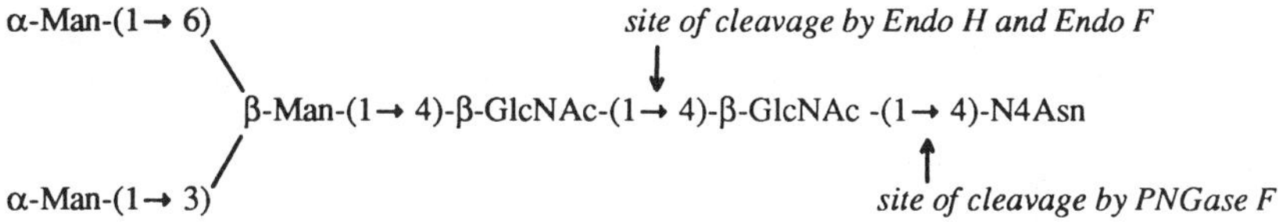

```
α-Man-(1→ 6)                                site of cleavage by Endo H and Endo F
            \                                    ↓
             β-Man-(1→ 4)-β-GlcNAc-(1→ 4)-β-GlcNAc -(1→ 4)-N4Asn
            /                                                  ↑
α-Man-(1→ 3)                                         site of cleavage by PNGase F
```

3

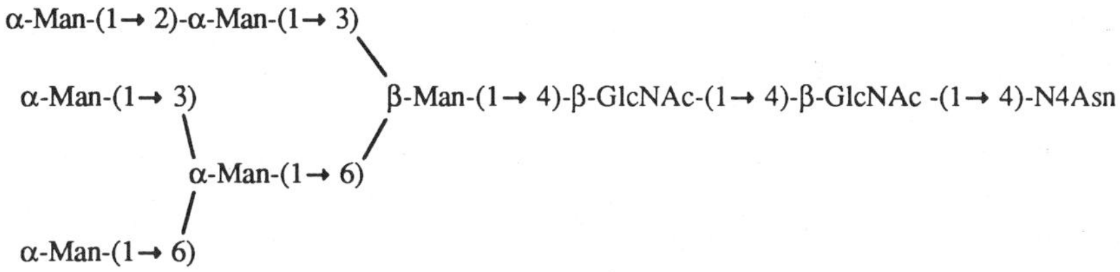

```
α-Man-(1→ 2)-α-Man-(1→ 3)
                         \
 α-Man-(1→ 3)             β-Man-(1→ 4)-β-GlcNAc-(1→ 4)-β-GlcNAc -(1→ 4)-N4Asn
            \            /
             α-Man-(1→ 6)
            /
 α-Man-(1→ 6)
```

4

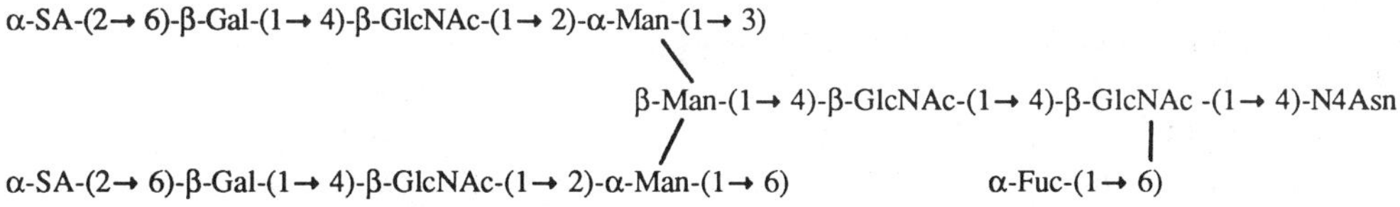

```
α-SA-(2→ 6)-β-Gal-(1→ 4)-β-GlcNAc-(1→ 2)-α-Man-(1→ 3)
                                              \
                                               β-Man-(1→ 4)-β-GlcNAc-(1→ 4)-β-GlcNAc -(1→ 4)-N4Asn
                                              /                                   |
α-SA-(2→ 6)-β-Gal-(1→ 4)-β-GlcNAc-(1→ 2)-α-Man-(1→ 6)                      α-Fuc-(1→ 6)
```

5

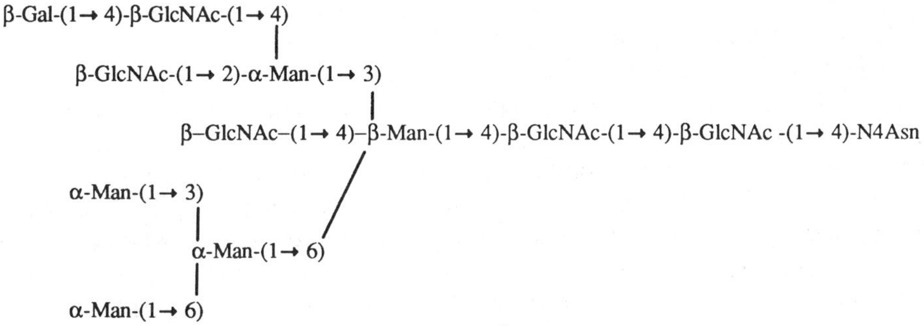

```
β-Gal-(1→ 4)-β-GlcNAc-(1→ 4)
                          |
     β-GlcNAc-(1→ 2)-α-Man-(1→ 3)
                              |
             β–GlcNAc–(1→ 4)–β-Man-(1→ 4)-β-GlcNAc-(1→ 4)-β-GlcNAc -(1→ 4)-N4Asn
                              /
 α-Man-(1→ 3)                /
          |                 /
          α-Man-(1→ 6)
          |
 α-Man-(1→ 6)
```

6

Scheme 2

different physical and biochemical properties. For example, ovalbumin contains one site of *N*-glycosylation, yet a variety of oligosaccharide structures have been found appended to this glycoprotein [3]. From a biological standpoint, it is the glycoform population—specifically, the weighted average of the activity and the quantity of each glycoform—that determines the activity of a given glycoprotein in vivo.

Thus, to determine the structure of a cell surface oligosaccharide attached to a glycoprotein requires purification of the target protein, followed by cleavage of the covalently bound carbohydrate chain from the protein. As discussed in more detail in other chapters of this book, the reagents used for selective cleavage may be chemical or enzymic. Isolation of the oligosaccharide chain from a glycoprotein is usually accomplished by one of the following methods: treatment with $NaOH-NaBH_4$, hydrazinolysis, or proteolysis. The first of these is used, somewhat specifically, to liberate *O*-linked chains from serine or threonine residues, whereas hydrazinolysis is used to cleave *N*-linked oligosaccharides. In contrast with these two chemical methods that liberate the native oligosaccharide ($NaOH-NaBH_4$ treatment gives an oligosaccharide alditol), enzyme-catalyzed proteolysis yields glycopeptides. All of these methods have their limitations, as discussed previously by Dill et al. [4].

Recently, *N*-linked oligosaccharides have been isolated by treatment of the intact glycoprotein with the specific endoglycosidases, endoglycosidase H (*Streptomyces plicatus*) and endoglycosidase F (*Flavobacterium meningosepticum*), or with *N*-glycosidase (PNGase) F [5]. Endo H and F cleave the di-*N*-acetylchitobiose moiety in the core of Asn-linked oligosaccharide chains (see Scheme 2), whereas PNGase F is a peptide *N*-glycosidase that specifically hydrolyzes *N*-acetylglucosaminylasparagine linkages in *N*-linked glycoproteins (see Scheme 2). Treatment of a glycoprotein with these and related enzymes liberates the oligosaccharide under mild conditions, thereby minimizing potential chemical perturbations during isolation. The use of endoglycosidases is likely to replace, or at least complement, the more traditional chemically based methodologies.

Regardless of the method chosen to liberate the bound oligosaccharide chain(s), the free oligosaccharides require purification, usually by some form of high-pressure liquid chromatography (HPLC) [6]; several systems have been developed for this purpose, including those equipped with pulsed amperometric detection (PAD) to permit the detection of small ($<10^{-10}$g) amounts of material. Sample purity is a prerequisite to a structure determination by any method, including nuclear magnetic resonance (NMR) spectroscopy; conflicting reported oligosaccharide structures in the past are likely to have derived from the use of impure, heterogeneous samples in the analysis.

The choice of method to evaluate the molecular structure of a cell surface oligosaccharide depends highly on the available quantity of sample. NMR methods require substantially more sample for analysis than do gas–liquid chromatography or mass spectrometry. Typically, a 0.5–5 mM solution (0.5 ml) of a pure oligosaccharide is required to conduct 500 MHz ^{1}H NMR experiments in the fourier-transform (FT) mode in two dimensions (2D) in a reasonable time frame, although less material (0.05 mM) is needed for single-dimension (1D) experiments [7]; ^{13}C NMR spectroscopic studies require in excess of ten times more sample (unless reverse detection is used; see later) than is needed for ^{1}H studies.

I. WHY USE NUCLEAR MAGNETIC RESONANCE SPECTROSCOPY TO ASSESS OLIGOSACCHARIDE STRUCTURE?

Several experimental methods may be applied to determine the chemical structure of an isolated oligosaccharide. Wet chemical methods (e.g., methylation analysis by gas chromatography–mass spectrometry (GCMS)) have been used traditionally to determine primary structure [8], although the reductive cleavage method has been introduced recently to assist in primary structure elucidation [9,10]. The analysis of isolated, *intact* oligosaccharides by mass spectrometry promises to provide information on primary structure rapidly and with high sensitivity [11,12]; for this, ion or atom liquid matrix sputtering ionization techniques and higher mass, high-performance tandem mass spectrometry show particular promise. Most importantly, these approaches require considerably less sample than is required to conduct an analysis by NMR spectroscopy. Thus, when sample quantities are limited, the investigator must rely on these more sensitive methods to determine structure. What, then, are the incentives to developing NMR spectroscopy as a tool to determine oligosaccharide structure?

An advantage of the NMR approach commonly cited is its nondestructive nature; that is, the sample may be recovered unadulterated after completion of the analysis. Although this is an attractive feature of the method, it should be recognized that the alternate, more sensitive methods identified in the foregoing, although destructive to the sample, usually require only a small investment of sample for a successful analysis. Thus, sample noninvasiveness is not the primary reason for developing NMR strategies to assess oligosaccharide structure.

The main incentive for the use of the NMR method lies in the fact that NMR yields a more complete picture of oligosaccharide structure and behavior in solution. In addition to determining primary structure, NMR provides information on the conformation and molecular dynamics (motional characteristics) of the molecule in the solution state. Knowledge of this higher-order structure is essential to understanding how these molecules mediate biological recognition on cell surfaces. Thus, although NMR spectroscopy may not be the method of choice to determine oligosaccharide primary structure, it is clearly preferred when molecular conformation and dynamics are the focus of attention.

The analytical reader may realize that the molecular behavior of an isolated oligosaccharide chain from a glycoprotein hardly mimics the behavior of the chain when bound to a protein. For example, the molecular motion of a freely tumbling oligosaccharide will be substantially different from that of the same oligosaccharide anchored to a protein backbone. It is less clear whether molecular conformation/internal dynamics is affected when a free oligosaccharide is covalently attached to a protein. Thus, appropriate caution must be exercised when extrapolating NMR-derived conformation and dynamics results on isolated oligosaccharides to similar structures linked to a protein anchor.

It should be noted that ^{13}C NMR spectra obtained on intact glycoproteins [13] in solution show that those signals arising from the sugar component are considerably sharper than those arising from the polypeptide backbone. This observation suggests that the oligosaccharides of at least some intact glycoproteins possess motional characteristics other than those conferred to them by the tumbling of the protein; that is, local (segmental) motion is experienced by the oligosaccharide chain. This local motion may play a role in the thermodynamics and energetics of specific binding and recognition in vivo.

II. SAMPLE PREPARATION FOR NUCLEAR MAGNETIC RESONANCE STUDIES

Sample preparation is an important component of a successful NMR investigation. For high-resolution studies at high fields (>300 MHz), only premium quality NMR tubes should be employed, to ensure optimal line shape and line width. Typically 5- and 10-mm internal diameter (ID) NMR tubes are used for ^{1}H and ^{13}C studies, respectively, although some ^{13}C studies are conducted with 5-mm tubes (smaller sample tubes may be used if the NMR spectrometer is equipped with a microprobe) when sample quantities are limited.

The choice of solvent is a key variable in sample preparation. For oligosaccharide studies for which biological relevance is essential (i.e., conformation and dynamic studies), the use of ^1H_2O or ^2H_2O is preferred. When oligosaccharide primary structure is the sole aim, nonaqueous solvents (e.g., DMSO-d_6) may be employed; nonaqueous solvents are commonly used in this context when the spectral dispersion is found to be improved in these solvents relative to that found in $^1H_2O/^2H_2O$. If dimethyl sulfoxide (DMSO)-d_6 is used, care must be taken to exchange the sample in ^2H_2O by repeated evaporation or lyophilization from ^2H_2O before dissolving the sample in DMSO-d_6 [14]. This exchange substitutes the exchangeable protons in the sample with ^2H and thereby eliminates their signals in ^1H NMR spectra. In the absence of this exchange, these exchangeable ^1H signals will contribute additional unnecessary complexity to the spectra. In some cases, however, the observation of hydroxyl protons is of value in carbohydrate structure determinations [15].

Normally NMR studies in aqueous solution are conducted in ^2H_2O solvent, but in some cases ^1H_2O is desirable. Studies with ^1H FT-NMR that are conducted in ^1H_2O are complicated by the dynamic range problem [for brief discussion, see 16] caused by the presence of an intense solvent signal in the midst of much weaker solute signals. Unless the solvent signal is reduced in magnitude, the detection of these weak solute signals cannot be achieved. Several methods have been developed for solvent signal suppression [17]. These methods include simple presaturation by coherent homonuclear decoupling [18], the use of partially relaxed spectra [19], the jump-and-return method [20], multipulse methods (e.g., the Redfield 2-1-4 pulse sequence [21], and pulse-shaping techniques [22].

Care should also be exercised to exclude contaminating paramagnetic agents from the sample solution. Paramagnetic metals may be removed by treating the sample solution with a chelating resin, or by adding a chelating agent (e.g., ethylenediaminetetraacetic acid; EDTA) in low concentrations to the sample solution. Oxygen is removed by degassing the sample with argon before the analysis.

III. FUNDAMENTAL STRUCTURAL CHARACTERISTICS OF OLIGOSACCHARIDES

Oligosaccharides are composed of monosaccharides linked to one another by *O*-glycosidic bonds. To solve the complete structure of a simple oligosaccharide, the following questions must be addressed: (1) What monosaccharide units are present in the structure, and what are their ring forms (i.e., pyranose or furanose)? (2) How are the monosaccharide units linked to one another? (3) What are the anomeric configurations of each glycosidically linked monosaccharide unit? (4) What are the preferred conformations of the constituent monomers, and the preferred torsion angles about each *O*-glycoside bond of the structure? (5) What are the preferred molecular torsion angles describing hydroxymethyl (if present) conformation in the structure? (6) What are the overall or local motional/dynamical properties of the molecule? Question (1), although sometimes solvable by NMR, is probably best addressed by chemical or enzymic hydrolysis of the oligosaccharide, followed by reduction and analysis of the alditol acetates by gas chromatography–mass spectrometry [9]; in some cases, a comparison of the ^1H or ^{13}C NMR spectrum of an unknown intact oligosaccharide with spectral data found in appropriate data bases [23a] may provide a means to assess composition; two review articles have appeared recently in which ^{13}C chemical shift data have been tabulated for monosaccharides [23b] and oligosaccharides [23c]. Questions (2) and (3) may be addressed by techniques in addition to NMR spectroscopy [e.g., methylation analysis (2) and hydrolysis by specific exo- or endoglycosidases (3)]. Questions (4–6) are best addressed through the assessment of specific NMR parameters (Table 1).

Nuclear magnetic resonance spectroscopy provides an array of physical parameters that may be brought to bear on the problem of oligosaccharide structure determination. Why is it, then, that such structure determinations are not—at least at present—considered straightforward? The complexity of the problem arises primarily because ^{1}H NMR spectra of complex oligosaccharides contain a multitude of resonances clustered over a limited range of chemical shifts (typically between 3 and 4 ppm) (Fig. 1). The significant signal overlap makes it difficult to locate the resonances of individual nuclei and assign these resonances to specific nuclei in the oligosaccharide structure. Furthermore, in ^{1}H spectra, the limited chemical shift range and the presence of homonuclear ^{1}H-^{1}H spin-coupling often conspire to create non–first-order spectra that cannot be easily interpreted to yield valid ^{1}H-^{1}H spin-couplings without assistance from spectral simulation by computer. Thus, NMR analyses of oligosaccharides are best conducted on high-field spectrometers (>400 MHz) to maximize spectral dispersion and minimize higher-order effects.

Useful information can, however, be obtained from one-dimensional ^{1}H NMR spectra of oligosaccharides, despite the often encountered complications caused by resonance overlap. The use of "structural-reporter groups," first described by Vliegenthart and co-workers [24], permits an interpretation, if only partial, in terms of primary structure. This interpretation is possible because specific types of protons resonate in characteristic regions of the spectrum; this chemical shift information can be used in conjunction with vicinal ^{1}H-^{1}H coupling constants ($^3J_{HH}$) (e.g., $^3J_{H1,H2}$), signal line-widths, and spectral integrations to identify the kinds of monosaccharide units present and their relative abundances. In *N*-linked oligosaccharides, several reporter groups have been identified: (1) anomeric protons, (2) mannose H-2 and H-3 protons, (3) sialic acid H-3 protons, (4) fucose H-5 and CH_3 protons, (5) galactose H-3 and H-4 protons, and (6) aminosugar *N*-acetyl CH_3 protons (see Fig. 1) [24].

Because the anomeric protons of oligosaccharides resonate within a distinct region of the ^{1}H spectrum (typically between 4.5 and 5.5 ppm), measurement of anomeric proton chemical shifts and $^3J_{H1,H2}$ is often straightforward and gives essential information on the monosaccharide type. For example, α-D-Gal, α-D-GlcNAc, and α-L-Fuc residues give $^3J_{H1,H2}$ of ~2–3 Hz (H-1 and H-2 are axial-equatorial; dihedral angle ~60°), whereas $^3J_{H1,H2}$ of ~7–9 Hz is observed in the corresponding β-anomers (H-1 and H-2 are diaxial; dihedral angle ~180°). Mannose residues pose a problem since $^3J_{H1,H2}$ is small in both anomers (α,

Table 1 NMR Parameters and Their Use in Oligosaccharide Structure Determination

Information	Relevant NMR parameter
Linkage sites	Chemical shift, NOE, spin coupling (^{13}C–^{1}H)
Anomeric configuration	Chemical shift, spin-couplings (^{1}H–^{1}H, ^{13}C–^{1}H)
Monomer conformation	Spin-couplings (^{1}H–^{1}H, ^{13}C–^{1}H, ^{13}C–^{13}C)
O-Glycoside conformation	Nuclear spin relaxation (T_1, T_2), NOE, spin-coupling (^{13}C–^{1}H, ^{13}C–^{13}C)
Hydroxymethyl conformation	Spin-couplings (^{1}H–^{1}H, ^{13}C–^{1}H)
Motional/dynamical properties	Nuclear spin relaxation (T_1, T_2), NOE

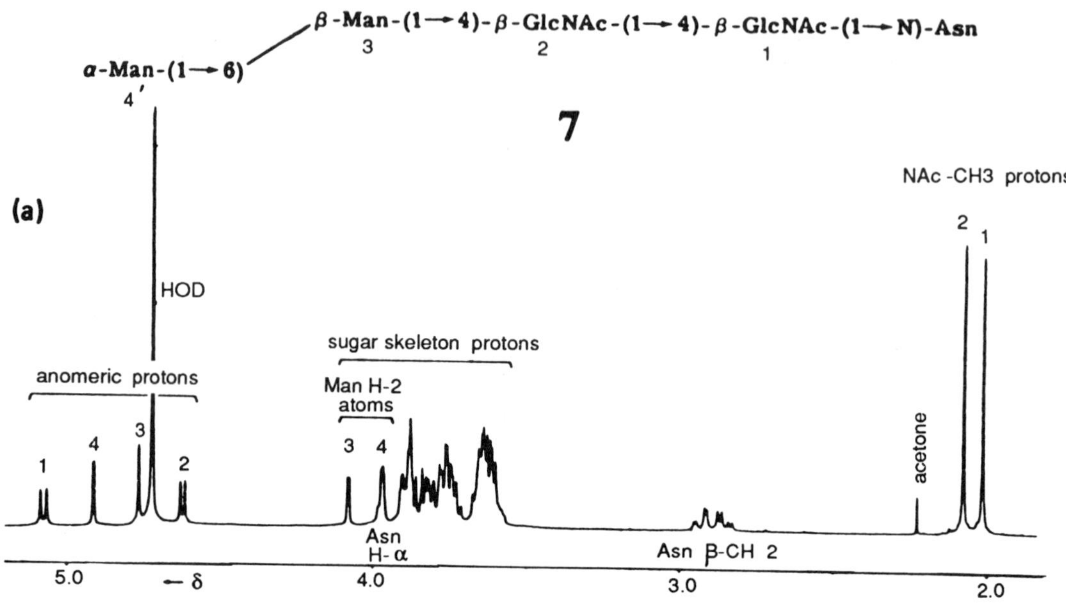

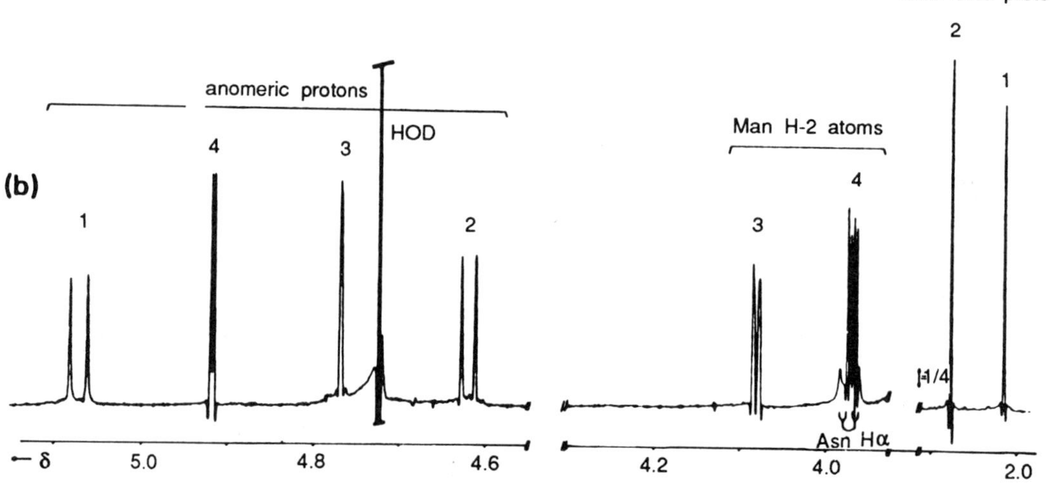

Figure 1 (a) The 500-MHz ^{1}H NMR spectrum of compound **7**. (b) The structural-reporter-group regions of the resolution-enhanced 500-MHz ^{1}H NMR spectrum of **7**. The bold numbers in the spectra refer to the corresponding residues in the indicated structure. The relative intensity scale of the expanded, *N*-acetyl-proton region differs from that of the other parts of the spectrum, as indicated. (From Ref. 24.)

~1.6 Hz; β, ~0.8 Hz; the dihedral angles between H-1 and H-2 are ~60° in both anomers), but chemical shift can sometimes assist in distinguishing these anomers [24].

Significant improvement in the interpretation of ^{1}H and ^{13}C NMR spectra of complex oligosaccharides derives from the application of two-dimensional (2D) NMR methods, which have been stimulated, in large part, by dramatic advances in computer technology and in the construction of high-field, superconducting magnets over the past decade. A

Table 2 Two-Dimensional NMR Methods Commonly Used in Oligosaccharide Structure Determinations

2D method	F1 Axis	F2 Axis	1D Analogue
J-Resolved spectroscopy			
homonuclear 2-DJ	J_{HH}	$\delta(^1H)$	None
Correlated spectroscopy			
COSY, relayed COSY	$\delta(^1H)$	$\delta(^1H)$	Homonuclear decoupling
DQF-COSY, TQF-COSY	$\delta(^1H)$	$\delta(^1H)$	Homonuclear decoupling
HOHAHA (TOCSY)	$\delta(^1H)$	$\delta(^1H)$	1-D HOHAHA
NOESY	$\delta(^1H)$	$\delta(^1H)$	1-D NOE
HETCOR, HMQC, HMBC	$\delta(X)$	$\delta(^1H)$	Selective ^1H-decoupling

multitude of 2D NMR pulse sequences (Table 2) have been developed and applied to carbohydrate-related systems; for example, a recent issue of the *Journal of Carbohydrate Chemistry* [25] was devoted to this subject, and several excellent books and reviews have appeared recently on the theory and applications of 2D NMR spectroscopy [26–30]. The 2D methods simplify complex spectra by increasing the spectral dispersion to facilitate a complete assignment of signals. Many of these experiments have one-dimensional analogues (see Table 2) that are applied in lieu of the full 2D version for reasons of speed and convenience [31].

There are two fundamental types of 2D NMR spectroscopy: J-resolved spectroscopy and correlated spectroscopy (see Table 2). Spectra of the first type contain two frequency axes (F1 and F2, respectively), one containing spin coupling (J) information and the other containing information on chemical shift (δ). In correlated spectroscopy, both frequency axes contain chemical shift (δ) information. The third dimension in both types of 2D spectroscopy is signal intensity, which in a typical 2D display is represented in the form of contours. J-Spectroscopy and correlated spectroscopy may be performed in either the homonuclear (i.e., couplings between similar nuclear spins) or heteronuclear (i.e., coupling between different nuclear spins) modes.

Correlated 2D NMR spectroscopy may be divided into two groups based on the mechanism of interaction producing the observed signals. The first group, which includes COSY and its descendants, HOHAHA, HETCOR, and HMQC, is based on scalar couplings through coherent transfer of transverse magnetization, and reveals *through-bond* connectivities. The second group is based on dipole couplings through incoherent transfer of magnetization and provides *through-space* connectivities; the NOESY experiment is representative of this group. In this review, spin-echo correlated spectroscopy (SECSY) is not discussed, as this method is not commonly practiced today, although there are some advantages of this method in oligosaccharide NMR as discussed by Koerner et al. [32].

IV. SALIENT FEATURES OF TWO-DIMENSIONAL NUCLEAR MAGNETIC RESONANCE SPECTROSCOPY AS APPLIED TO OLIGOSACCHARIDES

In this section, the fundamental features of each 2D NMR experiment are described, and examples of each method as applied to oligosaccharides are given. The reader is referred to other recent reviews [32–37] on oligosaccharide NMR for additional treatments of this subject. No attempt has been made to describe the individual pulse sequences and their implementation, and the reader is referred to recent literature dealing with these matters [26–30].

A. Correlated Spectroscopy

Correlated spectroscopy (COSY) [38] provides information on spin- coupling networks within the constituent residues of the oligosaccharide through the observation of cross-peaks that correlate mutually coupled signals in the 1D spectrum (which lies along the diagonal) (Fig. 2). The aforenoted structural reporter groups serve as convenient "initiation points" in the analysis; in practice, the anomeric protons, which are frequently well resolved and have characteristic chemical shifts, are commonly exploited. Within a typical aldohexopyranosyl ring, the coupling network is unidirectional, that is, H-1 couples to H-2, H-2 couples to H-1 and H-3, H-3 couples to H-2 and H-4, and so forth; the absence of coupling between H-1 and H-5 confers this unidirectionality and, thereby, notably simplifies the interpretation of COSY spectra of carbohydrates.

The success of the COSY experiment as applied to oligosaccharides in its simplest form is often compromised by two problems: residual spectral overlap or crowding that defies analysis, and the presence of no or small scalar couplings between vicinally related protons that prevents the detection of cross-peaks. A common example of the latter is the small coupling typically observed between H-4 and H-5 of galactopyranosyl (and fucopyranosyl) rings. This small coupling gives rise to a weak cross-peak (oftentimes no cross-peak is observed) and, thereby, obviates the determination of a complete set of coupling correlations within the ring (i.e., H-5, H-6, and H-6′ cannot be correlated). A potentially more severe problem is encountered in mannose residues in larger oligosaccharides in which the small $^3J_{H1,H2}$ (see foregoing) may be exceeded by natural resonance line-widths, and weak (or no) cross-peaks between H-1 and H-2 are observed.

B. Double-Quantum Filtered Correlated Spectroscopy

The basic COSY sequence has been modified to select for multiple quantum coherence transfers to simplify (edit) complex COSY spectra. DQF-COSY [39] (Fig. 3) is the most common modified form and offers some advantages over the basic COSY experiment. The introduction of a double-quantum filter (DQF) generates COSY spectra in which both the cross-peak and diagonal multiplets possess an antiphase structure and, thus, the cross-peaks close to the diagonal may be more easily observed and analyzed. Furthermore, DQF-COSY spectra suppress the detection of spin-isolated protons (i.e., uncoupled singlets) such as those arising from solvent or isolated methyl groups. The suppression of these signals is useful, especially when they occur in crowded regions of the spectrum.

C. Triple-Quantum Filtered Correlated Spectroscopy

In some cases COSY employing a triple-quantum filter (TQF-COSY) [40] can help to further edit COSY spectra of oligosaccharides. This technique is applied often to assist in the detection and assignment of the H-6 and H-6′ protons of constituent aldohexopyranosyl rings containing an exocyclic hydroxymethyl group (Fig. 4) (isolated hydroxymethyl

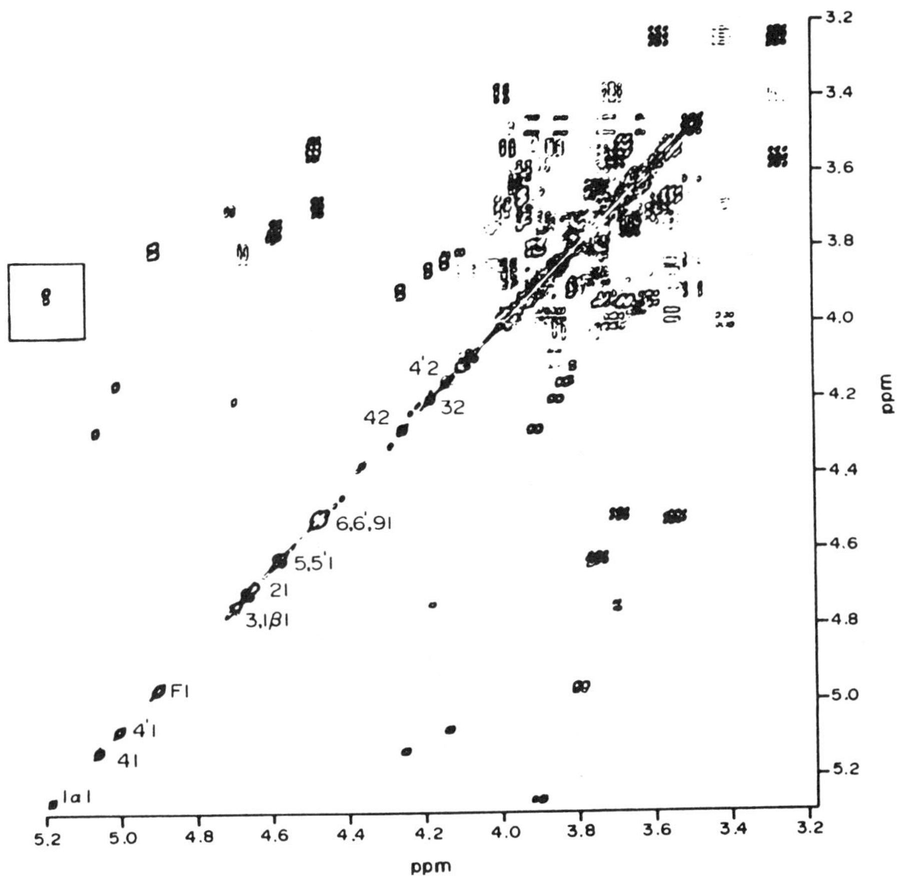

Figure 2 The homonuclear ^{1}H-^{1}H COSY spectrum of the oligosaccharide:

```
β-Gal-(1→4)-β-GlcNAc (1→2)-α-Man-(1→6)          α-Fuc-(1→6)
                                        |              |
                     β-GlcNAc-(1→4)-β-Man-(1→4)-β-GlcNAc-(1→4)-GlcNAc
                                        |
β-Gal-(1→4)-β-GlcNAc-(1→2)-α-Man-(1→3)
```

(From Ref. 34.)

groups, such as those appended to C-2 of 2-ketoses, will not be detected by TQF-COSY). From the selection rules of multiple-quantum coherence, only n-spin systems with n-1 resolved couplings will pass an n quantum filter [40b]. An additional coupling is required so that the spin system will evolve to generate a cross-peak in the TQF-COSY experiment. Thus, all cross-peaks will be edited from a conventional COSY map except those correlating the H-5, H-6, and H-6′ protons. A notable problem with TQF-COSY is its inherent low sensitivity compared with COSY.

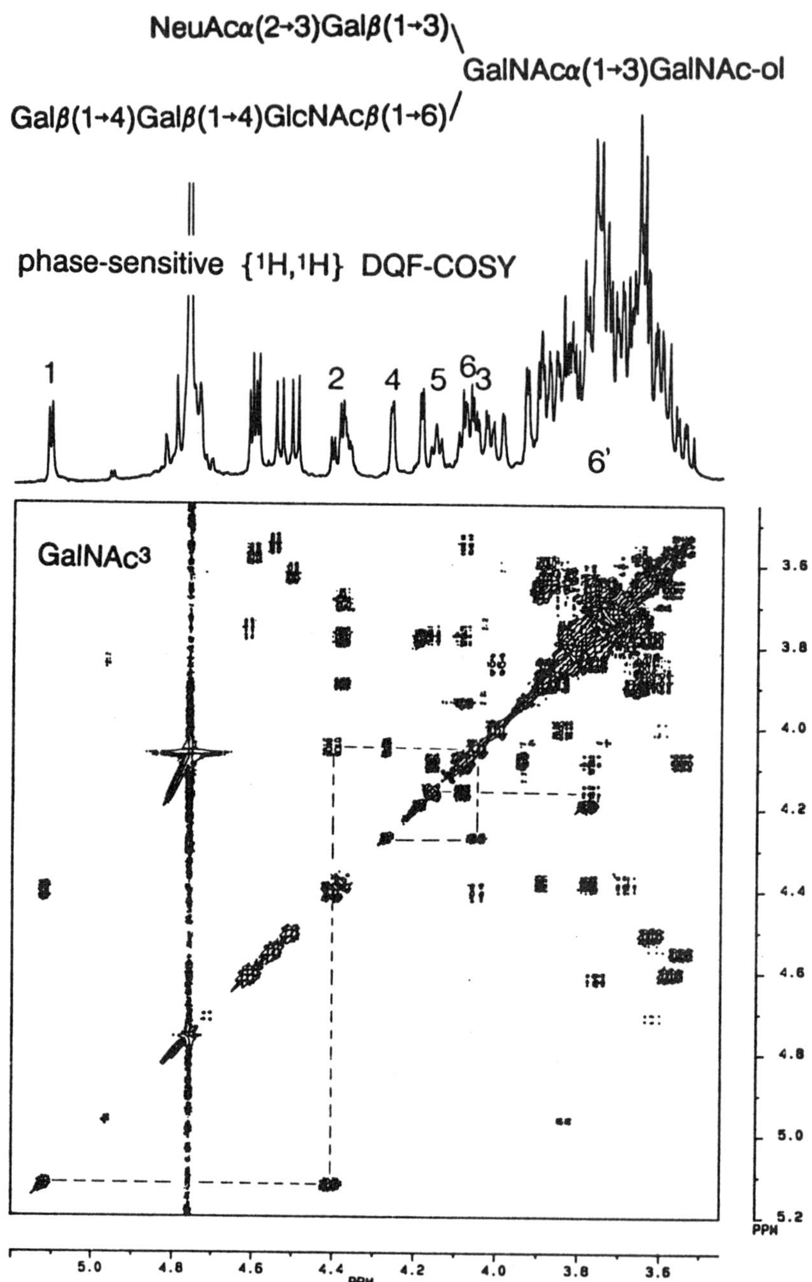

Figure 3 Section of the 2D{^{1}H,^{1}H} DQF-COSY spectrum of a heptasaccharide isolated from the salivary mucins of the Chinese swiftlet (genus *Collocalia*). The structure of the compound and the conventional 1D ^{1}H spectrum are shown on top of the contour map. Dashed lines connect signals from protons (1–6 and 6′) of the branching GalNAc^3 residue. The sample contained 4 mg of the heptasaccharide in 0.4 ml ^2H_2O at p^{2}H 6 and 25°C. The COSY experiment was performed in the phase-sensitive mode (TPPI) at 500 MHz on a Bruker AM-500. Data matrix: 512 × 2048; 64 scans per t_1 value. Sine-bell windows were used in both dimensions. Total measuring time: 4 hr. (From Ref. 37).

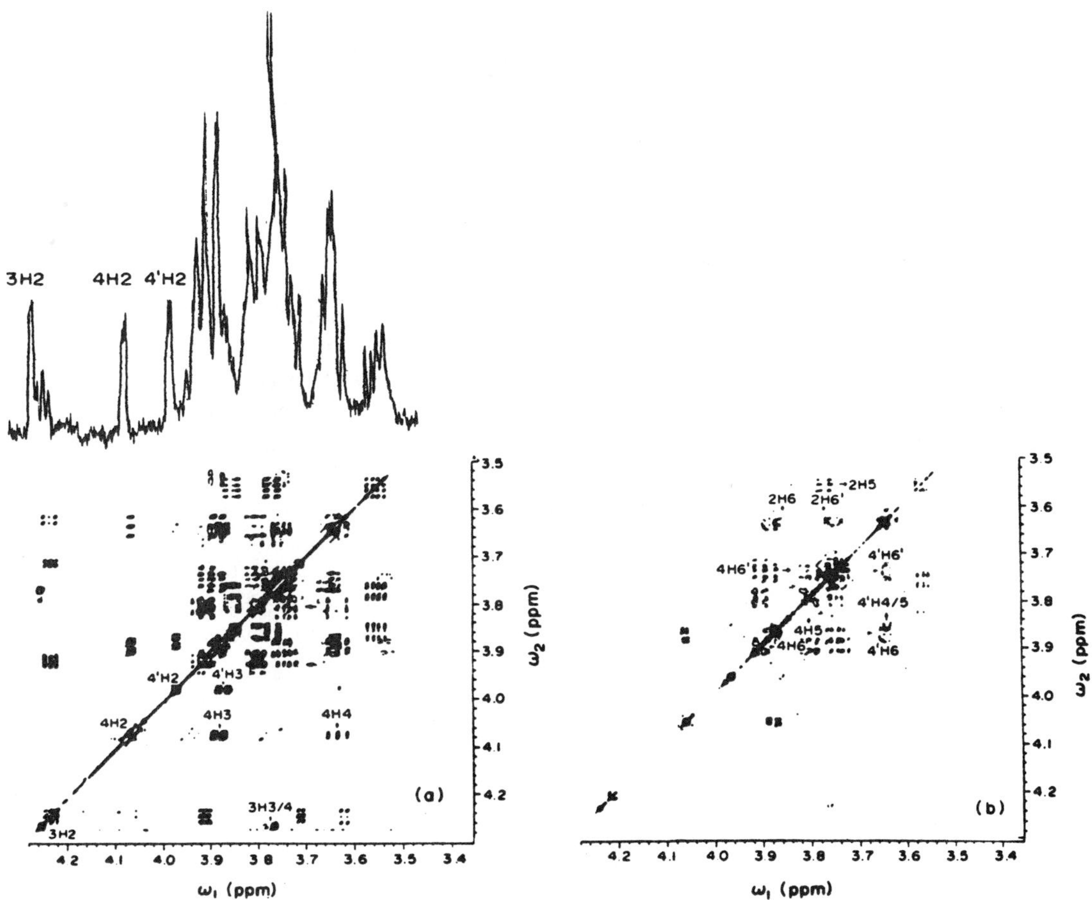

Figure 4 Illustration of COSY (left) vs TQF-COSY (right) for the pentasaccharide:

```
        4'
α-Man-(1→6)
              \  3                  2                    1
               β-Man-(1→4)-β-GlcNAc-(1→4)-β-GlcNAc
              /
α-Man-(1→3)
        4
```

All cross-peaks are purged in TQF-COSY, except those correlating the C-5 proton with C-6 protons. An exception is the cross-peak correlating the C-2 proton with the C-3 proton (labeled 4H3 in the COSY spectrum) of residue 4′, owing to the resolved coupling between the C-2 and C-4 protons of this residue. (From Ref. 40a).

D. Relayed Correlation Spectroscopy

Relayed correlation spectroscopy (RELAY-COSY) [41] provides correlations between the anomeric protons of oligosaccharides and other intraresidue protons that are not directly spin-coupled to them. For example, in an aldohexopyranosyl moiety, cross-peaks between

H-1 and H-2, H-3, H-4, and H-5 can be observed, provided that the three delays, t1, t2, and t3, in the pulse sequence are chosen correctly (Fig. 5). The success of RELAY-COSY depends on the fact that long-range ($>^3J$) coupling between pyranosyl ring protons is effectively zero. The presence of strong coupling, however, as found in ^1H spectra that are non–first-order, can interfere with the transfer efficiency of the experiment, resulting in the detection of weak cross-peaks.

E. Homonuclear Hartmann-Hahn Spectroscopy

Homonuclear Hartmann–Hahn (HOHAHA) spectroscopy [42, 43] is related to total correlation spectroscopy (TOCSY) introduced by Braunschweiler and Ernst [44]. During the mixing time of the pulse sequence, coherence transfer is induced between all spins in a given spin system, resulting in a 2D spectrum analogous to that obtained from multiple-step relayed COSY. In contrast to RELAY-COSY, however, spin propagation throughout the spin system is less sensitive to the magnitudes of $^3J_{HH}$, although in large oligosaccharides coherence transfer through small couplings (e.g., the small $^3J_{H4,H5}$ of aldopyranosyl rings having the *galacto* configuration) can be a problem. To generate a suitable tailored Hamiltonian during the mixing time for coherence transfer, a spin-locking pulse is usually employed with different degrees of compensation for resonance offset [29]. Like RELAY-COSY, the anomeric protons are used to "spy" on the remaining protons within each

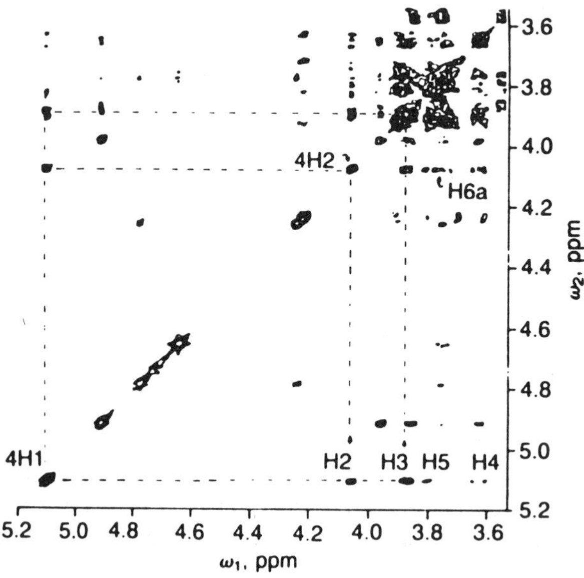

Figure 5 Contour plot of the 2D multistep relayed correlation spectrum (3.6–5.2 ppm) of the pentasaccharide Manα(1→3)[Manα(1→6)]Manβ(1→4)GlcNAcβ(1→4)GlcNAc (units labeled 4, 4′, 3, 2, 1), recorded at 470 MHz. The broken lines correlate mannose 4H1 with 4H2. In addition, mannose 4H1 can be correlated directly with mannose 4H3, as shown. The correlations with mannose 4H4 and 4H5 are determined similarly. The correlation between mannose 4H2 and 4H6 is also illustrated. (From Ref. 41.)

constituent residue of the oligosaccharide, thereby facilitating their identification and assignment (Fig. 6). A slice through a HOHAHA spectrum at each anomeric proton along the diagonal yields a ^{1}H subspectrum containing all scalar-coupled protons within that residue.

The one-dimensional version of HOHAHA is often useful and more time-efficient than the 2D version when the anomeric protons are well resolved. The 1D version is performed by applying a selective 180° pulse at each anomeric proton resonance, followed by the application of the spin-locking pulse and signal detection [45]. This method produces subspectra composed of each spin system detected via each anomeric proton (essentially ^{1}H spectra of each monosaccharide residue in the oligosaccharide). The 1D analogue of HOHAHA has found application in circumventing the small coupling problem, as described by Inagaki et al. [46].

F. Nuclear Overhauser Effect Spectroscopy

Although NOESY [47] may be used as an assignment aid, its principal use in oligosaccharide structure determination has been in the assessment of molecular conformation (i.e., 3D structure). Cross-peaks are observed in 2D NOESY spectra between proton pairs that are close in space (i.e., typically less than 5 Å). Thus, NOESY cross-peaks are observed between intraresidue proton pairs; for example, the internuclear distance between H-1 and H-5 (and H-3) of β-D-galactopyranosyl residues is about 2.5 Å, and thus a strong cross-peak is observed between these protons in the NOESY spectrum; in the α-configuration, only a nuclear Overhauser effect (NOE) between H-1 and H-2 is observed. In general, 1,3-diaxial and 1,2-eq-ax proton pairs in pyranosyl rings will produce intraresidue NOESY cross-peaks. More importantly, however, NOESY cross-peaks may be observed between interresidue proton pairs. Thus, in some cases, NOESY is valuable in determining and confirming linkage sites and *O*-glycoside conformation in oligosaccharides (Fig. 7).

The 2D NOE method is superior to the standard one-dimensional steady-state NOE (ss-NOE) method [48–50]. The latter experiment involves the selective saturation of an anomeric proton for a given period, followed by signal detection in the absence of saturation. A similar spectrum is obtained with the saturation frequency off-resonance, and the two data sets are subtracted to yield an NOE difference spectrum. This difference spectrum contains signals from protons that experience either a positive or negative NOE with the anomeric proton that was saturated, thus suggesting spatial proximity. Although the ss-NOE experiment has been used quantitatively to assess *O*-glycoside bond conformation [49,50], the method suffers from significant limitations. These include errors introduced by the assumption that the oligosaccharide tumbles in solution isotropically, and errors caused by the well-known "three-spin effect" [48]. Even though motional assumptions still plague 2D NOE measurements, the proper choice of mixing time in NOESY, such that the initial rate approximation is not violated, will minimize errors caused by three-spin effects.

The magnitude of an NOE depends not only on ^{1}H-^{1}H internuclear distance, but also on the rotational correlation time of the molecule. For conventional Overhauser effects, the NOE may vary from +0.5 (a 50% enhancement in signal intensity) in the extreme narrowing limit (i.e., small molecules), to –1.0 (signal intensity is inverted) in the spin diffusion limit (i.e., large molecules) [48]. Thus, in larger oligosaccharides, small or no NOEs are observed, thereby limiting the usefulness of the method. In contrast, Overhauser effects in the rotating frame (ROESY) [51a–c] increase monotonically in a positive direction with increasing correlation time and provide a potential solution to this problem. This method

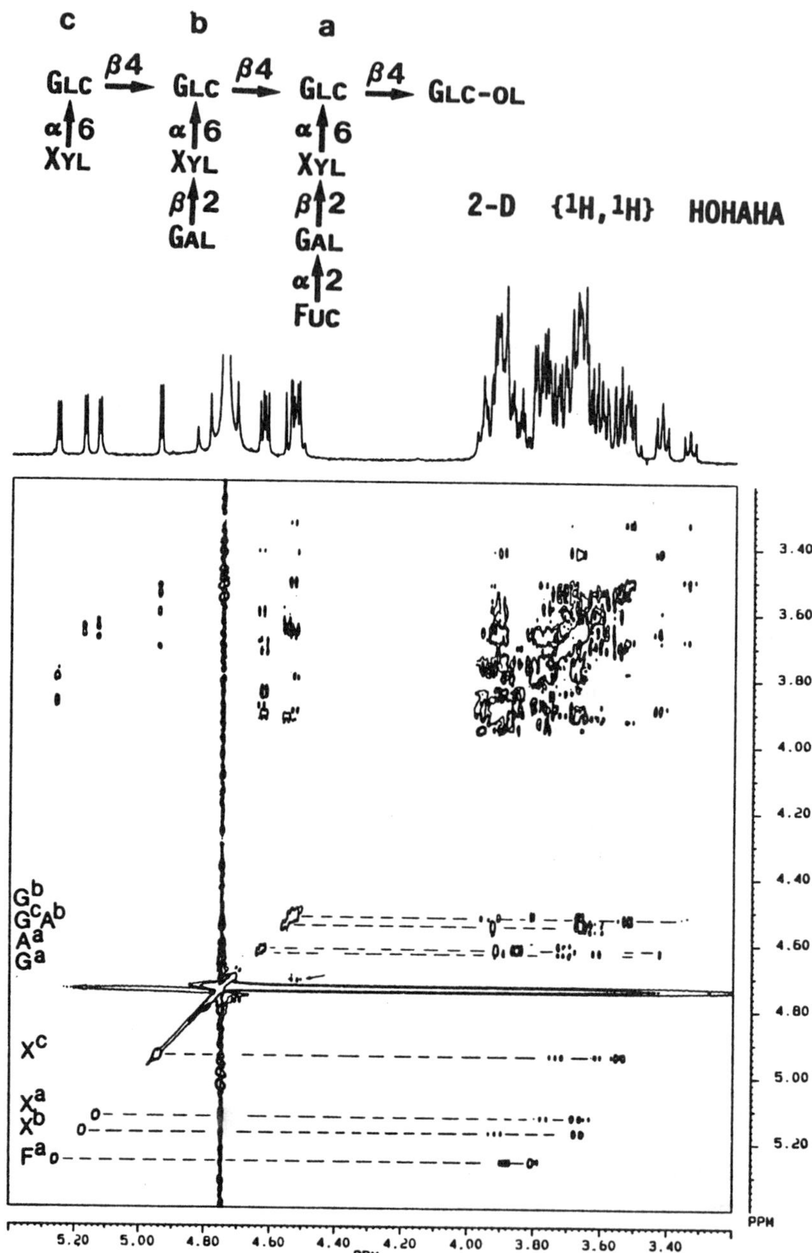

Figure 6 The 2D HOHAHA spectrum of a decasaccharide isolated from the xyloglucan of rapeseed hulls. The structure of the compound and the conventional 1D ^{1}H spectrum are shown on top of the contour map. Horizontal, dashed lines connect signals from protons of a particular glycosyl residue. The sample contained 4 mg of the decasaccharide in 0.4 ml of ^2H_2O at p^{2}H 7 and 27°C. The HOHAHA experiment was performed in the phase-sensitive mode (TPPI) at 500 MHz on a Bruker AM-500. Data matrix: 256 x 2048; 80 scans per t_1 value. Spin locking was achieved by the MLEV-17 sequence; the mixing time was 200 msec. Sine-bell window multiplication was applied in both dimensions before transformation of the data. Total measuring time: 6 hr. (From Ref. 37.)

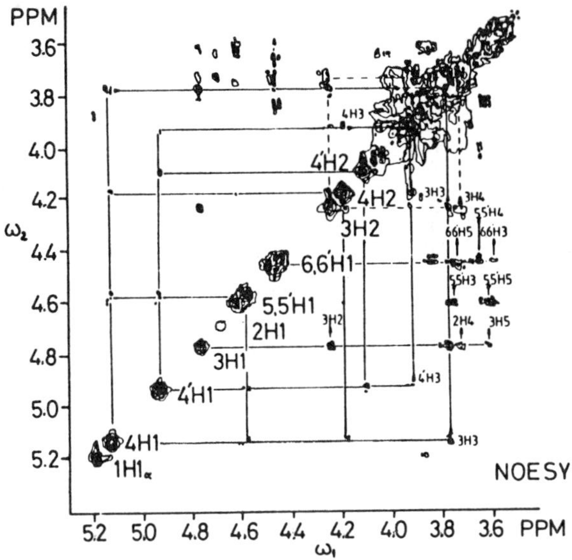

Figure 7 Two-dimensional NOE contour plot of the reduced biantennary oligosaccharide derived
6 5 4
from human serum transferrin, NeuNAcα(2→6)Galβ(1→4)GlcNAcβ(1→2)Manα(1→3)
6′ 5′ 4′ 3′ 2′ 1′
[NeuNAcα(2→6)Galβ(1→4)GlcNAcβ(1→2)Manα(1→6)]Manβ(1→4)GlcNAcβ(1→4)GlcNAc, recorded at 470 MHz. The region at ~4.7 ppm, corresponding to the strong H^2HO signal has been removed. The strong peaks on the diagonal correspond to the normal 1D spectrum. Cross-peaks showing proton–proton through-space connectivities appear symmetrically with the diagonal. Several of these have been connected with solid lines. The dashed lines show cross-peaks resulting from magnetization transfer between the tightly coupled H-3 and H-4 protons of Man-**3**. (From Ref. 51d).

may also be useful in distinguishing direct and indirect (three-spin) NOEs in oligosaccharides. However, caution must be exercised in the interpretation of ROESY data, as discussed by Homans [34].

G. Homonuclear J-Resolved Spectroscopy

Homonuclear 2D J-spectroscopy (HOMO 2DJ) [52] is useful in resolving overlapping multiplets in ^1H NMR spectra of oligosaccharides (Fig. 8). In addition, improved resolution is commonly observed in these spectra relative to that found in the 1D spectrum, since field inhomogeneity effects are canceled in the former by the refocusing 180° pulse. As a consequence, unresolved couplings, particularly long-range (e.g., $^4J_{HH}$) couplings, may be observed and measured in relatively small oligosaccharides [53a] and glycosphingolipids [54]. Although a valuable method in studies of small- to intermediate-sized oligosaccharides, the utility of homonuclear 2DJ spectroscopy declines with increasing molecular size, and the technique is considered to be of limited value in studies of more complex structures. Furthermore, in complex spectra containing overlapping mutually coupled

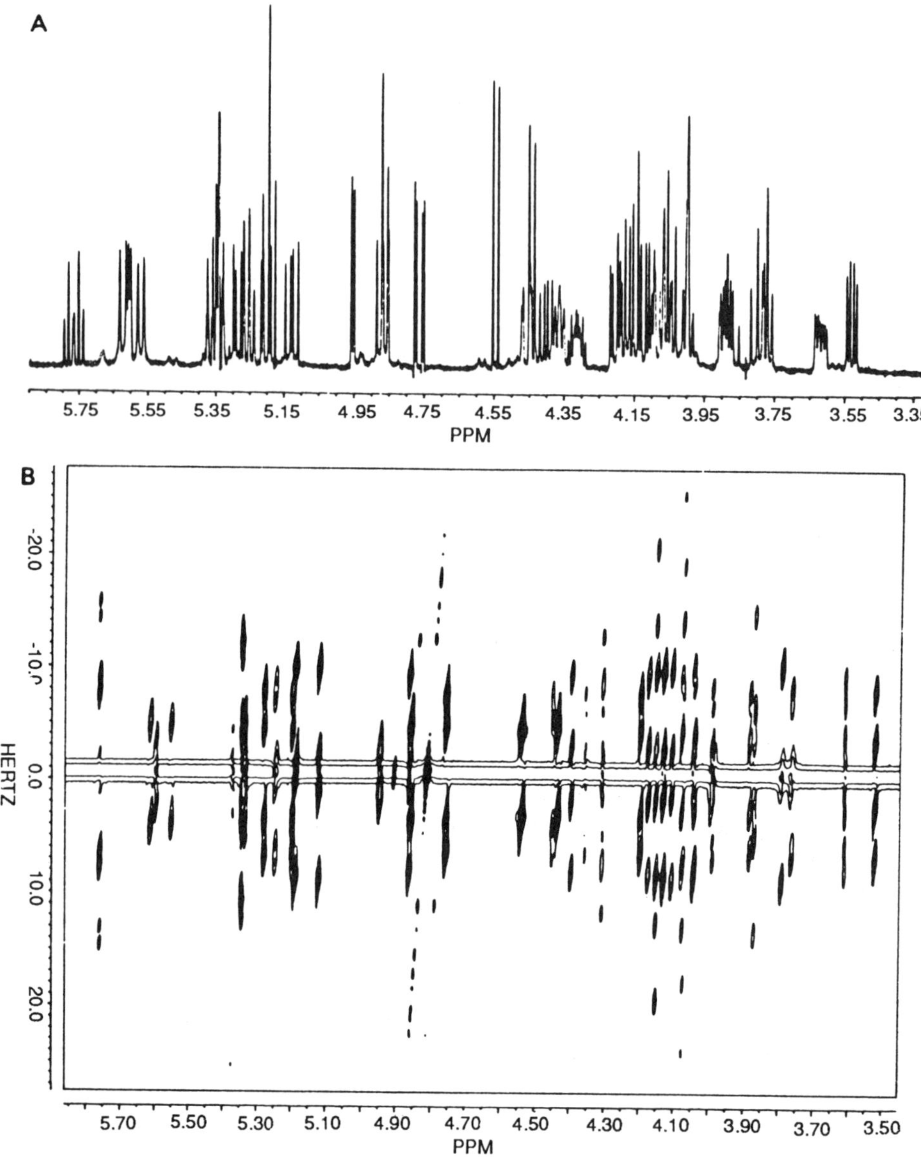

Figure 8 The partial 500 MHz ^1H NMR spectra of a 15-mM solution of the peracetylated globotetraosylceramide (globoside) in C^2HCl_3 at 303°K. (a) Resolution-enhanced 1D spectrum. (b) Contour plot of the 2D *J*-resolved spectrum; each of the multiplets, which are viewed here from the top, can be plotted individually in the conventional form (not shown here). (From Ref. 53b.)

signals, second-order effects cause distortions in the multiplet patterns and prevent the use of cross sections to observe individual multiplets and extract the desired ^1H-^1H couplings.

H. Heteronuclear ^{13}C–^1H Shift-Correlated Spectroscopy

The HETCOR experiment, as originally proposed [55], permits correlations between carbon and proton signals of the oligosaccharide. The method is based on coherent transfer of transverse magnetization between scalar-coupled protons and other nuclei having smaller gyromagnetic ratios (usually ^{13}C in oligosaccharides). The HETCOR experiment is normally optimized for the detection of directly bonded protons ($^1J_{CH} \approx$ 150–170 Hz), and carbon detection is employed. This experiment is used to assist in the assignment of carbon signals from assigned ^1H signals, or vice versa. Practically speaking, HETCOR suffers from poor sensitivity, and it is not employed unless relatively large quantities of oligosaccharide are available.

The problem of low sensitivity of the HETCOR experiment has spawned the development of indirect detection methods to obtain ^{13}C–^1H shift correlations; this experiment is referred to as heteronuclear multiple quantum coherence or HMQC [56] (Fig. 9). Instead of observing the less sensitive ^{13}C nucleus, as in the HETCOR experiment, the more sensitive ^1H is observed in HMQC. The resulting sensitivity enhancement, thus, may be attributed to the larger gyromagnetic ratio of ^1H and the shorter T_1 (spin-lattice or longitudinal relaxation time) of ^1H relative to that of ^{13}C. There are several variants of inverse (reverse) detection, but all rely on the generation of multiple-quantum coherence between coupled ^1H and ^{13}C energy levels.

In addition to one-bond ^{13}C–^1H correlations that facilitate ^{13}C or ^1H signal assignments, long-range ^{13}C–^1H correlation maps have been found to be invaluable in identifying linkage sites by transglycosidic ^{13}C–^1H couplings ($^3J_{COCH}$). *This approach is more reliable than NOESY in assigning linkage sites in oligosaccharides.* Like the one-bond correlation experiment, the long-range experiment may be performed in either direct or indirect modes, with the latter having much improved sensitivity. Bax and co-workers [57,58] have developed the HMBC method (heteronuclear multiple-bond correlation) for long-range correlations (Fig. 10); this method suppresses the detection of directly bonded ^{13}C–^1H pairs by inserting a low-pass J-filter in the pulse sequence.

It should be noted that one-dimensional analogues of the HETCOR and HMQC methods have been described for use in identifying linkage position in oligosaccharides; these include selective INEPT [59] and decoupled SPT [59]. The 2D DEPT has also been applied by Batta and Liptak [60] to identify interglycosidic linkages in oligosaccharides.

V. THREE-DIMENSIONAL NUCLEAR MAGNETIC RESONANCE SPECTROSCOPY OF OLIGOSACCHARIDES

Three-dimensional NMR spectroscopy (3D NMR) has been introduced recently to resolve the resonance overlap observed in 2D NMR spectra of complex molecules. In the homonuclear mode, 3D NMR combines two ^1H 2D experiments (e.g., COSY–COSY) in which the detection period of the first experiment serves as the evolution period of the second experiment [61].

At present, only a few studies have been reported in which 3D NMR has been applied to carbohydrates. Vuister et al. [62a] have used 3D NOE–HOHAHA spectroscopy to study the structure of a diantennary *N*-linked oligosaccharide, while Fesik et al. [62b] have used

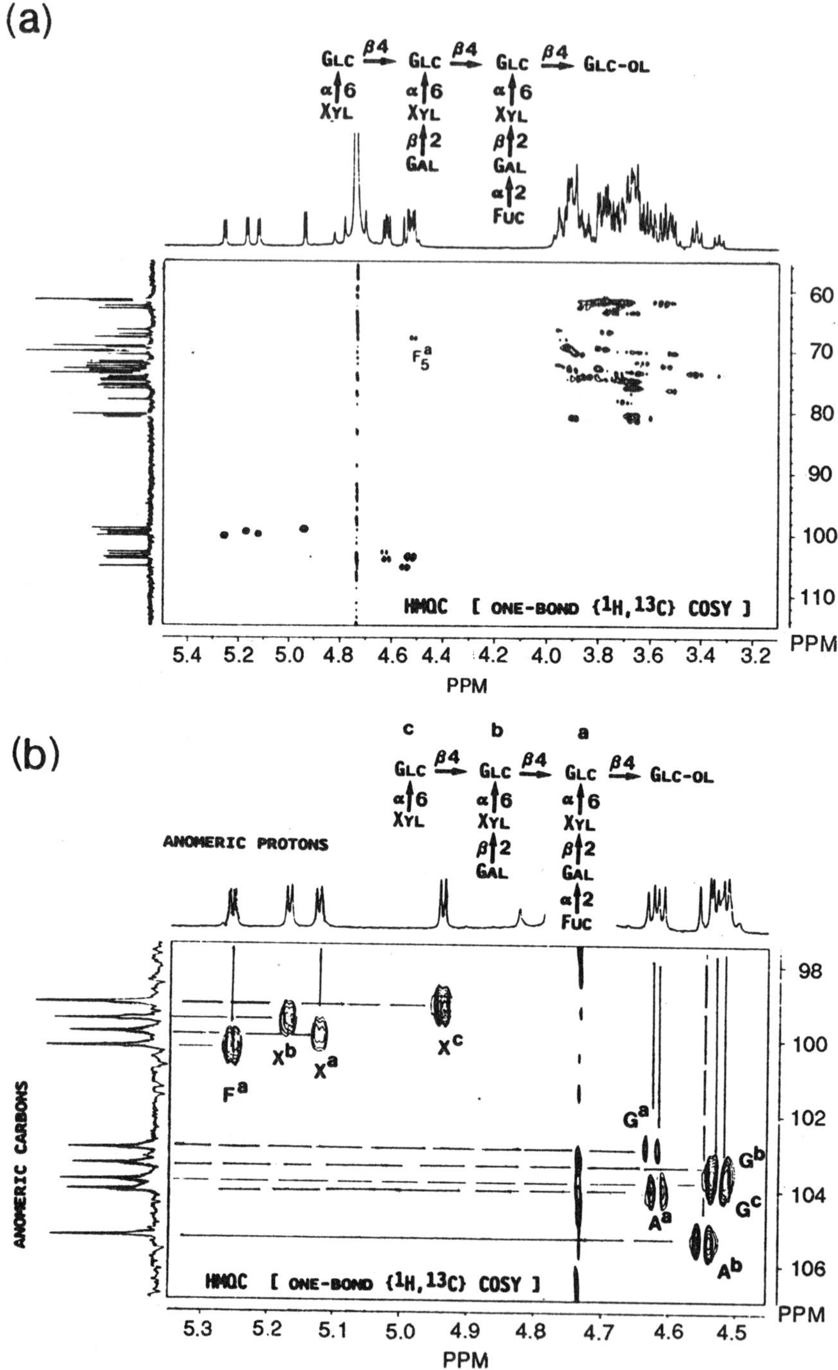

Figure 9 (a) The HMQC spectrum of the decasaccharide isolated from the xyloglucan of rapeseed hulls. The conventional 1D ^{1}H and ^{13}C spectra are shown along the horizontal and vertical axes, respectively. The sample was the same as described in the legend to Figure 6. Data matrix: 128 × 2048; 128 scans per t_1 value. A 24-Hz gaussian broadening was used in the t_1 dimension; a squared sine bell window shifted over $\pi/3$ was applied in the t_2 dimension. Total measuring time: 5 hr. (b) Anomeric region of the HMQC spectrum. Dashed lines indicate the one-bond connectivities between the H-1 and C-1 atoms. Symbols for residues are as in Figure 6. (From Ref. 37.)

(a)

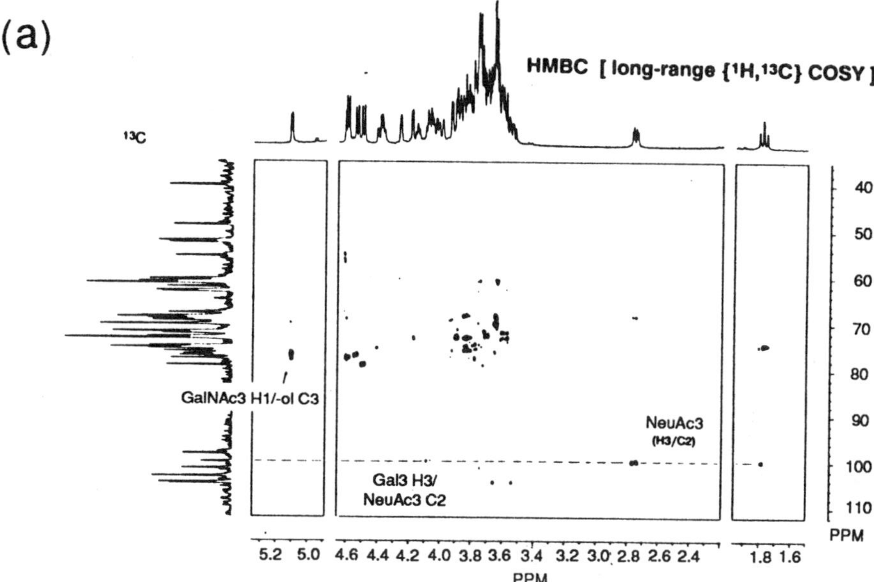

(b)

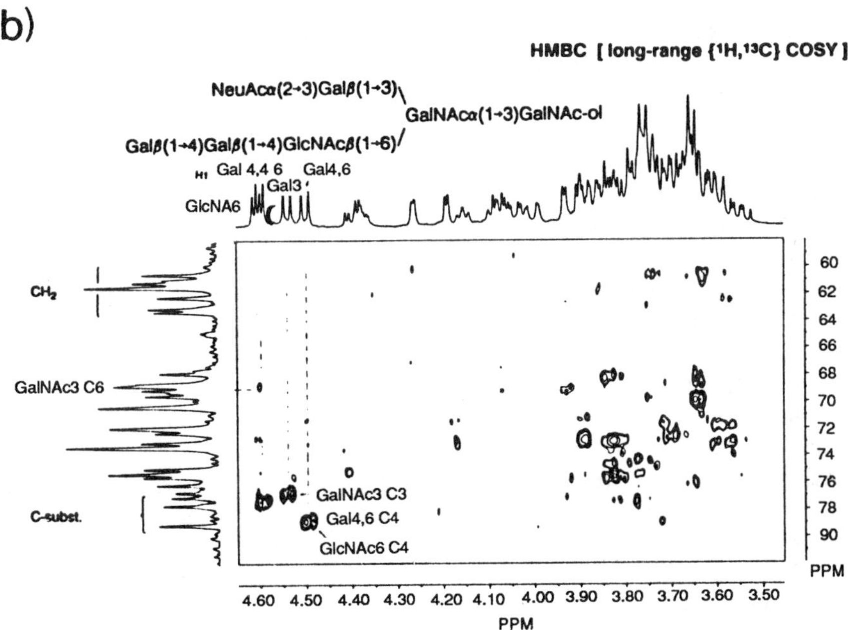

Figure 10 (a) The HMBC spectrum of the heptasaccharide isolated from the salivary mucins of the Chinese swiftlet (see structure in Figure 3). The conventional 1D ^{1}H and ^{13}C spectra are shown along the horizontal and vertical axes, respectively. The ^{1}H spectrum had been assigned by COSY (see Fig. 3) and HOHAHA, and the ^{13}C spectrum was subsequently assigned by HMQC. The sample was the same as described in the legend to Figure 3. Data matrix: 128 × 2048; 128 scans per t_1 value. A 12-Hz gaussian broadening was used in the t_1 dimension; a squared sine bell window shifted over $\pi/3$ was applied in the t_2 dimension. Total measuring time: 16 hr. (b) Anomeric region of the HMBC spectrum. Connectivities arising from $^{3}J_{CH}$ couplings across the glycosidic bonds are indicated. (From Ref. 37.)

heteronuclear 3D HMQC–COSY to assist in the study of an aminoglycoside. These methods, and related 3D methods that are likely to develop in the coming years, will be important in deciphering NMR spectra of complex oligosaccharides containing ten or more residues, situations in which 2D methods can be compromised. The routine implementation of 3D methods, however, will likely follow the standard incorporation of large-capacity disks (> 1 gigabyte), faster computers (for data processing), and computer software for pulse sequencing and data display, on commercial spectrometers.

Most recently, NMR spectroscopy in four dimensions has been reported [62c,d], but this method has not yet been applied to carbohydrate structural problems.

VI. SPIN-COUPLING CONSTANTS IN CARBOHYDRATES

A wide array of multidimensional NMR methods are frequently employed to assign the signals in complex ^{1}H and ^{13}C spectra and extract precise chemical shifts and spin-spin coupling constants (J) for use in the determination of molecular structure and conformation. Spin-spin coupling constants may be homonuclear (e.g., ^{1}H–^{1}H, ^{13}C–^{13}C) or heteronuclear (e.g., ^{13}C–^{1}H), and they occur typically across one-bond (^{1}J), two bonds, (^{2}J, geminal coupling), or three bonds (^{3}J, vicinal coupling). Of these three coupling types, three-bond couplings are most valuable to assess molecular conformation because of their well-documented sensitivity to molecular dihedral angles (Karplus relationships). In some cases, longer-range couplings (e.g., $^4J_{HH}$) may be observed, but the geometric requirements necessary to produce these couplings are high, and their magnitudes are usually small (<2 Hz). In addition to their application to structure elucidation, knowledge of spin couplings, especially ^{13}C–^{1}H couplings, is important in the proper choice of delay times in some 2D experiments (e.g., HETCOR, HMQC).

If we restrict our attention to the common monosaccharides found in animal glycoproteins, it is noted that the monosaccharide units are present as pyranosyl rings having configurations that confer significant conformational rigidity to the ring. In other words, pyranosyl rings having the D-*gluco*, D-*galacto*, or D-*manno* configurations highly prefer the $^4C_1$ chair conformation as opposed to other conformations (e.g., $^1C_4$, skew-boat) [63]. Therefore, observed couplings within these residues will reflect the presence of a single conformation, and the problem of conformational heterogeneity, like that experienced by furanosyl rings [64] and some pyranosyl rings (e.g., those having the *ido* configuration [65]) may be ignored.

As a consequence of the foregoing considerations, two general types of $^3J_{HH}$ are observed in the aldopyranosyl rings of most biologically important oligosaccharides. For vicinal protons that are oriented axial–equatorial (ax–eq) or equatorial–equatorial (eq–eq), the observed $^3J_{HH}$ between them will be ~2–3 Hz, whereas for an axial–axial (ax–ax) orientation, $^3J_{HH}$ will be ~7–9 Hz. Thus, for example, the α-*gluco* and β-*gluco* configurations may be distinguished based on $^3J_{H1,H2}$ (α, ~2–3 Hz; β, ~7–9 Hz). These differences result from the known dependence of $^3J_{HH}$ on dihedral angle in which a large coupling is observed between two eclipsed or anti protons, and minimal coupling is observed when the dihedral angle between the two protons is approximately orthogonal [66]. A more precise Karplus equation for $^3J_{HH}$ has been developed that takes substituent geometry and electronegatives into account [67].

Hydroxymethyl conformation in aldohexopyranosyl rings may be assessed by evaluating $^3J_{H5,H6R}$ and $^3J_{H5,H6S}$. Three staggered conformations about the C5-C6 bond are

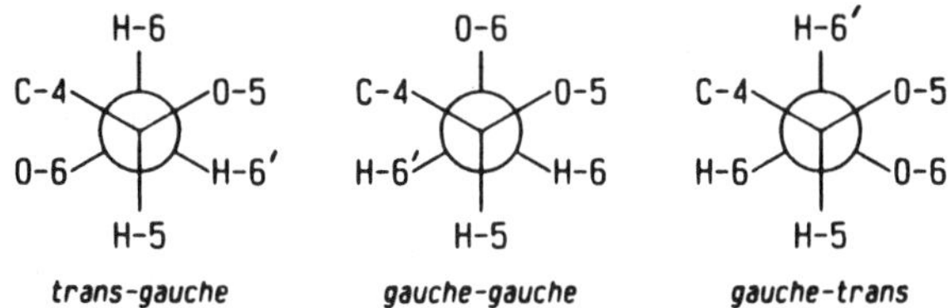

Figure 11 Newman projections for three C-5—C-6 rotamers. The *trans-gauche* rotamer is energetically unfavorable for 4,6-linked β-Glc residues.

possible and are denoted as the *gg, gt,* and *tg* conformers (Fig. 11). To evaluate the relative populations of these forms requires stereochemical assignments of the H-6 and H-6′ signals. These assignments may be achieved by selective deuteration [68] or by the concerted use of $^3J_{HH}$ and $^3J_{CH}$ [69]. In the absence of these stereochemical assignments, only the *gg* and *gt+tg* populations can be evaluated. Several methods have been proposed to assess hydroxymethyl conformation in carbohydrates based on $^3J_{HH}$, and their limitations have been discussed elsewhere [70]. Hydroxymethyl conformation is of critical importance to conformational studies of oligosaccharides containing [1→6]-linkages [71].

The ^{13}C–^{1}H couplings in carbohydrates have been studied in some detail in recent years [reviewed in 72a]. In aldopyranosyl rings, one-bond ^{13}C–^{1}H coupling between C-1 and H-1 ($^1J_{C1,H1}$) has been shown to be sensitive to anomeric configuration; $^1J_{C1,H1}$ is ~160 Hz when the C1-H1 bond is axial (i.e., β-D-aldopyranosyl rings in the $^4C_1$ conformation), whereas $^1J_{C1,H1}$ is ~170 Hz when the C1-H1 bond is equatorial (i.e., α-D-aldopyranosyl rings in the $^4C_1$ conformation) [72b]. Thus, if these one-bond couplings can be measured, either through gated-decoupled 1D ^{13}C spectra or from heteronuclear 2D J-spectra [72c], they may be used to establish or confirm anomeric configuration [65].

Two-bond ^{13}C–^{1}H coupling ($^2J_{CH}$) has been shown to be sensitive to the nature and orientation of the substituents appended to the carbons along the coupling pathway. Perlin and co-workers [73,74] and Bock and Pedersen [75] have proposed rules to interpret $^2J_{CH}$ in terms of carbohydrate structure, but in general, these couplings have not been used routinely in carbohydrate structure analysis, despite the claim that they may prove valuable for this [74].

In contrast with $^2J_{CH}$, three-bond ^{13}C-^{1}H couplings ($^3J_{CH}$) have received more attention in the conformational analysis of carbohydrates. There are two types of $^3J_{CH}$ of importance to carbohydrate structure determination, $^3J_{CCCH}$ and $^3J_{COCH}$ (Fig. 12). The first of these describes coupling pathways within each monosaccharide residue of the oligosaccharide

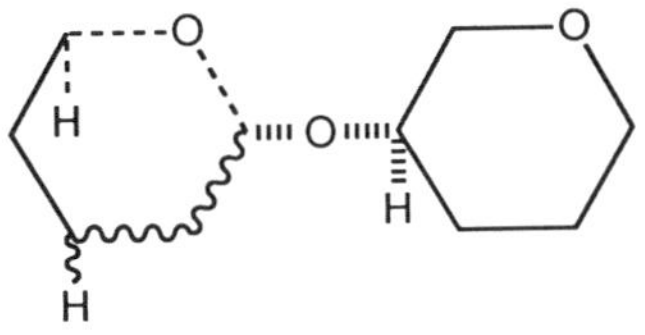

Figure 12 Vicinal ^{13}C–^{1}H coupling pathways involving an anomeric carbon of a monosaccharide glycosidically linked to another residue: intraresidue CCCH(~~), intraresidue COCH(- - -), and interresidue COCH (IIIII) coupling pathways.

chain, and thus provides information on monomer structure and conformation. The second, $^3J_{COCH}$, describes coupling pathways within a given residue (e.g., coupling between C-1 and H-5 in an aldopyranosyl ring), or coupling pathways *between* residues (e.g., coupling across the glycoside bond); the latter couplings are, therefore, of value in assessing *O*-glycoside bond conformation in oligosaccharides and are the targets of interest in the HMBC 2D experiment discussed earlier, although they are normally not measured accurately by this approach. Bax and Freeman [76a] have described an NMR approach to measure long-range ^{13}C–^1H coupling constants with high precision in unlabeled molecules. The Karplus curves relating $^3J_{CCCH}$ and $^3J_{COCH}$ to molecular dihedral angles have a conserved shape, but differ in amplitude [74a,76b,77].

The ^{13}C–^{13}C coupling constants are not commonly used in oligosaccharide analysis, in part because they are difficult to measure. NMR methods, such as INADEQUATE [78,79] in one- and two-dimensions, have been devised to assess these couplings in natural abundance compounds, but large quantities of material are required to conduct this analysis, making it impractical for most oligosaccharide studies. The ^{13}C–^{13}C couplings are best measured in ^{13}C-labeled compounds [69,80–84], but obtaining the latter is oftentimes difficult; however, recent advances in oligosaccharide chemical synthesis [85,86], and in the availability of specific glycosyltransferases (by genetic engineering) for enzymic preparations, may make labeled oligosaccharides more accessible in the future. When these couplings are available, however, they can provide information complementary to $^3J_{COCH}$ on *O*-glycoside bond conformation [69]; for example, $^3J_{CCOC}$ or $^3J_{COCC}$ across the glycosidic bond give information on the phi ϕ and psi ψ angles, respectively, defining the two linkage torsions (Fig. 13). In comparison with $^3J_{HH}$ and $^3J_{CH}$, however, the Karplus relationships for $^3J_{CC}$ are not as well established and are very sensitive to local structure, and these facts, in addition to the difficulty of their measurement, have at present restricted their use to relatively simple systems.

It should be appreciated that the interpretation of vicinal spin-coupling constants, and indeed any NMR parameter, in terms of molecular conformation depends on the conformational model invoked in the analysis. If the molecule under consideration is conformationally rigid (e.g., the ring conformation of β-D-glucopyranose), then the analysis is relatively straightforward if appropriate Karplus curves are available. However, when conformational flexibility exists, the values of the observed couplings will be averaged, and erroneous conclusions may be drawn if these couplings are interpreted in terms of a single conformation, as discussed previously by Jardetzky [87a]. This problem has been discussed by Cumming and Carver [87b] in the context of oligosaccharide conformation, and should be borne in mind when NMR parameters pertaining to linkage conformation are under consideration.

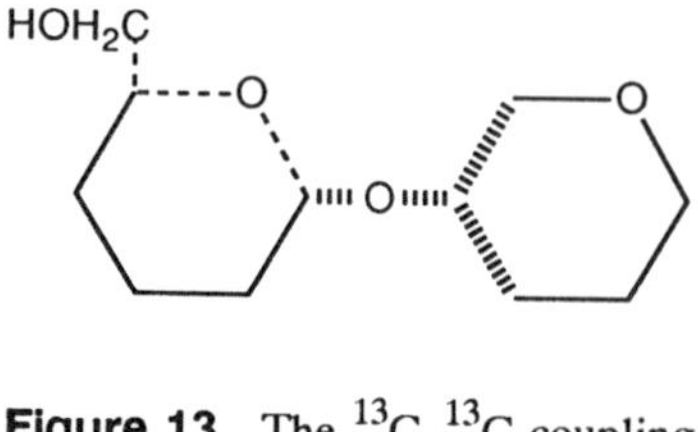

Figure 13 The ^{13}C–^{13}C coupling pathways involving an anomeric carbon of a monosaccharide glycosidically linked to another residue: intraresidue COCC (—) and interresidue COCC (ıııı) coupling pathways.

VII. FUNDAMENTAL ASPECTS OF NUCLEAR SPIN RELAXATION IN CARBOHYDRATES

Nuclear spin relaxation in carbohydrates has been discussed recently in some detail by Dais and Perlin [88]. The dissipation of energy by an excited nucleus may be achieved by transferring this energy to the environment (an enthalpic process), or by mutual exchange of this energy between other molecules in solution (an entropic process). The rates at which energy is dissipated by these two processes are called the spin-lattice (longitudinal; $1/T_1$) and spin-spin (transverse; $1/T_2$) relaxation rates, respectively, where T_1 and T_2 are the spin-lattice and spin-spin relaxation times, respectively. The T_1 values are commonly measured experimentally by the inversion-recovery method [89], and T_2 is measured by the Carr–Purcell–Meiboom–Gill (CPMG) method [90]. The efficiency of these relaxation processes depends, among other factors, on the motional properties of the molecule or the field strength at which the measurement is being determined, or on both.

For protons and carbons of simple monosaccharides in dilute solution in the absence of paramagnetic agents, $T_1 = T_2$ at current field strengths, both are dominated by the dipole–dipole mechanism of relaxation [91], and internuclear distances between dipoles become a prime determinant of T_1 and T_2. Both T_1 and T_2 will be inversely proportional to the inverse r^6 distance(s) between the dipoles. Protonated carbons relax primarily through their attached protons, since the C—H bond length (~1.08 Å) is much shorter than other C—H internuclear distances in the molecule. Since monosaccharides tumble isotropically in solution (i.e., there is no preferred axis of rotation), T_1 of the *ring* carbons may be used to determine the correlation time, τ_r (a measure of the tumbling rate of the molecule), from the equation [92]

$$\frac{1}{T_{1,C}} = \frac{N\gamma_H^2\gamma_C^2\hbar^2}{r_{CH}^6}\tau_r \qquad [1]$$

where N is the number of attached protons, γ_H and γ_C are the magnetogyric ratios of ^1H and ^{13}C, respectively, and $\hbar$ is Planck's constant divided by 2π. From Eq. [1], it might be expected that the T_1 of a singly protonated ring carbon of an aldohexopyranosyl ring will be twofold larger than the T_1 of an exocyclic unsubstituted hydroxymethyl carbon, since the latter is bonded to a twofold greater number of protons. In practice, however, this twofold difference is not realized (it is usually less than twofold), since the motional properties of the —CH_2OH carbon will be determined not only by the motional properties of the molecule, but also by the local (segmental) motion caused by rotation about the exocyclic C5-C6 bond [93].

For small carbohydrates, T_1 and T_2 are inversely proportional to τ_r; that is, T_1 and T_2 increase (relaxation is less efficient) as the molecule tumbles more rapidly (τ_r decreases). Thus, decreasing molecule size or increasing temperature, both of which decrease τ_r, will result in longer relaxation times [94]. Decreasing τ_r by increasing temperature often improves spectral resolution, since T_2 is inversely related to resonance line-width.

A given proton I in a monosaccharide will relax predominantly through other nearby protons according to the equation [89,91]

$$\frac{1}{T_{1,I}} = \frac{3\gamma_H^4\hbar^2\tau_{IS}}{2}\Sigma r_{IS}^{-6} \qquad [2]$$

HO CH2OH O HO OCH3 HO

Structure 8

where γ_H is the magnetogyric ratio of ^1H, and Σr_{IS}^{-6} is the sum of all internuclear ^1H–^1H distances to the inverse sixth power in the molecule involving proton I. Thus, for example, H1 of methyl β-D-galactopyranoside **8** will relax primarily through H-3 and H-5, as the latter are closest in space to H-1 (an internuclear distance of ~ 2.6 Å). As shown in Table 3, H-2 has the largest T_1 of the ring protons of **8**, which may be explained by noting that this proton is relatively isolated in the molecule, having no nearby protons through which to relax.

In oligosaccharides, protonated carbons relax predominantly through the directly bonded protons, as in monosaccharides. However, the protons of a given residue of an oligosaccharide may not only relax through other protons in the same residue (intraresidue relaxation), *but also potentially through protons in other residues (interresidue) relaxation.* The efficiency of the latter process will depend on the conformational properties of the oligosaccharide. This fact is at the heart of the NOE methods (see foregoing) that are used to determine oligosaccharide structure—NOEs are observed at protons that contribute to the dipolar relaxation of the proton being perturbed (saturated). It should be noted that oligosaccharides may not tumble isotropically, and anisotropic behavior should increase as the oligosaccharide assumes a more extended conformation. In these cases, Eqs. [1] and [2] are invalid, and must be modified [95] to account for a preferred rotational axis.

Spin-lattice relaxation times have been used to determine intra- and interresidue ^1H–^1H distances in oligosaccharides through the use of selective deuteration. This method (DESERT; *de*uterium *s*ubstitution *e*ffects on *r*elaxation *t*imes) [96,97] is based on the fact that deuterium substitution affects the relaxation efficiency of nearby protons in the molecule. By measuring the T_1 of a given proton in the presence and absence of deuterium

Table 3 Proton T_1 Values[a] for Methyl β-D-Galactopyranoside (Structure **8**) at 26° in ^2H_2O

		$T_1/T_{1,H\text{-}1}$	
Proton	T1 (s)	Experimental	Calculated[b]
H-1	1.5	1.0	1.0
H-2	6.5	4.2	4.2
H-3	1.9	1.2	1.3
H-4	1.8	1.2	1.3
H-5	1.1	0.73	0.94
H-6	1.0	0.67	0.36
H-6′	0.97	0.63	0.40

[a]Nonselective; 50 mM glycoside.
[b]By using Eq. [2].

at a specific site in the molecule, the internuclear distance between this proton and the site of deuteration can be determined. This approach has been used to evaluate *O*-glycoside bond conformation in [1→4]- and [1→6]-linked disaccharides [71,98].

VIII. THEORETICAL METHODS TO EVALUATE OLIGOSACCHARIDE CONFORMATION

Spin-spin coupling constants derived from COSY, HOHAHA, or homonuclear J-spectroscopy, and NOE determinations by 1D methods, NOESY, or ROESY, provide a wealth of information about the conformational properties of oligosaccharides. This information is often supplemented by energy calculations using several theoretical methods. These latter methods include ab initio calculations [99], semiempirical methods [reviewed in 100], molecular mechanics methods [100], and HSEA (hard-sphere) [101,102] calculations. In principle, the ab initio method should yield the best estimate of the potential surface for an isolated (gas-phase) molecule, as only a limited number of approximations are employed in these computations. However, ab initio calculations require a significant amount of computer time to conduct (especially when the larger basis sets are used, e.g., 6-31G*) and, consequently, have currently been restricted to studies of monosaccharides [103]. On the other extreme are HSEA calculations, first introduced by Lemieux and co-workers [101a,b], that treat each monosaccharide ring of an oligosaccharide as a rigid geometry and assess total energies as a function of the two *O*-glycoside torsion angles, phi (ϕ) and psi (ψ). The HSEA method determines energies by evaluating the nonbonded van der Waals interactions in the molecule (i.e., charges are ignored). This method has been refined [102] to include the torsional potential from the "exoanomeric effect" [104], the latter being an important factor regulating conformation about the glycosidic linkage. Despite its simplicity, HSEA calculations have been used successfully to accurately predict conformations consistent with those determined by crystallographic or NMR data.

It is important to recognize that these theoretical methods suffer from two limitations. These methods have as their aim to determine the global energy minimum of the molecule, but, in practice, it is not feasible to assess the full potential energy surface to ensure that the global minimum has indeed been located. Consequently, the minimization algorithms employed by these methods may converge on a geometry that does not represent the global minimum, but rather represents a local minimum in the potential surface. Initiating calculations at different starting geometries is often used to test for convergence at a local minimum, but this procedure is not unequivocal.

In addition to the local minimum problem, the effect of solvent is often not treated explicitly in most present-day energy calculations of oligosaccharides. One possible outcome of neglecting solvent contributions may be erroneous conclusions that overestimate the importance of intramolecular hydrogen bonding in stabilizing particular geometries, as competition with solvent water may effectively eliminate these interactions in solution.

The interactive use of NMR parameters and theoretical methods has proved to be a powerful tool in studies of proteins [105] and nucleic acids [106]. Distance geometry algorithms such as DISGEO [107], DISMAN [108], and DSPACE [109] use NOE or J values, or a combination thereof, derived from NMR spectroscopy to limit the conformational options available to the molecule. The addition of experimental NMR parameters to theoretical calculations is advantageous, as the former provide constraints to the conformational space and, thereby, improve the likelihood that a predicted preferred conformation actually lies at the global minimum. This approach has not yet been as successful in 3D determinations of oligosaccharides owing to the usually limited number of measurable NOEs in these molecules, although some success has been achieved recently by combining NOE and J values with a full classical Hamiltonian [110].

In addition to theoretical methods, described in the foregoing, that yield a relatively static picture of molecular structure, molecular dynamics (MD) simulations [111] provide useful information on the time-dependent molecular fluctuations of molecules. These motions are believed to play a role in mediating both chemical and biological reactions. In contrast with their applications to protein and nucleic acids, little work has been done as yet on the use of MD simulations in the carbohydrates. Brady and co-workers have studied monosaccharides (D-glucose) [112] and simple disaccharides (maltose, sucrose) [113,114], Yan and Bush [115] have applied MD methods to blood group oligosaccharides, and Homans et al. [116] have performed MD simulations on mannose-containing disaccharides. It is likely that MD methods will play an important role in assessing the effect of primary structure on the molecular motions of oligosaccharides in the future.

ACKNOWLEDGMENTS

The author would like to thank Dr. R. Bretthauer (University of Notre Dame), Dr. L. Lerner (University of Wisconsin—Madison), and Dr. C. A. Bush (University of Maryland—Baltimore County) for their helpful discussions and critical reading of this review. I am also thankful to several authors and publishers who have given permission to reproduce their work in this article.

REFERENCES

1. (a)Lote, C. J., and J. B. Weiss (1971). FEBS Lett. 16:81–85. (b) Weiss, J. B., Lote, C. J., and Bobinski, H. (1971). *Nature, 234*:25–26.
2. Kato, Y., and Spiro, R. G. (1989). *J. Biol. Chem. 264*:3364.
3. (a) Yamashita, K., Tachibana, Y., and Kobata, A. (1978). *J. Biol. Chem. 253:3862–3869. (b) Tai, T., Yamashita, K., Ogata-Arakawa, M., Koide, N., Maramatsu, T., Iwashita, S., Inoue, Y., and Kobata, A. (1975). J. Biol. Chem.* 250:8569–8575.
4. Dill, K., Berman, E., and Pavia, A. A. (1985). *Adv. Carbohydr. Chem. Biochem. 43*:1–49.
5. Thotakura, N. R., and Bahl, O. P. (1987). *Methods Enzymol. 138*:350–359.
6. Hicks, K. B. (1988). *Adv. Carbohydr. Chem. Biochem. 46*:17–72.
7. Vliegenthart, J. F. G., van Halbeek, H., and Dorland, L. (1981). *Pure Appl. Chem. 53*:45–77.
8. Rauvala, H., Finne, J., Krusius, T., Karkkainen, J., and Jarnefelt, J. (1981). *Adv. Carbohydr. Chem. Biochem. 38*:389–416.
9. Rolf, D., and Gray, G. R. (1982). *J. Am. Chem. Soc. 104*:3539–3541.
10. (a) Rolf, D., and Gray, G. R. (1986). *Carbohydr. Res. 152*:343–349. (b) Bennek, J. A., Rice, M. J., and Gray, G. R. (1986). *Carbohydr. Res. 157*:125–137.
11. Burlingame, A. L., Maltby, D., Russell, D. H., and Holland, P. T. (1988). *Anal. Chem. Fund. Rev. 60*:249R–342R.
12. Dell, A. (1987). *Adv. Carbohydr. Chem. Biochem. 45*:19.
13. (a) Dill, K., and Allerhand, A. (1979). *J. Biol. Chem. 254*:4524–4531. (b) Berman, E., Allerhand, A., and De Vries, A. L. (1980). *J. Biol. Chem. 255*:4407–4410. (c) Berman, E., Walters, D. E., and Allerhand, A. (1981). *J. Biol. Chem. 256*:3853–3857. (d) Rao, B. N. N., and Bush, C. A. (1987). *Biopolymers 26*:1227–1244.
14. Rao, B. N. N ., and Bush, C. A. (1988). *Carbohydr. Res. 180*:111.

15. (a) Dais, P., and Perlin, A. S. (1982). *Can. J. Chem. 60*:1648. (b) Adams, B., and Lerner, L. (1990). Div. of Carbohydr. Chem., American Chemical Society 199th National Meeting, Boston, April 22–27, 1990, Abstr. 93.
16. Derome, A. (1987). *Modern NMR Techniques For Chemistry Research*. Pergamon Press, Oxford, pp. 58–61.
17. (a) Sklenar, V. (1990). In: *NMR in Agriculture and Biochemistry*, (J. Finley, S. Schmidt, and A. S. Serianni, eds.). Plenum Press, Oxford, p. 63. (b) Hore, P. J. (1989). *Methods Enzymol. 176*:65–77.
18. (a) Schaefer, J. (1972). *J. Magn. Reson. 6*:670. (b) Jesson, J. P., Meakin, P., and Kneissel, G. (1973). *J. Am. Chem. Soc. 95*:618. (c) Bleich, H. E., and Glasel, J. A. (1975). *J. Magn. Reson. 18*:401. (d) Campbell, I. D., Dobson, C. M., and Ratcliffe, R. G. (1977). *J. Magn. Reson. 27*:455.
19. (a) Patt, S., and Sykes, B. D. (1972). *J. Chem. Phys. 56*:3182. (b) Benz, F. W., Feeney, J., and Roberts, G. C. K. (1972). *J. Magn. Reson. 8*:114. (c) Gupta, R. K., and Gupta, P. (1979). *J. Magn. Reson. 34*:657.
20. Plateau, P., and Gueron, M. (1982). *J. Am. Chem. Soc. 104*:7310.
21. Redfield, A. G., Kunz, S. D., and Ralph, E. K. (1975). *J. Magn. Reson. 19*:275.
22. Starcuk, Z., Pucek, L., and Starcuk, Z., Jr. (1988). *J. Magn. Reson. 80*:352.
23. (a) The CarbBank Carbohydrate database has been established to facilitate these searches, and information about this database may be obtained by contacting the Complex Carbohydrate Research Center, University of Georgia, 220 Riverbend Road, Athens, GA 30602. (b) Bock, K., and Pedersen, C. (1983). *Adv. Carbohydr. Chem. Biochem. 41*:27. (c) Bock, C., Pedersen, C., and Pedersen, H. (1984). *Adv. Carbohydr. Chem. Biochem. 42*:193.
24. Vliegenthart, J. F. G., Dorland, L., and van Halbeek, H. (1983). *Adv. Carbohydr. Chem. Biochem. 41*:209–374.
25. *J. Carbohydr. Chem.*, Vol. 3, 1984.
26. Schraml, J., and Bellama, J. M. (1988). *Two-Dimensional NMR Spectroscopy*. John Wiley & Sons, New York.
27. Martin, G. E., and Zektzer, A. S. (1988). *Two-Dimensional NMR Methods for Establishing Molecular Connectivity*. VCH Publishers, New York.
28. Brey, W. S., ed. (1988). *Pulse Methods in 1D and 2D Liquid-Phase NMR*. Academic Press, San Diego.
29. Ernst, R. R., Bodenhausen, G., and Wokaun, A. (1987). *Principles of Nuclear Magnetic Resonance in One and Two Dimensions*. Clarendon Press, Oxford.
30. (a) Croasmun, W. R., and Carlson, R. M. K., eds. (1987). *Two-Dimensional NMR Spectroscopy—Applications For Chemists and Biochemists*. VCH Publishers, New York. (b) H. Kessler, Gehrke, M., and Griesinger, C. (1988). *Angew. Chem. Int. Ed. Engl. 27*:490.
31. van Halbeek, H., and Poppe, L. Div. Carbohydr. Chem., American Chemical Society 199th National Meeting, Boston, April 22–27, 1990, Abstr. 64.
32. Koerner, T. A. W., Prestegard, J. H., and Yu, R. K. (1987). *Methods Enzymol. 138*:38–59.
33. Dabrowski, J. (1987). Application of two-dimensional NMR methods in the structural analysis of oligosaccharides and other complex carbohydrates. In *Two-Dimensional NMR Spectroscopy—Applications For Chemists and Biochemists* (W. R. Croasmun and R. M. K. Carlson, eds.). VCH Publishers, New York, pp. 349–386.
34. Homans, S. W. (1990). *Prog. NMR Spectrosc. 22*:55–81.
35. Bush, C. A. (1988). *Bull. Magn. Reson. 10*:73.
36. Sweeley, C. C., and H. A. Nunez. (1985). *Annu. Rev. Biochem. 54*:756–801.
37. van Halbeek, H. (1990). *Frontiers of NMR in Molecular Biology*. Alan R. Liss, New York, p. 195.
38. This experiment was first proposed by Jeener (Jeener, J.) (1971). Ampere International Summer School II, Basko Polje, Yugoslavia.
39. Piantini, U., Sorensen, O. W., and Ernst, R. R. (1982). *J. Am. Chem. Soc. 104*:6800.

40. (a) Homans, S. W., Dwek, R. A., Boyd, J., Mahmoudian, M., Richards, W. G., and Rademacher, T. W. (1986). *Biochemistry 25*:6342. (b) Braunschweiler, L., Bodenhausen, G., and Ernst, R. R. (1983). *Mol. Phys. 48*:535.
41. Homans, S. W., Dwek, R. A., Fernandes, D. L., and Rademacher, T. W. (1984). *Proc. Natl. Acad. Sci. USA 81*:6286.
42. Davis, D. G., and Bax, A. (1985). *J. Am. Chem. Soc. 107*:2820 (1985).
43. Homans, S. W., Dwek, R. A., Boyd, J., Soffe, N. and Rademacher, T. W. (1987). *Proc. Natl. Acad. Sci. USA 84*:1202.
44. Braunschweiler, L., and Ernst, R. R. (1983). *J. Magn. Reson. 53*:521.
45. Davis, D. G., and Bax, A. (1985). *J. Am. Chem. Soc. 107*:7197.
46. Inagaki, F., Shimada, I., Kohda, D., Suzuki, A., and Bax, A. (1989). *J. Magn. Reson. 81*:186.
47. Macura, S., and Ernst, R. R. (1980). *Mol. Phys. 41*:95.
48. Noggle, J. H., and Schirmer, R. E. (1971). *The Nuclear Overhauser Effect—Chemical Applications*. Academic Press, New York.
49. (a) Brisson, J.-R., and Carver, J. (1983). *Biochemistry 22*:3671. (b) Brisson, J.-R., and Carver, J. (1983). *Biochemistry 22*:3680. (c) Homans, S. W., Dwek, R. A., Fernandes, D. L., and Rademacher, T. W. (1982). *FEBS Lett. 150*:503.
50. Paulsen, H., Peters, T., Sinnwell, V., Lebuhn, R., and Meyer, B. (1984). *Liebigs Ann. Chem. 5*:951.
51. (a) Bothner-By, A. A., Stephens, R. L., Lee, J. T., Warren, C. D., and Jeanloz, R. W. (1984). *J. Am. Chem. Soc. 106*:811. (b) Bax, A., and Davis, D. G. (1985). *J. Magn. Reson. 63*:207. (c) Bax, A., Sklenar, V., and Summers, M. F. (1986). *J. Magn. Reson. 70*:327. (d) Homans, S. W., Dwek, R. A., Fernandes, D. L., and Rademacher, T. W. (1983). *FEBS Lett. 164*:231.
52. (a) Aue, W. P., Karhan, J., and Ernst, R. R. (1976). *J. Chem. Phys. 64*:4226. (b) Bax, A., Mehlkopf, A. F., and Smidt, J. (1980). *J. Magn. Reson. 40*:213.
53. (a) Hall, L. D., Sukumar, S., and Sullivan, G. R. (1979). *J. Chem. Soc. Chem. Commun.* p. 292. (b) Egge, H., Dabrowski, J. and Hanfland, P. (1984). *Pure Appl. Chem. 56*:807.
54. Yamada, A., Dabrowski, J., Hanfland, P., and Egge, H. (1980). *Biochem. Biophys. Acta 618*:473.
55. Bodenhausen, G., and Freeman, R. (1977). *J. Magn. Reson. 28*:471.
56. Bax, A., and Subramanian, S. (1986). *J. Magn. Reson, 67*:565–569.
57. Bax, A., and Summers, M. F. (1986). *J. Am. Chem. Soc. 108*:2093–2094.
58. Summers, M. F., Marzilli, L. G., and Bax, A. (1986). *J. Am. Chem. Soc. 108*:4285–4294.
59. Bax, A., Egan, W., and Kovac, P. (1984). *J. Carbohydr. Chem. 3*:593–611.
60. Batta, G., and Liptak, A. (1985). *J. Chem. Soc. Chem. Commun.* pp. 368–370.
61. (a) Vuister, G. W., and Boelens, R. (1987). *J. Magn. Reson. 73*:328. (b) Griesinger, G., Sorensen, O. W., and Ernst, R. R. (1987). *J. Am. Chem. Soc. 109*:7227.
62. (a) Vuister, G. W., de Waard, P., Boelens, R., Vliegenthart, J. F. G., and Kaptein, R. (1989). *J. Am. Chem. Soc. 111*:772–774. (b) Fesik, S. W., Gampe, R. T., Jr., and Zuiderweg, E. R. P. (1989). *J. Am. Chem. Soc. 111*:770–772. (c) Kay, L. E., Clore, G. M., Bax, A., and Gronenborn, A. M. (1990). *Science 249*:411. (d) Clore, G. M., and Gronenborn, A. M. (1991). *Science 252*:1390.
63. Angyal, S. J. (1969). *Angew. Chem. Int. Ed. Engl. 8*:157.
64. Serianni, A. S., and Barker, R. (1984). *J. Org. Chem. 49*:3292.
65. Snyder, J. R., and Serianni, A. S. (1986). *J. Org. Chem. 51*:2694.
66. (a) Karplus, M. (1959). *J. Chem. Phys. 30*:11. (b) Jardetzky, O., and Roberts, G. C. K. (1981). *NMR in Molecular Biology*. Academic Press, New York, pp. 37–43.
67. Haasnoot, C. A. G., deLeeuw, F. A. A. M., and Altona, C. (1980). *Tetrahedron 36*:2783.
68. (a) Nishida, Y., Ohrui, H., and Meguro, H. (1984). *Tetrahedron Lett. 25*:1575. (b) Ohrui, H., Nishida, Y., Watanabe, M., Hori, H., and Meguro, H. (1985). *Tetrahedron Lett. 26*:3251.
69. Hayes, M. L., Serianni, A. S., and Barker, R. (1982). *Carbohydr. Res. 100*:87.
70. (a) Kline, P. C., and Serianni, A. S. (1988). *Magn. Reson. Chem. 26*:120. (b) Kline, P. C., and Serianni, A. S. (1990). *Magn. Reson. Chem. 28*:324.
71. (a) Nishida, Y., Hori, H., Ohrui, H., Meguro, H., and Uzawa, J. (1988). *Tetrahedron Lett. 29*:4461. (b) Hori, H., Nishida, Y., Ohrui, H., Meguro, H., and Uzawa, J. (1988). *Tetrahedron Lett. 29*:4457.

72. Barker, R., and Walker, T. E. (1980). *Methods Carbohydr. Chem. 8*:151. (b) Bock, K., and Pedersen, C. (1977). *Acta Chem. Scand.* Ser. [B] *31*:354. (c) Bodenhausen, G., Freeman, R., and Turner, D. L. (1976). *J. Chem. Phys. 65*:839.
73. Schwarcz, J. A., and Perlin, A. S. (1972). *Can. J. Chem. 50*:3667.
74. Schwarcz, J. A., Cyr, N., and Perlin, A. S. (1975). *Can. J. Chem. 53*:1872.
75. Bock, K., and Pedersen, C. (1977). *Acta Chem. Scand.* [B] *31*:354.
76. (a) Bax, A., and Freeman, R. (1982). *J. Am. Chem. Soc. 104*:1099. (b) Perlin, A. S., and Hamer, G. K. (1979). In *Carbon-13 NMR in Polymer Science* (W. M. Pasika, ed.). American Chemical Society, Washington, D.C., p. 123.
77. Tvaroska, I., Hricovini, M., and Petrakova, E. (1989). *Carbohydr. Res. 189*:359.
78. Bax, A., Freeman, R., and Kempsell, S. P. (1980). *J. Am. Chem. Soc. 102*:4849.
79. (a) Bax, A., Freeman, R., Frenkiel, T. A., and Levitt, M. H. (1981). *J. Magn. Reson. 43*:478. (b) Mareci, T. H., and Freeman, R. (1982). *J. Magn. Reson. 48*:158.
80. Excoffier, G., Gagnaire, D. Y., and Taravel, F. R. (1974). *Carbohydr. Res. 56*:229.
81. Gagnaire, D. Y., Nardin, R., Taravel, F. R., and Vignon, M. R. (1977). *Nouv. J. Chim. 1*:423.
82. Nunez, H. A., and Barker, R. (1980). *Biochemistry 19*:489.
83. Rosevear, P. R., Nunez, H. A., and Barker, R. (1982). *Biochemistry 21*:1421.
84. King-Morris, M. J., and Serianni, A. S. (1987). *J. Am. Chem. Soc. 109*:3501.
85. Schmidt, R. R. (1986). *Angew. Chem. Int. Ed. Engl. 25*:212.
86. (a) Mootoo, D. R., Konradsson, P., and Fraser-Reid, B. (1989). *J. Am. Chem. Soc. 111*:8540. (b) Halcomb, R. L., and Danishefsky, S. J. (1989). *J. Am. Chem. Soc. 111*:6661. (c) Barrett, A. G. M., Bezuidenhoudt, B. C. B., Gasiecki, A. F., Howell, A. R., and Russell, M. A. (1989). *J. Am. Chem. Soc. 111*:1392.
87. (a) Jardetzky, O. (1980). *Biochem. Biophys. Acta 621*:227. (b) Cumming, D. A., and Carver, J. P. (1987). *Biochemistry 26*:6664.
88. Dais, P., and Perlin, A. S. (1987). *Adv. Carbohydr. Chem. Biochem. 45*:125.
89. Vold, R. L., Waugh, J. S., Klein, M. P., and Phelps, D. E. (1968). *J. Chem. Phys. 48*:3831.
90. Meiboom, S., and Gill, P. (1958). *Rev. Sci. Instrum. 29*:688.
91. Berry, J. M., Hall, L. D., Welder, D. G., and Wong, K. F. (1979). In *Anomeric Effect—Origin and Consequences* (W. A. Szarek and D. Horton, eds.). American Chemical Society, Washington, D.C., pp. 30–49.
92. Lyerla, J. R., and Levy, G. C. (1974). In *Topics in Carbon-13 NMR Spectroscopy*, Vol. 1 (G. C. Levy, ed.), John Wiley & Sons, New York, pp. 79–148.
93. Wu, G. D., Serianni, A. S., and Barker, R. (1983). *J. Org. Chem. 48*:1750.
94. Serianni, A. S., and Barker, R. (1982). *J. Magn. Reson. 49*:335–340.
95. Werbelow, L. G., and Grant, D. M. (1977). *Adv. Magn. Reson. 9*:190.
96. Akasaka, K., Imoto, T., Shibata, S., and Hatano, H. (1975). *J. Magn. Reson. 18*:328.
97. Zens, A. P., Williams, T. J., Wisowaty, J. C., Risher, R. R., Dunlap, R. B., Bryson, T. A., and Ellis, P. D. (1975). *J. Am. Chem. Soc. 97*:2850.
98. Kline, P. C., Huang, S.-G., Hayes, M. L., Barker, R., and Serianni, A. S. (1990). *Can. J. Chem. 68*:2171.
99. Hehre, W. J., Radom, L., Schleyer, P., and Pople, J. A. (1986). *Ab Initio Molecular Orbital Theory*. John Wiley & Sons, New York.
100. Clark, T. (1985). *A Handbook of Computational Chemistry—A Practical Guide to Chemical Structure and Energy Calculations*. John Wiley & Sons, New York.
101. (a) Lemieux, R. U., Bock, K., Delbaere, L. T., Koto, S., and Rao, V. S. (1980). *Can. J. Chem. 58*:631. (b) Thogersen, H., Lemieux, R. U., Bock, K., and Meyer, B. (1982). *Can. J. Chem. 60*:44.
102. Bock, K. (1983). *Pure Appl. Chem. 55*:605.
103. (a) Melberg, S., Rasmussen, K., Scordamaglia, R., and Tosi, C. (1979). *Carbohydr. Res. 76*:23. (b) Serianni, A. S., and Chipman, D. M. (1987). *J. Am. Chem. Soc. 109*:5297. (c) Garrett, E. C., and Serianni, A. S. (1990). *Computer Modeling of Carbohydrate Molecules* (A. D. French and

J. W. Brady, eds.). American Chemical Society, Washington, D.C., p. 91. (d) Garrett, E. C., and Serianni, A. S. (1990). *Carbohydr. Res. 206*:183.
104. Lemieux, R. U. (1971). *Pure Appl. Chem. 25*:527.
105. Wüthrich, K. (1986). *NMR of Proteins and Nucleic Acids*. John Wiley & Sons, New York.
106. Clore, G. M., Brunger, A. T., Karplus, M., and Gronenborn, A. M. (1986). *J. Mol. Biol. 191*:523.
107. Havel, T. F., and Wüthrich, K. (1985). *J. Mol. Biol. 186*:281.
108. Braun, W., and Go, N. (1985). *J. Mol. Biol. 186*:611.
109. This program can be obtained from Hare Research, Inc., 14810 216th Avenue NE, Woodinville, WA 98072.
110. Homans, S. W., Edge, C. J., Ferguson, M. A. J., Dwek, R. A., and Rademacher, T. W. (1989). *Biochemistry 28*:2881.
111. Brooks, B. R., Bruccoleri, R. E., Olafson, B. D., States, D. J., Swaminathan, S., and Karplus, M. (1983). *J. Comput. Chem. 4*:187.
112. Brady, J. W. (1986). *J. Am. Chem. Soc. 108*:8153.
113. Ha, S. N., Madsen, L. J., and Brady, J. W. (1988). *Biopolymers 27*:1927.
114. Tran, V. H., and Brady, J. W. (1990). *Biopolymers 29*:977.
115. Yan, Z.-Y., and Bush, C. A. (1990). *Biopolymers 29*:799.
116. Homans, S. W., Pastore, A., Dwek, R. A., and Rademacher, T. W. (1987). *Biochemistry 26*:6649.

5

Mass Spectrometry of Carbohydrates

Roger A. Laine *Louisiana State University and A&M College, and The LSU Agricultural Center, Baton Rouge, Louisiana*

Three major sets of compounds constitute most of the biological chemicals in living organisms: proteins, lipids, and carbohydrates. All three of these as well as the nucleic acids have subclasses that, in function, are structural, nutrient, or bioactive. Collagen and elastin are structural proteins; seed and yolk proteins provide nutrients to the developing embryo. Endorphins, on the other hand, are small molecular weight peptides that have a high level of biological activity that is based on further functions of a receptor protein. Lipids' structural representatives are membrane phospholipids and cholesterol; lipid storage compounds include triglycerides, whereas the prostaglandins and steroid hormones have high biological activity, based on recognition by a receptor protein either at the cell surface or in the cytoplasm or nucleus. This receptor protein then mediates the cascade of activities necessary to cause a change in function.

Complex carbohydrates also comprise a set of substances that includes fibrous structural compounds, such as cellulose and chitin, and energy storage compounds, such as sucrose, starch, and glycogen. Another structural set of carbohydrates comprises the common *N*-linked oligosaccharides associated with eucaryotic (and archebacterial) glycoproteins, either secreted or on the surface of cells. The *N*-linked oligosaccharides, and others that are *O*-linked to serine, threonine, or hydroxyproline, may make structural contributions to the whole glycoprotein by assisting in or maintaining polypeptide folding, and may confer physical characteristics such as enhanced solubility or protease resistance.

A third and rapidly emerging category for complex carbohydrates are receptor–ligand pairs of a bioactive oligosaccharide sequence with a protein. Examples are in the ABO blood group antigens and the antibodies that recognize these structures; the influenza hemaglutinin, which recognizes sialic acid; bacterial adhesion proteins that recognize and bind to glycoproteins or glycolipids; bacterial toxins, such as cholera toxin, that bind to certain gangliosides; sperm–egg interactions that may involve terminal sugars on glycoproteins; and surface-expressed glycosyltransferases. These areas will not be reviewed herein. The heparin pentasaccharide that inhibits the clotting mechanism by binding to antithrombin III is another example, as is the most common bioactive saccharide: sucrose, which stimulates the sweetness receptor, beginning a neural response—regarded as pleasant by most of us—probably originating from evolution toward consumption of sucrose-rich fruits. The possibility of development of pharmaceuticals is exemplified by the fact that structural analogues of sucrose include saccharin, sucralose, and the highly active sweetener dipeptide aspartame.

Some extremely promising areas for carbohydrate–receptor interactions include lymphocyte trafficking receptors such as ELAM-1, gp90mel, GP140, all of which have been shown to contain C-lectinlike domains, with their counterpart carbohydrate ligands either on the lymphocyte or vascular endothelium. Expression of the lectin or the ligand is transient, and affects the extravasation of lymphocytes and neutrophils that play a large role in the inflammatory response and in disorders such as rheumatoid arthritis. Other viral receptors include rotavirus, coronavirus, and herpesviruses. Glycosaminoglycans have bioactive sequences within the polymer; for example, dermatan sulfate sequences that inhibit the clotting cascade, heparin sequences that bind to fibroblast growth factor, other heparin sequences that inhibit smooth-muscle cell growth and may have many other biological activities related to the state of cellular growth, angiogenesis, and other proliferative processes. The literature in this area is burgeoning, and three companies have set up on the last 3 years to seek pharmaceuticals from complex carbohydrates: Glycomed, Glytek (Cytel), and Glycogen.

I. STRUCTURAL DIVERSITY OF CARBOHYDRATES

A substantial and complex difference, however, exists in the intricacy and detail of chemical characterization of structures of complex carbohydrates when compared with the other biochemical classes.

Complex carbohydrates constitute the most complicated set of biological polymers for structural characterization. Molecular parameters of ring size, anomeric configuration, position of linkage between monomers, and sequence of sugar epimers or sugar types all contribute toward five to six orders of magnitude higher structural diversity in a short sequence than found in proteins. All of these stereochemical and isomeric parameters can confer biological activity, such as antigenicity, viral or bacterial receptor activity, lymphocyte trafficking specificity, and other interesting and important functions. Branching adds another dimension that can produce three further orders of magnitude of possible molecular permutations.

Most of the differences are given by stereochemical or positional isomers, producing a very large possible set of compounds of the same mass and relative polarity, making separation science the most important and difficult aspect of analysis. Additional biological substitutions with phosphates, sulfates, phospono, acyl groups, ether groups, pyruvate ketals, and other moieties contribute another unimaginable number of isomers from which a single structure must be discerned.

Compounds as large as 18 sugars in the *N*-linked system are biosynthesized into an exact structure, without benefit of a polymeric template. These structures are built up by processive glycosyltransferases, each enzyme possessing a binding site that recognizes the previous structure and precisely adds the next monomer, with the proper anomericity, position of linkage, and ring size. Postassembly modification can take place at the oligomer of polymer stage. The template is thus carried in the binding and recognition sites of a whole series of glycosyltransferases. Little wonder that carbohydrates are enormously conserved in genetic families. A whole new enzyme must be evolved by extensive modification of the binding site to transfer a sugar to a different position or add a new sugar. Obviously, a single DNA base change is not going to bring about structural addition mutations in sugars; deletions, yes.

Most biological activity of carbohydrates has been found—as related to the binding-site size of a protein receptor—in short saccharide sequences of six or fewer, often part of a mono-, di-, or trisaccharide structure.

The number of possible structures in an otherwise unsubstituted linear polymer of six different D-hexoses A-B-C-D-E-F is as follows:

$$\text{“S”} = 6! \times 2^6_a \times 2^6_r \rightarrow 4^5 = 3{,}019{,}898{,}880$$

The first term (6!) represents permutations of six linear moieties; the second (2^6_a), anomeric configuration; the third (2^6_r), pyranose or furanose ring size; and the fourth (4^5), position of linkage. Adding the variable of *L*-sugars would increase this number by a factor of 64.

If one allows branching of the sequence, placing sugars A-B, for example, on the internal saccharides D, E, or F, increases the number of possibilities another 100-fold or more. Naturally occurring substituents (acetates, sulfates, phosphates, or others) make the actual, possible number . . . mind boggling.

```
A              A              A              A
 \              \              \              \
B–C–D–E–F      B–C–D–E–F      B–C–D–E–F      B–C–D–E–F

               A–B            A–B            A–B
                 \              \              \
                 C–D–E–F      C–D–E–F      C–D–E–F

               A–B–C       A–B–C
                   \           \
                   D–E–F     D–E–F

A            A–B                  A–B–C–D
 \             \                        \
  D–E–F        D–E–F                    E–F      and many others
     /            /
   B–C           C
```

Each structure, representing one linkage position around the ring of each of the possible branch points of these examples of branching configurations, has about 1.5 billion permutations.

A. Structure Cannot Be Inferred From Chromatographic Mobility

The most any single chromatographic or electrophoretic system can provide is about 600 windows of resolved individual structures, usually far fewer. One can quickly see that no set of chromatographic and electrophoretic systems can hope to provide resolution for such a large number of compounds of the same mass and polarity; therefore, one cannot rely on chromatographic or electrophoretic mobility behavior alone. If a compound is judged pure, only analysis can prove it.

Also, no single analytical system can provide the necessary resolution of data points to discern among even a fraction of this number. Biology has fortunately not confounded us with this many structures, but this complex nature of carbohydrates dictates that there exists the chemical problem of establishing an exact structure among the huge number possible.

In principle, given 5 mmol of pure oligosaccharide of fewer than ten sugars, high-field two-dimensional (2D) or three-dimensional (3D) NMR in a series of experiments including COSY, NOESY, and TOCSY, can be used to gain complete structural analysis, including spatial conformation. These considerations are covered in Chapter 4 in this volume. However, given the extreme number of possibilities in the foregoing example of a hexasaccharide, imagine the data resolution necessary to discern among all of these compounds, probably impossible without 3D-slicing methods.

The reality of biological systems gives the common result that unknown pure substances with biological activity, isolated from complex biological systems, usually yield samples, in the multi-picomole to 100-nmol range, that are available for structural determination. Iodine-125–labeled binding proteins can detect picomoles of active substance on a thin-layer plate. In this usual sample range, therefore, we must employ a strategy of chemical, enzymatic, and mass spectrometric methods to establish structures. Although DNA sequencing and peptide sequencing can be performed at femtomole to picomole levels, we must begin with 100 nmol to hope for a complete structure. This level is probably attainable in only a handful of laboratories in the world.

Details of separation science using thin-layer chromatography, high-performance liquid chromatography, column chromatography, gel filtration, capillary zone electrophoresis, supercritical fluid chromatography, ion exchange, reversed-phase, ion-pairing, and others can be found in the extensive literature on complex carbohydrates and will not be reviewed herein.

II. MASS SPECTROMETRY

In fact, a rich compendium of the most recent mass spectrometric methods for complex carbohydrates has recently been published in a volume of *Methods in Enzymology* [1]. The new volume contains a set of 11 chapters on the latest in mass spectrometry of carbohydrates, and the details of the experiments are included in those chapters.

Thus, for this chapter, instead of repeating parts of the current and extensive treatise in *Methods in Enzymology*, I will concentrate on illustrating the details of structure determination of an example from an *N*-linked oligosaccharide of a glycoprotein, and outline the strategy for combining wet chemistry and enzymology with mass spectrometry to establish the complete carbohydrate structure, explaining the benefit and limitations of the use of mass spectrometry in the analysis.

A. Strategy

Let us assume a pure sample of 50–100 nmol of a biantennary *N*-linked structure of the following configuration:

```
NANA(α2→6)Gal(β1→4)GlcNAc(β1→2)
                               \
                        Man(α1→6)           Fucose(α1–6)
                                 \                      \
                                  Man(β1→4)GlcNAc(β1→4)GlcNAc
                                 /
                        Man(α1→3)
                               /
NANA(α2→3)Gal(β1→3)GlcNAc(β1–2)
```

For isolation of compounds from glycoproteins, refer to Akira Kobata's work using hydrazinolysis [2], and references therein. See references to *N*-glycanase (PNGase) and other enzymatic methods in previous volumes of *Methods in Enzymology* dealing with complex carbohydrates, edited by Victor Ginsburg.

Note that the sialic acids (NANA) and the galactoses (Gal) have different linkage positions to the next sugar on the two antennary arms in this hypothetical structure. Determining the branch location of these groups is difficult in biantennary structures. Luckily, the branch to the mannose (man) is not 2,4 instead of 3,6. After perusing the chapter the reader can work this alternate structure out as an exercise. Determination for higher-branched structures (triantennary, tetrantennary, and so on), can be extremely difficult.

With 50–100 nmol, 2D NMR would be difficult or impossible in 1991; 5 μmol is closer to what is needed. Perhaps the NMR analysis will be limited to discerning the number and type of anomeric protons. The literature reviewed in Chapter 4 contains a reasonable database to suggest that the chemical shift of the anomeric protons can be used to confirm the presence of a given structure and linkage. The number and anomer-type linkages can usually be discerned from 100 nmol. Some chemical shifts may fall within windows for certain compounds, but remember the large number of possible structures and the small number of available windows in any high-resolution system.

B. Sugar Composition and Molecular Size

In the ensuing discussion, the following is an important analytical axiom:

> A known authentic standard compound with a similar structure should be carried through all of the procedures to establish molar response ratios for the entire analysis.

The sugar composition and molecular size are related structural determinations. Mass spectrometry alone will give molecular mass and perhaps a few informative fragments indicating sequence of sugar type and branching. For example, in this case, the data would indicate five hexoses (162-u each as internal units), four *N*-acetylhexosamines (203-u each), one methylpentose (163-u as a terminal unit) and two *N*-acetylneuraminic acids (308-u each as terminal units); add water: the mass should be 2402 u.

1. *Molecular Size*

Fast-atom bombardment (FAB; liquid secondary ion mass spectrometry; LSIMS) would usually utilize about 100 pmol to 1 nmol of material to obtain a relative molecular mass for these types of compounds. Limits of detection for electron multiplier detectors are in the 10- to 50-pmol range. Prior permethylation of the sample has some distinct advantages, as shown by Dell [3]. For example, the derivatized compound can also be used to determine linkage position, as described later. Cesium ion guns are more efficient than FAB ion guns for producing molecular ions in these instruments.

Some new instruments manufactured by the four major mass spectrometer companies have "array detectors." These instruments are about 100 times more sensitive than scanning detectors that use an exit slit and electron multiplier. The limits of detection for array detectors are in the range of 50–500 fmol applied to the probe.

Most biochemists will not have local access to a high-mass instrument equipped with FAB or cesium gun source. A few array detectors are installed in the field for those who have vanishingly small amounts of material. A collaborative effort with a mass spectrometry laboratory experienced in handling carbohydrates is important for efficient analysis (see [1] for a recent set of chapters by carbohydrate mass spectrometrists).

The mass spectrometry experiments would ordinarily not discern the stereochemical identity of the individual sugars (the hexoses, mannose and galactose, for example, have the same mass). I say "not ordinarily" because there are a few observations by a few researchers that suggest that, in the future, soft ionization plus collisional-activated dissociation in tandem mass spectrometry might provide a more sensitive physical probe for stereochemistry than previously believed. None of this is available for practical use and probably will not be for several years.

2. *Sugar Composition*

By using about 250 pmol of the sample, the total sugar composition could be accurately determined after acid hydrolysis by a modern instrument manufactured by Dionex and employed early [4] in the laboratory of Y. C. Lee at Johns Hopkins with the collaboration of Reid Townsend, now at UCSF and Mike Hardy, now at Dionex. With use of an anion-exchange column at high pH, the monosaccharides could be separated in two different gradients, one for the sialic acid and uronic acids and another for the neutral and amino-sugars. This is a new and useful technology, with high resolution for oligosaccharides.

Alternatively, gas chromatography–mass spectrometry (GCMS) could be utilized

1. By performing methanolysis on the sample and preparing the trimethylsilyl derivatives, gas chromatography could be used to determine sugar composition [5], and chemical ionization (CI) GCMS could be used to increase the sensitivity and selectivity of this analysis [6]. The limits of detection are about 1–5 nmol on a quadupole instrument and perhaps five times less on an ion trap mass spectrometer.

2. By preparing alditol acetates after acid hydrolysis, followed by chemical ionization (CI) GCMS to increase sensitivity and selectivity [7].

3. Computed "mass chromatography" increases the apparent signal/noise ratio by at least tenfold over gas chromatography alone [7]. Mass chromatography records the complete mass spectrum each second or less during a gas-chromatographic run, and stores the data. Then the computer is asked to plot an individual ion to selectively plot the sugar peaks in a mixture. (In methane chemical ionization, m/z 375, for example [7], representing the MH-60 ion, is usually the largest ion in the mass spectrum of alditol acetates from hexoses, whereas 374 is the most intense ion for alditol acetates from hexosamines. These ions can be used to selectively plot the chromatographic location and integration related to these structures while elmininating signal from all background except that at m/z 374 and 375.) These techniques are usable in the 1–10 nmol range of sample size with realistic individual injections of one-tenth of the sample. Ion trap mass spectrometers are about five times more sensitive than quadrupoles, owing to ion storage.

4. Another technique to increase sensitivity would be to collect only selected ion-monitoring integration data, using the sugar ions of interest (see your mass spectrometrist to explain this). The danger in this is missing an important, but unknown, component sugar of a different mass than usually expected.

After any of the foregoing procedures the analyst will have the following information:

The sugar ratio is two galactose, three mannose, four hexosamines, one fucose, and two sialic acid. (The hexosamines may not give integral numbers in the analysis and will need to be compared with numbers obtained in the authentic standard.)

The FABMS determined mass of the compound (2402 u) confirms the foregoing sugar ratios.

What the analyst does **not** know after FABMS and composition includes

The complete internal sequence of the sugars
Branching points
Position of linkage
Anomeric configuration
Ring size of constituents

In other words, lacking other information, such as history of analysis of similar compounds in nature, biological source, antibody or lectin binding, or other information, most of the structural parameters of an oligosaccharide remains unknown after FABMS and sugar composition analysis.

Clearly, claiming structure from the mobility of a peak eluting from a gel filtration column is hopeful, to say the least. Furthermore, claims of structure from a molecular ion and a few fragments is not scientifically sound.

Some information on the sequence of the sugars will be available in the FAB mass spectrum of the intact compound [7], the ABEE or ABOE [8,9], or the permethyl derivative [3]. These ions, recognized by the investigations of several groups of researchers have been catalogued according to their origins [3,9,10].

Ionization–energy-induced cleavages or collision-induced dissociation (CID) of Compound I would yield ions representing losses of the terminal sugars, principally with charge retention on the amino sugars, with positive ion detection, but ions originating from both

```
          The a arm                                       The c branch

    NANA(α2→6)Gal(β1→4)GlcNAc(β1–2)
                                  \
                             Man(α1→6)            Fucose (α1–6)
                                    \                         \
                                     Man(β1→4)GlcNAc(β1→4)GlcNAc
                                    /
                             Man(α1→3)            the core
                                  /
    NANA(α2→3)Gal(β1→3)GlcNAc(β1–2)

          The b arm
```

Compound I

the reducing and nonreducing end can be apparent in the spectrum, with either positive or negative ion detection [3,8–11].

For the purpose of simplifying the principal information currently available from an FAB mass spectrum, the following outline of the structure and the principal ions expected are ordinarily depicted according to the assignments of Costello [10]. However, I find these assignments a little difficult, owing to the possible losses of ions from any of several branches, and therefore, have modified the Costello fragment ion assignment system as follows:

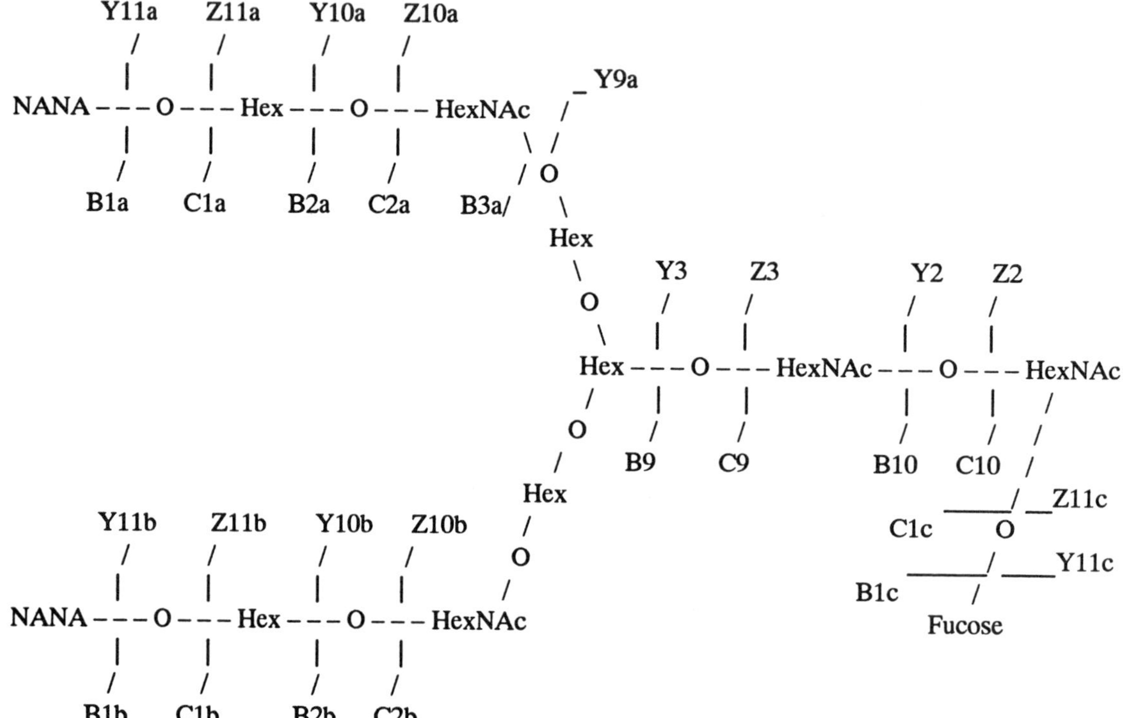

Scheme 1

Identification in Scheme 1 of C9a and Z9a, as well as the B8 and C8, B4b, C4b, Y9b, and Z9b were omitted to eliminate crowding.

FRAGMENT IONS. These general descriptions of the types of ions to expect in the spectra will assume values possible with either negative or positive FAB, with appropriate proton transfers.

N-Acetylneuraminic acid (sialic acid). The loss of this moiety from either arm on the nonreducing side of the glycosidic oxygen with negative ion detection would produce a reducing-end ion with a loss of mass 292 or 291 with proton transfer (the Y11a and Y11b ions). Charge retention on the nonreducing side of the bond would produce an ion of mass 290, 291, or 292, with proton transfer (B1a and B1b ions), with a two-proton shift and 308 on the reducing end side of the glycosidic oxygen (the C1a and C1b ions) [10]. Since

NANA can be lost from either arm, the respective ions would be expected to be equally likely within the bounds of this experiment.

Loss of two sialic acids, however, would be more rare, unless they were tandemly linked, since it would require two bond-scission events. We have reported that in low-energy CID, some spectral differences occur in position of linkage of the terminal sugar [12] when attached to aminosugars, but work with penultimate hexoses has not yet been reported. Further development of this area is needed. Costello and Domon [10,11], and Gillece-Castro and Burlingame [9], using CID, have noted some through-the-ring cleavages that can give evidence for position of linkages. Yields of these ions are low, however, and close in intensity to background or other, as yet unidentified, ions. Research progress in this area would greatly benefit the state-of-the-art analytical scheme described below.

NANA (α2–6)Gal (B2a, C2a)

and

NANA (α2–3)Gal (B2b, C2b)

These two fragments (the B2 and C2 ions) can be expected from either the upper 1→6 linked (a) arm or the lower 1→3 linked (b) arm as the molecule is depicted (see Compound I). Both fragments have the same mass (308 + 162 = 470, ± 1,2 for the C2 on the reducing side of the galactose's glycosidic oxygen and 454 ± 1,2, or 16 fewer for the B2 ions on the nonreducing side); hence, it will be difficult or impossible to discern between them or to discern their origin in the unknown structure. Although the mass of the disaccharide equals an *N*-acetylneuraminic acid and a hexose, the identity of the hexose will not be known from this experiment.

What will be clear is that terminal sialic acid is connected to a penultimate hexose, and that owing to lack of a disialo ion, sialic acid is probably not connected to another sialic acid or other sugar, unless there is one connected to the inner core of the oligosaccharide. Other ions described later, however, will eliminate these possibilities.

The absence of ions in the m/z 179 range along with the absence of ions showing a loss of this moiety will indicate, but not prove, that the oligosaccharide does not contain a terminal hexose.

FUCOSE. Loss of 146 or 163 ± 1,2 for the proximal fucose [9], a Y11c or Z11c ion, will identify the presence of fucose, but not its location. Cleavage of the bond between the two reducing glucosamines would yield, for example, an m/z 599 ion, with the very useful ABOE derivative [8,9] showing the fucose location on the proximal glucosamine, but not identifying its position of linkage.

NANA (α2→6)Gal(β1→4)GlcNAc (B3a, C3a)

and

NANA (α2→3)Gal(β1→3)GlcNAc (B3b, C3b)

These trisaccharide ions will be of the same mass (308 + 162 + 203 = m/z 673 ± 1,2 from the C3 657 ± 1,2 from the B3, and will be produced from the two possible locations on the *a* or *b* arm 2-linked mannose. The mass of the B3 fragments in permethylated oligosaccharides will be at m/z 825, ± 1,2 [3,10,11]. The reducing end fragments Y8a, Z8a, Y8b, Z8b will show the same mass for the loss of the trisaccharide from either arm.

NANA (α2→6)Gal(β1→4)GlcNAc(β1–2)Man (B4a, C4a)

and

NANA (α2→3)Gal(β1→3)GlcNAc(β1–2)Man (B4b, C4b)

These tetrasaccharide ions of identical mass would be produced from cleavage at the glyosidic oxygen of the 2-substituted mannose in each arm. The former structure would be produced from arm a and the latter from arm b. Incremental mass for the pair of ions at each side of the glycosidic oxygen, corresponding to one added hexose (673 + 162) would be expected at m/z 735 for the C4 and 719 for the B4 ± 1,2). Since the mass is identical, the ion produced from an unknown would simply represent cleavage from either arms a or b at the 6- and 3-positions of the branching mannose. An additional ion could be expected between carbons 5 and 6 of the branching mannose, yielding an ion 14-mass units higher than the C4a ion at m/z 749, ± 1,2.

LOCATION OF THE A,B BRANCH-ARM JUNCTURE. Because two glycosidic bonds need to be broken to produce a pentasaccharide from the unknown, ions representing a terminal pentasaccharide from either end will either be absent or will be at extremely low intensity, a principle first described for saccharides by Watanabe et al. [13].

```
NANA(α2→6)Gal(β1→4)GlcNAc(β1→2)Man(α1→6)
                                        \
                                         Man (B9, C9)
                                        /
NANA(α2→3)Gal(β1→3)GlcNAc(β1→2)Man(α1→3)
```

Presence of nonasaccharide B9, C9 ions one hexose (162 u) larger than two of the B4, C4 tetrasaccharides above added together, and the absence of B5, C5—or B6, C6—sugar fragments strongly suggests the presence of two equal-length branches of four sugars each. If the branches had been unequal at three and five sugars long, a pentasaccharide B5 fragment would have appeared in the spectrum. Similarly, if the branches had be two and six sugars long, B6, C6 ions would have appeared in the spectrum. The use of a reducing terminal label, such as ABOE, can help greatly with assignment of ions [9].

```
NANA(α2→6)Gal(β1→4)GlcNAc(β1→2)Man(α1→6)
                                        \
                                         Man (β1–4)GlcNAc
                                        /
NANA(α2→3)Gal(β1→3)GlcNAc(β1→2)Man(α1→3)
                        (B10, C10)
```

An additional set of B10, C10 ions, an increment of one hexosamine (+203), and a set of Y2, Z2 ions indicate that the next sugars in line are two hexosamines. The Y2, Z2 ions were already used for identification of the location of the *c* branch.

FAST ATOM BOMBARDMENT MASS SPECTROMETRY OF HIGH MANNOSE OLIGOSACCHARIDES. Burlingame and Gillece-Castro have examined a number of ABOE derivatives of *N*-linked high mannose oligosaccharides for which the spectra and interpretation have been recently reviewed [9].

From the sugar composition and mass spectra, so far, the stsructure can be predicted to be as follows:

```
NANA–Hex–GlcNAc
                \
                 Hex
                   \
                    Hex–HexNAc–HexNAc–Fuc
                   /
                 Hex
                /
NANA–Hex–GlcNAc
```

Essentially no linkage position or anomeric information has been obtained.

3. *Enzyme-Assisted Sequence Analysis: Ring Size and Anomeric Configuration in One Assay*

This step in the analysis is probably the most important nonspectroscopic microanalytical technique now available for saccharide analysis.

Since it is known from the literature that sialic acids usually are α-linked, an enzyme, neuraminidase, available from several microbial sources, can be employed to remove the sialic acids from the molecule, revealing the number of sialic acids and exposing penultimate or branching locations in the molecule. If the sialic acid is not removed by the enzymes, several possibilities exist, all interesting. One is that the sialic acid is substituted with an acetyl or other group on one of the free hydroxyls, or it exists in a lactone form. The sialic acid may be a branch in the molecule, and the enzyme chosen has specificity only for terminal linear sugars. When the next sugar is identified as a newly exposed terminal sugar by methylation linkage analysis, described later, it too can be removed by an anomeric-specific, ring size-specific glycosidase (β-galactopyranosidase, for example) establishing the sequence of sugars, identifying the ring size, and discerning the anomeric configuration all at once. The neuraminic acid can also be removed by specific mild hydrolytic conditions (0.1 N HCl, 1 hr, 80–100°C, for the purpose, but the information of the anomeric linkage type and substitutions is lost.

4. *Methylation Linkage Analysis*

The next important step is to catalogue the linkage position of each sugar in the compound. A 20-year-old method with a few improvements is still the only sure way to obtain this information [1]. Bengt Lindberg's methylation analysis system originally employed the Hakomori methylation [14], but many workers are now turning to the Cicanu and Kerek [15] method as simpler. Steps are as follows:

1. Methylate the pure saccharide [14,15]
2. Hydrolyze [16]
3. Reduce to the alditols with sodium borodeuteride [16]
4. Acetylate [3,16]
5. Run gas liquid chromatography–mass spectrometry, preferably with chemical ionization [16,17].

The resulting chemical entities appear as follows:

```
            CH2–O–CO–CH3
               |
3HC–OC–O–CH
               |
      3HC–O–CH
               |
            HC–O–CH3
               |
            HC–O–CO–CH3
               |
            CH2–O–CH3
```

(from 2-substituted mannopyranose)

These methods have recently been reviewed by Hellerquist [1,16] in detail. The advantages of chemical ionization (CI) in detection of the products has been published [17].

The resulting compounds are methylated on the free hydroxyls and acetylated on the following hydroxyls:

1. The reduced anomeric carbon (carbon-1, above)
2. The ring hydroxyl which was taken up in the acetal linkage (carbon-5 for a hexopyranose as shown above)
3. The position of linkage, if any (carbon-2 in the example given above).

Data expected from the unknown would give the following *O*-methyl alditols:

1 mol of 1,3,5-tri-*O*-acetylgalactitol (from 3-substituted galactopyranose)
1 mol of 1,5,6-tri-*O*-acetylgalactitol (from 6-substituted galactopyranose)
1 mol of 1,3,5-tri-*O*-acetyl-2-*N*-methyl-*N*-acetylglucosaminitol (from 3-substituted *N*-acetylglucopyranosamine)
2 mol of 1,4,5-tri-*O*-acetyl, 2-*N*-methyl-*N*-acetylglucosaminitol (from 4-substituted *N*-acetylglucosamine, ring size unknown)
1 mol of 1,4,5,6-tetra-*O*-acetyl, 2-*N*-methyl-*N*-acetylglucosaminitol (from 4,6-disubstituted *N*-acetylglucosamine)
1 mol of 1,3,5,6-tetra-*O*-acetylmannitol (from 3,6-disubstituted mannopyranose)
2 mol of 1,2,5-tri-*O*-acetylmannitol (from 2-substituted mannopyranose)
2 mol of 2,4,6,7,8,9-penta-*O*-acetyl-5-*N*-methyl-*N*-acetylneuraminic acid ketol (from terminal NANA-pyranose)
1 mol of 1,5-di-*O*-acetylfucitol (from terminal fucopyranose)

The total is 12 stoichiometric sugars. The ring size is now known for most of the compounds owing to the methylation at the 4-position. The only compounds for which this is unknown are the 3,4-substituted GlcNAcs. This ring-size information will come from later methylation analysis. The analyst must be certain to compare this with the results from a known, authentic standard treated in the same way in the same sample amount to ensure accurate molar response ratios. Extreme care must be taken not to evaporate the samples for different lengths of time, owing to differential volatility of the methylated derivatives.

5. Combined Enzyme/FABMS/Methylation Linkage Analysis for Sequence Determination

Neuraminidase treatment

2 mol of 1,5-di-*O*-acetylgalactosaminitol (from newly revealed terminal galactopyranose)

1 mol of 1,3,5-tri-*O*-acetyl-2-*N*-methyl-*N*-acetylglucosaminitol (from 3-substituted *N*-acetylglucosamine)
2 mol of 1,4,5-tri-*O*-acetyl, 2-*N*-methyl-*N*-acetylglucosaminitol (from 4-substituted *N*-acetylglucosamine)
1 mol of 1,4,5,6-tetra-*O*-acetyl, 2-*N*-methyl-*N*-acetylglucosaminitol (from 4,6-disubstituted *N*-acetylglucosamine)
1 mol of 1,3,5,6-tetra-*O*-acetylmannitol (from 3,6-disubstituted mannose)
2 mol of 1,2,5-tri-*O*-acetylmannitol (from 2-substituted mannose)
2 mol of 2,4,6,7,8,9-penta-*O*-acetyl-5-*N*-methyl-*N*-acetylneuraminic acid ketol (from terminal NANA)
1 mol of 1,5-di-*O*-acetylfucitol (from terminal fucose)

The foregoing list contains no substituted galactoses, and notice the 2 mol of 1,5-di-*O*-acetylgalctitol (from terminal galactose).

After the neuraminidase treatment, the sample should be tested by FABMS to see how many sialic acids were removed and whether the sample was completely hydrolyzed: 1 nmol needed. If the results show all sialic acids removed, the investigator may ignore the presence of terminal sialic acid in the sample and may not need to perform cleanup after neuraminidase before methylation. However, on this and subsequent steps, removal of the monosaccharides by microgel permeation desalting from the hydrolytic mixture before methylation will eliminate problems with identification of new terminal sugars.

The foregoing data shows that neuraminic acid was terminal and unsubstituted, and that 1 mol was glycosidically linked to the 3-position of a penultimate galactose, whereas 1 mol was linked to the 6-position of a penultimate galactose. Whether the 3- and 6-linked galactoses were on the same molecule or on a mixture of molecules is not clear at this point.

Mass spectrometry: At each step the resulting oligosaccharide, after enzyme digestion, can be analyzed by FABMS to confirm the stoichiometry of loss of sugars and to assess completion of the digestion. Thus, loss of 2 × 291 (± 1,2) will show two sialic acids cleaved by neuraminidase.

From the methylation data, it is clear that galactose is the next sugar to which NANA is substituted in the direction of the reducing end after the terminal NANA, and that one or more of the terminal sequences in the molecule is NANA-Gal, with two different substitutions present. This also shows that galactose occurs nowhere else in the molecule, by the absence of other substituted galactoses. This information is gleaned from the gas chromatographic behaviors, which differ for galactose and mannose derivatives.

ENIGMA. If the compound in question were not a single component, this result would be indistinguishable from the result in which a 1:1 mixture existed that comprised two different branched molecules, one containing two sialic acids linked to the 3-position of the penultimate galactose and the other containing two silaic acids bound to the 6-position of galactose, each molecule having the same galactose substitution on both branch arms. Another possibility is a 1:1:1 ratio of compound I with the other two possibilities just mentioned. Other ratios of mixtures of various conceivable proportions of either substitution in a series of molecules (worth contemplating) could be consistent with the data.

```
Gal(β1→X)GlcNAc(β1→2)Man(1→3,6)                Fucose(1→4,6)
                                  \                         \
                                   Man(1→4)GlcNAc(1→4,6)GlcNAc
                                  /
Gal(β1→X)GlcNAc(β1→2)Man(1→3,6)
```

Compound II

6. Galactosidases/FABMS/Methylation

Knowing that galactose is the next sugar in line, the investigator returns to the neuraminidase-treated original compound and attempts digestion with α- and β-galactopyranosidases. β-Galactosidase is found to remove the terminal galactopyranoses, after checking the compound with FABMS, and the resulting compound is methylated and gives the following data:

2 mol of 1,5-di-*O*-acetyl-2-*N*-methyl-*N*-acetylglucosaminitol
1 mol of 1,4,5-tri-*O*-acetyl, 2-*N*-methyl-*N*-acetylglucosaminitol (from 4-substituted *N*-acetylglucosamine)
1 mol of 1,4,5,6-tetra-*O*-acetyl, 2-*N*-methyl-*N*-acetylglucosaminitol (from 4,6-disubstituted *N*-acetylglucosamine)
1 mol of 1,3,5,6-tetra-*O*-acetylmannitol (from 3,6-disubstituted mannose)
2 mol of 1,2,5-tri-*O*-acetylmannitol (from 2-substituted mannose)
2 mol of 2,4,6,7,8,9-penta-*O*-acetyl-5-*N*-methyl-*N*-acetylneuraminic acid ketol (from terminal NANA)
1 mol of 1,5-di-*O*-acetylfucitol (from terminal fucose)

The foregoing list contains no galactoses and notice the 2 mol of 1,5-di-*O*-acetyl-2-*N*-methyl-*N*-acetylglucosaminitol (from terminal *N*-acetylglucosamine), showing that removal of the galactoses caused two GlcNAc moieties to become nonreducing terminal. This data confirms the branching nature of the molecule and indicates that the branch point has not yet been reached. Also, the pyranoside ring size of two of the GlcNAc moieties has now been proved. Does the reader see how?

Other significant observations at this point include the loss of

1 mol of 1,4,5-tri-*O*-acetyl-2-*N*-methyl-*N*-acetylglucosaminitol from 3-substituted GlcNAc, and
1 mol of 1,3,5-tri-*O*-acetyl-2-*N*-methyl-*N*-acetylglucosaminitol (from 4 substituted *N*-acetylglucosamine)

This indicates that, in the sample, each of the galactoses was connected to a different position on two separate *N*-acetylglucosamines, the 3-position on one and the 4-position on the other one. It is not clear to which arm these were attached, nor at this time is it clear that they are on the same molecule.

```
GlcNAc(β1→2)Man(1→3,6)                Fucose(1→4,6)
                       \                         \
                        Man(1→4)GlcNAc(1→4,6)GlcNAc
                       /
GlcNAc(β1→2)Man(1→3,6)
```

Compound III

7. β-Hexosaminidase/FABMS/Methylation

A good guess of the use of β-hexosaminidase from jack beans was next found by the investigator to remove both of the new terminal glucosamines, producing an hexasaccharide as determined by FABMS from the neuraminidase–galactosidase-treated original material. Methylation linkage analysis produces the following result:

1 mol of 1,4,5-tri-*O*-acetyl, 2-*N*-methyl-*N*-acetylglucosamine (from 4-substituted *N*-acetylglucosamine)
1 mol of 1,4,5,6-tetra-*O*-acetyl, 2-*N*-methyl-*N*-acetylglucosamine (from 4,6-disubstituted *N*-acetylglucosamine)
2 mol of 1,5-di-*O*-acetylmannose (from terminal mannose)
1 mol of 1,5-di-*O*-acetylfucose (from terminal fucose)
1 mol of 1,2,4,5-tetra-*O*-acetylmannose

The use of FABMS of this material shows a hexose trisaccharide fragment B3 and C3 ions; a 1-fucose, 1-hexosamine fragment, Y2, Z2; and a 1-fucose, 2-hexosamine fragment (Y3, Z3 ions), but not a 2-hexose disaccharide fragment, indicating that the branch point will be reached upon the next glycosidase digestion. The hexoses are identifiable as mannose by GLC or Dionex analysis, and, thus, the compound appears to be as follows by a combination of the methylation and FABMS data:

```
Man(1→3,6)              Fucose(1→4,6)
          \                          \
           Man(1→4)GlcNAc(1→4,6)GlcNAc
          /
Man(1→3,6)
```

Compound IV

Fragments from FABMS or CID could be useful to clarify the structure of each of these intermediate steps.

8. Mannosidases/FABMS/Methylation

Treatment with α- and β-mannosidase shows two α-mannopyranoses as the terminal sugar, and methylation analysis of the resulting FABMS-identified tetrasaccharide product gives the following:

1 mol of 1,4,5,6-tetra-*O*-acetyl, 2-*N*-methyl-*N*-acetylglucosamine (from 4-6-disubstituted *N*-acetylglucosamine)
1 mol of 1,5-di-*O*-acetylfucose (from terminal fucose)
1 mol of 1,5-di-*O*-acetyl-2-*N*-methyl-*N*-acetylglucosaminitol (from terminal *N*-acetylglucosamine)
1 mol of 1,5-di-*O*-acetylmannitol (from terminal mannopyranose

Obviously, we have reached the first branch point in the decasaccharide with the conversion of disubstituted mannose to terminal mannose, and have shown that the two glucosamines are inserted to the 3- and 6-positions of the branching mannose.

Man(1→4)GlcNAc(1→4,6)GlcNAc(4,6←1)Fucose

Compound V

The mannosidases are tried again and this time the β-mannosidase works to remove 1 mol of mannose from the sample, resulting in the following methylation linkage analysis:

1 mol of 1,5-di-*O*-acetylfucose (from terminal fucose)
1 mol of 1,4,5,6-tetra-*O*-acetyl, 2-*N*-methyl-*N*-acetylglucosamine (from 6-substituted *N*-acetylglucosamine)
1 mol of 1.5-di-*O*-acetyl, 2-*N*-methyl-*N*-acetylglucosamine (from terminal hexosamine)

GlcNAc(1→4,6)GlcNAc(4,6←1)Fucose

Compound VI

9. β-Hexosaminidase

Another round of β-hexosaminidase, removes one *N*-acetylglucosamine from the original product, and methylation analysis shows the following:

1 mol of 1,5-di-*O*-acetylfucose (from terminal fucose)
1 mol of 1,5,6-tri-*O*-acetyl-2-*N*-methyl-*N*-acetylglucosaminitol (from 6-substituted *N*-acetylglucosamine)

This completes the branching structure by showing the location of the final branch point, which is the reducing-end *N*-acetylglucosamine, containing terminal fucose substituted in its 6-position and *N*-acetylglucosamine substituted in the 4-position, since the 4,6-disubstituted GlcNAc has now been replaced with a 6-substituted GlcNAc.

C. Fucose Location

By using a combination of FAB and methylation analysis, the location of the fucose on the reducing-end GlcNAc could be determined at any time during the analysis, but not its position of insertion on the GlcNAc. However, the fucosidase treatment could be performed at any time, followed by methylation linkage analysis, showing the conversion of the 4,6-disubstituted GlcNAc to a 4-substituted GlcNAc. The FABMS B10,C10 converted to B9,C9 ions and Y2→Y1 or Z2→Z1 ions could then be used to determine that this was lost from the reducing end and also show that the reducing end was connected to the next sugar (another GlcNAc from FABMS Y3, Z3 ions) through the 4-position.

D. Still-To-Be-Determined Structural Parameters to Obtain a Complete Characterization

The remaining questions are the locations of the various substituted galactoses and *N*-acetylglucosamines in the two arms of the biantennary structure (Scheme 2).

```
        a arm

NANA(α2→6)Gal(β1→4)GlcNAc(β1–2)
                              \
                         Man(α1→6)            Fucose(α1–6)
                                 \                       \
                                  Man(β1→4)GlcNAc(β1→4)GlcNAc
                                 /
                         Man(α1→3)
                              /
NANA(α2→3)Gal(β1→3)GlcNAc(β1–2)

        b arm
```

Scheme 2

1. Periodate Cleavage

Periodate cleavage of this molecule would cleave the last two carbons from the sialic acids, destroy the 4-linked galactose in the *a* arm by oxidation between hydroxyls 2 and 3, destroy the terminal fucose in the same manner, destroy the 2-linked mannoses by oxidation between the 3- and 4-hydroxyls and preserve the remaining sugars, none of which have oxidizable vicinal hydroxyls.

```
                         H     H
 =                       |     |
   IO4             R-- C -- C -- R       (cleaved by periodate)
                         |     |
                        HO    OH
```

Reduction of the resulting dialdehydes and mild acid gives the resulting products:

GlcNAc

Man(β1→4)GlcNAc(β1→4)GlcNAc

*NANA(α2→3)Gal(β1→3)GlcNAc (*NANA has lost carbons 8 and 9)

Separation and enzyme–methylation analysis of the products obviously gives some very useful information, but nothing about which branches contain the linkages in question.

Confirmation could come by methylation analysis of the resulting "b" arm *NANA(α2→3)Gal(B1→3)GlcNAc fragment. The neuraminic acid left on the b arm would have its last two carbons removed and probably would not respond to neuraminidase treatment after deacetylation. Mild acid, however, could remove this sialic acid remnant, if not removed in the Smith degradation. This fragment could be treated with enzyme–methylation analysis to confirm the sequence using β-galactosidase, after which β-hexosaminidase would reveal an α-mannose, and treatment with α-mannosidase would, after methylation linkage analysis, confirm that the 3-branched b arm carried the 3-linked galactose and 3-linked GlcNAc.

Acetolysis of the parent compound would preferentially remove sugars glycosidically linked to the primary hydroxyls on hexopyranoses, including the 6-linked sialic acid and the 6-linked mannose on the a arm. Methylation analysis could confirm this supposition by showing that the resulting heptasaccharide contains a 3-substituted mannose, a 3-substituted galactose, and a 3-substituted GlcNAc, showing that the 3-substituted galactose and GlcNAc are in the b arm and, by inference, that the 6-linked Gal and 4-linked GlcNAc are in the *a* arm.

Thus, the final structure:

```
NANA(α2→6)Gal(β1→4)GlcNAc(β1–2)
                              \
                         Man(α1→6)          Fucose(α1–6)
                                  \                    \
                                   Man(β1→4)GlcNAc(β1→4)GlcNAc
                                  /
                         Man(α1→3)
                              /
NANA(α2→3)Gal(β1→3)GlcNAc(β1–2)
```

III. RECENT ADVANCES AND FUTURE PROMISE OF NOVEL METHODS

Tandem mass spectrometry may yield stereochemical and isomer-sensitive spectra of small complex oligosaccharides. In addition to work already mentioned earlier by Domon and Costello [10,11], Gillece-Castro and Burlingame [9], and Laine et al. [12] in this area, Richter and his colleagues have shown that the epimer forms of nonreducing-end sugars give specific ion ratios in tandem mass spectrometry studies after FABMS [17]. Garozzo et al. have obtained information for linkage position and identification of reducing ends of saccharides by negative ion FABMS-CID-MS [18]. Although the literature in this area, since 1988, comprises only a handful of papers, progress in this area may significantly enhance the usefulness of mass spectrometry in reducing the difficult and time-consuming steps of total structure determination.

REFERENCES

1. Laine, R. A. (1990). Glycoconjugates, overview and strategy. *Methods Enzymol. 193*:539–553 (J. A. McCloskey, ed.).
2. Takamoto, M., Endo, T., Isemura, M., Kochibe, N., and Kobata, A. (1989). Structures of asparagine-linked oligosaccharides of human placental fibronectin. *J. Biochem. 105*:42–50.
3. Dell, A. (1990). Preparation and desorption mass spectrometry of permethyl and peracetyl derivatives of oligosaccharides. *Methods Enzymol. 193*:647–660 (J. A. McCloskey, ed.).
4. Hardy, M. R., Townsend, R. R., and Lee, Y. C. (1988). Monosaccharide analysis of glycoconjugates by anion exchange chromatography with pulsed amperometric detection. *Anal. Biochem. 170*:54–62.
5. Laine, R. A., Esselman, W. J., and Sweeley, C. C. (1972). Gas liquid chromatography of carbohydrates. *Methods Enzymol. 28*:159–167.
6. Ashraf, J., Butterfield, D. A., Jarnefelt, J., and Laine, R. A. (1980). *J. Lipid. Res. 21*:1137.
7. Laine, R. A. (1981). Chemical ionization GC-mass spectrometry of carbohydrates. *Anal. Biochem. 116*:383.
8. Poulter, L., and Burlingame, A. L. (1990). Desorption mass spectrometry of oligosaccharides coupled with hydrophobic chromophores. *Methods Enzymol. 193*:661–688 (J. A. McCloskey, ed.).
9. Gillece-Castro, B. L., and Burlingame, A. L. (1990). Oligosaccharide characterization with high energy collision-induced dissociation mass spectrometry. *Methods Enzymol. 193*:689–712 (J. A. McCloskey, ed.)
10. Domon, B., and Costello, C. E. (1988). *Biochemistry 27*:1534.
11. Costello, C. E., and Vath, J. E. (1990). Tandem mass spectrometry of glycolipids. *Methods Enzymol. 193*:738–768 (J. A. McCloskey, ed.).
12. Laine, R. A., Pamidimukkala, K. M., French, A. D., Hall, R.W., Abbas, S. A., Jain, R. K., and Matta, E. L. (1988). Linkage analysis of carbohydrates by tandem mass spectrometry and molecular modelling. *J. Am. Chem. Soc. 110*:6931; Laine, R. A. (1989). *Methods Enzymol. 179*:157–169 (V. Ginsberg, ed.). Laine, R. A., Yoon, E. S., Mahier, T. J. (1991). *Biol. Mass Spectrom. 20*:505–514.
13. Watanabe, K., Laine, R. A., and Hakomori, S.-I. (1975). *Biochemistry 14*:2725.
14. Hakomori, S.-I. (1964). Simple method for methylation of sugars. *J. Biochem. 55*:205.
15. Ciucanu, I., and Kerek, F. (1984). *Carbohydr. Res. 131*:209.
16. Hellerqvist, C. G. (1990). Linkage analysis by the Lindberg method. *Methods Enzymol. 193*:xxx.
17. Muller, D. R., Domon, B. M., Blum, W., Rashdorf, F., and Richter, W. J. (1988). *Biomed. Environ. Mass Spectrom. 15*:441; Muller, D. R., Domon, B. M., and Richter, W. J. (1989). *Spectrosc. Int. J. 7*:11; Richter, W. J., Muller, D. R., and Domon, B. M. (1990). Tandem mass spectrometry in the structural characterization of oligosaccharide residues in glyconjugates. *Methods Enzymol. 193*:607–622 (J. A. McCloskey, ed.).
18. Garozzo, D., Giuffrida, M., Impallomeni, G., Ballistrere, A., and Montando, G. (1990). Determination of linkage position and identification of the reducing end in linear oligosaccharides by negative ion fast atom bombardment mass spectrometry. *Anal. Chem. 62*:279-286.

6

Synthetic Glycoconjugates

Y. C. Lee and Reiko T. Lee *Johns Hopkins University, Baltimore, Maryland*

The interest in the glycoconjugates has risen tremendously in the past decade. The progress in the areas of synthetic glycoconjugates also has kept up with this upsurge in the growing interest. The needs for synthetic glycoconjugates in the studies involving carbohydrates may be worth emphasizing. First, there is the traditional need for confirmation of structure by synthesis. The structure of a glycoconjugate is usually elucidated by a combination of different methods: chemical, enzymic, immunological, and physical. Each of the methods has its inherent weakness, and it is highly desirable to confirm the results obtained with one method by other methods.

There is still a need for structural confirmation by organic synthesis, whenever possible. An example is the establishment of the position of sulfation in the oligosaccharide chain in bovine lutropin by synthesis of a sulfated oligosaccharide [1]. The second reason for requiring synthetic glycoconjugates is because, in a large number of cases, the materials isolated from natural sources are not of a sufficiently high purity or quantity. It is becoming more and more apparent that subtle differences in structural features can bring about great differences in biological activity [2]. A minor contaminant, with a much greater biological activity than the major constituent, can jeopardize a proper interpretation of carbohydrate structural–functional relationships. Although synthetic methods are not infallible in this, at least the products from chemical synthesis are likely to contain types of contaminants different from those obtained from nature.

Another reason for synthesizing glycoconjugates is also related to probing the structural–functional relationships. Chemical synthesis can often provide not only the exactly identical sugar chains found in nature, but also many variants with which structural requirement of binding, recognition, and so on, can be probed in detail. The "fraudulent" oligosaccharides used by Lemieux et al. [e.g., 3,4] exemplify this aspect of synthetic glycoconjugates. One can also include in this category those synthetic glycoconjugates that are quite far removed from the natural products (e.g., carbohydrates attached to solid matrices [5] and unnatural substances, such as poly-D-lysine [6,7]).

Chemical synthesis of glycoconjugates is still a laborious task because of the inherent structural complexities of carbohydrates. It requires much greater effort than syntheses of peptides and nucleic acids, which have been almost totally automated. In view of this, synthetic products that do not strictly imitate the natural structure, but contain the essential elements needed for biological activities, can be useful. An example for this type of "shorthand synthesis" is a two-branched structure (*biantennary*) containing lactose or *N*-acetylgalactosamine (GalNAc) at the terminal of branches, which are made of 6-aminohexanol and amino acids [8,9] (see Diagram 3).

The recent progress in synthetic glycoconjugates is apparent in the areas of technique for oligosaccharide synthesis as well as in the methods of conjugating oligosaccharides to noncarbohydrate materials.

I. BASIC STRATEGIES FOR CARBOHYDRATE SYNTHESIS

In planning synthesis of carbohydrate compounds to be used in biological studies, several factors must be carefully considered. In assembling monosaccharide units, one must decide on the anomeric configuration of the glycosidic linkage, the mode of activation of the anomeric carbon, appropriate regiospecific or regioselective protective groups for glycosylation (Diagram 1). For large oligosaccharides, the order of building the chain or chains (i.e., stepwise addition or block synthesis) is an important factor. If the target compound is a glycoside, the choice of aglycon must also be considered. These strategies are closely interrelated and are governed by the ultimate purpose of utilization as well as practicality of synthesis. Unfortunately, each synthetic endeavor usually is a unique project, and there are no "standard rules" that can apply to all oligosaccharide syntheses. Several important aspects for consideration in planning the synthesis are discussed in the following.

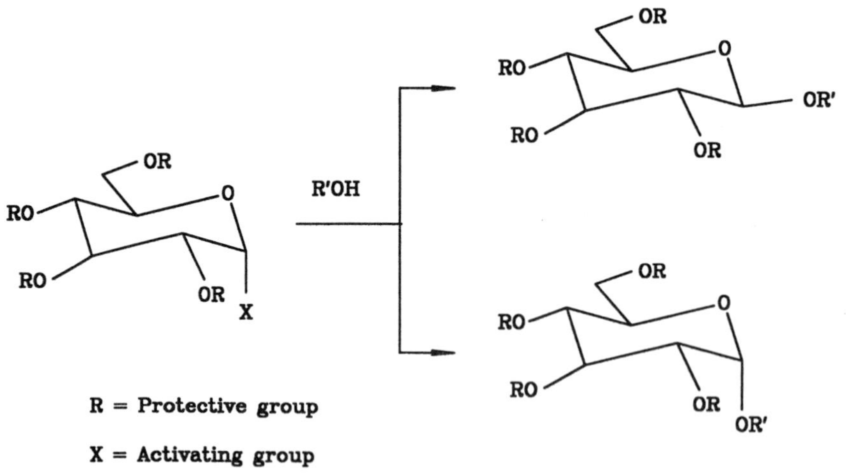

OR
RO
O
RO
OR
X
R'OH
OR
RO
O
OR'
RO
OR
OR
RO
O
RO
OR
OR'

R = Protective group

X = Activating group

Diagram 1

A. "Protective" Aglycon Versus "Functional" Aglycon

One important aspect of synthesis of glycoconjugates that has often escaped judicious consideration is the choice of aglycon. In the past, synthesis of oligosaccharides, especially those closely imitating the natural products (e.g., heparin pentasaccharide fragment [10] or fragments of *N*-glycosides [e.g., 11,12]) was designed to produce oligosaccharides with reducing termini. The terminal sugar residue may contain a temporary aglycon (e.g., benzyl group), to protect the reducing terminus during the synthesis, which is removed at the end of the synthesis to generate the reducing oligosaccharides. Alternatively, some oligosaccharides were synthesized with a persistent but nonfunctional aglycon, such as methyl group [e.g., 13–15], in which case conversion or replacement of the methyl group to another functional group will be inconvenient or impractical.

In numerous instances, the biological recognition of carbohydrate moieties is more dramatically manifested when the oligosaccharides are "clustered" [16,17]. In other cases, studies of binding involving carbohydrate moieties of glycolipids have been greatly facilitated by conjugating the oligosaccharide groups to proteins [18–20]. Therefore, it would be wise to incorporate a functionally active or activatable group into the aglycon so that the painstakingly synthesized oligosaccharides can be further modified or attached to a carrier. For example, if the oligosaccharides are synthesized with an aglycon such as the 8-methoxycarbonyloctyl group ($MeOCO(CH_2)_7CH_2$-) [21], then the oligosaccharides can be easily coupled to proteins or other amino-containing materials after conversion of the methyl ester to acyl azide. Usually, such an aglycon will not hinder the synthetic operations. The trimethylsilylethyl group has been championed as an aglycon [22] for its inertness during various synthetic manipulations and ease of conversion to another functional group. Examples of "convertible" aglycon (in contrast with mere "protective" aglycons) are shown in Table 1.

B. Nature of the Atom Linked to Anomeric Carbon

Another matter to consider is the choice among *C*-glycosides, *N*-glycosides, *O*-glycosides, and *S*-glycosides (and possibly other atoms). The most common practice is to prepare *O*-glycosides, because this is the type nature provides most abundantly, and most synthetic schemes of glycoside formation have been developed for *O*-glycosides. However, nature also provides ample examples of *N*-glycosides, not only in nucleosides, but in one of the common carbohydrate to peptide linkages, *N*-acetylglucosamine-asparagine

Table 1 Examples of "Convertible" Aglycons

Aglycon	Convertion to	Ref.
$-OCH_2CH{=}CH_2$	$-(CH_2)_2CH_2{-}S(CH_2)NH_2$	198
$-O(CH_2)_8COOMe$	$-O(CH_2)_8CON_3$	21
$-O(CH_2)_6NHCOOCH_2Ph$	$-O(CH_2)_6NH_2$	199
$-O(CH_2)_6NHCOCF_3$	$-O(CH_2)_6NH_2$	202
$-SCH_2CN$	$-SCH_2C({=}N)OMe$	113
$-SCH_2CONHCH_2CH(OMe)_2$	$-SCH_2CONHCH_2CHO$	135
$-OCH_2CH_2Br$	$-OCH_2CH_2OR$	146
$-OCH_2CH(CH_2Br)_2$	$-OCH(CH_2OR)_2$	147
$-O(CH_2)_2SiMe_3$	$-OCOCH_3$	22

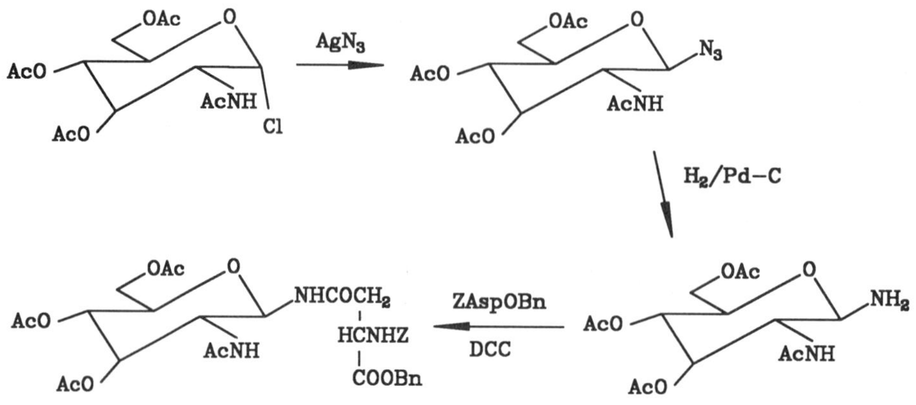

Diagram 2

(GlcNAc-Asn). If the intended synthesis is the direct imitation of *N*-glycosides in glycoproteins, then the aglycon should be an asparagine derivative (in which the β-amide *N* is the atom glycosylated). A rather recent review on synthesis of this type of compounds has appeared [23]. A most frequently used synthetic scheme for GlcNAc-Asn fragment is shown in Diagram 2. *N*-Glycosides have been used in a number of cases as an affinant for isolation of glycosidases and lectins. In these syntheses, glycosylamine is prepared by a direct action of methanolic ammonia on reducing sugars, and the amino group is acylated (for example, with *N*-protected 6-aminohexanoic acid) to stabilize the *N*-glycosidic bond [24].

On the other hand, *S*-glycosides are rarer in nature than *O*- or *N*-glycosides. In some synthetic designs [25,26], *S*-glycosides were chosen for their expected resistance to glycosidases. Most glycosidases do not cleave *S*-glycosides (or cleave only slowly), and the use of *S*-glycosides may provide advantages when the glycosides must be used over a prolonged period in biological studies. For some carbohydrate-binding systems, *S*-glycosides may have greater affinities than the corresponding *O*-glycosides. For example, Largent et al. [27] found that the alveolar macrophages attached to the gel layers containing *S*-glycosides much better than to the similar gel layers containing *O*-glycosides. Inhibition experiments with simple synthetic *O*- and *S*-glycosides showed that, in this case, the cell surface receptors indeed have greater affinity for *S*-glycosides than the corresponding *O*-glycosides.

One advantage of *S*-glycosides is that the thioglycosidic linkage can be readily and selectively cleaved by mercuric ion [28]. Thus, *S*-glycosides can be analyzed in the presence of *O*-glycosides or, if necessary, released for reexamination without affecting the coexisting *O*-glycosides. Additionally, thioglycosides can be utilized for the formation of *O*-glycosides (this aspect will be discussed later in Section II.A.7). One can design a thioglycoside that contains a "convertible" functional group, so that in addition to being a glycosylating agent, this thioglycoside can be attached to proteins or other compounds (by the convertible functional group). For example, *p*-nitrophenyl thioglycosides can be reduced to *p*-aminophenyl compound, which can be converted to either the diazonium salt and coupled to a phenolic group in proteins [26] or converted to isothiocyanate to react with amino groups [29]. The same *p*-nitrophenyl thioglycoside can be used as a glycosylating agent to prepare *O*-glycosides with suitable catalysts [30].

There are disadvantages to the *S*-glycosides as well. In preparation of antibodies containing *S*-glycosides, it appears that the glycosidic sulfur atom is strongly antigenic [Pazur, J., personal communication]. Antibodies made against mannose-*S*-bovine serum albumin (Man-*S*-BSA) did not cross-react with those made against Man-*O*-BSA. This cautions against the use of *S*-glycosides in antigen production.

Although the synthesis of *C*-glycosides has been attracting considerable attention [e.g., 31,32], *C*-glycosides have not been fully utilized in glycoconjugate research, except for the triazene derivatives used as "suicide" substrates [e.g., 33]. From the standpoint of resistance against both chemicals and enzymes, *C*-glycosides should be the most preferred. The glycosyl cyanides [31,34] can be readily converted to either amino or carboxyl derivatives for attachment to proteins or other materials. Active development in this area is expected in the future.

Although the discussion on the choice of *O*-, *C*-, *N*-, or *S*-glycosides in the preceding paragraphs was mostly for the glycosides of monosaccharides, it is conceivable to prepare oligosaccharides that possess glycosidic linkages resistant to exoglycosidases (e.g., *S*-glycosides) only at the nonreducing terminal position. Such oligosaccharide derivatives should be useful for studies of endoglycosidases, without being concerned about exoglycosidase contamination.

C. "Shorthand" Synthesis

Although it has been reported that antidextran antibodies may recognize residues as long as seven sugars [35], many times, the range of recognition by a carbohydrate-binding protein is much smaller [36]. For instance, hepatic sugar receptors from vertebrates bind bovine serum albumin bearing only monosaccharide derivatives as well as, or better than, any natural products, indicating dispensability of the sugars of the inner sequence. One can obtain a potent ligand for such receptors and lectins without synthesizing complete or substantially similar analogues of the natural product [37]. For example, for the mammalian hepatic Gal/GalNAc receptor, simple divalent galactose (Gal) and *N*-acetylgalactosamine (GalNAc) compounds, such as shown in Diagram 3, showed K_d of 0.3 μM and 3 nM, respectively, for the receptor–ligand complex [8,9]. The latter is as good a ligand as asialoorosomucoid (possessing more than 15 exposed galactose residues). The high affinity of these divalent glycosides is apparently due to simultaneous binding of two galactose (or GalNAc) residues to the adjacent sugar-combining sites, which are probably spatially

$$
\begin{array}{l}
\quad\; CH_2CONH(CH_2)_6\text{-O-Lac} \\
\quad\;\, | \\
RNHCHCONH(CH_2)_6\text{-O-Lac}
\end{array}
$$

$$
\begin{array}{l}
\quad\; CH_2CONHCH_2CONH(CH_2)_6\text{-O-GalNAc} \\
\quad\;\, | \\
RNHCHCONHCH_2CONH(CH_2)_6\text{-O-GalNAc}
\end{array}
$$

$$R = \text{Tyr}$$

Diagram 3 Divalent Ligands for Gal/CalNAc Lectin

separated by about 20 Å [38]. Long and flexible aglycons allow these compounds to assume an appropriate conformation to position the galactose residues about 20Å apart for simultaneous binding of both galactose residues.

Once the basic parameters for optimal binding of ligands are known, one can then devise compounds that will satisfy them without laborious synthetic mimicking of the carbohydrate structure in question. In this type of approach, here called *shorthand* synthesis, only the structural features necessary for biological activities are constructed by the most expedient methods. This allows simpler synthetic procedure to achieve the same biological effect.

D. "Fraudulent" Oligosaccharides

Oligosaccharides or their glycosides can be designed to be similar to, but different from, the naturally occurring structures. This is elegantly exploited by Lemieux et al. [3,4] in their studies of antibody and lectin binding. In addition to the usual replacement of OH with H or F, a technique such as introduction of 6-C-Me at the critical (1,6)-linkage [39] was used to fix two distinct conformations normally associated with a 1,6-linkage. These studies led the authors to conclude that both initial polar interaction and subsequent stabilization by hydrophobic interaction participate during binding of carbohydrates by proteins [4].

With the use of a series of fluorinated D-galactose oligosaccharides, Glaudemans et al. [36] studied binding specificities of anti-β-(1,6)-D-galactopyran antibodies. They concluded that there are four monosaccharide-combining sites in these antibodies, and the strongest interaction appears to be at the HO-groups of C-2 and C-3.

II. TECHNIQUES FOR OLIGOSACCHARIDE SYNTHESIS

This topic has been reviewed by several workers [40–43], and readers should refer to these reviews and the original articles contained therein for more detailed information.

A. Activation of the Anomeric Carbon and Conditions for Glycosylation

The anomeric carbon must be activated for glycosylation. The most important consideration in choosing an activation method among several available methods is the expected outcome of anomeric configuration of the product. However, for certain difficult glycosylations (e.g., a bulky aglycon with a hindered secondary hydroxyl group), a trial-and-error approach may be the only choice.

1. *Koenigs–Knorr Method*

Until recently, the Koenigs–Knorr method [42] was the most widely used for glycoside synthesis. In its most basic form, a sugar residue is totally *O*-acylated (usually *O*-acetylated), and the reactive acyloxy group at the anomeric carbon is replaced with a bromine (e.g., by reaction with HBr in acetic acid) or chlorine atom. The "acetohalo sugar" thus formed is used as the glycosylating agent, using heavy-metal salts [e.g., Ag_2CO_3, $Hg(CN)_2$] as catalysts (Diagram 4).

This method suffers from several disadvantages [41]. The acetohalo sugars are not particularly stable and must be prepared relatively fresh. The yield of glycosylation can be very low, especially when hindered or unreactive hydroxyl groups are to be glycosylated. The stereochemistry of the glycosides formed usually favors a 1,2-*trans*-configuration,

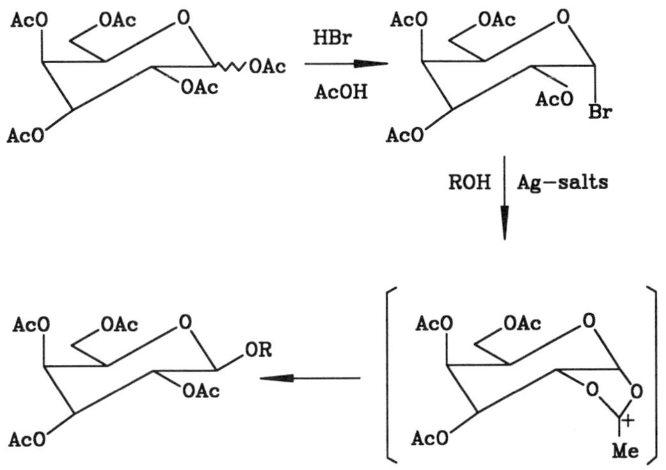

Diagram 4

although with some catalysts [notably $Hg(CN)_2$], a considerable amount of 1,2-*cis*-glycosides may be formed.

With acetohalo or similar 1-halo-sugar peracetates, anomeric configuration of the product glycoside is strongly influenced by the catalyst used in the reaction [40]. Silver salicylate and silver silicate, which are not soluble in most organic solvents used for glycosylation, promote formation of 1,2-*cis*-glycosides. Use of an insoluble catalyst can also result in regioselectivity. For example, in the presence of silver silicate, per-*O*-acetylated 2-deoxy-2-phthalmiido-D-glycopyranosyl bromide treated with a lactose derivative with open 3′- and 4′-OHs to yield a β(1″,4′)-linked product, whereas the same reactants in the presence of silver triflate yielded predominantly β(1″,3′)-linked product [44]. The effect of substituents on the anomeric ratio of the products was studied by van Boeckel et al. [45].

2. *Orthoester Method*

The major reason for the propensity of 1,2-*trans*-glycosides in the Koenigs–Knorr type reaction is the formation of an orthoester intermediate (see Diagram 4). Kochetkov and his co-workers developed a method that used orthoesters as glycosylating agents [46]. The main advantage of this method is a higher degree of stereoselectivity (1,2-*trans*) of the glycoside formed. In addition, many orthoesters are crystalline and have a longer shelf-life than the acetohalo sugars. Glycosylation with an orthoester is usually carried out at high temperature with acid catalysts, such as *p*-toluenesulfonic acid or $HgBr_2$.

3. *Glycosylation with 1-O-acetate by Lewis-Acid Catalysis*

Peracetates of sugars can glycosylate in the presence of a Lewis acid catalyst under mild conditions. Among the catalysts found useful for such syntheses ar BF_3 [47,48] and $SnCl_4$ [49,50]. An example for such a reaction is shown in Diagram 5. Paulsen and Paal [51] reported that oligosaccharide peracetate can be used as a glycosylating agent in the presence of trimethylsilytrifluoromethanesulfonate (TMS-triflate).

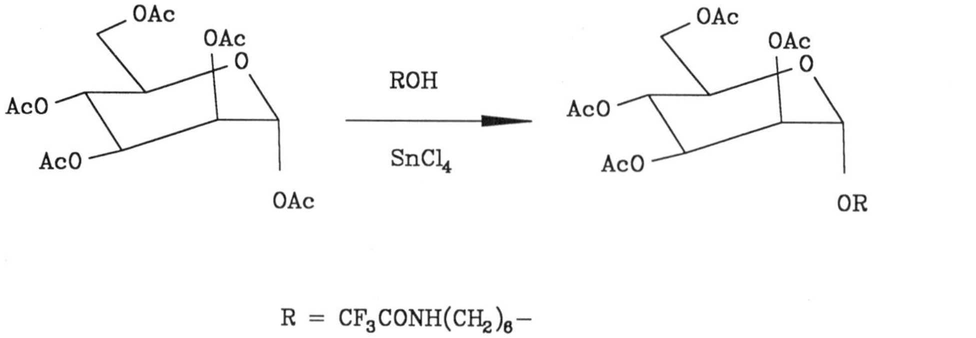

OAc OAc O AcO AcO OAc — ROH / $SnCl_4$ → OAc OAc O AcO AcO OR

R = $CF_3CONH(CH_2)_6-$

Diagram 5

4. *Oxazoline Method (for 2-Acetamido-2-deoxy Sugars)*

This method for 2-acetamido-2-deoxy sugars is analogous to the orthoester method described in the foregoing. Oxazoline derivatives of amino sugars can be prepared from peracylated 1-halo derivatives [52] or directly from peracylated amino sugars [53,54], as shown in Diagram 6. The latter method requires that the 1-*O*-acetyl group be in 1,2-*trans*-orientation. However, the use of $SnCl_4$ allows a 1-*O*-acetate in 1,2-*cis*-orientation [55] to form an oxazoline. In one report [54], trimethylsilyltrifluoromethanesulfonate (TMS-triflate) was used for the formation of an oxazoline derivative. Successful glycosylation with oligosaccharides isolated from human urine has been reported [56]. This method, as does the orthoester method, has advantages over the 1-halo derivatives in a cleaner stereochemistry of the glycoside formed (favoring 1,2-*trans*).

5. *Halide-Catalyzed Glycosylation for 1,2-CIS-Glycosides*

One of the more difficult problems in glycoside synthesis is the formation of 1,2-*cis*-glycosides, which are frequently encountered in nature. This problem was first addressed by the halide-catalyzed glycosylation method [57]. In this method, 1-halo sugars (usually in 1,2-*cis*-configuration), the hydroxyl groups of which are protected with a

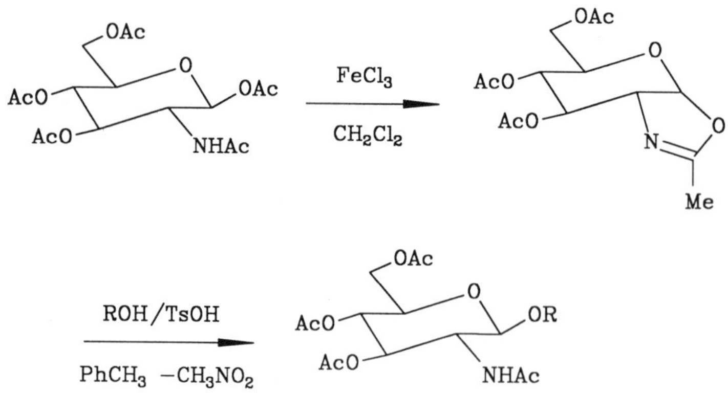

OAc O AcO AcO OAc NHAc — $FeCl_3$ / CH_2Cl_2 → OAc O AcO AcO N O Me

— ROH/TsOH / $PhCH_3$ $-CH_3NO_2$ → OAc O AcO AcO OR NHAc

Diagram 6

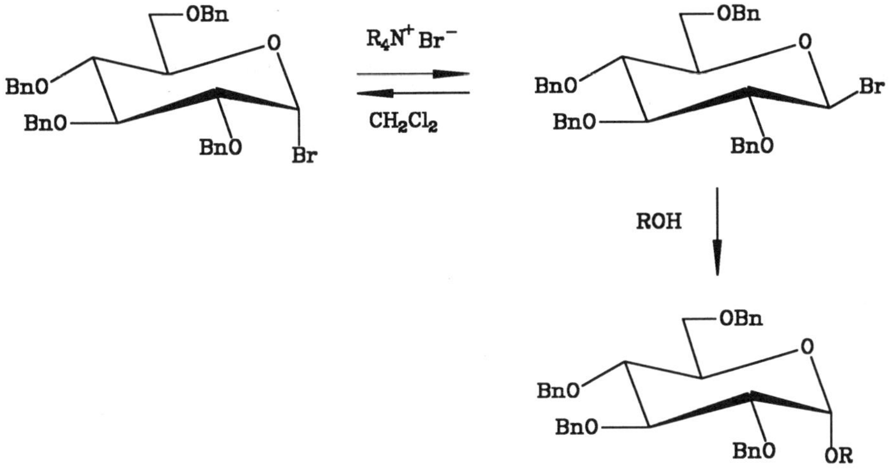

Diagram 7

"nonparticipating" group, such as benzyl ether, are reacted with excess tetraalkylammonium halide to produce in situ 1,2-*trans*-halide in equilibrium with 1,2-*cis*-halide. 1,2-*trans*-Glycosyl halide, although unstable, is very reactive and reacts faster with alcohols to yield 1,2-*cis*-glycosides (Diagram 7). Although this method is useful, it usually takes a relatively long reaction time (e.g., several days).

Efficient formation of a 1,2-*cis*-glycosidic bond requires a nonparticipating group at C-2. A most frequently used functional group for this purpose is benzyl ether. For example, β-D-mannosylation was accomplished with a 2-*O*-benzyl-α-1-bromo- derivative of D-mannopyranose [12]. In the case of amino sugars, a 2-azido-2-deoxy- group can be used as the nonparticipating group (see later) for the same purpose.

6. *Imidate Method [41]*

In 1978, Sinaÿ's group [58,59] demonstrated that sugars modified with acetimidate at the anomeric OH group can be used for glycosylation, yielding 1,2-*cis*-glycosides. Grundler and Schmidt [60] improved this technique by using trichloroacetimidate (Diagram 8). The efficiency of this reaction, together with its mild reaction condition, makes this method quite popular in recent years.

7. *Glycosylation by Thioglycoside [30]*

The use of thioglycosides in the presence of mercuric salts as *O*-glycosylating agents was first proposed by Ferrier [61] in the 1970s. Newer reagents for the activation of thioglycosides include methyl trifluoromethanesulfonate [62] and dimethyl(methylthio) sulfonium trifluoromethanesulfonate [63]. The latter catalyst was used with thioglycosides of mono-, di-, and trisaccharides to yield *O*-glycosides α-linked to serine [64]. Lönn reported that the use of sulfuryl chloride/trifluoromethylsulfonic acid [65] led to higher yields of the product. The use of NO_2BF_4 for activating thioglycosides offers great promise because it does not generate strong acid [66]. The use of sulfenate esters as glycosyl acceptors in *O*-glycoside synthesis from thioglycosides has been reported [67]. Diagram 9 illustrates the underlying principle for using thioglycoside as a glycosylating agent.

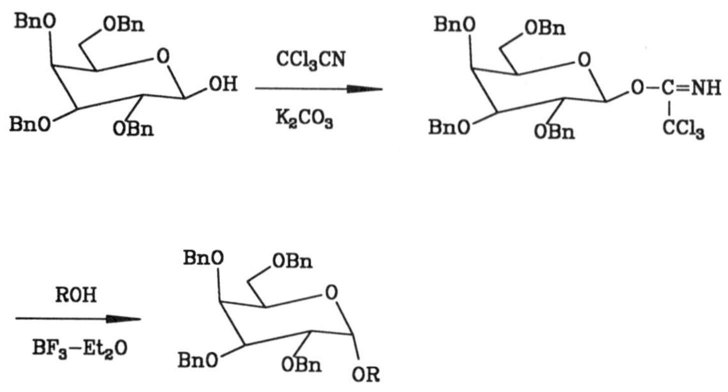

Diagram 8

8. *Further Modification of Reducing Oligosaccharides for Glycosylation*

In a number of cases, oligosaccharides are obtained from natural sources as reducing sugars. Examples include oligosaccharides generated by the action of *endo*-β-hexosaminidase [68] or glycopeptidase [69] and oligosaccharides isolable from urines of some patients [54,56]. *O*-Glycosides in glycoproteins are often obtained as reduced oligosaccharides [70]. In addition, some large synthetic oligosaccharides, such as a sulfated pentasaccharide fragment active in binding to antithrombin III [10,71], and octa- [12], nona-, and undecasaccharides [11] of the *N*-glycoside type, have been obtained as reducing sugars. There are methods for attaching the reducing and reduced oligosaccharides to proteins or other amino compounds either directly or after some modification.

When an oligosaccharide is fully acetylated, trimethylsilyltrifluoromethanesulfonic acid [51] can be used to perform glycosylation as mentioned earlier. The same reagent causes formation of an oxazoline derivative, if the reducing terminus is GlcNAc. This has been used for synthesis of lipid-linked oligosaccharides by Warren et al. [56].

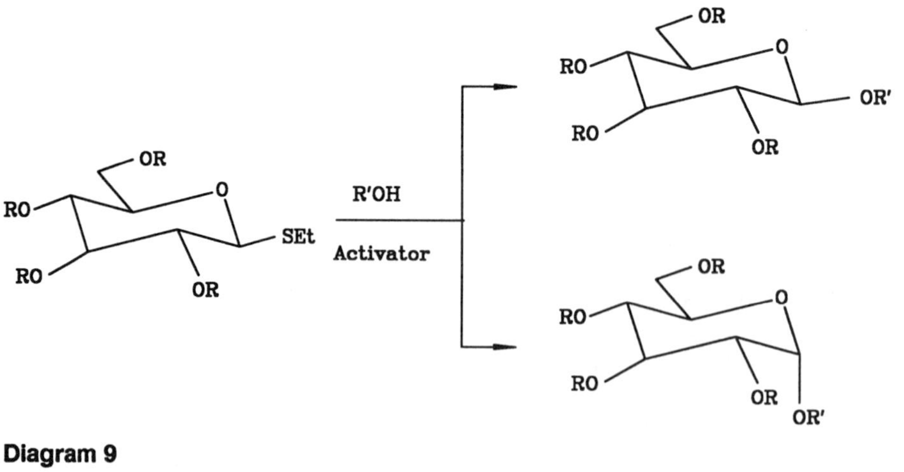

Diagram 9

B. Selective Protection of Hydroxyl and Amino Groups

There are numerous protective groups that can be used during glycoside synthesis [72,73]. Each protective group has its characteristic stability and susceptibility toward different reagents (for topics on selective deprotection, see Haines [74]). Sometimes, a protective group influences both the reactivity and stereochemistry of glycosylation of the neighboring hydroxyl group (such as 2-*O*-benzyl group mentioned in Sect. III.A.5). A judicious choice of protective groups is necessary for a successful oligosaccharide synthesis.

Table 2 lists protective groups used in actual synthesis of relatively complex oligosaccharides of *N*-glycoside type and a glycolipod (G_{M1}). The structure of compounds listed in

Table 2 Comparison of Protective Groups and Activation Methods

	Compounds[a]			
Protective Groups	I [75]	II [11]	III [12]	IV [76]
	For OH-groups			
Ester				
Acetyl	+	+	+	+
p-Nitrobenzoyl	+	–	–	–
Orthoester	+	+	+	–
Ether				
Benzyl	+	+	+	+
Allyl	+	+	+	+
1,6-Anhydro	–	–	+	–
Acetal/ketal				
Benzylidene	–	+	–	–
	For NH_2-groups			
Phthalimido	+	+	+	+
Azido	–	–	+	–
	For COOH-group			
Methyl ester	+	–	–	+
	For glycosylation			
Activation				
-Cl	+	+	–	–
-Br	–	+	+	+
-O-C(=NH)CCl_3	+	–	–	+
Tributyltin	+	–	–	–
Catalyst				
AgOTf	+	+	+	+
BF_3-Et_2O	+	–	–	+
$(Bu)_4NBr$	–	+	–	–

[a]See Diagram 10 for structures.

I Sialylated biantennary undecasaccharide of *N*-glycoside [75].

```
NeuAcα(2,6)Galβ(1,4)GlcNAcβ(1,2)Manα(1,6)
                                         \
                                          Manβ(1,4)GlcNAcβ(1,4)GlcNAc
                                         /
NeuACα(2,6)Galβ(1,4)GlcNAcβ(1,2)Manα(1,3)
```

II Tetraantennary undecasaccharide of *N*-glycoside [11].

```
Galβ(1,4)GlcNAcβ(1,6)
                    \
Galβ(1,4)GlcNAcβ(1,2)Manα(1,6)
                             \
                              Man
                             /
Galβ(1,4)GlcNAcβ(1,2)Manα(1,3)
                    /
Galβ(1,4)GlcNAcβ(1,4)
```

III Biantennary octasaccharide of *N*-glycoside [12].

```
Galβ(1,4)GlcNAcβ(1,2)Manα(1,6)
                             \
                              Manβ(1,4)GlcNAc
                             /
Galβ(1,4)GlcNAcβ(1,2)Manα(1,3)
```

IV Ganglioside GM_1 [76].

```
        Galβ(1,3)GalNAcα(1,4)Galβ(1,4)Glcβ→ceramide
                            /
              NeuAcα(2,3)
```

Diagram 10

Table 2 are shown in Diagram 10. The number of protective groups, as well as the number of methods used for activation, is not as large as one might imagine. Different investigators use different techniques for synthesis of an identical compound. Table 3 was compiled to illustrate how different groups of investigators have used different approaches to synthesize the same heparin pentasaccharide fragment (Diagram 11).

1. *Esters [72,73]*

ACETYL GROUP. The acetyl group is the most widely used protective group for the OH function because of its ease of formation as well as ease of removal. When a totally unmasked sugar is to be per-*O*-acetylated, the sugar can be dissolved in pyridine and treated with acetic anhydride at room temperature overnight, or it can be heated with acetic anhydride with acid (e.g., $HClO_4$) or base (sodium acetate) catalysts. If one stirs the reaction mixture into ice water, a crystalline acetate usually results. When a partially protected sugar is to be *O*-acetylated, the choice of reaction conditions may be more limited. For example, the presence of a benzylidene group precludes acetylation at a high

Table 3 Comparison of Techniques Used in Heparin Fragment Syntheses[a]

	I [79]	II [71]	III [200]
Starting materials	D-Glc, D-GlcN	D-Glc	D-Glc, cellobiose
Protection of HO-			
For regeneration	Bn	Bn	Bn
For sulfation	Ac *t*-Bu-orthoester	Ac	Ac Bz
For glycosylation	$ClCH_2CO$- Allyl	$MeCO(CH_2)_2CO$-	$ClCH_2CO$-
Glycosylation			
Activation	1-Br Orthoester	1-Br	1-Br
Catalysts	AgOTf Ag_2CO_3	AgOTf $Hg(CN)_2/HgBr_2$	AgOTf
Generation of IdoUA			
Oxidation	CrO_3/H_2SO_4	CrO_3/H_2SO_4	CrO_3/H_2SO_4
Inversion	TfO at C-5	MsO at C-5	Hydroboration at C5-C6
Sulfation	SO_3-NMe_3	SO_3-NMe_3	SO_3-NMe_3

[a]Bracketed numbers denote research groups.

temperature under acidic conditions. A selective *O*-acetylation can be performed with a much less reactive reagent (such as acetylimidazole).

Although not normally used for simple acetylation, acetyl chloride can be useful in acetylation with concomitant chlorination of anomeric carbon. For example, *N*-acetylglucosamine can be treated with acetyl chloride to yield an acetochloro derivative [77]. In the first stage of the reaction, 4 mol of HCl are produced as a consequence of total *O*-acetylation, and the HCl thus produced reacts with the 1-*O*-acetyl group to yield the

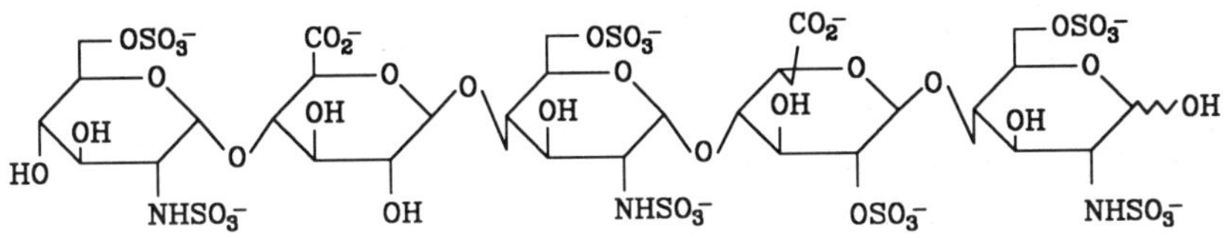

Diagram 11

1-chloro- derivative. A similar approach has been used for production of a GalNAc counterpart [48] as well as per-*O*-acetyl-2-chloro- derivative of sialic acid (NeuAc) [78].

As mentioned in Section II.A.2, 2-*O*-acetyl group participates in the Koenigs–Knorr-type reaction usually resulting in 1,2-*trans*-glycoside formation through an orthoester intermediate.

Removal of *O*-acetyl groups is most efficiently performed with a catalytic amount of sodium methoxide in dry methanol. The use of triethylamine [48] can be advantageous sometimes because of its volatility. Aqueous alkali can be used for de-*O*-acetylation, if the reducing group is securely masked (such as in the benzyl or methyl glycoside).

BENZOYL GROUP. The benzoyl ester is formed by reactions of hydroxyl groups with benzoyl chloride. Benzoyl derivatives tend to be more easily crystallized than acetyl derivatives. The same methods for de-*O*-acetylation can be used for removal of benzoyl groups, except that benzoates cannot be removed by evaporation.

Benzoylation of free sugars with limiting amounts of benzoyl chloride can result in selective protection of more reactive groups. For example, benzoylation of the methyl glycoside of mannose or glucose results in the formation of 2,3,6-tri-*O*-benzoate, leaving the 4-OH free for other reactions.

p-NITROBENZOYL GROUP. The *p*-nitrobenzoyl group is often used at the anomeric carbon prior to formation of glycosyl halide. Being a better leaving group, the *p*-nitrobenzoyloxy group can be replaced with halide under milder conditions. This is an important consideration when other protective groups are either acid- or halide-sensitive.

CHLOROACETYL GROUP. A unique feature of this group is that it can be removed specifically, for example, by heating with thiourea in alcohol [examples of the use of this protecting group are found in Refs. 53,79].

ORTHOESTER GROUP. 1,2-Orthoester derivatives (also see Sect. II.A.2) are resistant to alkaline conditions that remove *O*-benzyl derivative of D-mannose, which eventually would serve as the glycosyl acceptor at C-2 [80].

2. *Ethers [72,73]*

BENZYL ETHERS. Benzyl ether is the most useful ether-type protective group, because of its ease of removal by simple catalytic hydrogenolysis or brief HBr treatment. Formation of a benzyl ether requires benzyl halide (usually chloride or bromide), with a suitable base (KOH, NaH). A novel reductive cleavage (using $NaCNBH_3$) of 4,6-benzylidene ring results in formation of 6-*O*-benzyl-derivative, freeing the 4-OH group (see later discussion). Mild benzylation can be accomplished by using benzyl trichloroacetimidate [81,82], as shown in Diagram 12. Regioselective benzylation can be accomplished by means of dibutyltin [83] (see later). In addition to hydrogenolysis and HBr treatment, benzyl ethers can also be removed by acetolysis [72], resulting in *O*-acetates. Conditions for selective acetolysis to remove 6-*O*-benzyl group have been reported [84].

ALLYL ETHERS. The protective allyl ether group, originally proposed by Gigg and co-workers [e.g., 85], are usually formed by reaction of OH-groups with allyl halide under alkaline conditions. A method using allyl trichloroacetimidate under acidic conditions was reported [82]. Regioselective allylation using dibutyltin is known [86].

Allyl ethers are normally stable toward acidic conditions, but upon isomerization with *t*BuOK or $(Ph_3P)_3RhCl$ [85], the compound becomes acid labile and can be cleaved with as mild an acid reagent as $HgCl_2$ (Diagram 13). Another useful method for cleavage of allyl ether is by means of Pd-C (without hydrogen) in the presence of trace acid catalyst [87].

R–OH

$$\text{R–OH} \xrightarrow{\text{PhCH}_2\text{–O–C(=NH)–CCl}_3} \text{R–O–CH}_2\text{Ph}$$

Diagram 12

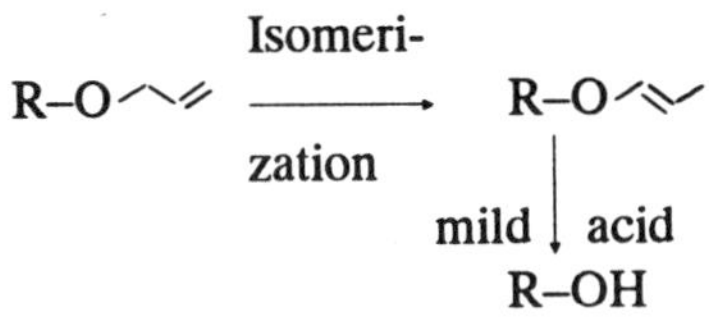

$$\text{R–O–CH}_2\text{CH=CH}_2 \xrightarrow[\text{zation}]{\text{Isomeri-}} \text{R–O–CH=CHCH}_3 \xrightarrow{\text{mild acid}} \text{R–OH}$$

Diagram 13

TRIPHENYLMETHYL (TRITYL) ETHERS [72]. The usefulness of this protective group lies in its high selectivity for the primary OH groups. With a controlled amount of reagent (trityl chloride), tritylation usually occurs only at the primary hydroxyl group. Detritylation is accomplished under extremely mild acidic conditions (e.g., 30 sec with HBr in acetic acid or 5 min in 50% trifluoroacetic acid).

SILYL ETHERS. The trimethylsilyl (TMS) group has been used for protection of OH-groups. Although this is a standard method for "volatilization" of sugars in gas chromatography, it is not as widely used in synthetic carbohydrate chemistry. However, cyclic silyl ethers have been useful in masking 1,3-diols (Diagram 14). Ichikawa et al. [88] found that 1,3-dichloro-1,1,3,3-tetraisopropyldisiloxane reacts with 1,6-anhydro-4′6′-benzylidene-maltose and -cellobiose differently, the former yielding 2′,3′-disilyl ether (Diagram 15), and the latter yielding 2,3-disilyl ether, respectively, in 60–64% yield (Diagram 16).

Removal of the cyclic silylether is accomplished under mildly acidic conditions, such as tetrabutylammonium fluoride in tetrahydrofuran or 0.2 M HCl in methanol.

```
|                                 |
CH–OH                             CH–O–Si–(iPr)2
|                                 |      \
CHR  +  (ClSi(iPr)2)2O  →         CHR      O
|                                 |      /
CH–OH                             CH–O–Si–(iPr)2
|                                 |
```

Diagram 14

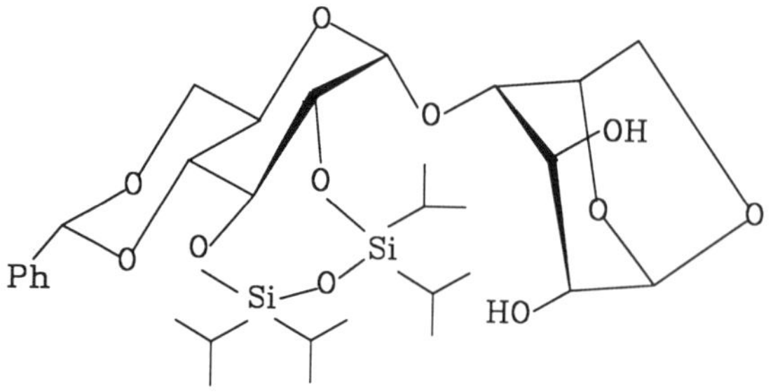

Diagram 15

3. *Acetals and Ketals*

ISOPROPYLIDENE DERIVATIVE. The isopropylidene group, also known as "acetonide," is quite widely used. Formation of an isopropylidene derivative (Diagram 17) can be carried out by shaking the sugar with acetone in the presence of acid catalysts (e.g., H_2SO_4, $ZnCl_2$, cation-exchange resin). Alternatively, transketalization with (2,2-dimethyloxy)propane can be used. The isopropylidene group masks two OH-groups simultaneously. In this process, a five-membered ring is generally formed (i.e., reaction with 1,2-diol) [73], but the outcome of the reaction is intricately controlled by the stereochemistry of the parent sugar as well as by the reagent used. Transketalization tends to form kinetically controlled products, such as the less stable 4,6-isopropylidene derivative, whereas direct reaction with acetone usually results in thermodynamically more stable products. Removal of isopropylidene group is by means of mild acid hydrolysis. Conditions used range from 50% acetic acid to dilute HCl. Treatment with 80–90% trifluoroacetic acid at room temperature usually gives satisfactory results.

BENZYLIDENE DERIVATIVE. Reaction of benzaldehyde with polyhydroxyl compounds results in the formation of benzylidene derivatives. Like isopropylidene derivatives described in the foregoing, benzylidene derivatives can also be made by transacetalization with benzaldehyde dimethyl acetal. Normally, six-membered ring benzylidene derivatives involving 1,3-diols are favored. Benzilydene groups can also be removed either by mild acid hydrolysis (e.g., 90% trifluoroacetic acid) or by catalytic hydrogenolysis. Benzylidene groups can be opened either reductively or oxidatively. Reductive cleavage with $NaCNBH_3$ leads to 6-*O*-benzyl derivatives [89], providing free

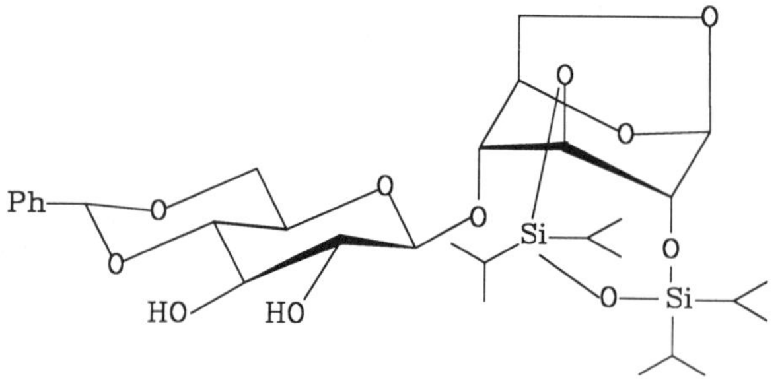

Diagram 16

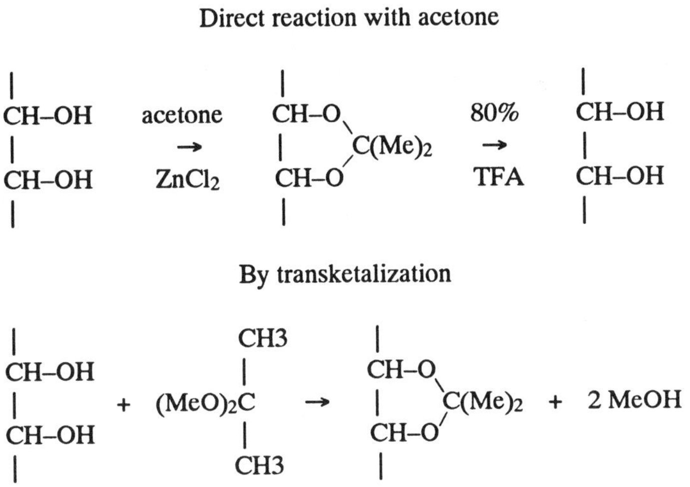

Direct reaction with acetone

```
|                    |                             |
CH–OH    acetone    CH–O               80%       CH–OH
|          →        |    \C(Me)2        →        |
CH–OH    ZnCl2      CH–O/              TFA       CH–OH
|                   |                            |
```

By transketalization

```
|                CH3        |
CH–OH            |          CH–O
|      +  (MeO)2C     →     |    \C(Me)2  +  2 MeOH
CH–OH            |          CH–O/
|                CH3        |
```

Diagram 17

4-OH. Oxidative opening of 4,6-*O*-benzylidene derivative with *N*-bromosuccinimide yields 4-*O*-benzoyl-6-bromo-6-deoxy-derivative [90,91] (Diagram 18). One report [92] describes oxidative cleavage of 3,4-*O*-benzylidene derivatives under photochemical conditions. Generally speaking, the reaction favors formation of derivatives with axial *O*-benzoyl and equatorial HO groups.

4. *Activation of the Hydroxyl Group*

Cyclic stannylene derivatives [93] have been useful in activation of hydroxyl groups. The use of tributyltin is more for "activation" of a certain hydroxyl group than for simple

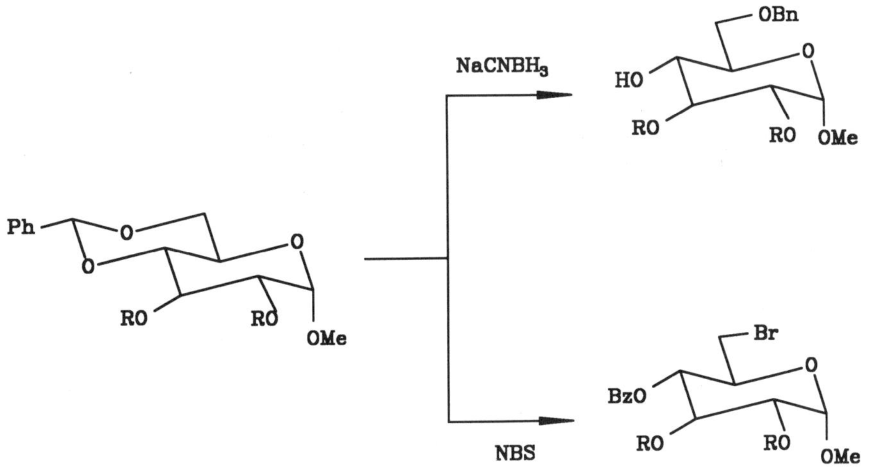

Diagram 18

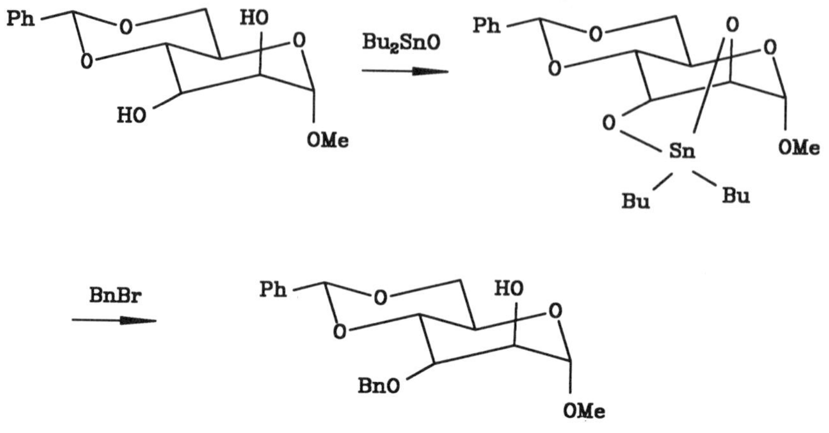

Diagram 19

protection per se. Bis(tributyltin)oxide reacts readily with free sugars to form cyclic coordinate derivatives [94]. Dibutyltin derivatives [83] also have been used in a similar fashion. When the cyclic tin derivatives are benzylated, the reaction occurs selectively at the equatorial hydroxyl component of the cyclic ether (Diagram 19). Primary hydroxyl group readily reacts with the tributyltin oxide, and benzylation of the resultant tin ether yields 6-*O*-benzyl derivative (Diagram 20).

The trityl group, useful in selective protection of the primary hydroxyl group, also finds its use as an activating group in glycosylation [42]. The trityl group is believed to be a better leaving group than a proton in such a reaction and, thus, promotes the glycosylation reaction with glycosyl halide at the site of the tritylated hydroxyl group.

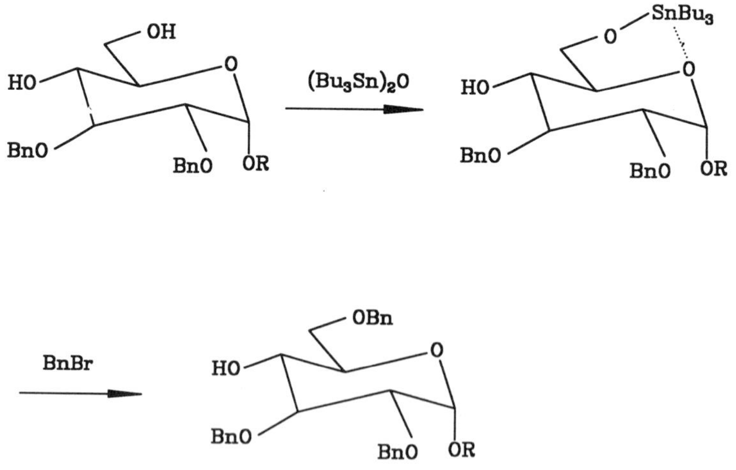

Diagram 20

5. *Protection of the Amino Group*

In the common amino sugars (2-amino-2-deoxy-hexoses) found in glycoproteins, the 2-amino- group is always *N*-acetylated, and the synthesis of glycosides frequently starts with 2-acetamido-2-deoxy-D-glucose or -D- galactose. Preparation of "acetochloro" derivatives from these sugars is well documented [48,77].

The amino group is also frequently protected by formation of a phthalimido group (Diagram 21A). The 2-deoxy-2-phthalimido derivative is readily prepared by reacting the amino sugar with phthalic anhydride, followed by acetic anhydride treatment. The resultant product is fully *O*-acetylated and is converted to the 1-halo derivative for glycosylation.

One of the advantages of the phthalimido group over the acetamido group is that 1-halo-2-deoxy-2-phthalimido derivatives are more readily obtainable in crystalline form and the phthalimido group is easily removed by treatment with ethanolic hydrazine. The phthalimido group, similar to the acetamido group, also appears to participate in the glycosylation reactions [95], and 1,2-*cis*-glycosides can be prepared only from the β-halo derivative under kinetic control.

An alternative method for masking the 2-amino group is to use 2-azido-2-deoxy derivatives (see Diagram 21B). As in the synthesis of *N*-lactosamine, the 2-azido derivative is generated from nonnitrogenous sugars. Unlike the 2-acetamido group, the 2-azido group does not participate in the glycosylation reaction and, with a suitable catalyst, the 1-halo-2-azido derivative can lead to formation of 1,2-*cis*-glycoside (e.g., α-glycoside of GalNAc to serine [96] and β-glycoside of ManNAc [97]). At the final stage of synthesis, the 2-azido group can be readily hydrogenolyzed to a 2-amino group, which is acetylated to obtain the desired product.

C. Phosphorylation and Sulfation

Some oligosaccharide chains in glycoconjugates are phosphorylated or sulfated. Lysosomal enzymes are targeted for the lysosomal compartment by recognition of mannose 6-phosphate residues on the *N*-glycosides [98]. Hindsgaul and his co-workers have synthesized mono- or dibranched mannose oligosaccharides having phosphate groups on different positions [99,100]. These compounds would be valuable in deciphering the exact binding mode during the process of mannose 6-phosphate recognition by its receptors. Similarly, the synthetic sulfated oligosaccharide [1], not only establishes the position of the

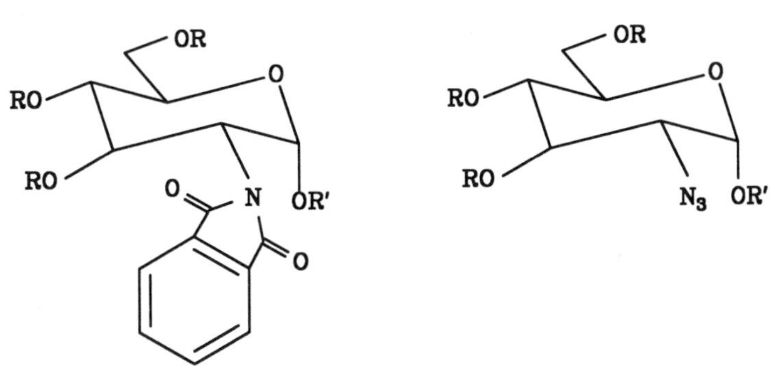

A B

Diagram 21

sulfate group, but promises to play an important role in understanding the mode of binding of some pituitary hormones.

D. Enzymatic Synthesis

Enzymes, stereospecific biological catalysts, are used in organic synthesis more and more frequently. Synthesis of simple carbohydrate derivatives and glycosides has been reviewed [101]. In the areas of synthesis of glycoconjugates, both glycosyltransferases and glycosidases have been used to prepare glycosides and oligosaccharides.

1. *Reversal of Glycosidase Reaction*

A promising result was reported [102] for the synthesis of mannose oligosaccharides from mannose by the reversal of α-mannosidase reaction, in which a total yield of disaccharides was 37% (w/w) of the total oligosaccharides formed. This is a much improved result, compared with a similar effort using the reversal of galactosidase for synthesis of Galβ(1,6)GalNAc [103].

2. *Glycosyltransferases*

Glycosyltransferases, the function of which is to transfer a sugar moiety from an appropriate sugar nucleotide to a suitable sugar or other acceptor compound, have been used for preparation of oligosaccharides or neoglycoproteins. Several examples are mentioned in Section IV.B.

There are also examples of di- and trisaccharide syntheses [104–106]. An interesting example by Palcic et al. [107] uses 6-substituted GlcNAc as an acceptor for bovine galactosyltransferase. The enzyme can tolerate an acceptor having α-L-fucose and methyl ester of NeuAc (but not free NeuAc) on the 6-hydroxyl of the GlcNAc residue.

III. SYNTHESIS OF GLYCOPEPTIDES

The synthesis of glycopeptides has been covered in a review by Garg and Jeanloz [23]. Newer examples of *O*-glycoside-containing glycopeptides include partial structures of asialoglycophorin A (i.e., a tetrapeptide containing four disaccharide side chains [108] and a hexapeptide with two disaccharide chains [109]). Bencomo and Sinaÿ [110,111] synthesized M and N active glycopeptides containing Galβ(1,3)GalNAc side chains. A glycopeptide related to nephritogenoside has been prepared [112] and coupled to bovine serum albumin.

IV. NEOGLYCOCONJUGATES

The term *neoglycoprotein* was first introduced in 1976 [113] to indicate a group of proteins that have been modified with carbohydrate derivatives. The usage has spread to the area of glycolipids [e.g., 114]. The compound to which the synthetic carbohydrates is conjugated can be a peptide, a lipid, or even some other type of compound. A collective term of *neoglycoconjugates* is suggested to designate all these compounds derivatized with mono- or oligosaccharides.

A. Synthetic Oligosaccharides

In recent years, synthetic chemists have been successful in preparing oligosaccharides of considerable complexity. At present, the synthetic oligosaccharides composed of the largest number of monosaccharide units are undecasaccharides: a nonsialylated tetra-antennary *N*-glycoside analogue [11] and sialylated biantennary structure [75]. Some aspects of the synthesis of these oligosaccharides have been reviewed in the earlier section.

The successful syntheses of these compounds depended on the advent of newer methods of glycosylation, not only for the efficiency of glycosylation, but also for higher anomeric selectivity, as well as improved methods for selective protection and clever application of "persistent" and "temporary" blocking groups.

B. Neoglycoproteins

There have been reviews on the neoglycoproteins covering the development up to about 1981 [7,25,26,115]. A more recent review on synthetic glycoconjugates has also appeared [116]. In this review, only special methods and newer applications will be discussed.

1. *Enzymatic Addition of Glycosyl Units and Glycopeptides (see also Sect. II.D)*

Enzymic addition of a glycosyl unit is very specific both for anomeric configuration and for the position of attachment. The enzymatic method is especially suited for preparing a small amount of sample when chemical modification may be impractical.

Berman et al. [117,118] used a galactosyltransferase to incorporate [^{13}C]galactose into *N*-acetylglucosamine (to form *N*-acetyllactosamine) or into ovalbumin for binding studies by CMR. Paulson et al. [119,120] desialylated red blood cells with a sialidase of wide specificity, and then resialylated using specific (e.g., α-2,3- and α-2,6-) sialyltransferases to produce "synthetic glycoconjugates" containing specific sialyl linkages, which allowed them to probe viral-binding specificities.

Hill and co-workers [121–124] serially added galactose and L-fucose to GlcNAc-AI-BSA [26] with a galactosyltransferase and two different L-fucosyltransferases, respectively, to produce L-fucose-containing neoglycoproteins for their studies of fucose-binding proteins (Diagram 22).

Glycopeptides have been attached to proteins by the action of transglutaminase by Yan and Wold [125]. The β-carboxyl of glutamine in the protein (in which all the amino groups have been masked) and α-amino group of glycopeptides are conjugated by this reaction. In β-casein modified with ovalbumin asparagine-oligosaccharides, four glutamine sites were readily modified by this enzymatic reaction.

$(GlcNAc)_n$–BSA
↓
$[Gal\beta(1,4)GlcNAc]_n$–BSA → $[L\text{–}Fuc\alpha(1,2)Gal\beta(1,4)GlcNAc]_n$–BSA
↓
$\{Gal\beta(1,4)[L\text{–}Fuc\alpha(1,3)]GlcNAc\}_n$–BSA

Diagram 22

2. *Glycoconjugates of Noncovalent Attachment*

A novel concept for attaching glycopeptides by means of noncovalent linkages has been promoted by Wold and his co-workers [126]. In their approach, glycopeptides are biotinylated, either directly or with a spacer arm. The biotinylated glycopeptides can then be bound very tightly, albeit noncovalently, to avidin or streptavidan. The dissociation constant of biotin–avidin complex is believed to be in the picomolar range, and these complexes have been shown to survive assay conditions in many studies [127–129]. An advantage of the avidin–(biotinyl glycopeptide) is that the orientation of sugar moieties can be estimated from the known geometry of combining sites of the avidin. This may provide an interesting handle for probing the relative orientation of combining sites of lectins and receptors. In another application, Asn-GlcNAc$_2$, Man$_5$ or its N^{α}-(6-aminohexanoyl) derivative was biotinylated and bound to avidin to study the effect of the peptide backbone on the accessibility of the sugar chains. Since the inner core of the glycan chains are "buried" in the binding site, these derivatives provide opportunities to study the effect of the protein matrix on glycan processing [127,130,131].

An alternative approach, based on the same principle, is to use enzyme–coenzyme complexing. A successful example is aspartate aminotransferase complexed with phosphopyridoxylated asparagine-oligosaccharides derived from ovalbumin [132]. These new types of neoglycoproteins, containing "high mannose" structures, have been used in studies of binding by alveolar macrophages [129].

3. *Application of Reductive Amination*

Modification of proteins with reductive amination [133,134] (Diagram 23) is one of the most gentle methods, as the reaction proceeds smoothly at neutral pH and room temperature. In addition, the modified amino groups remain positively charged; thus, there appears to be little change in the overall protein conformation. A refined scheme for preparing 1-thioglycosides possessing an ω-aldehydo group in the aglycon has been reported [135,136] (Diagram 24).

Many oligosaccharides can be prepared from *N*-glycosides of glycoproteins by means of an endo-β-hexosaminidase [68] or glycopeptidase [69]. These oligosaccharides can be directly attached to proteins by reductive amination with sodium cyanoborohydride [133,134] or pyridine borane [137] (see Diagram 23). In the process, the reducing terminus (usually GlcNAc) will become acyclic, which may cause difficulties in antigenicity or in abberation in the conformation of the rest of carbohydrate chain. Although a direct reductive amination with a reducing oligosaccharide is a slow process, owing to a low concentration of the free carbonyl form of the reducing terminus in aqueous solutions, the method is useful, nonetheless, because it is simple and requires little manipulation.

If the reducing terminus is to be sacrificed, as is inevitable by the direct reductive amination, it is possible to improve both the reaction rate and the yield of conjugation by first performing a limited periodate oxidation of the reducing terminus, to create free aldehydo group(s), before beginning the reductive amination. Oligosaccharide chains from *O*-glycosides are usually obtained by β-elimination, accompanied by reduction of the released terminus and, thus, are quite suited for this approach. This method is successfully applied in the modification of hyaluronic acid [138] with 1,6-diaminohexane and the production of neoglycolipids as will be described later. It is noteworthy that antibodies made against polylysine derivatized with cellulose oligosaccharides by reductive amination did not recognize the secondary amino group, although the acyclic portion of the oligosaccharides was recognized [139].

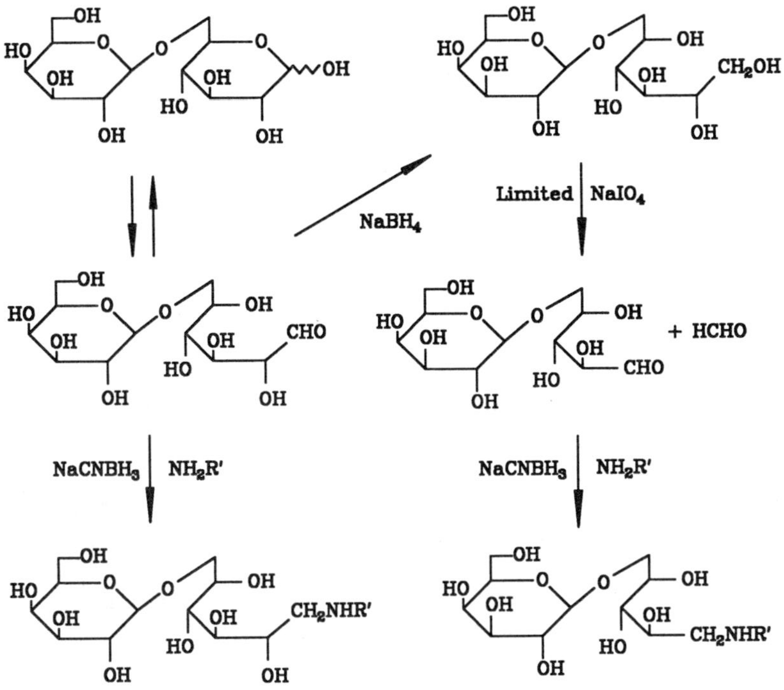

Diagram 23

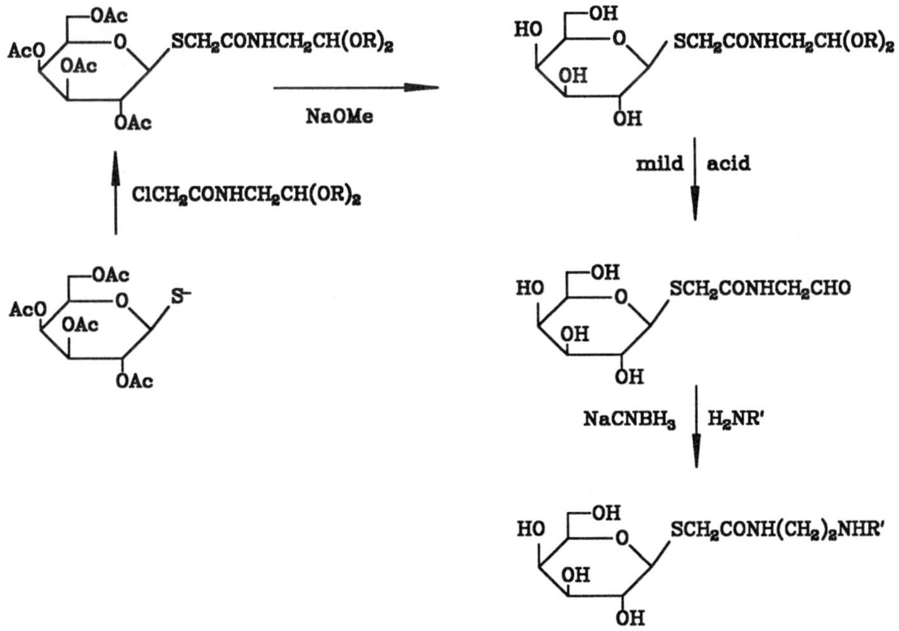

Diagram 24

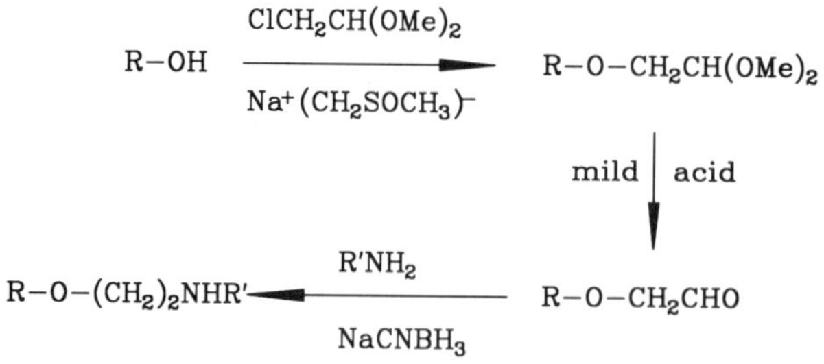

$$\text{R-OH} \xrightarrow[\text{Na}^+(\text{CH}_2\text{SOCH}_3)^-]{\text{ClCH}_2\text{CH(OMe)}_2} \text{R-O-CH}_2\text{CH(OMe)}_2$$

$$\text{R-O-CH}_2\text{CH(OMe)}_2 \xrightarrow{\text{mild acid}} \text{R-O-CH}_2\text{CHO}$$

$$\text{R-O-CH}_2\text{CHO} \xrightarrow[\text{NaCNBH}_3]{\text{R'NH}_2} \text{R-O-(CH}_2)_2\text{NHR'}$$

Diagram 25

An interesting method of attaching an aldehydic function to polysaccharides has been devised [140] (Diagram 25). A polysaccharide is alkylated under conventional conditions with chloroacetaldehyde dimethyl acetal. Deacetalization of the product yields a polysaccharide derivative containing aldehydic groups for protein conjugation. This method should be applicable to oligosaccharides as well, although there is little control on the precise position of alkylation or the extent of modification. Some glycosaminoglycans, such as heparin, that possess free or potentially free amino groups can be treated with nitrous acid, and the 2,5-anhydro-D-mannose residue produced thereby can be used for conjugation to proteins [141] (Diagram 26).

4. *Bifunctional Coupling Agents*

When a carbohydrate material that contains a free amino group, such as a glycopeptide, is to be coupled to protein, a bifunctional reagent can be used. As early as 1971, Rogers and Kornfeld [142] used toluene-2,4-diisocyanate to couple fetuin glycopeptides to proteins.

Lysogangliosides are obtained by complete removal of acyl groups followed by selective re-*N*-acetylation of amino groups on the sugars, leaving the amino group on the sphingosine available. Tiemeyer et al. [20] modified lysogangliosides with a homobifunctional reagent, such as bis(sulfosuccinimidyl)suberate, and the resultant ganglioside

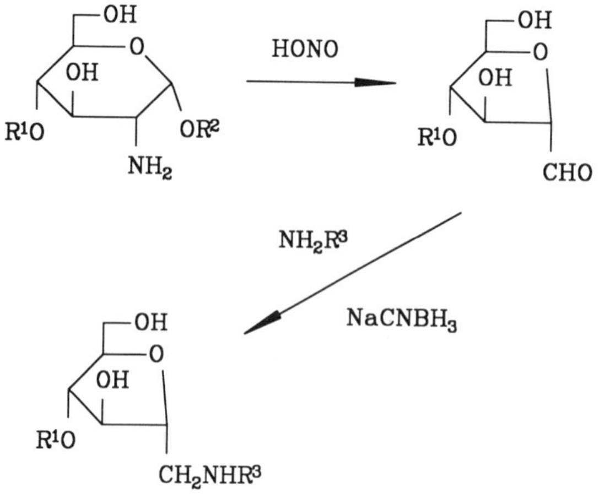

OH
O
OH
R^1O
OR^2
NH_2
HONO
OH
O
OH
R^1O
CHO
NH_2R^3
$NaCNBH_3$
OH
O
OH
R^1O
CH_2NHR^3

Diagram 26

$$(MeO)_2CHCH_2NHCO(CH_2)_4CO\text{–}NHNH_2 \longrightarrow (MeO)_2CHCH_2NHCO(CH_2)_4CO\text{–}N_3$$

$$\xrightarrow{RNH_2\ (glycopeptide)} (MeO)_2CHCH_2NHCO(CH_2)_4CO\text{–}NHR \xrightarrow{mild\ acid}$$

$$O{=}CHCH_2NHCO(CH_2)_4CO\text{–}NHR \xrightarrow[C_5H_5N\text{–}borane]{R'NH_2\ (protein)}$$

$$R'NH\text{–}CH_2CH_2NHCO(CH_2)_4CO\text{–}NHR$$

Diagram 27

derivative was coupled to bovine serum albumin. Oxidation of the double bond in the sphinganine in the sphingolipids also produced oligosaccharide derivatives containing an ω-aldehydo group which can be used for coupling to proteins by an extender arm (1,6-diaminohexane) in a similar way [143]. These neoglycoproteins were used as probes for ganglioside-binding receptors in neural tissues.

A heterobifunctional reagent has been prepared that utilizes mild conditions of reductive amination for coupling glycopeptides to proteins [Lee, Y. C., et al., unpublished results], as shown in Diagram 27. This reagent can convert a glycopeptide quantitatively into the acetal derivative which, after conversion to aldehyde, was then coupled to bovine serum albumin in very high yield (88%).

C. Nondegradable Glycoconjugates

Neoglycoconjugates can be designed to withstand the normal degradative processes in lysosomal compartment [7]. For example, poly-D-lysine derivatives modified with a 1-thio-*D*-mannoside (Diagram 28) were efficiently taken up by alveolar macrophages, and

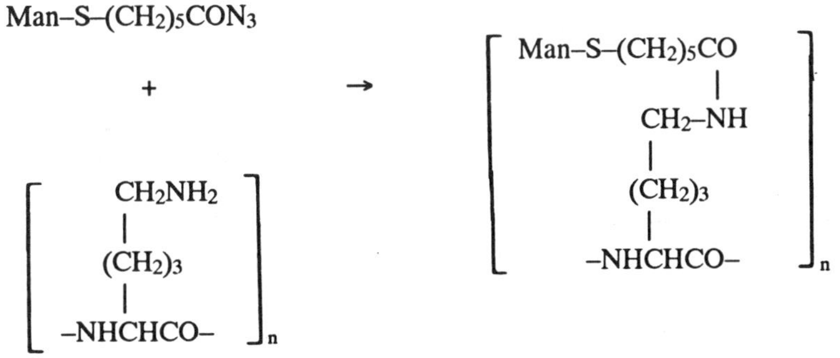

$$Man\text{–}S\text{–}(CH_2)_5CON_3 \quad + \quad \left[\begin{array}{c} CH_2NH_2 \\ | \\ (CH_2)_3 \\ | \\ \text{–}NHCHCO\text{–} \end{array}\right]_n \rightarrow \left[\begin{array}{c} Man\text{–}S\text{–}(CH_2)_5CO \\ | \\ CH_2\text{–}NH \\ | \\ (CH_2)_3 \\ | \\ \text{–}NHCHCO\text{–} \end{array}\right]_n$$

Poly-lysine Man-poly-lysine

Diagram 28

accumulated in the lysosomal compartment without being degraded [6]. Analogous derivatives based on poly-L-lysine were degraded normally.

D. Cluster Glycosides

Clustering of sugar residues often has tremendous enhancing effect on biological activities. This is most dramatically demonstrated for the mammalian hepatic receptors for Gal/GalNAc [37].

As mentioned in Section I.C, structurally simple, yet extremely potent, cluster glycosides [144] have been synthesized for these hepatic receptors [8,9]. The unique features of these cluster glycosides (see Diagram 3) are that the branching is provided by a dicarboxylic amino acid (aspartic acid), instead of the branching sugar, and flexibility is provided by aminohexyl aglycon. This type of shorthand synthesis may find wider applications in other sugar-binding systems.

E. Neoglycolipids

1. *Aliphatic Neoglycolipids*

Simple neoglycolipids based on the reaction of oligosaccharide lactones and aliphatic amines have been described [145]. The 2-bromoethyl group was used as an aglycon for eventual neoglycolipid synthesis by Magnusson and co-workers [86,146]. They further devised an interesting scheme that provides two fatty acyl residues per molecule [147] by using 3-bromo-2-(bromomethyl)propyl glycosides as an intermediate. From this intermediate, a diverse range of compounds can be prepared (Diagram 29).

2. *Steroid Glycosides*

Kempen et al. [148] attached a clustered tris-Gal-glycoside [149] to cholesterol by a spacer of succinyl glycine. When the product, Tris-Gal-Chol (Diagram 30), was incorporated into low-density lipoprotein (LDL) or high-density lipoprotein (HDL), they were both rapidly taken up by the liver. Interestingly, when Tris-Gal-Chol was incorporated into LDL, most of the uptake occurred in the Kupfer cells [150], and when it was incorporated into HDL,

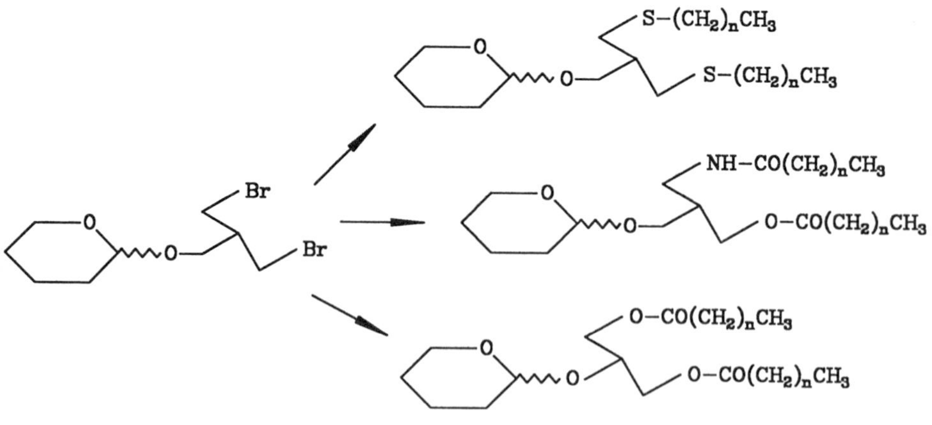

Diagram 29

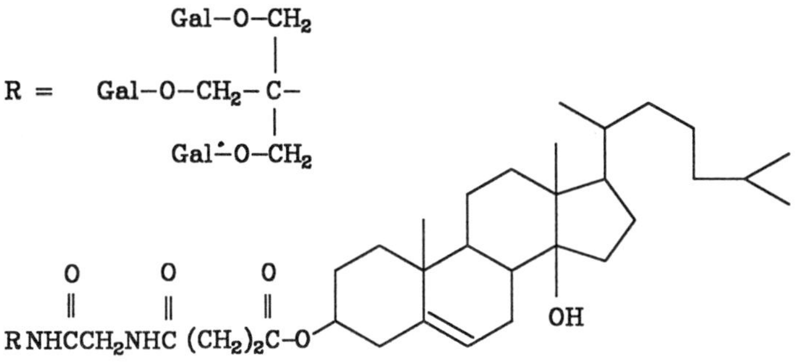

$$R = \begin{array}{l} \text{Gal–O–CH}_2 \\ \text{Gal–O–CH}_2\text{–C–} \\ \text{Gal–O–CH}_2 \end{array}$$

$$\text{RNHCCH}_2\text{NHC(CH}_2)_2\text{C–O–}$$ (cholesterol, OH)

Diagram 30

the uptake was mainly in the parenchymal cells [151]. When Tris-Gal-Chol, which is water soluble, was infused into a rat, the level of serum cholesterol decreased as much as 70%. The decrease of cholesterol level was shown to be sugar-specific and dose-dependent [152].

3. *Neoglycolipids Containing Phosphatidylenthanolamine*

Tang et al. [153] reported a method in which reduced oligosaccharides are partially oxidized at the glycitol moiety (by virtue of its greater susceptibility to periodate oxidation), and the resulting aldehydic compounds were conjugated to phosphatidylethanolamine by reductive amination (Diagram 31). These neoglycolipids are then coated on the bottom of plastic dishes to screen antigens [114]. Many new antigens to the known antibodies have been detected by this method.

F. Neoglycoproteins Versus Neoglycolipids

The choice between neoglycoproteins and neoglicolipids as probes for biological interactions involving carbohydrates is an interesting one. The antigenic determinants of the leprosy bacillus (*Mycobacterium leprae*) are phenolic glycolipids. However, Chatterjee et al. [154] found that neoglycoproteins containing the determinant disaccharide, (3,6-Me_2-Glc)β(1,4)(2,3-Me_2-L-Rha), served better for serological assays for leprosy than the natural glycolipids. The neoglycoprotein-containing trisaccharide (3,6-Me_2Glc)β(1,4)(2,3-Me_2-L-Rha)α(1,2)(3-Me-L-Rha), was even more effective. Tiemeyer et al. [20] and Yasuda et al. [143] also found that neoglycoproteins obtained by coupling the carbohydrate portion of gangliosides were more effective than gangliosides in detection of ganglioside receptors in nerve tissues. On the other hand, Feizi's group [114,153], chose to prepare neoglycolipids using oligosaccharides derived from glycoproteins for screening antitumor antigens. For this, the hydrophobic nature of neoglycolipids was utilized to coat the surface of culture plates.

Neoglycoproteins have been coated on colloidal gold for electronmicroscopic examination of sugar-binding proteins in the cells [155]. To the best of our knowledge, there has not been a report on using neoglycolipids on colloidal gold as electron microscopic probes.

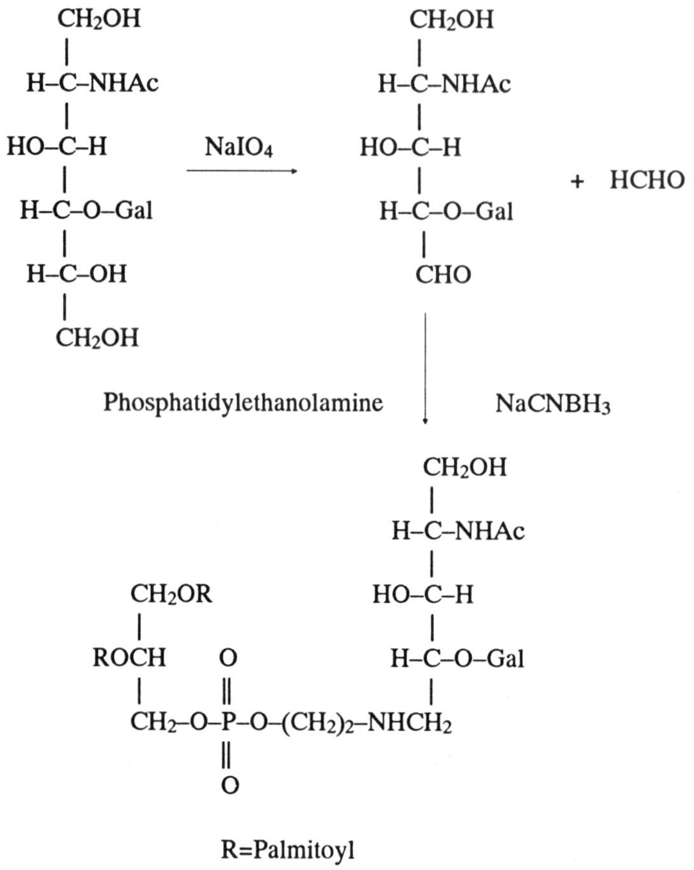

CH_2OH
|
H–C–NHAc
|
HO–C–H
|
H–C–O–Gal
|
H–C–OH
|
CH_2OH

$\xrightarrow{NaIO_4}$

CH_2OH
|
H–C–NHAc
|
HO–C–H
|
H–C–O–Gal
|
CHO

+ HCHO

Phosphatidylethanolamine ↓ $NaCNBH_3$

CH_2OH
|
H–C–NHAc
|
HO–C–H
|
H–C–O–Gal
|
CH_2OR
|
ROCH O
| ||
CH_2–O–P–O–$(CH_2)_2$–$NHCH_2$
||
O

R=Palmitoyl

Diagram 31

V. APPLICATION

In this section, only a limited number of examples from a diverse use of the synthetic glycoconjugates in biochemical and biological studies will be presented. The monosaccharide derivatives that are used to probe carbohydrate metabolism (e.g., triazine derivatives for inhibition of glycosidases) will not be included.

A. Binding Specificity and Mechanism

The binding specificity of hepatic carbohydrate receptors has been extensively studied with synthetic ligands. Lee and his co-workers [26] used BSA-based neoglycoproteins having different sugars at various degrees of modification to establish sugar-binding specificities of the hepatic lectins of the rabbit (Gal/GalNAc) and the chicken (GlcNAc). Stahl and co-workers [156] also used the same type of neoglycoproteins to establish the binding specificity of rat and rabbit alveolar macrophages (Man, L-Fuc). A surprising result was that the macrophages bound L-Fuc-BSA as well as Man-BSA. However, this is explainable on the basis of L-fucose and D-mannose having a similar overall structural feature [26]. The result implies that the nature of substituents at C-1 and C-5 (for both sugars) are not important for binding. This is borne out by the facts that *S*- or *O*-mannosides bind equally effectively, and that the presence of C-6 in the D-mannose residue is not essential for the binding.

Use of neoglycoproteins played a prominent role in establishing the specificity of L-fucose-binding lectin of the mammalian liver by Hill and co-workers [123,124]. A broad specificity of this lectin, as well as the presence of more abundant Gal/GalNAc lectin and Man/GlcNAc lectin in the liver, made it quite difficult to study this lectin with natural glycoproteins. With use of a series of BSA-based neoglycoproteins, each containing one specific mono- or oligosaccharide at a comparable sugar density, they established that the L-Fuc-lectin recognizes L-fucose as well as D-galactose residue.

There are two important aspects to these studies. First, it was established that, at least for some animal lectins, the binding requires mainly nonreducing terminal sugars. For instance, the BSA derivatives containing only glycosides of galactose, *N*-acetylgalactosamine, lactose, and *N*-acetyllactosamine possessed comparable binding affinity toward the mammalian hepatic Gal/GalNAc lectin, and the affinity surpassed even those obtainable from the natural sources. The second important fact is that the linear increase in the sugar density on the protein brings about logarithmic increase in binding potency [17,26]. This phenomenon is called the *clustering effect*, and it appears to manifest in the carbohydrate-binding systems of hepatocytes and alveolar macrophages.

The first small clustered glycosides [149] prepared to probe this phenomenon were rather simple (Diagram 32), but the effect of bundling the glycosides together was quite evident. The bundling of sugars in bis- and tris-glycosides enhanced the binding affinity much more than that accountable by simple arithmetic increase in the net sugar concentration [157]. Interestingly, the glycoside cluster effect is greatly reduced once the receptor protein has been "solubilized" by detergent extraction. A much more refined study [37,38] on the effect of glycoside clustering was carried out with a series of oligosaccharides that are analogues of naturally occurring *N*-glycosides [11,158] (Diagram 33). In this study, doubling and tripling the number of branches (each carrying a nonreducing terminal galactose) resulted in an increase in binding affinity of as much as 1000- and 1 million-fold, respectively (Table 4). Conformational analysis of these oligosaccharides revealed that the intergalactose distance in the most tightly bound triantennary oligosaccharide (estimated K_d=ca. 1 nM) are about 15–25 Å [38]. Studies that used several shorthand synthetic cluster glycosides (see Diagram 3) showed that glycoconjugates capable of fulfilling the aforementioned spacial arrangement of Gal/GalNAc residues are good ligands for binding, and otherwise, poor ligands [9]. Interestingly, the only tetraantennary oligosaccharide tested revealed no greater increase in affinity than can be accounted for by statistical increase in the availability of the terminal galactose residues.

```
Glyc–OCH2                 Glyc–OCH2                Glyc–OCH2
          \                         \                        \
Glyc–OCH2–C–NHR         Glyc–OCH2–C–NHR            HOCH2–C–NHR
          /                         /                        /
Glyc–OCH2                     HOCH2                    HOCH2

     Tris                      Bis                      Mono
```

Diagram 32

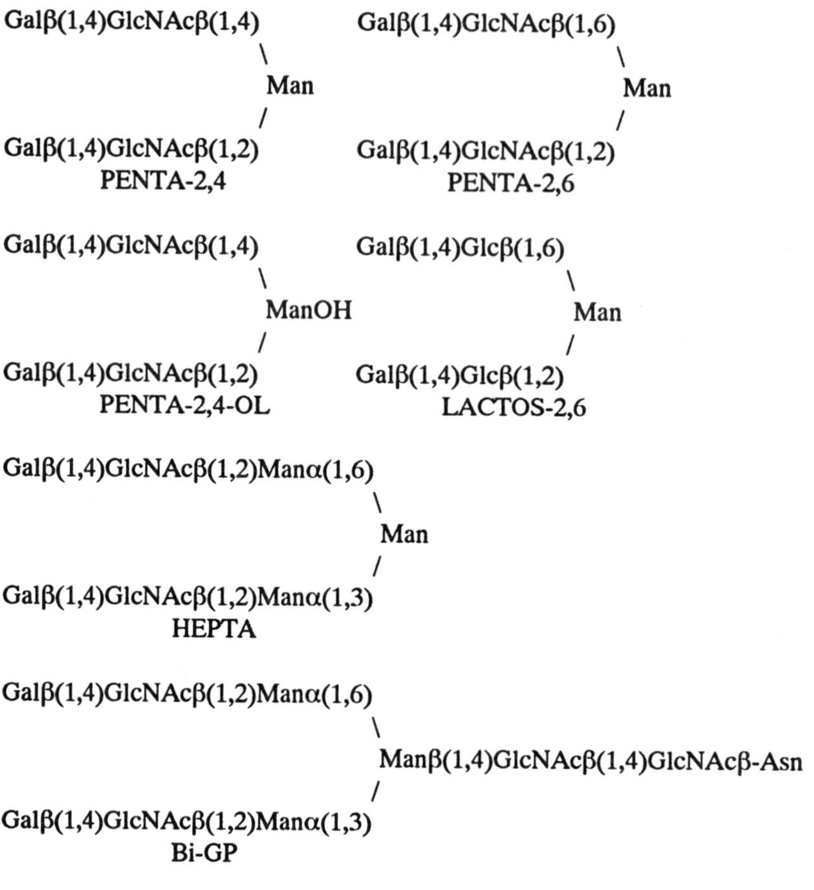

Diagram 33 (1)

With the same series of oligosaccharides, Crowley et al. [159] reported that the cluster requirement for *Datura* lectin is different from that of mammalian hepatic lectins. It favors the biantennary structure ("penta-2,6") over other tri- or tetravalent compounds (see Table 4). Inhibition of hemagglutination by a *Phaseolus* lectin by these oligosaccharides is also different. "Hepta" (a biantennary structure) was the most potent inhibitor, but "nona II" (a triantennary structure) was only marginally better than hepta. More surprisingly, "nona I," which is one of the two best ligands in the hepatocyte system is not even as good as the monovalent ligand, suggesting serious stereochemical interference by the additional branches. These data may be useful in understanding subunit organization of these carbohydrate-binding proteins.

Lemieux's group has been vigorously probing the molecular recognition of carbohydrates by various lectins and anti–blood-group antibodies using synthetic antigens [e.g., 21,160–165]. Their group meticulously synthesized a series of antigenic determinant oligosaccharides (mostly in the form of 8-carboxymethyloctanol glycosides [21]) and closely related analogues (referred to as *fraudulent* antigens). For example, to probe the nature of binding of lectin I of *Ulex europaeus*, they synthesized oligosaccharides αLFuc(1,2)βDGal(1,4)βDGlcNAc, βDGal(1,4)[αLFuc(1,3)]βDGlcNAc, and

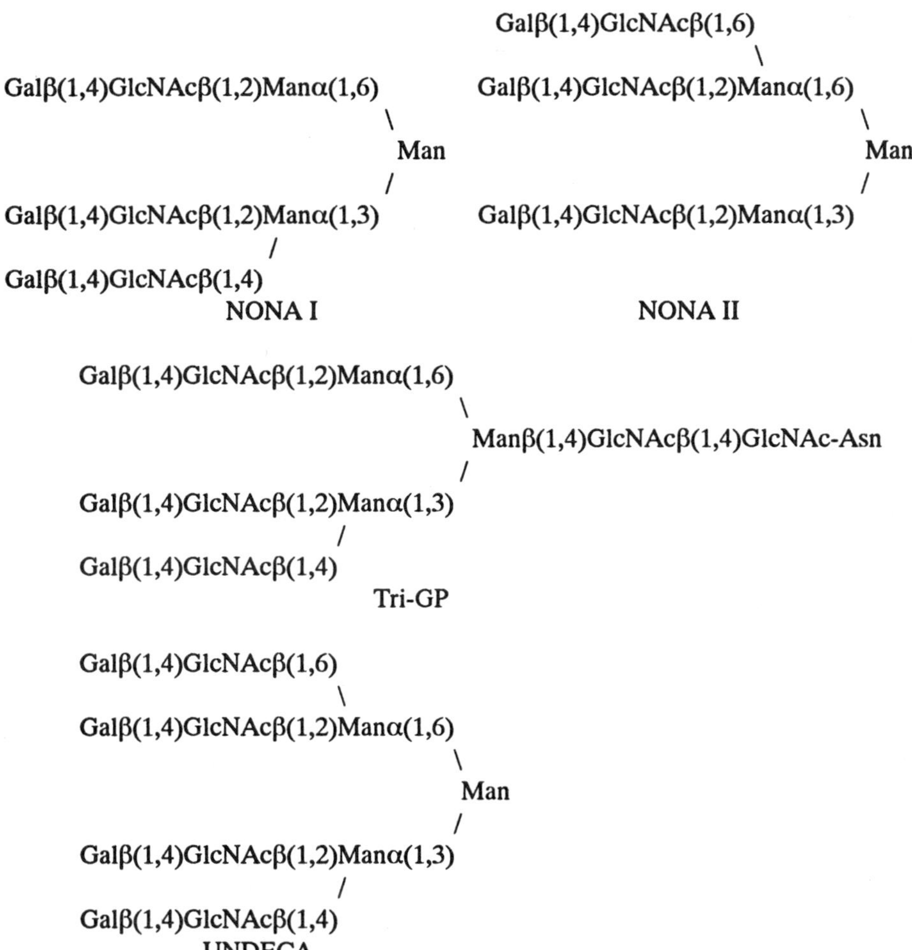

DIAGRAM 33 (2)

αLFuc(1,2)βDGal(1,4)[αLFuc(1,3)]βDGlcNAc, as well as analogous derivatives containing 6-deoxy-GlcNAc, 4′-epimer of Gal, and a 5′-nor-homolog of Gal [166]. The results from these studies revealed that a cluster involving OH-4, OH-3, and OH-2 of one of the αFuc's along with OH-3 of the βDGal provides polar interaction with lectin. Other examples of neoglycoproteins used in surveying of sugar binding specificities are demonstration of sugar binding in mammalian cell nuclei [167] and nucleoli [155], and the GlcNAc-binding activity in thyroid [168].

As mentioned in V.B.i, Paulson et al. [119,120] took advantage of specificity of sialidase and sialyltransferases in establishing binding specificity of viral receptors.

B. Neoglyconjugates as Affinant and Antigen

Neoglycoconjugates can be used not only as analytical tools but also as effective affinants for isolation of carbohydrate binding proteins. Isolation of L-fucose binding proteins

Table 4 [I_{50}] of Animal and Plant Lectins

Compounds[a]	Hepatocytes[b]	Datura lectin[c]	Phaseolus lectin[d]
		(mM)	
Gal-GlcNAc-Man	0.82		>18
Gal-GlcNAc		1.2	
		(μM)	
Hepta	20.4	450	5.5
Penta-2,6	2.8	2.5	>11
Penta-2,4	0.25	64	>11
		(nM)	
Nona II	111	4600	4
Nona I	2.4	120,000	>27
Undeca	1.3	6400	

[a]For the structures, see Diagram 33.
[b]Reference ligand = triantennary glycopeptide [37].
[c]Reference ligand = asialofetuin [159].
[d]Hammarström et al. [201].

[121,122], hepatic lectins [25,26], and chicken serum and egg yolk mannose-binding IgG's [169] are but a few examples. Specific blood group antibodies have been isolated with definitive oligosaccharide-derivatives of BSA attached to glass beads [21,161].

Neoglycoproteins have served as good antigens. Generally speaking, small oligosaccharides are not effective immunogens, and production of antibodies against small oligosaccharides is benefited by using neoglycoproteins carrying the target oligosaccharides. Excellent reports on the methodology and strategy have appeared [21,161,170]. Antibodies against β-D-glucosyl, β-D-galactosyl, and β-lactosyl residues have been made with neoglycoproteins containing glucose, galactose, and lactose [171].

C. Targeting

Targeting of Gal-human serum albumin (HSA) bearing a radiotracer (^{99m}Tc) for radiological examination of patients has been reported [172–174]. The great affinity of Gal-HSA for hepatic tissue makes this approach more specific than direct application of technetium compounds [175]. Neoglycoprotein was chosen over the derivatives of natural glycoproteins, such as ceruloplasmin, for several practical reasons. Preclinical and some clinical tests indicated that ^{99m}Tc-labeled HSA is nontoxic, and it is a promising noninvasive method for assessment of liver function [174].

The BSA-based neoglycoproteins were used to demonstrate that in vitro homing of hematopoietic stem cells is mediated by a recognition of galactose and mannose residues [176]. When neoglycoproteins are conjugated with drugs, a tenfold better specificity in cytotoxicity in cultured human embryonal carcinoma cells was observed [177]. Cawley et al. [178] reported that conjugation of diphtheria toxin, or its subunit A, to asialofetuin resulted in 600- to 1800-fold increase in toxicity toward rat hepatocytes.

D. Synthetic Substrates for Glycosyltransferase

Some glycosyltransferases have stringent structural requirements [179], and the measurement and expression of the glycosyltransferase activities can be misleading unless definitive substrates are used. The use of synthetic compounds as substrates and as reference compounds for products can alleviate this problem. Synthetic *p*-nitrophenyl glycosides of βGlcNAc(1,6)βGal and βGlcNAc(1,6)αMan were useful in critically evaluating the galactosyltransferase activities in human ovarian cancer patients [180]. The acceptor specificity of the GlcNAc-phosphotransferase for lysosomal enzyme targeting was examined with synthetic mannose oligosaccharides [181]. Activity of GnT V (a GlcNAc-transferase leading to the βGlcNAc(1,4)αMan(1,3) sequence in the *N*-glycoside) has been assayed with reverse-phase HPLC using a synthetic trisaccharide with a lipophilic aglycon [182]. In an even simpler application of glycosides with hydrophobic aglycons, a C_{18} cartridge (normally used for sample preparation for HPLC) was used for rapid analysis of glycosyltransferase activity [183].

Conformational preference of a *N*-acetylglucosaminyltransferase was probed with synthetic oligosaccharides. Srivastava et al. [184] synthesized isomeric derivatives of a trisaccharide, βGlcNAc(1,2)αMan(1,6)βGlc, by replacing either the *pro-R* or *pro-S* hydrogen on the C-6 of the glucose residue with a methyl group. One prefers the *gt* conformation at the αMan(1,6)-linkage, whereas the other favors the *gg* conformation [also see 39]. The isomer existing preferentially in the *gt* conformation was twice as active, whereas that favoring the *gg* conformation was only half as active as the parent compound. Such results support the importance of conformational structure of substrates for some glycosyltransferase [179].

E. Cell Biology

1. Cell Adhesion

Sugar-imbedded polyacrylamide gel layers have been used extensively in cell adhesion studies [e.g., 17,185]. Rat and chicken hepatocytes specifically adhered to the gel layers containing galactose and *N*-acetylglucosamine, respectively. Cell culture dishes, coated with lactose-bearing polystyrene, supported both the adhesion and viability of rat liver cells better than the dishes without such coating [186]. Similarly, rabbit alveolar macrophages which contain mannose-specific lectins on the surface adhered to mannose-bearing gel layers [27].

2. Decompaction

It was hypothesized that compaction during embryogenesis is related to the X-hapten, βGal(1,4)[αFuc(1,3)]βGlcNAc. Lacto-*N*-fucopentaose, βGal(1,4)[αFuc(1,3)]βGlcNAc (1,3)βGal(1,4)Glc, conjugated to the α- and ε-amino groups of lysyllysine by reductive amination was found to be effective in decompaction of a preimplanted mouse embryo, whereas free lacto-*N*-fucopentaose was not effective [16], demonstrating the importance of clustering glycosides.

3. Undegradable Neoglycoconjugates

The use of neoglycoproteins, especially undegradable ligands (Man_n-poly-D-lysine), in the studies of rabbit alveolar macrophages led to the discovery that the mannose-containing ligands stimulate the secretion of lysosomal enzymes by alveolar macrophages [187]. Man_n-poly-D-lysine was taken up as well as Man_n-poly-L-lysine by the macrophages, but unlike the latter, the poly-D-lysine derivative accumulated in the lysosomal compartment. Upon administering Man_n-BSA at this point, Man_n-poly-D-lysine that had accumulated in the lysosomal compartment was secreted together with lysosomal enzymes. The stimulation of enzyme secretion by mannose-containing ligands is not affected by preincubation with cycloheximide, indicating that the event is not likely to be caused by shunting of the newly synthesized lysosomal enzymes to the outside of cells. Man_n-poly-D-lysine was also found to be useful in detecting rabbit alveolar macrophage endosomes containing cathepsin D-like protease [188].

4. *Activation of Macrophages*

It is well known that muramyldipeptide (MDP) can activate macrophages. However, when MDP was conjugated to Man–BSA, and the conjugate was administered to rat or mice alveolar macrophages, tumoricidal activity increased [189] and resulted in metastasis eradication [190]. Presumably the MDP–Man–BSA conjugate was effectively taken up by the cells through the mannose receptors, and the MDP generated upon degradation of the conjugate effectively stimulated the macrophages.

5. *Neoglycoproteins as Probes in Flow Cytometry*

Neoglycoproteins can be readily labeled with fluorescent compounds, and the fluorescent neoglycoproteins can be used for quantitation of binding, uptake, and degradation of ligands [155,167,192].

6. Enzyme-Based Neoglycoproteins

Lactoperoxidase was modified with 2′-imino-2′-(methoxylethyl)-1-thio-D-mannopyranoside [17], and the mannose-modified lactoperoxidase was used to affinity-radiolabel the alveolar macrophage cell surface mannose receptor. This provided a means for identifying the mannose receptor protein as a single polypeptide band [193]. Similarly, *N*-acetylgalactosamine-modified lactoperoxidase has been useful in labeling the Gal/GalNAc-specific hepatic lectin of rat liver [194]. Galactose-modified lactoperoxidase was also useful in demonstrating two populations of endocytic vesicles [195].

IV. CONCLUSION

It is clear that synthetic oligosaccharides and glycoconjugates, either as full replicas of natural structures or as their abbreviated or fraudulent representation, will continue to be useful in studies involving carbohydrate recognition in biological systems. As methods for synthesizing more and more complex structures are developed, even more intricate features of carbohydrate recognition can be studied by synthetic compounds.

It is expected that the enzymatic method will be used more frequently, because it provides a very specific technique for addition of glycosyl residues on anomeric and positional isomers. Modern genetic engineering may provide a sufficient quantity of different glycosyltransferases in suitable form for this purpose. For instance, glycosyltransferases are generally membrane bound and insoluble without detergent, but they can be obtained in a soluble form by removing a part of the NH_2-terminal peptide sequence [196]. Site-specific mutagenesis may provide glycosyltransferases of "custom-tailored" specificity.

It is expected that chemical methods will continue to be refined and efficiency of synthesizing more complex oligosaccharides will be improved. Combinations of chemical and enzymatic methods that utilize the strength of each method, such as demonstrated by Sebasan and Paulson [197], should prove to be effective.

ACKNOWLEDGMENT

The authors thank Dr. Y. Ichikawa for reading the manuscript and offering valuable suggestions. We are also greatly indebted to the investigators who kindly sent us reprints and preprints that were extremely useful in preparation of this chapter.

REFERENCES*

1. Vandana, Hindsgaul, O., and Baenziger, J. U. (1987). Synthesis of oligosaccharide structures unique to pituitary glycoprotein hormones. *Can. J. Chem. 65*:1645–1652.
2. Townsend, R. R., Hardy, M. R., Wong, T. C., and Lee, Y. C. (1986). Binding of *N*-linked bovine fetuin glycopeptides to isolated rabbit hepatocytes: Gal/GalNAc hepatic lectin discrimination between Galβ(1,4)GlcNAc and Galβ(1,3)GlcNAc in a triantennary structure. *Biochemistry 25*:5716–5725.
3. Lemieux, R. U. (1982). The binding of carbohydrate structures with antibodies and lectins. In *Frontiers in Chemistry* (K. J. Laidler, ed.). Pergamon Press, Oxford, pp. 3–24.
4. Lemieux, R. U. (1985). The hydrated polar-group "gate" effect on the specificity and strength of the binding of oligosaccharides by protein receptor sites. *Proc. Int. Symp. Med. Chem. 1*: 329–351.
5. Schnaar, R. (1984). Immobilized glycoconjugates for cell recognition studies. *Anal. Biochem. 143*:1–13.
6. Hoppe, C. A., and Lee, Y. C. (1984). Accumulation of a nondegradable mannose ligand within rabbit alveolar macrophages. Receptor reutilization is independent of ligand degradation. *Biochemistry 23*:1723–1730.
7. Lee, Y. C., Hoppe, C. A., Kawaguchi, K., and Lin, P. (1987). Nondegradable carbohydrate ligands. *Methods Enzymol. 138*:404–409.
8. Lee, R. T., and Lee, Y. C. (1987). Preparation of cluster glycosides of *N*-acetylgalactosamine that have subnanomolar binding constants towards the mammalian hepatic Gal/GalNAc-specific receptor. *Glycoconjugate J. 4*:316–328.
9. Lee, R. T., Lin, P., and Lee, Y. C. (1984). New synthetic cluster ligands for galactose/*N*-acetylgalactosamine-specific lectin of mammalian liver. *Biochemistry 23*:4255–4261.
10. Petitou, M., Duchaussoy, P., Lederman, I., Choay, J., Sinaÿ, P., Jacquinet, J.-C., and Torri, G. (1986). Synthesis of heparin fragments. A chemical synthesis of the pentasaccharide *O*-(2-deoxy-2-sulfamido-6-*O*-sulfo-α-D-glucopyranosyl)-(1,4)-*O*-(β-D-glucopyranosyluronic acid)-(1,4)-*O*-(2-deoxy-2-sulfamido-3,6-di-*O*-sulfo-α-D-glucopyranosyl)-(1,4)-*O*-(2-*O*-sulfo-α-L-idopyranosyluronic acid)-(1,4)-2-deoxy-2-sulfamido-6-*O*-sulfo-D-glucopyranose decasodiuim salt. A heparin fragment having high affinity for antithrombin III. *Carbohydr. Res. 147*:221–236.
11. Lönn, H ., and Lönngren, J. (1983). Synthesis of a nona- and an undecasaccharide that form part of the complex type of carbohydrate moiety of glycoproteins. *Carbohydr. Res. 120*:17–24.
12. Paulsen, H., and Lebuhn, R. (1984). Synthese von pentasaccharid- und octasaccharaid-Sequenzen der Kohlenhydrat-kette von *N*-glycoproteinen. *Carbohydr. Res. 125*:21–45.
13. Gomtsyan, A. R., Byramova, N. E., Backinwosky, L. V., and Kochetkov, N. K. (1985). Synthesis of a branched pentasaccharide fragment of the O-antigen of *Shigella flexneri* serotype 5b. *Carbohydr. Res. 138*:C1–C4.
14. Abbas, S. A., Kohata, K., and Matta, K. L. (1987). Synthetic mucin fragments. Methyl 6-*O*-(2-acetamido-2-deoxy-β-D-glucopyranosyl)-β-D-galactopyranoside, methyl 3,4-di-*O*-(2-acetamido-2-deoxy-β-D-glucopyranosyl)-β-D-galactopyranoside, and methyl-*O*-(2-acetamido-2-

*The original manuscript was completed in 1987, and the literature surveyed covers up to 1987.

deoxy-β-D-glucopyranosyl)-(1,3)-*O*-β-D-galactopyranosyl-(1,3)-*O*-(2-acetamido-2-deoxy-β-D-glucopyranosyl)-(1,3)-β-D-galactopyranoside. *Carbohydr. Res. 161*:39–47.

15. Wessel, H.-P., Iversen, T., and Bundle, D. R. (1984). Synthesis of the trisaccharide moiety of gangliotriosylceramide (asialo GM_2). *Carbohydr. Res. 130*:5–21.
16. Fenderson, B. A., Zehavi, U., and Hakomori, S. (1984). A multivalent lacto-*N*-fucopentaose III-lysyllysine conjugate decompacts preimplantation mouse embryos, while the free oligosaccharide is ineffective. *J. Exp. Med. 160*:1591–1596.
17. Krantz, M. J., Holtzman, N. A., Stowell, C. P., and Lee, Y. C. (1976). Attachment of thioglycosides to proteins: Enhancement of liver membrane binding. *Biochemistry 15*:3963–3968.
18. Chatterjee, D., Cho, S.-N., and Brennan, P. J. (1986). Chemical synthesis and seroreactivity of *O*-(3,6-di-*O*-methyl-β-D-glucopyranosyl)-(1,4)-*O*-(2,3-di-*O*-methyl-α-L-rhamnopyranosyl)-(1,9)-oxynonanoyl-bovine serum albumin—the leprosy-specific, natural disaccharide-octyl-neoglycoprotein. *Carbohydr. Res. 156*:39–56.
19. Chatterjee, D., Cho, S.-N., Stewart, C., Douglas, J. T., Fujiwara, T., and Brennan, P. J. (1988). Synthesis and immunoreactivity of neoglycoproteins containing the trisaccharide unit of phenolic glycolipid I of *Mycobacterium leprae*. *Carbohydr. Res. 183*:241–260.
20. Tiemeyer, M., Yasuda, Y., and Schnaar, R. L. (1989). Ganglioside receptors on rat brain membranes. *J. Biol. Chem. 265*:1675–1985.
21. Lemieux, R. U., Eaker, D. A., and Bundle, D. R. (1977). A methodology for the production of carbohydrate-specific antibody. *Can. J. Biochem. 55*:507–512.
22. Jansson, K., Frejd, T., Kihlberg, J., and Magnusson, G. (1986). Boron trifluoride etherate as an effective reagent for the stereoselective one-pot conversion of acetylated 2-trimethylsilylethyl glycosides into sugar 1,2-*trans*-acetate. *Tetrahedron Lett. 27*:753–756.
23. Garg, H. G., and Jeanloz, R. W. (1985). Synthetic *N*- and *O*-glycosyl derivatives of L-asparagine, L-serine, and L-threonine. *Adv. Carbohydr. Chem. Biochem. 43*:135–201.
24. Gordon, J. A., Blumberg, S., Lis, H., and Sharon, N. (1972). Purification of soybean agglutinin by affinity chromatography on Sepharose-*N*-ε-aminocaproyl-β-D-galactopyranosylamine. *FEBS Lett. 24*:193–196.
25. Stowell, C. P., and Lee, Y. C. (1980). Neoglycoproteins. The preparation and application of synthetic glycoproteins. *Adv. Carbohydr. Chem. Biochem. 37*:225–281.
26. Lee, Y. C., and Lee, R. T. Neoglycoproteins as probes for binding and cellular uptake of glycoconjugates. In *The Glycoconjugates* (M. I. Horowitz, ed.). Academic press, New York, pp. 85–104.
27. Largent, B. L., Walton, K. M., Hoppe, C. A., Lee, Y. C., and Schnaar, R. L. (1984). Carbohydrate-specific adhesion of alveolar macrophages to mannose-derivatized surfaces. *J. Biol. Chem. 259*:1764–1769.
28. Krantz, M. J., and Lee, Y. C. (1976). Quantitative hydrolysis of thioglycosides. *Anal. Biochem. 71*:318–321.
29. McBroom, C. R., Samanen, C. H., and Goldstein, I. J. (1972). Carbohydrate antigens: Coupling of carbohydrates to proteins by diazonium and phenylisothiocyanate. *Methods Enzymol. 28*: 212–219.
30. Fuegedi, P., Garegg, P. J., Lönn, H., and Norberg, T. (1987). Thioglycosides as glycosylating agents in oligosaccharide synthesis. *Glycoconjugate J. 4*:97–108.
31. Myers, R. W., and Lee, Y. C. (1986). Improved preparation of some per-*O*-acetylated aldo-hexopyranosyl cyanides. *Carbohydr. Res. 154*:145–163.
32. Wu, T. C., Geokjian, P. G., and Kishi, Y. (1987). Preferred conformation of *C*-glycosides. 1. Conformational similarity of glycosides and corresponding *C*-glycosides. *J. Org. Chem. 52*: 4819–4823.
33. Docherty, P. A., and Aronson, N. N., Jr. (1987). α-D-Mannopyranosylmethyl-*p*-nitro-phenyltriazene inhibition of rat liver α-D-mannosidases. *Biochim. Biophys. Acta 914*:283–288.
34. Myers, R. W., and Lee, Y. C. (1984). Synthesis and characterization of some anomeric pairs of per-*O*-acetylated aldohexopyranosyl cyanides (per-*O*-acetylated 2,6-anhydroheptononitriles). On

the reaction of per-*O*-acetylaldohexopyranosyl bromides with mercuric cyanide in nitromethane. *Carbohydr. Res. 132*:61–82.
35. Kabat, E. A., and Mayer, M. M. (1961). Inhibition reactions. In *Experimental Immunochemistry*. Charles C. Thomas, Springfield, Ill., pp. 241–267.
36. Glaudemans, C. P. J., and Kovac, P. (1988). Deoxyfluoro carbohydrates as a tool for probing the binding subsites of monoclonal anti-saccharide antibodies. *ACS Symp. Ser. 374*:78–108.
37. Lee, Y. C., Townsend, R. R., Hardy, M. R., Lönngren, J., Arnarp, J., Haraldsson, M., and Lönn, H. (1983). Binding of synthetic oligosaccharides to the hepatic Gal/GalNAc lectin. *J. Biol. Chem. 258*:199–202.
38. Lee, Y. C., Townsend, R. R., Hardy, M. R., Lönngren, J., and Bock, K. (1984). Binding of synthetic clustered ligands to the Gal/GalNAc lectin on isolated rabbit hepatocytes. In *Biochemical and Biophysical Studies of Proteins and Nucleic Acids* (T. B. Lo, T. Y. Liu, and C. H. Li, eds.). Elsevier, New York, pp. 349–360.
39. Lemieux, R. U., Wong, T. C., and Thøgersen, H. (1982). The synthesis and conformational properties of the diastereoisomeric βDGal(1,4)βDGlcNAc(1,6)6-C-CH_3-D-Gal trisaccharides. *Can. J. Chem. 60*:81–86.
40. Paulsen, H. (1982). Advances in selective chemical syntheses of complex oligosaccharides. *Angew. Chem. Int. Ed. Engl. 21*:155–173.
41. Schmidt, R. R. (1986). New methods for the synthesis of glycosides and oligosaccharides—are there alternatives to the Koenigs–Knorr method? *Angew. Chem. Int. Ed. Engl. 25*:212–235.
42. Bochkev, A. F., and Zaikov, G. E. (1979). *Chemistry of the* O-*Glycosidic Bond: Formation and Cleavage*. Pergamon Press, Oxford.
43. Flowers, H. M. (1987). Chemical synthesis of oligosaccharides. *Methods Enzymol. 138*:359–404.
44. Paulsen, H., Hadamczyk, D., Kutschker, W., and Buensch, A. (1985). Regioselektive glycosylierung an 3′-OH oder 4′-OH der Lactose durch einsatz unterschiedlicher Katalysatorsysteme. *Liebigs Ann. Chem.* 129–141.
45. van Boeckel, C. A. A., Beetz, T., and van Aelst, S. F. (1984). Substituent effects on carbohydrate coupling reactions promoted by insoluble silver salts. *Tetrahedron 40*:4097–4107.
46. Kochetkov, N. K., and Bochkov, A. F. (1972). Synthesis of oligosaccharides by the orthoester method. *Methods Carbohydr. Chem. 6*:480–486.
47. Dahmen, J., Frejd, T., Magnusson, G., and Noori, G. (1983). Boron trifluoride etherate-induced glycosidation: Formation of alkyl glycosides and thioglycosides of 2-deoxy-2-phthalimidoglycopyranoses. *Carbohydr. Res. 114*:328–330.
48. Lee, R. T., Wong, T. C., and Lee, Y. C. (1986). Synthesis of 6′-aminohexyl-2-acetamido-2-deoxy-D-galactoside isomers and a unique isomerization catalyzed by ion exchange resin. *J. Carbohydr. Chem. 5*:343–357.
49. Hanessian, S., and Banoub, J. (1977). Chemistry of the glycosidic linkage. *O*-Glycosylations catalyzed by stannic chloride, in the D-ribofuranose and D-glucopyranose series. *Carbohydr. Res. 59*:261–267.
50. Banoub, J., and Bundle, D. R. (1979). Stannic tetrachloride catalysed glycosylation of 8-ethoxycarbonyloctanol by cellobiose, lactose, and maltose octaacetates; synthesis of α- and β-glycosidic linkages. *Can. J. Chem. 57*:2085–2090.
51. Paulsen, H., and Paal, M. (1984). Lewissäure-katalysierte Synthesen von di-und Trisaccharid-Sequenzen der *O*- und *N*-Glycoproteine. Anwendung von trimethylsilyl Trifluoromethanesulfonat. *Carbohydr. Res. 135*:53–69.
52. Lemieux, R. U., and Driguez, H. (1975). The chemical synthesis of 2-acetamido-2-deoxy-4-*O*-(α-L-fucopyranosyl)-3-*O*-(β-D-galactopyranosyl)-D-glucose. The Lewis a blood-group antigenic determinant. *J. Am. Chem. Soc. 97*:4063–4069.
53. Matta, K. L., and Barlow, J. J. (1976). A facile synthesis of 2-methyl-[3,4-di-*O*-acetyl-6-*O*-(chloroacetyl)-1,2-dideoxy-α-D-glucopyrano]-[2′,1′:4,5]-2-oxazoline. *Carbohydr. Res. 51*: 215–222.

54. Nakabayashi, S., Warren, C. D., and Jeanloz, R. W. (1986). A new procedure for the preparation of oligosaccharide oxazolines. *Carbohydr. Res. 150*:C7–C10.
55. Srivastava, V. K. (1982). A facile synthesis of 2-methyl-(3,4,6-tri-*O*-acetyl-1,2-dideoxy-α-D-glucopyrano)-[2,1-*d*]-2-oxazoline. *Carbohydr. Res. 103*:286–292.
56. Warren, C. D., Nakabayashi, S., and Jeanloz, R. W. (1987). Synthesis of the tetrasaccharide lipid intermediate P^1-dolychyl P^2-[*O*-α-D-mannopyranosyl-(1,6)-*O*-β-*D*-mannopyranosyl-(1,4)-*O*-(2-acetamido-2-deoxy-β-D-glucopyranosyl-(1,4)-2-acetamido-2-deoxy-α-D-glucopyranosyl]diphosphate. *Carbohydr. Res. 169*:221–233.
57. Lemieux, R. U., Hendriks, K. B., Stick, R. V., and James, K. (1975). Halide ion catalyzed glycosidation reactions. Syntheses of α-linked disaccharides. *J. Am. Chem. Soc. 97*:4056–4062.
58. Pougny, J.-R., Jacquinet, J.-C., Nassr, M., Duchet, D., Milat, M.-L., and Sinaÿ, P. (1978). A novel synthesis of 1,2-*cis*-disaccharides. *J. Am. Chem. Soc. 99*:6762–6763.
59. Pougny, J.-R., Nassr, M. A. M., Naulet, N., and Sinaÿ, P. (1978). A novel glucosidation reaction. Application to the synthesis of α-linked disaccharides. *Nouveau J. Chim. 2*:389–395.
60. Grundler, G., and Schmidt, R. R. (1985). Anwendung des Trichloracetimidatverfahrens auf 2-Desoxy-2-phthalimido-D-glucose-derivate. Synthese von Oligosacchariden der "Core-region" von *O*-Glycoproteinen des Mucin typs. *Carbohydr. Res. 135*:203–218.
61. Ferrier, R. J., Hay, R. W., and Vethaviyasar, N. (1973). A potentially versatile synthesis of glycosides. *Carbohydr. Res. 27*:55–61.
62. Lönn, H. (1987). Glycosylation using a thioglycoside and methyl trifluoromethanesulfonate. A new and efficient method for *cis* and *trans* glycoside formation. *J. Carbohydr. Chem. 6*:301–306.
63. Fuegedi, P., and Garegg, P. J. (1986). A novel promoter for the efficient construction of 1,2-*trans* linkages in glycoside synthesis, using thioglycosides as glycosyl donors. *Carbohydr. Res. 149*:C9–C12.
64. Paulsen, H., Rauwald, W. and Weichert, U. (1988). Glycosidierung mit Thioglycosiden von Oligosacchariden zu Segmenten von *O*-Glycoproteinen. *Liebigs Ann. Chem. 1988*:75–86.
65. Lönn, H. (1987). Sulfurylchloride/trifluoromethanesulfonic acid. A novel promoter system for glycoside synthesis using thioglycosides as glycosyl donors. *Glycoconjugate J. 4*:117–118.
66. Pozsgay, V., and Jennings, H. J. (1987). A new method for the synthesis of *O*-glycosides from *S*-glycosides. *J. Org. Chem. 52*:4635–4637.
67. Ito, Y., and Ogawa, T. (1987). Sulfenate esters as glycosyl acceptors: A novel approach to *O*-glycosides from thioglycosides and sulfenate esters. *Tetrahedron Lett. 28*:4701–4704.
68. Trimble, R. B., Trumbly, R., and Maley, F. (1987). Endo-β-*N*-acetylglucosaminidase H from *Streptomyces plicatus. Methods Enzymol. 138*:763–770.
69. Tarentino, A. L., and Plummer, T. H., Jr. (1987). Peptide-N^4-(*N*-acetyl-β-glucosaminyl)asparagine amidase and endo-β-*N*-acetylglucosaminidase from *Flavobacterium meningosepticum. Methods Enzymol. 138*:770–778.
70. Spiro, R. G. (1972). Study of the carbohydrates of glycoproteins. *Methods Enzymol. 28*:3–42.
71. van Boeckel, C. A. A., Beetz, T., Vos, J. N., de Jong, A. J. M., van Aelst, S. F., van den Bosch, R. H., Mertens, J. M. R., and van der Vlugt, A. F. (1985). Synthesis of a pentasaccharide corresponding to the antithrombin III binding fragment of heparin. *J. Carbohydr. Chem. 4*: 293–321.
72. Greene, T. W. (1981). Protection for the hydroxyl group, including 1,2- and 1,3-diols. In *Protective Groups in Organic Chemistry*. Wiley-Interscience Publication, New York, pp. 10–86.
73. Binkley, R. W. (1988). Protective groups. In *Modern Carbohydrate Chemistry*. Marcel Dekker, New York, pp. 113–165.
74. Haines, A. H. (1981). The selective removal of protecting groups in carbohydrate chemistry. *Adv. Carbohydr. Chem. Biochem. 39*:13–70.
75. Ogawa, T., Sugimoto, M., Kitajima, T., Sadozai, K. K., and Nukada, T. (1986). Total synthesis of a undecasaccharide: A typical carbohydrate sequence for the complex type of glycan chains of a glycoprotein. *Tetrahedron Lett. 27*:5739–5742.

76. Sugimoto, M., Numata, M., Koike, K., Nakahara, Y., and Ogawa, T. (1986). Total synthesis of gangliosides G_{M1} and G_{M2}. *Carbohydr. Res. 156*:C1–C5.
77. Horton, D. (1972). 2-Acetamido-3,4,6-tri-*O*-acetyl-2-deoxy-α-D-glucopyranosyl chloride. *Methods Carbohydr. Chem. 6*:282–285.
78. Kuhn, R., Lutz, P., and MacDonald, D. L. (1966). Synthese anomerer Sialin saure-methylketoside. *Chem. Ber. 99*:611–617.
79. Petitou, M., Duchaussoy, P., Lederman, I., Choay, J., Jacquinet, J.-C., Sinaÿ, P., and Torri, G. (1987). Synthesis of heparin fragments: A methyl α-pentaoside with high affinity for antithrombin III. *Carbohydr. Res. 167*:67–75.
80. Ogawa, T., and Sasajima, K. (1981). Synthesis of a model of an inner chain of cell-wall proteoheteroglycan isolated from *Piricularia oryzae*: Branched D-mannopentaosides. *Carbohydr. Res. 93*:67–81.
81. Iversen, T., and Bundle, D. R. (1981). Benzyl trichloroacetimidate, a versatile reagent for acid-catalyzed benzylation of hydroxy-groups. *J. Chem. Soc. Chem. Commun.* pp. 1240–1241.
82. Wessel, H.-P., Iversen, T., and Bundle, D. R. (1985). Acid-catalyzed benzylation and allylation by alkyl trichloroacetimidates. *J. Chem. Soc. Perkin Trans. I* pp. 2247–2250.
83. David, S., Thieffry, A., and Veyrieres, A. (1981). A mild procedure for the regiospecific benzylation and allylation of polyhydroxy-compounds via their stannylene derivatives in non-polar solvents. *J. Chem. Soc. Perkin Trans. I* pp. 1796–1801.
84. Eby, P., Sondheimer, S. J., and Schuerch, C. (1979). Selective acetolysis of primary benzyl ethers. *Carbohydr. Res. 73*:273–276.
85. Gent, P. A., and Gigg, R. (1974). The allyl ether as a protecting group in carbohydrate chemistry. Isomerisations with tristriphenylphosphine-rhodium(I) chloride. *J. Chem. Soc. Chem. Commun.* pp. 277–278.
86. Dahmen, J., Frejd, T., Groenberg, G., Lave, T., Magnusson, G., and Noori, G. (1983). 2-Bromoethyl glycosides: Applications in the synthesis of spacer-arm glycosides. *Carbohydr. Res. 118*:292–301.
87. Boss, R., and Scheffold, R. (1976). Cleavage of allyl ethers with Pd/C. *Angew. Chem. Int. Ed. Engl. 15*:558–559.
88. Ichikawa, Y., Manaka, A., and Kuzuhara, H. (1985). Discrimination between the 2,3- and the 2′,3′-hydroxyl groups of maltose and cellobiose through their specific protection. *Carbohydr. Res. 138*:55–64.
89. Garegg, P., Haltberg, H., and Wallin, S. (1982). A novel reductive ring-opening of carbohydrate benzylidene acetals. *Carbohydr. Res. 108*:97–101.
90. Hullar, T. L., and Siskin, S. B. (1970). The facile bromination by *N*-bromosuccinmide of benzylidene acetals of carbohydrates. Application to the synthesis of 2,6-imino carbohydrates (substituted 2,5-oxazabicyclo-2.2.2.octanes) *J. Org. Chem. 35*:225–228.
91. Hanessian, S., and Plessas, N. R. (1969). The reaction of *O*-benzylidene sugars with *N*-bromosuccinimide. II. Scope and synthetic utility in the methyl 4,6-*O*-benzylidenehexopyranoside series. *J. Org. Chem. 34*:1035–1044.
92. Binkley, R. W., Goewey, G. S., and Johnston, J. C. (1984). Regioselective ring opening of selected benzylidene acetals. A photochemically initiated reaction for partial deprotection of carbohydrates. *J. Org. Chem. 49*:992–996.
93. Ogawa, T., and Matsui, M. (1977). A new approach to regioselective acylation of polyhydroxy compounds. *Carbohydr. Res. 56*:C1–C6.
94. Ogawa, T., and Matsui, M. (1978). Regioselective enhancement of the nucleophilicity of hydroxyl groups through trialkylstannylation: A route to partial alkylation of polyhydroxy compounds. *Carbohydr. Res. 62*:C1–C4.
95. Lemieux, R. U., Takeda, T., and Chung, T. Y. (1976). Synthesis of 2-amino-2-deoxy-β-D-glycopyranosides. Properties and use of 2-deoxy-2-phthalimidoglycosyl halide. In *Synthetic Methods for Carbohydrates* H. S. El Khadem, ed.). *ACS Symp. Ser. 39*:90–115.

96. Paulsen, H., and Hoelck, J.-P. (1982). Synthese der Glycopeptide *O*-β-D-galactopyranosyl-(1,3)-*O*-(2-acetamido-2-deoxy-α-D-galactopyranosyl)-(1,3)-L-serin und -L-threonin. *Carbohydr. Res. 109*:89–107.
97. Paulsen, H., and Lorentzen, J. P. (1984). Darstellung von β-glycosidisch verknuepften Sacchariden der 2-Amino-2-deoxy-D-mannose und der 2-Amino-2-deoxy-D-mannuronsäure. *Carbohydr. Res. 133*:C1–C4.
98. Kornfeld, S. (1987). Trafficking of lysosomal enzymes. *FASEB J. 1*:462–468.
99. Srivastava, O. P., and Hindsgaul, O. (1987). Synthesis of 6′-*O*-phosphorylated *O*-α-D-mannopyranosyl-(1,3)- and -(1,6)-α-D-mannopyranosides. *Carbohydr. Res. 161*:324–329.
100. Srivastava, O. P., and Hindsgaul, O. (1987). Synthesis of phosphorylated trimannosides corresponding to end groups of the high-mannose chains of lysosomal enzymes. *Carbohydr. Res. 161*:195–210.
101. Whitesides, G. M., and Wong, C.-H. (1985). Enzymes as catalysts in synthetic organic chemistry. *Angew, Chem. Int. Ed. Engl. 24*:617–638.
102. Johansson, E., Hedbys, L., Larsson, P.-O., Mosbach, K., Gunnausson, A., and Svensson, S. (1986). Synthesis of mannose oligosaccharides via reversal of the α-mannosidase reaction. *Biotechnol. Lett. 8*:421–424.
103. Hedbys, L., Larsson, P.-O., and Mosbach, K. (1984). Synthesis of the disaccharide 6-*O*-β-D-galactopyranosyl-2-acetamido-2-deoxy-D-galactose using immobilized β-galactosidase. *Biochem. Biophys. Res. Commun. 123*:8–15.
104. Augé, C., David, S., Mathieu, C., and Gautheron, C. (1984). Synthesis with immobilized enzymes of two trisaccharides, one of them active as the determinant of a stage antigen. *Tetrahedron Lett. 25*:1467–1470.
105. Rosevear, P. R., Nuenz, H. A., and Barker, R. (1982). Synthesis and solution conformation of the type 2 blood group oligosaccharide αFuc(1,2)βDGal(1,4)βDGlcNAc. *Biochemistry 21*:1421–1431.
106. Wong, C.-H., Haynie, S. L., and Whitesides, G. M. (1982). Enzyme-catalyzed synthesis of *N*-acetyllactosamine with in situ regeneration of uridine 5′-diphosphate glucose and uridine 5′-diphosphate galactose. *J. Org. Chem. 47*:5416–5418.
107. Palcic, M., Srivastava, O. P., and Hindsgaul, O. (1987). Transfer of D-galactosyl groups to 6-*O*-substituted 2-acetamido-2-deoxy-D-glucose residues by use of bovine D-galactosyltransferase. *Carbohydr. Res. 149*:315–324.
108. Paulsen, H., and Schultz, M. (1986). Synthese der Tetrapeptidsequenz 12 bis 15 des Asialoglycophorin A mit vier Disaccharidseitenketten. *Liebigs Ann. Chem.* pp. 1435–1447.
109. Paulsen, H., and Schultz, M. (1987). Synthese der Hexapaptidsequenz 43 bis 48 des Asialoglycophorin mit zwei Disaccharidseitenketten. *Carbohydr. Res. 159*:37–52.
110. Bencomo, V. V., and Sinaÿ, P. (1984). Synthesis of M and N active glycopeptides. Part of the *N*-terminal region of human glycophorin A. *Glycoconjugate J. 1*:5–8.
111. Bencomo, V. V., Jacquinet, J.-C., and Sinaÿ, P. (1982). The synthesis of derivatives of *O*-β-D-galactopyranosyl-(1,3)-*O*-(2-acetamido-2-deoxy-α-D-galactopyranosyl)-L-serine and L-threonine. *Carbohydr. Res. 110*:C9–C11.
112. Takeda, T., Sawaki, M., Ogihara, Y., and Shibata, S. (1985). Synthesis of *N*-glycopeptides and a neoglycoprotein. *Carbohydr. Res. 139*:133–140.
113. Lee, Y. C., Stowell, C. P., and Krantz, M. J. (1976). 2-Imino-2-methoxyethyl 1-thioglycosides: New reagents for attaching sugars to proteins. *Biochemistry 15*:3956–3963.
114. Tang, P. W., and Feizi, T. (1987). Neoglycolipid micro-immunoassays applied to the oligosaccharides of human milk galactosyltransferase detect blood-group related antigens on both *O*- and *N*-linked chains. *Carbohydr. Res. 161*:133–143.
115. Aplin, J. D., and Wriston, J. C., Jr. (1981). Preparation, properties, and applications of carbohydrate conjugates of proteins and lipids. *Crit. Rev. Biochem. 10*:259–306.
116. Magnusson, G. (1986). Synthesis of neo-glycoconjugates. In *Protein–Carbohydrate Interactions in Biological Systems* (D. A. Lark, ed.). Academic Press,, New York, pp. 215–223.

117. Berman, E., Brown, J. H., Lis, H., and Sharon, N. (1985). Binding of [1-^{13}C]galactose-labeled *N*-acetyllactosamine to *Erythrina crystagalli* agglutinin as studied by ^{13}C-NMR. *Eur. J. Biochem. 152*:447–451.
118. Berman, E., Lis, H., and James, T. L. (1986). Binding of [1-^{13}C]galactose-enriched hen ovalbumin to *Erythrina crystagalli* agglutinin as studied by ^{13}C-NMR spectroscopy. *Eur. J. Biochem. 161*:589–594.
119. Paulson, J. C., Rogers, G. N., Carroll, S. M., Higa, H. H., Pritchett, T., Milks, G., and Sabesan, S. (1984). Selection of influenza virus variants based in sialyloligosaccharide receptor specificity. *Pure Appl. Chem. 56*:797–805.
120. Paulson, J. C., and Rogers, G. N. (1987). Resisalylated erythrocytes for assessment of the specificity of sialyloligosaccharide binding proteins. *Methods Enzymol. 138*:162–168.
121. Lehrman, M. A., and Hill, R. L. (1986). The binding of fucose-containing glycoproteins by hepatic lectins. Purification of a fucose-binding lectin from rat liver. *J. Biol. Chem. 261*: 7419–7425.
122. Haltiwanger, R. S., Lehrman, M. A., Eckhardt, A. E., and Hill, R. L. (1986). The distribution and localization of the fucose-binding lectin in rat tissues and the identification of a high affinity form of the mannose/*N*-acetylglucosamine-binding lectin in rat liver. *J. Biol. Chem. 261*:7433–7439.
123. Lehrman, M. A., Haltiwanger, R. S., and Hill, R. L. (1986). The binding of fucose-containing glycoproteins by hepatic lectins. The binding specificity of the rat liver fucose lectin. *J. Biol. Chem. 261*:7426–7432.
124. Lehrman, M. R., Pizzo, S. V., Imber, M. J., and Hill, R. J. (1986). The binding of fucose-containing glycoproteins by hepatic lectins. Re-examination of the clearance from blood and the binding to membrane receptors and pure lectins. *J. Biol. Chem. 261*:7412–7418.
125. Yan, S.-C. B., and Wold, F. (1984). Neoglycoproteins: In vitro introduction of glycosyl units at glutamines in β-casein using transglutaminase. *Biochemistry 23*:3759–3765.
126. Chen, V. J., and Wold, F. (1986). Neoglycoproteins: Preparation and properties of complexes of biotinylated asparagine-oligosaccharides with avidin and streptavidin. *Biochemistry 25*: 939–944.
127. Shao, M.-C., and Wold, F. (1987). The use of avidin-biotinylglycan as the model for in vitro glycoprotein processing. *J. Biol. Chem. 262*:2968–2972.
128. Shao, M.-C., Chin, C. C. Q., Caprioli, R. M., and Wold, F. (1987). The regulation of glycan processing in glycoproteins. The effect of avidin on individual steps in the processing of biotinylated glycan derivatives. *J. Biol. Chem. 262*:2973–2979.
129. Ohsumi, Y., Chen, V. J., Yan, S.-C. B., Wold, F., and Lee, Y. C. (1988). Interaction between new neoglycoproteins and the D-Man/L-Fuc receptor of rabbit alveolar macrophages. *Glycoconjugate J. 5*:99–106.
130. Yet, M.-G., Shao, M.-C., and Wold, F. (1988). Effects of the protein matrix on glycan processing in glycoproteins. *FASEB J. 2*:22–31.
131. Chen, V. J., and Wold, F. (1984). Neoglycoproteins: Preparation of noncovalent glycoproteins through high-affinity protein-(glycosyl) ligand complexes *Biochemistry 23*:3306–3311.
133. Gray, G. R. (1974). The direct coupling oligosaccharides to proteins and derivatized gels. *Arch. Biochem. Biophys. 163*:426–428.
134. Gray, G. R. (1978). Antibodies to carbohydrates: Preparation of antigens by coupling carbohydrates to proteins by reductive amination with cyanoborohydride. *Methods Enzymol. 50*:155–160.
135. Lee, R. T., and Lee, Y. C. (1979). Preparation of 1-thioglycosides having ω-aldehydo aglycons useful for attachment to proteins by reductive amination. *Carbohydr. Res. 77*:149–156.
136. Lee, R. T., and Lee, Y. C. (1980). Preparation and some biochemical properties of neoglycoproteins produced by reductive amination of thioglycosides containing an ω-aldehydoaglycon. *Biochemistry 19*:156–163.
137. Cabacungan, J. C., Ahmed, A. I., and Feeney, R. E. (1982). Amine boranes as alternative reducing agents for reductive alkylation of proteins. *Anal. Biochem. 124*:272–278.

138. Raja, R. H., LeBoeuf, R. D., Stone, G. W., and Weigel, P. H. (1984). Preparation of alkylamine and ^{125}I-radiolabeled derivatives of hyaluronic acid uniquely modified at the reducing end. *Anal. Biochem. 139*:168–177.

139. Danielson, S. J., and Gray, G. R. (1986). Immunochemical characterization of polylysine conjugates containing reductively aminated cellulose oligosaccharides. *Glycoconjugate J. 3*:363–377.

140. Bøgwald, J., Seljelid, R., and Hoffman, J. (1986). Coupling of polysaccharides activated by means of chloroacetaldehyde dimethyl acetal to amines or proteins by reductive amination. *Carbohydr. Res. 148*:101–107.

141. Hoffman, J., Larm, O., and Scholander, E. (1983). A new method for covalent coupling of heparin and other glycosaminoglycans to substances containing primary amino groups. *Carbohydr. Res. 117*:328–331.

142. Rogers, J. C., and Kornfeld, S. (1971). Hepatic uptake of proteins coupled to fetuin glycopeptide. *Biochem. Biophys. Res. Commun. 45*:622–629.

143. Yasuda, Y., Tiemeyer, M., Blackburn, C. C., and Schnaar, R. L. (1987). Neuronal recognition of gangliosides: Evidence for a brain ganglioside receptor. In *New Trends in Ganglioside Research: Neurochemical and Neuroregenerative Aspects* (Ledeen, Tettamanati, Yu, Hogan, and Yates (eds.). Springer-Verlag, New York, pp. 229–243.

144. Lee, R. T., and Lee, Y. C. (1987). Cluster glycosides. *Methods Enzymol. 133*:424–429.

145. Williams, T. J., Plessas, N. R., and Goldstein, I. J. (1979). A new class of model glycolipids: Synthesis, characterization, and interaction with lectins. *Arch. Biochem. Biophys. 195*:145–151.

146. Dahmen, J., Frejd, T., Groenberg, G., Lave, T., Magnusson, G., and Noori, G. (1983). 2-Bromoethyl glycosides: Synthesis and characterization. *Carbohydr. Res. 116*:303–307.

147. Ansari, A. A., Frejd, T., and Magnusson, G. (1987). 3-Bromo-2-bromomethylpropyl glycosides in the preparation of double-chain bis-sulfide neo-glycolipids *Carbohydr. Res. 161*:225–233.

148. Kempen, H. J. M., Hoes, C., van Boom, J. H., Spanjer, H. H., de Lange, J., Lanagendoen, A., and van Berkel, C. T. J. (1984). A water-soluble cholesteryl-containing trigalactoside: Synthesis, properties and use in directing lipid-containing particles to the liver. *J. Med. Chem. 27*:1306–1313.

149. Lee, Y. C. (1978). Synthesis of some cluster glycosides suitable for attachment to proteins or solid matrices. *Carbohydr. Res. 67*:509–514.

150. van Berkel, T. J. C., Kruijt, J. K., Spanjer, H. H., Nagelkerke, J. F., Harkes, L., and Kempen, H. J. M. (1985). The effect of a water-soluble tris-galactoside-terminated cholesterol derivative on the fate of low density lipoproteins and liposomes. *J. Biol. Chem. 260*:2694–2699.

151. van Berkel, T. J. C., Kruijt, J. K., and Kempen, H. J. M. (1985). Specific targeting of high density lipoproteins to liver hepatocytes by incorporation of a tris-galactoside-terminated cholesterol derivatives. *J. Biol. Chem. 260*:12203–12207.

152. Kempen, H. J. M., Kuipers, F., van Berkel, T. J. C., and Vonk, R. J. (1987). Effect of infusion of "tris-galactosyl-cholesterol" on plasma cholesterol, clearance of lipoprotein cholesteryl esters, and biliary secretion in the rat. *J. Lipids Res. 28*:659–666.

153. Tang, P. W., Gooi, H. C., Hardy, M., Lee, Y. C., and Feizi, T. (1985). Novel approach to the study of the antigenicities and receptor functions of carbohydrate chains of glycoproteins. *Biochem. Biophys. Res. Commun. 132*:474–480.

154. Chatterjee, D., Cho, S.-N., Brennan, P. J., and Aspinall, G. O. (1986). Chemical synthesis and seroreactivity of *O*-(3,6-di-*O*methyl-β-D-glucopyranosyl)-(1,4)-*O*-(2,3-di-*O*-methyl-α-L-rhamnopyranosyl)-(1,9)-oxynonanoyl-bovine serum albumin—the leprosy-specific, natural disaccharide-octyl-neoglycoprotein *Carbohydr. Res. 156*:39–56.

155. Bourgeois, C. A., Seve, A. P., Monsigny, M., and Hubert, J. (1987). Detection of sugar-binding sites in the fibrillar and the granular components of the nucleolus: An experimental study in cultured mammalian cells. *Exp. Cell Res. 172*:365–376.

156. Stahl, P. D., Schlesinger, P. H., Sigardson, E., Rodman, J. S., and Lee, Y. C. (1980). Receptor-mediated pinocytosis of mannose glycoconjugates by macrophages: Characterization and evidence for receptor recycling. *Cell 19*:207–215.

157. Connolly, D. T., Townsend, R. R., Kawaguchi, K., Bell, W. R., and Lee, Y. C. (1982). Binding and endocytosis of cluster glycosides by rabbit hepatocytes. Evidence for a short-circuit pathway that does not lead to degradation. *J. Biol. Chem.* *257*:939–945.
158. Baumann, H., Lönn, H., and Lönngren, J. (1983). Synthesis of a hexasaccharide that forms part of an alveolar glycoprotein. *Carbohydr. Res.* *114*:317–321.
159. Crowley, J. F., Goldstein, I. J., Arnarp, J., and Lönngren, J. (1984). Carbohydrate binding studies on the lectin from *Datura stramonium* seeds. *Arch. Biochem. Biophys.* *231*:524–533.
160. Hindsgaul, O., Norberg, T., Le Pendu, J., and Lemieux, R. U. (1982). Synthesis of type 2 human blood-group antigenic determinants. The H, X, and Y haptens and variations of the H type 2 determinant as probes for the combining site of the lectin I of *Ulex europaeus*. *Carbohydr. Res.* *109*:109–142.
161. Lemieux, R. U., Baker, D. A., Weinstein, W. M., and Switzer, C. M. (1981). Artificial antigens. Antibody preparations for the localization of Lewis determinants in tissues. *Biochemistry* *20*:199–205.
162. Lemieux, R. U., Venot, A. P., Spohr, U., Bird, P., Mandal, G., Morishima, N., Hindsgaul, O., and Bundle, D. R. (1985). Molecular recognition. V. The binding of the B human blood group determinant by hybridoma monoclonal antibodies. *Can. J. Chem.* *63*:2664–2668.
163. Sabesan, S., and Lemieux, R. U. (1984). Synthesis of tri- and tetrasaccharide haptens related to the *asialo* forms of the gangliosides Gm_2 and Gm_1. *Can. J. Chem.* *62*:644–654.
164. Spohr, U., Hindsgaul, O., and Lemieux, R. U. (1985). Molecular recognition. II. The binding of the Lewis b and Y human blood group determinants by the lectin IV of *Griffonia simplifolia*. *Can. J. Chem.* *63*:2644–2652.
165. Spohr, U., Morishima, N., Hindsgaul, O., and Lemieux, R. U. (1985). Molecular recognition. IV. The binding of the Lewis b human blood group determinant by a hybridoma monoclonal antibody. *Can. J. Chem.* *63*:2659–2663.
166. Hindsgaul, O., Khare, D. P., Bach, M., and Lemieux, R. U. (1985). Molecular recognition. III. The binding of the H-type 2 human blood group determinant by the lectin I of *Ulex europaeus*. *Can. J. Chem.* *63*:2653–2658.
167. Seve, A.-P., Hubert, J., Bouvier, D., Bourgeois, C., Midoux, P., Roche, A.-C., and Monsigny, M. (1986). Analysis of sugar-binding sites in mammalian cell nuclei by quantitative flow microfluorometry. *Proc. Natl. Acad. Sci. USA* *83*:5997–6001.
168. Miquelis, R., Alquier, C., and Monsigny, M. (1987). The *N*-acetylglucosamine-specific receptor of the thyroid. Binding characteristics, partial characterization, and potential role. *J. Biol. Chem.* *262*:15291–15298.
169. Wang, K.-Y., Hoppe, C. A., Datta, P. K., Fogelstrom, A., and Lee, Y. C. (1986). Identification of the major mannose-binding proteins from chicken egg yolk and chicken serum as immunoglobulins. *Proc. Natl. Acad. Sci. USA* *83*:9670–9674.
170. Lemieux, R. U., Bundle, D. R., and Baker, D. A. (1975). The properties of a "synthetic" antigen related to the human blood-group Lewis a. *J. Am. Chem. Soc.* *97*:4076–4083.
171. Pazur, J. H. (1982). Purification and properties of antibodies having specificity for the carbohydrate moieties of synthetic glycoconjugates. *Carbohydr. Res.* *107*:243–254.
172. Vera, D. R., Krohn, K. A., Stadalnik, R. C., and Scheibe, P. O. (1984). Tc-99-m-galactosyl-neoglycoalbumin: In vivo characterization of receptor-mediated binding to hepatocytes. *Nucl. Med.* *151*:191–196.
173. Vera, D. R., Krohn, K. A., Stadalnik, R. C., and Scheibe, P. O. (1984). Tc-99m galactosyl-neoglcoalbumin: In vitro characterization of receptor-mediated binding. *J. Nucl. Med.* *25*: 779–787.
174. Stadalnik, R. C., Vera, D. R., Woodle, E. S., Trudeau, W. L., Porter, B. A., Ward, R. E., Krohn, K. A., and O'Grady, L. F. (1985). Technetium-99m NGA functional hepatic imaging: Preliminary clinical experience. *J. Nucl. Med.* *26*:1233–1242.
175. Stadalnik, R. C., Vera, D. R., and Krohn, K. A. (1986). Receptor-binding radiopharmaceuticals: Experimental and clinical aspects. In *Nuclear Medicine Annual 1986*, (L. M. Freeman and H. S. Weissman, eds.). Raven Press, New York, pp. 105–139.

176. Aizawa, S., and Tavassoli, M. (1987). In vitro homing of hemopoietic stem cells is mediated by a recognition system with galactosyl and mannosyl specificities. *Proc. Natl. Acad. Sci. USA 84*:4485–4489.
177. Gabius, H.-J., Bokemeyer, C., Hellmann, T., and Schmoll, H.-J. (1987). Targeting of neoglycoprotein–drug conjugates to cultured human embryonal carcinoma cells. *J. Cancer Res. Clin. Oncol. 113*:126–130.
178. Cawley, D. B., Simpson, D. L., and Herschman, H. R. (1981). Asialoglycoprotein receptor mediates the toxic effects of an asialofetuin–diphtheria toxin fragment A conjugate on cultured rat hepatocytes. *Proc. Natl. Acad. Sci. USA 78*:3383–3387.
179. Schachter, H. (1986). Biosynthetic controls that determine the branching and microheterogeneity of protein-bound oligosaccharides. *Biochem. Cell Biol. 64*:163–181.
180. Madiyalakan, R., Piskorz, C. F., Piver, M. S., and Matta, K. L. (1987). Serum β-(1,4)-galactosyltransferase activity with synthetic low molecular weight acceptor in human ovarian cancer. *Eur. J. Cancer Clin. Oncol. 23*:901–906.
181. Madiyalakan, R., Chowdhary, M. S., Rana, S. S., and Matta, K. L. (1986). Lysosomal-enzyme targeting: The phosphorylation of synthetic D-mannosyl saccharides by UDP-*N*-acetylglucosamine:lysosomal-enzyme *N*-acetylglucosamine phosphotransferase from rat-liver microsomes and fibroblasts. *Carbohydr. Res. 152*:183–194.
182. Pierce, M., Arango, J., Tahir, S. H., and Hindsgaul, O. (1987). Activity of UDP-GlcNAc: α-mannose β(1,6)*N*-acetylglucosaminyltransferase (GnT V) in cultured cells using a synthetic trisaccharide acceptor. *Biochem. Biophys. Res. Commun. 146*:679–684.
183. Palcic, M. M., Heerze, L. D., Pierce, M., and Hindsgaul, O. (1988). The use of hydrophobic synthetic glycosides as acceptors in glycosyltransferase assays. *Glycoconjugate J. 5*:59–63.
184. Srivastava, O. P., Hindsgaul, O., Shoreibah, M., and Pierce, M. (1988). Recognition of oligosaccharide substrates by *N*-acetylglucosaminyltransferase-V. *Carbohydr. Res. 179*:137–161.
185. Pless, D., Lee, Y. C., Roseman, S., and Schnaar, R. L. (1983). Specific cell adhesion to immobilized glycoproteins demonstrated using new reagents for protein and glycoprotein immobilization. *J. Biol. Chem. 25S*:2340–2349.
186. Kobayashi, A., Akaike, T., Kobayashi, K., and Sumitomo, H. (1986). Enhanced adhesion and survival efficienty of liver cells in culture dishes coated with a lactose-carrying styrene homopolymer. *Makromol. Chem. Rapid Commun. 7*:645–650.
187. Ohsumi, Y., and Lee, Y. C. (1987). Mannose-receptor ligands stimulate secretion of lysosomal enzymes from rabbit alveolar macrophages. *J. Biol. Chem. 262*:7955–7962.
188. Diment, S., and Stahl, P. (1985). Macrophage endosomes contain proteases which degrade endocytosed protein ligands. *J. Biol. Chem. 260*:15311–15317.
189. Monsigny, M., Roche, A.-C., and Bailly, P. (1984). Tumoricidal activation of murine alveolar macrophages by muramyldipeptide substituted mannosylated serum albumin. *Biochem. Biophys. Res. Commun. 121*:579–584.
190. Roche, A.-C., Bailly, P., and Monsigny, M. (1985). Macrophage activation of MDP bound to neoglycoproteins: Metastasis eradication in mice. *Invasion Metastasis 5*:218–232.
191. Midoux, P., Roche, A.-C., and Monsigny, M. (1986). Estimation of the degradation of endocytosed material by flow cytofluorometry using two neoglycoproteins containing different numbers of fluorescein molecules. *Biol. Cell 58*:221–226.
192. Midoux, P., Roche, A.-C., and Monsigny, M. (1987). Quantification of the binding, uptake, and degradation of fluoresceinylated neoglycoproteins by flow cytometry. *Cytometry 8*:327–334.
193. Wileman, T. E., Lennartz, M. R., and Stahl, P. D. (1986). Identification of the macrophage mannose receptor as a 175-kDa membrane protein. *Proc. Natl. Acad. Sci. USA 83*:2501–2505.
194. Lee, R. T., and Lee, Y. C. (1987). Affinity labeling of the galactose/*N*-acetylgalactosamine-specific receptor of rat hepatocytes: Preferential labeling of one of the subunits. *Biochemistry 26*:6320–6329.
195. Baenziger, J. U., and Fiete, D. (1986). Separation of two populations of endocytic vesicles involved in receptor–ligand sorting in rat hepatocytes. *J. Biol. Chem. 281*:7445–7454.

196. Weinstein, J., Lee, E. U., McEntee, K., Lai, P.-H., and Paulson, J. C. (1987). Primary structure of β-galactoside α-2,6-sialyltransferase. Conversion of membrane-bound enzyme to soluble forms by cleavage of the NH_2-terminal signal anchor. *J. Biol. Chem.* *262*:17735–17743.
197. Sabesan, S., and Paulson, J. C. (1986). Combined chemical and enzymatic synthesis of sialyloligosaccharides and characterization by 500-MHz ^{1}H and ^{13}C NMR spectroscopy. *Biochemistry* *108*:2068–2080.
198. Lee, R. T., and Lee, Y. C. (1974). Synthesis of 3-(2-aminoethylthio)propyl glycosides. *Carbohydr. Res.* *37*:193–201.
199. Chipowsky, S., and Lee, Y. C. (1973). Synthesis of 1-thioaldosides having an amino group at the aglycon terminal. *Carbohydr. Res.* *31*:339–346.
200. Ichikawa, Y., Monden, R., and Kuzuhara, H. (1988). Synthesis of methyl glycoside derivatives of tri- and pentasaccharides related to the antithrombin II-binding sequence of heparin, employing cellobiose as a key starting material. *Carbohydr. Res.* *172*:37–64.
201. Hammarström, S., Hammarström, M.-L., Sundblad, G., Arnarp, J., and Lönngren, J. (1982). Mitogenic leukoagglutinin from *Phaseolus vulgaris* binds to a pentasacchride unit in *N*-acetyllactosamine-type glycoprotein glycans. *Proc. Natl. Acad. Sci. USA* *79*:1611–1615.
202. Weigel, P. H., Naoi, M., Roseman, S., and Lee, Y. C. (1979). Preparation of 6-aminohexyl D-aldopyranosides. *Carbohydr. Res.* *70*:83–91.
203. Dahmen, J., Frejd, T., Magnusson, G., and Noori, G. (1982). Preparation and application of 2-bromoethyl glycosides: Synthesis of spacer-arm glycosides and agglutination inhibitor *Carbohydr. Res.* *111*:C1–C4.

7

Proteoglycans: Structure, Synthesis, Function

V. P. Bhavanandan *The Milton S. Hershey Medical Center, The Pennsylvania State University, Hershey, Pennsylvania*

E. A. Davidson *Georgetown University School of Medicine, Washington, D.C.*

Proteoglycans belong to the class of macromolecules referred to as glycoconjugates. They differ from other members of the glycoconjugates (i.e., glycoproteins and mucins) in having one or more unbranched polysaccharide chains (rather than oligosaccharides) covalently linked to protein. Additional characteristic features that distinguish proteoglycans from glycoproteins and mucins are their repeating disaccharide structure and the high negative charge of their carbohydrate moiety at physiological conditions. The linear polysaccharide chains of proteoglycans, together with hyaluronic acid are referred to as *glycosaminoglycans*. Historically, these polysaccharides were named *muco*polysaccharides by Karl Meyer (1938) because of their ability to yield viscous, slimy, mucinlike solutions. The nomenclature committees would prefer that old terminology be dropped from use to prevent confusion, but this is not always practical. For example, even though the term glycosaminoglycans has been widely accepted, the diseases involving inborn errors of metabolism of these macromolecules are still referred to as mucopolysaccharidoses.

The term glycosaminoglycans (GAGs) was coined by Jeanloz (1960) to indicate that they are glycans (polysaccharides) that contain glycosamines. Each glycosaminoglycan is a polymer of a disaccharide composed of a hexosamine (glycosamine) and a nonnitrogenous sugar, which is usually an uronic acid, except in keratan sulfate, in which it is D-galactose. The hexosamine is either D-galactosamine or D-glucosamine, both of which are present mostly in the *N*-acetylated form except in heparan sulfate and heparin, wherein some hexosamine residues are *N*-sulfated. The glycosaminoglycans can be divided into two subgroups, namely, the galactosaminoglycans (chondroitins and dermatan sulfate) and glucosaminoglycans (hyaluronic acid, keratan sulfate, heparan sulfate, and heparin). The glycosidic bond between the hexosamine and the nonnitrogenous sugar may be in either the α- or β-configuration. The glycosaminoglycans, except hyaluronic acid, contain ester sulfate groups, mostly on the hydroxyl groups of the saccharides, but sometimes also on the amine groups as mentioned earlier. The major structural features of the glycosaminoglycans are summarized in Table 1 and Figure 1. It must be stressed that the disaccharide structures of the GAGs depicted in Figure 1 are idealized repeating units and wide variations occur in the charge density (degree of sulfation and distribution of sulfates) and in the ratio of glucuronic acid to iduronic acid. These variations together with variations in the polysaccharide chain length are responsible for the high degree of heterogeneity in this class of macromolecules. The biosynthetic explanation of this inherent heterogeneity and its functional relevance are discussed in the following. Although microheterogeneity in a structural molecule, such as the cartilage proteoglycan, may be inconsequential, this may not be true for heparin or for an informational molecule, such as cell surface heparan sulfate. Thus, the antithrombin III binding to heparin and heparan sulfate is very specific and is mediated by a defined pentasaccharide structure (Atha et al., 1984).

The glycosaminoglycans, with the exception of keratan sulfate and hyaluronic acid, are linked to protein by a core tetrasaccharide (Roden and Smith, 1966) consisting of glucuronic acid, two galactoses, and xylose; the trisaccharide component is illustrated (Fig. 2). The linkages between keratan sulfates and protein are the same as those in glycoproteins and mucins (see Table 1); hyaluronic acid apparently exists as the free polysaccharide.

Proteoglycans are exclusively animal products, and it is now clear that they are present in all animal tissues, although they are particularly abundant in connective tissues, where they are the major component of the ground substances surrounding collagen fibrils. Proteoglycans are probably essential components for holding large masses of organized cells together in tissues of multicellular organisms. In addition to the long-known occurrence of proteoglycans in the extracellular matrix, they are also found on the surface of cells, both as extrinsic molecules and as integral components of the plasma membrane (HööK et al., 1984) and intracellularly in the nucleus (Stoddart, 1979), mitochondria (Kozutsumi et al., 1979), and in various granules.

In the following sections, first the occurrence, primary structure, and specific functions of the five structural classes of glycosaminoglycans, defined by the saccharide composition of their repeated units are briefly discussed. This is followed by discussion of the structure and function of selected, reasonably well-characterized proteoglycans. Next the biosynthesis and assembly of the proteoglycans is dealt with, and, finally, the current research on the molecular biology of these molecules and possible future directions of research in this field are discussed.

A comprehensive review of all the information available on proteoglycans falls outside the scope of this chapter. However, a number of excellent overviews and reviews (Carney and Muir, 1988; Chakrabarti and Park, 1980; Comper and Laurent, 1978; Fransson, 1987; Gallagher et al., 1986; Hardingham, 1981; Hascall, 1981; Hassell et al., 1986; Heinegård and Sommarin, 1987; Höök et al., 1984; Iozzo, 1985; Kolset and Gallagher, 1990; Kuettner and Kimura, 1985; Poole, 1986; Ruoslahti, 1988) on proteoglycans and glycosaminoglycans have appeared, in the last 10 years, that provide additional information.

I. GLYCOSAMINOGLYCANS

A. Hyaluronic Acid

The macromolecule, hyaluronic acid, was first isolated from the vitreous fluid of the eye by Meyer and Palmer (1934). It has since been found widely distributed in various animal tissues, such as cartilage, skin, umbilical cord, cock's comb, and in the synovial fluid of joints. It is believed to be ubiquitous in connective tissues. Hyaluronic acid is the only

Table 1 Structural Details of Glycosaminoglycans and Proteoglycans

Name	Repeating dissaccharide unit: Amino sugar	Repeating dissaccharide unit: Nonnitrogenous sugar	Positions of *O*-sulfate	Linkage: Uronide	Linkage: Hexosaminidic	Other sugars	Linkage to protein
Hyaluronic acid	D-GlcNAc	D-GlcUA		$\beta1\rightarrow3$	$\beta1\rightarrow4$	?	?
Chondroitin	D-GalNAc	D-GlcUA		$\beta1\rightarrow3$	$\beta1\rightarrow4$	D-Gal D-Xyl	Gal$\beta1\rightarrow$3Gal$\beta1\rightarrow$4Xyl$\beta\rightarrow$Ser
Chondroitin-4-sulfate	D-GalNAc	D-GlcUA	4 of GalNAc	$\beta1\rightarrow3$	$\beta1\rightarrow4$	D-Gal D-Xyl	Gal$\beta1\rightarrow$3Gal$\beta1\rightarrow$4Xyl$\beta\rightarrow$Ser
Chrondroitin-6-sulfate	D-GalNAc	D-GlcUA	6 of GalNAc	$\beta1\rightarrow3$	$\beta1\rightarrow4$	D-Gal D-Xyl	Gal$\beta1\rightarrow$3Gal$\beta1\rightarrow$4Xyl$\beta\rightarrow$Ser
Dermatan sulfate	D-GalNAc	L-IdUA or D-GlcUA	4 of GalNAc	$\alpha1\rightarrow3$ $\beta1\rightarrow3$	$\beta1\rightarrow4$	D-Gal D-Xyl	Gal$\beta1\rightarrow$3Gal$\beta1\rightarrow$4Xyl$\beta\rightarrow$Ser
Heparan sulfate	D-GlcNAc or D-$GlcNSO_3$	D-GlcUA or L-IdUA	6 of GlcNAc 2 of IdUA	$\beta1\rightarrow4$ $\alpha1\rightarrow4$	$\beta1\rightarrow4$	D-Gal D-Xyl	Gal$\beta1\rightarrow$3Gal$\beta1\rightarrow$4Xyl$\beta\rightarrow$Ser
Heparin	D-$GlcNSO_3$ or D-GlcNAc	L-IdUA or D-GlcUA	6 of GlcNAc 3 of GlcNAc 2 of IdUA	$\alpha1\rightarrow4$ $\beta1\rightarrow4$	$\beta1\rightarrow4$	D-Gal D-Xyl	Gal$\beta1\rightarrow$3Gal$\beta1\rightarrow$4Xyl$\beta\rightarrow$Ser
Keratan sulfate I	D-GlcNAc	D-Gal	6 of GlcNAc 6 of Gal	Galactosidic $\beta1\rightarrow4$	$\beta1\rightarrow3$	Sialic acid L-Fuc D-Man	GlcNAc$\beta\rightarrow$Asn
Keratan sulfate II	D-GlcNAc	D-Gal	6 of GlcNAc 6 of Gal	$\beta1\rightarrow4$	$\beta1\rightarrow3$	Sialic acid L-Fuc D-Man D-GalNAc	GalNAc$\alpha\rightarrow$Ser/Thr

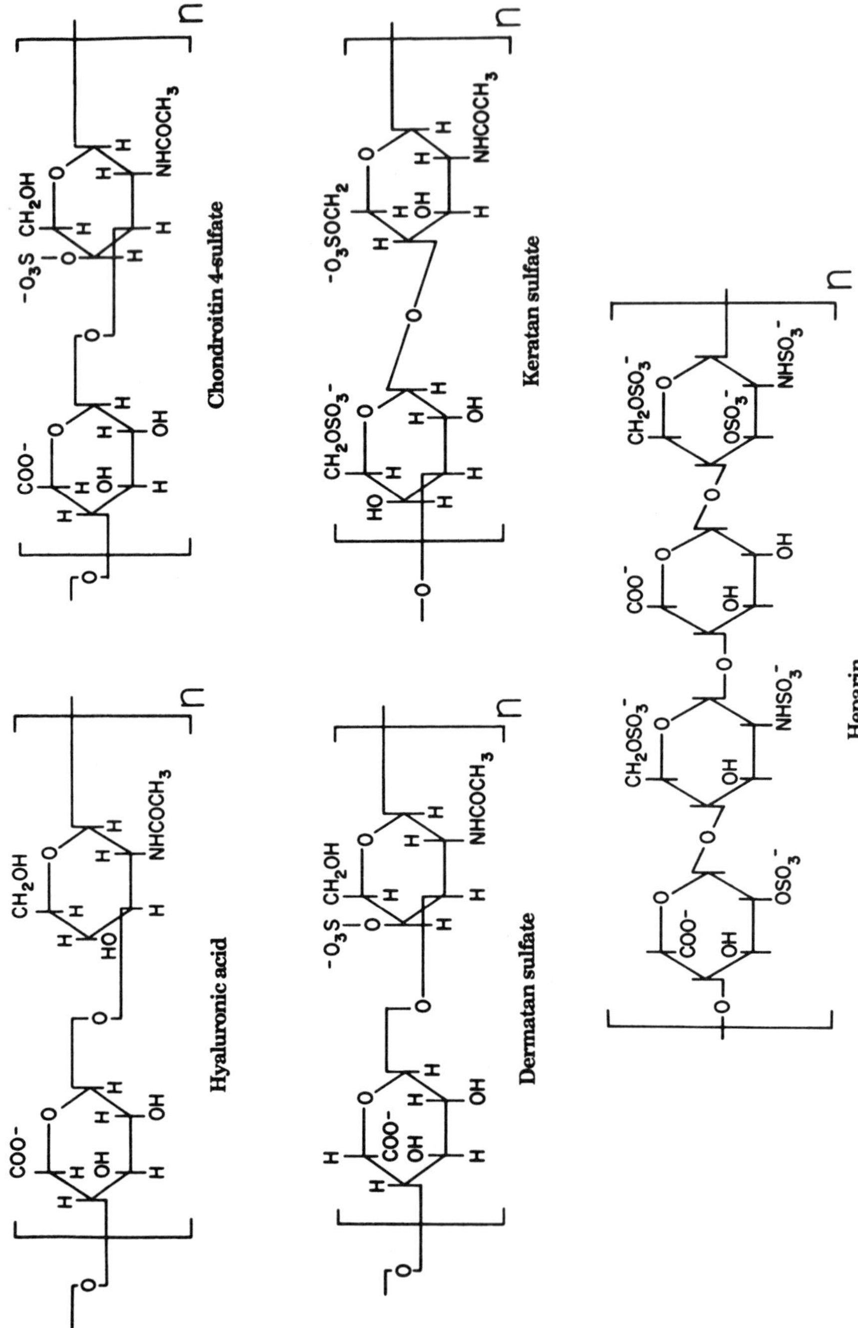

Figure 1 Idealized repeating units of glycosaminoglycans.

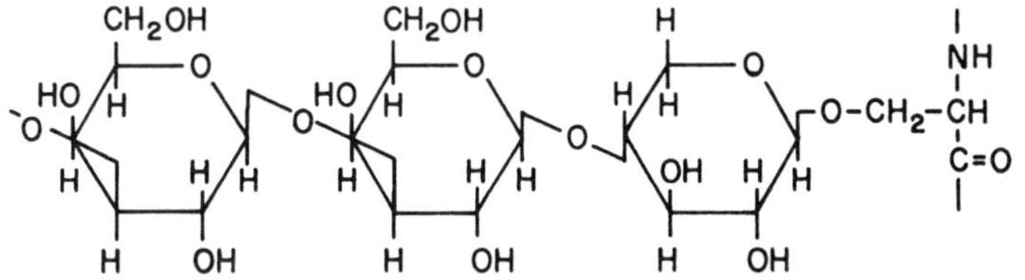

Figure 2 Linkage region: trisaccharide structure linking the chondroitins, dermatan sulfate, heparan sulfate, and heparin to serine of core proteins.

glycosaminoglycan that is also found in bacteria as components of capsules (Kendall et al., 1937). This may be of evolutionary significance, since hyaluronic acid could be the earlier form of all GAGs. It is made up of repeating disaccharide units of *N*-acetyl-D-glucosamine (GlcNAc) and D-glucuronic acid (see Table 1) linked β1→4 and β1→3, respectively. The polysaccharide chains are very large, and relative molecular masses (M_r) of 10^5 to 10^7 have been recorded, depending on the source (Laurent, 1970). Since this is the only nonsulfated GAG, its primary structure is homogeneous, and the heterogeneity of hyaluronic acid from a particular source is solely due to variations in chain length. All indications are that it is not covalently linked to a protein. However, because of the technical limitations of methods to detect a single reducing end of a large polysaccharide (a linear hyaluronic acid of M_r 10^6 will have a total of about 5000 monosaccharides) and because of problems in detecting a few amino acids in the presence of an excess of saccharides, one cannot firmly exclude the presence of a covalently linked small peptide. The role of a peptide as a primer for the initiation of the synthesis of glycogen has now been firmly established (Lomako et al., 1988), and a similar situation may exist for hyaluronic acid (Mikuni-Takagaki and Toole, 1981). However, inhibitors of protein synthesis (puromycin, cycloheximide) do not inhibit biosynthesis of hyaluronic acid (Matalon and Dorfman, 1968), and the investigations of Mason et al. (1982) and Prehm (1983) failed to detect the presence of covalently linked protein in newly synthesized hyaluronate. A wide variety of proteins possessing the ability to bind to hyaluronic acid have been identified; however, the functional relevance of the interactions between many of these proteins and hyaluronic acid is unclear. Some of these interactions may be artifactual, as suggested by the strong interaction of a serum-derived 85-kd protein with hyaluronic acid (Yoneda et al., 1990).

The physical properties and functions of hyaluronic acid are based on its ability to form stable aqueous gels. The gel state of the solution is influenced by pH and by the counterions sodium, potassium, and calcium. The molecule is considered to be a double helix (Arnott et al., 1983) and is heavily hydrated (Sheehan and Atkins, 1983). Because of the very large hydrodynamic domain of hyaluronic acid, even at low concentrations of about 0.1% (w/v), individual molecules begin to overlap (Comper and Laurent, 1978). One molecule of synovial fluid hyaluronate of M_r 8×10^6 in solution is estimated to occupy 10^3 times more space than the unhydrated molecule (Ogston and Stanier, 1951). In a superficial way, it is not difficult to relate the viscoelastic properties of hyaluronic acid solutions to the functions postulated for the molecule. The roles assigned to it include lubrication of synovial surfaces and other similar interfaces of solid and liquid, shock absorber of tissues and cells against mechanical stress and, in cartilage, a role in the assembly of proteoglycan aggregates. Hyaluronic acid is a constituent of the pericellular coat of unfertilized eggs (McClean and Rowlands, 1942) and a variety of other cells (Goldberg and Toole, 1984) where it may function in cell–cell interaction and cell adhesion (Underhill and Dorfman, 1978; Mikuni-Takagaki and Toole, 1980; Turley and Roth, 1980).

B. Chondroitin and Chondroitin Sulfates

Structurally, chondroitin differs from hyaluronic acid in the substitution of *N*-acetylgalactosamine (GalNAc) for GlcNAc and in its lower M_r (2–5 × 10^4). It was isolated from bovine cornea (Davidson and Meyer, 1954) and is believed to be the precursor of chondroitin sulfates (Sugumaran and Silbert, 1989). This polysaccharide could also be prepared from chondroitin sulfates by chemical desulfation. The chondroitin 4-sulfate (previously called chondroitin sulfate A) and chondroitin 6-sulfate (called chondroitin sulfate C) are commonly found forms having sulfate groups predominantly on the C-4 or C-6 position of GalNAc, respectively. These are the glycosaminoglycans that are most widely distributed in animal tissues. Chondroitin 4-sulfate (*chondrin schwefelsäure*) was isolated almost a century ago and was the first member of these macromolecules to be studied. In certain instances, it appears that the same chondroitin sulfate chain may have sulfate on one or the other isomeric position (Seno et al., 1975). All current information indicates that even a preparation of chondroitin sulfate with a sulfate/disaccharide ratio of 1.0 is likely to have an uneven distribution of sulfate ester and, therefore, have unsulfated and disulfated disaccharides. In one study, the region closer to the protein linkage region was found to have fewer sulfate groups than the rest of the chain (Wasteson and Lindahl, 1971).

Several undersulfated and oversulfated chondroitin sulfates have been isolated. In the oversulfated chondroitin sulfates from shark cartilage (named chondroitin sulfate D) and king crab cartilage the excess sulfate groups are located at the C-2 and C-3 positions of the glucuronic acid (Suzuki, 1960; Seno et al., 1974a). On the other hand, chondroitin sulfate E from squid cartilage was shown to contain primarily 4,6-disulfated galactosamine residues (Suzuki et al., 1968). There is hardly any information on what factors control the amount and position of sulfate on the glycosaminoglycan chains. Certainly, the substrate specificities of the sulfotransferases and the regulation of these transferases are two important factors; however, this area of research has been so far neglected. The biological implications of the position of the sulfate as well as the intrachain variation in the position of the sulfate are also unknown.

C. Dermatan Sulfate

Dermatan sulfate was first isolated from pig skin by Meyer and Chaffee (1941) and named chondroitin sulfate B because of the recognition that it is an isomer of chondroitin sulfates A and C. Structurally, dermatan sulfate has a variable proportion of α-L-iduronic acid in place of its 5-epimer, β-D-glucuronic acid. In having variable proportions of the uronic acid epimers and in the presence of some sulfate on the C-2 position of iduronic acid, dermatan sulfate resembles the heparan sulfate/heparin class of glycosaminoglycans. Because of this additional sulfate, the degree of sulfation of dermatan sulfate is generally higher than that of the chondroitin sulfates. In fact, certain preparations of dermatan sulfate show some anticoagulant activity, and one other previous name for this glycosaminoglycan is β-heparin (Marbet and Winterstein, 1951). The distribution of dermatan sulfate differs somewhat from that of the chondroitin sulfates in that it occurs in high concentrations in fibrous connective tissues, such as skin, intestine, blood vessels, tendon, and heart valves. From the foregoing discussion, it is clear that, even though dermatan sulfate is a galactosaminoglycan like chondroitin, it has sufficient difference to be considered as a class by itself.

D. Heparan Sulfate and Heparin

The structurally related glycosaminoglycans, heparan sulfate and heparin, consist of repeating disaccharide units of D-glucosamine and either β-D-glucuronic acid or α-L-iduronic acid. The linkage between the hexosamine and uronic acid is uniformly 1→4, rather than alternating 1→4 and 1→3 as in the other glycosaminoglycans. Other unique features of these glycosaminoglycans are the α-D-linkage between the hexosamine and the uronic acid and the presence of *N*-sulfated hexosamines (sulfamido glucosamine) in addition to *N*-acetylglucosamine. The distinction between heparan sulfate and heparin is arbitrary and is based on both structural and functional features. Structurally, heparan sulfate has fewer iduronic acid residues (40–50% versus 80%) and fewer sulfate ester groups (0.5–1.5 versus 1.5–2.5 per disaccharide) than heparin. In heparin, the major portion of glucosamine residues are *N*-sulfated and, in addition, most iduronic acid and glucosamine residues are *O*-sulfated at the C-2 and C-6 positions, respectively. Some glucosamine residues are also *O*-sulfated at the C-3 position, believed to be critical for the anticoagulant activity of heparin (Lindahl et al., 1980). Therefore, certain disaccharide units of heparin may have as many as four sulfate groups, in addition to the carboxyl group, making heparin the most highly negatively charged molecule in the animal. Heparan sulfate contains blocks of *N*-acetylglucosamine residues and, accordingly, deaminative cleavage by nitrous acid which attacks the *N*-sulfated glucosamine residues, generates fragments of much larger size than would be expected from a more even distribution (Linker and Hovingh, 1973). As with the chondroitins, the factors regulating the amounts and distribution of sulfate groups and the ratio of the two uronic acids are unknown.

The differences in the distribution and function of heparin and heparan sulfate are more striking than the differences in their structural features. Heparin is primarily present in mast cell granules as proteoglycans and is released into the extracellular space as free chains. In contrast, heparan sulfate occurs widely in a variety of cells and is generally present on the cell surface and in the extracellular matrix as proteoglycans of various molecular mass. However, intracellular forms of heparan sulfate are also known that are of relatively low molecular mass (Kraemer, 1971; Yanagishita and Hascall, 1984) and are believed to be degradation products of the extracellular variety. In addition to being a potent blood anticoagulant, heparin also exhibits marked antilipemic activity. Because of its ubiquitous cell surface location, heparan sulfate has been implicated in a number of biological processes, such as cell–cell recognition, cell–cell and cell–matrix adhesion, extracellular matrix organization, control of differentiation and cell growth, growth factor receptor and so on. In addition to vertebrate tissues, it is also present in invertebrates (Nader et al., 1984). Recent studies by Nader et al. (1988) suggest that the key structural features of heparan sulfate have been maintained throughout evolution. Both quantitative and qualitative changes in cell surface heparan sulfate have been noted as a result of malignant transformation (Underhill and Keller, 1975; Winterbourne and Mora, 1981; Bhavanandan, 1981). However, the actual correlation between these alterations and neoplastic transformation remains to be elucidated. Heparan sulfate is also found as proteoglycans in specialized intercellular matrices such as basement membrane.

E. Keratan Sulfate

Keratan sulfate, also first isolated by Karl Meyer (Meyer et al., 1953), is most unusual in that it has features common to both proteoglycan/GAGs and glycoproteins. The repeating sulfated disaccharide structure is its major claim for inclusion with the glycosaminoglycans, even though the uronic acid in the disaccharide is replaced by galactose. The sulfate groups are located on varying proportions of *N*-acetylglucosamine and galactose on the C-6 position (Bhavanandan and Meyer, 1966). Sulfate stoichiometry is somewhat variable and can exceed one per disaccharide unit. The unsulfated repeating unit (*N*-acetyllactosamine) is present in glycoproteins referred to as polylactosaminoglycans. Thus the anion transport glycoprotein of human erythrocytes (erythroglycan) (Järnefelt et al., 1978) could be considered a desulfated keratan sulfate proteoglycan. The M_r of the saccharide chains of corneal keratan sulfate ranges from a relatively low 4000, to about 22,000 (Stuhlsatz et al., 1981). Some structural features of keratan sulfate similar to those of glycoproteins are the presence of sialic acids, fucose (Bray et al., 1967), and mannose (Bhavanandan and Meyer, 1968). Branch points on the linear chain consisting of galactose are also indicated by the presence of an excess of galactose over glucosamine (Gregory and Roden, 1961; Bhavanandan and Meyer, 1967, 1968). The most striking resemblance to glycoprotein is the manner in which the keratan sulfate is linked to protein. Two types of keratan sulfate are known that can be distinguished based on the linkage to protein. Keratan sulfate I, so far found only in cornea, is linked to protein by GlcNAc→asparagine (Seno et al., 1965). It has been suggested that the mode of biosynthesis of corneal keratan sulfate proteoglycan is similar to that of the serum-type glycoproteins, since tunicamycin inhibits its synthesis (Hart and Lennarz, 1978). Studies on the biosynthesis of keratan sulfate I are of interest, because patients with macular corneal dystrophy fail to synthesize mature keratan sulfate proteoglycan (Hassell et al., 1980). Keratan sulfate II is the second type and has been found so far in two issues of skeletal origin (cartilage and intervertebral disks). Keratan sulfate II chains are linked to protein by *O*-glycosidic linkages between *N*-acetylgalactosamine and serine or threonine (Anderson et al., 1964). Keratan sulfate II cooccurs with chondroitin sulfate in the proteoglycan of cartilage. The limited distribution of keratan sulfate in contrast to the other GAGs is of interest. However, the significance of this cannot be ascertained until the biosynthetic relationship of keratan sulfates and polylactosaminoglycans is better understood. As mentioned previously, these two glycoconjugates may belong to a single family, bridging the gap between proteoglycans and glycoproteins. The full characterization of the keratan sulfatelike glycosaminoglycan isolated from other tissues [for example from rat brain by Vitello et al. (1978)] will be helpful.

The heterogeneity of keratan sulfate is high owing to variations in chain length, in the amount and distribution of sulfate groups, and in the presence of minor sugars (sialic acid, fucose, mannose).

II. PROTEOGLYCANS

A. Core Proteins

Investigations of the core proteins of proteoglycans have lagged behind in the past, primarily because of their inaccessibility to biochemical analysis and to difficulties in obtaining undegraded forms by chemical or enzymatic deglycosylation. More recently, molecular biological techniques have been utilized to successfully clone the cDNAs and, thereby, determine the sequence of the core proteins of several proteoglycans (Ruoslahti, 1988). In addition to providing information on sequence homologies and other relationships between these primary gene products, these investigations, in the long run, would also yield material for full physicochemical characterization of these proteins. It can be hoped that information on core protein families would provide a rigorous basis for the classification of the parent proteoglycans. Certain proteins have been identified that appear to occur as "part-time" proteoglycans in that they exist both as unsubstituted proteins and as proteoglycans (Fransson, 1987).

Currently, the proteoglycans are mainly distinguished either on the basis of the predominant glycosaminoglycan side chain (e.g., chondroitin sulfate proteoglycan, dermatan sulfate proteoglycan) or by the tissue of origin (cornea proteoglycan, platelet proteoglycan). The size of the proteoglycan (large, small) and its topographical distribution (extracellular, intracellular, cell surface) also may be used in describing these molecules. Since it appears that the type of proteoglycans is tissue-specific, rather than species-specific (Toledo and Dietrich, 1977), the animal of origin is normally not included in describing the proteoglycan. However, until the universality of this is established, it will be best to be cautious in making firm conclusions. For the purpose of this article, the proteoglycans will be referred to mainly by their tissue of origin. When necessary, other characteristics such as the predominant glycosaminoglycan present, the localization in the tissue (e.g., cell surface versus basement membrane), and overall properties or structural features will be mentioned. Only a few selected, reasonably well-characterized and representative proteoglycans are dealt within the following.

B. Cartilage Proteoglycans

The large, aggregating chondroitin sulfate-rich proteoglycan from cartilage is the most extensively investigated member of this class of glycoconjugates. This is partly because it is present in relatively high amounts, sometimes accounting for about 10% of the dry weight of cartilage. In addition to the large proteoglycans that have the ability to aggregate in the presence of hyaluronic acid, other smaller nonaggregating proteoglycans are also present in cartilage (Heinegärd et al., 1981). An additional reason for the special interest in the cartilage proteoglycans is because of the destruction of articular cartilages in arthritic diseases. Age-related changes of the proteoglycans are also known; for example, the overall keratan sulfate (KS) concentration of cartilage appears to vary directly with age (Kaplan and Meyer, 1959; Seno et al., 1965; Bray et al., 1967; Brandt and Muir, 1969; Roughley and Mort, 1986).

Early (before the 1950s) research on chondroitin sulfate (CS) and other glycosaminoglycans was done on the basis that these polysaccharides existed free or were associated in the tissue with proteins by noncovalent, primarily ionic, linkages. The method utilized to purify GAGs included proteolysis and alkali extraction, both of which inadvertently removed the protein component of the macromolecules. However, studies by many investigators during the early 1950s indicated that chondroitin sulfate chains were covalently attached to protein. Finally, the work of Muir (1958), Gregory et al. (1964), and Anderson et al. (1965) firmly established that chondroitin sulfates were indeed attached via xylose residues to serine of the core protein. As the result of intensive research over a period of about 20 years, an idealized "bottle-brush" model for the structure of the large aggregating proteoglycan of cartilage was developed (Fig. 3). In this model, the proteoglycan structure contains a protein core with three distinct domains (Hardingham,

1981). These are (1) the globular NH_2-terminal region that is low in GAGs, but has oligosaccharides, and is involved in interaction with hyaluronic acid and link glycoproteins; (2) a keratan sulfate-rich region (containing 30–60 chains per proteoglycan monomer); (3) a long extended region that is rich in serine, glycine, and proline and contains most of the chondroitin sulfate chains (50–100 per proteoglycan monomer). It is essential to stress that there is no sharp demarcation and that the regions occur with mixtures of the GAGs and saccharides.

Both *N*- and *O*-linked oligosaccharide units are present in the mature proteoglycan (Thonar and Sweet, 1979; De Luca et al., 1980). The *N*-linked structures appear to be

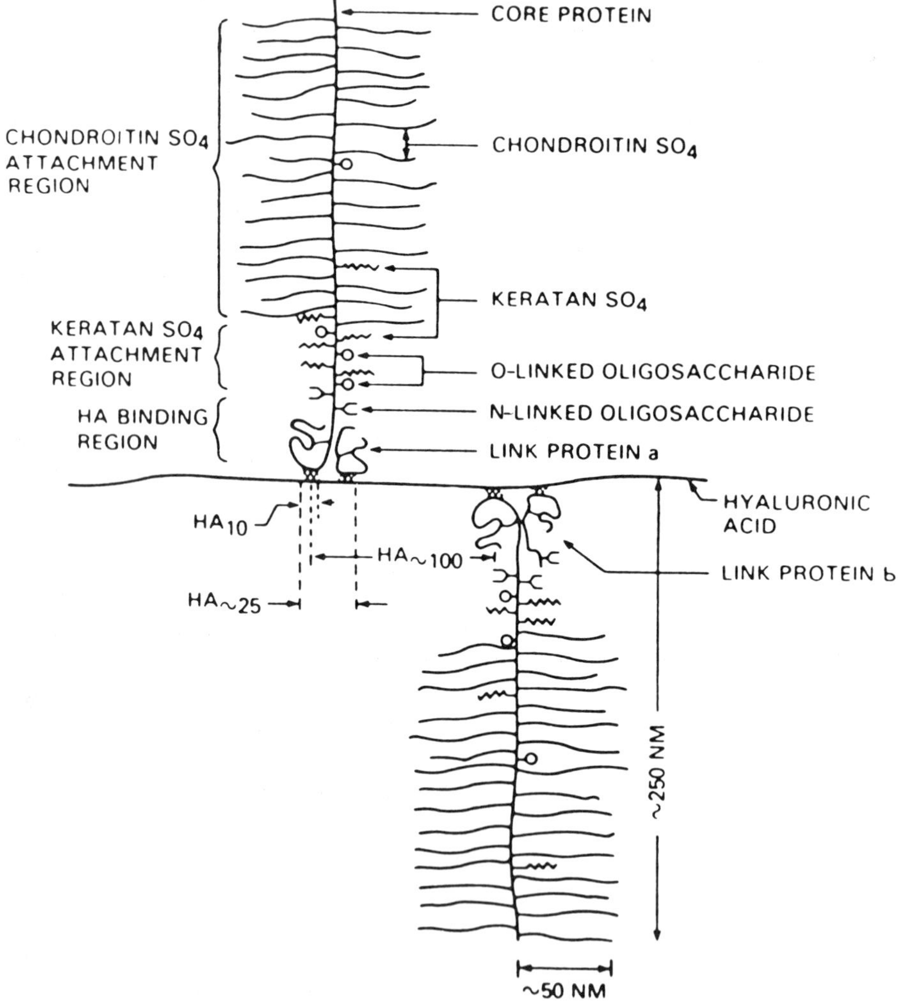

Figure 3 Structural model of cartilage protein aggregates showing interaction of proteoglycan monomer units with hyaluronic acid and link proteins. [From Hascall, V. C. (1981). Proteoglycans: Structure and function. In: *Biology of Carbohydrates*, Vol. 1 (V. Ginsburg and P. Robbins, eds.). Reprinted with permission of John Wiley & Sons, Inc., New York, pp. 1–49.]

located exclusively in the NH_2-terminal globular region of the protein and are of the complex triantennary type; up to 15 units may be present. Saccharide heterogeneity exists in this group, with about half of the internal *N*-acetylglucosamine residues fucosylated (Nilsson et al., 1982). A number of *O*-linked units are present, distributed more or less randomly outside of the hyaluronic acid binding region. Structures of these oligosaccharides, determined for this and other proteoglycans, indicate a broad variety in the di- to hexa- range; all are linked to the protein via *N*-acetylgalactosamine units (Nilsson et al., 1982; Santer et al., 1982).

The results of electron microscopic examination of proteoglycan monomers and aggregates are in general agreement with these models (Rosenberg et al., 1970). By molecular biological techniques, the amino acid sequences of the core proteins of this class of proteoglycans have been deduced within the last 5 years. The amino acid sequence of the core proteins of proteoglycans from rat chondrosarcoma (Doege et al., 1987), chick sternal cartilage (Tanaka et al., 1988), and bovine articular cartilage (Oldberg et al., 1987) appears to be conserved, having about 80% homology. An unexpected finding, with implications for function of the proteoglycans, is that the COOH-terminal segment of chicken cartilage proteoglycan (Sai et al., 1986) and of rat cartilage proteoglycan (Halberg et al., 1988) have striking sequence homology with chicken hepatic lectin.

Early determination of the core protein M_r, inferential in nature, gave values of about 2×10^5. Cell-free translation of mRNA indicates that the primary product has an M_r of about 350,000. The extensive processing that occurs following translation makes difficult an accurate measure of the core protein size of the mature proteoglycan, but working values somewhat in excess of 200,000 seem reasonable (Hascall and Sajdera, 1970; Hascall and Riolo, 1972). The hyaluronic acid-binding region begins near the NH_2-terminus in a globular domain of about 50,000 M_r (Heinegard and Hascall, 1974a; Hascall and Heinegard, 1974); a second globular region precedes the GAG-rich region that stretches for about 1500 amino acids, beginning at about residue 750 (Matthews, 1971; Heinegard and Hascall, 1974b). There is also a sparsely or nonglycosylated COOH-terminal globular domain of about 250 residues. Within the GAG-rich region, up to 80% of the seryl residues may carry GAG chains. Minimalist numerology calculates an M_r of 2×10^6 for the final product. A recently described modification is the presence of several, three to four, phosphate ester residues, presumably on hydroxy amino acids (Peters et al., 1981; Oegema et al., 1984; Anderson and Schwartz, 1984). The protein core of the aggregating proteoglycan is initially synthesized in the chondrocytes as a precursor of approximately 370 kd (Kimura et al., 1981). This protein is then glycosylated with oligosaccharides (both *N*- and *O*-linked) and glycosaminoglycans (chondroitin sulfate and keratan sulfate) before being secreted into the extracellular matrix. Processing of the core protein occurs either intracellularly during glycosylation or extracellularly, since the average size of the core proteins of secreted proteoglycan is about 200 kd, but the details of the processing steps are not known. Once secreted, the proteoglycans assemble into aggregates by interacting with hyaluronic acid. The reversible interaction with hyaluronic acid is stabilized by the link glycoproteins of 46 kd and 51 kd. The primary structure of one of the link proteins (43 ± 2 kd) from rat chondrosarcoma has been determined (Neame et al., 1986). Interestingly, the amino acid sequence of the link protein was very similar to that of the hyaluronic acid-binding region of the proteoglycan core protein (Neame et al., 1987). However, there are differences in their glycosylation patterns, and the two proteins are immunologically distinct (Caterson et al., 1979). Choi et al. (1985) were able to separate the two link proteins by WGA affinity chromatography and showed that, although the 51-kd link protein had a pronounced effect on the size of the aggregate formed, it alone did not effectively stabilize the aggregate. It was concluded that the 46-kd protein was essential to stabilize the proteoglycan aggregates.

Given the enormous structural diversity present in the final product, it is unlikely that a completely defined structure can be elucidated for any cartilage proteoglycan. Relatively simple-appearing problems may not yield easily. Thus, we do not know if a given serine residue is always glycosylated or, if it is, what type of substituent is allowed or indeed programmed. The control of KS/CS content as well as distribution is an intriguing regulatory question, especially since there is no evidence for multiple genes. Control of the distribution and structural choice of the GalNAc-linked oligosaccharide units is also unknown and likely to remain so. Since the biological functions of the cartilage proteoglycans appear to be primarily physical rather than informational (contrast structures such as heparin), the variations implied by the uncertainties noted in the foregoing are probably tolerable.

Cartilage is a highly specialized tissue, subject to considerable load stress under normal conditions. Biological properties range from strictly support functions (e.g., nasal septum) to nearly friction-free surface behavior in certain joints. Largely acellular, the characteristic physical properties are due to the combination of proteoglycan aggregates and collagen fibers that represent nearly the entire organic component of the extracellular matrix. The concentration of the proteoglycan ranges upward of 5% and the mechanical entrapment of these macromolecules in the nonelastic collagen mesh restricts their tendency to shrink or swell owing to solvent movement occasioned by the compressive loads exerted on the tissue. The physical properties of cartilage may be viewed within a cyclic set of events initiated by compression and solvent loss, increased expansion pressure owing to the hydrophilicity of the proteoglycan, abatement of load, followed by solvent uptake to reestablish the starting condition. Even in non–load-bearing tissues, resiliency and restoration of the initial dimensions follow distortion and characteristic features, again modulated by the proteoglycan–collagen network (Maroudas, 1968, 1975; Comper and Laurent, 1978; Myers et al., 1984).

An additional function for the proteoglycan in cartilage may relate to its intrinsic ion-binding capacity. A mature cartilage proteoglycan monomer of a nominal 2×10^6 M_r will have an aggregate charge of approximately –10,000. The maintenance of electrical neutrality requires the presence of appropriate counterions, and it has long been known that the entire and considerable ion-binding capacity of this tissue can be accounted for by its proteoglycan content. Furthermore, the relevant functional groups of the GAG chains, combined with their physical restriction (i.e., attachment to a common polypeptide core), serve to impart ion selectivity as well. Accordingly, divalent cations, especially calcium, are preferentially bound. Under normal conditions, the high affinity of the proteoglycan for calcium serves to prevent mineralization in the form of calcium phosphate or hydroxy apatite precipitation. Degradation of the proteoglycan could result in loss of this property and resultant foci of abnormal calcification; controlled management of the same phenomenon may be involved in normal bone deposition (Woodward and Davidson, 1968).

A specific role may also be envisioned for the link proteins present in cartilage. The physical distortions associated with normal tissue function may result in transient displacements that affect the interaction between the hyaluronate-binding domain of the proteoglycan and the hyaluronic acid ligand. Stabilization of this interaction or facilitation of proper orientation may require these auxiliary proteins.

Additional small cartilage proteoglycans containing chondroitin sulfate, dermatan sulfate, or keratan sulfate chains and referred to as decorin, biglycan, and fibromodulin, have also been reported. The cDNA clone corresponding to human bone biglycan was sequenced and found to code for a protein containing 12 tandem repeats of 24 amino acid residues. Furthermore, the protein sequence of biglycan showed homology with that of decorin (Fisher et al., 1989). The deficient expression of the gene coding for decorin in skin fibroblast cultures of individuals with Marfan's syndrome, an autosomal dominant connective tissue disorder, is an illustration of the importance of proteoglycans as structural components of connective tissue (Pulkkinen et al., 1990).

C. Cornea Proteoglycans

Two populations of proteoglycans, one containing keratan sulfate and the other containing dermatan sulfate, have been identified in studies carried out primarily with bovine (Axelsson and Heinegard, 1978; Keller et al., 1981) and monkey cornea (Nakazawa et al., 1983). In contrast with the large cartilage proteoglycans, these have much smaller hydrodynamic size, which is thought to be essential for their interaction with the precisely packed collagen fibrils and for the maintenance of optical transparency of the cornea (Hassell et al., 1986). The keratan sulfate proteoglycan that is considered to be unique to cornea is estimated to have a core protein of M_r about 53,000 and one to three keratan sulfate I chains in addition to several oligosaccharides, all linked to asparagine. Studies on corneal proteoglycans of embryonic chicken by Midura and Hascall (1989) also revealed a similar keratan sulfate proteoglycan having an overall size of about 100 kd and core protein size of 51 kd. Recent work shows that corneal explant cultures, derived from patients with macular corneal dystrophy, synthesize lactosaminoglycan glycoproteins instead of keratan sulfate proteoglycan (Midura et al., 1990). Interestingly, the core protein of the glycoprotein has hydrophobicity and molecular mass similar to the core protein of normal corneal keratan sulfate proteoglycan. It appears that the corneal clouding in this genetic disorder is the result of undersulfation of keratan sulfate I proteoglycan. The local precipitates causing the corneal clouding could be due to the poor solubility of the undersulfated proteoglycan. In this context, the observation that a major portion of keratan sulfate I became insoluble on desulfation (Bhavanandan and Meyer, 1968) could be of relevance. The basis (enzymic defect) for the undersulfation is not known, but it is considered unlikely that the undersulfation is due to the lack of a functional sulfate transferase (Hascall, 1981).

The dermatan sulfate proteoglycan from monkey cornea has an M_r of about 150,000 consisting of one or two polysaccharide chains of about 55 kd on a core protein of about 40 kd. In contrast with the dermatan sulfate from other sources, corneal dermatan sulfate is glucuronic acid-rich, considerably less sulfated (less than one residue per disaccharide), and is of higher molecular mass. The core proteins of the dermatan proteoglycan and keratan proteoglycan from cornea are distinct, as indicated by the difference in their size and by the lack of immunological cross-reactivity (Hassell et al., 1986). That the dermatan sulfate and keratan sulfate might reside on separate gene products is not surprising, since the protein signal for the initiation of these polysaccharide chains must be distinct.

It has been noted that the dermatan sulfate proteoglycan synthesized by macular dystrophy cornea has more sulfate, which might be an attempt by the tissue to compensate for the undersulfation of the keratan sulfate proteoglycan (Nakazawa et al., 1984).

D. Mast Cell (Heparin) Proteoglycan

Rat mast cell proteoglycan is an interesting molecule, having a fairly small core protein of M_r 20,000 and a repeating Ser-Gly sequence. About two-thirds of the serine residues are substituted by high molecular mass (M_r 80,000–100,000) heparin chains to yield a densely substituted molecule having an overall M_r of about 10^6 for the proteoglycan (Robinson et al., 1978). The smaller, biologically active heparin molecules are apparently produced by the action of mastocytosomal endoglycosidases before secretion from the mast cells. Heparin-containing proteoglycan molecules are also present in the intracellular granules of leukocytes (Tantravahi et al., 1986). Thus, most of the commercial porcine or bovine heparin are apparently endoglycosidase-mediated degradation products of heparin proteoglycans. The native rat mast cell proteoglycan exhibits lower anticoagulant activity, when compared with commercial heparin consisting of short polysaccharide chains.

E. Hepatic Proteoglycans

Compared with cartilage, cornea, or skin, the amount of proteoglycans and GAGs in organs such as liver, brain, and kidney, is low. However, these organs have a variety of proteoglycans and, in liver, heparan sulfate is the primary GAG associated with hepatocytes. The hepatocyte heparan sulfate proteoglycans are present both intracellularly (cytosol, nucleus, mitochondria) and extracellular on the cell surface. Two species of cell surface heparan sulfate proteoglycans have been identified, one an extrinsic molecule that can be released by treatment with heparin, and the other an intrinsic molecule anchored to the lipid bilayer (Kjellen et al., 1980, 1981). The latter was extracted by detergents from rat liver membranes and shown to have a core protein of M_r ~35 kd, with four to five heparan sulfate chains of M_r 14 kd. Ishihara et al. (1987) have shown that this proteoglycan is anchored by phosphoinositol to the lipid bilayer. The precise functions of the various species of proteoglycans in the hepatocytes have not been established; proposed functions include cell adhesion, growth control, receptors, and control of ion flux. There is good evidence to support the cell attachment–cell adhesion function of cell surface heparan sulfate proteoglycan. This function is mediated by the ability of this molecule to self-interact (Fransson et al., 1983) as well as to interact with matrix macromolecules such as fibronectin, laminin, and collagen. Chiarugi and Vannucchi (1976) proposed that cell surface heparan sulfate could function as a control element in eukaryotic cell growth.

F. Basement Membrane Proteoglycans

Earlier studies by Kanwar and Farquhar (1979) demonstrated the presence of heparan sulfate proteoglycan in glomerular basement membrane. The removal of this molecule by treatment with heparitinase increased permeability of the membrane to ferritin, indicating the importance of heparan sulfate proteoglycans to selective filtration of macromolecules (Kanwar et al., 1980). Furthermore, it has been noted that in human congenital nephrotic syndrome there is a defect in the production of heparan sulfate in the glomerular basement membrane (Vernier et al., 1983). The glomerular proteoglycan has heparan sulfate chains that are clustered in a very limited segment of the polypeptide and *N*-sulfated glucosamine is confined to regions of the polysaccharide remote from the linkage end. Its M_r ranges from 130 to 250 kd, and the core protein constitutes about 70% by weight of the molecule

(Parthasarathy and Spiro, 1984). An unusual feature of the core protein is the large number of half-cystine residues, about 6.1 mol%.

Hassel et al. (1980) isolated a heparan sulfate proteoglycan, which was much larger (~750 kd) than the glomerular proteoglycan, from the basement membrane of the Engelbreth–Holm–Swarm sarcoma. About 50% by weight of this molecule was estimated to be heparan sulfate chains of an average 70-kd size. Antibody prepared against this proteoglycan reacted with basement membranes in various tissues (skin, kidney, cornea), suggesting that most animal basement membranes contain heparan sulfate proteoglycans. A partial amino acid sequence of the core protein of the basement membrane heparan sulfate proteoglycan has been recently deduced from cDNA clones. The deduced sequence revealed two different domains, one a cysteine-rich region with sequence homology to laminin, and the other, with looping structures, homologous to the neural cell adhesion molecule (N-CAM) and other proteins of the immunoglobulin gene super family (Noonan et al., 1988).

G. Cell Surface Proteoglycans

Historically, histochemists have noted the presence of a dense, polyanionic "fuzz" on animal cell surfaces, the chemical nature of which was not entirely understood. From current information, it is clear that this fuzz is composed of various glycoconjugates and, hence, is now referred to as the *glycocalyx.* Depending on the tissue and cell type, the glycocalyx may be composed of membrane glycolipids, intrinsic and extrinsic (sialo)glycoproteins, mucins, hyaluronic acid, and proteoglycans. The work by Kraemer (1971) and others in the early 1970s suggested that proteoglycans may be a ubiquitous component of most animal cell membranes or cell surfaces. Studies, mostly on cultured cells, have substantiated this suggestion, and it can now be stated with certainty that all animal cells, with the exception of erythrocytes, produce cell surface-associated glycosaminoglycans/proteoglycans. As discussed in the foregoing for hepatocytes, two types of cell surface proteoglycans are known: namely, intrinsic and extrinsic types. Presently, the evidence for most of the intrinsic membrane proteoglycans is based primarily on indirect evidence, such as extractability with detergents, interaction with hydrophobic gel matrices (Octyl-Sepharose), and so on. In hepatocytes (Ishihara et al., 1987) and Schwann cells (Carey and Stahl, 1990), direct evidence has been obtained for phospholipid anchoring of proteoglycans. Confirmatory evidence for this as well as direct evidence for possible anchoring of proteoglycan to the lipid bilayer by a hydrophobic intramembranous peptide domain has yet to be obtained.

Concerning the extrinsic proteoglycans of the cell surface, it is currently difficult to make a clear distinction in a functional sense between those that are in the process of being secreted and, therefore, transiently present on the surface, from those that are true cell surface components. There is evidence for the association of glycosaminoglycans with receptors. Thus, in hepatocytes, heparan sulfate proteoglycans are bound to cell surface receptors (about 4×10^6 per cell are estimated to be present; Kjellen et al., 1980), but this binding is not a prelude to endocytosis and degradation as it is in fibroblasts (Prinz et al., 1978). These proteoglycans remain associated with the cell surface after trypsinization (Kjellen et al., 1980), but are releasable by heparin and, to a lesser extent, by dermatan sulfate. Specific-binding sites (receptors) for hyaluronic acid have been detected in SV40-transformed 3T3 cells by Underhill and Toole (1980). The nature of the cell surface proteoglycans of hepatocytes was briefly discussed earlier. A few other examples of cell surface proteoglycans are given in the following.

1. *Human Melanoma Cell Surface Proteoglycans*

Although most of the other cell surface proteoglycans investigated are of the heparan sulfate type, the one from melanoma cells is a chondroitin sulfate proteoglycan. This was discovered as a result of a search for monoclonal antibodies specific for human melanoma (Bumol and Reisfeld, 1982). The epitope recognized by the specific antibody turned out to be the core protein of a cell surface chondroitin sulfate proteoglycan. This molecule has been characterized as consisting of a core protein of about 250 kd, in addition to chondroitin sulfate chains and asparagine-linked complex oligosaccharides (Reisfeld et al., 1985). The antibody for this proteoglycan interfered with the melanoma cell adhesion and spreading, suggesting a role for this molecule in cell–matrix interaction and attachment.

2. *Ovarian Granulosa Cell Proteoglycans*

Studies with rat ovarian granulosa cells have shown the synthesis of two cell surface proteoglycans, one with dermatan sulfate chains (DS-PG-II) and one with heparan sulfate chains (HS-PG-I) (Yanagishita and Hascall, 1987). Both proteoglycans have a similar core protein of about 230 kd containing four to five asparagine-linked and 20 to 30 serine- or threonine-linked oligosaccharides. Although the former has four to five dermatan sulfate chains, the latter has five to six heparan sulfate chains. These proteoglycans appear to be intrinsic membrane components, but the nature of anchorage is not yet elucidated. Both proteoglycans are metabolized in a similar manner, with about 30% of the truncated proteoglycan being released into the medium and the balance internalized and degraded in lysosomes by two distinct degradation pathways (Yanagishita and Hascall, 1984).

These cells also synthesize another dermatan sulfate proteoglycan (DS-PG-I), which is secreted into the medium in vitro and into the follicular fluid in vivo. This molecule is unusual and interesting. It is very large, with a hydrodynamic size estimated to be similar to that of the cartilage monomer proteoglycan, M_r 2–3 × 10^6. The core protein is about 400 kd and is substituted with about 20 dermatan sulfate chains, 20–30 asparagine-linked and 200–300 serine- or threonine-linked oligosaccharides. The core protein lacks the binding sites for hyaluronic acid characteristic of the large cartilage proteoglycan. In its very large size and in the presence of several hundred *O*-linked saccharides, this molecule resembles epithelial mucins, rather than proteoglycans. Analysis of the distribution of the dermatan sulfate chains and the oligosaccharide would be interesting and would also confirm that this is indeed a hybrid proteoglycan–mucin molecule and not two molecules copurifying on DEAE-Sephacel and Sepharose CL-4B for unknown reasons. This molecule is considered to be a specialized proteoglycan characteristic of the ovarian granulosa cells involved in follicle development and ovulation (Yanagishita and Hascall, 1984).

3. *Human Platelet Proteoglycan*

The presence of glycosaminoglycans associated with the surface membranes and α-granules of platelets has been known for some years (Barber et al., 1972). A chondroitin sulfate proteoglycan has been purified from human platelets by Okayama et al. (1986) and characterized as a 136-kd molecule containing four chondroitin 4-sulfate chains of 28 kd. Alliel et al. (1988) have recently determined the complete amino acid sequence of the core protein by a combination of amino acid sequence analysis and cDNA sequencing. The core protein has an M_r of 14,641 and consists of 131 amino acids. The core has eight Ser–Gly repeats and a high degree of homology to the core protein of the rat yolk-sac tumor proteoglycan (pg19). It is postulated that this proteoglycan is the source of the inhibitor of complement subcomponent Clq previously detected in human serum (Silvestri et al., 1981). The platelet proteoglycan is also believed to play a role in packaging of platelet factor 4 in α-granules and in maintaining the granular structure.

4. *Mammary Epithelial Cell Proteoglycans*

The cell surface proteoglycan (named *syndecan*) of normal mouse mammary epithelial cells is unusual in containing both chondroitin sulfate and heparan sulfate. The core protein consists of a hydrophobic domain that is presumably intercalated into the plasma membrane. The GAG-rich domain is released intact from the cells by trypsin and is similar to proteoglycan found in the culture medium. Treatment of this trypsin-released proteoglycan with chondroitinase and heparitinase produced a core protein of M_r 53 kd. The intact proteoglycan is estimated to be about 250 kd and contain one or two chondroitin sulfate chains (M_r ~17 kd) and up to four heparan sulfate chains (M_r ~36 kd) (Rapraeger et al., 1985). The cDNA core protein of syndecan has been recently cloned and sequenced (Saunders et al., 1989).

This molecule is considered to be important for the interaction and attachment of these epithelial cells to the matrix, since it binds fibrillar collagen and fibronectin. Sanderson and Bernfield (1988) recently showed that syndecan from simple and stratified epithelia have identical core proteins, but differ in GAG content. The proteoglycan from uterus (simple) epithelia has an M_r of ~160 kd and that from vaginal (stratified) epithelia has an M_r of ~92 kd. They conclude that this cell surface proteoglycan shows a tissue-specific structural polymorphism caused by distinct posttranslational modifications.

The proteoglycans of a normal human breast cell line (HBL-100) and of a malignant human breast cell line (MDA-MB-231) have been purified and partially characterized (Gowda et al., 1986a,b). Both cell lines secreted approximately 70% of the synthesized proteoglycans, which were composed of 20% heparan sulfate and 80% chrondroitin sulfate proteoglycans. The MDA cell line secreted heparan sulfate proteoglycans of two different hydrodynamic sizes, in contrast with the HBL cell line, which secreted molecules of only one size. Of the plasma membrane-associated proteoglycans, 80 and 50% were of the heparan sulfate species in the HBL and MDA cells, respectively. There was no evidence for a proteoglycan containing both heparan sulfate and chondroitin sulfate, similar to that found in the mouse system. The heparan sulfate as well as chondroitin sulfate proteoglycans of both cell lines contain a series of *O*-linked oligosaccharides ranging in size from di- to hexasaccharides. The structures of two neutral and seven acidic oligosaccharides, including a novel sulfated sialyl oligosaccharide have been established (Gowda et al., 1986c).

H. Intracellular Proteoglycans

In addition to the extracellular matrix and cell surface proteoglycans, the presence of intracellular proteoglycans associated with granules and certain organelles is now well established.

Of the granule-associated proteoglycans, the best known is the heparin proteoglycan in mast cells, discussed earlier. Intracellular granule proteoglycans are also present in platelets, leukocytes, and basophils (Metcalfe and Austen, 1979). Two proteoglycans, one containing predominantly chondroitin sulfate and the other with dermatan sulfate, have been purified from the chromaffin granules of bovine adrenal medulla by Kiang et al. (1982). Whether all granule proteoglycans have primary mediator functions (e.g., mast cell heparin) or are merely structural matrix components of the granules is not known. Usually the intracellular granule proteoglycans are released by certain stimulants (e.g., IgE binding).

A well-characterized intracellular granule proteoglycan is serglycin, which is present in rat yolk-sac tumor cells. This proteoglycan is secreted by these cells and is considered to function by preventing cell adhesion to matrix through its binding to fibronectin and collagen (Brennan et al., 1984). The complete sequence of the core protein has been deduced from cloned cDNA. The sequence contains 104 amino acids, of which about one-half constitute a central region of uninterrupted Ser-Gly repeats. About 14 chondroitin sulfate chains are attached in this region. There are no potential sites for asparagine-glycosylation in the sequence (Bourdon et al., 1985).

A proteoglycan (clock protein) similar to that of serglycin is present in *Drosophila melanogaster*, which is involved in the control of the circadian rhythms (Reddy et al., 1986). Recent findings indicate that various hematopoietic cells also synthesize proteins similar to serglycin (Kolset and Gallagher, 1990).

I. Nuclear Proteoglycans

Kinoshita (1971, 1974) showed that sea urchin embryos synthesize a heparinlike protein complex in the cytoplasm, which is subsequently transferred to the nucleus and preferentially located in the template-active portion of the chromatin. The heparinlike molecule was suspected to induce changes in embryonic nuclei, such as loosening of chromatin and the initiation of RNA synthesis. We presented the first evidence for the presence of GAGs associated with mammalian cell nuclei (Bhavanandan and Davidson, 1975). Soon afterward the presence of nuclear proteoglycans in human fibroblasts (Fromme et al., 1976), HeLa cells (Stein et al., 1975), rat brain (Margolis et al., 1976), and normal and regenerating rat liver and rat ascites hepatoma (Furukawa and Terayama, 1977, 1979) was documented, thereby substantiating and extending our findings. The elucidation of the function of the proteoglycans in the nucleus has been much more challenging. The nuclear proteoglycans can be implicated in both structural and regulatory functions. For example, GAGs are able to increase template availability and increase the net synthesis of DNA in isolated nuclei and chromatin (Smith and Cook, 1977). Kinoshita and Yoshii (1979) provided evidence that nuclear proteoglycans are essential for the normal development of sea urchin embryos. When proteoglycan synthesis was inhibited by addition of aryl-β-D-xylosides, embryonic development did not proceed beyond the blastula stage. This arrest in development could be overcome by the administration of exogenous proteoglycans.

Furukawa and Bhavanandan (1982) found that GAGs could be classified into three groups, depending on their influence on DNA synthesis in normal and cancer cell nuclei. Hyaluronic acid, dermatan sulfate, keratan sulfate, and desulfated GAGs have no effect on DNA synthesis. Chondroitin sulfates, heparan sulfate, *N*-desulfated heparin, and chondroitin sulfate proteoglycan cause inhibition of DNA synthesis in both normal and cancer cell nuclei. Highly sulfated GAGs (heparin and chondroitin sulfate H) at low concentration stimulated DNA synthesis in normal cell nuclei, but inhibited DNA synthesis in cancer cell nuclei. At high concentration they inhibited DNA synthesis in both normal and cancer cell nuclei. These effects were not simply due to electrostatic interactions, and subtle changes in the degree of sulfation, size of the GAG, and nature of saccharide units can drastically change the influence of these molecules on DNA synthesis (Furukawa and Bhavanandan, 1983). Studies by Fedarko and Conrad (1986) showed that in rat hepatocytes, the nuclear pool represented 5–6% of the total cell-associated heparan sulfate and that the nuclear species had a high proportion of glucuronic acid and glucuronic acid 2-sulfate. Their work also suggested that the extracellular pool may serve as the precursor for the nuclear heparan sulfate (Ishihara et al., 1986).

III. PROTEOGLYCAN BIOSYNTHESIS

A. Overview of the Problem

The assembly of a molecule as complex as a cartilage proteoglycan requires the cooperative efforts of both a large number of enzymes and intercellular transport systems. Over and above the regulatory controls that are exerted at the transcriptional level (hormonal factors, tissue-specific activators or inhibitors), individual steps are also subject to allosteric control. Further complexity is provided by the numerous translocations involved as the growing macromolecule moves from the rough endoplasmic reticulum (RER), through the Golgi stack, to the cell surface. It is no surprise, therefore, that much of the current information is descriptive and that major questions concerning the biosynthetic pathway remain unanswered (Lennarz, 1980; Hascall and Kimura, 1982; Poole, 1986; Evered and Whelan, 1986; Hassell et al., 1986; Wright and Mecham, 1987; Carney and Muir, 1988).

The timing has been analyzed in a series of pulse–chase experiments employing either amino acid or sulfate as the reporter precursors. Detection of mature, extracellular product is either by means of antibodies directed against the core protein or by the chemical and physical properties of the proteoglycan. Intermediates in the assembly process (i.e., migration of biosynthetic intermediates through the organelles of the secretory pathway) can be visualized by autoradiographic methods or, occasionally, by fractionation of cellular homogenates.

The generally accepted temporal sequence requires about 45 min from initiation of translation to expression at the exterior surface. The overall order of events is less clear. Protein synthesis is initiated in the usual manner and addition of *N*-linked saccharide occurs cotranslationally (Parodi and Leloir, 1979; Hubbard and Ivatt, 1986). These latter components may play a role in trafficking, since correct processing appears to be necessary for final expression (Farquhar, 1985). Xylosylation obviously must precede glycosaminoglycan chain synthesis, but the exact locus of the xylosyltransferase (distal ER or *cis*-Golgi) is not entirely clear (Robinson et al., 1966; Baker et al., 1972; Hoffman et al., 1984). The addition of the other sugars involved in the core saccharide as well as any *O*-linked (galactosamine-initiated) units apparently begins in the *cis*-Golgi. Assembly of the repeat domain of the glycosaminoglycan chain, as well as its essentially concurrent sulfation, appears rapid and is likely to be processive; this is a median to *trans*-Golgi event (Lohmander et al., 1983; Geetha-Habib et al., 1984; Ratcliffe et al., 1985). Termination signals remain unclear, as do controls over polysaccharide numbers and sulfate content. The extensive saccharide chain processing required for heparin, heparan sulfate, and, to a lesser extent, dermatan sulfate, suggests a specialized trafficking pathway, allowing access to the relevant enzymes. The role of specific amino acid sequences in overall control remains under study; clearly identified are a glycine-rich domain associated with xylosylation (Heinegard et al., 1986; Sai et al., 1986; Bourdon et al., 1987; Oldberg et al., 1987; Alliel et al., 1988). The solution of these problems for molecules such as type IX collagen remains a major challenge (Bruckner et al., 1985; Vaughan et al., 1985; Van De Rest et al., 1985; Huber et al., 1986; Konomi et al., 1986; McCormick et al., 1987).

The initiation of proteoglycan synthesis (i.e., transcription of the appropriate message and translation of the core protein) is clearly subject to a broad spectrum of regulatory controls. Global effects, such as cell density (Bassleer et al., 1987; Uhlin-Hansen and Kolset, 1988), nature of the growth matrix (Keller et al., 1987; Needham et al., 1988), and developmental status (Kato and Gospodarowicz, 1985; Hutchison and Yasin, 1986; Noonan et al., 1986; Tian et al., 1986; Venn and Mason, 1986; Webber et al., 1986; Miller et al., 1987; Kolset et al., 1988; Plaas et al., 1988), represent only the iceberg tip in an ever-increasing series of descriptive reports. Putative modulators include (but are not limited to) retinoic acid (Campbell and Handley, 1987), insulin (Breton et al., 1988), insulinlike growth factor (McQuillan et al., 1986), fetal calf serum (McQuillan et al., 1986b), estradiol 17-β (MacKintosh and Mason, 1988), interleukin-1 (Langris et al., 1987), transforming growth factor-β (Bassols and Massagu'e, 1988; O'Keefe et al., 1988; Rosen et al., 1988), parathyroid hormone (Kato et al., 1988), cAMP (forskolin; Malemud et al., 1986), somatomedin-C (O'Keefe et al., 1988), thyroid-stimulating hormone (Shishiba et al., 1988), glucocorticoids (Kato and Gospodarowicz, 1985), oxygen tension (Humphries et al., 1986a), DC current (Okihana and Shimomura, 1988), and cartilage-inducing factor (Seyedin et al., 1987). Regrettably, none of these studies provide information on control of expression or specificity of the effect at the molecular level. Somewhat more focused are studies involving use of tunicamycin, monensin, ammonium chloride, and related inhibitors of glycosylation (Hart and Lennarz, 1978; Stevens and Austen, 1982; Lohmander et al., 1983; Burditt et al., 1985; Harper et al., 1986; Humphries et al., 1986b; Spiro et al., 1986; Yanagishita, 1986; Seyedin et al., 1987; Spiro et al., 1989), or sulfate deprivation (Silbert et al., 1986; Humphries et al., 1988).

It can be reasoned that both tissue and maturation specificities involve signals active at the transcriptional level, but since the molecular biology of this field is only in the beginning stages, little definitive data is available.

Sequence information derived from cloned cDNA indicates the presence of a typical signal domain as well as glycine-rich regions, presumably the site of GAG chain attachment. The relatively long transit time of the protein core from the RER to the *trans*-Golgi or outer surface suggests that rate control of synthesis may also be exerted by the complex path this macromolecule must traverse to be fully outfitted. Recent results indicate that a certain level of processing of the *N*-linked saccharide units must occur before mature product is made. Thus, processing of *N*-linked saccharides through the stage of glucose removal appears necessary both for cell surface expression and glycosaminoglycan chain synthesis. Mannose trimming appears not to be essential, however, for either process. These data indicate that correct trafficking signals for proteoglycan assembly involve early glycosylation events, which thereby can serve as regulatory loci (Elbein, 1987).

The late-stage processing involved in the synthesis of heparin or heparan sulfate proteoglycans (uronic acid epimerization, de-*N*-acetylation, various sulfations) may require yet another set of targeting cues to direct the precursor macromolecule to those compartments that possess the requisite enzymes. Future studies with cell mutants defective in synthesizing active antithrombin heparan sulfate can be expected to provide information on how the specific saccharide sequences are generated (DeAgostini et al., 1990). Currently, other than recognition of the simplest instructions (Asn-X-Ser/Thr for *N*-linked saccharide addition—necessary, but clearly not sufficient), little detail is known regarding the cellular pathway involved or how it is chosen.

B. Monosaccharide Components

Despite extensive attempts, no evidence has been found to support any mechanism of glycosaminoglycan chain synthesis other than alternative saccharide addition from nucleotide sugar precursors. The availability of these donors to the requisite glycosyltransferases involves a membrane transport system in the Golgi apparatus, the control elements of which are not known (Corey et al., 1980; Fleischer, 1981; Sommers and Hirschberg, 1982; Deutscher et al., 1984; Schwarz et al., 1984; Perez and Hirschberg, 1985; Capasso and Hirschberg, 1987). The de novo synthesis of the activated sugars as well as the sugars themselves is subject to regulation.

1. *Amino Sugar Biosynthesis*

In mammalian systems, at least, all relevant amino sugars are derived from glucosamine which, in turn, is formed from fructose 6-phosphate (Leloir and Cardini, 1953):

$$\text{Fructose 6-PO}_4 + \text{L-glutamine} \rightarrow \text{glucosamine 6-phosphate} + \text{L-glutamate}$$

This reaction is essentially irreversible and has some unusual features. Neither pyridoxal phosphate nor iron are involved in the transformation. The conversion of glutamine to glutamate and ammonia requires the stoichiometric participation of water, whereas the simple amination of the keto function of fructose 6-PO_4 would generate a water molecule. Partial mechanistic studies have suggested a role for an enzyme sulfhydryl, possible displacing the amide nitrogen to form a transient thioester. The binding of the substrates is ordered with the fructose 6-PO_4 first on. This indicates that the furanoside ring is opened first, probably followed by hydrogen migration from C-1 to C-2 and attack at C-2 by the amide group which, in turn, is activated by the neighboring sulfur. It is conceivable that the fructose oxygen ends up in glutamate, but the requisite ^{18}O-studies have not been performed. The mammalian enzyme is large (350,000 M_r) and is subject to negative allosteric regulation by UDP-GlcNAc; the bacterial enzyme is not so regulated (Gryder and Pogell, 1960; Kornfeld et al., 1964; Winterburn and Phelps, 1971; Chmara et al., 1985).

Galactosamine is formed at the nucleotide level by 4-epimerization of UDP-GlcNAc, a reaction exactly analogous to that catalyzed by UDP-glucose 4-epimerase.

Although sulfated nucleotidyl sugars have been identified, these appear to play no role in glycosaminoglycan synthesis (Suzuki and Strominger, 1960).

2. *Uronic Acids and Xylose*

Glucuronic acid is formed at the nucleotide level by the oxidation of UDP-glucose:

$$\text{UDP-glucose} + 2\text{NAD}^+ \rightleftharpoons \text{UDP-GlcUA} + 2\text{NADH} + 2\text{H}^+$$

Although early work did not indicate any intermediate in this conversion, more directed studies showed that the initial step involved formation of the 6-aldehydo sugar, which then added to a key sulfhydryl on the enzyme to form a thio adduct that, in turn, is oxidized to the thioester before hydrolytic cleavage (Nelsestuen and Kirkwood, 1971; Ridley et al., 1975). The latter part of this mechanism is analogous to that of glyceraldehyde 3-phosphate dehydrogenase; UDP-xylose is a negative regulator.

The key initiator for glycosaminoglycan chain synthesis, xylose, is formed from UDP-GlcUA by decarboxylation (Bean and Hassid, 1956; Smith et al., 1958; Feingold et al., 1960; Ankel and Feingold, 1965):

$$\text{UDP GlcUA} \rightleftharpoons \text{UDP-Xyl} + CO_2$$

C. Sulfate Activation

The free energy of hydrolysis of the mixed anhydride present in phosphoadenosine phosphosulfate (PAPS) is approximately –17 kcal, twice that of ATP. It is not surprising, therefore, that sulfate activation is a two-stage process (Robbins and Lippman, 1957, 1958):

$$\text{ATP} + SO_4 \rightleftharpoons \text{AMP-}SO_4 + \text{P-Pi}$$

$$\text{AMP-}SO_4 + \text{ATP} \rightarrow \text{PAPS} + \text{ADP}$$

The equilibrium constant for the first reaction strongly favors ATP formation and, as in other systems, the hydrolysis of pyrophosphate helps drive the synthesis. Although generally considered a soluble enzyme system, the mode of access of PAPS to the proper face of the Golgi compartment involves a transport system. Limitations on sulfate availability can affect glycosaminoglycan synthesis (Silbert, 1978; Ito et al., 1982; Humphries et al., 1986b; Tyree et al., 1986).

D. Assembly Sequence

Early studies in cell-free systems showed an mRNA-driven product recognized by antibodies against chondroitinase-digested proteoglycan to have an M_r of about 350,000 (Treadwell et al., 1980; Vertel et al., 1984). It appears that the precursors do not undergo extensive processing, since recent deglycosylation studies indicate that the M_r of the intact core protein of mature cartilage proteoglycan has an M_r of approximately 370,000 (Campbell et al., 1990). In cultured cells, newly synthesized material has a half-life of 45–90 min before translocation to the Golgi (Kimura et al., 1980). Upon entrance to this region, *O*-linked glycosylation, sulfation, and secretion are rapid, probably totaling less than 10 min; intermediates are not detectable under normal conditions, consistent with the hypothesis that GAG chain synthesis and subsequent sulfation are largely, if not completely processive events (Sugumaran and Silbert, 1989).

Following initiation of translation on the RER, *N*-glycosylation occurs as a cotranslational event. Processing of the $GlcNAc_2Man_9Glc_3$, at least to the stage of glucose removal, is a prerequisite to translocation to that section of the Golgi stack at which GAG chain synthesis takes place. The *N*-linked oligosaccharides, present at least in the cartilage proteoglycan, appear to have the same core structure as those in other glycoproteins. This indicates their cotranslational addition by the dolichol pathway with standard processing, the latter an essential step. Synthesis of the final complex structures is completed in the Golgi. Assembly of the core oligosaccharide (Gal-Gal-Xyl→Ser) is followed by alternate transfer of the two constituent saccharides of the unit in question. The latter stage may be processive; sulfation appears to be concurrent or nearly so with elongation and is likely to be processive as well. Described relatively recently for proteoglycans, and as yet of uncertain significance, are a wide variety of *O*-linked (Ser/Thr) oligosaccharides (Nilsson et al., 1982; Kiang et al., 1982; Yanagishita and Hascall, 1983). The addition of these galactosamine-keyed *O*-linked saccharides, as in other systems, appears to require both *cis*-

and *trans*-Golgi elements. Although some controversy exists over the locus of the initial transfer event (GalNAc), completion of these structures is also a Golgi function. As in other examples of serine or threonine glycosylation, control of substrate selection and achieved final structures remain unknown.

All saccharide transfer reactions utilize nucleotidyl sugars and, hence, requires their transport across the organellar membrane. The exact requirements for these systems remain undefined, and they may serve as kinetic controls.

The key glycosyl transfer for GAG chain synthesis, that of the xylosyl unit, is probably a *cis*-Golgi event (Robinson et al., 1966). Other than the apparent preference of the enzyme for seryl residues in glycine-rich regions, little is known about structural controls. Certainly not all serines in multiple-substituted proteoglycans bear GAG chains, whereas for those proteoglycans with one or two chains, unique sequences directing glycosylation have not yet been identified.

Although all naturally occurring proteoglycans have their GAG chains attached by serine residues, replacement of a key serine by threonine in decorin through site-directed mutagenesis resulted in 90% reduction in GAG chain synthesis, but not complete oblation. In addition, one of the flanking glycine residues from the characteristics Gly-Ser-Gly sequence could be removed without inactivating chain initiation. These results suggest that the signal for attachment of GAG chains, or at least xylosyl residues, to the core protein is protein conformation, rather than primary amino acid sequence (Mann et al., 1990).

The completion of the core trisaccharide involves two discrete galactosylations, probably in the *cis*-Golgi (Helting and Roden, 1969). Given the ability of synthetic xylosides to serve as initiators for chondroitin sulfate GAG chain formation (Roden, 1980), it is unlikely that the core protein is involved at this or subsequent steps. These artificial substrates appear relatively ineffective with heparin or heparan sulfate (Hopwood and Dorfman, 1977; Johnston and Keller, 1978; Robinson and Lindahl, 1981; Stevens and Austen, 1982).

As has been previously mentioned, chain elongation occurs by sequential monosaccharide addition from nucleotide sugars. Intermediates (e.g., partially completed chains) are not found. One possible mechanism would involve transient immobilization of the growing macromolecule at the relevant membrane site. The chains are finished as nonsulfated precursors that are rapidly esterified. Given the high free-energy change involved in sulfate transfer from PAPS to the sugar acceptor (–14 kcal/mol), it is conceivable that some of this is transduced into molecular motion, thereby facilitating the highly processive nature of sulfation.

Transport to the cell surface and insertion/secretion follow standard pathways.

Recent results have shown that the xylosyl residues at the initiation site tend to be phosphorylated in the 2-position (Oegema et al., 1984; Fransson et al., 1985). The possible regulatory nature of this modification is unknown, as is the mechanism of addition of the phosphate ester groups. Since phosphorylation frequency may be 80% or greater, a functional role seems possible. It should be noted, however, that GAG chains synthesized on artificial initiators (i.e., xylosides) do not appear to undergo phosphorylation of the terminal pentose.

E. Termination Signals

Chain initiation, at least for the glycosaminoglycan, appears to be relatively simplistic. Construction of the core saccharide requires only an available xylose. Termination and size control generally are, on the surface, more complex. Known stop signals for saccharide chains include fucose and sialic acid (most cases, but not all), neither of which are present. There is little to indicate that any relevant information is provided by the protein sequence. At least one report has suggested that sulfation at a nonreducing terminus may halt chain elongation (Otsu et al., 1985).

Thus, terminal residues on chondroitin sulfate chains may be sulfated in either or both the 4- and 6-positions (Telser et al., 1966; Roden et al., 1972; Silbert, 1978). It is generally agreed that 4-sulfation of a GalNAc results in lack of acceptor activity for glucuronsyl residues, but that 6-sulfation has no such effect. Since sulfation must temporally follow chain synthesis (or elongation), the role of site-specific sulfation in termination events remains inconclusive.

If termination of chain elongation is related in any way to sulfation, then it follows that the latter event is concurrent. This is hard to visualize in a physical sense and is contraindicated by recent data showing that unesterified chains of the (presumably) correct size can be formed.

The relatively low polydispersity exhibited by cartilage proteoglycan chondroitin sulfate chains may arise from purely kinetic controls. Thus, the presence of multiple sites on a given polypeptide competing for the same sugar nucleotide pool will result in a relatively narrow distribution of chain lengths. The overall determination of saccharide number in any chain, however, may be governed, in turn, by transit time (i.e., access of the glycosyl transferase complex to the growing macromolecule), nucleotidyl sugar, or PAPS concentrations, their transport, and other as yet undetermined factors. Although mutant cell lines with global defects in proteoglycan synthesis have been described, no reports have been made on cells with partial blocks in chain elongation.

F. Regulatory Aspects

Given the numerous descriptive reports, it seems clear that factors that influence differentiation status and cell division will also exert an effect on proteoglycan synthesis. In this, as in most other related cases, specificity remains obscure. Although the cartilage proteoglycan is certainly the best studied of the group, its overall multiply substituted structure may be in a distinct minority. Virtually all cells make proteoglycans, and most seem to have one or only a very few associated glycosaminoglycan chains.

REFERENCES

Alliel, P. M., P'erin, J. P., Maillet, P., Bonnet, F., Rosa, J. P., and Jolles, P. (1988). Complete amino acid sequence of a human platelet proteoglycan. *FEBS Lett. 236*:123–126.

Anderson, B., Hoffman, P., and Meyer, K. (1964). The *O*-serine linkage in peptides of chondroitin 4- or 6-sulfate. *J. Biol. Chem. 240*:156–167.

Anderson, B., Seno, N., Sampson, P., Riley, J. G., Hoffman, P., and Meyer, K. (1964). Threonine and serine linkages in mucopolysaccharides and glycoproteins. *J. Biol. Chem. 239*:PC2716–PC2717.

Anderson, R. S., and Schwartz, E. R. (1984). Phosphorylation of proteoglycans from human articular cartilage. *Arthritis Rheum. 27*:58–71.

Ankel, H., and Feingold, D. S. (1965). Biosynthesis of uridine diphosphate D-xylose I. Uridine diphosphate glucuronate carboxylase of wheat germ. *Biochemistry 4*:2468–2475.

Arnott, S., Mitra, A. K., and Raghunathan, S. (1983). Hyaluronic acid double helix. *J. Mol. Biol. 169*:861–872.

Atha, D. H., Stephens, A. W., and Rosenberg, R. D. (1984). Evaluation of critical groups required for the binding of heparin to antithrombin. *Proc. Natl. Acad. Sci. USA 81*:1030–1034.

Axelsson, I., and Heinegard, D. (1978). Characterization of the keratan sulfate proteoglycans from bovine corneal stroma. *Biochem. J. 169*:517–530.

Baker, J. R., Roden, L., and Stoolmiller, A. C. (1972). Biosynthesis of chondroitin sulphate proteoglycan. Xylosyl transfer to Smith-degraded cartilage proteoglycan and other exogenous acceptors. *J. Biol. Chem. 247*:3838–3847.

Barber, A. J., Käser-Glanzmann, R., Jakabova, M., and Lüscher, E. F. (1972). Characterization of a chondroitin 4-sulfate proteoglycan carrier for heparin neutralizing activity (platelet factor 4) released from human blood platelets. *Biochim. Biophys. Acta 286*:312–319.

Bassleer, C., Gysen, P., Bassleer, R., and Franchimont, P. (1987). Proteoglycans synthesized by human chondrocytes cultivated in clusters. *Am. J. Med. 83*:25–28.

Bassols, A., and Massagu'e, J. (1988). Transforming growth factor beta regulates the expression and structure of extracellular matrix chondroitin/dermatan sulfate proteoglycans. *J. Biol. Chem. 263*:3030–3045.

Bean, R. C., and Hassid, W. Z. (1956). Carbohydrate oxidase from a red alga, *Iridophycus flaccidum. J. Biol. Chem. 218*:425–436.

Bhavanandan, V. P. (1981). glycosaminoglycans of cultured human fetal uveal melanocytes and comparison with those produced by cultured human melanoma cells. *Biochemistry 20*:5595–5602.

Bhavanandan, V. P., and Davidson, E. A. (1975). Mucopolysaccharides associated with nuclei of cultured mammalian cells. *Proc. Natl. Acad. Sci. USA 72*:2032–2036.

Bhavanandan, V. P., and Meyer, K. (1966). Mucopolysaccharides: *N*-Acetyl glucosamine- and galactose-6-sulfates from keratosulfate. *Science 151*:1404–1405.

Bhavanandan, V. P., and Meyer, K. (1967). Studies on keratosulfates: Methylation and partial acid hydrolysis on bovine corneal keratosulfate. *J. Biol. Chem. 242*:4352–4359.

Bhavanandan, V. P., and Meyer, K. (1968). Studies on keratosulfates: Methylation, desulfation, and acid hydrolysis studies on old human rib cartilage keratosulfate. *J. Biol. Chem. 243*:1052–1059.

Bourdon, M. A., Krusius, T., Campbell, S., Schwartz, N. B., and Ruoslahti, E. (1987). Identification and synthesis of a recognition signal for the attachment of glycosaminoglycans to proteins. *Proc. Natl. Acad. Sci. USA 84*:3194–3198.

Bourdon, M. A., Oldberg, A., Pierschbacher, M., and Ruoslahti, E. (1985). Molecular cloning and sequence analysis of a chondroitin sulfate proteoglycan cDNA. *Proc. Natl. Acad. Sci. USA 82*:1321–1325.

Brandt, K. D., and Muir, H. (1969). Characterisation of proteinpolysaccharides of articular cartilage from immature pigs. *Biochem. J. 114*:871–876.

Bray, B. A., Lieberman, R., and Meyer, K. (1967). Structure of human skeletal keratosulfate: The linkage region. *J. Biol. Chem. 242*:3373–3380.

Brennan, M. J., Oldberg, A., Hayman, E. G., and Ruoslahti, E. (1984). Effect of a proteoglycan produced by rat tumor cells on their adhesion to fibronectin-collagen substrata. *Cancer Res. 43*:4302–4307.

Breton, M., Berrou, E., Deudon, E., Brahimi-Horn, M. C., and Picard, J. (1988). Effect of insulin on sulfated proteoglycan synthesis in cultured smooth muscle cells from pig aorta. *Exp. Cell. Res. 177*:212–220.

Bruckner, P., Vaughan, L., and Winterhalter, K. H. (1985). Type IX collagen from sternal cartilage of chicken embryo contains covalently bound glycosaminoglycans. *Proc. Natl. Acad. Sci. USA 82*:2608–2612.

Bumol, T. F., and Reisfeld, R. A. (1982). Unique glycoprotein–proteoglycan complex defined by monoclonal antibody on human melanoma cells. *Proc. Natl. Acad. Sci. USA 79*:1245–1249.

Burditt, L. J., Ratcliffe, A., Fryer, P. R., and Hardingham, T. E. (1985). The intracellular localisation of proteoglycans and their accumulation in chondrocytes treated with monensin. *Biochim. Biophys. Acta. 844*:247–255.

Campbell, M. A., and Handley, C. J. (1987). The effect of retinoic acid on proteoglycan biosynthesis in bovine articular cartilage cultures. *Arch. Biochem. Biophys. 253*:462–474.

Campbell, S. C., Krueger, R. C., and Schwartz, N. B. (1990). Deglycosylation of chondroitin sulfate proteoglycan and derived peptides. *Biochemistry 29*:907–914.

Capasso, J. M., and Hirschberg, C. B. (1987). Mechanisms of glycosylation and sulfation in the Golgi apparatus: Evidence for nucleotide sugar/nucleoside monophosphate and nucleotide sulfate/nucleoside monophosphate antiports in the Golgi apparatus membrane. *Proc. Natl. Acad. Sci. USA 81*:7051–7055.

Carey, D. J., and Stahl, R. C. (1990). Identification of a lipid-anchored heparan sulfate proteoglycan in Schwann cells. *J. Cell Biol. 111*:2053–2062.

Carney, S. L., and Muir, H. (1988). The structure and function of cartilage proteoglycans. *Physiol. Rev. 68*:858–910.

Caterson, B., Baker, J. R., Levitt, D., and Paslay, J. W. (1979). Radioimmunoassay of the link proteins associated with bovine nasal cartilage proteoglycan. *J. Biol. Chem. 254*:9369–9372.

Chakrabarti, B., and Park, J. W. (1980). Glycosaminoglycans: Structure and interaction. *CRC Crit. Rev. Biochem. 8*:225–313.

Chiarugi, V. P., and Vannucchi, S. (1976). Surface heparan sulfate as a control element in eukaryotic cells: A working model. *J. Theor. Biol. 61*:459–475.

Chmara, H., Andruszkiewicz, R., and Borowski, E. (1985). Inactivation of glucosamine 6-phosphate synthetase from *Salmonella typhimurium* Lt. 2 by fumaroyl diaminopropanoic acid derivatives, a novel group of glutamine analogs. *Biochim. Biophys. Acta 870*:357–367.

Choi, H. U., Tang, L. H., Johnson, T. L., and Rosenberg, L. (1985). Proteoglycans from bovine nasal and articular cartilages: Fractionation of the link proteins by wheat germ agglutinin affinity chromatography. *J. Biol. Chem. 260*:13370–13376.

Comper, W. D., and Laurent, T. C. (1978). Physiological function of connective tissue polysaccharides. *Physiol. Rev. 58*:255–315.

Corey, D. J., Sommers, L. W., and Hirschberg, C. B. (1980). CMP-*N*-acetylneuraminic acid: Isolation from and penetration into mouse liver microsomes. *Cell 19*:597–605.

Davidson, E. A., and Meyer, K. (1954). Chondroitin, a new mucopolysaccharide. *J. Biol. Chem. 211*:605–611.

DeAgostini, A. L.,Lau, H. K., Leone, C., Youssoufian, H., and Rosenberg, R. D. (1990). Cell mutants defective in synthesizing a heparan sulfate proteoglycan with regions of defined monosaccharide sequence. *Proc. Natl. Acad. Sci. USA 87*:9784–9788.

De Luca, S., Lohmander, L. S., Nilsson, B., Hascall, V. C., and Caplan, A. I. (1980). Proteoglycans from chick limb bud chondrocyte cultures. Keratan sulfate and oligosaccharides which contain mannose and sialic acid. *J. Biol. Chem. 255*:6077–6083.

Deutscher, S. L., Nuwayhid, N., Stanley, P., Briles, E. I., and Hirschberg, C. B. (1984). Translocation across Golgi vesicle membranes: A CHO glycosylation mutant deficient in CMP-sialic acid transport. *Cell 39*:295–299.

Doege, K., Sasaki, M., Horigan, E., Hassell, J. R., and Yamada, Y. (1987). Complete primary structure of the rat cartilage proteoglycan core protein deduced from cDNA clones. *J. Biol. Chem. 262*:17757–17767.

Elbein, A. (1987). Inhibitors of the biosynthesis and processing of *N*-linked oligosaccharide chains. *Annu. Rev. Biochem. 56*:497–534.

Evered, D., and Whelan, W. J., eds. (1986). Functions of proteoglycans. *Ciba Found. Symp. 124*:1–299 (Review).

Farquhar, M. G. (1985). Progress in unraveling pathways of Golgi traffic. *Annu. Rev. Cell. Biol. 1*:447–448.

Fedarko, N. S., and Conrad, H. E. (1986). A unique heparan sulfate in the nuclei of hepatocytes: Structural changes with the growth state of the cells. *J. Cell Biol. 102*:587–599.

Feingold, D. S., Neufeld, E. F., and Hassid, W. Z. (1960). The 4-epimerization and decarboxylation of uridine diphosphate D-glucuronic acid by extracts from *Phaseolus aureus* seedlings. *J. Biol. Chem. 235*:910–913.

Fisher, L. W., Termine, J. D., and Young, M. F. (1989). Deduced protein sequence of bone small proteoglycan I (biglycan) shows homology with proteoglycan II (decorin) and several nonconnective tissue proteins in a variety of species. *J. Biol. Chem. 264*:4571–4576.

Fleischer, B. (1981). The nucleotide content of rat liver Golgi vesicles. *Arch. Biochem. Biophys. 212*:602–610.

Fransson, L.-A. (1987). Structure and function of cell-associated proteoglycans. *TIBS 12*:406–411.

Fransson, L.-A., Carlstedt, I., Cöster, L., and Malmström, A. (1983). Proteoheparan sulfate from human skin fibroblasts: Evidence of self-interaction via the heparan sulfate side chains. *J. Biol. Chem. 258*:14342–14345.

Fransson, L.-A., Silverberg, I., and Carlstedt, I. (1985). Structure of the heparan sulfate–protein linkage region. *J. Biol. Chem. 260*:14722–14726.

Fromme, H. G., Buddecke, E., von Figura, K., and Kresse, H. (1976). Localization of sulfated glycosaminoglycans within cell nuclei by high-resolution autoradiography. *Exp. Cell Res. 102*:445–449.

Furukawa, K., and Bhavanandan, V. P. (1982). Influence of anionic polysaccharides on DNA synthesis in isolated nuclei and by DNA polymerase a; correlation of observed effects with properties of the polysaccharides. *Biochim. Biophys. Acta 740*:466–475.

Furukawa, K., and Bhavanandan, V. P. (1983). Influence of glycosaminoglycans on endogenous DNA synthesis in isolated normal and cancer cell nuclei. *Biochim. Biophys. Acta 697*:344–352.

Furukawa, K., and Terayama, H. (1977). Isolation and identification of glycosaminoglycans associated with purified nuclei from rat liver. *Biochim. Biophys. Acta 499*:278–279.

Furukawa, K., and Terayama, H. (1979). Patterns of glycosaminoglycans and glycoproteins associated with nuclei of regenerating liver of rat. *Biochim. Biophys. Acta 585*:574–588.

Gallagher, J. T., Lyon, M., and Steward, W. P. (1986). Structure and function of heparan sulphate proteoglycans [Review]. *Biochem. J. 236*:313–325.

Geetha-Habib, M., Campbell, S. C., and Schwartz, N. B. (1984). Subcellular localization of the synthesis and glycosylation of chondroitin sulfate proteoglycan core protein. *J. Biol. Chem. 259*:7300–7310.

Goldberg, R. L., and Toole, B. P. (1984). Pericellular coat of chick embryo chondrocytes: Structural role of hyaluronate. *J. Cell. Biol. 99*:2114–2121.

Gowda, D. C., Bhavanandan, V. P., and Davidson, E. A. (1986a).Isolation and characterization of proteoglycans secreted by normal and malignant human mammary epithelial cells. *J. Biol. Chem. 261*:4926–4934.

Gowda, D. C., Bhavanandan, V. P., and Davidson, E. A. (1986b). Isolation and characterization of membrane-associated proteoglycans from normal and malignant human mammary epithelial cells. *Glycoconjugate J. 3*:55–70.

Gowda, D. C., Bhavanandan, V. P., and Davidson, E. A. (1986c). Structure of *O*-linked oligosaccharides present in the proteoglycans secreted by human mammary epithelial cells. *J. Biol. Chem. 261*:4935–4939.

Gregory, J. D., and Roden, L. (1961). Isolation of keratosulfate from chondromucoprotein of bovine nasal septa. *Biochem. Biophys. Res. Commun. 5*:430–434.

Gregory, J. D., Laurent, T. C., and Roden, L. (1964). Enzymatic degradation of chondromucoprotein. *J. Biol. Chem. 239*:3312–3320.

Gryder, R. M., and Pogell, B. M. (1960). Further studies on glucosamine 6-phosphate synthesis by rat liver enzymes. *J. Biol. Chem. 236*:558–562.

Halberg, D. F., Proulx, G., Doege, K., Yamada, Y., and Drickamer, K. (1988). A segment of the cartilage proteoglycan core protein has lectin-like activity. *J. Biol. Chem. 263*:9486–9490.

Hardingham, T. (1981). Proteoglycans: Their structure, interactions and molecular organization in cartilage. *Biochem. Soc. Trans. 9*:489–497.

Harper, J. R., Quaranta, V., and Reisfeld, R. A. (1986). Ammonium chloride interferes with a distinct step in the biosynthesis and cell surface expression of human melanoma-type chondroitin sulfate proteoglycan. *J. Biol. Chem. 261*:3600–3606.

Hart, G. W., and Lennarz, W. J. (1978). Effects of tunicamycin on the biosynthesis of glycosaminoglycans by embryonic chick cornea. *J. Biol. Chem. 253*:5795–5801.

Hascall, V. C. (1981). Proteoglycans: Structure and function. In *Biology of Carbohydrates*, Vol. 1 (V. Ginsburg and P. Robbins, eds.). John Wiley & Sons, New York, pp. 1–49.

Hascall, V. C., and Heinegard, D. (1974). Aggregation of cartilage proteoglycans. I. The role of hyaluronic acid. *J. Biol. Chem. 249*:4232–4241.

Hascall, V. C., and Kimura, J. H. (1982). Proteoglycans: Isolation and characterization [Review]. *Methods Enzymol. 82A*:769–800.

Hascall, V. C., and Riolo, R. L. (1972). Characteristics of the protein-keratan sulfate core and of keratan sulphate prepared from bovine nasal cartilage proteoglycan. *J. Biol. Chem. 247*:4529–4538.

Hascall, V. C., and Sajdera, S. W. (1970). Physical properties and polydispersity of proteoglycans from bovine nasal cartilage. *J. Biol. Chem. 245*:4920–4930.

Hassell, J. R., Kimura, J. H., and Hascall, V. C. (1986). Proteoglycan core protein families. *Annu. Rev. Biochem. 55*:539–567.

Hassell, J. R., Robey, P. G., Barrach, H.-J., Wilczek, J., Rennard, S. I., and Martin, G. R. (1980). Isolation of a heparan sulfate-containing proteoglycan from basement membrane. *Proc. Natl. Acad. Sci. USA 77*:4494–4498.

Heinegård, D., Frazen, A., Hedbom, E., and Sommarin, Y. (1986). Common structures of the core proteins of interstitial proteoglycans. In *Functions of the Proteoglycans* (D. Evered and J. Whelan, eds.), *Ciba Found. Symp. 124*:69–88.

Heinegård, D., and Hascall, V. C. (1974a). Aggregation of cartilage proteoglycans. III. Characteristics of the proteins isolated from trypsin digests of aggregate. *J. Biol. Chem. 249*:4250–4256.

Heinegård, D., and Hascall, V. C. (1974b). Characterisation of chondroitin sulfate isolated from trypsin-chymotrypsin digests of cartilage proteoglycans. *Arch. Biochem. Biophys. 165*:427–441.

Heinegård, D., Paulsson, M., Inerot, S., and Carlström, C. (1981). A novel low molecular-weight chondroitin sulfate proteoglycan isolated from cartilage. *Biochem. J. 197*:355–366.

Heinegård, D., and Sommarin, Y. (1987). Proteoglycans: An overview. *Methods Enzymol. 144*: 305–319.

Hetling, T., and Roden, L. (1969). Biosynthesis of chondroitin sulfate. I. Galactosyl transfer in the formation of the carbohydrate–protein linkage region. *J. Biol. Chem. 244*:2799–2805.

Hoffman, H., Schwartz, N., Roden, L., and Prockop, D. (1984). Location of xylosyl transferase in the cisternae of the rough endoplasmic reticulum of embryonic cartilage cells. *Connect. Tissue Res. 12*:151–164.

Höök, M., Kjellen, L., Johansson, S., and Robinson, J. (1984). Cell surface glycosaminoglycans. *Annu. Rev. Biochem. 53*:847–869.

Hopwood, J. J., and Dorfman, A. (1977). Glycosaminoglycan synthesis by cultured human skin fibroblasts after transformation with simian virus 40. *J. Biol. Chem. 252*:4777–4785.

Hubbard, C. S., and Ivatt, R. J. (1986). Synthesis and processing of asparagine-linked oligosaccharides. *Annu. Rev. Biochem. 50*:4568–4576.

Huber, S., Van Der Rest, M., Bruckner, P., Rodriguez, E., Winterhalter, K. H., and Vaughan, L. (1986). Identification of the type IX collagen polypeptide chains. *J. Biol. Chem. 261*:5965–5968.

Humphries, D. E., Lee, S. L., Fanburg, B. L., and Silbert, J. E. (1986a). Effects of hypoxia and hyperoxia on proteoglycan production by bovine pulmonary artery endothelial cells. *J. Cell Physiol. 126*:249–253.

Humphries, D. E., Silbert, C. K., and Silbert, J. E. (1986b). Glycosaminoglycan production by bovine aortic endothelial cells cultured in sulfate-depleted medium. *J. Biol. Chem. 261*:9122–9127.

Humphries, D. E., Silbert, C. K., and Silbert, J. E. (1988). Sulphation by cultured cells. *Biochem. J. 252*:305–308.

Hutchison, C. J., and Yasin, R. (1986). Developmental changes in sulphation of chondroitin sulphate proteoglycan during myogenesis of human muscle cultures. *Dev. Biol. 115*:78–83.

Iozzo, R. V. (1985). Proteoglycans: Structure, function and role in neoplasia. *Lab. Invest. 53*:373–396.

Ishihara, M., Fedarko, N. S., and Conrad, H. E. (1986). Transport of heparan sulfate into the nuclei of hepatocytes. *J. Biol. Chem.* *261*:13575–13580.

Ishihara, M., Fedarko, N. S., and Conrad, H. E. (1987). Involvement of phosphatidyl inositol and insulin in the coordinate regulation of proteoheparan sulfate metabolism and hepatocyte growth. *J. Biol. Chem.* *262*:4708–4716.

Ito, K., Kimata, K., Sobue, M., and Suzuki, S. (1982). Altered proteoglycan synthesis by epiphyseal cartilages in culture at low $SO_4$2-concentration. *J. Biol. Chem.* *257*:917–923.

Järnefelt, J., Rush, J., Li, Y-T., and Laine, R. A. (1978). Erythroglycan, a high molecular weight glycopeptide with a repeating structure [galactosyl 1→4 2-deoxy-2-acetamido-glycosyl 1→3] comprising more than one-third of the protein-bound carbohydrate of human erythrocyte stroma. *J. Biol. Chem.* *253*:8006–8009.

Jeanloz, R. W. (1960). The nomenclature of mucopolysaccharides. *Arthritis Rheum.* *3*:233–237.

Johnston, L. S., and Keller, J. M. (1978). The effect of β-xylosides on heparan sulfate synthesis by SV40-transformed swiss mouse 3T3 Cells. *J. Biol. Chem.* *254*:2575–2578.

Kanwar, Y. S., and Farquhar, M. G. (1979). Isolation of heparan sulfate from glomerular basement membranes. *Proc. Natl. Acad. Sci. USA* *76*:4493–4497.

Kanwar, Y. S., Linker, A., and Farquhar, M. G. (1980). Increased permeability of the glomerular basement membrane to ferritin after removal of heparan sulfate by enzyme digestion. *J. Cell Biol.* *86*:688–693.

Kaplan, D., and Meyer, K. (1959). Ageing of human cartilage. *Nature* *183*:1267–1268.

Kato, Y., and Gospodarowicz, D. (1985). Stimulation by glucocorticoid of the synthesis of cartilage-matrix proteoglycans produced by rabbit costal chondrocytes in vitro. *J. Biol. Chem.* *260*:2364–2373.

Kato, Y., Koike, T., Iwamoto, M., Kinoshita, M., Sato, K., Hiraki, Y., and Suzuki, F. (1988). Effects of limited exposure of rabbit chondrocyte cultures to parathyroid hormone and dibutyryl adenosine 3′,5′-monophosphate. *Endocrinology* *122*:1991–1997.

Keller, R., Pratt, B. M., Furthmayr, H., and Madri, J. A. (1987). Aortic endothelial cell proteoheparan sulfate. II. Modulation by extracellular matrix. *Am. J. Pathol.* *128*:299–306.

Keller, R., Stein, T., Stuhlsatz, H. W., Greiling, H., Ohst, E., Müller, E., and Scharf, H.-D. (1981). Studies on the characterization of the linkage-region between polysaccharide chain and core protein in bovine corneal proteokeratan sulfate. *Hoppe-Seyler Z. Physiol. Chem.* *362*:327–336.

Kendall, F. E., Heidelberger, M., and Dawson, M. H. (1937). A serologically inactive polysaccharide elaborated by mucoid strains of group A hemolytic streptococcus. *J. Biol. Chem.* *118*: 61–69.

Kiang, W.-L., Krusius, T., Finne, J., Margolis, R. U., and Margolis, R. K. (1982). Glycoproteins and proteoglycans of the chromaffin granule matrix. *J. Biol. Chem.* *257*:1651–1659.

Kimura, J. A., Thonar, E. J.-M., Hascall, V. C., Reiner, A., and Poole, A. R. (1981). Identification of core protein, an intermediate in proteoglycan biosynthesis in cultured chondrocytes from the swarm rat chondrosarcoma. *J. Biol. Chem.* *256*:7890–7897.

Kimura, J. H., Lohmander, L. S., and Hascall, V. C. (1980). Studies on the biosynthesis of cartilage proteoglycan in model system of cultured chondrocytes from the swarm rat chondrosarcoma. *J. Biol. Chem.* *255*:7134–7139.

Kinoshita, S. (1971). Heparin as a possible initiator of genomic RNA synthesis in early development of sea urchin embryos. *Exp. Cell Res.* *64*:403–411.

Kinoshita, S. (1974). Some observations on a protein–mucopolysaccharide complex found in sea urchin embryos. *Exp. Cell Res.* *85*:31–40.

Kinoshita, S., and Yoshii, K. (1979). The role of proteoglycan synthesis in the development of sea urchins. *Exp. Cell Res.* *124*:361–369.

Kjellen, J., Oldberg, A., and Höök, M. (1980). Cell surface heparan sulfate: Mechanisms of proteoglycan-cell association. *J. Biol. Chem.* *255*:10407–10413.

Kjellen, L., Pettersson, I., and Höök, M. (1981). Cell surface heparan sulfate: An intercalated membrane proteoglycan. *Proc. Natl. Acad. Sci. USA* *78*:5371–5375.

Kolset, S. O., and Gallagher, J. T. (1990). Proteoglycans of haemopoietic cells. *Biochim. Biophys. Acta 1032*:191–211.

Kolset, S. O., Ivhed, I., Overtan, A., and Nilsson, K. (1988). Differentiation-associated changes in the expression of chondroitin sulfate proteoglycan in induced U-937 cells. *Cancer Res. 48*: 6103–6108.

Konomi, H., Seyer, J. M., Yoshifumi, N., and Olsen, B. R. (1986). Peptide-specific antibodies identify the a_2 chain as the proteoglycan subunit of type IX collagen. *J. Biol. Chem. 261*:6742–6746.

Kornfeld, S., Kornfeld, R., Neufeld, E. F., and O'Brien, P. J. (1964). The feedback control of sugar nucleotide biosynthesis in liver. *Proc. Natl. Acad. Sci. USA 52*:371–379.

Kozutsumi, Y., Itoh, N., Fujimoto, K., Kawasaki, T., and Yamashina, I. (1979). Isolation and characterization of mucopolysaccharides from rat liver mitochondria. *J. Biochem. (Tokyo) 86*:1049–1054.

Kraemer, P. M. (1971). Heparan sulfates of cultured cells. I. Membrane-associated and cell-sap species in chinese hamster cells. *Biochemistry 10*:1437–1445 and 1445–1451.

Kuettner, K. E., and Kimura, J. H. (1985). Proteoglycans: An overview. *J. Cell. Biochem. 27*:327–336.

Langris, M., Daireauz, M., Jouis, V., Bocquet, J., and Loyau, G. (1987). Interleukin-1-like factor (mononuclear cell factor) modulates proteoglycan synthesis in cultured human synovial cells. *Biochim. Biophys. Acta. 929*:34–42.

Laurent, T. C. (1970). The structure of hyaluronic acid. In *Chemistry and Molecular Biology of the Intercellular Matrix*, Vol. 2 (E. A. Balazs, ed.). Academic Press, New York, pp. 703–732.

Leloir, L. F., and Cardini, C. E. (1953). The biosynthesis of glucosamine. *Biochim. Biophys. Acta 12*:15–22.

Lennarz, W. T., ed. (1980). *The Biochemistry of Glycoproteins and Proteoglycans* [Review]. Plenum Press, New York.

Lindahl, U., Bäckström, G., Thunberg, L., and Leder, I. G. (1980). Evidence for a 3-*O*-sulfated D-glucosamine residue in the antithrombin-binding sequence of heparin. *Proc. Natl. Acad. Sci. USA 77*:6551–6555.

Linker, A., and Hovingh, P. (1973). The heparitin sulfates (heparan sulfates). *Carbohydr. Res. 29*:41–62.

Lohmander, L. S., Fellini, S. A., Kimura, J. H., Stevens, R. L., and Hascall, V. C. (1983). Formation of proteoglycan aggregates in rat chondrosarcoma chondrocyte cultures treated with tunicamycin. *J. Biol. Chem. 258*:12280–12286.

Lohmander, L. S., Hascall, V. C., Yanagishita, M., Kuettner, K. E., and Kimura, J. H. (1986). Post-translational events in proteoglycan synthesis: Kinetics of synthesis of chondroitin sulfate and oligosaccharides on the core protein. *Arch. Biochem. Biophys. 250*:211–227.

Lomako, J., Lomako, W. M., and Whelan, W. J. (1988). A self-glucosylating protein is the primer for rabbit muscle glycogen bisynthesis. *FASEB J. 2*:3097–3103.

MacKintosh, D., and Mason, R. M. (1988). Pharmacological actions of 17 beta-oestradiol on articular cartilage chondrocytes and chondrosarcoma chondrocytes in the absence of oestrogen receptors. *Biochim. Biophys. Acta 964*:295–302.

Malemud, C. J., Mills, T. M., Shuckett, R., and Papay, R. S. (1986). Stimulation of sulfated proteoglycan synthesis by forskolin in monolayer cultures of rabbit articular chondrocytes. *J. Cell Physiol. 129*:51–59.

Mann, D. M., Yamaguchi, Y., Bourdon, M. A., and Ruoslahti, E. (1990). Analysis of glycosaminoglycan substitution in decorin by site-directed mutagenesis. *J. Biol. Chem. 265*:5317–5323.

Marbet, R., and Winterstein, A. (1951). β-Heparin, ein neueer, Blutgerinnungshemmender Mucoitinschwefel säureester. *Helv. Chim. Acta 34*:2311–2320.

Margolis, R. K., Crockett, C. P., Kiang, W. L., and Margolis, R. U. (1976). Glycosaminoglycans and glycoproteins associated with rat brain nuclei. *Biochim. Biophys. Acta 451*:465–469.

Maroudas, A. (1968). Physiochemical properties of cartilage in the light of ion-exchange theory. *Biophys. J. 8*:575–595.

Maroudas, A. (1975). Biophysical chemistry of cartilaginous tissue with special reference to solute and fluid transport. *Biorheology 12*:233–248.

Mason, R. M., d'Arville, C., Kimura, J. H., and Hascall, V. C. (1982). Absence of covalently linked core protein from newly synthesized hyaluronate. *Biochem. J. 207*:445–457.

Matalon, R., and Dorfman, A. (1968). The structure of acid mucopolysaccharides produced by Hurler fibroblasts in tissue culture. *Proc. Natl. Acad. Sci. USA 60*:179–185.

Mathews, M. B. (1971). Comparative biochemistry of chondroitin sulfate-proteins of cartilage and notochord. *Biochem. J. 125*:37–46.

McClean, D., and Rowlands, I. W. (1942). Role of hyaluronidase in fertilization. *Nature 150*:627–628.

McCormick, D., Van Der Rest, M., Goodship, J., Lozano, G., Ninomiya, Y., and Olsen, B. R. (1987). Structure of the glycosaminoglycan domain in the type IX collagen-proteoglycan. *Proc. Natl. Acad. Sci. USA 84*:4044–4048.

McQuillan, D. J., Handley, C. J., Campbell, M. A., Bolis, S., Milway, V. E., and Herington, A. C. (1986a). Stimulation of proteoglycan biosynthesis by serum and insulin-like growth factor-I in cultured bovine articular cartilage. *Biochem. J. 240*:423–430.

McQuillan, D. J., Handley, C. J., and Robinson, H. C. (1986b). Control of proteoglycan biosynthesis. Further studies on the effect of serum on cultured bovine articular cartilage. *Biochem. J. 237*: 741–747.

Metcalfe, D. D., and Austen, K. F. (1979). Structure and function of intracellular proteoglycans. *Monogr. Allergy 14*:236–248.

Meyer, K. (1938). The chemistry and biology of mucopolysaccharides and glycoproteins. *Cold Spring Harbor Symp. Quant. Biol. 6*:91–102.

Meyer, K., and Chaffee, E. (1941). The mucopolysaccharides of skin. *J. Biol. Chem. 138*: 491–499.

Meyer, K., Linker, A., Davidson, E. A., and Weissmann, B. (1953). The mucopolysaccharides of bovine cornea. *J. Biol. Chem. 205*:611–616.

Meyer, K., and Palmer, J. W. (1934). The polysaccharide of the vitreous humor. *J. Biol. Chem. 107*:629–634.

Midura, R. J., and Hascall, V. C. (1989). Analysis of the proteoglycans synthesized by corneal explants from embryonic chicken. *J. Biol. Chem. 264*:1423–1430.

Midura, R. J., Hascall, V. C., MacCallum, D. K., Meyer, R. F., Thonar, E. J.-M. A., Hassell, J. R., Smith, C. F., and Klintworth, G. K. (1990). Proteoglycan biosynthesis by human corneas from patients with types 1 and 2 macular corneal dystrophy. *J. Biol. Chem. 265*:15947–15955.

Mikuni-Takagaki, Y., and Toole, B. P. (1980). Cell-substratum attachment and cell surface hyaluronate of Rous sarcoma virus-transformed chondrocytes. *J. Cell Biol. 85*:481–488.

Mikuni-Takagaki, Y., and Toole, B. P. (1981). Hyaluronate-protein complex of Rous sarcoma virus-transformed chick embryo fibroblasts. *J. Biol. Chem. 256*:8463–8469.

Miller, R. R., Rao, J. S., and Festoff, B. W. (1987). Proteoglycan synthesis by primary chick skeletal muscle during in vitro myogenesis. *J. Cell Physiol. 133*:258–266.

Muir, H. (1958). The nature of the link between protein and carbohydrate of a chondroitin sulphate complex from hyaline cartilage. *Biochem. J. 69*:195–204.

Myers, E. R., Armstrong, C. G., and Mow, V. C. (1984). Swelling pressure and collagen tension. In *Connective Tissue Matrix* (D. W. L. Hukins, ed.). Macmillan, London, pp. 161–168.

Nader, H. B., Ferreira, T. M. P. C., Paiva, J. F., Medeiros, M. G. L., Jeronimo, S. M. B., Paiva, V. M. P., and Dietrich, C. P. (1984). The isolation and structural studies of heparan sulfates and chondroitin sulfates from three species of molluscs. *J. Biol. Chem. 259*:1431–1435.

Nader, H. B., Ferreira, T. M. P. C., Toma, L., Chavante, S. F., Dietrich, C. P., Casu, B., and Torri, G. (1988). Maintenance of heparan sulfate structure throughout evolution: Chemical and enzymatic degradation, and ^{13}C-N.M.R.—spectral evidence. *Carbohydr. Res. 184*:292–300.

Nakazawa, K., Hassell, J. R., Hascall, V. C., Lohmander, S., Newsome, D. A., and Krachmer, J. (1984). Defective processing of keratan sulfate in macular corneal dystrophy. *J. Biol. Chem. 259*:13751–13757.

Nakazawa, K., Newsome, D. A., Nilsson, B., Hascall, V. C., and Hassell, J. R. (1983). Purification of keratan sulfate proteoglycan from monkey cornea. *J. Biol. Chem. 258*:6051–6055.

Neame, P. J., Christner, J. E., and Baker, J. R. (1986). The primary structure of link protein from rat chondrosarcoma proteoglycan aggregate. *J. Biol. Chem. 261*:3519–3535.

Neame, P. J., Christner, J. E., and Baker, J. R. 91987). Cartilage proteoglycan aggregates. The link protein and proteoglycan amino-terminal globular domains have similar structures. *J. Biol. Chem. 262*:17768–17778.

Needham, L. K., Adler, R., and Hewitt, A. T. (1988). Proteoglycan synthesis in flat cell-free cultures of chick embryo retinal neurons and photoreceptors. *Dev. Biol. 129*:304–314.

Nelsestuen, G. L., and Kirkwood, S. (1971). The mechanism of action of uridine diphosphoglucae dehydrogenase. *J. Biol. Chem. 246*:3828–3834.

Nilsson, B., De Luca, S., Lohmander, S., and Hascall, V. C. (1982). Structures of *N*-linked and *O*-linked oligosaccharides on proteoglycan monomer isolated from the Swarm rat chondrosarcoma. *J. Biol. Chem. 257*:10920–10927.

Noonan, D. M., Horigan, E. A., Ledbetter, S. R., Vogeli, G., Sasaki, M., Yamada, Y., and Hassell, J. R. (1988). Identification of cDNA clones encoding different domains of the basement membrane heparan sulfate proteoglycan. *J. Biol. Chem. 263*:16379–16387.

Noonan, D. M., Malemud, D. J., and Przybylski, R. J. (1986). Biosynthesis of heparan sulfate proteoglycans of developing chick breast skeletal muscle in vitro. *Exp. Cell. Res. 166*:327–339.

O'Keefe, R. J., Puzas, J. E., Brand, J. S., and Rosier, R. N. (1988). Effects of transforming growth factor-beta on matrix synthesis by chick growth plate chondrocytes. *Endocrinology 122*:2953–2961.

Oegema, T. R., Jr., Kraft, E. L., Jourdian, G. W., and Van Valen, T. R. (1984). Phosphorylation of chondroitin sulfate in proteoglycans from the Swarm rat chondrosarcoma. *J. Biol. Chem. 259*:1720–1726.

Ogston, A. G., and Stanier, J. E. (1951). The dimensions of the particle of hyaluronic acid complex in synovial fluid. *Biochem. J. 49*:585–590.

Okayama, M., Oguri, K., Fujiwara, Y., Nakanishi, H., Yonekura, H., Kondo, T., and Ui, N. (1986). Purification and characterization of human platelet proteoglycan. *Biochem. J. 233*:73–81.

Okihana, H., and Shimomura, Y. (1988). Effect of direct current on cultured growth cartilage cells in vitro. *J. Orthop. Res. 6*:690–694.

Oldberg, Å., Antonsson, P., and Heinegard, D. (1987). The partial amino acid sequence of bovine cartilage proteoglycan, deduced from a cDNA clone, contains numerous Ser-Gly sequences arranged in homologous repeats. *Biochem. J. 243*:255–259.

Otsu, K., Inoue, H., Tsuzuki, Y., Yonekura, H., Nakanishi, Y., and Suzuki, S. (1985). A distinct terminal structure in newly synthesized chondroitin sulphate chains. *Biochem. J. 227*:37–48.

Parodi, A. J., and Leloir, L. F. (1979). The role of lipid intermediates in the glycosylation of proteins in the eukaryotic cell. *Biochem. J. 245*:763–772.

Parthasarathy, N., and Spiro, R. G. (1984). Isolation and characterization of the heparin sulfate proteoglycan of the bovine glomerular basement membrane. *J. Biol. Chem. 259*:12749–12755.

Perez, M., and Hirschberg, C. B. (1985). Translocation of UDP-*N*-acetylglucosamine into vesicles derived from rat liver rough endoplasmic reticulum and Golgi apparatus. *J. Biol. Chem. 260*:4671–4678.

Peters, C., Schwermann, J., Schmidt, A. S., and Buddecke, E. (1981). Phosphate ester groups in proteoglycans from bovine nasal cartilage. *Biochim. Biophys. Acta 673*:270–278.

Plaas, A. H., Sandy, J. D., and Kimura, J. H. (1988). Biosynthesis of cartilage proteoglycan and link protein by articular chondrocytes from immature and mature rabbits. *J. Biol. Chem. 263*:7560–7566.

Poole, A. R. (1986). Proteoglycans in health and disease: Structures and functions [Review]. *Biochem. J. 236*:1–14.

Prehm, P. (1983). Synthesis of hyaluronate in differentiated teratocarcinoma cells. Mechanism of chain growth. *Biochem. J. 211*:191–198.

Prinz, R., Schwermann, J., Buddecke, E., and von Figura, K. (1978). Endocytosis of sulphated proteoglycans by cultured skin fibroblasts. *Biochem. J. 176*:671–676.

Pulkkinen, L., Kainulainen, K., Krusius, T., Mäkinen, P., Schollin, J., Gustavsson, K.-H., and Peltonen, L. (1990). Deficient expression of the gene coding for decorin in a lethal form of Marfan syndrome. *J. Biol. Chem. 265*:17780–17785.

Rapraeger, A., Jalkanen, M., Endo, E., Koda, J., and Bernfield, M. (1985). Cell surface proteoglycan from mouse mammary epithelial cells bears chondroitin and heparan sulfate glycosaminoglycans. *J. Biol. Chem. 260*:11046–11052.

Ratcliffe, A., Fryer, P. R., and Hardingham, T. E. (1985). Proteoglycan biosynthesis in chondrocytes: Protein A: gold localization of proteoglycan, chondroitin sulfate and link protein within Golgi sub-compartments. *J. Cell Biol. 101*:2355–2365.

Reddy, P., Jacquier, A. C., Abovich, N., Petersen, G., and Rosbash, M. (1986). The period clock locus of *D. melanogaster* codes for a proteoglycan. *Cell 46*:53–61.

Reisfeld, R. A., Schulz, G., Cheresh, D. A., and Harper, J. R. (1985). Molecular and biological profiles of two unique antigens associated with human melanoma. In *Molecular Biology of Tumor Cells* (B. Wahren, ed.). Raven Press, New York, pp. 169–181.

Ridley, W., Houchins, J. P., and Kirkwood, S. (1975). Mechanism of action of uridine diphosphoglucose dehydrogenase. *J. Biol. Chem. 250*:8761–8767.

Robbins, P. W., and Lippman, F. (1957). Isolation and identification of active sulfate. *J. Biol. Chem. 229*:837–851.

Robbins, P. W., and Lippman, F. (1958). Separation of the two enzymatic phases in active sulfate synthesis. *J. Biol. Chem. 233*:681–685.

Robinson, H. C., and Lindahl, U. (1981). Effect of cycloheximide, β-D-xylosides and β-D-galactosides on heparin biosynthesis in mouse mastocytoma. *Biochem. J. 194*:575–586.

Robinson, H. C., Horner, A. A., Höök, M., Ogren, S., and Lindahl, U. (1978). A proteoglycan form of heparin and its degradation to single-chain molecules. *J. Biol. Chem. 253*:6687–6693.

Robinson, H. C., Telser, A., and Dorfman, A. (1966). Studies on the biosynthesis of the linkage region of chondroitin sulphate–protein complex. *Proc. Natl. Acad. Sci. USA 56*:1859–1866.

Roden, L. (1980). In *The Biochemistry of Glycoproteins and Proteoglycans* (Lennarz, W. J. ed.). Plenum Press, New York, pp. 267–371.

Roden, L., Baker, J. R., Helting, T., Schwartz, N. B., Stoolmiller, A. C., Yamagata, S., and Yamagata, T. (1972). Biosynthesis of chondroitin sulfate. *Methods Enzymol. 28*:638–676.

Roden, L., and Smith, R. (1966). Structure of the neutral trisaccharide of the chondroitin-4-sulphate protein linkage region. *J. Biol. Chem. 241*:5949–5954.

Rosen, D. M., Stempien, S. A., Thompson, A. Y., and Seyedin, S. M. (1988). Transforming growth factor-beta modulates the expression of osteoblast and chondroblast phenotypes in vitro. *J. Cell Physiol. 134*:337–346.

Rosenberg, L., Hellmann, W., and Kleinschmidt, A. K. (1970). Macromolecular models of protein polysaccharides from bovine nasal cartilage based on electron microscopic studies. *J. Biol. Chem. 245*:4123–4130.

Roughley, P. J., and Mort, J. S. (1986). Ageing and the aggregating proteoglycans of human articular cartilage. *Clin. Sci. 71*:331–344.

Ruoslahti, E. (1988). Structure and biology of proteoglycans. *Annu. Rev. Cell Biol. 4*:229–255.

Sai, S., Tanaka, T., Kosher, R. A., and Tanzer, M. L. (1986). Cloning and sequence analysis of a partial cDNA for chicken cartilage proteoglycan core protein. *Proc. Natl. Acad. Sci. USA 83*:5081–5085.

Sandell, L. J. (1987). Molecular biology of proteoglycans, in *Biology of Proteoglycans* (T. N. Wight and R. P. Mecham, eds.). Academic Press, New York, pp. 27–57.

Sanderson, R. D., and Bernfield, M. (1988). Molecular polymorphism of a cell surface proteoglycan: Distinct structures on simple and stratified epithelia. *Proc. Natl. Acad. Sci. USA 85*:9562–9566.

Santer, V., White, R. J., and Roughley, P. J. (1982). *O*-Linked oligosaccharides of human articular cartilage proteoglycan. *Biochim. Biophys. Acta 716*:277–282.

Saunders, S., Jalkanen, M., O'Farrell, S., and Bernfeld, M. (1989). Molecular cloning of syndecan, an integral membrane proteoglycan. *J. Cell Biol. 108*:1547–1556.

Schwarz, J. K., Capasso, J. M., and Hirschberg, C. B. (1984). Translocation of adenosine 3′-phosphate 5′-phosphosulfate into rat liver Golgi vesicles. *J. Biol. Chem. 259*:3554–3559.

Seno, M., Akiyama, F., and Anno, K. (1974b). Substrate specificity of chondrosulfatases from *Proteus vulgaris* for sulphated tetrasaccharides. *Biochim. Biophys. Acta 362*:290–298.

Seno, H., Anno, K., Yaegashi, Y., and Okuyama, T. (1975). Microheterogeneity of chondroitin sulfates from various cartilages. *Connect. Tissue Res. 3*:87–96.

Seno, N., Meyer, K., Anderson, B., and Hoffman, B. (1965). Variations in keratosulfates. *J. Biol. Chem. 240*:1005–1010.

Seno, N., Yamashiro, S., and Anno, K. (1974a). Isolation and characterization of a new disaccharide disulfate: 2-Acetoamido-2-deoxy-3-*O*-(2- or 3-*O*-sulfo-β-D-glucopyranosyluronic acid)-4-*O*-sulfo-D-galactose. *Biochim. Biophys. Acta 343*:423–426.

Seyedin, S. M., Thomas, T. C., Thompson, A. Y., Rosen, D. M., and Piez, K. A. (1987). Purification and characterization of two cartilage-inducing factors from bovine demineralized bone. *Proc. Natl. Acad. Sci. USA 84*:7977–7981.

Sheehan, J. K., and Atkins, E. D. T. (1983). X-ray fibre diffraction study of conformational changes in hyaluronate induced in the presence of sodium, potassium and calcium cations. *Int. J. Biol. Macromol. 5*:215–221.

Shishiba, Y., Yanagishita, M., Hascall, V. C., Takeuchi, Y., and Yokoi, N. (1988). Characterization of proteoglycans synthesized by rat thyroid cells in culture and their response to thyroid-stimulating hormone. *J. Biol. Chem. 263*:1745–1754.

Silbert, J. E. (1978). Biosynthesis of chondroitin sulphate. Chain termination. *J. Biol. Chem. 253*:6888–6892.

Silbert, J. E., Palmer, M. E., Humphries, D. E., and Silbert, C. K. (1986). Formation of dermatan sulfate by cultured human skin fibroblasts. *J. Biol. Chem. 261*:13397–13400.

Silvestri, L., Baker, J. R., Roden, L., and Stroud, R. M. (1981). The Clq inhibitor in serum is a chondroitin 4-sulfate proteoglycan. *J. Biol. Chem. 256*:7383–7387.

Smith, E. E., Mills, G. T., Bernheimer, H. P., and Austrian, R. (1958). The formation of a uridine pyrophosphoglucuronic acid from uridine pyrophosphoglucose by extracts of a noncapsulated strain of pneumococcus. *Biochim. Biophys. Acta 28*:211–212.

Smith, M. R., and Cook, R. T. (1977). Mechanisms of polyanion stimulation of nuclear DNA synthesis. *Exp. Cell Res. 110*:15–23.

Sommers, L. W., and Hirschberg, C. B. (1982). Transport of sugar nucleotides into rat liver Golgi. *J. Biol. Chem. 257*:10811–10817.

Spiro, R. C., Casteel, H. E., Laufer, D. M., Reisfeld, R. A., and Harper, J. R. (1989). Post-translational addition of chondroitin sulfate glycosaminoglycans. *J. Biol. Chem. 264*:1779–1786.

Spiro, R. C., Parsons, W. G., Perry, S. K., Caulfield, J. P., Hein, A., Reisfeld, R. A., Harper, J. R., Austen, K. F., and Stevens, R. L. (1986). Inhibition of post-translational modification and surface expression of a melanoma-associated chondroitin sulfate proteoglycan by diethylcarbamazine or ammonium chloride. *J. Biol. Chem. 261*:5121–5129.

Stein, G. S., Roberts, R. M., Davis, J. L., Head, W. J., Stein, J. L., Thrall, C. L., van Veen, J., and Welch, D. W. (1975). Are glycoproteins and glycosaminoglycans components of the eukaryotic genome? *Nature 258*:639–641.

Stevens, R. L., and Austen, K. F. (1982). Effect of *p*-nitrophenyl-β-D-xyloside on proteoglycan and glycosaminoglycan biosynthesis in rat serosal mast cell cultures. *J. Biol. Chem. 257*:253–259.

Stoddart, R. W. (1979). Nuclear glycoconjugates and their relation to malignancy. *Biol. Rev. 54*:199–235.

Stuhlsatz, H. W., Hirtzel, F., Keller, R., Cosma, S., and Greiling, H. (1981). Studies on the polydisperity and heterogeneity of proteokeratan sulfate from calf and porcine cornea. *Hoppe-Seyler Z. Pysiol. Chem. 362*:841–852.

Sugumaran, G., and Silbert, J. E. (1989). Biosynthesis of chondroitin sulfate: Organization of sulfation. *J. Biol. Chem. 264*:3864–3868.

Suzuki, S. (1960). Isolation of novel disaccharides from chondroitin sulfates. *J. Biol. Chem. 235*:3580–3588.

Suzuki, S., Saito, H., Yamagata, T., Anno, K., Seno, N., Kawai, Y., and Furuhashi, T. (1968). Formation of three types of disulfated disaccharides from chondroitin sulfates by chondroitinase digestion. *J. Biol. Chem. 243*:1543–1550.

Suzuki, S., and Strominger, J. L. (1960). Enzymatic sulfation of mucopolysaccharide in hen oviduct. *J. Biol. Chem. 235*:257–266.

Tanaka, T., Har-El, R., and Tanzer, M. L. (1988). Partial structure of the gene for chicken cartilage proteoglycan core protein. *J. Biol. Chem. 263*:15831–15835.

Tantravahi, R. U., Stevens, R. L., Austen, K. F., and Weis, J. H. (1986). A single gene in mast cells encodes the core peptides of heparin and chondroitin sulfate proteoglycans. *Proc. Natl. Acad. Sci. USA 83*:9207–9210.

Telser, A., Robinson, H. C., and Dorfman, A. (1966). The biosynthesis of chondroitin sulfate. *Arch. Biochem. Biophys. 116*:458–465.

Thonar, E. J.-M. A., and Sweet, M. B. E. (1979). An oligosaccharide component in proteoglycans of articular cartilage. *Biochim. Biophys. Acta 584*:353–357.

Tian, M. Y., Yanagishita, M., Hascall, V. C., and Reddi, A. H. (1986). Biosynthesis and fate of proteoglycans in cartilage and bone during development and mineralization. *Arch. Biochem. Biophys. 247*:221–232.

Toledo, O. M. S., and Dietrich, C. P. (1977). Tissue specific distribution of sulfated mucopolysaccharides in mammals. *Biochim. Biophys. Acta 498*:114–122.

Treadwell, B. V., Mankin, D. P., Ho, P. K., and Mankin, H. J. (1980). Cell-free synthesis of cartilage proteins: Partial identification of proteoglycan core and link proteins. *Biochemistry 19*:2269–2275.

Turley, E. A., and Roth, S. (1980). Interactions between the carbohydrate chains of hyaluronate and chondroitin sulfate. *Nature 283*:268–271.

Tyree, B., Hassell, J. R., and Hascall, V. C. (1986). Altered synthesis of heparan sulfate proteoglycans at low sulfate concentration. *Arch. Biochem. Biophys. 250*:202–210.

Uhlin-Hansen, L., and Kolset, S. O. (1988). Cell density-dependent expression of chondroitin sulfate proteoglycan in cultured human monocytes. *J. Biol. Chem. 263*:2526–2531.

Underhill, C. B., and Dorfman, A. (1978). The role of hyaluronic acid in intercellular adhesion of cultured mouse cells. *Exp. Cell Res. 117*:155–164.

Underhill, C. B., and Keller, J. M. (1975). A transformation-dependent difference in the heparan sulfate associated with the cell surface. *Biochem. Biophys. Res. Commun. 63*:448–454.

Underhill, C. B., and Toole, B. P. (1980). Physical characteristics of hyaluronate binding to the surface of simian virus 40-transformed 3T3 cells. *J. Biol. Chem. 255*:4544–4549.

Van De Rest, M., Mayne, R., Ninomiya, Y., Seidah, N. G., Chretien, M., and Olsen, B. R. (1985). The structure of type IX collagen. *J. Biol. Chem. 260*:221–225.

Vaughan, L., Winterhalter, K. H., and Bruckner, P. (1985). Proteoglycan Lt from chicken embryo sternum identified as type IX collagen. *J. Biol. Chem. 260*:4758–4763.

Venn, G., and Mason, R. M. (1986). Changes in mouse invertebral-disc proteoglycan synthesis with age. Hereditary kyphoscoliosis is associated with elevated synthesis. *Biochem. J. 234*:475–479.

Vernier, R. L., Klein, D. J., Sisson, S. P., Mahan, J. D., Oegema, T. R., and Brown, D. M. (1983). Heparan sulfate-rich anionic sites in the human glomerular basement membrane. Decreased concentration in congenital nephrotic syndrome. *N. Engl. J. Med. 309*:1001–1008.

Vertel, B. M., Upholt, W. B., and Dorfman, A. (1984). Cell-free translation of messenger RNA for chondroitin sulfate proteoglycan core protein in rat cartilage. *Biochem. J. 217*:259–263.

Vitello, L., Breen, M., Weinstein, H. G., Sittig, R. A., and Blacik, L. J. (1978). Keratan sulfate-like glycosaminoglycan in the cerebral cortex of the brain and its variation with age. *Biochim. Biophys. Acta 539*:305–314.

Wasteson, Å., and Lindahl, U. (1971). The distribution of sulfate residues in the chondroitin sulfate chain. *Biochem. J. 125*:903–908.

Webber, R. J., Zitaglio, T., and Hough, A. J. (1986). In vitro cell proliferation and proteoglycan synthesis of rabbit meniscal fibrochondrocytes as a function of age and sex. *Arthritis Rheum. 29*:1010–1016.

Winterbourne, D. J., and Mora, P. T. (1981). Cells selected for high tumorigenicity or transformed by simian virus 40 synthesize heparan sulfate with reduced degree of sulfation. *J. Biol. Chem. 256*:4310–4320.

Winterburn, P. J., and Phelps, C. F. (1971). Studies on the control of hexosamine biosynthesis by glucosamine synthetase. *Biochem. J. 121*:711–720.

Woodward, C., and Davidson, E. A. (1968). Structure–function relationships of protein polysaccharide complexes: Specific ion-binding properties. *Proc. Natl. Acad. Sci. USA 60*:201–205.

Yanagishita, M. (1986). Tunicamycin inhibits proteoglycan synthesis in rat ovarian granulosa cells in culture. *Arch. Biochem. Biophys. 251*:287–298.

Yanagishita, M., and Hascall, V. C. (1983). Characterization of low buoyant density dermatan sulfate proteoglycans synthesized by rat ovarian granulosa cells in culture. *J. Biol. Chem. 258*:12847–12856.

Yanagishita, M., and Hascall, V. C. (1984). Proteoglycans synthesized by rat ovarian granulosa cells in culture. Isolation, fractionation, and characterization of proteoglycans associated with cell layer. *J. Biol. Chem. 259*:10260–10269; 10270–10283.

Yanagishita, M., and Hascall, V. C. (1987). Proteoglycan metabolism by rat ovarian granulosa cells in vitro. In *Biology of Extracellular Matrix*, Vol. 2. *Biology of Proteoglycans* (T. N. Wight and R. P. Mecham, eds.). Academic Press, New York, pp. 105–128.

Yoneda, M., Suzuki, S., and Kimata, K. (1990). Hyaluronic acid associated with the surfaces of cultured fibroblasts is linked to a serum-derived 85-kDa protein. *J. Biol. Chem. 265*:5247–5257.

8

Glycolipids: Structure, Synthesis, Functions

Samar K. Kundu *Abbott Laboratories, Abbott Park, Illinois*

Glycolipids are ubiquitous membrane components of mammalian cells. The majority of these lipids are assumed to be localized at the outer leaflet of the plasma membrane, with their hydrophilic sugar chain protruding on the external surface [1–3]. Because of the presence of glycolipids at the cell surface and their unique structural diversity in different tissues and cell types of normal and diseased states, there appears to be an outburst of interest among scientists to examine the possible functions of these molecules.

The present chapter is not intended to be a review in the traditional sense with a complete catalog of literature. I will select classic and recent literature and cover general aspects of glycolipid structural diversity, biosynthesis, catabolism, and chemical synthesis, and present-day knowledge and speculations of the role of glycolipids in biological functions. A number of reviews on various aspects of glycolipids are found in the literature [1–17].

I. STRUCTURES

Glycolipids are made of a hydrophobic ceramide (*N*-acylsphingosine) moiety (Fig. 1) and one or more sugars linked glycosidically to the terminal primary hydroxyl group of sphingosine.

Although glycolipids show heterogeneity in both the oligosaccharide and ceramide (Cer) portions, they are characterized and identified on the basis of their carbohydrate structures. The major types of sugars that are found in animals, including humans, are glucose (Glc), galactose (Gal), fucose (Fuc), *N*-acetylgalactosamine (GalNAc), and *N*-acetylglucosamine (GlcNAc). Other sugars have also been identified in rare instances [1]. Glycolipids containing one or more sialic acids [*N*-acetyl-neuraminic acid (NeuAc) or *N*-glycolylneuraminic acid (NeuGc) or its derivatives] attached to a neutral sugar(s) by an α-ketosidic bond are known as *gangliosides*. Most glycolipids, including gangliosides, are derived from glucosylceramide or galactosylceramide by sequential addition of other monosaccharides.

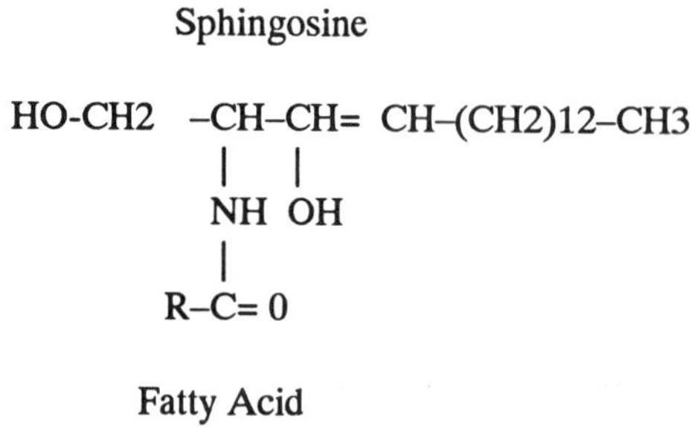

Sphingosine

HO-CH2 –CH–CH= CH–(CH2)12–CH3
| |
NH OH
|
R–C= 0

Fatty Acid

Figure 1 Structure of ceramide (*N*-acylsphingosine).

Table 1 shows the major basic structures of the glycolipid series. Substitutions at different sugars lead to many diversified structures. My intention is not to cover the listing of all glycolipids, most of which have been reviewed recently [1,2,4–7,10–12]. The structures and the major sources of glycolipids are listed in Table 2.

The list covers all major basic types of glycolipids. More complex polyglycosylceramides with 18–22 sugar residues are reported to be present in hog gastric mucosa [30] and in the glycoprotein fraction of human erythrocyte membrane [31]. The hog mucosa glycolipids seemed to have a blood group A and H activity (see Table 2). Polyglycosylceramides from human erythrocytes contain branched, repeated *N*-acetyllactosamine (i.e., Galβ1-4GlcNAcβ1-3[Gal]β1-4GlcNAcβ1-6]Galβ1-4GlcNAcβ1-R) and chitobiose structure (GlcNAcβ1-4GlcNAc) [32]. The structures of the polyglycosylceramide are yet to be confirmed.

II. CERAMIDES IN GLYCOLIPIDS

Although glycolipids are usually classified on the basis of oligosaccharide structure, wide variation in the ceramide composition have been reported. The sphingosine and the fatty acid vary in chain length, in degree of unsaturation, and in substitutions by hydroxyl groups.

Sphingosine (4-sphingenine; 2-D-amino-4-octadecene-1,3-diol) is the preponderant long-chain base in the ceramic components of most mammalian glycolipids. Others that are often found in glycolipids include dihydrosphingosine (sphinganine), and 4-eicosphingenine (C_{20} analogue of sphingosine). A number of long-chain analogues in variable amounts have been also found in glycolipids. In general, gangliosides from brain contain a mixture of C_{18} and C_{20} long-chain bases, whereas other glycolipids from the same source do not show any detectable C_{20} long-chain base. Interestingly, the proportion of C_{20} long-chain base is a function of development. This base is virtually absent in neonatal rat and human brains, but increases to adult level during the early stages of maturation [33]. Also, it is interesting that the proportion of C_{20} long-chain base depends on the number of sialic acid moieties in brain gangliosides. Trisialogangliosides contain 70% of the C_{20} homologue, compared with 30% of the monosialo species [34]. Extraneural glycolipids, including gangliosides, contain C_{18} long-chain bases as major components, although small amounts of trihydroxy and branched-chain bases (C_{18} as well as C_{20}) have been detected.

Table 1 Major Basic Glycolipid Structures

Series	Structure	Name of sugar moiety	Abbreviation
Lacto series	[a]Galβ1-4GlcNAcβ1-3Galβ1-4Glcβ1-1-Cer	Lacto-neo-tetraosyl	LcnOSe$_4$
	Galβ1-3GlcNAcβ1-3Galβ1-4Glcβ1-1-Cer	Lactotetraosyl	LcOSe$_4$
	[b]Galβ1-4GlcNAcβ1-3Galβ1-4GlcNAcβ1-3Galβ1-4Glc1-1-Cer	Lacto-nor-hexaosyl	LcOSe$_6$
	[b]Galβ1-4GlcNAcβ1 \\ $^6_3$Galβ1-4GlcNAcβ1-3Galβ1-4Glcβ1-1Cer / [d]Galβ1-4GalNAcβ1	Lacto-iso-octaosyl	LciOSe$_8$
Globo series	[a]GalNAcβ1-3Galα1-4Galβ1-4Glcβ1-1-Cer	Globotetraosyl	GbOSe$_4$
	GalNAcβ1-3Galα1-3Galβ1-4Glcβ1-1-Cer	Globo-neo-tetraosyl	GbnOSe$_4$
	GalNAcα1-3GalNAcβ1-3Galα1-4Galβ1-4Glcβ1-1-Cer	Globopentaosyl	GbOSe$_5$
Ganglio series	[c]Galβ1-3GalNAcβ1-4Gal[b]β1-4Glcβ1-1Cer	Gangliotetraosyl	GgOSe$_4$
	GalNAcβ1-4Gal[d]β1-4Glcβ1-1Cer	Gangliotriaosyl	GgOSe$_3$
	GalNAcβ1-4Gal[d]β1-3GalNAcβ1-4Gal[d]β1-4Glcβ1-1Cer	Gangliopentaosyl	GgOSe$_5$
Gala series	[b]Galα1-4Galβ1-1-Cer	Galabiosyl	GalOSe$_2$
Isoglobo series	Galα1-3Galβ1-4Glcβ1-1-Cer	Isoglobotriaosyl	GbiOSe$_3$
	GalNAcβ1-3Galα1-3Galβ1-4Glcβ1-1-Cer	Isoglobotetraosyl	GbiOSe$_4$
Muco series	[b]Galβ1-3Gal[e]β1-4Glcβ1-1-Cer	Mucotriaosyl	MuOSe$_3$
	[b]Galβ1-3Galβ1-3Galβ1-4Glcβ1-1-Cer	Mucotetraosyl	MuOSe$_4$

The abbreviations are in accordance to IUPAC-IUB commission, 1977. Substitution sites in core structures.
[a]2-3Sialosyl, 2-6 sialosyl, Galβ1-3,Galα1-3,fucosylα1-2
[b]Fucosylβ1-2 or 2-3 sialosyl
[c]2-3 Sialosyl, 2-8 sialosyl 2-3 sialosyl, fucosyl 1-2, or fucosyl 1-3
[d]2-3 Sialosyl, 2-8 sialosyl 2-3 sialosyl
[e]GalNAcα1-3Galβ1-6

Table 2 Structure and Major Sources of Gycolipids

Structure	Designation/IUPAC-IUB[a]	Symbol[b]	Major source	Ref.
Ganglio series				
NeuAcα2-3Gal-Cer	I^3-NeuAc-Gal-Cer	G_{M4}	Human brain, avian brain	6,11
R_1 ≠ Galβ1-4Glc-Cer				
R_1 = NeuAcα2-3	II^3NeuAc-Lac-Cer	G_{M3}	Extraneural tissue,[c] erythrocytes, plasma	6,11,18,19
R_1 = NeuAcα2-8NeuAcα2-3	II^3NeuAc$_2$-Lac-Cer	G_{D3}	Extraneural tissue,[c] erythrocytes, plasma	6,11,18,19
R_1 = NeuAcα2-8NeuAcα2-8NeuAcα2-3	II^3NeuAc$_3$-Lac-Cer	G_{T3}	Fish brain	6,11
GalNAcβ1-4Galβ1-4Glc-Cer (with R_1 on Gal)				
R_1 = H	GgOSe$_3$	Asialo G_{M2}	Guinea pig erythrocytes; spleen, liver of Tay-Sachs patients	20,21
R_1 = NeuAcα2-3	II^3NeuAc-GgOSe$_3$-Cer	G_{M2}	Tay-Sachs brain	6,11,21
R_1 = NeuAcα2-8NeuAcα2-3	II^3NeuAc-GgOSe$_3$-Cer	G_{D2}	Human brain	6,11,21
R_1 = NeuAcα2-8NeuAcα2-8NeuAcα2-3	II^3NeuAc$_2$-GgOSe$_3$-Cer	G_{T2}	Pig adipose tissue, fish brain	11,22
Galβ1-3GalNAcβ1-4Galβ1-4Glc-Cer (with R_1 on terminal Gal, R_2 on inner Gal)				
R_1 = H; R_2 = H	GgOSe$_4$	AsG_{M1}	Brain of GM1-gangliosidosis	21
R_1 = H; R_2 = NeuAcα2-3	II^3NeuAc-GgOSe$_4$-Cer	G_{M1a}	Tumor	23
R_1 = NeuAcα2-3; R_2 = H	IV^3NeuAc-GgOSe$_4$-Cer	G_{M1b}	Human brain[c], GM1-gangliosidosis	6,10,11
R_1= Fucα1-2; R_2 = NeuAcα2-3	IV^2Fuc-II^3NeuAc-GgOSe$_4$-Cer	Fuc G_{M1}	Pig adipose, bovine thyroid*	11

R$_1$ = Fucα1-3; R$_2$ = NeuAcα2-3	IV^{3}Fuc-II^{3}NeuAc-GgOSe$_4$-Cer	Fuc G$_{M1}$	Pig adipose	11
R$_1$ = Fucα1-2; R$_2$ = NeuAcα2-8NeuAcα2-3	IV^{2}Fuc-II^{3}NeuAc$_2$-GgOSe$_4$-Cer	Fuc-G$_{D1b}$	Human brain, pig cerebellum	11
R$_1$ = Galα1-3; R$_2$ = NeuAcα2-3	IV^{3}Gal-II^{3}NeuAc-GgOSe$_4$-Cer	Gal-G$_{M1b}$	Frog fat body	11
R$_1$ = Galβ1-3Galα1-3; R$_2$ = NeuAcα2-3	IV^{3}Gal$_2$-II^{3}NeuAc-GgOSe$_4$-Cer	Gal$_2$-G$_{M1b}$	Frog fat body	11
R$_1$ = Galα1-3Galβ1-3Galα1-3; R$_2$ = NeuAc	IV^{3}Gal$_3$-II^{3}NeuAc-GgOSe$_4$-Cer	Gal$_3$-G$_{M1b}$	Frog fat body	11
R$_1$ = NeuAcα2-3; R$_2$ = NeuAcα2-3	IV^{3}NeuAc-II^3-NeuAc-GgOSe$_4$-Cer	G$_{D1a}$	Human brain[c]	6,10,11
R$_1$ = H; R$_2$ = NeuAcα2-8NeuAcα2-3	II^{3}NeuAc$_2$-GgOSe$_4$-Cer	G$_{D1b}$	Human brain[c]	6,10,11
R$_1$ = H; R$_2$ = NeuAcα2-8NeuAcα2-8NeuAcα2-3	II^{3}NeuAc$_3$-GgOSe$_4$-Cer	G$_{T1c}$	Fish brain	6,10,11
R$_1$ = NeuAcα2-3; R$_2$ = NeuAcα2-8NeuAcα2-3	IV^{3}NeuAc-II^{3}NeuAc$_2$-GgOSe$_4$-Cer	G$_{T1b}$	Human brain	6,10,11
R$_1$ = NeuAcα2-8NeuAcα2-3; R$_2$ = NeuAcα2-3	IV^{3}NeuAc$_2$-II^{3}NeuAc-GgOSe$_4$-Cer	G$_{T1a}$	Human brain	6,10,11
R$_1$ = NeuAcα2-8NeuAcα2-3; R$_2$ = NeuAcα2-8NeuAcα2-3	IV^{3}NeuAc$_2$-II^{3}NeuAc$_2$-GgOSe$_4$-Cer	G$_{Q1b}$	Human, bovine, chicken brain	6,11
R$_1$ = NeuAcα2-3; R$_2$ = NeuAcα2-8NeuAcα2-8NeuAcα2-3	IV^{3}NeuAc-II^{3}NeuAc$_3$-GgOSe$_4$-Cer	G$_{Q1c}$	Fish brain	6,10,11
R$_1$ = NeuAcα2-8NeuAcα2-3; R$_2$ = NeuAcα2-8NeuAcα2-8NeuAcα2-3	IV^{3}NeuAc$_2$-II^{3}NeuAc$_3$-GgOSe$_4$-Cer	G$_{P1}$	Fish brain	6,10,11
GalNAcβ1-4Galβ1-3GalNAcβ1-4Galβ1-4Glc-Cer (R$_1$ on GalNAc at 3GalNAc; R$_2$ on Gal at 4Galβ1-4Glc)				
R$_1$ = H; R$_2$ = NeuAcα2-3	IV^{4}GalNAc-II^{3}NeuAc-GgOSe$_4$-Cer	GalNAc-G$_{M1a}$	Human brain	11
R$_1$ = NeuAcα2-3; R$_2$ = H	IV^{3}NeuAc-IV^{3}GalNAc-GgOSe$_4$Cer	GalNAc-G$_{M1b}$	Tay-Sachs brain	11
NeuAcα2 \ $^6_3$Galβ1-3GalNAcβ1-3Galα1-4Galβ1-4Glc-Cer / NeuAcα2			Human erythrocytes	18

[a]The designation is in accordance with the recommendations of IUPAC-IUB Commission on Biochemical Nomenclature. *Hoppe-Seylers Z. Physiol. Chem. 358*:617 (1977); *J. Lipid Res. 19*:4 (1978).
[b]The symbols are based on the Svennerholm system for gangliosides; for neutral glycolipids, trivial names are also given.
[c]In nonhuman samples, other types of sialosyl group, *N*-glycolylneuraminic acid (NeuGc), are known to exist [see Refs. 6, 10, 100].

Table 2 (Continued)

Structure	Designation/IUPAC-IUB[a]	Symbol[b]	Major source	Ref.
Lacto series				
Galβ1-4GlcNAcβ1-3Galβ1-4Glc-Cer | | R_1 R_2				
R_1 = H; R_2 = H	LcnOSe$_4$-Cer	Paragloboside	Human erythrocytes	18
R_1 = Fucα1-2; R_2 = H	IV^2-Gal-LcnOSe$_4$-Cer	H_1 (type 2)	Human erythrocytes	1,24
R_1 = Galα1-4; R_2 = H	IV^{4}Gal-LcnOSe$_4$-Cer	P_1	Human erythrocytes	24
R_1 = Galα1-3; R_2 = H	IV^{3}Gal-LcnOSe$_4$-Cer		Human erythrocytes	1,24
R_1 = Galβ1-3; R_2 = H	IV^{3}Gal-LcnOSe$_4$-Cer		Human erythrocytes	1,24
R_1 = NeuAcα2-3; R_2 = H	IV^{3}NeuAc-LcnOSe$_4$-Cer	2-3SPG	Human erythrocytes	1,24
R_1 = NeuAcα2-6; R_2 = H	IV^{3}NeuAc-LcnOSe$_4$-Cer	2-6SPG	Human erythrocytes	1,24
R_1 = NeuAcα2-8NeuAcα2-3; R_2 = H	IV^{3}NeuAc$_2$-LcnOSe$_4$-Cer		Human kidney, human erythrocytes	1,11, 20,24
R_1 = NeuAcα2-3; R_2 = Fucα1-3	IV^{3}NeuAc-III^{3}Fuc-LcnOSe$_4$-Cer		Human kidney	11
NeuAcα2-3Galβ1-3Galβ1-4GlcNAcβ1-3Galβ1-4Glc-Cer	IV^3[NeuAcα2-3Galβ1-]-]LcnOSe$_4$Cer		Human, bovine erythrocytes	10,11, 24
NeuAcα2-3GalNAcβ1-3Galβ1-4GlcNAcβ1-3Galβ1-4Glc-Cer	IV^3[NeuAcα2-3GalNAcβ1-]-LcnOSe$_4$Cer		Human erythrocytes	1,11,24
GalNAcα1 \\ $^3_2$Galβ1-4GlcNAcβ1-3Galβ1-4Glc-Cer / Fucα1		A_1(type 2)	Human erythrocytes	1,24
Galα1 \\ $^3_2$Galβ1-4GlcNAcβ1-3Galβ1-4Glc-Cer / Fucα1		B_1(type 2)	Human erythrocytes	1,24

Structure	Name	Abbreviation	Source	Ref.
Galβ1-3GlcNAcβ1-3Galβ1-4Glc-Cer	LcOSe$_4$-Cer			
Galα1 \\ $^{3}_{2}$Galβ1-3GlcNAcβ1-3Galβ1-4Glc-Cer / Fucα1		B$_1$ (type 1)	Human erythrocytes	1,24
GalNAcα1 \\ $^{3}_{2}$Galβ1-3GlcNAcβ1-3Galβ1-4Glc-Cer / Fucα1		B1 (type 1)	Human erythrocytes	1,24
Galβ1-4GlcNAcβ1-3Galβ1-4GlcNAcβ1-3Galβ1-4Glc-Cer | R$_1$				
R$_1$ = H	LcOnSe$_6$-Cer		Human erythrocytes	1,11,20,24
R$_1$ = NeuAcα2-3	VI^{3}NeuAc-LcnOSe$_6$-Cer	2-3SNH	Human erythrocytes[c]	1,11,20,24
R$_1$ = NeuAcα2-6	VI^{6}NeuAc-LcnOSe$_6$-Cer	2-6SNH	Human erythrocytes	1,24
R$_1$ = Fucα1-2	VI^2-Fuc-LcnOSe$_6$-Cer	H$_2$ (type 2)	Human erythrocytes	1,24
Galα1 \\ $^{3}_{2}$Galβ1-4GlcNAcβ1-3Galβ1-4GlcNAcβ1-3Galβ1-4Glc-Cer / Fucα1		B$_2$ (type 2)	Human erythrocytes	1,24
Galβ1-3GlcNAcβ1 \\ $^{6}_{3}$Galβ1-3GlcNAcβ1-3Galβ1-4Glc-Cer / Galβ1-3GlcNAcβ1	LcOSe$_8$-Cer		Rat small intestine	2

Table 2 (Continued)

Structure	Designation/IUPAC-IUB[a]	Symbol[b]	Major source	Ref.
R_1-Galβ1-4GlcNAcβ1 \\ ${}^6_3$Galβ1-4GlcNAcβ1-3Galβ1-4Glc-Cer / R_2-Galβ1-4GlcNAcβ1				
R_1 = H; R_2 = H	LcOnSe$_8$-Cer		Human erythrocytes	26
R_1 = Fucα1-2; R_2 = Fucα1-2		H_3 (type 2)	Human erythrocytes	1,24
R_1 = Fucα1-2; R_2 = NeuAcα2-3			Human erythrocytes	1,24
R_1 = NeuAcα2-3; R_2 = NeuAcα2-3	VIII[3]NeuAc$_2$-LcOnSe$_8$-Cer		Human erythrocytes	20
GalNAcα1 \\ ${}^3_2$Galβ1-4GlcNAcβ1-]2 ${}^6_3$ Galβ1-4GlcNAcβ1-3Galβ1-4Glc-Cer / Fucα1		A_3 (type 2)	Human erythrocytes	24
Globo Series				
Galα1-4Galβ1-4Glc-Cer	GbOSe$_3$-Cer	P[k]	Human erythrocytes	20,25
GalNAcβ1-3Galα1-4Galβ1-4Glc-Cer	GbOSe$_4$-Cer	P (globoside)	Human erythrocytes	20
GalNAcα1-3GalNAcβ1-3Galα1-4Galβ1-4Glc-Cer	GbOSe$_5$-Cer	Forssman	Sheep, goat erythrocytes	20
GalNAcβ1-3GalNAcβ1-3Galα1-4Galβ1-4Glc-Cer		Para-Forssman	Human erythrocytes	27
NeuAcα2-3GalNAcβ1-3Galα1-4Galβ1-4Glc-Cer	IV[3]NeuAc-GbOSe$_4$-Cer		Human muscles in amyotropic lateral sclerosis	28,29
NeuAcα2-8NeuAc2-3Galβ1-3Galα1-4Galβ1-4Glc-Cer	IV[3]NeuAc$_2$-GgOSe$_4$-Cer		Chicken muscle	29
NeuAcα2-3Galβ1-3GalNAcβ1-3Galα1-4Galβ1-4Glc-Cer	IV[3][NeuAcα2-3Galβ1-]GbOSe$_4$-Cer		Chicken muscle	29

The fatty acid components of glycolipids are usually more complex than the long-chain bases. They range in complexity from C_{14} to C_{24} in chain length. In general, fatty acids of shorter chain lengths (C_{16} and C_{18}) are present to some degree in most mammalian glycolipids. In some cases, longer chain fatty acids (C_{22} and C_{24}) are preponderant. The major gangliosides of the mammalian brain contain 70–90% of $C_{18:20}$ acids [35,36], but the minor species G_{M3} and G_{D3} contain less of $C_{18:0}$ and relatively more $C_{16:0}$ and $C_{16:1}$ acids [36]. Gangliosides from extraneural sources usually show significant amounts of C_{18}, C_{22}, and C_{24} fatty acids in addition to C_{16} fatty acids. The fatty acid components also show significant variation between organs and species. The preponderant fatty acid components in human plasma gangliosides are $C_{16:0}$, $C_{16:1}$, $C_{18:0}$, and $C_{18:1}$ [19]. Similarly, it was observed that the mono- and diglycosylceramides of porcine erythrocytes contain more than 60% of C_{16} and C_{18} fatty acids, as compared with tetraglycosylceramides from the same organ, which contain C_{22} and C_{24} as the major species (90%) [37]. Kannagi et al. [38] presented data on the relationship between fatty acid and oligosaccharide structures of a number of glycolipids. It was suggested that human erythrocyte monosialosylgangliosides that possess a terminal NeuAcα2-6Gal structure contain short-chain fatty acids, compared with gangliosides with a terminal NeuAcα2-3Gal structure that had a preponderance of long-chain fatty acids. Two monosialogangliosides that have the same carbohydrate structure, NeuAcα2-3Galβ1-4GlcNAcβ1-3Galβ1-4GlcNAcβ1-4Glc-Cer, but differ in fatty acids, have been recently characterized by Kundu et al. [18]. One of the gangliosides contains C_{22}, C_{23}, and C_{24} fatty acids as the major components, whereas the other contains exclusively C_{14}, C_{15}, C_{16}, and C_{18} fatty acids. These data indicate that the relationship, as hypothesized by Kannagi et al., may not be as complete as suggested by this group, but it does not detract from the conceptual importance of this proposal. It is interesting that brain gangliosides of the ganglio series have a preponderance of C_{16} and C_{18} fatty acids, and most gangliosides of the lacto and globo series contain mostly C_{22} and C_{24} fatty acids [18,19]. Two gangliosides having the same carbohydrate structure, NeuAcα2-3Galβ1-4Glc-Cer (G_{M3}) isolated from human plasma also show a diversity in fatty acid composition [19].

Neutral glycolipids isolated from extraneural organs show a wide diversity in fatty acid composition. In general, galactosylceramides from mammalian brain contain large proportions of 2-hydroxy fatty acids, but such acids are encountered only rarely in extraneural glycolipids. Glucosylceramides from extraneural sources usually do not contain 2-hydroxy fatty acids. Glucosyl-, lactosyl-, and lactotriaosylceramides from extraneural sources usually contain C_{16} to C_{20} as major fatty acid components, compared with globo and lacto series that contain a higher proportion of fatty acids with chain lengths between C_{20} and C_{24} [19].

Because of the possibilities of a great variety of naturally occurring fatty acids and long-chain bases, the number of combinations in ceramide structure could be enormous. Recent advances in chromatographic techniques would definitely enhance separation of molecular species of glycolipids (i.e., with the same carbohydrate structure, but different ceramide components). Several reviews on various aspects of ceramide structures are found in the literature [2,5,11,21,39–41], and the methods used in analysis of ceramides are described in published papers [18,19,42–48].

III. DISTRIBUTION OF GLYCOLIPIDS

Glycolipids are widely distributed in all living organisms. The brain and other nervous tissues usually have a high content of glycolipid, compared with extraneural tissues. Under

Table 3 Percentage Distribution of Major Glycolipids in Human Tissues[a]

Sample	CMH[b]	CDH[c]	LcOSe3	LcOSe4	Ref.
Plasma	45	27	9	18	50
Erythrocytes	8	14	9	67	50
Liver	15–20	65–75	7–10	18	51, 52
Spleen	10–15	50–60	12–18	11–12	51, 52
Kidney	5	11	28	55	53

[a]Percent distribution is expressed for the total neutral glycolipid fraction only. Values for plasma, erythrocytes, and kidney have been calculated from the literature cited.
[b]CMH, ceramide monohexoside: Glc-Cer + Gal-Cer.
[c]CDH, ceramide dihexoside: Gal-Glc-Cer + Gal-Gal-Cer.

normal conditions, the brain and nervous tissues contain only monohexosyl ceramides, cerebrosides (Gal-Cer), and sulfatide ($3SO_3$-Gal-Cer). Glucosylceramide and other neutral glycolipids are rarely present in the nervous system under normal states. On the other hand, in extraneural tissues and in erythrocytes, glucosylceramide and other longer-chain oligoglycosylceramides are preponderant. In the kidneys and intestines, both galactosylceramide and glucosylceramide are present in almost equal amounts. The presence of both types of monohexaosylceramides is reported in newborn and fetal human brains [49]. The distribution of glycolipids differs in each tissue or cell type within the same species. Table 3 shows the percentage distribution of the major glycolipids in extraneural human samples.

The distribution of glycolipids is very much species-dependent. Table 4 shows an example of distribution of the major neutral glycolipids in erythrocytes from different species.

The distribution of gangliosides are similarly organ- and species-dependent (e.g., the major gangliosides in human erythrocytes are G_{M3} and 2-3NeuAc SPG, whereas in bovine erythrocytes the major gangliosides are G_{M3} and 2-3NeuGc SPG [6,10,11]. *N*-Acetylneuraminic acid (NeuAc) is considered to be the major type of sialic acid in the human brain. *N*-Glycolylneuraminic acid (NeuGc) is a minor species in brain gangliosides of some mammals, but it is widely distributed in extraneural tissues. Several di- and trisialogangliosides of nonhuman tissues contain two different types of sialic acid in the same molecule [10,11]. In general, the major brain gangliosides of warm-blooded animals contain one to three sialic acids in a molecule, whereas in cold-blooded animals, such as

Table 4 Major Neutral Glycolipids in Mammalian Erythrocytes

Species	Major glycolipid	Ref.
Human	Globoside (GbOSe4)	54, 55
Sheep/goat	Forssman (GbOSe5)	56–59
Bovine	$IV^3\alpha$Gal-LcnOSe4	60–62
Rabbit	$IV^3\alpha$Gal-LcnOSe4 $IV^3\alpha$Gal-LcOSe4	63–65
Guinea pig	As G_{M2} (GgOSe3)	66

Table 5 Comparison of Ganglioside Contents in Human Brain, Erythrocytes, and Plasma

Tissue	Lipid-bound sialic acid	Ref.
Brain gray matter	880 mg/g fresh tissue	11
Brain white matter	227 mg/g fresh tissue	11
Erythrocytes	7.5 mg/mL packed cells	67
Plasma	2.0 mg/mL	19

fish brain, the major gangliosides contain three to five sialic acids in a molecule [10,11]. The concentrations of gangliosides in different organs within the same species differ appreciably. Table 5 shows the ganglioside content in the human brain, erythrocytes, and plasma.

The glycolipid content and distribution depend primarily on the type of tissues used. The glycolipid pattern also depends on the developmental stage of a tissue or an organ and on whether the tissue is normal or diseased [1,11,21] (see later).

IV. ISOLATION AND PURIFICATION OF GLYCOLIPIDS

Several methods for the isolation and purification of glycolipids are described in the literature [11,67–72]. One of the most commonly used methods for the isolation of gangliosides was first introduced by Folch et al. [73]. It is based on the special property of these compounds to partition from the organic solvent in which they are extracted into an aqueous phase. However, this partition procedure cannot be used to quantitatively isolate all species of gangliosides from the aqueous phase because the less polar gangliosides could not be totally extracted from the organic solvent phase [47,72]. Moreover, neutral glycolipids with more than four oligosaccharide residues tend to partition into the aqueous phase, thereby creating more problems in purification of the gangliosides during later steps. This is a serious problem if one desires quantitative isolation of neutral glycolipids and gangliosides from animal tissues or cells. To overcome these disadvantages, anion exchangers such as DEAE–Sephadex [47,72], DEAE–Sepharose [74], Spherosil–DEAE-dextran [75], DEAE–Sephacel [76], DEAE–CPG [67,72], and DEAE–silica [67,72] have been used. All of the DEAE supports basically utilize the same principle. Gangliosides, sulfatides, and acidic phospholipids bind with the anion exchanger, whereas the neutral and zweitterionic lipids do not bind. Gangliosides can be easily eluted from the anion exchanger with a salt solution in quantitative yield [47,67,72]. Although any of the anion exchangers can be used, DEAE–silica definitely enjoys superiority over the others because of its faster flow rate, rapid equilibration, easier regeneration, low cost, lower susceptibility to microbial attack, and physical and dimensional stability. The isolation and purification of neutral glycolipids and gangliosides using DEAE-silica gel has been elaborately described [72]. Figure 2 shows a flowchart for purification of glycolipids. In brief, tissues or cells are extracted with a mixture of chloroform–methanol (1:1, v/v). For dried samples, 10% water in the chloroform–methanol mixture is necessary to facilitate extraction. The extract containing the total lipid mixture is then fractionated on a DEAE–matrix (acetate form) column. Acidic lipids (gangliosides and phospholipids) and sulfatides bind to the DEAE–matrix, whereas the neutral and zweitterionic phospholipids elute out and do not bind (fraction 1). The acidic lipids bound to the column are eluted with 0.2 M potassium acetate in methanol (fraction 2). Fraction 1 is acetylated with acetic anhydride and pyridine, and the acetylated neutral glycolipid fraction is separated from other lipids as described [72,77]. The acetylated glycolipids are then deacetylated with 0.2 M potassium hydroxide in methanol or Dowex 50 (H^+ form) [72]. As an illustration, a thin-layer chromatogram of the total neutral glycolipid fraction from human erythrocytes is shown in Figure 3 (for purification of each component, see later discussion). The acidic lipid fraction from the DEAE column (fraction 2) is hydrolyzed with 0.1 N sodium hydroxide in methanol and dialyzed against water at 4°C [47,71].

If the amount of starting tissue is small, it is advisable to avoid the dialysis step to minimize loss of samples. For handling small amounts of samples, the use of hydrophobic-interaction chromatography on a C_{18}–silica cartridge or column is recommended [72,78]. Lipid samples bind to C_{18}–silica in aqueous solution, whereas other nonlipid impurities pass through. The bound lipids are eluted with chloroform–methanol. The dialyzed sample, or the C_{18}–silica eluted sample containing gangliosides, fatty acids, and sulfatide is subjected to silica gel column chromatography [72]. Sulfatide and other impurities can be removed with a mixture of chloroform–methanol (85:15 v/v); total gangliosides are eluated out by chloroform–methanol (1:1, v/v). As an illustration, thin-layer chromatographic patterns of the total gangliosides from human brain white matter, human erythrocytes, and human plasma are shown in Figure 4. The total ganglioside mixture, if desired, can be separated according to the number of sialic acid residues by using a concave gradient elution on a DEAE–matrix (acetate form). The gradient is carried out with a three-chamber system, methanol in the first mixing chamber, 0.2 M ammonium acetate in methanol in the second, and 0.5 M ammonium acetate in methanol in the third [72]. Figures 5 and 6 show the elution profiles of the total gangliosides from beef brain and human erythrocytes according to the number of sialic acid residues. Figures 7 and 8 show individual ganglioside species profiles of beef brain and human erythrocytes. Beef brain gangliosides contain approximately 20–21% monosialo-, 53–55% disialo-, 20–22% trisialo-, and 4–5% tetrasialo-species [72], whereas human erythrocyte gangliosides contain approximately 84–86% monosialo- and 14–16% disialo-species [72].

Purification of individual glycolipids, including gangliosides, are usually carried out on the basis of the number of carbohydrate moieties in a molecule. Preparative thin-layer chromatography can be done with small amounts of samples [11,17,71]. Column chromatographies using varying proportions of chloroform–methanol and water on different silica-based matrices are described in the literature [24,67]. Separation using these procedures often leads to poor resolution, and repeated rechromatographies are required to isolate pure glycolipids. In recent years, use of porous silica gel as a chromatographic support has eliminated many of the problems. Therefore, the use of porous silica gel, such as Iatrobeads or other porous silica gels that are commercially available, is recommended [72]. A linear gradient containing varying proportions of isopropranol–hexane–water has been successfully used for the isolation of neutral glycolipids and gangliosides from human erythrocytes and human plasma, and many minor components that previously could not be separated have now been identified [1,18,19,24]. Figure 9 shows a thin-layer chromatogram of purified gangliosides from beef brain by a high-performance liquid chromatographic (HPLC) procedure [79–81]. Figures 10 and 11 show thin-layer chromatograms of purified monosialo- and disialogangliosides from human erythrocytes by a similar procedure [80].

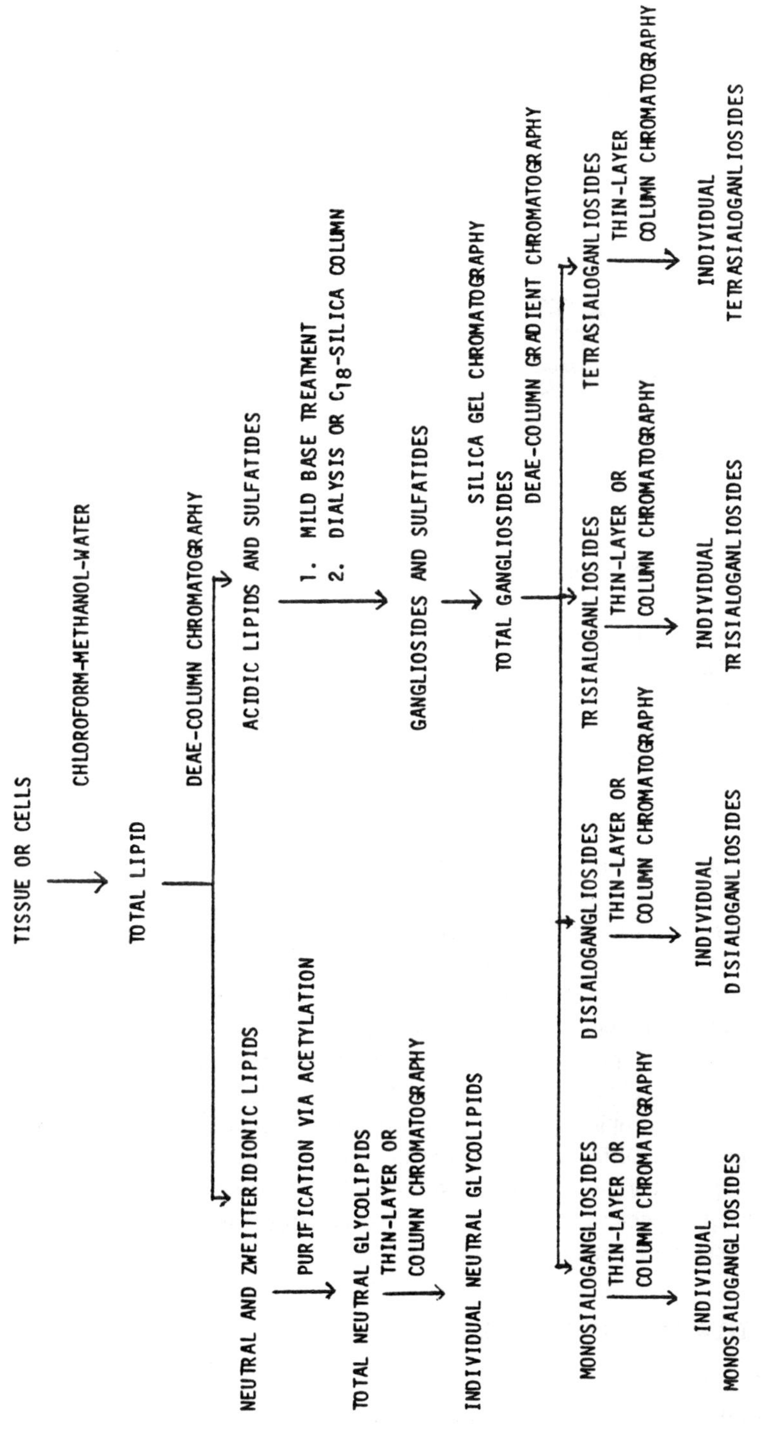

Figure 2 Flowchart for separation and purification of glycolipids.

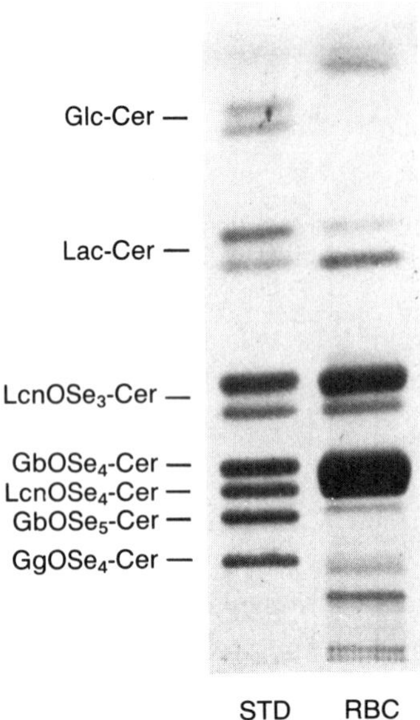

Figure 3 Thin-layer chromatogram of neutral glycolipids of human erythrocytes on a precoated Merck high-performance silica gel 60,200-μm-thick plate. The solvent was chloroform–methanol–water, 60:30:5 (v/v/v), and the glycolipids were detected with α-naphthol–sulfuric acid [17]. Lane STD contains a standard mixture of neutral glycolipids; lane RBC contains neutral glycolipids of normal human erythrocytes.

It is definitely preferable to use a HPLC procedure for purification of glycolipids, although good separation of individual glycolipids can also be achieved in a low-pressure or even with a no-pressure (gravity) column, if one uses porous silica gel and isopropanol–hexane–water mixtures as the gradient, as in HPLC.

The glycolipids thus purified are homogeneous in terms of carbohydrate structures. If one desires to purify individual glycolipids on the basis of molecular species, hydrophobic-interaction chromatography on a porous silica gel matrix that is coated with octadecyl (C_{18}), octyl (C_8), or phenyl groups, can be adopted. Advantage can be taken of the differences in the hydrophobic binding of different molecular species of a glycolipid to separate components. Different proportions of methanol and water can be used to separate the molecular species of glycolipid. This reversed-phase elution procedure has been used to separate molecular species of brain gangliosides G_{M4}, G_{M3}, G_{M2}, G_{M1}, G_{D1a}, and G_{D1b} on a C_{18}-silica column with methanol as an eluant [82–85].

V. QUANTITATION OF GLYCOLIPIDS

The amounts of glycolipid samples isolated from animal sources are occasionally so small that quantitation by a conventional weighing method is not possible. Therefore, quantitation is done from an aliquot of a glycolipid solution in which it is stored. A neutral glycolipid fraction can be quantitated colorimetrically by estimation of the total hexose content with the phenol–sulfuric acid reaction [86]. The amount of glycolipid can also be

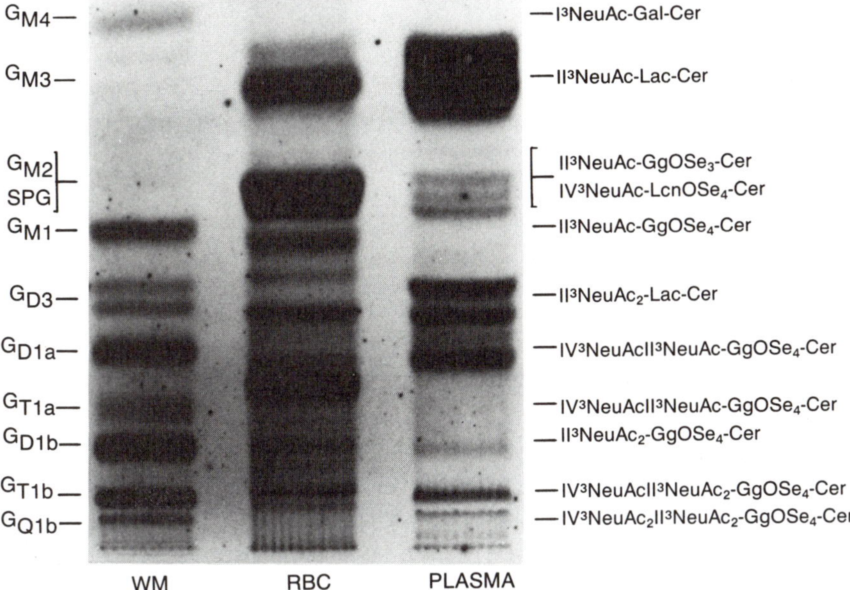

Figure 4 Thin-layer chromatogram of gangliosides of human brain white matter, human erythrocytes, and human plasma on a precoated Merck high-performance silica gel 60,200-μm-thick plate. The solvent was chloroform–methanol–water, 55:45:10 (v/v/v), with 0.02% $CaCl_2 \cdot 2H_2O$ (w/v), and the gangliosides were detected with resorcinol–hydrochloric acid [88]. Lane WM contains gangliosides of human brain white matter; lane RBC contains gangliosides of human erythrocytes; lane PLASMA contains gangliosides of human plasma.

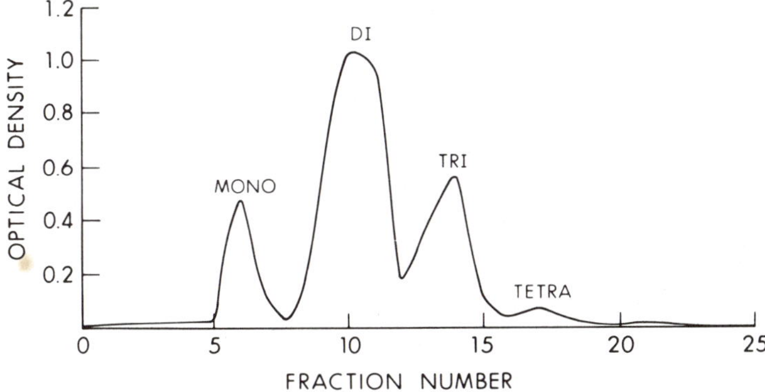

Figure 5 Elution profile of the total gangliosides from beef brain on DEAE-silica gel (acetate form). Sample applied: 40 mg ganglioside mixture; bed dimension: 2 × 60 cm; flow rate: 1.5 mL/min. Eluants: methanol (chamber 1); 0.2 M ammonium acetate in methanol (chamber 2); and 0.5 M ammonium acetate in methanol (chamber 3), 200 mL of each connected to each other through a gradient mixer. Fractions of 15-mL effluents were collected and 500 μL aliquots were used for sialic acid assay with resorcinol reagent [98]. MONO, DI, TRI, and TETRA denote the numbers of sialic acid residues in the ganglioside fractions.

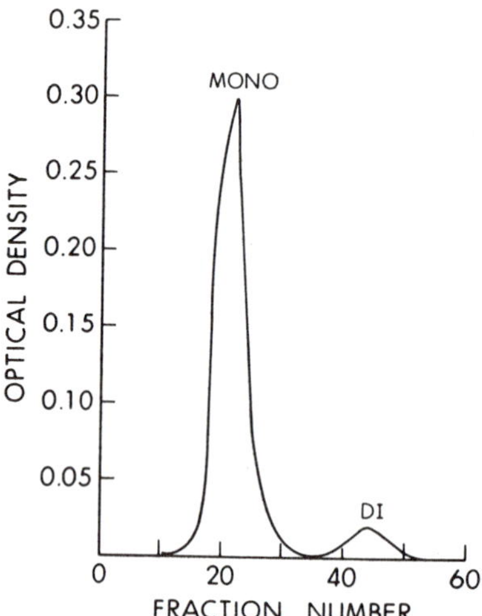

Figure 6 Elution profile of the total gangliosides from human erythrocytes on DEAE-silica gel (acetate form). Sample applied: 21 mg ganglioside mixture. Other details are as described in Figure 5.

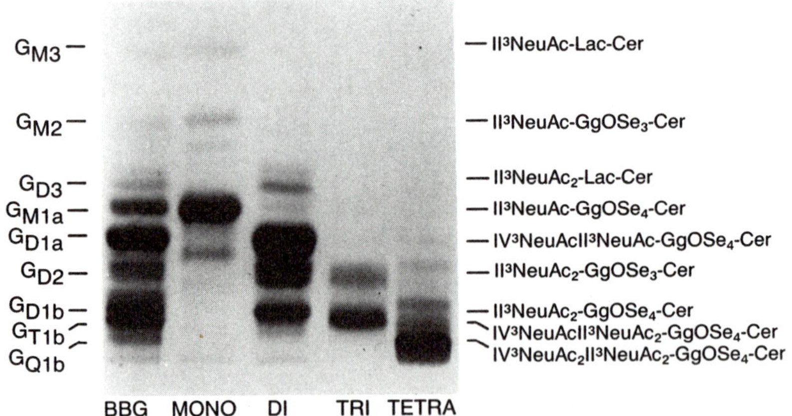

Figure 7 Thin-layer chromatogram of individual ganglioside species of beef brain separated by DEAE-gradient chromatography (see Figure 5). A precoated Merck silica gel 60,250-μm-thick plate was used. The solvent was chloroform–methanol–2.5 N ammonium hydroxide, and the gangliosides were detected with resorcinol–hydrochloric acid [88]. Lane BBG contains a standard beef brain ganglioside mixture; lane MONO contains monosialo-species; lane DI contains disialo-species; lane TRI contains trisialo-species; lane TETRA contains tetrasialo-species.

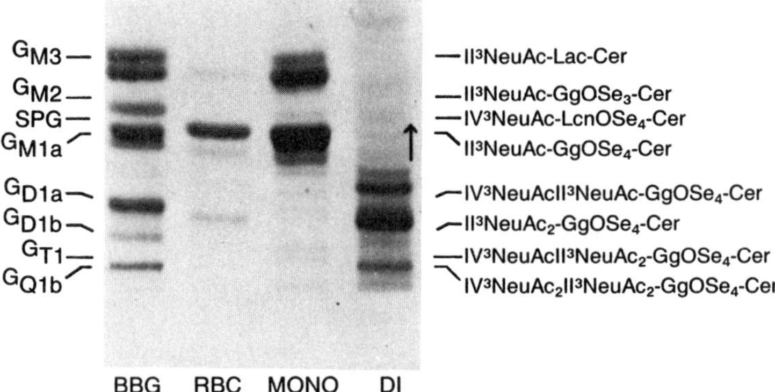

Figure 8 Thin layer chromatogram of individual ganglioside species of human erythrocytes, separated by DEAE-gradient chromatography (see Figure 6). A precoated Merck silica gel 60,250-μm-thick plate was used. The solvent was chloroform–methanol–water, 60:40:9 (v/v/v), with 0.02% $CaCl_2 \cdot 2H_2O$ (w/v), and the gangliosides were detected with resorcinol–hydrochloric acid [88]. Lane BBG contains a standard beef brain ganglioside mixture; lane RBC contains total ganglioside mixture from human erythrocytes; lane MONO contains monosialo-species; lane DI contains disialo-species. All bands were purple with resorcinol spray, except lane DI. The faint bands in this lane starting from the arrow are yellow and, presumably, are nonganglioside contaminants.

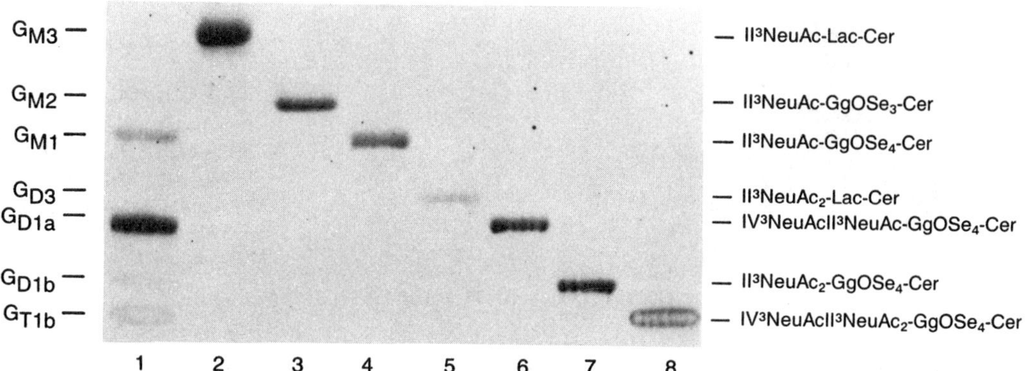

Figure 9 Thin-layer chromatogram of purified gangliosides of beef brain, separated by high-performance liquid chromatography [80]. Lane 1 contains total ganglioside mixture of beef brain; lanes 2–8 are purified gangliosides of beef brain. Other details are as described in Figure 2.

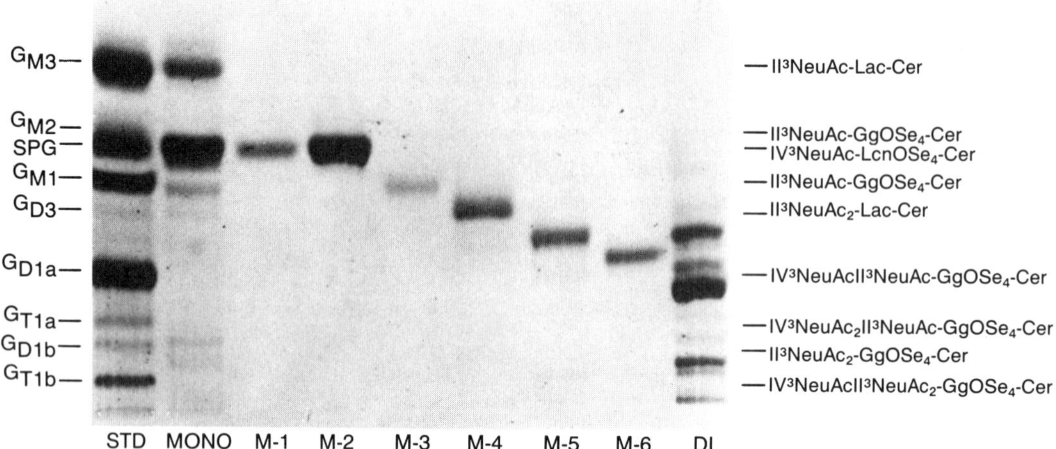

Figure 10 Thin-layer chromatogram of purified monosialogangliosides of human erythrocytes, separated by high-performance liquid chromatography [80]. Lane STD contains a standard mixture of G_{M1}, G_{M2}, SPG, and total beef brain gangliosides. Lane MONO contains total monosialo-species of human erythrocytes (see Fig. 6); lanes M-1 to M-6 are purified monosialogangliosides of human erythrocytes; lane DI contains total disialo-species of human erythrocytes (see Fig. 6). Other details are as described in Figure 2.

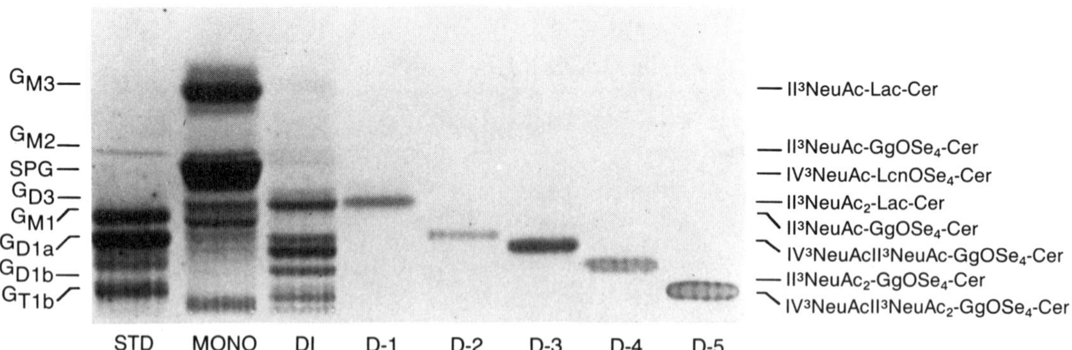

Figure 11 Thin-layer chromatogram of purified disialogangliosides of human erythrocytes, separated by high-performance liquid chromatography [80]. Lane STD contains a standard beef brain ganglioside mixture; lanes MONO and DI contain monosialo- and disialo-species of human erythrocytes (see Fig. 6); lanes D-1 to D-5 are purified disialogangliosides of human erythrocytes. Other details are as described in Figure 7.

determined by hydrolysis in 90% acetic acid containing 0.5 N sulfuric acid, reduction with sodium borohydride, acetylation, and analysis of galactitol or glucitol acetate by gas chromatography [18]. The later procedure, although accurate, is very time-consuming and is not generally used for quantitation of the total glycolipid mixture, but it is used for a purified individual glycolipid.

The total amount of ganglioside in a given sample is based on the quantitative estimation of the sialic acid moiety. Although a number of colorimetric methods are described in the literature [11,42,87], the resorcinol method, developed by Svennerholm, in 1957, is still in use [88]. The chromagen is extracted with *n*-butylacetate–butanol (85:15, v/v) [89], and the absorption is measured at 580 nm. The resorcinol method requires 5–30 μg of free or bound sialic acid. The resorcinol method has been modified by oxidation with periodate before reaction with resorcinol to increase the sensitivity [90]. The periodate–resorcinol method suffers from a serious disadvantage for some polysialogangliosides because the internal sialic acid residue linked to the basic oligosaccharide chain remains unaffected by mild periodate treatment. In contrast with these procedures, which measure both free and bound sialic acid, the method based on thiobarbituric acid [91,92] detects only free sialic acid and, thus, requires a hydrolysis step. This method requires about 3 μg of free sialic acid. A micromodification [93] has been used to assay as little as 25 ng of sialic acid. The most reliable and accurate method for estimation of sialic acid, developed by Yu and Ledeen [94], is based on gas chromatography. The technique involves mild methanolysis (0.05 N hydrochloric acid in methanol) of a ganglioside sample at 80°C for 1 h and estimation of the methyl ketoside methylesters of sialic acid by gas chromatography. This method distinguishes two major types of sialic acid—NeuAc and NeuGc—and as little as 0.3 μg of sialic acid can be analyzed accurately.

Quantitation of individual neutral glycolipids in a mixture can be done by running a thin-layer chromatogram with a 0.3% solution of α-naphthol in 95% ethanol, followed by spraying with 50% aqueous sulfuric acid and heating at 100°C for 5 min after which each band is quantitated by densitometric scanning [95,96]. The thin-layer plate can alternatively by sprayed with 0.2% solution of orcinol in sulfuric acid–water (3:1, v/v) and heated at 100°C for 10 min and then quantitated densitometrically as before. The densitometric methods all suffer from difficulty in obtaining reproducible results. Isolation of individual glycolipids by preparative thin-layer chromatography and analysis by gas chromatography is potentially accurate, but its use is rather limited because of the poor recovery of the long-chain glycolipids from the silica gel plate. Acetylated glycolipids can be easily prepared [17,77] and are much less polar than the parent glycolipids; their recovery from the silica gel plate is quantitative and can be used for quantitation of individual glycolipids. Analysis of neutral glycolipids by HPLC from perbenzoylated derivatives [82] or from the *O*-acetyl-*N*-*p*-nitrobenzoyl derivative [97] is the most accurate method for quantitation of individual glycolipids.

Quantitation of individual gangliosides separated by thin-layer chromatography can be carried out by a simple and elegant method developed by Suzuki [98] in which the gangliosides are visualized on the plate by exposure to iodine vapor and are scraped into centrifuge tubes. Sialic acid is then determined directly with the resorcinol reagent without eluting the ganglioside from the silica gel. In a modification of this method, gangliosides are weakly visualized by heating the plate lightly, spraying with resorcinol reagent [99,100], and assaying the scrapings, as in the Suzuki method. Densitometric scanning methods have been developed and are being used routinely in many laboratories [101–103] for quantitation of individual gangliosides. As little as 29 ng of sialic acid could be detected with a good signal/noise ratio. This lower limit is roughly 20 times more sensitive than the resorcinol assay in test tubes.

VI. STRUCTURE DETERMINATION OF GLYCOLIPIDS

Elucidation of the structure of an isolated glycolipid sample requires determination of (1) carbohydrate composition, (2) sugar sequence, (3) glycosidic substitution sites, (4) stereochemistry of the glycosidic bond, (5) sphingosine bases, and (6) fatty acid composition.

A. Carbohydrate Composition

Compositional analysis of hexoses and hexosamines were carried out for many years by colorimetric procedures [104]. Since most of these methods do not differentiate individual sugars and have low sensitivity, the preferred methods are based on gas chromatographic analysis of individual sugar components. The assay is done by analyzing the methylglycosides or alditol acetates. The widely used method of Sweeley and co-workers [50,105] involves methanolysis of the intact glycolipid sample with 0.75 N hydrochloric acid in methanol and converting the methylglycosides formed to volatile trimethylsilyl ethers [106]. If the methanolysate contains amino sugars, they must be *N*-acylated with acetic anhydride in presence of pyridine [107]. The methyl glycosides can also be analyzed as their trifuroacetyl derivative [108], for which the *N*-acetylation step is unnecessary. The problem associated with this gas chromatographic method is that methanolysis produces two or more methyl glycosides for each sugar component because of anomerization, which often makes quantitation difficult. This problem has been minimized by gas chromatographic analysis of the sugar components as hexitol or hexosaminitol acetates. The procedures involved in this assay are liberation of monosaccharides from a glycolipid sample by acetolysis, followed by hydrolysis in 90% acetic acid in 0.5 N sulfuric acid; reduction of the monosaccharides with sodium borohydride; acetylation of alditols with acetic anhydride; and analysis as alditol acetates by gas chromatography [105,109]. Each sugar component thus gives rise to a single peak, and fairly good recovery of hexosaminitol acetates can be obtained. This method, however, cannot be used to determine sialic acids [110].

B. Sugar Sequence

The carbohydrate sequence was originally studied by partial hydrolysis of glycolipids and analysis of either the liberated sugar components or the partially hydrolyzed glycolipid fragments. Reaction of glycolipids with acetic acid–acetic anhydride–sulfuric acid mixture was used to generate a family of acetylated oligosaccharides that were then identified [111]. Treatment of intact polysialogangliosides with 5.6 N formic acid is used for partial removal of sialic acid, and the family of fragmented gangliosides with intact oligosaccharide chains thus obtained, are identified by thin-layer chromatographic analysis with known standards [112,113].

The most reliable and specific method for determining the carbohydrate sequence in glycolipids is the use of specific glycosidases. Exoglycosidases cleave the monosaccharide units from the nonreducing end of the oligosaccharide chain in glycolipids, whereas endoglycosidases do not recognize monosaccharide residues, but do recognize oligosaccharides as the specific aglycon. The isolation and properties of many of the exoglycosidases from plant, animal, and bacterial sources have been reviewed [114] and most of the exoglycosidases are commercially available. The use of exo- and endoglycosidases has been reviewed [115]. It is important to note that not all exoglycosidases from various sources behave similarly in terms of hydrolysis of a specific monosaccharide from a glycolipid. Several of the exoglycosidases that are commonly used in the hydrolysis of glycolipids are commercially available. These are α-galactosidase from coffee beans (Sigma Chemical Company, St. Louis, Missouri) and ficin [114]; β-galactosidase from jack bean meal (Sigma) and from *Charonia lampas* (Sigma); β-*N*-acetylhexasaminidase from jack bean meal [114] and from *C. lampas* (Miles Biochemicals, Elkhardt, Indiana); β-*N*-acetylglucosaminidase from beef kidney (Boehringer Manheim Biochemicals, Indianapolis, Indiana); α-fucosidase from *C. lampas* (Sigma) and from beef kidney (Sigma); neuraminidase from *Vibrio cholerae* (Behring Diagnostics, LaJolla, California) and from *Arthrobacter ureafaciens* (Boehringer Manheim). These exoglycosidases have been widely used to establish sugar sequence and anomeric configuration by sequential degradation of the parent glycolipid to a fragmented glycolipid of known and established carbohydrate structure [18,21,116].

Neuraminidase from *V. cholerae* is often used to establish the position of a sialic acid residues in a ganglioside molecule. Gangliosides containing terminal sialic acid residues can be cleaved easily with *V. cholerae* neuraminidase [11], whereas gangliosides containing internal sialic acid, such as G_{M1}, remain unaffected. Neuraminidase from *A. ureafaciens*, on the other hand, removes all sialic acid residues from a ganglioside, irrespective of the position, and results in the formation of an asialo derivative of the parent ganglioside [11].

The most commonly used endoglycosidase used for hydrolysis of the oligosaccharide chain in glycolipids is endo-β-galactosidase from *Escherichia freundii* [117]. This enzyme specifically cleaves all glycolipids of the lacto and lactoneo series with the common structure R-GlcNAcβ-3Galβ-4Glc (or GlcNAc) groups, in which R represents either hydrogen or sugars. Further substitution of the galactose residue in the essential trisaccharide core structure greatly reduces the susceptibility of the oligosaccharide chains to the enzyme. On the other hand, substitution of a sialosyl group at the terminal galactosyl residue of lactotetraosyl structure enhances the hydrolyzability of the internal Galβ1-Glc linkage. β-Galactosyl linkages of the globo and ganglio series remain unaffected by this enzyme [118].

Mass spectrometry is increasingly becoming a useful tool for elucidating carbohydrate sequences in glycolipids. The greatest advantage of mass spectrometry is that only microgram amounts are required. Early studies employing trimethylsilyl [119] and acetyl [120] derivatives are being replaced by the use of permethylated glycolipids [121–124]. Electron impact mode is used for structural analysis of permethylated glycolipids. Karlsson and co-workers [121,122] discovered that more structural information on a glycolipid can be obtained by analyzing permethylated, permethylated-reduced, and permethylated–reduced-silylated glycolipids. It is possible to determine not only the carbohydrate sequences, but also substitution sites and sphingosine bases and fatty acid distribution in a glycolipid [86,121–125]. The structure of a new ganglioside with ten sugar residues has been elucidated [18] with the Karlsson technique. Chemical ionization mass spectrometry with permethylated glycolipids has also been successfully used to elucidate glycolipid structure. This method provides simpler fragmentation patterns and also provides more distinct information about their molecular weights than electron impact mass spectrometry [126]. Electron impact as well as chemical ionization mass spectrometric methods both require derivatization of the parent glycolipid, and the ion intensities in the high-mass region are very weak, which is often difficult to detect. Newer ionization methods have been developed in which derivatization is not necessary and have been successfully applied for glycolipid structure analysis. The methods employed include field desorption, fast-atom bombardment, and secondary ion mass spectrometry [126–128].

C. Glycosidic Substitution

Glycosidic substitution sites have been traditionally determined by periodate oxidation and permethylation. Periodate is an oxidizing agent, cleaves carbon–carbon bonds between vicinal diols to produce aldehydes. The aldehydes are reduced with sodium borohydride, and the resulting fragments are identified following liberation by acid hydrolysis. Glucose substituted at the C-4 hydroxyl position but not at the C-2 or C-3 hydroxyl position, would give rise to erythritol. On the other hand, if galactose is similarly substituted, the hydrolysis product would be threitol. These two compounds can be easily identified by gas chromatography [129]. Glucose or galactose substituted at the C-2 hydroxyl yields glycerol, as do unsubstituted sugars at a terminal position, whereas C-3–substituted sugars would block oxidation. Unsubstituted fucose produces 1,2-dihydroxypropane and glyceraldehyde. It is to be noted that a glucose residue linked to ceramide in glycolipids tends to oxidize sluggishly, even though it is substituted at the C-4 hydroxyl [130]. Conformational effects [131] and micellar structure have been attributed for this effect. Sialic acid at a terminal position yields the seven-carbon analogues (*N*-acetyheptulosaminic acid), whereas a substitution at C-8 hydroxyl remains unaffected. Thus, periodate oxidation has been useful in identification of the NeuAcα-8NeuAc linkage.

Permethylation is still the most widely used method for determining substitution sites and can be used on a microscale. Permethylation of glycolipids can be done according to the method developed by Hakomori [132,133]. This method uses reaction of a glycolipid with excess methyl iodide in the presence of dimethylsulfinylcarbanion formed by reaction of sodium hydride with dimethylsulfoxide. Hydrolysis of the permethylated derivative, conversion of the substituted monosaccharides into partially methylated alditol acetates, and identification by gas chromatography [134] or gas chromatography–mass spectrometry [135,136] then follow. Hexosamines require a modified procedure to ensure adequate yields [65]. The procedure involves sequential acetolysis and hydrolysis. An alternative procedure utilizing acetylated methylglycoside derivatives can also be used in revealing substitution positions on galactosamine [133] and glucosamine [137].

D. Stereochemistry of the Glycosidic Linkage

The standard method for determining the anomeric configuration of a glycolipid has been the use of specific exoglycosidases described earlier under sugar sequence. In recent years, high-resolution nuclear magnetic resonance (NMR) spectroscopy has been successfully used to elucidate the structures of a number of glycolipids [138–140]. Although NMR methods have an added advantage of being nondestructive, the amount of sample needed, even with a high-resolution (200–300 MHz instruments) is quite high (over 200 μg), which is often difficult to purify. It is possible, however, to elucidate the structure of a glycolipid by ^{1}H-NMR. The information that can be deduced from an NMR spectrum is as follows: (1) The number of sugar residues can be determined by integrating the H-1 signals. The number of amino sugar units is given by the integration of the methyl signals of the acetamido groups. (2) The anomeric configuration of a given sugar unit follows from its

$J_{1,2}$-coupling constant, a value of about 3–4 Hz is observed for α-anomers and about 7–9 Hz for the β-anomers. The possible confusion between different hexose and hexosamine units can be unambiguously excluded by establishing the relationship with the corresponding H-2 resonance by the aid of spin-decoupling difference spectroscopy. (3) The sites of substitution of sugar residues can be deduced by studying the chemical shift of different H-signals.

In addition to conventional ^{1}H-NMR, other methods such as J-resolved two-dimensional ^{1}H-NMR [141], two-dimensional nuclear Overhauser ^{1}H-NMR [142], homonuclear two-dimensional spin-echo J-correlated ^{1}H-NMR [143] and ^{13}C-NMR [144] have been used for structure elucidation of glycolipids.

VII. INDIRECT METHODS FOR CHARACTERIZATION OF GLYCOLIPIDS

A. Thin-Layer Chromatography

Thin-layer chromatography (TLC) is routinely used for resolution of glycolipid mixtures into individual components for pattern distribution. It is often used for preliminary identification of glycolipids by comparison with known standards. One-dimensional TLC is most commonly used for such purpose, which has been reviewed [11,17,71]. For neutral glycolipids, the most effective developing solvents constitute a mixture of chloroform–methanol–water in various amounts. For gangliosides, both neutral and basic solvents are used. The neutral solvent mixtures comprise various amounts of chloroform–methanol–water containing 0.02% calcium chloride. The basic solvent mixtures usually comprised various amounts of chloroform–methanol–2.5 N ammonium hydroxide.

As an example, the ganglioside profiles of human brain white matter, beef brain, and human erythrocytes in neutral and basic solvent mixtures are depicted in Figure 2. Two-dimensional TLC is also used in resolution and fingerprinting of complex glycolipid mixtures [11,22].

B. High-Performance Liquid Chromatography

High-performance liquid chromatography is beginning to show much promise for resolution of glycolipid mixtures into individual components. Although glycolipid mixtures have been separated into individual components without derivatization [82], the sensitivity of detection is quite low. Thus, methods have been developed to analyze glycolipids as their perbenzoylated [82] or *O*-acetyl-*N*-*p*-nitrobenzoylated [97] derivatives. The sensitivity of individual components can be as low as 50 pmol, and complete analysis can be done within 1 h. However, the derivatized glycolipid samples can be used for analysis purpose only. Hydrolysis of the derivatized glycolipids to the native forms is difficult to achieve.

C. Immunostaining of Glycolipids

In recent years, numerous monoclonal and polyclonal antibodies against defined carbohydrate structures have been identified [145–149]. These antibodies can be used to identify intact glycolipids of their hydrolysis products. Glycolipid mixtures can be separated into individual components by thin-layer chromatography with use of a suitable developing solvent system. Individual glycolipids can be detected by binding with specific anti-carbohydrate antibody. If binding occurs to a glycolipid on a plate, the bound anti-carbohydrate antibody can be detected by tagging with ^{125}I-labeled protein A or

^{125}I-anti-immunoglobulin antibody, which can label the sandwiched anticarbohydrate antibody [19,150–151]. The bound antibody to glycolipid can also be detected by enzyme-labeled anti-immunoglobulin antibody [152,153].

VIII. CHEMICAL SYNTHESIS OF GLYCOLIPIDS

Complete chemical synthesis of glycolipids identical with natural materials was performed primarily by Shapiro and co-workers [154,155]. The scheme used by this group is outlined in Scheme 1. The synthetic racemic erythrooxazoline (I) is treated with dilute acid (HCl or H_2SO_4) to give racemic *O*-benzoyl amine salts (II), which were resolved through crystalline salts with chiral tartaric acids. The chiral derivatives were acylated with long-chain acyl chlorides to give chiral ceramide derivatives (III). Glycosidation of (III) with acetobromo sugar in the presence of $Hg(CN)_2$ (Helferich modification of Koenigs–Knorr reaction) gives the glycoside (IV) which is de-*O*-acylated to produce glycolipids (V).

A number of new methods have been developed in recent years for the synthesis of complex oligosaccharides [152,154]. In addition to the oxazoline method, the phthalimide

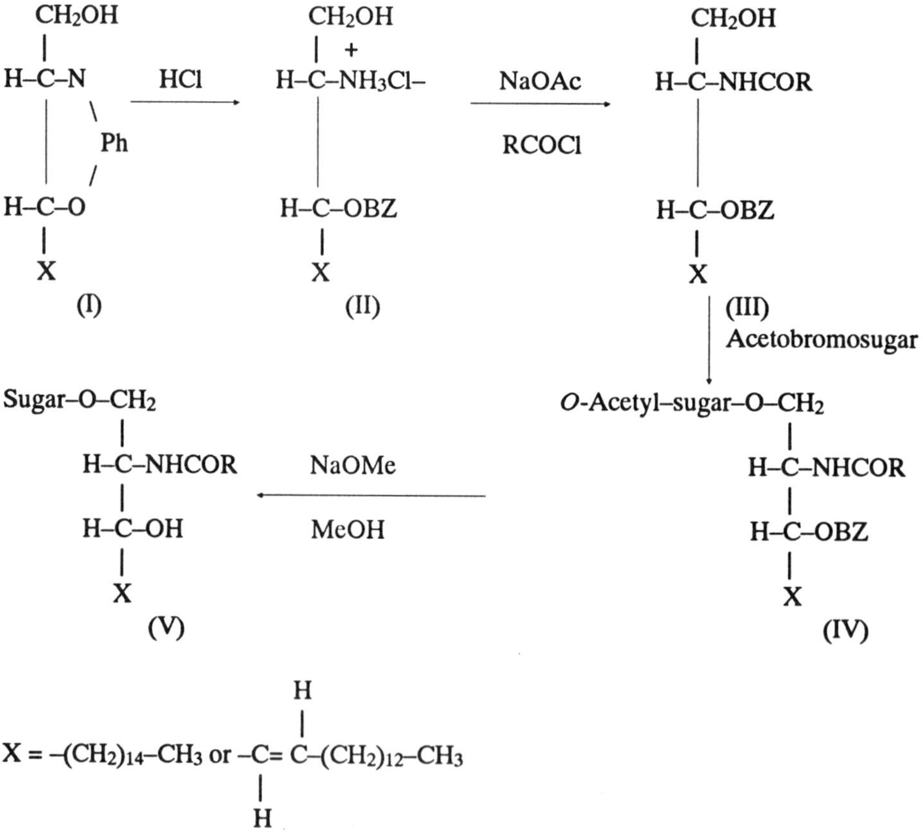

```
CH2OH                        CH2OH                               CH2OH
  |                            | +                                 |
H–C–N         HCl          H–C–NH3Cl–          NaOAc           H–C–NHCOR
  |   \      ------->          |            ----------->           |
  |    Ph                      |               RCOCl               |
  |   /                        |                                   |
H–C–O                      H–C–OBZ                             H–C–OBZ
  |                            |                                   |
  X                            X                                   X
 (I)                         (II)                                  | (III)
                                                                   | Acetobromosugar
                                                                   v
Sugar–O–CH2                                          O-Acetyl–sugar–O–CH2
     |                                                                |
   H–C–NHCOR          NaOMe                                       H–C–NHCOR
     |           <-------------                                       |
   H–C–OH             MeOH                                        H–C–OBZ
     |                                                                |
     X                                                                X
    (V)                                                             (IV)

                                H
                                |
X = –(CH2)14–CH3 or –C= C–(CH2)12–CH3
                     |
                     H
```

R = $-(CH_2)_{14}-CH_3$, $-(CH_2)_{16}-CH_3$, $-(CH_2)_{22}-CH_3$, or $-(CH_2)_{21}-CHOH-$
Sugar = galactose, glucose or lactose

Scheme 1

method has been an effective method for β-glycoside synthesis. A phthalimide group at C-2 is sterically demanding and favors the β-form. 2-Azidopyranosyl halides, if sufficiently reactive, can also be converted into β-glycosides. Several glycolipids that have been chemically synthesized are P^k antigen (GbOSe$_3$Cer) [155,156], gangliotriosylceramide (GgOSe$_3$Cer) [154,157], G_{M4} [158], G_{M3} [159,160], G_{M2}, and G_{M1} [161]. Many other oligosaccharide structures that constitute the backbone of glycolipids have also been synthesized [162]. These include gangliotriosyl (GbOSe$_4$) [163,164], lactoneotetraosyl (LcnOSe$_4$) [165], Lewis B blood determinant [166], Forssman (GbOSe$_5$) [167,168] and i antigen [β(1-3) dimer of lactosamine; 169]. Kabat and co-workers used several synthetic oligosaccharides and glycosides to elegantly establish that the combining site of anti-I antibody (blood group I) was Galβ1-4GlcNAcβ1-6Galβ1- [170]. The usefulness of synthetic oligosaccharides and their derivatives in determining the combining sites of antibodies and lectins and in understanding the conformational properties has been reviewed [171].

IX. BIOSYNTHESIS OF GLYCOLIPIDS

It is generally accepted that glycolipids are synthesized by sequential addition of a specific monosaccharide from sugar nucleotides. The addition occurs to the ceramide portion at the nonreducing end of the growing oligosaccharide chain. In 1970, Roseman [172] proposed that monosaccharides are transferred from the nucleotide derivatives to appropriate acceptors and that each transfer is catalyzed by a specific glycosyltransferase that is a part of a multienzyme complex. The exact localization of all glycosyltransferase is unclear, particularly for gangliosides, for which the assembly of neutral or sialylated oligosaccharide precursors may be at one locus and final sialylation at another. It is believed that glycosylation of glycolipids takes place in the Golgi apparatus or in the endoplasmic reticulum of cells, or both [173–176], from which the newly synthesized compounds move to their final destination, the plasma membrane. Earlier reports that glycosyltransferases could be found on the cell surface [177,178], where they were thought to be involved in *cis* or *trans* (cell-to-cell) glycosylation reactions [179,180], seem open to question [181–183]. The incorporation kinetics of radioactive galactose into the total and surface pools of gangliosides do not support such a mechanism [184]. In addition, it was observed that ganglioside synthesis was not reduced in cells treated with neuraminidase, even though the levels of surface gangliosides were extensively altered [183]. Although glycolipids seem to be synthesized intracellularly, their mechanism of transport to the cell surface is not yet clear.

A membrane flow model [185] is now generally accepted as the mechanism of intracellular translocation of many compounds. It seems likely that the major pool of cellular glycolipids is localized in the plasma membrane with their oligosaccharide chain exposed to the external environment. The remaining glycolipids, presumably located within the cell being degraded by lysosomes, are synthesized in the Golgi apparatus or are being transported to the plasma membrane.

The proposed pathways for biosynthesis of the ganglio series of gangliosides are presented in Figure 12. The specificities of different glycosyltransferases are described in the literature [10,185–190]. Each biosynthetic step involves specific glycosyltransferase (βGalT, β-galactosyltransferase; βGlcT, β-glucosyltransferase; βGalNAcT, β-*N*-acetylgalactosaminyltransferase; αSAT, α-sialosyltransferase) isolated from developing animal brains.

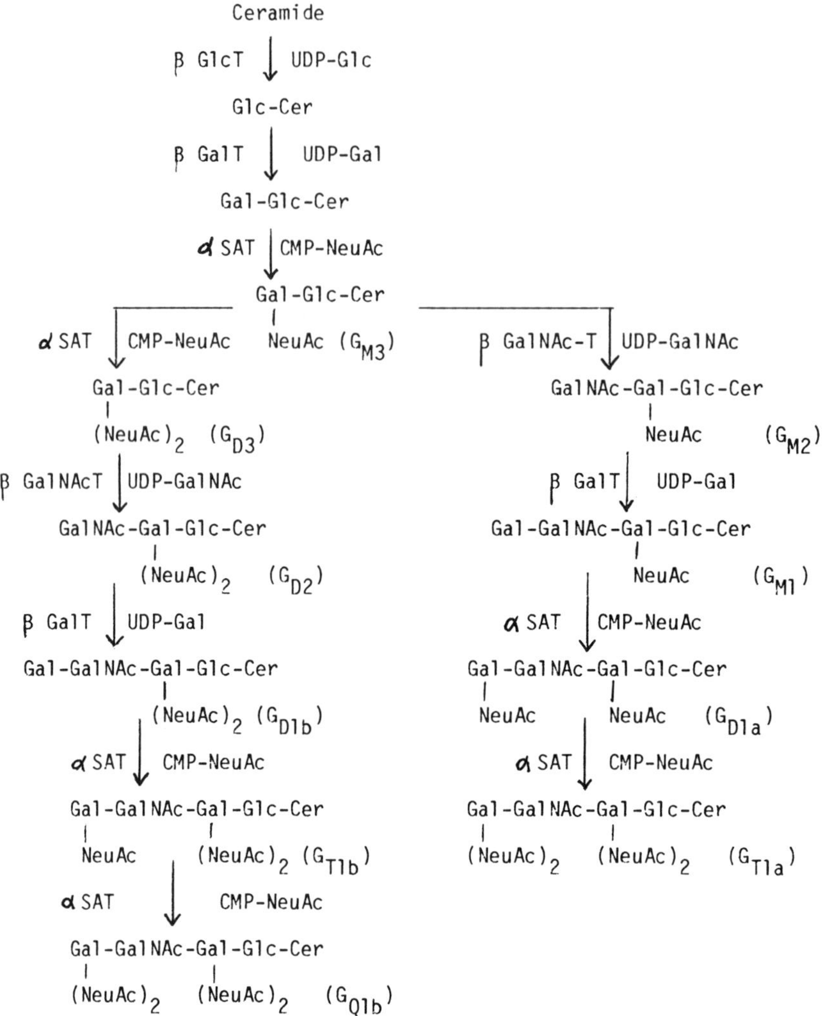

Figure 12 Proposed biosynthetic pathways of brain gangliosides.

The proposed pathways for biosynthesis of blood group-active glycolipids of the lacto series are presented in Figure 13. The stepwise biosynthesis of this group of glycolipids was conducted in chemically differentiated cultured cells [186–189]. The core structure, $LcnOSe_4$-Cer (Galβ1-4GlcNAcβ1-3Galβ1-4Glc-Cer), synthesized by stepwise addition of βGlcNAc and β-Gal to lactosylceramide, gives rise to different blood group-active compounds.

The proposed pathways for the globo series of glycolipids are presented in Figure 14. Specific glycosyltransferases needed in each biosynthetic step have been isolated from developing animal brain and chemically differentiated cultured cells [190,191].

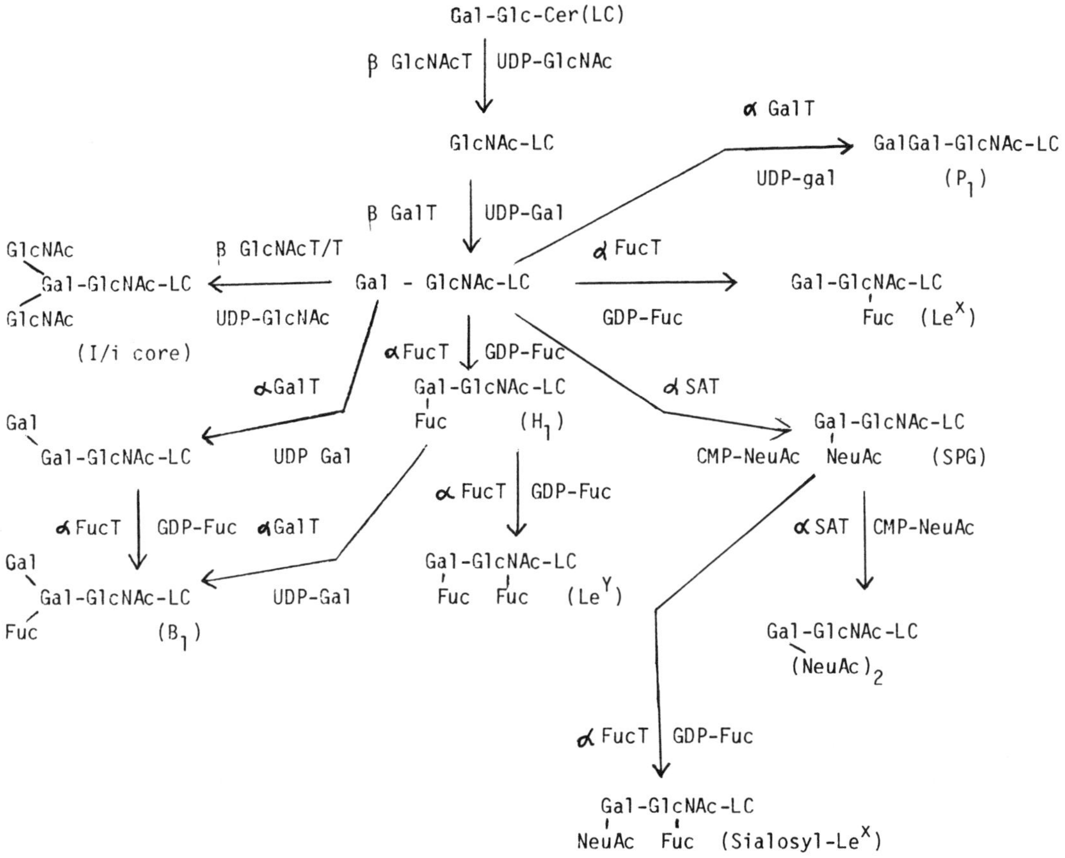

Gal-Glc-Cer(LC)
β GlcNAcT
UDP-GlcNAc
GlcNAc-LC
α GalT
UDP-gal
GalGal-GlcNAc-LC
(P1)
β GalT
UDP-Gal
GlcNAc
Gal-GlcNAc-LC
GlcNAc
(I/i core)
β GlcNAcT/T
UDP-GlcNAc
Gal - GlcNAc-LC
α FucT
GDP-Fuc
Gal-GlcNAc-LC
Fuc (LeX)
αGalT
Gal
Gal-GlcNAc-LC
UDP Gal
αFucT
GDP-Fuc
Gal-GlcNAc-LC
Fuc (H1)
α SAT
Gal-GlcNAc-LC
CMP-NeuAc
NeuAc (SPG)
α FucT
GDP-Fuc
αGalT
Gal
Gal-GlcNAc-LC
Fuc (B1)
UDP-Gal
α FucT
GDP-Fuc
Gal-GlcNAc-LC
Fuc Fuc (LeY)
α SAT
CMP-NeuAc
Gal-GlcNAc-LC
(NeuAc)2
α FucT
GDP-Fuc
Gal-GlcNAc-LC
NeuAc Fuc (Sialosyl-LeX)

Figure 13 Proposed biosynthetic pathways of blood group-active glycolipids.

Gal-Glc-Cer
αGalT
UDP-Gal
Gal-Gal-Glc-Cer
(Pk)
βGalNAcT
UDP-GalNAc
GalNAc-Gal-Gal-Glc-Cer
(P)
αGalNAcT
UDP-GalNAc
GalNAc-GalNAc-Gal-Gal-Glc-Cer
(Forssman)

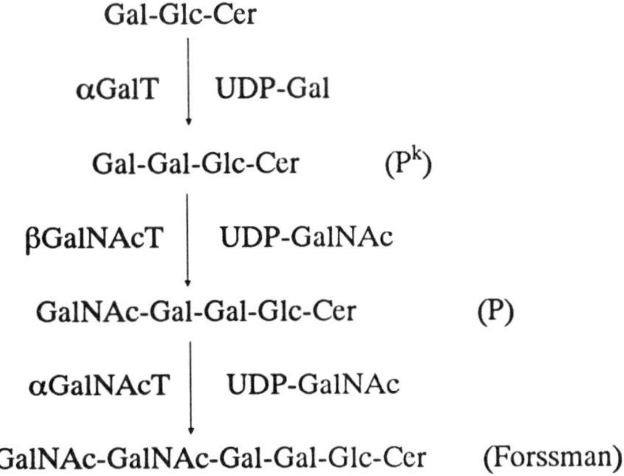

Figure 14 Proposed biosynthetic pathways of the globo series of glycolipids.

X. CATABOLISM OF GLYCOLIPIDS AND STORAGE DISEASES

Glycolipids are catabolized in a stepwise fashion by exoglycosidases. Evidence indicates that lysosomes are the principal site of glycolipid catabolism and that endoglycosidases are not involved in the catabolic process [192]. Virtually all of the catabolic enzymes for glycolipids show acidic pH optima. In addition, many occur in particulate fractions and show enhancement of an activity by detergents. In recent years, several individual glycosides have been isolated in varying degrees of purity, and the degradative pathways of many of the complex glycolipids have been proposed [21,192]. Figure 15 depicts the catabolism of trisialoganglioside G_{T1a} by sequential action of eight exoglycosidases. On the other hand, the major trisialo ganglioside G_{T1b} is believed to metabolize primarily to G_{D1b},

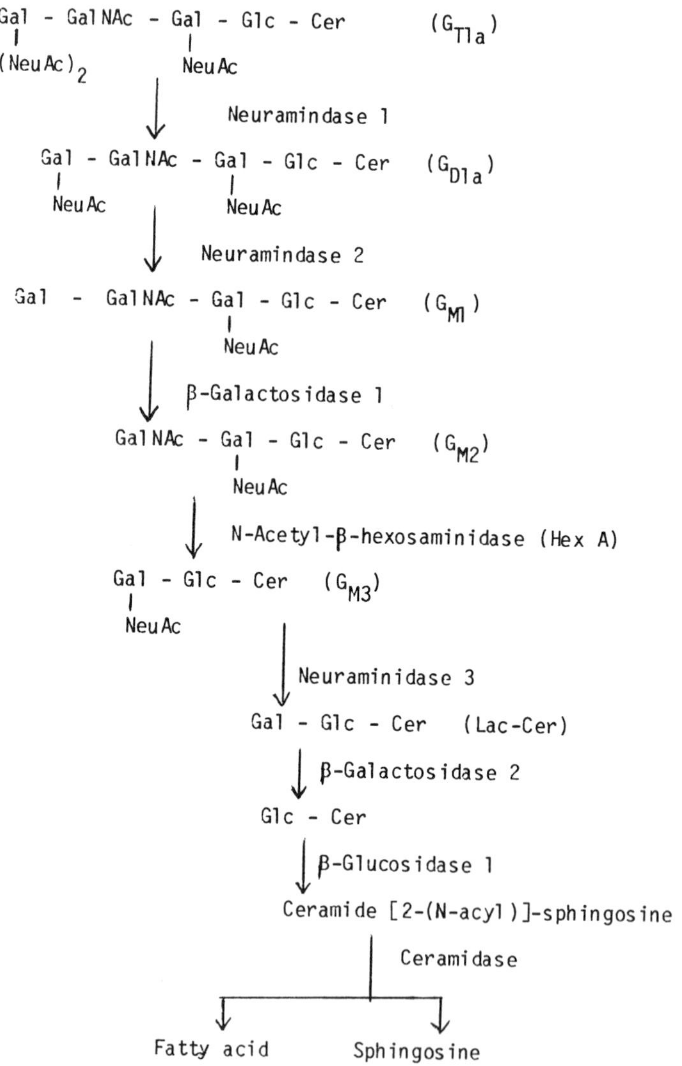

Figure 15 Proposed pathways for the catabolism of G_{T1a} ganglioside.

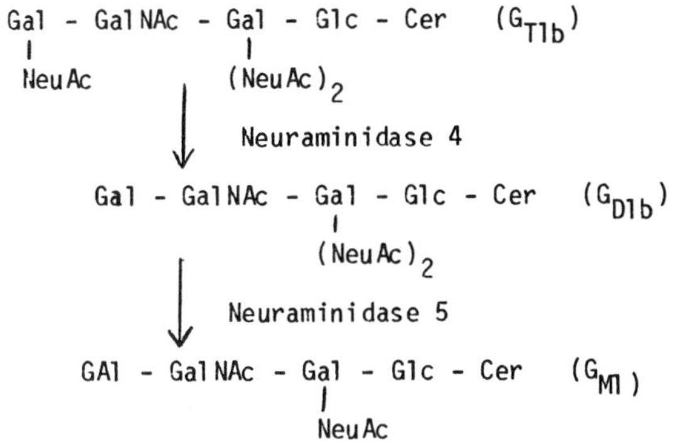

```
Gal - GalNAc - Gal - Glc - Cer    (GT1b)
 |              |
NeuAc          (NeuAc)2
         |
         ↓  Neuraminidase 4
     Gal - GalNAc - Gal - Glc - Cer    (GD1b)
                     |
                    (NeuAc)2
         |
         ↓  Neuraminidase 5
    GAl - GalNAc - Gal - Glc - Cer    (GM1)
                    |
                   NeuAc
```

Figure 16 Proposed pathways for the catabolism of G_{T1b} ganglioside.

rather than to G_{D1a} because of the greater activity of NeuAc bound to the terminal galactose (Fig. 16). G_{D1b} then metabolizes to G_{M1} and, subsequently, degrades as shown in Figure 16. This rate differential was demonstrated for the enzyme of *V. cholerae* [193] and human brain neuraminidases [194]. The sialic acid residue in G_{M1} remains resistant to neuraminidase until the terminal galactose and galactosamine moieties have been removed. Nonreactivity of the sialic acid in G_{M1} appears as a steric block of *N*-acetylgalactosamine bound to the C-4 hydroxyl of galactose, to which the sialic acid is attached [21].

The catabolic pathways of blood group antigens are depicted in Figure 17. The pathways described in Figure 17 represent normal metabolic pathways. A disorder in lipid metabolism that results from an inherited deficiency of any of the exoglycosidases causes accumulation of a particular glycolipid. Currently, at least ten distinct disorders of lipid metabolism are well established. These diseases are collectively called lysosomal storage diseases [for review see Ref. 195–197]. Table 6 shows the most common inherited metabolic disorders. The major glycolipids that accumulates because of a defect or deficiency in the catabolic enzyme or activator proteins [198–201] for each specific disorder are listed. The Sphingolipid storage diseases, except Fabry's disease, are inherited as autosomal recessive genetic mutations.

XI. FUNCTIONS OF GLYCOLIPIDS

The direct function of a glycolipid in mammalian tissue or cells is still an open question. However, glycolipids have been implicated in various biological phenomena. This review covers a brief summary of the present-day knowledge of some of the possible functions in different biological processes.

A. Cell Surface Receptors

Glycolipids, particularly gangliosides, interact with many bioactive factors by inhibiting or interfering with the physiological effects of these factors or cells. These effects have been interpreted as being receptors, at which glycolipids are involved with different biofactors.

The most extensively studied receptor function of glycolipids is the interaction of ganglioside G_{M1} with cholera toxin, which has been reviewed [202]. This toxin is produced by *V. cholerae*. Van Heyningen and co-workers [202] were the first to show that brain

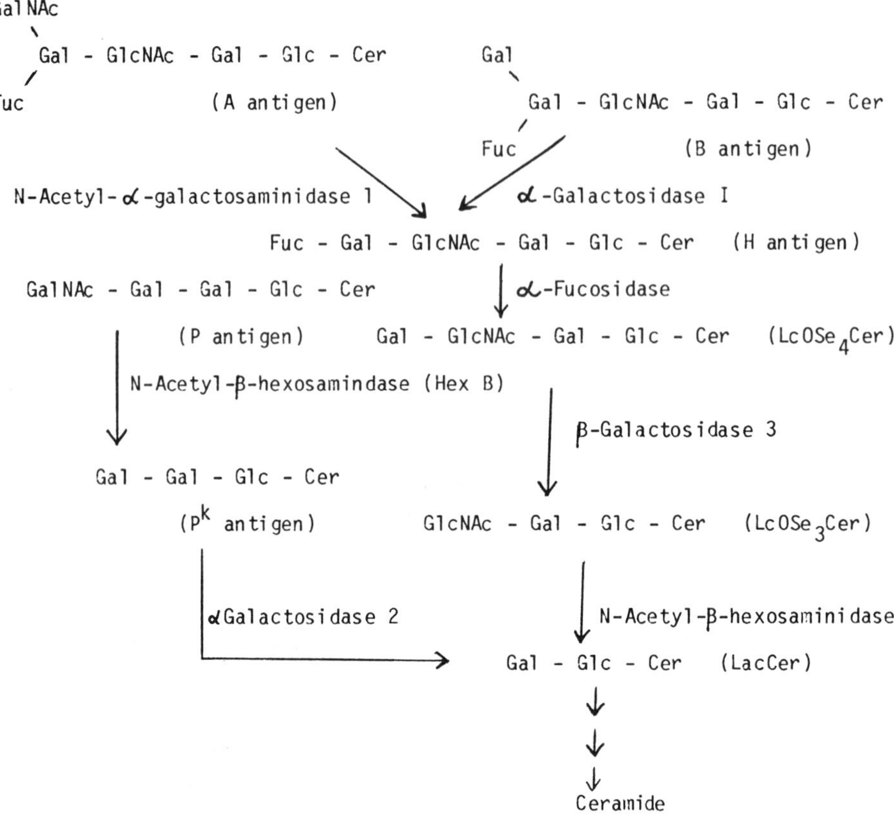

Figure 17 Proposed pathways for the catabolism of blood group glycolipids.

gangliosides bind cholera toxin and block its physiological effect. It was later demonstrated that ganglioside G_{M1} was the most effective inhibitor [204,205]. The evidence to substantiate specific binding was based upon measurement of biological response resulting in cAMP formation and included various experimental approaches including the blockage of biological effect by exogenous gangliosides and the lack of response of G_{M1}-deficient SV40-transformed cells to cholera toxin. When gangliosides other than G_{M1} were taken up in the cell membrane to substitute for G_{M1}, no cAMP formation was noted if cells were exposed to cholera toxin. Further evidence was inferred by the observation that endogenous and exogenous G_{M1} is protected by cholera toxin from the action of galactose oxidase or periodate followed by sodium borohydride. The binding of G_{M1} results in a conformational change of the toxin, resulting in a perturbation of the cellular membrane. This, in turn, is thought to facilitate dissociation and penetration of the A subunits, which triggers activation of adenylate cyclase. Although G_{M1} seems to be the best-characterized receptor for cholera toxin for many cells, this may not be true for all cells. G_{M1} is absent in fat cells and adrenal cells, but they still respond to cholera toxin [208]. The binding of cholera toxin to mouse LY and human KB-3 cell membranes is trypsin-sensitive [209]. Glycoproteins of rat microvillus membranes, prepared by *Ricinus communis* lectin-Sepharose chromatography, were shown to react with cholera toxin [210]. Similarly, some proteins from Balb/C 3T3

Table 6 Major Glycolipid Storage Diseases

Disease	Major glycolipid accumulated	Enzyme defect
Gaucher's	Glcβ1-1-Cer	β-Glucosidase
Krabbe's	Galβ1-1-Cer	β-Galactosidase
Metachromatic leukodystrophy	$3SO_3H$-Galβ1-1 Cer	Arylsulfatase A
Fabry's	Galα1-4 Galβ1-4Glc-Cer	α-Galactosidase
Fucosidosis	Fucα1-2Galβ1-4GlcNAcβ1-3Galβ1-4Glc-Cer (H_1) Fucα1-2Galβ1-3GalNAcβ1-4Galβ1-4Glc-Cer (Fuc-G_{M1}) NeuAc	α-Fucosidase
G_{M1}-gangliosidosis	Galβ1-3GalNAcβ1-4Galβ1-4Glc-Cer (G_{M1}) NeuAc	β-Galactosidase
Tay-Sachs, type II (Sandhoff, O variant)	GalNAcβ1-4Galβ1-4Glc-Cer (G_{M2}) NeuAc GalNAcβ1-4Galβ1-4Glc-Cer (as G_{M2})	*N*-Acetyl-β-hexosaminidases A and B
Tay-Sachs, type I (classic, B variant)	G_{M2}	*N*-Acetyl-β-hexosaminidase A
G_{M2}-gangliosidosis, type III (juvenile) and type IV (adult, AB variant)	G_{M2}	Partial *N*-Acetyl-β-hexosaminidase A (deficiency in activator protein)

cells seem to react with cholera toxin [211]. Thus, G_{M1} could be the major receptor for some cells, but proteins that carry a structure similar to G_{M1} [212] could also be receptors.

Tetanus toxin, produced by *Clostridium tetani*, binds specifically with gangliosides of the ganglio series that carry two nonterminal sialic acid residues, G_{D1b} and G_{T1b} [213,214]. The binding is more specific than other gangliosides, G_{M2}, G_{M1}, or G_{D1a} [215]. Although several other studies also have indicated similar specificity for tetanus toxin, others imply that the original concept may not be true. It was claimed by Zimmerman and Gifferetti [216] that differentiating mouse neuroblastoma C 1300 cells pretreated with neuraminidase and β-galactosidase still fixed tetanus toxin (direct immunofluorescence) by a mechanism unrelated to gangliosides. Stoeckel et al. [217] indicated that tetanus toxin is internalized at the peripheral nerve and transported retrogradely. Ganglioside G_{T1b} reduced this axonal transport by only 50%. It was also demonstrated that G_{D1b}, but not its free oligosaccharide, binds to tetanus toxin, indicating a requirement of the lipid moiety in the binding process [218]. It seems apparent that gangliosides may not be the functional receptors for tetanus toxin. Additional experiments are required to elucidate the role of gangliosides with the binding and action of tetanus toxin. By using ganglioside-deficient cells that do not bind or respond to tetanus toxin, one might be able to demonstrate that treatment of the cells with G_{T1b} and G_{D1b} elicited toxin binding and action. One obstacle is that the mechanisms of toxin action and its subsequent biochemical effects have not been resolved.

Botulinum toxin was also reported to bind specifically to G_{T1b} and G_{Q1b} [219,220]. G_{T1a}, a positional isomer of G_{T1b} containing two sialic acid residues at the terminal end, is about 136 times less effective than GT1b, which contains one sialic acid residue at the terminal end. From this, it can be inferred that a disialosyl group at the internal galactose and a mono- or disialosyl group at the terminal galactose are essential for fixation of toxin action.

Other toxins that seem to bind to gangliosides include staphylococcal α-toxin, which binds to 2-3-sialosyllactoneotetraosylceramide [221]; gonococcal pilli protein, which binds to G_{M1} [222]; hemolysins of *Vibrio parahaemolyticus*, which bind to G_{T1b} [223]; *Escherichia coli* toxin, which binds to G_{M1} and appears to be a clone of cholera toxin [202,224,225].

A conspicuous parallelism appears to exist for the binding of toxin and that of thyroid-stimulating hormone, thyrotropin, chronionic gonodotropin and luteinizing hormone, the common gangliosides being G_{T1b} and G_{D1b} [226–228]. Gangliosides G_{M2} and G_{T1b} inhibit both mouse and human interferon activity [229,230]. Mouse fibroblast interferon is absorbed on a ganglioside–Sepharose column and eluted by sialosyllactose [229,231]. These nonspecific interactions of gangliosides with hormones and interferon strongly suggest that gangliosides are not the real receptors for glycoprotein hormones and interferons. A real receptor may be an amphipathic polypeptide associated with the ganglioside fraction [232]. Gangliosides may be the cofactor for the real receptor and may regulate hormone binding to the receptor. The analogy was recently described for expression of the Paul–Bunnel antigen, which is a ganglioside polypeptide [232]. The interaction of G_{T1b} and G_{D1b} with the hormones seems to be a cross-reaction between a real receptor with a structure similar to that of G_{T1b} and G_{D1b}. A ganglioside with an unidentified structure was more inhibiting of thyrotropin binding among a larger number of thyroid gangliosides [228]. Most work that was carried out in these studies utilized gangliosides that may not have been pure because they are often contaminated with peptides having similar mobility in thin-layer chromatography [233]. Careful studies are needed to identify receptor molecules among candidate glycolipids, since the purity of the test lipid is of utmost importance in such work.

Woolley and Gommi [234] observed that the serotonin sensitivity of neuraminidase-treated fungal preparations can be restored by adding a G_{D3} ganglioside. Further studies have disputed the finding that gangliosides may be the receptor for serotonin [215,235]. A hydrophobic protein or a proteolipid is more likely to be responsible for the high-affinity binding of serotonin.

Sendai virus has been shown to bind with gangliosides containing a common terminal sequence NeuAcα2-8NeuAcα2-3-Galβ1-3GalNAc-R which is present in G_{T1a}, G_{Q1b}, and G_{P1c} [236,237]. However, these polysialogangliosides are absent in various cell lines that are susceptible to Sendai virus-mediated cell fusion [1]. Binding of Sendai virus by HeLa cells or with gangliosides and proteins isolated from HeLa cells and bovine brain gangliosides, indicated that only the HeLa proteins, but not the gangliosides, inhibit virus-induced hemagglutination [238]. However, it has been claimed [239] that gangliosides, such as G_{D1a}, G_{T1b}, and G_{Q1b}, function as receptors for Sendai virus by conferring susceptibility to infection when they are incorporated into receptor-deficient cells. The endogenous gangliosides of the three commonly used host cells for Sendai virus, MDBK, HeLa, and MDCK cells, were shown to have G_{M3} (major), G_{M1}, G_{D1a}, and other polysialogangliosides of the ganglio series [240]. Suzuki et al. [241] claimed that a sialosylglycoprotein fraction isolated from hog erythrocytes is responsible for activities with Sendai virus. The concept that ganglioside is the real receptor will undoubtedly remain open to controversy until the mechanisms of viral infection and its subsequent biochemical effects are clearly understood.

Kleinman and co-workers [242] discovered that gangliosides G_{T1b} and G_{D1b}, but not G_{M1} and G_{M3}, inhibit cell adhesion on a fibronectin–collagen layer absorbed on plastic surfaces. However, cell adhesion on a collagen or lectin layer is also inhibited by these gangliosides. It is known that various fibroblasts that have a high content of fibronectin lack higher gangliosides [243]. Thus, it seems that gangliosides may not be the receptor of fibronectin.

Antigen or mitogen stimulates lymphocytes and release a variety of soluble mediators, lymphokines. The lymphokines then affect the functional properties of macrophages, neutrophils, and other cells. Macrophage migration inhibitory factor (MIF), one of the most extensively studied lymphokines, renders macrophages cytotoxic for tumor cells. The MIF activity can be eliminated by treatment of guinea pig macrophages with liposomes containing a glycolipid from guinea pig macrophages, but not with liposomes containing glycolipid of neutrophils and brain tissues [244–246]. Of the various oligosaccharides, only α-L-fucosyl oligosaccharides inhibit MIF activity [247]. Fucose-binding lectins of *Lotus tetragonolobus* and *Ulex europeus* agglutinin, but not other lectins, interfere with the response of macrophages to MIF [245]. A fucolipid with the structure Galα1-3(Fucα1-2)Galβ1-3GalNAcβ1-4Galβ1-4Glc-Cer, isolated from rat granuloma and peritoneal macrophages, inhibits MIF activity [1]. On the other hand, a fucolipid with a lactoglycosyl structure does not inhibit the activity. On this basis, it can be inferred that MIF interacts with a fucosylated glycolipid containing a ganglio series structure Fucα1-2, 1-3, or 1-6 Galβ1-3GalNAcβ-R, but not with fucosylated glycolipid containing the lacto series structure Fucα1-2, 1-3, or 1-6Galβ1-4GlcNAcβ1-R [248]. From these results, it can be inferred that the receptor for MIF is a fucosylated glycolipid of the ganglio series or its analogue.

Exogenous addition of gangliosides (G_{M3} and its lyso derivative) has caused a decrease in the tyrosine protein kinase activity of the epidermal growth factor (EGF) receptor and an inhibition of EGF-stimulated growth [249]. It was hypothesized that the EGF receptor is physiologically regulated by gangliosides. This hypothesis was tested with a mutant Chinese hamster ovary (CHO) cell line that has a defect in the biosynthesis of gangliosides [250]. An inverse correlation between the level of ganglioside expression and signal transduction by the EGF receptor was observed that is consistent with the hypothesis that the function of the EGF receptor is physiologically regulated by gangliosides. Interestingly, a ganglioside, de-*N*-acetyl G_{M3} (II^3NeuNH2LacCer) isolated from A431 cells and B16 melanoma was shown to enhance the kinase activity of EGF receptor and an enhancement of EGF-stimulated growth [251]. These properties of de-*N*-acetyl G_{M3} are in striking contrast with those of G_{M3} and its lyso derivative as discussed in the foregoing. These data indicate that de-*N*-acetylation at the sialic acid moiety of G_{M3} is an important mechanism for modulation of EGF-dependent cell growth. This is antagonistic to that of G_{M3}-dependent modulation of receptor function.

From the results of the interaction of gangliosides with various biofactors, as described in the foregoing, it can be inferred that some of the interaction seem ionic and nonspecific. The binding of a ganglioside should not be interpreted as the presence of a receptor. However, it is definitely possible that gangliosides regulate or modulate the receptor function.

B. Cellular Antigens

It has been clearly demonstrated in recent years that many blood group antigens are glycolipids. The study of glycolipids not only offered a clear chemical structure of blood group A, B, H. I, i, P_1, P, and P^k antigens, but also delineate their structural polymorphism and their immunological significance. This has been reviewed [24,24,252–256]. These blood group determinants are carried by glycolipids as well as by a few specific membrane proteins. Four polymorphic components of blood group A, designated by A_1, A_2, A_3, and A_4 have been isolated from A erythrocytes [24]. They all contain the type 2 backbone structure found in glycoproteins (i.e., the subterminal disaccharide is Galβ1-4GlcNAc). The structural requirement for blood group A activity is the presence of GalNAcα1-3[Fucα1-2]Galβ1-3GlcNAc. The polymorphic forms differ from each other in the chain length or branching (see Table 2). Blood group H-active polymorphic forms, H_1, H_2, H_3, and H_4 have been isolated from human O erythrocytes [24] and also contain a type 2 chain backbone. They differ from each other in chain length or branching, as in A-active determinants. For H activity, the structural requirement is Galα1-3[Fucα1-4]Galβ1-4GlcNAc. Blood group B-active glycolipids have not as yet been thoroughly investigated. Two structures, ceramide hexasaccharide (B_1) and ceramide octasaccharide (B_2) have been isolated from human B erythrocytes [24,255,256]. Blood group ABH glycolipids isolated from erythrocytes belong to the type 2 chain. Blood group ABH-active glycolipids with a type 1 chain containing Galβ1-3GlcNAc subterminal disaccharide have also been isolated from animal tissues, including intestinal mucosa, intestine, and bovine serum [9,24]. The type 2 chain is also found in animal tissues. The distribution of type 1 and type 2 chains among glycolipids is species-dependent and organ-dependent; this is reviewed by Hakomori [24]. Blood group ABH-active glycolipids containing the type 1 chain could have a straight or branched carbohydrate structure, ranging from 6 to 20 sugar residues.

The blood group I and i antigens are precursors for ABH antigens. The mechanism of incomplete synthesis of ABH determinants that create the Ii structure is not known. The reactivity of sialosyllacto-*N-nor*-hexaosylceramide and their derivatives with six monoclonal anti-i antibodies results in a clear conclusion that the minimum structural requirement for i activity was the two repeating *N*-acetyllactosamine (Galβ1-4GlcNAcβ1-3) structures (i.e., lacto-*N*-nor-hexaosyl). The blood group I activity against anti-I auto-antibodies varies in specificity, but most of these antibodies react with a determinant present in the branched lacto-*N-nor*-octaosyl- structure that is found in glycolipids and glycoproteins (see Table 2) [24,252]. The I antibodies are capable of reacting with this structure, even when the terminal nonreducing galactosyl residues are substituted by sialic acid, fucose, or other sugars [24].

The human erythrocyte P blood group system contains two common antigens, P_1 and P and the rare P^k antigen. These three antigens have been identified as glycolipids, and not glycoproteins, by hemagglutination inhibition studies and chemical analysis [25]. The P_1 antigen contains a lactoneotetraosylceramide to which Galα1-4 is linked to the terminal galactose residue. The P antigen is present in all normal human erythrocytes and has been identified as globotetraosylceramide; the P^k antigen is a precursor of P antigen and is identified as globotriaosylceramide. The immunochemistry and genetics of blood group P system has been reviewed [25].

C. Interaction Antigens

Some antibodies appear to be directed against interaction products of the I and P blood group systems (e.g., IP_1 and IP) [25]. As discussed earlier, the requirement of I specificity is a branched lacto-*nor*-octaosylceramide and for P_1, the structured requirement is Galα1-4 linked to lactoneotetraosylceramide. Anti-IP_1 probably reacts with a branched glycolipid that has this basic structure plus a terminal Galα1-4 residue that would create a P_1 determinant. P_1 activity was detected in a larger glycolipid that has not yet been identified [25]. However, it can be hypothesized that a glycolipid structure with dual specificity, which would react with anti-IP_1, can be isolated from human erythrocytes.

In the Lutheran blood group system, the In(Lu) phenotype, which is inherited as a dominant autosomal trait, the expression of the Lutheran, P_1, i, and Auberger antigens is suppressed [25]. Both P_1 and i antigens contain the lactoneotetraosyl structure, and it seems likely that the addition of a terminal Galα1-4 residue to the lacto-*N-nor*-hexaosyl structure would create a P_1 determinant. We have detected Lu_b activity in the human erythrocyte gangliosides, most of which contain the lactoneotetraosyl structure [18,24]. It can be hypothesized that the Lutheran, P_1 and Ii antigens contain certain common structures, namely, lactoneotetraosylceramide or a homologous larger structure with repeating *N*-acetyllactosamine units or a branching structure built on this backbone. The simultaneous depression of the Lutheran, P_1, and i antigens could result from modification of this common backbone structure by addition of an extra sugar moiety, such as sialic acid or fucose, or by a block in biosynthesis of a common intermediate structure. The later possibility is excluded because excessive phenotype Lu_{a-b-} have normal P_1 and i antigens. Our analysis indicated that this recessive phenotype contains normal quantities of lactoneotetraosylceramide. Moreover, the ABH antigens share the same backbone, and they are normal in both recessive Lu_{a-b-} and the In(Lu) phenotypes. It seems likely that In(Lu) phenotypes result from the action of a glycosyltransferase that adds an extraneous sugar moiety shared by the Lutheran, P_1, i, and Auberger antigens [25].

D. Acquired Cellular Antigens

It has been generally assumed that cellular antigens are synthesized by the cells in which they are detected, but more recent studies have indicated that glycolipids are acquired by cells [257,258]. A number of erythrocyte blood group antigens, including Le^a, and Le^b, and others that resemble the Lewis antigens of cattle, sheep, and pigs, are acquired from plasma and are not synthesized in situ [9,13,257,258]. Relatively little is known about the mechanism, extent, specificity, or reversibility of glycolipid transfer in vivo. Glycolipids of human plasma are, within experimental error, all associated with lipoproteins [259–262], and all of the glycolipids were found to be in the same relative proportions in the major classes of lipoproteins. Approximately 60% of glycolipids are associated with low-density lipoproteins (LDL), 25–30% with high-density lipoproteins (HDL), and 12–14% with very low-density lipoproteins (VLDL). Recent studies indicated that transfer of glycolipids between vesicles and cell membranes is facilitated in the presence of transfer proteins isolated from liver [263,264]. Glycolipids can be transferred from a donar HDL to a lipoprotein-deficient acceptor in the presence of an exchange factor isolated from human plasma [265]. Similarly, ω-pyrenyl-labeled glycolipids can be transferred from the donar HDL to unlabeled lymphocytes in the presence of the plasma protein factor, but $[^3H]G_{M3}$ was not transferred from HDL to LDL or to lymphocytes at a measurable rate in the absence of the plasma protein factor [265]. The reversible exchange of glycolipid between HDL and LDL in vitro has been reported [266,267]. Most of the data showing alteration in lymphocyte reactivity with glycolipid vesicles has very little biological import. The vesicles used may cause changes in cell membrane cholesterol and phospholipids. The lymphocyte changes, therefore, are not primarily due to acquisition of glycolipid. Any manipulation that perturbs the membrane composition or organization can activate or inhibit lymphocyte response. Several lines of evidence have indicated that transfer of glycolipids from lipoproteins to cells in vivo is a selective and complex process that differs markedly from the procedures that were used to demonstrate glycolipid transfer in vitro [13,267]. However, it is known that there is a marked difference between the glycolipid compositions of plasma, erythrocytes, and lymphocytes [19,266].

E. Tumor-Associated Glycolipids

Glycolipids of cell surface membranes alter their chemical composition and organization in tumors in vivo and upon oncogenic transformation in vitro. The tumor-associated glycolipids are immunologically distinct for tumor cells, but in many cases, the same antigens are present, in small quantity, in normal homologous cells. Several reviews have appeared in recent years on this subject [1,12,14,268–271]. The changes associated with the oncogenic transformation can be classified into four major types: (1) incomplete synthesis of the sugar chain owing to a block of a single or multiple glycosyltransferase(s) which results in an overall "simplification" of the glycolipid structures; (2) synthesis of a new single or multiple glycolipid(s), with the same core structure, that is unique to the tumor, such as incompatible blood group antigens and new structures that are absent in normal homologous cells; (3) a shift in the glycolipid synthesis from one series of a glycolipid to another (e.g., ganglio series to globo series or lacto series); (4) an organizational change of glycolipids on the cell surface that alters their antigenic expression. This change may be due to a change in glycolipid environment, to change in the hydrophobicity of the ceramide moiety, or to a combination of both factors.

I have not attempted to provide a comprehensive review on this subject, but rather to provide recent findings on human cancer antigens that have been defined by specific monoclonal antibodies [145,146,269].

1. Ganglio Series

Human Melanoma. Ganglioside G_{D3} was identified as melanoma antigen, as defined by a human melanoma-specific IgG antibody [272] and by an IgM antibody [273]. This antigen contains a much higher proportion of long-chain fatty acids (C_{22} and C_{24}), compared with ganglioside G_{D3} isolated from brain [274]. Although G_{D3} is chemically detected in various tissues, the monoclonal antibody shows high specificity for human melanoma cells and tissues. A monoclonal antibody with higher specificity for human melanoma has also been identified [275,276]. This antibody is directed against the G_{D3} ganglioside containing 9-*O*-acetyl sialic acid. An anti-G_{D3} IgG3 antibody has been used in clinical trials in two independent laboratories. The results showed a regression of melanoma, without causing damage of normal tissue [277,278].

A monoclonal antibody against G_{M2} (KM531) was evaluated with the ganglioside fraction of human gastric mucosa [279]. The monoclonal antibody binds to G_{M2} as well as with the CAD antigen [GalNAcβ1-4(NeuAcα2-3)Galβ1-4GlcNAcβ1-3Galβ1-4Glc-Cer] having the same terminal determinant. A decrease in CAD antigen and an increase in G_{M2} have been associated with gastric mucosa.

Oncofetal Antigens. Two human monoclonal antibodies against oncofetal antigens (OFA-1-1 and OFA-1-2) have been produced by transforming B-lymphocytes of melanoma patients with Epstein–Barr virus [280]. The anti-OFA-1-1 reacts with a variety of human cancer cells: melanoma, brain tumors, breast cancers, and other types of tumors. The OFA-1-1 was identified as G_{M2} [281]; OFA-1-2 was identified as ganglioside G_{D2}.

The anti-OFA-1-2 defines an antigen on neuroectodermal tumors [282]. A melanoma-associated antibody (AH) detected in the serum of patients was found to be specific for G_{D2} [283]. Although G_{M2} and G_{D2} are detected in normal cells as minor components or are cryptic; they are highly exposed and accumulate in tumor cells, when other longer-chain gangliosides of the ganglio series are deleted. An iodine-131–labeled monoclonal antibody against G_{D2} has been used in the diagnosis and therapy of human neuroblastoma [284].

A specific monoclonal antibody was raised against the fucosyl-G_{M1} ganglioside isolated from human small-cell lung carcinoma [285,286]. This ganglioside contains phytosphingosine exclusively, compared with other sources that contain the same carbohydrate structure [287]. Fucosyl-G_{M1} has been identified as a marker for human small-cell carcinoma [285–287].

2. *Globo Series*

Burkitt's Lymphoma Antigen. This antigen, as defined by a rat monoclonal IgM antibody [288], was identified as globotriaosylceramide ($GbOSe_3$-Cer). It is specifically expressed on Burkitt's lymphoma cells, irrespective of whether or not they possess the Epstein–Barr virus genome [289]. This antigen was previously identified as the blood group P^k antigen [25]. Normal human erythrocytes contain a moderate quantity of $GbOSe_3$-Cer (5–10 μg/10 mg protein), whereas, in blood group P^k erythrocytes, an increase of approximately fivefold in $GbOSe_3$-Cer is observed [12,25]. Burkitt's lymphoma cells contained large quantity of $GbOSe_3$-Cer (200–800 μg/10 mg protein). P^k erythrocytes react very well with anti-P^k antibodies, but normal erythrocytes do not react with them [25]. The rat monoclonal IgM antibody does not react with normal erythrocytes. This may imply that the small quantity of $GbOSe_3$-Cer in various normal cells is cryptic. Recently, it has been documented [290] that Gbo_3Se_3-Cer massively accumulates in human testicular tumors and may possibly be used as a marker, especially for seminomas.

Forssman Glycolipid Antigen. Humans have been considered as a Forssman-negative species. More recent studies indicate that, in 20–30% of the population, small quantities of the Forssman glycolipid are contained in gastrointestinal tissue; however, in most of the population (70–80%) this tissue does not contain the Forssman glycolipid [291]. In human gastrointestinal cancer specimens derived from Forssman- negative tissue, the Forssman glycolipid was found to be present [292]. Various specimens of squamous carcinoma and adenocarcinoma of the lung and of fetal lung contain 60–90 times more Forssman glycolipid than the normal counterparts [293]. Forssman glycolipid has also been detected in various human lung, breast, and gastric cancer cell lines by indirect immunofluorescence [294] and by immunostaining of thin-layer chromatograms of total glycolipids [295]. In most lung cancers, enhanced α-GalNAc transferase activity was observed [296] and, therefore, it was not surprising that Forssman glycolipid appears in Forssman-negative tissues. However, the amount of Forssman glycolipid in various human cancers is quite low, and most anti-Forssman antibodies could not stain the Forssman antigen in tumor tissue, indicating that antigenicity in these tumors is of minor importance.

Galactosylgloboside Antigen. Galactosylgloboside (Galβ1-3-$GbOSe_4$-Cer) has been identified in patients with primary lung cancer [297]. Immunological methods, using monoclonal antibodies J309 and D579, and chemical methods were used to identify this antigen. The implications of the findings for the biology of lung cancer and the presence or absence of this glycolipid in other tissues await further confirmation for its use as a new marker.

Sialosylglobotetraosylceramide Antigen. The only ganglioside of the globo series identified as a tumor-associated glycolipid is sialosylglobotetraosylceramide ($IV^3NeuAcGbOSe_4Cer$) isolated from human teratocarcinoma [298]. It is also found as a minor component in muscle biopsy specimens of amyotropic lateral sclerosis, identified by immunostaining of the neuraminidase-treated total glycolipids in the monospecific anti-$GbOSe_4Cer$ IgG antibody [28].

3. Lacto Series

Sialosyl-Le^a Antigen. A number of monoclonal antibodies directed against colorectal adenocarinomas have been discussed in the literature [299]. Among these antibodies, N-19-9, was reported to be highly specific for colorectal adenocarcinoma as well as for gastric and pancreatic cancer. The binding of the antibody to carcinoma tissue extract was inhibited by serum from colon adenocarcinoma patients, but not with serum from patients with bowel diseases or from normal subjects [300]. The specific antigen defined by this monoclonal antibody was identified as sialosyl-Le^a ($IV^3NeuAcIII^4FucLcOSe_4Cer$) [301]. The increased antigen level in sera of patients with cancers, particularly pancreatic cancer, provides a useful tool in the diagnosis of pancreatic cancer. In serum, this antigen seems to be expressed on glycoproteins [302].

Disialosyl-Le^a Antigen. This antigen, $III^6IV^3NeuAc_2III^4FucLcOSe_4Cer$, present in human adenocarcinomas [303], was strongly immunogenic and its concentrations in normal subjects and noncancerous patients are extremely low. The monoclonal antibody FH7 recognizes this antigen and seems to be cross-reactive with other sialosyl-Le^a structures. This antibody detects circulating antigens from a variety of human cancers, including gastric, colonic, lung, and pancreatic cancers.

Sialosyl-Lc_4 Antigen. This antigen, $IV^3NeuAcLcOSe_4Cer$, is found in human teratocarcinoma PAI [304] and in colon cancer tissues [305]. A monoclonal antibody CA50, which was produced by immunization of mice with colon cancer tissue, appears to recognize this antigen [306]. In normal tissue, this antigen, which possesses a type 1 chain (Galβ1-3GlcNAcβ1-) structure, seemed to be absent. However, CA50 cross-reacts with several sialosyl-Le^a structures.

4. Lactone Series

Sialosyl-nLc_4 Antigen. This antigen, $IV^3NeuAcnLcOSe_4Cer$, is found in human glioma tissues of grades III and IV. Normal brain tissue lacks this ganglioside. Radioimmunoassay with a specific monoclonal antibody to this ganglioside was used to identify it as a marker ganglioside for human malignant gliomas [307].

Le^x, Dimeric-Le^x and Trimeric-Le^x Antigens. In recent years, numerous monoclonal antibodies have been directed against a particular family of structures that contain the x structure (Galβ1-4[Fucα1-3]GlcNAcβ1-3Gal-) [1,72,145,146]. Monoclonal antibodies that exhibit this specificity have been elicited by immunization with murine teratocarcinomas [308–310], carcinomas of the pancreas, lung, colon, and stomach [150], and myeloid leukemia cells and granulocytes [301,311,312]. Antibodies to embryonic antigen SSEA-1 have been identified as directed against the x determinant [309]. All these antibodies detect a series of glycolipids in normal granulocytes [313], erythrocytes [310], normal colonic mucosa, and normal liver [314]. These glycolipids all contain an unbranched type 2 chain with Fucα1-3 linkage at the penultimate GlcNAc residue. Cancer tissues, particularly adenocarcinomas, show a large accumulation of lactoneofucopentaosylceramide ($III^3FucLcnOSe_4Cer$), difucosyllacto*nor*hexaosylceramide ($III^3V^3Fuc_2LcnOSe_6Cer$), and trifucosyllacto*nor*octaosylceramide ($III^3V^3VII^3Fuc_3LcnOSe_8Cer$), compared with normal liver, which virtually lacks these compounds [315]. Two monoclonal antibodies (FH4 and AcFH12) have been produced that react with only di- or trifucosylated chains [316]. Because of the large accumulation of the di- and trifuxosylated-Le^x glycolipid in adenocarcinomas, these antibodies may provide a useful tool in immunodiagnosis.

Recently, a monoclonal antibody (624H12), recognizing the sugar sequence ($III^3V^3Fuc_2LcnOSe_6Cer$) has been used for early detection of lung cancer, which had not been possible by conventional diagnostic techniques [317]. A monoclonal antibody (FH6), which is specific for sialosyl-dimeric Le^x (SSEA-1 antigen) has been utilized to implicate increased expression of this antigen in liver metastases of human colorectal carcinoma [318]. Sialosyl-SSEA-1 antigen has also been recognized as a tumor marker with a high specificity for adenocarcinoma of the reproductive organs [319–321]. A cutoff level of 50 U/mL of sialosyl-SSEA-1 antigen was suggested for diagnosis of ovarian tumors [321]. Sialosyl-SSEA-1 antigen has also been identified as a marker for human natural killer cells and immature lymphoid cells [322].

Sialosyl-Fuc-Lcn_6 Antigen. This antigen ($III^3FucVI^6NeuAcLcnOSe_6Cer$), was found, in large amounts, in some cases of human colon and liver adenocarcinoma, but it was absent in normal colonic mucosa and normal liver specimens [323]. A monoclonal antibody IB9 recognizes this cancer-associated antigen as a sialosylα2-6–containing ganglioside.

The mechanism and biological implications of glycolipid changes are still open to question. Whatever, the mechanism is, the accumulation of neoglycolipid in a specific tumor can have an excellent potential for the diagnosis and immunotherapy of human cancer [324–326].

F. Immunomodulation by Glycolipids

It was reported that liposomes containing G_{M1} or G_{D1b} ganglioside inhibited the antisheep erythrocyte plaque-forming response (PFC) by mouse spleen cells [327]. Gangliosides G_{M1}, G_{D1a}, G_{D1b}, as well mixed gangliosides from bovine brain were also found to inhibit lipopolysaccharide-induced activation of mouse spleen B lymphocytes [328]. Brain gangliosides also had a suppressive effect on mitogenic response of mouse splenocytes [329] and thymocytes [330] when activated with either lectins or lipopolysaccharides. The modulatory or suppressive effects of B and T lymphocytes were also reported for gangliosides isolated from the supernatant of cultured cells [331–333] and human leukemic lymphocytes [334]. It has been reported [333] that a disialoganglioside fraction could prevent the development of tolerance in immature B cells. It was concluded that glycolipids released by antigen-stimulated T cells modulate the differentiation of B cells into antibody-secreting cells and that glycolipid can protect B cells from tolerance induced either by antigen overload or hapten presented on an unrelated carrier. The evidence for the dependence of the suppressive effect on ganglioside structure is still contradictory. Most of the studies were performed with a mixture of bovine or human brain gangliosides, but some studies were performed with individual gangliosides. Polysialogangliosides were more effective in inhibiting concanavalin A (ConA)-induced proliferation than monosialogangliosides [330], and high concentrations of sialic acid or monoglycosyl neutral glycolipid had no inhibitory effect. On the other hand, Whisler and Yates [335] noted differences in the ability of individual gangliosides to inhibit Con A-induced proliferation, compared with mitogen-induced lymphocyte activation, but there was no difference between monosialo- and polysialogangliosides. Gangliosides incorporated into liposomes containing lecithin and cholesterol were more effective inhibitors than dispersions of pure gangliosides [335].

The data indicate that inhibition occurs during the early phases of activation. There appears to be a minimal period during which the cells must be in contact with the gangliosides before inhibitory effect is established. It has been noted that Con A also enhances some stages of ganglioside biosynthesis in lymphocytes [14]. The enhancement of ganglioside synthesis appears to be associated with cell division, since it becomes visible within a few hours of lymphocyte activation.

G. Glycolipids in Central and Peripheral Nerves

In the central nervous system (CNS), the gangliosides compose up to 5–10%, which exceeds the protein-bound fraction. Virtually all cellular and subcellular fractions of the brain contain gangliosides to one degree or another, but most brain gangliosides are located in the neurons. For a long time it was believed that gangliosides were localized in the nerve endings, but more recent work in different laboratories [6,10] suggests that gangliosides are distributed over a large part of the neuronal surface. If the neuronal plasma membrane is viewed as a continuum, individual lipids inserted into the perikaryal membrane would be expected to diffuse laterally into the adjoining plasma membranes of the processes. Such mechanism would allow equilibrium of surface molecules between the neuron and its processes, whereas the more remote axonal and synaptic regions would be expected to receive gangliosides by rapid axonal transport [336]. The hypothesis that gangliosides are distributed over the major part of the neuronal surface will probably be validated with improved methods for isolating perikaryon, plasma membranes, and processes.

The peripheral nervous system (PNS) also contains large amounts of gangliosides. Myelin isolated from PNS has the same lipid composition as that of CNS myelin, but they vary in the distribution patterns [6,10,11]. However, sialosylgalactosylceramide (G_{M4}) is absent from peripheral nerves and seems to be a specific marker for CNS myelin [45,47]. There is increasing body of evidence that suggests that CNS and PNS gangliosides play an important functional role in a variety of cellular events [1,2,4,6–8,10,12–14], as well as in a number of neurological and neuromuscular functions [1,337–342].

The involvement of glycolipids in in vitro development of neuromuscular junctions in mixed cultures of skeletal muscle and spinal cord was suggested by Obata and co-workers [343]. Ganglioside G_{M1} and globotetrasylceramide were added to the mixed cultures and the development of the neuromuscular junction was measured electrophysiologically by quantifying the end-plate potential and counting the number of myotubes formed. The major finding was that, at low concentrations of glycolipids (8–63 μM), stimulation of the development of neuromuscular junction occurred, whereas at high glycolipid concentrations (0.25–0.5 mM) inhibition was noted. It is not known, however, whether the glycolipids are involved in this process. Nothing is known about in vivo involvement. However, it was suggested that endogenous G_{M1} may be important in neuromuscular formation, since its blockage by cholera toxin or the B-subunit of the toxin inhibited neuromuscular formation [344].

The indication that gangliosides possess neuritrogenic properties came from the work of Purpura and co-workers [345,346]. It was shown that mature neurons in both G_{M2} and G_{M1}-gangliosides produced new processes, including aberrant secondary neurites with occasional synapses from regions of the cell that were laden with stored ganglioside.

Subsequently, it was shown that the neuroblastoma N-2A cell line underwent prolific neuritogenesis upon addition of single or mixed gangliosides to the culture medium [347,348]. Primary neuronal cultures, derived from various parts of the nervous system, were also reported to undergo neuritogenesis in the presence of exogenous gangliosides. Extensive studies on the effect of exogenous gangliosides in neuronal differentiation and neuritogenesis have been performed and have been reviewed [339,349]. It is also reported that antibodies to gangliosides inhibit neurite outgrowth from regenerating retinal explants of the goldfish [350].

The indication that exogenous gangliosides are capable of nerve repair in vivo was demonstrated by Ceccarelli et al. [351]. Subsequently, Gorio and co-workers [352] demonstrated a similar effect. Rats treated daily with 5 mg/kg of gangliosides showed more rapid recovery of functional reinnervation of sympathetic nerves after transecting the extensor digitorium muscle. The effect was shown to be mediated through stimulation of motor neuron sprouting, rather than by elongation. Sprouting, which is a response of the nerve to damage, must be a functional regulator of the nerve generation by intrinsic influences. A number of studies [337–341] have shown that gangliosides are also active in the recovery process of the CNS. Gangliosides administered peripherally were shown to modulate serotonin receptor function in normal brain [353]. Studies conducted on the recovery process in the hippocampus after brain lesions, by ganglioside injection, indicated that the cholinergic parameters in the hippocampus were improved [354].

The regenerating optic system of the goldfish was reported to respond to ganglioside treatment [355]. Low doses showed a stimulating effect, whereas higher doses of gangliosides were inhibitory. Antibodies to gangliosides showed inhibition in the recovery of the axons as expected [356]. It has been shown that the G_{M1} ganglioside is an agent with a strong potential for enhancing CNS repair [357]; also, local application of ganglioside to the site of nerve injury stimulated axonal sprouting much more rapidly than intraperitoneal injection [357].

No clear picture has yet emerged concerning the mode of action of gangliosides in these various systems. The questions still remain unanswered: How are the gangliosides adhering to cells in vitro (i.e., are they inserted within the plasma membrane or are they physically adsorbed to the cell surface, or are they subjected to endocytosis)?

Concerning the nature of cell–ganglioside association, it is known that gangliosides added to the culture medium are taken up by the cells [358–364]. The binding of gangliosides to the cells seems to be heterogeneous in affinity. A portion of the ganglioside that is attached to the surface of the membrane is easily removed by washing, whereas the remainder becomes more tightly bound [358,363]. The ganglioside either binds to membrane proteins or becomes anchored in hydrophobic parts of the membrane [358,364].

The current belief is that some amount of gangliosides may be inserted into the cell membrane, but a major portion of the ganglioside is not associated with the cell in a normal manner. It is not clear, however, which portion of the cell-associated ganglioside is responsible for the biological effects. However, gangliosides added exogenously to cells and subcellular preparations actually behave like endogenous gangliosides and are known to change overall membrane dynamics, as indicated by modifications of Na^+, K^+-ATPase activity, phosphodiesterase activity, adenylate cyclase activity, and cell surface sialylation processes [365,366]. Such modification contributes substantially to cell behavior. Gangliosides have been shown to be therapeutically effective in the treatment of diabetic neuropathy [338,364,362]. Diabetic neuropathy in humans involves electrophysiological and morphological alterations of peripheral nerves. The nerve conduction velocity is reduced, and axonal degeneration is detectable. Accordingly, this was considered to be a neuroaxonal disorder. This pathological picture suggests that an ideal drug for treatment of diabetic neuropathy should be such that it should improve conduction velocity and induce replacement of degenerated axons by stimulating sprout formation from the remaining axons. Gangliosides have been reported in the treatment of diabetes-induced peripheral neuropathy of animal models, such as the mutant mouse C57 BL/KS (db/db) [368], in which diabetes was induced by injection with alloxan or streptozocin. Clinical trials in humans are underway to determine the possible effectiveness of ganglioside therapy for diabetic neuropathy [340,369].

To study the function of a specific glycolipid, specific antiglycolipid antibody has been used to observe functional changes. Rapport and co-workers [370,371] showed that injection of anti-G_{M1} antibodies into the sensorimotor cortex of the rat induced recurrent epileptic seizure activity; absorption of antibody with G_{M1} ganglioside completely abolished this activity. It was also determined that antibodies to G_{M1} injected into the ventricle of rats and mice immediately induced memory loss (inhibition of learning) and also interfered with normal dentric growth in neonatal rats [372,373]. Injection of anti-G_{M1} antibodies into the periaqueductal gray (PAG) matter of rats was also shown to block morphine-induced analgesia [374]. The selection of PAG was made because PAG contains a high concentration of morphine receptors, and a small amount of morphine injected in this region induces analgesia. These observations suggest that gangliosides should be an important focus of attention for examining neurochemical correlates of processes and that antiganglioside antibodies could be useful as reagents to study such functions.

Gangliosides have also been used in the therapy of alcoholic neuropathies [375,376], stroke patients [377], experimental CS_2 neuropathy in the rat [378], and ouabain-induced retinopathy in the rabbit [379]. Systemetic administration of gangliosides has been shown to be pharmacologically effective in reducing injury and facilitating recovery after CNS damage in various animal paradigms [380]. More recently, it has been documented [381] that ganglioside therapy following CNS ischemia in animals and humans causes reductions in the extent of injury (acute phase) and enhanced neurological recovery (long-term effects). The pharmacologic effects of ganglioside G_{M1} treatment to enhance recovery after CNS injury have been recently reviewed [382]. The key mechanism through which G_{M1} exerts its clinical effects is by enhancing the neuroplasticity of the CNS. The neuropharmacologic effects of G_{M1} in humans have been confirmed in controlled clinical trials in head and spinal cord injury and in stroke patients [382]. Undoubtedly, this area of research is expanding very rapidly, and clinical trials are in progress in different sites with different neuropathies. One can only hope that the beneficial effects of ganglioside therapy may prove useful sometime in the near future.

ACKNOWLEDGMENT

The author sincerely thanks Ms Rose Cote and Donna Weichbrodt in preparation of this manuscript.

REFERENCES

1. Hakomori, S.-I. (1981). Glycosphingolipids in cellular interaction, differentiation and oncogenesis. *Annu. Rev. Biochem.* *50*:733–764.

2. Karlsson, K.-A. (1982). Glycosphingolipids and surface membranes. In *Biological Membranes*, Vol. 4 (D. Chapman, ed.). Academic Press, New York, pp. 1–74.
3. Thompson, T. E., and Tillack, T. E. (1985). Organization of glycosphingolipids in bilayers and plasma membranes of mammalian cells. *Annu. Rev. Biophys. Chem. 14*:361–386.
4. Ando, S. (1983). Gangliosides in the nervous system. *Neurochem. Int. 5*:507–537.
5. Sweeley, C. C., and Siddiqui, B. (1977). Chemistry of mammalian glycolipids. In *Biochemistry of Mammalian Glycoproteins and Glycolipids* (M. I. Horowitz and W. Pigman, eds.). Academic Press, New York, pp. 459–485.
6. Ledeen, R. W. (1978). Ganglioside structures and distribution: Are they localized at the nerve ending? *J. Supramol. Struct. 8*:1–17.
7. Yamakawa, T., and Nagai, Y. (1978). Glycolipids at the cell surface and their biological functions. *Trends Biochem. Sci. 3*:128–131.
8. Fishman, P. H., and Brady, R. O. (1976). Biosynthesis and function of gangliosides. *Science 194*:906–915.
9. Marcus, D. M., and Schwarting, G. A. (1976). Immunochemical properties of glycolipids and phospholipids. *Adv. Immunol. 23*:203–240.
10. Wiegandt, H. (1982). The gangliosides. *Adv. Neurochem. 4*:149–223.
11. Ledeen, R. W., and Yu, R. K. (1982). Gangliosides: Structure, isolation and analysis. *Methods Enzymol. 83*:139–191.
12. Hakomori, S.-I. (1984). Tumor associated antigens. *Annu. Rev. Immunol. 2*:103–126.
13. Marcus, D. M. (1984). A review of the immunogenic and immunomodulatory properties of glycosphingolipids. *Mol. Immunol. 21*:1083–1091.
14. Dyatlovitskaya, E. V., and Bergelson, L. D. (1987). Glycosphingolipids and antitumor immunity. *Biochim. Biophys. Acta 907*:125–143.
15. Hakomori, S.-I. (1986). Glycosphingolipids. *Sci. Am. 254*:44–53.
16. Curatolo, W. (1987). The physical properties of glycolipids. *Biochim. Biophys. Acta. 906*: 111–136.
17. Kundu, S. K. (1981). Thin-layer chromatography of gangliosides and neutral glycosphingolipids. *Methods Enzymol. 72*:185–204.
18. Kundu, S. K., Marcus, D. M., Pascher, I., and Samuelsson, B. E. (1983). New gangliosides from human erythrocytes. *J. Biol. Chem. 258*:13857–13866.
19. Kundu, S. K., Diego, I., Osovitz, S., and Marcus, D. M. (1985). Glycosphingolipids of human plasma. *Arch. Biochem. Biophys. 238*:388–400.
20. Yamakawa, Y. (1983). Glycolipids of the red blood cells. In *Red Blood Cells of Domestic Mammals* (N. S. Agar and P. G. Board, eds.). Elsevier Science Publisher, Amsterdam, pp. 37–53.
21. Ledeen, R. W., and Yu, R. (1973). Structure and enzymic degradation of sphingolipids. In *Lysosomes and Storage Diseases*. Academic Press, New York, pp. 105–145.
22. Ohashi, M., and Yamakawa, T. (1977). Isolation and characterization of glycosphingolipids in pig adipose tissue. *J. Biochem. 81*:1675–1690.
23. Hirabayashi, Y., Taki, T., and Matsumoto, M. (1979). Tumor ganglioside—natural occurrence of G_{M1b}. *FEBS Lett. 100*:253–257.
24. Hakomori, S. I. (1981). Blood group ABH and Ii antigens of human erythrocytes: Chemistry, polymorphism and thin developmental change. *Semin. Hematol. 18*:39–62.
25. Marcus, D. M., Kundu, S. K., and Suzuki, A. (1981). The P blood group system: Recent program in immunochemistry and genetics. *Semin. Hematol. 18*:63–71.
26. Okada, Y., Kannagi, R., Levery, S. B., and Hakomori, S.-I. (1984). Glycolipid antigens with blood group I and i specificities from human adult and umbilical cord erythrocytes. *J. Immunol. 133*:835–842.
27. Ando, S., Kon, K., Nagai, Y., and Yamakawa, T. (1982). A novel pentaglycosylceramide containing di-β-*N*-acetylgalactosaminyl residue (para-Forssman glycolipid) isolated from human erythrocyte membrane. In *New Vistas in Glycolipid Research* (A. Makita, S. Handa, T. Taketomi, and Y. Nagai, eds.). Plenum Press, New York, pp. 71–81.

28. Kundu, S. K., Harati, Y., and Misra, L. K. (1984). Sialosylglobotetraosylceramide: A marker for amyotropic lateral sclerosis. *Biochem. Biophys. Res. Commun. 118*:82–89.
29. Hogan, E. L., Happel, R. D., and Chien, J.-L. (1982). Membrane glycolipids in chicken muscular dystrophy. In *New Vistas in Glycolipid Research* (A. Makita, S. Handa, T. Taketomi, and Y. Nagai, eds.). Plenum Press, New York, pp. 273–290.
30. Slomiany, A., and Slomiany, B. L. (1980). Structure of the ceramide octadecahexoside isolated from gastric mucosa. *Biochem. Biophys. Res. Commun. 93*:770–775.
31. Gardas, A. (1976). A structural study on a macro-glycolipid containing 22 sugars isolated from human erythrocytes. *Eur. J. Biochem. 68*:177–183.
32. Zdaebske, E., and Koscielak, J. (1978). Studies on the structure and I-blood-group activity of poly(glycosyl)ceramides. *Eur. J . Biochem. 91*:517–525.
33. Rosenberg, A., and Stern, N. (1966). Changes in sphingosine and fatty acid components of the gangliosides in developing rat and human brain. *J. Lipid Res. 7*:122–128.
34. Schengrund, C. L., and Garrigan, O. W. (1969). A comparative study of gangliosides from the brain of various species. *Lipids 4*:488–492.
35. Sambasivarao, K., and McCluer, R. H. (1964). Lipid components of gangliosides. *J. Lipid Res. 5*:103–110.
36. Ledeen, R., Salsman, K., and Carbrera, M. (1968). Gangliosides in subacute sclerosing leukoencephalitis: Isolation and fatty acid composition of nine fractions. *J. Lipid Res. 9*: 129–136.
37. Coles, E., and Foote, J. L. (1970). Fatty acids of glycophingolipids from pig blood fractions. *J. Lipid Res. 11*:433–438.
38. Kannagi, R., Nudelman, E., and Hakomori, S. (1982). Possible role of ceramide in defining structure and function of membrane glycolipids. *Proc. Natl. Acad. Sci. USA 79*:3470–3474.
39. Karlsson, K. A. (1970). Sphingolipid long chain bases. *Lipids 5*:878–891.
40. Karlsson, K. A. (1970). On the chemistry and occurrence of sphingolipid long chain bases. *Chem. Phys. Lipids 5*:6–43.
41. Martensson, E. (1969). Glycosphingolipids of animal tissue. *Progr. Chem. Fats Other Lipids 10*:367–407.
42. Yang, H. J., and Hakomori, S. (1971). A sphingolipid having a novel type of ceramide and a lacto-*N*-fucopentose. III. *J. Biol. Chem. 246*:1192–1200.
43. Ledeen, R., and Salsman, K. (1970). Fatty acid and long-chain lose composition of adrenal medulla gangliosides. *Lipids 5*:751–756.
44. Vance, D. E., Krivit, W., and Sweeley, C. C. (1969). Concentrations of glycosyl ceramides in plasma and red cells in Fabry's disease, a glycolipid lipidosis. *J. Lipid Res. 10*:188–192.
45. Fong, J. W., Ledeen, R. W., Kundu, S. K., and Brostoff, S. W. (1976). Gangliosides of peripheral nerve myelin. *J. Neurochem. 26*:157–162.
46. Yu, R. K., and Ledeen, R. W. (1972). Gangliosides of human, bovine and rabbit plasma. *J. Lipid Res. 13*:680–686.
47. Ledeen, R. W., Yu, R. K., and Eng, L. F. (1973). Gangliosides of human myelin: Sialosylgalactosylceramide (G_7) as a major component. *J. Neurochem. 21*:829–839.
48. Rosenfelder, G., Chang, J. Y., and Braun, D. G. (1983). Sphingosine determination at the picomole level using dimethylaminoazobenzene sulphonyl chloride. *J. Chromatogr. 272*:21–27.
49. Svennerholm, L. (1964). The gangliosides. *J. Lipid Res. 5*:145–165.
50. Vance, D. E., and Sweeley, C. C. (1967). Quantitative determination of the neutral glycosyl ceramides in human blood. *J. Lipid Res. 8*:621–630.
51. Svennerholm, E., and Svennerholm, L. (1963). The separation of neutral blood-serum glycolipids by thin-layer chromatography. *Biochim. Biophys. Acta. 70*:432–441.
52. Svennerholm, L. (1964). The distribution of lipids in the human nervous system. 1. Analytical procedure. Lipids of foetal and newborn brain. *J. Neurochem. 11*:839–853.
53. Martensson, E. (1966). Sulfatides of human kidney. Isolation, identification, and fatty acid composition. *Biochim. Biophys. Acta 116*:521–531.

54. Yamakawa, T., Nishimura, S., and Kamimura, M. (1965). The chemistry of the lipids of posthemolytic residue or stroma of erythrocytes. 8. Further studies on human red cell glycolipids. *Jpn. J. Exp. Med.* *35*:201–207.
55. Hakomori, S., Siddiqui, B., Li, Y. T., Li, S. C., and Hellerquist, C. G. (1971). Anomeric structures of globoside and ceramide trihexoside of human erythrocytes and hamster fibroblasts. *J. Biol. Chem.* *246*:2271–2277.
56. Ando, S., and Yamakawa, T. (1970). On the oligosaccharide of Forssman-active glycolipid. *Chem. Phys. Lipids.* *5*:91–95.
57. Siddiqui, B., and Hakomori, S. (1971). A revised structure for the Forssman glycolipid hapten. *J. Biol. Chem.* *246*:5766–5769.
58. Taketomi, T., and Kawamura, N. (1972). Preparation of lysohematoside (neuraminylgalactosyl-glycosyl-sphingosine) from hematoside of equine erythrocyte and its chemical and hemolytic properties. *J. Biochem.* *72*:799–806.
59. Taketomi, T., Hara, A., Kawamura, N., and Hayashi, A. (1974). Immunochemical studies of lipids. IV. Chemical modification of Forssman globoside and immunological activities. *J. Biochem.* *75*:192–199.
60. Kuhn, R., and Wiegandt, H. (1964). Weitere ganglioside aus Menschenhirn. *Z. Naturforsch,* *19b*:256–264.
61. Uemura, K., Yuzawa, M., and Taketomi, T. (1978). Characterization of major glycolipids in bovine erythrocyte membrane. *J. Biochem.* *83*:463–471.
62. Chien, J. L., Li, S.-C., Laine, R. A., and Li, Y. T. (1978). Characterization of gangliosides from bovine erythrocyte membranes. *J. Biochem.* *253*:4031–4035.
63. Yamakawa, T., Irie, R., and Iwanaga, M. (1960). The chemistry of lipid of post hemolytic residue or stroma of erythrocytes. IX. Silicic acid chromatography of mammalian stroma glycolipids. *J. Biochem.* *48*:490–507.
64. Eto, T., Ichikawa, T., Nishimura, K., Ando, S., and Yamakawa, T. (1968). Chemistry of lipid of the posthemolytic residue or stroma of erythrocytes. XVI. Occurrence of ceramide pentasaccharide in the membrane of erythrocytes and reticulocytes of rabbit. *J. Biochem.* *64*: 205–213.
65. Stellner, K., Saito, H., and Hakomori, S. (1973). Determination of amino sugar linkages in glycolipids by methylation. Amino sugar linkages of ceramide pentasaccharides of rabbit erythrocytes and of Forssman antigen. *Arch. Biochem. Biophys.* *155*:464–472.
66. Seyma, Y., and Yamakawa, T. (1974). Chemical structure of glycolipid of guinea pig red blood cell membrane. *J. Biochem.* *75*:837–842.
67. Kundu, S. K., Chakravarty, S. K., Roy, S. K., and Roy, A. K. (1979). DEAE-silica gel and DEAE-controlled porous glass as ion-exchangers for isolation of glycolipids. *J. Chromatogr.* *170*:65–72.
68. Rouser, R., Kritchevsky, G., Yamamoto, A., Simon, G., Galli, G., and Bauman, A. J. (1969). Diethylaminoethyl cellulose column chromatographic procedures for phospholipids, glycolipids and pigments. *Methods Enzymol.* *14*:272–317.
69. Renkonen, O., and Varo, P. (1956). In *Lipid Chromatographic Analysis*, Vol. 1 (G. V. Marinetti, ed.). Marcel Dekker, New York, pp. 41–61.
70. Hakomori, S., and Siddiqui, B. (1974). Isolation and characterization of glycosphingolipid from animal cells and their membranes. *Methods Enzymol.* *32*:345–369.
71. Ledeen, R. W., and Yu, R. K. (1978). Methods for isolation and analysis of gangliosides. *Res. Methods Neurochem.* *4*:371–410.
72. Kundu, S. K. (1981). DEAE-silica gel and DEAE-controlled porous glass as ion-exchangers for the isolation of glycolipids. *Methods Enzymol.* *72*:174–185.
73. Folch, J., Less, M., and Sloane Slanley, G. H. (1957). A simple method for the isolation and purification of total lipids from animal tissues. *J. Biol. Chem.* *226*:497–509.
74. Iwamori, M., and Nagai, Y. (1978). A new chromatographic approach to the resolution of individual gangliosides. *Biochim. Biophys. Acta* *528*:257–267.

75. Fredman, P., Nilsson, O., Tayot, J. L., and Svennerholm, L. (1980). Separation of gangliosides on a new type of anion-exchange resin. *Biochim. Biophys. Acta 618*:42–50.
76. Itoh, T., Li, Y.-T., Li, S.-C., and Yu, R. K. (1981). Isolation and characterization of a novel monosialosylpentahexaosyl ceramide from Tay-Sachs brain. *J. Biol. Chem. 256*:165–169.
77. Saito, T., and Hakomori, S. (1971). Quantitative isolation of total glycosphingolipid from animal cells. *J. Lipid Res. 12*:257–259.
78. Kundu, S. K., and Suzuki, A. (1981). A simple micromethod for the isolation of gangliosides by reversed-phase chromatography. *J. Chromatogr. 224*:249–256.
79. Kannagi, R., Watanabe, K., and Hakomori, S. (1987). Isolation and purification of glycosphingolipids by high-performance liquid chromatography. *Methods Enzymol. 138*:3–12.
80. Kundu, S. K., and Scott, D. D. (1982). Rapid separation of gangliosides by high-performance liquid chromatography. *J. Chromatogr. 232*:19–27.
81. Hakomori, S. (1978). Isolation of blood group ABH glycolipids from human erythrocyte membranes. *Methods Enzymol. 50*:207–211.
82. McCluer, R. H., Ullman, M. D., and Jungalwala, F. B. (1986). HPLC of glycosphingolipids and phospholipids. *Adv. Chromatogr. 25*:309–353.
83. Kadowaki, H., Evans, J. E., and McCluer, R. H. (1984). Separation of brian monosialoganglioside molecular species by high-performance liquid chromatography. *J. Lipid Res. 25*: 1132–1139.
84. Sonnino, S., Ghidoni, R., Gazzotti, G., Krischner, G., Gall, G., and Tettamont, G. (1984). High performance liquid chromatography preparation of the molecular species of G_{M1} and G_{D1a} gangliosides with homogeneous long chain base composition. *J. Lipid Res. 25*:620–629.
85. Sonnino, S., Kirschner, G., Ghidoni, R., Acquotti, D., and Tettamonti, G. (1985). Preparation of G_{M1} ganglioside molecular species having homogeneous fatty acid and long chain moieties. *J. Lipid Res. 26*:248–257.
86. Dubois, M. K., Gilles, K. A., Hamilton, J. K., Rebers, P. A., and Smith, F. (1956). Colorimetric method for determination of sugars and related substances. *Anal. Chem. 28*:350–356.
87. Watanabe, K., Powell, M., and Hakomori, S. (1979). Isolation and characterization of gangliosides with a new sialosyl linkage and core structure. *J. Biol. Chem. 254*:8223–8229.
88. Svennerholm, L. (1957). Quantitative estimation of sialic acids. II. A colorimetric resorcinol–hydrochloric acid method. *Biochim. Biophys. Acta 24*:604–611.
89. Miettinen, T. A., and Takki-Luukkainen, I. T. (1959). Use of butylacetate in determination of sialic acid. *Acta Chem. Scand. 13*:856–858.
90. Jourdian, G. W., Dean, L., and Roseman, S. (1971). The sialic acids. XI. A periodate–resorcinol method for the quantitative estimation of free sialic acids and their glycosides. *J. Biol. Chem. 246*:430–435.
91. Warren, L. (1959). The thiobarbituric acid assay of sialic acids. *J. Biol. Chem. 234*:1971–1976.
92. Aminoff, D. (1959). The determination of free sialic acid in the presence of the bound compound. *Virology 7*:355–357.
93. Hahn, H., Hellman, B., Lernmark, A., Sehlin, J., and Tuljedal, T. (1974). The pancreatic B-cell recognition of insulin secretagogues. *J. Biol. Chem. 249*:5275–5284.
94. Yu, R. K., and Ledeen, R. W. (1970). Gas-liquid chromatography assay of lipid-bound sialic acids: Measurement of gangliosides in brain of several species. *J. Lipid Res. 11*:506–512.
95. Robert, J., and Rebel, G. (1975). Quantitation of neutral glycolipids by thin-layer chromatography on precoated plates. *J. Chromatogr. 110*:393–397.
96. Friedrich, V. L., and Hauser, A. (1973). Biosynthesis of psychosine and levels of cerebrosides in the central and peripheral nervous systems of quaking mice. *J. Neurochem. 20*:1131–1141.
97. Suzuki, A., Kundu, S. K., and Marcus, D. M. (1980). An improved technique for separation of neutral glycosphingolipids by high-performance liquid chromatography. *J. Lipid Res. 21*: 473–477.
98. Suzuki, K. (1964). A simple and accurate micromethod for quantitative determination of ganglioside patterns. *Life Sci. 3*:1227–1223.

99. MacMillan, V. H., and Wherrett, J. R. (1969). A modified procedure for the analysis of mixtures of tissue gangliosides. *J. Neurochem. 16*:1621–1624.
100. Yates, A. J., and Thompson, D. (1977). An improved assay of gangliosides separated by thin-layer chromatography. *J. Lipid Res. 18*:660–663.
101. Ando, S., Chang, N.-C., and Yu, R. K. (1978). High-performance thin-layer chromatography and densitometric determination of brain ganglioside compositions of several species. *Anal. Biochem. 89*:437–450.
102. Brady, R. O., Barek, C., and Bradley, R. M. (1969). Composition and synthesis of gangliosides in rat hepatocyte and hepatoma cell lines. *J. Biol. Chem. 244:6552-6554.*
103. Smid, F., and Reinisova, J. (1973). A densitometric method for the determination of gangliosides after their separation by thin-layer chromatography and detection with resorcinol reagent. *J. Chromatogr. 86*:200–204.
104. Dische, Z. (1962). General color reactions. *Methods Carbohydr. Chem. 1*:477–497.
105. Laine, R., Esselman, W. J., and Sweeley, C. C. (1972). Gas-liquid chromatography of carbohydrates. *Meth. Enzymol. 18*:159–167.
106. Carter, H. E., and Gaver, R. C. (1967). Improved reagent for trimethylsilylation of sphingolipid bases. *J. Lipid Res. 8*:391–395.
107. Etchinson, J. R., and Holland, J. J. (1975). A procedure for the rapid quantitative *N*-acetylation of amino sugar methyl glycosides. *Anal. Biochem. 66*:87–92.
108. Ando, S., and Yamakawa, Y. (1971). Application of trifluoracetyl derivatives to sugar and lipid chemistry. I. Gas chromatography analysis of common constituents of glycolipids. *J. Biochem. 70*:335–340.
109. Sawardeker, J. S., Sloneker, H. J., and Jeanes, A. (1965). Quantitative determination of monosaccharides as their alditol acetates by gas-liquid chromatography. *Anal. Chem. 37*:1602–1604.
110. Metz, J., Ebert, W., and Weicker, H. (1971). Quantitative determination of neutral and amino sugars by gas-liquid chromatography. *Chromatographia 4*:345–350.
111. Kuhn, R., and Wiegandt, H. (1963). Die Konstitutin der Gangliotetraose and des Gangliosides G_1. *Chem. Ber. 96*:866–880.
112. Ishizuka, I., and Wiegandt, H. (1972). An isomer of trisialoganglioside and the structure of tetra- and pentasialoganglioside from fish brain. *Biochim. Biophys. Acta 260*:279–289.
113. Ando, S., and Yu, R. K. (1979). Isolation and characterization of two isomers of brain tetrasialogengliosides. *J. Biol. Chem. 254*:12224–12229.
114. Ginsburg, V., ed. (1972). Complex carbohydrates, Part B. *Methods Enzymol. 18*:699–873.
115. Kobata, A. (1979). Use of endo- and exoglycosidases for structural studies of glycoconjugates. *Anal. Biochem. 100*:1–14.
116. Hakomori, S., and Watanabe, K. (1976). Blood group glycolipids of human erythrocytes. In *Glycolipid Methodology* (L. A. Witting, ed.). American Oil Chemist's Society, Champaign, Ill., pp. 13–47.
117. Fakuda, M. N., and Matsumura, G. (1976). Endo-β-galactosidase of *Escherichia freundii. J. Biol. Chem. 251*:6218–6225.
118. Fakuda, M. N., Watanabe, K., and Hakomori, S. I. (1978). Release of oligosaccharides from various glycosphingolipids by endo-β-galactosidase. *J. Biol. Chem. 253*:6814–6819.
119. Dawson, G., and Sweeley, C. C. (1971). Mass spectrometry of neutral, mono-, and disialoglycosphingolipids. *J. Lipid Res. 12*:56–64.
120. Markay, S. P., and Wenger, D. A. (1974). Mass spectra of complex molecules. I. Chemical ionization of sphingolipids. *Chem. Phys. Lipids 12*:182–200.
121. Karlsson, K. A. (1973). Carbohydrate composition and sequence analysis of cell surface components by mass spectrometry. Characterization of the major monosialoganglioside of brain. *FEBS Lett. 32*:317–320.
122. Karlsson, K. A. (1974). Carbohydrate composition and sequence analysis of a derivative of brain disialyganglioside by mass spectrometry, with molecular weight ions at m/e 2245. Potential use in the specific microanalysis of cell surface components. *Biochemistry 13*: 3643–3647.

123. Ledeen, R. W., Kundu, S. K., Price, H. C., and Fong, J. W. (1974). Mass spectra of permethyl derivatives of glycosphingolipids. *Chem. Phys. Lipids 13*:429–446.
124. Karlsson, K. A. (1980). Structural fingerprinting of gangliosides and other glycoconjugates by mass spectrometry. In *Structure and Function of Gangliosides* (L. Svennerholm, P. Mandel, H. Dreyfus, and P. F. Urban, eds.). Plenum Press, New York, pp. 47–61.
125. Watanabe, K., Powell, M., and Hakomori, S. (1978). Isolation and characterization of a novel fucoganglioside of human erythrocyte membranes. *J. Biol. Chem. 253*:1049–1056.
126. Ariga, T., Yu, R. K., and Miyatake, T. (1984). Characterization of gangliosides by direct inlet chemical ionization mass spectrometry. *J. Lipid Res. 25*:1096–1101.
127. Handa, S., and Kushi, Y. (1984). Application of field desorption and secondary ion mass spectrometry for glycolipid analysis. In *Ganglioside Structure, Function and Biomedical Potential* (R. W. Ledeen, R. K. Yu, M. M. Rapport and K. Suzuki, eds.). Plenum Press, New York, pp. 65–73.
128. Arita, M., Iwamori, M., Higuchi, T., and Nagai, Y. (1984). Positive and negative ion fast atom bombardment mass spectrometry of glycosphingolipids. Discrimination of the positional isomers of gangliosides with sialic acids. *J. Biochem. 95*:971–981.
129. Ledeen, R., and Salsman, K. (1965). Structure of the Tay-Sachs' ganglioside. *Biochemistry 4*:2225–2233.
130. Johnson, G. A., and McCluer, R. H. (1964). Periodate oxidation studies of human brain gangliosides. *Biochim. Biophys. Acta 84*:587–597.
131. Mayer, H., Framberg, K., and Wackesser, J. (1974). 6-*O*-methyl-D-glucosamine in lipopolysaccharides of *Rhodopseudomonas palustris* strains. *Eur. J. Biochem. 44*:181–187.
132. Hakomori, S. (1964). Rapid permethylation of glycolipid and pentasaccharide catalyzed by methylsulfinylcarbanion in dimethyl sulfoxide. *J. Biochem. 55*:205–211.
133. Kundu, S. K., Ledeen, R. W., and Gorin, P. A. (1975). Determination of position of substitution on 2-acetamido-2-deoxy-D-galactosyl residues in glycolipids. *Carbohydr. Res. 39*: 179–191.
134. Bjorndal, H., Hellerqvist, C. G., and Svensson, S. (1967). Gas-liquid chromatography of partially methylated alditols as their acetates. *Acta Chem. Scand. 21*:1801–1804.
135. Bjorndal, H., Hellerqvist, C. G., Lindberg, B., and Svensson, S. (1970). Gas-liquid chromatography and mass spectrometry in methylation analysis of polysaccharides. *Angew. Chem. Int. Ed. Engl. 9*:610–619.
136. Hellerqvist, C. G., Lindberg, B., Pilotti, A., and Lindberg, A. A. (1971). Structural studies of the O specific side chains of the cell wall polysaccharide from *Salmonella senftenberg*. *Carbohydr. Res. 16*:297–302.
137. Kundu, S. K., Ledeen, R. W., and Gorin, P. A. J. (1975). Determination of position of substitution on 2-acetamido-2-deoxy-D-glycosyl residues in glycolipids. *Carbohydr. Res. 39*:329–348.
138. Dabrowski, J., Hanfland, P., and Egge, H. (1982). Analysis of glycosphingolipids by high-resolution proton nuclear magnetic resonance spectroscopy. *Methods Enzymol. 83*: 69–86.
139. Falk, K. E., Karlsson, K. A., and Samuelsson, B. E. (1979). Proton nuclear magnetic resonance analysis of anomeric structure of glycosphingolipids. *Arch. Biochem. Biophys. 192*:191–202, and references cited therein.
140. Gasa, S., Mitsuyama, T., and Makita, A. (1983). Proton nuclear magnetic resonance of neutral and acidic glycosphingolipids. *J. Lipid Res. 24*:174–182.
141. Yamada, A., Dabrowski, J., Hanfland, P., and Egge, H. (1980). Preliminary results of J-resolved, two-dimensional H-NMR studies on glycosphingolipids. *Biochim. Biophys. Acta 618*:473–479.
142. Koerner, T. A. W., Prestegard, U. H., Demou, P. C., and Yu, R. K. (1983). High-resolution proton NMR studies of gangliosides. 2. Use of two-dimensional nuclear Overhauser effect spectroscopy and silylation shifts for determination of oligosaccharide sequence and linkage sites. *Biochemistry 22*:2687–2690.

143. Koerner, T. A. W., Prestegard, J. H., Demou, P. C., and Yu, R. K. (1983). High-resolution proton NMR studies of gangliosides. 1. use of homonuclear two-dimensional spin-echo J-correlated spectroscopy for determination of residue composition and anomeric configurations. *Biochemistry 22*:2676–2687.
144. Sillerud, L. O., Yu, R. K., and Schafer, D. E. (1982). Assignment of the carbon-13 nuclear magnetic resonance spectra of gangliosides G_{M4}, G_{M3}, G_{M1}, D_{D1a}, G_{D1b} and G_{T1b}. *Biochemistry 21*:1260–1271, and references cited therein.
145. Magnani, J. L. (1986). Carbohydrate sequences detected by murine monoclonal antibodies. *Chem. Phys. Lipids 42*:65–74.
146. Hakomori, S. (1983). Tumor-associated glycolipid antigens defined by monoclonal antibodies. *Bull. Cancer 70*:118–126.
147. Marcus, D. M. (1976). Applications of immunological techniques to the study of glycosphingolipids. In *Glycolipid Methodology* (L. A. Witting, ed.). American Oil Chemists' Society Press, Champaign, Ill., pp. 233–245.
148. Kundu, S. K., and Roy, S. K. (1976). Aminopropyl silica gel as a solid support for the preparation of glycolipid immunoadsorbent and purification of antibodies. *J. Lipid Res. 20*:825–833.
149. Marcus, D. M., and Kundu, S. K. (1980). Preparation and properties of antibodies to gangliosides. In *CNRS International Symposium on Structure and Function of Gangliosides*, Vol. 125 (L. Svennerholm, P. Mandel, H. Dreyfus, and P. F. Urban, eds.). Plenum Press, New York, pp. 321–326.
150. Brockhaus, M., Magnani, J. L., Herlyn, M. D., Steplewski, Z., Koprowski, H., and Ginsburg, V. (1982). Monoclonal antibodies directed against the sugar sequences of lacto-*N*-fucopentaose III are obtained from mice immunized with human tumors. *Arch. Biochem. Biophys. 217*:647–651.
151. Kundu, S. K., Pleatman, M. A., Redwine, W. A., Boyd, A. E., and Marcus, D. M. (1983). Binding of monoclonal antibody A_2B_5 to gangliosides. *Biochem. Biophys. Res. Commun. 116*:836–848.
152. Kundu, S. K., Misra, L. K., and Luthra, M. G. (1982). Muscle glycolipids in inherited muscular dystrophy of chickens. *FEBS Lett. 150*:359–364.
153. Anderson, B., Slota, J., Kundu, S., Patrick, J., Manderino, G., Rittenhouse, H., and Tomita, J. (1987). Characterization of monoclonal antibodies to paragloboside (PG) and sialosyl-PG (2,6-SPG) and an improved chromatogram binding assay for rapidly identifying antibodies to tumor antigens. *J. Cell. Biochem. S110*:157.
154. Shapiro, D., Acher, A. J., and Robinson, Y. (1973). Studies in the ganglioside series. VIII. Total synthesis of Tay-Sachs globoside. *Chem. Phys. Lipids 10*:28–36, and references cited therein.
155. Shapiro, D., and Acher, A. J. (1978). Total synthesis of ceramide trihexoside accumulating with Fabry's disease. *Chem. Phys. Lipids 22*:197–206.
156. Koike, K., Suginoto, M., Sato, S., Ito, Y., Nakahara, Y., and Ogawa, T. (1984). Total synthesis of globotriaosyl-E and Z-ceramides and isoglobotriaosyl-E-ceramides. *Carbohydr. Res. 63*: 15–25.
157. Wessel, H. P., Iverson, T., and Bundle, D. R. (1984). Synthesis of the trisaccharide moiety of gangliotriosylceramide (asialo G_{M2}). *Carbohydr. Res. 130*:5–21.
158. Namata, M., Sugimoto, M., Koike, K., and Ogawa, T. (1987). Total synthesis of sialosylcerebroside, G_{M4}. Carbohydr. Res. 163:209–225.
159. Gigg, R. (1980). Synthesis of glycolipids. *Chem. Phys. Lipids 26*:287–404.
160. Shapiro, D. (1973). Approaches to the synthesis of gangliosides. *Int. Cong. Pure Appl. Chem. 2*:153–166.
161. Sugimoto, M., Numata, M., Koike, K., Nakahara, Y., and Ogawa, T. (1986). Total synthesis of gangliosides G_{M1} and G_{M2}. *Carbohydr. Res. 156*:C1–C5.
162. Paulsen, H. (1982). Advances in selective chemical synthesis of complex oligosaccharides. *Angew, Chem. Int. Ed. Engl. 21*:155–173.
163. Sebesan, S., and Lemieux, R. (1984). Synthesis of tri- and tetrasaccharide haptens related to the asialo forms of the gangliosides G_{M2} and G_{M1}. *Can. J. Chem. 62*:644–653.

164. Paulsen, H., and Bunsch, A. (1976). Synthese der Tetrasaccharid-kettedes P-antigen-globoside. Eine β-D-glycosidysynthese fur 2-amino-2-desoxyzucker. *Carbohydr. Res. 101*:21–30.
165. Ponpipom, M. M., Bugianese, R. L., and Shen, T. Y. (1978). Synthesis of paragloboside analogs. *Tetrahedron Lett.* 1717–1720.
166. Rana, S., Barlow, J. J., and Matta, K. L. (1981). The chemical synthesis of *O*-α-L-fucopyranosyl (1-2)-*O*-β-D-galactopyranosyl-(1-3)-*O*-[α-L-fucopyranosyl-(1-4)-2-acetamido-2-deoxy-D-glucopyranose, the Lewis b blood-group antigenic determinant. *Carbohydr. Res. 96*: 231–241.
167. Paulsen, H., and Bunsch, A. (1982). Synthesis of pentasaccharide chain of Forssman antigens. *Carbohydr. Res. 100*:143–167.
168. Nunomuro, S., Mori, M., Ito, Y., and Ogawa, T. (1989). A total synthesis of Forssman glycolipid, globopentaosylceramide, IV^3GalNAcαGb4Cer, *Tetrahedron Lett. 30*:6713–6716.
169. Veyrieres, A. (1981). Blood group Ii-active oligosaccharides—synthesis of a tetrasaccharide, a β-(1-3)dimer of *N*-acetyl-lactosamine. *J. Chem. Soc. Perkin 1*:1626–1629.
170. Kabat, E. A., Liao, J., and Lemieux, R. J. (1978). Immunochemical studies on blood groups. LXVIII. The combining site of anti-I Ma (group I). *Immunochemistry 15*:727–731.
171. Lemieux, R. U. (1982). The binding of carbohydrate structures with antibodies and lectins. In *IUPAC Frontiers of Chemistry* (K. J. Laidler, ed.). Pergamon Press, New York, pp. 3–24.
172. Roseman, S. (1970). The synthesis of complex carbohydrates by multiglycosyltransferase systems and their potential function in intercellular adhesion. *Chem. Phys. Lipids 5*:270–297.
173. Keenan, T. W., Morre, D. J., and Basu, S. (1974). Concentration of glycosphingolipid glycosyltransferase in Golgi apparatus from rat liver. *J. Biol. Chem. 251*:310–315.
174. Fleischer, B. (1977). Localization of some glycolipid glycosylating enzymes in the Golgi apparatus of rat kidney. *J. Supramol. Struct. 7*:79–89.
175. Pacuszka, T., Duffard, R. O., Nishimura, R. N., Brady, R. O., and Fishman, P. H. (1978). Biosynthesis of bovine thyroid gangliosides. *J. Biol. Chem. 253*:5839–5846.
176. Eppler, C. M., Morre, D. J., and Keenan, T. W. (1980). Ganglioside biosynthesis in rat liver: Characterization of cytidine-5-monophospho-*n*-acetylneuraminic acid:hematoside (G_{M3}) sialosyltransferase. *Biochim. Biophys. Acta 619*:318–331.
177. Patt, L. M., and Grimes, W. J. (1974). Cell surface glycolipid and glycoprotein by glycosyltransferases of normal and transformed cells. *J. Biol. Chem. 249*:4157–4165.
178. Yogeeswaran, G., Laine, R. A., and Hakomori, S. (1974). Mechanism of cell contact-dependent glycolipid synthesis: Further studies with glycolipid glass-complex. *Biochem. Biophys. Res. Commun. 59*:591–599.
179. Bosman, H. B. (1972). Cell surface glycosyltransferases and acceptors in normal and RNA- and DNA-virus transformed fibroblasts. *Biochem. Biophys. Res. Commun. 48*:523–529.
180. Roth, S., and White, D. (1971). Intracellular contact and cell-surface galactosyl transferase activity. *Proc. Natl. Acad. Sci. USA 48*:523–529.
181. Keenan, T. W., and Morre, D. J. (1973). Glycosyltransferases: Do they exist on the surface membrane of mammalian cells? *FEBS Lett. 37*:124–228.
182. Deppert, W., and Walter, G. (1978). Cell surface glycosyltransferases—do they exist? *J. Supramol. Struct. 8*:19–37.
183. Miller-Podraja, H., Bradley, R. M., and Fishman, P. H. (1982). Biosynthesis and localization of gangliosides in cultured cells. *Biochemistry 21*:3260–3265.
184. Miller-Podraja, H., and Fishman, P. H. (1982). Translocation of new gangliosides to cell surface. *Biochemistry 21*:3265–3270.
185. Morre, D. J., Kartenback, J., and Franke, W. W. (1979). Membrane flow and interconversions among endomembranes. *Biochim. Biophys. Acta 559*:71–152.
186. Basu, S., and Basu, M. (1982). Expression of glycosphingolipid glycosyltransferases in development and transformation. In *The Glycoconjugates*. Academic Press, New York, pp. 265–282.

187. Basu, S., Basu, J., De, T., Kyle, J. W., Das, K. K., and Schaeper, R. J. (1986). Biosynthesis of gangliosides and blood group glycolipids using solubilized glycosyltransferases. In *Enzymes of Lipid Metabolism II* (L. Freysz, H. Dreyfus, R. Massarelli, and S. Gatt, eds.). Plenum Press, New York, pp. 233–243.
188. Dawson, G. (1974). Glycolipid biosynthesis. In *The Glycoconjugates*, Vol. 2. Academic Press, New York, pp. 256–283.
189. Shur, B. D. (1982). Cell surface glycosyltransferase activities during fertilization and early embryogenesis. In *The Glycoconjugates*, Vol. 3. Academic Press, New York, pp. 146–185.
190. Chien, J. L., Williams, T., and Basu, S. (1973). Biosynthesis of a globoside-type glycosphingolipid by a *N*-acetylgalactosaminyltransferase from embryonic chicken brain. *J. Biol. Chem.* *248*:1778–1785.
191. Basu, S., Basu, M., Higashi, H., and Evans, C. H. (1982). Biosynthesis and characterization of globoside and Forssman glycosphingolipids in guinea pig tumor cells. In *New Vistas in Glycolipid Research* (A. Makita, S. Handa, T. Taketomi, and Y. Nagai, eds.). Plenum Press, New York, pp. 131–137.
192. Dawson, G. (1978). Glycolipid catabolism. In *The Glycoconjugates*, Vol. 2. Academic Press, New York, pp. 285–336.
193. Svennerholm, L. (1963). Chromatographic separation of human brain gangliosides. *J. Neurochem.* *10*:613–623.
194. Ohman, R., Rosenberg, A., and Svennerholm, L. (1970). Human brain sialidase. *Biochemistry* *9*:3774–3782.
195. Brady, R. O. (1978). Sphingolipidoses. *Annu. Rev. Biochem.* *47*:687–713.
196. Brady, R. O. (1973). Hereditary diseases—causes, cures and problems. *Angew. Chem. Int.* *12*:1–11.
197. Brady, R. O. (1973). Inborn errors of lipid metabolism. *Adv. Enzymol.* *38*:294–315.
198. Li, S. C., and Li, Y. T. (1976). An activator stimulating the enzymic hydrolysis of sphingolipids. *J. Biol. Chem.* *251*:1159–1163.
199. Li, S. C., Wan, C. C., Mazotte, M. Y., and Li, Y. T. (1979). Requirements of an activator for the hydrolysis of sphingoglycolipids by glycosidases of human liver. *Carbohydr. Res.* *34*: 189–193.
200. Li, S. C., Nakumura, T., Ogamo, A., and Li, Y. T. (1979). Evidence for the presence of two separate protein activators for the enzymic hydrolysis of G_{M1} and G_{M2} gangliosides. *J. Biol. Chem.* *254*:592–595.
201. Metz, R. J., and Radin, N. S. (1980). Glucosylceramide uptake protein from spleen cytosol. *J. Biol. Chem.* *255*:4463–4467.
202. Fishman, P. H. (1982). Role of membrane gangliosides in the binding and action of bacterial toxins. *J.Membr. Biol.* *69*:85–97.
203. Van Heyningen, W. E., Carpenter, C. C. L., Pierce, N. F., and Greendugh, W. B. (1971). Deactivation of cholera toxin by ganglioside. *J. Infect. Dis.* *124*:415–418.
204. Cuatrecases, P. (1973). Gangliosides and membrane receptors for cholera toxin. *Biochemistry* *12*:3558–3566.
205. Holmgren, J., Lonroth, L., Mansson, J. E., and Svennerholm, L. (1973). Tissue receptor for cholera exotoxin: Postulated structure from studies with G_{M1} ganglioside and related glycolipids. *Infect. Immun.* *8*:208–214.
206. Moss, J., Osborne, J. C., Fishman, P. H., Brewer, H. B., Vaughan, M., and Brady, R. O. (1977). Effect of gangliosides and substrate analogues on the hydrolysis of nicotinamide adenine dinucleotide by choleragen. *Proc. Natl. Acad. Sci. USA* *74*:74–78.
207. Fishman, P. H., Moss, J., and Osborne, J. C., Jr. (1978). Interaction of choleragen with the oligosaccharide of ganglioside G_{M1}. Evidence for multiple oligosaccharide binding sites. *Biochemistry* *17*:711–716.
208. Kanfer, J. N., Carter, T. P., and Katzman, H. M. (1976). Lipolytic action of cholera toxin on fat cells. Re-examination of the concept implicating G_{M1} ganglioside as the native membrane receptor. *J. Biol. Chem.* *251*:7610–7619.

209. Grollman, E. F., Lee, G., Lamos, S., Lazo, P. S., Kabach, H. R., Friedman, R. M., and Kohn, L. D. (1978). Relationships of the structure and function of the interferon receptor to hormone receptors and establishment of the antiviral state. *Cancer Res. 38*:4172–4185.
210. Marita, A., Tsao, D., and Kim, Y. S. (1980). Identification of cholera toxin binding glycoproteins in rat in micro villus membranes. *J. Biol. Chem. 255*:2549–2553.
211. Critchley, D. R., Ansell, S., Perkins, R., Dilks, S., and Ingram, S. (1979). Isolation of cholera toxin receptors from a mouse fibroblast and lymphoid cell line by immune precipitation. *J. Supramol. Struct. 12*:273–291.
212. Tonegawa, Y., and Hakomori, S. (1977). Ganglioprotein and globoprotein: The glycoproteins reacting with anti-ganglioside and anti-globoside antibodies and the ganglioprotein change associated with transformation. *Biochem. Biophys. Res. Commun. 76*:9–17.
213. Van Heyningen, W. E., and Mellanby, J. (1968). The effect of cerebroside and other lipids on the fixation of tetanus toxin by gangliosides. *J. Gen. Microbiol. 52*:447–454.
214. Lee, G., Consiglio, E., Habig, W., Dryer, S., Hardgree, C., and Kohn, L. D. (1978). Structure and functional studies of receptors for thyrotropin and tetanus toxin: Lipid modulation of effects binding to the glycoprotein receptor component. *Biochem. Biophys. Res. Commun. 83*: 313–320.
215. Van Heyningen, W. E. (1974). Gangliosides as membranes receptors for tetanus toxin, cholera toxin and serotonin. *Nature 249*:415–417.
216. Zimmerman, J. M., and Gifferetti, J. C. (1977). Interaction of tetanus toxin and toxoid with cultured neuroblastoma cells. *Naunyn Schmiedebergs Arch. Pharmacol. 296*:271–277.
217. Stoeekel, D., Schwab, M., and Thoenen, H. (1977). Role of gangliosides in the uptake and retrograde axonal transport of cholera and tetanus toxin as compared to nerve growth factor and wheat germ agglutinin. *Brain Res. 132*:273–285.
218. Helting, T. B., Zwisler, O., and Weigandt, H. (1977). Structure of tetanus toxin. II. Toxin binding to gangliosides. *J. Biol. Chem. 252*:194–198.
219. Simpson, L. L., and Rapport, M. M. (1971). Ganglioside interaction with botulinum toxin. *J. Neurochem. 18*:1341–1343.
220. Kitamura, M., Iwamori, M., and Nagai, Y. (1980). Interaction between *Clostridium botulinum* neurotoxin and gangliosides. *Biochim. Biophys. Acta 628*:328–335.
221. Kato, J., and Naiki, M. (1976). Ganglioside and rabbit erythrocyte membrane receptor for staphylococcal alpha-toxin. *Infect. Immun. 13*:289–291.
222. Buchanan, T. M., Pearce, W. A., and Chen, K. C. S. (1978). In *The Immunobiology of* Niesseria gonorrhoeaa (G. Brooks, ed.). American Society for Microbiology, Washington, D.C., pp. 242–249.
223. Takeda, J., Takeda, T., Honda, T., and Miwatani, T. (1978). Inactivation of the biological activities of the thermostable direct hemolysin of *Vibrio parahaemolyticus* by ganglioside G_{T1}. *Infect. Immun. 14*:1–5.
224. Dallas, W. S., and Falkow, S. T. (1979). The molecular nature of heat-labile enterotoxin (LT) of *Escherichia coli. Nature 277*:406–407.
225. Donta, S. T., and Viner, J. P. (1975). Inhibition of the stereodogenic effects of cholera toxin and heat-labile *Escherichia coli* enterotoxins by G_{M1} ganglioside; evidence for a similar receptor for the two toxins. *Infect. Immun. 11*:982–985.
226. Mullin, B. R., Fishman, P. H., Lee, G., Aloj, S. M., Ledley, F. D., Winand, R. J., Kohn, L. D.,and Brady, R. O. (1976). Thyrotropin–ganglioside interactions and their relationship to the structure and function of thyrotropin receptors. *Proc. Natl. Acad. Sci. USA 73*: 842–846.
227. Lee, G., Aloj, S. M., Brady, R. O., and Kohn, L. D. (1976). The structure and function of glycoprotein in hormone receptors—ganglioside interactions with human chorionic gonodotropin. *Biochem. Biophys. Res. Commun. 73*:370–377.
228. Lee, G., Aloj, S. M., and Kohn, L. D. (1977). The structure and function of glycoprotein hormone receptors—ganglioside interactions with luteinizing hormone. *Biochem. Biophys. Res. Commun. 77*:434–441.

229. Besancon, F., Ankal, H., and Basu, S. (1976). Specificity and reversibility of interferon ganglioside interaction. *Nature 259*:576–578.
230. Vengries, V. E., Reylonds, F. H., Hallenberg, M. D., and Pitha, P. M. (1976). Interferon action; role of membrane gangliosides. *Virology 72*:486–493.
231. Ankel, J., Krishnamurti, C., Besancon, F., Stefanos, S., and Falcoft, E. (1980). Mouse fibroblast (type 1) and immune (type II) interferons. Pronounced differences in affinity for gangliosides and in antiviral and antigrowth effects on mouse leukemia L-1210 R cells. *Proc. Natl. Acad. Sci. USA 77*:2528–2532.
232. Karlsson, K. A. (1977). Aspects on structure and function of sphingolipids in cell surface membranes. In *Structure of Biological Membranes* (S. Abrahamson and I. Pascher, eds.). Plenum Press, New York, pp. 245–274.
233. Byrne, M.C., Sbaschnig-Agler, M., Aquino, D. A., Scolafani, J. R., and Ledeen, R. W. (1985). Procedure for isolation of gangliosides in high yield and purity; simultaneous isolation of neutral glycosphingolipids. *Anal. Biochem. 148*:163–173.
234. Wooley, D. W., and Gommi, B. W. (1965). Serotonin receptor. VII. Activities of various pure gangliosides as the receptor. *Proc. Natl. Acad. Sci. USA 53*:959–963.
235. De Robertis, E., Lapetina, E. G., and de Plaza, S. F. (1976). Subcellular distribution and possible role of gangliosides in the CNS. In *Ganglioside Function: Biochemical and Pharmacological Implications*, Vol. 71 (G. Porcellati, B. Ceccarelli, and G. Tettamonti, eds.). Academic Press, New York, pp. 105–121.
236. Haywood, A. M. (1974). Characterization of Sendai virus receptors in a model membrane. *J. Mol. Biol. 83*:427–436.
237. Holmgren, J., Svennerholm, L., Elwing, H., Fredman, P., and Strannegrad, O. (1980). Sendai virus receptor: Proposed recognition structure based on binding to plastic adsorbed gangliosides. *Proc. Natl. Acad. Sci. USA 77*:1947–1950.
238. Wu, P. S., Ledeen, R. W., Udem, S., and Isaacson, Y. A. (1980). Nature of the Sendai virus receptor: Glycoprotein versus ganglioside. *J. Virol. 33*:304–310.
239. Markwell, M. A. K., Svennerholm, L., and Paulson, J. C. (1981). Specific ganglioside function as host cell receptor for Sendai virus. *Proc. Natl. Acad. Sci. USA 78*:5406–5410.
240. Markwell, M. A. K., Fredman, P., and Svennerholm, L. (1984). Receptor ganglioside content of three hosts for Sendai virus MDBK, HeLa and MDCK cells. *Biochim. Biophys. Acta 775*:7–16.
241. Suzuki, Y., Suzuki, N., Suzuki, T., and Matsumato, M. (1982). Major sialoglycolipids of hog erythrocytes—isolation and involvement in Sendai virus receptor. *Proc. Jpn. Conf. Biochem. Lipids 24*:69–72.
242. Kleinman, H. K., Martin, G. R., and Fishman, P. (1979). Ganglioside inhibition of fibronectin-mediated cell adhesion to collagen. *Proc. Natl. Acad. Sci. USA 76*:3367–3371.
243. Rauvala, H., Carter, W. G., and Hakomori, S. (1980). Studies of cell on cell adhesion and recognition. 1. Extent and specificities of cell adhesion triggered by carbohydrate reactive proteins, lectins and by fibronectin. *J. Cell Biol. 88*:127–137.
244. Higgins, T. J., Sabatino, A. P., Remold, H. O., and David, J. R. (1978). Possible role of macrophage glycolipids as receptors for migration inhibitory factor (MIF). *J. Immunol. 121*:880–886.
245. Higgins, T. J., Liu, D. Y., Remold, H. G., and David, J. R. (1980). Further characterization of the putative glycolipid receptor for MIF. Role of fucose associated with an acidic glycolipid. *Biochem. Biophys. Res. Commun. 93*:1259–1265.
246. Liu, D. Y., Petschek, K. D., Remold, H. G., and David, J. R. (1982). Isolation of a guinea pig macrophage glycolipid with the properties of a putative migration inhibitory factor receptor. *J. Biol. Chem. 257*:159–162.
247. Poste, G., Allen, H., and Matta, K. L. (1979). Cell surface receptors for lymphokines. II. Studies on the carbohydrate composition of the MIF receptor on macrophages using synthetic saccharides and plant lectins. *Cell. Immunol. 44*:89–98.
248. Miura, T., Handa, S., and Yamakawa, T. (1979). Specific inhibition of macrophage migration inhibition factor by fucosylated glycolipid RM. *J. Biochem. 86*:773–776.

249. Bremer, E. G., Schlessinger, J., and Hakomori, S. (1986). Ganglioside mediated modulation of cell growth. Specific effects of G_{M3} tyrosine phosphorylation of epidermal growth factor. *J. Biol. Chem. 261*:2434–2440.

250. Weis, F. M., and Davis, R. J. (1990). Regulation of epidermal growth factor receptor signal transduction. Role of ganglioside. *J. Biol. Chem. 265*:12059–12066.

251. Hanai, N., Dohi, T., Nores, G. A., and Hakomori, S. (1988). A novel ganglioside, de-*N*-acetyl G_{M3} (II^3NeuNH2LacCer), acting as a strong promoter for epidermal growth factor receptor kinase and as a stimulator for cell growth. *J. Biol. Chem. 263*:6296–6301.

252. Kabat, E. K. (1982). Philip Levine Award Lecture. Contributions of quantitative immunochemistry to knowledge of blood group A, B, H, Le, I and i antigens. *Am. J. Clin. Pathol. 78*:281–292.

253. Egge, H., and Hanfland, P. (1981). Immunochemistry of the Lewis-blood group system: Mass spectrometric analysis of permethylated Le^a, Le^b and H-type 1 (Le^{dH}) blood-group active and related glycosphingolipids from human plasma. *Arch. Biochem. 210*:396–404.

254. Hanfland, P., and Graham, H. A. (1981). Immunochemistry of the Lewis-blood group system; partial characterization of Le^a, Le^b and H type (Le^{dH})-blood group active glycosphingolipids from human plasma. *Arch. Biochem. Biophys. 210*:383–395.

255. Hanfland, P. (1975). Characterization of B and H blood-group active glycosphingolipids from human B erythrocyte membranes. *Chem. Phys. Lipids 15*:105–108.

256. Koscielak, J., Piasek, A., and Gorniak, H. (1973). Structures of fucose-containing glycolipid with H and B blood group activity and of sialic acid and glucosamine-containing glycolipid of human erythrocyte membrane. *Eur. J. Biochem. 37*:214–225.

257. Marcus, D. M., and Cass, L. E. (1969). Glycosphingolipids with Lewis-blood group activity: Uptake by human erythrocytes. *Science 164*:553–555.

258. Crookston, M. C. (1980). Blood group antigens acquired from the plasma. In *Immunobiology of the Erythrocytes* (S. G. Sandler, I. Nusbacher, and M. S. Schanfield, eds.). Alan R. Liss, New York, *Prog. Clin. Biol. Res. 43*:35–65.

259. Dawson, G., Kruski, A. W., and Scanu, A. M. (1976). Distribution of glycosphingolipids in the serum lipoproteins of normal human subjects and patients with hypo and hyperlipidemias. *J. Lipid Res. 17*:125–131.

260. Chatterjee, S., and Kwiterovich, P. O. (1976). Glycosphingolipids of human plasma lipoproteins. *Lipids 11*:462–466.

261. Clarke, J. T. R. (1981). The glycosphingolipids of human plasma lipoproteins. *Can. J. Biochem. 59*:412–417.

262. van den Bergh, F. A. J. T. M., and Tager, J. M. (1976). Localization of neutral glycosphingolipids in human plasma. *Biochim. Biophys. Acta 441*:391–402.

263. Metz, R. J., and Radin, N. A. (1982). Purification and properties of cerebroside transfer protein. *J. Biol. Chem. 257*:12901–12907.

264. Bloj, B., and Zilversmidt, D. B. (1981). Accelerated transfer of neutral glycosphingolipid and ganglioside G_{M1} by a purified lipid transfer protein. *J. Biol. Chem. 256*:5988–5991.

265. Via, D. P., Massey, J. B., Kundu, S. K., Marcus, D. M., and Pownall, H. J. (1982). Spontaneous and plasma factor-mediated transfer of (ω-pyrenyl) alkanoylglycolipids between model and native high density lipoproteins. *Fed. Proc. 41*:684.

266. Loeb, J., and Dawson, G. (1982). Reversible exchange of glycosphingolipids between human high and low density lipoproteins. *J. Biol. Chem. 257*:11982–11987.

267. Loeb, J., and Dawson, G. (1983). High density lipoprotein exchange reactions. *Mol. Cell. Biochem. 52*:161–176.

268. Hakomori, S. (1986). Tumor-associated glycolipid antigens, their metabolism and organization. *Chem. Phys. Lipids 42*:209–233.

269. Sell, S. (1990). Cancer associated carbohydrates identified by monoclonal antibodies. *Hum. Pathol. 21*:1003–1019.

270. Radin, N. S., and Inokuchi, J. (1988). Glucosphingolipids as sites of action in the chemotherapy of cancer. *Biochem. Pharmacol. 37*:2879–2886.

271. Herlyn, M., Menrad, A., and Koprowski, H. (1990). Structure, function and clinical significance of human tumor-antigens. *JNCI 82*:1883–1889.
272. Dippold, W. G., Lloyd, K. O., Li, L. T., Ikeda, H., Oettgen, H. F., and Old, L. J. (1980). Cell-surface antigens of human malignant melanoma. Defination of six antigenic systems with mouse monoclonal antibodies. *Proc. Natl. Acad. Sci. USA 77*:6114–6118.
273. Yeh, M. I., Hellstrom, I., Abe, K., Hakomori, S., and Hellstrom, K. E. (1982). A cell-surface antigen which is present in the ganglioside fraction and shared by human melanomas. *Int. J. Cancer 29*:269–275.
274. Nudelman, E., Hakomori, S., Kannagi, R., Levery, S., Yeh, M. Y., Hellstrom, K. E., and Hellstrom, I. (1982). Characterization of human melanoma-associated ganglioside antigen defined by monoclonal antibody 4.2. *J. Biol. Chem. 257*:12752–12756.
275. Cheresh, D. A., Reisfeld, R. A., and Varki, A. P. (1984). *O*-Acetylation of disialoganglioside G_{D3} by human melanoma cell creates a unique antigenic determinant. *Science 225*:844–846.
276. Cheresh, D. A., Varki, A. P., Varki, N. M., Stallcup, W. B., Levine, J., and Reisfeld, R. A. (1984). A monoclonal antibody recognizes an *O*-acylated sialic acid in a human melanoma-associated ganglioside. *J. Biol. Chem. 259*:7453–7459.
277. Houghton, A. N., Mintzer, D., Cordon-Cardo, C., Wett, S., Fliegel, B., Vadhan, S., Carswell, L., Melamed, M. R., Oettgen, H. F., and Old, L. J. (1985). Mouse monoclonal IgG3 antibody detecting G_{D3} ganglioside: Phase 1 trial in patients with malignant melanoma. *Proc. Natl. Acad. Sci. USA 82*:1242–1246.
278. Herberman, R. B., Morgan, A. C., Reisfeld, R., and Ortaldo, J. R. (1986). In *Monoclonal Antibodies and Cancer Therapy* (R. A. Reisfeld and S. Sell, eds.). Alan R. Liss, New York.
279. Dohi, T., Ohta, S., Hanai, N., and Oshima, M. (1990). Sialosylpentaosylceramide detected with anti G_{M2} monoclonal antibody. Structural characterization and complementary expression with G_{M2} in gastric cancer and normal gastric mucosa. *J. Biol. Chem. 265*:7880–7885.
280. Irie, R. F., Sze, L. L., and Saxton, R. E. (1982). Human antibody to OFA-1, a tumor antigen, produced in vitro by Epstein–Barr virus-transformed human B-lymphoid cell lines. *Proc. Natl. Acad. Sci. USA 79*:5666–5671.
281. Tai, T., Paulson, J. C., Cahan, L. D., and Irie, R. F. (1983). Ganglioside G_{M2} as a human tumor antigen (OFA-1-1). *Proc. Natl. Acad. Sci. USA 80*:5392–5396.
282. Cahan, L. D., Irie, R. E., Singh, R., Casidenti, A., and Paulsen, J. C. (1982). Identification of human neuroectodermal tumor antigen [1A-1-2] as ganglioside G_{D2}. *Proc. Natl. Acad. Sci. USA 79*:7629–7633.
283. Watanabe, T., Pukel, C. S., Takeyama, H., Lloyd, K. O., Shiku, H., Li, L. T. C., Trabassos, L. R., Oettgen, H. F., and Old, L. J. (1982). Human melanoma AH is an autoantigen ganglioside related to G_{D2}. *J. Exp. Med. 156*:1884–1889.
284. Cheung, N. K., and Miraldi, F. D. (1988). Iodine-131 labeled G_{D2} monoclonal antibody in the diagnosis of human neuroblastoma. *Progr. Clin. Biol. Res. 271*:595–604.
285. Fredman, P., Brezicka, F. T., Holmgren, J., Lindholm, L., Nilsson, O., and Svennerholm, L. (1986). Binding specificity of monoclonal antibodies to ganglioside, Fuc-G_{M1}. *Biochim. Biophys. Acta 875*:316–323.
286. Boltovskaya, M. N. (1988). Biochemical and immunochemical analysis of gangliosides of human small cell lung carcinoma; production of monoclonal antibodies against a unique marker of small cell lung carcinoma, ganglioside Fuc-G_{M1}. *Biotechnicol. Appl. Biochem. 10*:273–286.
287. Nilsson, O., Mansson, J. E., Brezicka, T., Holmgren, J., Lindholm, L., Sorenson, S., Yngvason, F., and Svennerholm, L. (1984). Fuc-G_{M1}-A ganglioside associated with small cell lung carcinomas. *Glyconjugates 2*:623–624.
288. Wiels, J., Fellous, M., and Tursz, T. (1981). Monoclonal antibody against a Burkitt lymphoma associated antigen. *Proc. Natl. Acad. Sci. USA 78*:6485–6488.
289. Nudelman, E., Kannagi, R., Hakomori, S., Parsons, S., Lipinski, M., Wiels, J., Fellous, M., and Tursz, M. (1983). A glycolipid antigen associated with Burkitt lymphoma defined by a monoclonal antibody. *Science 220*:509–511.

290. Ohyama, C., Fukushi, Y., Satoh, S., Orikasa, S., Nudelman, E., Stroud, M., and Hakomori, S. (1990). Changes in glycolipid expression in human testicular tumor. *Int. J. Cancer 45*:1040–1044.
291. Buchbinder, L. (1935). Heterophile phenomena in immunology. *Arch. Pathol. 19*:841–880.
292. Hakomori, S., Wang, S. H., and Young, W. W. (1977). Isoantigenic expression of Forssman glycolipid in human gastric and colonic mucosa; its possible identity with "A-like antigen" in human cancer. *Proc. Natl. Acad. Sci. USA 74*:3023–3027.
293. Yoda, Y., Ishibashi, T., and Makita, A. (1980). Isolation, characterization and biosynthesis of Forssman antigen in human lung and lung carcinoma. *J. Biochem. 88*:1887–1890.
294. Mori, R., Sudo, T., and Kano, K. (1983). Expression of heterophile Forssman antigens in cultured malignant cell lines. *JNCI 70*:811–814.
295. Mori, E., Mori, T., Sanai, Y., and Nagai, Y. (1982). Radioimmuno-thin-layer chromatographic detection of Forssman antigen in human carcinoma cell lines. *Biochem. Biophys. Res. Commun. 108*:920–926.
296. Taniguchi, N., Yokosawa, N., Narita, J., Mitsuyama, T., and Makita, A. (1981). Expression of Forssman antigen synthesis and degradation in human lung cancer. *JNCI 67*:577–583.
297. Schrump, D. S., Furukawa, K., Yamaguchi, A., Lloyd, K. O., and Old, L. J. (1988). Recognition of galactosyl-globoside by monoclonal antibodies from patients with primary lung cancer. *Proc. Natl. Acad. Sci. USA 85*:4441–4448.
298. Schwarting, G. A., Carroll, P. G., and DeWolf, W. C. (1983). Fucosyl-globoside and sialosyl-globoside are new glycolipids isolated from human teratocarcinoma cells. *Biochem. Biophys. Res. Commun. 112*:935–940.
299. Koprowski, J., Steplewski, Z., Mitchell, K., Herlyn, M., Herlyn, D., and Fuhrer, P. (1979). Colorectal carcinoma antigens detected by hybridoma antibodies. *Somatic Cell Genet. 5*:957–972.
300. Koprowski, H., Herlyn, M., Steplewski, Z., and Sears, H. F. (1981). Specific antigen in serum of patients with colon carcinoma. *Science 212*:53–72.
301. Magnani, J. L., Nilsson, G., Brockhaus, M., Zopf, D., Steplewski, Z., Koprowski, H., and Ginsburg, V. (1982). A monoclonal antibody-defined antigen associated with gastrointestinal cancer is a ganglioside containing sialylated lacto *N*-fucopentaose II. *J. Biol. Chem. 257*:14365–14369.
302. Magnani, J. L., Steplowski, Z., Koprowski, H., and Ginsburg, V. (1983). Identification of the gastrointestinal and pancreatic cancer-associated antigen detected by monoclonal antibody 19-9 in the sera of patients as mucin. *Cancer Res. 43*:5489–5492.
303. Nudelman, E., Fukushi, Y., Levery, S. B., Higuchi, T., and Hakomori, S. (1986). Novel fucolipids of human adenocarcinoma: Disialosyl Le^a antigen ($III^4FucIII^6NeuAcIV^3NeuAcLc_4$) of human colonic adenocarcinoma and the monoclonal antibody (FH7) defining this structure. *J. Biol. Chem. 261*:5486–5495.
304. Rettig, W. J., Cordon-Cardo, C., Ng, T. S. C., Oettgen, H. F., Old, L. J., and Hoyd, K. O. (1985). High-molecular-weight glycoproteins of human teratocarcinoma defined by monoclonal antibodies to carbohydrate determinants. *Cancer Res. 45*:815–821.
305. Fakuda, M. N., Bothner, B., Lloyd, K. L., Rettig, W. J., Tiller, P. R., and Dell, A. A. (1986). Structures of glycosphingolipids isolated from embryonal carcinoma cells. The presence of mono- and disialosyl glycolipids with blood group type 1 sequence. *J. Biol. Chem. 261*:5145–5153.
306. Nilsson, O., Mansson, J. L., Lindholm, L., Holmgren, J., and Svennerholm, L. (1985). Sialosyllactotetraosylceramide, a novel ganglioside antigen detected in human carcinomas by a monoclonal antibody. *FEBS Lett. 182*:398–402.
307. Fredman, P., von Holst, H., Collins, V. P., Granholm, L., and Svennerholm, L. (1988). Sialosylactoneotetrasylceramide, a ganglioside marker for human malignant gliomas. *J. Neurochem. 50*:912–919.
308. Solter, D., and Knowles, B. B. (1978). Monoclonal antibody defining a stage-specific embryonic antigen (SSEA-1). *Proc. Natl. Acad. Sci. USA 75*:5565–5569.

309. Gooi, H. C., Feizi, T., Kapadia, A., Knowles, B. B., Solter, D., and Evans, J. M. (1981). Stage-specific embryonic antigen involves 1-3-fucosylated type 2 blood group chains. *Nature 292*:156–158.
310. Kannagi, R., Nudelman, E., Levery, S. B., and Hakomori, S. (1984). A series of human erythrocyte glycosphingolipids reacting to the monoclonal antibody directed to a developmentally regulated antigen SSEA-1. *J. Biol. Chem. 257*:14865–14874.
311. Skubitz, K. M., Pessano, S., Bottero, L., Ferrero, D., Rovera, G., and August, A. J. (1983). Human granulocyte surface molecules identified by murine monoclonal antibodies. *J. Immunol. 131*:1882–1888.
312. Urdal, D. G., Brentnall, T. A., Bernstein, I. D., and Hakomori, S. (1983). A granulocyte reactive monoclonal antibody IG 10 identifies the Galβ1-4(Fuc 1-3)GlcNAc (X determinant) expressed in HL-60 cells on both glycolipid and glycoprotein molecules. *Blood 62*:1022–1026.
313. Gooi, J., Thorpe, S. J., Hounsell, E. I., Rumpold, H., Kraft, D., Forster, O., and Feizi, T. (1983). Marker of peripheral blood granulocytes and monocytes of man recognized by two monoclonal antibodies VE P8 and VE P9 involves the trisaccharide 3-fucosyl-*N*-acetylactosamine. *Eur. J. Immunol. 13*:306–312.
314. Hakomori, S., Nudelman, E., Kannagi, R., and Levery, S. B. (1982). The common structure in fucosyl-lactosaminolipids accumulating in human adenocarcinomas and its absence in normal tissue. *Biochem. Biophys. Res. Commun. 109*:36–44.
315. Hakomori, S., Nudelman, E., Levery, S. B., and Kannagi, R. (1984). Novel fucolipids accumulating in human adenocarcinoma I. Glycolipids with di- or trifucosylated type 2 chain. *J. Biol. Chem. 259*:4672–4680.
316. Fukushi, Y., Hakomori, S., Nudelman, E., and Cochran, N. (1984). Novel fucolipids accumulating in human adenocarcinoma. II. Selective isolation of hybridoma antibodies that differentially recognize mono-, di- and trifucosylated type 2 chains. *J. Biol. Chem. 259*:4681–4685.
317. Kyogashima, M., Mulshine, J., Linnoila, R. I., Jensen, S., Magnani, J. L., Nudelman, E., Hakomori, S., and Ginsberg, V. (1989). Antibody 624H12, which detects lung cancer at early stages, recognizes a sugar sequence in the glycosphingolipid difucosylneolactonorhexaosylceramide ($V^3FucIII^3FucnLc_6Cer$). *Arch. Biochem. Biophys. 275*:309–314.
318. Hoff, S. D., Matsushita, Y., Ota, D. M., Cleary, K. R., Yamori, T., Hakomori, S., and Irimura, T. (1989). Increased expression of sialyl-dimeric Le_x antigen in liver metastases of human colorectal carcinoma. *Cancer Res. 49*:6883–6888.
319. Suzuki, M., Ohwada, M., and Tamada, T. (1990). Clinical value of sialyl-SSEA-1 antigen in patients with ovarian cancer. *Gynecol. Oncol. 36*:371–375.
320. Iwanari, O., Miyoko, J., Date, Y., Nakayama, S., Kijima, S., Moriyama, M., Karina, K., Endoh, J., and Kitao, M. (1990). Clinical evaluations of the tumor marker sialyl-SSEA-1 antigen for clinical gynecological disease. *Gynecol. Obstet. Invest. 29*:214–218.
321. Iwanari, O., Miyoko, J., Date, Y., Nakayamas, S., Kijima, S., Moriyana, M., Karino, K., Endoh, J., and Kitao, M. (1990). Differential diagnosis of ovarian cancer, benign ovarian tumor and endometriosis by a combination assay of serum sialyl SSEA-1 antigen and CA125 levels. *Gynecol. Obstet. Invest. 29*:71–74.
322. Ohmori, K., Yoneda, T., Ishikara, G., Shigeta, K., Hirashima, K., Kanai, M., Itai, S. (1989). Sialyl SSEA-1 antigen as a carbohydrate marker of natural killer cells and immature lymphoid cells. *Blood 74*:255–261.
323. Hakomori, S., Nudelman, I., Levery, S. B., and Patterson, S. J. (1983). Human cancer associated gangliosides defined by a monoclonal antibody (IB9) directed to sialosyl 2-6 galactosyl residue. A preliminary note. *Biochem. Biophys. Res. Commun. 113*:791–798.
324. Steplewski, Z., Koprowski, H., and Thurin, M. (1988). Monoclonal antibodies against glycolipid antigens and pharmaceutical compositions containing them for diagnosis and treatment of adenocarcinomas. European Patent 285059A2.
325. Thurin, M., Koprowski, H., Herlyn, M., and Steplewski, Z. (1988). Monoclonal antibodies against melanoma-associated ganglioside antigens, hybrid cell lines producing these antibodies, and their use in diagnosis and therapy of tumors. European Patent 280209A2.

326. Nudelman, E. D., Levery, S. B., Stroud, M. R., Salyan, M. E. K., and Hakomori, S. (1989). Unbranched ceramide polysaccharide tumor antigens for antibody production and antitumor vaccines. European Patent 344955A2.
327. Miller, H. C., and Esselman, W. J. (1975). Modulation of the immune response by antigen reactive lymphocytes after cultivation with gangliosides. *J. Immunol. 67*:839–843.
328. Ryan, J. L., and Shintzky, M. (1979). Possible role for glycosphingolipids in the control of immune responses. *Eur. J. Immunol. 9*:171–175.
329. Krishnaraj, R., Lengle, E., and Kemp, R. G. (1982). Murine leukemia. Proposed role of gangliosides in immune suppression. *Eur. J. Cancer Clin. Oncol. 18*:89–98.
330. Lengle, E. E., Krishnaraj, R., and Kemp, R. R. (1979). Inhibition of lectin-induced mitogenic response of thymocytes by glycolipids. *Cancer Res. 39*:817–822.
331. Ladisch, S., Gillard, B., Wong, C., and Ulsh, L. (1983). Shedding and immunoregulatory activity of YAC-1 lymphoma cell ganglioside. *Cancer Res. 43*:3808–3813.
332. Freimuth, W. W., Miller, H. C., and Esselman, W. J. (1979). Soluble factors containing Thy-1 antigen shed from lymphoblastoid cells modulate in vitro plaque-forming cell response. *J. Immunol. 123*:201–207.
333. Miller, H. C., Chancey, W. G., Klinman, N. R., and Esselman, W. J. (1982). Regulation of B cell tolerance by murine gangliosides. *Cell. Immunol. 67*:390–395.
334. Gonwa, T. A., Westrick, M. A., and Macher, B. A. (1984). Inhibition of mitogen- and antigen-induced lymphocyte activation by human leukemia cell gangliosides. *Cancer Res. 44*:3467–3470.
335. Whisler, R. L., and Yates, A. J. (1980). Regulation of lymphocyte responses by human gangliosides. *J. Immunol. 125*:2106–2111.
336. Ledeen, R. W., Skirvanek, J. A., Tirri, L. J., Margolis, R. K., and Margolis, R. V. (1976). *Adv. Exp.Med. Biol. 17*:83–93.
337. Rapport, M. M., and Gorio, A., eds. (1981). *Gangliosides in Neurological and Neuromuscular Function, Development and Repair*. Raven Press, New York.
338. Gorio, A., Karpiac, S. E., and Davis, J. N. (1986). Ganglioside enhancement of neuronal differentiation, plasticity, and repair. *CRC Crit. Rev. Clin. Neurobiol. 2*:241–296.
339. Ledeen, R. W. (1984). Biology of gangliosides neuritogenic and neuronotrophic properties. *J. Neurochem. Res. 12*:147–159.
340. Ledeen, R. W., Yu, R. K., Rapport, M. M., and Suzuki, K., eds. (1984). Ganglioside structure, function and biomedical potential. Plenum Press, New York.
341. Porcellati, G., Ceccarelli, B., and Tettamonti, G., eds. (1976). Ganglioside function. Plenum Press, New York.
342. Svennerholm, L., Mandel, P., Dreyfus, H., and Urban, P. F., eds. (1980). *CNRS International Symposium on Structure and Function of Gangliosides*. Plenum Press, New York.
343. Obata, K., Oide, M., and Handa, S. (1977). Effects of glycolipids on in vitro development of neuromuscular junction. *Nature 266*:369–371.
344. Obata, K., and Handa, S. (1979). Development of neuromuscular transmission in culture and its facilitation by glycolipids. In *Integrative Control Functions of the Brain* (M. Ito, ed.). Kodanska, Elsevier, Tokyo, Amsterdam, pp. 5–15.
345. Purpura, D. P., and Baker, H. J. (1977). Meganeurities and other aberrant processes of neurons in feline G_{M1} gangliosidosis: A Golgi study. *Brain Res. 116*:1–21.
346. Walkey, S. V., Wurzelmann, S., and Purpura, D. P. (1981). Ultrastructure of neurites and meganeurites of cortical pyramidal neurons in feline gangliosidosis as revealed by the combined Golgi-EM technique. *Brain Res. 211*:393–398.
347. Roisen, F. J., Bartfeld, H., Nagele, R., and Yorke, G. (1981). Ganglioside stimulation of axonal sprouting in vitro. *Science 214*:577–578.
348. Dimpfel, W., Moller, W., and Mengs, U. (1981). Ganglioside-induced neurite formation in cultured neuroblastoma cells. In *Gangliosides in Neurological and Neuromuscular Function, Development and Repair* (M. M. Rapport and A. Gorio, eds.). Raven Press, New York, pp. 119–134.

349. Cannella, M. S., Oderfeld-Nowak, B., Gradkowska, M., Skup, M., Garofalo, L., Cuello, A. C., and Ledeen, R. W. (1990). Derivatives of ganglioside G_{M1} as a neuronotrophic agents; comparison of in vivo and in vitro effects. *Brain Res. 513*:286–294.
350. Spirman, N., Sela, B. N., and Schwartz, M. (1982). Antiganglioside antibodies inhibit neuritic outgrowth from regenerating goldfish retinal explants. *J. Neurochem. 39*:874–877.
351. Ceccarelli, B., Aporti, F., and Finesso, M. (1976). Effects of brain gangliosides on functional recovery in experimental regeneration and reinnervation. In *Ganglioside Function* (G. Porcellati, B. Ceccarelli, and G. Tettamonti, eds.) *Adv. Exp. Med. Biol. 71*:275–293.
352. Gorio, A., Carmignoto, G., Facci, L., and Finesso, M. (1980). Motor nerve sprouting induced by ganglioside treatment. Possible implications for ganglioside on neuronal growth. *Brain Res. 197*:236–241.
353. Agnati, L. F., Benfanati, F., Battistini, N., Cavicchioli, L., Fuxe, K., and Toffano, G. (1983). Selective modulation of ^{3}H-spiperone labeled 5-HT receptors by subchronic treatment with the ganglioside G_{M1} in the rat. *Acta Physiol. Scand. 117*:311–314.
354. Wojcik, M., Ulas, J., and Oderfeld-Nowak, B. (1982). The stimulating effect of ganglioside injections on the recovery of choline acetylcholinesterase activities in the hippocampus of the rat after septal lesions. *Neuroscience 1*:495–499.
355. Grafstein, B., and Yip, H. K. (1982). Techniques for improving axonal regeneration: Assay in goldfish optic nerve. In *Nervous System Regeneration* (A. M. Giuffride-Stella, B. Haber, G. Hashim, and J. R. Perez-Polo, eds.). Alan R. Liss, New York, pp. 105–118.
356. Sparrow, J. R., Grafstein, B., Mc Guinness, C., and Schwartz, M. (1984). Antibodies to gangliosides inhibit goldfish optic nerve regeneration in vivo. *J. Neurosci. Res. 12*:223–243.
357. Sparrow, J. R., and Grafstein, B. (1982). Sciatic nerve regeneration in ganglioside-treated rats. *Exp. Neurol. 77*:230–235.
358. Radsak, K., Schwarzman, G., and Wiegandt, H. (1982). Studies on the cell association of exogenously added sialoglycolipids. *Hoppe Seylers Z. Physiol. Chem. 363*:263–272.
359. Sharom, F. J., and Grant, C. M. W. (1978). A model for ganglioside behavior in cell membranes. *Biochim. Biophys. Acta 507*:280–293.
360. Moss, J., Fishman, P. H., Manganiello, V. S., Vaugan, M., and Brady, R. O. (1976). Functional incorporation of ganglioside into intact cells: Induction of choleragen responsiveness. *Proc. Natl. Acad. Sci. USA 73*:1034–1037.
361. Toffano, G., Benvengnu, D., Bonetti, A. C., Facci, L., Leon, A., Orlando, P., Ghidoni, R., and Tettamonti, G. (1980). Interactions of G_{M1} ganglioside with crude rat brain neuronal membranes. *J. Neurochem. 35*:861–866.
362. Kanda, S., Inone, K., Nojima, S., Utsumi, H., and Wiegandt, H. (1982). Incorporation of a ganglioside and a spin-labeled ganglioside analogue into cell and liposomal membranes. *J. Biochem. 91*:2095–2098.
363. Schwarzman, G., Hoffmann-Bleihauer, P., Schubert, J., Sandhoff, K., and Marsh, D. (1983). Incorporation of ganglioside analogues into fibroblast cell membranes. A spin-label study. *Biochemistry 22*:5041–5048.
364. Facci, L., Leon, A., Toffano, G., Sonnino, S., Ghidoni, R., and Tettamonti, G. (1984). Promotion of neuritrogenesis in mouse neuroblastoma cells by exogenous gangliosides. Relationship between the effect and the cell association of ganglioside G_{M1}. *J. Neurochem. 42*:299–305.
365. Leon, A., Facci, L., Toffano, G., Sonnino, S., and Tettamonti, G. (1981). Activation of (Na^+, K^+)-ATPase by nanomolar concentrations of G_{M1} ganglioside. *J. Neurochem. 37*: 350–357.
366. Leon, A., Tettamonti, G., and Toffano, G. (1981). Changes in functional properties of neuronal membranes by insertion of exogenous ganglioside. In *Gangliosides in Neurological and Neuromuscular Function, Development and Repair* (M. M. Rapport and A. Gorio, eds.). Raven Press, New York, pp. 45–54.
367. Gorio, A., Aporti, F., and Norido, F. (1981). Ganglioside treatment in experimental diabetic neuropathy. In *Gangliosides in Neurological and Neuromuscular Function. Development and Repair* (M. M. Rapport and A. Gorio, eds.). Raven Press, New York, pp. 259–266.

368. Gorio, A., Aporti, F., Di Gregorio, F., Schiavinato, A., Siliprando, R., and Vitradello, M. (1984). Ganglioside treatment of genetic and alloxan-induced diabetic neuropathy. In *Ganglioside Structure, Function and Biomedical Potential* (R. W. Ledeen, R. K. Yu, M. M. Rapport, and K. Suzuki, eds.). Plenum Press, New York, pp. 549–564.
369. Crepaldi, G., et al. (1983). Ganglioside treatment in diabetic peripheral neuropathy. A multicenter trial. *Acta Diabetol. Lat. 20*:265–276.
370. Karpiak, S. E., Mahadik, S. P., Graf, L., and Rapport, M. M. (1981). An immunological model of epilepsy: Seizures induced by antibodies to G_{M1} ganglioside. *Epilepsia 22*:189–196.
371. Rapport, M. M. (1981). Specificity of antiganglioside serum in the perturbation of CNS functions. In *Gangliosides in Neurological and Neuromuscular Function, Development and Repair* (M. M. Rapport and A. Gorio, eds.). Raven Press, New York, pp. 91–97.
372. Karpiac, S. D., Graf, L., and Rapport, M. M. (1978). Antibodies to G_{M1} ganglioside inhibit a learned avoidance response. *Brain Res. 151*:637–640.
373. Karpiac, S. E., Vilim, F., and Mahadik, S. P. (1984). Gangliosides accelerate rat neonatal learning and levels of cortical acetylcholinesterase. *Dev. Neurosci. 6*:127–135.
374. Karpiac, S. E. (1982). Antibodies to G_{M1} ganglioside inhibit morphine analgesia. *Pharm. Biochem. Behav. 16*:611–613.
375. Mamoli, B., Brunner, G., Mader, R., and Schanda, H. (1980). Effects of cerebral gangliosides in the alcoholic polyneuropathies. *Eur. Neurol. 19*:320–326.
376. Massarotti, M. (1983). Ganglioside treatment of alcoholic neuropathies: Experimental and clinical aspects. *Pharm. Biochem. Behav. 18*:51–54.
377. Bassi, S., Albizzati, M. G., Sbachi, M., Frattola, L., and Massarotti, M. (1984). Double-blind evaluation of monosialoganglioside (G_{M1}) therapy in stroke. *J. Neurosci. Res. 12*:493–498.
378. Maroni, M., Colombi, A., Gilioli, R., Rota, E., De Paschale, G., Castano, P., and Foa, V. (1981). Effects of ganglioside therapy on experimental CS_2 neuropathy. *Clin. Toxicol. 18*:1475–1484.
379. Norido, F., Canella, R., Zanoni, R., and Gorio, A. (1983). Monosialoganglioside (G_{M1}) treatment of ouabain-induced retinopathy in the rabbit. *Acta Neuropathol. 62*:46–50.
380. Mahadik, S. P., and Karpiak, S. K. (1988). Gangliosides in treatment of neural injury and disease. *Drug. Dev. Res. 15*:337–360.
381. Karpiak, S. E., and Mahadik, S. P. (1990). Ganglioside reduction of ischemic injury. *Crit. Rev. Neurobiol. 5*:221–237.
382. Samson, J. C. (1990). G_{M1} ganglioside treatment of CNS injury: Clinical evidence for improved recovery. *Drug Dev. Res. 19*:209–224.

9

The Biosynthesis of Serine (Threonine)-*N*-Acetylgalactosamine-Linked Carbohydrate Moieties

Harry Schachter and Inka Brockhausen *University of Toronto and The Hospital for Sick Children, Toronto, Ontario, Canada*

Glycoproteins are biopolymers that contain one or more carbohydrate chains linked covalently to a polypeptide backbone. The carbohydrate chains are classified according to the linkage between sugar and amino acid. This chapter will deal with the biosynthesis of oligosaccharides linked to polypeptide by serine (threonine)-*N*-acetylgalactosamine [Ser(Thr)-GalNAc]*O*-glycosidic (mucin-type) linkages (*O*-glycans). Recent reviews have considered various aspects of this topic [33,150,287,296,298,299,301,303]. Many glycoproteins contain both *N*- and *O*-linked glycans on the same polypeptide chain (Table 1).

O-Glycosidically linked *N*-acetylglucosamine (*O*-GlcNAc) does not come under the topic of Ser(Thr)-linked GalNAc oligosaccharides. These residues are enriched on many proteins localized to the cytoplasmic and nucleoplasmic compartments of the cell (e.g., numerous DNA-binding proteins, including several transcription factors, and nuclear pore proteins contain multiple *O*-GlcNAc residues) [2,130,133,166].

Mucins (mucous glycoproteins), the main class of glycoproteins containing *O*-glycans, are responsible for the gel-forming properties of mucus, the viscous fluid lining the epithelium of the gastrointestinal, respiratory, and genitourinary tracts. The function of the mucous gel is to protect and lubricate the mucous epithelium. Mucins are large molecules (usually over 1×10^6 in relative molecular mass; M_r) and contain from 50 to 80% or more by weight of carbohydrate. The oligosaccharides range in size from a single monosaccharide to some 20 residues, and there may be several hundred oligosaccharide chains per polypeptide molecule; these *O*-glycans usually exhibit a striking structural heterogeneity.

Structural information on the oligosaccharide moieties has been obtained by cleaving the Ser(Thr)-GalNAc linkage, usually by alkali-catalyzed β-elimination, and purifying the resultant mixture of oligosaccharides by electrophoresis, paper chromatography, gel filtration, thin layer chromatography, high-pressure liquid chromatography, affinity chromatography on lectin or antibody columns, or by other techniques. Recent advances in the determination of oligosaccharide fine structure have been applied to many *O*-glycosidic

Table 1 Glycoproteins with *O*-Glycosyl Oligosaccharides

Species	Glycoproteins	Linkage[a]	Core class[b]	Refs.
	Mucins			
Collocalia	Salivary mucin	O,N	1, 2, 5	125, 379
Cow	Submaxillary	O	1, 2, 3, 5	77, 138, 293, 294, 352, 354
Dog	Gastric	O	1	327
	Submaxillary	O	1	215
Goat	Submaxillary	O	1	87
Hen	Ovomucin[c]	O	2	338
Horse	Gastric	O	1, 2, 3	247
Human	Amniotic fluid	O	2	122, 128
	Bronchial (CF)[c]	O	1, 2, 3, 4	39, 200, 222, 368
	Bronchial[c]	O	1, 2, 3, 4	38, 183, 201, 202, 277, 364
	Cervical	O	1, 2	398–401
	Colonic	O	3	266, 267
	Gastric	O	1, 2	252, 331
	Milk[d]	O	1, 2	127, 129
	Ovarian cyst	O	1, 2, 5, 6	49, 85, 213, 214, 218, 234, 284, 347, 385, 386
	Salivary	O	1, 2	275, 276
Monkey	Cervical	O	1, 2, 3	134, 239–242
Mouse	Submandibular	O, N	1	372
Pig	Gastric	O	1, 2	81, 185, 328, 369
	Cowper's gland	O	1	226, 271
	Submaxillary	O	1	6, 7, 53, 295, 365
	Tracheal	O	1, 2	64, 271, 272
Rat	Colonic	O	3	329
	Duodenal	O	2	370
	Submandibular	O	1	343
	Sublingual	O	3	330
	Small intestinal	O	1, 3	57
Sheep	Gastric	O	1, 2, 4	151, 156, 383, 384
	Submaxillary	O	1, 2	142, 143, 288, 289, 362, 389

Table 1 (Continued)

Species	Glycoproteins	Linkage[a]	Core class[b]	Refs.
	Cancer cells			
Human	Rectal adenocarcinoma	O	1, 3, 5	199, 390
	Mammary epithelial MDA-MB-231 proteoglycan	O	1, 2	117
	Leukosialin from leukemic cell lines:			
	K562 (erythroid)	O	1	
	HL-60 (myeloid)	O	[1], 2	
	HSB-2 (T-lymphoid)	O	[1], 2	56
	Leukemic leukocytes:			
	AML	O, N	1, 2	
	CML	O, N	1, 2	102
	Embryonal carcinoma PA1 cell line[d]	O	1	210
	Choriocarcinoma hCG, β subunit	O, N	1, 2	5, 68
Mouse	Epiglycanin	O, N	1	360
Rat	Ascites hepatoma *Zajdela*	O	1, 2	243
	Ascites hepatoma AH 66 plasma membrane	O	1, 2	109
	Mammary carcinoma	O	2	159, 160
	Swarm rat chondrosarcoma proteoglycan	O, N	1, 2	250
	Secreted glycoproteins			
Calf	Fetuin	O, N	1, 2	88, 251, 336
Codfish	Antifreeze	O	1	141
Cow	Adrenal medulla chromaffin chromogranins	O, N	1	180
	Factor X	O, N	1	228
	kappa-Casein	O	1, 2	96, 290, 366, 367
	Plasma kininogen	O	1	92
	Plasminogen	O, N	1	220
Horse	Chorionic gonadotropin	O, N	1, 2	79
Human	Apolipoprotein CIII	O	1	356
	Fibronectin	O, N	1	194
	hCG β subunit	O, N	1, [2]	20, 68, 179
	Interleukin-2	O	1	69

Table 1 (Continued)

Species	Glycoproteins	Linkage[a]	Core class[b]	Refs.
	Secreted glycoproteins (continued)			
Human (continued)	kappa-Casein	O	1, 2, 6	97, 291, 371
	lambda Light chain	O, N	1	63
	Meconium	O	1, 2, 3, 4, 5, 6	52, 152, 153, 155
	Milk IgA	O, N	1, 2	258, 259
	Myeloma IgA	O, N	1	13
	Myeloma IgD	O, N	1	225
	Plasminogen	O, N	1	135, 220
	Plasma galactoprotein	O, N	1, 2	4
	Plasma α2HS glycoprotein, B chain	O	1	112
	Seminal plasma glycoprotein	O	1, 2	123, 124, 126
	Serum cholinesterase	O	1	355
	Urinary glycopeptide	O	1, 2	144, 208, 211, 255
	Vitamin D-binding glycoprotein	O	1	373
Pig	Plasminogen	O, N	1	220
Sheep	kappa-Casein	O	1,2	333
	Membrane glycoproteins			
Cow	Glycophorin	O, N	1, 2	104, 107
Dog	Glycophorin	O, N	1	231, 391
Horse	Glycophorin	O, N	1	106
Human	Erythrocyte membrane	O, N	1	344
	Granulocytes	O, N	1, 2	102
	Glycophorin (MM variant)	O, N	1, 2	3, 212, 350
	Glycophorin	O, N	1	105, 110, 233, 269
	Glycophorin (Cad)	O, N	1	35, 137
	HLA-DR–associated invariant chain	O, N	1	217
	Platelet glycocalicin	O, N	1, 2	187, 188, 353
	Mammary epithelial HBL-100 proteoglycan	O	1, 2	117
	Transferrin receptor	O	1	83
Monkey	Glycophorin	O	1	232
Mouse	Glycophorin	O, N	1, 2	193
Pig	Glycophorin	O, N	1	178

Table 1 (Continued)

Species	Glycoproteins	Linkage[a]	Core class[b]	Refs.
	Membrane glycoproteins (continued)			
Rabbit	Glycophorin	O, N	1	103, 231
Rat	Erythrocyte	O, N	1, 2	89
	Other glycoproteins			
Herring	Egg glycoproteins	O	1	164
Rat	Brain glycoproteins	O	1	98
	Brain proteoglycan[e]	O, N	1	195
Salmon	Egg glycoproteins	O	1	324
Trout	Egg glycoproteins	O	1, 5	163, 175
Coronavirus	E1 glycoprotein	O	1	248
Friend leukemia virus	Glycoprotein 71	O, N	1	114
Epstein–Barr virus	Glycoprotein 350	O, N	1	313
Pseudorabies virus in insect line SF9		O	1	351
Mouse hepatitis virus	Glycoprotein E1	O	1	249

[a] *N,N*-glycosyl oligosaccharides; *O,O*-glycosyl oligosaccharides.
[b] The core classes 1–6 are defined in the text and synthesis of cores 1–4 is illustrated in Fig. 2. Brackets indicate a relatively small amount of material.
[c] Oligosaccharides may be sulfated.
[d] These glycoproteins have been reported to contain oligosaccharides with the core structure Galα1-3GalNAc-R.
[e] These glycoproteins have been reported to contain oligosaccharides with the core structure GlcNAcβ1-3Man-Ser(Thr).

oligosaccharides. As will be discussed in this chapter, this new structural information is an essential prerequisite to the understanding of the biosynthesis of these molecules.

I. BIOSYNTHETIC STUDIES WITH INTACT CELLS

Glycosylation of proteins can be studied in two general ways [296]. Radiolabeled monosaccharides can be administered to whole animals, tissue incubations, or cell cultures, and the incorporation of radioactivity into glycoproteins can be followed either by autoradiography or by chemical analysis of tissue fluids and subcellular fractions. Alternatively, cell-free preparations can be monitored for glycosyltransferase activities. All these procedures have been applied to the study of *O*-glycosyl oligosaccharides.

The autoradiographic approach has two serious disadvantages. First, the glycoprotein being labeled cannot be identified with certainty. Second, many radiolabeled monosaccharide precursors undergo extensive conversions to other monosaccharides. The first problem has been approached by using tissues that secrete predominantly mucous glycoproteins, such as the mucous epithelium of the respiratory, gastrointestinal, and genitourinary systems, and mucous acinar cells of exocrine glands, such as the submaxillary gland. The second problem can be overcome by using radiolabeled precursors that either do not convert to other radiolabeled sugars or convert to only a limited number of sugars (e.g., L-fucose, 2-[^{3}H]-D-mannose or *N*-acetyl-D-mannosamine).

Autoradiographic experiments have been carried out with radioactive amino acids, sulfate, glucose, galactose, glucosamine, mannose, fucose, and *N*-acetylmannosamine in a variety of mucus-secreting tissues, such as gastric mucous cells, intestinal goblet cells, and bronchial mucous glands [21,22,190–192,227,245,246]. Although it is clear from work with radioactive amino acid precursors that the polypeptide backbones of mucins must be assembled in the rough endoplasmic reticulum (RER), the addition of most of the carbohydrate and sulfate occurs after the protein has migrated to the Golgi apparatus. Radioautographic studies with radiolabeled *N*-acetylglucosamine as a precursor for *N*-acetylgalactosamine showed almost exclusive initial localization of the label to the Golgi apparatus and minimal labeling of the rough endoplasmic reticulum. Thus, even the very first sugar (GalNAc) appears to be incorporated into peptide primarily in the Golgi apparatus. This differs markedly from the mechanism of *N*-glycosyl oligosaccharide synthesis in which the asparagine (Asn)-GlcNAc linkage is synthesized in the rough endoplasmic reticulum by incorporation of a preformed oligosaccharide chain into the polypeptide. Biosynthetic studies on intact cells involving biochemical analyses of subcellular fractions are considered in detail in Section V.

II. SERINE(THREONINE)-*N*-ACETYLGALACTOSAMINE OLIGOSACCHARIDE STRUCTURE

Structural information is an essential prerequisite to an understanding of oligosaccharide biosynthesis. The sugars commonly found in mucin-type oligosaccharides are GalNAc, galactose (Gal), GlcNAc, sialic acid (SA), and fucose (Fuc), but mannose (Man) and glucose (Glc) have also been reported [87]. Gal, GlcNAc, and GalNAc may occur in a sulfated form and thereby contribute with sialic acid to the acidity of some mucin-type oligosaccharides.

The tremendous diversity of *O*-glycan structure makes detailed discussion a very difficult task. The procedure adopted in this chapter is to arrange as many structures as possible in biosynthetic pathways (Figs. 1–10). The same arrow direction (Fig. 1) is used for a particular glycosyltransferase in all these figures. Structures are referred to in the text by short names enclosed in braces { }; these names are defined in the figures. It is important to emphasize at the outset that most of the arrows in Figures 2–10 refer to hypothetical biosynthetic steps based on structures. Those steps that have, in fact, been tested by in vitro enzyme assays will be identified when they are discussed. There may be tissue and species

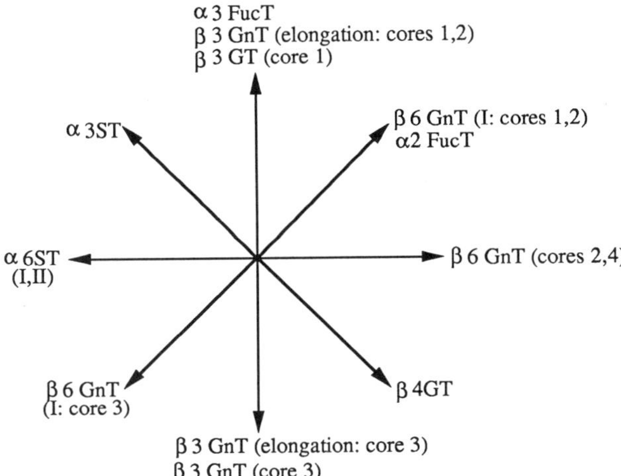

Figure 1 All arrows in Figs. 2–10 will follow the same directions, as indicated in the figure. In those cases for which more than one enzyme is represented by a single arrow, it will be possible to determine the enzyme from the structures in the figure. Abbreviations: S, sialic acid; G,D-galactose (Gal); Gn, *N*-acetyl-D-glucosamine (GlcNAc); Fuc, L-fucose; T, transferase.

variations for the different glycosyltransferases and it may not always be accurate to apply the schemes of Figures 2–10 to a particular situation.

Table 1 lists literature references to the *O*-glycan structures described in this chapter. Tables 2–4 attempt to assign as many of these structures as possible to various glycoproteins, but these lists are not complete. It is convenient to classify *O*-glycosyl oligosaccharides according to their core structures [154,303], and the chapter is organized on this basis.

A. Monosaccharide Chain

The simplest oligosaccharide of the Ser(Thr)-GalNAc class is obviously GalNAc. The glycoproteins that contain single GalNAc residues are listed in Table 2.

B. Disaccharide Chains

Four types of Ser(Thr)-GalNAc disaccharide chains have been described to date:

1. Sialylα2-6GalNAc-α-Ser(Thr)-R (Table 2)
2. Galβ1-3GalNAc-α-Ser(Thr)-R (Table 2)
3. GlcNAcβ1-3GalNAc-α-Ser(Thr)-R (Table 4)
4. GalNAcα1-3GalNAc-α-Ser(Thr)-R (Table 1)

C. Chains with More than Two Residues

Chains with more than two residues vary in size from trisaccharides to chains with 18 or more residues. The larger chains contain three distinct regions, a core, a backbone, and a nonreducing terminus.

Table 2 Glycoproteins with Core 1 *O*-Glycans[a]

Species	Glycoproteins	Oligosaccharide structures[b,c]															
		1	2	3	4	5	6	7	8	9	10	11	12	13	14	15	16
	Mucins																
Cow	Submaxillary	x	x	x	x												
Dog	Submaxillary				x				z		x						x
	Gastric									y							
Horse	Gastric	x							y							x	
Human	Ovarian cyst			x					y	y	z			x		x	
	Salivary			x		x										x	
	Bronchial			z					y		x		x	x	x	x	
	Cervical		x		x		x										
	Milk					x									x		
Monkey	Cervical		x		x			y									x
Pig	Submaxillary				x	x										x	x
	Gastric	x								y	y						
	Cowper's gland		x				x										
	Tracheal				x	x										x	
Rat	Salivary															x	x
	Small intestinal															x	
Sheep	Submaxillary	x	x	x												x	
	Gastric										z						

	Cancer cells							
Human	Mammary proteoglycan:			x	x	x	x	
	leukemic cell lines (leukosialin):							
	K562 (erythroid)	x	x	x	x	x	x	
	HL-60 (myeloid)					x		
	HSB-2 (T-lymphoid)					x		
	Leukemic leukocytes:							
	AML				x	x	x	
	CML					x	x	
	Choriocarcinoma hCG, β subunit			x	x	x	x	
Mouse								
	Epiglycanin	x		x		x		y
Rat								
	Swarm rat chondrosarcoma proteoglycan					x	x	
	Ascites hepatoma AH 66 membrane	x		x		x		
	Ascites hepatoma *Zajdela*					x	x	
	Secreted glycoproteins							
Calf								
	Fetuin					x	x	
Cow								
	kappa-Casein			x	x	x	x	x
	Adrenal medulla chromaffin chromogranins				x	x	x	
	Plasminogen					x	x	

Table 2 (continued)

Species	Glycoproteins	Oligosaccharide structures[b,c] 1	2	3	4	5	6	7	8	9	10	11	12	13	14	15	16
	Secreted Glycoproteins (continued)																
Horse	Chorionic gonadotropin					x	x										
Human	Normal hCG, β subunit		x	x	x	x	x										
	IgA	x				x					y						
	Plasminogen					x	x										
	Fibronectin																
	Amniotic fluid			x		x	x										
	Plasma					x											
	Seminal plasma glycoprotein			x					x		x	x	x	x	x	x	
	Meconium	x		x													
	kappa-Casein	x		x													
	Urinary glycopeptide					x	x										
Pig	Plasminogen					x	x										
	Membrane glycoproteins																
Cow	Glycophorin									y	y						
Dog	Glycophorin				x	x	x										
Horse	Glycophorin				x	x	x										

Human	Glycophorin				x	x	x				
	Normal granulocyte leukosialin					x	x				
Monkey	Glycophorin						x				
Mouse	Glycophorin					x	x				
Pig	Glycophorin				x						y
Rabbit	Glycophorin			x		x	x				
Rat	Erythrocyte				x	x	x				
Friend leukemia virus	Glycoprotein 71			x	x	x	x				
Pseudo-rabies virus in SF9 cells		x		x							

[a]The presence of various oligosaccharides in a particular glycoprotein is indicated as follows: x, isolated as such from the glycoprotein; y, not isolated as such from a particular glycoprotein, but a larger oligosaccharide containing the designated structure was isolated; z, a combination of x and y.

[b]The oligosaccharide structures are as follows:
1, GalNAc-α-Ser(Thr)-; 2, Sialylα2-6GalNAc-α-Ser(Thr)-; 3, Galβ1-3GalNAc-α-Ser(Thr)- (core 1).
The following structures are named in the short nomenclature defined in Figs. 2 and 3:
4, **{S6[1]}**; 5, **{S3[1]}**; 6, **{S6[S3][1]}**; 7, **{S6[Gn3][1]}**; 8, **{Gn3[1]}**; 9, **{Gn6[Gn3][1]}**; 10, **{G4Gn3[1]}**; 15, **{F2[1]}**; 16, **{S6[F2][1]}**.
The following structures have not been defined in the figures:
11, Fucα1-2Galβ1-4GlcNAcβ1-3Galβ1-3GalNAc-α-Ser(Thr)- (H determinant)
12, Galβ1-4(Fucα1-3)GlcNAcβ1-3Galβ1-3GalNAc-α-Ser(Thr)- (X determinant)
13, Fucα1-2Galβ1-4(Fucα1-3)GlcNAcβ1-3Galβ1-3GalNAc-α-Ser(Thr)- (Y determinant)
14, Sialylα2-3Galβ1-4(Fucα1-3)GlcNAcβ1-3Galβ1-3GalNAc-α-Ser(Thr)- (sialylated X determinant)

[c]Oligosaccharides are arranged in groups according to the biosynthetic schemes in the various figures: 1–3 (Fig. 2); 4–6 (Fig. 2); 7 (Fig. 2); 8–9 (Fig. 2); 10–14 (Fig. 3 and 4); 15–16 (Fig. 3).

Table 3 Glycoproteins with Core 2 *O*-Glycans

Species	Glycoproteins	Oligosaccharide structures[f]																	
		1	2	3	4	5	6	7	8	9	10	11	12	13	14	15	16	17	18
	Mucins																		
Avian	Salivary					y													
	Ovomucin[c]			z		y													
Cow	Submaxillary		x															x	
Horse	Gastric			y										y				z	y
	Chorionic gonadotropin					x	x	x											
Human	Ovarian cyst	x	x	z					y	y		y	y	z	x	y		x	z
	Salivary			y		x							y		z	x		y	
	Bronchial		x	z		x	x				z			x	x	z		z	x
	Gastric									y	x								
	Cervical			z		z	z							x	z	x			
	Milk[a]			x		x								x	x				
	Amniotic fluid																x		
Monkey	Cervical[b]			y			y											x	
Pig	Gastric[d]			y							z	y							
	Tracheal		x	x	x	x		x		y	y	y	y	x				x	x
Sheep	Submaxillary		x																
	Gastric			y						y	z								

	Cancer cells													
Human	Mammary proteoglycan[c]:		x		x	x	x							
	Leukosialin from leukemic cell lines:													
	HL-60 (myeloid)				x	x	x							x
	HSB-2 (T-lymphoid)				x	x	x							x
	Leukemic leukocytes:													
	AML				x	x	x							x
	CML				x	x	x							x
	Choriocarcinoma hCG, β subunit		x		x	x	x							
Rat	Swarm rat chondrosarcoma proteoglycan						x							
	Ascites hepatoma AH 66 membrane	x	x	x	x	x	x							
	Mammary carcinoma[c]		x		x	x	x							
	Ascites hepatoma *Zajdela*				x	x	x							
	Secreted glycoproteins													
Calf	Fetuin						x							
Cow	kappa-Casein	x	x		x									x

Table 3 (continued)

Species	Glycoproteins	Oligosaccharide structures[f]																	
		1	2	3	4	5	6	7	8	9	10	11	12	13	14	15	16	17	18
	Secreted Glycoproteins (continued)																		
Human	IgA[e]			x		x					x	x						x	
	Plasma galactoprotein							x											
	Seminal plasma glycoprotein		x	x		x								x	x	x			
	Meconium[a,d]	x	x	x						x	x							x	
	kappa-Casein	x	z	x						z									
	Normal hCG, β subunit			x		x	x	x											
	Urinary glycopeptide							x											
	Membrane glycoproteins																		
Cow	Glycophorin										y								
Human	Glycophorin				z	z													
	Platelet glycocalicin							x											
	Normal granulocyte leukosialin					x	x	x	y		y				x				
	Mammary proteo-glycan[c]			x		x	x	x											
Mouse	Glycophorin							x											
Rat	Erythrocyte							x											

[a]Oligosaccharides may contain the GlcNAcβ1-6Galβ- structure.
[b]Oligosaccharides may contain GalNAcβ3 or 4Galβ-.
[c]Oligosaccharides may be sulfated.
[d]Oligosaccharides may contain GlcNAcα1-4Galβ-.
[e]A great variety of structures were found including the SSEA-1, I, i, and Le[a] determinants (see Table 5).
[f]The presence of various oligosaccharides in a particular glycoprotein is indicated as follows: x, isolated as such from the glycoprotein; y, not isolated as such from a particular glycoprotein, but a larger oligosaccharide containing the designated structure was isolated; z, a combination of x and y.

The oligosaccharide structures are as follows:

1, GlcNAcβ1-6GalNAc-α-Ser(Thr)-; 2, GlcNAcβ1-6(Galβ1-3)GalNAc-α-Ser(Thr)- (core 2).

The following structures are named in the short nomenclature defined in Figs. 2, 3, 5–7:

3, **{G4[2]}**; 4, **{S3[2]}**; 5, **{G4[S3][2]}**; 6, **{S3G4[2]}**; 7, **{S3G4[S3][2]}**; 9, **{G4[Gn3][2]}**; 10, **{G4[G4Gn3][2]}** and **{G4[Gn3G4Gn3][2]}**; 11, **{Gn6[Gn3][2]}**, **{G4[Gn6][Gn3][2]}**; **{G4[Gn6][G4Gn3][2]}**; and **{G4[G4Gn6][G4Gn3][2]}**; 12, **{G4[Gn6[Gn3]G4Gn3][2]}**, **{G4[Gn6[G4Gn3]G4Gn3][2]}**; and **{G4[G4Gn6[G4Gn3]G4Gn3][2]}**; 17, **{G4[F2][2]}** and **{F2[2]}**; 18, **{F2G4[F2][2]}**.

The following structures have not been defined in the figures:

8, GlcNAcβ1-3Galβ1-4GlcNAcβ1-6(Galβ1-3)GalNAc-α-Ser(Thr)-; 13, Fucα1-2Galβ1-4GlcNAcβ1-6(Galβ1-3)GalNAc-α-Ser(Thr)- and Fucα1-2Galβ1-4GlcNAcβ1-6(Sialylα2-3Galβ1-3)GalNAc-α-Ser(Thr)- (H determinants); 14, Galβ1-4(Fucα1-3)GlcNAcβ1-6(Galβ1-3)GalNAc-α-Ser(Thr)- and Galβ1-4(Fucα1-3)GlcNAcβ1-6(Sialylα2-3Galβ1-3)GalNAc-α-Ser(Thr)- (X determinants); 15, Fucα1-2Galβ1-4(Fucα1-3)GlcNAcβ1-6(Galβ1-3)GalNAc-α-Ser(Thr)- and Fucα1-2Galβ1-4(Fucα1-3)GlcNAcβ1-6(Sialylα2-3Galβ1-3)GalNAc-α-Ser(Thr)- (Y determinants); 16, Sialylα2-3Galβ1-4(Fucα1-3)GlcNAcβ1-6(Sialylα2-3Galβ1-3)GalNAc-α-Ser(Thr)- (sialylated X determinant).

The oligosaccharides are arranged in groups according to the biosynthetic schemes in the figures: 1 (Fig. 2); 2, 3 (Fig. 5); 4–7 (Fig. 5); 9, 10 (Fig. 6); 11 (Fig. 6); 12 (Fig. 6); 13–16 (Fig. 4 and 7); 17, 18 (Fig. 7).

Table 4 Glycoproteins with Core 3 *O*-Glycans

Species	Glycoproteins	Oligosaccharide structures[b]												
		1	2	3	4	5	6	7	8	9	10	11	12	13
	Mucins													
Cow	Submaxillary	x				x	x			x	x			
Horse	Gastric		x											
Human	Colonic	x	x	x	x	x	x	z	z					
	Bronchial[a]	z	z	y		x	z							x
Monkey	Cervical						y							
Rat	Salivary						y							
	Colon		x			x	z	z	z					
	Secreted glycoproteins													
Human	Meconium[a]	x	x				x			x		x	x	

[a]Structures may be sulfated.

[b]The presence of various oligosaccharides in a particular glycoprotein is indicated as follows: x, isolated as such from the glycoprotein; y, not isolated as such from a particular glycoprotein, but a larger oligosaccharide containing the designated structure was isolated; z, a combination of x and y.

The oligosaccharide structures are as follows:
1, GlcNAcβ1-3GalNAc-α-Ser(Thr)- (core 3).
The following structures are named in the short nomenclature defined in Fig. 8:
2, **{G4[3]}**; 3, **{Gn3G4[3]}** and **{G4Gn3G4[3]}**; 4, **{Gn6[Gn3]G4[3]}**, **{Gn6[G4Gn3]G4[3]}**, and **{G4Gn6[G4Gn3]G4[3]}**; 5, **{S6[3]}**; 6, **{S6[G4][3]}**; 7, **{S6[Gn3G4][3]}** and **{S6[G4Gn3G4][3]}**; 8, **{S6[Gn6[Gn3]G4][3]}**, **{S6[Gn6[G4Gn3]G4][3]}**, and **{S6[G4Gn6[G4Gn3]G4][3]}**.

The following structures have not been defined in Fig. 8:
9, Sialylα2-6[Galβ1-4(Fucα1-3)GlcNAcβ1-3]GalNAc-α-Ser(Thr)-; 10, Sialylα2-6[Fucα1-2Galβ1-4GlcNAcβ1-3]GalNAc-α-Ser(Thr)-; 11, Galβ1-3GlcNAcβ1-3Galβ1-4GlcNAcβ1-3GalNAc-α-Ser(Thr)-; 12, Galβ1-3GlcNAcβ1-3Galβ1-3GlcNAcβ1-3GalNAc-α-Ser(Thr)-; 13, Sialylα2-6[Fucα1-2Galβ1-4(Fucα1-3)GlcNAcβ1-3]GalNAc-α-Ser(Thr)-.
The oligosaccharides are arranged in groups according to the biosynthetic scheme in Fig. 8 (i.e., 1–4 and 5–8).

1. Core Classes

There are at least six core classes of the Ser(Thr)-GalNAc type. We have named these classes as follows:

Core class 1: Galβ1-3GalNAc-α-Ser(Thr)-R (Tables 1, 2)
Core class 2: GlcNAcβ1-6(Galβ1-3)GalNAc-α-Ser(Thr)-R (Tables 1, 3)
Core class 3: GlcNAcβ1-3GalNAc-α-Ser(Thr)-R (Tables 1, 4)
Core class 4: GlcNAcβ1-6(GlcNAcβ1-3)GalNAc-α-Ser(Thr)-R (Table 1)
Core class 5: GalNAcα1-3-GalNAc-α-Ser(Thr)-R (Table 1)
Core class 6: GlcNAcβ1-6GalNAc-α-Ser(Thr)-R (Table 1)

These classes must not be confused with the oligosaccharide-typing system commonly used for the backbone structures (see following section).

2. Backbone Structures

The backbone of the Ser(Thr)-GalNAc oligosaccharide is formed by elongation of the core. Elongation usually involves addition of galactose and *N*-acetylglucosamine residues in β-linkages. The following moieties are commonly found in the backbone structure [95,154]:

1. Galβ1-3GlcNAcβ-, usually referred to as a type 1 structure.
2. Galβ1-4GlcNAcβ-, usually referred to as a type 2 structure.
3. (Galβ1-4GlcNAcβ1-3-)$_n$, a linear sequence of repeating type 2 disaccharides (also referred to as polylactosaminoglycans), is an antigenic determinant recognized by anti-blood group i antibodies.
4. Galβ1-4GlcNAcβ1-6
```
Galβ1-4GlcNAcβ1-6
                 \
                  Galβ-
                 /
Galβ1-4GlcNAcβ1-3
```
   is a branched structure that has been shown to be an antigenic determinant recognized by several anti-blood–group I antibodies.
5. GlcNAcβ1-6Galβ-, a structural unit found only in human glycoproteins to date.

In addition to type 1 and 2 structures mentioned, the blood group determinant nomenclature [121] recognizes two other structural types [i.e., Galβ1-3GalNAcα- (type 3) and Galβ1-3GalNAcβ- (type 4)].

3. Nonreducing Termini

The larger Ser(Thr)-GalNAc oligosaccharides are frequently terminated by antigenic determinants such as the human blood groups ABH, Le^a, and Le^b determinants, and the X (or mouse stage-specific embryonic antigen SSEA-1) and Y determinants (Table 5). These carbohydrate determinants are all due to α-linked sugars. Other sugars have also been found in terminal nonreducing positions, such as GlcNAcα1-4 [154] and GalNAcβ1-4/3 [240]. Feizi [95] has pointed out that the incorporation of a branch into the blood group i determinant is essential for blood group I activity, but blocks expression of blood group i activity. Similarly, the presence of the H determinant blocks expression of determinants i, I, and SSEA-1, and the presence of either the blood group A or B determinant blocks the H determinant. Thus, internal antigenic determinants are blocked as new antigenic determinants are added to the growing oligosaccharide chain. In addition, sulfate groups may be added onto terminal or internal sugars.

III. FUNCTIONS OF *O*-LINKED OLIGOSACCHARIDES

The lack of specific inhibitors for *O*-glycosylation has hindered studies on the functions of *O*-glycans. However, several interesting reports have appeared in the literature that suggest the importance of these molecules.

The *O*-glycans contribute to the chemical and biological properties of a glycoprotein. They have been found on numerous glycoproteins with different biological functions. There are several studies indicating important roles in receptor–ligand interactions, virus infectivity, fertilization, in the control of the immune system, and during cell differentiation. *O*-Glycan structures often show significant alterations in disease, for example, cancer and inflammatory diseases. Variations in oligosaccharide conformation may be important not only for interactions of carbohydrates with glycosyltransferases (see Sect. XI.B) but also in other ligand–receptor systems involving protein–carbohydrate interactions.

Differences in *O*-glycan structures have been reported between cytotoxic T lymphocytes and T-helper cells [209] suggesting different roles of *O*-glycans in these cell types. The stable cell surface expression of human interleukin-2 receptor and the major antigen envelope protein of Epstein–Barr virus all depend on normal *O*-glycosylation [189].

The *O*-glycans may inhibit proteolytic degradation of proteins or can contribute to the conformation of a protein or its stability. The activity of human granulocyte colony-stimulating factor is greatly decreased upon removal of its *O*-glycan chain SAα2-3Galβ1-3(SAα2-6)GalNAc- [253]. The role of this *O*-glycan was suggested to be either stabilization of protein conformation or inhibition of polymerization, which would deactivate the factor.

It has also been shown that the receptor for low-density lipoproteins (LDL) requires *O*-glycans for proper stability and function [182,189,396]. Kingsley et al. [182] selected a mutant Chinese hamster ovary cell line (complementation group *ldlD*) with a defective LDL receptor owing to incomplete glycosylation. The mutant line lacks UDP-Gal/UDP-GalNAc 4-epimerase and therefore can make neither UDP-Gal nor UDP-GalNAc. The defect can be completely corrected by exogenous galactose and *N*-acetylgalactosamine. Mutant cells remain LDL receptor-defective by the addition of exogenous galactose to the medium, but the addition of *N*-acetylgalactosamine to the medium restores normal *O*-glycosylation and overcomes the LDL receptor defect. Absence of *O*-glycosylation of human recombinant decay-accelerating factor expressed in this mutant ldlD cell line caused increased proteolytic degradation of the factor upon reaching the cell surface [278]. *O*-Glycans may, therefore, regulate the level of cell surface glycoprotein expression.

It has been shown that the glycoprotein ZP3 on the mouse egg's zona pellucida serves as a sperm receptor [37,100]. Removal of *O*-linked oligosaccharides from ZP3 destroyed its sperm receptor activity, whereas removal of *N*-linked oligosaccharides had no effect. *O*-Glycans isolated from ZP3 by reductive β-elimination were shown to possess sperm receptor activity. An α-linked galactose residue at the nonreducing terminus of these *O*-glycans is essential for binding of sperm to the zona pellucida. The authors conclude that the mouse sperm bind to eggs by *O*-linked oligosaccharides present on ZP3.

IV. GLYCOSYLTRANSFERASES

The incorporation of *N*-acetylgalactosamine into the polypeptide chain and subsequent elongation depends on the availability and relative activity of various glycosyltransferases. The glycosyltransferases are a large group of enzymes catalyzing the following general reaction:

$$\text{Nucleotide sugar + acceptor} \longrightarrow \text{sugar–acceptor + nucleotide}$$

The glycosyltransferases have been extensively reviewed [33,287,296,298,301,332]. They are membrane-bound enzymes and require detergent for optimum activity. The application of affinity chromatography to detergent-solubilized enzyme preparations has resulted in the purification of many of these enzymes [33]. Several of the glycosyltransferases acting on *O*-glycans (i.e., the pig gastric core 2 and blood group I β6-GlcNAc-transferases and the pig colonic core 3 and elongation β3-GlcNAc-transferases, all discussed later) are inhibited at Triton X-100 concentrations above 0.5% [42,43]; this has made it difficult to solubilize and purify any of these enzymes [unpublished data].

One of the most interesting features of these enzymes is their ability to discriminate between different oligosaccharide structures. This substrate specificity directs the synthetic pathway along various routes and is a major factor in the control of oligosaccharide synthesis. The study of highly purified glycosyltransferases has provided evidence for the hypothesis that every sugar–sugar linkage requires a separate transferase for its synthesis. This one linkage–one glycosyltransferase hypothesis appears to hold true in most instances, although at least one exception is now known (i.e., the Lewis blood group-dependent fucosyltransferase catalyzes the synthesis of both α1-3 and α1-4 linkages) [268]. It has long been known that identical linkages in different complex carbohydrates may be synthesized by the same glycosyltransferase (e.g., the determinants for the human blood groups A, B, H, Le^a, and Le^b are synthesized by the same transferases whether they occur in glycosphingolipids, *N*-glycosyl oligosaccharides, *O*-glycosyl oligosaccharides or milk oligosaccharides with lactose at the reducing terminus). Examples have appeared in the literature in which a particular linkage is made by more than one glycosyltransferase (see later).

The following discussion will deal with the glycosyltransferases that are involved in Ser(Thr)-GalNAc oligosaccharide synthesis.

V. INITIATION OF OLIGOSACCHARIDE SYNTHESIS

All Ser(Thr)-GalNAc oligosaccharides appear to be initiated by the incorporation of a single *N*-acetylgalactosamine residue from UDP-GalNAc into the polypeptide chain. This mechanism of initiation differs markedly from the complex process required for synthesis of the asparagine-*N*-acetylglucosamine (Asn-GlcNAc) linkage [341]. The latter involves preassembly of a large oligosaccharide as a dolichol pyrophosphate oligosaccharide and transfer of the entire oligosaccharide to the polypeptide chain in the rough endoplasmic reticulum. In the *O*-glycosylation process, individual sugar chains are transferred, apparently in the Golgi apparatus.

The UDP-GalNAc:polypeptide α-*N*-acetylgalactosaminyltransferase (polypeptide GalNAc-transferase; EC 2.4.1.41) was first described by McGuire and Roseman [224] in sheep submaxillary glands. It has subsequently been found in many other tissues [296,337]. It has been purified or partially purified from ascites hepatoma AH 66 cells (48,100-fold,

Table 5 Antigenic Determinants

Structure of determinant	Name of antigen
Type 1[a]	
Fucα1-2Galβ1-3GlcNAc-	Blood group H, type 1
GalNAcα1-3(Fucα1-2)Galβ1-3GlcNAc-	Blood group A, type 1
Galα1-3(Fucα1-2)Galβ1-3GlcNAc-	Blood group B, type 1
Galβ1-3(Fucα1-4)GlcNAc-	Blood group Lewis[a] (Le^a)
Fucα1-2Galβ1-3(Fucα1-4)GlcNAc-	Blood group Lewis[b] (Le^b)
Type 2[b]	
Fucα1-2Galβ1-4GlcNAc-	Blood group H, type 2
GalNAcα1-3(Fucα1-2)Galβ1-4GlcNAc-	Blood group A, type 2
Galα1-3(Fucα1-2)Galβ1-4GlcNAc-	Blood group B , type 2
Galβ1-4(Fucα1-3)GlcNAc-	Determinant X (Le^x), Stage-specific embryonic antigen-1 (SSEA-1)
Fucα1-2Galβ1-4(Fucα1-3)GlcNAc-	Determinant Y (Le^y)
Polylactosaminoglycan	
Galβ1-4GlcNAcβ1-3Galβ1-4GlcNAcβ1-3-R	Blood group i
Galβ1-4GlcNAcβ1-6 \ Galβ1-4GlcNAc-R / Galβ1-4GlcNAcβ1-3	Blood group I
GalNAcα1-3GalNAcβ1-3Gal-	Forssman antigen

[a]Determinants containing the sequence Galβ1-3GlcNAc are called type 1.
[b]Determinants containing the sequence Galβ1-4GlcNAc are called type 2.

4% yield) [342], from bovine colostrum (516,000-fold, 18.9% yield) [91], from mouse lymphoma BW5147 cells (2300-fold, 8–10% yield) [91], and from lactating bovine mammary glands (7460-fold, 0.23% yield) [345]. Purification of all these preparations was achieved by affinity chromatography on columns of carbohydrate-free bovine submaxillary apomucin coupled to cyanogen bromide-activated Sepharose 4B or 6B. Both uridine diphosphate (UDP) and Mn^{2+} were required for binding of enzyme to the affinity column. The purified hepatoma enzyme gave a single band on sodium dodecyl sulfate–polyacrylamide gel electrophoresis (SDS-PAGE) at an M_r of 56,000, whereas the bovine colostrum enzyme showed an M_r of 70,000. Divalent cation is essential for activity, with Mn^{2+} being the most effective cation.

The ability of various compounds to accept *N*-acetylgalactosamine has been studied by many laboratories with both crude and purified enzyme preparations [40,73,74,120, 142,143,158,224,296,311,342,345,397]. Effective acceptors are apomucins made from ovine or bovine submaxillary glands, kappa-casein, bovine myelin A1 protein (myelin basic protein), carbohydrate-free antifreeze glycoprotein from Antarctic fish, and immature viral glycoproteins from herpes simplex virus type, or peptides of a certain length and sequence. Native mucins, asialomucins or pronase-digested apomucins are ineffective acceptors.

Large tryptic peptides from apomucin were effective substrates [142,143], but no unique primary sequences were identified adjacent to the *O*-glycosidically substituted serine and threonine residues. More recently [40,397], synthetic polypeptides have been tested as acceptors for the GalNAc-transferase. The smallest effective peptide is Thr-Pro-Pro-Pro with a blocked NH_2-terminus; all peptides with this sequence are effective acceptors. At least three proline residues are essential.

The amino acid sequences of many *O*-glycosylated glycoproteins and mucins have now been determined, either by purification of the glycoprotein and sequencing of the peptide, or by gene cloning. The amino acid sequences near *O*-glycosylation sites in most glycoproteins contain proline residues. These proline residues are apparently required for polypeptide GalNAc-transferase action either because the proline residues are directly recognized by the enzyme or because they cause a conformational change in the polypeptide chain that is essential for enzyme action. A small number of glycoproteins, such as glycophorin, carry *O*-glycosyl oligosaccharides in regions devoid of nearby proline residues. It is possible that sites of *O*-glycosylation lacking proline residues in the adjacent primary sequence may either adopt the required conformation even in the absence of these residues, or proline residues required for enzyme binding may be brought into the region of *O*-glycosylation by three-dimensional factors. The presence of other oligosaccharide chains may influence the *O*-glycosylation of serine or threonine residues [342]. It has been suggested that disulfide bond formation near the site of *O*-glycosylation may control the glycosylation of rabbit IgG at the hinge region [158].

The polypeptide GalNAc-transferase does not act on low molecular mass sugars. It is also ineffective toward a variety of different glycolipids and can be readily differentiated from the α3-GalNAc-transferase responsible for human blood group A determinant synthesis, from globoside α3-GalNAc-transferase, from globotriaosylceramide β3-GalNAc-transferase and from blood group Cad β4-GalNAc-transferase.

The *O*-glycosylation by polypeptide GalNAc-transferase from porcine or bovine submaxillary glands and hen oviduct is not mediated by a dolichol or lipid intermediate [12,132]. It is of interest that *O*-mannosylation of serine and threonine residues during fungal glycoprotein biosynthesis does require dolichol monophosphate mannose [348].

Histochemical studies using monoclonal antibodies recognizing polypeptide GalNAc-transferase showed that the enzyme is localized at neuromuscular junctions in the rat skeletal muscle. Expression of the enzyme appears to be regulated by synaptic interactions [308].

Various attempts have been made to determine the subcellular location of the polypeptide GalNAc-transferase. The enzyme has been reported to be in the smooth-surfaced microsome fraction of rat small intestinal mucosa [181], brain [184], hen oviduct [132], and HeLa cells [119]. Abeijon and Hirschberg [1] found that the specific activity of UDP-GalNAc:polypeptide GalNAc-transferase was enriched 37-fold relative to homogenate in a rat liver Golgi preparation; virtually no enzyme activity was detected in membranes from the rough and smooth endoplasmic reticulum. Sealed right-side-out Golgi vesicles were able to translocate UDP-GalNAc into their lumen; this translocation was saturable, indicating that a transport protein was involved.

Both autoradiography and biochemical methods have been used to assess incorporation of labeled sugars into the glycoproteins of intact cells (see Sect. I). Some studies have indicated that N-acetylgalactosamine incorporation is a relatively early event. For example, Strous [339] has reported the isolation from rat stomach polyribosomes of peptidyl-tRNA containing alkali-labile *N*-acetylgalactosamine. This type of experiment is technically very difficult, and it is impossible to rule out small amounts of nonnascent glycopeptide. Johnson and Heath [167] isolated fetuin peptidyl-tRNA from fetal calf liver and analyzed for *O*-glycans released after β-elimination in the presence of NaB^3H_4. They found that < 1.3% of the potential *O*-glycan chains of fetuin had been initiated. Rough microsomes were used to program the cell-free synthesis of fetuin; < 1% of the radioactive fetuin product was bound by GalNAc-specific lectins. These results show that *N*-acetylgalactosamine addition is primarily a posttranslational event, but that some of its incorporation occurs very early.

Studies on the biosynthesis of glycophorin A in K562 cells [172,173] suggested that *N*-acetylgalactosamine was incorporated early in the biosynthetic pathway, whereas galactose was added later. Piller et al. [265] studied the maturation of *N*- and *O*-linked oligosaccharides of leukosialin in K562 cells by radiolabeling, immunoprecipitation, and lectin binding. They found that *N*-glycosylation occurs in the rough endoplasmic reticulum; initiation of *O*-glycosylation follows in early Golgi compartments, followed rapidly by galactosylation and sialylation. Cummings et al. [75] studied the biosynthesis of the low-density lipoprotein (LDL) receptor in normal human fibroblasts and in a human epidermoid carcinoma cell line (A-431). Pulse-labeling studies indicated a precursor form of the receptor (about 124,000 Da) that contained a single *N*-glycan of the high mannose type and several neutral *O*-glycans. This precursor was converted to a mature form of the receptor (about 176,000 Da) in which the *N*-glycan had been processed to the complex type (containing sialylated antennae), and the *O*-glycans had become sialylated. The conclusion was that N-acetylgalactosamine incorporation into threonine or serine must occur before, or at the same time as, α2-mannosidase I-catalyzed processing of high mannose *N*-glycans, in the *cis*-Golgi or earlier [116].

Other studies indicate that *O*-glycosylation is initiated later in the biosynthetic pathway. *O*-Glycosylation of human chorionic gonadotrophin (hCG) by a human choriocarcinoma cell line was shown to occur shortly before secretion, presumably in the Golgi [131].

Three classes of virus, all of which bud through intracellular membranes, rather than through the plasma membrane, have been shown to contain glycoproteins with *O*-linked glycans (i.e., herpes simplex virus type 1 HSV-1 [254], vaccinia virus [322], and coronavirus [248]). *N*-Acetylgalactosamine incorporation into E1 of murine hepatitis virus A 59 occurred only in the Golgi-enriched fraction and not in the rough microsome fraction. Monensin, an ionophore believed to interfere with movement within the Golgi apparatus, completely blocked *N*-acetylgalactosamine incorporation into E1. The authors concluded that this incorporation was a posttranslational event occurring in the Golgi apparatus. A similar study was carried out with HSV-1 [168]. Several HSV-1 glycoproteins contain both *N*- and *O*-linked glycans. These glycoproteins are first synthesized as precursor forms with no detectable *O*-glycans but with *N*-glycans that are completely in the high mannose form. On maturation, the *N*-glycans are processed completely to the complex type,and sialylated *O*-glycans appear. Monensin prevents maturation (i.e., *N*-glycans remain high mannose, and *O*-glycans are not incorporated). The conclusion is that *N*-acetylgalactosamine incorporation into polypeptide is a relatively late event, probably occurring in the Golgi apparatus. In contrast, the HSV-1 glycoprotein C appeared to be *O*-glycosylated before its entry into the Golgi, and the subsequent addition of sugars occurred much later [310].

Roth [281,282] has used *Helix pomatia* lectin labeled with particles of colloidal gold to study electron micrographs of intestinal goblet cells from chicks and rats. The lectin locates terminal α-linked *N*-acetylgalactosamine residues. No particles were detected in the rough endoplasmic reticulum. Label was confined to several cisternae at the *cis*-Golgi side and one or two cisternae at the *trans*-Golgi side with minimal labeling of the medial-Golgi. The interpretation suggested was that label at the *cis*-Golgi was due to incorporation of a single *N*-acetylgalactosamine into polypeptide, and that the label observed in the *trans*-Golgi was due to GalNAc at the nonreducing termini of large *O*-glycans. Deschuyteneer et al. [82] studied the subcellular localization of porcine submaxillary gland apomucin (antibody) and *N*-acetylgalactosamine (*H. pomatia* lectin) in porcine submaxillary gland by cytochemical methods. They concluded that apomucin is first glycosylated primarily in the *cis*-Golgi apparatus.

In summary, therefore, the evidence now available indicates that the bulk of *O*-glycosylation occurs after release of nascent peptide from the polyribosomes, at a stage when the peptide has moved to the smooth endoplasmic reticulum or Golgi apparatus region of the cell. There is, however, also some very early *O*-glycan initiation.

Studies on the biosynthesis of sialomucins have suggested a hypothesis to resolve these conflicting reports. Sialomucins are abundant on the surfaces of epithelial cells, lymphocytes [56,102], and certain tumor cells [58,109,113,136,159,203,243,321,360]. These sialomucins are large, highly glycosylated glycoproteins, rich in sialylated *O*-glycans (see Tables 1–3). It has been suggested that these sialic acid-rich cell surfaces play an important role in metastatic potential and in the escape of tumors from immune destruction [58]. Carraway's group [58,334,335] has studied the biosynthesis of asialoglycoprotein-1 (ASGP-1) from rat mammary adenocarcinoma ascites cells using radioactive leucine, threonine, galactose, and glucosamine as precursors. They have concluded that *O*-glycosylation is initiated throughout the endomembrane system. *O*-Linked carbohydrate is detected either cotranslationally or shortly after translation, and further carbohydrate initiation occurs over almost the entire period of transit of ASGP-1 from the site of polypeptide synthesis in the rough endoplasmic reticulum to the plasma membrane. The amount of early initiation may depend on the size of the molecule being synthesized. This continuous model of *O*-glycan initiation may explain the discrepancies observed between different tissues. It is also possible that the intracellular location of the polypeptide GalNAc-transferase or its activity distribution varies among different cells.

VI. ASSEMBLY OF OLIGOSACCHARIDES WITH CORE CLASS 1

A. Synthesis of the Core

The enzyme that routes the biosynthetic pathway toward synthesis of core class 1 oligosaccharides is UDP-Gal:GalNAc-R β3-galactosyltransferase (EC 2.4.1.122). The enzyme catalyzes the following reaction:

$$\text{UDP-Gal} + \text{GalNAc-R} \longrightarrow \text{Gal}\beta 1\text{-3GalNAc-R} + \text{UDP}$$

The β3-galactosyltransferase was first described in porcine and ovine submaxillary glands [300] and has since been reported in rat pancreas, liver, stomach, and intestine; human erythrocytes, leukocytes, platelets, serum, and trachea; porcine stomach, colon, and Cowper's gland; baby hamster kidney cells; and canine trachea [8–10,15,23,41,44, 54,61,65,115,139,140,226,279,280,296,323,340,357]. The β3-Gal-transferase acts not only on mucins, but is required for the synthesis of core classes 1 and 2 on other glycoproteins (see Table 1). The enzyme is membrane bound and requires detergent and Mn^{2+} for activity.

The enzyme has been purified (2000-fold, 20% yield) from detergent extracts of swine trachea [226]. The enzyme bound to an affinity column prepared by using ionic forces to attach asialo Cowper's gland mucin (see Tables 1 and 2) to DEAE-cellulose.

Furukawa and Roth [111] have reported the purification from chick embryo liver of UDP-Gal:GlcNAc-R β4-Gal-transferase (EC 2.4.1.38 or EC 2.4.1.90) and UDP-Gal:GalNAc-R β3-Gal-transferase (EC 2.4.1.122). The two enzymes behaved similarly on several affinity columns, but were eventually separated from each other by affinity chromatography using the respective acceptors as ligands. The β3-Gal-transferase was purified 4300-fold with a yield of 5%,

Several laboratories have reported on the substrate specificity of the β3-Gal-transferase from various sources. The most effective acceptors are mucins containing terminal GalNAcα-Ser(Thr) residues (e.g., asialo ovine submaxillary, asialo porcine submaxillary, or asialo porcine Cowper's gland mucins). The enzyme can transfer galactose to free *N*-acetylgalactosamine or to glycosides of GalNAc but the K_m for these acceptors (100–180 mM) is much higher than for mucins (90 μM for the pure enzyme from pig trachea). Mucins in which the Ser(Thr)-linked GalNAc residue is substituted by one or more sugar residues and glycoproteins lacking the Ser(Thr)-GalNAc moiety are not effective as acceptors. Interestingly, substitution of the Ser(Thr)-linked GalNAc by a sialylα2-6-residue (e.g., native ovine submaxillary, porcine submaxillary, or porcine Cowper's gland mucins) prevents enzyme action. This serves as an important control in the biosynthesis of mucins [223,300]. If the CMP-NAN:GalNAc-R α6-sialyltransferase I (see later) acts first, the synthetic pathway stops at the disaccharide sialyl-α2-6-GalNAc-R. Mucins rich in this disaccharide (such as ovine submaxillary mucin or porcine Cowper's gland mucin) are made in tissues that have a high α6-sialyltransferase I activity relative to β3-Gal-transferase. On the other hand, mucins that have larger oligosaccharides with core class 1 (such as porcine submaxillary mucin) are made in tissues in which the β3-Gal-transferase level is high relative to the α6-sialyltransferase I level. Prior action of the β3-Gal-transferase does not prevent subsequent action of the α6-sialyltransferase I (Fig. 2). The synthesis of {**S6[1]**} (see Fig. 2), therefore, obeys what we have called the *three-before-six* rule (i.e., carbon-3 must be substituted before carbon-6 can be substituted). Other examples of this rule will be pointed out as they occur. Unpublished results from our laboratory indicate that the purified β3-Gal-transferase from rat liver requires the 3- and 4-hydroxyl, but not the 6-hydroxyl of GalNAc-R substrate. This is compatible with the finding that core class 6 (GlcNAcβ1-6GalNAc-R) oligosaccharide is a substrate for the enzyme (see Sect. IX).

We have purified the enzyme from rat liver 15,000-fold using 5-Hg-UDP-GlcNAc-thiopropyl-Sepharose as an affinity column [unpublished]. The rat liver enzyme acts readily on various synthetic and glycoprotein-derived GalNAc-peptides containing either four or more amino acids, or protective groups, to eliminate charges near the glycosylation site. The activity is strongly influenced by the composition and sequence of the peptide moiety, the attachment position of GalNAc, and the presence of other GalNAc residues. Six amino acids as well as the GalNAc residue can be accommodated in the binding site [44].

The subcellular location of the β3-Gal-transferase has been studied in rat pancreas [279,280], rat liver [8–10], mouse lymphoma cells [90], and porcine Cowper's gland [226]. The consensus is that the enzyme is greatly enriched in the Golgi apparatus. Andersson and Eriksson [9,10], in their studies on rat liver, found the Golgi apparatus to be enriched in the β3-Gal-transferase, but also found appreciable amounts of this enzyme in the endoplasmic reticulum. The Golgi and endoplasmic reticulum enzymes differed in their susceptibility to detergent, and the authors suggested that the two locations may have different enzymes.

Elhammer and Kornfeld [90] subfractionated mouse lymphoma BW 5147 cell homogenates on sucrose gradients. The β3-Gal-transferase migrated in the same fractions as UDP-Gal:GlcNAc β4-Gal-transferase, an enzyme previously localized to the *trans*-Golgi [283].

The β3-Gal-transferase can be measured in membranes prepared from normal human erythrocytes but erythrocytes from patients with a rare disease called permanent mixed-field polyagglutinability lack this enzyme [23,60,139]. Glycophorin A from these defective erythrocytes contains *O*-glycosyl oligosaccharides with reduced amounts of sialic acid and galactose (i.e., the *O*-glycosyl oligosaccharides are GalNAc and presumably sialylα2-6GalNAc). The presence of terminal *N*-acetylgalactosamine residues on these red cells explains why they react with antibodies to the T_n antigenic determinant. A human T-lymphoblastoid cell line, Jurkat, was found to specifically lack the core 1 β3-Gal-transferase [264]. Glycoproteins synthesized in these cells contain mainly GalNAc- and lack core 1 structures. Jurkat cells therefore represent a model for functional studies of core 1 oligosaccharides.

B. Elongation of the Core: Synthesis of the Blood Group Ii Backbone

When the Galβ1-3GalNAc core is synthesized (see Fig. 2), elongation can proceed along a variety of different pathways (Figs. 2–4). Elongation almost always involves the addition of galactose and *N*-acetylglucosamine residues in β1-3, β1-4, and β1-6 linkages. The repeating sequence $(\text{Gal}\beta 1\text{-}4\text{GlcNAc}\beta 1\text{-}3)_n$ often serves in *O*-glycans as an intermediary sequence between core and nonreducing terminus. This repeating sequence is also found in some glycosphingolipids and *N*-glycans and has been identified as the antigenic determinant for an interesting developmentally regulated blood group antigen on human erythrocytes named i [95]. *N*-Acetylglucosamine residues can be inserted in β1-6 linkage to a galactose residue in this repeating sequence to create a branched structure Galβ1-4GlcNAcβ1-3(Galβ1-4GlcNAcβ1-6)Gal. This branched structure has been associated with human blood group I antigenic activity [95] (see Table 5).

1. *UDP-GlcNAc:Galβ1-3GalNAc-R (GlcNAc to Gal) β3-GlcNAc- Transferase*

N-Acetylglucosamine addition in β1-3 linkage to core 1 to form {**Gn3[1]**} and elongated derivatives of {**Gn3[1]**} (see Figs. 2–4) occurs during the synthesis of many mucins and glycoproteins, as indicated in Table 2. Data from our laboratory [45,47] indicate that porcine gastric membrane preparations contain a UDP-GlcNAc:Galβ1-3GalNAc-R (GlcNAc to Gal)β3-GlcNAc-transferase (EC 2.4.1.146), where R is polypeptide from mucin or antifreeze glycoprotein. The enzyme also elongates core 2 oligosaccharides and is discussed further in Section VII.B.

2. *UDP-GlcNAc:Galβ1-4GlcNAc-R (GlcNAc to Gal) β3-GlcNAc-Transferase*

Several reports have appeared [17,18,149,162,261,273,285,361,363,394,402–404] indicating the incorporation of *N*-acetylglucosamine into various Galβ1-4GlcNAc-terminated acceptors to form the linear i structure, GlcNAcβ1-3Galβ1-4GlcNAc-. The enzyme, UDP-GlcNAc:Galβ1-4GlcNAc-R (GlcNAc to Gal)β3-GlcNAc-transferase (EC 2.4.1.149), is different from the β3-GlcNAc-transferase described in the foregoing section, which elongates the mucin core Galβ1-3GalNAc [45] because of different tissue distributions, although the two enzymes share requirements for divalent cation, UDP-GlcNAc, and terminal Galβ- in the acceptor.

3. UDP-GlcNAc:GlcNAcβ1-3Galβ-R (GlcNAc to Gal) β6-GlcNAc-Transferase

In vitro enzymatic synthesis of the branched I antigenic determinant, Galβ1-4GlcNAcβ1-6(Galβ1-4GlcNAcβ1-3)Gal-, has been achieved by Piller et al. [262] using GlcNAcβ1-3Galβ1-4Glc-β-methyl as acceptor for β6-GlcNAc-transferase activity. Pig gastric mucosal extracts have been shown to incorporate *N*-acetylglucosamine in β1-6 linkage to galactose using GlcNAcβ1-3Galβ1-3GalNAc-α-benzyl as substrate to form GlcNAcβ1-6(GlcNAcβ1-3)Galβ1-3GalNAc-α-benzyl ({**Gn3[1]**} to {**Gn6[Gn3][1]**}; see Fig. 2) [43] (see Sect. VII.B and XI for further discussion) (i.e., the pig stomach β6-GlcNAc-transferase obeys the three-before-six rule). No GlcNAcβ1-3Galβ1-3(GlcNAcβ1-6)GalNAc-α-benzyl was detected in these experiments, indicating that elongation of core 1 by the β3-GlcNAc-transferase prevents addition of GlcNAc in β1-6 linkage to GalNAc to form a core 2 structure (i.e., {**Gn3[1]**} cannot be converted to {**Gn3[2]**}, as indicated in Fig. 2).

There have been reports on in vitro synthesis of the linear structure GlcNAcβ1-6Galβ1-4Glc(NAc)- [17,186,363,395,403,404] by a different β6-GlcNAc-transferase (EC 2.4.1.150) in Novikoff ascites tumor cells, human serum, human ovarian tissue, and mouse T-lymphoma cells. This enzyme does not obey the three-before-six rule. Novikoff ascites tumor cells appear to contain both β6-GlcNAc-transferase activities [186].

4. UDP-Gal:GlcNAc-R β4-Galactosyltransferase

The enzyme UDP-Gal:GlcNAc β4-galactosyltransferase (EC 2.4.1.38 or EC 2.4.1.90) has been purified from bovine milk and from various other sources and is one of the most thoroughly studied glycosyltransferases [33,301]. In combination with α-lactalbumin, it becomes lactose synthetase, the enzyme responsible for the synthesis of the lactose secreted in milk. This same enzyme appears to be capable of the synthesis of the Galβ1-4GlcNAc moiety found in the poly-*N*-acetyllactosaminoglycans with blood group i antigenic activity [177,304,305]. It is possible that the enzyme activity is regulated by phosphorylation of the enzyme protein [48].

In an interesting study, Blanken et al. [36] showed that pure bovine colostrum UDP-Gal:GlcNAc β4-Gal-transferase could act on the synthetic branched trisaccharide GlcNAcβ1-3(GlcNAcβ1-6)Gal to form the pentasaccharide Galβ1-4GlcNAcβ1-3(Galβ1-4GlcNAcβ1-6)Gal. This pentasaccharide shows blood group I activity with some anti-I antibodies. The galactosyltransferase showed a marked preference for the β1-6–linked GlcNAc terminus relative to the β1-3–linked GlcNAc, although with time both *N*-acetylglucosamine residues were galactosylated.

The genes for the bovine, human, and mouse UPD-Gal:GlcNAc β4-Gal-transferase have been cloned [11,76,161,221,237,238,314,317]. The human gene is localized to chromosome 9 [86,315], and the mouse gene [145,316] maps to chromosome 4. There is no homology to other glycosyltransferases, including the cloned α3-Gal-transferase [174,205]. The enzyme has the structure of a type 2 membrane protein, similar to other cloned glycosyltransferases [256], with a short NH_2-terminal end directed toward the cytoplasm, a membrane-bound region serving as a signal-anchor, a stem region that has the potential to be *O*-glycosylated because of the presence of proline, threonine, and serine residues, and a COOH-terminal region containing the active site directed into the lumen of the Golgi. The enzyme exists as two different membrane-bound forms, a long and a short peptide; the longer form contains an additional NH_2-terminal peptide sequence and is probably transcribed from the same gene utilizing a different upstream promoter [286]. With the exception of male germ cells of the mouse, which contain only the long form, most organs investigated contain both forms of the enzyme [318]. The

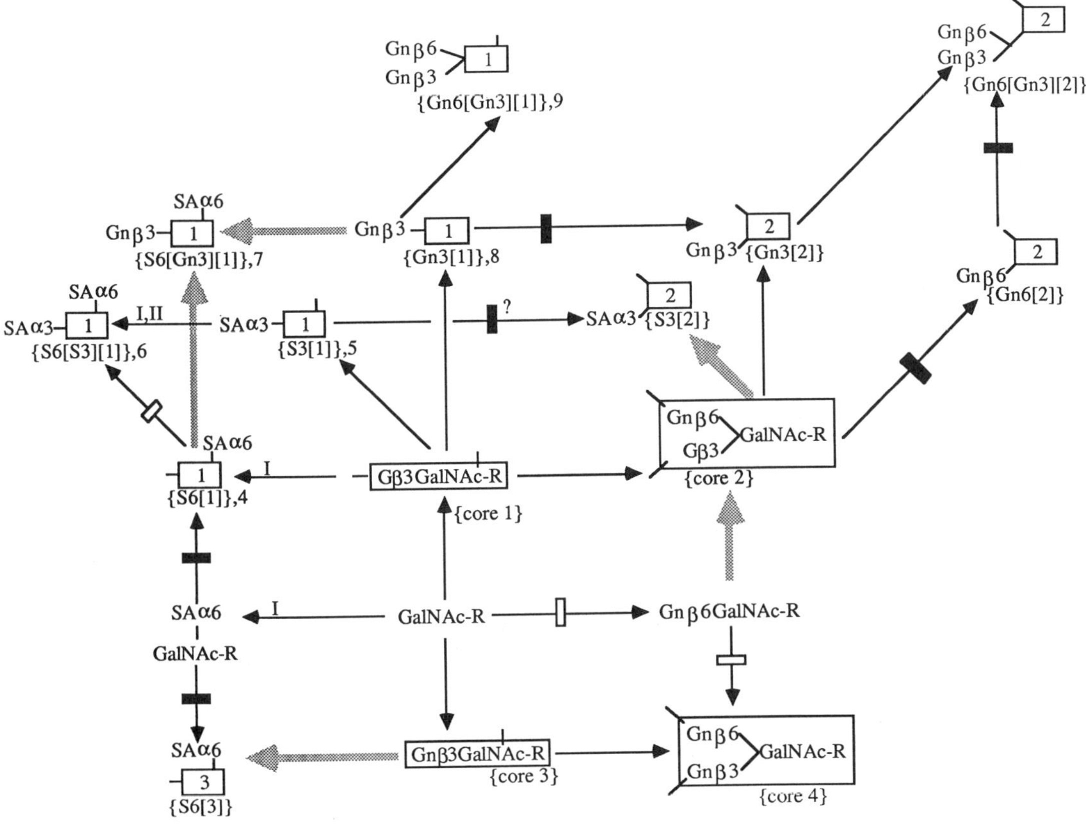

Figure 2 Assembly of Ser(Thr)-GalNAc oligosaccharides, showing synthesis of the sialylα2-6GalNAc disaccharide, core classes 1–4, and 6 (see text) and some commonly occurring derivatives of core classes 1 and 2. The numbered rectangular boxes represent the core structures. The names in braces { } are defined in the figure and are used to refer to structures in the text. There are two α6-sialyltransferases designated I and II (see text). All the reactions in the figure except the steps shown by discontinuous arrows have been studied in cell-free preparations. Arrows blocked with filled rectangles indicate reactions that do not take place and arrows blocked with open rectangles indicate slow reactions. Structures **{S3[1]}** and **{S6[S3][1]}** are commonly found on sialomucins of normal circulating cells (erythrocyte glycophorins and leukocyte leukosialins) and of tumor cells, but also occur on secreted glycoproteins and on some mucins (see Table 2).

The conversion of GalNAc-R to GlcNAcβ1-6GalNAc-R could not be demonstrated by us; however, the GlcNAcβ1-6GalNAc- structure has been isolated from some human glycoproteins (see Table 3) and the enzyme activity has been reported in human ovarian tissue [395]. Abbreviations: GalNAc-R, Ser(Thr)-linked GalNAc residue; SA or S, sialic acid; G,D-galactose (Gal); GA, *N*-acetyl-D-galactosamine (GalNAc); Gn, *N*-acetyl-D-glucosamine (GlcNAc).

galactosyltransferase may be located in the Golgi as well as on the cell surface of sperm and other normal and cancer cells. The mouse sperm β4-Gal-transferase apparently plays a role in gamete recognition [325].

5. *UDP-Gal:GlcNAcβ1-3GalNAc-R β3-Gal-Transferase*

More recently, a UDP-Gal:GlcNAc β3-Gal-transferase has been highly purified from pig trachea [319,320]. This enzyme is clearly different from the previously studied UDP-Gal:GlcNAc β4-Gal-transferase and UDP-Gal:GalNAc-R β3-Gal-transferase. The physiological function of this enzyme appears to be the elongation of Ser(Thr)-GalNAc oligosaccharides and glycolipids, since the enzyme works best with the GlcNAcβ1-3Gal(NAc)-R acceptors and does not work on *N*-glycosyl oligosaccharides terminating in *N*-acetylglucosamine.

C. Terminal Glycosylation

A common feature of all Ser(Thr)-GalNAc oligosaccharides is the presence of nonreducing terminal α-linked sugars. The incorporation of a sugar in α-linkage may mask underlying antigenic determinants and, usually, prevents further sugar additions, thereby inhibiting chain growth. Addition of a sialic acid in α2-6 linkage to the GalNAc residue of Galβ1-3GalNAc-Ser(Thr)-R does not prevent further sugar additions to the galactose residue, although the sialyl residue usually remains unsubstituted. Sialyl residues can be substituted by another sialyl residue in α2-8 linkage in both gangliosides and glycoproteins [99].

The most common α-linked terminal sugars are sialic acid and L-fucose, but terminal galactose, *N*-acetylgalactosamine, and *N*-acetylglucosamine residues also occur in α-linkage. Frequently, the α-linked sugars, either alone (e.g., blood group H determinant) or in combination (e.g., blood group Le^b determinant), constitute an antigenic determinant (see Table 5). The synthesis of the human ABO and Lewis determinants has been thoroughly reviewed elsewhere [33,302,374,387,388] and only a relatively brief description will be given here. *N*-Acetylgalactosamine can also be added in β1-4 linkage to form the blood group Cad determinant (see later in Sect. VI.C.6).

1. *CMP-Sialic Acid:GalNAc-R α6-Sialyltransferase I (EC 2.4.99.3)*

This enzyme was mentioned briefly in section VI.A. It was pointed out that UDP-Gal:GalNAc-R β3-Gal-transferase could not act after the α6-sialyltransferase (see Fig. 2) (i.e., the incorporation of a sialic acid in α2-6 linkage into GalNAc-mucin prevented further sugar addition and accounted for the relatively large amounts of sialylα2-6GalNAc- disaccharides in many mucins, such as ovine submaxillary and porcine Cowper's gland mucins). Mucins with larger oligosaccharides containing the sialylα2-6GalNAc moiety, such as porcine submaxillary mucin, are made in tissues containing relatively more of the β3-Gal-transferase than the α6-sialyltransferase. The α6-sialyltransferase I catalyzes the following reaction:

$$\text{CMP-sialic acid} + \text{R}_1\text{-GalNAc-Ser(Thr)-R}_2 \longrightarrow$$

$$\text{R}_1\text{-(sialyl}\alpha\text{2-6)GalNAc-Ser(Thr)-R}_2 + \text{CMP}$$

where R_1 can be H-, Galβ1-3-, or sialylα2-3Galβ1-3-, and where the sialyl residue can be either *N*-acetylneuraminic (NeuNAc) acid or *N*-glycolylneuraminic acid (NeuNGc). The β3-Gal-transferase and α6-sialyltransferase I both compete for the same substrate, GalNAc-R (see Fig. 2). The sialyltransferase, first described in sheep submaxillary glands

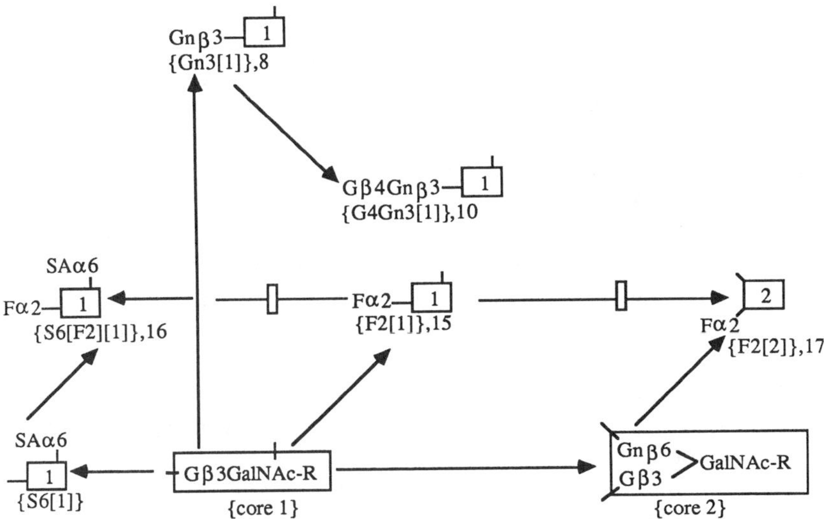

Figure 3 Synthesis of some oligosaccharides with class 1 core. These reactions, or analogous reactions with different oligosaccharides having similar nonreducing termini, have been studied in vitro. The numbered rectangular boxes represent the core structures. The names in braces { } are defined in the figure and are used to refer to structures in the text. Arrows blocked with open rectangles indicate slow reactions. Fucosylated core 1 *O*-glycans occur primarily on mucins (see Table 2; references in Table 1). Abbreviations: as in the legends to Figs. 1 and 2.

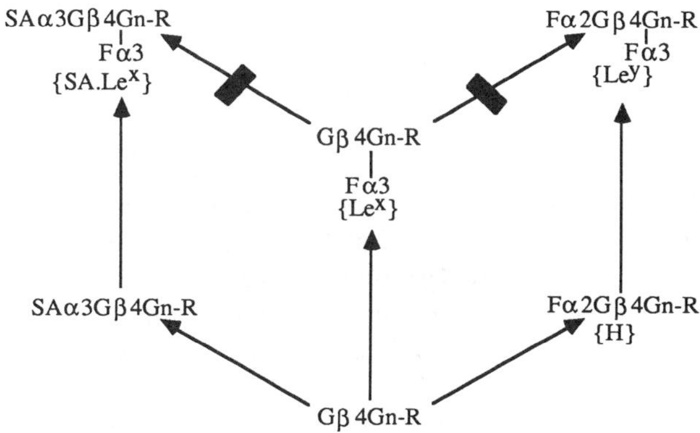

Figure 4 Synthesis of determinants H, X, Y, and sialylated X (see Table 5). All the reactions in the figure, or analogous reactions, have been studied in cell-free preparations. Arrows blocked with filled rectangles indicate reactions that do not take place. These reactions require a Galβ1-4GlcNAc-R starting terminus and *O*-glycans have been reported carrying these determinants based on **{G4Gn3[1]}** (see Fig. 3), on **{G4[2]}** and **{G4[S3][2]}** (see Fig. 7), and on **{G4[4]}** and **{G4[G4][4]}** (see Fig. 9). Abbreviations: as in the legends to Figs. 1 and 2.

[55], has also been found in porcine, bovine, and canine submaxillary glands, and has been highly purified from porcine submaxillary glands [33,288,289]. Some of the glycoproteins that carry the sialylα2-6GalNAc moiety, either alone or as part of a larger core 1 or core 3 oligosaccharide, are shown in Tables 2 and 4.

Purification of α6-sialyltransferase I (EC 2.4.99.3) [288,289] was complicated by the presence in porcine submaxillary gland of another enzyme, CMP-sialic acid:Galβ1-3GalNAc-R α3-sialyltransferase (EC 2.4.99.4; see next section). Both enzymes were purified to homogeneity from Triton X-100 extracts of submaxillary gland. Key steps in the purifications were repeated adsorptions to CDP-hexanolamine-Sepharose columns. The final preparation of α6-sialyltransferase I was obtained in 2% yield with a purification factor of 117,000, had a specific activity of 44.6 units/mg, and was free of the α3-sialyltransferase. The enzyme has no requirement for cation.

As indicated in Figure 3, α6-sialyltransferase I acts poorly on {**F2[1]**} and, therefore, the preferential pathway towards {**S6[F2][1]**}, a component of several mucins (see Table 2), involves the prior action of α6-sialyltransferase I to make {**S6[1]**} followed by α2-fucosyltransferase [31,33]. Pigs and other species with the appropriate genotype make mucins carrying the blood group A determinant (see Table 5). Thus, porcine submaxillary mucin from A^+-pigs contains appreciable amounts of GalNAcα1-3(Fucα1-2)Galβ1-3(±sialylα2-6)GalNAc-R. Since α6-sialyltransferase I is totally inactive toward GalNAcα1-3(Fucα1-2)Galβ1-3GalNAc-R, the blood group A-active pentasaccharide must be synthesized by the action of the blood group A-dependent α3-GalNAc-transferase (discussed later) on {**S6[F2][1]**} (Fig. 3) [31,33].

2. *CMP-Sialic Acid:Galβ1-3GalNAc-R (Sialyl to Gal) α3-Sialyltransferase (EC 2.4.99.4)*

Porcine submaxillary glands contain two distinct sialyltransferases acting on *O*-glycans, the α6-sialyltransferase I discussed in the previous section and an α3-sialyltransferase catalyzing the following reaction:

CMP-sialic acid + Galβ1-3GalNAc-R ⟶ sialylα2-3Galβ1-3GalNAc-R + CMP

Both enzymes have been purified to homogeneity. The final preparation of the α3-sialyltransferase from porcine submaxillary glands [274,288,289] was obtained in 5% yield and had a purification factor of 92,000 and a specific activity of 10.6 units/mg protein. The purified enzyme can form the sialylα2-3Gal bond with a variety of galactosyl-terminal acceptors, namely, Galβ1-3GalNAc-Ser(Thr)-proteins (e.g., various asialomucins, antifreeze glycoprotein, asialofetuin), Galβ1-3GalNAc ($K_m = 0.2$ mM), Galβ1-3GlcNAc ($K_m = 85$ mM), Galβ1-4Glc ($K_m = 130$ mM) and ganglioside G_{M1} (to form ganglioside G_{D1a}).

These two enzymes are different from two other sialyltransferases, which act on *N*-glycans and that have both been purified to homogeneity and cloned (i.e., CMP-sialic acid:Galβ1-4GlcNAc-R α6-sialyltransferase, EC 2.4.99.1 [33]; and CMP-sialic acid: Galβ1-3(4)GlcNAc-R α3-sialyltransferase, EC 2.4.99.6 [375,376]). The α6-sialyltransferases acting on *N*- and *O*-glycans synthesize different linkages (i.e., sialylα2-6Gal and sialylα2-6GalNAc, respectively) and, therefore, it is not surprising that the enzymes are different. The α3-sialyltransferases, however, both synthesize the sialylα2-3Gal moiety, but yet they are quite distinct enzymes; evidently these enzymes recognize not only the terminal galactosyl residue of the acceptor, but also the penultimate residue(s). The sialylα2-3Galβ1-4GlcNAc-R structures occurring in *O*-glycans are probably synthesized by the α3-sialyltransferase that acts on *N*-glycans.

The action of the highly purified CMP-sialic acid:Galβ1-3GalNAc-R α3-sialyltransferase on ganglioside G_{M1} implies that the same enzyme acts on both *O*-glycans (EC 2.4.99.4) and gangliosides (EC 2.4.99.2). The sialyltransferase does not, however, act on lactosylceramide, indicating the existence of a third α3-sialyltransferase responsible for synthesis of sialylα2-3Galβ1-4Glc-ceramide (EC 2.4.99.9). The CMP-sialic acid:Galβ1-3GalNAc-R α3-sialyltransferase has been reported in ovine submaxillary glands [26] and must be present in many other tissues to account for the synthesis of **{S3[1]}**, **{S6[S3][1]}**, **{S3[2]}**, and derivatives of **{S3[2]}** (Figs. 2 and 5; see Tables 2 and 3).

As shown in Figure 2, the synthesis of **{S6[S3][1]}** proceeds by the three-before-six rule (i.e., addition of sialic acid in α2-6 linkage to **[S3[1]}**, since α3-sialyltransferase acts poorly on **{S6[1]}** [31.33]). The **{S3[1]}** and **{S6[S3][1]}** structures are found in a large variety of *O*-glycans present on mucins and other secreted glycoproteins, as well as on cell membrane glycoproteins from both tumors and normal cells (see Table 2). It is interesting that both Sendai virus-mediated erythrocyte hemagglutination and infection of bovine kidney cells can be prevented by treating the cells with neuraminidase and restored by treating the neuraminidase-treated cells with CMP-NAN and the α3-sialyltransferase; other sialyltransferases are totally ineffective, implying that the receptor for Sendai virus is the sialylα2-3Gal moiety [219]. This moiety appears to be a receptor for polyomavirus, Newcastle disease virus, an equine influenza A virus, and a human influenza A2 virus, although hemagglutination by the human influenza virus was restored by sialyltransferases other than the α3-sialyltransferase [50, 257].

The CMP-sialic acid:Galβ1-3GalNAc-R α3-sialyltransferase was found in granulocytes from both normal and chronic (CML) and acute (AML) myelogenous leukemia patients, but the activity in CML cells was 2.8 times higher than in the normal cells, explaining the decreased reactivity of CML granulocytes with peanut agglutinin [14]. The α3-sialyltransferase also appears to be induced in leukemia-derived cell lines during differentiation [176]. The human enzyme has been tentatively localized to chromosome 11 [80].

3. *CMP-Sialic Acid:Sialylα2-3Galβ1-3GalNAc-R (Sialyl to GalNAc) α6-Sialyltransferase II (EC 2.4.99.7)*

The synthesis of the tetrasaccharide **{S6[S3][1]}** both in fetal calf liver [359] and in salivary glands [31,33] proceeds by the three-before-six rule (i.e., α3-sialyltransferase acts before α6-sialyltransferase; see Fig. 2). Porcine liver, and probably also rat, human, and canine liver, contain a CMP-NAN:Galβ1-3GalNAc-R α3-sialyltransferase similar to the salivary gland enzyme discussed in the preceding section [24,25,358,377]. However, whereas highly purified porcine submaxillary gland α6-sialyltransferase I can act on GalNAc-protein, Galβ1-3GalNAc-protein and sialylα2-3Galβ1-3GalNAc-protein (as indicated in Fig. 2), the α6-sialyltransferase in fetal calf liver microsomes can act only on the latter acceptor. In addition, the pure salivary gland α6-sialyltransferase I cannot act on the *p*-nitrophenyl derivatives of GalNAc, Galβ1-3GalNAc, and sialylα2-3Galβ1-3GalNAc, whereas fetal calf liver microsomes can incorporate sialic acid in α2-6 linkage to the *N*-acetylgalactosamine residue of sialylα2-3Galβ1-3GalNAc-R, where R can be either protein or phenyl derivatives. The α3-sialyltransferase and α6-sialyltransferase II are also present in brain tissue [19]. Inhibition studies with *N*-ethylmaleimide showed that only the α6-sialyltransferase II contains sulfhydryl groups, near the nucleotide–sugar-binding site, that were protected by binding to CMP-sialic acid.

It is clear that there are at least two distinct α6-sialyltransferases involved in synthesis of *O*-glycans. Both enzymes make the sialylα2-6GalNAc linkage. One enzyme (α6-sialyltransferase I, EC 2.4.99.3) is probably involved in mucin synthesis by porcine and ovine submaxillary glands, is absent from fetal and adult mammalian livers, and can act on R_1-GalNAc-R_2, where R_1 = H-, Galβ1-3-, or sialylα2-3Galβ1-3-, and R_2 = protein. It cannot act on acceptors in which R_2 = *p*-nitrophenyl. The other sialyltransferase (α6-sialyltransferase II, EC 2.4.99.7) is more widely distributed, being present in both fetal calf liver and porcine and ovine submaxillary glands [27]. This enzyme is probably involved in making the **{S6[S3][1]}** tetrasaccharide of mucins and nonmucin glycoproteins such as fetuin (secreted by the liver into the plasma) and glycophorin (present on the human red blood cell membrane) (see Table 2). This enzyme acts only on sialylα2-3Galβ1-3GalNAc-R, but R can be either a protein or *p*-nitrophenyl. It is not known which α6-sialyltransferases are involved in sialylation of core 3 (see Fig. 8) and core 5.

4. GDP-Fuc:β-Galactoside α2- (EC 2.4.1.69), α3- (EC 2.4.1.152), and α3/4- (EC 2.4.1.65) Fucosyltransferases

Table 5 lists some of the antigenic determinants that contain fucose. These determinants occur at the nonreducing ends of *O*-glycans, of *N*-glycans of the poly-*N*-acetyllactosaminoglycan type, and of glycosphingolipids. There are at least four separate fucosyltransferases acting on acceptors with a β-galactoside nonreducing terminus [33,171,268,287,296,301,302,374]: (1) the human blood group H-dependent GDP-L-Fuc:β-galactoside α2-fucosyltransferase (EC 2.4.1.69); (2) the human secretory gene Se-dependent GDP-L-Fuc:β-galactoside α2-fucosyltransferase; (3) the human blood group Lewis-dependent GDP-L-Fuc:Galβ1-3(4)GlcNAc-R (Fuc to GlcNAc) α4(3)-fucosyltransferase (EC 2.4.1.65); and (4) GDP-L-Fuc:Galβ1-4GlcNAc-R (Fuc to GlcNAc) α3-fucosyltransferase (EC 2.4.1.152). All four enzymes have an absolute requirement for a terminal galactosyl residue and, therefore, are different from the GDP-L-Fuc:β-*N*-acetylglucosaminide (Fuc to Asn-linked GlcNAc) α6-fucosyltransferase acting on *N*-glycans [216].

The blood group H enzyme attaches fucose in α1-2 linkage to the terminal galactosyl residue of any β-D-galactoside, both large and small, irrespective of the subterminal sugars. The blood group H determinant occurs on both type 1 and type 2 chains (see Table 5) at the nonreducing termini of *N*- and *O*-glycans and of glycosphingolipids, and directly on the *O*-glycan core 1, 2, 3, and 4 structures to form **{F2[1]}**, **{S6[F2][1]}**, and **{F2[2]}** (see Fig. 3 and Tables 2–4) and other structures that are not listed.

Phenyl β-D-galactoside is a useful specific substrate for the enzyme, since it is inactive with the other known fucosyltransferases. Structures like Galβ1-3(Fucα1-4)GlcNAc-R (the Le^a determinant) or Galβ1-4(Fucα1-3)GlcNAc-R (the X-determinant; see Fig. 4) are not acceptors. The enzyme occurs in the serum and bone marrow (erythrocytes and platelets but not leukocytes) of all humans with the *H* gene irrespective of secretor status. The changes in α2-L-fucosyltransferase levels, previously reported in leukemia, have shown to correlate with platelet counts and are not diagnostic markers for leukemia [326]. Rare individuals with the *hh* genotype (Bombay O_h) lack the enzyme. A similar enzyme occurs in other species (e.g., dog, pig, rat, and cow). The enzyme has been purified 124,000-fold to homogeneity from pig submaxillary glands [30,32,33] using repeated affinity chromatography on GDP-hexanolamine-agarose. The final preparation was obtained in 6% yield with a specific activity of 20 U/mg protein. The H-dependent α2-fucosyltransferase has also been purified 10 million-fold from human serum. An extremely effective step in the purification procedure, achieving separation from α3-fucosyltransferase, Se-gene-encoded α2-fucosyltransferase, and other proteins, was the binding to the cation-exchange column S-Sepharose [292]. The native enzyme consists of several subunits and has an M_r of 148,000, as determined by gel filtration. The human *H*-dependent enzyme has been cloned and expressed in several mammalian systems and has been localized to chromosome 19 [94,204,270].

The presence of α2-fucosyltransferase activity, similar to the blood group H-dependent enzyme, in the sera of individuals with the 'para-Bombay' phenotype who secrete H-substances (*Se* genotype), but have H-negative erythrocytes (*hh* genotype) [207], led to the concept that the *Se* gene controls the appearance of a second α2-fucosyltransferase distinct from the *H*-dependent enzyme [197,206]. This enzyme accounts for the presence of α2-fucosyltransferase activity in secretory tissues of *Se* individuals (i.e., human milk, submaxillary glands, and gastric mucosa). The *H* gene is not expressed in secretory tissues, and the *Se* gene is totally suppressed in hematopoietic tissues. Individuals with the rare Bombay O_h phenotype, who are lacking H antigens in both erythrocytes and secretory tissues, have both the *hh* and *sese* genotypes. Although both the *H*- and *Se*-dependent α2-fucosyltransferases can utilize both type 1 and type 2 acceptors, the *Se*-enzyme prefers type 1 substrates, whereas the *H*-enzyme prefers type 2 [28,29].

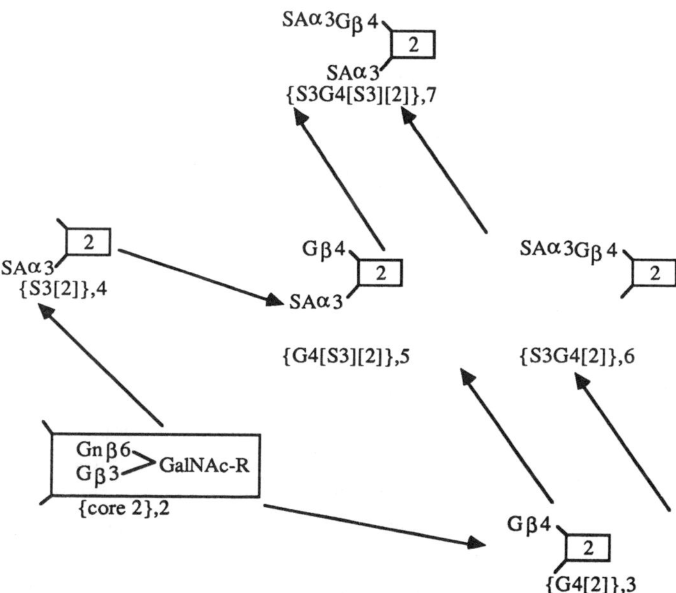

Figure 5 Assembly of some sialylated core 2 *O*-glycans. The numbered rectangular boxes represent the core structures. The names in braces { } are defined in the figure and are used to refer to structures in the text. The reactions in the figure have not been studied in cell-free preparations. These sialylated *O*-glycans have been found on mucins, on tumor cells, and on both secreted and membrane-bound glycoproteins (see Table 3). Glycans have been isolated from monkey cervical mucin with the Cad, H, A, and B determinants (see Table 5) attached to **{S3G4[2]}**. Abbreviations: as in the legends to Figs. 1 and 2.

The blood group Lewis enzyme attaches fucose in α1-4 linkage to the subterminal *N*-acetylglucosamine of Galβ1-3GlcNAc-R (type 1) structures (see Table 5). The enzyme has been found in the submaxillary glands, gastric mucosa, cervical epithelium, and milk of individuals with the Lewis (*Le*) gene. The enzyme is not present in human serum, irrespective of the Lewis genotype. The best substrates for specific assay of the enzyme are milk oligosaccharides, like lacto-*N*-fucopentaose I (Fucα1-2Galβ1-3GlcNAcβ1-3Galβ1-4Glc) that are not acceptors for the other known fucosyltransferases. The enzyme has been purified over 500,000-fold to apparent homogeneity from human milk by affinity chromatography on GDP-hexanolamine-agarose [33,268]. The highly purified enzyme carries out all three of the following reactions:

GDP-Fuc + Galβ1-3GlcNAc-R ⟶ Galβ1-3(Fucα1-4)GlcNAc-R + GDP

GDP-Fuc + Galβ1-4GlcNAc-R ⟶ Galβ1-4(Fucα1-3)GlcNAc-R + GDP

GDP-Fuc + Galβ1-4Glc ⟶ Galβ1-4(Fucα1-3)Glc + GDP

The K_m values for these three substrates are, respectively, 1.9, 1.6, and 59 mM. The α1-3- and α1-4-activities copurify, are activated to the same extent by various cations, and are inactivated at identical rates by various procedures. Kinetic analysis indicates that a single active site is involved in both α1-3- and α1-4-Fuc incorporation. The human α3/4-fucosyltransferase cDNA has been cloned by gene transfer into mammalian cells and screening enzyme product in expressing cells with anti SSEA-1 antibodies [196]. The gene for this enzyme, like the H-dependent α2-Fuc-transferase, was localized to chromosome 19. The cloned enzyme carried out both α1-3 and α1-4 Fuc incorporation. This is a rare example of a single glycosyltransferase catalyzing sugar attachment to two different carbon atoms at appreciable rates and, therefore, is an exception to the one linkage–one glycosyltransferase rule. Enzyme activity is enhanced by the presence of a Fucα1-2-residue on the terminal galactose of the substrate. The enzyme is inactive toward substrates with the structure Galβ1-3GalNAc-Ser(Thr), but *O*-glycans with class 1 core structures may carry the Lewis determinant at Galβ1-3GlcNAc termini (e.g., in pig gastric mucin and human ovarian cyst mucin).

Saliva from individuals with the *Le* gene can catalyze the three reactions shown [171]. Saliva from *Le*-negative individuals, however, can catalyze only the attachment of fucose in α1-3 linkage to *N*-acetylglucosamine. There is, therefore, a fourth enzyme, GDP-L-Fuc:Galβ1-4GlcNAc (Fuc to GlcNAc) α3-fucosyltransferase, which is different from the *H*-, *Se*-, and *Le*-dependent fucosyltransferases. This enzyme and the *Le*-dependent enzyme can be separated by isoelectric focusing of human saliva from *Le*-positive individuals. Human serum contains the non–*Le*-dependent α3-fucosyltransferase but lacks completely the *Le*-dependent α3/4-fucosyltransferase even in *Le*-positive individuals. The non–*Le*-dependent α3-fucosyltransferase in serum and saliva acts only on Galβ1-4GlcNAc-R and cannot add fucose in α1-4 linkage to Galβ1-3GlcNAc nor in α-3 linkage to Galβ1-4Glc. Pig liver contains a GDP-L-Fuc:β-galactoside fucosyltransferase that is highly active toward Galβ1-4GlcNAc, but not toward Galβ1-3GlcNAc, Galβ1-6GlcNAc, nor Galβ1-4Glc [165]. It thus resembles the human non–*Le*-dependent α3-fucosyltransferase in substrate specificity.

The GDP-L-Fuc:Galβ1-4GlcNAc (Fuc to GlcNAc) α3-fucosyltransferase has been purified about 2000-fold from mouse F9 embryonal carcinoma cells [230]. The pure enzyme showed an M_r of 65,000 on SDS-PAGE and did not contain detectable GDP-L-Fuc:β-galactoside α2-fucosyltransferase nor GDP-L-Fuc:Galβ1-3(4)GlcNAc-R (Fuc to

GlcNAc) α4(3)-fucosyltransferase. The enzyme has also been purified from human neuroblastoma cells [101] and from human milk [93]. The milk enzyme has an M_r of 44,000 Da.

The structure Galβ1-4(Fucα1-3)GlcNAc (see Fig. 4) has been called the X or the SSEA-1 determinant (mouse stage-specific embryonic antigen, see Table 5) and occurs in complex *N*-glycans (e.g., in human serum α1-acid glycoprotein), in glycosphingolipids, and in *O*-glycans with core classes 1–4 (see Tables 2–4). The Y determinant (see Fig. 4 and Table 5) has also been detected in *O*-glycans with core classes 1–4 (see Tables 2–4). The sequences sialylα2-3Galβ1-4GlcNAc- and sialylα2-3Galβ1-4(Fucα1-3)GlcNAc- (sialylated X; see Fig. 4) usually occur in *N*-glycans and glycosphingolipids, but have also been detected in *O*-glycans of core classes 1–4 (see Tables 2–4). Sialylated X is of special interest as a tumor-associated antigen [148,108].

The pattern of synthesis of the sialylα2-3Galβ1-4GlcNAc-, H, X, Y, and sialylated X determinants is shown in Figure 4. The starting point is a glycan with a type 2 disaccharide at its nonreducing terminus (Galβ1-4GlcNAc-R). Such termini are provided by {**G4Gn3[1]**} (see Fig. 3), {**G4[S3][2]**} and {**G4[2]**} (see Fig. 7), and {**G4[4]**} and {**G4[G4][4]**} (see Fig. 9). Analogous pathways can be drawn for synthesis of the Le^a and Le^b determinants, using a type 1 starting point (Galβ1-3GlcNAc-R). It is interesting to note from Figure 4 that neither the α3-sialyltransferase nor the α2-fucosyltransferase can act on the Galβ1-4(Fucα1-3)GlcNAc-terminus and that, therefore, both of these enzymes must act before the α3-fucosyltransferase. Both the Le-independent α3-fucosyltransferase and the Le-dependent α3/4 fucosyltransferase from normal human tissues have been shown to act on NeuAcα2-3Galβ1-4GlcNAc and on Fucα1-2Galβ1-4GlcNAc [169], but enzyme from leukemic granulocytes could not use NeuAcα2-3Galβ1-4GlcNAc as an acceptor [170]. There may, in fact, be two different Le-independent α3-fucosyltransferases, only one of which can act on sialylated substrates [51,157,229].

Expression of GDP-L-Fuc:Galβ1-4GlcNAc-R (Fuc to GlcNAc) α3-fucosyltransferase (but not of CMP-sialic acid:Galβ1-4GlcNAc-R α2-3 sialyltransferase), and of the X, Y, and sialylated X determinants, has been localized to human chromosome 11 by use of human –mouse myeloid cell hybrids [80,349]. This finding suggests that the expression of α3-fucosyltransferase (and not of α2-3 sialyltransferase) controls the expression of the sialylated X determinant and is independent of the α3/4-Fuc-transferase (on chromosome 19). However, the expression of X and sialylated X determinants in human colonic adenocarcinoma is due to neither the α3-fucosyltransferase nor the α2-3 sialyltransferase, but rather, to a precursor enzyme, UDP-GlcNAc:Galβ1-4GlcNAc-R (GlcNAc to Gal) β3-GlcNAc-transferase (EC 2.4.1.149) discussed earlier [146–148].

5. Blood Group A- and B-Dependent α3-GalNAc- and α3-Gal- Transferases

The determinants for the human blood groups A and B (see Table 5) occur on *N*- and *O*-glycans and on glycosphingolipids. These determinants have also been found in other species. The enzymes responsible for the synthesis of these determinants are, respectively, UDP-GalNAc:Fucα1-2Gal-R α3-N-acetylgalactosaminyltransferase (EC 2.4.1.40) and UDP-Gal:Fucα1-2Gal-R α3-galactosyltransferase (EC 2.4.1.37) [33,287,302,374]. Both enzymes have a very similar substrate specificity (i.e., they require the *H* determinant (see Table 5) at the nonreducing terminus of the substrate). The enzymes can, however, act on the disaccharide Fucα1-2Gal, as well as on larger oligosaccharides, protein-bound *N*- and *O*-glycans, and glycosphingolipids, provided there is Fucα1-2Gal at the terminus. In some species (e.g., rabbit bone marrow [16]) analogous glycosyltransferases do not require the presence of the α2-linked fucose residue.

Much of the early structural work on the human blood group antigens was carried out on the water-soluble ovarian cyst mucins [284,374,386]. Other glycoproteins with core class 1 *O*-glycans carrying either A or B blood group determinants are porcine submaxillary and gastric mucins, rat intestinal and salivary mucins, and ovine and equine gastric mucins (see Table 1 for references).

The A-dependent GalNAc-transferase has been purified from human serum [235,378], from porcine submaxillary glands [306,307], and from human lung [67] and gastric and duodenal scrapings [244]. The porcine submaxillary gland enzyme was purified 38,000-fold to homogeneity by repeated chromatography of Triton X-100 extracts on UDP-hexanolamine-agarose. The purified enzyme was obtained in 13% yield and had a specific activity of 30 U/mg protein. The human gut enzyme was purified 125,000-fold on UDP-hexanolamine-agarose and octyl-Sepharose and showed an M_r of 40,000 Da [244]. The B-dependent α3-Gal-transferase has been purified 400,000-fold from human serum [236]. The A and B enzymes have overlapping substrate specificities under certain conditions in vitro [118,392,393] (i.e., the A enzyme may transfer galactose from UDP-Gal to 2′-fucosyllactose and the B enzyme may transfer GalNAc from UDP-GalNAc to the same acceptor). These activities do not, however, appear to be functional under physiological conditions in vivo, although they may account for the anomalous expression of blood group antigenic activity by certain human tumors. These activities are examples of a single glycosyltransferase catalyzing different reactions and, therefore, are exceptions to the one linkage–one glycosyltransferase rule.

The genes for the human blood group A and B transferases have been cloned based on peptide sequences from the human A enzyme [387,388]. Interestingly, the *A* and *B* genes show a small number of base pair substitutions; the enzyme proteins, therefore, differ only in four amino acids. This apparently changes the specific recognition of UDP-GalNAc by the A enzyme to UDP-Gal by the B enzyme [388]. The *O* gene does not express either the A or B enzymes, but is structurally identical with the *A* and *B* genes, except for a base pair deletion causing a shift in the reading frame; this leads to the production of an entirely different protein, which lacks A or B activity [387].

The blood group A-dependent α3-GalNAc-transferase appears to be different from two GalNAc-transferases synthesizing the Forssman antigenic terminus GalNAcα1-3GalNAcβ1-3Gal- (see Table 5) on glycosphingolipids. The sequence of GalNAcα1-3GalNAcβ1-3Gal- has been reported on a core 1 *O*-glycan isolated from canine gastric mucin [327]. It is not known whether the Forssman GalNAc-transferases responsible for synthesis of the glycolipids can also act on *O*-glycans.

6. *Blood Group Cad-Dependent UDP-GalNAc:sialylα2-3Galβ-R (GalNAc to Gal) β4-GalNAc-Transferase*

Cad is a rare blood group carbohydrate antigen carried by both sialoglycoproteins and gangliosides of human erythrocytes. The major *O*-glycan (71%), isolated from Cad glycophorin A by reductive β-elimination, is the pentasaccharide GalNAcβ1-4(NeuAcα2-3)Galβ1-3(NeuAcα2-6)GalNAc-ol [35,137], whereas glycophorin A from Cad-negative red blood cells has primarily the tetrasaccharide structure **{S6[S3][1]}** (see Fig. 2). The Cad antigen is serologically related to the Sd^a blood group antigen, but the structure responsible for Sd^a reactivity on the red cell surface is not known [34].

Conzelmann and Kornfeld [71] reported the presence of *O*-glycans with GalNAcβ1-4Gal-termini in a mouse cytotoxic T-lymphocyte line. A mutant line resistant to the lectin from *Vicia villosa* lacked terminal GalNAcβ1-4-residues. The parent line contains a UDP-GalNAc:sialylα2-3Galβ-R (GalNAc to Gal) β4-GalNAc-transferase, which is absent in the *V. villosa* lectin-resistant line [72]. This enzyme transferred GalNAc in β-linkage to both *O*-glycans of human glycophorin A and *N*-glycans of Tamm–Horsfall glycoprotein, probably synthesizing the GalNAcβ1-4(NeuAcα2-3)Galβ-terminus on both the *N*- and *O*-glycans. The presence of the sialyl residue is essential for β4-GalNAc-transferase activity. Conzelmann and Bron [70] devised several clever techniques for assaying the UDP-GalNAc:sialylα2-3Galβ-R β4-GalNAc-transferase in crude cell and tissue extracts. They used biotinylated human glycophorin A as acceptor and purified the product by adsorption to avidin–agarose. Mouse intestinal tissue and most cytotoxic murine T lymphocytes were found to have high levels of the transferase, as expected from their reactivity with the *V. villosa* lectin; noncytolytic T-cell lines generally had low enzyme levels. The enzyme is also active in the cecum and the colon, but not the ileum, of the rat intestine, and it appears to undergo specific patterns of expression during cell differentiation and postnatal development [78].

The UDP-GalNAc:sialylα2-3Galβ-R (GalNAc to Gal) β4-GalNAc-transferase has been studied in human kidney [260], guinea-pig kidney [309], and human plasma [346]. All three enzymes can utilize NeuAcα2-3Galβ1-4Glc and NeuAcα2-3Galβ1-4GlcNAc-glycoprotein (fetuin and Tamm–Horsfall glycoprotein) as acceptors, but they do not work on ganglioside G_{M3} (NeuAcα2-3Galβ1-4Glc-ceramide). The human kidney and plasma enzymes have been shown to act on sialosylparagloboside (NeuAcα2-3Galβ1-4GlcNAcβ1-3Galβ1-4Glc-ceramide). The guinea pig kidney enzyme can incorporate GalNAc into native glycophorin A (i.e., NeuAcα2-3Galβ1-3 (± NeuAcα2-6)GalNAc-protein), but the human plasma enzyme acts poorly on glycophorin A, whereas the human kidney enzyme cannot use glycophorin at all. This difference is not understood but may be related to the fact that the human enzymes act poorly on the NeuAcα2-3Galβ1-3GalNAc-terminus. It is interesting to note that the rare Cad determinant is GalNAcβ1-4(NeuAcα2-3)Galβ1-3(NeuAcα2-6)GalNAc-Ser(Thr)-X on glycophorin A, whereas the commonly found Sd^a determinant is GalNAcβ1-4(NeuAcα2-3)Galβ1-4GlcNAcβ1-3Gal- attached to an *N*-glycan of urinary Tamm–Horsfall glycoprotein; the human plasma and kidney enzymes described in the foregoing, therefore, may not be involved in Cad determinants synthesis, but rather, in Sd^a determinant synthesis. Recent reports [84,312] indicate the presence in urine from Sd(a+) donors of a UDP-GalNAc:sialylα2-3Galβ-R (GalNAc to Gal) β4-GalNAc-transferase; urine from an Sd(a-) donor had no detectable activity.

VII. ASSEMBLY OF OLIGOSACCHARIDES WITH CORE CLASS 2

A. Synthesis of the Core: UDP-GlcNAc:Galβ1-3GalNAc-R (GlcNAc to GalNAc) β6-*N*-Acetylglucosaminyltransferase (Ec 2.4.1.102)

The GlcNAcβ1-6(Galβ1-3)GalNAc-R core is synthesized from Galβ1-3GalNAc-R by an enzyme described in canine submaxillary glands [380,381], rabbit intestine [382], bovine trachea [66], porcine gastric mucosa and rat colon [42,43,46], human granulocytes [41], and other tissues. The enzyme catalyzes the following reaction:

UDP-GlcNAc + Galβ1-3GalNAc-R ⟶
GlcNAcβ1-6(Galβ1-3)GalNAc-R + UDP

O-Glycans with core class 2 have been described in many glycoproteins (see Tables 1 and 3). Core class 2 structures were not found in canine submaxillary mucin [215], although canine submaxillary glands are very rich in the β6-GlcNAc-transferase.

Like all β6-GlcNAc-transferases, the core 2 β6-GlcNAc-transferase is active in the absence of Mn^{2+} and in the presence of EDTA. The enzyme obeys the three-before-six rule (i.e., it requires a 3-substituted GalNAc terminus on its substrate). The enzyme will work on the disaccharide Galβ1-3GalNAc-R (R = H, benzyl, *o*-, or *p*-nitrophenyl, methyl) as well as on Galβ1-3GalNAc-Ser(Thr)-R, where R is derived from mucin or antifreeze glycoprotein. The enzyme is totally inactive on GalNAc-Ser(Thr)-mucin, indicating that the UDP-Gal:GalNAc-mucin β3-Gal-transferase must act before the β6-GlcNAc-transferase, as indicated in Figure 2. However, an alternative pathway to core 2 synthesis is possible by the action of core 1 β3-Gal-transferase on core 6 (see Sect. IX) [unpublished]. The presence of a sialic acid residue (e.g., in native calf fetuin, Sialylα2-3Galβ1-3GalNAc) inhibits enzyme action [380]; the presence of a fucose residue (i.e., Fucα1-2Galβ1-3GalNAc-R; {**F2[1]**}; see Fig. 3), makes β6-GlcNAc-transferase action less favorable. The tetrasaccharide {**F2[2]**} (see Fig. 3) and the pentasaccharide {**G4[F2][2]**} (see Fig. 7) have been reported in several *O*-glycans (see Table 3); these structures are probably synthesized by the addition of Fucα1-2 and Galβ1-4 residues after the completion of core 2, as indicated in Figures 3 and 7. Substitution of core 1 makes conversion to core 2 unlikely (see Figs. 2 and 3).

The core 2 β6-GlcNAc-transferase was fourfold increased over normal values in granulocytes from CML patients and 18-fold in leukocytes from AML patient [41a]. The AML and CML leukocytes represent different stages of differentiation, whereas normal granulocytes are more mature cells. CaCo-2 human adenocarcinoma cells that spontaneously differentiate in culture show an increase of the β6-GlcNAc-transferase upon differentiation [41b]. The enzyme activity increases also when human lymphocytes are activated by anti CD3 antibodies and interleukin-2 [263]. Therefore, it appears that the β6-GlcNAc-transferase is specifically induced during the process of cell differentiation.

B. Elongation of the Core: Synthesis of the Ii Backbone

The elongation of the class 2 core is very similar to the process described earlier for class 1 core (see Sect. VI.B) (i.e., the process involves addition of *N*-acetylglucosaminyl and galactosyl residues in β1-3, β1-4 and β1-6 linkages). The process is more complex than for class 1 cores because elongation can occur both at the galactosyl and at the *N*-acetylglucosaminyl residues of the class 2 core. The branched structure GlcNAcβ1-6(GlcNAcβ1-3)Gal- is usually attached to the galactosyl moiety of core 2 as indicated in Figures 2 and 6, but has also been reported on the GlcNAc moiety (e.g., in human glycophorin [3]).

We have found in porcine gastric mucosa [45,47] an elongation enzyme, UDP-GlcNAc:R_1β1-6(Galβ1-3)GalNAc-R_2 (GlcNAc to Gal) β3-*N*-acetylglucosaminyltransferase (EC 2.4.1.146) that catalyzes the reaction

$$\text{UDP-GlcNAc} + R_1\beta1\text{-}6(\text{Gal}\beta1\text{-}3)\text{GalNAc-}\alpha\text{-}R_2 \longrightarrow R_1\beta1\text{-}6(\text{GlcNAc}\beta1\text{-}3\text{Gal}\beta1\text{-}3)\text{GalNAc-}\alpha\text{-}R_2 + \text{UDP}$$

When R_1 is GlcNAc, R_2 can be *o*-nitrophenyl, benzyl, or polypeptide from mucin or antifreeze glycoprotein. When R_1 is H, however, R_2 must be derived from mucin or antifreeze glycoprotein. The enzyme shows a strict requirement for divalent cation (Mn^{2+}

or Co^{2+}) and is markedly stimulated by addition of Triton X-100, suggesting that, like most glycosyltransferases, the elongation β3-GlcNAc-transferase is membrane bound. Since canine submaxillary glands are rich in the UDP-GlcNAc:Galβ1-3GalNAc-R (GlcNAc to GalNAc) β6-GlcNAc-transferase responsible for core 2 synthesis (see Sect. VII.A) [380,381], but lack the elongation β3-GlcNAc-transferase activity, the β3- and β6-enzymes must be different [47].

The same elongation β3-GlcNAc-transferase is believed to act on both core 1 (see Sect. VI.B.1) and core 2 oligosaccharides. The metal ion requirements, tissue distribution, and substrate specificity of this enzyme differ from those of the β3-GlcNAc-transferase acting on Galβ1-4GlcNAc-R or Galβ1-4Glc-R to produce blood group i determinants (EC 2.4.1.149; see Sect. VI.B.2), and from those of the β3-GlcNAc-transferase that converts GalNAc-Ser(Thr) to GlcNAcβ1-3GalNAc-Ser(Thr) (core 3, EC 2.4.1.147; see sect. VIII) [42,46]. Thus, there are at least three different β3-GlcNAc-transferases acting on *O*-glycans.

Derivatives containing the structures {**Gn3[2]**} and {**Gn6[Gn3][2]**} (see Figs. 2 and 6) have been found in many core 2 *O*-glycans (see Table 3). We have shown [43] that porcine gastric mucosal extracts can transfer GlcNAc in β1-6 linkage to both core 1 and core 2 derivatives, as follows (R = benzyl):

GlcNAcβ1-3Galβ1-3GalNAc-α-R ⟶ GlcNAcβ1-6(GlcNAcβ1-3)Galβ1-3GalNAc-α-R
{**Gn3[1]**} {**Gn6[Gn3][1]**}

GlcNAcβ1-6(GlcNAcβ1-3Galβ1-3)GalNAc-α-R ⟶
{**Gn3[2]**}
GlcNAcβ1-6[GlcNAcβ1-6(GlcNAcβ1-3)Galβ1-3]GalNAc-α-R
{**Gn6[Gn3][2]**}

Competition data suggest that the β6-GlcNAc-transferase activities that convert GlcNAcβ1-3Galβ1-4Glc(NAc)-R to GlcNAcβ1-6(GlcNAcβ1-3)Galβ1-4Glc(NAc)-R, core 1 to core 2 (see Sect. VII.A; EC 2.4.1.102) and core 3 to core 4 (see Section VIII; EC 2.4.1.148) may all be due to a single enzyme in pig gastric mucosa. The core 2 β6-GlcNAc-transferase from CML and AML cells, however, cannot make either core 4 or the GlcNAcβ1-6(GlcNAcβ1-3)Galβ- branch [41], suggesting that the enzyme in leukocytes is different from the enzyme in mucus-secreting tissues. Enzyme purification is, however, required to establish this.

C. Addition of α-Linked Sugars

The termination of core class 2 oligosaccharides probably follows the same rules as outlined in Section VI.C for class 1 oligosaccharides, although this has not been studied. Addition of α-linked sialyl, fucosyl, *N*-acetylgalactosaminyl and galactosyl residues can occur on oligosaccharides attached both to the galactosyl and *N*-acetylglucosaminyl residues of the core. Some of the glycoproteins containing sialylated (see Fig. 5) and fucosylated (see Figs. 3 and 7) core 2 oligosaccharides are listed in Table 3. The Lewis determinant (see Table 5) has been found on core 2 oligosaccharides in human submaxillary and ovarian cyst mucins and other glycoproteins. The X and Y determinants (see Table 5) have also been found on core 2 oligosaccharides as indicated in Table 3; the synthesis of these determinants is indicated in Figure 4 using either {**G4[S3][2]**} or {**G4[2]**} as starting points (see Fig. 7).

Core class 2 oligosaccharides terminated with blood group A or B determinants have been described in ovine and equine gastric mucin, human ovarian cyst mucin, and porcine tracheal mucin. Finally, there have also been reports of termination by α-linked *N*-acetylglucosaminyl residues in rat and porcine gastric and duodenal mucins, and in human ovarian cyst mucin and meconium.

VIII. ASSEMBLY OF OLIGOSACCHARIDES WITH CORE CLASSES 3 AND 4

The glycoproteins that contain oligosaccharides of core classes 3 and 4 are listed in Table 1. These cores are synthesized by two GlcNAc-transferases that have been studied in rat colon mucosal extracts [42,46] and that catalyze the following reactions in the presence of Triton X-100 and Mn^{2+}:

UDP-GlcNAc + GalNAc-α-R ⟶ GlcNAcβ1-3GalNAc-αR + UDP

UDP-GlcNAc + GlcNAcβ1-3GalNAc-α-R ⟶
GlcNAcβ1-6(GlcNAcβ1-3)GalNAc-α-R + UDP

R can be phenyl, benzyl, or polypeptide derived from mucin. In rat colonic mucosa the β6-GlcNAc-transferase making core 4 (EC 2.4.1.148) proceeds at an appreciably faster rate than the β3-GlcNAc-transferase making core 3 (EC 2.4.1.147). The core 4 enzyme can be readily assayed with GlcNAcβ1-3GalNAc-α-R and does not require Mn^{2+}. The tissue distribution (Table 6) and kinetic properties of the core 3 β3-GlcNAc-transferase are markedly different from those of the β3-GlcNAc-transferase that elongates core classes 1 and 2. These two enzymes are, therefore, different. The core 2 and 4 β6-GlcNAc-transferase activities, however, may be due to the same enzyme in mucin-secreting tissues, as they occur in parallel (see Table 6). The core 4 activity is absent from lymphoid cells [41] and from CaCo-2 human colonic adenocarcinoma cells [41b]. These tissues are rich in core 2 activity indicating that there are at least two different core 2 β6-GlcNAc-transferases.

The β3-Gal-transferase reported by Sheares and Carlson [319] has a preference for core class 3, but also acts on core class 2. No other elongation or termination reaction for core class 3 and 4 oligosaccharides has been studied in vitro. In general, most core 3 (Fig. 8) and 4 oligosaccharides show features similar to core class 1 and 2 oligosaccharides. Glycoproteins with sialylated core class 3 oligosaccharides are listed in Table 4. The sialylα2-6GalNAc-mucin is not an acceptor for the core 3 β3-GlcNAc-transferase (see Fig. 2) [42]; therefore, the synthesis of the sialylated core 3 structure {**S6[3]**} [i.e., GlcNAcβ1-3(sialylα2-6)GalNAc-mucin] probably occurs through the action of α6-sialyltransferase on GlcNAcβ1-3GalNAc-mucin (see Fig. 2). The latter reaction has not, however, been demonstrated in vitro. Fucosylated core class 3 oligosaccharides occur in rat colonic, monkey cervical, and human bronchial and colonic mucins. Blood group A determinants are found in rat colonic mucin.

Core class 4 oligosaccharides have been described only in human bronchial and ovine gastric mucins and human meconium (see Table 1), although the core 4 β6-GlcNAc-transferase is a commonly found enzyme (see Table 6). The reason for this may be the limiting activities of the core 3 β3-GlcNAc-transferase (see Table 6), or the presence of a highly active α6-sialyltransferase competing for common substrates with the core 4 β6-GlcNAc-transferase. Some of the sialylated and fucosylated core 4 oligosaccharides found in human bronchial mucin are shown in Figures 9 and 10. The core 4 oligosaccharides containing the

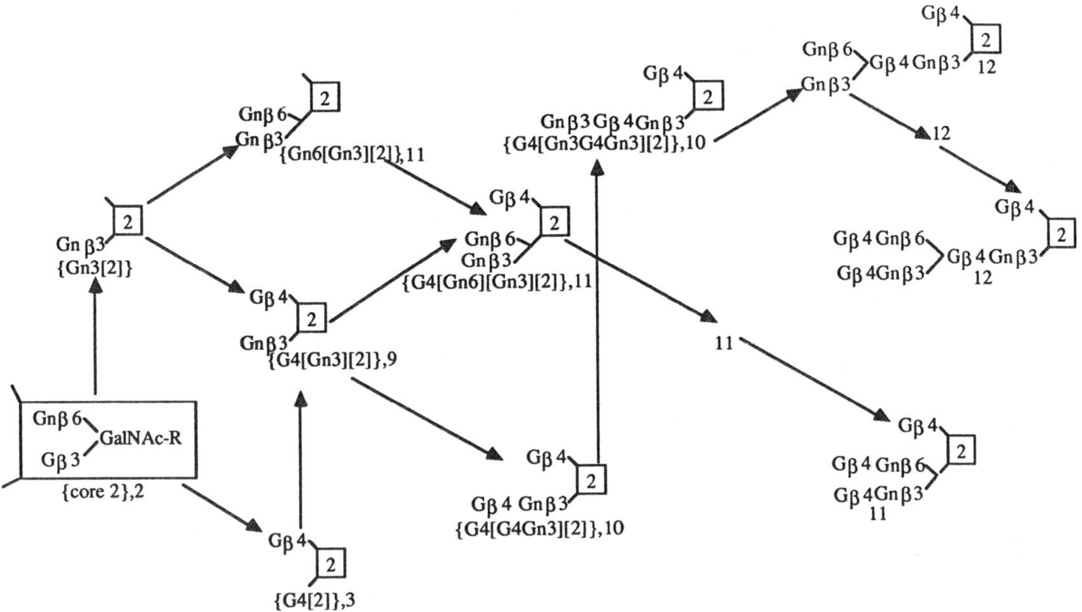

Figure 6 Assembly of some core 2 *O*-glycans. The numbered rectangular boxes represent the core structures. The names in braces { } are defined in the figure and are used to refer to structures in the text. Only the conversions of core 2 to **{Gn3[2]}** and **{Gn6[Gn3][2]}** were studied in cell-free preparations. These structures are based on fucosylated and sialylated *O*-glycans isolated from porcine gastric and tracheal mucins and human ovarian cyst mucin (see Table 3; see Table 1 for references). Some of these glycans have been shown to carry the blood group A and H determinants (see Table 5). Abbreviations: as in the legends to Figs. 1 and 2.

X and Y determinants are probably synthesized by the pathway shown in Figure 4 using **{G4[4]}** and **{G4[G4][4]}** as starting points.

IX. ASSEMBLY OF OLIGOSACCHARIDES WITH CORE CLASSES 5 AND 6

The core class 5 structure GalNAcα1-3GalNAc-Ser(Thr)-R has been reported on human meconium and ovarian cyst mucins, and dog and collocalia salivary mucins (see Table 1). Kurosaka et al. [198] have reported that detergent extracts of human intestinal cancerous tissues contain a UDP-GalNAc:GalNAc-mucin α-*N*-acetylgalactosaminyltransferase capable of synthesizing this core structure in vitro. The enzyme acts on asialo-bovine submaxillary mucin in the presence of Mn^{2+}, but the linkage position in the GalNAcα-GalNAc-enzyme product has yet to be demonstrated. Nothing is known about the elongation reactions for core 5.

Core class 6 structure GlcNAcβ1-6GalNAc-Ser(Thr)-R has been reported only on human glycoproteins (see Table 3), suggesting that human tissues may contain a β6-GlcNAc-transferase which acts directly on GalNAc-R (see Fig. 2) and does not obey the three-before-six rule (i.e., does not require the prior synthesis of the Galβ1-3GalNAc- or GlcNAcβ1-3GalNAc-sequence). However, it is important to rule out degradation of oligosaccharides. An activity synthesizing the GlcNAcβ6-GalNAc-linkage, and not requiring prior 3-substitution of GalNAc-R, has been reported in human ovarian tissue [395]. The reason for the low abundance of core 6 in *O*-glycans may be the activity of the core 1 β3-Gal-transferase that converts core 6 to core 2 [unpublished]. It is not known if the enzyme that synthesizes core 6 also synthesizes the GlcNAcβ1-6 Galβ-linkage found in human glycoproteins; this enzyme was discussed in Section VI.B.3. Elongated core 6 structures are similar to those for cores 1 and 2, but the reactions have not been studied.

X. SULFATION OF *O*-GLYCANS

Sulfate may be linked to various positions at terminal or internal residues of *O*-linked oligosaccharides (see Tables 1–3). A portion of probably most mucins is sulfated; structural analysis, however, has been difficult in the past owing to the lability of the sulfate–sugar linkages. The degree of sulfation has been found to be altered in diseases such as cancer, cystic fibrosis, and inflammation, but the role of sulfate is not known. Rat gastric mucins can be sulfated by a Golgi-localized sulfotransferase; the rat gastric enzyme adds sulfate to the *N*-acetylglucosaminyl residue of the elongated core 1 structure to form SO_3H-6GlcNAcβ1-3Galβ1-3GalNAc- [59]. Radiolabeling studies showed that sulfation of rat adenocarcinoma cell surface mucin is a relatively late event and probably occurs in the same compartment as β4-galactosylation, but before sialylation [159].

XI. CONTROL OF SYNTHESIS

The control of protein and nucleic acid synthesis depends primarily on the assembly of exact copies on templates. A template biosynthetic mechanism is feasible, however, only with linear structures like proteins and nucleic acids. One of the most characteristic and

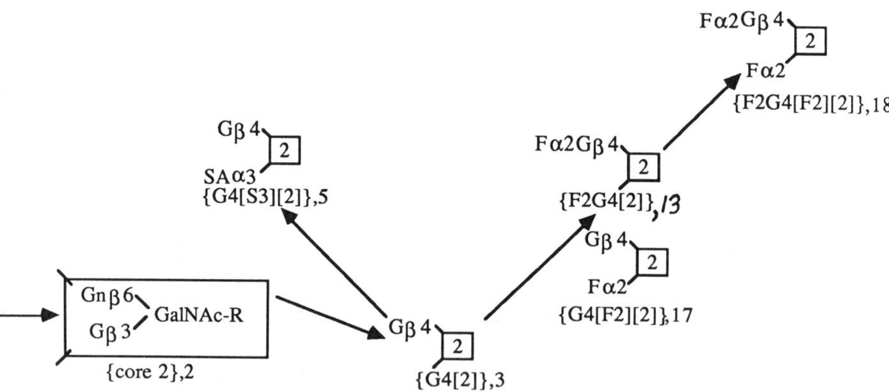

Figure 7 Assembly of some core 2 *O*-glycans. The numbered rectangular boxes represent the core structures. The names in braces { } are defined in the figure and are used to refer to structures in the text. Only the reaction from core 2 to **{G4[2]}**, but none of the remaining reactions in the figure, has been studied in vitro. These structures are based on *O*-glycans isolated from human ovarian cyst and bronchial mucins (references in Table 1). Glycans have been isolated from several glycoproteins (see Table 3) with the X and Y determinants (see Table 5) on **{G4[S3][2]}** and **{G4[2]}** (see Fig. 4). Abbreviations: as in the legends to Figs. 1 and 2.

potentially very important features of complex carbohydrates is their branched structure. The control of the synthesis of such a structure cannot be by a template mechanism. Rather, control is exerted through a variety of factors [297]. Obviously, the genes must control complex carbohydrate synthesis, but they do so by being structural genes for the glycosyltransferases and for protein backbones of the glycoproteins, and by determining the structure of the endomembrane system on which glycosylation takes place.

There are numerous situations that show fluctuations in glycosyltransferase activities. These include normal biological events, such as cellular differentiation, lymphocyte activation, maturation, and growth. Many pathological conditions, such as cancer, inflammatory disease, and cystic fibrosis, show changes in *O*-glycan structures, as well as in glycosyltransferase activities. The expression of enzymes also varies among species, organs, and cell types. Although several genes for glycosyltransferases acting on *O*-glycans have been cloned, we do not yet know how these genes are controlled.

The major known factors controlling synthesis are the substrate specificities and availability of a large number of glycosyltransferases. However, many other factors, such as nucleotide–sugar and acceptor availability, cations, and subcellular membrane organization, undoubtedly influence the final product. Also the rate of transit through the endomembrane system may be faster than the ability of the glycosylation apparatus to add sugars, thereby resulting in incomplete chains. Various questions arise that are difficult to answer at the present stage of knowledge. We do not understand why *O*-glycosyl oligosaccharides present such a bewildering variety of structures, even on a single glycoprotein. What determines the core class? What determines the degree of elongation? Why are some structures branched and others linear? Why does branching occur at several different points? Why are some structures terminated by α-linked sugars and others are not? What factors determine which α-linked sugars terminate a structure and the points of addition of the terminating sugars?

O-Glycan structures and the glycosyltransferases involved in their synthesis often show organ and species specificity. Mucins contain the greatest diversity of core 1 and core 2 oligosaccharides, whereas normal secreted glycoproteins and membrane glycoproteins, and particularly glycoproteins from cancer tissues, show a more restricted *O*-glycan distribution and shorter oligosaccharide chains (mainly the sialylated oligosaccharides in columns 4–6 of Table 2 and columns 4–7 of Table 3). Blood group H oligosaccharides (in columns 15–16 of Table 2 and 17–18 of Table 3) occur almost exclusively on mucins.

A. Competition for a Common Substrate

It appears that glycosyltransferase availability and specificity can answer some of the questions raised. For example, ovine submaxillary glands make the sialylα2-6GalNAc disaccharide because of a preponderance of α6-sialyltransferase I over β3-galactosyltransferase, whereas, in porcine submaxillary glands, the reverse situation applies, and larger structures with core class 1 (Galβ1-3GalNAc) are made. This type of control involves a simple competition between two or more glycosyltransferases for a common substrate, and such control points have also been described in the assembly of *N*-glycosyl oligosaccharides [297]. The route taken by the synthetic pathway at a competition point is presumably dictated primarily by the relative activities of the competing transferases. Table 6, for example, shows the variation in *O*-glycan GlcNAc-transferase activities between different tissues and species.

Table 6 Tissue and Species Distribution for *O*-glycan GlcNAc-Transferases[a]

		GlcNAc-transferase (GnT) activity (nmol/h/mg protein)				
Species	Tissue	Core 3 β3-GnT	Core 4 β6-GnT	Core 2 β6-GnT	core2/core 4 ratio	Elongation β3-GnT
Dog	Submaxillary	0	17.8	114	6.4	0
Human	Colon	5.5	10.3	24.0	2.3	0–9.6
	Serum					0
	Ovary	1.8	8.1	17.2	2.1	
	Granulocytes	0	0	0.3		0.4
	AML cells	0	0	5.2		0
	CML cells	0	0	1.2		0
Monkey	Colon	2.4	4.7	11.9	2.5	
	Stomach	0.4	9.5	25.9	2.7	
Pig	Colon	20.5	13.4	51.2	3.8	14
	Stomach	0.8	167	334	2.0	9
	Submaxillary	0	0	0		0
Rat	Colon	19.7	108	135	1.3	17
	Stomach	0.4	25.2	29.6	1.2	1
	Submaxillary	0	4.1	4.4	1.1	0
	Small intestine	3.0	38.6	48.6	1.3	1
	Liver	<0.6	<0.6	<0.6		<0.4
Sheep	Stomach	0.6	16.9	21.3	1.3	

[a]The enzymes are described in the text. The various activities were not assayed under identical conditions.
Sources: Refs. 41–43, 47, 297, 395.

At least four different enzymes compete for GalNAc-R (see Fig. 2); depending on the tissue, the preponderant core type synthesized may be sialylα2-6GalNAc, core 1 or core 3. At least five enzymes compete for core 1 (see Figs. 2 and 3; i.e., α6-sialyltransferase I, α3-sialyltransferase, the elongation β3-GlcNAc-transferase, and α2-fucosyltransferase all compete with the core 2 β6-GlcNAc-transferase); core 2 synthesis is reduced if substitution of core 1 occurs. Since nonmucinous tissues, such as liver, lack α6-sialyltransferase I, it is the ratio of either α3-sialyltransferase or the elongation β3-GlcNAc-transferase or the α2-fucosyltransferase to core 2 β6-GlcNAc-transferase that determines whether or not core 2 is made in these tissues.

It has been shown [5,68] that human urinary chorionic gonadotrophin (hCG) from normal pregnant women and from women with hydatiform moles had almost entirely core 1 *O*-glycans (see Table 2), whereas hCG from choriocarcinoma patients' urine was very rich in core 2 *O*-glycans (see Table 3), mainly {S3G4[S3][2]} (see Fig. 5). Patients with invasive moles had intermediate levels of these core 2 *O*-glycans. {S3G4[S3][2]} is found in most of the tumors and tumor lines listed in Table 3. The suggested explanation is an

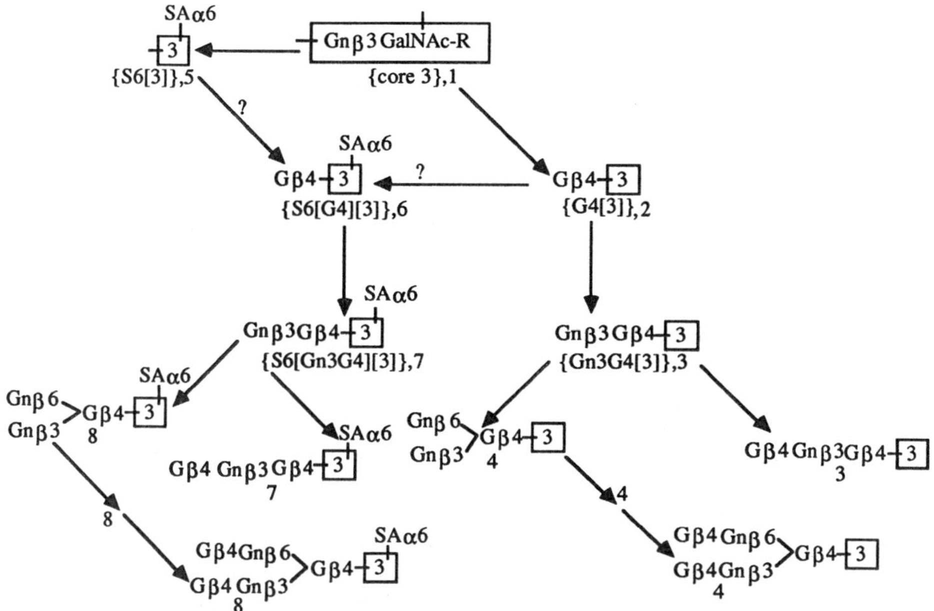

Figure 8 Assembly of some core 3 *O*-glycans. The numbered rectangular boxes represent the core structures. The names in braces { } are defined in the figure and are used to refer to structures in the text. Only the conversion of core 3 to **{S6[3]}**, but none of the remaining reactions in the figure, has been studied in cell-free preparations. These structures are based on sialylated *O*-glycans isolated from human and rat colonic mucin (see Table 4; references in Table 1). Some of the human glycans carry the blood group A determinant (see Table 5). It is not known whether the nonsialylated structures on the right side of the figure can be converted to the sialylated structures on the left side of the figure. Abbreviations: as in the legends to Figs. 1 and 2.

increase in the ratio of core 2 β6-GlcNAc-transferase to CMP-sialic acid:Galβ1-3GalNAc-R (sialyl to Gal) α3-sialyltransferase (see Fig. 2).

Other competition points are (1) sialylation versus elongation of core 2 (see Fig. 2); (2) decision as to whether to synthesize the X, Y, or sialylated X determinants (see Fig. 4); and (3) decisions frequently have to be made between two apparently similar termini in a single oligosaccharide (e.g., competition for the two termini of **{G4[2]}** by α3-sialyltransferase; see Fig. 5). Figures 6–10 show other types of competition points.

B. Substrate Specificity

An important type of control [297] involves the turning on or the turning off of a particular reaction by the insertion of a key glycosyl residue. The following are examples of reactions that can be turned off by this mechanism: (1) the insertion of a sialyl residue in α2-6 linkage into GalNAc to make sialylα2-6GalNAc prevents both core 1 (β3-Gal-transferase) and core 3 (β3-GlcNAc-transferase) synthesis (see Fig. 2); (2) the substitution of core 1 by sialyl, *N*-acetylglucosaminyl, or fucosyl residues prevents or retards core 2 synthesis (see Figs. 2 and 3); (3) the three-before-six rule; that is, several reaction sequences

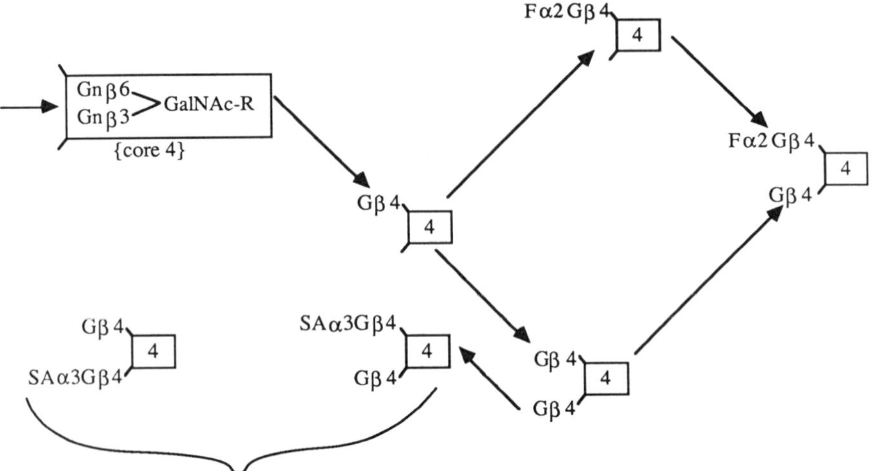

Figure 9 Assembly of some core 4 *O*-glycans. The numbered rectangular boxes represent the core structures. None of these reactions have been studied in cell-free preparations. These structures are based primarily on an extensive series of *O*-glycans isolated from human bronchial mucin (references in Table 1). Glycans have been reported with determinants Y and sialylated X (see Table 5) on **{G4[4]}** and **{G4[G4][4]}** (see Fig. 4). Abbreviations: as in the legends to Figs. 1 and 2.

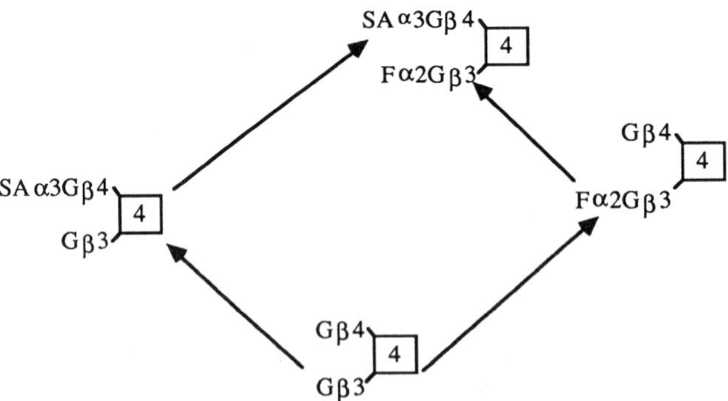

Figure 10 Assembly of some core 4 *O*-glycans. The numbered rectangular boxes represent the core structures. None of these reactions has been studied in cell-free preparations. These structures are based on *O*-glycans from human bronchial mucin (references in Table 1). Abbreviations: as in the legends to Figs. 1 and 2.

require substitution of carbon-3 before carbon-6 (i.e., carbon-3 substitution is prevented by carbon-6 substitution, for example, synthesis of core 4, **{S6[1]}**, **{S6[3]}**, **{S6[S3][1]}**, and GlcNAcβ1-6(GlcNAcβ1-3)Gal-; see Fig. 2); (4) substitution of the penultimate GlcNAc of Galβ1-4GlcNAc-R by Fucα1-3 prevents the action of both α3-sialyltransferase and H-dependent α2-fucosyltransferase (see Fig. 4), as well as of A-dependent α3-GalNAc-transferase and B-dependent α3-Gal-transferase; and (5) substitution of the GlcNAcβ1-3Gal- terminus to form Galβ1-4GlcNAcβ1-3Gal- prevents addition of GlcNAc in β1-6-linkage to the internal galactose [i.e., the GlcNAcβ1-6(GalNAcβ1-3)Gal-moiety is completed before addition of further galactosyl residues, as indicated in Figs. 2, 6, and 8 [262]].

Alternatively, the addition of a particular residue may be essential for enzyme action; for example, (1) addition of GlcNAc in β1-6 linkage to GalNAc to form core 2 or core 4 structures cannot proceed in most tissues unless galactose or *N*-acetylglucosamine, respectively, are first inserted in β1-3 linkage (the three-before-six rule; see Fig. 2); and (2) addition of *N*-acetylglucosamine in β1-6 linkage to galactose to form GlcNAcβ1-6(GlcNAcβ1-3)Gal- cannot proceed in most tissues unless *N*-acetylglucosamine is first inserted in β1-3 linkage (the three-before-six rule; see Figs. 2, 6, and 8).

There is good evidence to suggest that the interaction of a carbohydrate-binding protein, such as a glycosyltransferase, lectin, or antibody, with its carbohydrate ligand is dependent on the three-dimensional conformation of the oligosaccharide substrate [62,150,297]. Furthermore, although oligosaccharides may exist in several different conformations in free solution, it is clear from x-ray crystallographic studies of glycoproteins that interaction with protein can stabilize oligosaccharide structure into a single conformation that may or may not be a suitable substrate for a particular glycosyltransferase. This is an important concept that may play a role in the control of biosynthetic pathways.

The initiation of *O*-glycan biosynthesis and the synthesis of the cores depend on the peptide structure of the substrate. We have shown that in rat colon the two enzymes acting on GalNAc-R, the core 1 β3-Gal-transferase and the core 3 β3-GlcNAc-transferase, are sensitive to changes in the peptide moiety of glycopeptide substrates as well as to the presence of additional *N*-acetylgalactosaminyl residues in the substrate [44]. This means that the peptide plays an important role in determining the core structures made at each glycosylation site.

XII. CONCLUDING REMARKS

Many more enzymes involved in the synthesis of *O*-glycans have to be identified and their roles and control mechanisms have to be elucidated. The refinement of structural analyses and gene cloning have recently advanced our knowledge of *O*-glycan structures and their biosynthesis. Research in these areas will advance rapidly, but other areas of research are indispensable, such as cell biology and the search for specific enzyme inhibitors. We hope that this will resolve many of the reported discrepancies and open questions.

ACKNOWLEDGMENTS

This work was supported by the Medical Research Council of Canada and by the Canadian Cystic Fibrosis Foundation.

REFERENCES

1. Abeijon, C., and Hirschberg, C. B. (1987). Subcellular site of synthesis of the *N*-acetylgalactosamine (α1-0) serine (or threonine) linkage in rat liver. *J. Biol. Chem. 262*:4153–4159.
2. Abeijon, C., and Hirschberg, C. B. (1988). Intrinsic membrane glycoproteins with cytosol-oriented sugars in the endoplasmic reticulum. *Proc. Natl. Acad. Sci. USA 85*:1010–1014.
3. Adamany, A. M., Blumenfeld, O. O., Sabo, B., and McCreary, J. (1983). A carbohydrate structural variant of MM glycoprotein (glycophorin A). *J. Biol. Chem. 258*:11537–11545.
4. Akiyama, K., Simons, E. R., Bernasconi, P., Schmid, K., van Halbeek, H., Vliegenthart, J. F. G., Haupt, U., and Schwick, H. G. (1984). The structure of the carbohydrate units of human plasma galactoglycoprotein determined by 500-megahertz ^{1}H NMR spectroscopy. *J. Biol. Chem. 259*:7151–7154.
5. Amano, J., Nishimura, R., Mochizuki, M., and Kobata, A. (1988). Comparative study of the mucin-type sugar chains of human chorionic gonadotropin present in the urine of patients with trophoblastic diseases and healthy pregnant women. *J. Biol. Chem. 263*:1157–1165.
6. Aminoff, D., Baig, M. M., and Gathmann, W. D. (1979). Glycoproteins and blood group activity. Oligosaccharides of A^+ hog submaxillary glycoproteins. *J. Biol. Chem. 254*:1788–1793.
7. Aminoff, D., Gathmann, W. D., and Baig, M. M. (1979). Glycoproteins and blood group activity. Isolation and characterization of oligosaccharides of H^+ hog submaxillary glycoprotein and their comparison to those found in A^+ and A^-H^- glycoproteins. *J. Biol. Chem. 254*:8909–8913.
8. Andersson, G. N., and Eriksson, L. C. (1979). Characterization of UDP-galactosyl:asialomucin transferase activity in the Golgi system of rat liver. *Biochim. Biophys. Acta 570*:239–247.
9. Andersson, G. N., and Eriksson, L. C. (1980). Studies on the latency of UDP-galactose: asialomucin galactosyltransferase activity in microsomal and Golgi subfractions from rat liver. *Biochim. Biophys. Acta 600*:571–576.
10. Andersson, G. N., and Eriksson, L. C. (1981). Endogenous localization of UDP-galactose: asialomucin galactosyltransferase activity in rat liver endoplasmic reticulum and Golgi apparatus. *J. Biol. Chem. 256*:9633–9639.
11. Appert, H. E., Rutherford, T. J., Tarr, G. E., Wiest, J. S., Thomford, N. R., and McCorquodale, D. J. (1986). Isolation of a cDNA coding for human galactosyltransferase. *Biochem. Biophys. Res. Commun. 139*:163–168.
12. Babczinski, P. (1980). Evidence against the participation of lipid intermediates in the in vitro biosynthesis of serine(threonine)-*N*-acetyl-D-galactosamine linkages in submaxillary mucin. *FEBS Lett. 117*:207–211.
13. Baenziger, J., and Kornfeld, S. (1974). Structure of the carbohydrate units of IgA1 immuno globulin. II. Structure of the *O*-glycosidically linked oligosaccharide units. *J. Biol. Chem. 249*:7270–7281.
14. Baker, M. A., Kanani, A., Brockhausen, I., Schachter, H., Hindenburg, A., and Taub, R. N. (1987). Presence of cytidine 5′-monophospho-*N*-acetylneuraminic acid:Galβ1-3GalNAc-R α(2-3)-sialyltransferase in normal human leukocytes and increased activity of this enzyme in granulocytes from chronic myelogenous leukemia patients. *Cancer Res. 47*:2763–2766.
15. Baker, A. P., and Munro, J. R. (1971). Multiglycosyltransferase system of canine respiratory tissue. Uridine diphosphate galactose:mucin galactosyltransferase. *J. Biol. Chem. 246*:4358–4362.
16. Basu, M., and Basu, S. (1973). Enzymatic synthesis of a blood group B-related pentaglycosylceramide by an α-galactosyltransferase from rabbit bone marrow. *J. Biol. Chem. 248*:1700–1706.
17. Basu, M., and Basu, S. (1984). Biosynthesis in vitro of Ii core glycosphingolipids from neolactotetraosylceramide by β1-3 and β1-6-*N*-acetylglucosaminyltransferases from mouse T-lymphoma. *J. Biol. Chem. 259*:12557–12562.
18. Basu, M., Chon, H., and Basu, S. (1982). Biosynthesis in vitro of core structure LM1 and Ii glycosphingolipids in mouse lymphomas. Annual Meeting Society for Complex Carbohydrates, Abstr. 63.

19. Baubichon-Cortray, H., Broquet, P., George, P., and Louisot, P. (1989). Different reactivity of two brain sialyltransferases towards sulfhydryl reagents. Evidence for a thiol group involved in the nucleotide-sugar binding site of the NeuAcα2-3Galβ1-3GalNAc α(2-6)sialyltransferase. *Glycoconjugate J. 6*:115–127.
20. Bedi, G. S., Reddy, M. S., Shah, R. H., and Bahl, O. P. (1979). Carbohydrate structure of gonadotropins. Annual Meeting of the Society for Complex Carbohydrates, Abstr. 2.
21. Bennett, G. C. (1977). Goblet cells and secretion of mucus. In *Mucus Secretions and Cystic Fibrosis*. (G. G. Forstner, ed.). *Probl. Paediatr. 19*:66–76.
22. Bennett, G. C., Leblond, C. P., and Haddad, A. (1974). Migration of glycoprotein from the Golgi apparatus to the surface of various cell types as shown by radioautography after labeled fucose injection into rats. *J. Cell Biol. 60*:258–285.
23. Berger, E. G., and Kozdrowski, I. (1978). Permanent mixed-field polyagglutinable erythrocytes lack galactosyltransferase activity. *FEBS Lett. 93*:105–108.
24. Bergh, M. L. E., Hooghwinkel, G. J. M., and van den Eijnden, D. H. (1981). Specificity of porcine liver Galβ(1-3)GalNAc-R α(2-3)sialyltransferase. Sialylation of mucin-type acceptors and ganglioside GM1 in vitro. *Biochim. Biophys. Acta 660*:161–169.
25. Bergh, M. L. E., Koppen, P., and van den Eijnden, D. H. (1981). High pressure liquid chromatography of sialic acid-containing oligosaccharides. *Carbohydr. Res. 94*:225–229.
26. Bergh, M. L. E., Koppen, P. L., and van den Eijnden, D. H. (1982). Specificity of ovine submaxillary gland sialyltransferase. Applications of high-pressure liquid chromatography in the identification of sialooligosaccharide products. *Biochem. J. 201*:411–415.
27. Bergh, M. L. E., and van den Eijnden, D. H. (1983). Aglycon specificity of fetal calf liver and ovine and porcine submaxillary gland α-*N*-acetylgalactosaminide α2-6-sialyltransferase. *Eur. J. Biochem. 136*:113–118.
28. Betteridge, A., and Watkins, W. M. (1985). Acceptor substrate specificities of human α-2-L-fucosyltransferases from different tissues. *Biochem. Soc. Trans. 13*:1126–1127.
29. Betteridge, A,, and Watkins, W. M. (1985). Variant forms of alpha-2-L-fucosyltransferase in human submaxillary glands from blood group ABH "secretor" and "non-secretor" individuals. *Glycoconjugate J. 2*:61–78.
30. Beyer, T. A., and Hill, R. L. (1980). Enzymatic properties of the β-galactoside α1→2 fucosyltransferase from porcine submaxillary gland. *J. Biol. Chem. 255*:5373–5379.
31. Beyer, T. A., Rearick, J. I., Paulson, J. C., Prieels, J. P., Sadler, J. E., and Hill, R. L. (1979). Biosynthesis of mammalian glycoproteins. Glycosylation pathways in the synthesis of the non-reducing terminal sequences. *J. Biol. Chem. 254*:12531–12541.
32. Beyer, T. A., Sadler, J. E., and Hill, R. L. (1980). Purification to homogeneity of the H blood group β-galactoside α1→2 fucosyltransferase from porcine submaxillary gland. *J. Biol. Chem. 255*:5364–5372.
33. Beyer, T. A., Sadler, J. E., Rearick, J. I., Paulson, J. C., and Hill, R. L. (1981). Glycosyltransferases and their use in assessing oligosaccharide structure and structure-function relationships. *Adv. Enzymol. 52*:23–175.
34. Blanchard, D., Capon, C., Leroy, Y., Cartron, J. P., and Fournet, B. (1985). Comparative study of glycophorin A derived *O*-glycans from human Cad, Sd(a+) and Sd(a-) erythrocytes. *Biochem. J. 232*:813–818.
35. Blanchard, D., Cartron, J. P., Fournet, B., Montreuil, J., van Halbeek, H., and Vliegenthart, J. F. G. (1983). Primary source of the oligosaccharide determinant of blood group Cad specificity. *J. Biol. Chem. 258*:7691–7695.
36. Blanken, W. M., Hooghwinkel, G. J. M., and van den Eijnden, D. H. (1982). Biosynthesis of blood-group I and i substances. Specificity of bovine colostrum β-*N*-acetyl-D-glucosaminide β1-4-galactosyltransferase. *Eur. J. Biochem. 127*:547–552.
37. Bleil, J. D., and Wassarman, P. M. (1988). Galactose at the non-reducing terminus of *O*-linked oligosaccharides of mouse egg zona pellucida glycoprotein ZP3 is essential for the glycoprotein's sperm receptor activity. *Proc. Natl. Acad. Sci. USA 85*:6778–6782.

38. Breg, J., van Halbeek, H., Vliegenthart, J. F. G., Klein, A., Lamblin, G., and Roussel, P. (1988). Primary structure of neutral oligosaccharides derived from respiratory mucus glycoproteins of a patient suffering from bronchiectasis, determined by combination of 500-MHz proton NMR spectroscopy and quantitative sugar analysis. 2. Structure of 19 oligosaccharides having the GlcNAcβ(1-3)GalNAc-ol core (type 3) or the GlcNAcβ(1-3)[GlcNAcβ(1-6)]GalNAc-ol core (type 4). *Eur. J. Biochem. 171*:643–654.

39. Breg, J., van Halbeek, H., Vliegenthart, J. F. G., Lamblin, G., Houvenaghel, M. C., and Roussel, P. (1987). Structure of sialyl-oligosaccharides isolated from bronchial mucus glycoproteins of patients (blood group O) suffering from cystic fibrosis. *Eur. J. Biochem. 168*:57–68.

40. Briand, J. P., Andrews, S. P., Cahill, E., Conway, N. A., and Young, J. D. (1981). Investigation of the requirements for *O*-glycosylation by bovine submaxillary gland UDP-*N*-acetylgalactosamine:polypeptide *N*-acetylgalactosamine transferase using synthetic peptide substrates. *J. Biol. Chem. 256*:12205–12207.

41a. Brockhausen, I., Kuhns, W., Schachter, H., Matta, K. L., Sutherland, D. R., and Baker, M. A. (1991). Biosynthesis of *O*-glycans in leukocytes from normal donors and from patients with leukemia: Increase in *O*-glycan core 2 UDP-GlcNAc: Galβ3GalNAcα-R (GlcNAc to GalNAc) β(1-6)-*N*-acetylglucosaminyltransferase in leukemic cells. *Cancer Res. 51*:1257–1263.

41b. Brockhausen, I., Romero, P. A., Herscovics, A. (1991). *Cancer Res. 51*:3136–3142.

42. Brockhausen, I., Matta, K. L., Orr, J., and Schachter, H. (1985). Mucin synthesis. VI. UDP-GlcNAc:GalNAc-R β3-N-acetylglucosaminyltransferase and UDP-GlcNAc:GlcNAcβ1-3GalNAc-R (GlcNAc to GalNAc) β6-*N*-acetylglucosaminyltransferase from pig and rat colon mucosa. *Biochemistry 24*:1866–1874.

43. Brockhausen, I., Matta, K. L., Orr, J., Schachter, H., Koenderman, A. H. L., and van den Eijnden, D. H. (1986). Mucin synthesis. VII. Conversion of R1-β1-3Gal-R2 to R1-β1-3(GlcNAcβ1-6)Gal-R2 and of R1-β1-3GalNAc-R2 to R1-β1-3(GlcNAcβ1-6)GalNAc-R_2 by a β6-*N*-acetylglucosaminyltransferase in pig gastric mucosa. *Eur. J. Biochem. 157*:463–474.

44. Brockhausen, I., Möller, G., Merz, G., Adermann, K., and Paulsen, H. (1990). Control of mucin synthesis: The peptide portion of synthetic *O*-glycopeptide substrates influences the activity of *O*-glycan core 1 UDP-galactose:*N*-acetyl-α-galactosaminyl-R β3-galactosyltransferase. *Biochemistry 29*:10206–10212.

45. Brockhausen, I., Orr, J., and Schachter, H. (1984). Mucin synthesis. V. The action of pig gastric mucosal UDP-GlcNAc:Galβ1-3(R1)GalNAc-R2 (GlcNAc to Gal) β3-*N*-acetylglucosaminyltransferase on high molecular weight substrates. *Can. J. Biochem. Cell Biol. 62*:1081–1090.

46. Brockhausen, I., Rachaman, E. S., Matta, K. L., and Schachter, H. (1983). Mucin synthesis. IV. The separation by high performance liquid chromatography of phenyl, benzyl and *ortho*-nitrophenyl oligosaccharide glycosides. Analysis of substrates and products for four *N*-acetyl-D-glucosaminyl-transferases involved in mucin synthesis. *Carbohydr. Res. 120*:3–16.

47. Brockhausen, I., Williams, D., Matta, K. L., Orr, J., and Schachter, H. (1983). Mucin synthesis. III. UDP-GlcNAc:Galβ1-3(GlcNAcβ1-6)GalNAc-R (GlcNAc to Gal) β3-*N*-acetylglucosaminyltransferase, an enzyme in porcine gastric mucosa involved in the elongation of mucin-type oligosaccharides. *Can. J. Biochem. Cell Biol. 61*:1322–1333.

48. Bunnell, B. A., Adams, D. E., and Kidd, V. J. (1990). Transient expression of a p58 protein kinase cDNA enhances mammalian glycosyltransferase activity. *Biochem. Biophys. Res. Comm. 171*:196–203.

49. Bush, C. A., Panitch, M. M., Dua, V. K., and Rohr, T. E. (1985). Carbon nuclear magnetic resonance spectra of oligosaccharides isolated from human milk and ovarian cyst mucin. *Anal. Biochem. 145*:124–136.

50. Cahan, L. D., and Paulson, J. C. (1980). Polyoma virus adsorbs to specific sialyloligosaccharide receptors on erythrocytes. *Virology 103*:505–509.

51. Campbell, C., and Stanley, P. (1984). The Chinese Hamster ovary glycosylation mutants LEC11 and LEC12 express two novel GDP-fucose:*N*-acetylglucosaminide 3-α-L-fucosyltransferase enzymes. *J. Biol. Chem. 259*:11208–11214.

52. Capon, C., Leroy, Y., Wieruszeski, J.-M., Ricart, G., Strecker, G., Montreuil, J., and Fournet, B. (1989). Structures of *O*-glycosidically linked oligosaccharides isolated from human meconium glycoproteins. *Eur. J. Biochem. 182*:139–152.
53. Carlson, D. M., and Blackwell, C. (1968). Structures and immunochemical properties of oligosaccharides isolated from pig submaxillary mucins. *J. Biol. Chem. 243*:616–626.
54. Carlson, D. M., David, J., and Rutter, W. J. (1973). Galactosyltransferase activities in pancreas, liver, and gut of the developing rat. *Arch. Biochem. Biophys. 157*:605–612.
55. Carlson, D. M., McGuire, E. J., Jourdian, G. W., and Roseman, S. (1973). The sialic acids. XVI. Isolation of a mucin sialyltransferase from sheep submaxillary gland. *J. Biol. Chem. 248*:5763–5773.
56. Carlsson, S. R., Sasaki, H., and Fukuda, M. (1986). Structural variations of *O*-linked oligosaccharides present in leukosialin isolated from erythroid, myeloid, and T-lymphoid cell lines. *J. Biol. Chem. 261*:12787–12795.
57. Carlsson, H. E., Sundblad, G., Hammarstrom, A., and Lonngren, J. (1978). Structure of some oligosaccharides derived from rat-intestinal glycoproteins. *Carbohydr. Res. 64*:181–188.
58. Carraway, K. L., and Spielman, J. (1986). Structural and functional aspects of tumor cell sialomucins. *Mol. Cell. Biochem. 72*:109–120.
59. Carter, S. R., Slomiany, A., Gwozdzinski, K., Liau, Y. H., and Slomiany, B. L. (1988). Enzymatic sulfation of mucus glycoprotein in gastric mucosa. Effect of ethanol. *J. Biol. Chem. 263*:11977–11984.
60. Cartron, J., Andrev, J., Cartron, J., Bird, G. W. G., Salmon, C., and Gerbal, A. (1978). Demonstration of T-transferase deficiency in Tn-polyagglutinate blood samples. *Eur. J. Biochem. 92*:111–119.
61. Cartron, J. P., and Nurden, A. T. (1979). Galactosyltransferase and membrane glycoprotein abnormality in human platelets from Tn-syndrome donors. *Nature 282*:621–623.
62. Carver, J. P., and Brisson, J. (1984). The three-dimensional structure of *N*-linked oligosaccharides. In *Biology of Carbohydrates*, Vol. 2 (V. Ginsburg and P. W. Robbins, eds.). John Wiley & Sons, New York, pp. 289–331.
63. Chandrasekaran, E. V., Mendicino, A., Garver, F. A., and Mendicino, J. (1981). Structures of sialylated *O*-glycosidically and *N*-glycosidically linked oligosaccharides in a monoclonal immunoglobulin light chain. *J. Biol. Chem. 256*:1549–1555.
64. Chandrasekaran, E. V., Rana, S. S,, Davila, M., and Mendicino, J. (1984). Structures of the oligosaccharide chains in swine trachea mucin glycoproteins. *J. Biol. Chem. 259*:12908–12914.
65. Cheng, P., and Bona, S. J. (1982). Mucin biosynthesis. Characterization of UDP-galactose:α-*N*-Acetylgalactosaminide β3-galactosyltransferase from human tracheal epithelium. *J. Biol. Chem. 257*:6251–6258.
66. Cheng, P., Wingert, W. E., Little, M. R., and Wei, R. (1985). Mucin biosynthesis. Properties of a bovine tracheal mucin β-6-*N*-acetylglucosaminyltransferase. *Biochem. J. 227*:405–412.
67. Clausen, H., White, T., Takio, K., Titani, K., Stroud, M., Holmes, E., Karkov, J., Thim, L., and Hakomori, S. (1990). Isolation to homogeneity and partial characterization of a histo-blood group A defined Fucalpha1→2Galapha1→3-*N*-acetylgalactosaminyltransferase from human lung tissue. *J. Biol. Chem. 265*:1139–1145.
68. Cole, L. A. (1987). The *O*-linked oligosaccharide structures are striking different on pregnancy and choriocarcinoma hCG. *J. Clin. Endocrinol. Metab. 65*:811–813.
69. Conradt, H. S., Ausmeier, M., Dittmar, K. E. J., Hauser, H., and Lindenmaier, W. (1986). Secretion of glycosylated human interleukin-2 by recombinant mammalian cell lines. *Carbohydr. Res. 149*:443–450.
70. Conzelmann, A., and Bron, C. (1987). Expression of UDP-*N*-acetylgalactosamine:β-galactose β1,4-*N*-acetylgalactosaminyltransferase in functionally defined T-cell clones. *Biochem. J. 242*:817–824.
71. Conzelmann, A., and Kornfeld, S. (1984). β-Linked *N*-acetylgalactosamine residues present at the non-reducing termini of *O*-linked oligosaccharides of a cloned murine cytotoxic T

lymphocyte line are absent in a *Vicia villosa* lectin-resistant mutant cell line. *J. Biol. Chem. 259*:12528–12535.

72. Conzelmann, A., and Kornfeld, S. (1984). A murine cytotoxic T lymphocyte cell line resistant to *Vicia villosa* lectin is deficient in UDP-GalNAc:β-galactose β1,4-*N*-acetylgalactosaminyltransferase. *J. Biol. Chem. 259*:12536–12542.
73. Cruz, T. F., and Moscarello, M. A. (1983). Enzymic glycosylation of myelin basic protein: Identification of the major sites. *Biochim. Biophys. Acta 760*:403–410.
74. Cruz, T. F., Wood, D. D., and Moscarello, M. A. (1984). Identification of threonine-95 as the major site of glycosylation in normal human myelin basic protein. *Biochem. J. 220*: 849–852.
75. Cummings, R. D., Kornfeld, S., Schneider, W. J., Hobgood, K. K., Tollelshaug, H., Brown, M. S., and Goldstein, J. L. (1983). Biosynthesis of *N*- and *O*-linked oligosaccharides of the low density lipoprotein receptor. *J. Biol. Chem. 258*:15261–15273.
76. D'Agostaro, G., Bendiak, B., and Tropak, M. (1989). Cloning of cDNA encoding the membrane-bound form of bovine β1,4-galactosyltransferase. *Eur. J. Biochem. 183*:211–217.
77. D'Arcy, S. M., Donoghue, C. M., Koeleman, C. A. M., van den Eijnden, D. H., and Savage, A. V. (1989). Determination of the structure of a novel acidic oligosaccharide with blood-group activity isolated from bovine submaxillary-gland. *Biochem. J. 260*:389–393.
78. Dal'olio, F., Malagolini, N., di Stefano, G., Ciambella, M., and Serafini-Cessi, F. (1990). Postnatal development of rat colonic epithelial cells is associated with changes in the expression of the β1,4-*N*-acetylgalactosaminyltransferase involved in the synthesis of Sda antigen and of α2,6-sialyltransferase activity towards *N*-acetyllactosamine. *Biochem. J. 270*: 519–524.
79. Damm, J. B., Hard, K., Kamerling, J. P., van Demen, G. W. K., and Vliegenthart, J. F. G. (1990). Structure determination of the major *N*- and *O*-linked carbohydrate chains of the β subunit from equine chorionic gonadotropin. *Eur. J. Biochem. 189*:175–813.
80. de Heij, H. T., Tetteroo, P. A. T., van Kessel, A. H. M. G., Schoenmaker, E., Visser, F. J., and van den Eijnden, D. H. (1988). Specific expression of a myeloid-associated CMP-NeuAc:Galβ1-3GalNAcα-R α2-3-sialyltransferase and the sialyl-X determinant in myeloid human–mouse cell hybrids containing human chromosome 11. *Cancer Res. 48*:1489–1493.
81. Derevitskaya, V. A., Arbatsky, N. P., and Kochetkov, N. K. (1978). The structure of carbohydrate chains of blood-group substance. Isolation and elucidation of the structure of higher oligosaccharides from blood-group substance H. *Eur. J. Biochem. 86*:423–437.
82. Deschuyteneer, M., Eckhardt, A., Roth, J., and Hill, R. L. (1988). The subcellular localization of apomucin and nonreducing terminal *N*-acetylgalactosamine in porcine submaxillary glands. *J. Biol. Chem. 263*:2452–2459.
83. Do, S.-I., Enns, C., and Cummings, R. D. (1990). Human transferrin receptor contains *O*-linked oligosaccharides. *J. Biol. Chem. 265*:114–125.
84. Donald, A. S. R., Soh, C. P. C., Yates, A. D., Feeney, J., Morgan, W. T. J., and Watkins, W. M. (1987). Structure, biosynthesis, and genetics of the Sda antigen. *Biochem. Soc. Trans. 15*:606–608.
85. Dua, V. K., Rao, B. N. N., Wu, S., Dube, V. E., and Bush, C. A. (1986). Characterization of the oligosaccharide alditols from ovarian cyst mucin glycoproteins of blood group A using high pressure liquid chromatography (HPLC) and high field proton NMR spectroscopy. *J. Biol. Chem. 261*:1599–1608.
86. Duncan, A. M. V., McCorquodale, M. M., Morgan, C., Rutherford, T. J., Appert, H. E., and McCorquodale, D. J. (1986). Chromosomal localization of the gene for a human galactosyltransferase (GT-1). *Biochem. Biophys. Res. Commun. 141*:1185–1188.
87. Dutta, B., and Rao, C. V. N. (1982). Structures of carbohydrate chains of glycoprotein isolated from goat submaxillary mucin. *Biochim. Biophys. Acta 701*:72–85.
88. Edge, A. S. B., and Spiro, R. G. (1987). Presence of an *O*-glycosidically linked hexasaccharide in fetuin. *J. Biol. Chem. 262*:16135–16141.

89. Edge, A. S. B., van Langenhove, A., Reinhold, V., and Weber, P. (1986). Characterization of *O*-glycosidically linked oligosaccharides of rat erythrocyte membrane sialoglycoproteins. *Biochemistry 25*:8017–8024.
90. Elhammer, A., and Kornfeld, S. (1984). Two enzymes involved in the synthesis of *O*-linked oligosaccharides are localized on membranes of different densities in mouse lymphoma BW5147 cells. *J. Cell Biol. 98*:327–331.
91. Elhammer, A., and Kornfeld, S. (1986). Purification and characterization of UDP-*N*-acetylgalactosamine:polypeptide *N*-acetylgalactosaminyltransferase from bovine colostrum and murine lymphoma BW5147 cells. *J. Biol. Chem. 261*:5249–5255.
92. Endo, Y., Yamashita, K., Han, Y. N., Iwanaga, S., and Kobata, A. (1977). The carbohydrate structure of a glycopeptide released by the action of plasma kallikrein on bovine plasma high-molecular-weight kininogen. *J. Biochem. 82*:545–550.
93. Eppenberger-Castori, S., Lötscher, H., and Finne, J. (1989). Purification of the *N*-acetylglucosaminide α(1-3/4)-fucosyltransferase of human milk. *Glycoconjugate J. 6*:101–114.
94. Ernst, L. K., Rajan, V. P., Larsen, R. D., Ruff, M. M., and Lowe, J. B. (1989). Stable expression of blood group H determinants and GDP-L-fucose:β-D-galactoside 2-α-L-fucosyltransferase in mouse cells after transfection with human DNA. *J. Biol. Chem. 264*:3436–3447.
95. Feizi, T. (1982). Antigenicities of mucins—their relevance to tumour associated and stage specific embryonic antigens. In *Mucus in Health and Disease*, Vol. II (E. N. Chantler, J. B. Elder, and M. Elstein, ed.). Plenum Press, New York, pp. 29–37.
96. Fiat, A., Chevan, J., Jolles, P., de Waard, P., Vliegenthart, J. F. G., Piller, F., and Cartron, J. (1988). Structural variability of the neutral carbohydrate moiety of cow colostrum kappa-casein as a function of time after parturition. Identification of a tetrasaccharide with blood group I specificity. *Eur. J. Biochem. 173*:253–259.
97. Fiat, A. M., Jolles, J., Aubert, J. P., Loucheux-Lefebvre, M. H., and Jolles, P. (1980). Localisation and importance of the sugar part of human casein. *Eur. J. Biochem. 111*:333–339.
98. Finne, J. (1975). Structure of the *O*-glycosidically linked carbohydrate units of rat brain glycoproteins. *Biochim. Biophys. Acta 412*:317–325.
99. Finne, J., Krusius, T., and Rauvala, H. (1977). Occurrence of disialosyl groups in glycoproteins. *Biochem. Biophys. Res. Comm. 74*:405–410.
100. Florman, H. M,, and Wassarman, P. M. (1985). *O*-linked oligosaccharides of mouse egg ZP3 account for its sperm receptor activity. *Cell 41*:313–324.
101. Foster, C. S., and Glick, M. C. (1987). *N*-Acetyl-β-D-glucosaminide α1-3-fucosyltransferase purified from human neuroblastoma. *Biochem. Soc. Trans. 15*:398–399.
102. Fukuda, M., Carlsson, S. R., Klock, J. C., and Dell, A. (1986). Structures of *O*-linked oligosaccharides isolated from normal granulocytes, chronic myelogenous leukemia cells, and acute myelogenous leukemia cells. *J. Biol. Chem. 261*:12796–12806.
103. Fukuda, K., Honma, K., Manabe, H., Utsumi, H., and Hamada, A. (1987). Alkali-labile oligosaccharide units of a sialoglycoprotein from rabbit erythrocyte membranes. *Biochim. Biophys. Acta 926*:132–138.
104. Fukuda, K., Kawashima, I., Tomita, M., and Hamada, A. (1982). Structural studies of the acidic oligosaccharide units from bovine glycophorin. *Biochim. Biophys. Acta 717*: 278–288.
105. Fukuda, M., Lauffenburger, M., Sasaki, H., Rogers, M. E., and Dell, A. (1987). Structures of novel sialylated *O*-linked oligosaccharides isolated from human erythrocyte glycophorins. *J. Biol. Chem. 262*:11952–11957.
106. Fukuda, K., Tomita, M., and Hamada, A. (1980). Isolation and characterization of alkali-labile oligosaccharide units from horse glycophorin. *J. Biochem. (Tokyo) 87*:687–693.
107. Fukuda, K., Tomita, M., and Hamada, A. (1981). Isolation and structural studies of the neutral oligosaccharide units from bovine glycophorin. *Biochim. Biophys. Acta 677*:462–470.
108. Fukushima, K., Hirota, M,, Terasaki, P. I., Wakisaka, A., Toagshi, H., Chia, D., Suyama, N., Fukushi, Y., Nudelman, E., and Hakomori, S. (1984). Characterization of sialylated Lewis-X as a new tumor-associated antigen. *Cancer Res. 44*:5279–5285.

109. Funakoshi, I., and Yamashina, I. (1982). Structures of *O*-glycosidically linked sugar units from plasma membranes of an ascites hepatoma, AH 66. *J. Biol. Chem.* *257*:3782–3787.

110. Furthmayr, H. (1978). Structural comparison of glycophorins and immunochemical analysis of genetic variants. *Nature* *271*:519–524.

111. Furukawa, K., and Roth, S. (1985). Co-purification of galactosyltransferases from chick-embryo liver. *Biochem. J.* *227*:573–582.

112. Gejyo, F., Chang, J., Burgi, W., Schmid, K., Offner, G. D., Troxler, R. F., van Halbeek, H., Dorland, L., Gerwig, G. J ., and Vliegenthart, J. F. G. (1983). Characterization of the B-chain of human plasma alpha2HS-glycoprotein. The complete amino acid sequence and primary structure of its heteroglycan. *J. Biol. Chem.* *258*:4966–4971.

113. Gendler, S. J., Lancaster, C. A., Taylor-Papadimitriou, J., Duhig, T., Peat, N., Burchell, J., Pemberton, L., Lalani, E.-N., and Wilson, D. (1990). Molecular cloning and expression of human tumor-associated polymorphic epithelial mucin. *J. Biol. Chem.* *265*:15286–15293.

114. Geyer, R., Dabrowski, J., Dabrowski, U., Linder, D., Schlüter, M., Schott, H.-H., and Stirm, S. (1990). Oligosaccharides at individual glycosylation sites in glycoprotein 71 of Friend leukemia virus. *Eur. J. Biochem.* *187*:95–110.

115. Gleeson, P. A., Feeney, J., Mills, G., and Hughes, R. C. (1984). Galactosyltransferases of BHK cells. Characterization of two oligosaccharide products synthesized using bovine asialo submaxillary gland mucin as acceptor. *Eur. J. Biochem.* *144*:143–150.

116. Goldberg, D. E., and Kornfeld, S. (1983). Evidence for extensive subcellular organization of asparagine-linked oligosaccharide processing and lysosomal enzyme phosphorylation. *J. Biol. Chem.* *258*:3159–3165.

117. Gowda, D. C., Bhavanandan, V. P., and Davidson, E. A. (1986). Structures of *O*-linked oligosaccharides present in the proteoglycans secreted by human mammary epithelial cells. *J. Biol. Chem.* *261*:4935–4939.

118. Greenwell, P., Yates, A. D., and Watkins, W. M. (1986). UDP-*N*-acetyl-D-galactosamine as a donor substrate for the glycosyltransferase encoded by the *B* gene at the human blood group *ABO* locus. *Carbohydr. Res.* *149*:149–170.

119. Hagopian, A., Bosmann, H. B., and Eylar, E. H. (1968). Glycoprotein biosynthesis: The localization of polypeptidyl:*N*-acetylgalactosaminyl, collagen:glucosyl, and glycoprotein:galactosyl transferases in HeLa cell membrane fractions. *Arch. Biochem. Biophys.* *128*:387–396.

120. Hagopian, A., Westall, F. C., Whitehead, J. S., and Eylar, E. H. (1971). Glycosylation of the A1 protein from myelin by a polypeptide *N*-acetylgalactosaminyltransferase. Identification of the acceptor sequence. *J. Biol. Chem.* *246*:2519–2523.

121. Hakomori, S., Clausen, H., and Levery, S. B. (1987). A new series of blood group A and H antigens expressed in human erythrocytes and the incompatible A antigens expressed in tumours of blood group O and B individuals. *Biochem. Soc. Trans.* *15*:593–596.

122. Hanisch, F. G., Egge, H., Peter-Katalinic, J., and Uhlenbruck, G. (1986). Primary structure of a major sialyl-oligosaccharide alditol from human amniotic mucins expressing the tumor-associated sialyl-X antigenic determinant. *FEBS Lett.* *200*:42–46.

123. Hanisch, F. G., Egge, H., Peter-Katalinic, J., and Uhlenbruck, G. (1986). Structure of neutral oligosaccharides derived from mucus glycoproteins of human seminal plasma. *Eur. J. Biochem.* *155*:239–247.

124. Hanisch, F. G., Egge, H., Peter-Katalinic, J., Uhlenbruck, G., Dienst, C., and Fangmann, R. (1985). Primary structures and Lewis blood group dependent expression of major sialylated saccharides from mucus glycoproteins of human seminal plasma. *Eur. J. Biochem.* *152*: 343–351.

125. Hanisch, F. G., and Uhlenbruck, G. (1984). Structural studies on *O*- and *N*-glycosidically linked carbohydrate chains on *Collocalia* mucin. *Hoppe-Seyler Z. Physiol. Chem.* *365*:119–128.

126. Hanisch, F. G., Uhlenbruck, G., and Dienst, C. (1984). Structure of tumor-associated carbohydrate antigen Ca 19-9 on human seminal-plasma glycoproteins from healthy donors. *Eur. J. Biochem.* *144*:467–474.

127. Hanisch, F. G., Uhlenbruck, G., Dienst, C., Stottrop, M., and Hippauf, E. (1985). Ca 125 and Ca 19-9: Two cancer-associated sialylsaccharide antigens on a mucus glycoprotein from human milk. *Eur. J. Biochem. 149*:323–330.
128. Hanisch, F.-G., Uhlenbruck, G., Peter-Katalinic, J., and Egge, H. (1988). Structural studies on oncofetal carbohydrate antigens (CA 19-9, CA 50, and CA 125) carried by *O*-linked sialyloligosaccharides on human amniotic mucins. *Carbohydr. Res. 178*:29–47.
129. Hanisch, F.-G., Uhlenbruck, G., Peter-Katalinic, J., Egge, H., Dabrowski, J., and Dabrowski, U. (1989). Structures of neutral *O*-linked polylactosaminoglycans on human skim milk mucins. *J. Biol. Chem. 264*:872–883.
130. Hanover, J. A., Cohen, C. K., Willingham, M. C., and Park, M. K. (1987). *O*-Linked *N*-acetylglucosamine is attached to proteins of the nuclear pore. Evidence for cytoplasmic and nucleoplasmic glycoproteins. *J. Biol. Chem. 262*:9887–9894.
131. Hanover, J. A., Elting, J., Mintz, G. R., and Lennarz, W. J. (1982). Temporal aspects of the *N*- and *O*-glycosylation of human chorionic gonadotropin. *J. Biol. Chem. 257*:10172–10177.
132. Hanover, J. A., Lennarz, W. J., and Young, J. D. (1980). Synthesis of *N*- and *O*-linked glycopeptides in oviduct membrane preparations. *J. Biol. Chem. 255*:6713–6716.
133. Hart, G. W., Holt, G. D., and Haltiwanger, R. S. (1988). Nuclear and cytoplasmic glycosylation: Novel saccharide linkages in unexpected places. *TIBS 13*:380–384.
134. Hatcher, V. B., Schwarzmann, G. O. H., Jeanloz, R. W., and McArthur, J. W. (1977). Purification, properties, and partial structure elucidation of a high-molecular-weight glycoprotein from cervical mucus of the bonnet monkey (*Macaca radiata*). *Biochemistry 16*:1518–1524.
135. Hayes, M. L., and Castellino, F. J. (1979). Carbohydrate of the human plasminogen variants. III. Structure of the *O*-glycosidically linked oligosaccharide units. *J. Biol. Chem. 254*:8777–8780.
136. Heffernan, M., Yousefi, S., and Dennis, J. W. (1989). Molecular characterization of P2B/LAMP-1, a major protein target of a metastasis-associated oligosaccharide structure. *Cancer Res. 49*:6077–6084.
137. Herkt, F., Parente, J. P., Leroy, Y., Fourne, B., Blanchard, D., Cartron, J. P., van Halbeek, H., and Vliegenthart, J. F. G. (1985). Structure determination of oligosaccharides isolated from Cad erythrocyte membranes by permethylation analysis and 500-MHz proton NMR spectroscopy. *Eur. J. Biochem. 146*:125–129.
138. Herp, A., Wu, A. M., and Moschera, J. (1979). Current concepts of the structure and nature of mammalian salivary mucous glycoproteins. *Mol. Cell. Biochem. 23*:27–43.
139. Hesford, F. J., and Berger, E. G. (1981). Human erythrocyte galactosyltransferase. Characterization, membrane association and sidedness of active site. *Biochim. Biophys. Acta 649*: 709–716.
140. Hesford, F. J., Berger, E. G., and van den Eijnden, D. H. (1981). Identification of the product formed by human erythrocyte galactosyltransferase. *Biochim. Biophys. Acta 659*: 302–311.
141. Hew, C. L., Slaughter, D., Fletcher, G. L., and Joshi, S. B. (1981). Antifreeze glycoproteins in the plasma of Newfoundland Atlantic cod (*Gadis morhua*). *Can. J. Zool. 59*:2186–2192.
142. Hill, H. D., Reynolds, J. A., and Hill, R. L. (1977). Purification, composition, molecular weight and subunit structure of ovine submaxillary mucin. *J. Biol. Chem. 252*:3791–3798.
143. Hill, H. D., Schwyzer, M., Steinman, M. H., and Hill, R. L. (1977). Ovine submaxillary mucin: Primary structure and peptide substrates of UDP-*N*-acetylgalactosamine:mucin transferase. *J. Biol. Chem. 252*:3799–3804.
144. Hirabayashi, Y., Matsumoto, Y., Matsumoto, M., Toida, T., Iida, N., Matsubara, T., Kanzaki, T., Yokota, M., and Ishizuka, I. (1990). Isolation and characterization of major urinary amino acid *O*-glycosides and a dipeptide *O*-glycoside from a new lysosomal storage disorder (Kanzaki disease). *J. Biol. Chem. 265*:1693–1701.
145. Hollis, G. F., Douglas, J. G., Shaper, N. L., Shaper, J. H., Stafford-Hollis, J. M., Evans, R. J., and Kirsch, I. R. (1989). Genomic structure of murine β1,4-galactosyltransferase. *Biochem. Biophys. Res. Commun. 162*:1069–1075.

146. Holmes, E. H. (1988). Characterization of a β1-3*N*-acetylglucosaminyltransferase associated with synthesis of type 1 and type 2 lacto-series tumor-associated antigens from the human colonic adenocarcinoma cell line SW403. *Arch. Biochem. Biophys. 260*:461–468.
147. Holmes, E. H., Hakomori, S., and Ostrander, G. K. (1987). Synthesis of type 1 and 2 lacto series glycolipid antigens in human colonic adenocarcinoma and derived cell lines is due to activation of a normally unexpressed β1-3*N*-acetylglucosaminyltransferase. *J. Biol. Chem. 262*:15649–15658.
148. Holmes, E. H., Ostrander, G. K., Clausen, H., and Graem, N. (1987). Oncofetal expression of Le^x carbohydrate antigens in human colonic adenocarcinomas. Regulation through type 2 core chain synthesis rather than fucosylation. *J. Biol. Chem. 262*:11331–11338.
149. Hosomi, O., Takeya, A., and Kogure, T. (1989). Separation into two major forms of β(1-3)*N*-acetylglucosaminyltransferase from human serum. *Jpn. J. Med. Sci. Biol. 42*:77–82.
150. Hounsell, E. F. (1987). Tate and Lyle Lecture. Structural and conformational characterization of carbohydrate differentiation antigens. *Chem. Soc. Rev. 16*:161–185.
151. Hounsell, E. F., Fukuda, M., Powell, M. E., Feizi, T., and Hakomori, S. (1980). A new *O*-glycosidically linked tri-hexosamine core structure in sheep gastric mucin: A preliminary note. *Biochem. Biophys. Res. Commun. 92*:1143–1150.
152. Hounsell, E. F., Lawson, A. M., Feeney, J., Cashmore, G. C., Kane, D. P., Stoll, M., and Feizi, T. (1988). Identification of a novel oligosaccharide backbone structure with a galactose residue monosubstituted at C-6 in human foetal gastrointestinal mucins. *Biochem. J. 256*:397–401.
153. Hounsell, E. F., Lawson, A. M., Feeney, J., Gooi, H. C., Pickering, N. J., Stoll, M. S., Lui, S. C., and Feizi, T. (1985). Structural analysis of the *O*-glycosidically linked core-region oligosaccharides of human meconium glycoproteins which express oncofoetal antigens. *Eur. J. Biochem. 148*:367–377.
154. Hounsell, E. F., Lawson, A. M., and Feizi, T. (1982). Structural and antigenic diversity in mucin carbohydrate chains. In *Mucus in Health and Disease*, Vol. II (E. N. Chanatler, J. B. Elder, and M. Elstein, ed.). Plenum Press, New York, pp. 39–41.
155. Hounsell, E. F., Lawson, A. M., Stoll, M. S., Kane, D. P., Chasmore, G. C., Carruthers, R. A., Feeney, J., and Feizi, T. (1989). Characterization by mass spectrometry and 500 MHz proton nuclear magnetic resonance spectroscopy of penta- and hexasaccharide chains of human foetal gastrointestinal mucins (meconium glycoproteins). *Eur. J. Biochem. 186*:597–610.
156. Hounsell, E. F., Wood, E., Feizi, T., Fukuda, M., Powell, M. E., and Hakomori, S. (1981). Structural analysis of hexa- to octasaccharide fractions isolated from sheep gastric glycoproteins having blood-group I and i activities. *Carbohydr. Res. 90*:283–307.
157. Howard, D. R., Fukuda, M., Fukuda, M. N., and Stanley, P. (1987). The GDP-fucose: *N*-acetylglucosaminide 3-α-L-fucosyltransferase of LEC11 and LEC 12 Chinese hamster ovary mutants exhibit novel specificities for glycolipid substrates. *J. Biol. Chem. 262*:16830–16837.
158. Hughes, R. C., Bradbury, A. F., and Smyth, D. G. (1988). Substrate recognition by UDP-*N*-acetyl-α-D-galactosamine:polypeptide *N*-acetyl-α-D-galactosaminyltransferase. Effects of chain length and disulfide bonding of synthetic peptide substrates. *Carbohydr. Res. 178*:259–269.
159. Hull, S. R., and Carraway, K. L. (1989). Sulfation of the tumor cell surface sialomucin of the 13762 rat mammary adenocarcinoma. *J. Cell. Biochem. 40*:67–81.
160. Hull, S. R., Laine, R. A., Kaizu, T., Rodriguez, J., and Carraway, K. L. (1984). Structures of the *O*-linked oligosaccharides of the major cell surface sialoglycoprotein of MAT-B1 and MAT-C1 ascites sublines of the 13762 rat mammary adenocarcinoma. *J. Biol. Chem. 259*:4866–4877.
161. Humphreys-Beher, M. G., Bunnell, B., van Tuinen, P., Ledbetter, D. H., and Kidd, V. J. (1986). Molecular cloning and chromosomal localization of human 4-β-galactosyltransferase. *Proc. Natl. Acad. Sci. USA 83*:8918–8922.
162. Humphreys-Beher, M., and Carlson, D. M. (1982). Synthesis of a GlcNAc-Gal linkage on asialo-orosomucoid. Annual Meeting of the Society for Complex Carbohydrates, Abstr. 25.

163. Inoue, S., Iwasaki, M., and Matsumura, G. (1981). Novel carbohydrate structures in trout egg glycoprotein. Occurrence of a neuraminidase-resistant *N*-glycolylneuraminosyl-(2→3)-*N*-acetylgalactosamine linkage. *Biochem. Biophys. Res. Commun. 102*:1295–1301.
164. Iwasaki, M., and Inoue, S. (1981). Structures of *O*-glycosidically linked carbohydrate units of herring egg sialoglycoprotein. *J. Biochem. (Tokyo) 89*:1067–1074.
165. Jabbal, I., and Schachter, H. (1971). Pork liver guanosine diphosphate-L-fucose glycoprotein fucosyltransferases. *J. Biol. Chem. 246*:5154–5161.
166. Jackson, S. P., and Tjian, R. (1988). *O*-Glycosylation of eukaryotic transcription factors: Implications for mechanisms of transcriptional regulation. *Cell 55*:125–133.
167. Johnson, W. V., and Heath, E. C. (1986). Evidence for posttranslational *O*-glycosylation of fetuin. *Biochemistry 25*:5518–5525.
168. Johnson, D. C., and Spear, P. G. (1983). *O*-Linked oligsaccharides are acquired by herpes simplex virus glycoproteins in the Golgi apparatus. *Cell 32*:987–997.
169. Johnson, P. H., and Watkins, W. M. (1985). Sialyl compounds as acceptor substrates for the human α-3- and α-3/4-L-fucosyltransferases. *Biochem. Soc. Trans. 13*:1119–1120.
170. Johnson, P. H., and Watkins, W. M. (1987). Sialyl compounds as acceptor substrates for fucosyltransferases in normal and leukemic human granulocytes. *Biochem. Soc. Trans. 15*:396.
171. Johnson, P. H., Yates, A. D., and Watkins, W. M. (1981). Human salivary fucosyltransferases. Evidence for two distinct α-3-L-fucosyltransferase activities one of which is associated with the Lewis blood group *Le* gene. *Biochem. Biophys. Res. Commun. 100*:1611–1618.
172. Jokinen, M., Andersson, L. C., and Gahmberg, C. G. (1985). Biosynthesis of the major red cell sialoglycoprotein, glycophorin A. *O*-Glycosylation. *J. Biol. Chem. 260*:11314–11321.
173. Jokinen, M., Gahmberg, C. G., and Andersson, L. C. (1979). Biosynthesis of the major human red cell sialoglycoprotein, glycophorin A, in a continuous cell line. *Nature 279*:604–607.
174. Joziasse, D. H., Shaper, J. H., van den Eijnden, D. H., van Tunen, A. J., and Shaper, N. L. (1990). Bovine α1-3-galactosyltransferase: Isolation and characterization of a cDNA clone. Identification of homologous sequences in human genomic DNA. *J. Biol. Chem. 264*:14290–14297.
175. Kanamori, A., Inoue, S., Iwasaki, M., Kitajima, K., Kawai, G., Yokoyama, S., and Inoue, Y. (1990). Deaminated neuraminic acid-rich glycoprotein of rainbow trout egg vitelline envelope. *J. Biol. Chem. 265*:21811–21819.
176. Kanani, A., Sutherland, D. R., Fibach, E., Matta, K. L., Hindenburg, A., Brockhausen, I., Kuhns, W., Taub, R. N., van den Eijnden, D. H., and Baker, M. A. (1990). Human leukemic myeloblasts and myeloblastoid cells contain the enzyme cytidine 5′-monophosphate-*N*-acetylneuraminic acid:Galβ1-3GalNAcα(2-3)-sialyltransferase. *Cancer Res. 50*:5003–5007.
177. Kaur, K. J., Turco, S. J., and Laine, R. A. (1982). Erythroglycan can be elongated by bovine milk UDP-galactose:D-glucose 4-β-galactosyltransferase. *Biochem. Int. 4*:345–351.
178. Kawashima, I., Fukuda, K., Tomita, M., and Hamada, A. (1982). Isolation and characterization of alkali-labile oligosaccharide units from porcine erythrocyte glycophorin. *J. Biochem. (Tokyo) 91*:865–872.
179. Kessler, M. J., Mise, T., Ghai, R. D., and Bahl, O. P. (1979). Structure and location of the *O*-glycosidic carbohydrate units of human chorionic gonadotropin. *J. Biol. Chem. 254*: 7909–7914.
180. Kiang, W. L., Krusius, T., Finne, J., Margolis, R. U., and Margolis, R. K. (1982). Glycoproteins and proteoglycans of the chromaffin granule matrix. *J. Biol. Chem. 257*:1651–1659.
181. Kim, Y. S., Perdomo, J., and Nordberg, J. (1971). Glycoprotein biosynthesis in small intestinal mucosa. I. A study of glycosyltransferases in microsomal subfractions. *J. Biol. Chem. 246*: 5466–5476.
182. Kingsley, D. M., Kozarsky, K. F., Hobbie, L., and Krieger, M. (1986). Reversible defects in *O*-linked glycosylation and LDL receptor expression in a UDP-Gal/UDP-GalNAc 4-epimerase deficient mutant. *Cell 44*:749–759.
183. Klein, A., Lamblin, G., Lhermitte, M., Roussel, P., Breg, J., van Halbeek, H., and Vliegenthart, J. F. G. (1988). Primary structure of neutral oligosaccharides derived from respiratory mucus

glycoproteins of a patient suffering from bronchiectasis, determined by combination of 500-MHz proton NMR spectroscopy and quantitative sugar analysis. 1. Structure of 16 oligosaccharides having the Galβ(1-3)GalNAc-ol core (type 1) or the Galβ(1-3)[GlcNAcβ(1-6)]GalNAc-ol core (type 2). *Eur. J. Biochem. 171*:631–642.

184. Ko, G. K. W., and Raghupathy, E. (1972). Glycoprotein biosynthesis in the developing rat brain. II. Microsomal galactosaminyltransferase utilizing endogenous and exogenous protein acceptors. *Biochim. Biophys. Acta 264*:129–143.
185. Kochetkov, N. K., Derevitskaya, V. A., and Arbatsky, N. P. (1976). The structure of pentasaccharides and hexasaccharides from blood group substance H. *Eur. J. Biochem. 67*: 129–136.
186. Koenderman, A. H. L., Koppen, P. L., and van den Eijnden, D. H. (1987). Biosynthesis of polylactosaminoglycans. Novikoff ascites tumor cells contain two UDP-GlcNAc:β-galactoside β1-6-*N*-acetylglucosaminyltransferase activities. *Eur. J. Biochem. 166*:199–208.
187. Korrell, S. A. M., Clemetson, K. J., van Halbeek, H., Kamerling, J. P., Sixma, J. J., and Vliegenthart, J. F. G. (1984). Structural studies of the *O*-linked carbohydrate chains of human platelet glycocalicin. *Eur. J. Biochem. 140*:571–576.
188. Korrell, S. A. M., Clemetson, K. J., van Halbeek, H., Kamerling, J. P., Sixma, J. J., and Vliegenthart, J. F. G. (1985). The structure of a fucose-containing *O*-glycosidic carbohydrate chain of human platelet glycocalicin. *Glycoconjugate J. 2*:229–234.
189. Kozarsky, K., Kingsley, D., and Krieger, M. (1988). Use of a mutant cell line to study the kinetics and function of *O*-linked glycosylation of low density lipoprotein receptors. *Proc. Natl. Acad. Sci. USA 85*:4335–4339.
190. Kramer, M. F., and Geuze, J. J. (1977). Glycoprotein transport in the surface mucous cells of the rat stomach. *J. Cell Biol. 73*:533–547.
191. Kramer, M. F., and Geuze, J. J. (1980). Comparison of various methods to localize a source of radioactivity in ultrastructural autoradiographs. The site of [^{3}H]-galactose incorporation in surface mucous cells of the rat stomach. *J. Histochem. Cytochem. 28*:381–387.
192. Kramer, M. F., Geuze, F. F., and Strous, G. J. A. M. (1978). Site of synthesis, intracellular transport, and secretion of glycoprotein in exocrine cells. In *Respiratory Tract Mucus. CIBA Found. Symp. 54*:25–51.
193. Krotkiewski, H., Lisowska, E., Angel, A.-S., and Nilsson, B. (1988). Structural analysis by fast-atom-bombardment mass spectrometry of the mixture of alditols derived from the *O*-linked oligosaccharides of murine glycophorins. *Carbohydr. Res. 184*:27–38.
194. Krusius, T., Fukuda, M., Dell, A., and Ruoslahti, E. (1985). Structure of the carbohydrate units of human amniotic fluid fibronectin. *J. Biol. Chem. 260*:4110–4116.
195. Krusius, T., Reinhold, V. N., Margolis, R. K., and Margolis, R. U. (1987). Structural studies on sialylated and sulphated *O*-glycosidic mannose-linked oligosaccharides in the chondroitin sulphate proteoglycan of brain. *Biochem. J. 245*:229–234.
196. Kukowska-Latallo, J. F., Larsen, R. D., Nair, R. P., and Lowe, J. B. (1990). A cloned human cDNA determines expression of a mouse stage-specific embryonic antigen and the Lewis blood group α(1,3/1,4)fucosyltransferase. *Genes Deve. 4*:1288–1303.
197. Kumazaki, T., and Yoshida, A. (1984). Biochemical evidence that secretor gene *Se* is a structural gene encoding a specific fucosyltransferase. *Proc. Natl. Acad. Sci. USA 81*: 4193–4197.
198. Kurosaka, A., Funakoshi, I., Matsuyama, M., Nagayo, T., and Yamashina, I. (1985). UDP-GalNAc:GalNAc-mucin α-*N*-acetylgalactosamine transferase activity in human intestinal cancerous tissues. *FEBS Lett. 190*:259–262.
199. Kurosaka, A., Nakajima, H., Funakoshi, J., Matsuyama, M., Nagayo, T., and Yamashina, I. (1983). Structures of the major oligosaccharides from a human rectal adenocarcinoma glycoprotein. *J. Biol. Chem. 258*:11594–11598.
200. Lamblin, G., Boersma, A., Klein, A., Roussel, P., van Halbeek, H ., and Vliegenthart, J. F. G . (1984). Primary structure determination of five sialylated oligosaccharides derived from bronchial mucus glycoproteins of patients suffering from cystic fibrosis. The occurrence of the

NeuAcα(2-3)Galβ(1-4)[Fucα(1-3)]GlcNAcβ(1-*) structural element revealed by 500-MHz ^{1}H NMR spectroscopy. *J. Biol. Chem.* *259*:9051–9058.

201. Lamblin, G., Boersma, A., Lhermitte, M., Roussel, P., Mutsaers, J. H. G. M., van Halbeek, H., and Vliegenthart, J. F. G. (1984). Further characterization, by a combined high-performance liquid chromatography/proton NMR approach, of the heterogeneity displayed by the neutral carbohydrate chains of human bronchial mucins. *Eur. J. Biochem.* *143*: 227–236.
202. Lamblin, G., Lhermitte, M., Boersma, A., Roussel, P., and Reinhold, V. (1980). Oligosaccharides of human bronchial glycoproteins. Neutral di- and trisaccharides isolated from a patient suffering from chronic bronchitis. *J. Biol. Chem.* *255*:4595–4598.
203. Lan, M. S., Batra, S. K., Qi, W.-N., Metzgar, R. S., and Hollingsorth, M. A. (1990). Cloning and sequencing of a pancreatic tumor mucin cDNA. *J. Biol. Chem.* *265*:15294–15299.
204. Larsen, R. D., Ernst, L. K., Nair, R. P., and Lowe, J. B. (1990). Molecular cloning, sequence and expression of a human GDP-L-fucose:β-D-galactoside 2-α-L-fucosyltransferase cDNA that can form the H blood group antigen. *Proc. Natl. Acad. Sci. USA* *87*:6674–6678.
205. Larsen, R. D., Rajan, V. P., Ruff, M. M., Kukowska-Latallo, J., Cummings, R. D., and Lowe, J. B. (1989). Isolation of a cDNA encoding a murine UDPgalactose:β-D-galactosyl-1,4-*N*-acetyl-D-glucosaminide α-1,3-galactosyltransferase: Expression cloning by gene transfer. *Proc. Natl. Acad. Sci. USA* *86*:8227–8231.
206. Le Pendu, J., Cartron, J. P., Lemieux, R. U., and Oriol, R. (1985). The presence of at least two different H-blood-group-related β-D-Gal α-2-L-fucosyltransferases in human serum and the genetics of blood group H substances. *Am. J. Hum. Genet.* *37*:749–760.
207. Le Pendu, J., Cartron, J. P., Oriol, R., and Lemieux, R. U. (1983). Two α-2-L-fucosyltransferases in human serum under distinct genetic control. In *Seventh International Symposium on Glycoconjugates*, p. 748.
208. Lecat, D., Lemonnier, M., Derappe, C., Lhermitte, M., van Halbeek, H., Dorland, L., and Vliegenthart, J. F. G. (1984). The structure of sialylglycopeptides of the *O*-glycosidic type, isolated from sialidosis (mucolipidosis I) urine. *Eur. J. Biochem.* *140*:415–420.
209. Lefrancois, L., Puddington, L., Machamer, C. E., and Bevan, M. J. (1985). Acquisition of cytotoxic T lymphocyte-specific carbohydrate differentiation antigens. *J. Exp. Med.* *162*:1275–1293.
210. Leppanen, A., Korvuo, A., Puro, K., and Renkonen, O. (1986). Glycoproteins of human teratocarcinoma cells (PA1) carry both anomers of *O*-glycosyl-linked D-galactopyranosyl-(1-3)-2-acetamido-2-deoxy-α-D-galactopyranosyl group. *Carbohydr. Res.* *153*:87–95.
211. Linden, H.-U., Klein, R. A., Egge, H., Peter-Katalinic, J., Dabrowski, J., and Schindler, D. (1989). Isolation and structural characterization of sialic acid-containing glycopeptides of the *O*-glycosidic type from the urine of two patients with an hereditary deficiency in α-*N*-acetylgalactosaminidase activity. *Biol. Chem. Hoppe-Seyler* *370*:661–672.
212. Lisowska, E., Duk, M., and Dahr, W. (1980). Comparison of alkali-labile oligosaccharide chains of M and N blood-group glycopeptides from human erythrocyte membrane. *Carbohydr. Res.* *79*:103–113.
213. Lloyd, K. O., and Kabat, E. A. (1968). Immunochemical studies on blood groups. XLI. Proposed structures for the carbohydrate portions of blood group A, B, H, Lewis^a, and Lewis^b substances. *Proc. Natl. Acad. Sci. USA* *61*:1470–1477.
214. Lloyd, K. O., Kabat, E. A., and Licerio, E. (1968). Immunochemical studies on blood groups. XXXVIII. Structures and activities of oligosaccharides produced by alkaline degradation of blood-group Lewis^a substance. Proposed structure of the carbohydrate chains of human blood-group A, B, H, Le^a, and Le^b substances. *Biochemistry* *7*:2976–2990.
215. Lombart, C. G., and Winzler, R. J. (1974). Isolation and characterization of oligosaccharides from canine submaxillary mucin. *Eur. J. Biochem.* *49*:77–86.
216. Longmore, G. D., and Schachter, H. (1982). Control of glycoprotein synthesis. VI. Product identification and substrate specificity studies of the GDP-L-fucose:2-acetamido-2-deoxy-β-D-glucoside (Fuc to Asn-linked GlcNAc) 6-α-L-fucosyltransfrase in a Golgi-rich fraction from porcine liver. *Carbohydr. Res.* *100*:365–392.

217. Machamer, C. E., and Cresswell, P. (1984). Monensin prevents terminal glycosylation of the *N*- and *O*-linked oligosaccharides of the HLA-DR-associated invariant chain and inhibits its dissociation from the alpha-beta chain complex. *Proc. Natl. Acad. Sci. USA 81*:1287–1291.

218. Maisonrouge-McAuliffe, F., and Kabat, E. A. (1976). Immunochemical studies on blood groups. LXV. Structures and immunochemical properties of oligosaccharides from two fractions of blood group substance from human ovarian cyst fluid differing in B, I, and i activities and reactivity towards concanavalin A. *Arch. Biochem. Biophys. 175*:90–113.

219. Markwell, M. A. K., and Paulson, J. C. (1980). Sendai virus utilizes specific sialyloligosaccharides as host cell receptor determinants. *Proc. Natl. Acad. Sci. USA 77*:5693–5697.

220. Marti, T., Schaller, J., Rickli, E. E., Schmid, K., Kamerling, J. P., Gerwig, G. J., van Halbeek, H., and Vliegenthart, J. F. G. (1988). The *N*- and *O*-linked carbohydrate chains of human, bovine, and porcine plasminogen. Species specificity in relation to sialylation and fucosylation patterns. *Eur. J. Biochem. 173*:57–63.

221. Masri, K. A., Appert, H. E., and Fukuda, M. N. (1988). Identification of the full-length coding sequence for human galactosyltransferase (β-*N*-acetylglucosaminide:β1,4-galactosyltransferase). *Biochem. Biophys. Res. Commun. 157*:657–663.

222. Mawhinney, T. P., Adelstein, E., Morris, D. A., Mawhinney, A. M., and Barbero, G. J. (1987). Structure determination of five sulfated oligosaccharides derived from tracheobronchial mucus glycoproteins. *J. Biol. Chem. 262*:2994–3001.

223. McGuire, E. J. (1970). Biosynthesis of submaxillary mucins. In *Blood and Tissue Antigens* (D. Aminoff, ed.). Academic Press, New York, pp. 461–478.

224. McGuire, E. J., and Roseman, S. (1967). Enzymatic synthesis of the protein-hexosamine linkage in sheep submaxillary mucin. *J. Biol. Chem. 242*:3745–3747.

225. Mellis, S. J., and Baenziger, J. U. (1983). Structures of the *O*-glycosidically linked oligosaccharides of human IgD. *J. Biol. Chem. 258*:11557–11563.

226. Mendicino, J., Sivakami, S., Davila, M., and Chandrasekaran, E. V. (1982). Purification and properties of UDP-Gal:*N*-acetylgalactosaminide mucin:β1,3-galactosyltransferase from swine trachea mucosa. *J. Biol. Chem. 257*:3987–3994.

227. Meyrick, B., and Reid, L. (1975). In vitro incorporation of (^{3}H)threonine and (^{3}H)glucose by the mucous and serous cells of the human bronchial submucosal gland. A quantitative electron microscope study. *J. Cell Biol. 67*:320–344.

228. Mizuochi, T., Yamashita, K., Fujikawa, K., Titani, K., and Kobata, A. (1980). The structures of the carbohydrate moieties of bovine blood coagulation factor X. *J. Biol. Chem. 255*:3526–3531.

229. Mollicone, R., Gibaud, A., Francois, A., Ratcliffe, M., and Oriol, R. (1990). Acceptor specificity and tissue distribution of three human α-3-fucosyltransferases. *Eur. J. Biochem. 191*:169–176.

230. Muramatsu, H., Kamada, Y., and Muramatsu, T. (1986). Purification and properties of *N*-acetylglucosaminide α1-3-fucosyltransferase from embryonal carcinoma cells. *Eur. J. Biochem. 157*:71–75.

231. Murayama, J., Fukuda, K., Yamashita, T., and Hamada, A. (1983). Structural studies on sugar chains in animal glycophorins. In *Seventh International Symposium on Glycoconjugates*, p. 184.

232. Murayama, J., Manabe, H., Fukuda, K., Utsumi, H., and Hamada, A. (1989). Structure of the major *O*-glycosidic oligosaccharide of monkey erythrocyte glycophorin. *Glycoconjugate J. 6*:499–510.

233. Murayama, J., Tomita, M., and Hamada, A. (1982). Primary structure of horse erythrocyte glycophorin HA. Its amino acid sequence has a unique homology with those of human and porcine erythrocyte glycophorins. *J. Membr. Biol. 64*:205–215.

234. Mutsaers, J. H. G. M., van Halbeek, H., Vliegenthart, J. F. G., Wu, A. M., and Kabat, E. A. (1986). Typing of core and backbone domains of mucin-type oligosaccharides from human ovarian cyst glycoproteins by 500 MHz proton NMR spectroscopy. *Eur. J. Biochem. 157*: 139–146.

235. Nagai, M., Dave, V., Kaplan, B. E., and Yoshida, A. (1978). Human blood group glycosyltransferases. I. Purification of *N*-acetylgalactosaminyltransferase. *J. Biol. Chem. 253*: 377–379.
236. Nagai, M., Dave, V., Meunsch, H., and Yoshida, A. (1978). Human blood group glycosyltransferases. II. Purification of galactosyltransferase. *J. Biol. Chem. 253*:380–381.
237. Nakazawa, K., Ando, T., Kimura, T., and Narimatsu, H. (1988). Cloning and sequencing of a full-length cDNA of mouse *N*-acetylglucosamine (β1-4)galactosyltransferase. *J. Biochem. (Tokyo) 104*:165–168.
238. Narimatsu, H., Sinha, S., Brew, K., Okayama, H., and Qasba, P. K. (1986). Cloning and sequencing of cDNA of bovine *N*-acetylglucosamine (β1-4)galactosyltransferase. *Proc. Natl. Acad. Sci. USA 83*:4720–4724.
239. Nasir-Ud-Din. (1987). Structure of acidic oligosaccharides isolated from pronase-treated glycoprotein of bonnet monkey. (*Macaca radiata*) cervical mucus. *Carbohydr. Res. 159*: 95–107.
240. Nasir-ud-Din, Hussain, S. A., Jeanloz, R. W., and Walker-Nasir, E. (1990). Studies on cervical glycoproteins. Isolation and characterization of neutral oligosaccharides from pronase-treated glycoproteins of bonnet monkey (*Macacca radiata*). *Carbohydr. Res. 205*: 444–452.
241. Nasir-Ud-Din, Jeanloz, R. W., Lamblin, G., Roussel, P., van Halbeek, H., Mutsaers, J. H. G. M., and Vliegenthart, J. F. G. (1986). Structure of sialyloligosaccharides isolated from bonnet monkey (*Macaca radiata*) cervical mucus glycoproteins exhibiting multiple blood group activities. *J. Biol. Chem. 261*:1992–1997.
242. Nasir-Ud-Din, Walker-Nasir, E., and Malghani, M. A. K. (1983). Bonnet monkey cervical mucus: Isolation and characterization of oligosaccharides from the pronase-treated periovulatory phase glycoprotein. In *Seventh International Symposium on Glycoconjugates*. p. 603.
243. Nato, F., Goulut, C., Bourillon, R., van Halbeek, H., and Vliegenthart, J. F. G. (1986). The structure of *O*-glycosidic oligosaccharide chains of the major *Zajdela* hepatoma ascites cell membrane glycoprotein. *Eur. J. Biochem. 159*:303–308.
244. Navaratnam, N., Findlay, J. B. C., Keen, J. N., and Watkins, W. M. (1990). Purification, properties and partial amino acid sequence of the blood-group-A-gene-associated α-3-*N*-acetylgalactosaminyltransferase from human gut mucosal tissue. *Biochem. J. 271*:93–98.
245. Neutra, M., and Leblond, C. P. (1966). Radioautographic comparison of the uptake of galactose-[^{3}H] and glucose-[^{3}H] in the Golgi region of various cells secreting glycoproteins or mucopolysaccharides. *J. Cell Biol. 30*:137–150.
246. Neutra, M., and Leblond, C. P. (1966). Synthesis of the carbohydrate of mucus in the Golgi complex as shown by electron microscope radioautography of goblet cells from rats injected with glucose-H^3. *J. Cell Biol. 30*:119–136.
247. Newman, W., and Kabat, E. A. (1976). Immunochemical studies on blood groups. Structures and immunochemical properties of nine oligosaccharides from B-active and non-B-active blood group substances of horse gastric mucosae. *Arch. Biochem. Biophys. 172*: 535–550.
248. Niemann, H., Boschek, B., Evans, D., Rosing, M., Tamura, T., and Klenk, H. (1982). Post-translational glycosylation of coronavirus glycoprotein E1—inhibition by monensin. *EMBO J. 1*:1499–1504.
249. Niemann, H., Geyer, R., Klenk, H. D., Linder, D., Stirm, S., and Wirth, M. (1984). The carbohydrates of mouse hepatitis virus (MHV) A59: Structures of the *O*-glycosidically linked oligosaccharides of glycoprotein E1. *EMBO J. 3*:665–670.
250. Nilsson, B., De Luca, S., Lohmander, S., and Hascall, V. C. (1982). Structures of *N*-linked and *O*-linked oligosaccharides on proteoglycan monomer isolated from the swarm rat chondrosarcoma. *J. Biol. Chem. 257*:10920–10927.
251. Nilsson, B., Norden, N. E., and Svensson, S. (1979). Structural studies on the carbohydrate portion of fetuin. *J. Biol. Chem. 254*:4545–4553.

252. Oates, M. D. G., Rosbottom, A. C., and Schrager, J. (1974). Further investigations into the structure of human gastric mucin: The structural configuration of the oligosaccharide chains. *Carbohydr. Res. 34*:115–137.

253. Oh-eda, M., Hasegawa, M., Hattori, K., Kuboniwa, H., Kojima, T., Orita, T., Tomonou, K., Yamazaki, T., and Ochi, N. (1990). *O*-Linked sugar chain of human granulocyte colony-stimulating factor protects it against polymerization and denaturation allowing it to retain its biological activity. *J. Biol. Chem. 265*:11432–11435.

254. Olofsson, S., Blomberg, J., and Lycke, E. (1981). *O*-Glycosidic carbohydrate–peptide linkages of herpes simplex virus glycoproteins. *Arch. Virol. 70*:321–329.

255. Parkkinen, J., and Finne, J. (1983). Isolation and structural characterization of five major sialyloligosaccharides and a sialylglycopeptide from normal human urine. *Eur. J. Biochem. 136*:355–361.

256. Paulson, J. C., and Colley, K. J. (1989). Structure, localization, and control of cell type-specific glycosylation. *J. Biol. Chem. 264*:17615–17618.

257. Paulson, J. C., Sadler, J. E., and Hill, R. L. (1979). Restoration of specific myxovirus receptors to asialoerythrocytes by incorporation of sialic acid with pure sialyltransferases. *J. Biol. Chem. 254*:2120–2124.

258. Pierce-Crétel, A., Decottignies, J.-P., Wieruszeski, J.-M., Strecker, G., Montreuil, J., and Spik, G. (1989). Primary structure of twenty-three neutral and monosialylated oligosaccharides *O*-glycosidically linked to the human secretory immunoglobulin A hinge region determined by a combination of permethylation analysis and 400 MHz ^{1}H-NMR spectroscopy. *Eur. J. Biochem. 182*:457–476.

259. Pierce-Cretel, A., Pamblanco, M., Strecker, G., Montreuil, J., and Spik, G. (1981). Heterogeneity of the glycans *O*-glycosidically linked to the hinge region of the secretory immunoglobulins from human milk. *Eur. J. Biochem. 114*:169–178.

260. Piller, F., Blanchard, D., Huet, M., and Cartron, J. (1986). Identification of a α-NeuAc-(2-3)-β-D-galactopyranosyl *N*-acetyl-β-D-galactosaminyltransferase in human kidney. *Carbohydr. Res. 149*:171–184.

261. Piller, F., and Cartron, J. (1983). UDP-GlcNAc:Galβ1-4Glc(NAc) β1-3*N*-acetylglucosaminyltransferase. Identification and characterization in human serum. *J. Biol. Chem. 258*:12293–12299.

262. Piller, F., Cartron, J. P., Maranduba, A., Veyrieres, A., Leroy, Y., and Fournet, B. (1984). Biosynthesis of blood group I antigens. Identification of a UDP-GlcNAc:GlcNAcβ1-3Gal(-R) β1-6 (GlcNAc to Gal) *N*-acetylglucosaminyltransferase in hog gastric mucosa. *J. Biol. Chem. 259*:13385–13390.

263. Piller, F., Piller, V., Fox, R. I., and Fukuda, M. (1988). Human T-lymphocyte activation is associated with changes in *O*-glycan biosynthesis. *J. Biol. Chem. 263*:15146–15150.

264. Piller, V., Piller, F., and Fukuda, M. (1990). Biosynthesis of truncated *O*-glycans in the T cell line Jurkat. Localization of *O*-glycan initiation. *J. Biol. Chem. 265*:9264–9271.

265. Piller, V., Piller, F., Klier, F. G., and Fukuda, M. (1989). *O*-glycosylation of leukosialin in K562 cells. Evidence for initiation and elongation in early Golgi compartments. *Eur. J. Biochem. 183*:123–135.

266. Podolsky, D. K. (1985). Oligosaccharide structures of human colonic mucin. *J. Biol. Chem. 260*:8262–8271.

267. Podolsky, D. K. (1985). Oligosaccharide structures of isolated human colonic mucin species. *J. Biol. Chem. 260*:15510–15515.

268. Prieels, J., Monnom, D., Dolmans, M., Beyer, T. A., and Hill, R. L. (1981). Copurification of the Lewis blood group *N*-acetylglucosaminide α1-4-fucosyltransferase and *N*-acetylglucosaminide α1-3-fucosyltransferase from human milk. *J. Biol. Chem. 256*:10456–10463.

269. Prohaska, R., Koerner, T. A. W., Armitage, I. M., and Furthmayr, H. (1981). Chemical and C-13 NMR studies of the blood group M and N active sialoglycopeptides from human glycophorin A. *J. Biol. Chem. 256*:5781–5791.

270. Rajan, V. P., Larsen, R. D., Ajmera, S., Ernst, L. K., and Lowe, J. B. (1989). A cloned human DNA restriction fragment determines expression of a GDP-L-fucose:β-D-galactoside 2-α-L-fucosyltransferase in transfected cells. Evidence for isolation and transfer of the human blood group locus. *J. Biol. Chem. 264*:11158–11167.
271. Rana, S. S., Chandrasekaran, E. V., Kennedy, J., and Mendicino, J. (1984). Purification and structures of oligosaccharide chains in swine trachea and Cowper's gland mucin glycoproteins. *J. Biol. Chem. 259*:12899–12907.
272. Rana, S. S., Chandrasekaran, E. V., and Mendicino, J. (1987). Structures of the sialylated oligosaccharide chains in swine trachea mucin glycoproteins. *J. Biol. Chem. 262*:3654–3659.
273. Rearick, J. I., Cummings, R., and Kornfeld, S. (1981). Specific assay for UDP-GlcNAc: β-galactoside *N*-acetylglucosaminyltransferase. Annual Meeting of The Society for Complex Carbohydrates, Abstr. 63.
274. Rearick, J. I., Sadler, J. E., Paulson, J. C., and Hill, R. L. (1979). Enzymatic characterization of β-D-galactoside α2-3-sialyltransferase from porcine submaxillary gland. *J. Biol. Chem. 254*:4444–4451.
275. Reddy, M. S., Prakobphol, A., Levine, M. J., and Tabak, L. A. (1982). Structures of the *O*-glycosidic units of a lower molecular weight salivary mucin. Annual Meeting of The Society for Complex Carbohydrates, Abstr. 29.
276. Reddy, M. S., Shah, R. H., and Bahl, O. P. (1979). Structures of the carbohydrate units of mucins from normal and fibrocystic human submaxillary secretions. Annual Meeting of The Society for Complex Carbohydrates, Abstr. 40.
277. Reddy, M. S., Tabak, L. A., and Levine, M. J. (1983). Characterization of oligosaccharides from human tracheobronchial mucin. Annual Meeting of The Society for Complex Carbohydrates, Abstr. 18.
278. Reddy, P., Caras, I., and Krieger, M. (1989). Effects of *O*-linked glycosylation on the cell surface expression and stability of decay-accelerating factor, a glycosphingolipid-anchored membrane protein. *J. Biol. Chem. 264*:17329–17336.
279. Ronzio, R. A. (1973). Glycoprotein synthesis in the adult rat pancreas. I. Subcellular distributions of uridine diphosphate galactose:glycoprotein galactosyltransferase and thiamine pyrophosphate phosphohydrolase. *Biochim. Biophys. Acta 313*:286–295.
280. Ronzio, R. A. (1973). Glycoprotein synthesis in the adult rat pancreas. II. Characterization of Golgi-rich fractions. *Arch. Biochem. Biophys. 159*:777–784.
281. Roth, J. (1984). Cytochemical localization of terminal *N*-acetyl-D-galactosamine residues in cellular compartments of intestinal goblet cells: Implications for the topology of *O*-glycosylation. *J. Cell Biol. 98*:399–406.
282. Roth, J. (1987). Subcellular organization of glycosylation in mammalian cells. *Biochim. Biophys. Acta 906*:405–436.
283. Roth, J., and Berger, E. G. (1982). Immunocytochemical localization of galactosyltransferase in HeLa cells: Codistribution with thiamine pyrophosphatase in trans-Golgi cisternae. *J. Cell Biol. 92*:223–229.
284. Rovis, L., Anderson, B., Kabat, E. A., Gruezo, F., and Liao, J. (1973). Structures of oligosaccharides produced by base-borohydride degradation of human ovarian cyst blood group H, Le^b and Le^a active glycoproteins. *Biochemistry 12*:5340–5354.
285. Russin, T. Z., Laine, R. A., and Turco, S. J. (1981). Cell-free biosynthesis of erythroglycan in a microsomal fraction from K-562 cells. *Biochem. J. 197*:327–332.
286. Russo, R. N., Shaper, N. L., and Shaper, J. (1990). Bovine β1-4-galactosyltransferase: Two sets of mRNA transcripts encode two forms of the protein with different amino-terminal domains. In vitro translation experiments demonstrate that both the short and the long forms of the enzyme are type II membrane-bound glycoproteins. *J. Biol. Chem. 265*: 3324–3331.
287. Sadler, J. E. (1984). Biosynthesis of glycoproteins: Formation of *O*-linked oligosaccharides. In *Biology of Carbohydrates*, Vol. 2 (V. Ginsburg and P. W. Robbins, eds.). John Wiley & Sons, New York, pp. 199–288.

288. Sadler, J. E., Rearick, J. I., and Hill, R. L. (1979). Purification to homogeneity and enzymatic characterization of an α-*N*-acetylgalactosaminide α2-6-sialyltransferase from porcine submaxillary glands. *J. Biol. Chem. 254*:5934–5941.

289. Sadler, J. E., Rearick, J. I., Paulson, J. C., and Hill, R. L. (1979). Purification to homogeneity of a β-galactoside α2-3-sialyltransferase and partial purification of an α-*N*-acetylgalactosaminide α2-6-sialyltransferase from porcine submaxillary glands. *J. Biol. Chem. 254*:4434–4443.

290. Saito, T., Itoh, T., and Adachi, S. (1981). The chemical structure of a tetrasaccharide containing *N*-acetylglucosamine obtained from bovine colostrum kappa-casein. *Biochim. Biophys. Acta 673*:487–494.

291. Saito, T., Itoh, T., and Adachi, S. (1988). Chemical structure of neutral sugar chains isolated from human mature milk kappa-casein. *Biochim. Biophys. Acta 964*:213–220.

292. Sarnesto, A., Köhlin, T., Thurin, J., and Blaszczyk-Thurin, M. (1990). Purification of *H*-gene-encoded β-galactoside α1-2-fucosyltransferase from human serum. *J. Biol. Chem. 265*:15067–15075.

293. Savage, A. V., Donoghue, C. M., D'Arcy, S. M., Koeleman, C. A. M., and van den Eijnden, D. H. (1990). Structure determination of five sialylated trisaccharides with core types 1, 3, or 5 isolated from bovine submaxillary mucin. *Eur. J. Biochem. 192*:427–432.

294. Savage, A. V., Donoghue, C. M., Koeleman, C. A. M., and van den Eijnden, D. H. (1990). Structural characterization of sialylated tetrasaccharides and pentasaccharides with blood group H and Le^x activity isolated from bovine submaxillary mucin. *Eur. J. Biochem. 193*: 837–843.

295. Savage, A. V., Koppen, P. L., Schiphorst, W. E. C. M., Trippelvitz, L. A. W., van Halbeek, H., Vliegenthart, J. F. G., and van den Eijnden, D. H. (1986). Porcine submaxillary mucin contains α2-3 and α2-6-linked *N*-acetyl- and *N*-glycolyl-neuraminic acid. *Eur. J. Biochem. 160*: 123–129.

296. Schachter, H. (1978). Glycoprotein biosynthesis. In *The Glycoconjugates*, Vol. 2 (W. Pigman and M. I. Horowitz, eds.). Academic Press, New York, pp. 87–181.

297. Schachter, H. (1986). Biosynthetic controls that determine the branching and microheterogeneity of protein-bound oligosaccharides. *Biochem. Cell Biol. 64*:163–181.

298. Schachter, H., and Brockhausen, I. (1989). The biosynthesis of branched *O*-glycans. In *Symposia of the Society for Experimental Biology. Mucus and Related Topics*, No. 43 (E. Chantler and N. A. Ratcliffe, eds.). Cambridge, Engl., pp. 1–26.

299. Schachter, H., Brockhausen, I., and Hull, E. (1989). HPLC assays for the *N*-acetylglucosaminyltransferases involved in *N*- and *O*-glycan synthesis. *Methods Enzymol. Complex Carbohydr.* (Part F) *179*:351–397.

300. Schachter, H., McGuire, E. J., and Roseman, S. (1971). Sialic acids. XIII. A uridine diphosphate D-galactose:mucin galactosyltransferase from porcine submaxillary gland. *J. Biol. Chem. 246*: 5321–5328.

301. Schachter, H., and Roseman, S. (1980). Mammalian glycosyltransferases: Their role in the synthesis and function of complex carbohydrates and glycolipids. In *Biochemistry of Glycoproteins and Proteoglycans* (W. J. Lennarz, ed.). Plenum Press, New York, pp. 85–160.

302. Schachter, H., and Tilley, C. A. (1978). The biosynthesis of human blood group substances. In *Biochemistry of Carbohydrates* (D. J. Manners, ed.). University Park Press, Baltimore, pp. 209–246.

303. Schachter, H., and Williams, D. (1982). Biosynthesis of mucus glycoproteins. In *Mucus in Health and Disease*, Vol. II (E. N. Chantler, J. B. Elder, and M. Elstein, ed.). Plenum Press, New York, pp. 3–28.

304. Schenkel-Brunner, H. (1973). Incorporation of galactose into blood groups (ABH) precursor substance by lactose synthetase from human milk. Effects on cross-reactivity with anti-Type-14 pneumococcus serum. *Eur. J. Biochem. 33*:30–35.

305. Schenkel-Brunner, H., Kabat, E. A., and Liao, J. (1979). Biosynthesis of a blood-group-I determinant reacting with anti-I Ma serum (group 1). *Eur. J. Biochem. 98*:573–575.

306. Schwyzer, M., and Hill, R. L. (1977). Porcine A blood group-specific *N*-acetylgalactosaminyltransferase. I. Purification from porcine submaxillary glands. *J. Biol. Chem. 252*: 2338–2345.
307. Schwyzer, M., and Hill, R. L. (1977). Porcine A blood group-specific *N*-acetylgalactosaminyltransferase. II. Enzymatic properties. *J. Biol. Chem. 252*:2346–2355.
308. Scott, L. J. C., Balsamo, J., Sanes, J. R., and Lilien, J. (1990). Synaptic localization and neural regulation of an *N*-acetylgalactosaminyltransferase in skeletal muscle. *J. Neurosci. 10*: 346–350.
309. Serafini-Cessi, F., Dall'Olio, F., and Malagolini, N. (1986). Characterization of *N*-acetyl-β-D-galactosaminyltransferase from guinea-pig kidney involved in the biosynthesis of Sd^a antigen associated with Tamm–Horsfall glycoprotein. *Carbohydr. Res. 151*:65–76.
310. Serafini-Cessi, F., Dall'olio, F., Malagolini, N., and Campadelli-Fiume, G. (1989). Temporal aspects of *O*-glycosylation of glycoprotein C from herpes simplex virus type-1. *Biochem. J. 262*:479–484.
311. Serafini-Cessi, F., Dall'Olio, F., Scannavini, M., Costanzo, F., and Campadelli-Fiume, G. (1983). *N*-Acetylgalactosaminyltransferase activity involved in *O*-glycosylation of herpes simplex virus type 1 glycoproteins. *J. Virol. 48*:325–329.
312. Serafini-Cessi, F., Malagolini, N., and Dall'Olio, F. (1988). Characterization and partial purification of β-*N*-acetylgalactosaminyltransferase from urine of Sd(a+) individuals. *Arch. Biochem. Biophys. 266*:573–582.
313. Serafini-Cessi, F., Malagolini, N., Nanni, M., Dall'olio, F., Campadelli-Fiume, G., Tanner, J., and Kieff, E. (1989). Characterization of *N*- and *O*-linked oligosaccharides of glycoprotein 350 from Epstein–Barr virus. *Virology 170*:1–10.
314. Shaper, N. L., Hollis, G. F., Douglas, J. G., Kirsch, I. R., and Shaper, J. H. (1988). Characterization of the full length cDNA for murine β-1,4-galactosyltransferase. Novel features at the 5'-end predict two translational start sites at two in-frame AUGs. *J. Biol. Chem. 263*:10420–10428.
315. Shaper, N. L., Shaper, J. H., Bertness, V., Chang, H., Kirsch, I. R., and Hollis, G. F. (1986). The human galactosyltransferase gene is on chromosome 9 at band p13. *Somat. Cell. Mol. Genet. 12*:633–636.
316. Shaper, N. L., Shaper, J. H., Hollis, G. F., Chang, H., Kirsch, I. R., and Kozak, C. A. (1987). The gene for galactosyltransferase maps to mouse chromosome 4. *Cytogenet. Cell. Genet. 44*:18–21.
317. Shaper, N. L., Shaper, J. H., Meuth, J. L., Fox, J. L., Chang, H., Kirsch, I. R., and Hollis, G. F. (1986). Bovine galactosyltransferase: A clone identified by direct immunological screening of a cDNA expression library. *Proc. Natl. Acad. Sci. USA 83*:1573–1577.
318. Shaper, N. L., Wright, W. W., and Shaper, J. H. (1990). Murine β1,4-galactosyltransferase: Both the amounts and structure of the mRNA are regulated during spermatogenesis. *Proc. Natl. Acad. Sci. USA 87*:791–795.
319. Sheares, B. T., and Carlson, D. M. (1983). Characterization of UDP-galactose:2-acetamido-2-deoxy-D-glucose 3beta-galactosyltransferase from pig trachea. *J. Biol. Chem. 258*:9893–9898.
320. Sheares, B. T., Lau, J. Y. T., and Carlson, D. M. (1982). Biosynthesis of galactosyl-β1,3-*N*-acetylglucosamine. *J. Biol. Chem. 257*:599–602.
321. Sherblom, A. P., Huggins, J. W., Chestnut, R. W., Buck, R. L., Ownby, C. L., Dermer, G. B., and Carraway, K. L. (1980). Cell surface properties of ascites sublines of the 13762 rat mammary adenocarcinoma. Relationship of the major sialoglycoprotein to xenotransplantability. *Exp. Cell Res. 126*:417–426.
322. Shida, H., and Dales, S. (1981). Biogenesis of vaccinia: Carbohydrate of the hemagglutinin molecules. *Virology 111*:56–72.
323. Shier, W. T., and Roloson, G. J. (1977). Preparation and galactosyltransferase acceptor activity of derivatives of antifreeze glycoproteins of an Antarctic fish. *Can. J. Biochem. 55*:886–893.
324. Shimamura, M., Endo, T., Inoue, Y., and Inoue, S. (1983). A novel neutral oligosaccharide chain found in polysialoglycoproteins isolated from Pacific salmon eggs. Structural studies by

secondary ion mass spectrometry, proton nuclear magnetic resonance spectroscopy, and chemical methods. *Biochemistry 22*:959–963.

325. Shur, B. D., and Neely, C. A. (1988). Plasma membrane association, purification, and partial characterization of mouse sperm β1,4-galactosyltransferase. *J. Biol. Chem. 263*:17706–17714.

326. Skacel, P. O., and Watkins, W. M. (1988). Significance of altered α-2-L-fucosyltransferase levels in serum of leukemic patients. *Cancer Res. 48*:3998–4001.

327. Slomiany, B. L., Banas-Gruszka, Z., Zdebska, E., and Slomiany, A. (1982). Characterization of the Forssman-active oligosaccharides from dog gastric mucus glycoprotein isolated with the use of a monoclonal antibody. *J. Biol. Chem. 257*:9561–9565.

328. Slomiany, B. L., and Meyer, K. (1973). Oligosaccharides produced by acetolysis of blood group active (A+H) sulfated glycoproteins from hog gastric mucin. *J. Biol. Chem. 248*:2290–2295.

329. Slomiany, B. L., Murty, V. L. N., and Slomiany, A. (1980). Isolation and characterization of oligosaccharides from rat colonic mucus glycoprotein. *J. Biol. Chem. 255*:9719–9723.

330. Slomiany, A., and Slomiany, B. L. (1978). Structures of the acidic oligosaccharides isolated from rat sublingual glycoprotein. *J. Biol. Chem. 253*:7301–7306.

331. Slomiany, B. L., Zdebska, E., and Slomiany, A. (1984). Structural characterization of neutral oligosaccharides of human $H^{+}Le^{b+}$ gastric mucin. *J. Biol. Chem. 259*:2863–2869.

332. Snider, M. D. (1984). Biosynthesis of glycoproteins: Formation of *N*-linked oligosaccharides. In *Biology of Carbohydrates*, Vol. 2 (V. Ginsburg and P. W. Robbins, eds.), John Wiley & Sons, New York, pp. 163–198.

333. Soulier, S., Sarfati, R. S., and Szabo, L. (1980). Structure of the asialyl oligosaccharide chains of casein isolated from ovine colostrum. *Eur. J. Biochem. 108*:465–472.

334. Spielman, J., Hull, S. R., Sheng, Z., Kanterman, R., Bright, A., and Carraway, K. L. (1988). Biosynthesis of a tumor cell surface sialomucin. Maturation and effects of monensin. *J. Biol. Chem. 263*:9621–9629.

335. Spielman, J., Rockley, N. L., and Carraway, K. L. (1987). Temporal aspects of *O*-glycosylation and cell surface expression of ascites sialoglycoprotein-1, the major cell surface sialomucin of 13762 mammary ascites tumor cells. *J. Biol. Chem. 262*:269–275.

336. Spiro, R. G., and Bhoyroo, V. D. (1974). Structure of the *O*-glycosidically linked carbohydrate units of fetuin. *J. Biol. Chem. 249*:5704–5717.

337. Stojanovic, D., Vischer, P., and Hughes, R. C. (1984). Glycosyltransferases of baby hamster kidney cells and ricin-resistant mutants. *O*-Glycan biosynthesis. *Eur. J. Biochem. 138*:551–562.

338. Strecker, G., Wieruszeski, J., Martel, C., and Montreuil, J. (1987). Determination of the structure of sulfated tetra- and pentasaccharides obtained by alkaline borohydride degradation of hen ovomucin. A fast atom bombardment–mass spectrometric and ^{1}H-NMR spectroscopic study. *Glycoconjugate J. 4*:329–337.

339. Strous, G. J. M. (1979). Initial glycosylation of proteins with acetylgalactosaminylserine linkages. *Proc. Natl. Acad. Sci. USA 76*:2694–2698.

340. Strous, G. J. M., Hendriks, H. G. C. J. M., and Kramer, M. F. (1980). Role of galactosyl-transferases in rat gastric epithelial glycoprotein synthesis. *Biochim. Biophys. Acta 613*:381–391.

341. Struck, D. K., and Lennarz, W. J. (1980). The function of saccharide-lipids in synthesis of glycoproteins. In *The Biochemistry of Glycoproteins and Proteoglycans* (W. J. Lennarz, ed.). Plenum Press, New York, pp. 35–83.

342. Sugiura, M., Kawasaki, T., and Yamashina, I. (1982). Purification and characterization of UDP-GalNAc:polypeptide *N*-acetylgalactosamine transferase from an ascites hepatoma, AH 66. *J. Biol. Chem. 257*:9501–9507.

343. Tabak, L. A., Dickson, L., Reddy, M. S., Levine, M. J., Kuatt, B. L., and Baum, B. J. (1982). Isolation and partial characterization of a mucin-glycoprotein from rat submandibular glands. Annual Meeting of the Society for Complex Carbohydrates, Abstr. 27.

344. Takasaki, S., Yamashita, K., and Kobata, A. (1978). The sugar chain structures of ABO blood group active glycoproteins obtained from human erythrocyte membrane. *J. Biol. Chem. 253*: 6086–6091.

345. Takeuchi, M., Yoshikawa, M., Sasaki, R., and Chiba, H. (1985). Purification and characterization of UDP-*N*-acetylgalactosamine:kappa-casein polypeptide *N*-acetylgalactosaminyltransferase from mammary gland of lactating cow. *Agric. Biol. Chem. 49*: 1059–1069.
346. Takeya, A., Hosomi, O., and Kogure, T. (1987). Identification and characterization of UDP-GalNAc:NeuAcα2-3Galβ1-4Glc(NAc) β1-4(GalNAc to Gal) *N*-acetylgalactosaminyltransferase in human blood plasma. *J. Biochem. (Tokyo) 101*:251–259.
347. Tanaka, M., Anderson, B., and Dube, V. E. (1982). A new I-active sequence in oligosaccharides of ovarian cyst glycoprotein. Annual Meeting of the Society for Complex Carbohydrates, Abstr. 10.
348. Tanner, W., and Lehle, L. (1987). Protein glycosylation in yeast. *Biochim. Biophys. Acta 906*:81–99.
349. Tetteroo, P. A. T., de Heij, H. T., van den Eijnden, D. H., Visser, F. J., Schoenmaker, E., and van Kessel, A. H. M. G. (1987). A GDP-fucose:[Galβ1-4]GlcNAc α1-3-fucosyltransferase activity is correlated with the presence of human chromosome 11 and the expression of the Le^x, Le^y, and sialyl-Le^x antigens in human–mouse cell hybrids. *J. Biol. Chem. 262*: 15984–15989.
350. Thomas, D. B., and Winzler, R. J. (1969). Structural studies on human erythrocyte glycoproteins. *J. Biol. Chem. 244*:5943–5946.
351. Thomsen, D. R., Post, L. E., and Elhammer, A. P. (1990). Structure of *O*-glycosidically linked oligosaccharides synthesized by the insect cell line Sf9. *J. Cell. Biochem. 43*:67–79.
352. Tsuji, T., and Osawa, T. (1986). Carbohydrate structures of bovine submaxillary mucin. *Carbohydr. Res. 151*:391–402.
353. Tsuji, T., Tsunehisa, S., Watanabe, Y., Yamamoto, K., Tohyama, H., and Osawa, T. (1983). The carbohydrate moiety of human platelet glycocalicin. *J. Biol. Chem. 258*:6335–6339.
354. Tsuji, T., Yamamoto, K., Konami, Y., Irimura, T., and Osawa, T. (1982). Separation of acidic oligosaccharides by liquid chromatography. Application to analysis of sugar chains of glycoproteins. *Carbohydr. Res. 109*:259–269.
355. Uhlenbruck, G., Haupt, H., Reese, J., and Steinhausen, G. C. (1977). Serum cholinesterase as a model glycoprotein. *J. Clin. Chem. Clin. Biochem. 15*:561–564.
356. Vaith, P., Assmann, G., and Uhlenbruck, G. (1978). Characterization of the oligosaccharide side chain of apolipoprotein C-III from human plasma very low density lipoproteins. *Biochim. Biophys. Acta 541*:234–240.
357. van den Eijnden, D. H., Barneveld, R. A., and Schiphorst, W. E. C. M. (1979). Structure of the disaccharide chain of galactosyl-*N*-acetylgalactosaminyl-protein synthesized in vitro. *Eur. J. Biochem. 95*:629–637.
358. van den Eijnden, D. H ., Bergh, M. L. E., Dieleman, B., and Schiphorst, W. E. C. M. (1981). Specificity of sialyltransferase: Sialylation of ovine submaxillary mucin in vitro. *Hoppe-Seyler Z. Physiol. Chem. 362*:113–124.
359. van den Eijnden, D. H., Bergh, M. L. E., Joziasse, D. H., Blanken, W. M., and Koppen, P. L. (1982). Application of high pressure liquid chromatography (HPLC) in the study of glycosyltransferases. *Eleventh International Carbohydrate Symposium*. Abstr. IV-21.
360. van den Eijnden, D. H., Evans, N. A., Codington, J. F., Reinhold, V., Silber, C., and Jeanloz, R. W. (1979). Chemical structure of epiglycanin, the major glycoprotein of the TA3-Ha ascites cell. The carbohydrate chains. *J. Biol. Chem. 254*:12153–12159.
361. van den Eijnden, D. H., Koenderman, A. H. L., and Schiphorst, W. E. C. M. (1988). Biosynthesis of blood group i-active polylactosaminoglycans. Partial purification and properties of an UDP-GlcNAc:*N*-acetyllactosaminide β1→3-*N*-acetylglucosaminyltransferase from Novikoff tumor cell ascites fluid. *J. Biol. Chem. 263*:12461–12471.
362. van den Eijnden, D. H., Schiphorst, W. E. C. M., and Berger, E. G. (1983). Specific detection of *N*-acetylglucosamine-containing oligosaccharide chains on ovine submaxillary asialomucin. *Biochim. Biophys. Acta 755*:32–39.

363. van den Eijnden, D. H., Winterwerp, H., Smeeman, P., and Schiphorst, W. E. C. M. (1983). Novikoff ascites tumour cells contain *N*-acetyllactosaminide β1,3 and 1,6 GlcNAc-transferase activity. *J. Biol. Chem. 258*:3435–3437.

364. van Halbeek, H., Breg, J., Vliegenthart, J. F. G., Klein, A., Lamblin, G., and Roussel, P. (1988). Isolation and structural characterization of low-molecular-mass monosialyl oligosaccharides derived from respiratory-mucus glycoproteins of a patient suffering from bronchiectasis. *Eur. J. Biochem. 177*:443–460.

365. van Halbeek, H., Dorland, L., Haverkamp, J., Veldink, G. A., Vliegenthart, J. F. G., Fournet, B., Ricart, G., Montreuil, J., Gathmann, W. D., and Aminoff, D. (1981). Structure determination of oligosaccharides isolated from A^+, H^+ and A-H-hog-submaxillary-gland mucin glycoproteins, by 360-MHz ^{1}H-NMR spectroscopy, permethylation analysis and mass spectrometry. *Eur. J. Biochem. 118*:487–495.

366. van Halbeek, H., Dorland, L., Vliegenthart, J. F. G., Fiat, A. M., and Jolles, P. (1980). A 360-MHz ^{1}H-NMR study of three oligosaccharides isolated from cow kappa-casein. *Biochim. Biophys. Acta 623*:295–300.

367. van Halbeek, H., Dorland, L., Vliegenthart, J. F. G., Fiat, A. M., and Jolles, P. (1981). Structural characterization of a novel acidic oligosaccharide unit derived from cow colostrum kappa-casein. *FEBS Lett. 133*:45–50.

368. van Halbeek, H., Dorland, L., Vliegenthart, J. F. G., Hull, W. E., Lamblin, G., Lhermitte, M., Boersma, A., and Roussel, P. (1982). Primary-structure determination of fourteen neutral oligosaccharides derived from bronchial-mucus glycoproteins of patients suffering from cystic fibrosis, employing 500-MHz ^{1}H-NMR spectroscopy. *Eur. J. Biochem. 127*: 7–20.

369. van Halbeek, H., Dorland, L., Vliegenthart, J. F. G., Kochetkov, N. K., Arbatsky, N. P., and Derevitskaya, V. A. (1982). Characterization of the primary structure and the microheterogeneity of the carbohydrate chains of porcine blood-group H substance by 500 MHz ^{1}H-NMR spectroscopy. *Eur. J. Biochem. 127*:21–29.

370. van Halbeek, H., Gerwig, G. J., Vliegenthart, J. F. G., Smits, H. L., van Kerkhof, P. J. M., and Kramer, M. F. (1983). Terminal alpha(1-4)-linked *N*-acetylglucosamine: A characteristic constituent of duodenal-gland mucous glycoproteins in rat and pig. A high resolution proton NMR study. *Biochim. Biophys. Acta 747*:107–116.

371. van Halbeek, H., Vliegenthart, J. F. G., Fiat, A., and Jolles, P. (1985). Isolation and structural characterization of the smaller-size oligosaccharides from desialylated human kappa-casein. Establishment of a novel type of core for a mucin-type carbohydrate chain. *FEBS Lett. 187*: 81–88.

372. van Nieuw-Amerongen, A., Oderkerk, C. H., Roukema, P. A., Wolf, J. H., Lisman, J. J. W., and Overdijk, B. (1983). Murine submandibular mucin (MSM): A mucin carrying *N*- and *O*-glycosidically bound carbohydrate chains. *Carbohydr. Res. 115*:C1–C5.

373. Viau, M., Constans, J., Debray, H., and Montreuil, J. (1983). Isolation and characterization of the *O*-glycan chain of the human vitamin-D binding protein. *Biochem. Biophys. Res. Commun. 117*:324–331.

374. Watkins, W. M. (1980). Biochemistry and genetics of the ABO, Lewis and P blood group systems. *Adv. Hum. Genet. 10*:1–385.

375. Weinstein, J., DeSouza-e-Silva, U., and Paulson, J. C. (1982). Purification of a Galβ1-4GlcNAc α2-6-sialyltransferase and a Galβ1-3(4)GlcNAc α2-3-sialyltransferase to homogeneity from rat liver. *J. Biol. Chem. 257*:13835–13844.

376. Weinstein, J., DeSouza-e-Silva, U., and Paulson, J. C. (1982). Sialylation of glycoprotein oligosaccharides *N*-linked to asparagine. Enzymatic characterization of a Galβ1-3(4)GlcNAc α2-3-sialyltransferase and a Galβ1-4GlcNAc α2-6-sialyltransferase from rat liver. *J. Biol. Chem. 257*:13845–13853.

377. Wetmore, S., Mahley, R. W., Brown, W. V., and Schachter, H. (1974). Incorporation of sialic acid into sialidase-treated apolipoprotein of human very low density lipoprotein by a pork liver sialyltransferase. *Can. J. Biochem. 52*:655–664.

378. Whitehead, J. S., Bella, A., Jr., and Kim, Y. S. (1974). An *N*-acetylgalactosaminyltransferase from human blood group A plasma I. Purification and agarose binding properties. *J. Biol. Chem. 249*:3442–3447.

379. Wieruszeski, J. M., Michalski, J. C., Montreuil, J., Strecker, G., Peter-Katalinic, J., Egge, H., van Halbeek, H., Mutsaers, J. H. G. M., and Vliegenthart, J. F. G. (1987). Structure of the monosialyl oligosaccharides derived from salivary gland mucin glycoproteins of the Chinese swiftlet (genus *Collocalia*). Characterization of novel types of extended core structure, Galβ(1-3)[GlcNAcβ(1-6)]GalNAcα(1-3)GalNAc-ol, and of chain termination, $[Gal\alpha(1\text{-}4)]_{0\text{-}1}[Gal\beta(1\text{-}4)]_2GlcNAc\beta(1\text{-})$. *J. Biol. Chem. 262*:6650–6657.

380. Williams, D., Longmore, G. D., Matta, K. L., and Schachter, H. (1980). Mucin synthesis. II. Substrate specificity and product identification studies on canine submaxillary gland UDP-GlcNAc:Galβ1-3GalNAc (GlcNAc to GalNAc) β6-*N*-acetylglucosaminyltransferase. *J. Biol. Chem. 255*:11253–11261.

381. Williams, D., and Schachter, H. (1980). Mucin synthesis. I. Detection in canine submaxillary glands of an *N*-acetylglucosaminyltransferase which acts on mucin substrates. *J. Biol. Chem. 255*:11247–11252.

382. Wingert, W. E., and Cheng, P. (1984). Mucin biosynthesis: Characterization of rabbit small intestinal UDP-*N*-acetylglucosamine:galactose β-3-*N*-acetylgalactosaminide (*N*-acetylglucosamine to *N*-acetylgalactosamine) β6-*N*-acetylglucosaminyltransferase. *Biochemistry 23*: 690–697.

383. Wood, E., Hounsel, E. F., and Feizi, T. (1981). Preparative affinity chromatography of sheep gastric mucins having blood-group Ii activity, and release of antigenically active oligosaccharides by alkaline-borohydride degradation. *Carbohydr. Res. 90*:269–282.

384. Wood, E., Hounsell, E. F., Langhorne, J., and Feizi, T. (1980). Sheep gastric mucins as a source of blood-group-I and -i antigens. *Biochem. J. 187*:711–718.

385. Wu, A. M., Kabat, E. A., Nilsson, B., Zopf, D. A., Gruezo, F. G., and Liao, J. (1984). Immunochemical studies on blood groups. LXXI. Purification and characterization of radioactive ^{3}H-reduced di- to hexasaccharides produced by alkaline beta-elimination-borohydride ^{3}H reduction of Smith degraded blood group A active glycoproteins. *J. Biol. Chem. 259*: 7178–7186.

386. Wu, A. M., Kabat, E. A., Pereira, M. E. A., Gruezo, F. G., and Liao, J. (1982). Immunochemical studies on blood groups. The internal structure and immunological properties of water-soluble human blood group A substance studied by Smith degradation, liberation, and fractionation of oligosaccharides and reaction with lectins. *Arch. Biochem. Biophys. 215*: 390–404.

387. Yamamoto, F.-I., Clausen, H., White, T., Marken, J., and Hakomori, S. I. (1990). Molecular genetic basis of the histo-blood group ABO system. *Nature 345*:229–233.

388. Yamamoto, F.-I., and Hakomori, S.-I. (1990). Sugar-nucleotide donor specificity of histo-blood group A and B transferases is based on amino acid substitutions. *J. Biol. Chem. 265*:19257–19262.

389. Yamamoto, M., and Yosizawa, Z. (1978). Glycopeptides isolated from ovine submaxillary mucin. *J. Biochem. (Tokyo) 83*:1159–1164.

390. Yamashina, I., Funakoshi, I., Kawasaki, T., Kurosaka, A., Sugiura, M., and Fukui, S. (1983). Alterations of glycoproteins and glycosaminoglycans associated with malignant transformation. In *Membrane Alterations in Cancer*. (A. Makita, S. Tsuiki, S. Fujii, and L. Warren, eds.), *GANN Monogr. Cancer Res. 29*:149–156.

391. Yamashita, T., Murayama, J., Utsumi, H., and Hamada, A. (1985). Structural studies of *O*-glycosidic oligosaccharide units of dog erythrocyte glycophorin. *Biochim. Biophys. Acta 839*:26–31.

392. Yates, A. D., Feeney, J., Donald, A. S. R., and Watkins, W. M. (1984). Characterization of a blood group A-active tetrasaccharide synthesized by a blood group B gene-specified glycosyltransferase. *Carbohydr. Res. 130*:251–260.

393. Yates, A. D., and Watkins, W. M. (1982). The biosynthesis of blood group B determinants by the blood group A gene-specified α-3-GalNAc-transferase. *Biochem. Biophys. Res. Commun. 109*:958–965.

394. Yates, A. D., and Watkins, W. M. (1983). Enzymes involved in the biosynthesis of glycoconjugates. A UDP-2-acetamido-2-deoxy-D-glucose:β-D-galactopyranosyl-(1→4)-saccharide (1→3)-2-acetamido-2-deoxy-β-D-glucopyranosyltransferase in human serum. *Carbohydr. Res. 120*:251–268.

395. Yazawa, S., Abbas, S. A., Madiyalakan, R., Barlow, J. J., and Matta, K. L. (1986). *N*-Acetyl-β-D-glucosaminyltransferases related to the synthesis of mucin-type glycoproteins in human ovarian tissue. *Carbohydr. Res. 149*:241–252.

396. Yoshimura, A., Yoshida, T., Seguchi, T., Waki, M., Ono, M., and Kuwano, M. (1987). Low binding capacity and altered *O*-linked glycosylation of low density lipoprotein receptor in a monensin-resistant mutant of Chinese hamster ovary cells. *J. Biol. Chem. 262*:13299–13308.

397. Young, J. D., Tsuchiya, D., Sandlin, D. E., and Holroyde, M. J. (1979). Enzymic *O*-glycosylation of synthetic peptides from sequences in basic myelin protein. *Biochemistry 18*:4444–4448.

398. Yurewicz, E. C., Matsuura, F., and Moghissi, K. S. (1982). Sialylated oligosaccharides of human cervical mucin. Annual Meeting of the Society for Complex Carbohydrates, Abstr. 30.

399. Yurewicz, E. C., Matsuura, F., and Moghissi, K. S. (1982). Structural characterization of neutral oligosaccharides of human midcycle cervical mucin. *J. Biol. Chem. 257*:2314–2322.

400. Yurewicz, E. C., Matsuura, F., and Moghissi, K. S. (1987). Structural studies of sialylated oligosaccharides of human midcycle cervical mucin. *J. Biol. Chem. 262*:4733–4739.

401. Yurewicz, E. C., and Moghissi, K. S. (1981). Purification of human midcycle cervical mucin and characterization of its oligosaccharides with respect to size, composition, and microheterogeneity. *J. Biol. Chem. 256*:11895–11904.

402. Ziderman, D., Gompertz, S., Smith, Z. G., and Watkins, W. M. (1967). Glycosyltransferases in mammalian gastric mucosal linings. *Biochem. Biophys. Res. Commun. 29*:56–61.

403. Zielenski, J., and Koscielak, J. (1983). The occurrence of two novel GlcNAc-transferase activities in human serum. *FEBS Lett. 158*:164–168.

404. Zielenski, J., and Koscielak, J. (1983). Sera of i subjects have the capacity to synthesize the branched GlcNAcβ(1-6)[GlcNAc(β1-3)]Gal..... structure. *FEBS Lett. 163*:114–118.

10

Synthesis of Asparagine-Linked Oligosaccharides: Pathways, Genetics, and Metabolic Regulation

Richard D. Cummings *The University of Georgia, Athens, Georgia*

The pathway of biosynthesis of asparagine (Asn)-linked oligosaccharides in glycoproteins has been studied intensely during the last decade. Initial steps in this pathway, which appear to be generally conserved among eukaryotes, involve the synthesis of a large-sized precursor oligosaccharide attached to lipid in pyrophosphate linkage and the transfer of that preformed carbohydrate unit to asparagine residues in newly formed polypeptides. The metabolic fate and function of the initial glycoprotein-bound oligosaccharide, however, differs remarkably and importantly among different cell types, organisms, and different glycoproteins within the same cell.

In recent years, several excellent reviews on the biosynthetic pathway for Asn-linked oligosaccharides have appeared [Kornfeld and Kornfeld, 1980, 1985; Struck and Lennarz, 1980; Snider, 1984; Fuhrmann et al., 1985; Schachter, 1986; Lennarz, 1987; Kaplan et al., 1987; and Krag, 1985]. Because of the large scope of the field, this present chapter cannot comprehensively cover this research area, and it will not attempt to recapitulate all that is contained in the previous reviews. Rather, this chapter will focus on certain new information about several important aspects of the biosynthetic pathway for Asn-linked oligosaccharides, including newly discovered steps and enzymes, as well as some of the genetic and metabolic factors that appear to contribute to the regulation and control of biosynthesis.

I. STRUCTURALLY DISTINCT ASPARAGINE-LINKED OLIGOSACCHARIDES

Glycoproteins containing Asn-linked oligosaccharides are found in cell secretions, serum, and other body fluids, and as membrane-bound constituents of the plasma membrane, Golgi apparatus, lysosome, endoplasmic reticulum, and the nuclear envelope outer membrane. In recent years, numerous studies have been done on a wide variety of animal and plant cell glycoproteins, and the overwhelming conclusion is that there are perhaps thousands of structurally different Asn-linked oligosaccharides. This diversity in the structures of the

Asn-linked oligosaccharides is central to hypotheses that these chains are important in biological functions, including cellular recognition, cellular adhesion, and protein targeting.

Although there are many different Asn-linked oligosaccharides, there are certain features common to virtually all members of this class. These common features are illustrated in the oligosaccharide structures shown in Figure 1 and include (1) the presence of a common core pentasaccharide containing two α-linked mannosyl residues and one mannosyl residue in β-linkage to an N,N′-diacetylchitobiosyl unit, (2) the β-linkage of this chitobiosyl unit to the asparagine amide nitrogen, and (3) the presence of the oligosaccharide on asparagine residues in the consensus sequence -Asn-X-Ser/Thr-. The presence of the common core structure initially suggested the now well-accepted fact that Asn-linked oligosaccharides are synthesized by a common pathway.

The Asn-linked oligosaccharides can be divided into three main classes that are biosynthetically related, but that differ in their prominent structural features. The three classes are high mannose type, hybrid type, and complex type. High mannose-type Asn-linked oligosaccharides contain mannose as the predominant constituent and usually only mannose or glucose residues are present in the nonreducing termini. Hybrid-type Asn-linked oligosaccharides contain structural features similar to high mannose-type chains, but, in addition, contain monosaccharides other than mannose or glucose at the nonreducing termini. Finally, the complex-type oligosaccharides contain only three mannose residues and, usually, many other monosaccharides in addition to mannose. In the complex-type chains, none of the mannose residues are present at the nonreducing termini. In vertebrates, the complex-type chains often contain the disaccharide *N*-acetyllactosamine (Galβ1,4GlcNAc), and they are sometimes called *N*-acetyllactosamine-type chains.

Although this simple classification has a certain logical quality, it tends to diminish the most distinguishing feature of Asn-linked oligosaccharides, namely, their tremendous structural diversity. Examples of this variety of the Asn-linked oligosaccharides are shown in Figure 1. High mannose-type chains vary in the number of mannose residues, the position of the mannose residues, and in the degree to which these mannose residues are phosphorylated or sulfated. Hybrid-type oligosaccharides may be phosphorylated or sulfated, or both, on the mannose residues and contain sialic acids and other monosaccharides on other branches.

Complex-type Asn-linked oligosaccharides can be highly branched, depending on the substitution pattern of *N*-acetylglucosamine (GlcNAc) residues to the mannose residues in the core, to generate structures given the trivial names of bi-, tri-, tetraantennary, and so on. Even pentaantennary complex-type Asn-linked oligosaccharides have been described [Paz-Parente et al., 1982, 1983]. The outer branches of many complex-type Asn-linked oligosaccharides in animal cell glycoproteins contain the repeating disaccharide units of $[3Gal\beta1,4GlcNAc\beta]_n$ (termed poly-*N*-acetyllactosamine), where $n = 2$ to approximately 50 [Järnefelt et al., 1978; Fukuda et al., 1984; Cummings and Kornfeld, 1984; Carlsson et al., 1988; Arumugham et al., 1986; for review see Fukuda, 1985]. Because of their repetitious character, poly-*N*-acetyllactosamine chains allow an almost infinite variety of possible structures.

Asparagine-linked oligosaccharides may be sulfated [Baenziger and Green, 1988; Spiro and Bhoyroo, 1988]; highly fucosylated and carry blood group antigens [Tsuji et al., 1981; Krusius et al., 1978, Järnefelt et al., 1978; Cummings et al., 1985;

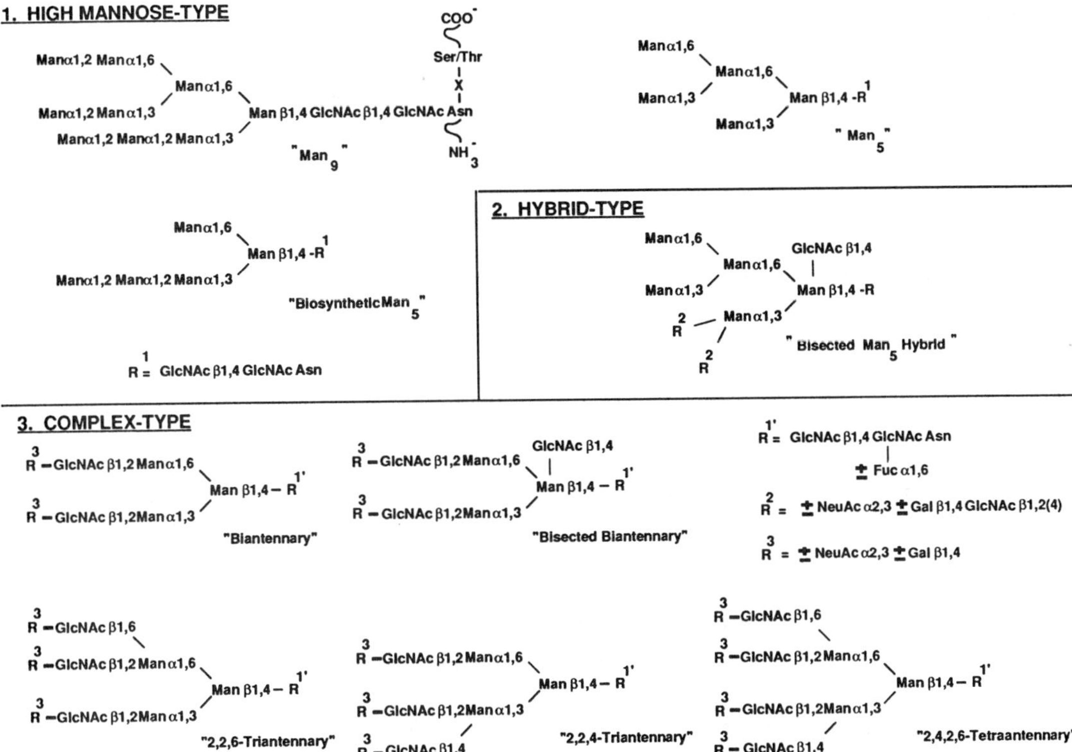

Figure 1 Diversity of structural features of Asn-linked oligosaccharides.

and Finne et al., 1989]; and phosphorylated [reviewed by Kornfeld, 1987]. For example, corneal keratan sulfate consists of a complex-type Asn-linked oligosaccharide with poly-*N*-acetyllactosamine chains containing sulfate SO_4-6-GlcNAc and SO_4-6-Gal [Roden, 1980]. Sulfated glycosaminoglycanlike oligosaccharides and other types of sulfated structures have also been found in complex-type Asn-linked oligosaccharides [Freeze and Wolgast, 1986; Van Kuik et al., 1987; Sundblad et al., 1988a,b; Kato and Spiro, 1989]. Some neural glycoproteins, including neuronal cell adhesion molecule (N-CAM), contain polysialic acid in complex-type Asn-linked oligosaccharides [Rothbard et al., 1982; Finne and Finne, 1983]. These polysialic acid-containing oligosaccharides may have a wider distribution than first believed; they appear to occur, to some extent, in a variety of cells and tissues [Lipinski et al., 1987; Livingston et al., 1987, 1988; Katajima et al., 1988].

Finally, unusual types of Asn-linked oligosaccharides are known in glycoproteins from some more primitive organisms: for example, the methylphosphate on the mannose residues of the cellular slime mold *Dictyostelium discoideum* [Gabel et al., 1984], bisecting high mannose chains, also in *D. discoideum* [Couso et al., 1987], xylose in plant glycoproteins [Sturm et al., 1987], and unusual sialic acid derivatives [Corfield and Schauer, 1982; Kammerling et al., 1988; Hanaoka et al., 1989; Diaz et al., 1989]. This list is undoubtedly incomplete, but it does suggest that a large number of structurally different Asn-linked oligosaccharides exist in eukaryotic glycoproteins.

4. POLY-N-ACETYLLACTOSAMINE (LACTOSAMINOGLYCAN)

R^3 = ± NeuAc α2,3(6) [Gal β1,4 GlcNAc β1,3]$_n$ Gal β1,4
("Linear", Type 2 Chain, i-antigen)

Gal β1,4 GlcNAc β1,6
R^3 = ± NeuAc α2,3(6) [Gal β1,4 GlcNAc β1,3]$_n$ Gal β1,4
(Branched, Type 2 Chain, I-Antigen)

5. POLY-SIALIC ACID

NeuAc α2,8 NeuAc α2,8 NeuAc α2,8 NeuAc α2,8 NeuAc α2,3 Gal –

6. BLOOD GROUP ANTIGENS AND HIGH FUCOSYLATION

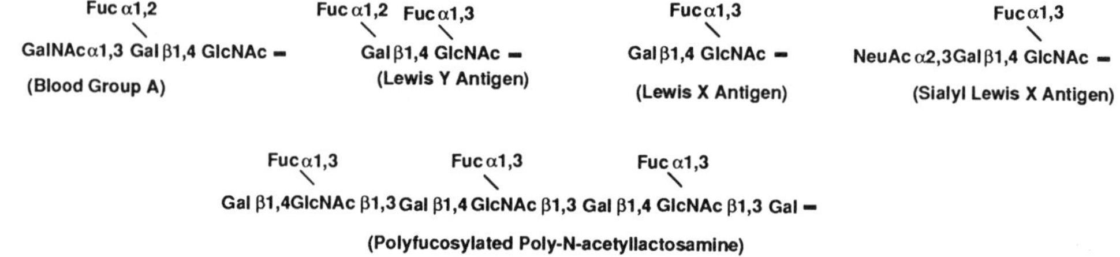

7. SULFATED N-LINKED OLIGOSACCHARIDES

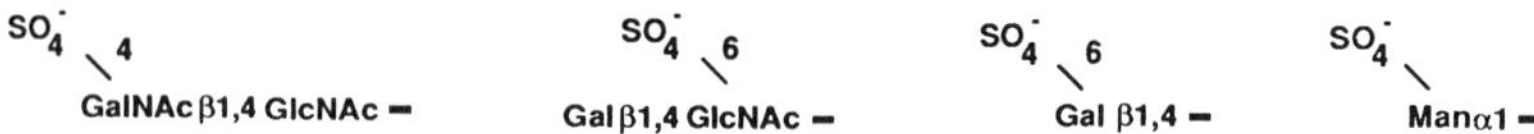

Figure 1 (continued)

II. INTRACELLULAR BIOSYNTHESIS OF ASPARAGINE-LINKED OLIGOSACCHARIDES

The biosynthesis of Asn-linked oligosaccharides is separated both spatially and temporally into discrete enzymatic steps, which are generally categorized as early-stage processing, middle-stage processing, and late-stage processing. These stages involve the endoplasmic reticulum (ER), different stacks of the Golgi apparatus, and the *trans*-Golgi network [Kornfeld and Kornfeld, 1985; Griffiths and Simons, 1986]. The broad features of this scheme are idealized in Figure 2.

Biosynthesis of Asn-linked oligosaccharides is initiated by the cotranslational transfer of oligosaccharide by the oligosaccharyltransferase from the dolichol-linked oligosaccharide $Glc_{0\text{-}3}Man_{5\text{-}9}GlcNAc_2$-$(PO_4)_2$-dolichol to asparagine residues in nascent polypeptides in the rough endoplasmic reticulum (RER) [Struck and Lennarz, 1980; Hubbard and Ivatt, 1981]. Numerous studies have verified that the oligosaccharide is added to asparagine residues in the tripeptide, consensus sequence -Asn-X-Ser/Thr-. There is evidence that the OH-group of the serine/threonine (Ser/Thr) residues participates in the enzymatic transfer of the oligosaccharide [Bause and Legler, 1981; Bause, 1983]. It should be noted, however,

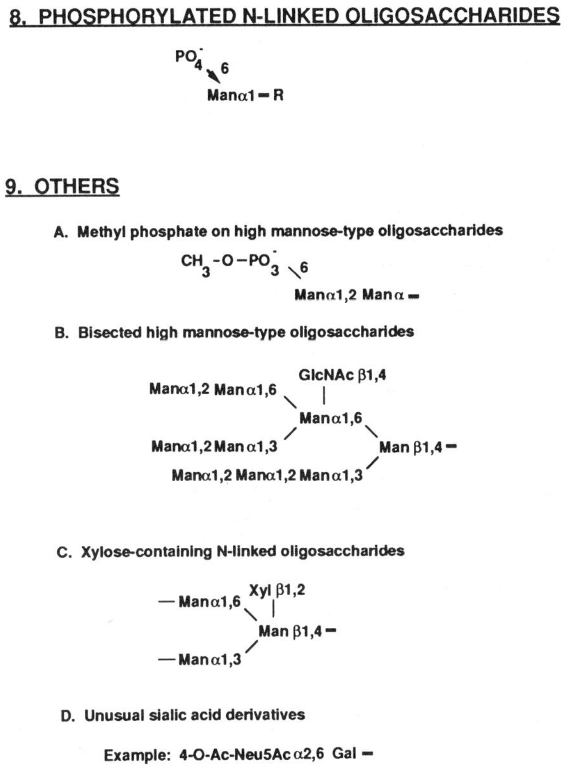

8. PHOSPHORYLATED N-LINKED OLIGOSACCHARIDES

PO_4^- → 6
Manα1 – R

9. OTHERS

A. Methyl phosphate on high mannose-type oligosaccharides

CH_3-O-PO_3^- ↘6
Manα1,2 Manα –

B. Bisected high mannose-type oligosaccharides

```
                       GlcNAc β1,4
Manα1,2 Manα1,6  \        |
                    Manα1,6 \
                  /          \
Manα1,2Manα1,3 /              Man β1,4 –
                               /
  Manα1,2 Manα1,2 Manα1,3  /
```

C. Xylose-containing N-linked oligosaccharides

```
              Xyl β1,2
— Manα1,6 \   |
            Man β1,4 –
          /
— Manα1,3
```

D. Unusual sialic acid derivatives

Example: 4-O-Ac-Neu5Ac α2,6 Gal –

Figure 1 (continued)

that fewer than half of the known tripeptide sequences in secreted glycoproteins are glycosylated, and the reason may be differences in accessibility of the oligosaccharyltransferase to the asparagine amide.

After transfer to protein, the oligosaccharide is subject to trimming or removal of glycoside residues by a number of glycosidases. This aspect of the pathway is termed early-stage processing [Kornfeld and Kornfeld, 1985]. Much of this early-stage processing occurs in the ER, transitional ER, and *cis*-Golgi apparatus. Middle-stage processing occurs in both the *cis*- and medial-Golgi apparatus and involves elongation of the processed Asn-linked oligosaccharides by addition of other glycoside residues. Finally, late-stage processing occurs, largely in the *trans*-Golgi apparatus and *trans*-Golgi network and involves the elongation of the Asn-linked oligosaccharides with additional glycosides and other constituents.

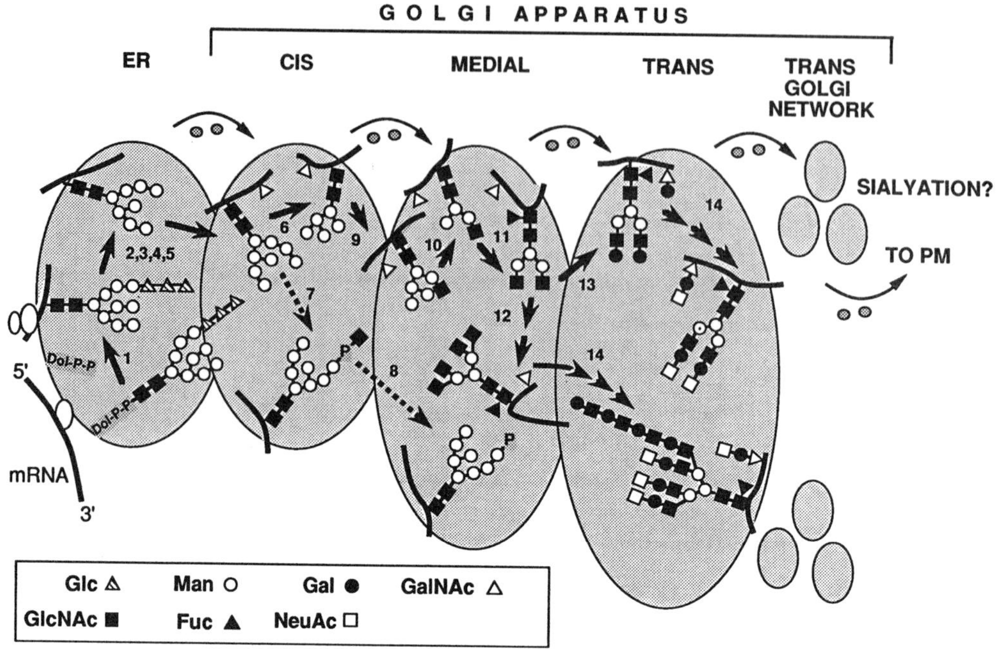

Figure 2 Idealized pathway for the biosynthesis of Asn-linked oligosaccharides. Biosynthesis begins in the rough endoplasmic reticulum (RER) and proceeds through the *cis*-, medial-, and *trans*-Golgi apparatus. Some of the enzymes and steps in the process are numbered as follows: 1. The oligosaccharyl transferase transfers oligosaccharide from dolichol-linked oligosaccharide to nascent polypeptide; 2. glucosidases I and II; 3. glucosidase II; 4. reglucosylation; 5. ER-mannosidase; 6. mannosidase I; 7. lysosomal enzyme *N*-acetylglucosamine phosphotransferase and addition of GlcNAc-1-phosphate (P = PO_4); 8. *N*-acetylglucosamine-1-phosphodiester α-*N*-acetylglucosaminidase; 9. *N*-acetylglucosaminyltransferase I; 10. mannosidase II; 11. *N*-acetylglucosaminyltransferase II; 12. *N*-acetylglucosaminyltransferases IV and V; 13. galactosyltransferase; 14. terminal glycosylation and synthesis of poly-*N*-acetyllactosamine structures by sialyltransferases, *N*-acetylglucosaminyltransferases, and galactosyltransferases.

III. DIFFERENCES IN FORMATION OF DOLICHOL-LINKED OLIGOSACCHARIDES

The principal pathway for biosynthesis of the dolichol-linked oligosaccharides in various organisms are shown in Figure 3. The large, dolichol-linked oligosaccharide is synthesized by enzymes in the ER membrane. The major steps in this synthesis have been reviewed [Lennarz, 1987; Kaplan et al., 1987; Hirschberg and Snider, 1987]. The assembly of the lipid-linked oligosaccharide is initiated by the enzyme UDP-GlcNAc:dolichol phosphate *N*-acetylglucosamine-1-phosphotransferase, which catalyzes the formation of Dol-P-P-GlcNAc. This enzyme is highly inhibited by the antibiotic tunicamycin, and inhibition eventually results in the complete lack of assembled lipid-linked oligosaccharides and in the cessation of Asn-linked oligosaccharide biosynthesis.

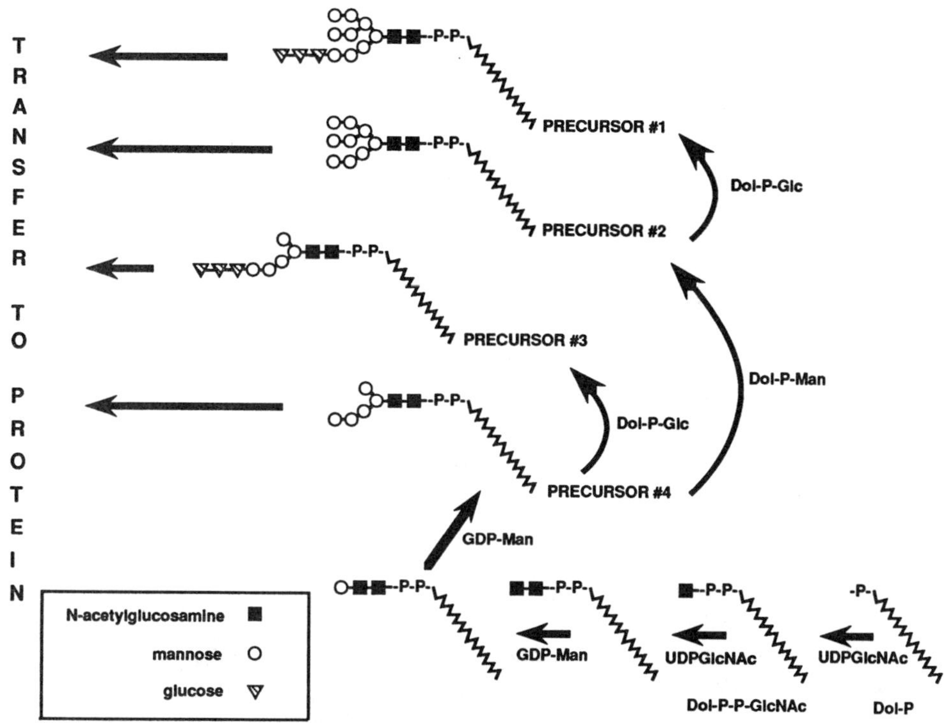

Figure 3 Pathway of biosynthesis of lipid-linked oligosaccharides.

Some animal cells, selected for resistance to tunicamycin, develop elevated levels of this enzyme activity [Criscuolo and Krag, 1982], and it was proposed that this was due to increases in copy number for the structural gene for the enzyme. A genetic element containing the putative structural gene for tunicamycin-resistance was recently identified in *Saccharomyces cerevisiae* [Rine et al., 1983; Barnes et al., 1984]. Lehrman et al. (1988) isolated and identified a segment of the gene for the dolichol phosphate-dependent *N*-acetylglucosamine-1-phosphotransferase in Chinese hamster ovary cells, which were selected for increased resistance to tunicamycin, and they were found to have an elevated enzyme activity. The gene segment for the enzyme in these cells has 92% amino acid sequence homology to the *ALG-7* gene from yeast [Barnes et al., 1984], which appears to be the yeast gene for the enzyme.

Higher animals and plants synthesize the large precursor $Glc_3Man_9GlcNAc_2$-P-P-Dol (shown in Fig. 3 as precursor 1). The detailed structure of the $Glc_3Man_9GlcNAc_2$ moiety of the lipid-linked oligosaccharide is shown in Figure 4. The presence of glucose residues on the dolichol-linked oligosaccharide, however, is not a universal phenomenon. For example, trypanosomatids synthesize dolichol-linked oligosaccharides devoid of glucose (Glc) and containing six to eight mannose (Man) residues, depending on the species [Parodi et al., 1981, 1983; Parodi and Quesada-Allue, 1982; Previato et al., 1986]. These altered precursors are indicated as precursor 2 in Figure 3.

In mammalian cells, it appears that glucose residues are essential components of the dolichol-linked oligosaccharide, whereas certain mannose residues may be dispensable.

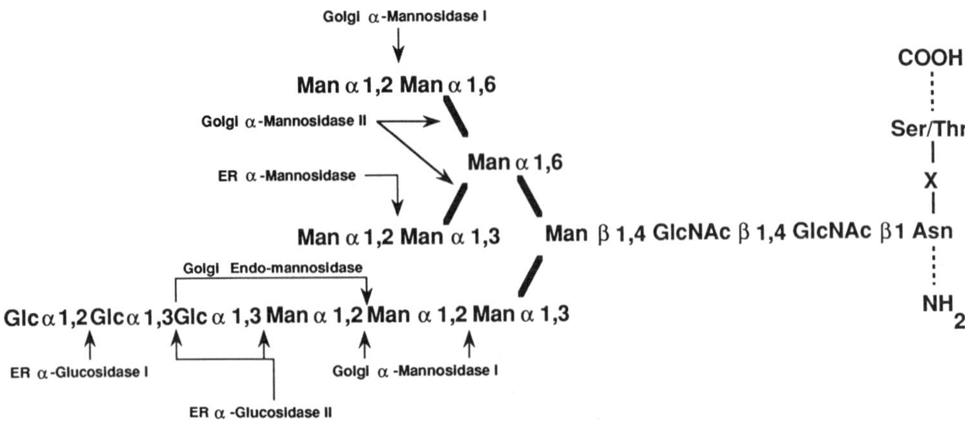

Figure 4 Glycosidases involved in processing of Asn-linked oligosaccharides.

For example, a mouse lymphoma cell line, Thy-1$^-$, class E, selected for lack of expression of Thy antigen, synthesizes a glucosylated, truncated lipid-linked oligosaccharide, which contains glucose (precursor 3, Fig. 3) [Gabel and Kornfeld, 1982]. The Thy-1$^-$, class E cells do not synthesize dolichol-phosphate-mannose (Dol-P-Man), the essential donor for the final four mannose residues in $Glc_3Man_9GlcNAc_2$-P-P-dolichol [Chapman et al., 1980]. Similar mutants of Chinese hamster ovary (CHO) cells were obtained by screening for reduced mannose-6-phosphate receptor levels [Stanley, 1984; Robbin et al., 1981]. Although these cells grow in culture, it should be noted that aberrant glycosylation of glycoproteins caused by addition of truncated lipid-linked oligosaccharides may alter normal subunit association of some glycoproteins [Hearing et al., 1989]. The lack of Dol-P-Man is lethal in yeast, as discussed later.

Recently, Lehrman and Zeng (1989) derived a mutant CHO cell line resistant to the lethal effects of two inhibitors of processing glycosidases, castanospermine and swainsonine, inhibitors of glucosidase I/II and mannosidase II, respectively. The mutant cells, designated PIR for processing inhibitor resistance, fail to synthesize significant levels of glucose-containing lipid-linked oligosaccharides and, instead, accumulate the lipid-linked oligosaccharide, $Man_5GlcNAc_2$-P-P-Dol (precursor 4). It is possible that PIR cells use this nonglucosylated lipid-linked oligosaccharide as the direct donor for Asn-linked oligosaccharides. Direct proof that precursor 4 is used will require analysis of the newly synthesized glycoproteins to determine if they contain any glucose residues. As pointed out by Lehrman and Zeng (1989), PIR cells are slightly sensitive to castanospermine, which suggests that some glucosylation of the lipid-linked oligosaccharide may occur.

The gene for Dol-P-Man synthase in yeast has been cloned by Orlean et al. (1988) and found to be active in *Escherichia coli*. Sequence analysis predicts a protein of 30 kd, with no potential *N*-glycosylation sites, which is in agreement with the relative molecular mass (M_r) of the purified yeast enzyme [Haselbeck and Tanner, 1982]. Because of the lack of a cleavage site for signal peptidase, the orientation of the enzyme in the ER membrane has not yet been determined. The gene for Dol-P-Man synthetase appears to be essential for yeast growth, not because of the role of the enzyme in *N*-linked glycosylation, but perhaps because Dol-P-Man is the donor for *O*-linked mannose residues in yeast glycoproteins. Deficiencies in Dol-P-Man are not lethal to mammalian cells in culture, as discussed earlier.

IV. TRANSFER OF OLIGOSACCHARIDE TO ASPARAGINE BY OLIGOSACCHARYLTRANSFERASE

The glycosylation of asparagine residues in nascent polypeptides in the lumen of the REF depends on several critical factors. Some of the more obvious factors include (1) the presence of the consensus tripeptide sequence -Asn-X-Ser/Thr-; (2) the availability of the dolichol-linked oligosaccharide $Glc_3Man_9GlcNAc_2$-P-P-Dol; (3) and an oligosaccharyl transferase and the catalytic transfer of the oligosaccharide [Kiely et al., 1976; Hart et al., 1979; Gabel et al., 1980; Struck and Lennarz, 1980; see also Kaplan et al., 1987]. Wieland et al. (1987) have reported that a synthetic tripeptide containing the consensus sequences and blocked NH_2- and COOH-termini could be *N*-glycosylated by intact cells and then secreted, indicating that the compound entered the ER of the treated cells and contained all the information necessary for in vivo recognition and glycosylation. However, Geetha-Habib et al. (1990) found the synthetic tripeptide acceptor microinjected into *Xenopus* oocytes was glycosylated, but not secreted, and appeared to be degraded in an intracellular acidic compartment. Interestingly, neither Wieland et al. (1987), nor Geetha-Habib et al. (1990) found that the glycosylated tripeptide was efficiently processed to mature complex-type chains. This indicates that perhaps the glycosylated tripeptide never gained access to the full Golgi apparatus or was inaccessible to enzymes in that organelle.

A clear understanding of the biochemical mechanisms underlying the transfer of the lipid-linked oligosaccharide has been hampered by difficulties encountered in purifying the oligosaccharyl transferase. However, Lennarz and colleagues identified a 57-kd protein that could be photoaffinity labeled using an [^{125}I]-labeled probe that is a derivative of the tripeptide Asn-Lys-Thr [Welply et al., 1986; Kaplan et al., 1988]. An unusual finding in these studies was the observation that the 57-kd protein specifically interacted with the photoaffinity probe, but the protein had no oligosaccharyltransferase activity. Kaplan et al. (1988) hypothesized that the 57-kd protein is a catalytically inactive component of the oligosaccharyltransferase multisubunit complex responsible for recognizing and binding to the Asn-glycosylation site, and that another polypeptide in the complex contains enzymatic activity. It was subsequently demonstrated, using antibodies to the 57-kd protein, that it resides in the lumen of ER, and that it is insensitive to proteases in intact microsomes [Geetha-Habib et al., 1988]. In contrast, the overall oligosaccharyltransferase activity of the sealed microsomes was sensitive to trypsin, suggesting that at least one critical component of the presumed enzyme complex traverses the ER membrane. Antibodies to the 57-kd protein were used to identify and clone the gene for the protein from a cDNA expression library in λgt11 [Geetha-Habib et al., 1988].

The 57-kd component of the oligosaccharyltransferase has high-sequence similarity to three other 57-kd proteins known to be residents of the ER lumen, namely, protein disulfide isomerase, the β-subunit of prolyl hydroxylase, and thyroid hormone-binding protein [Geetha-Habib et al., 1988]. The sequences of all these proteins from one organism are not yet available, and it is possible that they are, in fact, all identical. The sequence similarity of the oligosaccharyl-binding protein to these other enzymes raises the possibility that one single polypeptide contains all the activities, but that remains to be proven.

The oligosaccharyltransferase in vertebrate cells appears to be selective for the structure of the dolichol-linked oligosaccharide it transfers, preferring as a substrate the fully glycosylated $Glc_3Man_9GlcNAc_2$-P-P-Dol [Kornfeld and Kornfeld, 1985; Elbein, 1987] (see Fig. 3). In contrast, some protozoans, such as trypanosomatids, do not synthesize glucosylated dolichol-linked oligosaccharides [Engel and Parodi, 1985; Parodi et al., 1981, 1984a], and the oligosaccharyltransferase utilized nonglycosylated dolichol-linked oligosaccharides just as efficiently as it does $Glc_3Man_9GlcNAc_2$-P-P-Dol [Bosch et al., 1988]. Because the trypanosomatids do not synthesize glucosylated lipid-linked oligosaccharides, they might be expected to lack certain α-glucosidase–processing enzymes, especially α-glucosidase I. Bosch et al. (1988) demonstrated this to be true, but they found that the organisms have some glucosidase II activity. This enzyme may be in the Golgi apparatus in this organism and may be involved in removing the glucose residues added transiently to the glycoproteins in that organelle.

V. TRANSFER OF SUGAR TO ASPARAGINE RESIDUES IN THE TRIPEPTIDE SEQUENCE -ASN-X-SER/THR BY LOWER ORGANISMS

Halobacteria synthesize glycoproteins and glycosaminoglycans that contain glycosylated asparagine residues that are substituted *N*-glycosidically with glucose [Wieland et al., 1983]. Paul et al. (1986) and Paul and Wieland (1987) analyzed the structure of a sulfated glycosaminoglycan synthesized by these organisms and found that the polysaccharide chain was attached through *N*-acetylgalactosamine to asparagine in the tripeptide sequence -Asn-Ala-Ser-. Other evidence indicates that transfer of both the glycosaminoglycan and the glucose residues involve dolichollike lipid-linked oligosaccharides and transfer of sugar to asparagine residues at the surface membrane of the halobacteria. The glycosylation of asparagine in the consensus tripeptide in these organisms further supports the likely involvement of the OH-group on serine or threonine in the consensus tripeptide in the reaction mechanism, as proposed by Marshall (1974) and experimentally supported by Bause and Legler (1981). As pointed out by Paul et al. (19860, the glycosylation of asparagine residues in the surface membrane of halobacteria resembles, in some way, the *N*-glycosylation of proteins in the animal cell ER.

VI. EARLY-STAGE ENZYMATIC ACTION AND PROCESSING OF THE TRANSFERRED OLIGOSACCHARIDE

The oligosaccharide that is initially transferred to protein, $Glc_3Man_9GlcNAc_2$Asn-R, is a substrate for a variety of "processing" glycosidases in the ER and Golgi apparatus. The early-stage processing is initiated in the ER by a set of at least three different enzymes, designated α-glucosidase I, α-glucosidase II, and ER-α-mannosidase (see Fig. 4). Glucosidase I removes the single, terminal α1,2-linked glucose residue and then glucosidase II removes the two remaining α1,3-linked glucose residues. Bischoff and Kornfeld (1983) described an ER-mannosidase, which is sensitive to the effects of the inhibitor deoxymannojirimycin, and which efficiently removes a single mannose residue from $Man_9GlcNAc_2$-R to generate $Man_8GlcNAc_2$-R. Complementary DNA for the rat liver ER α-mannosidase has recently been cloned and sequenced [Bischoff et al., 1990]. The protein is predicted to contain 1040 amino acids and to lack a classic signal sequence or definable transmembrane domain. Interestingly, the ER α-mannosidase has significant sequence homology to the yeast vacuolar α-mannosidase [Yoshihisa and Anraku, 1989].

However, there may be other mannosidase activities in the ER, based on the studies of Rizzolo and Kornfeld (1988). They found that the high mannose-type Asn-linked oligosaccharides on a hybrid protein, unable to exit the ER, were processed by an α-mannosidase activity sensitive to deoxymannojirimycin. In addition, structural analyses of the $Man_8GlcNAc_2$-Asn on the hybrid protein indicated that it has the same structure as that produced by the ER α-mannosidase, thereby raising the possibility that the ER contains multiple mannosidases, which may be compartmentalized. In yeast, the transfer of the $Glc_3Man_9GlcNAc_2$- to asparagine residues and subsequent removal of the three glucose residues occurs, as in vertebrates. Only a single α-linked mannose residue is removed in yeast, however, to generate a specific isomer of $Man_8GlcNAc_2$- in which the middle mannose residue, linked α1,2 to Manα1,3-R, is removed [Byrd et al., 1982; Trimble and Atkinson, 1986] (see Fig. 4). Removal of this mannose residue occurs by the action of a specific mannosidase in the ER [Jelinek-Kelly and Herscovics, 1988] that appears to be equivalent to the one in vertebrates. The $Man_8GlcNAc_2$- (and $Man_9GlcNAc_2$-) are substrates for an enzyme unique in yeast, an α1,6-mannosyltransferase, that utilizes guanosine diphosphate (GDP)-mannose and adds a mannose to the outer α1,6-linked mannosyl residue [Romero and Hersovics, 1989].

The functions of mannose processing in the ER is not known. The Golgi α-mannosidase I can function in vitro to convert $Man_9GlcNAc_2$ to $Man_5GlcNAc_2$. Perhaps action of the ER mannosidase alters the conformation of the glycoprotein, making it more susceptible to the Golgi α-mannosidase I. Alternatively, the combination of the ER and Golgi α-mannosidases may be a type of fine-tuning mechanism by cells to produce surface glycoproteins with particular assortments of high mannose-type chains [Bischoff et al., 1990].

Further processing of the mannose chains of Asn-linked oligosaccharides is obligatory for the formation of complex-type chains. The processing occurs by the action of both α-mannosidase I, which is active in the *cis*-Golgi compartments, and α-mannosidase II, which is active in the medial-Golgi compartments. Recently, Baron and Garoff (1990) reported that the peptide sequence of an antigen recognized by a medial-Golgi–specific monoclonal antibody [Burke et al., 1982] is α-mannosidase II. Not all *N*-linked oligosaccharides on glycoproteins that enter the Golgi apparatus, however, are recognized by the Golgi α-mannosidase I. Many mature surface and secreted glycoproteins, such as transferrin receptor and IgM, contain high mannose-type chains that are relatively unprocessed. It has been proposed that the high mannose-type chains at some sites might be sterically inaccessible to α-mannosidase I [Hsieh et al., 1983b; Trimble et al., 1983; Pollack and Atkinson, 1983; Forsee et al., 1989]. However, the precise factors controlling recognition of substrates by α-mannosidase I in vitro remain uncertain.

Mannose-processing is important in the biosynthetic pathway of phosphorylation of the high mannose-type chain in lysosomal enzymes [Lazzarino and Gabel, 1988]. Depending on the site of attachment of phosphate residues, the accessibility of the mannose residues to the action of mannosidases can be dramatically affected. For example, addition of phosphate by the phosphotransferase to specific mannose residues can limit the action of mannosidase I, but allow action of GlcNAc-transferase I on these oligosaccharides, to generate phosphorylated or sialylated hybrid-type molecules [Varki and Kornfeld, 1983]. Lazzarino and Gabel (1989) have postulated that, following the action of the ER mannosidase, initial phosphorylation of high mannose-type chains occurs on mannose residues of the α1,6-branch. These monophosphorylated oligosaccharides have several possible fates. The Golgi mannosidase I may or may not act on these monophosphorylated oligosaccharides. It appears that if it does remove additional mannose residues, then it is unlikely that the oligosaccharides become further phosphorylated, and they tend to either remain monophosphorylated high mannose-type chains or become substrates for GlcNAc-transferase I and, subsequently, phosphorylated, hybrid-type oligosaccharides. On the other hand, the $Man_8GlcNAc_2$-Asn oligosaccharides may be phosphorylated on the α1,3-branch to the core β-mannose residue to become mature diphosphorylated high mannose-type chains. The distinctive phosphorylation patterns suggest different possible functions. In this light, it is interesting that different phosphorylated oligosaccharides differ in their binding affinity to the two mannose-6-phosphate receptors [Kornfeld, 1989].

VII. INVOLVEMENT OF ENDOGLYCOSIDASES IN EARLY-STAGE PROCESSING OF ASPARAGINE-LINKED OLIGOSACCHARIDES

There is evidence that the removal of glucose residues from glucosylated high mannose-type Asn-linked oligosaccharides during biosynthesis does not occur exclusively in the ER. Lubas and Spiro (1987) first described the presence in the rat liver Golgi apparatus of an endo-α-D-mannosidase capable of removing the disaccharide unit Glcα1,3Man from the glucosylated high mannose-type chain $Glc_1Man_9GlcNAc$-R (see Fig. 4). The enzyme has a near-neutral pH optimum and was enriched in Golgi membranes 69-fold over ER membranes. The inhibitors 1-deoxynojirimycin, 1-deoxymannojirimycin, and swainsonine, which inhibits the processing enzymes glucosidase II, α-mannosidase I, and α-mannosidase II, respectively, are without effect on the endomannosidase [Lubas and Spiro, 1988]. The discovery of this enzyme suggests the possibility that even in the presence of inhibitors of early-stage processing, it is possible to have processing of *N*-linked chains to complex-types, because this enzyme could allow alternative processing of high mannose chains without the action of glucosidase II and α-mannosidase I.

Recent studies indicate that the endomannosidase can also remove the tri- and tetrasaccharides, Glcα1,3Glcα1,3Man and Glcα1,2Glcα1,3Glcα1,3Man [Lubas and Spiro, 1988]. The precise role of the endomannosidase in *N*-linked processing is currently obscure, but there are several possibilities. The enzyme could be a type of "backup" to the ER glucosidases (i.e., completing removal of glucose residues in glycoproteins that escaped action by the ER glucosidases). Additionally, the enzyme could be specific, and preferentially process glucosylated high mannose-type oligosaccharides, in particular glycoproteins in which the glucose residues are refractory to the ER glucosidases. Alternatively, the enzyme could be important in removing those glucose residues added transiently to high mannose-type oligosaccharides in the Golgi apparatus, as discussed later.

VIII. REGLYCOSYLATION OF OLIGOSACCHARIDES AFTER TRANSFER TO PROTEINS

Parodi and colleagues [Parodi et al., 1983, 1984b] were the first to describe the presence in plant and animal cells of a glucosyltransferase capable of adding glucose to the high mannose-type Asn-linked oligosaccharides of newly synthesized glycoproteins. Their experiments suggested that the glucosyltransferase utilizes uridine diphosphate (UDP) glucose as the donor and is contained within the ER (see Fig. 2). Parodi et al. (1984) suggested, as have others [Hoflack et al., 1981], that terminal glucose residues are required to prevent extensive degradation of the high mannose-type oligosaccharides in the ER by α-mannosidases. Although this is possible, it should be recalled that the ER contains a specific mannosidase activity, which works principally on only one α-linked mannose residue, as discussed earlier.

Recently, Suh et al. (1989) have found that the high mannose-type Asn-linked oligosaccharides in the G protein from a temperature-sensitive vesicular stomatitis virus (VSV), is monoglucosylated at the nonpermissive temperature in which the G protein is retained in the ER. Their structural analyses indicated that the oligosaccharides contained no fewer than eight mannose residues and only one glucose residue. The results of pulse-chase radiolabeling experiments in the absence of protein synthesis strongly suggested that the oligosaccharides on the G protein were deglucosylated and reglucosylated in a posttranslational process. These results are interesting, since electron microscopic evidence indicates that at the nonpermissive temperature the mutant VSV G protein is a resident of the ER [Bergmann and Singer, 1983a,b]. Considering the previous observations by Lodish and Kong (1984) that glucose removal is required for many glycoproteins to exit the ER, the possibility is raised that a glucose-binding protein might be involved in glycoprotein retention in the ER. Currently, however, there is no direct evidence for such a binding protein. Suh et al. (1989) speculated that the presence of glucose on the ER proteins may signal them for degradation and that reglucosylation may be a part of the machinery for protein degradation in the ER. Glucosylated high mannose-type oligosaccharides may also be involved in early aspects of protein folding [Schlesinger et al., 1984]. Clearly, the precise role of reglucosylation will require a better understanding of both the glucosyltransferase and its specificity and the state of glucosylation of nonaberrant ER proteins.

In a related earlier study, Rizzolo and Kornfeld (1988) demonstrated that a hybrid protein, derived from a gene containing fused coding domains for a region of the influenza virus hemagglutinin and rat growth hormone, does not exit the ER and appears to be retained in the smooth cisterna of the ER. Approximately one-half of the high mannose-type Asn-linked oligosaccharides of the protein were glucosylated. In light of these findings, it is noteworthy that Strous et al. (1987) isolated glucosidase II from rat cell lines and found that it was a soluble, glycoprotein and, presumably, contains glucosylated high mannose-type Asn-linked oligosaccharides. Because this enzyme is a resident ER protein, but is not an intrinsic membrane glycoprotein, the possibility is raised that retention of glycoproteins in the ER may be through protein–carbohydrate interactions.

IX. PARTICIPATION OF ORGANELLES AND THEIR ENZYMES IN THE BIOSYNTHESIS OF ASPARAGINE-LINKED OLIGOSACCHARIDES

Although cytoplasmically oriented polypeptides have been found to be glycosylated, the Asn-linked oligosaccharides have not yet been found in the cytoplasmic compartment, but only in polypeptide domains within the lumen of the outer nuclear envelope, ER, Golgi apparatus, lysosome, and intracellular vesicles involved in membrane trafficking, and, of course, the plasma membrane. As one might predict, most of the enzyme machinery involved in Asn-linked oligosaccharide biosynthesis is found within the lumen of the ER and Golgi apparatus.

The overall biosynthetic pathway for Asn-linked oligosaccharides in animal and plant cells involves many proteins. (1) Each monosaccharide unit in a mature Asn-linked oligosaccharides is derived from a high-energy donor, the sugar nucleotides. The synthesis of these sugar nucleotide donors, such as UDPGlcNAc, UDPGal, GDPMan, GDPFuc, UDPGlc, UDPGalNAc, and UDPXyl occurs in the cytoplasm, and for many of the sugar nucleotides, this synthesis requires several enzymatic steps. (2) The addition of each monosaccharide unit in an Asn-linked oligosaccharides is catalyzed by glycosyltransferases, most of which are located in the lumen of the ER and Golgi apparatus. In general, each enzyme is specific for the monosaccharide unit transferred and the structure of the acceptor oligosaccharide or aglycone. Most of these enzymes are capable of transferring a monosaccharide to make only one specific linkage, but there appear to be exceptions to this, as discussed later. In addition, all modifications of the monosaccharide residues in Asn-linked oligosaccharides, such as phosphorylation, sulfation, and acetylation, require additional enzymes. Considering the large array of Asn-linked oligosaccharides shown in Figure 1, it is apparent that many enzymes are involved in the biosynthesis of Asn-linked oligosaccharides. (3) The removal of monosaccharide residues during biosynthesis of the Asn-linked oligosaccharides requires specific glycosidases. Some of these glycosidases are highly specific for the structure of the oligosaccharide they act upon. (4) The utilization by the glycosyltransferases in the lumen of the ER and Golgi apparatus of sugar nucleotides requires specific transport of sugar nucleotides across these membranes. Many of the transport systems have been identified and have been shown to involve specific polypeptide components in the proper membranes [Hirschberg and Snider, 1987]. None of these transport systems has yet been fully biochemically characterized, owing to the difficulty in isolating them. The foregoing considerations, just for the biosynthesis of Asn-linked oligosaccharides, clearly indicate the degree to which many aspects of cellular structure and metabolism are committed to glycoprotein and overall glycoconjugate biosynthesis.

X. FACTORS THAT REGULATE THE BIOSYNTHESIS OF ASPARAGINE-LINKED OLIGOSACCHARIDES

The difficulty in understanding the regulation of the pathways for Asn-linked oligosaccharide biosynthesis is probably due to the many factors involved in determining the structures. Some factors that contribute to the regulation of biosynthesis are listed in Table 1. Although each of these factors could be reviewed individually, some comments on recent developments related to the determinants that influence protein glycosylation are included here. The consensus sequence -Asn-X-Ser/Thr- is a necessary, but not sufficient, requirement for addition of Asn-linked oligosaccharides [Kornfeld and Kornfeld, 1985]. The availability of dolichol-linked intermediates may affect the efficiency of asparagine glycosylation [Stoll and Krag, 1988; Stoll et al., 1988]. Protein folding during synthesis of nascent chains may also be important in allowing access of the oligosaccharyltransferase to the asparagine glycosylation sites [Struck and Lennarz, 1980]. The structures of oligosaccharides at individual glycosylation sites in glycoproteins also appear to be greatly influenced by the site to which they are attached. Some examples of glycoproteins in which these observations have been made include human IgD [Mellis and Baenziger, 1983], the murine major histocompatibility antigen [Swiedler et al., 1983], Sindbis virus glycoproteins [Hsieh et al., 1983a; Hubbard, 1988], Friend murine leukemia virus glycoproteins [Schluter et al., 1984], the hemagglutinin of influenza virus [Keil et al., 1985], murine Ia antigen [Cowan et al., 1985], erythropoietin [Dube et al., 1988], vesicular stomatitis virus G protein [Hubbard, 1987], recombinant human tissue plasminogen activator [Spellman et al., 1989], ovomucoid and fetuin [Yet et al., 1988a], and neoglycoproteins [Shao and Wold, 1988, 1989].

The protein conformation may limit access of glycosidases and glycosyltransferases to the sugar chains [reviewed by Yet et al., 1988b]. The processing at individual glycosylation sites may also be affected by oligomerization of proteins. For example, subunit association

Table 1 Some Factors Regulating the Structures of Asn-Linked Oligosaccharides

Primary structure of the glycoprotein
Intracellular location of a glycoprotein
Types and activity levels of glycosidases in the cell
Types and activity levels of glycosyltransferases in the cell
Physical accessibility of oligosaccharides to enzymes
Oligomerization of subunits of a glycoprotein
Specificity of glycosyltransferase for certain glycoprotein substrates
Association of a glycoprotein with trafficking receptors
Transit time of a glycoprotein in ER and other organelles
Recycling of membrane-bound glycoproteins
Availability of dolichol and dolichol-linked oligosaccharides

affects the site-specific glycosylation of the common α-chain from Mac-1 and LFA-1 [Dahms and Hart, 1986] and the common α-subunit of glycoprotein hormones [Green et al., 1986; Kobata, 1988]. Alternatively, glycosylation of specific asparagine residues may be critical for subunit association, as observed during studies on the biosynthesis of human chorionic gonadotropin [Matzuk and Boime, 1988a,b; Matzuk et al., 1989].

It is clear that there are cell type-specific influences on the structures of Asn-linked oligosaccharides present in animal glycoproteins [Hsieh et al., 1983a,b; Hubbard, 1987, 1988; Sheares and Robbins, 1987; Parekh et al., 1987; Kagawa et al., 1988]. The basis for this is a result of many cell type–specific factors. The array of glycosyltransferases and glycosidases and the levels of their activities in cells greatly influence the types of oligosaccharides synthesized [Schachter et al., 1983; Kornfeld and Kornfeld, 1985; Schachter, 1986]. The expression of glycosyltransferases involved in late-stage processing of Asn-linked oligosaccharides is tissue- and cell-specific, related to the differentiation state of the cells, and inducible by a number of agents [Cummings and Mattox, 1988; Wang et al., 1989; Muramatsu, 1988; Paulson et al., 1989; Symington et al., 1989; Dutt et al., 1988].

Viral and chemical transformations of cells have been found to cause cells to synthesize glycoproteins that generally contain complex-type Asn-linked oligosaccharides with increased degrees of branching and altered sialylation [Hubbard, 1987; Yamashita et al., 1985; Pierce and Arango, 1986; Warren et al., 1978; Smets and van Beek, 1984; Reading and Hutchins, 1985; Passaniti and Hart, 1988]. These results suggest that expression of glycosyltransferases acting on Asn-linked oligosaccharides is altered in transformed animal cells. Recent findings support these predictions. For example, the activity of *N*-acetylglucosaminyltransferase V, an enzyme involved in generating the 2,6-triantennary complex chain [Cummings et al., 1982] is elevated after transformation of animal cells by certain viruses [Yamashita et al., 1985; Dennis et al., 1987; Marchase et al., 1988; Arango and Pierce, 1988; Dennis and Laferte, 1988]. Other examples in which glycosylation of proteins is altered upon malignant transformation, during disease conditions, or upon cellular differentiation are discussed by Rademacher et al. (1988).

Although animals share common pathways for early-stage biosynthesis and processing of Asn-linked oligosaccharides, the enzymes involved in late-stage processing in certain organisms (invertebrates) are different from those in mammalian cells. For example, adult males of *Schistosoma mansoni*, an helminthic human parasite, synthesize high mannose-type Asn-linked oligosaccharides with the typical $Man_9GlcNAc_2$-Asn- structure found in mammalian cells [Nyame et al., 1988]. The complex-type Asn-linked oligosaccharides, however, are devoid of sialic acid (as are the organisms in general) and contain terminal unsulfated, β-linked *N*-acetylgalactosamine, an unusual terminal sequence [Nyame et al., 1989]. Insect cells synthesize the normal dolichol-linked oligosaccharide $Glc_3Man_9GlcNAc_2$-P-P-Dol and process the oligosaccharide to the core $Man_3GlcNAc_2$-Asn [Butters et al., 1981; Hsieh and Robbins, 1984]. The organisms lack sialyltransferases, galactosyltransferases, and many other enzymes found to be involved in late-stage processing in vertebrate cells. *Dictyostelium discoideum* cells synthesize unusual "bisected" high mannose-type Asn-linked oligosaccharides [Couso et al., 1987] and have been found to contain a unique bisecting *N*-acetylglucosaminyltransferase, specific for high mannose-type chains [Sharkey and Kornfeld, 1989].

In addition, the final structures of Asn-linked oligosaccharides may be partly determined by enzyme competition. There are many glycosyltransferases that compete for similar acceptors, thereby generating the possibility for heterogeneity in structure [Elices and Goldstein, 1988, 1989; Lee et al., 1989; Brockhausen et al., 1989; Merkle and Cummings, 1987; Smith et al., 1990]. This competition between enzymes for similar substrates may, however, be greatly affected by other nearby determinants, such as the accessibility of enzymes to the oligosaccharides and perhaps, even, peptide determinants that influence transient associations of glycosyltransferases with each other and with their glycoprotein substrates.

XI. GLYCOSYLTRANSFERASES SPECIFIC FOR THE GLYCOPROTEIN SUBSTRATES

One obvious way in which cells could regulate the types and structures of Asn-linked oligosaccharides is to have glycosyltransferases that are specific for the glycoprotein substrates. Although most of the glycosyltransferases studied do not appear to be highly specific in this way, two different glycosyltransferases have recently been discovered that are very specific for their glycoprotein substrates.

The first example of glycoprotein-specific *N*-linked oligosaccharide biosynthesis was the formation of phosphorylated high mannose-type oligosaccharides in lysosomal hydrolases [Kornfeld, 1987; von Figura and Hasilik, 1986]. The discovery of this pathway was aided by the study of a human disorder, termed I-cell disease, in which the sera of affected individuals contain relatively large amounts of secreted lysosomal hydrolases. The above-normal rate of secretion was found to be due to lack of a "recognition marker" [Hickman and Neufeld, 1972; Neufeld et al., 1975], which was subsequently identified as a mannose-6-phosphate moiety [Creek and Sly, 1984]. The normal biosynthesis of lysosomal enzymes was shown to involve a phosphotransferase that adds GlcNAc-1-PO_4 from UDPGlcNAc to the mannose residues in the high mannose-type Asn-linked oligosaccharides of lysosomal hydrolases during their traverse of the Golgi apparatus (see Fig. 2) [Hasilik et al., 1981; Reitman and Kornfeld, 1981a,b]. The enzyme can add GlcNAc-1-phosphate to any of five outer mannose residues [Kornfeld, 1987]. The phosphotransferase was found to be specific for lysosomal enzymes and to have altered activity in I-cell fibroblasts [von Figura and Hasilik, 1986]. After addition of GlcNAc-1-PO_4, the terminal *N*-acetylglucosaminyl residue is removed by a specific *N*-acetylglucosamine-1-phosphodiester α-*N*-acetylglucosaminidase [Varki and Kornfeld, 1980], and the terminal mannose-6-phosphate moiety is bound by the mannose-6-phosphate receptor. The receptor allows segregation of the lysosomal enzymes from the secretory pathway and promotes targeting of the enzymes to lysosomes [Kornfeld, 1987].

Although the specific molecular recognition between the phosphotransferase and lysosomal hydrolases is not well understood, there is ample evidence that it involves both a recognition of a peptide determinant in the lysosomal hydrolases and the high mannose-type Asn-linked oligosaccharides of the protein [Reitman and Kornfeld, 1981a,b; Lang et al., 1984]. These observations explain the specific occurrence of mannose-6-phosphate residues on lysosomal hydrolases, in spite of the fact that acid hydrolases and virtually all glycoprotein entering the Golgi apparatus contain similar high mannose-type Asn-linked oligosaccharides. Recent experiments, however, suggest that some glycoproteins other than lysosomal hydrolases contain the mannose-6-phosphate moiety [Purchio et al., 1988; Faust et al., 1987; Lee and Nathan, 1988; Herzog et al., 1987; Todderud and Carpenter, 1988]. Phosphorylation of these proteins may involve phosphotransferases other than the one specific for lysosomal enzymes, or these other phosphorylated glycoproteins may contain recognition determinants for the known phosphotransferase.

The lysosomal hydrolases, however, are not the only glycoproteins known to contain unique Asn-linked oligosaccharides. The pituitary glycoprotein hormones in higher animals contain relatively small amounts of galactose and sialic acid and have unusual Asn-linked oligosaccharides with terminal 4-sulfated, β-linked *N*-acetylgalactosaminyl residues in the sequence SO_4GalNAcβ1,2-Manα-R [Pierce and Parsons, 1981; Green et al., 1986; Baenziger and Green, 1988; Green and Baenziger, 1988]. The occurrence of the unusual disaccharide sequence GalNAcβ1,4GlcNAc-R led Smith and Baenziger (1988) to examine pituitary membranes for the presence of a specific glycosyltransferase. They identified a *N*-acetylgalactosaminyltransferase in bovine pituitary membranes that specifically recognizes the complex-type Asn-linked oligosaccharides in α-subunit of pituitary hormones. Interestingly, structural studies on the Asn-linked oligosaccharides of the glycoprotein hormones suggest that the *N*-acetylgalactosaminyltransferase and β1,4-galactosyltransferase compete for the terminal, nonreducing *N*-acetylglucosaminyl residues on these hormones during biosynthesis [Smith and Baenziger, 1988]. Galactosylation allows subsequent sialylation, whereas the addition of *N*-acetylgalactosamine allows for only sulfation on that specific branch. The association of β-subunits with α-subunits appears to modulate the glycosylation in specific ways, depending on the glycoprotein hormone.

The *N*-acetylgalactosaminyltransferase may then have at least dual recognition domains [Smith and Baenziger, 1988]. One domain may recognize the α-subunit of pituitary hormones and the other domain may recognize the oligosaccharide chain. Since the enzyme has not been purified, it is possible that these two recognition domains reside in different subunits; the analogy, however, between this system and the phosphotransferase system involved in lysosomal enzyme biosynthesis is obvious. Both enzymes must recognize some feature of the protein to be glycosylated, as well as some structural features of the specific Asn-linked oligosaccharides of the polypeptide. The conceptual difference between these enzymes resides in the complexity of the recognition signal for the polypeptide. The recognition marker for *N*-acetylgalactosaminyltransferase may be simple, since the enzyme is required to recognize only the single α-subunit of these hormones. The recognition marker for the lysosomal phosphotransferase may be more complex, since it appears to require specific secondary or tertiary structures relatively unique to the acid hydrolases [Kornfeld, 1987].

The discoveries of the α-subunit-specific *N*-acetylgalactosaminyltransferase and lysosomal enzyme-specific phosphotransferase raise the possibility that other glycosyltransferases might be polypeptide-specific. For both lysosomal enzymes and pituitary glycoprotein hormones, their unusual structural features led to the discovery of glycoprotein-specific glycosyltransferases. Undoubtedly, such investigations should begin with a careful understanding of the structures of the Asn-linked oligosaccharides in different glycoprotein species. A very recent example of how the finding of unusual types of Asn-linked oligosaccharides may lead to glycoprotein-specific enzymes, involves the study the glycosylation of human IgG. The complex-type Asn-linked oligosaccharides of human IgG are deficient in galactose residues [Kornfeld et al., 1971] and more recent studies indicate that the deficiency is pronounced in patients with rheumatoid arthritis [Rademacher et al., 1986; Parekh et al., 1985]. The altered glycosylation may be due to a defective IgG-specific galactosyltransferase in human B-cells [Furukawa et al., 1987; Axford et al., 1987], although a clear understanding of this presumed enzyme and pathway awaits further studies.

Recently, oligosaccharides derived from several sources, including animal cell surfaces, laminin, and erythrocyte band 3 glycoprotein, have been shown to contain the repeating disaccharide sequence [3Galβ1,4GlcNAcβ1]$_n$ or poly-*N*-acetyllactosamine sequence. This sequence occurs, however, on selected surface glycoproteins, such as the lysosome-associated membrane glycoprotein (LAMPs) [Lippincott-Schwartz and Fambrough, 1986; Mane et al., 1989; Carlsson et al., 1988; Howe et al., 1988]. Interestingly, in Chinese hamster ovary cells, virtually all of the poly-*N*-acetyllactosamine sequences synthesized by these cells are contained in LAMPs, and other glycoproteins made by these cells lack these sequences [Do et al., 1990].

These studies imply that poly-*N*-acetyllactosamine addition to glycoproteins may be highly selective and may involve recognition of peptide determinants shared by LAMPs, laminin, and others that are recognized by some critical glycosyltransferase in the biosynthetic pathway. The critical enzyme could be *N*-acetylglucosaminyltransferase V, since it has been shown earlier that in the absence of this enzyme, cells synthesize drastically reduced levels of poly-*N*-acetyllactosamine chains [Cummings et al., 1982; Cummings and Kornfeld, 1984]. It is known that the so-called extension *N*-acetylglucosaminyltransferase responsible for elongating poly-*N*-acetyllactosamine sequences preferentially recognizes the tri- and tetraantennary branch of complex-type *N*-linked chains containing the product of the *N*-acetylglucosaminyltransferase V [van den Eijnden et al., 1988]. Thus, *N*-acetylglucosaminyltransferase V might initiate poly-*N*-acetyllactosamine synthesis by synthesizing a preferred substrate for the extension *N*-acetylglucosaminyltransferase. Support for this hypothesis has come from studies demonstrating a strong correlation between elevation of *N*-acetylglucosaminyltransferase V activity and poly-*N*-acetyllactosamine synthesis in transformed animal cell lines [Takasai et al., 1980; Yamashita et al., 1984, 1985; Pierce and Arango, 1986; Dennis and Laferte, 1989; Debray et al., 1986; Dennis et al., 1989; Yousefi et al., 1991]. This hypothesis, though interesting, conflicts with certain findings that poly-*N*-acetyllactosamine chains are found on biantennary structures of erythrocyte band 3 glycoprotein [Tsuji et al., 1981]. Several explanations are possible. There may be several different extension *N*-acetylglucosaminyltransferases, the specificity and expression of which are tissue-specific and developmentally regulated. Alternatively, it is possible that the extension *N*-acetylglucosaminyltransferase prefers the more highly branched oligosaccharides because of steric constraints on its recognition of substrates. If that were so, then band 3 glycoprotein oligosaccharides might be inherently more accessible to the extension *N*-acetylglucosaminyltransferase.

XII. STRUCTURAL REMODELING OF ASPARAGINE-LINKED OLIGOSACCHARIDES IN CELL SURFACE GLYCOPROTEINS BY RECYCLING TO PARTS OF THE GOLGI

The structures of newly synthesized, membrane-bound glycoproteins as they exit the *trans*-Golgi network can be altered if these molecules reenter the Golgi. A class of cell surface glycoproteins that do appear to reenter elements of the Golgi apparatus are surface receptors for a variety of ligands. Recent studies have shown that mature mannose-6-phosphate receptors at the cell surface can recycle to the *trans*-Golgi apparatus, as evidenced by the fact that these mature receptors can acquire new sialic acid residues [Goda and Pfeffer, 1988; Jin et al., 1989]. Similar studies had been conducted on the transferrin receptor that also demonstrated that the receptor desialylated at the cell surface could reacquire sialic acid residues [Regoeczi et al., 1982; Evans et al., 1983; Snider and Rogers, 1985].

The precise compartments to which mature surface glycoproteins recycle are not known. Kreisel et al. (1988) obtained evidence for the readdition of both fucose and sialic acid to the rat hepatocyte dipeptidylpeptidase IV during recycling. Addition of core fucose to complex-type Asn-linked oligosaccharides most likely occurs in the medial-Golgi [Goldberg and Kornfeld, 1983]. Kreisel et al. (1988) speculate that the refucosylation of dipeptidylpeptidase IV occurs in this core region, and thus could argue that during recycling the enzyme reached the medial-Golgi apparatus. Unfortunately, the precise location of the fucosyltransferases in the Golgi apparatus of the rat hepatocytes is unknown; hence, the results are now somewhat equivocal. All of these studies provide additional information about the recycling pathway for surface receptors, and indicate that these receptors can reenter some aspects of the secretory pathway and probably mix with internal, newly synthesized receptors [Griffiths et al., 1988].

XIII. CLONED AND EXPRESSED GENES FOR EUKARYOTIC GLYCOSYLTRANSFERASES IN HETEROLOGOUS CELL TYPES

An understanding of the factors that regulate the biosynthesis of Asn-linked oligosaccharides will clearly require the cloning and expression of the genes for the glycosyltransferases and glycosidases. Until recently, the progress in cloning the genes for mammalian glycosyltransferases had been slow because of the difficulties in obtaining purified enzymes in sufficient quantities. The genes for a few animal cell glycosyltransferases involved in late-stage processing of complex-type Asn-linked oligosaccharides have recently been cloned and the predicted amino acid sequences determined (Fig. 5).

Several groups have obtained animal genes for UDPGal:GlcNAc β1,4-galactosyltransferase, the enzyme responsible for lactose synthesis and partially responsible for synthesizing *N*-acetyllactosamine-type structures in complex-type Asn-linked oligosaccharides [Appert et al., 1986; Naramatsu et al., 1986; Shaper et al., 1986, 1988; Masri et al., 1988]. The murine galactosyltransferase gene [Shaper et al., 1988] codes for a polypeptide of 399 amino acids with a subunit M_r of 44.4 kd. This protein is 82 and 92% similar at the amino acid level to the bovine and partial human sequences, respectively. Masri et al. (1988) obtained a full-length cDNA clone for the human β-1,4-galactosyltransferase and found that it was highly homologous with the bovine enzyme. Interestingly, Shaper et al. found a second in-frame AUG codon in the cloned cDNA, suggesting the possibility of a

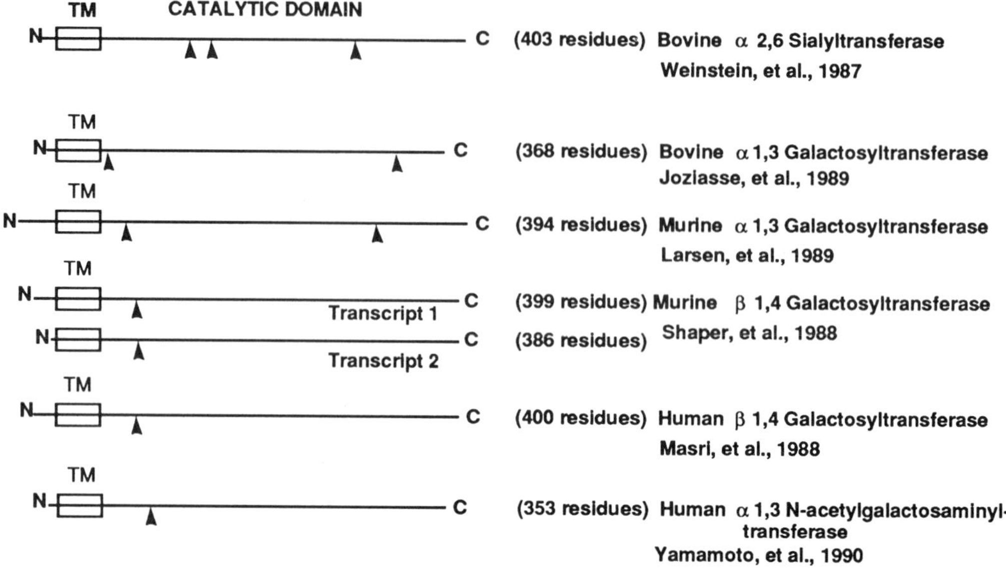

Figure 5 Comparison of the domain structures of some animal glycosyltransferases.

second transcript, coding for a smaller protein with 386 amino acids. It was hypothesized that initiation at the second AUG could generate a protein lacking a signal peptide, which could presumably retain its catalytic domain in the cytoplasm. Nevertheless, direct evidence for a cytoplasmically active galactosyltransferase has not yet been obtained. Similar evidence for dual initiation sites in the human β1,4-galactosyltransferase gene was obtained by Masri et al. (1988). Weinstein et al. (1987) cloned the cDNA for rat liver CMPNeuAc:Galβ1,4GlcNAc α2,6-sialyltransferase and found that it codes for a polypeptide of 403 residues, with a subunit M_r of 46.7 kd. Recently, two groups, Larsen et al. (1989) and Joziasse et al. (1989), have obtained the sequence for a third type of glycosyltransferase, the UDPGal:β-D-galactoside α1,3-galactosyltransferase. This enzyme is expressed in numerous murine tissues, and the level of enzyme activity is influenced by a number of factors [Maddox et al., 1982; Cummings and Mattox, 1988]. The murine α1,3-galactosyltransferase gene displays significant homology with the human blood group A and B gene, trivially named the A-type α1,3-*N*-acetylgalactosaminyltransferase and B-type α1,3-galactosyltransferase, respectively [Yamamoto et al., 1990a,b]. Even more recently, Kumar et al. (1990) have cloned and sequenced the cDNA for the human *N*-acetylglucosaminyltransferase I.

It is interesting that the genes for these different types of glycosyltransferase, such as the β1,4-galactosyltransferase, the α2,3-sialyltransferase, and the α1,3-galactosyltransferase are similar in size, topology, and in their presumed intracellular location in the *trans*-Golgi apparatus of mammalian cells. However, except for the foregoing observed relationship between the α1,3-glycosyltransferases, there is little, if any, sequence homology among these proteins. These observations suggest several hypotheses. Consider the observation that there is an apparent conservation in size of the enzymes. This may be related to the fact they all utilize *N*-linked oligosaccharide substrates and, in some cases, *O*-linked oligosaccharides and glycosphingolipid substrates. Thus, the enzymes may have to be a certain length and dimension to extend the enzyme active site into the lumen far enough to be flexible and gain access to many different regions of glycoconjugate substrates.

The absence of large homologous domains suggest that the determinants that "target" these enzymes to become residents of the *trans*-Golgi apparatus are not simply due to binding of a peptide determinant to some type of residence-specifying receptor. There may be lipid-binding domains (i.e., the transmembrane regions) that help specify the final destination of a protein by a receptor-independent–type mechanism. Considering that the secretory apparatus has gradients of lipids and density, it is probably worth contemplating the possibility that newly synthesized glycosyltransferases "spontaneously" sort to correct compartments. With use of these cloned glycosyltransferases genes, it should now be possible to explore the determinants that establish membrane localization.

The absence of sequence homology indicates that there is little recent evolutionary relationship between most of these enzymes. It is important to recall that certain glycosylation processes, such as sialylation and blood group antigen formation, are late-stage processing events not present in many lower vertebrates and invertebrates and probably absent in plants. However, some of events in early- and middle-stage processing, occur in both plants and animals, suggesting that the enzymes might be similar between plants and animals. In light of this possibility, it is interesting that pea seedlings contain *N*-acetylglucosaminyltransferase I and the α1,2-fucosyltransferase activities (the blood group H glycosyltransferase) [Palcic and Hindsgaul, personal communication]. This raises the possibility that some animal cell glycosyltransferases genes might be related to those in organisms from other phyla. Availability of cloned genes from animals will permit assessment of this possibility.

Clearly, it will be necessary to obtain the sequences for a large number of glycosyltransferases before we can begin to understand all of the factors that regulate their expression and localization in the different stacks of the Golgi apparatus. The slow progress in this area has recently been overcome by several investigators who have developed promising and novel approaches that enable the rapid isolation and cloning of glycosyltransferase genes. These approaches rely on direct screening for cells expressing an enzyme under consideration by using lectins or antibodies that bind to the surface-localized product of the enzyme. This genetic expression approach has been recently described [Ernst et al., 1989; Rajan et al., 1989; Lowe, 1989; Ripka et al., 1989a,b], and was used successfully by Larsen et al. (1989) in cloning the murine α1,3-galactosyltransferase. Cummings and Mattox (1988) originally reported that the mouse teratocarcinoma cell line F9 contained abundant levels of the enzyme UDPGal:β-D-galactosidase α1,3-galactosyltransferase and that retinoic acid-induced differentiation results in elevated levels of enzyme activity. Larsen et al. (1989) then prepared a cDNA expression library in a mammalian expression vector using mRNA from the retinoic acid-differentiated F9 cells. When COS-1 cells were transfected with this cDNA library, some cells expressed the functional α1,3-galactosyltransferase and were identified by their ability to adhere to the immobilized plant lectin from *Griffonia simplicifolia* I-B_4. Nontransfected COS-1 cells do not express this enzyme and do not bind to the immobilized lectin. The plasmids isolated from the positive, transfected COS-1 cells were amplified, and the transfection and selection was repeated until a purified, cloned plasmid was obtained. Northern blot analysis, using the cloned α1,3-galactosyltransferase gene, demonstrated that the transcript for the α1,3-galactosyltransferase is elevated severalfold upon retinoic acid-induced differentiation of F9 cells [Larsen et al., 1989].

This approach should be generally useful for the cloning and isolation of virtually all glycosyltransferases involved in the processing of Asn-linked oligosaccharides. The ability to isolate the cloned genes for animal cell glycosyltransferases will permit studies of their expression and regulation in a variety of tissues and cell types. Furthermore, it is possible that genetic expression of these enzymes in heterologous cell types not normally expressing them will give new insights into the factors that determine Golgi localization and oligosaccharide structure. Recently, Lee et al. (1989) expressed the α2,6-sialyltransferase in CHO cells, which normally lack this activity, but instead have the α2,3-sialyltransferase. Their results indicated that the cloned gene coded for an active enzyme that caused the formation of α2,6-linked sialic acid residues on surface glycoconjugates. In a related study, Smith et al. (1990) expressed the murine α1,3-galactosyltransferase in CHO cells and demonstrated that the terminal α1,3-galactosyl residues were added in lieu of terminal α2,3-sialic acid residues, demonstrating enzyme competition. Future experiments of this type and the expression of glycosyltransferase genes in heterologous cell types should dramatically increase our knowledge of the roles of glycosyltransferases in directing the biosynthesis of specific carbohydrate structures on glycoproteins and the functions of cell surface glycoprotein oligosaccharides.

ACKNOWLEDGMENTS

The research of the author in relation to the work described in this article was supported by research grants from the National Cancer Institute (CA37626), National Institute of Allergy and Infectious Diseases (AI26725) and National Eye Institute (EY05971). The author thanks Dr. Roberta Merkle, Dr. David Smith, and Sandra Cummings for critical reading of the manuscript.

REFERENCES

Appert, H. E., Rutherford, T. J., Tarr, G. E., Wiest, J. S., Thomrod, N. R., and McCorquodale, D. J. (1986). *Biochem. Biophys. Res. Commun. 139*:163–168.

Arango, J., and Pierce, M. (1988). *J. Cell. Biochem. 37*:225–231.

Arumugham, R. G., Hsieh, T. C.-Y., Tanzer, M. L., and Laine, R. A. (1986). *Biochim. Biophys. Acta 883*:112–126.

Axford, J. S., MacKensie, L., Lydyard, P. M., Hay, F. C., Isenberg, D. A., and Roitt, I. M. (1987). *Lancet 2*:1486–1488.

Baenziger, J. U., and Green, E. D. (1988). *Biochim. Biophys. Acta 947*:287–306.

Baron, M. D., and Garoff, H. (1990). *J. Biol. Chem. 265*:19928–19931.

Barnes, G., Hansen, W., Holcomb, C., and Rine, J. (1984). *Mol. Cell. Bio. 4*:2381–2388.

Bause, E. (1983). *Biochem. J. 209*:331–336.

Bause, E., and Legler, G. (1981). *Biochem. J. 195*:639–644.

Bergmann, J. E., and Singer, S. J. (1983a). *J. Cell Biol. 97*:1777–1787.

Bergmann, J. E., and Singer, S. J. (1983b). *Proc. Natl. Acad. Sci. USA 78*:1746–1750.

Bischoff, J., and Kornfeld, R. (1983). *J. Biol. Chem. 258*:7907–7910.

Bischoff, J., Moremen, K., and Lodish, H. F. (1990). *J. Biol. Chem. 265*:17110–17117.

Bosch, M., Trombetta, S., Engstrom, U., and Parodi, A. J. (1988). *J. Biol. Chem. 263*:17360–17365.

Brockhausen, I., Hull, E., Hindsgaul, O., Schachter, H., Shah, R. N., Michnick, S. W., and Carver, J. P. (1989). *J. Biol. Chem.* *264*:11211–11221.

Butters, T. D., Hughes, R. C., and Vischer, P. (1981). *Biochim. Biophys. Acta 640*:672–686.

Byrd, J. C., Tarentino, A. L., Maley, F., Atkinson, P. H., and Trimble, R. B. (1982). *J. Biol. Chem.* *257*:14657–14666.

Carlsson, S. R., Roth, J., Piller, F., and Fukuda, M. (1988). *J. Biol. Chem.* *263*:18911–18919.

Chapman, A., Fujimoto, K., and Kornfeld, S. (1980). *J. Biol. Chem.* *255*:4441–4445.

Corfield, A. P., and Schauer, R. (1982). In *Sialic Acids Chemistry, Metabolism and Function* (R. Schauer, ed.). Springer-Verlag, Vienna, pp. 6–50.

Couso, R., van Halbeek, H., Reinhold, V., and Kornfeld, S. (1987). *J. Biol. Chem.* *262*:4521–4527.

Cowan, E. P., Cummings, R. D., Lee, D. R., Schwartz, B. D., and Cullen, S. E. (1985). *Mol. Immunol.* *22*:135–143.

Creek, K. E., and Sly, W. S. (1984). In *Lysosomes in Biology and Pathology* (J. T. Dingle, R. T. Dean, and W. Sly, eds.). Elsevier, Amsterdam, pp. 63–82.

Criscuolo, B., and Krag, S. (1982). *J. Cell Biol.* *94*:586–591.

Cummings, R. D., and Kornfeld, S. (1984). *J. Biol. Chem.* *259*:6253–6260.

Cummings, R. D., Mangelsdorf, S., and Carpenter, G. (1985). *J. Biol. Chem.* *260*:11944–11952.

Cummings, R. D., and Mattox, S. (1988). *J. Biol. Chem.* *263*:511–519.

Cummings, R. D., Trowbridge, I. S., and Kornfeld, S. (1982). *J. Biol. Chem.* *257*:13421–13427.

Dahms, N. M., and Hart, G. W. (1986). *J. Biol. Chem.* *261*:13186–13196.

Dahms, N. M., and Kornfeld, S. (1989). *J. Biol. Chem.* *264*:11458–11467.

Debray, H., Qin, Z., Palannoy, P., Montreuil, J., Dus, D., Radjikowski, C., Christensen, B., and Kieler, J. (1986). *Int. J. Cancer 37*:607–611.

Dennis, J., and Laferte, S. (1989). *Cancer Res.* *49*:945–951.

Dennis, J. W., Kosh, K., Bryce, D.-M., and Brietman, M. L. (1989). *Oncogene 4*:853–860.

Dennis, J. W., and Laferte, S. (1988). In *Altered Glycosylation in Tumor Cells.* UCLA Symposia on Molecular and Cellular Biology (C. L. Reading, S. Hakomori, and D. M. Marcus, eds.). Alan R. Liss, New York, pp. 257–267.

Dennis, J. W., Laferte, S., Waghorne, C., Breitman, M. L., and Kerbel, R. S. (1987). *Science* *236*:582–585.

Diaz, S., Higa, H. H., Hayes, B. K., and Varki, A. (1989). *J. Biol. Chem.* *264*:19416–19426.

Do, K.-Y., Smith, D. F., and Cummings, R. D. (1990). *Biochem. Biophys. Res. Commun.* *173*:1123–1128.

Dube, S., Fisher, J. W., and Powell, J. S. (1988). *J. Biol. Chem.* *263*:17516–17521.

Dutt, A., Tang, J.-P., and Carson, D. D. (1988). *J. Biol. Chem.* *263*:2270–2279.

Elbein, A. D. (1987). *Annu. Rev. Biochem.* *56*:497–534.

Elices, M. J., and Goldstein, I. J. (1988). *J. Biol. Chem.* *263*:3354–3362.

Elices, M. J., and Goldstein, I. J. (1989). *J. Biol. Chem.* *260*:1375–1380.

Engel, J. C., and Parodi, A. J. (1985). *J. Biol. Chem.* *260*:10105–10110.

Ernst, L. K., Rajan, V. P., Larsen, R. D., Ruff, M. M., and Lowe, J. B. (1989). *J. Biol. Chem.* *264*:3436–3447.

Evans, W. H., Flint, N., Debanne, M. T., and Regoeczi, E. (1983). In *Structural Carbohydrates of the Liver* (H. Popper, W. Reutter, F. Gudat, and E. Kottgen, eds.). MTP Press, Lancaster, Engl., pp. 301–311.

Faust, P. L., Chirgwin, J. M., and Kornfeld, S. (1987). *J. Cell Biol.* *105*:1947–1955.

Finne, J., Breimer, M. E., Hansson, G. C., Karlsson, K-A., Leffier, H., Vliegenthart, J. F. G., and van Halbeek, H. (1989). *J. Biol. Chem.* *264*:5720–5735.

Finne, J., and Finne, U. (1983). *Biochem. Biophys. Res. Commun.* *112*:482–487.

Forsee, W. T., Palmer, C. F., and Schutzback, J. S. (1989). *J. Biol. Chem.* *264*:3869–3876.

Freeze, H. H., and Wolgast, D. (1986). *J. Biol. Chem.* *261*:127–134.

Fuhrmann, U., Bause, E., and Ploegh, H. L. (1985). *Biochim. Biophys. Acta 825*:95–100.

Fukuda, M. (1985). *Biochim. Biophys. Acta 780*:119–150.

Fukuda, M., Dell, A., Oates, J. E., and Fukuda, M. N. (1984). *J. Biol. Chem. 259*:8260–8273.

Furukawa, K., Matsuta, K., Takeuchi, F., et al. (1987). The Ninth International Symposium on Glycoconjugates, Lille, France, 6–11 July, E56.

Gabel, C. A., Costello, C. E., Reinhold, V. N., Kurz, L., and Kornfeld, S. (1984). *J. Biol. Chem. 259*:13762–13769.

Gabel, C. A., and Kornfeld, S. (1982). *J. Biol. Chem. 257*:10605–10612.

Geetha-Habib, M., Noiva, R., Kaplan, H. A., and Lennarz, W. J. (1988). *Cell 54*:1053–1060.

Geetha-Habib, M., Park, H. R., and Lennarz, W. J. (1990). *J. Biol. Chem. 265*:13655–13660.

Glabe, C. G., Hanover, J. A., and Lennarz, W. J. (1980). *J. Biol. Chem. 255*:9236–2\9242.

Goldberg, D., and Kornfeld, S. (1983). *J. Biol. Chem. 258*:3159–3165.

Goda, Y., and Pfeffer, S. R. (1988). *Cell 55*:309–320.

Green, E. D., Adelt, G., and Baenziger, J. U. (1988). *J. Biol. Chem. 263*:18253–18268.

Green, E. D., and Baenziger, J. U. (1988). *J. Biol. Chem. 263*:25–35.

Green, E. D., Boime, I., and Baezinger, J. U. (1986). *Mol. Cell. Biochem. 72*:81–100.

Griffiths, G., Hoflack, B., Simons, K., Mellman, I., and Kornfeld, S. (1988). *Cell 52*:329–341.

Griffiths, G., and Simons, K. (1986). *Science 234*:438–443.

Hanaoka, K., Pritchett, T. J., Takasaki, S., Kochibe, N., Sabesan, S., Paulson, J. C., and Kobata, A. (1989). *J. Biol. Chem. 264*:9842–9849.

Hart, G. W., Brew, K., Grant, G. A., Bradshaw, R. A., and Lennarx, W. J. (1979). *J. Biol. Chem. 254*:9747–9753.

Haselbeck, A., and Tanner, W. (1984). *FEMS Microbiol. Lett. 21*:305–308.

Hasilik, A., Waheed, A., and von Figura, K. (1981). *Biochem. Biophys. Res. Commun. 98*: 761–767.

Hearing, J., Gething, M.-J., and Sambrook, J. (1989). *J. Cell Biol. 108*:355–365.

Herzog, V., Neumuller, W., and Holzmann, B. (1987). *EMBO J . 6*:555–560.

Hickman, S., and Neufeld, E. F. (1972). *Biochem. Biophys. Res. Commun. 49*:992–999.

Hirschberg, C. B., and Snider, M. D. (1987). *Annu. Rev. Biochem. 56*:63–88.

Hoflack, B., Cacan, R., and Verbert, A. (1981). *Eur. J. Biochem. 117*:285–290.

Howe, C. L., Granger, B. L., Hull, M., Green, S. A., Gabel, C. A., Helenius, A., and Mellman, I. (1988). *Proc. Natl. Acad. Sci. USA 85*:7577–7581.

Hsieh, P., and Robbins, P. W. (1984). *J. Biol. Chem. 259*:2375–2382.

Hsieh, P., Rosner, M. R., and Robbins, P. W. (1983b). *J. Biol. Chem. 258*:2548–2554.

Hsieh, P., Rosner, M. R., and Robbins, P. W. (1983a). *J. Biol. Chem. 258*:2555–2561.

Hubbard, S. C. (1988). *J. Biol. Chem. 263*:19303–19317.

Hubbard, S. C. (1987). *J. Biol. Chem. 262*:16403–16411.

Hubbard, S. C., and Ivatt, R. J. (1981). *Annu. Rev. Biochem. 50*:555–583.

Järnefelt, J., Rush, J., Li, Y. T., and Laine, R. A. (1978). *J. Biol. Chem. 253*:8006–8009.

Jelinek-Kelly, S., and Herscovics, A. (1988). *J. Biol. Chem. 263*:14757–14763.

Jin, M., Sahagian, G. G., and Snider, M. D. (1989). *J. Biol. Chem. 264*:7675–7680.

Joziasse, D. H., Shaper, J. H., van den Eijinken, D. H., van Tunen, A. J., and Shaper, N. L. (1989). *J. Biol. Chem. 264*:14290–14297.

Kagawa, Y., Takasaki, S., Utsumi, J., Hosoi, K., Shimizu, H., Kochibe, N., and Kobata, A. (1988). *J. Biol. Chem. 263*:17508–17515.

Kammerling, J. P., Damm, J. B. L., Hard, K. J., and Vliegenthart, J. F. G. (1988). In *Sialic Acids: Proceedings of the Japanese–German Symposium on Sialic Acids* (R. Schauer and T. Yamakawa, eds.), Barble Mende, Kiel, FRG, pp. 74–75.

Kaplan, H. A., Naider, F., and Lennarz, W. J. (1988). *J. Biol. Chem. 263*:7814–7820.

Kaplan, H. A., Welpy, J. K., and Lennarz, W. J. (1987). *Biochim. Biophys. Acta 906*:161–173.

Katajima, K., Inoue, S., Inoue, Y., and Troy, F. A. (1988). *J. Biol. Chem. 263*:18269–18276.

Kato, Y., and Spiro, R. G. (19898). *J. Biol. Chem. 254*:3364–3371.

Keil, W., Geger, R., Dabroski, J., Dabroski, U., Niemann, H., Stirm, S., and Klenk, H.-D. (1985). *EMBO J. 4*:2711–2720.

Kiely, M. L., McKnight, G. S., and Schimke, R. T. (1976). *J. Biol. Chem. 251*:5490–5495.
Kobata, A. (1988). *J. Cell. Biochem. 37*:79–90.
Kornfeld, R., Keller, J., Baenziger, J., and Kornfeld, S. (1971). *J. Biol. Chem. 246*:3259–
Kornfeld, R., and Kornfeld, S. (1980). In *The Biochemistry of Glycoproteins and Proteoglycans* (W. J. Lennarz, ed.). Plenum Press, New York, pp. 1–34.
Kornfeld, R., and Kornfeld, S. (1985). *Annu. Rev. Biochem. 54*:631–664.
Kornfeld, S. (1987). *FASEB J. 1*:462–468.
Kornfeld, S. (1989). *J. Biol. Chem. 264*:12115–12118.
Krag, S. S. (1985). *Curr. Top. Membr. Transp. 24*:181–249.
Kreisel, W., Hanski, C., Tran-Thi, T.-A., Katz, N., Deckers, K., Reutter, W., and Gerok, W. (1988). *J. Biol. Chem. 263*:11736–11742.
Krusius, T., Finne, J., and Rauvala, H. (1978). *Eur. J. Biochem. 92*:289–300.
Kumar, R., Yang, J., Larsen, R. D., and Stanley, P. (1990). *Proc. Natl. Acad. Sci. USA 87*:9948–9952.
Lang, L., Reitman, M. L., Tang, J., Roberts, R. M., and Kornfeld, S. (1984). *J. Biol. Chem. 259*:14663–14671.
Larsen, R. D., Rajan, V. P., Ruff, M. M., Kukowska-Latallo, J., Cummings, R. D., and Lowe, J. B. (1989). *Proc. Natl. Acad. Sci. USA 86*:8227–8231.
Lazzarino, D. A., and Gabel, C. A. (1989). *J. Biol. Chem. 264*:5015–5023.
Lazzarino, D. A., and Gabel, C. A. (1988). *J. Biol. Chem. 263*:10118–10126.
Lee, E. U., Roth, J., and Paulson, J. C. (1989). *J. Biol. Chem. 264*:13848–13855.
Lee, S. J., and Nathan, D. (1988). *J. Biol. Chem. 263*:3521–3527.
Lehrman, M. A., and Zeng, Y. (1989). *J. Biol. Chem. 264*:1584–1593.
Lehrman, M. A., Zhu, X., and Khounlo, S. (1988). *J. Biol. Chem. 263*:19796–19803.
Lennarz, W. J. (1987). *Biochemistry 26*:7205–7210.
Lipinski, M., Hirsch, M.-R., Deagostini-Bazin, H., Yamada, O., Tursz, T., and Goridis, C. (1987). *Int. J. Cancer 40*:81–86.
Lippincott-Schwartz, J., and Fambrough, D. M. (1986). *J. Cell Biol. 102*:1593–1605.
Livingston, B. D., Jacobs, J. L., Glick, M. C., and Troy, F. A. (1988). *J. Biol. Chem. 263*:9443–9448.
Livingston, B. D., Jacobs, J., Shaw, G. W., Glick, M. C., and Troy, F. A. (1987). *Fed. Proc. 46*: 2151.
Lodish, H. F., and Kong, N. (1984). *J. Cell. Biol. 98*:1720–1729.
Lowe, J. (1989). *J. Biol. Chem. 264*:3436–3447.
Lubas, W. A., and Spiro, R. G. (1987). *J. Biol. Chem. 262*:3775–3781.
Lubas, W. A., and Spiro, R. G. (1988). *J. Biol. Chem. 263*:3990–3998.
Maddox, D. E., Shibata, S., and Goldstein, I. (1982). *Proc. Natl. Acad. Sci. USA 79*:166–170.
Mane, S. M., Mazella, L., Bainton, D. F., Holt, V. K., Cha, Y., Hildreath, J. E. K., and August, T. J. (1989). *Arch. Biochem. Biophys. 268*:360–378.
Marchase, R. B., Kidd, V. J., Rivera, A. A., and Humphreys-Beher, M. G. (1988). *J. Cell. Biochem. 36*:453–465.
Marshall, R. D. (1974). *Biochem. Soc. Symp. 40*:17–26.
Masri, KlA., Appert, H. E., and Fukuda, M. N. (1988).*Biochem. Biophys. Res. Commun. 157*:657–663.
Matzuk, M. M., and Boime, I. (1988a). *J. Biol. Chem. 263*:17106–17111.
Matsuk, M. M., and Boime, I. (1988b). *J. Cell. Biol. 106*:1049–1059.
Matzuk, M. M., Keene, J. L., and Boime, L. (1989). *J. Biol. Chem. 264*:2409–2414.
Mellis, S. J., and Baenziger, J. U. (1983). *J. Biol. Chem. 258*:11546–11556.
Merkle, R. K., and Cummings, R. D. (1987). *J. Biol. Chem. 262*:8179–8189.
Mizouchi, T., Loveless, R. W., Lawson, A. M., Chai, W., Lachmann, P. J., Childs, R. A., Thiel, S., and Feizi, T. (1989). *J. Biol. Chem. 264*:13834–13839.
Moskaug, J.-O., Sandvig, K., and Olsnes, S. (1989). *J. Biol. Chem. 264*:11367–11372.
Muramatsu, T. (1988). *J. Cell. Biochem. 36*:1–14.
Naramatsu, H., Sinha, S., Brew, K., Okayama, H., and Qasba, P. K. (1986). *Proc. Natl. Acad. Sci. USA 83*:4720–4724.

Neufeld, E. F., Lim, T. W., and Shapiro, L. J. (1975). *Annu. Rev. Biochem. 44*:357–376.

Nyame, K., Cummings, R. D., and Damian, R. T. (1988). *Mol. Biochem. Parasitol. 28*:265–274.

Nyame, K., Smith, D. F., Damian, R. T., and Cummings, R. D. (1989). *J. Biol. Chem. 264*: 3235–3243.

Orlean, P., Albright, C., and Robbins, P. W. (1988). *J. Biol. Chem. 263*:17499–17507.

Parekh, R. B., Dwek, R. A., Sutton, B. J., et al. (1985). *Nature 316*:452–457.

Parekh, R. B., Tse, A. G. D., Dwek, R. A., Williams, A. F., and Rademacher, T. W. (1987). *EMBO J. 6*:1233–1244.

Parodi, A. J., Lederkremer, G. Z., and Mendelzon, D. H. (1983). *J. Biol. Chem. 258*:5589–5595.

Parodi, A. J., Martin-Barrientos, J., and Engel, J. C. (1984a). *Biochem. Biophys. Res. Commun. 118*:1–7.

Parodi, A. J., Mendelson, D. H., Lederkremer, G. Z., and Martin-Barrientos, J. (1984b). *J. Biol. Chem. 259*:6351–6357.

Parodi, A. J., Quesada-Allue, L. A., and Cazzulo, J. J. (1981). *Proc. Natl. Acad. Sci. USA 78*:6201–6205.

Parodi, A. J., and Quesada-Allue, (1982). *J. Biol. Chem. 257*:7637–7640.

Passanti, A., and Hart, G. W. (1988). *J. Biol. Chem. 263*:7591–7603.

Paul, G., Lottspeich, F., and Wieland, F. (1986). *J. Biol. Chem. 261*:1020–1024.

Paul, G., and Wieland, F. (1987). *J. Biol. Chem. 262*:9587–9593.

Paulson, J. C., Weinstein, ?., and Schauer, A. (1989). *J. Biol. Chem. 264*:10931–10934.

Paz-Parente, J., Strecker, G., Leroy, Y., Montreuil, J., Fournet, B., van Halbeek, H., Dorland, L., and Vliegenthart, J. F. G. (1983). *FEBS Lett. 152*:145–152.

Paz-Parente, J., Wieruszeski, J.-M., Strecker, G., Montreuil, J., Fournet, B., van Halbeek, H., Dorland, L., and Vliegenthart, J. F. G. (1982). *J. Biol. Chem. 257*:13173–13176.

Pierce, J. G., and Parsons, T. F. (1981). *Annu. Rev. Biochem. 50*:465–495.

Pierce, M., and Arango, J. (1986). *J. Biol. Chem. 261*:10772–10777.

Pollack, L., and Atkinson, P. H. (1983). *J. Cell Biol. 97*:293–299.

Previato, J. O., Mendelzon, D. H., and Parodi, A. J. (1986). *Mol. Biochem. Parasitol. 18*:343–353.

Purchio, A. F., Cooper, J. A., Brunnert, A. M., Lioubin, M. N., Gentry, L. E., Kovacina, K. S., Roth, R. A., and Marquardt, H. (1988). *J. Biol. Chem. 263*:14211–14215.

Rademacher, T. W., Homans, S. W., Parekh, R. B., and Dwek, R. A. (1986). *Biochem. Soc. Symp. 51*:131–148.

Rademacher, T. W., Parekh, R. B., and Dwek, R. A. (1988). *Annu. Rev. Biochem. 57*:785–838.

Rajan, V. P., Larsen, R. D., Ajmera, S., Ernst, L. K., and Lowe, J. B. (1989). *J. Biol. Chem. 264*:11158–11167.

Reading, C. L., and Hutchins, J. T. (1985). *Cancer Metastasis Res. 4*:221–260.

Regoeczi, E., Chindemi, P. A., Debanne, M. T., and Charlwood, P. A. (1982). *Proc. Natl. Acad. Sci. USA 79*:2226–2230.

Reitman, M. L., and Kornfeld, S. (1981a). *J. Biol. Chem. 256*:4275–4281.

Reitman, M. L., and Kornfeld, S. (1981b). *J. Biol. Chem. 256*:11977–11980.

Rine, J., Hansen, W., Hardeman, E., and Davis, R. (1983). *Proc. Natl. Acad. Sci. USA 80*:6750–6754.

Ripka, J., Pierce, M., and Fregien, N. (1989b). Cotransformation of Lec-1 CHO cells with *N*-acetylglucosaminyltransferase I activity and a selectable marker. *J. Cell. Biochem.* (in press).

Ripka, J., Pierce, M., and Fregien, N. (1989a). DNA-mediated transformation of *N*-acetylglucosaminyltransferase I activity into an enzyme-deficient cell line. *Biochem. Biophys. Res. Commun. 159*:554–560.

Rizzolo, L. J., and Kornfeld, R. (1988). *J. Biol. Chem. 263*:9520–9525.

Robbins, A. R., Myerowita, Youle, R. J., Murray, G. J., and Neville, D. M., Jr. (1981). *J. Biol. Chem. 256*:10618–10622.

Roden, L. (1980). In *The Biochemistry of Glycoproteins and Proteoglycans* (W. J. Lennarz, ed.). Plenum Press, New York, pp. 267–372.

Romero, P., and Herscovics, A. (1989). *J. Biol. Chem.* *264*:1946–1950.

Rothbard, J. B., Brackenberry, R., Cunningham, B. A., and Edelman, G. M. (1982). *J. Biol. Chem.* *257*:11064–11069.

Schachter, H. (1986). *Biochem. Cell Biol.* *64*:163–181.

Schachter, H., Narasimhan, S., Gleeson, P., and Vella, G. (1983). *Can. J. Biochem.* *61*:1049–1066.

Schluter, M., Linder, D., Geyer, R., Hunsmann, G., Schneider, J., and Stirm, S. (1984). *FEBS Lett.* *169*:194–198.

Schlesinger, S., Malfer, C., and Schlesinger, M. J. (1984). *J. Biol. Chem.* *259*:7597–7601.

Shao, M-C, and Wold, F. (1989). *J. Biol. Chem.* *264*:6245–6251.

Shao, M-C, and Wold, F. (1988). *J. Biol. Chem.* *263*:5771–5774.

Shaper, N. L., Hollis, G. F., Douglas, J. G., Kirsch, I. R., and Shaper, J. H. (1988). *J. Biol. Chem.* *263*:10420–10428.

Shaper, N. L., Shaper, J. H., Meuth, J. L., Fox, J. L., Chang, H., Kirsch, I. R., and Hollis, G. F. (1986). *Proc. Natl. Acad. Sci. USA* *83*:1573–1577.

Sharkey, D. J., and Kornfeld, R. (1989). *J. Biol. Chem.* *264*:10411–10419.

Sheares, B. T., and Robbins, P. W. (1987). *Proc. Natl. Acad. Sci. USA* *84*:3570–3574.

Smets, L. A., and van Beek, W. P. (1984). *Biochim. Biophys. Acta* *738*:237–249.

Smith, D. F., Larsen, R. D., Mattox, S., Lowe, J. B., and Cummings, R. D. (1990). *J. Biol. Chem.* *265*:6225–6234.

Smith, P. L., and Baenziger, J. U. (1988). *Science* *242*:930–933.

Snider, M. D. (1984). In *Biology of Carbohydrates* (V. Ginsburg and P. W. Robbins, eds.). Wiley-Intersciences, New York, pp. 163–198.

Snider, M. D., and Rogers O. C. (1985). *J. Cell Biol.* *100*:826–834.

Spellman, M. W., Basa, L. J., Leonard, C. K., Chakel, J. A., and O'Connor, J. V. (1989). *J. Biol. Chem.* *264*:14100–14111.

Spiro, R. G., and Bhoyroo, V. D. (1988). *J. Biol. Chem.* *263*:14361–14358.

Stanley, P. (1984). *Annu. Rev. Genet.* *18*:525–552.

Stoll, J., and Krag, S. (1988). *J. Biol. Chem.* *263*:10766–10773.

Stoll, J., Rosenwald, A. G., and Krag, S. S. (1988). *J. Biol. Chem.* *263*:10774–10782.

Strous, G. J., Van Kerkhof, P., Brok, R., Roth, J., and Brada, D. (1987). *J. Biol. Chem.* *262*:3620–3625.

Struck, D. K., and Lennarz, W. J. (1980). In *The Biochemistry of Glycoproteins and Proteoglycans* (W. J. Lennarz, ed.). Plenum Press, New York, pp. 35–84.

Sturm, A., Van Juik, J. A., Vliegenthart, J. F. G., and Chrispeels, M. J. (1987). *J. Biol. Chem.* *262*:13392–13403.

Suh, K., Bergmann, J. E., and Gabel, C. A. (1989). *J. Cell Biol.* *108*:811–819.

Sundblad, G., Holojda, S., Roux, L., Varki, A., and Freeze, H. H. (1988a). *J. Biol. Chem.* *263*:8890–8896.

Sundblad, G., Kajiji, S., Quaranta, V., Freeze, H. H., and Varki, A. (1988b). *J. Biol. Chem.* *263*:8897–8903.

Swiedler, S. J., Hart, G. W., Tarentino, A. L., Plummer, T. H., and Freed, J. H. (1983). *J. Biol. Chem.* *258*:11515–11523.

Symington, B. E., Symington, F. W., and Rohrschneider, L. R. (1989). *J. Biol. Chem.* *264*:13258–13266.

Takasaki, S., Ikehira, H., and Kobata, A. (1980). *Biochem. Biophys. Res. Commun.* *92*:735–742.

Todderud, G., and Carpenter, G. (1988). *J. Biol. Chem.* *263*:17893–17896.

Trimble, R. B., and Atkinson, P. H. (1986). *J. Biol. Chem.* *261*:9815–9824.

Trimble, R. B., Maley, F., and Chu, F. K. (1983). *J. Biol. Chem.* *258*:2562–2561.

Tsuji, T., Irimura, T., and Osawa, T. (1981). *J. Biochem.* *256*:10497–10502.

Van Kuik, J. A., Berg, J., Kolsteeg, C. E. M., Kamerling, J. P., and Vliegenthart, J. F. G. (1987). *FEBS Lett.* *221*:150–154.

van den Eijnden, D. H., Koenderman, A. H. L., and Schiphorst, W. E. C. M. (1988). *J. Biol. Chem.* *263*:12461–12471.

Varki, A., and Kornfeld, S. (1980). *J. Biol. Chem. 255*:8398–8401.

Varki, A., and Kornfeld, S. (1983). *J. Biol. Chem. 258*:2808–2818.

von Figura, K., and Hasilik, A. (1986). *Annu. Rev. Biochem. 55*:167–193.

Wang, X.-C., O'Hanlon, T. P., and Lau, J. T. Y. (1989). *J. Biol. Chem. 264*:1864–1859.

Warren, L., Buck, C. A., and Tuszynski, G. P. (1978). *Biochim. Biophys. Acta 516*:97–127.

Weinstein, J., Lee, E. U., McEntee, K., Lai, P-H, and Paulson, J. C. (1987). *J. Biol. Chem. 262*:17735–17743.

Welply, J. K., Kaplan, H. A., Shenbagamurthi, P., Naider, F., and Lennarz, W. J. (1986). *Arch. Biochem. Biophys. 246*:808–819.

Wieland, F., Heitzer, R., and Schaefer, W. (1983). *Proc. Natl. Acad. Sci. USA 80*:5470–5474.

Wieland, F. T., Gleason, M. L., Serafini, T. A., and Rothman, J. E. (1987). *Cell 50*:289–300.

Yamamoto, F., Clausen, H., White, T., Marken, J., and Hakomori, S. (1990b). *Nature 345*:229–233.

Yamamoto, F., Marken, J., Tsuji, T., White, T., Clausen, H., and Hakomori, S. (1990a). *J. Biol. Chem. 265*:1146–1151.

Yamashita, K., Ohkura, T., Tachibana, Y., Takasaki, S., and Kobata, A. (1984). *J. Biol. Chem. 259*:10834–10840.

Yamashita, K., Tachibana, Y., Ohkura, T., and Kobata, A. (1985). *J. Biol. Chem. 260*:3963–3969.

Yet, M.-G., Chin, C. C. Q., and Wold, F. (1988a). *J. Biol. Chem. 263*:111–117.

Yet, M.-G., Shao, M. C., and Wold, F. (1988b). *FASEB J. 2*:22–31.

Yoshihisa, T., and Anraku, Y. (1989). *Biochem. Biophys. Res. Commun. 163*:908–915.

Yousefi, S., Higgins, E., Daoling, Z., Pollex-Kruger, A., Hindsgaul, O., and Dennis, J. W. (1991). *J. Biol. Chem. 266*:1772–1782.

11

Recent Advances in the Development of Potential Inhibitors of Glycosyltransferases

Shaheer H. Khan and Khushi L. Matta *Roswell Park Cancer Institute, Buffalo, New York*

The glycoconjugates (glycoproteins and glycolipids) found on cell surface membranes, in which the oligosaccharide residues are exposed, play important roles in intercellular recognition and interaction [1–5]; in the control of cell growth and, thereby, in tumor formation [6,7]; in the interactions with biological factors, such as hormones, enzymes, lectins, and antibodies [1,2,8,9]; in the binding of bacteriotoxins and viruses [1,2,10–14] to host cells, to mention but a few phenomena. Over the past decade, a considerable amount of evidence has accumulated indicating that a large number of glycoconjugates become altered during normal and abnormal cellular development [15–18]. Among the abnormal alterations is a characteristic shift toward the synthesis and expression of larger, high relative molecular mass (M_r), asparagine (Asn)-linked oligosaccharides following neoplastic transformation [19–23]. Such changes have been detected in both rodent [19–21] and human tumor cells [23] when they are transformed by chemical mutagens [19], by oncogenic viruses [20], or by transection with DNA obtained from neoplastic cells [21]. Several reports have indicated that the change in size is related to an increase of sialic acid content in the oligosaccharide structures [22–24]. Other commonly observed changes include the occurrence of polyfucosylated, repeated *N*-acetyllactosamine chains [25–27] and increased branching of the trimannosyl core of *N*-linked structures [28,29].

These aberrant glycosylation processes result in consistent changes in cell surface carbohydrate chains and are known to accompany the development of human melanoma, neuroblastoma, and colorectal, gastric, and pancreatic carcinoma [30,31].

The new structures that appear at the cell surface during these processes are referred to as *tumor-associated antigens*. Consequently, the majority of tumor-associated antigens are glycoconjugates in which carbohydrate moieties are altered [8,16,31–37]. The functional significance, if any, of these structures is still unclear, but these aberrant oligosaccharide structures are already attracting a great deal of interest as potential tumor markers. Thus, there is much interest in compounds that can prevent the aberrant glycosylation of glycoconjugates that accompanies cancers.

The biosynthesis of *N*-linked glycoproteins is complex and involves several metabolic pathways, such as lipid intermediates, nucleotide sugar substrates, glycosyltransferases, processing by specific glycosidases, and packaging of end products [38]. Therefore, many potential sites exist for both the regulation and inhibition of protein glycosylation, and the inhibition of any of the pathways involved may affect the final product [38–40]. Several recent reviews [38,41–46] have extensively covered various aspects of the inhibition of biosynthesis of the oligosaccharide chains of *N*-linked glycoproteins. However, inhibitors of glycosyltransferases have not been covered in any detail. This may be partly because the search for these inhibitors has only recently begun. Therefore, it is timely that the literature on potential inhibitors of glycosyltransferases should be reviewed.

The descriptor *glycosyltransferase* defines [47,48] a class enzymes that catalyze the transfer of glycosyl residues from a donor to an acceptor substrate. The donor molecules are usually nucleotide diphosphate sugars that donate, in a stereospecific manner, glycosyl residues during the biosynthesis of glycoproteins, glycolipids, and polysaccharides as follows:

Sugar nucleotide + HO-acceptor → sugar-*O*-acceptor + nucleotide

As we have just discussed, consistent changes in cell surface carbohydrate chains have long been known to accompany malignant behavior, but it is only comparatively recently that these changes have been indicated to correlate reproducibly with alteration in glycosyltransferase expression [49–51]. In addition, several laboratories, including ours, have reported that elevations of various serum glycosyltransferase activities correlate with the presence of malignancy in humans [52–61; see Ref. 32 for review]. Clearly, potential specific inhibitors of these glycosyltransferases represent a new approach to the regulation of uncontrolled growth and may have some potential as cancer chemotherapeutic agents. Several reports have appeared supporting the concept that inhibitors of protein glycosylation possess certain antitumor properties [38,43,62–67]. Therefore, a number of sugar nucleotide analogues have been synthesized and tested as potential inhibitors of glycosyltransferases. Several of these compounds have also been evaluated as antitumor and antimetastatic agents and may have potential as chemotherapeutic agents.

I. DEVELOPMENT OF SIALYLTRANSFERASE INHIBITORS

Sialic acid (*N*-acetylneuraminic acid; NANA, NeuNAc) generally occupies nonreducing terminal positions in the carbohydrate chains of glycoproteins and glycolipids and is considered to play important roles in the various biological phenomena involving cell surfaces. Interest in sialic acid and sialyltransferase has been increased owing to observations that neoplastic transformation and differentiation of cells are accompanied by alterations in the composition and metabolism of cell surface sialoglycoconjugates [68,69]. In addition, several reports have appeared that indicate that there is a correlation between cell surface sialic acid or sialyltransferase activity and the growth [70] or metastatic potential of tumor cells [71–75].

Alterations in sialyltransferase activities have been reported in the serum or plasma of patients with malignant disease, in various transformed cell lines, and in tumor-bearing animal systems [70,76–78]. Such alterations in serum sialyltransferase activity are thought to be the result of both an increased production of the enzyme and its release, perhaps through cell surface shedding from the metastasizing mammary tumor cells [70]. Thus, since sialylation of cell surface glycoconjugates plays a crucial role in malignancy, it becomes highly desirable to develop specific inhibitors for sialyltransferase. It should be emphasized here that sialyltransferase is the enzyme that catalyzes the transfer of sialic acid from the precursor, cytidine monophosphate (CMP)-sialic acid, to the terminal cell surface sugar on asialoglycoprotein. Previous reports from the laboratory of Bernacki et al. have indicated that CMP, which is a by-product following the transfer of NANA from CMP-NANA, inhibits the transfer of NANA from CMP-NANA to appropriate acceptor substrates using rat liver microsomes [79], L1210 cells [43], or serum [80], as a source of sialyltransferase. Therefore, several CMP, CMP-NANA analogues [81] and sialic acid–nucleoside conjugates [82–84] have been synthesized and evaluated for inhibitory activity against a sialyltransferase (EC 2.4.99.1) system.

A. Synthesis

1. *Synthesis of Cytidine Monophosphate and Cytidine Monophosphate-Sialic Acid Analogues*

These potential sialyltransferase inhibitors have been designed and synthesized by Korytnyk and co-workers [81]. For the synthesis of CMP-NANA analogues, namely 5′-(*cis*-4-*N*-acetylcyclohexyl)cytidylic acid hydrochloride (**1**) and 5′-(*trans*-4-*N*-acetylcyclohexyl)cytidylic acid hydrochloride (**2**), two approaches have been applied. Both routes required the preparation of a common intermediate, *N*-benzoyl-2′,3′-di-*O*-benzoylcytidine 5′-β-cyanoethylphosphate, which was, in turn, obtained by condensing *N*-benzoyl-2′3′-di-*O*-benzoylcytidine with β-cyanoethylphosphate in the presence of 2,4,6-triisopropylbenzenesulfonyl chloride (TPS). This intermediate could be coupled either with *cis*- or *trans*-(4-hydroxycyclohexyl)acetamide in the presence of TPS, resulting in the production of either the *cis* **1** or *trans* **2** derivative, after removal of protecting benzoyl groups. This approach provides a respectable yield (39%) for the *trans*-isomer **2**, whereas in the case of the *cis*-isomer **1**, the yield was discouragingly low (about 3%). The low yield in the latter was attributed [81] to unfavorable steric interactions from axial orientation of the hydroxyl group of *cis*-(4-hydroxycyclohexyl)acetamide. An alternate approach for other synthesis of **1** involved the use of benzenesulfonyltetrazole (BST), instead of TPS, as the condensing reagent, which resulted in a significant improvement in yield (29%).

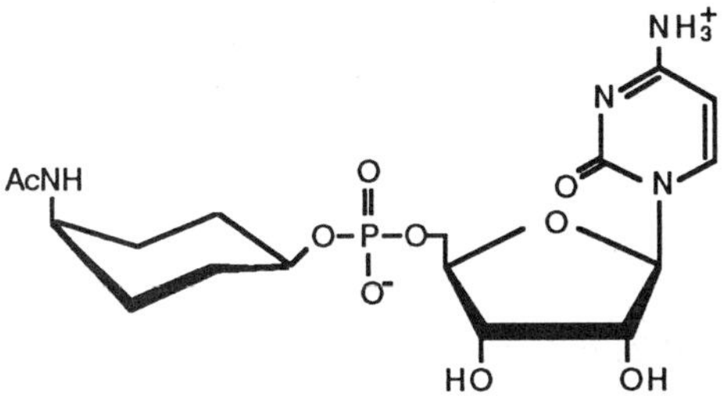

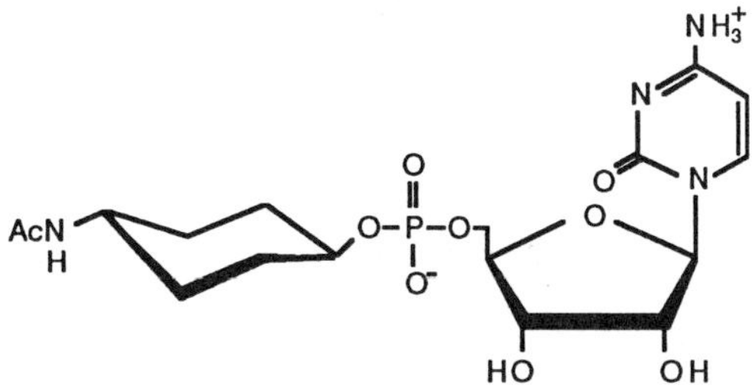

Cytidine 5′-fluorophosphate (5′F-CMP; **3**) was prepared by reacting cytidine 5′-phosphate with 2,4-dinitrofluorobenzene (Sanger's reagent) in the presence of tri-*N*-butylamine in dimethylformamide.

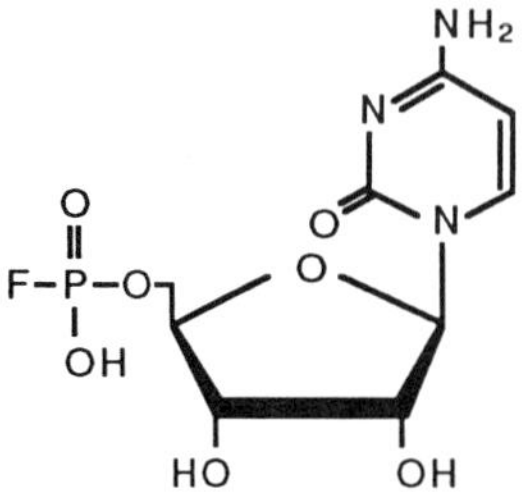

3

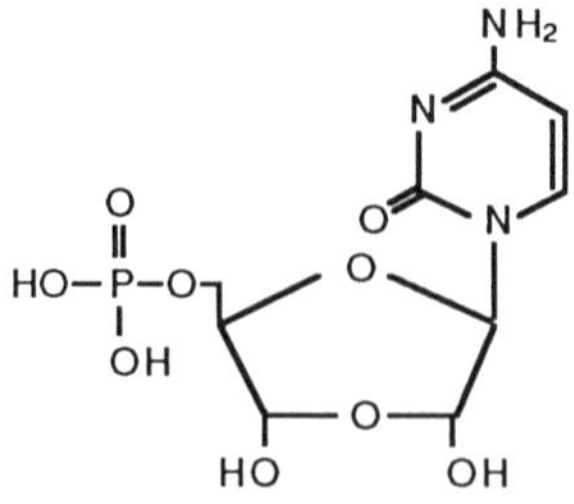

4

Cytidine 5′-monophosphate 2′,3′-dialdehyde (**4**) was obtained by the sodium periodate oxidation of cytidine 5′-monophosphate.

2. *Synthesis of Sialic Acid–Nucleoside Conjugates*

The straightforward synthesis of nucleoside derivatives of *N*-acetylneuraminic acid, namely, 5-fluoro-2′,3′-isopropylidene-5′-*O*-(methyl-5-acetamido-4,7,8,9-tetra-*O*-acetyl-2,3,5-trideoxy-D-glycero-α-D-galacto-2-nonulopyranosate)uridine (**5**) and 2′,3′-di-*O*-acetyl-5′ (methyl-5-acetamido-4,7,8,9-tetra-*O*-acetyl-2,3,5-trideoxy-D-glycero-α-D-galacto-2-nonulopyranosate)iosine (**6**) have been described by Kijima et al. [82]. The condensation of methyl-5-acetamido-4,7,8,9-tetra-*O*-acetyl-2-chloro-2,3,5-trideoxy-D-glycero-α-D-galacto-2-nonulopyranosate with 5-fluoro-2′,3′-isopropylidene uridine and 2′,3′-di-*O*-acetylinosine in acetonitrile and in the presence of mercuric bromide and mercuric cyanide resulted in the preparation of compounds **5** and **6**, respectively.

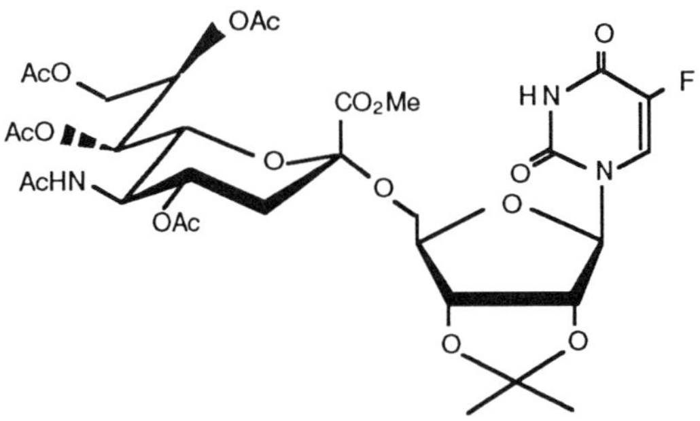

5

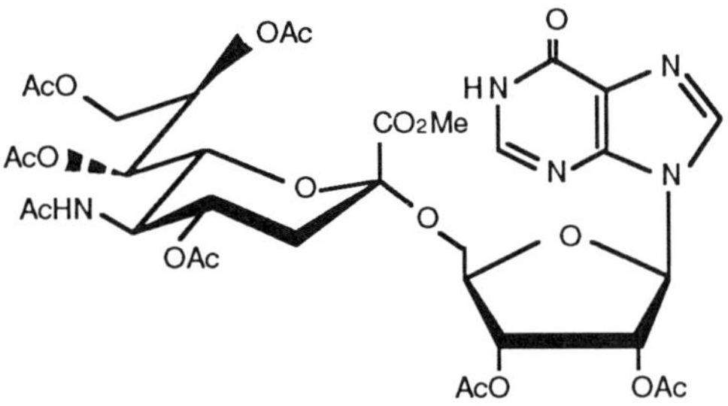

6

B. Inhibition of Sialyltransferase Activity

1. *Inhibition of Sialyltransferase Activity by Cytidine Monophosphate and Cytidine Monophosphate-Sialic Acid Analogues*

The CMP and CMP-NANA analogues were evaluated (Table 1) as inhibitors of the ectosialyltransferase of L1210 cells and of human serum cytidine 5′-monophosphate (CMP)-*N*-acetylneuraminic acid:glycoprotein sialyltransferase (EC 2.4.99.1) [38,43, 80,81]. Cytidine and *N*-acetylneuraminic acid had no inhibitory activity on the ecto-sialyltransferase system, whereas CMP (1.25 mM) showed 72% inhibition. 5′-F-CMP **3** was the most potent inhibitor, being almost as active as its parent compound CMP, and resulted in 74% inhibition at 1.25 mM. Cytidine monophosphate and 5′-F-CMP **3** were found to be competitive inhibitors of human serum sialyltransferase, with K_i values of 50 and 70 μM, respectively. Ribodialdehyde CMP (**4**) also demonstrated a 46 and 55% inhibition of ectosialyltransferase and human sialyltransferase, respectively, at a concentration of 1.25 mM.

Although this compound was less active than CMP or 5′-F-CMP **3**, it showed considerably higher activity than CMP-NANA analogues, such as 5′-(*cis*-4-*N*-acetylcyclohexyl)cytidylic acid hydrochloride (**1**) and 5′-(*trans*-4-*N*-acetylcyclohexyl)cytidylic acid hydrochloride (**2**). Neither of the CMP-NANA analogues (**1** and **2**) exhibited any significant inhibitory activity in either enzyme system. The extent of inhibition was found to be greater with human serum sialyltransferase than with the L1210 ectosialyltransferase system, suggesting different substrate specificities for the two enzyme systems [81].

The reduced activity of the two CMP-NANA analogues, **1** and **2**, versus a higher inhibitory activity of the two CMP analogues, **3** and **4**, was explained by Korytnyk and associates [81]. These authors have suggested that these transferases have a regulatory site that is sensitive to CMP, the end product of the reaction they catalyze. It is conceivable that CMP-NANA analogues may not affect this site because of steric reasons. They do, however, bind to the catalytic site, although to a small extent, since the *N*-acetylcyclohexanol group is not closely related to the sialic acid residue for faithful recognition by the active site of the enzyme.

2. *Inhibition of Sialyltransferase Activity by Sialic Acid–Nucleoside Conjugates*

The two sialic acid–nucleoside conjugates, **5** and **6**, inhibited sialyltransferase (EC 2.4.99.1) on the lymphocyte surface, which resulted in decreased incorporation of sialic acid into endogenous cellular acceptors or into exogenous desialylated glycoconjugates [83]. The type of inhibition was determined to be competitive, and the apparent K_i values

Table 1 Inhibition of Sialyltransferases by CMP and C MP-NANA Analogues

Compound	Concentration (mM)	Ectosialyltransferase activity (% inhibition)	Human serum ST activity (% inhibition)
CMP	0.125	18	85
	1.25	72	100
1	0.125	00	00
	1.25	02	20
2	0.125	00	00
	1.25	15	28
3	0.125	19	76
	1.25	74	98
4	0.125	00	00
	1.25	46	55

Source: Refs. 38 and 81.

for compounds **5** and **6** were 2.3 and 2.5 mM, respectively. Inhibition studies have also suggested that the inhibitory effect of compounds **5** and **6** is specific for the sialylation of glycoproteins containing *O*-linked saccharides [83].

C. Antitumor Activity of Cytidine Monophosphate and Cytidine Monophosphate-Sialic Acid

The CMP and CMP-NANA analogues were also evaluated as antitumor agents in a variety of tumor models [81]. Among the various compounds that have been tested, the only compound that demonstrated significant antitumor activity was CMP-ribodialdehyde (**4**). Compound **4** was administered interperitoneally at doses of 75, 100, or 150 mg/kg per day for 5 days, and this prolonged the life span of L1210 tumor-bearing mice, with 33% of the mice being long-term survivors. The antileukemic effects of compound **4** may be due to the inhibition of sialyltransferase, which is elevated in the sera [76]. In addition, it probably affects DNA biosynthesis and other metabolic pathways, as postulated by Korytnyk et al. [81].

D. Antimetastatic Activity of Sialic Acid–Nucleoside Conjugates

In an interesting study, Kijima-Suda et al. [84] have evaluated the antimetastatic activity of sialic acid–nucleoside analogue **5** against murine colon adenocarcinoma-26 sublines of high- (NL-17) and low- (NL-44) metastatic potential. It was shown that the inhibition of sialyltransferase by this compound resulted in a significant prevention of lung metastasis and prolongation of survival in the mouse. This compound, which is an inhibitor of sialyltransferase, decreased platelet-aggregating activity, which might have resulted in a decrease of metastatic potential of tumor cells. It is also conceivable that this antimetastatic effect may be due to modification of tumor cell surface properties by affecting the sialic acid content of the cells as a result of inhibition of sialyltransferase by compound **5**.

II. DEVELOPMENT OF GALACTOSYLTRANSFERASE INHIBITORS

Recent efforts toward producing galactosyltransferase inhibitors have concentrated on the design of unreactive sugar nuclotide analogues [48,85,86]. The design of these compounds was based on the consideration that, during the biosynthesis of glycoproteins, both the ester linkage between sugar and phosphate and the anhydride linkage of the pyrophosphate are potential targets for enzymic cleavage [85]. It was hypothesized [85] that replacement of the O-P or C-O linkage with a more stable C-P bond should provide nucleoside diphosphate–sugar analogues that are more resistant to enzymatic degradation and, thus, have enhanced potential to selectively inhibit glycoprotein synthesis. Such compounds may prove valuable as substrate analogues for the various glycosyltransferases covalently attached to the outer surface of the plasma membrane.

Realizing the possible impact of such an approach, Vaghefi et al. [85,86] synthesized and tested the biological activity of certain analogues of naturally occurring nucleotide diphosphate sugars, such as uridine diphosphate galactose (UDPGal), adenosine diphosphate glucose (ADPGlc), and guanosine diphosphate mannose (GDPMan). These analogues, which do not readily penetrate cellular membranes, can selectively inhibit glycosyltransferases located at the outer surface of the tumor cell plasma membrane. The inhibitory properties observed may be exploited to learn more about the role of carbohydrate moieties in glycoconjugates.

A. Synthesis

1. Synthesis of Nucleotide Diphosphate Analogues

Vaghefi et al. [85] have described the synthesis of adenosine 5′-phosphoric α-D-glucopyranosylphosphonic anhydride (7), guanosine 5′-phosphoric α-D-mannopyranosylphosphonic anhydride (8), and uridine 5′-phosphoric α-D-galactopyranosylphosphonic anhydride (9).

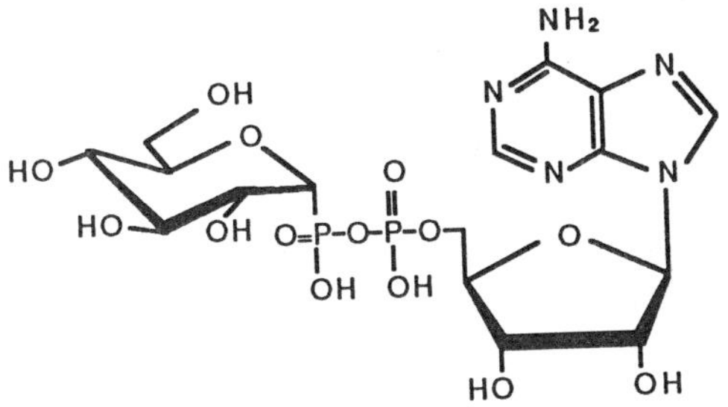

7

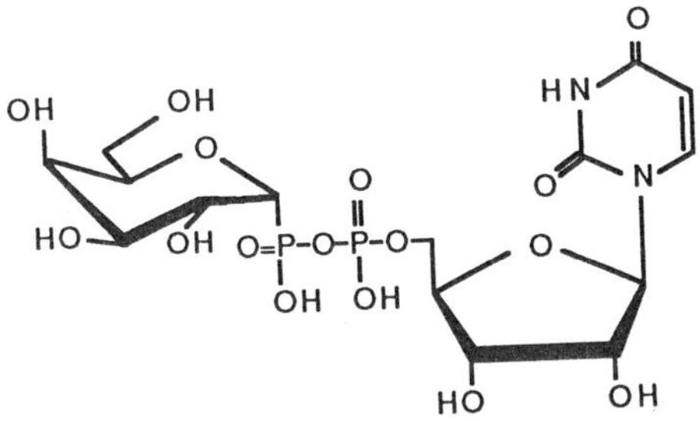

8

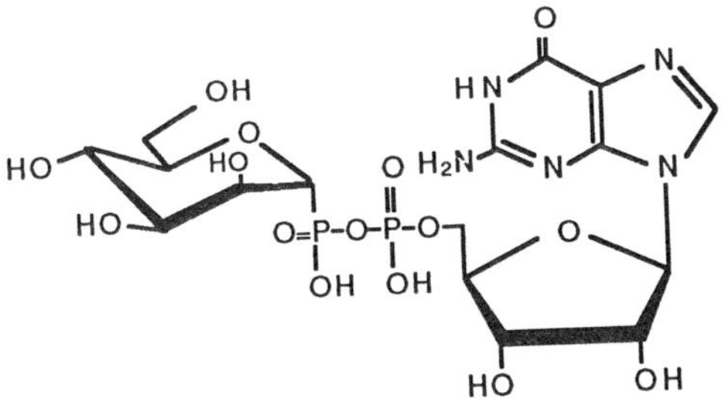

9

The synthetic schemes for these compounds required the preparation of an α-D-glycopryanosylphosphonate of each sugar that could be coupled with the corresponding nucleoside 5′-monosphosphate to give the final compound. The synthesis of compound **7** is summarized in the following: other members of this class of compounds (**8** and **9**) can also be synthesized by following the same procedure. Methyl-2,3,4,6-tetra-*O*-benzyl-α-D-glucopyranoside was acetolyzed to give 2,3,4,6-tetra-*O*-benzyl-1-*O*-acetyl-α-D-glucopyranose. This compound was condensed with tris(trimethylsilyl)phosphite in the presence of trimethylsilyl trifluoromethanesulfonate to afford trimethylsilyl-[2,3,4,6-tetra-*O*-benzyl-α-D-glucopyranosyl]phosphonic acetic anhydride. Catalytic hydrogenation followed by alkaline hydrolysis provided α-D-glucopyranosylphosphonate, which was coupled with adenosine 5′-phosphoric di-*n*-butylphosphinothioic anhydride in dry pyridine and in the presence of silver nitrate to give adenosine 5′-phosphoric α-D-glycopyranosylphosphonic anhydride (**7**).

2. *Synthesis of Nucleoside Methylenediphosphonate Sugars*

The syntheses of adenosine 5′-[(α-D-glucopyranosylhydroxyphosphinyl-methyl]phosphorate (**10**), uridine 5′-[(α-D-galactopyranosylhydroxyphosphinyl)methylphosphonate (**11**), and guanosine 5′-[(α-D-mannopyranosylhydroxyphosphinyl)methyl]phosphonate (**12**) have been described by Vaghefi and associates [86].

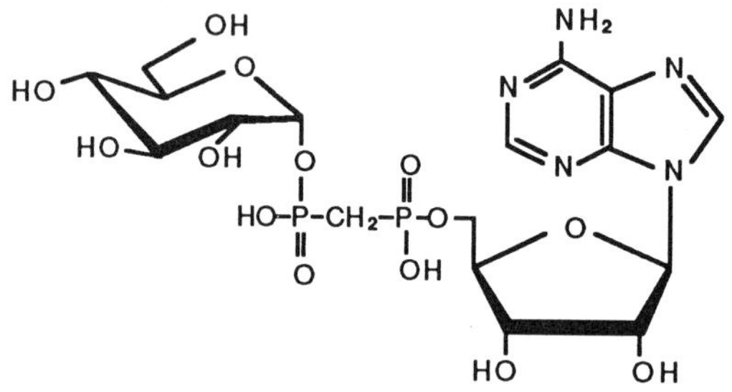

10

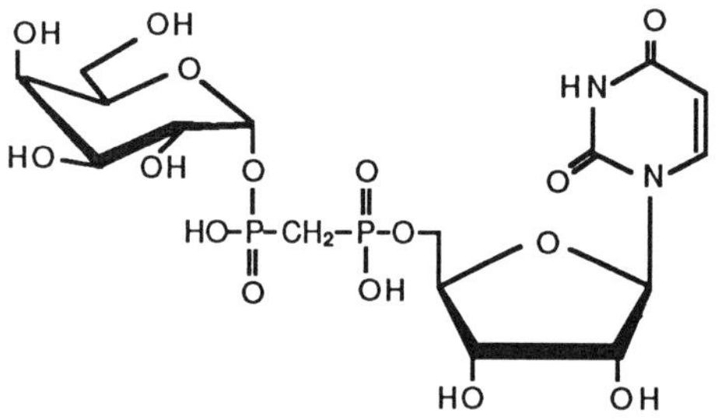

11

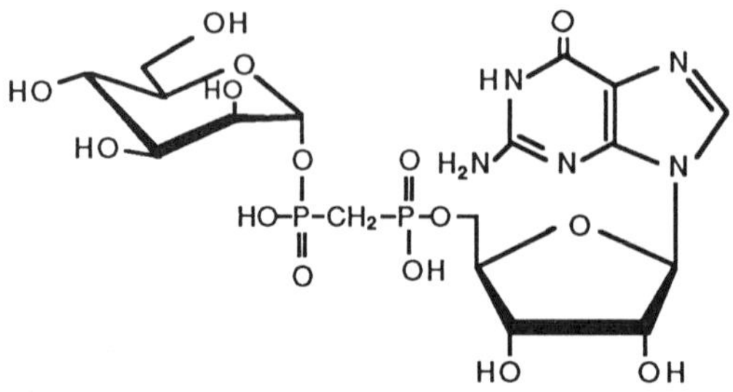

12

This class of compounds is similar to those described in Section II.A.1, except that the oxygen bridge between the two phosphorus atoms has been replaced by a methylene moiety. Here we will briefly discuss the synthesis of compound **10**, other analogues (**11** and **12**) can also be prepared through a similar sequence of chemical reactions. [(Diphenoxyphosphinyl)methyl]phosphonic acid, prepared by two different routes, was coupled with 1,2,3,4,6-penta-*O*-acetyl-β-D-glycopyranose at 170°C. Under these conditions, the thermodynamically controlled α-anomer, 2,3,4,6-tetra-*O*-acetyl-α-D-glycopyranosyl 1 [[(diphenoxyphosphinyl)methyl]phosphonate] was obtained. Catalytic hydrogenation afforded 2,3,4,6-tetra-*O*-acetyl-α-D-glucopyranosyl 1-7-(methylenediphosphonate). The triethylammonium salt of this compound was condensed with 2′,3′-di-*O*-acetyladenosine in dry pyridine and in the presence of 1-(mesitylene-2-sulfonyl)-3-nitro-1,2,4-triazole to give the acetylated intermediate which was *O*-deacetylated to furnish adenosine 5′-[(α-D-glucopyranosylhydroxyphosphinyl)methyl]phosphonate (**10**).

B. Inhibition of Galactosyltransferase Activity

1. *Inhibition of Galactosyltransferase Activity by Nucleoside Diphosphate Sugar Analogues*

Analogues **7–9** were tested as inhibitors of specific glycoprotein galactosyltransferase (EC 2.4.1.38) activity secreted by L1210 leukemia cells grown in mice peritoneal ascites [85]. The UDPGal analogue, uridine 5′-phosphoric α-D-galactopyranosylphosphonic anhydride (**8**) showed the greatest inhibitory activity against galactosyltransferase at concentrations of 20 and 50 μg/mL. At 20 μg/mL (35 μM), enzyme activity was decreased by 50%, whereas a concentration of 50 μg/mL (89 μM) resulted in 58% inhibition. The type of enzyme inhibition caused by **8** was competitive, and the K_i for this inhibitor was 165 μM. Since galactosyltransferases are present at the surface of metastatic cells, specific inhibitors of these enzymes are highly desirable because they might prevent metastasis.

2. *Inhibition of Galactosyltransferase Activity by Nucleoside Methylenediphosphonate Sugars*

Compounds **10–12** were evaluated as potential inhibitors of galactosyltransferase (EC 2.4.1.38) [86]. The adenosine methylenediphosphonate glucose analogue **10** did not demonstrate any significant inhibitory effect at concentrations up to 50 μg/mL, whereas the uridine galactose derivative **11** reduced galactosyltransferase activity by over 40% at concentrations of 1 μg/mL. The type of inhibition caused by **11** was competitive, and the K_i value was near 97 μM. It has been suggested that, in natural products of this type, the bond between sugar and phosphate is cleaved by the enzyme, but Vaghefi et al. [86] have proposed that inhibition of the enzyme by **11** might be due to another mechanism.

C. Antitumor Activity

1. *Antitumor Activity of Nucleoside Diphosphate Sugar Analogues*

Analogues **7**, **8**, and **9** were all evaluated for antitumor activity in vitro against L1210 leukemia, human B-lymphoblastic leukemia (WI-L2), and human T-lymphoblastic leukemia (CEM) cells [85]. Compounds **7** and **9** showed slight cytotoxicity against WI-L2 and CEM, but were inactive against L1210. Compound **8** showed no cytotoxicity against any of the foregoing cell lines. The low cytotoxicity of these analogues may be due to their ionic strength, which prevents the molecule from penetrating the cell membrane.

2. Antitumor Activity of Nucleoside Methylenediphosphonate Sugars

The in vitro antitumor activities of **10**, **11**, and **12** were measured against L1210 leukemia, human B-lymphoblastic leukemia (WW-L2), and human T-lymphoblastic leukemia (CEM) [86]. None of these compounds showed any antitumor activity against any of the cell lines.

III. DEVELOPMENT OF FUCOSYLTRANSFERASE INHIBITORS

Hindsgaul et al. [87] have convincingly argued that, although unreactive sugar nucleotide analogues can be invaluable tools in understanding the mechanism of action of glycosyltransferases, this class of compounds may not act as specific inhibitors of glycosyltransferases because any given sugar nucleotide can act as a biosynthetic glycosyl donor for several, often competing glycosyltransferases. An interesting and reasonable hypothesis has been put forward by these authors [87] describing a possible role for a new class of mechanism-based *bisubstrate analogue* inhibitors for α-(1→2)-L-fucosyltransferase. This enzyme catalyzes the reaction:

GDP-Fuc + β-Gal-R → α-Fuc-(1→2)-β-Gal-R + GDP

where R represents the rest of the blood group-active molecule [48]. The design of the bisubstrate analogue inhibitor is based on the concept that the specificity of a glycosyltransferase is determined by its ability to recognize both the oligosaccharide acceptor substrate as well as the sugar nucleotide donor. Therefore, for the inhibitor to be specific, it should contain structural features common to both the donor and acceptor molecules involved in the glycosylation process. The strategy for the production of such type of inhibitors involves the covalent attachment of an acceptor specific for a glycosyltransferase to the nucleotide portion of the donor sugar nucleotide through the acceptor hydroxyl group to which the enzyme normally transfers a glycosyl residue. The resultant bisubstrate analogue satisfied the criteria put forth for specific glycosyltransferase inhibitors. The enzyme α-(1→2)-fucosyltransferase was selected to test this inhibitor, because this enzyme utilizes the simple molecule, phenyl α-D-galactopyranoside as an acceptor substrate, which simplified the chemistry involved.

A. Synthesis of Bisubstrate Analogue

The elegant synthesis of phenyl-2-*O*-[2-phosphonoethyl]-β-D-galactopyranoside guanosine-5′-phosphate anhydride (**13**) as its disodium salt was recently accomplished by Hindsgaul and associates [87]. To the best of our knowledge, this is the first advancement of a mechanism-based bisubstrate analogue inhibitor for a glycosyltransferase.

Compound **13** was obtained by coupling GMP-morpholidate with phenyl 2-*O*-(2-phosphonoethyl)-β-D-galactopyranoside. The latter compound was obtained from 2-*O*-acetyl-3,4,6-tri-*O*-benzyl-α-D-galactopyranosyl bromide through a succession of several high-yielding chemical steps.

B. Inhibition of Fucosyltransferase Activity by a Bisubstrate Analogue Inhibitor

Compound **13** is the first example of a mechanism-based bisubstrate analogue inhibitor for a glycosyltransferase that binds to both the acceptor and donor recognition sites of

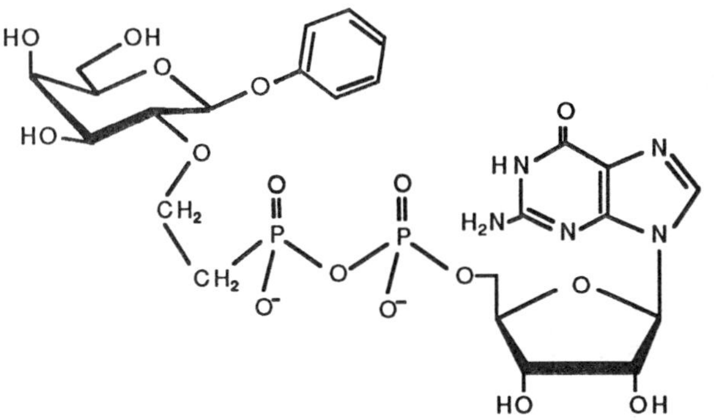

13

the enzyme. This compound was tested as an inhibitor of β-galactoside-α-(1→2)-L-fucosyltransferase; both the membrane-bound and soluble forms of the enzyme were assayed [87]. The pattern of inhibition was competitive for GDP-fucose, phenyl β-D-galactopyranoside, and the alternative acceptor β-D-Galp-(1→3)-β-D-GlcpNAc-*O*-$(CH_2)_7$-COOMe. The K_i values were in the range of 2.3–16 μM.

IV. DEVELOPMENT OF *N*-ACETYLGLUCOSAMINYLTRANSFERASE V INHIBITORS

N-Acetylglucosaminyltransferase V (GnT-V, EC 2.4.1.155) is thought to be responsible for the transfer of a 2-acetamido-2-deoxy-β-D-glucopyranosyl group to O-6 of the (1→6)-linked α-D-mannosyl residue that forms part of the trimannopyranosyl core of asparagine-linked *N*-glycans [88]. This enzyme has attracted a great deal of interest as a potential tumor marker because of its increased expression in cells transformed by tumor viruses [89–92] or oncogenes [93]. Furthermore, recent work of Dennis and co-workers [94,95] has suggested that an increase in intracellular activity of GnT-V is directly related to the metastatic potential of certain tumor cell lines. In addition, these authors [96] have also reported that an increased expression of GnT-V activity and resulting cell surface structures are associated with a number of human breast carcinomas, when compared with nonmalignant tissues. For the past few years, we have been actively involved with the synthesis of acceptor substrates for GnT-V [97–99]. Now we have turned our attention toward the design, synthesis, and biological evaluation of potential inhibitors for this particular enzyme. Our strategy in considering the design of inhibitors for the GnT-V is focused on defining the specific acceptor substrate for the enzyme and then chemically synthesizing this acceptor, but with the active hydroxyl group masked where the enzyme would have transferred the glycosyl residue. On the basis of this rationale, we have envisioned β-D-GlcpNAc-(1→2)-6-*O*-Me-α-D-Manp-(1→6)-Glcp-$OC_6H_4NO_2$ (**14**) as a potential inhibitor for GnT-V. Very recently, Palcic et al. [100] have reported the synthesis of a similar trisaccharide, β-D-GlcpNAc-(1→2)-α-D-6-deoxy-Manp-(1→6)-β-D-Glcp-*O*-$(CH_2)_7CH_3$ (**15**) and have shown it to be a competitive inhibitor for GnT-V.

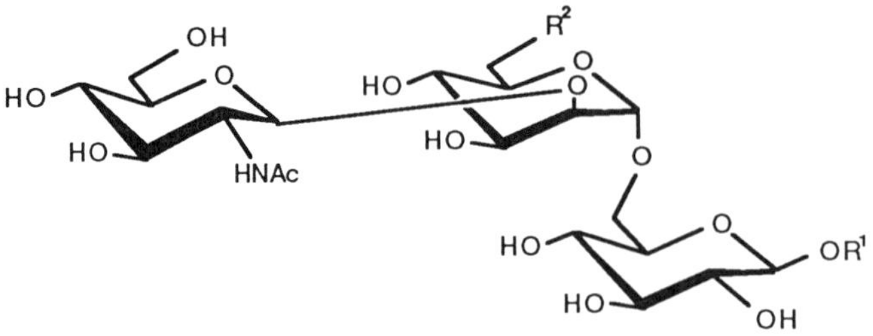

14 $R^1 = C_6H_4NO_2$ -(4), $R^2 = OCH_3$
15 $R^1 = (CH_2)_7CH_3$, $R^2 = H$

A. Synthesis

1. Proposed Synthesis of the Methylated Analogue

A general and convenient method for the synthesis of oligosaccharides containing β-D-GlcpNAc-(1→2)-α-D-Manp as a terminal unit has been developed and applied in our laboratory for the synthesis of potential substrates for GnT-V [97–99]. The proposed synthesis of compound **14** can also be effectively accomplished by using similar methodology.

2. Synthesis of Deoxygenated Analogue

n-Octyl-2-acetamido-2-deoxy-β-D-glucopyranosyl-(1→2)-*O*-6-deoxy-α-D-mannopyrano syl-(1→6)-β-D-glucopyranoside (**15**) is the latest addition [100] to the family of potential inhibitors of glycosyltransferases. The synthesis of this compound was achieved by condensation of 3,4,6-tri-*O*-acetyl-2-deoxy-2-phthalimido-β-D-glucopyranosyl bromide with *n*-octyl-3,4-di-*O*-benzyl-6-deoxy-α-D-mannopyranosyl-(1→6)-2,3,4-tri-*O*-benzyl-β-D-glu copyranoside in dichloromethane and in the presence of silver trifluoromethansulfonate and *sym*-collidine, followed by removal of protecting groups. The latter disaccharide acceptor was prepared by coupling of *n*-octyl-2,3,4-tri-*O*-benzyl-β-D-glucopyranoside with 2-*O*-acetyl-3,4-di-*O*-benzyl-6-deoxy-α-D-mannopyranosyl chloride in the presence of silver trifluoromethanesulfonate and *N,N,N′,N′*-tetramethylurea, followed by removal of 2′-*O*-acetyl group.

B. Inhibition of *N*-Acetylglucosaminyltransferase V Activity by a Deoxygenated Analogue

The deoxygenated analogue **15** was evaluated [100] as an inhibitor of GnT-V from baby hamster kidney (BHK) cells, BHK cells transformed with Rous sarcoma virus (RS-BHK), and a lectin-resistant BHK cell line LP3.3. Compound **15** was an inhibitor for GnT-V activity, and the mode of inhibition was determined to be competitive relative to the acceptor, β-D-GlcpNAc-(1→2)-α-D-Manp-(1→6)-β-D-Manp-*O*-$(CH_2)_7COOCH_3$ for the enzymes from these three sources. The K_i values for the inhibitor were in the range of 60–77 μM.

V. CONCLUSIONS

A common feature observed in cells following oncogenic transformation is the alteration of cell surface glycoconjugates. There is a growing consensus that the metabolic basis for these alterations involves primarily the activation of specific glycosyltransferases and entails their biosynthetic pathways. Thus, it may be therapeutically advantageous to inhibit these glycosyltransferases. Concerted efforts in this direction from various laboratories have met with some degree of success.

The inhibition of various glycosyltransferases has been studied using nucleoside and nucleotide–sugar analogues. The analogues 5′-F-CMP, **3**, and CMP-ribodialdehyde, **4**, have been found to inhibit both human serum sialyltransferase and L1210 ectosialyltransferase activity. Compound **4** also showed significant antileukemic effects. Two sialic acid–nucleoside conjugates, **5** and **6**, competitively inhibited sialyltransferase. Inhibition of this enzyme by compound **5** also resulted in a significant prevention of lung metastasis and the prolongation of life span in the mouse.

Inhibition of galactosyltransferase has been studied using unreactive sugar–nucleotide analogues. Among the various analogues tested, only UDP-Gal derivatives **8** and **11** demonstrated inhibitory activity.

A new approach in the inhibition of glycosyltransferases involving the use of a bisubstrate analogue inhibitor **13** has been applied for the first time. This mechanism-based inhibitor was found to be a competitive inhibitor of α-(1→2)-L-fucosyltransferase. This strategy opens a new area for development of inhibitors of various glycosyltransferases and should aid in the design and evaluation of more active and specific inhibitors. These inhibitors may find a potential application in cancer chemotherapy, in view of recent studies that have provided evidence that fucosyltransferases are increased in the sera or saliva of patients with a wide variety of cancers.

A deoxygenated substrate analogue **15**, has also been synthesized and tested as an inhibitor of GnT-V. Analogue **15** inhibited this enzyme in a competitive fashion. Studies toward the development of specific inhibitors for GnT-V are also in progress in our laboratory.

It is only relatively recently that glycobiologists have come to appreciate the crucial role played by glycosyltransferases in causing structural alterations of the carbohydrate moieties of cell surface glycoconjugates found in malignancy. The search for potential inhibitors of these enzymes is also at an early stage, but continued research toward this goal should lead to the development of more potent inhibitors that could be developed into effective chemotherapeutic agents.

ACKNOWLEDGMENTS

The authors would like to thank Drs. H. Allen, A. Sarkar, and P. R. Ramakrishnan for helpful suggestions in preparing this manuscript. We acknowledge Mr. C. Piskorz for excellent editorial help, Mrs. M. Vallina for secretarial help, and Ms. P. J. Pajak for art work. Research from our laboratory cited in this review was supported in part by grants from National Cancer Institute (CA 35329) and American Cancer Society (CH 419) awarded to K. L. Matta.

REFERENCES

1. Montreuil, J. (1980). Primary structure of glycoprotein glycans: Basis for the molecular biology of glycoproteins. *Adv. Carbohydr. Chem. Biochem.* *37*:157–223.
2. Schachter, H. (1984). Glycoproteins: Their structure, biosynthesis and possible clinical implications. *Clin. Biochem.* *17*:3–14.

3. Sharon, N., and Lis, H. (1989). Lectins as cell recognition molecules. *Science 246*:227–234.
4. Sharon, N. (1975). In *Complex Carbohydrates: Their Chemistry, Biosynthesis, and Function.* Addison-Wesley Publishing, Reading, Mass., pp. 177–188.
5. Glick, M. C., and Flowers, H. (1978). Surface membranes. In *The Glycoconjugates*, Vol. 2 (M. I. Horowitz and W. Pigman, eds.). Academic Press, New York, pp. 337–384.
6. Dennis, J. W., and Laferté, L. (1985). Recognition of asparagine-linked oligosaccharides on tumor cells by natural killer cells. *Cancer Res. 45*:6034–6040.
7. Goldhirsch, A., Berger, E., Müller, O., Maibach, R., Misteli, S., Buser, K., and Brunner, K. (1988). Ovarian cancer and tumor markers: Sialic acid, galactosyltransferase and CA-125. *Oncology 45*:281–286.
8. Hakomori, S.-I. (1984). Tumor-associated antigens. *Annu. Rev. Immunol. 2*:103–126.
9. Sharon, N. (1986). Lectins as molecules and as tools. *Annu. Rev. Biochem. 55*:35–67.
10. Horowitz, M. (1978). Immunological aspects. In *The Glycoconjugates*, Vol. 2 (M. I. Horowitz and W. Pigman, eds.). Academic Press, New York, pp. 387–436.
11. Boch, K., Breimer, M. E., Brignole, A., Hansson, G. C., Karlsson, K.-A., Larson, G., Leffler, H., Samuelsson, B. E., Strömberg, N., Edén, C. S., and Thurin, J. (1985). Specificity of binding of a strain of uropathogenic *Escherichia coli* to Galα1→4Gal-containing glycosphingolipids. *J. Biol. Chem. 260*:8545–8551.
12. Krivan, H. C., Roberts, D. D., and Gingburg, V. (1988). Many pulmonary pathogenic bacteria bind specifically to the carbohydrate sequence GalNAcβ1→4Gal found in some glycolipids. *Proc. Natl. Acad. Sci. USA 85*:5157–6161.
13. Pritchett, T. J., Brossmer, R., Rose, U., and Paulson, J. C. (1987). Recognition of monovalent sialosides by influenza virus H3 hemagglutinin. *Virology 160*:502–506.
14. Gruters, R. A., Neefjes, J. J., Tersmette, M., de Goede, R. E. Y., Tulp, A., Huisman, H. G., Miedema, F., and Ploegh, H. L. (1987). Interference with HIV-induced syncytium formation and viral infectivity by inhibitors of trimming glucosidase. *Nature 330*:74–77.
15. Smets, L. A., and Van Beek, W. P. (1984). Carbohydrates of the tumor cell surface. *Biochem. Biophys. Acta 738*:237–249.
16. Hakomori, S.-I. (1981). Glycosphingolipids in cellular interaction, differentiation, and oncogenesis. *Annu. Rev. Biochem. 50*:733–764.
17. Hakomori, S.-I. (1984). Glycosphingolipids as differentiation-dependent, tumor-associated markers and as regulators of cell proliferation. *Trends Biochem. Sci. 9*:453–458.
18. Feizi, T., and Childs, R. A. (1985). Carbohydrate structures of glycoproteins and glycolipids as differentiation antigens, tumour-associated antigens, and components of receptor systems. *Trends Biochem. Sci. 10*:24–29.
19. Warren, L., Buck, C. A., and Tusgynski, G. P. (1978). Glycopeptide changes and malignant transformation. A possible role for carbohydrate in malignant behavior. *Biochem. Biophys. Acta 516*:97–127.
20. Santer, U. V., and Glick, M. C. (1979). Partial structure of a membrane glycoprotein from virus transformed hamster cells. *Biochemistry 18*:2533–2540.
21. Collard, J. G., Van Beek, W. P., Janssen, J. W. G., and Schijven, J. F. (1985). Transfection by human oncogenes. Concomitant induction of tumorigenicity and tumor-associated membrane alterations. *Int. J. Cancer 35*:207–214.
22. Warren, L., Fuhrer, J. P., and Buck, C. A. (1972). Surface glycoproteins of normal and transformed cells: A difference determined by sialic acid and a growth-dependent sialyltransferase. *Proc. Natl. Acad. Sci. USA 69*:1838–1842.
23. Van Beek, W. P., Smets, L. A., and Emmelot, P. (1975). Changed surface glycoprotein as a marker of malignancy in human leukaemic cells. *Nature 253*:457–460.
24. Hunt, L. A., and Wright, S. E. (1985). Both acidic-type and neutral-type asparaginyl-oligosaccharides of host–cell glycoproteins are altered in Rous-sarcoma-virus-transformed chick-embryo fibroblasts. *Biochem. J. 229*:441–451.

25. Hakomori, S.-I., Nudelman, E., Levery, S. B., and Kannagi, R. (1984). Novel fucolipids accumulating in human adenocarcinoma. I. Glycolipids with di- or trifucosylated type 2 chain. *J. Biol. Chem. 259*:4672–4680.
26. Fukushi, Y., Hakomori, S.-I., Nudelman, E., and Cochran, N. (1984). Novel fucolipids accumulating in human adenocarcinoma. II. Selective isolation of hybridoma antibodies that differentially recognize mono-, di-, and trifucosylated type 2 chain. *J. Biol. Chem. 259*: 4681–4685.
27. Holmes, E. H., Ostrander, G. K., and Hakomori, S.-I. (1985). Enzymatic basis for the accumulation of glycolipids with X and dimeric X determinants in human lung cancer cells (NCI-H69). *J. Biol. Chem. 260*:7619–7627.
28. Takasai, S., Ikehira, H., and Kobata, A. (1980). Increase of asparagine-linked oligosaccharides with branched outer chains caused by cell transformation. *Biochem. Biophys. Res. Commun. 92*:735–742.
29. Debray, H. D., Qin, Z., Delannoy, P., Montreuil, J., Dus, D., Radzikowski, C., Christensen, B., and Kieler, J. (1986). Altered glycosylation of membrane glycoproteins in human uroepithelial cell lines. *Int. J. Cancer. 37*:607–611.
30. Cheresh, D. A., Reisfeld, R. A., and Varki, A. P. (1984). *O*-Acetylation of disialoganglioside GD_3 by human melanoma cells creates a unique antigenic determinant. *Science 225*: 844–846.
31. Hakomori, S.-I. (1985). Aberrant glycosylation in cancer cell membranes as focused on glycolipids: Overview and perspectives. *Cancer Res. 45*:2405–2414.
32. Alhadeff, J. A. (1989). Malignant cell glycoproteins and glycolipids. *CRC Crit. Rev. Oncol. Hematol. 9*:37–107.
33. Reading, C. L., and Hutchins, J. T. (1985). Carbohydrate structure in tumor immunity. *Cancer Metastasis Rev. 4*:221–260.
34. Feizi, T. (1984). Monoclonal antibodies reveal saccharide structures of glycoproteins and glycolipids as differentiation and tumor-associated antigens. *Biochem. Soc. Trans. 12*: 545–549.
35. Hakomori, S.-I. (1986). Tumor-associated antigens, their metabolism and organisation. *Chem. Phys. Lipids 42*:209–233.
36. Hakomori, S.-I. (1989). Aberrant glycosylation in tumors and tumor-associated carbohydrate antigens. *Adv. Cancer Res. 53*:257–331.
37. Fukuda, M. (1985). Cell surface glycoconjugates as oncodifferentiation markers in hematopoietic cells. *Biochem. Biophys. Acta 780*:119–150.
38. Bernacki, R. J., and Kortnyk, W. (1982). Development of membrane sugar and nucleotide sugar analogs as potential inhibitors or modifiers of cellular glycoconjugates. In *The Glycoconjugates*, Vol. 4 (M. I. Horowitz, ed.). Academic Press, New York, pp. 245–263.
39. Bernacki, R. J., Sharma, M., Porter, N. K., Rustum, Y., Paul, B., and Korytnyk, W. (1977). Biochemical characteristics, metabolism, and antitumor activity of several acetylated hexosamines. *J. Supramol. Struct. 7*:235–250.
40. Schachter, H. (1978). Glycoprotein biosynthesis. In *The Glycoconjugates*, Vol. 2 (M. I. Horowitz and W. Pigman, eds.). Academic Press, New York, pp. 87–181.
41. Schwarz, R. T., and Datema, R. (1982). Inhibition of lipid-dependent glycosylation. In *The Glycoconjugates*, Vol. 3 (M. I. Horowitz, ed.). Academic Press, New York, pp. 47–79.
42. Schwarz, R. T., and Datema, R. (1982). The lipid pathway of protein glycosylation and its inhibitors: The biological significance of protein-bound carbohydrates. *Adv. Carbohydr. Chem. Biochem. 40*:287–379.
43. Bernacki, R. J., Porter, C., Korytnyk, W., and Mihich, E. (1978). Plasma membrane as a site for chemotherapeutic intervention. *Adv. Enzyme Regul. 16*:217–237.
44. Kornfeld, R., and Kornfeld, S. (1985). Assembly of asparagine-linked oligosaccharides. *Annu. Rev. Biochem. 54*:631–664.

45. Elbein, A. D. (1984). Inhibitors of the biosynthesis and processing of *N*-linked oligosaccharides. *Crit. Rev. Biochem. 16*:21–47.
46. Elbein, A. D. (1987). Inhibitors of the biosynthesis and processing of *N*-linked oligosaccharide chains. *Annu. Rev. Biochem. 56*:497–534.
47. Watkins, W. M. (1986). Glycosyltransferases. Early history, development and future prospects. *Carbohydr. Res. 149*:1–12.
48. Beyer, T. A., Sadler, J. E., Rearick, J. I., Paulson, J. C., and Hill, R. L. (1981). Glycosyltransferases and their use in assessing oligosaccharide structure and structure–function relationships. *Adv. Enzymol. 52*:23–175.
49. Holmes, E. H., and Hakomori, S.-I. (1983). Enzymatic basis for changes in fucoganglioside during chemical carcinogenesis. *J. Biol. Chem. 258*:3706–3713.
50. Weiser, M. M., Klohs, W. D., Podolsky, D. K., and Wilson, J. R. (1982). Glycosyltransferases in cancer. In *The Glycoconjugates*, Vol. 4 (M. I. Horowitz, ed.). Academic Press, New York, pp. 301–333.
51. Narasimhan, S., Schachter, H., and Rajalakshmi, S. (1988). Expression of *N*-acetylglucosaminyltransferase III in hepatic nodules during rat liver carcinogenesis promoted by orotic acid. *J. Biol. Chem. 263*:1273–1281.
52. Madiyalakan, R., Yazawa, S., Abbas, S. A., Barlow, J. J., and Matta, K. L. (1986). Use of *N*-acetyl-2′-*O*-methyllactosamine as a specific acceptor for the determination of α-L-(1→3)-fucosyltransferase in human serum. *Anal. Biochem. 152*:22–28.
53. Madiyalakan, R., Yazawa, S., Abbas, S. A., Barlow, J. J., and Matta, K. L. (1986). Elevated serum α(1→3)-L-fucosyltransferase activity with synthetic low molecular weight acceptor in human ovarian cancer. *Cancer Lett. 30*:201–205.
54. Yazawa, S., Madiyalakan, R., Piver, M. S., and Matta, K. L. (1986). Elevated activities of blood group Le gene dependent α(1→3)-L-fucosyltransferase in human saliva of Lewis negative patients with epithelial ovarian cancer. *Cancer lett. 32*:165–169.
55. Yazawa, S., Madiyalakan, R., Izawa, H., Asao, T., Furukawa, K., and Matta, K. L. (1988). Cancer-associated elevation of α(1→3)-L-fucosyltransferase activity in human serum. *Cancer 62*:516–520.
56. Madiyalakan, R., Yazawa, S., and Matta, K. L. (1988). Characterization of plasma α(1→3)-L-fucosyltransferase from an ovarian cancer patient with the aid of synthetic substrates. *Indian J. Biochem. Biophys. 25*:32–35.
57. Yazawa, S., Asao, T., Nagamachi, Y., Abbas, S. A., and Matta, K. L. (1989). Tumor-related elevation of serum (α1→3)-L-fucosyltransferase activity in gastric cancer. *J. Cancer Res. Clin. Oncol. 115*:451–455.
58. Yazawa, S., Asao, T., Nagamachi, Y., and Matta, K. L. (1989). Elevated activities of serum α(1→3)-L-fucosyltransferase in human cancer. *J. Tumor Marker Oncol. 4*:355–362.
59. Madiyalakan, M., Piskorz, C. F., Piver, M. S., and Matta, K. L. (1987). Serum β-(1→4)-galactosyltransferase activity with synthetic low molecular weight acceptor in human ovarian cancer. *Eur. J. Cancer Clin. Oncol. 23*:901–906.
60. Yazawa, S., Asao, T., and Izawa, H., Miyamoto, Y., and Matta, K. L. (1988). The presence of CA19-9 in serum and saliva from Lewis blood-group negative cancer patients. *Jpn. J. Cancer Res. 79*:538–543.
61. Madiyalakan, R., Di Cioccio, R. A., and Matta, K. L. (1986). Serum α-L-(1→6)-fucosyltransferase activity in patients with liver cancer. *IRCS Med. Sci. 14*:373.
62. Knowles, R. W., and Person, S. (1976). Effects of 2-deoxyglucose, glucosamine, and mannose on cell fusion and the glycoproteins of herpes simplex virus. *J. Virol. 18*:644–651.
63. Büchsel, R., Hassels-Vischer, B., Tauber, R., and Reutter, W. (1980). 2-Deoxy-D-galactose impairs the fucosylation of glycoproteins of rat liver and Morris hepatoma. *Eur. J. Biochem. 111*:445–453.
64. Schmidt, M. F. G., Schwartz, R. T., and Ludwig, H. (1976). Fluorosugars inhibit biological properties of different enveloped viruses. *J. Virol. 18*:819–823.

65. Bessell, E. M., Courtenay, V. D., Foster, A. B., Jones, M., and Westwood, J. H. (1973). Some in vivo and in vitro antitumor effects of the deoxyfluoro-D-glucopyranoses. *Eur. J. Cancer* *9*:463–470.
66. Schmidt, M. F. G., Biely, P., Karátky, Z., and Schwartz, R. T. (1978). Metabolism of 2-deoxy-2-fluoro-D-[^{3}H]glucose and 2-deoxy-2-fluoro-D-[^{3}H]mannose in yeast and chick-embryo cells. *Eur. J. Biochem.* *87*:55–68.
67. Morin, M. J., and Bernacki, R. J. (1983). Biochemical effects and therapeutic potential of tunicamycin in murine L1210 leukemia. *Cancer Res.* *43*:1669–1674.
68. Saat, Y. A., Krishnaraj, R., and Kemp, R. G. (1981). Glycolipid sialyltransferases in normal and neoplastic murine thymocytes. *Biochem. Biophys. Acta* *678*:213–220.
69. Augener, W., Brittinger, G., Abel, A. C., and Goldblum, N. (1980). Sequential expression of fucosyltransferase and *N*-acetylneuraminyltransferase activities in human leukemic cells arrested at different stages of mutation. *Cancer Biochem. Biophys.* *5*:33–39.
70. Bernacki, R. J. (1977). Concomitant elevations in serum sialyltransferase activity and sialic acid in rats with metastasizing mammary tumors. *Science* *195*:577–580.
71. Yogeeswaran, G., and Salk, P. (1981). Metastatic potential is positively correlated with cell surface sialylation of cultured murine tumor cell lines. *Science* *212*:1514–1516.
72. Dennis, J. W., Waller, C., Timpl, R., and Schirrmacher, V. (1982). Sialic acid on metastatic tumour cells reduces cell attachment to fibronectin and collagen type IV. *Nature* *300*:274–276.
73. Fogel, M., Altevogt, P., and Schirrmacher, V. (1983). Metastatic potential severely altered by changes in tumor cell adhesiveness and cell-surface sialylation. *J. Exp. Med.* *157*:371–376.
74. Evans, I. M., Hilf, R., Murphy, M., and Bosmann, H. B. (1980). Correlation of serum, tumor, and liver serum glycoprotein: *N*-Acetylneuraminic acid transferase activity with growth of the R3230AC mammary tumor in rats and relationship of the serum activity to tumor burden. *Cancer Res.* *40*:3103–3111.
75. Skipski, V. P., Carter, S. P., Terebus-Kekish, O. I., Podlaski, F. J., Jr., Peterson, R. H. F., and Stock, C. C. (1981). Ganglioside profiles of metastases and of metastasizing and non-metastasizing rat primary mammary carcinomas. *JNCI* *67*:1251–1258.
76. Bosmann, H. B., Spataro, A. C., Myers, M. W., Bernacki, R. J., Hillman, M. J., and Caputi, S. E. (1975). Serum and host liver activities of glycosidases and sialyltransferases in animals bearing transplantable tumors. *Res. Commun. Chem. Pathol. Pharmacol.* *12*:499–512.
77. Kessel, D., and Allen, J. (1975). Elevated plasma sialyltransferase in the cancer patient. *Cancer Res.* *35*:670–672.
78. Ip, C., and Dao, T. (1978). Alterations in serum glycosyltransferase and 5′-nucleotidase in breast cancer patients. *Cancer Res.* *38*:723–728.
79. Bernacki, R. J. (1975). Regulation of rat liver glycoprotein: *N*-Acetylneuraminic acid transferase activity by pyrimidine nucleotides. *Eur. J. Biochem.* *58*:477–481.
80. Klohs, W. D., Bernacki, R. J., and Korytnyk, W. (1979). Effects of nucleotides and nucleotide:analogs on human serum sialyltransferase. *Cancer Res.* *39*:1231–1238.
81. Korytnyk, W., Angelino, N., Klohs, W., and Bernacki, R. J. (1980). CMP and CMP-sugar analogs as inhibitors of sialic acid incorporation into glycoconjugates. *Eur. J. Med. Chem.* *15*:77–84.
82. Kijima, I., Ezawa, K., Toyoshima, S., Furuhata, K., Ogura, H., and Osawa, T. (1982). Induction of suppressor T cells by neuraminic acid derivatives. *Chem. Pharm. Bull.* *30*:3278–3283.
83. Kijima-Suda, I., Toyoshima, S., Itoh, M., Furuhata, K., Ogura, H., and Osawa, T. (1985). Inhibition of sialyltransferases of murine lymphocytes by disaccharide nucleotides. *Chem. Pharm. Bull.* *33*:730–739.
84. Kijima-Suda, I., Miyamoto, Y., Toyoshima, S., Itoh, M., and Osawa, T. (1986). Inhibition of experimental pulmonary metastasis of mouse colon adenocarcinoma 26 sublines by a sialic acid:nucleoside conjugate having sialyltransferase inhibiting activity. *Cancer Res.* *46*: 858–862.
85. Vaghefi, M. M., Bernacki, R. J., Dalley, N. K., Wilson, B. E., and Robins, R. K. (1987). Synthesis of glycopyranosylphosphonate analogs of certain natural nucleoside diphosphate sugars as potential inhibitors of glycosyltransferases. *J. Med. Chem.* *30*:1383–1391.

86. Vaghefi, M. M., Bernacki, R. J., Hennen, W. J., and Robins, R. K. (1987). Synthesis of certain nucleoside methylenediphosphonate sugars as potential inhibitors of glycosyltransferases. *J. Med. Chem. 30*:1391–1399.
87. Palcic, M. M., Heerze, L. D., Srivastava, O. P., and Hindsgaul, O. (1989). A bisubstrate analog inhibitor for α(1→2)-fucosyltransferase. *J. Biol. Chem. 264*:17174–17181.
88. Cummings, R. D., Trowbridge, I. S., and Kornfeld, S. (1983). A mouse lymphoma cell line resistant to the leukoagglutinating lectin from *Phaseolus vulgaris* is deficient in UDP-GlcNAc:α-D-mannoside β1,6 *N*-acetylglucosaminyltransferase. *J. Biol. Chem. 257*:13421–13427.
89. Yamashita, K., Ohkura, T., Tachibana, Y., Takasuki, S., and Kobata, A. (1984). A comparative study of the oligosaccharides released from baby hamster kidney cells and their polyoma transformant by hydrazinolysis. *J. Biol. C hem. 259*:10834–10840.
90. Yamashita, K., Tachibana, Y., Ohkura, T., and Kobata, A. (1985). A enzymatic basis for the structural changes of asparagine-linked sugar chains of membrane glycoproteins of baby hamster kidney cells induced by polyoma transaction. *J. Biol. Chem. 260*:3963–3969.
91. Pierce, M., and Arango, J. (1986). Rouse sarcoma virus-transformed baby hamster kidney cells express higher levels of asparagine-linked tri- and tetraantennary glycopeptides containing [GlcNAc-β(1,6)Man-α(1,6)Man] and poly-*N*-acetyllactosamine sequences than baby hamster kidney cells. *J. Biol. Chem. 261*:10772–10777.
92. Arango, J., and Pierce, M. (1988). Comparison of *N*-acetylglucosaminyltransferase V activities in Rous sarcoma-transformed baby hamster kidney (RS-BHK) and BHK cells. *J. Cell. Biochem. 37*:199–205.
93. Dennis, J. W., Kosh, K., and Breitman, M. L. (1989). Oncogenes conferring metastatic potential induce increased branching of Asn-linked oligosaccharides in rat2 fibroblasts. *Oncogene 4*: 853–860.
94. Dennis, J. W., Laferté, S., Waghorne, C., Breitman, M. L., and Kerbel, R. S. (1987). β1-6 Branching of Asn-linked oligosaccharide is directly associated with metastasis. *Science 236*: 582–585.
95. Dennis, J. W. (1988). Asn-linked oligosaccharide processing and malignant potential. *Cancer Surv. 7*:572–595.
96. Dennis, J. W., and Laferté, S. (1989). Oncodevelopmental expression of GlcNAcβ1-6Manα1-6Manβ1- branched asparagine-linked oligosaccharides in murine tissues and human breast carcinomas. *Cancer Res. 49*:945–950.
97. Khan, S. H., Abbas, S. A., and Matta, K. L. (1989). Synthesis of some oligosaccharides containing the *O*-(2-acetamido-2-deoxy-β-D-glucopyranosyl)-*O*-α-D-mannopyranosyl unit. Potential substrates for UDP-GlcNAc:α-D-mannopyranosyl-(1→6)-*N*-acetyl-β-D-glucosaminyltransferase (GnT-V). *Carbohydr. Res. 193*:125–139.
98. Khan, S. H., Abbas, S. A., and Matta, K. L. (1990). Synthesis of 4-nitrophenyl *O*-(2-acetamido-2-deoxy-β-D-glucopyranosyl)-(1→2)-*O*-(4-*O*-methyl-α-D-mannopyranosyl)-(1→6)-β-D-gluco pyranoside. A potential specific acceptor substrate for *N*-acetylglucosaminyltransferase-V (GnT-V). *Carbohydr. Res. 205*:385–397.
99. Khan, S. H. (1990). Studies in oligosaccharide synthesis. PhD dissertation. State University of New York at Buffalo, Buffalo, New York, pp. 59–61.
100. Palcic, M. M., Ripka, J., Kaur, K. J., Shoreibah, M., Hindsgaul, O., and Pierce, M. (1990). Regulation of *N*-acetylglucosaminyltransferase V activity. *J. Biol. Chem. 265*:6759–6769.

12

Function of Carbohydrate Moieties: Membrane and Nonsecretory Glycoproteins

Minoru Fukuda *The La Jolla Cancer Research Foundation, La Jolla, California*

Our knowledge of the mechanisms of inheritance in biological systems has advanced greatly in the past three decades. We now know that the information for the development of a mature organism is encoded within the chemical structure of deoxyribonucleic acid (DNA). This information is translated into amino acid sequences, which define how polypeptides fold and eventually form units of multiple polypeptides. The next question we need to address is how such information is organized to specify more complex structures, such as organelles and, ultimately, how these different levels of organization lead to development of the human body. This will require an understanding of the sequences of reactions that determine development and differentiation.

To answer this question, it is important to try to understand how the expression of a particular protein is regulated at the level of DNAs, at the level of interaction between DNA and protein, and at the level of effector molecules, such as hormones. However, this approach reveals only one aspect of the regulation that a mature organism must achieve. The regulation of these events is the result not only of expression of particular proteins, but also the result of interaction between these molecules in the same cell or among different cells. In this chapter, we would like to emphasize the significance of the glycosylation of proteins in molecular interactions. For this, we will first summarize the current knowledge on the function of mannose-6-phosphate structures for lysosomal enzyme targeting; then outline several examples of the functional significance of polylactosaminoglycans; and, finally, we describe several examples that demonstrate the functions of carbohydrates.

I. MANNOSE-6-PHOSPHATE, A MARKER FOR LYSOSOMAL ENZYMES

In the past decade our knowledge of the carbohydrate structures attached to protein has vastly advanced. The carbohydrate moieties can be classified into two groups according to the mode of linkage to an amino acid: *N*-linked saccharides and *O*-linked saccharides. The *N*-linked saccharides are further classified into at least three subgroups: the complex-type, hybrid-type, and high mannose-type (see Chap. 2 for these structures).

In the same period, our knowledge on the biosynthetic pathways used to from these various arrays of carbohydrates has greatly advanced. In particular, the clarification of processing of *N*-linked saccharides during the biosynthesis of glycoproteins has brought us new insight that describes the formation of these three different *N*-glycan groups into one unified series of reactions (Fig. 1) [1]. A slight variation of this scheme also includes the pathway leading to the formation of structures containing mannose-6-phosphate residues, which are specific markers for lysosomal enzymes.

As shown in Figure 1, plasma membrane proteins, as well as lysosomal enzymes and secretory proteins, are synthesized in the rough endoplastic reticulum (RER). All

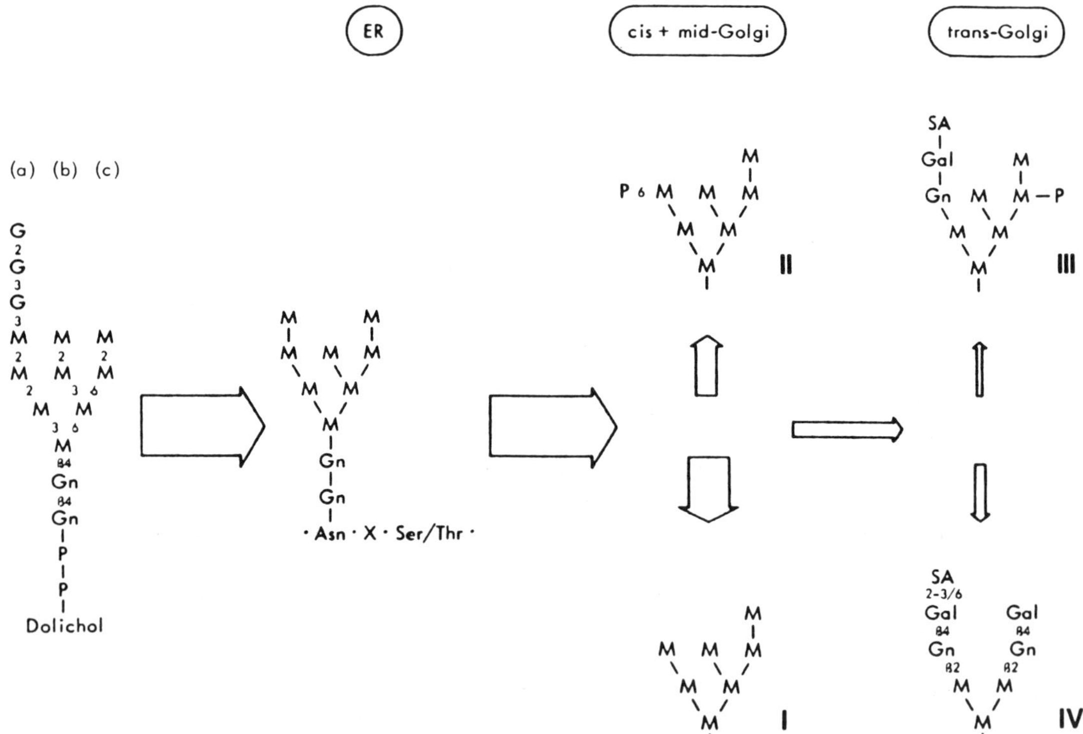

Figure 1 Main stages in the processing of oligosaccharides in lysosomal enzymes. The organelles to which the processing is localized are indicated at the top of the figure. Four typical structures that have been found in lysosomal enzymes, as formed in different parts of the Golgi complex, are shown: I, high mannose; II, phosphorylated high mannose; III, phosphorylated hybrid; IV, complex of oligosaccharide. Hybrid oligosaccharides without phosphate groups have also been found. The arrows indicate the approximate relative abundance of the four oligosaccharide types. The symbols are G, glycose; Gal, galactose; Gn, *N*-acetylglucosamine; M, mannose; SA, *N*-acetylneuraminic acid; P, phosphate. Single numbers indicate the positions of α-anomerically linked sugars. (From Ref. 1.)

of the three different kinds of proteins enter into the luminal side of RER. These proteins then undergo cotranslational glycosylation of asparagine (Asn) residues present in Asn-X-Ser/Thr sequences. This is initiated by en bloc transfer of a large glucose-containing high mannose saccharide from a lipid intermediate to the nascent polypeptide. This large saccharide is then gradually trimmed down and then modified into the sialic acid- containing saccharides typically found in secretory and membrane glycoproteins. Some of the saccharides, however, are not extensively trimmed and, instead, remain as high mannose saccharides. On the other hand, high mannose saccharides on lysosomal enzymes acquire phosphate residues by addition of *N*-acetylglucosamine 1-phospho-diester to mannose (Man) residues, followed by removal of *N*-acetylglucosamine (GluNAc) [2]. The resulting mannose-6-phosphate serves as a recognition marker that leads to high-affinity binding of newly synthesized lysosomal enzymes to mannose-6-phosphate receptors found in the Golgi [1,3,4].

Thus, it is critical for the phosphotransferase to transfer *N*-acetylglucosamine-1-phosphate to the high mannose saccharides attached to nascent lysosomal enzymes. The phosphotransferase specifically recognizes lysosomal enzymes and phosphorylates high mannose saccharides that are present only in lysosomal enzymes [5]. If this recognition marker is deficient, it leads to the multiple deficiency of lysosomal enzymes, as found in I-cell disease. This discovery, made by Hickman and Neufeld [6], actually provided the basis for many subsequent studies that eventually led to the recognition marker and its receptor. By examining these processes, we can develop two interesting concepts on the significance of processing and modification of carbohydrate moieties.

First, it is most likely that the processing of *N*-linked saccharides has evolved so that it provides a single unified system that can generate an array of saccharides with potentially different functions. By having the same precursor saccharide in differently final predetermined proteins, all can undergo similar modifications within the same apparatus until they are sorted to their final destinations. It is only with the addition of GlcNAc-1-phosphate or *N*-acetylglucosamine that proteins with different destinations are glycosylated differently (see Fig. 1).

The second interesting carbohydrate modification that derives from the addition of mannose-6-phosphate to lysosomal enzymes is one of biological economy. Lysosomes contain more than 30 different hydrolytic enzymes, each with quite different structures. If cells needed to develop a different marker to target each of these enzymes or to the lysosomes, the number of markers would equal the number of lysosomal enzymes. Instead, by adding mannose-6-phosphate with a specific marker to the saccharides attached to lysosomal enzymes, this can all be accomplished with a single specific marker. Targeting of various proteins to lysosomes is accomplished in only one series of reactions. This clever innovation is also economical, since only one common receptor for mannose-6-phosphate is necessary for targeting. Furthermore, this process can ignore the diversity of protein structures of the various lysosomal enzymes. The specificity required in this system is found in the phosphotransferase, which recognizes a common peptide conformation found on various lysosomal enzymes.

These two points illustrate the general importance of protein glycosylation. It provides a unified biosynthetic pathway for entirely different proteins and a unified specific marker for specific kinds of proteins that are structurally different in their peptide portions.

II. POSSIBLE ROLES OF POLYLACTOSAMINOGLYCAN

A. Structural Characteristics of Polylactosaminoglycans

During the past decade, a novel type of asparagine-linked carbohydrate chain has been gaining prominence. This type of carbohydrate chain contains essentially the same core structure as other asparagine-linked carbohydrate chains and has a characteristic repeating unit composed of *N*-acetyllactosamine (Galβ1→4GlcNAcβ1→3). This type of repeating chain is also called a *polylactosaminoglycan* [7]. Some of the characteristics and historical background on polylactosaminoglycan have been presented previously [7,8], and three recent findings on polylactosaminoglycan structure will be discussed here.

First, we have systematically analyzed the carbohydrate chains of whole human erythrocytes to determine what minimum number of *N*-acetyllactosaminyl units is required to accept fucose at C-2 of galactose and form the ABH blood group antigen. This recent analysis revealed that only those side chains containing two or more *N*-acetyllactosamino units can be substrates for α1→2-fucosyltransferase. Thus, the presence of *N*-acetyllactosaminyl repeats convert the simple *N*-linked saccharides into species that express various antigenic markers. Human erythrocyte glycolipids also contain polylactosamine units, and it became apparent that the foregoing rule also applies to them: the smallest glycolipid containing α1→2-linked fucose is Galβ1→4GlcNAcβ1→3Galβ1→4Glcβ1→Cer [9], and fucose is not added to Galβ1→4Glcβ→Cer.

The second important finding is that the addition of polylactosaminyl units occurs preferentially on certain side chains. In human erythrocytes, the activities of *N*-acetylglucosaminyltransferase IV or V is low; hence, most band 3 carbohydrates contain biantennary saccharides [10,11]. as we have previously shown, the side chain arising from α-mannose attached to C-6 of β-mannose (mannose C-6 side) contains *N*-acetyllactosamine repeats, whereas that from C-3 of β-mannose contains a minimum amount of *N*-acetyllactosamine repeats (Fig. 2).

When glycopeptides from other sources are analyzed, similar but slightly different features are observed. Many cells contain *N*-acetylglucosaminyltransferase IV and V so that a large proportion of *N*-linked saccharides of those cells have triantennary or tetraantennary saccharides. In these cells, polylactosaminyl extension is found preferentially on the side chain attached to C-6 of α-mannose, which is, in turn, bound to C-6 of β-mannose. The next preferred side chain is linked to C-2 of α-mannose on the C-6 side (Fig. 3). In addition to these side chains, a small proportion of polylactosaminyl extension is present in the side chain arising from C-4 of α-mannose of the C-3 side [12–14]. This preferential extension of polylactosamine (i.e., *branch specificity*) can be explained by the specificity of β1→3 *N*-acetylglucosaminyltransferase, which forms the *N*-acetyllactosamine extension [15].

It is noteworthy that branch specificity can also be found in two sialyltransferases. It was discovered in various cell types that α2→3-sialyltransferase prefers the same side chains as β1→3 *N*-acetylglucosaminyltransferase, *extension enzyme* [8,10–14]. On the other hand, α2→6-sialyltransferase has almost the opposite branch specificity, and the most preferential site for this enzyme is the side chain arising from C-2 of α-mannose of the C-3 side [16]. Thus, α2→3 sialyltransferase and α2→6 sialyltransferase act complementarily in different side chains.

The preferential extension of *N*-acetyllactosamine in certain side chains is important to understand various other observations. For example, leukophytohemagglutinin (PHA-L) bind preferentially to saccharides containing the side chain arising from C-6 of α-mannose

[17]. Since the same side chain is preferentially extended by *N*-acetyllactosamine repeats, a PHA-L-resistant cell line lacks polylactosaminyl repeats as well as the side chain arising from C-6 of α-mannose. This is why PHA-L-resistant mutant cells lack both a branch arising from C-6 of mannose on C-6 side and polylactosaminyl repeats [18].

Third, important progress in structural studies on polylactosaminoglycan is that tomato lectin preferentially binds to polylactosaminoglycan. Previously, it has been shown that *Datura stramonium* lectin [12] and pokeweed mitogen [19] bind to polylactosaminoglycan. However, in both cases, these lectins also bind to complex-type *N*-linked saccharides when they have the GlcNAcβ1→6(GlcNacβ1→2)Manα1→6 structures [19,20]. This latter structure is usually present in tetraantennary saccharides; thus, these lectins bind to tetraantennary *N*-linked saccharides, regardless of whether or not the side chains have a *N*-acetyllactosaminyl repeat. Merkle and Cummings discovered that, on the other hand, tomato lectin, binds preferentially to polylactosaminoglycan, but does not bind to usual *N*-linked saccharides, even when they are tetraantennary [21]. We have followed up this report and found that tomato lectin is superior to *D. stramonium* lectin in specific isolation of polylactosaminoglycan. Tomato lectin will probably play a critical role in the studies on polylactosaminoglycan, and further studies on its specificity will be important to expand its usefulness.

B. The Requirement of Polylactosaminoglycan for Proper Folding of Band 3

Congenital dyserythropoietic anemias (CDA) are genetic erythropathies accompanied by bone marrow erythroid hyperplasia and erythroblastic polynuclearity [22]. In all types of CDAs, membrane abnormalities are prominent features on electron microscopic examination [23]. The type II CDAs are also called HEMPAS (*h*ereditary *e*rythroblastic *m*ultinuclearity associated with a *p*ositive *a*cidified *s*erum test) because the erythrocytes of those patients are lysed by acidified serum obtained from 30% of normal individuals [24].

When the erythrocytes from HEMPAS patients were analyzed by sodium dodecyl sulfated–polyacrylamide gel electrophoresis (SDS–PAGE), it was noted that the band 3 glycoprotein from HEMPAS erythrocytes migrates faster than that of normal erythrocytes [25–29]. This lower relative molecular mass (M_r) of band 3 is not due to proteolytic processing of native band 3, but is caused by a defect in glycosylation. In particular, it was suggested that polylactosaminoglycan is missing in HEMPAS band 3, since it was not affected by the treatment of endo-β-galactosidase, an enzyme specific for the lactosaminyl linkage [30]. On the other hand, HEMPAS erythrocytes accumulated a substantial amount of glycolipids that contained polylactosaminyl repeats, polylactosaminoglycolipids [31].

The carbohydrate structure of HEMPAS band 3 was determined and was found to be a truncated structure, as shown in Figure 2. This *N*-linked saccharide lacks entirely the extension of polylactosamine in the side chain arising from C-6 of α-mannose that is linked to C-6 of β-mannose. It was discovered that this lack of a polylactosaminyl side chain was due to the reduced activity of *N*-acetylglucosaminyltransferase II, which is the first enzyme necessary for *N*-acetyllactosamine formation in the C-6 side chain [32]. Because the first *N*-acetyllactosamine is absent, poly-*N*-acetyllactosamine is not formed in this carbohydrate. As shown in the previous section, poly-*N*-acetyllactosamine is exclusively formed in the side chain arising from α-mannose linked to C-6 of β-mannose in human band 3. In this defect, it is possible that cells containing triantennary or tetraantennary carbohydrate structures may escape from the damage caused by the absence of *N*-acetylglucosaminyltransferase II. Since only the extension of the side chain that arises from C-2 of

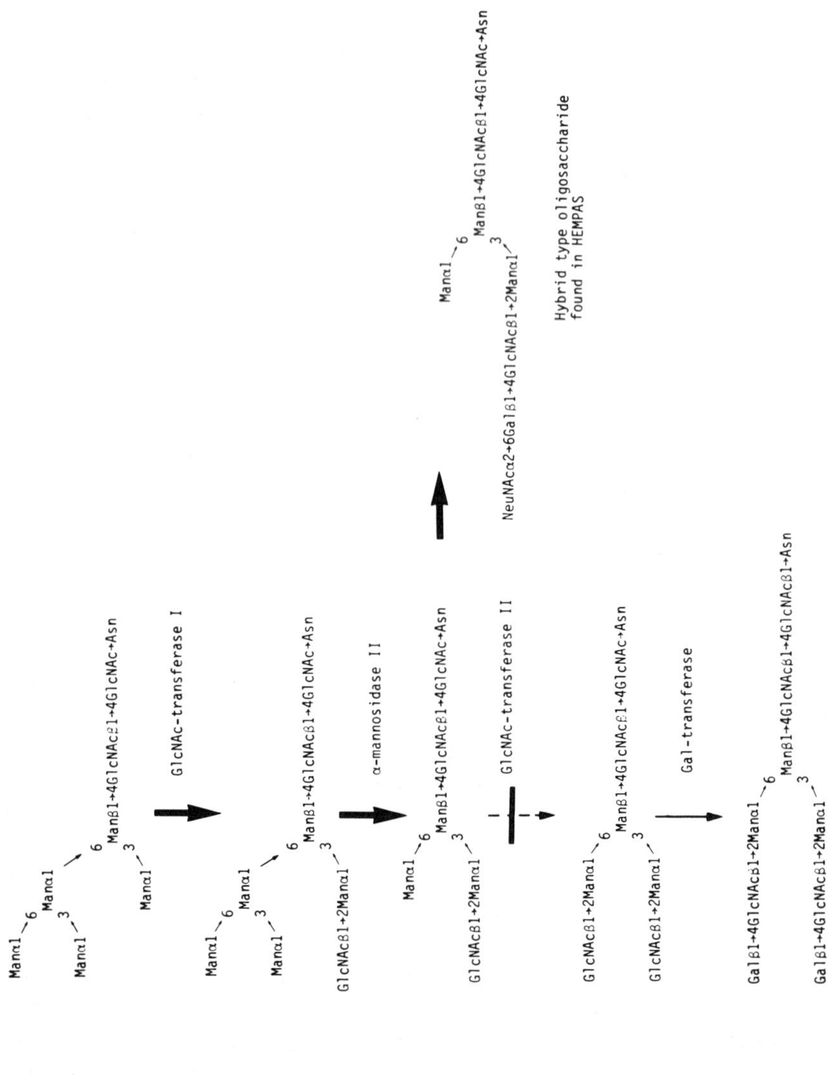

GlcNAc-transferase I
α-mannosidase II
GlcNAc-transferase II
Gal-transferase
Genetic defect in HEMPAS
Hybrid type oligosaccharide found in HEMPAS

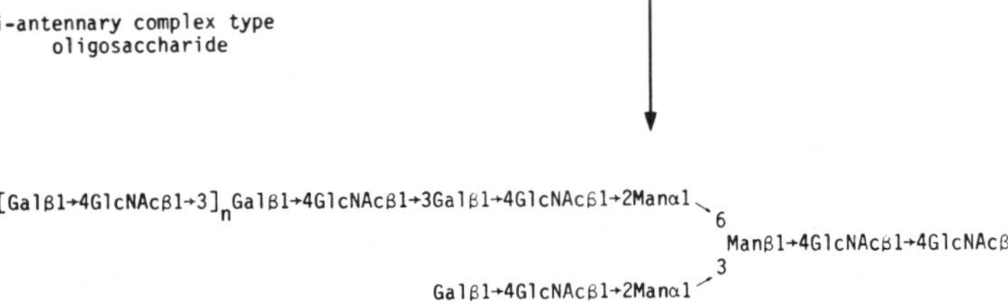

Figure 2 Proposed scheme for biosynthesis of *N*-linked oligosaccharides and the genetic defect in HEMPAS. A most likely course of polylactosaminoglycan synthesis is shown. Biosynthesis of *N*-linked oligosaccharides on proteins is according to Hubbard and Ivatt [100] and Kornfeld [101]. *N*-Acetylglucosaminyltransferase II is involved in the formation of the Manα1→6Manβ1→ arm to which the lactosaminyl repeats would attach. A genetic defect in *N*-acetylglucosaminyltransferase II blocks the normal biosynthesis of complex-type oligosaccharides and results in the accumulation of the trimannosyl hybrid oligosaccharide. In the other instance, the weak activity of *N*-acetylglucosaminyltransferase II may cause significant delay in biosynthesis of the Manα1→6Man arm, resulting in the decreased glycosylation of lactosaminoglycans. (From Ref. 31.)

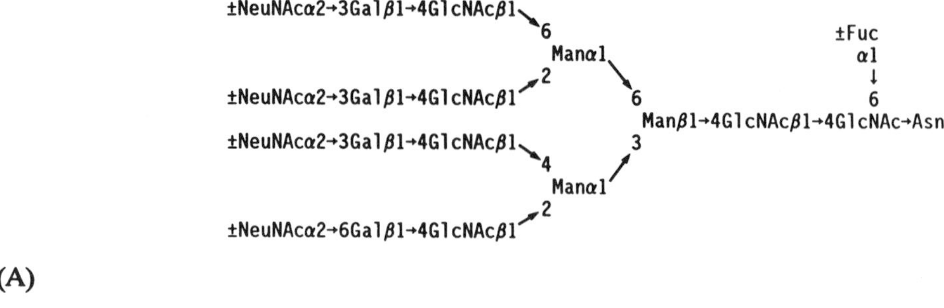

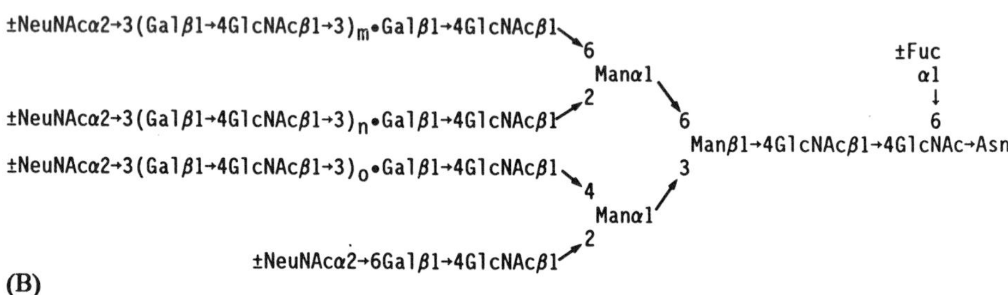

Figure 3 Most probable structures of tetraantennary saccharides (A) and tetraantennary saccharides with *N*-acetyllactosaminyl repeats (B). (A) α2-3 Sialylation takes place preferentially at the side chain arising from C-6 and then from the side chain at C-2 of 2,6-substituted mannose and C-4 of 2,4-substituted mannose. α2→6 Sialylation takes preferentially at the side chain arising from C-2 of 2,4-substituted mannose on the C-3 side [16]. (B) The distribution of *N*-acetyllactosaminyl repeats is as follows in recombinant erythropoietin: the saccharides with one *N*-acetyllactosaminyl repeat (m=1, n=0), and o=0, in 90% of the molecules, and o=1 and m=n=0 in 10% of the molecules); the saccharides with two *N*-acetyllactosaminyl repeats (m=n=1 and o=0 in 60% of the molecules, and m=o=1 and n=0 in 40% of the molecules); and the saccharides with three *N*-acetyllactosaminyl repeats (m+n+o=3). (Adapted from Ref. 14.)

α-mannose on the C-6 side is inhibited in this kind of HEMPAS, polylactosamine may be found in other side chains in these cells. In these cells, defects in glycoproteins containing polylactosaminoglycan are probably fewer than found in erythrocytes [see also Ref. 32]. This conclusion is consistent with the fact that HEMPAS erythrocytes contain a large amount of polylactosaminyl lipids [31], since glycolipid synthesis should not be impaired by this defect.

Erythroid cells from HEMPAS patients show morphological abnormalities. Many erythroblasts have more than one nucleus and often show blobs, clefts, myelin figures, and vacuoles. Erythrocytes very often are found with a second membrane underneath the plasma membrane, a so-called double membrane. This second membrane is apparently different from the plasma membrane, since band 3 and glycophorin A are absent from it. Furthermore, intracellular membranous structures at blobs and clefts are also lacking band 3 and glycophorin A. These results indicate that these abnormal membranes are probably part of the endoplasmic reticulum or derived from an intracellular organelle [33].

When the distribution of band 3 was examined, intriguing results were observed. First, it appears that band 3 at the plasma membrane is more clustered in HEMPAS erythrocytes than in normal erythrocytes. Second, band 3 proteins are clustered as a large mass in the cytoplasm of HEMPAS erythrocytes, whereas normal erythrocytes completely lack such structures. These clusterings can be observed only in band 3 distribution, but not in glycophorin A distribution [33]. These results strongly suggest that the absence of polylactosaminoglycan in HEMPAS band 3 results in the clustering of band 3. This clustering of band 3 then may cause the abnormality in membrane organization of HEMPAS erythrocytes.

In typical patients with HEMPAS, the membrane abnormality is relatively restricted to erythrocytes, probably because only erythrocytes lack polylactosaminoglycan. In some variants of these patients, however, the abnormality is extended to other blood cells, such as granulocytes and platelets. One of those variants was recently analyzed in detail [34]. Interestingly, polylactosamine formation in this patient was absent in both glycoproteins (band 3 and band 4.5) and glycolipids (polyglycosylceramide). It was discovered that this abnormality is probably caused by a defect in galactosyltransferase. The HEMPAS variant contains a comparable amount of soluble galactosyltransferase, but expresses a significantly reduced level of membrane-bound galactosyltransferase. Thus, it is possible that a genetic defect in galactosyltransferase results in the increased amount of the soluble form, probably by mutation at the transmembrane portion or presumed protease-catalyzed cleavage site [34]. Recently, cDNA sequences for galactosyltransferase were determined, and the deduced amino acid sequence indicates that a galactosyltransferase contains only one hydrophobic transmembrane portion, which could act as a combination of signal and anchor for membrane insertion and is not cleaved after processing [35,36]. If mutation brings about a basic amino acid in proximity of the NH_2-terminal of the hydrophobic segment, the hydrophobic portion could be cleaved as a signal peptide [37]. If this were to take place, the only hydrophobic domain would become a leader peptide, which is cleaved after processing, leading to expression of only a soluble form of the enzyme. It will be interesting to know if such is true in this HEMPAS variant.

These results, as a whole, provide clear examples of genetic disease in which the defect of glycosyltransferases is the primary cause for diseases. We expect more diseases that are due to defective in glycosyltransferase(s) will be discovered by analyzing various HEMPAS patients.

C. Lysosomal Membrane Glycoproteins as Major Carriers for Polylactosaminoglycans

The distribution of polylactosaminoglycan is unique not only among branches, but also among proteins. In human erythrocytes, polylactosaminoglycans are attached to band 3 and band 4.5 (which includes glucose transporter), but not to glycophorins [see Ref. 8], although both band 3 and glycophorins are synthesized in erythroid cells. On the other hand, little is known about the protein carriers for polylactosaminoglycan in nucleated cells. We have shown previously that only membrane glycoproteins contain polylactosaminoglycan, and secreted glycoproteins contain no detectable amount of polylactosaminoglycan in PA-1 human teratocarcinoma cells [38]. Furthermore, only a limited number of cellular glycoproteins contain polylactosaminoglycan [38]. These results strongly suggest that the nature of a protein determines whether it is modified by polylactosaminoglycan.

Our previous studies showed that glycoproteins with $M_r \sim 120{,}000$ are major carriers for polylactosaminoglycans in granulocytic cells [39]. To characterize the molecules that carry polylactosaminoglycan in nucleated cells, we isolated these glycoproteins and found them to be lysosomal membrane glycoproteins. Two related glycoproteins, termed lamp-1 and lamp-2 were found. Chemical characterization of purified glycoproteins was done [40], and the amino acid sequences of the glycoproteins were deduced from cDNAs [41]. These data showed several interesting aspects of the glycoproteins. First, a large portion of the lamp molecule resides in the lumen of lysosomes. This portion of the molecule can be roughly divided into two domains, and the two domains are separated by a hingelike structure enriched with proline and serine (lamp-1) or proline and threonine (lamp-2). Each domain contains 7–11 *N*-linked saccharides, and some of them are polylactosaminoglycans. The high number and large size of carbohydrate moieties, as a whole, probably serves to protect lysosomal membrane glycoproteins from degradation by lysosomal proteases (Fig. 4). In fact, the lysosomal glycoproteins were degraded faster when *N*-glycan synthesis was inhibited with tunicamycin [42].

Why, then, do the lysosomal membrane glycoproteins lamp-1 and lamp-2 contain polylactosaminoglycan? Our recent experiments provide some insights. We discovered that the distance of glycosylation sites from the membrane may affect the mode of glycosylation [43]. When human chorionic gonadotropin α-chain was fused with the transmembrane plus the cytoplasmic segment of vesicular stomatitis virus glycoprotein (G), the fused protein acquired polylactosaminoglycan, whereas the parent α-chain and G protein express only typical *N*-linked saccharides. We proposed that by bringing the glycosylation sites close to the transmembrane portion in the chimeric protein, the signal to acquire polylactosaminoglycan was formed [43]. In fact, band 3 glycoprotein [10,11], glucose transporter

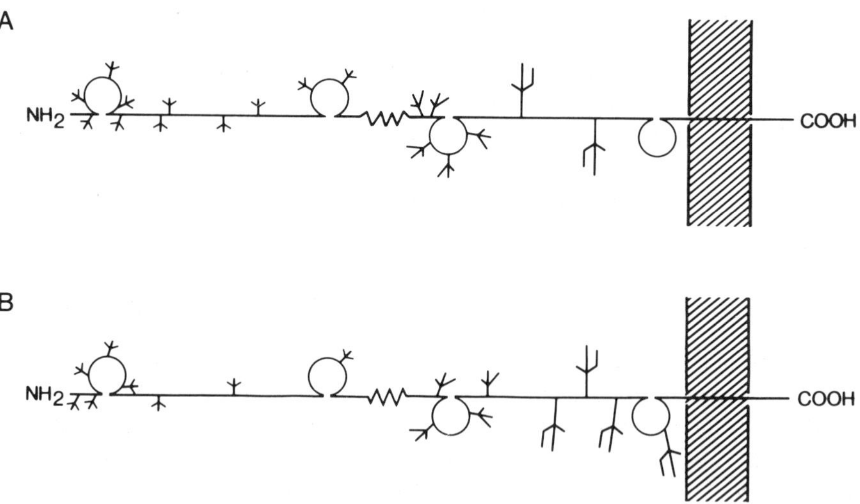

Figure 4 Depicted structure of human lamp-1 (A) and lamp-2 (B). *N*-linked carbohydrate moieties and hingelike structures are indicated by ⺅ and ⩘, respectively. It is assumed that the *N*-glycans attached closely to transmembrane are polylactosaminoglycan, and those at the NH_2-terminal end are typical complex *N*-linked saccharides. *N*-Glycans in the third loop may be of intermediate size. Each presumptive loop is made by a disulfide bond. Cysteine residues connected to each other are tentatively assigned. Most of the molecules reside in the luminal side of lysosomes. The lipid bilayer is indicated by a shaded area. Although this figure does not include *O*-glycan attachment sites, it is likely that they are clustered around the hingelike structures as shown in IgA α_1-chain and IgD δ-chain. (From Ref. 41.)

[44], and influenza B virus NB glycoprotein [45], which all contain polylactosaminoglycan, have *N*-glycosylation sites close to the transmembrane portions [45–47]. Furthermore, lamp-2 contains more polylactosaminoglycan than lamp-1, as expected from the fact that lamp-2 contains more glycosylation sites close to the transmembrane portion than does lamp-1. It will be more interesting if this rule can be applied to other glycoproteins in general.

D. Polylactosaminoglycan Determination of Metastatic Capability in Tumor Cells

Warren and his co-workers discovered that virally transformed neoplastic cells contain more glycopeptides with higher relative molecular masses than the parent untransformed cells [48]. It was originally thought that the basis of this difference was that transformed cells contain more sialic acid, which results in a higher molecular mass. However, further studies by Kobata and Muramatsu and their colleagues found two interesting observations: (1) transformed cells contain more tri- or tetraantennary saccharides than the parent untransformed cells, and (2) transformed cells contain more polylactosaminyl repeats than the parent untransformed cells [49,50]. As described in the foregoing, polylactosaminyl repeats are built up mainly on side chains that are present in tri- or tetraantennary saccharides. Therefore, it is likely that these two distinct features of transformed cells are closely related.

In several systems metastatic tumor cells contain more polylactosaminyl repeats than nonmetastatic counterparts. Furthermore, the metastatic capability of tumor cells can be significantly reduced when the cells are treated with tunicamycin [51], swainsonine [52,53], or castanospermine [54] in animal models. Since swainsonine and castanospermine inhibit the conversion of high mannose-type saccharides into the complex-type [55], these drugs, at the same time, inhibit the formation of polylactosaminyl repeats. Thus, these combined results apparently support the idea that polylactosaminoglycan is somehow involved in the metastatic process of tumor cells.

It has also been shown, by Dennis et al. [17], that metastatic tumor cells contain polylactosaminoglycan mainly in glycoproteins of 130 kd, and the content of polylactosamine, or the GlcNAcβ1→6Manα1→6 Man branch in the glycoprotein, is substantially reduced when nonmetastatic mutant tumor cells were analyzed. This glycoprotein(s) (Gp130) has now been identified as lysosomal membrane glycoprotein-1 [56]. These results, therefore, suggest that metastatic tumor cells express more polylactosaminoglycan on lamp-1 (and possibly lamp-2) than do nonmetastatic counterparts. Why, then, are these two seemingly unrelated properties connected? It has been reported that lysosomal enzymes may be involved in the invasion by tumor cells of the surrounding tissue or penetration of the endothelial membrane [57–59]. Metastatic tumor cells secrete more lysosomal enzymes than their nonmetastatic counterparts. It is likely that this secretion takes place by the fusion of whole lysosomes with the plasma membrane. This fusion may be facilitated by lysosomal membrane glycoproteins, since these glycoproteins are found in endosomes and plasma membranes, as well as lysosomes. It can be speculated that lysosomal membrane glycoproteins with increased amounts of polylactosaminoglycan could more efficiently bring lysosomal contents to be secreted than those with a normal amount of polylactosaminoglycan. The inhibition of polylactosaminoglycan biosynthesis then results in fusion inhibition of lysosomes with the plasma membrane, rendering the tumor cells no longer invasive. To study this hypothesis, it will be interesting to see if the cell surface expression of lysosomal membrane glycoproteins is related to the invasion capability of metastatic tumor cells.

E. Polylactosaminoglycan in Cell–Cell Interaction

Polylactosaminoglycan and its derivatives have been shown, in several cases, to play some role in cell–cell interaction. Gooi et al. [60] showed that the SSEA-1 antigen expressed on mouse preimplanting embryo is actually a Galβ1→4(Fucα1→F3)GalNAcβ1→3 structure. Bird and Kimber [61] and Fenderson et al. [62] exogenously added oligosaccharides having this structure to developing mouse early embryos. Strikingly, these oligosaccharides inhibited morula compaction or induced decompaction of morulae [62]. These results were interpreted as follows: surface saccharides with Galβ1→(Fucα1→3)GlcNAc structure are recognized by some molecules that are present in opponent cells. If such molecules are saturated with competitive saccharides, such recognition will not take place, which results in defective development. Although it is not clear what kind of molecules recognize such saccharides, one of those could be a glycosyltransferase, which uses such saccharides as substrates [see, for example, Ref. 63].

Although it is not yet entirely clear, polylactosaminyl carbohydrates appear to be involved in the adhesion of sperm to eggs during fertilization. According to Wassarman et al. [64], this polylactosaminyl carbohydrate is apparently carried by *O*-linked saccharides.

As a counterpart of this observation, Tsuji et al. [65] discovered that monoclonal antibody, obtained from an infertile woman, reacts with the polylactosaminyl carbohydrates on sperm. This monoclonal antibody inactivates human sperm mobility in the presence of complement, which results in infertility. It is interesting that polylactosaminyl structure is present in human sperm, whereas a similar structure is also detected in murine eggs. It is possible that both egg and sperm express different polylactosaminyl structures, which play different roles in egg–sperm interaction.

III. CELL SURFACE CARBOHYDRATES IN CELL–CELL INTERACTION

A. Possible Role of Cell Surface Carbohydrates in the Immune System

Cell surface carbohydrates, polylactosaminoglycan in particular, may be involved in the immune system. The T-200 glycoprotein probably participates in the recognition of target cells by natural killer cells [66]. More recently, it has become apparent that target cells also bind to natural killer cells, probably through binding to polylactosaminoglycan in T-200 (LCA) molecules. This is an interesting new finding, since T-200 binding to target cells is modulated by putative binding proteins in target cells [67]. In relation to this finding, killer cells also may bind to carbohydrate determinants of target cells. A monoclonal antibody, LeoMel 3, inhibits the killing of human melanoma cells by anomalous killer cells. Fukuta et al. [68] showed that this monoclonal antibody actually binds to the carbohydrate chains in ganglioside G_{D2} and its related gangliosides. The results suggest that killer T lymphocytes may bind to target cells by recognition of carbohydrate determinants on the surface of the target cells.

More recently, we have shown that *O*-glycans attached to a leukocyte major sialoglycoprotein, leukosialin, change dramatically during T-cell activation or in hematological disorders. Resting human T lymphocytes express tetrasaccharides NeuNAcα2→3Galβ1→3(NeuNAcα2→6)GalNAc on leukosialin. After T lymphocytes are activated by anti-CD3 antibodies and interleukin-2, those activated T lymphocytes express more complex hexasaccharides, NeuNAcα2→3Galβ1→3(NeuNAcα2→3 Galβ1→4GlcNAcβ1→6)GalNAc [69] (Fig. 5). This change was also observed when T lymphocytes from patients with immune deficiencies, such as Wiskott–Aldrich syndrome, were analyzed. In the latter disease, T lymphocytes are continuously depleted from circulation, and patients eventually die of recurrent infections [70,71]. It is possible that T lymphocytes with hexasaccharides are no longer recognized by peripheral lymph nodes or the thymus, so that such T cells can no longer sustain their lives by resting in such organs. This hypothesis is consistent with the fact that activated T lymphocytes do not go back to homing organs. It will be interesting to investigate whether a lectin is present in thymocytes or other lymphoid organs and whether it binds specifically to the tetrasaccharides but *not* to the hexasaccharide.

Ishizaka and his colleagues [72] described factors, produced by T cells, that have affinity for IgE and enhance or suppress the response of IgE. The IgE-potentiating factor has both *N*-linked oligosaccharides and *O*-linked saccharides, whereas the IgE-suppressive factor has only *O*-linked oligosaccharides. Further experiments showed that

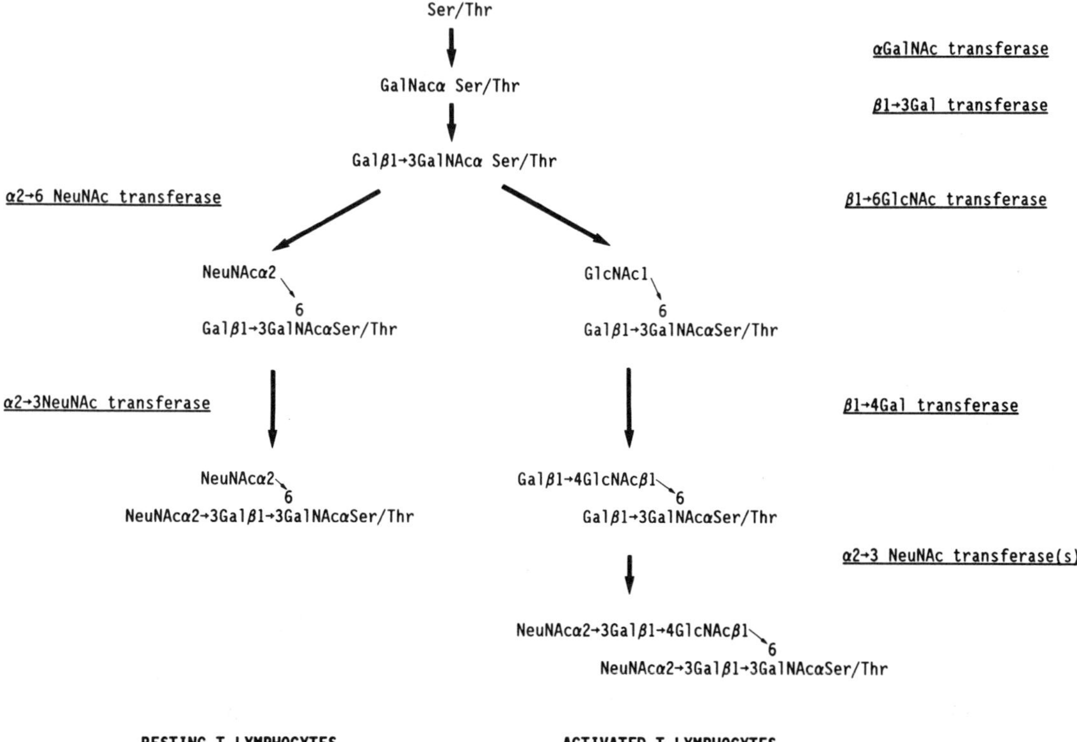

Figure 5 Biosynthesis of leukosialin-attached *O*-linked oligosaccharide in resting and activated T lymphocytes. The activities of six glycosyltransferases were measured. Resting T lymphocytes express no detectable amounts of Galβ1→3GalNAc β1→6GlcNAc (to GalNAc)-transferase, whereas activated T lymphocytes express a significant amount of the same enzyme. On the other hand, GalNAc-Ser(Thr)-α2→6 sialyltransferase is decreased after T lymphocytes are activated. (From Ref. 69.)

IgE-potentiating factor and IgE-suppressive factor are encoded by the same gene and that their biological activities are decided by a posttranslational glycosylation. Enhancing factor enhances the assembly of *N*-linked saccharides to the IgE-binding factor peptide and was found to be a kallikreinlike enzyme [73]. Glycosylation inhibition factor inhibits the same reaction and was found to be a fragment of phospholipase inhibitory protein [74]. Although it is not yet clear how these two factors can change the *N*-glycan synthesis of IgE-binding factors, it is attractive to think that the status of glycosylation in IgE-binding factors change the IgE response.

B. Homing of Lymphocytes and Focal Adhesion of Leukocytes to the Blood Vessel Lining: Possible Role of Lectinlike Structures in Cell Adhesion

To achieve lymphoid organ development and progression of the immune response, it is critical to interface between a lymphocyte's mobile circulating phase and its relatively sessile phase within a particular lymphoid organ. Recirculating lymphocytes specifically recognize and adhere to luminal walls of endothelia and migrate through highly specialized endothelium (high-walled endothelial venules; HEV) into the lymphoid organ parenchyma. Both T and B lymphocytes enter lymphoid organs through common HEV and, thereafter, eventually migrate to T-cell or B-cell domains of the organs. Recirculation of lymphocytes from the bloodstream to particular sites has been called *homing*. This homing is apparently mediated by cell surface structures of lymphocytes called *homing receptors*, which recognize and adhere to HEV of lymphoid organs.

It is apparent that there are two different kinds of lymphocytic migration from blood to lymphoid organs. Peripheral node lymphocytes bind preferentially to peripheral node HEV, whereas lymphocytes from Peyer's patch mucosa bind more preferentially to Peyer's patch HEV. The putative mouse lymphocyte-homing receptor for peripheral lymph node HEV is detected by monoclonal antibody MEL14 [75]. The protein sequence was recently deduced for this homing receptor, and the following interacting aspects were revealed [76,77]: First, this protein has three different functional domains that are tandemly connected. From the NH_2-terminal portion, a lectinlike domain, and epithelial growth factor (EGF)-like domain, and a domain of complement regulatory protein repeat are present. The presence of the lectinlike domain is consistent with the previous report that mannose-6-phosphate and its analogues inhibit binding to peripheral lymph node HEV [78]. However, lymphocyte binding to Peyer's patches is not inhibited by these carbohydrates. It is likely that this lectinlike domain actually binds to the carbohydrates of peripheral node HEV, which may be unique to these cells.

In parallel to these studies, the amino acid sequence of Peyer's patch-specific lymphocyte-homing receptor was deduced from cDNA sequence [79]. Interestingly, this receptor falls into an entirely different family of receptors, integrins. Since integrins are assumed to bind to the peptide moieties of ligand and not of carbohydrate, this is consistent with the previous results on specificity, indicating that no carbohydrate is involved in binding of lymphocytes to Peyer's patches.

A novel type of adhesion molecule seems to play substantial roles in various aspects of blood cell function. The inflammatory or immune cytokines, interleukin-1, tumor necrosis factor, and bacterial endotoxin, act directly on the lining of blood vessels and endothelial cells and induce the endothelial leukocyte adhesion molecule 1 (ELAM-1). This molecule is present only on stimulated endothelial cells and attracts neutrophils to inflamed sites, where the cells help clear the area by ingesting bacteria and other detritus. Bevilacqua et al.

[80] recently obtained the amino acid sequence of ELAM-1 by isolating cDNA coding for the molecule. The ELAM-1 is characterized as having a lectinlike domain, and EGF-like domain, and six repeats of complement regulatory protein. This structural feature is very similar to that of the peripheral lymph node homing receptor.

In addition to these molecules, which clearly play a major role in adhesion, another molecule has a strikingly similar structure. This protein, GMP-140, is found in the membranes of secretory granules in platelets and endothelial cells, which can be rapidly mobilized to the cell surface by thrombin and other fast-acting modulators. The amino acid sequence deduced from recently isolated cDNA indicates that GMP-140 contains an NH_2-terminal lectin domain, and EGF-like domain, and nine repetitive motifs of complement regulatory protein repeats [81]. Thus, these three newly described cell surface molecules appear to constitute a new gene family of adhesion molecules (Fig. 6). It will be crucial to know whether lectin domains actually play a role in the adhesion process and, at the same time, bind to the cell surface through carbohydrates.

Complementary DNA cloning of another homing receptor, however, revealed a new family of adhesion molecules. Stamenkovic et al. [82] and Goldstein et al. [83] reported cDNA and deduced amino acid sequence of a glycoprotein defined by the Hermes monoclonal antibodies. It has been shown that antibodies against distinct epitopes of this "homing receptor" can interfere with lymphocyte binding to lymph node, mucosal, or synovial HEV [84]. The amino acid sequence deduced from cDNA for this molecule indicates that this molecule contains a domain similar to the tandemly repeated domains in the cartilage link protein and proteoglycan monomer. Following this link proteinlike domain, the proximal extracellular domain contains *O*-glycosylation sites. This domain may also be important in the binding of homing receptor to mucosal vascular endothelial cells, since monoclonal antibody Hermes-3, which binds to this domain, inhibits the lymphocyte binding to endothelial cells.

Expression of this glycoprotein is not, however, restricted to hematopoietic cells. Furthermore, an immunologically and structurally related molecule isolated from a human fibrosarcoma line, termed CR/ECMR III, appears to bind to type I and type IV collagen. Thus, the Hermes/CD44 glycoprotein may function as multipurpose adhesion receptors [85].

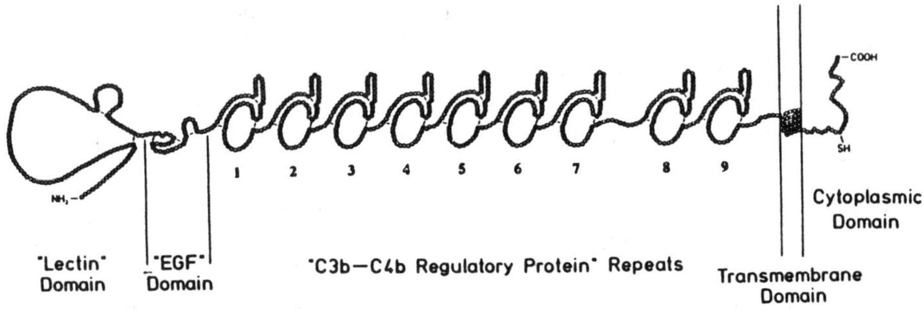

Figure 6 Proposed folding pattern of the domains in GMP-140. The formation of disulfide bridges in each domain was assumed to be the same as the pattern experimentally determined in similar regions of other proteins. The potential free sulfhydryl group in the cytoplasmic domain is indicated by SH. (From Ref. 81.)

These results now force us to reexamine some of the structural features of proteoglycan. In particular, it has been shown that a large cartilage proteoglycan [86] and a large fibroblast proteoglycan [87] contain a link proteinlike domain at the NH_2- and external portion and an EGF-like domain and lectinlike domain at their near COOH-terminal portions. It is tempting to assume that this variety of adhesion molecules, MEL-14, $gp90^{Hermes}$, and large proteoglycans are related to each other and evolved from a primordial binding protein [88,89]. It will be important to know how each domain functions in these different molecules.

C. Possible Roles of Animal Lectins

One of the earliest observations that indicate roles of carbohydrates was made in the clearance of plasma glycoproteins from the blood circulation after the sialic acid residues were removed from the glycoproteins [90]. Ashwell and his colleagues eventually discovered that this clearance is due to binding of asialoglycoproteins to liver through a galactose-binding lectin present in hepatocytes [91]. This discovery has led to a new concept of animal lectins because a number of carbohydrate-binding proteins have been discovered since. The amino acid sequences of animal lectins have been determined mainly by isolating and sequencing cDNAs coding for the lectins [92]. From comparing these sequences, it is now possible to infer the sequence that is probably responsible for its binding to carbohydrates. This data base was essential to define the lectinlike domain in adhesion molecules shown earlier. The function(s) of animal lectins has not yet been clarified in many cases. However, the following examples suggest that this family of molecules may play various roles in addition to a role in hepatic uptake. Stahl and his colleagues [93] provided a line of evidence that macrophages possess a cell surface protein that binds to high mannose-type *N*-linked saccharides of glycoproteins. These glycoproteins are eventually delivered to lysosomes, where they are degraded by lysosomal proteases. Interestingly, there is a homologous protein in the blood stream called serum mannan-binding protein (MBP). This serum MBP apparently binds to complement proteins. The binding of MBP to the erythrocytes fixed with mannan leads in hemolysis of sheep erythrocytes, which is dependent on complement. This complement activation is unique to serum MBP, and a similar MBP, isolated from liver, does not fix complement [94]. There are two possibilities for the mechanisms of complement fixation by serum MPB. One interesting possibility is that serum MBP mimics the role of C1q so that the binding of C1r and C1s is followed. This hypothesis is based on the fact that C1q and serum MBP have remarkably homologous structures. Another possibility is still that C1q is bound to serum MBP attached to sensitized sheep erythrocytes. It will be interesting to determine which is true.

Galactose-binding lectins, in general, bind to galactose terminals of glycoproteins, regardless of how the galactose residues are attached to the backbone of the carbohydrate chains. However, it became apparent that some galactose-binding lectins bind more preferentially to polylactosaminoglycan than to other saccharides [95–97]. This may imply some role for this kind of lectin, since the amount and structure of polylactosaminoglycan varies significantly, depending on the stage of cell differentiation and cell lineage (see Sect. II.A). It will be interesting to see if any correlation exists between the appearance of such lectins and the appearance of a particular polylactosaminoglycan that is a good ligand for the lectins.

IV. CONCLUDING REMARKS

As seen from the foregoing, there is some compelling evidence that carbohydrates play a key function in various biological processes. Although the demonstration of functional importance for carbohydrates has been demonstrated in somewhat isolated cases, those examples encourage us to look for functions of carbohydrates in many biological processes. In this context, it may be important that some of the apparent carbohydrate functions can be explained by their role in modulation of proteins that attach carbohydrates. Since carbohydrate structures are produced secondarily by a gene product—the glycosyltransferases—they are less rigidly controlled than the structures of proteins, which are primary products of genes. This provides some heterogeneity, which may be unfavorable in some instances. However, at the same time, glycosylation provides variety that may be important in modulating protein function. This modulation is one of the functions that is exerted by cell-specific or differentiation stage-specific glycosylation, differences that are widely observed in glycosylation.

To understand the structure and function of carbohydrates to the same extent as we understand those of proteins, two approaches should quickly be incorporated into this field. One approach is to chemically or enzymatically synthesize carbohydrates. Once a unique structure is elucidated, it is critical that we obtain that carbohydrate in large quantity. Having a large quantity of a given carbohydrate enables us to obtain antibodies against the carbohydrate, purify the specific antibodies by immunoadsorption, test the function of the carbohydrates by adding carbohydrates or its protein conjugates, or by adding the antibodies specific to the carbohydrate.

The second important approach is to isolate cDNA or genes encoding glycosyltransferases. This direction is immediately interfaced with the first approach, since production of a specific glycosyltransferase encoded by isolated genes is an extremely powerful means to produce a specific carbohydrate structure. It is likely that synthesis of carbohydrates by stepwise reaction of various glycosyltransferases will eventually replace chemical synthesis of saccharides, since the latter reaction usually lacks its specificity. The most recent success in producing a functional galactosyltransferase in bacteria promises this feasibility [98]. Isolation of the glycosyltransferase gene should broaden our capability beyond this immediate need. By expressing the glycosyltransferase gene in cells that lack that particular glycosyltransferase, we can test if the addition of the specific carbohydrate structure can change the function of cells. It is also possible that overproduction of one particular structure may lead to abnormal development of cells or organs. Furthermore, it is possible to inactivate the glycosyltransferase genes by targeting and interrupting the genes with homologous recombination [99]. These two approaches are possible with transgenic mice technology.

In the past decade, we have significantly progressed in our knowledge of the structures and biosynthetic pathways of carbohydrates attached to glycoproteins. By introducing the vast advancement of molecular biology into our field, we are now ready to regulate glycosylation and test the significance of glycosylation in biology.

ACKNOWLEDGMENT

The authors wish to thank Drs. M. N. Fukuda, H. Freeze, R. Cummings and M. A. Williams for useful discussion and critical reading of the manuscript. The authors also thank Ms Kaarin Soma for secretarial assistance, and Ms Henny Bierhuizen for arranging the references.

Our work and work by Dr. Michiko Fukuda's laboratory presented in this article are supported by R01 CA33000, CA33895, CA48737 (to MF) and DK37016 (to MNF).

REFERENCES

1. von Figura, K., and Hasilik, A. (1986). Lysosomal enzymes and their receptors. *Annu. Rev. Biochem. 55*:167–193.
2. Varki, A., and Kornfeld, S. (1980). Identification of a rat liver α-*N*-acetylglucosaminyl phosphodiesterase capable of removing "blocking" α-*N*-acetylglucosamine residues from phosphorylated high mannose oligosaccharides of lysosomal enzymes. *J. Biol. Chem. 255*:8398–8401.
3. Kornfeld, S. (1987). Trafficking of lysosomal enzymes. *FASEB J. 1*:462–468.
4. Brown, W. J., Goodhouse, J., and Farquhar, M. G. (1986). Mannose 6-phosphate receptors for lysosomal enzyme cycle between the Golgi complex and endosomes. *J. Cell Biol. 103*: 1235–1247.
5. Lang, L., Reitman, M., Tang, J., Roberts, R. M., and Kornfeld, S. (1984). Lysosomal enzyme phosphorylation. Recognition of a protein-dependent determinant allows specific phosphorylation of oligosaccharides present on lysosomal enzymes. *J. Biol. Chem. 259*:14663–14671.
6. Hickman, S., and Neufeld, E. F. (1972). A hypothesis for I-cell disease: Defective hydrolases that do not enter lysosomes. *Biochem. Biophys. Res. Commun. 49*:992–999.
7. Fukuda, M. (1985). Cell surface glycoconjugates as onco-differentiation markers in hematopoietic cells. *Biochem. Biophys. Acta 780*:119–150.
8. Fukuda, M., and Fukuda, M. N. (1984). Cell surface glycoproteins and carbohydrate antigens in development and differentiation of human erythroid cells. In *The Biology of Glycoproteins* (R. J. Ivatt, ed.). Plenum Press, New York, pp. 183–234.
9. Hakomori, S. (1981). Glycosphingolipids in cellular interaction, differentiation, and oncogenesis. *Annu. Rev. Biochem. 50*:733–764.
10. Fukuda, M., Dell, A., and Fukuda, M. N. (1984). Structure of fetal lactosaminoglycan, the carbohydrate moiety of band 3 isolated from umbilical cord erythrocytes. *J. Biol. Chem. 259*:4782–4791.
11. Fukuda, M., Dell, A., Oates, J. E., and Fukuda, M. N. (1984). Structure of branched lactosaminoglycan, the carbohydrate moiety of band 3 isolated from adult human erythrocytes. *J. Biol. Chem. 259*:8260–8273.
12. Cummings, R. D., and Kornfeld, S. (1984). The distribution of repeating [Galβ1,4GlcNAcβ1,3] sequences in asparagine-linked oligosaccharides of the mouse lymphoma cell lines BW5147 and PHA^R2.1. Binding of oligosaccharides containing these sequences to immobilized *Datura stramonium* agglutinin. *J. Biol. Chem. 259*:6253–6260.
13. Fukuda, M., Bothner, B., Ramasamooj, P., Dell, A., Tiller, P. R., Varki, A., and Klock, J. C. (1985). Structures of sialylated fucosyl polylactosaminoglycans isolated from chronic myelogenous leukemia cells. *J. Biol. Chem. 260*:12957–12967.
14. Sasaki, H., Bothner, B., Dell, A., and Fukuda, M. (1987). Carbohydrate structure of erythropoietin expressed in Chinese hamster ovary cells by a human erythropoietin cDNA. *J. Biol. Chem. 262*:12059–12076.
15. van den Eijnden, D. H., Koederman, A. H. L., and Shiphorst, W. E. C. M. (1988). Biosynthesis of blood group i-active polylactosaminoglycans. Partial purification and properties of an UDP-GlcNAc:N-acetyllactosaminide β1→3-*N*-acetylglucosaminlytransferase from Novikoff tumor cell ascites fluid. *J. Biol. Chem. 263*:12461–12471.
16. Joziasse, D. H., Schipherst, W. E. C. M., van den Eijnden, D. H., van Kuik, J. A., van Halbeek, H., and Vliegenthart, J. F. G. (1987). Branch specificity of bovine colostrum CMP-sialic acid:Galβ1→4GlcNac-R α2→6 sialyltransferase. Sialylation of bi-, tri-, and tetraantennary oligosaccharides and glycopeptides of the *N*-acetyllactosamine type. *J. Biol. Chem. 262*:2025–2033.

17. Dennis, J. W., Laferte, S., Waghorne, C., Breitman, M. L., and Kerbel, R. S. (1986). β1-6 Branching of Asn-linked oligosaccharides is directly associated with metastasis. *Science 236*: 582–585.
18. Cummings, R. D., and Kornfeld, S. (1982). Characterization of the structural determinants required for the high affinity interaction of asparagine-linked oligosaccharides with immobilized *Phaseolus vulgaris* leukoagglutinating and erythroagglutinating lectins. *J. Biol. Chem. 257*: 11230–11234.
19. Irimura, T., and Nicolson, G. L. (1983). Interaction of pokeweed mitogen with poly(*N*-acetyllactosamine)-type carbohydrate chains. *Carbohydr. Res. 120*:1887–195.
20. Crowley, J. F., Goldstein, I. J., Arnarp, J., and Lonngren, J. (1984). Carbohydrate binding studies on the lectin from *Datura stramonium* seeds. *Arch. Biochem. Biophys. 231*:524–533.
21. Merkle, R. K., and Cummings, R. D. (1987). Relationship of the terminal sequences to the length of poly-*N*-acetyllactosamine chains in asparagine-linked oligosaccharides from the mouse lymphoma cell line BW5147. Immobilized tomato lectin interacts with high affinity with glycopeptides containing long poly-*N*-acetyllactosamine chains. *J. Biol. C hem. 262*:8179–8189.
22. Heimpel, H., and Wendt, F. (1968). Congenital dyserythropoietic anaemia with karyorrhexis and multinuclearity of erythroblasts. *Helv. Med. Acta 34*:103–115.
23. Fresco, R. (1981). Electron microscopy in the diagnosis of the bone marrow disorders of the erythroid series. *Semin. Hematol. 18*:279–292.
24. Crookston, J. H., Crookston, M. C., Burnie, K. L., Francombe, W. H., Dacie, J. V., Davis, J. A., and Lewis, S. J. (1969). Hereditary erythroblastic multinuclearity associated with a positive acidified-serum test; a typical congenital dyserythropoietic anaemia. *Br. J. Haematol. 17*:11–26.
25. Anselstetter, V., Horstmann, H.-J., and Heimpel, H. (1977). Congenital dyserythropoietic anaemia, types I and II; aberrant pattern of erythrocyte membrane proteins in CDA II, as revealed by two-dimensional polyacrylamide gel electrophoresis. *Br. J. Haematol. 35*:209–215.
26. Harlow, R. W. H., and Lowenthal, R. M. (1982). Erythrocyte membrane proteins in an unusual case of congenital dyserythropoietic anaemia type II (CDA II). *Br. J. Haematol. 50*:35–41.
27. Baines, A. J., Banga, J. P. S., Gratzer, W. B., Linch, D. C., and Huehns, E. R. (1982). Red cell membrane protein anomalies in congenital dyserythropoietic anaemia, type II (HEMPAS). *Br. J. Haematol. 50*:563–574.
28. Mawby, W. J., Tanner, M. J. A., Anstee, D. J., and Clamp, J. R. (1983). Incomplete glycosylation of erythrocyte membrane proteins in congenital dyserythropoietic anaemia type II (CDA II). *Br. J. Haematol. 55*:357–368.
29. Scartezzini, P., Forni, G. L., Baldi, M., Izzo, C., and Sansone, G. (1982). Decreased glycosylation of band 3 and band 4.5 glycoproteins of erythrocyte membrane in congenital dyserythropoietic anaemia type II. *Br. J. Haematol. 51*:569–576.
30. Fukuda, M. N., Papayannopoulou, T., Gordon-Smith, E. C., Rochant, H., and Testa, U. (1984). Defect in glycosylation of erythrocyte membrane proteins in congenital dyserythropoietic anaemia type II (HEMPAS). *Br. J. Haematol. 56*:55–68.
31. Fukuda, M. N., Bothner, B., Scartezzini, P., and Dell, A. (1986). Isolation and characterization of poly-*N*-acetyllactosaminylceramides accumulated in the erythrocytes of congenital dyserythropoietic anemia type II patients. *Chem. Phys. Lipids 42*:185–197.
32. Fukuda, M. N., Dell, A., and Scartezzini, P. (1987). Primary defect of congenital dyserythropoietic anemia type II. Failure in glycosylation of erythrocyte lactosaminoglycan proteins caused by lowered *N*-acetylglucosaminyltransferase II. *J. Biol. Chem. 262*:7195–7206.
33. Fukuda, M. N., Klier, G., Yu, J., and Scartezzini, P. (1986). Anomalous clustering of underglycosylated band 3 in erythrocytes and their precursor cells in congenital dyserythropoietic anemia type II. *Blood 68*:521–529.
34. Fukuda, M. N., Masri, K. A., Dell, A., Thonar, E. J.-M., Klier, G., and Lowenthal, R. M. (1989). Defective glycosylation of erythrocyte membrane glycoconjugates in a variant of congenital dyserythropoietic anemia type II (HEMPAS); association of low level of membrane bound form of galactosyltransferase. *Blood 73*:1331–1339.

35. Masri, K. A., Appert, H. E., and Fukuda, M. N. (1988). Identification of the full-length coding sequence for human galactosyltransferase (β-*N*-acetylglucosaminide:β1→4-galactosyltransferase). *Biochem. Biophys. Res. Commun. 157*:657–663.
36. Shaper, N. L., Hollis, G. F., Douglas, J. G., Kirsch, I. R., and Shaper, J. H. (1988). Characterization of the full length cDNA for murine β1,4-galactosyltransferase. *J. Biol. Chem. 263*:10420–10428.
37. Szczesna-Skorupa, E., Browne, N., Mead, D., and Kemper, B. (1989). Positive charges at the NH_2 terminals convert the membrane-anchor signal peptide of cytochrome P-450 to a secretory signal peptide. *Proc. Natl. Acad. Sci. USA 85*:738–742.
38. Fukuda, M. N., Dell, A., Oates, J. E., and Fukuda, M. (1985). Embryonal lactosaminoglycan. The structure of branched lactosaminoglycans with novel disialosyl (sialylα2→-sialyl) terminals isolated from PA1 human embryonal carcinoma cells. *J. Biol. Chem. 260*:6623–6631.
39. Fukuda, M., Carlsson, S. R., Klock, J. C., and Dell, A. (1986). Structures of *O*-linked oligosaccharides isolated from normal granulocytes, chronic myelogenous leukemia cells, and acute myelogenous leukemia cells. *J. Biol. Chem. 261*:12796–12806.
40. Carlsson, S. R., Roth, J., Piller, F., and Fukuda, M. (1988). Isolation and characterization of human lysosomal membrane glycoproteins, h-lamp-1 and h-lamp-2. *J. Biol. Chem. 263*:18911–18919.
41. Fukuda, M., Viitala, J., Matteson, J., and Carlsson, S. R. (1988). Cloning of cDNAs encoding human lysosomal membrane glycoproteins, h-lamp-1 and h-lamp-2. *J. Biol. Chem. 263*:18920–18928.
42. Barrioccanal, J. G., Bonifacino, J. S., Yuan, L., and Sandoval, I. V. (1986). Biosynthesis, glycosylation, movement through the Golgi system, and transport to lysosomes, an *N*-linked carbohydrate-independent mechanism of three lysosomal integral membrane proteins. *J. Biol. Chem. 261*:16755–16763.
43. Fukuda, M., Guan, J.-L., and Rose, J. K. (1988). A membrane-anchored form but not the secretory form of human chorionic gonadotropin-A chain acquires polylactosaminoglycan. *J. Biol. Chem. 263*:5314–5318.
44. Gorya, F. R., Baldwin, S. A., and Lienhardt, G. E. (1979). The monosaccharide transport from human erythrocytes is heterogenously glycosylated. *Biochem. Biophys. Res. Commun. 91*: 955–961.
45. Williams, M. A., and Lamb, R. A. (1988). Polylactosaminoglycan modification of a small integral membrane glycoprotein, influenza B virus NB. *Mol. Cell. Biol. 8*:1186–1196.
46. Kopito, R. R., and Lodish, H. F. (1985). Primary structure and transmembrane orientation of the murine anion exchange protein. *Nature 316*:234–238.
47. Mueckler, M., Caruso, C., Baldwin, S. A., Panico, M., Blench, I., Morris, H. R., Allard, W. J., Lienhard, G. E., and Lodish, H. F. (1985). Sequence and structure of a human glucose transporter. *Science 229*:941–945.
48. Warren, L., Buck, C. A., and Tuszynski, G. P. (1978). Glycopeptide changes and malignant transformation. A possible role for carbohydrate in malignant behavior. *Biochim. Biophys. Acta 516*:97–127.
49. Ogata, S-I., Muramatsu, T., and Kobata, A. (1976). New structural characteristic of the large glycopeptides from transformed cells. *Nature 259*:580–582.
50. Yamashita, K., Ohkura, T., Tachibana, S., Takasaki, S., and Kobata, A. (1984). Comparative study of the oligosaccharides released from baby hamster kidney cells and their poloma transformant by hydrazinolysis. *J. Biol. Chem. 259*:10834–10840.
51. Irimura, T., Gonzalez, R., and Nicolson, G. L. (1981). Effects of tunicamycin on B16 metastatic melanoma cell surface glycoproteins and blood-borne arrest and survival properties. *Cancer Res. 41*:3411–3418.
52. Humphries, M. J., Matsumoto, K., White, S. L., and Olden, K. (1986). Oligosaccharide modification by swainsonine treatment inhibits pulmonary colonization by B16-F10 murine melanoma cells. Proc. Natl. Acad. Sci. USA 83:1752–1756.
53. Dennis, J. W. (1986). Effects of swainsonine and polyios: Polycytidylic acid on murine tumor cell growth and metastasis. *Cancer Res. 46*:5131–5136.

54. Humphries, M. J., Matsumoto, K., White, S. L., and Olden, K. (1986). Inhibition of experimental metastasis by castanospermine in mice: Blockage of two distinct stages of tumor colonization by oligosaccharide processing inhibitors. *Cancer Res.* *46*:5215–5222.
55. Elbein, A. D. (1987). Inhibitors of the biosynthesis and processing of *N*-linked oligosaccharide chains. *Annu. Rev. Biochem.* *56*:497–534.
56. Chen, J. W., Cha, Y., Yuksel, K. U., Gracy, R. W., and August, J. T. (1988). Isolation an sequencing of a cDNA clone encoding lysosomal membrane glycoprotein mouse LAMP-1. *J. Biol. Chem.* *263*:8754–8758.
57. Oosta, G. M., Farreau, L. V., Bealer, P. C., and Rosenburg, R. D. (1982). Purification and properties of human platelet heparitinase. *J. Biol. Chem.* *257*:11249–11255.
58. Sloane, B. F., Rozhin, J., Johnson, K., Taylor, H., Crissman, J., and Honn, K. V. (1986). Cathepsin B: Association with plasma membrane in metastatic tumors. *Proc. Natl. Acad. Sci. USA* *83*:2483–2487.
59. Niedbala, M. J., Madiyalakan, R., Matta, K., Crickard, K., Sharma, M., and Bernacki, R. J. (1987). Role of glycosidases in human ovarian carcinoma cell mediated degradation of subendothelial extracellular matrix. *Cancer Res.* *47*:4634–4641.
60. Gooi, H. C., Feizi, T., Kapadia, A., Knowles, B. B., Solter, D., and Evans, M. J. (1981). Stage-specific embryonic antigen involved $\alpha 1 \rightarrow 3$ fucosylated type 2 blood group chains. *Nature* *292*:156–158.
61. Bird, J. M., and Kimber, S. J. (1984). Oligosaccharides containing fucose linked α(1-3) and α(1-4) to *N*-acetylglucosamine cause decompaction of mouse morulae. *Dev. Biol.* *104*:449–460.
62. Fenderson, B. A., Zehavi, U., and Hakomori, S. (1984). A multivalent lacto-*N*-fucopentaose III-lysyllsine conjugate decompacts preimplantation mouse embryos, while the free oligosaccharide is ineffective. *J. Exp. Med.* *160*:1591–1596.
63. Bayna, E. V., Shaper, J. H ., and Shur, B. D. (1988). Temporally specific involvement of cell surface β1,4 galactosyltransferase during mouse embryo morula compaction. *Cell* *53*:145–157.
64. Wasserman, P. M. (1987). Early events in mammalian fertilization. *Annu. Rev. Cell Biol.* *3*:109–142.
65. Tsuji, Y., Clauson, H., Nudelman, E., Kaizu, T., Hakomori, S., and Isojima, S. (1988). Human sperm carbohydrate antigens defined by an antisperm human monoclonal antibody derived from infertile woman bearing antisperm antibodies in her serum. *J. Exp. Med.* *108*:343–356.
66. Newman, W. (1982). Selective blockade of human natural killer cells by a monoclonal antibody. *Proc. Natl. Acad. Sci. USA* *79*:3858–3862.
67. Gilbert, C. W., Zaroukian, M. H., and Esselman, W. J. (1988). Poly-*N*-acetyllactosamine structures on murine cell surface T200 glycoprotein participate in natural killer cell binding to YAC-1 targets. *J. Immunol.* *140*:2821–2828.
68. Fukuta, S., Werkmeister, J. A., Burns, G. F., Ginsburg, V., and Magnani, J. L. (1987). Monoclonal antibody Leo Mel 3, which inhibits killing of human melanoma cells by anomalous killer cells, binds to a sugar sequence in G_{D2} and several other gangliosides. *J. Biol. Chem.* *262*:4800–4803.
69. Piller, F., Piller, V., Fox, R. I., and Fukuda, M. (1988). Human T-lymphocyte activation is associated with changes in *O*-glycan biosynthesis. *J. Biol. Chem.* *263*:15146–15150.
70. Parkman, R., Remold-O'Donnell, E., Kenney, D. M., Perrine, S., and Rosen, F. S. (1981). Surface protein abnormalities in the lymphocytes and platelets from patients with Wiskott–Aldrich syndrome. *Lancet* *2*:1387–1389.
71. Remold-O'Donnell, E., Kenney, D. M., Parkman, R., Cairns, L., Savage, B., and Rosen, F. S. (1984). Characterization of a human lymphocyte surface sialoglycoprotein that is defective in Wiskott–Aldrich syndrome. *J. Exp. Med.* *159*:1705–1723.
72. Ishizaka, K. (1984). Regulation of IgE synthesis. *Annu. Rev. Immunol.* *2*:159–182.
73. Iwata, M., Fukutoni, Y., Hashimoto, T., Sato, Y., Sato, H., and Ishizaka, K. (1987). Augmentation of the antibody response by antigen-specific glycosylation-enhancing factor. *J. Immunol.* *138*:2561–2567.

74. Akasaki, M., Jardien, P., and Ishizaka, K. (1986). Immunosuppressive effects of glycosylation inhibiting factor on the IgE and IgG antibody response. *J. Immunol. 136*:3172–3177.
75. Gallatin, W. M., Weissman, I. L., and Butcher, E. C. (1983). A cell-surface molecule involved in organ-specific homing of lymphocytes. *Nature 304*:30–34.
76. Siegelman, M. H., van de Rijn, M., and Weissman, I. L. (1989). Mouse lymph node homing receptor cDNA clone encodes a glycoprotein revealing tandem interaction domains. *Science 243*:1165–1172.
77. Lasky, L. A., Singer, M. S., Yednock, T. A., Dowbonko, D., Fennie, C., Rodriguez, H., Nguyen, T., Stachel, S., and Rosen, S. (1989). Cloning of a lymphocyte receptor reveals a lectin domain. *Cell 56*:1045–1055.
78. Yednack, T. A., Stoolman, L. M., and Rosen, S. D. (1987). Phosphomannosyl-derivatized beads detect a receptor involved in lymphocyte homing. *J. Cell Biol. 104*:713–723.
79. Helzmann, B., McIntyre, B. W., and Weissman, I. L. (1989). Identification of a murine Peyer's patch-specific lymphocyte homing receptor on integrin molecule with an α-chain homologous to human VLA-4α. *Cell 56*:37–46.
80. Bevilacqua, M. P., Stengelin, S., Gimbrone, M. A., Jr., and Seed, B. (1989). Endothelial leukocyte adhesion molecule is an inducible receptor for neutrophils related to complement regulatory proteins and lectins. *Science 243*:1160–1165.
81. Johnston, G. I., Cook, R. G., and McEver, R. P. (1989). Cloning of GMP-140, a granule protein of platelets and endothelium: Sequence similarity to proteins involved in cell adhesion and inflammation. *Cell 56*:1033–1044.
82. Stamenkovic, I., Amiost, M., Pesadono, J. M., and Seed, B. (1989). A lymphocyte molecule implicated in lymph node homing is a number of the cartilage link protein family. *Cell 56*:1057–1069.
83. Goldstein, L. A., Zhou, D. F. H., Picker, L. J., Minty, C. N., Bargatze, R. F., Ding, J. F., and Butcher, E. C. (1989). A human lymphocyte homing receptor, the Hermes antigen, is related to cartilage proteoglycan core and link proteins. *Cell 56*:1063–1072.
84. Jalkanen, S., Bargatze, R. F., de los Toyos, J., and Butcher, E. C. (1987). Lymphocyte recognition of high endothelium: Antibodies to distinct epitopes of an 85–95 kd glycoprotein antigen differentially inhibit lymphocyte binding to lymph node, mucosal, or synovial endothelial cells. *J. Cell Biol. 105*:983–990.
85. Gallatin, W. H., Wagner, E. A., Hoffman, I. A., St. John, T., Butcher, E. C., and Carter, W. G. (1989). *Proc. Natl. Acad. Sci. USA 86*:4654–4658.
86. Doege, K., Sasaki, M., Horigan, E., Hassell, J. R., and Yamada, Y. (1987). Complete primary structure of the rat cartilage proteoglycan core protein deduced from cDNA clones. *J. Biol. Chem. 262*:17757–17767.
88. Ruoslahti, E. (1988). Structure and biology of proteoglycans. *Annu. Rev. Cell Biol. 4*:229–255.
89. Stoolman, L. M. (1989). Adhesion molecules controlling lymphocyte migration. *Cell 56*:907–910.
90. Morell, A. G., Irvine, K. A., Sternlieb, I., Scheinberg, I. H., and Ashwell, G. (1968). Physical and chemical studies on ceruloplasmin: V. Metabolic studies on sialic acid-free ceruloplasmin in vivo. *J. Biol. Chem. 243*:155–159.
91. Neufeld, E. F., and Ashwell, G. (1980). Carbohydrate recognition systems for receptor-mediated pinocytosis. In *The Biochemistry of Glycoproteins and Proteoglycans* (W. J. Lennarz, ed.). Plenum Press, New York, pp. 241–266.
92. Drickamer, K. (1988). Two distinct classes of carbohydrate-recognition domains in animal lectins. *J. Biol. Chem. 263*:9557–9560.
93. Lennartz, M. R., Cole, F. S., Shepard, V. L., Wileman, T. E., and Stahl, P. D. (1987). Isolation and characterization of a mannose-specific endocytosis receptor from human placenta. *J. Biol. Chem. 262*:9942–9944.
94. Ikeda, K., Sannoh, T., Kawasaki, N., Kawasaki, T., and Yamashina, I. (1987). Serum lectin with known structure activates complement through the classical pathway. *J. Biol. Chem. 262*: 7451–7454.

95. Leffler, H., and Barondes, S. H. (1986). Specificity of binding of three soluble rat lung lectins to substituted and unsubstituted mammalian β-galactosides. *J. Biol. Chem. 261*:10119–10126.
96. Merkle, R. K., and Cummings, R. D. (1989). Asparagine-linked oligosaccharides containing poly-*N*-acetyllactosamine chains are preferentially bound by immobilized colt heart agglutinin. *J. Biol. Chem. 263*:1043–1049.
97. Fukuda, M. N., Sasaki, H., Lopez, L., and Fukuda, M. (1989). Survival of recombinant erythropoietin on the circulation: The role of carbohydrates. *Blood 73*:84–89.
98. Aoki, D., Appert, H. E., Johnson, D., Wong, S. S., and Fukuda, M. N. (1990). Analysis of the substrate binding sites of human galactosyltransferase by protein engineering. *EMBO J. 9*:3171–3178.
99. Thomas, K. R., and Capecchi, M. R. (1987). Site-directed mutagenesis by gene targeting in mouse embryo-derived stem cells. *Cell 51*:503–512.
100. Hubbard, S. C., and Ivatt, R. J. (1981). Synthesis and processing of aspargine-linked oligosaccharide. *Annu. Rev. Biochem. 50*:555–583.
101. Kornfeld, S. (1982). Oligosaccharide processing during glycoprotein biosynthesis. In *The Glycoconjugates*, Vol. 3 (M. I. Horowitz, ed.). Academic Press, New York, pp. 3–23.

13

Function of the Carbohydrate Moieties of Secretory Glycoconjugates

Kenneth Olden *Howard University Cancer Center, Howard University Medical School, Washington, D.C. and National Institute of Environmental Health Sciences, Research Triangle Park, North Carolina*

Tet-Kin Yeo and Kiang-Teck Yeo *Howard University Cancer Center, Howard University Medical School, Washington, D.C. and Beth Israel Hospital and Harvard Medical School, Boston, Massachusetts*

The function of the carbohydrate moieties of glycoproteins is presently an area of intense investigation [Gibson et al., 1980; Olden et al., 1982a, 1985; Schwarz and Datema, 1982, 1984; von Figura and Hasilik, 1986; Datema et al., 1987]. Among the more established functions are included (1) cell–cell recognition and adhesion [Olden et al., 1982b; Raz and Lotan, 1987]; (2) modulation of physicochemical properties of proteins, such as solubility and tertiary conformation [Olden et al., 19897]; (3) regulation of proteolytic processing and stabilization against proteolysis [Olden et al., 1982a,b, 1985; Schwarz and Datema, 1982, 1984]; (4) intracellular sorting and pinocytotic uptake of lysosomal hydrolases [von Figura and Hasilik, 1986]; and (5) involvement in receptor-mediated biological reactions, such as activation of adenylate cyclase. Additionally, it has long been appreciated that the glycan moieties of glycoproteins are likely to encode important biological information that can subsequently be decoded by other protein molecules called lectins [Olden et al., 1982a; Sharon, 1984, 1987; Olden and Parent, 1987].

Asparagine- or *N*-linked oligosaccharides are among the common type of glycoproteins that are found in eukaryotic cells as secreted glycoproteins. The oligosaccharide portion of these glycoproteins may be either of the high-mannose, the hybrid-, or the complex-type structure [Kornfeld and Kornfeld, 1985; Montreuil, 1987]. Most of the recent surge of interest in glycoproteins has been to investigate the function of the complex-type oligosaccharide moieties because of their capacity for enormous structural diversity. Also, their synthesis involves an array of energy-dependent, enzymatic reactions that are subject to metabolic inhibition.

The biosynthesis of complex-type *N*-linked oligosaccharides has been well studied (see Chap. 10). The initial transfer of $Glc_3Man_9(GlcNAc)_2$ oligosaccharide from the lipid intermediate to the protein is followed by trimming of three glucose (Glc) and four mannose (Man) residues. After the addition of one *N*-acetylglucosamine (GlcNAc), Golgi mannosidase II removes two terminal mannosyl residues, to yield an oligosaccharide with the structure $GlcNAcMan_3(GlcNAc)_2$. The complex carbohydrate chain is then completed by the addition of various sugar residues, including galactose (Gal), *N*-acetylglucosamine, and sialic acid. The covalent attachment of carbohydrate chains represents a significant modification in the structure of the protein. The biological relevancy of this complex series of cotranslational and posttranslational events is still largely unknown; however, it is unlikely that some oligosaccharides would retain the high mannose-type structure, whereas others would be processed to the complex-type structure unless some selective advantage was conferred to the organism.

The function of the *N*-linked, complex-type oligosaccharides in the intracellular transport and secretion of glycoproteins has been extensively explored since our initial publication more than 10 years ago [Olden and Yamada, 1978]. Such studies have been stimulated by the discovery of inhibitors that block either the transfer of oligosaccharide chains from dolichol diphosphate to the asparagine (Asp) residues of nascent polypeptides, or the processing of the glucosylated intermediate to mature forms following transfer to the protein. Two other frequently used approaches involve isolation of cell mutants with defects in the biosynthesis of various intermediate glycan structures [Trowbridge et al., 1978] and, more recently, the use of site-directed mutagenesis to insert or delete glycosylation sites in cloned proteins [Machamer et al., 1985].

Eylar (1965) and Melchers (1973) were the first to propose that the attachment of carbohydrates to proteins served as a means for labeling the proteins for intracellular recognition and export. That the cellular glycosylation machinery can produce glycoproteins with a variety of different oligosaccharide structures is consistent with this view. However, their proposal was largely based on the general observation that most secreted cellular protein products contain covalently linked carbohydrates, meaning that most of the products processed by the intracellular transport pathway of eukaryotic cells are glycoproteins. Over the past 15 years, studies by Blobel and co-workers [see Blobel, 1975, 1980; Walter and Blobel, 1989], have elucidated the mechanism of cotranslational transport across the membrane of the rough endoplasmic reticulum (RER), a process mediated by specific amino acid sequences (signals) of the protein. However, less is known about the events that regulate (govern) the highly efficient unidirectional transport of secretory products through the membranous organelles that constitute the secretory pathway. The secretory pathway originates in the RER where secretory proteins are synthesized on ribosomes attached to these organelles. From their origin in the RER they move to the Golgi where they can be either concentrated and stored in secretory granules, or directly discharged by a mechanism of exocytosis [Palade, 1975]. As glycoproteins traverse the ER/Golgi organelles to the cell surface, the *N*-linked oligosaccharide moieties are processed (reconstructed or modified) by an array of compartment-specific reactions [Hubbard and Ivatt, 1981], for most of which the significance is currently unknown.

We postulated that the intracellular transport of secretory proteins occurs by both specific (regulated) and nonspecific, passive flow processes [Olden et al., 1982a]. According to our model, the regulated pathway involves organelle-associated receptors that selectively bind and sort the appropriate (glyco)protein substrate by recognizing structural features (signals; e.g., oligosaccharide moieties) of the exported product. The implications of regulated transport is that proteins, synthesized simultaneously in the RER, may be transported to the cell surface at different rates as a function of their relative binding affinities. In contrast, the nonspecific, passive-flow model of intracellular transport proposes that all secretory proteins are exported with similar kinetics. Since transport by the nonspecific pathway does not require active recognition, any protein can enter this more "primitive" pathway by default under the appropriate conditions.

I. VARIABILITY IN INTRACELLULAR TRANSPORT RATES

The most simplistic mechanism of intracellular transport is that all exported products move through the same secretory pathway at the same rate by a nonspecific, bulk flow or conveyor-belt–type mechanism [Palade, 1975]. The finding, using immunoelectron microscopy, that all types of secretory glycoproteins are localized in all intracellular ER and Golgi vesicles [Geuze et al., 1979] is consistent with this view. Furthermore, the compartmentalization of the enzymes required for the processing of the carbohydrate moiety suggests that all secretory glycoproteins take the same path to the cell surface. Even though these results indicate that all secretory proteins follow the same pathway, they need not imply that they do so at the same rate.

To this end, we investigated protein secretion in human hepatoma cell line HepG2. This cell line synthesizes and secretes most, if not all, of the proteins characteristic of human hepatocytes and in comparable relative amounts [Knowles et al., 1980]. Nine proteins were examined: namely, fibronectin, α-fetoprotein, α_1-proteinase inhibitor, transferrin, plasminogen, α_2-macroglobulin, ceruloplasmin, fibrinogen, and albumin [Parent et al., 1985]. To determine the rates of secretion of these proteins, HepG2 cultures were pulse-labeled with [^{35}S]methionine for 10 min at 37°C, washed with phosphate-buffered saline, and chased for varying intervals in medium containing excess (10 mM) unlabeled methionine. Specific proteins were quantitatively isolated from both the medium and cell lysate by immunoprecipitation with monospecific antibodies; we found that the proteins studied were secreted at characteristic rates suggestive of a specific or a receptor-mediated mechanism. For example, albumin, fibronectin, α-fetoprotein, and α_1-protease inhibitors were exported with a retention half-time of 30–40 min; ceruloplasmin, α_2-macroglobulin, and plasminogen with a retention half-time of 75–80 min; whereas transferrin and fibrinogen were secreted with an intracellular retention half-time of 110–120 min. Lodish and co-workers [Strous and Lodish, 1980; Lodish et al., 1983; Lodish and Kong, 1984], using the HepG2 system, also reported that serum proteins were not secreted from the cells at the same rate; albumin, α_1-protease inhibitor, and transferrin were exported at rates similar to those that we obtained, whereas half of newly synthesized complement C3 and α_1-antichymotrypsin were secreted by 75–80 min. The export of newly synthesized rat liver proteins is also asynchronous; for example, albumin is secreted more rapidly than transferrin [Morgan and Peters, 1971; Schreiber et al., 1979; Strous and Lodish, 1980; Ledford and Davis, 1983].

Similar results were obtained by Scheele and Tartakoff (1985), who investigated the intracellular transport of 12 nonglycosylated secretory proteins in the guinea pig exocrine pancreas. Major asynchrony was observed at four levels: (1) exit from the RER, (2) transport through the Golgi complex, (3) entry into storage granules, and (4) discharge from the cell. Trypsinogen, chymotrypsinogen 2, procarboxypeptidase A2, and lipase 2 were exported with rapid kinetics, whereas amylase and procarboxypeptidase B were exported with slower kinetics.

The finding that different secretory proteins are exported at variable rates is not consistent with the simple bulk-flow model for secretion [Palade, 1975; Lodish, 1988]. These results suggest intracellular transport is a selective process, and that proteins with similar kinetics may share structural determinants that are specific for their rate of export. To explain the existence of multiple secretory rates, we have proposed that structural determinants of the protein function as transport signals that are recognized by specific intracellular transport receptors [Olden et al., 1982a, 1985; Parent et al., 1985]. It is likely that the structural determinants responsible for the apparent specificity of intracellular transport are encoded in either the carbohydrate or protein moiety, or possibly in both. Specific receptors in the ER or Golgi could either selectively sort proteins into different secretory pathways, or differentially regulate their rate of transport along a common pathway. Currently, we do not know whether there is a correlation between the structure of the carbohydrate moieties of the glycoproteins and their kinetics of secretion, as their structures have not been determined on glycoproteins derived from the transformed HepG2 cells. Although a direct correlation may not exist, the results to be described in the following discussion strongly suggest that the oligosaccharide moiety of the glycoprotein plays a regulatory role in intracellular transport.

In summary, our findings that there are three discrete classes of secretory proteins in HepG2 cells strongly suggest that intracellular transport occurs by a highly regulated, selective process. The varied transport rates of the different hepatoma glycoproteins also suggest that, in an individual cell, there are either multiple pathways for protein export or one common pathway through which proteins traverse at different rates.

We further conducted experiments to determine precisely which segment of the ER–Golgi pathway contributed to the apparent differences in the rate of externalization among the secretory glycoproteins. It is possible to monitor intracellular movement of asparagine-linked glycoproteins by indirect measurement of different glycan forms, as glycosylation and subsequent processing are compartmentalized activities. For example, the addition of the oligosaccharide chain to the nascent polypeptide occurs cotranslationally in the RER, whereas processing is initiated in the ER and is completed in the *cis*-, *medial*-, or *trans*-compartments of the Golgi. Lodish and co-workers [Strous and Lodish, 1980; Lodish et al., 1983], based on rate of acquisition of resistance to endoglycosidase H, reported that differences of transport of secretory proteins in the hepatoma cells was due to variability in rates of movement from the site of synthesis in the RER to the Golgi; that is, retention in the ER was primarily responsible for the differences in overall rates of secretion. However, the acquisition of endoglycosidase H resistance does not monitor the entry of secretory glycoproteins into the Golgi because oligosaccharides become resistant to endoglycosidase H after being modified by Golgi α-mannosidase II [Tulsiani et al., 1982], an event believed to occur in the *medial*-Golgi compartment [Goldberg and Kornfeld, 1983; Dunphy and Rothman, 1985]. Therefore, secretory proteins may be present in the *cis*- and *medial*-Golgi for some interval of time before acquiring the endoglycosidase H-resistant structure. To circumvent this shortcoming, we employed three independent experimental approaches to determine the rate of transport of intracellular glycoproteins [Yeo et al., 1985]: cellular fractionation, acquisition of endoglycosidase H resistance, and binding to immobilized plant lectins. Specifically, we studied the average residency times of glycoproteins in the RER and Golgi.

With these three approaches, we observed that transferrin, with a cellular retention half-time of 110 min, exhibited a RER retention half-time of 45 min; ceruloplasmin, with a cellular retention half-time of 70 min, had a RER retention half-time of approximately 30 min; and α_1-protease inhibitor, with a cellular retention half-time of 30 min, showed an RER retention half-time of 10 min. Although there is clearly variability in transport rates from the ER to the Golgi, it cannot account for all the variability in overall secretion rates, since, on the average, the three glycoproteins reside in the RER for no more than two-fifths of their total intracellular residency times. The results summarized in Table 1 indicate that variability in rates of export of the liver secretory glycoproteins is due to differences in rates of transport within both the ER and Golgi. For example, the post RER retention half-times

Table 1 Schematic Representation of the Intracellular Transport of Human Liver Secretory Glycoproteins

Rough endoplasmic reticulum	Post-rough endoplasmic reticulum	Extracellular
α_1-Protease inhibitor		
——10 min——➤	——18 min——➤	Cellular $RT^1/2$ of 28 min
Ceruloplasmin		
——30 min——➤	——40 min——➤	Cellular $RT^1/2$ of 70 min
Transferrin		
——45 min——➤	——65 min——➤	Cellular $RT^1/2$ of 110 min

were 18 min, 40 min, and 65 min for α_1-protease inhibitor, ceruloplasmin and transferrin, respectively. Contrary to the suggestion of Lodish and co-workers (1983), all three proteins resided in the post-RER organelles longer than in the RER. Furthermore, transit of glycoproteins out of the RER occurred long before *medial*-Golgi modifications of the glycan occurred, as demonstrated by the time required to acquire endoglycosidase H resistance and wheat germ agglutinin binding. For example, half of transferrin became resistant to endoglycosidase H, a *medial*-Golgi modification, in 65 min; acquired wheat germ agglutinin (WGA)-binding activity, a *medial*-Golgi activity, in 75 min; and *Ricinus cummunis* agglutin I-binding activity, a *trans*-Golgi activity by 90 min, whereas the average time spent in the RER was only 45 min.

Therefore, we conclude that the asynchrony in rates of secretion of specific glycoproteins by human hepatoma cells is due to differences in rates of transport in both the ER and Golgi. The heterogeneity is not solely due to differences in transport rates from the RER to the Golgi, as previously suggested. The variability is probably a reflection of differential regulation by interaction with specific receptors in the ER and Golgi, as previously suggested [Olden et al., 1982a,b].

II. EFFECT OF GLYCOSYLATION ON INTRACELLULAR TRANSPORT

Do glycoproteins display a stringent dependency on glycosylation for intracellular transport and secretion? To best address this question, we and others have studied specific, and well-characterized molecules, such as the serum glycoproteins [Olden et al., 1982a, 1985], viral membrane glycoproteins [Gibson et al., 1980; Schwarz and Datema, 1982; Datema, 1987], muscle acetylcholine receptor [Prives and Olden, 1980], and immunoglobulins [Melchers, 1973; Weitzman et al., 1976; Eagon and Heath, 1977; Hickman et al., 1977].

The availability of specific inhibitors acting at different stages of the glycosylation process has permitted the synthesis of protein without oligosaccharide moieties, or protein with altered carbohydrate structures. Tunicamycin, swainsonine, 1-deoxynojirimycin, and bromoconduritol are the most commonly used agents for this purpose [Yamada and Olden, 1982; Schwarz and Datema, 1982; Olden et al., 1982a, 1985; Humphries and Olden, 1989]. Tunicamycin is a structural analogue of uridine diphosphate

(UDP)-2-acetamide-2-deoxyglucose, and blocks the initial step in the glycosylation of dolichol phosphate, an intermediate in the assembly of core oligosaccharides linked *N*-glycosidically to protein [Kuo and Lampen, 1976; Takatsuki et al., 1975; Tkacz and Lampen, 1975; Struck and Lennarz, 1977]. Swainsonine, an inhibitor of Golgi α-mannosidase II [Colegate et al., 1979; Elbein et al., 1981; Gross et al., 1983; Tulsiani and Touster, 1983], and 1-deoxynojirimycin and bromoconduritol, inhibitors of endoplasmic reticulum glucosidase I and II [Romero et al., 1983; Gross et al., 1983] or glucosidase II [Legler, 1977; Legler and Lotz, 1973], respectively, are the most useful for specifically modifying the structure of *N*-linked oligosaccharides. Important contributions to carbohydrate function have also been made using glycosylation-deficient mutants [Pouyssegur and Pastan, 1976, 1977; Trowbridge et al., 1978; Machamer et al., 1985].

A. Tunicamycin

We investigated the effect of tunicamycin treatment on the biosynthesis and secretion of fibronectin in chicken embryo fibroblasts [Olden and Yamada, 1978]. We chose this drug because the composition and chemical properties of the carbohydrate side chains associated with fibronectin are characteristic of the asparagine-linked type. Under the conditions of our experiment, we were able to block glycosylation almost completely (≥98%), with only a slight inhibition of protein synthesis (<15%). In general, the carbohydrate-depleted protein was secreted with normal kinetics, and there was no evidence of intracellular accumulation. The major effect of tunicamycin treatment (0.05 μg/mL) was to accelerate the rate of degradation of fibronectin by two to threefold [Olden and Yamada, 1978; Bernard et al., 1982]. Similarly, tunicamycin treatment had no apparent effect on the secretion on the carbohydrate-deficient protein species of procollagen [Olden and Yamada, 1978; Duksin and Bornstein, 1977], a glycoprotein containing both *N*- and *O*-linked oligosaccharides. Also, glycosylation was not required for the synthesis nor secretion of acetylcholine receptor [Prives and Olden, 1980]; adrenal corticotropic hormone (ACTH); and β-lipotropin endorphin precursor in the neurointermediate lobe of the African frog *Xenopus laevis* [Loh and Gainer, 1978, 1979]. But, like fibronectin, the carbohydrate-deficient protein precursor was rapidly degraded during intracellular transport; however, the nonglycosylated precursor, which did escape degradation, was processed to a set of typical peptides that were secreted normally. Additionally, glycosylation is apparently not required for the normal secretion of ovalbumin [Struck et al., 1978; Keller and Swank, 1978], nor apoprotein B of the low-density lipoprotein [Struck et al., 1978]. Likewise, the intracellular transport and externalization of most membrane glycoproteins examined were not impaired by inhibition of glycosylation (see Chap. 12). Again, the most frequent anomaly was loss of specificity for posttranslational proteolytic processing [Olden et al., 1985, 1987; Schwarz and Datema, 1982, 1984; Datema, 1987].

In contrast, there are several examples for which inhibition of glycosylation has been reported to inhibit intracellular transport and secretion. For example, treatment with tunicamycin prevented the secretion of invertase and acid phosphatase in yeast, and the carbohydrate-depleted proteins accumulated in unidentified intracellular membrane organelles [Onishi et al., 1979]. Hickman and co-workers (1977, 1978) reported that glycosylation was required for the efficient secretion of IgA, IgE, and for IgM, but not for IgG. The carbohydrate-depleted immunoglobulins accumulated in intracellular membranous organelles, reminiscent of ER/Golgi in mouse plasmacytoma. Perhaps carbohydrate-depletion of these glycoproteins altered their physiochemical properties, such as tertiary conformation or solubility [Olden et al., 1987]. In fact, Tarentino et al. (1974) have reported that the enzymatic removal of carbohydrates from IgM altered the solubility of the protein; however, the most prominent example is that of glycoprotein G of vesicular stomatitis virus (VSV). Gibson, Leavitt, and co-workers [Leavitt et al., 1977; Gibson et al., 1978, 1979] analyzed two strains of VSV (San Juan and Orsay), which have structurally different G proteins, to determine the effect of carbohydrate depletion on viral maturation and physical properties of the G protein. Tunicamycin treatment severely inhibited virus production at 38°C in both strains; however, at 30°C, nearly complete inhibition of virus yield was observed with VSV San Juan strain, whereas only partial inhibition was obtained for the VSV Orsay strain. Carbohydrate-depleted G protein derived from both strains was insoluble at 38°C, leading to aggregation.

It is apparent from the studies cited that glycosylation is required for the intracellular transport and secretion of some glycoproteins, but not others. But, except in a few instances, it is not known whether the effect of glycosylation is direct or indirect. It is generally assumed that inhibition of secretion is due to intracellular aggregation of insoluble, carbohydrate-depleted protein; however, this assumption is generally without experimental verification. Also, the finding that glycosylation is not obligatory for the secretion of all glycoproteins does not mean that the carbohydrate moiety does not play an important regulatory role.

To clear up some of the ambiguity associated with studies conducted by several different investigators in many different systems, we redirected our tunicamycin studies to the human hepatoma system because it provides the opportunity to examine the secretion of many glycoproteins simultaneously. We found that tunicamycin treatment of HepG2 cells markedly inhibited the secretion of α_2-macroglobulin, α_1-protease inhibitor, and ceruloplasmin, but had no effect on the intracellular transport of albumin, nor on the glycoproteins transferrin, fibrinogen, α-fetoprotein, fibronectin, and plasminogen [Bauer et al., 1985]. The most significant inhibitory effect was on the secretion of α_2-macroglobulin, with greater than 90% of the carbohydrate-depleted protein still in the cell in an undegraded form almost 7 hr after synthesis, whereas in untreated cells, 50% of newly synthesized α_2-macroglobulin was secreted by 75–80 min. Similarly, only 40–45% of non-glycosylated α_1-protease inhibitor was secreted by 7 hr, whereas 50% of the glycosylated species is secreted by 30–40 min. Also, half of newly synthesized ceruloplasmin is secreted in 165 min by tunicamycin-treated cells, compared with 75–80 min in untreated cells.

In summary, carbohydrate depletion inhibited the secretion of three serum glycoproteins: α_1-protease inhibitor, ceruloplasmin, and α_2-macroglobulin, but had no effect on the rate of secretion of albumin, α-fetoprotein, fibronectin, plasminogen, and transferrin. The possible relation among the three glycoproteins, the secretion of which was inhibited by tunicamycin treatment, is that all have at least one triantennary oligosaccharide chain of the complex type [Yamashita et al., 1981; Hodges et al., 1979; Morell et al., 1971; Baenziger and Fiete, 1980]. In contrast, the five glycoproteins, the secretion of which was not influenced by carbohydrate-depletion, lack triantennary, complex carbohydrate moieties [Spik et al., 1975; Yoshima et al., 1980; Takasaki et al., 1980; Townsend et al., 1982]. Again, one should not place too much emphasis in these possible correlations because the carbohydrate structural analyses were performed using glycoproteins from normal liver. However, the report of Bartalena and Robbins (1984) that the secretion of thyroxine-binding globulin, a glycoprotein containing triantennary complex oligosaccharides [Zinn et al., 1978], is also inhibited by tunicamycin treatment of HepG2 cells adds credence to this view.

It is currently unclear why glycosylation is crucial for the secretion of certain specific glycoproteins. Since aggregation has been demonstrated to be a major cause of the impairment in intracellular transport of carbohydrate-deficient proteins [Gibson et al., 1980; Olden et al., 1985, 1987]. Yeo et al. (1989) examined the solubility of the three inhibited glycoproteins using the following approaches: (1) Dounce homogenization of cells and incubation of the 100,000 × *g* particulate fraction in 0.1% Triton X-100; (2) subjection of the postnuclear fraction to four cycles of freeze–thaw, followed by centrifugation at 100,000 × *g*; and (3) sizing of the proteins present in the 100,000 × *g* supernatant fractions by Bio-Gel A-1.5 column chromotography. In the first two experiments, both glycosylated and nonglycosylated species of α_2-macroglobulin, ceruloplasmin, and α_1-protease inhibitor were recovered in the soluble fraction, similarly to transferrin, a glycoprotein the transport of which was not affected by tunicamycin treatment. In Bio-Gel chromatography of the 100,000 × *g* supernatant fraction derived from the freeze–thaw procedure, the carbohydrate-depleted protein did not elute earlier than the glycosylated species, as expected if significant aggregation had occurred. In fact, the carbohydrate-depleted proteins often eluted slightly later than their glycosylated counterparts owing to apparent decrease in molecular mass. These experiments strongly support the conclusion that tunicamycin-induced inhibition of the secretion of α_2-macroglobulin, α_1-protease inhibitor, and ceruloplasmin is not due to protein aggregation in the RER.

Yeo et al. (1989) also investigated the site within the ER–Golgi pathway where the carbohydrate-depleted species of α_i-protease inhibitor, α_2-macroglobulin, and ceruloplasmin accumulated. By using Percoll density centrifugation to separate subcellular organelles, it was found that the decrease in the rate of secretion of the three carbohydrate-depleted proteins was due to accumulation in the fraction corresponding to the RER.

Although these studies [Yeo et al., 1989] reveal no gross physical changes in the deglycosylated proteins, the secretion of which is inhibited, they do not exclude the possibility that some conformational alterations could have occurred that account for their impaired secretion. In fact, there are a number of studies showing that posttranslational modifications, such as the covalent attachment of fatty acids and carbohydrate moieties, play important roles in maintenance of the tertiary conformation of glycoproteins [see Olden et al., 1987 for review]. However, studies with RNase may be instructive, as it represents a class of glycoproteins that occur naturally in both nonglycosylated (RNase A) and glycosylated (RNase B) forms. Various physical measurements of bovine pancreatic RNase reveal no differences in the tertiary conformation between RNase A and RNase B with 10% carbohydrate content, suggesting that deglycosylation need not alter the physicochemical properties of the protein in all instances. Accordingly, the effect of carbohydrate-depletion on the intracellular transport of glycoproteins may vary depending on the primary structure of individual proteins. Alternatively, the intracellular transport of some glycoproteins may be more dependent on carbohydrate-mediated mechanisms.

B. Deoxynojirimycin

Deoxynojirimycin is an antibiotic derived from several bacterial and fungal sources, but can also be prepared by chemical synthesis. It inhibits RER glucosidase I and II, thus the removal of glucose from Glc_3-Man_9-$GlcNAc_2$. Parent et al. (1986) used 1-deoxynojirimycin to modify the structure of *N*-linked oligosaccharide moieties of secretory glycoproteins of HepG2 cells in culture. With use of a pulse-chase protocol, we found that deoxynojirimycin (1.25 mM) markedly reduced the rate of secretion of α_1-protease inhibitor, ceruloplasmin, and α_2-macroglobulin, but had no effect on the export of fibronectin, α-fetoprotein, and transferrin, nor on albumin, which lacks carbohydrate. For example, 50% of newly synthesized α_1-protease inhibitor, the glycoprotein most dramatically affected, was secreted in 27 min in control cultures versus 110 min in the drug-treated cultures. The inhibition of secretion was due to impairment in the ER transport mechanism(s); for example, the rate of acquisition of endoglycosidase resistance in the Golgi was greatly retarded, and Percoll gradient cell fractionation analysis revealed that specific carbohydrate-depleted proteins (α_1-protease inhibitor, α_2-macroglobulins, and ceruloplasmin) accumulated in the membrane fraction corresponding to the RER in the deoxynojirimycin-treated cultures. For instance, 50% of the newly synthesized glycosylated species of α_1-protease inhibitor was transferred from the RER to the Golgi–smooth membrane fraction by 10 min, whereas 70 min was required for the transfer of a comparable amount of the carbohydrate-depleted species. Apparently, the removal of the terminal glucose residues from the oligosaccharide moiety of secretory glycoproteins is required for transport of some glycoproteins from the RER to the Golgi. The small amounts of α_1-protease inhibitor, α_2-macroglobulin, and ceruloplasmin secreted by deoxynojirimycin-treated cells possessed fully processed complex-type carbohydrate moieties, suggesting that escape from drug inhibition was absolutely required for secretion. Why deoxynojirimycin treatment inhibits the intracellular transport of some glycoproteins, but not others, is still unclear. It is highly unlikely that accumulation in the RER is due to nonspecific aggregation or abnormal tertiary conformation, since glucosylated intermediates (e.g., $Glc_3Man_9GlcNAc_2$) are natural constituents of this organelle; whereas treatment with tunicamycin led to the generation of unnatural products. One possible explanation for the selective inhibition is that the oligosaccharide moieties of α_1-protease inhibitor, α_2-macroglobulin, and ceruloplasmin form part of the binding site for a receptor that regulates transport of these glycoproteins from the RER. However, our finding that export of the same three glycoproteins is sensitive to both tunicamycin and deoxynojirimycin treatment strongly suggests that the presence, rather than the removal, of a specific oligosaccharide structure on these glycoproteins is required for efficient transport from the RER. Because the oligosaccharide intermediates on glycoproteins transported from the RER are thought to be $Man_{8\text{-}9}GlcNAc_2$, we have proposed that a transport receptor with a binding specificity for this carbohydrate moiety is involved in the rapid transport of α_1-protease inhibitor, α_2-macroglobulin, and ceruloplasmin from the RER.

Our results are consistent with the hypothesis that *N*-linked oligosaccharides on fibronectin, α-fetoprotein, and transferrin are processed by glucosidases resistant to deoxynojirimycin or that there is another enzymatic pathway for processing oligosaccharides associated with these glycoproteins to the complex-type structure. It is highly unlikely that the oligosaccharide moiety is synthesized by a mechanism that does not involve the glycosylated, high mannose intermediate. Also, evidence that there exists an alternate pathway for processing $Glc_3Man_9GlcNAc_2$ is rapidly accumulating. For example, Lubas and Spiro (1987) discovered a novel enzyme in rat liver capable of converting $Glc_1Man_9GlcNAc_2$ to $Man_8GlcNac_2$ in a single step, yielding Glcα1-2Man disaccharide as a by-product. Thus a deficiency in glucosidase II activity may not lead to complete inhibition in the synthesis of complex oligosaccharides. Additionally, Krag (1979, 1985) and Krag and Robbins (1982) have shown that transfer of $Glc_3Man_9GlcNAc_3$ from dolichol to protein is not the only mechanism for *N*-glycosylation; they observed that Chinese hamster ovary cells synthesize low levels of lipid-linked $Glc_3Man_5GlcNAc_2$ in addition to the predominant $Glc_3Man_9GlcNAc_2$ species. The former species can be transferred to the asparagine residue of proteins and processed to complex-type oligosaccharides [Chapman et al., 1979, 1980; Rearick et al., 1981; Datema and Schwarz, 1981; Stoll et al., 1982]. Also, Yamashita et al. (1983) have reported that the *N*-linked oligosaccharide of ovomucoid is formed from the lipid-linked $Glc_3Man_5GlcNAc$ precursor in chicken oviduct, whereas the oligosaccharide moiety of ovalbumin is synthesized from the typical $Glc_3Man_9GlcNAc_3$ precursor. Their results show that the oligosaccharide moieties of different glycoproteins in the same cell can be synthesized by different mechanisms. Finally, Reitman et al. (1982) showed that a murine lymphoma cell line, containing less than 0.3% of the glucosidase II activity found in the parent cell line, still possessed 25% of the wild-type level of *N*-linked complex-type oligosaccharides. These are just some of the examples of possible alternate pathways for synthesis of *N*-linked complex type glycoproteins that might explain why the synthesis and intracellular transport of some glycoproteins is not sensitive to deoxynojirimycin or to other oligosaccharide-processing inhibitors (see later discussion) [for additional references see Tabas and Kornfeld, 1978; Tulsiani et al., 1982; Parodi et al., 1983; Huffaker and Robbins, 1983; Runge et al., 1984; Romero and Herscovics, 1986].

In summary, the deoxynojirimycin results are consistent with the view that *N*-linked oligosaccharide moieties of secretory glycoproteins play an important role in the intracellular transport of specific glycoproteins.

C. Bromoconduritol

We have used this inhibitor to further investigate the possible role of the oligosaccharide structure in the intracellular transport of secretory glycoproteins [Yeo et al., 1989]. Treatment of HepG2 cells with bromoconduritol (1 mM) inhibited the secretion of the same three glycoproteins (α_1-protease inhibitor, α_2-macroglobulin, and ceruloplasmin) inhibited by tunicamycin and deoxynojirimycin. But, unlike the other two drugs, bromoconduritol also inhibited the secretion of transferrin and α-fetoprotein. Again, we observed that glycoproteins containing glucosylated, high mannose-type oligosaccharides were not secreted. Since bromoconduritol is known to prevent the removal of the innermost glucose of *N*-linked oligosaccharides, our results represent the first demonstration that this glucose residue is crucial for efficient intracellular transport of some glycoproteins. As with deoxynojirimycin, the inhibition of the synthesis of the complex oligosaccharide was not complete, even at high concentrations of bromoconduritol. So, either glucosidase II is partially resistant to the drug, or alternative pathway(s) exist for processing the $Glc_1Man_9GlcNAc_2$ precursor. For example, a novel enzyme capable of trimming the innermost glucose residue has been described.

Precisely how bromoconduritol inhibits the secretion of specific glycoproteins is unclear. Again three possibilities exist, namely: (1) $GlcMan_9GlcNAc_2$ moiety may be part of the recognition site recognized by specific membrane receptors, the binding of which prevent export from the RER; (2) the $Man_9GlcNAc_2$ moiety may be part of the recognition site for transport receptors in the RER, the binding of which prevent export from the RER; or (3) the unlikely aggregation of conformationally altered glycoproteins containing $Glc_1Man_9GlcNAc_2$ structure. In any event, it is clear that the removal of all three glucose residues is necessary for normal secretion of several glycoproteins. That bromoconduritol inhibits the secretion of transferrin and α-fetoprotein, whereas deoxynojirimycin does not, is of great interest; it reveals the marked specificity in carbohydrate requirement for intracellular transport.

D. Swainsonine

Swainsonine is an indolizidine alkaloid isolated from various plant sources [Colegate et al., 1979; Molyneux and James, 1982]; chemical synthesis has also been achieved [J. Cha and D. Dime, unpublished]. The drug inhibits Golgi α-mannosidase II [Tulsiani et al., 1982], leading to the formation of hybrid-type oligosaccharides with one branch containing sugars commonly found in complex oligosaccharides (*N*-acetylglucosamine, galactose, sialic acid) and the other mannose-containing branch remains unsubstituted [Gross et al., 1983; Tulsiani and Touster, 1983; Arumugham and Tanzer, 1983]. Unlike complex oligosaccharides, the hybrid species is sensitive to the action of endoglycosidase [Elbein et al., 1981; Gross et al., 1983; Tulsiani and Touster, 1983; Arumugham and Tanzer, 1983].

In contrast with tunicamycin, deoxynojirimycin, and bromoconduritol, swainsonine treatment (1 μg/mL) of HepG2 cells accelerated the secretion of transferrin, ceruloplasmin, α_2-macroglobulin, and α_1-antitrypsin by decreasing lag period by 10–15 min relative to untreated cultures [Yeo et al., 1985a,b]. The enhanced secretion was specific for glycoproteins, as albumin, a nonglycoprotein, was not affected. The conversion of the high mannose precursor to the hybrid form in swainsonine-treated cells occurred more rapidly (by approximately 10 min) than the conversion to the complex form in control cells. Since both the hybrid and complex oligosaccharides are terminally sialylated by sialyltransferase in the *trans*-Golgi, this suggests that swainsonine enhances the rate of glycoprotein transfer from the RER to the *trans*-Golgi compartment.

E. Genetic Defects in the Biosynthesis of Oligosaccharides

The results obtained with inhibitors of various glycosylation reactions are analogous to known enzyme deficiency diseases resulting from impairment in secretion. For example, a genetic alteration at the α_1-antitrypsin locus leads to impairment in oligosaccharide processing and accumulation of enzymatically active protein in the RER [Carrell et al., 1982; Bathurst et al., 1984]. α_1-Antitrypsin (also referred to as α_1-protease inhibitor) is the major protease inhibitor in human plasma. It inhibits several serine proteases, including trypsin, chymotrypsin, elastase, and leukocyte proteases. α_1-Antitrypsin is a glycoprotein that consists of a single polypeptide chain with three *N*-linked complex-type carbohydrate moieties at positions 46, 83, and 247. The carbohydrate chains are typically biantennary, but under some conditions, the ones at positions 46 and 83 may be replaced by triantennary structures.

The fact that a genetic deficiency in this enzyme renders individuals susceptible to lung damage, and possibly death, has stimulated considerable interest in the molecular basis of the deficiency. In individuals homozygous for the *Z* allele of the protease inhibitor locus, the concentration of α_1-antitrypsin in the plasma is decreased to 15% of the normal level, and substantial amounts of the protein accumulates in the RER of the hepatocytes owing to impairment in secretion of the *Z* variant [Hercz et al., 1978; Hercz and Harpaz, 1980; Carrell et al., 1982; Bathurst et al., 1984]. That glycoproteins, other than α_1-antitrypsin, are normally processed to the complex-type structure indicates that the processing enzymes are present and functional. The amino acid substitution site is not near regions of carbohydrate attachment nor of proteolytic cleavage. Also, the native conformation of α_1-antitrypsin, isolated from the RER of the Z variant, must be normal with the active site intact, since the activity is comparable to the wild type [Bathurst et al., 1984]. This suggests that accumulation in the RER does not occur because of gross misfolding of the polypeptide chain.

As with oligosaccharide-processing inhibitors, the mechanism of inhibition of intracellular transport is unclear; however, the carbohydrate moiety accumulates as the glucosylated, high mannose-type precursor. It is also possible that the single amino acid substitution may alter a crucial amino acid sequence involved in intracellular transport. In any event, the α_1-antitrypsin deficiency could be prevented if the secretion block could be overcome. This phenomenon of intracellular accumulation of the mutant *Z* allele α_1-antitrypsin is not unique to liver cells, since expression of variant forms in macrophages or oocytes also lead to accumulation in the RER [Perlmutter et al., 1985; Verbanac and Heath, 1986].

A problem with using glycosylation or processing inhibitors to study secretion of a multiglycosylated protein is that the role of the individual oligosaccharide chains in secretion cannot be examined. This is an important consideration, as there might be site-specific intracellular functions of the various *N*-linked oligosaccharide chains on a given glycoprotein. To circumvent this problem, one can use oligonucleotide-directed mutagenesis. For example, using site-directed mutagenesis of the human chorionic gonadotropin-α chain glycosylation sites, Matzuk and Boime (1988) have shown that removal of a single oligosaccharide chain, at position 59, enhanced degradation of the protein, but had no effect on secretion, whereas loss of the second oligosaccharide chain, at position 78, alters the secretion of the dimer. Specific deletions in the carbohydrate residues of the μ-chain of IgM has also been shown to influence the rate of secretion [Sidman et al., 1981]. For example, mutant clone 482 secreted μ-protein at twice the rate of glycosylated μ under the same conditions.

III. CONCLUDING REMARKS

The studies cited in the foregoing indicate that intracellular transport of secretory glycoproteins occurs by a specific and highly regulated mechanism. Different secretory proteins are secreted at very different rates, and export from the RER does not occur until specific oligosaccharide modifications have occurred. Intracellular transport at different rates suggests that specific structural features of the transported proteins interact with receptors on the luminal surface of the ER–Golgi membranes. However, the precise nature of the putative signals that regulate intracellular transport is still far from understood; possibilities include covalently linked oligosaccharides, specific amino acid sequences, or conformational features of the polypeptide chain itself [reviewed by Pelham, 1989]. Also, it is possible that a single feature may not be operative for all secretory proteins or glycoproteins; in fact, there is no reason to limit our thinking to the existence of a single sorting signal. For example, some glycoproteins may use the oligosaccharide moiety exclusively, others a peptide sequence, and still others might use a combination of both during their intracellular passage. There might also exist primary and secondary transport signals, so that if one fails, the other would take over. In any event, the existence of multiple secretory rates suggests either that there are multiple secretory pathways, or that the secretory proteins traverse a single pathway at different rates. Both mechanisms imply that secretion occurs by a selective process, rather than by a nonspecific passive flow process.

We do not yet know what molecules regulate the differential rates of secretion. But, the data do suggest that this occurs by selective binding or retention to receptors by a mechanism possibly analogous to receptor-mediated endocytosis. Binding may be regulated by change in the ionic environment, as the pH of different segments of the exocytotic pathway varies considerably [Matlin, 1986]. Different proteins would enter the transport vesicles at rates commensurate with their binding affinities. For example, if the binding affinity of the glycoprotein is strong, the intracellular pool would be small, as newly synthesized glycoproteins in the pool would be exported rapidly. Conversely, if the binding affinity were weak, the intracellular pool would be large and would turnover relatively slowly.

The concept of specific transport receptors is an attractive model for the following reasons: (1) It explains the remarkable variability in transport kinetics, and (2) it provides a structural basis for sorting and substantial purification of glycoprotein products observed in living systems. Without such an active process, the efficiency of intracellular transport and protein composition or function of various cellular organelles would be lost.

ACKNOWLEDGMENTS

Our studies were supported by PHS grants R01/GM29804 and R01/CA34918. We thank Ms. Joyce Dempsey for typing the manuscript.

REFERENCES

Arumugham, R. G., and Tanzer, M. L. (1983). Abnormal glycosylation of human cellular fibronectin in the presence of swainsonine. *J. Biol. Chem. 258*:11883–11889.

Baenziger, J. U., and Fiete, D. (1980). Galactose and *N*-acetylgalactosamine specific endocytosis of glycopeptides by isolated rat hepatocytes. *Cell 22*:611–620.

Bartalena, L., and Robbins, J. (1984). Effect of tunicamycin and monensin on secretion of thyroxine-binding globulin by culture human hepatoma (HepG2) cells. *J. Biol. Chem. 259*:13610–13614.

Bathurst, I. C., Travis, J., George, P. M., and Carrell, R. W. (1984). Structural and functional characterization of the abnormal Z α_1-antitrypsin isolated from human liver. *FEBS Lett. 177*: 179–183.

Bauer, H. C., Parent, J. B., and Olden, K. (1985). Role of carbohydrate in glycoprotein secretion by human hepatoma cells. *Biochem. Biophys. Res. Commun. 128*:368–375.

Bernard, B. A., Yamada, K. M., and Olden, K. (1982). Carbohydrates selectively protect a specific domain of fibronectin against proteases. *J. Biol. Chem. 257*:8549–8554.

Blobel, G. (1980). Intracellular protein topogenesis. *Proc. Natl. Acad. Sci. USA 77*:1496–1500.

Blobel, G., and Dobberstein, B. (1975). Transfer of proteins across membranes. I. Presence of proteolytic processed and nonprocessed nascent immunoglobulin chains on membrane bound ribosomes. *J. Cell Biol. 67*:835–851.

Carrell, R. W., Jeppsson, J. O., Laurell, C. B., Brennan, S. O., Owen, M. C., Vaughan, L., and Boswell, D. R. (1982). Structure and variation of human α_1-antitrypsin. *Nature 298*:329–333.

Chapman, A., Fujimoto, K., and Kornfeld, S. (1980). The primary glycosylation defect in class E Thy-1 negative mutant mouse lymphoma cells is an inability to synthesize dolichol-P-mannose. *J. Biol. Chem. 255*:4441–4446.

Chapman, A., Li, E., and Kornfeld, S. (1979). The biosynthesis of the major lipid-linked oligosaccharide of Chinese hamster ovary cells occurs by the ordered addition of mannose residues. *J. Biol. Chem. 254*:10243–10249.

Colegate, S. M., Dorling, P. R., and Huxtable, C. R. (1979). A spectroscopic investigation of swainsonine: An α-mannosidase inhibitor isolated from *swainsona canescens. Aus. J. Chem. 32*:2257–2264.

Datema, R., Olofsson, S., and Romero, P. A. (1987). Inhibitors of protein glycosylation and glycoprotein processing in viral systems. *Pharmacol. Ther. 33*:221–286.

Datema, R., and Schwarz, R. T. (1981). Effect of energy depletion on the glycosylation of a viral glycoprotein. *J. Biol. Chem. 256*:11191–11198.

Duksin, D., and Bornstein, P. (1977). Changes in surface properties of normal and transformed cells caused by tunicamycin, an inhibitor of protein glycosylation. *Proc. Natl. Acad. Sci. USA 74*: 3433–3437.

Dunphy, W. G., and Rothman, J. E. (1985). Compartmental organization of the Golgi stack. *Cell 42*:13–21.

Eagon, P. K., and Heath, E. C. (1977). Glycoprotein synthesis in myeloma cells. Characterization of nonglycosylated immunoglobulin light chain secreted in presence of 2-deoxy-D-glucose. *J. Biol. Chem. 252*:2372–2383.

Elbein, A. D., Solf, R., Dorling, P. R., and Vosbeck, K. (1981). Swainsonine: An inhibitor of glycoprotein processing. *Proc. Natl. Acad. Sci. USA 78*:7393–7397.

Eylar, E. H. (1965). On the biological role of glycoproteins. *J. Theor. Biol. 10*:89–113.

Geuze, J. J., Slot, J. W., and Tokuyasu, K. T. (1979). Immunocytochemical localization of amylase and chymotrypsinogen in the exocrine pancreatic cell with special attention to the Golgi complex. *J. Cell Biol. 82*:697–707.

Gibson, R., Kornfeld, S., and Schlesinger, S. (1980). A role for oligosaccharides in glycoprotein biosynthesis. *Trends Biochem. Sci. 5*:290–293.

Gibson, R., Leavitt, R., Kornfeld, S., and Schlesinger, S. (1978). Synthesis and infectivity of vesicular stomatitis virus containing nonglycosylated G protein. *Cell 13*:671–679.

Gibson, R., Schlesinger, S., and Kornfeld, S. (1979). The nonglycosylated glycoprotein of VSV is temperature-sensitive and undergoes intracellular aggregation at elevated temperatures. *J. Biol. Chem. 254*:3600–3607.

Goldberg, D. E., and Kornfeld, S. (1983). Evidence for extensive subcellular organization of asparagine-linked processing and lysosomal enzyme phosphorylation. *J. Biol. Chem. 258*: 3159–3165.

Gross, V., Tran-Thi, T. A., Vosbeck, K., and Heinrich, P. C. (1983). Effect of swainsonine on the processing of the asparagine-linked carbohydrate chains of α_1-antitrypsin in rat hepatocytes. *J. Biol. Chem. 258*:4032–4036.

Hercz, A., and Harpaz, N. (1980). Characterization of the oligosaccharides of the liver Z variant α_1-antitrypsin. *Can. J. Biochem. 58*:644–648.

Hercz, A., Katona, E., Cutz, E., Wilson, J. R., and Barton, M. (1978). α_1-Antitrypsin: The presence of excess mannose in the Z variant isolated from liver. *Science 201*:1229–1232.

Hickman, S., and Kornfeld, S. (1978). Effect of tunicamycin on IgM, IgA and IgG secretion by mouse plasmacytoma cells. *J. Immunol. 121*:990–996.

Hickman, S., Kulczycki, A., Lynch, R. G., and Kornfeld, S. (1977). Studies of the mechanism of tunicamycin inhibition of IgA and IgE secretion by plasma cells. *J. Biol. Chem. 252*:4402–4408.

Hodges, L. C., Laine, R., and Chan, S. K. (1979). Structure of the oligosaccharide chains in human α_1-protease inhibitor. *J. Biol. Chem. 254*:8202–8212.

Hubbard, S. C., and Ivatt, R. (1981). Synthesis and processing of asparagine-linked oligosaccharides. *Annu. Rev. Biochem. 50*:555–583.

Huffaker, T. C., and Robbins, P. W. (1983). Yeast mutants deficient in protein glycosylation. *Proc. Natl. Acad. Sci. USA 80*:7466–7470.

Humphries, M. J., and Olden, K. (1989). Asparagine-linked oligosaccharides and tumor metastasis. *Pharmacol. Ther. 44*:85–105.

Keller, R. K., and Swank, G. D. (1978). Tunicamycin does not block ovalbumin secretion in the oviduct. *Biochem. Biophys. Res. Commun. 85*:762–768.

Knowles, B. B., Howe, C. D., and Aden, D. P. (1980). Human hepatocellular carcinoma cell lines secrete the major plasma proteins and hepatitis B surface antigen. *Science 209*:497–499.

Kornfeld, R., and Kornfeld, S. (1985). Assembly of asparagine-linked oligosaccharides. *Annu. Rev. Biochem. 54*:631–664.

Krag, S. S. (1985). Mechanisms and functional role of glycosylation in membrane protein synthesis. *Curr. Top. Membr. Transp. 24*:181–249.

Krag, S. S. (1979). A concanavalin A-resistant Chinese hamster ovary cell line is deficient in the synthesis of [^{3}H]glucosyl oligosaccharide-lipid. *J. Biol. Chem. 254*:9167–9177.

Krag, S. S., and Robbins, A. R. (1982). A Chinese hamster ovary cell mutant deficient in glucosylation of lipid-linked oligosaccharide synthesis lysosomal enzymes of altered structure and function. *J. Biol. Chem. 257*:8424–8431.

Kuo, S. C., and Lampen, J. O. (1976). Tunicamycin inhibition of [^{3}H]glucosamine incorporation into yeast glycoproteins: Binding of tunicamycin and interactions with phospholipids. *Arch. Biochem. Biophys. 172*:574–581.

Leavitt, R., Schlesinger, S., and Kornfeld, S. (1977). Impaired intracellular migration and altered solubility of nonglycosylated glycoproteins of vesicular stomatis virus and Sindbis virus. *J. Biol. Chem. 252*:9018–9023.

Ledford, B., and Davis, W. (1983). Kinetics of serum protein secretion by cultured hepatoma cells. *J. Biol. Chem. 258*:3304–3308.

Legler, G. (1977). Glucosidases. *Methods Enzymol. 46*:368–381.

Legler, G., and Lotz, W. (1973). Investigation of the mechanisms of action of glycoside splitting enzyme. VII. Functional groups on the active site of α-glucosidase. *Hoppe-Seyler Z. Physiol. Chem. 354*:243–254.

Lodish, H. (1988). Transport of secretory and membrane glycoproteins from the rough endoplasmic reticulum to the Golgi. *J. Biol. Chem. 263*:2107–2110.

Lodish, H. F., and Kong, N. (1984). Glucose removal from *N*-linked oligosaccharides is required for efficient maturation of certain secretory glycoproteins from the rough endoplasmic reticulum to the Golgi complex. *J. Biol. Chem. 98*:1720–1729.

Lodish, H. F., Kong, N., Snider, M., and Strous, G. J. A. M. (1983). Hepatoma secretory proteins migrate from the rough endoplasmic reticulum to Golgi at characteristic rates. *Nature 304*:80–83.

Loh, Y. P., and Gainer, H. (1978). The role of glycosylation in the biosynthesis, degradation and secretion of the ACTH-β-lipotropin common precursor and its peptide products. *FEBS Lett. 96*:269–272.

Loh, Y. P., and Gainer, H. (1979). The role of carbohydrate in the stabilization, processing, and packaging of the glycosylated adenocorticotropin–endorphin common precursor in toad pituitaries. *Endocrinology 105*:474–487.

Lubas, W. A., and Spiro, R. G. (1987). Golgi-endo-α-D-mannosidase from rat liver, a novel *N*-linked carbohydrate unit processing enzyme. *J. Biol. Chem. 262*:3775–3781.

Machamer, C. E., Florkiewicz, R. Z., and Rose, J. K. (1985). A single *N*-linked oligosaccharide at either of the two normal sites is sufficient for transport of vesicular stomatis virus G protein to the cell surface. *Mol. Cell. Biol. 5*:3074–3083.

Melchers, F. (1973). Biosynthesis, intracellular transport and secretion of immunoglobulins. Effect of 2-deoxy-D-glucose in tumor plasma cells producing and secreting immunoglobulin G_1. *Biochemistry 12*:1471–1476.

Molyneux, R. J., and James, L. F. (1982). Loco intoxication: Indolizidine alkaloids of spotted locoweed. *Science 216*:190–191.

Montreuil, J. (1987). Structure and conformation of glycoprotein glycans. In: *Vertebrate Lectins* (K. Olden and J. B. Parent, eds.). Van Nostrand Reinhold, New York, pp. 1–26.

Morell, A. G., Gregoriadis, G., and Schienberg, I. H. (1971). The role of sialic acid in determining the survival of glycoproteins in the circulation. *J. Biol. Chem. 246*:1461–1467.

Morgan, E. H., and Peters, T. (1971). Intracellular aspects of transferrin synthesis and secretion in the rat. *J. Biol. Chem. 246*:3508–3511.

Olden, K., Bernard, B. A., Humphries, M. J., et al. (1985). Function of glycoprotein glycans. *Trends Biochem. Sci. 10*:78–82.

Olden, K., Bernard, B. A., White, S. L., and Parent, J. B. (1982b). Function of the carbohydrate moieties of glycoproteins. *J. Cell. Biochem. 18*:313–335.

Olden, K., and Parent, J. B. (1987). *Vertebrate Lectins*. Van Nostrand Reinhold, New York.

Olden, K., Parent, J. B., and White, S. L. (1982a). Carbohydrate moieties of glycoproteins. A reevaluation of their function. *Biochim. Biophys. Acta 650*:209–232.

Olden, K., White, S. L., and Newton, S. A. (1987). Glycoproteins: Potential role of glycan moiety in regulation of physicochemical stability and proteolysis. In: *Thermotolerance* (K. J. Henle, ed.). CRC Press, Boca Raton, pp. 119–141.

Olden, K., and Yamada, K. M. (1978). Role of carbohydrates in protein secretion and turnover: Effects of tunicamycin on the major cell surface glycoprotein of chick embryo fibroblasts. *Cell 13*: 461–473.

Onishi, H. R., Tkacz, J. S., and Lampen, J. O. (1979). Glycoprotein nature of yeast alkaline phosphatase. Formation of active enzyme in the presence of tunicamycin. *J. Biol. Chem. 254*:11943–11952.

Palade, G. (1975). Intracellular aspects of protein secretion. *Science 189*:347–358.

Parent, J. B., Bauer, H. C., and Olden, K. (1985). Three secretory rates in human hepatoma cells. *Biochim. Biophys. Acta 846*:44–50.

Parent, J. B., Yeo, T. K., Yeo, K. T., and Olden, K. (1986). Differential effects of 1-deoxynojirimycin on the intracellular transport of secretory glycoproteins of human hepatoma cells in culture. *Mol. Cell. Biochem. 72*:21–33.

Parodi, A. J., Mendelzon, D. H., and Lederkremer, G. Z. (1983). Transient glycosylation of protein-bound $Man_9GlcNAc_2$, $Man_8GlcNAc_2$ and $Man_7GlcNAc_2$ in calf thyroid cells. A possible recognition signal in processing of glycoproteins. *J. Biol. Chem. 258*:8260–8265.

Pelham, H. R. B. (1989). Control of protein exit from endoplasmic reticulum. *Annu. Rev. Cell Biol. 5*:1–23.

Perlmutter, D. H., Kay, R. M., Cole, F. S., Rossing, T. H., van Thiel, D., and Cotten, H. R. (1985). The cellular defect in α_1-protease inhibitor deficiency is expressed in human monocytes and in *Xenopus* oocytes injected with human mRNA. *Proc. Natl. Acad. Sci. USA 82*:6918–6921.

Pouyssegur, J. M., and Pastan, I. (1976). Mutants of Balb/c3T3 fibroblasts defective in adhesiveness to substratum: Evidence for alteration in cell surface proteins. *Proc. Natl. Acad. Sci. USA 73*:544–548.

Pouyssegur, J. M., and Pastan, I. (1977). Mutants of mouse fibroblasts altered in the synthesis of cell surface glycoproteins. *J. Biol. Chem. 252*:1639–1646.

Prives, J. M., and Olden, K. (1980). Carbohydrate requirement for expression and stability of acetylcholine receptor on the surface of embryonic muscle cells in culture. *Proc. Natl. Acad. Sci. USA 77*:5263–5267.

Raz, A., and Lotan, R. (1987). Endogenous galactoside-binding lectins: A new class of functional tumor cell surface molecules related to metastasis. *Cancer Metastasis Rev. 6*:433–452.

Rearick, J. I., Chapman, A., and Kornfeld, S. (1981). Glucose starvation alters lipid-linked oligosaccharide biosynthesis in Chinese hamster ovary cells. *J. Biol. Chem. 256*:6255–6261.

Reitman, M. L., Trowbridge, I. S., and Kornfeld, S. (1982). A lectin-resistant mouse lymphoma cell line is deficient in glucosidase II, a glycoprotein processing enzyme. *J. Biol. Chem. 257*:10257–10263.

Romero, P. A., Datema, R., and Schwarz, R. T. (1983). *N*-Methyl-1-deoxynojirimycin, a novel inhibitor of glycoprotein processing and its effect of fowl plague virus maturation. *Virology 130*:238–242.

Romero, P. A., and Herscovics, A. (1986). Transfer of nonglucosylated oligosaccharide from lipid to protein in a mammalian cell. *J. Biol. Chem. 261*:15936–15940.

Runge, K. W., Huffaker, T. C., and Robbins, P. W. (1984). Two yeast mutations in glycosylation steps of the asparagine glycosylation pathway. *J. Biol. Chem. 259*:412–417.

Scheele, G., and Tarkakoff, A. (1985). Exit of nonglycosylated secretory proteins from the rough endoplasmic reticulum is asynchronous in the exocrine pancreas. *J. Biol. Chem. 360*: 926–931.

Schreiber, G., Dryburgh, H., Millership, A., et al. (1979). The synthesis and secretion of rat transferrin. *J. Biol. Chem. 254*:12013–12019.

Schwarz, R. T., and Datema, R. (1984). Inhibitors of trimming: New tools in glycoprotein research. *Trends Biochem. Sci. 9*:32–34.

Sharon, N. (1984). Surface carbohydrates and surface lectins are recognition determinants in phogocytosis. *Immunol. Today 5*:143–147.

Sharon, N. (1987). Lectins; an overview. In: *Vertebrate Lectins* (K. Olden and J. B. Parent, eds.). Van Nostrand Reinhold, New York, pp. 27–45.

Sidman, C., Potash, M. J., and Kohler, G. (1981). Roles of protein and carbohydrate in glycoprotein processing and secretion. Studies using mutants expressing IgM mu chains. *J. Biol. Chem. 256*:13180–131887.

Spik, G., Bayard, B., Fournet, B., Strecker, G., Bouquelet, S., and Montreuil, J. (1975). Studies on glycoconjugates. LXIV. Complete structure of carbohydrate units of human serotransferrin. *FEBS Lett. 50*:296–298.

Stoll, J., Robbins, A. R., and Krag, S. S. (1982). Mutant Chinese hamster ovary cells with altered mannose 6-phosphate receptor activity is unable to synthesize mannosylphosphoryldolichol. *Proc. Natl. Acad. Sci. USA 79*:2296–2300.

Strous, G. J. A. M., and Lodish, H. F. (1980). Intracellular transport of secretory and membrane proteins in human hepatoma cells injected by vesicular stomatitis virus. *Cell 22*:708–717.

Struck, D. K., Suita, P. B., Lane, M. D., and Lennarz, W. J. (1978). Effect of tunicamycin on the secretion of serum proteins by primary cultures of rat and chick hepatocytes. *J. Biol. Chem. 253*:5332–5337.

Struck, D. K., and Lennarz, W. J. (1977). Evidence for the participation of saccharide–lipids in the synthesis of the oligosaccharide chain of ovalbumin. *J. Biol. Chem. 252*:1007–1013.

Tabas, I., and Kornfeld, S. (1978). The synthesis of complex-type oligosaccharides. III. Identification of an α-D-mannosidase activity involved in a late stage of processing of complex-type oligosaccharides. *J. Biol. Chem. 252*:7779–7786.

Takasaki, S., Yamashita, K., Suzuki, K., and Kobata, A. (1980). Structural studies of the sugar chains of cold-insoluble globulin isolated from human plasma. *J. Biochem. 88*:1587–1594.

Takatsuki, A., Kohno, K., and Tamura, G. (1975). Inhibition of biosynthesis of polyisoprenol sugars in chick embryo microsomes by tunicamycin. *Agric. Biol. Chem. 39*:1089–2091.

Tarentino, A. L., Plummer, T. H., and Maley, F. (1974). The release of intact oligosaccharides from specific glycoproteins by endo-β-*N*-acetylglucosiminidase H. *J. Biol. Chem. 249*:818–826.

Tkacz, J. S., and Lampen, J. O. (1975). Tunicamycin inhibition of polyisoprenyl *N*-actylglucosaminyl pyrophosphate formation of calf liver microsomes. *Biochem. Biophys. Res. Commun. 65*:248–257.

Townsend, R. R., Hilliker, E., Li, Y. T., Laine, R. A., Bell, W. R., and Lee, Y. C. (1982). Carbohydrate structure of human fibrinogen. *J. Biol. Chem. 257*:9704–9710.

Trowbridge, I. S., Hyman, R., and Mazauskas, C. (1978). The synthesis and properties of T25 glycoprotein in Thy-1 negative mutant lymphoma. *Cell 14*:21–32.

Tulsiani, D. R. P., Harris, T. M., and Touster, O. (1982). Swainsonine inhibits the biosynthesis of complex glycoproteins by inhibition of Golgi mannosidase II. *J. Biol. Chem. 257*:7936–7939.

Tulsiani, D. R. P., and Touster, D. (1983). Swainsonine causes the production of hybrid glycoproteins by human skin fibroblasts and rat liver Golgi preparations. *J. Biol. Chem. 258*:7578–7585.

Verbanac, K. M., and Heath, E. C. (1986). Biosynthesis, processing, and secretion of M and Z variant human α_1-antitrypsin. *J. Biol. Chem. 261*:9979–9989.

von Figura, K., and Hasilik, A. (1986). Lysosomal enzymes and their receptors. *Annu. Rev. Biochem. 55*:167–193.

Walter, P., and Blobel, G. (1981). Translocation of protein across the endoplasmic reticulum. II. Signal recognition protein (SRP) mediates the selective binding to microsomal membranes of in vitro assembled polysomes synthesizing secretory proteins. *J. Cell Biol. 91*:551–556.

Weitzman, S., and Scharff, M. D. (1976). Mouse myeloma mutants blocked in the assembly, glycosylation and secretion of immunoglobulin. *J. Mol. Biol. 102*:237–252.

Yamada, K. M., and Olden, K. (1982). Actions of tunicamycin on vertebrate cells. In: *Tunicamycin* (G. Tamura, ed.). Japan Scientific Societies Press, Tokyo, pp. 119–144.

Yamashita, K., Kamerling, J. P., and Kobata, A. (1983). Structural studies of the sugar chains of hen ovomucoid. Evidence indicating that they are formed mainly by the alternate biosynthetic pathway of asparagine-linked sugar chains. *J. Biol. Chem. 258*:3099–3106.

Yamashita, K., Kamerling, J. P., and Kobata, A. (1983). Structural studies of the sugar chains of hen ovomucoid. Evidence indicating that they are formed mainly by the alternate biosynthetic pathway of asparagine-linked sugar chains. *J. Biol. Chem. 258*:3099–3106.

Yeo, K. T., Parent, J. B., Yeo, T. K., and Olden, K. (1985b). Variability in transport rates of secretory glycoproteins through the endoplasmic reticulum and Golgi in human hepatoma cells. *J. Biol. Chem. 260*:7896–7902.

Yeo, K. T., Yeo, T. K., and Olden, K. (1989). Bromoconduritol treatment delays intracellular transport of secretory glycoproteins in human hepatoma cell cultures. *Biochem. Biophys. Res. Commun. 161*:1013–1019.

Yeo, T. K., Yeo, K-T., and Olden, K. (1989). Accumulation of unglycosylated liver secretory glycoproteins in the rough endoplasmic reticulum. *Biochem. Biophys. Res. Commun. 160*:1421–1428.

Yeo, T. K., Yeo, K. T., Parent, J. B., and Olden, K. (1985a). Swainsonine treatment accelerates intracellular transport and secretion of glycoproteins in human hepatoma cells. *J. Biol. Chem. 260*:2565–2569.

Yoshima, H., Mizuochi, T., Ishie, and Kobata, A. (1980). Structure of the asparagine-linked sugar chains of α-fetoprotein purified from human ascites fluid. *Cancer Res. 40*:4276–4281.

Zinn, A. B., Marshall, J. S., and Carlson, D. M. (1978). Carbohydrate structures of thyroxine-binding globulin and their effects on hepatocyte membrane binding. *J. Biol. Chem. 252*:6768–6773.

14

Mechanisms and Control of Glycoconjugate Turnover

Paul H. Weigel *University of Texas Medical Branch, Galveston, Texas*

I. INTRODUCTION

A. Objectives and Nomenclature

The objective of this chapter is to discuss the role of carbohydrate in determining the mechanism of turnover and degradation in mammals of soluble glycoproteins. Length restrictions do not permit discussion of glycolipids or membrane-bound glycoproteins. Previous reviews have dealt extensively with the intracellular enzymes involved in the degradation of glycoconjugates [1,2]. In this chapter I will concentrate instead on the recognition of carbohydrate-containing molecules by cells and the intracellular processing that ultimately leads to their degradation. This discussion is from the perspective of the carbohydrate on the macromolecule and its influence on the fate of that molecule. There could be many receptors throughout the body that recognize and bind unique carbohydrate structures on a glycoprotein or that recognize a structural determinant that is a combination of amino acid sequence and carbohydrate sequence. These latter receptors could easily have escaped detection because the "ligands" used to discover the lectin receptors identified thus far may not have such determinants. In fact, the ligands generally used represent common readily available structures. Therefore, the number of cell surface molecules responsible for the recognition and turnover of different classes of glycoconjugates (Table 1) may be an underestimate and represent only the tip of the iceberg in terms of the future growth of this list.

There is very extensive literature, well reviewed by others, on the turnover and degradation of proteins directed by various features of the protein structure [3–7]. I have considered only situations in which the role of carbohydrate in the clearance and degradation of the molecule is direct. Only receptor systems that are known to recognize the carbohydrate portion of the molecule are discussed. The latter category of molecules has loosely come to be called *vertebrate lectins*. This, in some respects, is an unfortunate

Table 1 Known Mammalian Endocytic/Degradative Carbohydrate Receptors

Sugar specificity	Cell type
Gal/GalNAc	Liver parenchymal cells
Gal/GalNAc	Liver Kupffer and endothelial cells
Gal/GalNAc	Peritoneal macrophages
Man/GlcNAc	Liver Kupffer and endothelial cells
Man/GlcNAc	Macrophages in lung, bone marrow, peritoneum
Fuc	Liver Kupffer cells
Hyaluronan	Liver endothelial cells
CI-Man-6-P	Probably all cells
CD-Man-6-P	Probably all cells
Glucose-AGE	Peritoneal macrophages

parallel to the plant lectins, and the use of this nomenclature leads to a problem that should be addressed. Most of the plant lectins have had no function ascribed to them, and they are almost always soluble proteins. The vertebrate lectins so far fall into two broad categories: One group contains membrane-bound proteins, the other group contains soluble proteins.

Most of the membrane-bound vertebrate lectins have been localized in discrete cell populations, are found in plasma membranes, and have been shown to mediate the uptake and usually the subsequent degradation of an appropriate glycoconjugate molecule. Although this may not necessarily be the real function of these molecules, nonetheless, they are able to mediate this function in vivo and in a variety of in vitro systems. There is a major difference, therefore, in considering these molecules as lectins, compared with the sense that the plant molecules have been described as lectins. The use of the term *lectin* denotes a multivalent, nonimmune carbohydrate-binding protein with no subsequent function or known consequence of binding, whereas most of the membrane-bound vertebrate molecules are, in fact, *receptors*. That is, these proteins are not only able to bind a carbohydrate-containing ligand, but, as a consequence of binding, they are then able to execute some other function, such as endocytosis. As a criterion for their inclusion in this review, vertebrate lectins were considered only if there has been evidence for the cell surface localization of the lectin or the ability of the lectin to mediate endocytosis and degradation of a glycoconjugate bound by that lectin. Both of these findings make it reasonable to consider the "lectin" to be a receptor molecule, the function of which could be relevant in controlling the metabolism of the glycoconjugate.

The other group of vertebrate lectins, however, is composed predominantly of intracellular or extracellular soluble binding proteins, the role of which in the cellular metabolism of glycoconjugates is much less clear. These lectins are described in more detail in Chapter 2.

B. Sources of Confusion

Another unrelated problem in the field has been confusion in the literature concerning the number and nature of various lectinlike molecules in cells that are responsible for the uptake and degradation of glycoconjugates. This confusion arises from two sources.

In many cases, there are multiple lectins with different specificities in the same cell or in different cells in the same tissue. The source of confusion then resides in the complexity of the glycoconjugate being used. The larger, more complex oligosaccharides have many potential sugar recognition sites that can be bound by lectins or receptors of different specificities. Therefore, when one looks at the tissue clearance of particular glycoconjugates, especially if the glycoconjugate can be recognized by more than one lectin, the overall pattern may be very confusing. More than one receptor may recognize the same structure or different determinants on that same structure. This problem can be overcome by purifying the various cell types in the tissue and showing sugar-specific binding, uptake and degradation of the glycoconjugate(s).

A second source of confusion has only become apparent within the last few years and depends on the extent to which two different cell types in the same tissue intercommunicate and exchange macromolecules. The finding of an accumulation of a radiolabeled glycoprotein in cell type B cannot be taken as proof that such cells have the receptor system for that glycoprotein or that cell B was the original site of uptake. Uptake by cell type A and subsequent transfer, with or without first processing of the ligand, to cell type B must be considered. There are several interesting reports of this phenomenon. For example, liver endothelial cells (LECs) bind and internalize the soluble plasma glycoprotein ceruloplasmin. The cells then desialylate the protein and release it into the extracellular medium. The asialoceruloplasmin is then taken up by the parenchymal D-galactose (Gal) receptor system and degraded in hepatocytes [8]. Similarly, retinyl esters in chylomicron remnants are endocytosed by liver parenchymal cells, and retinol is subsequently transferred to stellate cells in the liver, where it is stored [9].

C. The Functional Significance of Macromolecular Carbohydrate

There is no consensus on the role of the carbohydrate in glycoproteins, glycolipids, and other macromolecular glycoconjugates. It seems clear that one cannot ascribe a singular function to, for example, complex *N*-linked oligosaccharides. A picture is emerging, however, that views the carbohydrate as playing several possible specific or nonspecific functional roles. These ideas have been reviewed by Baenziger [10] and West [11] and have been discussed in previous chapters. For example, in some cases, the thermal stability of a protein, usually a membrane-bound molecule, is much greater when it is glycosylated, although the exact structure of the oligosaccharide may not be critical. In other cases, specific oligosaccharide structures do appear to control or influence the subsequent metabolism of the protein. As another example, the presence of a phosphorylated high D-mannose (Man)-type oligosaccharide can determine whether the newly synthesized protein containing this structure will be targeted to lysosomes or be secreted from the cell. An enormous literature exists that implicates carbohydrate determinants as important in cell–cell and cell–matrix interactions during embryogenesis and development [12]. It is well beyond the scope of this chapter to review the evidence for functional roles of carbohydrate in glycoconjugates.

It is reasonable to expect that for the glycoconjugates in which the carbohydrate has an important specific functional role, some mechanism has evolved to regulate the amount or activity of these molecules by recognition of the carbohydrate portion of the molecule. It is presumably this type of function that is shared by the class of membrane-bound lectins described in the next section.

II. RECOGNITION OF SOLUBLE GLYCOCONJUGATES BY MEMBRANE-BOUND CARBOHYDRATE RECEPTORS

Probably there are specific receptors for most plasma and lymphatic proteins that recognize one or more structural features of the soluble protein and control its metabolism or its rate of degradation in the body. The nature and number of these determinants is now unknown, although there are many candidates, and this is an area of active investigation by many laboratories. One structural determinant that has clearly been shown to be important for controlling the turnover of soluble proteins is the nature of the carbohydrate or oligosaccharide on the proteins. The first recognition of this importance was through the pioneering work of Ashwell and Morell and their co-workers who discovered the asialoglycoprotein receptor, also referred to as the Gal receptor [13,14]. This is a class II receptor, according to the classification scheme (Table 2) initially proposed by Kaplan [15,16]. *Class I* receptors function to transmit information to the cell interior after binding the ligand. They are coupled to intracellular signal-transducing systems that generate second messengers within the cell. These do not require divalent cations for ligand binding, and the ligand and the receptor are usually degraded after they are internalized. The *class II* receptors, on the other hand, appear to function solely to internalize the ligand, to remove it from the extracellular fluid. They often require divalent cations, such as Ca^{+2}, to bind the ligand, and the ligand is usually degraded after internalization. In some cases the ligand undergoes *transcytosis* across the cell; that is, it is internalized from one surface of a polar cell, transferred across the cell, and externalized (secreted) into a different extracellular compartment. Class II receptors are reutilized and recycled. Since the discovery of the Gal receptor in the late 1960s, other members of this family of carbohydrate-specific endocytic receptors have been found.

At least eight different membrane carbohydrate receptors have been identified and characterized in several different cell types and different tissues (see Table 1). This field has expanded rapidly. The majority of these receptors have been found in just three cell

Table 2 Classification of Cell Surface Receptors

Class I receptors

Function to transmit information to the cell interior after binding ligand. They are coupled to one or more intracellular signal transduction pathways to generate second-messenger molecules.

Ligand binding is divalent cation-independent.

Both the ligand and receptor are usually degraded after internalization.

Examples: Receptors for insulin, glucagon, and epidermal growth factor

Class II receptors

The function of these receptors appears to be the internalization and removal of the ligand.

Ligand binding is usually divalent cation-dependent; Ca^{+2} is usually effective and presumably the important cation in vivo.

The ligand is either degraded or undergoes transcytosis after internalization.

The receptor is recycled back to the cell surface and is reutilized; it is not degraded after internalization.

Examples: Receptors for LDL, fibrin, and asialoglycoproteins.

Source: After Ref. 15.

types: macrophages, hepatocytes, and LECs. These cell types have numerous specialized functions, among which is the responsibility within the body of removing and degrading certain classes of molecules. For example, migratory macrophages have an active Fc receptor system for the recognition, removal, and degradation of antibody–antigen complexes, obviously an important function [17]. Hepatocytes in liver are the site of removal of excess low-density lipoprotein (LDL) from blood and, therefore, are involved in regulating its concentration, also an important function [18]. By inference then, the presence of many of the carbohydrate receptor systems discussed in this section in these cell types is an argument that these are important clearance systems for the removal of carbohydrate-containing molecules in vivo. The problems of identifying the endogenous in vivo ligand(s) for these receptors and deducing their true function(s) are discussed in a later section.

All of the proteins discussed in the following sections could be considered carbohydrate- binding proteins or lectins, but since they can mediate a known cellular function (endocytosis) after binding the ligand, they can more informatively and appropriately be called *receptors*. Carbohydrate-binding proteins that do not function in a known receptor-like way can be called *lectins*. Most of the carbohydrate-specific receptors that have been studied in enough detail, can be categorized as class II receptors. Additional information on some of these receptors, especially their history and initial characterization, can be found in earlier reviews [13,14].

A. General Structure of the Carbohydrate Receptors

The primary protein sequences of those receptors for which the structures have been determined all show a similar organization of distinct domains (Fig. 1). These domains have been predicted on the basis of the primary amino acid sequence and, in a few cases, verified experimentally [19]. The five proteins shown in Figure 1A are more closely related than the two D-mannose 6-phosphate (Man-6-P) receptors. In the former proteins, there is an amino-terminal cytoplasmic domain, a transmembrane-spanning domain, an extracellular "stalk" domain containing *N*-linked (and perhaps *O*-linked) oligosaccharides, and a COOH-terminal extracellular carbohydrate-binding domain (see Fig. 1A). The stalk domains may mediate subunit association and assembly into the oligomers needed for high-affinity recognition of multivalent oligosaccharide ligands. The Man-6-P receptors have their COOH-termini in the cytoplasm, their NH_2-termini are extracellular, and they have less well-defined external carbohydrate-binding domains (see Fig. 1B).

B. The Hepatocyte Galactosyl/*N*-Acetylgalactosaminyl Receptor

The initial work on the first mammalian carbohydrate-binding protein was biochemical, and this is reflected in the early jargon designating this protein as the "hepatic-binding protein," or even the "Ashwell binding protein" [13,14,20,21]. Although deservedly giving credit, the latter designation is an interesting misnomer. The term *hepatic-binding protein* also conveys less information than what is now known about this protein. As cell biologists began working in this field, appreciation for the function of the molecule at the cellular level increased and the "binding protein" gradually became a "receptor" [22]. It has since also been called the Gal or the Gal/*N*-acetyl-D-galactosamine (GalNAc) receptor. This nomenclature problem is also shared by the other glycoconjugate receptors and arises from the difficulty in not knowing the native ligand(s) for these receptor molecules. I will refer to this receptor as the hepatic Gal receptor.

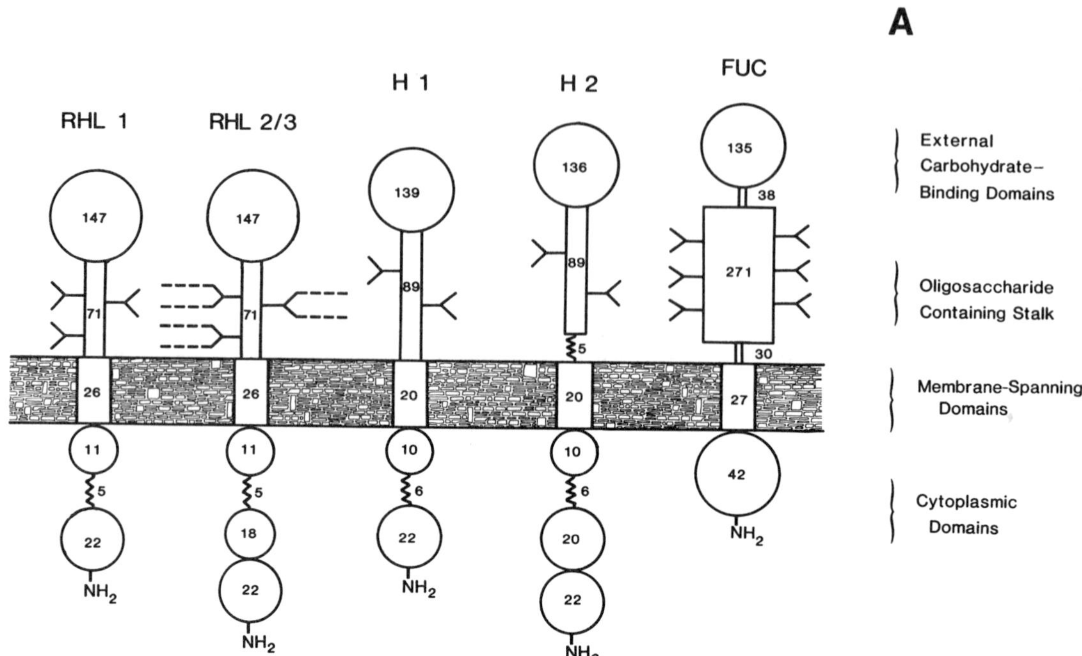

Figure 1 Structure of the carbohydrate receptor subunits. (A) The schematic representation illustrates the four major domains of five proteins for which the sequences have been deduced from cDNA clones or determined directly. Asparagine-linked oligosaccharide chains are indicated by the symbol Y. These proteins include the rat hepatic Gal receptor subunits RHL 1 and 2/3, the human hepatic Gal receptor subunits H1 and H2, and the rat Kupffer cell Fuc receptor. RHL 2 and 3 have identical polypeptide backbones, but differ in their oligosaccharides. RHL 3 has a greater mass of carbohydrate (indicated by the dashed lines), than RHL 2. These five proteins all have their NH_2-termini in the cytoplasm and their COOH-ends are extracellular. The numbers of amino acids in each domain is indicated. (B) The other two proteins shown, the cation-dependent and cation-independent Man-6-P receptors (MPR), have their COOH-termini inside the cell and their NH_2-termini outside the cell. The domain structure of the latter two proteins is also not as obviously related to the organization of the first five proteins, especially in the extracellular domains. Also note that only ~64% of the structure of the CI Man-6-P receptor is shown; the extracellular domains contain 2269 amino acids. In none of the above cases is the subunit composition and stoichiometry of the native receptor known. It is certain that the native receptors are oligomers of the subunits shown and not monomers. Not shown in this figure are the macrophage Man receptor [81,82] and the integral [106] and peripheral [102; also Cherayil et al. (1990). *Proc. Natl. Acad. Sci. USA 87*:7324–7328] membrane macrophage Gal receptors, the cDNAs of which have also been cloned.

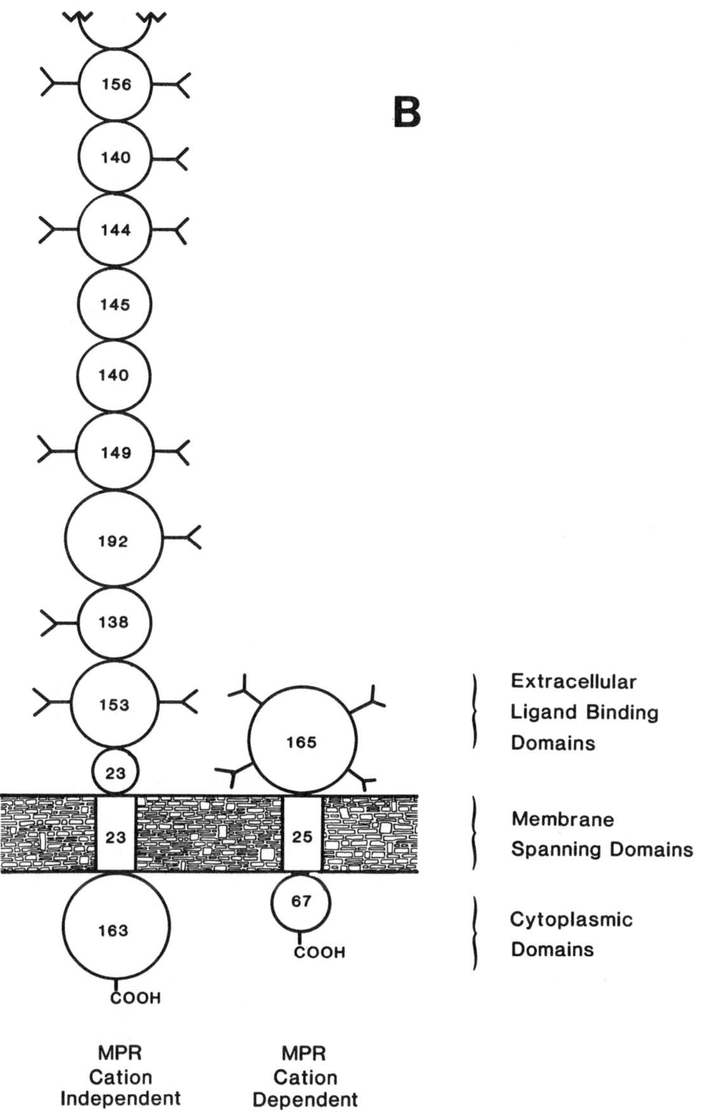

Figure 1 (continued)

1. General Characterization of the Gal Receptor

The interaction of the receptor with most ligands is of high affinity; the dissociation constants (K_d) are in the nanomolar range. Ligand dissociation occurs below pH 6, and Ca^{+2} is required for binding. Approximately 20 plasma glycoproteins, including some hormones, have been shown to be suitable natural ligands (when desialylated) for recognition by the Gal receptor. However, the ligand used most predominantly is asialoorosomucoid (ASOR). This is the desialylated form of orosomucoid or α_1-acid glycoprotein, which is a soluble plasma acute-phase protein. Neoglycoproteins, synthetic polymers, or solid surfaces containing a sufficient density of galactosyl groups will also bind to the receptor. The receptor appears to be present only in hepatocytes. Only 10–25% of the total cellular receptors are on the cell surface, 75–90% are inside the cell, and there are about a half million receptors per cell. This is the only receptor system that has been shown to recycle both in vivo and in vitro. There are two distinct receptor populations and two pathways for ligand dissociation and ligand degradation in hepatocytes [21]. Receptor recycling and intracellular processing are discussed in a later section.

2. *Specificity of the Galactosyl Receptor*

The most extensive work has been on receptors from rat, rabbit, and human. The receptor is specific for terminal clustered galactosyl or *N*-galactosaminyl residues. It requires two or three sugar residues close together to give efficient binding. Lee and co-workers [23] and Ashwell and co-workers [13] have done extensive studies utilizing saccharides of known structure to determine the sugar specificity and to map the sugar binding site(s) of the receptor. More is known about the effect of carbohydrate structure on recogniton by the Gal receptor than any of the other carbohydrate receptors. Potentially, five groups in a galactopyranoside can participate in the sugar–receptor binding interaction [23]. These include the aglycone, C2-OH_{eq} or C2-$NHAc_{eq}$, C3-OH_{eq}, C4-OH_{ax}, and C6-CH_2-groups. The appropriate group and stereoconformation must be present in at least four of these groups to observe significant binding.

Lee and co-workers have extensively studied the effect of oligosaccharide structure on the binding affinity of the Gal/GalNAc-terminal oligosaccharide to the rabbit receptor. The receptor–ligand affinity can vary over many orders of magnitude (K_d from 10^{-6} to 10^{-10} M), depending on the structure. The receptor interaction with a wide range of different oligosaccharide structures can best be described by a model, proposed by Hardy et al. [24], in which three individual monosaccharide-binding sites are arranged in space at the vertices of a triangle with sides of 15, 22, and 25 Å. This proposal represents a major contribution to understanding the molecular design of the carbohydrate receptors in general. These workers also proposed that clusters of receptors on the cell surface can present a lattice of monosaccharide-binding sites that will be able to recognize a great diversity of different oligosaccharide structures containing different numbers of branches, terminal sugars, and different spatial organization of those sugars. This proposal nicely explains the properties of the isolated receptor subunits (discussed later) and the broad specificity and affinity displayed by native receptors on cells for Gal/GalNAc-terminal glycoconjugates of different structure.

3. *Calcium Ion Requirement for Ligand Binding*

Oligosaccharide binding requires a relatively high Ca^{+2} concentration. The apparent K_d is ~1.5 mM and 3.3 Ca^{+2} were bound per polypeptide chain by the rabbit receptor [25]. It is unknown whether the ion is directly involved in the interaction between protein and sugar-binding site, or if an appropriate conformational change is induced by Ca^{+2} binding to a distal site. Very little ligand binding (<1%) occurs in the absence of Ca^{+2} or the presence of chelators, such as EGTA or EDTA. This fact is often used as the basis for most easily determining the nonspecific binding of radiolabeled ligands (i.e., binding that occurs in the presence of chelator is defined as *nonspecific*). In the presence of 2 mM Ca^{+2}, receptor–ligand complexes typically dissociate at 4°C with first-order rate constants of k_{off} ~10^{-5} s^{-1}. If chelator is added, however, dissociation of complexes is 10^2 to 10^3 times faster and is complete within minutes; it is, in fact, difficult to measure accurately. Presumably, the chelator can sequester only free Ca^{+2} and not bind to Ca^{+2} interacting with the receptor. This result can be interpreted to mean that Ca^{+2} bound to receptor–ligand complexes is in rapid dynamic equilibrium with free Ca^{+2}. At high Ca^{+2} concentrations, the rebinding of Ca^{+2} to the complex is sufficiently fast to prevent dissociation of the destabilized complex. In the presence of chelator, this Ca^{+2}-free unstable complex would then rapidly dissociate.

The role of Ca^{+2} in the binding and function of the Gal or other carbohydrate receptors is not known. Not all of the known carbohydrate receptors shown in Table 1 require Ca^{+2}. For example, the LEC hyaluronic acid (HA) receptor does not. Extracellular Ca^{+2} is also not necessary for receptor-mediated endocytosis nor apparently normal functioning of the coated pit pathway.

4. *Polypeptide Composition of the Galactosyl Receptor*

The first mammalian carbohydrate receptor for which the primary structure was determined was the Gal receptor. This receptor, like all of the members of this family for which the structure is now known, is an integral transmembrane glycoprotein. Given the primary amino acid sequence, determined directly on the protein or deduced from the cDNA sequence [19], the rat Gal receptor is composed of polypeptides that contain at least four domains (see Fig. 1A).

The size of the native Gal receptor in the membrane of a live hepatocyte is unknown. In nonionic detergents, the rat receptor is approximately 264 kd, as determined by gel filtration and sedimentation equilibrium analysis [25]. Radiation inactivation studies have determined that the binding domain for ASOR is approximately 148 kd in the presence of detergent and 105 kd in its absence [26]. In receptor preparations purified by ligand-affinity chromatography, one major band and two minor bands are identified on sodium dodecyl sulfate–polyacrylamide gel electrophoresis (SDS–PAGE). These three subunits have been designated rat hepatic lectins (RHL) 1, 2, and 3 [27]. The three separate subunits are, respectively, M_r = 41.5, 49, and 54 kd, and they are the products of two different genes [19,27]. The RHL 2 and RHL 3 proteins differ only in the type and extent of posttranslational carbohydrate modification and show considerable sequence homology with RHL 1 [19,28]. The stoichiometry and subunit composition of the native rat receptor is unknown. As analyzed by SDS–PAGE, the mole ratio for the RHL 1, 2, and 3 subunits in affinity-purified rat receptor preparations varies, respectively, from 2.5:1:1 to 8:1:1 [13,14,19]. The human hepatic Gal receptor also is the product of two genes and consists of two different polypeptides, H1 ($M_r \approx 46{,}000$) and H2 ($M_r \approx 50{,}000$) [29,30]. The smaller polypeptide H1 is severalfold more abundant than H2. Likewise, the rabbit receptor is composed of two polypeptides of 40,000 (R1) and 48,000 (R2) that are also isolated in a molar ratio of, respectively, 2:1 [13,14,20].

Apparently, because of the small size difference, early studies with the human Gal receptor assumed that only one type of subunit was present [20]. Subsequently, the presence of two different subunits was realized. Most recently, the H2 subunit has been shown to be two very closely related forms, designated H2a and H2b [31]. These two isoforms, which arise by alternative mRNA splicing, differ by the insertion in H2a of five additional amino acids immediately adjacent to the transmembrane domain on the exoplasmic side. This insertion in the stalk region may introduce an extra β turn in H2a, which could alter the relative positions of the carbohydrate-binding sites of H2a and H2b in the native receptor. Both mRNAs and proteins are expressed in transfected cells and, more importantly, both transcripts are found in HepG2 cells. Direct evidence for the presence of both proteins in these cells is lacking, but it is likely. This finding would then mean that three distinct subunits are present in the Gal receptor from both rat and human. As discussed later, the possibilities of Gal receptor isoforms or multiple receptor pathways, which has been shown in the rat, may well also be true in humans. Future detailed study of the Gal receptor from other mammalian species (such as rabbit) is needed to determine if this receptor is, in general, composed of three types of subunits.

5. *Subunit Interactions in the Galactosyl Receptor*

Halberg et al. [28] used 1,5-difluoro-2,4-dinitrobenzene to cross-link the rat Gal receptor in detergent solution and crude membranes in the absence of Ca^{+2}. The cross-linking of detergent-solubilized receptors resulted in the formation of separate polypeptide homo-oligomers, culminating with a hexamer. Each subunit is capable of binding Gal. In isolated rat liver microsomes, cross-linking resulted in formation of only dimers and trimers of RHL 1 or RHL 2/3. These workers concluded that RHL 1 and RHL 2/3 are present separately in independent molecular complexes (e.g., $[RHL\ 1]_6$ or $[RHL\ 2/3]_6$). The chicken and the rat receptors both show an interesting requirement for clusters of sugar residues for cell binding to synthetic culture surfaces [32] and for optimal binding and endocytosis of ligand by the receptor [33]. Therefore, some form of multimer arrangement of these hepatic carbohydrate-binding subunits is not unexpected.

From chemical affinity cross-linking studies using *N*-hydroxysuccinimide derivatives of ASOR, Herzig and Weigel [34], concluded that native receptors on intact hepatocytes are heterooligomers. These results do not exclude the possibility that homooligomers of RHL 1 and RHL 2/3 exist. In the absence of ligand, antiserum specific for RHL 1 coprecipitates RHL 2/3 in Triton X-100 extracts of whole cells [35]. The converse was also true. This indicates that heterooligomeric complexes exist. Furthermore, each of these antisera (which were a generous gift from Dr. Drickamer) inhibited identically and almost completely ^{125}I-labeled ASOR binding to hepatocytes [36], a result not predicted by a homooligomeric model for the subunits. Sawyer et al. [37] also concluded that the three RHL subunits can exist as a heteroligomer on the cell surface. They were able to immunoprecipitate both RHL 1 and RHL 2/3, from lactoperoxidase-labeled primary hepatocytes in culture, with antibody directed against unique peptide sequences of either the RHL 1 or the RHL 2/3 subunits. Bischoff et al. [30] reported similar results, supporting a hetero-oligomeric model for the human receptor. By using chemical cross-linking and a cross-immunoprecipitation approach with specific antisera, these investigators found that the H1 and H2 polypeptides associate to form an oligomeric complex in human hepatoma cells.

6. *Proposed Structure of the Galactosyl Receptor*

An understanding of the arrangement of the native rat Gal receptor is complicated by the unknown number of subunits per native receptor, the stoichiometry of the three RHL subunits in a native receptor, and the existence of two functionally distinct receptor populations [21]. We have proposed a model for the structure of the rat Gal receptor (Fig. 2) that is based on our data and the current literature [34]. We suggest that, in the rat, there are discrete types of Gal receptors present in the native state as heterohexamers composed of four RHL 1 subunits and two subunits of either RHL 2 or RHL 3. Each subunit would contribute a binding site for a single sugar residue (see Fig. 2A). Each hexamer would have two potential binding domains for tri- or tetraantennary oligosaccharide ligands. Each ligand-binding domain would be a heterotrimer composed of two RHL 1 subunits and one RHL 2 or RHL 3 subunit. These domains are arranged at the vertices of a right triangle of sides 15, 22, and 25 Å [24]. The RHL 2 and RHL 3

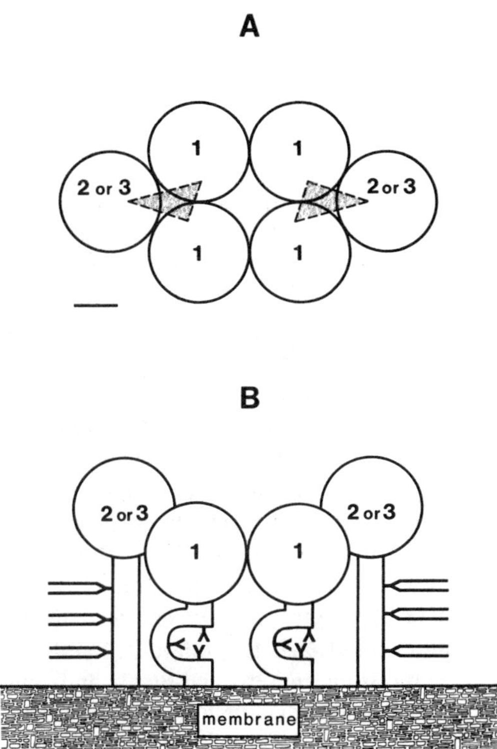

Figure 2 A model for the structure of the parenchymal Gal receptor. The proposed heterohexamer is viewed from above the plasma membrane in (A) and from the side, parallel to the membrane, in (B). The circles represent the COOH-terminal carbohydrate-binding domains shown in Figure 1A. The oligosaccharide chains are the Ys. The RHL 1 (rat), H1 (human), or R1 (rabbit) subunits are indicated by 1, the RHL 2, H2, or R2 subunits are indicated by 2; and the RHL 3 is indicated by 3. The shaded triangles denote oligosaccharide-binding sites. The vertices of the triangles are individual monosaccharide (Gal or GalNAc)-binding sites on each subunit. The bar represents 1.5 nm.

polypeptides have a substantially greater carbohydrate content than RHL 1 and migrate on SDS–PAGE as though they were larger by, respectively, about 8 kd and 14 kd. The RHL 2/3 polypeptide itself is only 18-amino acids (~1.5 kd) larger than RHL 1, which cannot account for the size difference of the mature proteins. Although the nature of the oligosaccharides on the three subunits is unknown, RHL 3 appears to have polylactosaminoglycan chains, since it is recognized by the lectin from *Datura stramonium* [28]. We propose that owing to the extra mass of their carbohydrate chains in the stalk region (see Fig. 1A), RHL 2 and RHL 3 will have their Gal-binding sites farther away from the plasma membrane than those of RHL 1 subunits in the heterohexamer (see Fig. 2B).

This model reconciles many experimental observations. (1) The structurally different rat Gal receptor populations $(RHL\ 1)_4(RHL\ 2)_2$, $(RHL\ 1)_4(RHL\ 3)_2$, and $(RHL\ 1)_4(RHL\ 2)(RHL\ 3)$ may explain the two functionally different pathways for ligand dissociation and degradation [21]. It is interesting that there are two forms of the human H2 subunit [29,31], which could also allow possible Gal receptor isotypes. (2) The reported M_r of the native rat receptor is 264 ± 16 kd [25]. The M_r of the three proposed heterohexamers would be close to this, at about 268, 279, and 288 kd. The predicted M_rs of homohexamers of RHL 1, 2, or 3 would not be as close, at 249, 294, or 324 kd. (3) Each of the three receptor subunits is able to bind Gal [19,28,34,38], and the ASOR-binding domain of the native receptor is 105 kd [26]. Since no single subunit approaches this size, several subunits must participate in binding to the asialoglycoprotein. We proposed that such a binding domain consists of the COOH-terminal domains from three subunits. If the cytoplasmic or transmembrane domains of each subunit can be irradiated without altering ASOR-binding ability, then the size of the remaining trimer ($M_r \approx 110$ kd) is in close agreement with the results of Steer et al. [26]. (4) The two complete trimeric sugar-binding sites can be simultaneously occupied by small glycopeptide ligands, but a large ligand, such as a glycoprotein, bound to one domain would sterically preclude binding of another ligand to the adjacent domain [24]. This would explain the inverse relationship between ligand size and the number of ligand-binding sites per cell. (5) In the presence of some nonionic detergents, there is a doubling of the valency of the affinity-purified receptor, even when immobilized on Sepharose [39]. We proposed that the hexamer can be distorted by the detergent to allow the second binding domain to also bind ligand. (6) The mole ratio of RHL 1, 2, and 3 in affinity-purified receptor preparations will be 4:1:1, a value very close to what most laboratories observe. (7) Expression of fully active Gal receptor (i.e., two complete ligand-binding sites, each built from three subunits) requires the cDNAs for both RHL 1 and RHL 2/3 [40]. (8) Immunoprecipitation of Gal receptor from cell extracts, with antiserum specific for either RHL 1 or RHL 2/3, coprecipitates the other subunit(s). (9) Chemical or photoaffinity labeling of native receptors labels all three subunits, but with different efficiencies, depending on the size of the ligand. (10) This same heterooligomer model also applies to the other mammalian Gal receptor systems (see Fig. 2).

C. The Mannosyl 6-Phosphate Receptor(s)

1. *General*

Those who have tried to follow the development of the Man-6-P receptor story over the previous decade might well liken the experience to a roller coaster ride full of sharp turns and surprises. This has been an exciting and dynamic field. The hunt for an acid hydrolase receptor began with the studies of Neufeld and co-workers in the late 1960s [41,42]. Corrective factors that could restore mucopolysaccharide catabolic function to otherwise defective cells were found to be lysosomal enzymes. Fibroblasts were able to endocytose (the process also referred to in the early literature as pinocytosis) these proteins in a receptor-mediated manner. The initial evidence, in 1974, that the recognition signal for uptake was a carbohydrate determinant was reported by Hickman, Shapiro and Neufeld (see 41, 42 for reviews). A flurry of activity then followed, during which time other laboratories made significant contributions including isolation of "the receptor" [for reviews, see 43–45].

Several hypotheses were advanced to explain the cellular need for the receptor, its location on the cell surface, and the observation that extracellular lysosomal enzymes can be internalized and packaged into functional lysosomes. The initial "secretion–recapture" model proposed by Neufeld and her co-workers [41,42] to explain the normal cellular mechanism for lysosomal enzyme packaging did not stand up to subsequent experimental tests. This hypothesis, however, stimulated a new wave of research. Sly and co-workers later proposed an alternative role for the receptor and postulated that the primary route for delivery of acid hydrolases to lysosomes involved an intracellular pathway [43]. The extracellular pathway involving recapture of secreted enzyme was considered a minor pathway representing "escape" of some lysosomal enzymes or Man-6-P receptor–enzyme complexes from the major intracellular pathway. For a brief period, the role of *the* Man-6-P receptor in cells seemed secure, although the exact itinerary of the receptor and the mechanisms by which lysosomal enzymes were correctly sorted were still obscure.

2. Identification of Two Receptors

The roller coaster veered again when investigators undertook to establish mutant cell lines defective in Man-6-P receptors. Robbins et al. [46] were the first to isolate Chinese hamster ovary (CHO) variant cells with altered receptors by selecting for resistance to Man-6-P–ricin conjugates. These mutants appeared to have reduced capacity to bind and internalize β-galactosidase, although at high concentrations of this ligand some mutants actually bound more than the parental line. Subsequently, Gabel et al. [47] also isolated mutants deficient in the Man-6-P receptor and showed that these cells used an alternative pathway to target lysosomal enzymes. This led the way for Kornfeld and his co-workers to identify, isolate, and characterize a second receptor [48].

The "classic" initial receptor is composed of very large subunits ($M_r \approx 275$ kd). The subunits are transmembrane glycoproteins (see Fig. 1B) and do not require a divalent cation for ligand binding. The native size of this receptor in intact cells is probably a dimer. The second "newer" Man-6-P receptor is composed of smaller transmembrane subunits (M_r = 46 kd) that require divalent cations, such as Ca^{+2}, for binding [44,48]. In the native state, this receptor is probably a dimer or trimer. The two receptors have very similar, although not identical, specificities for manno-oligosaccharides with two phosphomonoesters. The cation-independent receptor binds ligands with a higher affinity than the cation-dependent receptor. A possible third Man-6-P receptor or binding protein was also isolated by Distler and Jourdian [49]. This latter protein is also much smaller than the original receptor (~42 kd), but unlike the 46-kd protein, it is not dependent on divalent cations for ligand binding. This protein has not yet been as extensively studied, and it is not clear where it will fit into the evolving scheme.

Within the last few years, rapid progress has been made in characterizing the primary structure of the subunits of the two major Man-6-P receptors. Kornfeld adapted the nomenclature of designating the cation-dependence or cation-independence of those receptors by using a two-letter abbreviation. Thus, the 275-kd protein is referred to as the CI (cation-independent) Man-6-P receptor, and the 46-kd protein is referred to as the CD (cation-dependent) Man-6-P receptor. The early literature refers to the CI receptor as a 215-kd protein, based on its apparent migration in SDS–PAGE. Only when cDNA clones encoding the larger CI Man-6-P receptor were obtained and sequenced and the receptor protein sequence deduced was it recognized that the protein is expected to be 275 kd [50,51]. The mature protein from human or bovine is predicted to have 2451 [51] or 2455 [50] amino acids, respectively. The reason for the anomolous migration position is presumed to be due to the presence of carbohydrate, but this is not known.

The smaller CD Man-6-P receptor has been purified by Hoflack and Kornfeld [52] and by Stein et al. [53]. The cDNA clones for this receptor have also been isolated, and the protein sequence deduced [48,54]. The two receptor proteins are clearly different gene products, with many structural differences. They are also immunologically distinct. Nonetheless, they do share some sequence homology in the extracellular domains [48]. Both receptors are found in almost all cells examined, although there may be cell types in which one of the two receptors is predominant or absent.

3. Identity of the Mannosyl 6-Phosphate Receptor and the Insulinlike Growth Factor II Receptor

The most startling revelation from the cloning work was the finding that the deduced sequence of the human CI Man-6-P receptor is the same (99.8% identity at the nucleotide level and 99.4% identity to the amino acid sequence) as the human insulinlike growth factor II receptor [50,51,55,56]. The insulinlike growth factors are a family of proteins with sequence and structural homology to insulin. The identity of these two cell surface receptors, previously thought to be different unrelated molecules, has suddenly brought together two very different research fields. It will be very exciting to follow the future consequences of this unexpected merger.

Several interesting findings have already been reported. The type II insulinlike growth factor receptor has been found in serum in rats [57]. If it is true that both receptors are, in fact, the same, then the same protein should bind to each "ligand," and this is the case. The purified insulinlike growth factor II receptor also binds to pentamannosyl-6-phosphate immobilized on Sepharose [58]. Furthermore, the purified CI Man-6-P receptor binds to insulinlike growth factor II [59]. Neither ligand interferes with the binding of the other, suggesting that the binding domains for insulinlike growth factor II and Man-6-P reside in different regions of the protein. In addition, the binding of insulinlike growth factor, but not Man-6-P is coupled to G protein activation [60].

4. Function of the Mannosyl 6-Phosphate Receptor(s)

The discovery of two Man-6-P receptors and the identity of the insulinlike growth factor II and CI Man-6-P receptor means that virtually all of the previous studies to determine the functional role of *the* receptor need to be reexamined and probably reinterpreted. The field is again full of confusion and uncertainty, and many laboratories are working to sort out the respective function(s) of each Man-6-P receptor. The CI Man-6-P receptor appears to be primarily involved in endocytosis [61]. This is somewhat surprising, since most purely endocytic receptors, the class II receptors (see Table 2) that are not involved in signal transduction, are divalent cation-dependent. Many of these class II receptors require Ca^{+2} for binding their ligands and are usually recycling receptor systems involved in ligand clearance and degradation [15,16]. Another exception is the HA receptor of liver endothelial cells (discussed later).

A possible function for the endocytically active CI Man-6-P receptor would be the removal of extracellular lysosomal enzymes released into the circulation as a result of cell injury, disease, or death. Such endogenous ligands could clearly be detrimental to normal cells and to the organism because of their degradative activity, even at neutral pH. It is reasonable to believe that cells could have evolved several mechanisms (pathways) to deal with such an important, yet potentially dangerous, set of enzymes. Consistent with this possibility, Brown [62] has found that the CI Man-6-P receptors are concentrated intracellularly, particularly in the *trans*-Golgi region.

There is, as yet, no explanation for why these two apparently unrelated ligand-binding activities, one for a growth factor and one for an oligosaccharide, reside on the same cell surface receptor. The function(s) of the CI Man-6-P receptor will almost certainly go beyond its role as a carbohydrate receptor. One has the feeling that the roller coaster ride with this molecule is not over, but just coming to the top of another crest.

The vast majority of the CD Man-6-P receptors are located intracellularly, and only 3% are on the cell surface, at least in monocytes [53]. This presumably means that this receptor will be able to bind only glycoproteins containing Man-6-P within intracellular luminal compartments with a high Ca^{+2} concentration. Given what is currently known, this CD Man-6-P receptor is more likely to be involved in intracellular lysosomal enzyme sorting and packaging than the CI Man-6-P receptor [61]. In cells containing only the 46-kd CD Man-6-P receptor, antibodies to this receptor cause loss of intracellular lysosomal enzymes. In cells with both receptors, the same antibody does not induce this loss or secretion [61]. These latter investigators concluded that the CD receptor mediates transport of endogenous newly synthesized lysosomal enzymes and that the CI receptor is responsible for internalization of exogenous molecules containing Man-6-P.

D. The Liver Nonparenchymal Cell Mannosyl Receptor

1. *General*

Many investigators have shown that mannose-containing glycoproteins introduced into the blood are rapidly removed by the reticuloendothelial cells of the liver [63]. Actually, two major pathways for removal of mannose, fucose (Fuc), and *N*-acetyl-D-glycosamine (GlcNAc) glycoconjugates were discovered [64]. One pathway involves the liver and the other pathway is present in spleen, bone, and other elements of the reticuloendothelial system. I will discuss the two pathways separately, although they are clearly related and probably use the same receptor.

In 1978, the laboratories of Sly and of Stahl independently showed that hepatic reticuloendothelial cells have a Man/GlcNAc-specific binding and endocytic activity. This receptor activity was capable of removing glycoproteins from blood and mediating their degradation. Both Kupffer cells and LECs have this mannosyl receptor, based on several lines of evidence. Autoradiographic studies of captured glycoproteins in liver showed the receptor to be in endothelial and Kupffer cells, but not in hepatocytes [65]. After perfusion of intact liver with a radiolabeled glycoprotein, such as invertase (which contains about 50% mannose), the protein can be recovered in all the major cell types after subsequent dissociation and purification of cells [66]. For invertase, for example, the latter study found that 38%, 10%, and 52% of the recovered radiolabel was in parenchymal, Kupffer, and endothelial cells, respectively. It was virtually absent in stellate fat-storing cells. These workers showed that isolated hepatocytes expressed only about 10^3 binding sites per cell, which makes it unlikely that these cells have a large endocytic capacity in vivo. Rather, the 2% contamination of isolated hepatocytes with endothelial or Kupffer cells probably accounts for the invertase binding observed (at roughly 10^5 sites per cell on the nonparenchymal cells).

The result of the liver perfusion experiment in the foregoing study is very important because it emphasizes a point made earlier in the introduction. The finding of accumulation of a radiolabeled glycoprotein in cell type B cannot be taken as proof that such cells have the receptor system for that glycoprotein or that that cell was the original site of uptake. Uptake by cell type A and subsequent transfer, with or without first processing of the ligand, to cell type B must be considered. This appears to be likely in the foregoing case. Invertase taken up by endothelial or Kupffer cells may, in part, be transferred to hepatocytes. The latter cells, being larger, may have a greater degradative capability and share the burden of metabolizing and turning over these molecules. Recognition of this problem of intercellular catabolic communication is only recent, and it will have to be taken into account in future studies and in reinterpreting past studies.

2. Specificity of the Nonparenchymal Liver Cell Mannosyl Receptor

Maynard and Baenziger [67] concluded that the minimal structure required for recognition and endocytosis was a typical core $Man_3GlcNAc_2$-Asn structure [Manα1,6(Manα1,3) Manβ1,4GlcNAcβ1,4GlcNAcβAsn] substituted at the 1,6Man branch with either Manα1,6 or GlcNAcβ1,6. The latter residue of mannose or *N*-acetylglucosamine appears to be critical, since the core structure alone is not sufficient for uptake to occur. Binding of ligand is Ca^{+2}-dependent, and it is inhibited or reversed at low pH. Receptor recognition and uptake was not blocked by the presence of additional Man residues, such as found on high mannose-type oligosaccharides or the presence of additional *N*-acetylglucosaminyl residues, such as on tetraantennary complex oligosaccharides. However, addition of galactosyl residues to the latter tetraantennary structure blocked uptake [67]. This situation may represent a nice "conservation of cellular effort," since such galactosyl-terminal oligosaccharides are very efficiently recognized and removed by the hepatic Gal receptor anyway.

3. Isolation of the Mannosyl Receptor

The receptor has not yet been purified from liver. An apparently related protein was purified from rat cerebellum by Zanetta et al. [68]. This molecule termed "lectin R1" has specificity for mannose-rich glycoproteins and is localized in brain in neurons and, interestingly, in endothelial cells of the central nervous system. It is absent in glial cells. Antiserum to this purified brain lectin cross-reacted with a protein(s) in rat liver, which was localized in both endothelial and Kupffer cells. It is likely that the liver nonparenchymal Man receptor is essentially identical with the macrophage Man receptor, discussed in Section II.E, which has been purified.

4. Function of the Mannosyl Receptor

Because of its abundance, its localization in the reticuloendothelial cells of liver and the demonstration of cell surface binding and uptake of injected glycoproteins in vivo and in isolated cells, this receptor almost surely plays a role in the turnover and degradation of mannosyl glycoproteins. Isolated Kupffer or endothelial cells have about $0.5–1.5 \times 10^5$ surface receptors per cell and can internalize $13–80 \times 10^3$ molecules per cell per minute [67]. Although it has not been examined in these liver cells, it is safe to predict that there is a large intracellular pool of receptors relative to the surface number, and that this is a recycling receptor system. These features are typical characteristics of high-capacity endocytic class II recycling receptor systems.

The contribution of endothelial cells to the liver uptake in vivo of glycoproteins containing terminal *N*-acetylglucosamine was estimated to be three to seven times greater than that of the Kupffer cells [69]. In support of this, Magnusson and Berg [70] have shown, using highly purified rat LECs prepared by centrifugal elutriation, that the endocytic capacity of the Man receptor system on these cells is very great. This receptor has the highest endocytic rate constant (K_e = 4.12) of any yet measured [71]. This K_e value corresponds to a half-life of only 10 sec for the surface pool of receptor–ligand complexes; extremely rapid endocytosis indeed!

As with many of the carbohydrate receptor systems, the identification of function is a haunting problem. The native ligand(s) for this receptor are unknown. It is certain, however, that if a soluble or particulate glycoconjugate with the typical *N*-linked oligosaccharide core structure appears in the blood as a result of injury, infection, or normal glycoprotein processing, then its removal and degradation in the liver will be mediated by the nonparenchymal reticuloendothelial Man receptor.

E. The Macrophage Mannosyl/*N*-Acetylglucosaminyl Receptor

1. *General*

Numerous studies using cultured blood monocytes and macrophages from bone marrow, peritoneum, or lung have confirmed the presence of a second reticuloendothelial pathway outside of liver for mannosyl glycoconjugates [72]. Approximately 20% of the total cellular Man receptor content is on the cell surface and 80% is intracellular. There are about 3.6×10^5 receptors per cell. Alveolar macrophages, for example, can internalize Man-bovine serum albumin (BSA) or β-glucuronidase at a rate of 550 molecules per cell each second. Such a large endocytic capacity is typical for receptor systems, such as LEC hyaluronan (or hyaluronic acid; HA) receptor and the parenchymal Gal receptor (700 molecules per cell each second; Table 3), that can reutilize and recycle their receptors. The macrophage Man receptor is clearly a recycling receptor and has a cycle time of about 11 min [72,73]. Ligand binding is also Ca^{+2}-dependent and is disrupted at low pH (below pH 6), two characteristics of high-endocytic capacity class II receptor-recycling systems. The receptor has specificity for carbohydrate determinants containing Man, Fuc, or GlcNAc, but not Gal. Recognition of fucose ligands by the receptor had been observed earlier [74].

2. *Affinity and Specificity of the Mannosyl Receptor*

The specificity of the macrophage Man receptor appears to be identical with the liver nonparenchymal Man receptor discussed earlier. Most studies have been with alveolar macrophages from rabbits. Ohsumi et al. [75] used a series of related neoglycoprotein derivatives of β-casein containing asparagine (Asn) oligosaccharides (Man_5-$GlcNAc_2$-Asn) to show

Table 3 Endocytic/Degradative Capacity[a] of the Class II Carbohydrate-Specific Receptors

Receptor	Degradative rate (molecules/cell/hr)	Total receptor number/cell	Normalized rate (molecules/cell/hr/receptor)	Ref.
Galactosyl (liver parenchymal)	2,520,000	5.0×10^5	5.0	21
Mannosyl (macrophage)	1,980,000	3.6×10^5	5.5	72
Hyaluronan (liver endothelial)	900,00	1.8×10^5	5.0	129

[a]The maximum steady-state rates of ligand degradation in the Gal and Man receptor systems and endocytosis in the HA receptor system were normalized to the total cellular receptor activity (surface plus intracellular). The very close agreement of these normalized degradative rates suggests that the three receptor systems reflect the same fundamental cellular process. At steady state (see Fig. 3), for every ligand molecule degraded, there is a net cellular uptake of one ligand. Therefore, each receptor mediates the net uptake and degradation of one ligand every 12 min.

that the affinity of binding to the receptor increased over fourfold (to K_d = 253 nM) if three, rather than one, chains were present. The number of binding sites per cell did not change. A streptavidin–biotin conjugate with the same number of oligosaccharide chains had almost a tenfold higher affinity. This result confirmed an earlier suggestion that the orientation of the oligosaccharide extended from the protein matrix can affect the receptor–ligand interaction [76]. A series of Man-BSA neoglycoproteins bound with increasing affinity (41-fold) as the average number of Man residues per BSA molecule went from 5 to 43; K_d = 820 to 2 nM [75]. Here, the number of ligand molecules bound per cell also decreased from 130,000 to 42,700. This type of effect on affinity as the density of sugars is increased is typical for the carbohydrate receptors.

3. *Structure of the Mannosyl Receptor*

The Man receptor subunit was identified as a 175-kd protein by its selective iodination on alveolar macrophages, using a Man-lactoperoxidase neoglycoprotein [77]. The receptor has subsequently been purified from rabbit [78], human [79], and rat [80] alveolar macrophages. In each case, a single polypeptide of M_r 175–180 kd was obtained. Although its structure is unknown, it is reasonable to expect that the native receptor is oligomeric, like the other carbohydrate receptors.

Taylor et al. [81] and Ezekowitz et al. [82] cloned the cDNA for this receptor. Human placenta also contains the Man receptor and provides ample protein to determine the sequence of proteolytic fragments. Oligonucleotide probes were used to isolate cDNA clones, covering the entire mRNA-coding region, from a placental cDNA library. The encoded protein (minus signal sequence) is 1438 amino acids, with a predicted M_r of 164,135. The deduced protein has a single membrane-spanning domain (COOH-terminal cytoplasmic) and three extracellular domains [81]. The first NH_2-terminal domain is a unique cysteine-rich 139 amino acid segment. The second domain is similar to the type II repeats of fibronectin. The third domain, closest to the transmembrane domain, contains eight segments homologous with the C-type carbohydrate recognition domains of the hepatic Gal receptor and other Ca^{+2}-dependent animal lectins. As expected, deletion of the cytoplasmic tail gives a mutant receptor that can bind, but not internalize, ligand [82].

4. *Localization of the Mannosyl Receptor*

Antibody to the protein from rat lung was used in indirect immunofluorescence studies to show that this receptor is not in hepatocytes, but is in hepatic Kupffer and endothelial cells [83]. This result is consistent with the earlier clearance studies of Hubbard et al. [65] that showed accumulation by both cell types of mannose- or *N*-acetylglucosamine-containing ligands. Hill and his co-workers have played a major role in sorting out the distribution of the various carbohydrate-binding activities in liver. They have clearly shown that the same "macrophage" Man receptor can be isolated from rat liver Kupffer cells [84], and that antibodies to the Kupffer cell receptor colocalizes in these cells with antibody to the rat lung Man receptor [83]. It is now evident that the Man receptor is localized in liver, spleen, lung, and, surprisingly, in leg muscle. Presumably, resident and migratory macrophages throughout the body contain this receptor activity. Interestingly, fresh human blood monocytes and a number of macrophagelike cell lines do not express the Man receptor. This suggests that the Man receptor is expressed only in differentiated macrophages; the same result is true for the Gal receptor in hepatocytes.

5. *Function of the Mannosyl Receptor*

Mannose-containing ligands, such as high mannose-type oligosaccharides, mannose–neoglycoproteins, lysosomal glycosidases, and mannans are degraded after endocytosis mediated by this receptor. An important function seems likely for the Man receptor, considering its extensive activity and its location on an important cell type involved in inflammation, wound healing, antigen processing, and other surveillance functions within the organism.

Expression of Man receptor activity is modulated in cultured cells by lymphokines and anti-inflammatory steroids. Lee and co-workers have suggested that the receptor may facilitate the ability of macrophages to attack yeast or bacteria that contain Man in their cell walls. Hoppe and Lee [85] observed that exposure of macrophages to soluble monosaccharides such as Man, GlcNAc, or Fuc, stimulated an increase in the cell surface-binding activity of the receptor. This triggering response may be important in the ability of macrophages to deal with potential microbial pathogens. Subsequently, Ohsumi and Lee [86] reported that exposure of rabbit alveolar macrophages to mannose ligands at 37°C stimulated the cells to secrete lysosomal enzymes in a time- and dose-dependent manner. These workers proposed that occupancy of receptors on the cells may be a trigger for secretion of enzymes that will degrade a yeast or bacterial cell wall and enable the macrophage to kill and phagocytose the potential pathogen. Mannosyl receptors have also been shown to be comodulated with F_c receptors on macrophages bound to mannan-coated substrata [87]. This result suggests that the two receptors may be coordinated at least in some aspects of their function and implies a role for the Man receptor in some immunocompetent process(es).

F. The Kupffer Cell Fucosyl Receptor

1. *General*

The presence of one or more possible endocytic receptors in liver with specificity for fucose-containing glycoconjugates had been suspected since the late 1970s [63,74]. Several laboratories reported what appeared to be conflicting results using a variety of different fucose ligands. Hill and co-workers finally clarified this confusion in an extensive and important series of papers published in 1986. Their success in this effort hinged on the use of neoglycoproteins, the chemistry of which has been extensively developed and pioneered by Lee and his co-workers [88] and by Wold and his co-workers [89]. Depending on the oligosaccharide structure employed, liver homogenates can exhibit up to four different fucose-binding activities. Three of these lectins can function endocytically in intact liver in vivo or perfused liver in vitro to remove circulating glycoconjugates and, therefore, are receptors.

As discussed earlier, the reticuloendothelial cell receptors for Man/GlcNAc can also recognize fucose-containing glycoproteins, depending on the nature of the fucose linkage. Lehrman et al. [90] found that Fuc-BSA clearance by rat liver was not inhibited by ligands for the parenchymal Gal/GalNAc receptor or the nonparenchymal Man/GlcNAc receptor. The clearance of Fucα1,3(Galβ1,4)GlcNAc-BSA, however, was weakly inhibited by Fuc-BSA or Galβ1,4GlcNAc-BSA, but was effectively blocked by a mixture of both neoglycoproteins. This result was interpreted to reflect the presence of two receptors involved in the clearance of the branched fucosyl glycopeptide. The hepatocyte Gal/GalNAc receptor can apparently recognize this latter structure with one terminal galactosyl and one terminal fucosyl residue. Since the BSA conjugate had about 30 residues of fucose per molecule of BSA [90], it is likely that more than two sugars were recognized by each receptor. The effect of oligosaccharide number per BSA molecule was not examined in this study, but given the work of Baenziger and Fiete [91], one would expect a biantennate ligand, such as the Fuc1,3(Galβ1,4)GlcNAc to be bound, but not endocytosed, by the Gal/GalNAc receptor system in hepatocytes. A further complication in the earlier studies with liver extracts was the presence of the intracellular hepatocyte mannose-binding proteins (also referred to as the *core-specific lectin*) that can also bind to fucosylglycoconjugates [92]. In intact liver, this activity presumably need not be considered, since it is not found on the cell surface and appears to be a secreted plasma protein (see Sect. II.H). From all of the foregoing studies, Lehrman et al. [90] concluded that a fourth lectin was involved in the clearance of Fuc-BSA.

2. Structure of the Fucosyl Receptor

Lehrman and Hill [84] then went on to purify a fucose-binding lectin from rat liver by affinity chromatography on Fuc-BSA-Sepharose. This protein contained two polypeptides, of 88 and 77 kd, that were shown by peptide mapping to be distinct from the hepatocyte Gal/GalNAc and the reticuloendothelial cell Man/GlcNAc receptors. The authors concluded that, since the peptide maps for the 77- and 88-kd proteins were very similar, the 77-kd peptide was probably derived from the 88-kd peptide by proteolysis. This fucose-binding protein is, therefore, probably an oligomer of the 88-kd protein. *N*-Glycanase treatment of purified receptor, which removes all *N*-linked oligosaccharides, decreased the M_r of the two polypeptides on SDS–PAGE to 52,000 and 58,000. The Gal and Man/GlcNAc receptors mentioned previously have primary carbohydrate specificity for sugars other than fucose and seem to interact with fucose-containing glycoconjugates only when these other sugars are present in the appropriate position and linkages. It is, therefore, reasonable to consider that the binding and clearance activity toward Fuc-BSA reflects the presence of a discrete, genuine Fuc receptor.

This conclusion was confirmed when the cDNA for the rat Kupffer cell Fuc receptor was cloned and sequenced [93]. The deduced protein sequence codes for a protein of 550 amino acids, M_r 61,104, that is unique, but highly homologous to other hepatic carbohydrate receptors (see Fig. 1A). The protein is a transmembrane glycoprotein with an NH_2-terminal intracellular domain of 42 amino acids and a COOH-terminal carbohydrate-binding domain of 135 amino acids. A fourth domain between the membrane-spanning and carbohydrate-binding domains contains 18 continuous, homologous sequences of 14–18 residues each. Four of the six potential *N*-glycosylation sites are in this domain. This domain probably serves as a stalk, a spacer arm, to keep the carbohydrate-binding sites extended away from the membrane.

3. Specificity and Localization of the Fucosyl Receptor

Like most of the endocytic degradative class II receptors described previously, the Fuc receptor requires Ca^{+2} for binding and is pH-dependent [80]. Below pH 7, receptor–ligand complexes dissociate. Consistent with the oligomeric structure of the other glycoconjugate receptors characterized to date, the binding to ligand is very dependent on the fucose content. For example, at an average of 24 fucosyl groups per BSA molecule, little if any binding was detected, whereas maximal binding occurred at 40 fucoses per BSA [94]. Given competition studies with simple glycosides, the receptor sugar-binding site appears to be best fit by the C1 conformation of *N*-acetyl-D-galactosamine and will also bind D-galactose and D-mannose, but not *N*-acetyl-D-glucosamine.

The presence of Fuc receptor activity was screened for in small amounts of different rat tissues and was found only in liver [83]. Immunocytologic methods were used to show that the receptor is present only in Kupffer cells in the liver. Interestingly, this receptor was not found in macrophages from any other tissue source, making it the only other carbohydrate receptor, other than the hepatic Gal/GalNAc receptor and possibly the endothelial cell HA receptor, with a unique liver localization.

4. Function of the Fucosyl Receptor

Degradation of Fuc-BSA or other fucosyl-glycoconjugates in isolated Kupffer cells has not been characterized, but presumably occurs. A likely function of this receptor would be the removal and degradation of fucosyl-glycoconjugates from the circulation. As with many of the other carbohydrate receptors, the endogenous ligands are unknown and its function can only be surmised (see Sect. IV).

G. The Macrophage/Reticuloendothelial Cell Galactosyl Receptor(s)

1. General

This still developing story indicates that macrophages have two distinct Gal receptor systems. Although it is too premature to be sure, one system may be for particulate and one for soluble Gal-containing ligands.

In addition to the very endocytically active Gal/GalNAc receptor in liver hepatocytes, mammalian liver displays another recognition system specific for galactose-containing glycoconjugates. Kolb et al. [95] first demonstrated that Kupffer cells, the resident liver macrophages, can bind erythrocytes on which galactosyl residues have been exposed by neuraminidase treatment. These investigators and their co-workers have continued to characterize this receptor system, which is present also on endothelial cells. This Gal receptor on macrophages from liver and peritoneum can mediate the binding and phagocytosis of desialylated lymphocytes or erythrocytes [for a review see 96]. In an elegant study, using colloidal gold particles of different sizes, Schlepper–Schafer et al. [97] were able to show that the size of a galactose-containing ligand determined which of the two liver Gal receptors systems would recognize and bind it. The Kupffer and endothelial cell receptors bound particles ranging from 2.2 to 11.7 nm in diameter, whereas only particles of <7.8 nm were bound to hepatocytes. The two liver Gal receptors can be further distinguished by their different patterns of expression and independent modulation during embryonic development [98].

Peritoneal macrophages also endocytose soluble galactosyl-terminal glycoproteins, whereas macrophages from lung and lymph do not [99]. These latter cells may also have a Gal receptor, but its role (like the liver macrophage receptor) may be recognition of particulate ligands. It may not mediate efficient uptake of soluble ligands because it is either present in low amounts or not targeted to the coated pit pathway.

2. Purification and Characterization of the Galactosyl Receptors

Roos et al. [100] isolated a putative receptor from intact perfused liver and crude liver membranes by affinity chromatography on a galactose-containing matrix. Since they did not also use purified rat liver Kupffer cells, they could not rule out the possibility that the protein isolated was from a different cell type. However, antibodies to this purified protein differentially stained liver sinusoidal cells and gave a pattern different from that of antibody to the hepatic Gal receptor. An unusual characteristic of this receptor is that it does not behave like an integral (intrinsic) membrane protein. It is like a peripheral (extrinsic) membrane protein, since it is released in the absence of divalent cations [100]. After isolated liver macrophages were treated with EDTA, the lost surface galactose-binding activity was restored when the cells were incubated at 37°C. The cells, therefore, appear to have an intracellular reserve of this protein. The isolated protein had an apparent M_r of about 30,000. The significance of this protein not being an integral membrane protein, and yet apparently able to mediate phagocytosis or endocytosis of appropriate ligands, is unclear. Further studies on purified cell preparations will be required to determine its mode of action.

Kempka et al. [101] concluded that a membrane-associated form of C-reactive protein (a major acute-phase protein in rat and human) is the galactose-specific particle receptor on rat liver macrophages. Purified C-reactive protein (~30 kd) could functionally replace this receptor on EDTA-treated Kupffer cells. Another galactose-binding protein, the Mac-2 antigen on the surface of mouse peritoneal macrophages, was cloned [102] and shown to be identical with the 35-kd galactose-binding lectin of fibroblasts [103]. From the foregoing studies, it seems clear that macrophages contain peripheral membrane galactose-binding proteins on their surfaces that allow the uptake of large galactose-containing particles. However, it is not clear whether there are multiple galactose-binding proteins that can serve this function; perhaps different proteins are found on macrophages from different tissues. These extrinsic membrane proteins are distinct from another macrophage Gal receptor, which is an integral membrane protein.

Kawasaki et al. [99] isolated a galactose-binding protein from rat peritoneal macrophages by affinity chromatography of Triton X-100 extracts on ASOR-Sepharose. The Man/GlcNAc receptor was first removed by chromatography on mannan-Sepharose. The purified Gal receptor required Ca^{+2} for ligand binding and contained three polypeptides of M_r ~42, 60, and 65 kd. These were in apparent ratios similar to the hepatic Gal receptor; the latter two peptides were minor components. Antibody to the hepatic Gal receptor completely inhibits uptake of ASOR, but not of Man-BSA, by intact cells. The two liver membrane Gal receptors are, therefore, immunologically related and have similar subunit compositions. The authors estimate that ~50% of the receptors are on the macrophage surface. Considering the very similar ligand specificities between the two receptors [23,99,104], it is likely that at least the carbohydrate-binding domains are very homologous.

Kelm and Schauer [105] also isolated a Gal receptor from rat peritoneal macrophages and found three polypeptides by SDS–PAGE of M_r ~42, 56, and 68 kd. From several studies, they concluded that the latter two proteins are not part of the Gal receptor. The major 42-kd protein was selectively iodinated on intact cells by an ASOR–lactoperoxidase conjugate at 4°C. A photoaffinity ligand also labeled only this peptide. Furthermore, the two minor proteins were not readsorbed to galactose affinity matrix. However, since differential labeling is obtained in the hepatic Gal receptor as well, these two minor bands could still be part of the receptor. These and other questions should be answered when specific antibodies are developed to the major 42-kd protein.

Ii et al. [106] cloned the cDNA for the 42-kd Gal receptor from a rat peritoneal macrophage cDNA library. A unique portion of the cDNA for RHL 1 (with minimal homology to RHL 2/3) was used to screen for this macrophage cDNA. The deduced amino acid sequence indicates that this macrophage Gal "receptor" belongs to the C-type animal lectin family [107] and has 59% sequence homology with RHL 1. The macrophage protein is different from RHL 1 in several interesting ways. It has a shorter cytoplasmic tail and an insertion of 24 amino acids, which contain an RGD sequence, between the single transmembrane and carbohydrate-binding domains [106]. The deduced protein contains

306 amino acids, with a predicted M_r of 34,242. *N*-Glycosylation at two sites, identical with those in RHL 1, gives an $M_r \sim 42$ kd. Although there is no doubt that this protein is part of the macrophage Gal receptor, no evidence is available to show that this is the only component of the receptor. Functional Gal receptor activity has not been reconstituted in transfected cells. Therefore, like the hepatic Gal receptor, two cDNAs and multiple protein subunits may be required for expression of a functional receptor.

3. *Degradation of Carbohydrate Ligands*

Rat peritoneal macrophages can also take up and degrade soluble galactose-containing glycoproteins [104,108], such as ASOR, one of the plasma glycoproteins commonly used as a ligand for the liver parenchymal Gal/GalNAc receptor. Although only about 3×10^3 binding sites were detected per cell, the binding was Ca^{+2}-dependent, and specific for Gal or GalNAc, but not Fuc, Man, Glc, or GlcNAc. This low level of activity, if also present on Kupffer cells in liver, would not compete effectively with the parenchymal Gal/GalNAc receptor for the removal of soluble galactose-containing glycoconjugates from blood. However, even this low level of activity may be very important in allowing macrophages to remove and degrade soluble, as well as participate, galactosyl ligands from peripheral tissues outside the vasculature.

4. *Function of the Macrophage Galactosyl Receptor(s)*

A Gal receptor system(s) is present on resident liver and, at least, some migratory macrophages, and has been proposed to be involved in regulating the turnover and degradation of aged red blood cells [109,110]. It has long been recognized that the sialic acid content on the surface of erythrocytes decreases during the lifetime of these cells. Appearance of the "new" galactosyl groups thus exposed would provide a convenient carbohydrate determinant for a system needed to remove old or damaged cells from the body. A large amount of evidence supports this model, although it is not universally accepted. Desialylation of glycoproteins and gangliosides on the red blood cell surface would presumably occur spontaneously, reflecting the relative instability of the glycosidic bonds of the sialic acid family, or in response to injury or trauma. Neuraminidases, for example, are often released at a wound site or by some pathogenic organisms or parasites [111]. This functional role for the reticuloendothelial Gal receptor is very attractive. As expanded below in Section IV, it could be very important for mammals to remove efficiently both soluble and particulate glycoconjugates containing terminal galactose or *N*-acetylgalactosamine.

These macrophage Gal receptors have a degradative function, but like the macrophage Man receptor, the Gal receptors may also be involved in an immunologically related function. Intraperitoneal coadministration of the simple competing monosaccharide galactose with desialylated sheep red blood cells in mice resulted in an adjuvant activity [112]. The adjuvant activity was sugar-specific and concentration-dependent. Galactose, GalNAc, Man-6-P, and D-Fuc were effective stimulators of the immune response, whereas Glc, Man, Fructose (Fru), and L-Fuc were ineffective. The mechanism of this effect is unknown, but involvement of a macrophage Gal/GalNAc receptor is an interesting possibility.

H. The Liver Parenchymal Cell Mannosyl-Binding Proteins

1. *General*

Several groups have isolated and extensively characterized what has emerged as a family of related carbohydrate-binding proteins. Maynard and Baenziger [113] isolated a

Man/GlcNAc-specific lectin from rat liver that was distinct in specificity and molecular size from the Man/GlcNAc receptor of liver reticuloendothelial cells. This former protein was localized intracellularly, not on the cell surface, in rat hepatocytes. This molecule has since been referred to variously as the core-specific lectin, by Baenziger's laboratory; as the liver mannan-binding protein, by Yamashina and colleagues [114]; or as the mannose-binding protein, by Drickamer et al. [115]. For some time, this field was very confusing because this protein, which I will refer to as the mannose-binding protein (MBP), is the major liver protein isolated by affinity chromatography on a variety of different mannose-containing supports. The MBP is not, however, the same as the liver Man receptor, although understandably most investigators initially thought that it was.

The MBPs are included in this chapter (but not in Table 1), even though there is little evidence for cell surface localization or for their ability to mediate endocytosis and degradation of glycoconjugates. There are, nonetheless, reasons to believe this protein family may have important function(s). A small amount of the MBPs may be found at the plasma membrane, although the majority is soluble and, in fact, is found in plasma in significant amounts.

2. *Specificity of the Mannosyl-Binding Proteins*

The carbohydrate specificity of the soluble parenchymal MPBs is similar to that of the nonparenchymal Man receptor, but extensive studies with well-defined structures have not been done. Mori et al. [116] showed that the rat MBP bound Man_{33}-BSA almost as well as mannan; $GlcNAc_{43}$-BSA was bound even more effectively. Asialoagalacto- or ahexosamino-orosomucoid were not bound by the MBP. An additional form of this Man/GlcNAc lectin was reported by Haltiwanger et al. [83] to have a high affinity for Fuc-BSA. This subpopulation of the lectin had an identical M_r (32,000), antibody reactivity, and V-8 protease peptide map, compared with the "normal" form. Its significance or possible function is unknown.

3. *Structure of the Mannosyl-Binding Proteins*

The MBPs have been isolated from rat or human serum as well as from liver [117]. The intact binding protein is a hexamer composed of identical subunits with an M_r of 30,000 [19]. It is now clear that there are two distinct MBPs designated MBP-A and MBP-C. These two proteins are found in both liver and serum. The cDNAs for MBP-A [118] and MBP-C [113] have been isolated and sequenced and, together with primary protein sequence data [114], have revealed several interesting features [107].

The two proteins can be aligned with four gaps and 56% sequence identity. The proteins have a collagenous tail following an NH_2-terminal signal sequence. This segment of 18-glycine tripeptides contains hydroxylysine and hydroxyproline, as well as the characteristic collagen oligosaccharide Glc-Gal-HyLys [119]. The two MBPs are, therefore, members of a family of secreted plasma proteins that contain collagenlike sequences. This family also includes a form of acetylcholinesterase, the pulmonary surfactant proteins, and the complement protein C1q.

The two liver MPBs have extensive sequence homology with the following proteins: dog pulmonary surfactant apoprotein; an invertebrate soluble galactose-specific lectin; the family of mammalian hepatic membrane endocytic receptors discussed earlier (the rat and human Gal receptor subunits–RHL 1, 2, and 3, and H1 and H2; the Kupffer cell Fuc receptor; the macrophage Man receptor; and the macrophage integral membrane Gal receptor); and the chicken GlcNAc receptor (CHL). All of these proteins (coded for by 11 distinct genes) have a conserved 130- to 150-amino acid region, usually toward the COOH-terminal end, that forms the sugar-binding domain (see Fig. 1A). Even though the sugar specificities of these proteins are different (e.g., Gal, Man, or GlcNAc), the amino acid sequence similarities are extensive [19,107]. For example, the COOH-terminal end of the RHL-1 or CHL protein sequence can be aligned with either MBP-A or MBP-C, with five or four gaps, to give 23 and 32% sequence identity, respectively. Overall, 19 amino acids, particularly cysteine residues, are conserved in the carbohydrate recognition domains among these proteins. In fact, given the strong similarity, Drickamer predicted that dog pulmonary surfactant would have an as yet undiscovered carbohydrate-binding activity [19,107]. When tested, this protein did indeed show a Ca^{+2}-dependent binding toward matrices with Gal, Man, Glc, and Fuc, but not GalNAc or GlcNAc [120]. A similar MBP has been isolated from bovine serum and shown to be homologous to dog pulmonary surfactant and MBP-A and MBP-C [121].

4. *Functions of the Intracellular and Soluble Plasma Mannosyl-Binding Proteins*

At least two interesting functions are possible for the MBPs. On the basis of the intracellular localization in the Golgi apparatus and rough and smooth endoplasmic reticulum, Yamashina and colleagues proposed that the MBPs were involved in the transport of biosynthetic intermediates of glycoproteins [122]. Mori et al. [123] identified a set of endogenous ligands for the rat MBPs that included the hepatocyte proteins α_1-macroglobulin, α_1-antitrypsin, α_1-acid glycoprotein, and the lysosomal enzyme β-glucuronidase. The former three proteins are major plasma proteins synthesized and secreted by hepatocytes; they are also acute-phase proteins. All four endogenous ligands contained high mannose-type oligosaccharides ($Man_{8,9}GlcNAc_2$-) and had rapid turnover rates, with average half-lifes of 45 min. This finding makes it very likely that the MBPs are involved in the intracellular transport or processing of glycoproteins during their biosynthesis. If so, then smaller, but significant, amounts of the MBPs may be found in all cells.

The feeling that one or both of these MBPs is primarily a plasma protein has also led to the suggestion that they are involved in a primitive type of immune response directed against the mannose-containing glycoconjugates of microorganisms [114]. Ikeda et al. [124] showed that the plasma MBP (apparently MBP-A), but not the liver MBP, was able to activate the complement system in a sugar-dependent manner through the classic pathway. This is an exciting discovery and in a broad sense involves this protein with the recognition and eventual degradation of appropriate glycoconjugates, although by an indirect route—the clearance of antibody–antigen complexes and antibody-coated particles. The function of MBP-C, which is structurally distinct from MBP-A, is still unknown.

Two forms of MBPs have been reported in sera, but the relative amounts and how these concentrations change in disease, trauma, or aging, are also unknown. It is also unclear whether the extensive studies by Baenziger and co-workers on the biosynthesis and processing of core-specific lectin have, in fact, been on MBP-A or MBP. Baenziger believes that these studies have focused primarily on MBP-A (personal communication). Clearly, numerous interesting discoveries remain to be made in this lectin family.

I. The Liver Endothelial Cell Hyaluronan (Glycosaminoglycan) Receptor

1. *General*

The glycosaminoglycan, hyaluronan (HA), is a major constituent of the extracellular matrix and plays an important role in many developmental and regulatory processes, including cell adhesion, morphogenesis, wound healing, tumorigenesis, and angiogenesis [125]. Studies on the clearance of HA from the circulation have gained importance over the last decade because of the increasing therapeutic applications of HA in the treatment of arthritis, eye surgery, tendon repair, and the correction of depressed granulocyte function. The recent findings of elevated plasma HA levels in several disease conditions, such as liver cirrhosis, rheumatoid arthritis, and scleroderma, and the increased HA content of or production by many tumors, have also led to an increased interest in HA metabolism.

The elegant studies of Laurent and co-workers have shown that injected HA in rats and rabbits is rapidly cleared from the circulation ($t_{1/2}$ 2.5–4.5 min), and about 90% of the injected label is recovered in the liver [125]. A whole-body autoradiographic study on tissue uptake of circulating HA in mice also confirmed the accumulation of HA in the liver. Although both rat liver hepatocytes and liver endothelial cells (LECs) specifically bind HA, only endothelial cells are the major site of HA uptake and degradation [126]. A single class of high-affinity receptors for HA has been demonstrated on LECs. The HA receptor on these cells is different in both its function and sensitivity to pH from the fibroblast HA receptor identified by Underhill and Toole [125].

In LECs in suspension or culture, a large fraction of the total HA receptors (50–80%) are intracellular and have an affinity and specificity similar to that of cell surface HA receptors [127]. In cultured rat LECs there are 93,000 surface receptors per cell and about 87,000 intracellular receptors per cell.

2. *Affinity and Specificity of the Hyaluronan Receptor*

Cell surface and intracellular HA-binding sites have essentially the same affinity for ^{125}I-HA; $K_d = 5.8 \pm 2.8 \times 10^{-8}$ M [127]. Smedsrod et al. [128] reported a K_d of 6×10^{-11} M, using larger HA (M_r of 400,000) and about 10^4 receptors per cell. The affinity of the HA-receptor interaction is directly related, and the binding (B_{max}) is inversely related, to the chain length of HA. An almost identical pattern of competition with various saccharides was observed for binding at 4°C or endocytosis of HA by LEC at 37°C. Of the various compounds examined, only three—HA, chondroitin sulfate, and heparin—were efficient competitors of the binding and endocytosis of ^{125}I-HA in LEC [127,129]. Desulfated chondroitin and dextran sulfate also showed a significant ability to compete for endocytosis in these cells. Unrelated polyanions, however, such as DNA, RNA, and polygalacturonic acid, were not effective competitors for endocytosis of HA. Dextran, glucuronic acid, *N*-acetylglucosamine, inulin, *N,N'*-diacetylchitobiose, and haptoglobin were likewise unable to compete. Other investigators have also demonstrated that competition for HA endocytosis by LECs occurs using chondroitin sulfate, but not with heparin [128,130]. However, the size of the [^{3}H]HA used in those experiments was larger (M_r ~400,000 compared with ~44,000). This might suggest that heparin can compete effectively for endocytosis of small HA molecules, but not for very large HA molecules. McGary et al. [129] also concluded, as Laurent et al. proposed [130], that the LECs probably also function in vivo to remove these other glycosaminoglycans from the circulation by using the same receptor. Neither HA binding nor endocytosis requires Ca^{+2}; both processes occur in the presence of EDTA [131].

3. *Endocytic Capacity of the Hyaluronan Receptor*

The rate of ^{125}I-HA internalization by LECs at 37°C increases almost linearly with increasing concentrations until approximately 4×10^{-7} M [129]. A plateau was approached at HA concentrations greater than 3×10^{-6} M. The apparent K_m for the rate of internalization is about 0.4 × 10^{-6}M HA. At saturation, the maximal steady-state endocytic rate was 250 molecules per cell each second. Such a saturation effect is typical of a receptor-mediated endocytic process and would not occur if uptake was by a fluid-phase process. The maximum endocytic rate of 250 molecules of HA per LEC per second compares well with the maximum values of about 700 and 550 molecules per cell each second for the hepatocyte asialoglycoprotein receptor [21] and the macrophage mannose receptor [72], respectively. When normalized for the receptor content per cell, these three different receptor recycling systems are all essentially identical (see Table 3).

The accumulation of HA at 37°C by LECs follows kinetics typical of an endocytic internalization of receptor-bound ligand. The inability to strip cell-associated HA at 4°C suggested that almost all of the accumulated HA was intracellular. This was further substantiated by the ability of digitonin to permeabilize LECs and to release 87% of the HA accumulated after 90 min. The release of cell-associated HA after the digitonin permeabilization also indicates that dissociation of HA–receptor complexes occurs after endocytosis. Such a dissociation event is necessary for a system that degrades ligand and recycles the receptor. Unlike many of the other receptors characterized to date, the dissociation of HA from the endothelial HA receptor is *not* facilitated at low pH (e.g., pH 5.0). In fact, the amount of HA bound actually increases approximately 30-fold, at pH 5.0 [127]. However, washes at pH 9.0 remove ≥70% of the surface-bound or intracellular HA previously bound to intact or digitonin-permeabilized cells, respectively.

4. *Evidence for Receptor Recycling*

The large number of intracellular HA receptors suggests that they have a function. At least 50% of the intracellular HA receptors in LECs become occupied with HA of extracellular origin during continuous endocytosis at 37°C. However, this is an underestimate owing to the rapid rate of dissociation of HA (M_r = 44,000) from the receptor. At least 70% of the intracellular HA receptors in LECs probably function during endocytosis [129]. Since these cells mediate the clearance of HA in vivo, it is likely that the intracellular HA receptors are either involved in trafficking of the endocytosed HA (e.g., delivery to lysosomes) or are a reservoir of internal receptors involved in the continuous receptor recycling pathway.

When the cell surface and internal HA receptor number was compared with the amount of ^{125}I-HA processed per cell in 6 hr, it was clear that receptor reutilization must occur. Even if all the intracellular HA receptors were involved in the endocytic cycle (i.e., all cellular HA receptors are functional and functionally equivalent), then each HA receptor would have still been used about 13 times to account for the total amount of ^{125}I-HA processed. A similar situation occurred also in the absence of protein synthesis; each receptor was used six times. Experiments in the presence of cycloheximide show there is no loss of HA receptor activity even after 6 hr, a time when at least six rounds of endocytosis has occurred [129]. Therefore, there is no indication that HA receptors are degraded during continuous endocytosis.

Other recycling endocytic systems, such as the class II carbohydrate receptors and LDL receptors, utilize the coated pit pathway [16]. Hyperosmolarity induced with sucrose interrupts the coated pit pathway in hepatocytes [132,133]. The ability of 0.4 M sucrose to inhibit HA endocytosis to nonspecific levels [129] strongly suggests that a coated pit pathway is involved. Since binding of HA to LECs at 4°C is not affected by 0.4 M sucrose, the effect of sucrose is on endocytosis. Electron microscopic studies with colloidal gold probes, coated with chondroitin sulfate proteoglycans, also indicate that a coated pit pathway is involved [134].

5. *Degradation of Hyaluronic Acid by Liver Endothelial Cells*

Metabolically labeled HA is degraded by cultured LECs completely to water, acetate, and lactate, the end products of glycolytic metabolism [128]. There is only a 10–20 min lag before degradation products accumulate in the medium. Presumably all cells have a lysosomal hyaluronidase [125]. This endoglycosidase, together with the exoglycosidases β-*N*-acetyl-D-glucosaminidase and β-D-glucuronidase can account for the complete breakdown of HA or related glycosaminoglycans to their respective monosaccharides. It will be interesting to see whether the LECs possess an additional hyaluronidase activity or a greater amount of an enzyme common to most cells. The unique hydroxyphenylpropionyl derivative of HA used in our studies [127,129] is structurally similar to some of the "residualizing" labels developed by Baynes and co-workers [135] to localize the site of degradation of labeled molecules in tissues in vivo. A *residualizing label* is a structure that is covalently attached to a protein and that is not readily released from lysosomes after the protein has been degraded. It, therefore, remains behind as a "residue" to mark the location of its carrier protein. This uniquely modified HA derivative displays a greater lag (2–3 hr) before the release of radiolabeled degradation products into the medium and a much greater steady-state accumulation of intermediate degradation products inside the cell before they appear in the medium.

6. *Total Body Turnover of Hyaluronan*

From work by Fraser and his colleagues [113], it has recently been estimated that the total body turnover of HA in humans is as much as 4 g/day. Furthermore, it was discovered that lymph nodes actually remove about 80–90% of the HA from lymph before it is delivered into the blood [136]. Understandably, this was missed in studies following the fate of HA injected directly into the blood, thereby, bypassing the lymph system. The present model for HA and glycosaminoglycan metabolic turnover is that degradation of these molecules occurs in the extracellular matrices throughout tissues in the body. The smaller HA fragments (still very large by macromolecular standards; $M_r \sim 10^6$) then flow from the tissues into the collecting lymph system draining all tissues and are delivered from lymph into the blood. The HA content of lymph is much higher than the HA content of blood. Hyaluronic acid and other glycosaminoglycan chains not received by the lymph nodes are cleared by the LECs.

A future direction in this field will be the identification of the lymph node cell type(s) and receptor system that mediates HA clearance. This will certainly be a recycling class II receptor capable of efficient and high-capacity ligand processing. Conceivably this could be a different receptor from that in LECs and could be an addition to the list in Table 1.

A possible lung HA-binding activity has also been detected in alveolar macrophages, based on cross-reactivity with a monoclonal antibody to the BHK fibroblast 85-kd HA receptor [125,137]. The macrophage HA-binding protein is larger, at 99 kd. On the basis of the existing literature, it is reasonable to expect that macrophages possess the ability to endocytose and degrade extracellular HA.

7. *Function of the Hyaluronan/Glycosaminoglycan Receptor*

Considering the large endocytic capacity of LECs for HA and the large daily turnover of HA in the body, it is likely that removal of circulating HA is an important function. This is the only carbohydrate receptor for which the endogenous ligands are definitely known. The metabolic turnover of chondroitin sulfate or glycosaminoglycan chains other than HA recognized by this receptor has not been determined, but it is clear that this is a generic receptor able to mediate the clearance of this large polysaccharide family. The physiological consequence of not removing these molecules from the blood is unknown, but can be imagined. We have suggested that problems in the microcirculatory system would arise as the viscosity of blood increased with increasing HA concentration in the blood [125]. Human fibrinogen specifically binds to HA [138], and the presence of HA alters the ability of thrombin to induce fibrin clot formation in vitro [125]. Elevated levels of HA in the blood could, therefore, affect the normal dynamics of homeostasis and induce thromboses or bleeding [139].

J. The Macrophage Advanced Glycation End Product Receptor

1. General

Under physiological conditions (37°C, pH 7.4, 150 mM NaCl), reducing sugars readily react with the primary amino groups of proteins to form Schiff bases or imines. Although this adduct is reversible, it can also undergo a series of complex rearrangements and reactions in vitro to form stable, polymeric reaction products that are fluorescent and brown. This process is referred to as the Maillard or browning reaction [140] and results in the formation of posttranslationally modified proteins. These are *glycated*, rather than glycosylated, proteins to denote the nonenzymatic addition of the sugar to the protein. Since these molecules are technically glycoproteins, the receptor mediating their uptake and removal has been included in this section.

In extracellular fluids, such as blood, the most prevalent reducing sugar is glucose. The degree of glycation of circulating proteins such as albumin, or even the intracellular protein hemoglobin, correlates with the concentration of blood glucose. Protein glycation is increased in diabetes in proportion to the increase in blood glucose concentration. The reaction products, particularly those derived from glucose, have been termed advanced glycosylation end products (AGE) by Cerami and co-workers [141,142]. This designation is not entirely accurate and should more appropriately be called advanced glycation end products. This acronym calls attention to the fact that these glycation products are also correlated with age and probably account for at least some of the age-dependent chemical modification of proteins. These products include such compounds as *N*-fructoselysine and its oxidation products *N*-carboxymethylysine and 3(*N*-lysino)-lactic acid. Another rearrangement product 2-furoyl-4(5)-(2-furanyl)-1*H*-imidazole (FFI) is a condensation product of two lysine groups and two glucose molecules.

There is still controversy, however, about whether this condensation product would ever be formed in vivo [143]. In addition, the characterization of the AGE receptor has been complicated by the presence of multiple scavenger receptors for other abnormal modified proteins in the same cells. Both peritoneal macrophages and liver sinusoidal endothelial cells have more than one scavenger receptor for molecules like acetylated or oxidized low-density lipoprotein (LDL), or formaldehyde-treated albumin [126,144–146].

2. Affinity and Specificity of the Advanced Glycation End Product Receptor

Peritoneal macrophages are able to bind, internalize, and degrade glucosyl-modified proteins. Mouse macrophages have ~10^5 receptors per cell, with an affinity (K_a) of ~5×10^5 M^{-1} using AGE-BSA [141]. Binding was saturable and competed for with unlabeled AGE-BSA. The AGE-BSA binding to macrophages was not competed for by mannans or acetylated LDL, indicating that the AGE–protein receptor is different than the acetyl-LDL or Man receptors. These workers reported that formaldehyde-treated BSA, a ligand for the macrophage reticuloendothelial scavenger receptor [144], also did not compete for binding to the AGE receptor. However, polyanionic compounds, such as heparin, polyadenylic acid, polyinosinic acid, and fucoidin, were effective competitors. It remains to be determined whether ligand binding requires Ca^{+2} or other divalent cations. Radoff et al. [142] characterized the receptor–ligand interaction by competition studies using derivatives of FFI of known structure. The carbonyl group, furan ring(s), and the central imidazole structure were all important for receptor recognition.

In contrast, Takata et al. concluded that the AGE receptor is the same as the scavenger receptor [147]. This study employed an extensive series of FFI and aldehyde derivatives of BSA. The FFI-BSA was not specifically bound to either rat liver sinusoidal cells or rat peritoneal cells. The AGE-BSA and aldehyde-treated BSA were specifically bound by these cells, were cleared from plasma in vivo, and effectively competed with one another for binding or uptake. These workers conclude that AGE-modified proteins in vivo are the endogenous natural ligands for the macrophage scavenger receptor. The exact specificity of the AGE receptor and its relatedness to the scavenger receptor(s) must still be clarified.

3. Degradative Capacity of the Advanced Glycation End Product Receptor

At saturation, macrophages were able to process (i.e., endocytose or degrade) about 2×10^6 molecules of AGE-BSA per cell per hour [141]. Since there are ~10^5 receptors per cell on the cell surface, the cells processed an amount of ligand equivalent to their receptor content every 3 min. One can predict, therefore, that the AGE receptor will be a recycling class II receptor system with as many as, or more, intracellular receptors than cell surface receptors.

4. Structure of the Advanced Glycation End Product Receptor

Although the putative receptor has not yet been purified, it has been identified by cross-linking to ligand. ^{125}I-Labeled AGE-BSA was cross-linked to a murine macrophage-derived cell line, with a homobifunctional cross-linker [142]. The calculated M_r of the protein cross-linked to this ligand was 90 kd. Cross-linking was prevented by non-radiolabeled AGE-BSA but not BSA alone. Because ligand binding depends on the density of AGE groups per BSA, it is reasonable to assume that the native receptor is oligomeric.

5. Function of the Advanced Glycation End Product Receptor

It will be important to determine if this receptor is also present on other macrophage populations as well as on peritoneal-derived cells and liver sinusoidal cells. If other migratory and resident macrophages in the body have the AGE receptor, then it may have an important function; presumably, the removal of senescent proteins bearing the age-dependent modification of advanced glycation end products derived from Glc. A recent study by Patrick et al. [148], for example, documented an excellent correlation in the accumulation of carboxymethyllysine in the lens with age in humans. Because of the phagocytic capacity of macrophages, the AGE receptor would allow removal of "aged" cells as well as soluble proteins.

K. The Thyroid *N*-Acetylglucosaminyl Receptor

1. General

Consiglio et al. [149] identified a binding activity on cultured thyrocytes and thyroid membranes specific for thyroglobulin. From experiments with asialoagalactothyroglobulin, they suggested that the thyroid prohormone might be the endogenous ligand of a receptor specific for *N*-acetylglucosaminyl groups. Ligand binding requires Ca^{+2} and is blocked or reversed in the presence of EDTA. Unlike most of the carbohydrate receptors discussed in the foregoing, the pH optimum for binding was acidic. Only 10% of the binding activity was observed at pH 7, compared with pH 5. This behavior is similar to that of the LEC receptor for HA and glycosaminoglycans [127]. Although the GlcNAc-binding activity in thyrocytes does mediate uptake, it does not mediate degradation of the ligand.

2. *Specificity and Structure of the N-Acetylglucosaminyl Receptor*

Miquelis et al. [150] used neoglycoproteins to study this receptor further and confirmed the sugar specificity for GlcNAc. Ligands presenting highly clustered galactose residues were weakly recognized, whereas BSA neoglycoproteins containing mannose, fucose, glucose, or nonclustered galactose were not bound. With the use of GlcNAc-Sepharose these workers purified a detergent-soluble activity from porcine thyroid plasma membranes. The purified GlcNAc–thyroid receptor has a subunit size of 45 kd and appears to be a trimer. Fluorescein-labeled GlcNAc-BSA was used to show that the receptor is localized in pig thyroid, both on the cell surface and intracellularly. All follicle cells were labeled with the probe, and the apical surfaces of cells facing the follicular lumen were also labeled.

3. *Function of the N-Acetylglucosaminyl Receptor*

The effect of pH on receptor-binding activity is similar to that for the transferrin, polymeric IgA, and neonatal gut IgG receptors, which mediate *transcytosis* (i.e., the vectorial transport through a polar cell from one extracellular compartment to another) of their respective ligands. A similar function is, therefore, likely for the thyroid GlcNAc receptor. Miquelis et al. [150] suggested that the thyroid GlcNAc receptor may recycle incompletely processed thyroglobulin molecules and, perhaps, increase the efficiency of hormone production by the thyroid.

There is no evidence that GlcNAc-containing glycoproteins can be removed from plasma by this receptor. Interestingly, however, the thyroid GlcNAc receptor cross-reacts with polyclonal antibody raised against the liver parenchymal Gal receptor [149].

L. Other Putative Carbohydrate Receptors

1. *The Lewis Lung Carcinoma Glucosyl Receptor*

Although this receptor is in a cancer cell, it is included in this chapter because it has been demonstrated that it can mediate endocytosis and degradation of the carbohydrate ligands it binds [151]. Presumably, this same receptor is present, although perhaps not to the same extent, on a normal lung cell type. With use of neoglycoproteins, the receptor was shown to be somewhat specific for α-D-glucosyl residues [152]. The receptor also recognized α-L-fucose, α-L-rhamnose and α-D-mannose, but not β-D-GlcNAc, β-D-galactose, or α-D-galactose. Isolated cells at 4°C had 7.5×10^5 binding sites per cell for fluorescein-labeled α-D-Glc-BSA, with an apparent affinity (K_a) of $2 \times 10^6 M^{-1}$. Although this affinity is only moderately high, the number of receptors per cell is very large, suggesting that an important function may exist for this receptor. The receptor has not been extensively characterized or purified. The requirement for Ca^{+2} is also unreported.

2. *The Bone Marrow Galactosyl Receptor*

An "asialoglycoprotein receptor," was reported in mouse femoral bone marrow, using a frozen section assay with lymphocytes containing a high level of peanut agglutin-binding sites [153]. The use of the term *asialoglycoprotein receptor* for this activity is somewhat unfortunate, since it implies a functional similarity to the hepatic receptor, which has not yet been determined. This activity was not Ca^{+2}-dependent, was not blocked by antiserum to the rat hepatic Gal receptor, but was inhibited by asialofetuin or asialotransferrin. There is no evidence that cells in bone marrow are able to specifically internalize and degrade Gal ligands. This activity is likely to be the same Gal and Man recognition system involved in the "homing" of granulocyte–macrophage progenitor cells to hemopoietic stroma [154]. Like many other cell–cell or cell–matrix carbohydrate recognition systems, this interaction is probably involved in cell growth and differentiation, rather than glycoconjugate turnover. This activity might be the same as the macrophage Gal receptor(s), but since this is not yet known, it was not included in that section.

3. *The Macrophage β-Glucan Receptor*

A receptor specific for predominantly linear β1,3glucan polymers has been identified on human neutrophils, mouse bone marrow macrophages, and mouse resident and thioglycolate-stimulated peritoneal macrophages [155,156]. This is thought to be one of the three macrophage receptors capable of recognizing zymosan and heat-killed yeast. The other two are the Man receptor and the complement type 3 receptor. (Zymosan is a crude yeast cell wall preparation consisting chiefly of protein–carbohydrate complexes.) The glucan receptor almost exclusively mediates macrophage binding and phagocytosis of yeast, since it is inhibited up to 95% by soluble β-glucan polymers. The ingested yeast or zymosan particles and their associated glycoconjugates are then degraded. The efficiency of binding to receptor is increased with increasing polymer size and is not significantly affected by the presence of mannans. Although it has not yet been tested directly, this receptor, like the macrophage Man and Gal receptors discussed in the foregoing sections, probably will also mediate the endocytosis and degradation of soluble β-glucans. This receptor has not been extensively characterized or purified, but, presumably, plays an important role in host defense and protection against potential microbial pathogens.

III. MECHANISMS OF CELLULAR PROCESSING AND DEGRADATION OF INTERNALIZED SOLUBLE GLYCOPROTEINS

A. Introduction

In this section, I will give an overview of our present understanding of the cellular pathways by which glycoproteins are processed and degraded by carbohydrate receptor systems. Almost all work on the intracellular fate of internalized ligand and on receptor recycling has been done on the hepatic Gal receptor and the macrophage Man receptor. Therefore, I will discuss these two receptor systems, particularly the former, since it is one with which my laboratory has extensive experience. It is reasonable to believe that most, if not all, of the basic features that have been found for the Gal receptor also apply to all the other recycling carbohydrate receptors in other cell types. The Man-6-P "receptor" has also been extensively studied, but, since the recognition that there are actually two different receptors and the identity of one of these with the insulinlike growth factor II receptor, this field is now in a state of confusion. Consequently, most of the studies prior to 1987 will have to be reassessed and perhaps reinterpreted. I will, therefore, not include information on the processing of ligands containing Man-6-P.

There are four possible outcomes for processing of endocytosed receptor–ligand complexes [157,158]: (1) The receptor may recycle back to the cell surface while the ligand is degraded, as seen with the LDL or Gal receptor systems. (2) The receptor and ligand may both recycle, as seen with the transferrin receptor. (3) A third fate, for example, seen with the epidermal growth factor receptor complex, is that both the receptor and the ligand can be degraded. (4) Finally, both receptor and ligand may be transported across a polar cell and released into a different external compartment, as occurs during the transcytosis of IgA by secretory component. Where these outcomes have been determined for the membrane-bound carbohydrate binding receptors discussed in the foregoing, they are of the first type; they are recycling receptors, and their ligands are degraded. The thyroid GlcNAc receptor and the parenchymal MBPs are exceptions.

If one incubates isolated hepatocytes at 37°C with radiolabeled asialoglycoproteins, the cells very quickly take up the protein and, within a few hours, they reach a steady-state level of accumulation of intracellular radioactivity (Fig. 3). The cells can maintain that level for a very long time, during which they will continually be degrading the internalized glycoprotein and releasing the degradation products into the medium. The cells will continue to process the internalized asialoglycoprotein until virtually all of the extracellular protein has been degraded, at which time, the amount of cell-associated radioactivity will decline, until there is virtually none left in the cell. All of the class II recycling carbohydrate receptors process and metabolize glycoconjugates in this general manner. During the time in which the content of intracellular internalized glycoprotein and the rate of degradation

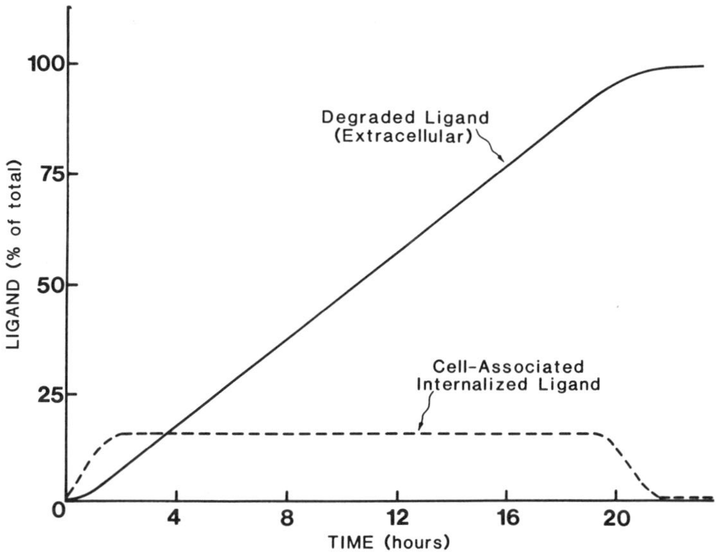

Figure 3 Cellular processing and degradation of extracellular glycoconjugates. The diagram depicts the intracellular accumulation (dashed line) of a ligand recognized by a class II recycling carbohydrate receptor in isolated cells. A steady-state amount of ligand is internalized within 1–2 hr; at this point, there is a dynamic balance of uptake and release of intact ligand, and the two opposing processes are equal. The rate of ligand degradation and the accumulation of extracellular breakdown products increases during the first hour to a final steady state (solid line). This rate of degradation continues until the extracellular concentration of ligand approaches the K_d for binding to the receptor. At this point, the intracellular ligand pool also decreases, until this and virtually all of the external ligand have been degraded. The figure is based on data obtained with isolated rat hepatocytes (0.2 pmol receptor per 10^6 cells) in the presence of 1.5 μg ASOR (38 pmol) per 10^6 cells.

are at a steady state, one can view these receptor systems as very efficient protein-degrading machines within the cell. The reason they can be so efficient is because the receptors are recycled and reutilized.

B. Glycoconjugate Binding and Internalization

1. *The Coated Pit Pathway*

The class II receptor systems that have been examined mediate ligand uptake by a coated pit pathway. *Coated pits* are discrete organelles found in all eukaryotic cells. They were first characterized and implicated in endocytosis by Roth and Porter [159]. Coated pits are approximately 100–200 nm in diameter and occupy about 1–2% of the cell surface. Most cells have hundreds to perhaps a few thousand coated pits on their surface. The structure, assembly, purification, and function of coated pits have been reviewed by others [160–163]. The coat appearance in transmission electron microscopy is primarily owing to the protein clathrin, which can self-assemble in the presence of associated minor proteins to form basketlike cages. Coated pits are continually internalized and reformed at the cell surface. One can speak of a coated pit cycle, in which the components of the pit are conserved and reused, even though some of the structures are transient [161,163]. Coated pit internalization, disassembly, and reassembly occur constitutively. The half-life for this cycle is on the order of minutes [164]. A simple, initial model to understand coated pit recycling and receptor recycling is shown in Figure 4. It illustrates the three major phases

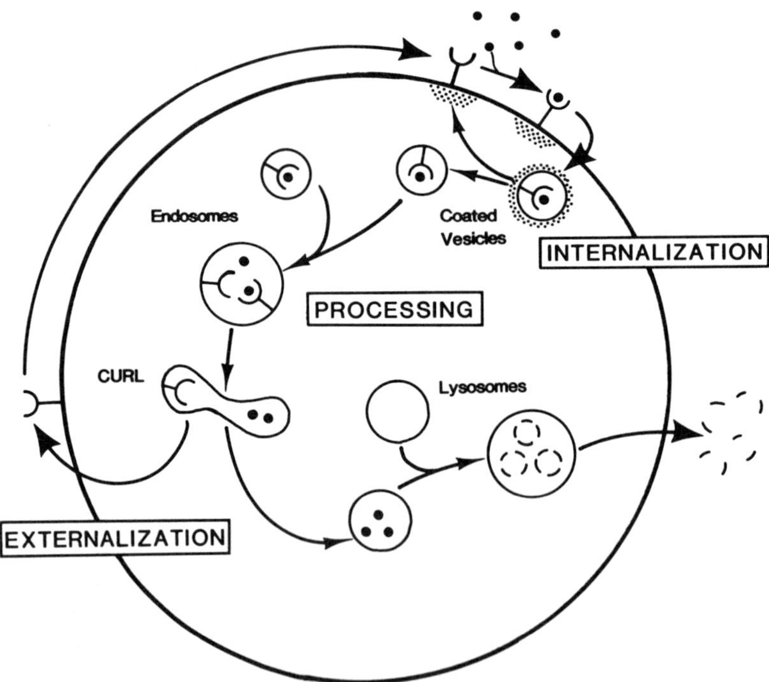

Figure 4 A "basic" model for receptor-mediated endocytosis and receptor recycling. A general scheme is depicted that can be broken down into three phases, as discussed in Section III.B.

for the process (indicated by the boxes); an internalization phase, an intracellular processing phase, and an externalization phase.

2. *Internalization*

The interactions between soluble glycoconjugate ligands and their membrane-bound receptors are generally of moderate to high affinity. Most investigators use ligands that, based on equilibrium binding analysis at 0–4°C, bind to receptor with dissociation constants (K_d) on the order of 10^{-10} to 10^{-8} M. In a few cases, such as the LEC HA receptor binding to large (M_r~10^6) HA, the affinity can be even greater (K_d~10^{-11} M). With K_d values in the nanomolar range, one can study ligand binding and processing by cells in vitro or tissues in vivo reasonably well, because the level of nonspecific interactions is usually acceptably low (≤10%), and the complex is relatively stable. Typically, when they have been measured, the on-rates (k_{on}) for ligand binding to receptor are on the order of $10^4 s^{-1} M^{-1}$ and the off-rates (k_{off}) for dissociation of receptor–ligand complexes are approximately $10^{-5} s^{-1}$. At 37°C, k_{on} should be even greater and, since the rate of internalization greatly exceeds the rate of dissociation (even at 37°C), endocytic capture of soluble carbohydrate ligands is very efficient, especially at low ligand concentrations.

In vivo clearance studies in rats, rabbits, or human (in the case of the HA receptor) indicate that carbohydrate ligands with high affinity for their receptors (K_a~$10^9 M^{-1}$) are efficiently and rapidly removed from the blood [13,14,88,125]. Typically, the receptor systems localized in the liver mediate clearance of circulating glycoconjugates with half-times on the order of 2–10 min. Because of the wide range of carbohydrate structures at least partially bound by the hepatic Gal, Fuc, Man/GlcNAc, and HA receptors, the liver is able to remove a very large variety of possible glycoconjugates. Anatomically as well, the liver is well designed as an effective filter. These features are hardly likely to be due to chance; therefore, we must consider that a significant general function of mammalian liver is the removal of various nonsialylated glycoconjugates. The rationale or teleologic basis for this function is discussed in Section IV.

3. *Receptor Targeting to Coated Pits*

After ligand binds to receptor on the cell surface, the receptor–ligand complex will diffuse in the plane of the plasma membrane and be captured by a coated pit [165]. In some cases, receptors are able to incorporate into a coated pit without having first bound the ligand. This ligand-independent targeting of receptors to coated pits must occur for receptors that recycle constitutively; that is recycle in the absence of ligand. Constitutive recycling is a general characteristic of class II receptors (see Table 2). In general, the class I receptors do not enter coated pits until they have bound their ligand; they then cluster and enter a coated pit.

The structural features of receptors that direct them to a coated pit are currently unknown. This area represents one of the most important problems now being examined by researchers in cell biology. No common primary protein sequence has been identified in the cytoplasmic domains of those coated pit receptors [163] for which structures are known (e.g., see Fig. 1). This list now includes the receptors for epidermal growth factor, insulin, LDL, the macrophage Man and Gal receptors, the two Man-6-P receptors (CD and CI), the chicken liver GlcNAc receptor, the liver Fuc receptor, and the rat and human hepatic Gal receptors. Despite that there is no consensus primary sequence among these receptors, they are all targeted to coated pits. These proteins must have a common feature of their secondary or tertiary structure that is necessary for incorporation into a coated pit.

The only common primary feature in the cytoplasmic domains of these receptors is the presence of a tyrosine or, occasionally, another aromatic residue, usually close to the membrane-spanning domain. Numerous studies with mutant receptors altered at this residue suggest that it is essential for endocytosis through the coated pit pathway [e.g., 166–169]. For example, mutation of one key aromatic acid in the transferrin [167,169] and LDL receptors [168] markedly reduces the rate of receptor internalization. Interestingly, mutation to a tyrosine in the hemagglutinin protein of influenza virus allowed this protein to then enter the coated pit pathway [166]. The tetrapeptide sequence NPXY (where X is any amino acid) was found in the cytoplasmic domains of ten endocytic receptors or cell surface proteins, including the LDL receptor [168]. Alteration of any of the residues in this sequence decreased internalization, suggesting its importance at least for LDL uptake [168]. Collawn et al. [169] proposed that a tight turn is the important structural feature required for entry into the coated pit pathway. They identified the tetrapeptide sequence YXRF in the cytoplasmic domain of the transferrin receptor as necessary for internalization. They also showed that, although the two sequences indicated are not the same, both favor a reverse-turn conformation or a closely related terminal turn of an α-helix [169]. Therefore, it now appears that entry into coated pits requires a common conformational feature of the receptor's cytoplasmic domain that provides the appropriate structural recognition.

After a receptor–ligand complex reaches a coated pit, it is internalized in a clathrin-coated vesicle, termed an *endosome* [165]. The vast majority of evidence indicates that a free internal vesicle buds off from the coated pit. The energy for this vesiculation process may come from the inherent thermodynamic driving force for the assembly of clathrin into a spherical basket structure [160–164]. This can occur in vitro under the appropriate conditions and could, therefore, occur as well in vitro, independently of ATP [21].

Many investigators studying receptor-mediated endocytosis have not clearly distinguished between measurements of continuous versus a single round of ligand uptake. One can measure a single wave of surface-bound ligand by binding radiolabeled ligand to cells at 4°C, at which temperature endocytosis does not occur; washing away the nonbound ligand, and then warming the cells to 37°C [21]. This protocol enables one to study individual steps in the processing pathways for ligand or receptor under conditions during which only one round of function is occurring. By using this protocol, Clarke and Weigel [170] showed that in severely ATP-depleted cells, the internalization phase of the overall endocytic process (see Fig. 4) still occurred normally in the hepatic Gal receptor system. The rate and extent of internalization of one round of surface-bound ligand was not affected by a 98% decrease in cellular ATP.

4. *Intracellular Processing*

The clathrin coat is removed from the internalized, coated vesicles by an ATP-dependent process [171], and the free clathrin can then go back to re-form coated pits on the cell surface [164]. Removal of clathrin from endosomes is an early event that occurs very soon after internalization and requires cytoplasmic proteins. The ATP requirement for clathrin uncoating is consistent with the idea that internalization per se is ATP-independent, but that energy must be fed into the coated pit cycle at other points to keep the cycle operating. After clathrin uncoating, numerous processing steps occur. Among these steps are the fusion of endosomal vesicles and the dissociation of ligand from receptor. Dissociation is an absolute requirement for a recycling receptor system to operate efficiently. There must also be a mechanism for segregating the ligand from the receptor and then delivering the ligand to a degradation compartment, such as lysosomes, where degradation can occur and degradation products will be released. In the externalization phase (see Fig. 4) the receptor is brought back to the cell surface in an active form to complete the cycle. The receptor is then ready to function again in another round of ligand uptake.

C. Multiple Pathways of Receptor-Mediated Endocytosis

Similar to any process, the real situation is usually more complicated than what one initially believes, and this is true for receptor-mediated endocytosis. Figure 5 shows our present working model for the hepatic Gal receptor system. We believe that this general two-pathway model also applies to most or all of the class II recycling receptor systems. Many investigators have not extensively pursued the details of possible multiple pathways. There is enough data from these other systems, however, to warrant the prediction that, when investigated, they will also be found to have two ligand-processing and receptor-recycling pathways.

Over the last few years, we have accumulated considerable evidence that two Gal receptor subpopulations and two ligand-processing endocytic pathways exist in rat hepatocytes [21]. A key point is that the pathways are parallel and not serial. Both isolated and primary cultures of rat hepatocytes endocytose desialylated glycoproteins through two functionally distinct Gal receptor subpopulations. The two receptor subsets, which we designate as state 1 and state 2 Gal receptors, operate in parallel to internalize, dissociate [172,173], and degrade [174,175] ASOR in two distinct pathways. Freshly isolated hepatocytes exhibit the state 1 pathway only, whereas hepatocytes equilibrated at 37°C following isolation exhibit both the state 1 and the state 2 pathways. Roughly half of the surface and intracellular Gal receptors are dedicated to each pathway. The state 1 pathway is a minor endocytic pathway. It represents no more than 20% of the glycoprotein-processing capability of the cells. The state 2 pathway is the major endocytic pathway.

The processing steps within each of these pathways comprise nine to ten discrete steps that can be identified experimentally (see Fig. 5). These involve processing of either the ligand or the receptor in the two different pathways. I will present the details of this model, some of the experimental evidence for it, and illustrate the kinds of experiments that one can do to examine the processing of internalized glycoproteins. Multiple receptor pathways and populations have also been postulated for the Man receptor system [176].

Table 4 is a summary of the characteristics that help define these two pathways. They include the following observations: In one pathway, but not in the other, the surface receptor activity can be modulated; that is, the activity can be lost and regained in a reversible way [177]. This occurs in the major state 2 pathway. In the same pathway, we can show that those receptors recycle constitutively, and that they undergo an inactivation–reactivation cycle that does not appear to occur in the state 1 pathway. *Diacytosis* (the return of internalized receptor–ligand complexes to the cell surface) occurs in the state 1, but not the state 2, pathway [178]. Dissociation of receptor–ligand complexes in these two pathways can be differentiated in at least three different ways. In the minor state 1 pathway, the rate of dissociation is very slow ($k = 0.014\ \text{min}^{-1}$) and in the major state 2 pathway it is 20-fold faster ($k = 0.28\ \text{min}^{-1}$). In the major pathway the receptor–ligand dissociation is blocked by the ionophore monensin, but still proceeds at 18°C.

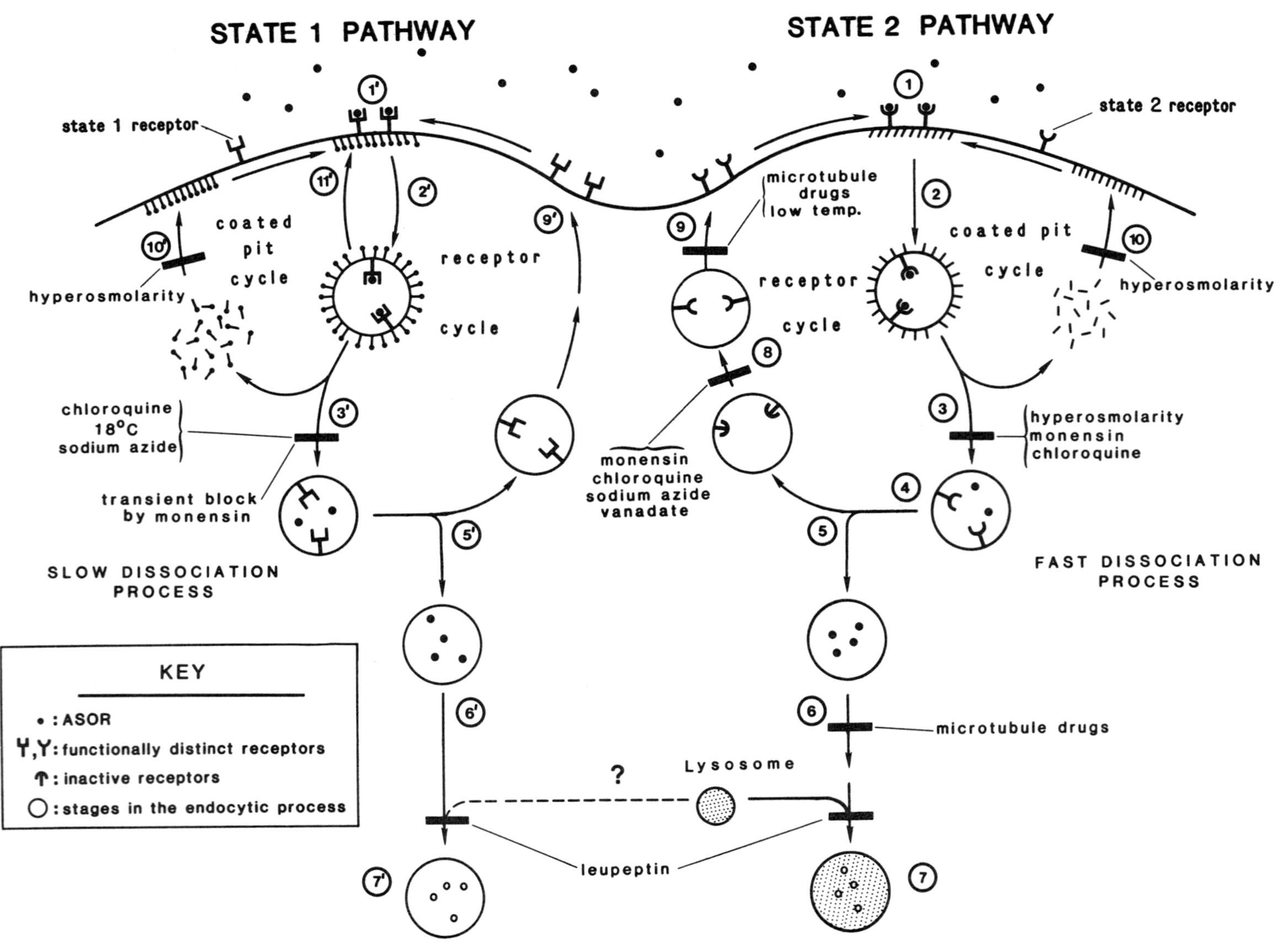
STATE 1 PATHWAY
STATE 2 PATHWAY
state 1 receptor
state 2 receptor
coated pit cycle
receptor cycle
hyperosmolarity
microtubule drugs low temp.
coated pit cycle
receptor cycle
hyperosmolarity
chloroquine 18°C sodium azide
transient block by monensin
monensin chloroquine sodium azide vanadate
hyperosmolarity monensin chloroquine
SLOW DISSOCIATION PROCESS
FAST DISSOCIATION PROCESS
microtubule drugs
?
Lysosome
leupeptin
1' 2' 3' 5' 6' 7' 9' 10' 11'
1 2 3 4 5 6 7 8 9 10
KEY
• : ASOR
Ⴤ,Y : functionally distinct receptors
↑ : inactive receptors
○ : stages in the endocytic process

Table 4 Characteristics of the Two Hepatic Gal Receptor Pathways[a]

Feature	The (minor) state 1 pathway	The (major) state 2 pathway	Ref.
Surface receptor modulation by temperature, monensin, ATP depletion, microtubule drugs, chloroquine	No	Yes	177,181,186
Receptors constitutively recycle	No	Yes	181,238
Receptors undergo an inactivation–reactivation cycle	No?	Yes	181,230,238
Diacytosis of internalized receptor–ligand complexes	Yes	No	178
Dissociation of receptor–ligand complexes	Slow	Fast	172,200
Ligand dissociation blocked by monensin	No	Yes	173,202
Ligand dissociation blocked at 18°C	Yes	No	172
Lag period before ligand degradation	No	Yes	174
Colchicine (1 μM) inhibits ligand degradation	No	Yes	175

[a]Two functionally distinct Gal receptor subpopulations and pathways are defined by nine characteristics. These features demonstrate differential effects on Gal receptor activity or function and ASOR processing in isolated rat hepatocytes in suspension and in culture. Some of these data have been reviewed previously [21].

Figure 5 The two-pathway model for endocytosis mediated by the hepatic Gal receptor system. The state 1 and state 2 Gal receptors and their respective parallel pathways are discussed in the text. The points of action of various inhibitors are indicated by the solid bars. The numbers denote the following stages in these pathways: 1,1′, ligand binding; 2,2′, receptor–ligand internalization; 3,3′, receptor–ligand dissociation; 4, state 2 Gal receptor inactivation; 5,5′, segregation of receptor and ligand; 6.6′, ligand delivery to a degradative compartment (lysosomes in the state 2 pathway, possibly an earlier endosome in the state 1 pathway); 7,7′, ligand degradation; 8, reactivation of state 2 Gal receptors; 9,9′, recycling of receptors to the cell surface; 10,10′, formation of new coated pits; 11′, diacytosis of receptor–ligand complexes in the state 1 pathway.

Just the opposite occurs in the state 1 pathway. In terms of ligand degradation, the state 2 pathway has a lag period before degradation products are released into the medium, whereas the state 1 pathway has no lag. In the major state 2 pathway, degradation also is 30 times more sensitive than in the minor state 1 pathway to inhibition by microtubule drugs, such as colchicine [175]. The following sections will discuss the dynamics of surface and intracellular receptors, ligand dissociation, ligand degradation, and then, summarize some of the recent data on the inactivation–reactivation cycle in the major state 2 pathway.

D. Modulation of Cell Surface Receptor Activity in the Absence of Ligand

It is generally thought that Gal receptors, like other class II receptors, migrate along a membrane pathway from the cell surface through the cell interior, then back to the plasma membrane [157]. Thus, at any particular moment, Gal receptors are present on the surface and the interior of hepatocytes, with 50–90% of all cellular receptors located intracellularly [13,14,21,179]. Maintenance of normal Gal receptor activity and content on the cell surface, however, is very sensitive to a variety of perturbations. In the absence of added ligand, hepatocytes lose surface receptor activity when treated with metabolic energy poisons [170,180,181], monensin [182,183], local anesthetics, weak bases [184,185], or after incubation either at low (24–37°C) temperatures [186] or with desialylated glycoproteins [187].

1. Evidence for Two Subsets of Active Surface Receptors

In virtually all of these cases, only about half of the surface Gal receptor activity was depleted, suggesting that surface Gal receptors may be divided into subsets of perturbant-sensitive and perturbant-insensitive receptors. McAbee et al. [177] characterized and compared the modulation of surface Gal receptor activity elicited by several inhibitors to determine if lost surface receptor activity constituted only one of the two subsets of Gal receptors. In the absence of ligand, 37°C-equilibrated hepatocytes lost 40–60% of their surface Gal receptor activity after treatment with monensin, chloroquine, colchicine, vanadate, or sodium azide. These cells, consistently retained residual, nonmodulatable Gal receptor activity, regardless of the perturbant concentration, length of incubation, or even when treated with multiple inhibitors, either simultaneously or in sequence. Moreover, these agents blocked the expression of state 2 receptor activity on freshly isolated cells incubated at 37°C, but did not decrease state 1 Gal receptor activity already expressed on these cells. Only the state 2 Gal receptor activity was sensitive to these inhibitors. Therefore, state 1 and state 2 Gal receptors represent the inhibitor-insensitive and inhibitor- sensitive receptor populations, respectively (Fig. 6).

The state 1 and state 2 Gal receptors may react differently to the actions of these inhibitors because the two receptor types exist in independent functional pathways. For instance, Gal receptors partition about equally into and out of coated membrane on the surfaces of hepatocytes both in vivo and in vitro [188]. Modulatable state 2 Gal receptors may constitutively be targeted to coated pits on the cell surface and recycle as they go "along for the ride" in these recycling endocytic structures. Nonmodulatable state 1 receptors may enter coated pits only when triggered by binding ligand. Hyperosmotic medium, which disrupts clathrin-coated pit formation and function [132,133], prevents these inhibitors from eliciting a loss of surface Gal receptor activity [189]. Thus, receptor

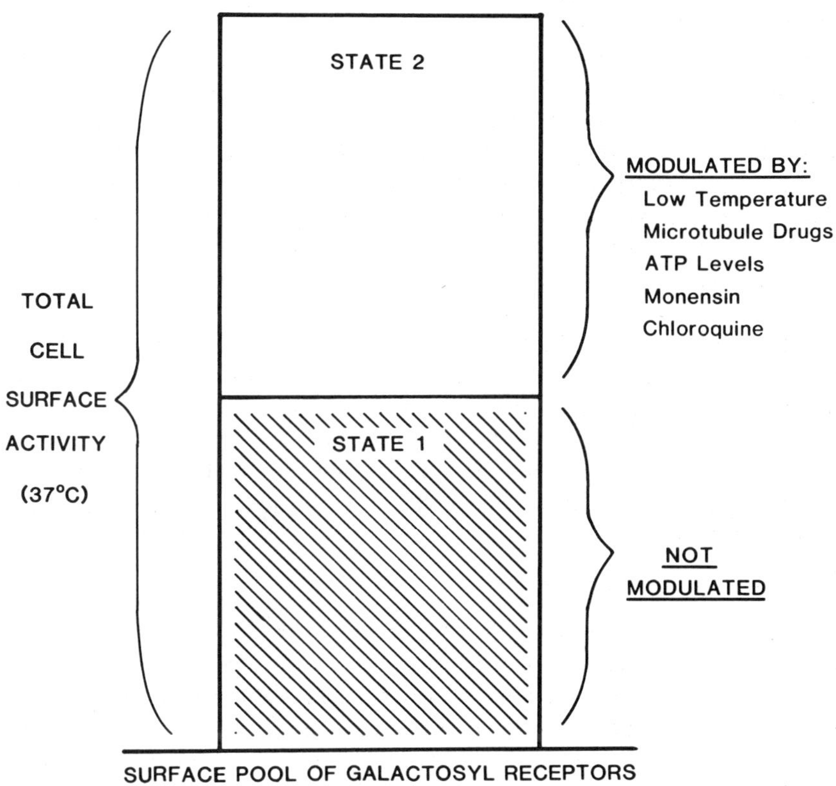

Figure 6 Modulation of "sensitive" cell surface receptor activity. The indicated inhibitors or treatments cause a loss of state 2, but not state 1 Gal receptor activity on the hepatocyte surface at 37°C, even in the absence of ligand. Sodium orthovanadate also selectively modulates the state 2 Gal receptor pool [259].

modulation apparently occurs by a clathrin-coated pit pathway, and nonmodulated receptors do not enter this pathway until they bind ligand. A similar explanation was postulated for the modulation of LDL receptors by monensin, which was only partial in the absence of ligand, but was complete when LDL was added [190].

It is important to note that the inhibitors discussed in the foregoing also induce similar patterns of surface receptor modulation in the absence of ligand in other carbohydrate receptor systems. Weak bases, such as chloroquine and NH_4Cl, reduce surface Man-6-P receptor activity to a similar extent [191] as well as Man receptors on alveolar macrophages [192]. Modulation of Man receptor activity on the surface of macrophages also occurs in the presence of phenylarsine oxide, calcium ionophores, or phorbol esters [193,194]. Surface receptor topography is also affected in other class II recycling receptor systems. Similar decreases in surface α_2-macroglobulin receptor activity were observed when alveolar macrophages were treated with methylamine [195]. Fibroblast LDL receptor activity is reduced by about 50% after monensin treatment in the absence of ligand [190].

Since these agents differentially affect the two subpopulations of Gal receptors [177], and have similar effects on other receptor systems, it is likely that other receptor systems also exhibit multiple receptor subsets and recycling pathways. Indeed, evidence has been reported that both the transferrin receptor system [196] and the Man receptor system [176] exhibit multiple-receptor pools and recycling pathways. Thus, differential sensitivities of migratory receptors to various perturbants may be diagnostic for the presence of multiple-receptor subsets and/or recycling pathways.

2. *Molecular Basis for Receptor Modulation*

The nature of the cell surface receptor activity reduction has been characterized in several instances: (1) Fiete et al. [183] showed that monensin-treated hepatocytes accumulated inactive Gal receptors on the cell surface, with only a minor loss of surface receptor protein. (2) Intracellular accumulation of surface Gal receptors was demonstrated on human hepatoma cells treated with weak bases [185]. It was not determined in this study whether these redistributed receptors were active or inactive. (3) We reported that ATP depletion caused an intracellular entrapment of surface state 2 Gal receptors [181]. In addition, these intracellularly accumulated Gal receptors were inactive.

The loss of surface receptor activity from inhibitor-treated cells, in the absence of ligand, is usually interpreted to mean that intracellular entrapment of receptors has occurred. Most investigators assume that modulation of receptor activity is evidence for constitutive recycling of receptors. This, however, has been demonstrated directly in only the two instances indicated in the foregoing. This explanation does not allow for the retention of inactive receptors on the cell surface [183].

E. Function of the Large Intracellular Receptor Pools

Active intracellular receptors can be detected by the protocol depicted in Figure 7A. It is a hallmark of class II recycling receptor systems that there are an equal number of or more receptors inside the cell than on the cell surface. This is true for the Gal receptor in rat hepatocytes [13,21,179], the LEC HA receptor [127], the macrophage Man receptor [194], and the CI Man-6-P receptor [44,45]. Although initially reported not to be true in human HepG2 cells, these cells, in fact, also have this typical distribution [179]. Likewise, the LDL receptor in fibroblasts is also now known to be intracellular [197].

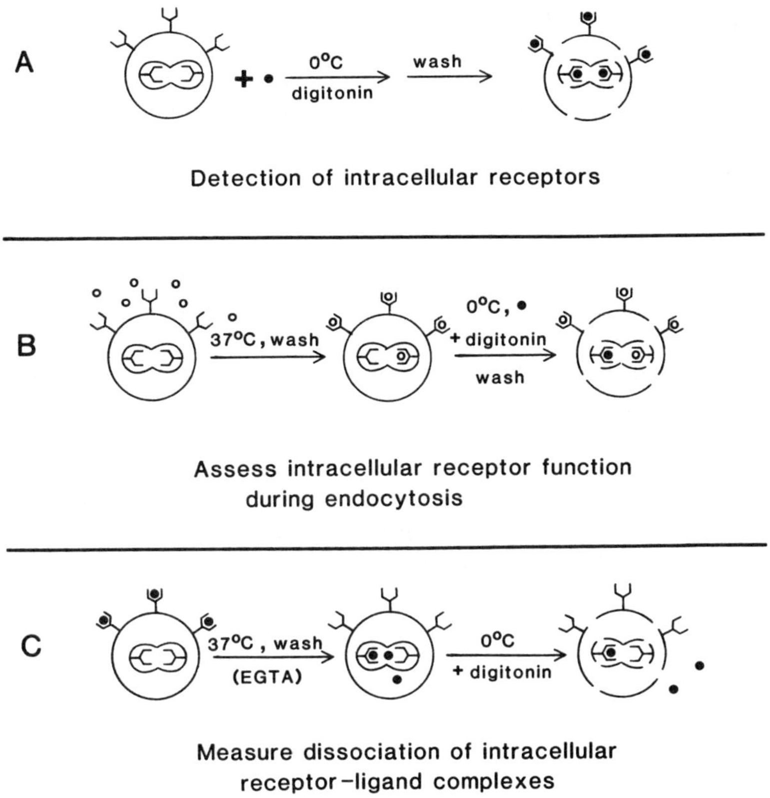

A — 0°C digitonin → wash →

Detection of intracellular receptors

B — 37°C, wash → 0°C, • + digitonin, wash →

Assess intracellular receptor function during endocytosis

C — 37°C, wash (EGTA) → 0°C + digitonin →

Measure dissociation of intracellular receptor–ligand complexes

Figure 7 Use of the permeabilizing detergent digitonin to assess intracellular receptor number or function and ligand dissociation. The three different protocols depict how digitonin has been used in the hepatic Gal receptor system [172,198,199] to characterize either (A) intracellular active receptor content, (B) intracellular receptor function, or (C) dissociation of internalized receptor–ligand complexes. Filled and open circles are radioactive and competing nonradioactive ligand, respectively.

There are two general possibilities for the function of the large pool of internal receptors: (1) They are part of the "train" of receptors moving through the recycling pathway (see Fig. 5). At any instant only a fraction of receptors in the entire pathway is on the cell surface, the remainder would be in intracellular compartments at various stages along the pathway. On the basis of Gal receptor inactivation studies, discussed in the following, we believe that the internal recycling receptor pool is at least the same size as the fraction of surface receptors that participate in recycling. This is the most likely function. (2) Alternatively, the intracellular receptors are "secondary carriers" involved in shuttling ligand from incoming recycling receptors to internal compartments along the pathway to lysosomes. This is an unlikely function, but difficult to rule out.

Attempts to determine whether the intracellular pool of Gal receptors in hepatocytes was functional were technically difficult and investigators initially concluded that these receptors were not functional. It was subsequently shown, by the protocol depicted in Figure 7B, that these intracellular receptors in isolated hepatocytes become occupied with ligand in a time- and concentration-dependent manner [198]. When cells have reached a steady-state rate of ligand uptake in the presence of excess ligand, at least 70% of the intracellular receptors and >90% of the cell surface receptors are bound to ligand. The remaining 30% that are unoccupied probably represent recycling receptors that have dissociated from the ligand, have passed the points of segregation or inactivation and reactivation, and are on the way back to the cell surface (see the externalization phase in Fig. 4; step 9/9′ in Fig. 5). A small fraction of active intracellular receptors (a small percentage) may represent molecules in the biosynthetic pathway. Virtually the same results were obtained with the HA receptor in LECs [129].

There can be little doubt that most intracellular carbohydrate receptors are functional. However, the question of whether all cellular receptors in a particular system are functionally equivalent has not been answered, nor even addressed in any receptor system. No one knows whether a given receptor will traverse any or all of the various cellular compartments that can contain that receptor. It is almost always tacitly assumed that all receptors are "created equal." This need not be true [21].

In calculating receptor-recycling times, many investigators fail to take into account the presence of the intracellular receptor pool and, consequently, underestimate the time needed for a given receptor to traverse the recycling pathway. What they often calculate is the time needed to replace the net number of endocytic surface receptors.

F. Dissociation of Internalized Receptor–Ligand Complexes

1. *The Kinetics of Dissociation*

To measure the rate of dissociation of receptor–ligand complexes, Weigel et al. [199] developed a general protocol (see Fig. 7C) involving the nonionic detergent, digitonin, which will permeabilize, but not solubilize, cells. Cells are allowed to take up radioactive ligand at 37°C, sampled at different times, chilled on ice, and washed with EGTA (or low pH) to remove surface-bound ligand. A portion of the cells is then washed with a buffer that contains digitonin at an optimal concentration. The control cells are washed in cold buffer without digitonin. The digitonin will put holes in membranes throughout the cell and allow any ligand that has already dissociated from the receptor, and therefore is free, to leave the cell. Free ligand can then be separated from the receptor-bound ligand, still associated with the cell ghost, by centrifugation.

The EGTA-resistant radioactivity associated with the cell is operationally defined as being endocytosed. Within 2–4 min after transfer to 37°C, virtually all of the surface-bound ligand is inside the cell (Fig. 8). This rapid endocytosis is characteristic of receptor–mediated uptake by the coated pit pathway. If parallel samples are washed with the digitonin-containing buffer, then one obtains the lower dashed curve in Figure 8. The difference between these two curves represents the pool of intracellular free or dissociated ligand. The area under the lower curve represents the receptor-bound ligand that has not yet dissociated, but that is inside the cell.

The surprising result obtained, when the kinetics of ligand dissociation from hepatic Gal receptors was measured, was that the increase in the intracellular free-ligand pool was biphasic [172]. For a few minutes, there is a very rapid rate of dissociation of internalized complexes, until roughly half of the initial surface receptor–ligand complexes are dissociated (see the shaded area in Fig. 8). However, the remaining complexes then dissociate only very slowly over a relatively long period. It takes 1.5–2 hr, in fact, to get complete dissociation of the complexes endocytosed during one synchronous wave of endocytosis.

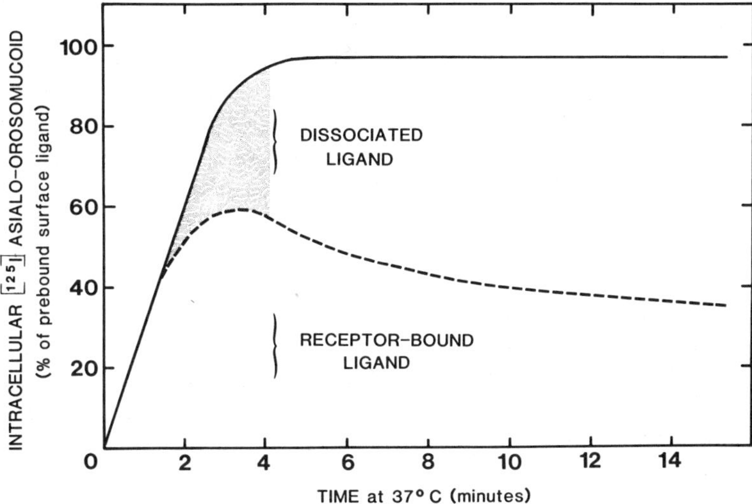

Figure 8 Biphasic dissociation of internalized receptor–ligand complexes. The solid line shows the time course for the internalization of one wave of surface-bound ^{125}I-ASOR by isolated rat hepatocytes at 37°C. Endocytosed ligand is determined by its resistance to release by EGTA at 4°C. The dashed line represents the radioactivity associated with cells that were first washed at 4°C with an EGTA-containing buffer and then with buffer containing digitonin to permeabilize the cells and allow dissociated ligand to leave the cell (See Fig. 7C). All intracellular complexes dissociate within 1–2 hr. The shaded area represents the rapidly dissociated ligand in the state 2 Gal receptor pathway. (Adapted from Ref. 172.)

The same result was obtained by Harford et al. [200], who used a different technique: detergent solubilization followed by selective precipitation.

When the same experiment was performed at 18°C, endocytosis of the surface-bound ligand still occurred, although it proceeded more slowly at this lower temperature [172]. Nonetheless, internalization went to completion, and dissociation of what corresponds to the rapidly dissociating intracellular complexes also occurred. At 18°C, however, the receptor-ligand complexes that were in the slowly dissociating compartment did not dissociate at all. The slow dissociation process occurs from 37°C down to 22°C, but below 18–20°C, this process is preferentially inhibited. Interestingly, those receptor–ligand complexes that would normally dissociate very slowly at 37°C are still able to return to the cell surface at 18°C [21]. This process, termed diacytosis (see step 11′ in Fig. 5) occurs in all class II recycling receptor systems, and it does not usually result in the release of free ligand into the medium. However, if EGTA is present in the medium, the ligand will be released from the receptor at the rate of about 7%/min of what is in this intracellular pool (corresponding to a $t_{1/2} \cong 10$ min). Despite that these receptor–ligand complexes continually go back and forth between the cell surface and this intracellular compartment, they do not enter the compartment where the fast dissociation process can occur, even at 18°C. This is one of the reasons for believing that the two ligand processing pathways are parallel, not serial, pathways or that perhaps the two pathways are in different hepatocyte subpopulations.

The kinetics of dissociation of Man-BSA after internalization by alveolar macrophages was determined by a detergent solubilization and precipitation assay [201]. The first-order rate constant for ligand dissociation was $k = 0.087\ \text{min}^{-1}$, considerably slower than the value ($k \geq 0.28\ \text{min}^{-1}$) obtained for the state 2 Gal receptor in hepatocytes [172,200]. However, since both endocytosis and dissociation were measured in the former study, the value obtained for dissociation must be an underestimate of the intrinsic dissociation rate. It is also interesting to note that the kinetics of dissociation were not measured to completion (only ~60%) in these studies, so that a biphasic dissociation curve might have been obtained. Other characteristics of the Man receptor, including modulation of only a portion of surface receptor activity [192] and diacytosis of internalized receptor–ligand complexes [176] are also common characteristics between this and the Gal receptor system.

Wileman et al. [201] showed, in a series of elegant experiments, that isolated endosomes containing Man-BSA–receptor complexes will dissociate ligand in the presence of ATP. This process was ATP-specific and was blocked by proton ionophores and inhibitors of the endosomal ATP-dependent proton pump. To date, this is the strongest evidence that receptor–ligand dissociation may be mediated by low pH in vivo.

2. *Dissociation Occurs Differently in the Two Cellular Pathways*

Another indication that there are two different asialoglycoprotein-processing pathways in hepatocytes was seen when the monensin sensitivity of ligand dissociation was examined [173]. A number of investigators have used monensin, which is an ionophore specific for Na^+ and H^+, to collapse intracellular pH and pNa gradients and to inhibit the dissociation of internalized receptor–ligand complexes. For example, Harford et al. [202] showed that monensin blocks ligand dissociation in the hepatic Gal receptor system. However, they studied periods of up to only about 45 min. When we examined longer times after internalization of surface-bound radioactive ligand, we found an interesting result [173]. In the control without monensin, the radioactivity in the cells decreases after about an hour because degradation occurs. In the presence of monensin, this degradation is inhibited. When the control cells are washed at 4°C in the presence of digitonin to assess the rate of dissociation, the rapid dissociation was complete in minutes, and the slow dissociation continued for several hours (as in Fig. 8).

In the presence of 20 μM monensin there was virtually no dissociation for the first 40 min, consistent with the results of Harford et al. [202]. However, after 40 minutes, ligand dissociation resumed and proceeded at a slow rate, similar to the control's slow rate [173]. When another addition of monensin was made, neither the onset of this delayed dissociation nor the rate was affected. Ligand dissociation in the presence of monensin continued until a plateau was reached; not all of the complexes dissociated. The receptor–ligand complexes that do not dissociate, correspond to those complexes that, in the control cells, would dissociate by the rapid state 2 pathway. The state 1 Gal receptor–ligand complexes were the ones that dissociated in the presence of monensin. We concluded from these studies that the partial inhibition by monensin is due to the introduction of a lag period in a step before dissociation of receptor–ligand complexes in the slow-dissociation state 1 pathway only. The fast-dissociation process in the state 2 Gal receptor pathway is completely blocked, whereas monensin delays, but does not stop completely, either a processing step or a translocation step in the state 1 Gal receptor pathway. These two kinetically distinguishable processes must reflect two different pathways by which ligand is processed, because one can proceed even when the other is inhibited.

3. *The Current Model*

In the current model for receptor–ligand dissociation [157], it is believed that acidification of the intraluminal volume within the endosome is the cause of dissociation. The acceptance of this model rests primarily on indirect correlative evidence. First, the complexes between ligands and most of both class I and class II receptors are sensitive to pH. Low pH, on the order of pH 5.5, will prevent ligand binding or will disrupt receptor–ligand complexes. Second, agents that disrupt or collapse pH gradients were found to block intracellular receptor–ligand dissociation in live cells.

It is clear that the intracellular compartments of the endocytic pathway become increasingly acidic, and an increasing pH gradient is established between the cytoplasm and the lumen of these vesicles [203–205]. There is a gradient in the magnitude of these pH gradients as one moves from endosomes near the cell periphery (pH ~6.5–7.0) to lysosomes deep in the cell interior (pH 5.0–5.5). Both endosomes and lysosomes contain ATP-dependent membrane-bound proton-translocating pumps, which are responsible for generating the pH gradients in vivo. Many investigators have shown (for example, with ligands containing pH-sensitive probes, such as fluorescein) that after endocytosis, the molecules sequentially pass through compartments of successively lower pH [165,179,206–209]. This has been demonstrated both with fluid-phase markers, which may or may not be internalized exclusively in coated pits, and with many different ligands specific for receptors that use the coated pit pathway, including the hepatic Gal receptor [206,207,209,210].

Although it is assumed that the receptors inside the cell have the same pH sensitivity as the cell surface receptors, this has rarely been tested directly. By using the trick of digitonin permeabilization (see Fig. 7), we were able to do this in isolated rat hepatocytes. We compared the pH sensitivity of receptor–ligand complexes, pre-formed on the cell surface at 4°C, with those complexes after internalization in live hepatocytes (Fig. 9). To our knowledge, such an experiment has not been done in any other receptor system. The initial cell surface receptor pool and the pool of internalized receptor–ligand complexes had identical sensitivity to decreasing pH. They both showed stable binding, until a threshold at about pH 6.2, and complete dissociation occurred at pH 5.4. This curve is typical of the pH sensitivity reported for cell surface receptors in other systems.

The structural basis for changes in ligand-binding affinity at low pH for the Gal or any other pH sensitive receptor is unknown, although several workers have shown physical changes in these proteins at low pH [14,16,20,179]. It remains to be determined whether the conformation of the ligand-binding COOH-terminal domains of these proteins (see Fig. 1A) is affected directly or if other regions of the proteins are affected.

4. *Ligand Dissociation Can Occur Without Acidification*

Although the role of pH in mediating ligand dissociation inside the cell is assumed to be direct, because low pH dissociates receptor–ligand complexes in vitro, there are numerous inconsistencies in this model.

One of the two hepatic Gal receptor pathways does not require low pH for ligand dissociation. Dissociation of state 1 Gal receptor–ligand complexes occurs in the absence of intracellular pH gradients under conditions (i.e., plus monensin) that block dissociation of complexes in the state 2 pathway.

The LEC HA receptor is a class II recycling receptor system that must depend on efficient ligand dissociation for rapid receptor recycling [129]. In this receptor system,

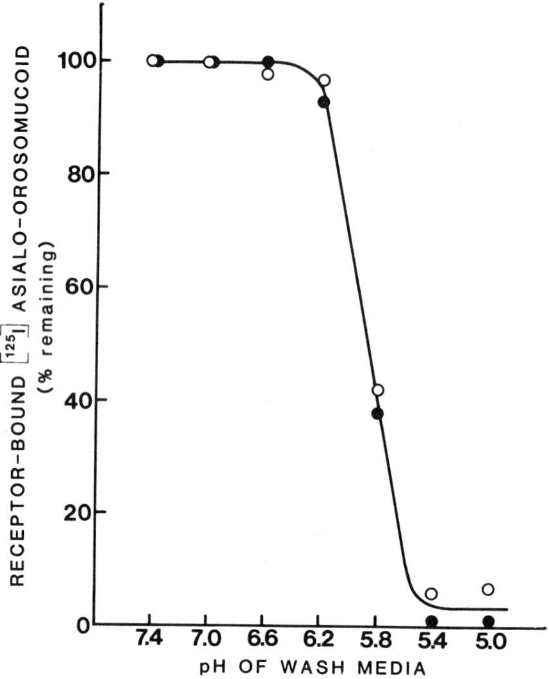

Figure 9 pH sensitivity of cell surface or internalized ASOR–Gal receptor complexes. Hepatocytes (2×10^6/mL) were allowed to bind ^{125}I-ASOR (1.5 μg/mL) at 4°C for 60 min, and unbound ligand was removed by centrifugation. Half of the cells were left at 4°C (open circles) and half were put at 37°C for 15 min (solid circles) to allow endocytosis to occur. Aliquots of cells were then washed, treated with 0.055% digitonin on ice for 10 min, washed again, and put through two 20-min incubations and washes in buffers at the indicated pHs. After a final wash, radioactivity in the cell samples was determined. In this experiment, the receptor–ligand complexes assessed after internalization are all in the state 1 Gal receptor pathway (see Fig. 5). MES was used as the buffer for pHs below 6.3 and HEPES for pHs above 6.3.

however, receptor binding to ligand is not decreased at low pH; in fact, it is increased [127]. In an acidic endosome, binding would, therefore, be even stronger, not weaker.

In the studies in which the kinetics of acidification have been measured, it usually takes about 5 min before internalized ligand is in an environment at pH 6 [203–209]. The compartment in which dissociation occurs has been estimated to be at pH 6–6.3 [209]. Yet, in the few cases for which the kinetics of receptor–ligand dissociation has been measured in live cells [172,200], dissociation of a complete wave of surface-bound ligand is essentially over by this time (for example, in the state 2 Gal receptor pathway). Even if the early (<5 min) endosome acidifies to pH 6, very little dissociation would be expected (see Fig. 9).

Although the affinities of receptors at pH 5.5 are dramatically less than at pH 7 (see Fig. 9), there may still be an interaction between ligand and receptor. An often overlooked feature of receptor-mediated endocytosis is that the ligand concentration within an early endosome can be greater by ~10,000-fold than in the extracellular medium. Consequently, even if the affinity of a receptor decreases to a K_d of 10^{-6} M from 10^{-9} M, the free ligand concentration in the endosome would still ensure that a large percentage of the receptors were still occupied inside the cell. For example, ASOR present at an external concentration of 1 nM would be internalized in hepatocytes at a steady-state rate of about 250 molecules per cell per second. This probably represents about 10–20 coated pits budding off per second to form intracellular coated vesicles, with an average diameter of 150 nm. The total volume occupied by 15 endosomes would therefore by $\sim 27 \times 10^6$ nm^3 (1 nm^3 = 10^{-27}m^3 = 10^{-21}cm^3 = 10^{-21} mL = 10^{-24} L). The concentration of ligand in this volume element is, therefore $[(250) \div (6.023 \times 10^{23})] \div (27 \times 10^{-18}\ \text{L}) \approx 1.6 \times 10^{-5}$ M. Under these conditions, the receptor affinity would have to decrease by about six orders of magnitude to ensure complete (>99%) and rapid dissociation in the acidified endosome. There is no evidence to support such dramatically great changes in receptor affinity at pH 5.5.

In a very small intracellular volume, such as an endosome, there are only ~10 protons at pH 5.5. It is difficult to conceive of a mechanism that will effectively employ so few protons to affect a larger number of receptor–ligand complexes.

Ligand binding by many class I receptors is also sensitive to low pH [15,16], even though these receptors are not recycled, but are usually degraded or down regulated. Therefore, there is no special correlation between the fact that a receptor can be recycled and that its affinity for ligand is decreased at lower pH.

5. *Alternative Reasons for Cellular pH Gradients*

There is an alternative explanation for the present data regarding the role of acidification in dissociation. Rather than directly mediating receptor–ligand dissociation, the pH gradient established in the endosomal pathway may provide energy for the vectorial movement of membranes or fluid between the plasma membrane and lysosomes. Drugs and inhibitors that collapse pH gradients and block ligand dissociation, thus, would do so indirectly by preventing translocation (or maturation) of early endosomes to the point in the intracellular pathway at which dissociation occurs. One reason to consider this alternative as a possible explanation is that the secretory pathway in which proteins are exocytosed also has a pH gradient from the beginning to the end of the pathway. The pH decreases in going from the cell interior to the cell perimeter [203]. Another reason is that Davoust et al. [211] found that endocytosis was stopped by lowering the external pH to 5.7. Under these conditions, the intracellular pH dropped to 6.2; the endosomal pH was not raised, but the cytoplasmic–endosomal lumen pH gradient was destroyed. Although receptor–ligand dissociation could still occur, endocytic traffic inside the cell stopped.

An attractive explanation to explain dissociation is a mechanism that inactivates the receptor, such that the ligand dissociates and then cannot rebind to the receptor. The high concentration of free ligand in the endosomal pathway would, therefore, not hinder the efficient segregation of receptor, which is to be recycled, from ligand, which is to be degraded. As discussed in the following section, we have shown that state 2 hepatic Gal receptors undergo a reversible inactivation and reactivation during their constitutive recycling itinerary.

G. Segregation of Ligand and Receptor: Intracellular Processing

1. *Identification of Endocytic Organelles*

Very little is known about the mechanism by which the dissociated ligand and unoccupied receptor are separated from one another and directed along their separate intracellular routes. Physically, this problem reduces to how the cell separates a fluid volume, the lumen of the endosome containing the now soluble ligand, from the surrounding membrane, containing the receptor. The cellular location in which segregation occurs is not well-defined.

By using the sophisticated combination of time-lapse video fluorescence microscopy and image intensification, Herman and Albertini [212] measured cytoplasmic organelle movements during endosome translocation. These workers showed that endosomal movement occurs in two stages: an initial saltatory motion at a mean velocity of 0.03 μm/sec; and a later continuous unidirectional movement toward the cell interior at a rate of 0.05 μm/sec. In this second phase, vesicles moved over distances of 5–40 μm until they fused with phase-dense structures, presumably lysosomes. Microtubule drugs blocked all phases of endosome movement. An interesting observation in this study was that lysosomes were initially throughout the cytoplasm and were also induced, by the presence of the ligands used, to undergo the same two phases of saltatory motion and centripetal intracellular translocation as did endosomes.

Pastan, Willingham and their coworkers characterized an endosomal compartment in fibroblasts that they designated a *receptosome* [165]. Although this terminology is not widely used, these workers helped focus attention on the problem of intracellular sorting. Others have noted, in a variety of endocytic systems, that after 15–30 min of exposure, ligands become localized in multivesicular bodies [165]. Geuze et al. [213] identified an organelle in parenchymal cells in intact liver that contained a central vesicle with tubular extensions. After internalization of asialoglycoproteins, the vesicular lumen contained free ligand, whereas the Gal receptor was predominantly concentrated in the tubular membranes. This organelle was called CURL; *c*ompartment of *u*ncoupling *r*eceptor and *l*igand. The different surface area/volume ratios of the vesicle and tubular portions of CURL would be well suited for an efficient separation of soluble ligand into the portion with the smaller ratio (vesicles) and the membrane-bound receptor into the portion with the larger ratio (tubules). These investigators proposed that the CURL vesicles transform into secondary lysosomes, in which the ligand is degraded, and the tubular vesicles are an intermediate in the pathway by which receptor is recycled back to the cell surface.

This important proposal has not been verified yet, nor has it been possible to purify the CURL organelle in vitro, although some isolated subcellular fractions do contain structures with this tubular–vesicular structure [214]. This type of compartment, however, has been identified in other cell types, including Vero cells [215], fibroblasts, reticulocytes, and macrophages [216]. Tubular–vesicular organelles probably occur in all cells. Many investigators believe, in fact, that in the liver cell, these various types of endosomes, transfer tubules, or others, exist as an intracellular continuous network, linking the various "parts" of the endocytic network. Only when the cell is treated with fixatives for electron microscopic (EM) studies or intentionally disrupted to generate subcellular fractions does fragmentation and vesiculation occur, which then results in the separation of apparently discrete "organelles." In support of this idea, Kessell et al. [217] have characterized a temperature-sensitive mutant in *Drosophila*, in which the endocytic membrane tubule network is reversibly disrupted by a temperature shift.

The terms receptosome, endosome, and CURL are used by various workers to designate the intracellular compartments that contain internalized ligand and receptor. There has been, and still is, confusion in this area, concerning nomenclature and the details of intracellular endocytic processing pathway. Several points are generally agreed on, however:

1. The endocytic pathway is acidic, as discussed earlier, and this has been demonstrated at the EM level [205] and in live cells by many investigators [203–208].
2. Receptor–ligand dissociation occurs early after endocytosis and is relatively rapid, whereas segregation of ligand and receptor occurs more slowly.
3. In the major endocytic (state 2) Gal receptor pathway, there is usually a lag of about 15 min before ligand degradation products are released into the medium. In this pathway microtubules are involved in delivering the segregated ligand to lysosomes [218]. The lag period represents the time needed to transport ligand from the cell surface to the cell interior, the so-called Golgi-associated endoplasmic reticulum-lysosome (GERL) region, where the bulk of the cellular lysosomes reside [212]. This distance can be on the order of 5 μm; roughly 25 times the diameter of an early endocytic vesicle.
4. When cells that have been endocytosing a ligand are disrupted by homogenization, cavitation, or other methods, at least three distinct subcellular fractions containing ligand can be separated by various centrifugation methods [219–221]. These can all be considered endosomal fractions, yet they must be derived from different organelles or different regions of a larger complex organelle. Most investigators have assumed that there is a sequential relation among the three endosomal compartments, since there is a time-dependent, apparently sequential appearance of ligand in the lumen of such vesicles. However, a major complication in interpreting these studies is the growing awareness that there are multiple endocytic pathways in most cells, even for the same ligand. If, as we believe for the hepatic Gal receptor, these are parallel, rather than serial (or branched) pathways, then the interpretations of subcellular fractionation studies of intracellular ligand translocation are even more difficult. Many earlier studies may need to be reinterpreted in this light, and the mechanism(s) of intracellular ligand and receptor processing may not be solved for some time. This general problem represents an important and difficult current question in cellular biology.
5. Different receptors can endocytose their respective ligands and enter the same endosome [165]. This has been shown in many different cell types and with many different receptor–ligand combinations. Although this result may have been initially surprising, it clearly presents no problem to the cell's ability to sort these molecules correctly. The endocytic-processing machinery is able to separate accurately ligands that are to be degraded from those that are being transcytosed across the cell. Likewise, receptors that are being recycled are directed along their appropriate pathway, while receptors that are to be down regulated or degraded are directed to a different "correct" pathway.

2. *In Vitro Reconstitution of Intracellular Processing*

In an effort to understand the molecular mechanisms of ligand-processing and receptor-recycling events that occur inside the cell after endocytosis, many investigators have developed cell-free in vitro systems [222]. These attempts generally fall into two categories: (1) reconstitution of endosomal vesicle fusion using vesicles isolated from two different cell samples that have endocytosed different ligands [223–225]; or (2) in vitro reconstitution in vesicles of a specific function, such as receptor–ligand dissociation, that would have occurred in the cells used had they not been disrupted for the experiment [201,226]. An extensive summary of this very active field [see 222] is beyond the scope of this review.

The difficulties in getting to the molecular level are enormous and reflected in the fact that few investigators have published more than one or two papers on the initial characterization of these systems. An exception is Rothman's group, which has succeeded in identifying and purifying factors participating in the early steps following endocytosis, particularly clathrin uncoating. In a series of elegant studies, Schmid, Rothman, and their co-workers have elucidated the mechanism by which a clathrin-uncoating ATPase dissociates and sequesters clathrin triskelions from coated vesicles [227]. The uncoating ATPase is present intracellularly in stoichiometric amounts relative to clathrin and is a member of the 70-kd family of heat-shock proteins [228].

Another exception is Stahl's group. Diaz et al. [224] have ingeniously used a two-ligand system targeted to the macrophage Man receptor. They used dinitrophenyl (DNP)-β-glucuronidase endocytosed in one group of J774E cells (a macrophage-derived cell line) and mannosylated anti-DNP IgG (a monoclonal antibody) in a second group of cells. Fusion events between vesicles (endosomes) prepared from the two groups of cells were detected by the recovery of enzyme activity in antibody–antigen complexes. Fusion was dependent on temperature, time, ATP, and the presence of both cytoplasmic and membrane proteins. Systems such as this have the potential to yield future details about the molecular events during intracellular processing.

These reconstitution systems, together with the elegant reconstitution of coated pit assembly by Moore et al. [164], may enable one to recreate in vitro the first few steps of the endocytic pathway (see 1–4 in Fig. 5). Unfortunately, some investigators have developed systems in which they cannot be sure that both ligands follow the same intracellular pathway. As many have done, they assume there is only one endocytic pathway for all ligands whether their uptake is receptor- or fluid-phase–mediated. Especially in the case of reconstitution between vesicles with a ligand taken up by a fluid-phase process and vesicles with a ligand taken up by a receptor-mediated process, the fusion events in vitro may not reflect an event that happens in vivo.

H. Degradation of Internalized Glycoprotein

Endocytosed glycoproteins are degraded to their constituent amino acids and monosaccharides, and the products are released into the extracellular medium [229]. They can be utilized for biosynthesis of macromolecules by the cell. In most receptor systems, when cells are exposed to the continuous presence of ligand, there is an apparent lag of about 15 min before these degradation products can be detected extracellularly. One hint that degradation proceeds by two different cellular pathways was the observation that the kinetics of release of acid-soluble degradation products at 37°C is actually biphasic [174]. Hepatocytes were allowed to internalize one wave of surface-bound radioactive ASOR, and the appearance of extracellular acid-soluble degradation products was measured. There was an immediate, significant, and reproducible release of degradation products that occurred with essentially no lag period. Then, after about 15 min, there was an onset of a much greater rate, until a final steady-state rate of degradation was obtained. It is very easy to overshadow and swamp out the slower initial rate of degradation if the cells are in the continuous presence of the ligand.

This immediate-onset degradation represents the minor state 1 Gal receptor pathway in which the ligand dissociates slowly. Hepatocytes can be treated so that they have on their surface just the state 1 Gal receptor activity or active receptors from both pathways. Radioactive ligand was bound to the two different cell suspensions at 4°C, the cells were washed, put at 37°, and the degradation and release of acid-soluble radioactivity was measured [174]. In the cells that had both receptor populations, a biphasic curve was obtained, whereas in the cells that had just the state 1 receptor population, there was a monotonic rate of degradation that had no lag and that extrapolated to time zero. Cells that have an increasingly greater amount of the surface state 2 Gal receptors show an increasing rate of degradation. The degradation curves become progressively more biphasic, and the increased rate appears only after a lag of about 15 min. In other words, one can change the amount of degradation in the state 2 receptor pathway by varying the amount of this receptor activity initially on the cell surface.

The rapid degradation of internalized glycoprotein, with only a small lag, has also been found in the Man receptor system in macrophages [226]. Here, endosomes were shown to contain proteases responsible for the nonlysosomal ligand degradation. This pathway may be analogous to the state 1 Gal receptor pathway found in hepatocytes.

There is also evidence for two degradation pathways for asialoglycoproteins based on how the cells respond to microtubule depolymerizing drugs, such as colchicine [218]. Numerous investigators have shown that microtubule drugs block ligand degradation in many different systems. These drugs block delivery of ligand, and not degradation per se. If hepatocytes are allowed to internalize and degrade excess radioactive ASOR at 37°C in the presence of colchicine, added at time zero, there is still a significant rate of degradation (about 10% of the control rate). If one adds the drug at different times (e.g., 10, 20, 40, 60 min) after the cells are exposed to the radioactive ligand, the extent of inhibition progressively decreases. It appears that the ligand can get past a block point and can then be degraded; a delivery, not a degradation step, is affected. Other microtubule destabilizing drugs, such as colcemide, podophylotoxin, and vinblastine sulfate, give similar results, whereas lumicolchicine has no affect.

Degradation in the two Gal receptor pathways is differentially inhibited by colchicine [175]. Hepatocytes with both Gal receptor populations or only the minor state 1 Gal receptors on the cell surface, containing surface-bound ligand, were warmed to 37°C, and the rate of release of acid-soluble degradation products was measured in the presence of increasing amounts of colchicine. A biphasic titration curve, with two inflection points, was obtained for the cells that had two active receptor populations. For the cells with just the state 1 pathway, only one inflection was observed. In the major state 2 pathway, the pathway in which the ligand dissociates very rapidly, but has a lag time (for delivery) before degradation occurs, inhibition occurs at 10^{-7} M colchicine. In the state 1 pathway, about a 30-fold greater amount of colchicine is required for a comparable degree of inhibition.

I. Constitutive Receptor Recycling and the Receptor Inactivation–Reactivation Cycle

1. Receptor Inactivation

Hepatic Gal receptors in the major state 2 pathway undergo constitutive recycling. They are able to go into and out of the cell even in the absence of ligand. During this constitutive-recycling process, these receptors are transiently inactivated and then reactivated inside the cell. We first considered this possibility several years ago when we and others observed that a number of different agents, such as metabolic energy poisons, would cause a loss of cell surface receptor activity at 37°C in the absence of ligand (see Fig. 6). Hepatocytes were incubated at 37°C, in the presence of different amounts of sodium azide without ligand, for 45 min and then chilled to 4°C and washed. Viable cells were isolated over discontinuous Percoll gradients, and radioactive ligand was then used to measure receptor activity, or an affinity purified polyclonal antireceptor antibody was used to measure receptor protein [181]. As a function of azide concentration, there was approximately a 50% loss of both receptor activity and protein from the cell surface. This suggested that there is constitutive receptor recycling, but that only a portion of receptors are involved.

When the azide-treated cells were permeabilized with digitonin (see Fig. 7A), we expected to recover the antibody-binding activity and the ligand-binding activity inside the cell. All of the antibody-reactive binding was recovered in the permeabilized cells; there was virtually no change, even at severe ATP depletion (>95%). However, as cellular ATP was depleted, there was a parallel decrease in the total cellular content of active receptors, as measured by the ligand-binding assay [181]. There was about a 50% loss of total cell Gal receptor activity over about 30 min. Again, there was a sensitive population of receptors, the state 2 receptor population, and an insensitive population of receptors, the state 1 receptors. The receptor activity loss could not be explained by the degradation of receptors, because they were still detected with the antibody. Furthermore, the receptor activity loss is reversible in the absence of protein synthesis. If cells are incubated with sodium azide, then washed, and kept at 37°C in the presence of cycloheximide, the cells recover their total receptor activity, ATP levels, and the normal distribution of receptor between the cell surface and the cell interior.

To determine the cellular site of receptor inactivation, the rate of change in ligand-binding capability on the cell surface and in the whole cell was measured. When azide was added to the cells, there was an immediate decrease in receptor activity on the cell surface [181]. However, no decrease was seen when the cells were permeabilized and the total cellular (i.e., both the surface and the intracellular) content of active receptors was measured. The apparently lost receptor activity from the surface was recovered inside the cell. For the first 2 or 3 min there was virtually no change in the total cellular receptor activity; only after about 3 min was there a decrease. The intracellular content of active receptors actually increased transiently before it decreased. The results clearly show that receptor inactivation is an intracellular event.

Inactivation of state 2 Gal receptors is likely due to a covalent modification, rather than to receptor occupancy by an endogenous ligand. To test this, cells were azide-treated at 37°C in the absence of ligand and then washed at 4°C in the presence or absence of EGTA, to expose cryptic Gal receptors. The cells were then assessed for their ability to bind radioactive ligand at 4°C (Fig. 10). There is still a loss of cellular Gal receptor activity, whether or not the cryptic, EGTA-sensitive receptors have been exposed before measuring receptor activity. Medh and Weigel [230] have reconstituted Gal receptor inactivation in digitonin-permeabilized hepatocytes. Only the state 2 Gal receptors were inactivated, and the process was ATP-dependent; no other cytosolic components appear to be required.

2. Receptor Recycling

A receptor cycle time is about 5–10 min [13,14,16,20,21,157,179]. Accordingly, the 30-min period to establish a new steady-state level of receptor activity in the presence of azide [181] suggests that more than one receptor cycle is required. The net amount of active receptor lost from the cell surface is only about half of the amount of receptor activity lost from the total cell [181]. More than one surface equivalent of receptor activity loss occurs in the total cell. The same results are obtained with a nitrogen atmosphere, although the kinetics are different because of the problems with gas exchange.

Our interpretation of these experiments is depicted in the cartoon in Figure 11. This model applies to only one of the two Gal receptor pathways in hepatocytes—the state 2 receptor pathway. Receptors in this pathway are able to enter the cell, through coated pits, even in the absence of ligand. Once inside the cell in an active form, they are then

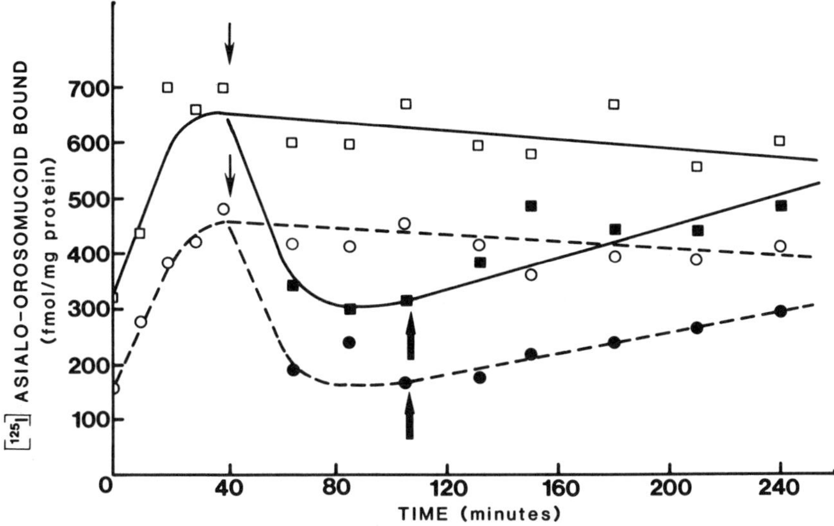

Figure 10 Effect of sodium azide on the cryptic Gal receptor pool. Freshly isolated hepatocytes (2 × 10^6 cells/mL) were incubated at 37°C. After 40 min, either sodium azide, to a final concentration of 10 mM (solid circles and squares), or PBS (open circles and squares) was added (small arrows) and the 37°C incubation was continued. After 100 min, the cells were collected by centrifugation, resuspended in fresh medium, and put at 37°C (large arrows). At the indicated times, samples of the cell suspensions were chilled on ice and EGTA, to 10 mM (solid and open squares), or PBS (solid and open circles) was added. After 10 min, the cells were collected, washed, and incubated in medium containing 1.5 μg/mL ^{125}I-ASOR for 60 min at 4°C. After washing away unbound ligand, the cells were assessed for bound radioactivity. Points represent the mean of duplicates.

inactivated by a mechanism that remains to be determined (see later discussion). Reactivation of receptors and externalization, getting them back to the cell surface, requires ATP. When cells are depleted of ATP by using metabolic energy poisons or nitrogen, then the receptors cannot be reactivated and reexternalized, and they accumulate inside the cell. At the start of the experiment, there is a substantial pool of active receptors still inside the cell beyond the block point(s)—the ATP-requiring step(s). The reason that more inactive receptors accumulate in the whole cell than the net number lost from the cell surface, may be that some of these receptors are externalized and then internalized and get trapped inside the cell at this block point. Some receptors may traverse through several cycles before being inactivated and trapped.

3. *Interpretation of Immunologicalization Studies*

An important point about the discovery of inactive intracellular receptors needs to be made concerning localization studies employing immunoelectron microscopy with polyclonal or monoclonal antibodies [16,20,157,179]. All studies to date, with any receptor system, would not have distinguished between active versus inactive receptors. It will be important to develop antibodies that are specific for one or the other receptor state, to assess their respective cellular localizations. Once the molecular basis for receptor inactivation is determined, synthetic peptides may permit the development of such antibodies.

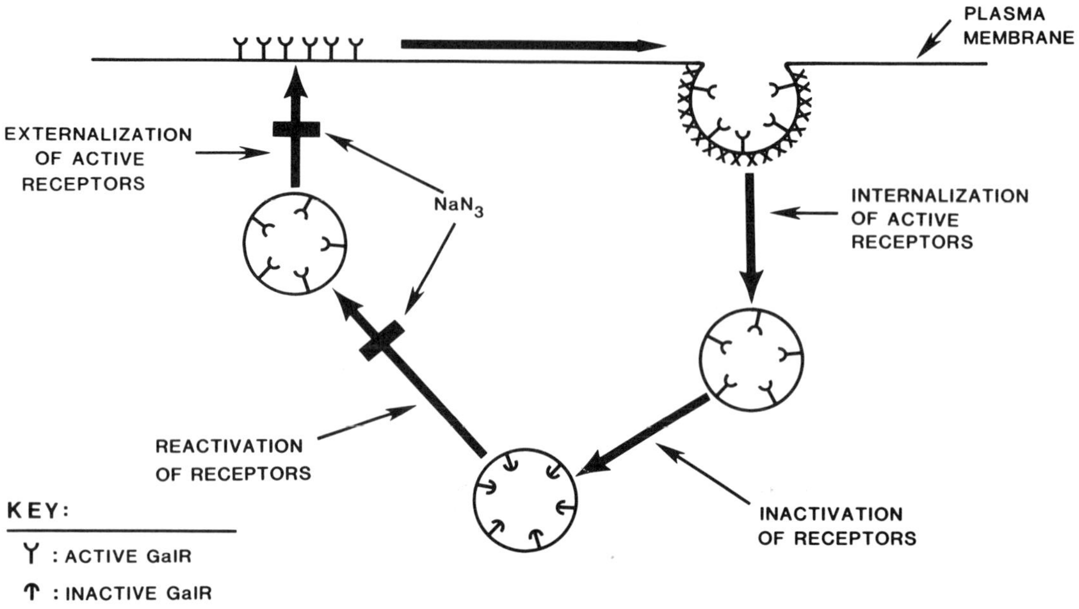

Figure 11 Constitutive recycling pathway for the hepatic state 2 Gal receptors (GalR). Gal receptors in the state 2 pathway enter the cell through coated pits, become transiently inactivated inside the cell and are then reactivated while still inside the cell and externalized back to the plasma membrane. The surface receptors then diffuse to coated pits and repeat the cycle. (From Ref. 181.)

4. Functional Significance and Generality of Receptor Inactivation

The activities of several class I receptors have been shown to be reversibly regulated. Vasopressin and angiotensin II receptors on rat hepatocytes are inactivated by a Ca^{+2}-dependent cytosolic protein. This inactivation is reversed by lowering the pH to 5.5, but not by EGTA [231]. The aromatic hydrocarbon receptor in Hepa-1 cells is reversibly inactivated in a time- and temperature-dependent manner in the presence of cytochalasin B. The extent of the loss of ligand-binding activity correlates with the depletion of cellular ATP. Reactivation is readily achieved by incubation of cells at 37°C, after removing cytochalasin B [232]. The activity of the pregnenolone receptor in guinea pig adrenocortical cytosol requires phosphorylation by a cytosolic kinase. Alkaline phosphatase treatment at pH 9.0 causes its reversible inactivation [233]. The calf uterus estradiol-17 β-receptor is also active in the phosphorylated form and inactive when it is dephosphorylated. A nuclear phosphatase and a cytosolic kinase are responsible for the reversible inactivation and reactivation process [234]. In addition to our reports on inactivation of the hepatic Gal receptor, evidence has also been presented by others for the presence or generation of inactive transferrin [235,236] and mannose [237] receptors.

Why would cells have a mechanism for inactivating receptors? There are several reasons why this might be a very valuable observation that probably applies to many other receptor systems as well. Probably the best reason is that *receptor inactivation would increase the efficiency of segregation of receptor and ligand after dissociation.* As discussed earlier (see Sect. III.F.4), when receptors bring ligand into the cell interior, the ligand is being highly concentrated (~10^4-fold) from the extracellular medium into the small volume of the endosome within the cell. Therefore, to get efficient separation and segregation of the ligand from the receptor, even at low pH, it would benefit the cell tremendously to have the receptor inactivated. This is an attractive general mechanism that would apply to all class II, and perhaps class I, receptor systems. Additionally, receptor inactivation may be an alternative mechanism to mediate receptor–ligand dissociation. Receptor inactivation during constitutive recycling could also provide a mechanism for cells to regulate their endocytic and recycling processes.

Another possible reason to inactivate the receptor may be to prevent it from binding to soluble or membrane-bound glycoproteins at stages of their biosynthesis when they have terminal Gal/GalNAc residues. Others have shown that receptors are present in the Golgi region [179], but it is not known if they are active. If they were, this might disrupt normal glycoprotein biosynthesis, processing, or translocation.

5. *Uncoupling of State 2 Galactosyl Receptor Inactivation and Receptor Redistribution*

The cellular process by which the state 2 Gal receptor population can be inactivated is distinct from the process by which these receptors become redistributed in the cell; that is, leave the cell surface and accumulate inside the cell. This conclusion is based on the finding that the two processes can be uncoupled. They can occur independently, depending on the perturbing agent to which the cells are exposed. McAbee et al. [238] examined a variety of treatments and inhibitors and followed Gal receptor activity and localization by measuring ASOR- or antireceptor antibody-binding capacity in intact and digitonin-permeabilized rat hepatocytes. The various agents fell into three categories (Table 5). Some agents, such as low temperature (24°–37°C) or microtubule drugs, do not cause receptor inactivation, but do lead to receptor redistribution. Other agents, such as monensin or chloroquine, cause receptor inactivation, but not redistribution. Agents that deplete cellular ATP cause both receptor redistribution and inactivation.

6. *The Role of Receptor Phosphorylation*

The rat and human hepatic Gal receptors and the chicken hepatic GlcNAc receptor are phosphoproteins [13,19,20,21,179]. All of the subunit polypeptides of these receptors have

Table 5 Effect of Modulating Agents on the Rat Liver Parenchymal State 2 Gal Receptor System[a]

Agent or treatment	Receptor inactivation	Receptor redistribution
Low temperature (24°–37°C)	No	Yes
Microtubule drugs	No	Yes
Monensin	Yes	No
Chloroquine	Yes	No
Sodium azide	Yes	Yes
N_2 atmosphere	Yes	Yes

[a]Cells treated at 37°C in the absence of ligand with the various agents show independent effects on receptor localization in the cell or on receptor inactivation. The agents fall into one of three categories. Receptor inactivation or redistribution from the cell surface to the cell interior are measured at 4°C using ligand or antireceptor antibody binding in the presence or absence of digitonin as described in Refs. 181 and 238.

one or more serine/threonine (Ser/Thr) or tyrosine residues in their cytoplasmic domains. This is also true for all the carbohydrate receptors for which sequences have been determined (see Fig. 1). Although it is not known if the Man or Fuc receptors are also phosphorylated in the live cell, it is reasonable to believe that they are. Most of the class I signal-transducing receptors that operate by either the phosphoinositide or the G-protein adenylate cyclase pathway are phosphorylated and dephosphorylated as part of the regulation of the receptor function. Often these receptors are also tyrosine kinases and modify themselves as well as other target intracellular proteins [239]. Whether the class II carbohydrate receptor cycle or function is regulated by phosphorylation and dephosphorylation of these receptors has not been determined. Addition of asialoglycoproteins to hepatocytes does not stimulate the turnover of the phosphoinositol lipids [240] nor, presumably, protein kinase C.

The reversible inactivation–reactivation cycle that the state 2 Gal receptors in rat hepatocytes undergo may be due to (de)phosphorylation. We believe that, under normal conditions in the presence or absence of ligand, these receptors are recycling and that at any instant only a small fraction of the state 2 receptors (probably ~10%) are inactivate. As shown in the studies cited earlier, the inactive state 2 Gal receptor pool can be increased by depleting intracellular ATP, and these inactive receptors are then trapped inside the cell. There was no substantial alteration in the pattern of human Gal receptor phosphylation, whether or not cells were internalizing ligand or if receptor recycling was abolished by primaquine [241]. Phorbol esters increase the phosphorylation of the human Gal receptor and alter its cellular distribution [242]. Phorbol esters stimulate the activity of protein kinase C and increase the phosphorylation of cellular Ser/Thr residues on target proteins [239]. The hyperphosphorylated human Gal receptor does not leave the cell interior [242]. The kinetics of redistribution and the finding of a transient decrease in affinity of cell surface receptors, indicated that the effect of the phorbol esters was complex and could involve several steps or sites. Hyperphosphorylation of receptors induced by agents that stimulate protein kinase C may be nonspecific and may not reflect a physiologically important activity.

Nonetheless, Gal receptor inactivation may be due to a specific phosphorylation or dephosphorylation. There is a dose- and time-dependent decrease in both surface and total cellular receptor activity when cells are exposed to the protein phosphatase inhibitor vanadate (Fig. 12). One would predict that vanadate should decrease the rate of dephosphorylation of receptors (or a regulatory protein that controls receptor activity), and therefore, the steady-state level of phosphoreceptor should increase. Since we observe that receptor activity is lost in the presence of vanadate, either the phosphorylated receptor could be inactive or a second protein is being phosphorylated and is regulating the activity of the receptor. Characterization of the relations between activity and phosphorylation in receptors isolated by ligand, versus by antibody-affinity chromatography, from normal and vanadate-treated cells should provide answers to some of these questions.

Active receptors isolated by ligand-binding affinity chromatography are phosphorylated [19]. Therefore, one needs to propose that there are either multiple phosphorylation sites on the cytoplasmic domains of these receptors or that other proteins are involved in the receptor inactivation. Multiple phosphorylation sites in the cytoplasmic domains are possible for all of the mammalian carbohydrate receptors. Consistent with this idea, we and others find that [^{32}P]phosphate incorporation into Gal receptors in hepatocytes is much slower (hours) than expected on the basis of the receptor cycle time (minutes). It is possible that there are both stable and dynamic phosphorylation sites in these proteins. This would

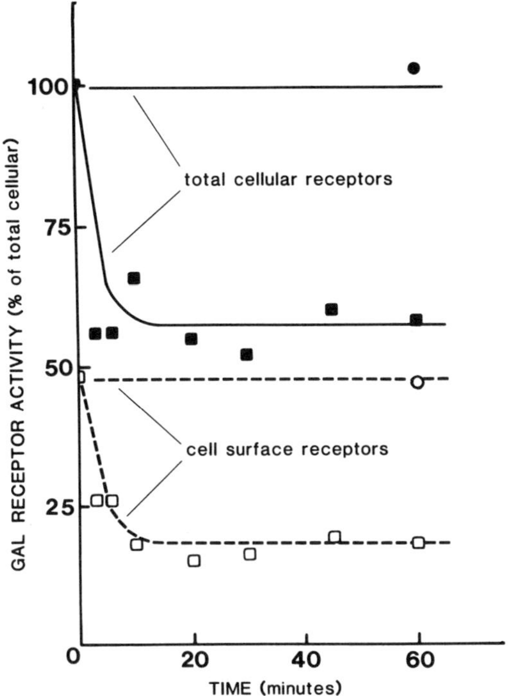

Figure 12 Effect of vanadate on the expression of surface and total cellular state 2 Gal receptor activity in isolated hepatocytes. Rat hepatocytes were incubated at 37°C in the absence of ligand. At the start of the experiment samples received medium alone (solid and open circles) or sodium vanadate to a final concentration of 8 mM (solid and open squares). Samples were chilled to 4°C and cell surface or total cellular ^{125}I-ASOR-binding ability was determined, respectively, in the absence (open circles and squares) or presence of digitonin (solid circles and squares). [Adapted from Ref. 259.]

explain some of the present conflicting results and confusion, but, unfortunately, would make the puzzle more difficult to solve.

J. Lessons from the Hepatic Galactosyl Receptor

We believe that a number of lessons can be learned from the Gal receptor system that will apply to the other class II carbohydrate receptor systems, and possibly to class I receptors as well.

1. Endocytic and degradative processes are more complicated than initially thought. There are probably multiple cellular pathways for ligand processing in all receptor systems. We believe that all class II receptor systems will show two functionally distinct subpopulations of receptors and two pathways, as described earlier for the state 1 and state 2 hepatic Gal receptor.
2. The surface receptors, the activity of which can be modulated by the agents such as shown in Figure 6, are constitutively recycling in the absence of ligand.
3. Constitutively recycling receptors undergo reversible inactivation–reactivation inside the cell as they traverse their recycling pathway.

IV. POSSIBLE FUNCTIONS OF THE ENDOCYTIC MEMBRANE-BOUND CARBOHYDRATE RECEPTORS

An intriguing idea for the function of the carbohydrate receptors, particularly for those located in liver, is that these are the last steps in a "housekeeping" pathway. That is, cells or tissues at other sites in the body may process a given family of glycoconjugates and convert them to soluble molecules containing the oligosaccharide recognition marker (like a zip code) needed to mail the molecule to the liver, where it will be completely degraded.

There are good candidate functions, as discussed in Section II, for the two Man-6-P receptors, the thyroid GlcNAc receptor, the Glc-AGE receptor, and the liver HA receptor. The likely endogenous ligands are known for these receptors. Several reasonable suggestions for the function of the macrophage Man receptor were also discussed (see Sect. II.D.). The possible functions of the other Gal and Fuc receptors are much less clear. They are generally assumed to remove "deleterious" agents [13–16,21,179], but hypotheses on the nature of these agents or the consequences of their not being removed have not been presented.

A. Previously Proposed Functions of the Hepatic Galactosyl Receptor

1. Glycoprotein Turnover

Ashwell and his co-workers initially proposed that the function of the hepatic Gal receptor was to remove spontaneously desialylated plasma glycoproteins [13]. These and other workers have shown that, after desialylation, a large number of different glycoproteins can be removed by the Gal receptor. They suggested that normal plasma glycoprotein turnover would involve loss of sialic acid, either spontaneously or because of low levels of circulating neuraminidases, and then clearance by the receptor in liver. Subsequent studies showed, however, that this was not the mechanism by which plasma proteins, such as orosomucoid, turned over metabolically [243,244].

2. Removal of Antibody–Antigen Complexes

Mammalian IgGs have galactosyl-terminal oligosaccharide chains in the hinge region of the protein, yet the molecule is not recognized and removed by the Gal receptor. It was suggested that formation of the IgG–antigen complex changed the conformation of the antibody so that this oligosaccharide chain was now recognized and the complex could be removed and degraded [21]. Although this is still an intriguing model, there currently is little evidence to support it.

3. Cell–Cell or Cell–Matric Adhesion

Cell–cell or cell–matrix adhesion can be mediated in vitro by the receptor [32] and is a possible function in vivo during metastasis [245]. Since cell surfaces and the extracellular matrix contain galactosyl-terminal molecules, an adhesive function for the receptor in vivo is certainly possible. It is more likely, though, that this adhesion is a fortuitous function expressed in vitro owing to the numerous surface receptors and their mobility.

4. Intracellular Routing of Glycoproteins

Intracellular routing is a possible function, especially since hepatocytes synthesize and secrete such large amounts of plasma glycoproteins. For example, the receptor could bind to the galactosyl-terminal oligosaccharides of newly made glycoproteins and shuttle them from the Golgi to the cell periphery. They could then be sialylated and released from the receptor and brought to the cell surface (exocytosed). Several observations support this idea. First, Gal receptors are found in the Golgi region [13,179] as well as other internal compartments between the Golgi and plasma membrane. Second, a small percentage of internalized asialoglycoproteins can be resialylated and returned intact to the extracellular medium [246]. This implies that Gal receptors can shuttle ligand to a compartment containing active sialyltransferases.

We believe that the state 1 Gal receptor population and pathway may be involved in this latter process, and that the state 1 pathway is also responsible for transcytosis in vivo. The function of the major state 2 Gal receptor pathway, we propose, is to remove circulating galactose-containing ligands. This removal is vitally important for the organism for the reasons developed in the following.

B. The Galactosyl Homeostasis Hypothesis

We propose that the major function of the several liver Gal/GalNAc receptors is, in fact, to remove circulating glycoconjugates. The reasons for this function are embodied in an idea referred to as the galactosyl homeostasis hypothesis. The basic tenet of this hypothesis (Fig. 13) is that cell surface and extracellular soluble glycoconjugates containing terminal Gal/GalNAc are necessary and important molecules during embryogenesis and organ development. Correct morphogenesis and organization of differentiated tissue relies, in as yet undetermined ways, on these molecules. After development is complete, the organization of cell–cell, cell–matrix, and intramatrix interactions and the exertion of normal growth control requires the continuous participation of galactosyl-specific lectins (either as binding proteins or true receptors) with galactosyl-glycoconjugates. Presumably, these molecules continuously turn over, being degraded and synthesized, and maintain a constant steady-state concentration of each interacting set of Gal/GalNAc molecules (ligands) and their respective specific lectins (receptors) in the particular tissue. Thus, there is a Gal/GalNAc homeostasis involved in regulating the expression of these molecules in tissues.

The introduction of soluble or particulate Gal/GalNAc glycoconjugates, for whatever reason, into the tissue will be deleterious, because these molecules will compete for binding to the endogenous matrix and cell surface galactosyl-specific lectins or receptors. The competing galactosyl-containing molecules would disrupt the normal pattern of organization, growth control, or function maintained by these interactions. This is why such soluble or particulate glycoconjugates generated by disease, trauma, cell death, or other reasons, are injurious to the organism. Mammals, therefore, have evolved three extremely efficient liver clearance systems for removing both soluble and particulate Gal/GalNAc-ligands: One, the parenchymal Gal receptor, is one of the most active endocytic and degradative systems known. It removes a wide range of soluble circulating galactosyl-glycoconjugates that would arise in the extracellular spaces in all tissues, the lymph, or the circulation. The other two systems, the macrophage Gal receptors present on resident and migratory cells remove a wide range of Gal/GalNAc-containing particles, including whole cells, matrix, and cellular debris. Unlike the liver system, which is fixed, the macrophage system is mobile and can perform a surveillance (search and destroy) role. A brief summary of the evidence in support of this function for these Gal receptors is presented next.

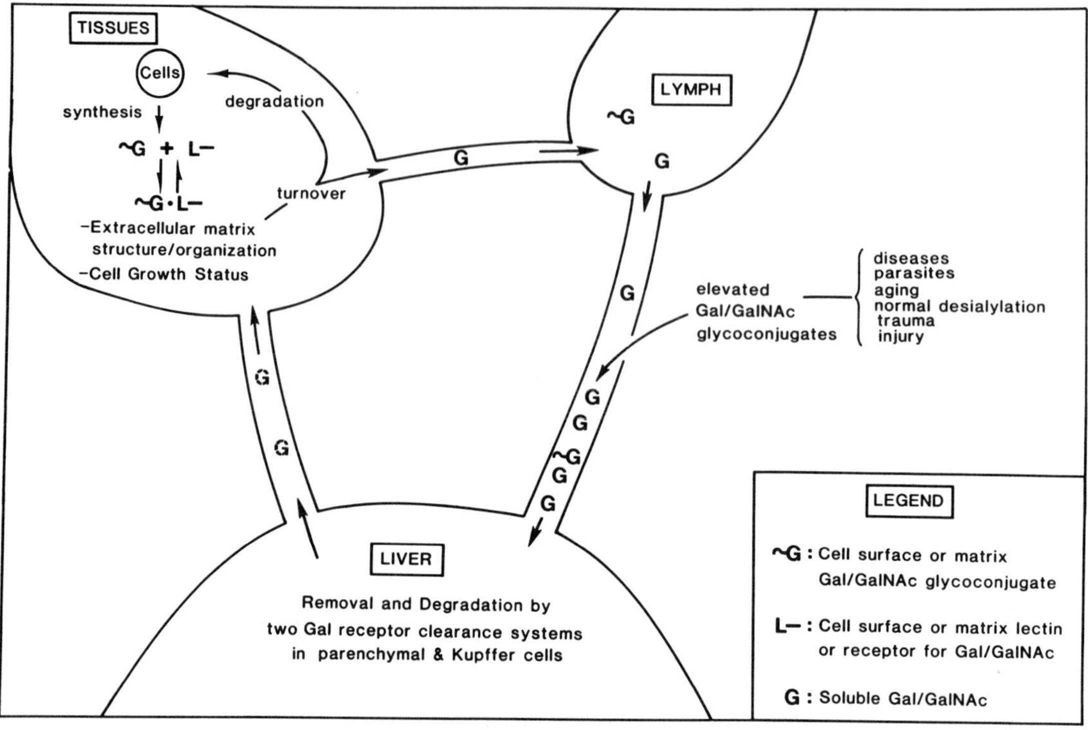

Figure 13 The galactosyl homeostasis hypothesis. The schematic diagram depicts the dynamic metabolic turnover of Gal/GalNAc-containing glycoconjugates and their removal by liver. Migratory macrophages throughout the tissues will also remove and degrade larger Gal/GalNAc-containing particles or cells. Cells synthesize and secrete the Gal/GalNAc lectins and glycoconjugates. They may also degrade some of these molecules as they turn over, but the bulk flows from the extracellular matrix to lymph and then the blood. The multiple Gal/GalNAc receptor systems then remove these glycoconjugates. The modified Gs leaving the liver represent molecules that might escape degradation in the liver if the receptor function was impaired or if the influx of Gal/GalNAc molecules (owing to diseases, and so forth) was so great that the liver's capacity for removal was saturated. These "escaped" molecules could then exert disruptive effects on cell growth or function or the structure or organization of the matrix.

1. Galactose Molecules and Galactosyl Lectins During Development

The developing embryo has a high content of Gal/GalNAc glycoconjugates, and during development, the hepatic Gal receptor is not expressed [98]. We propose that this is because there is no need to remove these molecules; they are part of the dynamic developmental process. In fact, the model predicts that their removal during development will be detrimental and would result in abnormal or abortive development. In support of this, a galactose-induced teratogenesis was noted in mice with defective galactokinase [247]. Litters of mothers receiving a high galactose diet showed a greater fetal loss, and 26% of the pups were malformed, compared with 0% for mothers on a control diet. Many investigators have shown that the various β-galactoside lectins are regulated during development. Their concentration and their distribution pattern changes as development of a tissue proceeds [248]. Cell migration during development can also depend on recognition and even modification of galactosyl-containing molecules in the matrix [249]. Cell surface galactosyltransferase activities may be functionally important during cell migration and for cell–cell recognition [250,251].

2. A Role for Galactose Molecules and Galactosyl Lectins in Differentiated Tissues

A large literature indicates that the extracellular matrix of normal adult tissues contains galactose-binding proteins termed *galaptins* (see Chap. 19). These soluble "lectins" are thought to be needed for the organization (and perhaps rigidity or cross-linking) of the matrix. Interestingly, a number of other important extracellular matrix molecules, such as the proteoglycan core and lung surfactant proteins [107], also have galactose-binding activity and protein domains that have sequence homology to galactosyl lectins. The endogenous ligands for these galaptins have not yet been identified, although one exception may be the major protein elastin. A galactose-binding protein was recently shown to be part of an elastin–receptor complex involved in elastin fiber assembly [252]. In some instances, the lectin–glycoconjugate interaction in the embryo may serve to block a glycoconjugate–receptor interaction that will eventually prevail in the adult tissue [248].

3. Cell Surfaces Contain Galactose/N-Acetylgalactosamine Glycoconjugates

Although eukaryotic cell surfaces are rich in sialylated glycolipids and glycoproteins, there is abundant evidence that cells also normally possess terminal Gal/GalNAc residues on their surfaces. For example, many pulmonary, urinary, or gastrointestinal pathogenic bacteria establish themselves in the host by binding to glycolipids containing GalNAc-β1,4-Gal [253]. The parasite *Trypanosoma cruzi* may facilitate its successful invasion of the host by virtue of a cell surface neuraminidase, specific for mammalian cell surface sialylconjugates. Normal persons have an endogenous circulating inhibitor that effectively stops trypanosome desialylation of mammalian cells [111].

4. Cellular Turnover Mediated by Carbohydrate Receptors

Although this chapter has concentrated on the turnover of macromolecular glycoconjugates, it is appropriate to mention instances in which the presence of cell surface galactosyl-glycoconjugates directs a cell to be destroyed (thus turning over all its constituents). This could be an important function of the macrophage Gal receptor(s). The example that has been studied the longest is the removal of aged red blood cells. It has been long recognized that as erythrocytes age they lose cell surface sialic acid and increase the number of exposed galactosyl and *N*-acetylgalactosaminyl residues [109,110]. One hypothesis centers around the major erythrocyte surface glycoprotein, the sialoglycoprotein, glycophorin A. Desialylation of this protein exposes multiple Gal-β1,3-GalNAc groups on the cell surface. This saccharide is recognized by the reticuloendothelial system, the macrophage Gal receptor, which mediates the binding, uptake, and degradation of these cells. Senescent erythrocytes can be phagocytosed by autologous blood monocytes [109] and by liver macrophages, the Kupffer cells [110]. Although the hepatic Gal receptor may mediate a static binding between hepatocytes and aged erythrocytes that might penetrate the sinusoidal cell layer, it cannot mediate endocytosis of such large "particles." Presumably, this latter receptor is locked into the coated pit pathway, whereas Kupffer (macrophage) cell Gal receptors are not. As mentioned in Section II.G.1, the liver Gal receptors in hepatocytes and macrophages have very different constraints on the size of ligands they can internalize, even though their sugar specificities appear almost identical.

Presumably, any cell type in the blood (or the tissues) that has a high enough cell surface content of galactosyl groups may be cleared by the macrophage receptor in liver (or by migratory cells). For example, bone marrow transplantation involving intravenous injection of pluripotent stem cells can be enhanced by the coadministration of asialoglycoproteins [254]. Although this observation was initially explained by the action of the hepatic Gal receptor, it is probably due to the function of the Kupffer cell Gal receptor. The normal homing system for bone marrow progenitor cells involves galactosyl groups on these cells. Under normal, physiological circumstances, liver sequestration of bone marrow stem cells probably does not occur, at least to the detriment of hematopoiesis.

5. *Galactosyl-glycoconjugates and Galactosyl Receptors Are Involved in Growth Control and Cancer*

Soluble galactosyl lectins are usually mitogenic when tested on eukaryotic cells in culture, which is further evidence that cell surfaces normally have some level of exposed terminal galactosyl-glycoconjugates. The phenomenon of contact inhibition of growth, displayed by cultured cells in continuous cell–cell contact, is generally considered to reflect a regulatory process of fundamental importance. Cell–cell contacts through plasma membrane glycoconjugates with terminal galactosyl residues have been shown to mediate this growth regulatory response in human fibroblasts [255]. The mechanism by which cell surface galactose is involved in regulating growth is unknown. Many tumors and tumor cell lines contain endogenous galactosyl lectins [256] that could be involved in tumor cell growth, adhesion, and metastasis [257]. Cultured Hodgkin's cells have a cell surface galactosyl lectin that is related immunologically to the hepatic Gal receptor [258]. This surface molecule has sialyltransferase activity.

6. *Summary of the Hypothesis*

Any Gal/GalNAc-terminal glycoconjugates in excess or in the "wrong" place in the mammalian body can disrupt normal cell growth control or cell–cell and cell–matrix organization (see Fig. 13). Removal of any excess or inappropriate Gal/GalNAc molecules by the two Gal receptor systems is necessary for the organism's health and possibly survival. This model leads to several predictions that can be tested experimentally. For example, if there is a dynamic total-body turnover of these molecules there should be elevated levels of Gal/GalNAc glycoconjugates in lymph, relative to blood. Also, chronically elevated plasma levels of Gal/GalNAc glycoconjugates should have pathological consequences. These and other related predictions can be tested to understand further the function of the liver parenchymal and macrophage Gal receptors.

ACKNOWLEDGMENTS

I thank Drs. D. McAbee and J. Yannariello-Brown for helpful discussions and Janet Oka for help preparing the figures and for her experimental acumen. I am very grateful to Betty Jackson and Lisa Raney for help preparing this manuscript. This work was supported by grants from the National Institutes of Health, GM 30218 and GM 35978.

REFERENCES

1. Conzelmann, E., and Sandhoff, K. (1987). Glycolipid and glycoprotein degradation. *Adv. Enzymol. 60*:89–216.

2. Kresse, H., and Glossl, J. (1987). Glycosaminoglycan degradation. *Adv. Enzymol. 60*:217–311.
3. Stadtman, E. R. (1990). Covalent modification reactions are marking steps in protein turnover. *Biochemistry 29*:6323–6331.
4. Bond, J. S., and Butler, P. E. (1987). Intracellular proteases. *Annu. Rev. Biochem. 56*:333–364.
5. Rechsteiner, M., Rogers, S., and Rote, K. (1987). Protein structure and intracellular stability. *Tends Biochem. Sci. 12*:390–394.
6. Khairallah, E. A., Bond, J. S., and Bird, J. W. C., eds. (1985). *Intracellular Protein Catabolism.* Alan R. Liss, New York.
7. Durand, P., and O'Brien, J. S., eds. (1982). *Genetic Errors of Glycoprotein Metabolism.* Springer-Verlag, New York.
8. Irie, S., and Tavassoli, M. (1986). Liver endothelium desialylates ceruloplasmin. *Biochem. Biophys. Res. Commun. 140*:94–100.
9. Blomhoff, R., Berg, T., and Norum, K. R. (1988). Transfer of retinol from parenchymal to stellate cells in liver is mediated by retinol-binding protein. *Proc. Natl. Acad. Sci. USA 85*:3455–3458.
10. Baenziger, J. U. (1985). The role of glycosylation in protein recognition. *Am. J. Physiol. 121*:382–391.
11. West, C. M. (1986). Current ideas on the significance of protein glycosylation. *Mol. Cell. Biochem. 72*:3–20.
12. Reading, C. L. (1984). Carbohydrate structure, biological recognition, and immune function. In *The Biology of Glycoproteins* (R. J. Ivatt, ed.). Plenum Press, New York, pp. 235–321.
13. Ashwell, G., and Harford, J. (1982). Carbohydrate-specific receptors of the liver. *Annu. Rev. Biochem. 51*:531–554.
14. Steer, C. J., and Ashwell, G. (1986). Hepatic membrane receptors for glycoproteins. *Prog. Liver Dis. 8*:99–123.
15. Kaplan, J. (1981). Polypeptide-binding membrane receptors: Analysis and classification. *Science 212*:14–20.
16. Anderson, R. G. W., and Kaplan, J. (1983). Receptor-mediated endocytosis. *Mod. Cell Biol. 1*:1–52.
17. Metzger, H., and Kinet, J.-P. (1988). How antibodies work: Focus on Fc receptors. *FASEB J. 2*:3–11.
18. Hofmann, S. L., Russell, D. W., Brown, M. S., Goldstein, J. L., and Hammer, R. E. (1988). Overexpression of low density lipoprotein (LDL) receptor eliminates LDL from plasma in transgenic mice. *Science 239*:1277–1281.
19. Drickamer, K. (1987). Membrane receptors that mediate glycoprotein endocytosis: Structure and biosynthesis. *Kidney Int. 32*:S167–S180.
20. Schwartz, A. L. (1984). The hepatic asialoglycoprotein receptor. *Crit. Rev. Biochem. 16*: 207–233.
21. Weigel, P. H. (1987). Receptor recycling and ligand processing mediated by hepatic galactosyl receptors: A two-pathway system. In *Vertebrate Lectins* (K. Olden and J. B. Parent, eds.). Van Nostrand Reinhold, New York, pp. 65–91.
22. Weigel, P. H. (1980). Characterization of the asialoglycoprotein receptor on isolated rat hepatocytes. *J. Biol. Chem. 255*:6111–6120.
23. Lee, R. T., Myers, R. W., and Lee, Y. C. (1982). Further studies on the binding characteristics of rabbit liver galactose/*N*-acetylgalactosamine-specific lectin. *Biochemistry 21*:6292–6298.
24. Hardy, M. R., Townsend, R. R., Parkhurst, S. M., and Lee, Y. C. (1985). Different modes of ligand binding to the hepatic galactose/*N*-acetylgalactosamine lectin on the surface of rabbit hepatocytes. *Biochemistry 24*:22–28.
25. Andersen, T. T., Freytag, J. W., and Hill, R. L. (1982). Physical studies of the rabbit hepatic galactoside-binding protein: Effects of calcium and ligands. *J. Biol. Chem. 257*:8036–8041.
26. Steer, C. J., Kempner, E. S., and Ashwell, G. (1981). Molecular size of the hepatic receptor for asialoglycoproteins determined in situ by radiation inactivation. *J. Biol. Chem. 256*:5851–5856.

27. Drickamer, K., Mamon, J. F., Binns, G., and Leung, J. O. (1984). Primary structure of the rat liver asialoglycoprotein receptor: Structural evidence for multiple polypeptide species. *J. Biol. Chem. 259*:770–778.
28. Halberg, D. F., Wager, R. E., Farrell, D. C., Hildreth, J., IV, Quesenberry, M. S., Loeb, J. A., Holland, E. C., and Drickamer, K. (1987). Major and minor forms of the rat liver asialoglycoprotein receptor are independent galactose-binding proteins: Primary structure and glycosylation heterogeneity of minor receptor forms. *J. Biol. Chem. 262*:9828–9838.
29. Spiess, M., and Lodish, H. F. (1985). Sequence of a second human asialoglycoprotein receptor: Conservation of two receptor genes during evolution. *Proc. Natl. Acad. Sci. USA 82*:6465–6569.
30. Bischoff, J., Libresco, S., Shia, M. A., and Lodish, H. F. (1988). The H1 and H2 polypeptides associate to form the asialoglycoprotein receptor in human hepatoma cells. *J. Cell. Biol. 106*:1067–1074.
31. Lederkremer, G. Z., and Lodish, H. F. (1991). An alternatively spliced miniexon alters the subcellular fate of the human asialoglycoprotein receptor H2 subunit. *J. Biol. Chem. 266*:1237–1244.
32. Weigel, P. H. (1980). Rat hepatocytes bind to synthetic galactoside surfaces via a patch of asialoglycoprotein receptors. *J. Cell Biol. 87*:855–861.
33. Kuhlenschmidt, T. B., Kuhlenschmidt, M. S., Roseman, S., and Lee, Y. C. (1984). Binding and endocytosis of glycoproteins by isolated chicken hepatocytes. *Biochemistry 23*:6437–6444.
34. Herzig, M. C. S., and Weigel, P. H. (1989). Synthesis and characterization of *N*-hydroxysuccinimide ester chemical affinity derivatives of asialoorosomucoid that covalently cross-link to galactosyl receptors on isolated rat hepatocytes. *Biochemistry 28*:600–610.
35. Herzig, M. C. S., and Weigel, P. H. (1990). Surface and internal galactosyl receptors are heterooligomers and retain this structure after ligand internalization or receptor modulation. *Biochemistry 29*:6437–6447.
36. Oka, J. A., Herzig, M. C. S., and Weigel, P. H. (1990). Functional galactosyl receptors on isolated rat hepatocytes are hetero-oligomers. *Biochem. Biophys. Res. Commun. 170*:1308–1313.
37. Sawyer, J. T., Sanford, J. P., and Doyle, D. (1988). Identification of a complex of the three forms of the rat liver asialoglycoprotein receptor. *J. Biol. Chem. 263*:10534–10538.
38. Lee, R. T., and Lee, Y. C. (1986). Preparation of a high-affinity photolabeling reagent for the Gal/GalNAc lectin of mammalian liver: Demonstration of galactose-combining sites on each subunit of rabbit hepatic lectin. *Biochemistry 25*:6835–6841.
39. Ray, D. A., Oka, J. A., and Weigel, P. H. (1986). Nonionic detergents increase the stoichiometry of ligand binding to the rat hepatic galactosyl receptor. *Biochemistry 25*:6097–6103.
40. McPhaul, M., and Berg, P. (1986). Formation of functional asialoglycoprotein receptor after transfection with cDNAs encoding the receptor proteins. *Proc. Natl. Acad. Sci. USA 83*:8863–8867.
41. Neufeld, E. F., Lim, T. W., and Shapiro, L. J. (1975). Inherited disorders of lysosomal metabolism. *Annu. Rev. Biochem. 44*:357–376.
42. Neufeld, E. F., and Ashwell, G. (1980). Carbohydrate recognition systems for receptor-mediated pinocytosis. In *The Biochemistry of Glycoproteins and Proteoglycans.* (W. J. Lennarz, ed.). Plenum Press, New York, pp. 241–266.
43. Sly, W. S., Fischer, H. D., Gonzalez-Noriega, A., Grubb, J. H., and Natowicz, M. (1981). Role of the 6-phosphomannosyl-enzyme receptor in intracellular transport and adsorptive pinocytosis of lysosomal enzymes. *Methods Cell Biol. 23*:191–214.
44. von Figura, K., and Hasilik, A. (1986). Lysosomal enzymes and their receptors. *Annu. Rev. Biochem. 55*:167–193.
45. Sahagian, G. G. (1984). The mannose 6-phosphate receptor: Function, biosynthesis and translocation. *Biol. Cell 51*:207–214.
46. Robbins, A. R., Myerowitz, R., Youle, R. J., Murray, G. J., and Neville, D. M., Jr. (1981). The mannose 6-phosphate receptor of Chinese hamster ovary cells. Isolation of mutants with altered receptors. *J. Biol. Chem. 256*:10618–10622.

47. Gabel, C. A., Goldberg, D. E., and Kornfeld, S. (1983). Identification and characterization of cells deficient in the mannose-6-phosphate receptor: Evidence for an alternate pathway for lysosomal enzyme targeting. *Proc. Natl. Acad. Sci. USA 80*:775–779.
48. Kornfeld, S. (1987). Trafficking of lysosomal enzymes. *FASEB J. 1*:462–468.
49. Distler, J. J., and Jourdian, G. W. (1987). Low molecular weight phosphomannosyl receptor from bovine testes. *Methods Enzymol. 138*:504–509.
50. Lobel, P., Dahms, N. M., and Kornfeld, S. (1988). Cloning and sequence analysis of the cation-independent mannose 6-phosphate receptor. *J. Biol. Chem. 263*:2563–2570.
51. Oshima, A., Nolan, C. M., Kyle, J. W., Grubb, J. H., and Sly, W. M. (1988). The human cation-independent mannose 6-phosphate receptor. Cloning and sequence of the full-length cDNA and expression of functional receptor in COS cells. *J. Biol. Chem. 263*:2553–2562.
52. Hoflack, B., and Kornfeld, S. (1985). Purification and characterization of a cation-dependent mannose 6-phosphate receptor from murine $P338D_1$ macrophages and bovine liver. *J. Biol. Chem. 260*:12008–12014.
53. Stein, M., Meyer, H. E., Hasilik, A., and von Figura, K. (1987). 46-kDa mannose 6-phosphate-specific receptor: Purification, subunit composition, chemical modification. *Biol. Chem. Hoppe-Seyler 368*:927–936.
54. Pohlmann, R., Nagel, G., Schmidt, B., Stein, M., Lorkowski, G., Krentler, C., Cully, J., Meyer, H. E., Grzeschik, K.-H., Mersmann, G., Hasilik, A., and von Figura, K. (1987). Cloning of a cDNA encoding the human cation-dependent mannose 6-phosphate-specific receptor. *Proc. Natl. Acad. Sci. USA 84*:5575–5579.
55. Roth, R. A. (1988). Structure of the receptor for insulin-like growth factor II: The puzzle amplified. *Science 239*:1269–1271.
56. Kiess, W., Blickenstaff, G. D., Sklar, M. M., Thomas, C. L., Nissley, S. P., and Sahagian, G. G. (1988). Biochemical evidence that the type II insulin-like growth factor receptor is identical to the cation-independent mannose 6-phosphate receptor. *J. Biol. Chem. 263*:9339–9344.
57. Kiess, W., Greenstein, L. A., White, R. M., Lee, L., Rechler, M. M., and Nissley, S. P. (1987). Type II insulin-like growth factor receptor is present in rat serum. *Proc. Natl. Acad. Sci. USA 84*:7720–7724.
58. MacDonald, R. G., Pfeffer, S. R., Coussens, L., Tepper, M. A., Brocklebank, C. M., Mole, J. E., Anderson, J. K., Chen, E., Czech, M. P., and Ullrich, A. (1988). A single receptor binds both insulin-like growth factor II and mannose-6 phosphate. *Science 239*:1134–1137.
59. Tong, P. Y., Tollefsen, S. E., and Kornfeld, S. (1988). The cation-independent mannose 6-phosphate receptor binds insulin-like growth factor II. *J. Biol. Chem. 263*:2585–2588.
60. Okamoto, T., Nishimoto, I., Murayama, Y., Ohkuni, Y., and Ogata, E. (1990). Insulin-like growth factor-II/mannose 6-phosphate receptor is incapable of activating GTP-binding proteins in response to mannose 6-phosphate, but capable in response to insulin-like growth factor-II. *Biochem. Biophys. Res. Comm. 168*:1201–1210.
61. Stein, M., Zijderhand-Bleekemolen, J. E., Geuze, H., Hasilik, A., and von Figura, K. (1987). M_r 46000 mannose 6-phosphate specific receptor: Its role in targeting of lysosomal enzymes. *EMBO J. 6*:2677–2681.
62. Brown, W. J. (1990). Cation-independent mannose 6-phosphate receptors are concentrated in *trans* Golgi elements in normal human and I-cell disease fibroblasts. *Eur. J. Cell Biol. 51*:201–210.
63. Stahl, P. D., and Schlesinger, P. H. (1980). Receptor-mediated pinocytosis of mannose/*N*-acetylglucosamine-terminated glycoproteins and lysosomal enzymes by macrophages. *Trends Biochem. Sci. 5*:194–196.
64. Schlesinger, P. H., Rodman, J. S., Doebber, T. W., Stahl, P. D., Lee, Y. C., Stowell, C. P., and Kuhlenschmidt, T. B. (1980). The role of extra-hepatic tissues in the receptor-mediated plasma clearance of glycoproteins terminated by mannose or *N*-acetylglucosamine. *Biochem. J. 192*:597–606.
65. Hubbard, A. L., Wilson, G., Ashwell, G., and Stukenbrok, H. (1979). An electron microscope autoradiographic study of the carbohydrate recognition systems in rat liver. *J. Cell Biol. 83*:47–64.

66. Tolleshaug, H., Berg, T., and Blomhoff, R. (1984). Uptake of mannose-terminated glycoproteins in isolated rat liver cells. *Biochem. J. 223*:151–160.
67. Maynard, Y., and Baenziger, J. U. (1981). Oligosaccharide specific endocytosis by isolated rat hepatic reticuloendothelial cells. *J. Biol. Chem. 256*:8063–8068.
68. Zanetta, J. P., Bingen, A., Dontenwill-Kieffer, M., Reeber, A., Meyer, A., and Vincendon, G. (1987). Cerebellar lectin "R1" is related to the receptor of circulating mannosyl glycoproteins of liver sinusoidal cells. *Cell. Mol. Biol. 33*:423–434.
69. Pranning-van Dalen, D. P., de Leeuw, A. M., Brouwer, A., and Knook, D. L. (1987). Rat liver endothelial cells have a greater capacity than Kupffer cells to endocytose *N*-acetylglucosamine- and mannose-terminated glycoproteins. *Hepatology 7*:672–679.
70. Magnusson, S., and Berg, T. (1989). Extremely rapid endocytosis mediated by the mannose receptor of sinusoidal endothelial rat liver cells. *Biochem. J. 257*:651–656.
71. Wiley, H. S., and Cunningham, D. D. (1982). The endocytotic rate constant: A cellular parameter for quantitating receptor-mediated endocytosis. *J. Biol. Chem. 257*:4222–4229.
72. Stahl, P. D., Wileman, T. E., Diment, S., and Shepherd, V. L. (1984). Mannose-specific oligosaccharide recognition by mononuclear phagocytes. *Biol. Cell 51*:215–218.
73. Lennartz, M. R., Cole, F. S., and Stahl, P. D. (1989). Biosynthesis and processing of the mannose receptor in human macrophages. *J. Biol. Chem. 264*:2385–2390.
74. Shepherd, V. L., Lee, Y. C., Schlesinger, P. H., and Stahl, P. D. (1981). L-Fucose-terminated glycoconjugates are recognized by pinocytosis receptors on macrophages. *Proc. Natl. Acad. Sci. USA 78*:1019–1022.
75. Ohsumi, Y., Chen, V. J., Yan, S.-C. B., Wold, F., and Lee, Y. C. (1988). Interaction between new neoglycoproteins and the D-Man/L-Fuc receptor of rabbit alveolar macrophages. *Glycoconjugate J. 5*:99–106.
76. Shao, M.-C., and Wold, F. (1988). The effect of the protein matrix on glycan processing in glycoproteins: Kinetic analysis of three rat liver Golgi enzymes. *J. Biol. Chem. 263*: 5771–5774.
77. Wileman, T. E., Lennartz, M. R., and Stahl, P. D. (1986). Identification of the macrophage mannose receptor as a 175-kDa membrane protein. *Proc. Natl. Acad. Sci. USA 83*:2501–2505.
78. Lennartz, M. R., Wileman, T. E., and Stahl, P. D. (1987). Isolation and characterization of a mannose-specific endocytosis receptor from rabbit alveolar macrophages. *Biochem. J. 245*: 705–711.
79. Stephenson, J. D., and Shepherd, V. L. (1987). Purification of the human alveolar macrophage mannose receptor. *Biochem. Biophys. Res. Commun. 148*:883–889.
80. Haltiwanger, R. S., and Hill, R. L. (1986). The isolation of a rat alveolar macrophage lectin. *J. Biol. Chem. 261*:7440–7444.
81. Taylor, M. E., Conary, J. T., Lennartz, M. R., Stahl, P. D., and Drickamer, K. (1990). Primary structure of the mannose receptor contains multiple motifs resembling carbohydrate-recognition domains. *J. Biol. Chem. 265*:12156–12162.
82. Ezekowitz, R. A. B., Sastry, K., Bailly, P., and Warner, A. (1990). Molecular characterization of the human macrophage mannose receptor: Demonstration of multiple carbohydrate recognition-like domains and phagocytosis of yeast in COS-1 cells. *J. Exp. Med. 172*:1785–1794.
83. Haltiwanger, R. S., Lehrman, M. A., Eckhardt, A. E., and Hill, R. L. (1986). The distribution and localization of the fucose-binding lectin in rat tissues and the identification of a high affinity form of the mannose/*N*-acetylglucosamine-binding lectin in rat liver. *J. Biol. Chem. 261*:7433–7439.
84. Lehrman, M. A., and Hill, R. L. (1986). The binding of fucose-containing glycoproteins by hepatic lectins: Purification of a fucose-binding lectin from rat liver. *J. Biol. Chem. 261*:7419–7425.
85. Hoppe, C. A., and Lee, Y. C. (1982). Stimulation of mannose-binding activity in the rabbit alveolar macrophage by simple sugars. *J. Biol. Chem. 257*:12831–12834.
86. Ohsumi, Y., and Lee, Y. C. (1987). Mannose-receptor ligands stimulate secretion of lysosomal enzymes from rabbit alveolar macrophages. *J. Biol. Chem. 262*:7955–7962.

87. Sung, S.-S. J., Nelson, R. S., and Silverstein, S. C. (1985). Mouse peritoneal macrophages plated on mannan- and horseradish peroxidase-coated substrates lose the ability to phagocytose by their Fc receptors. *J. Immunol. 134*:3712–3717.
88. Lee, Y. C., and Lee, R. T. (1982). Neoglycoproteins as probes for binding and cellular uptake of glycoconjugates. In *The Glycoconjugates*, (M. I. Horowitz, ed.) Vol. IV. Academic Press, New York, pp. 57–83.
89. Chen, V. J., and Wold, F. (1986). Neoglycoproteins: Preparation and properties of complexes of biotinylated asparagine-oligosaccharides with avidin and streptavidin. *Biochemistry 25*: 939–944.
90. Lehrman, M. A., Pizzo, S. V., Imber, M. J., and Hill, R. L. (1986). The binding of fucose-containing glycoproteins by hepatic lectins. *J. Biol. Chem. 261*:7412–7418.
91. Baenziger, J. U., and Fiete, D. (1980). Galactose and *N*-acetylgalactosamine-specific endocytosis of glycopeptides by isolated rat hepatocytes. *Cell 22*:611–620.
92. Brownell, M. D., Colley, K. J., and Baenziger, J. U. (1984). Synthesis, processing, and secretion of the core-specific lectin by rat hepatocytes and hepatoma cells. *J. Biol. Chem. 259*:3925–3932.
93. Hoyle, G. W., and Hill, R. L. (1988). Molecular cloning and sequencing of a cDNA for a carbohydrate binding receptor unique to rat Kupffer cells. *J. Biol. Chem. 263*:7487–7492.
94. Lehrman, M. A., Haltiwanger, R. S., and Hill, R. L. (1986). The binding of fucose-containing glycoproteins by hepatic lectins. The binding specificity of the rat liver fucose lectin. *J. Biol. Chem. 261*:7426–7432.
95. Kolb, H., Kriese, A., Kolb-Bachofen, V., and Kolb, H. A. (1978). Possible mechanism of entrapment of neuraminidase-treated lymphocytes in the liver. *Cell. Immunol. 40*:457–462.
96. Shepherd, V., Schlesinger, P., and Stahl, P. (1983). *Curr. Top. Membr. Transp. 18*:317–338.
97. Schlepper-Schafer, J., Hulsmann, D., Djovkar, A., Meyer, H. E., Herbertz, L., Kolb, H., and Kolb-Bachofen, V. (1986). Endocytosis via galactose receptors in vivo: Ligand size directs uptake by hepatocytes and/or liver macrophages. *Exp. Cell Res. 165*:494–506.
98. Dini, L., Conti-Devirgiliis, L., and Russo-Caia, S. (1987). The galactose-specific receptor system in rat liver during development. *Development 100*:13–22.
99. Kawasaki, T., Ii, M., Kozutsumi, Y., and Yamashina, I. (1986). Isolation and characterization of a receptor lectin specific for galactose/*N*-acetylgalactosamine from macrophages. *Carbohydr. Res. 151*:197–206.
100. Roos, P. H., Hartman, H.-J., Schlepper-Schafer, J., Kolb, H., and Kolb-Bachofen, V. (1985). Galactose-specific receptors on liver cells. II. Characterization of the purified receptor from macrophages reveals no structural relationship to the hepatocyte receptor. *Biochim. Biophys. Acta 847*:115–121.
101. Kempka, G., Roos, P. H., and Kolb-Bachofen, V. (1990). A membrane-associated form of C-reactive protein is the galactose-specific particle receptor on rat liver macrophages. *J. Immunol. 144*:1004–1009.
102. Cherayil, B. J., Weiner, S. J., and Pillai, S. (1989). The Mac-2 antigen is a galactose-specific lectin that binds IgE. *J. Exp. Med. 170*:1959–1972.
103. Jia, S., and Wang, J. L. (1988). Carbohydrate binding protein 35. Complementary DNA sequence reveals homology with proteins of the heterogeneous nuclear RNP. *J. Biol. Chem. 263*:6009–6011.
104. Lee, H., Kelm, S., Yoshino, T., and Schauer, R. (1988). Carbohydrate specificity of the galactose-recognizing receptor of rat peritoneal macrophages. *Biol. Chem. Hoppe-Seyler 369*:705–714.
105. Kelm, S., and Schauer, R. (1988). The galactose-recognizing system of rat peritoneal macrophages; identification and characterization of the receptor molecule. *Biol. Chem. Hoppe-Seyler 369*:693–704.
106. Ii, M., Kurata, H., Itoh, N., Yamashina, I., and Kawasaki, T. (1990). Molecular cloning and sequence analysis of cDNA encoding the macrophage lectin specific for galactose and *N*-acetylgalactosamine. *J. Biol. Chem. 265*:11295–11298.

107. Drickamer, K. (1988). Two distinct classes of carbohydrate-recognition domains in animal lectins. *J. Biol. Chem. 263*:9557–9560.
108. Kelm, S., and Schauer, R. (1986). The galactose-recognizing system of rat peritoneal macrophages. *Biol. Chem. Hoppe-Seyler 367*:989–998.
109. Aminoff, D. (1988). The role of sialoglycoconjugates in the aging and sequestration of red cells from circulation. *Blood Cells 14*:229–247.
110. Schlepper-Schafer, J., and Kolb-Bachofen, V. (1988). Red cell aging results in a change of cell surface carbohydrate epitopes allowing for recognition by galactose-specific receptors of rat liver macrophages. *Blood Cells 14*:259–269.
111. Prioli, R. P., Rosenberg, I., and Pereira, M. E. A. (1987). Specific inhibition of *Trypanosoma cruzi* neuraminidase by the human plasma glycoprotein "cruzin." *Proc. Natl. Acad. Sci. USA 84*:3097–3101.
112. Klerx, J. P. A. M., Molendijk, A. J., Van Dijk, H., Vloet, K. P., and Willers, J. M. N. (1986). Simple sugars with affinity for the macrophage asialoglycoprotein receptor are adjuvants for the humoral immune response to neuraminidase-treated sheep erythrocytes. *J. Immunol. 136*: 73–75.
113. Maynard, Y., and Baenziger, J. U. (1982). Characterization of a mannose and *N*-acetylglucosamine-specific lectin present in rat hepatocytes. *J. Biol. Chem. 257*:3788–3794.
114. Oka, S., Itoh, N., Kawasaki, T., and Yamashina, I. (1987). Primary structure of rat liver mannan-binding protein deduced from its cDNA sequence. *J. Biochem. 101*:135–144.
115. Drickamer, K., Dordahl, M. S., and Reynolds. L. (1986). Mannose-binding proteins isolated from rat liver contain carbohydrate-recognition domains linked to collagenous tails: Complete primary structures and homology with pulmonary surfactant apoprotein. *J. Biol. Chem. 261*:6878–6887.
116. Mori, K., Kawasaki, T., and Yamashina, I. (1983). Identification of the mannan-binding protein from rat livers as a hepatocyte protein distinct from the mannan receptor on sinusoidal cells. *Arch. Biochem. Biophys. 222*:542–552.
117. Taylor, M. E., and Summerfeld, J. A. (1987). Carbohydrate-binding proteins of human serum: Isolation of two mannose/fucose-specific lectins. *Biochim. Biophys. Acta 915*:60–67.
118. Drickamer, K., and McCreary, V. (1987). Exon structure of a mannose-binding protein gene reflects its evolutionary relationship to the asialoglycoprotein receptor and nonfibrillar collagens. *J. Biol. Chem. 262*:2582–2589.
119. Colley, K. J., and Baenziger, J. U. (1987). Identification of the post-translational modifications of the core-specific lectin. *J. Biol. Chem. 262*:10290–10295.
120. Haagsman, H. P., Hawgood, S., Sargeant, T., Buckley, D., White, R. T., Drickamer, K., and Benson, B. J. (1987). The major lung surfactant protein, SP 28-36, is a calcium-dependent, carbohydrate-binding protein. *J. Biol. Chem. 262*:13877–13880.
121. Young, N. M., and Leon, M. A. (1987). The carbohydrate specificity of conglutinin and its homology to proteins in the hepatic lectin family. *Biochem. Biophys. Res. Commun. 143*: 645–651.
122. Mori, K., Kawasaki, T., and Yamashina, I. (1984). Subcellular distribution of the mannan-binding protein and its endogenous inhibitors in rat liver. *Arch. Biochem. Biophys. 232*:223–233.
123. Mori, K., Kawasaki, T., and Yamashina, I. (1988). Isolation and characterization of endogenous ligands for liver mannan-binding protein. *Arch. Biochem. Biophys. 264*:647–656.
124. Ikeda, K., Sannoh, T., Kawasaki, N., Kawasaki, T., and Yamashina, I. (1987). Serum lectin with known structure activates complement through the classical pathway. *J. Biol. Chem. 262*:7451–7454.
125. Evered, D., and Whelan, J., eds. (1989). *The Biology of Hyaluronan*. John Wiley & Sons, London.
126. Smedsrod, B., Pertoft, H., Gustafson, S., and Laurent, T. C., (1990). Scavenger functions of the liver endothelial cell. *Biochem. J. 266*:313–327.

127. Raja, R. H., McGary, C. T., and Weigel, P. H. (1988). Affinity and distribution of surface and intracellular hyaluronic acid receptors in isolated rat liver endothelial cells. *J. Biol. Chem. 263*:16661–16668.
128. Smedsrod, B., Pertoft, H., Eriksson, S., Fraser, J. R. E., and Laurent, T. C. (1984). Studies in vitro on the uptake and degradation of sodium hyaluronate in rat liver endothelial cells. *Biochem. J. 223*:617–626.
129. McGary, C. T., Raja, R. H., and Weigel, P. H. (1989). Endocytosis of hyaluronic acid by rat liver endothelial cells: Evidence for receptor recycling. *Biochem. J. 257*:875–884.
130. Laurent, T. C., Fraser, J. R. E., Pertoft, H., and Smedsrod, B. (1986). Binding of hyaluronate and chondroitin sulphate to liver endothelial cells. *Biochem. J. 234*:653–658.
131. Yannariello-Brown, J., McGary, C. T., and Weigel, P. H. (1991). The endocytic hyaluronan receptor in rat liver sinusoidal endothelial cells is Ca^{+2}-independent and distinct from a Ca^{+2}-dependent hyaluronan binding activity. *J. Cell Biochem.* (in press).
132. Oka, J. A., and Weigel, P. H. (1988). Effects of hyperosmolarity on ligand processing and receptor recycling in the hepatic galactosyl receptor system. *J. Cell. Biochem. 36*:169–183.
133. Oka, J. A., Christensen, M. D., and Weigel, P. H. (1989). Hyperosmolarity inhibits galactosyl receptor-mediated but not fluid phase endocytosis in isolated rat hepatocytes. *J. Biol. Chem. 264*:12016–12024.
134. Smedsrod, B., Malmgren, M., Ericsson, J., and Laurent, T. C. (1988). Morphological studies on endocytosis of chondroitin sulphate proteoglycan by rat liver endothelial cells. *Cell Tissue Res. 253*:39–45.
135. Strobel, J. L., Baynes, J. W., and Thorpe, S. R. (1985). ^{125}I-Glycoconjugate labels for identifying sites of protein catabolism in vivo: Effects of structure and chemistry of coupling to protein on label entrapment in cells after protein degradation. *Arch. Biochem. Biophys. 240*:635–645.
136. Fraser, J. R. E., Kimpton, W. G., Laurent, C. T., Cahill, R. N. P., and Vakakis, N. (1988). Uptake and degradation of hyaluronan in lymphatic tissue. *Biochem. J. 256*:153–158.
137. Green, S. J., Tarone, G., and Underhill, C. B. (1988). Distribution of hyaluronate and hyaluronate receptors in the adult lung. *J. Cell Sci. 89*:145–156.
138. LeBoeuf, R. D., Raja, R. H., Fuller, G. M., and Weigel, P. H. (1986). Human fibrinogen specifically binds hyaluronic acid. *J. Biol. Chem. 261*:12586–12592.
139. Weigel, P. H., Fuller, G. M., and LeBoeuf, R. D. (1986). A model for the role of hyaluronic acid and fibrin in the early events during the inflammatory response and wound healing. *J. Theor. Biol. 119*:219–234.
140. Ahmed, M. U., Dunn, J. A., Walla, M. D., Thorpe, S. R., and Baynes, J. W. (1988). Oxidative degradation of glucose adducts to protein: Formation of 3-(*N'*-lysino)-lactic acid from model compounds and glycated proteins. *J. Biol. Chem. 263*:8816–8821.
141. Vlassara, H., Brownlee, M., and Cerami, A. (1986). Novel macrophage receptor for glucose-modified proteins is distinct from previously described scavenger receptors. *J. Exp. Med. 164*:1301–1309.
142. Radoff, S., Vlassara, H., and Cerami, A. (1988). Characterization of a solubilized cell surface binding protein on macrophages specific for proteins modified nonenzymatically by advanced glycosylated end products. *Arch. Biochem. Biophys. 263*:418–423.
143. Horiuchi, S., Shiga, M., Araki, N., Takata, K., Saitoh, M., and Morino, Y. (1988). Evidence against in vivo presence of 2-(2-furoyl)-4(5)-(2-furanyl)-1*H*-imidazole, a major fluorescent advanced end product generated by nonenzymatic glycosylation. *J. Biol. Chem. 263*:18821–18826.
144. Blomhoff, R., Eskild, W., and Berg, T. (1984). Endocytosis of formaldehyde-treated serum albumin via scavenger pathway in liver endothelial cells. *Biochem. J. 218*:81–86.
145. Horiuchi, S., Takata, K., Maeda, H., and Morino, Y. (1985). Scavenger function of sinusoidal liver cells. *J. Biol. Chem. 259*:53–56.
146. Nishikawa, K., Arai, H., and Inoue, K. (1990). Scavenger receptor-mediated uptake and metabolism of lipid vesicles containing acidic phospholipids by mouse peritoneal macrophages. *J. Biol. Chem. 265*:5226–5231.

147. Takata, K., Horiuchi, S., Araki, N., Shiga, M., Saitoh, M., and Morino, Y. (1988). Endocytic uptake of nonenzymatically glycosylated proteins is mediated by a scavenger receptor for aldehyde-modified proteins. *J. Biol. Chem. 263*:14819–14825.
148. Patrick, J. S., Thorpe, S. R., and Baynes, J. W. (1988). Oxidative degradation of glycated proteins: Age-dependent accumulation of carboxymethyllysine in lens proteins. *FASEB J. 2*:A1731.
149. Consiglio, E., Shifrin, S., Yavin, Z., Ambesi-Impiombato, F. S., Rall, J. E., Salvatore, G., and Kohn, L. D. (1981). Thyroglobulin interactions with thyroid membranes: Relationship between receptor recognition of *N*-acetylglucosamine residues and the iodine content of thyroglobulin preparations. *J. Biol. Chem. 256*:10592–10599.
150. Miquelis, R., Alquier, C., and Monsigny, M. (1987). The *N*-acetylglucosamine-specific receptor of the thyroid: Binding characteristics, partial characterization, and potential role. *J. Biol. Chem. 262*:15291–15298.
151. Midoux, P., Roche, A.-C., and Monsigny, M. (1986). Estimation of the degradation of endocytosed material by flow cytofluorometry using two neoglycoproteins containing different numbers of fluorescein molecules. *Biol. Cell 58*:221–226.
152. Roche, A. C., Barzilay, M., Midoux, P., Junqua, S., Sharon, N., and Monsigny, M. (1983). Sugar-specific endocytosis of glycoproteins by Lewis lung carcinoma cells. *J. Cell. Biochem. 22*:131–140.
153. Samlowski, W. E., Braaten, B. A., and Daynes, R. A. (1985). Characterization of the in vitro interaction of PNA^{hi} lymphocytes with the bone marrow and hepatic asialoglycoprotein receptors. *Cell. Immunol. 95*:1–14.
154. Aizawa, S., and Tavassoli, M. (1987). Interaction of murine granulocyte-macrophage progenitors and supporting stroma involves a recognition mechanism with galactosyl and mannosyl specificities. *J. Clin. Invest. 80*:1698–1705.
155. Czop, J. K., and Austen, K. F. (1985). A β-glucan inhibitable receptor on human monocytes: Its identity with the phagocytic receptor for particulate activators of the alternative complement pathway. *J. Immunol. 134*:2588–2593.
156. Goldman, R. (1988). Characteristics of the β-glucan receptor of murine macrophages. *Exp. Cell Res. 174*:481–490.
157. Goldstein, J. L., Brown, M. S., Anderson, R. G. W., Russell, D. W., and Schneider, W. J. (1985). Receptor-mediated endocytosis: Concepts emerging from the LDL receptor system. *Annu. Rev. Cell Biol. 1*:1–39.
158. van Deurs, B., Petersen, O. W., Olsnes, S., and Sandvig, K. (1989). The ways of endocytosis. *Int. Rev. Cytol. 117*:131–177.
159. Roth, T. F., and Porter, K. R. (1964). Yolk protein uptake in the oocyte of the mosquito *Aedes aegypti* L. *J. Cell Biol. 20*:313–332.
160. Salisbury, J. L., Condeelis, J. S., and Satir, P. (1983). Receptor-mediated endocytosis: Machinery and regulation of the clathrin-coated vesicle pathway. *Int. Rev. Exp. Pathol. 24*: 1–62.
161. Keen, J. H. (1990). Clathrin and associated assembly and disassembly proteins. *Annu. Rev. Biochem. 59*:415–438.
162. Steer, C. J., and Klausner, R. D. (1983). Clathrin-coated pits and coated vesicles: Functional and structural studies. *Hepatology 3*:437–454.
163. Pearse, B. M. F., and Robinson, M. S. (1990). clathrin, adaptors, and sorting. *Annu. Rev. Cell Biol. 6*:151–171.
164. Moore, M. S., Mahaffey, D. T., Brodsky, F. M., and Anderson, R. G. W. (1987). Assembly of clathrin-coated pits onto purified plasma membranes. *Science 263*:558–563.
165. Pastan, I., and Willingham, M. C., eds. (1985). *Endocytosis*. Plenum Press, New York.
166. Lazarovits, J., and Roth, M. (1988). A single amino acid change in the cytoplasmic domain allows the influenza virus hemagglutinin to be endocytosed through coated pits. *Cell 53*: 743–752.

167. McGraw, T. E., and Maxfield, F. R. (1990). Human transferrin receptor internalization is partially dependent upon an aromatic amino acid on the cytoplasmic domain. *Cell Regul.* *1*:369–377.
168. Chen, W.-J., Goldstein, J. L., and Brown, M. S. (1990). NPXY, a sequence often found in cytoplasmic tails, is required for coated pit-mediated internalization of the low density lipoprotein receptor. *J. Biol. Chem.* *265*:3116–3123.
169. Collawn, J. F., Stangel, M., Kuhn, L. A., Esekogwu, V., Jing, S., Trowbridge, I. S., and Tainer, J. A. (1990). Transferrin receptor internalization sequence YXRF implicates a tight turn as the structural recognition motif for endocytosis. *Cell 63*:1061–1072.
170. Clarke, B. L., and Weigel, P. H. (1985). Recycling of the asialoglycoprotein receptor in isolated rat hepatocytes: ATP depletion blocks receptor recycling but not a single round of endocytosis. *J. Biol. Chem.* *260*:128–133.
171. Schmid, S. L., and Rothman, J. E. (1985). Enzymatic dissociation of clathrin cages in a two-stage process. *J. Biol. Chem.* *260*:10044–10049.
172. Oka, J. A., and Weigel, P. H. (1983). Recycling of the asialoglycoprotein receptor in isolated rat hepatocytes: Dissociation of internalized ligand from receptor occurs in two kinetically and thermally distinguishable compartments. *J. Biol. Chem.* *258*:10253–10262.
173. Oka, J. A., and Weigel, P. H. (1987). Monensin inhibits ligand dissociation only transiently and partially and distinguishes two galactosyl receptor pathways in isolated rat hepatocytes. *J. Cell. Physiol.* *133*:243–252.
174. Clarke, B. L., Oka, J. A., and Weigel, P. H. (1987). Degradation of asialoglycoproteins mediated by the galactosyl receptor system in isolated hepatocytes: Evidence for two parallel pathways. *J. Biol. Chem.* *262*:17384–17392.
175. Clarke, B. L., and Weigel, P. H. (1989). Differential effects of leupeptin, monensin and colchicine on ligand degradation mediated by the two asialoglycoprotein receptor pathways in isolated rat hepatocytes. *Biochem. J.* *262*:277–284.
176. Tietze, C., Schlesinger, P., and Stahl, P. (1982). Mannose-specific endocytosis receptor of alveolar macrophages: Demonstration of two functionally distinct intracellular pools of receptor and their roles in receptor recycling. *J. Cell Biol.* *92*:417–424.
177. McAbee, D. D., Clarke, B. L., Oka, J. A., and Weigel, P. H. (1990). The surface activity of the same subpopulation of galactosyl receptors on isolated rat hepatocytes is modulated by colchicine, monensin, ATP depletion, and chloroquine. *J. Biol. Chem.* *265*:629–635.
178. Weigel, P. H., and Oka, J. A. (1984). Recycling of the hepatic asialoglycoprotein receptor in isolated rat hepatocytes: Receptor-ligand complexes in an intracellular slowly dissociating pool return to the cell surface prior to dissociation. *J. Biol. Chem.* *259*:1150–1154.
179. Breitfeld, P. P., Simmons, C. F., Jr., Strous, G. J. A. M., Geuze, H. J., and Schwartz, A. L. (1985). Cell biology of the asialoglycoprotein receptor system: A model of receptor-mediated endocytosis. *Int. Rev. Cytol.* *97*:47–95.
180. Tolleshaug, H., Kolset, S. O., and Berg, T. (1985). The influence of cellular ATP levels on receptor-mediated endocytosis and degradation of asialoglycoproteins in suspended hepatocytes. *Biochem. Pharmacol.* *34*:1639–1645.
181. McAbee, D. D., and Weigel, P. H. (1988). ATP-dependent inactivation and reactivation of constitutively recycling galactosyl receptors in isolated rat hepatocytes. *Biochemistry* *27*:2061–2069.
182. Berg, T., Blomhoff, R., Naess, L., Tolleshaug, H., and Drevon, C. A. (1983). Monensin inhibits receptor-mediated endocytosis of asialoglycoproteins in rat hepatocytes. *Exp. Cell Res.* *148*:319–330.
183. Fiete, D., Brownell, M. D., and Baenziger, J. U. (1983). Evidence for transmembrane modulation of the ligand-binding site of the hepatocyte galactose/*N*-acetylgalactosamine-specific receptor. *J. Biol. Chem.* *258*:817–823.
184. Tolleshaug, H., and Berg, T. (1977). Uptake and degradation of asialo-fetuin by isolated rat hepatocytes. *Acta. Biol. Med. Ger.* *36*:1753–1762.

185. Schwartz, A. L., Bolognesi, A., and Fridovich, S. E. (1984). Recycling of the asialoglycoprotein receptor and the effect of lysosomotropic amines in hepatoma cells. *J. Cell Biol. 98*:732–738.
186. Weigel, P. H., and Oka, J. A. (1983). The surface content of asialoglycoprotein receptors on isolated hepatocytes is reversibly modulated by changes in temperature. *J. Biol. Chem. 258*:5089–5094.
187. Steer, C. J., Weiss, P., Huber, B. E., Wirth, P. J., Thorgeirsson, S. S., and Ashwell, G. (1987). Ligand-induced modulation of the hepatic receptor for asialoglycoproteins in the human hepatoblastoma cell line, HepG2. *J. Biol. Chem. 262*:17524–17529.
188. Wall, D. A., and Hubbard, A. L. (1981). Galactose-specific recognition system of mammalian liver: Receptor distribution on the hepatocyte cell surface. *J. Cell Biol. 90*:687–696.
189. McAbee, D. D., Oka, J. A., and Weigel, P. H. (1989). Loss of surface galactosyl receptor activity on isolated rat hepatocytes induced by monensin or chloroquine requires receptor internalization via a clathrin coated pit pathway. *Biochem. Biophys. Res. Comm. 161*:261–266.
190. Basu, S. K., Goldstein, J. L., Anderson, R. G. W., and Brown, M. S. (1981). Monensin interrupts the recycling of low density lipoprotein receptors in human fibroblasts. *Cell 24*: 493–502.
191. Nolan, C. M., and Sly, W. S. (1987). Intracellular traffic of the mannose-6-phosphate receptor and its ligands. In *Immunobiology of Proteins and Peptides* (Z. Atassi, ed.). Plenum Press, New York, pp. 199–212.
192. Tietze, C., Schlesinger, P., and Stahl, P. (1980). Chloroquine and ammonium ions inhibit receptor-mediated endocytosis of mannose-glycoconjugates by macrophages: Apparent inhibition of receptor recycling. *Biochem. Biophys. Res. Commun. 93*:1–8.
193. Buys, S. S., Gren, L. H., and Kaplan, J. (1987). Phorbol esters and calcium ionophores inhibit internalization and accelerate recycling of receptors in macrophages. *J. Biol. Chem. 262*:12970–12976.
194. Kaplan, J., Ward, D. M., and Wiley, H. S. (1985). Phenylarsine oxide-induced increase in alveolar macrophage surface receptors: Evidence for fusion of internal receptor pools with the cell surface. *J. Cell Biol. 101*:121–129.
195. Kaplan, J., and Keogh, E. A. (1981). Analysis of the effect of amines on inhibition of receptor-mediated and fluid-phase pinocytosis in rabbit alveolar macrophages. *Cell 24*:925–932.
196. Stein, B. S., and Sussman, H. H. (1986). Demonstration of two distinct transferrin receptor recycling pathways and transferrin-independent receptor internalization in K562 cells. *J. Biol. Chem. 261*:10319–10331.
197. Pathak, R. K., Merkle, R. K., Cummings, R. D., Goldstein, J. L., Brown, M. S., and Anderson, R. G. W. (1988). Immunocytochemical localization of mutant low density lipoprotein receptors that fail to reach the Golgi complex. *J. Cell Biol. 106*:1831–1841.
198. Weigel, P. H., and Oka, J. A. (1983). The large intracellular pool of asialoglycoprotein receptors functions during the endocytosis of asialoglycoproteins by isolated rat hepatocytes. *J. Biol. Chem. 258*:5095–5102.
199. Weigel, P. H., Ray, D. A., and Oka, J. A. (1983). Quantitation of intracellular membrane-bound enzymes and receptors in digitonin-permeabilized cells. *Anal. Biochem. 133*:437–449.
200. Harford, J., Bridges, K., Ashwell, G., and Klausner, R. D. (1983). Intracellular dissociation of receptor-bound asialoglycoproteins in cultured hepatocytes: A pH-mediated nonlysosomal event. *J. Biol. Chem. 258*:3191–3197.
201. Wileman, T., Boshans, R., and Stahl, P. (1985). Uptake and transport of mannosylated ligands by alveolar macrophages: Studies on ATP-dependent receptor–ligand dissociation. *J. Biol. Chem. 260*:7387–7393.
202. Harford, J., Wolkoff, A. W., Ashwell, G., and Klausner, R. D. (1983). Monensin inhibits intracellular dissociation of asialoglycoproteins from their receptor. *J. Cell Biol. 96*:1824–1828.
203. Mellman, I., Fuchs, R., and Helenius, A. (1986). Acidification of the endocytic and exocytic pathways. *Annu. Rev. Biochem. 55*:663–700.
204. Yamashiro, D. J., and Maxfield, F. R. (1988). Regulation of endocytic processes by pH. *Trends Pharm. Sci. 9*:190–193.

205. Anderson, R. G. W., and Orci, L. (1988). A view of acidic intracellular compartments, *J. Cell Biol. 106*:539–543.
206. Yamashiro, D. J., and Maxfield, F. R. (1984). Acidification of endocytic compartments and the intracellular pathways of ligands and receptors. *J. Cell. Biochem. 26*:231–246.
207. Yamashiro, D. J., and Maxfield, F. R. (1987). Kinetics of endosome acidification in mutant and wild-type Chinese hamster ovary cells. *J. Cell Biol. 105*:2713–2721.
208. Sipe, D. M., and Murphy, R. F. (1987). High-resolution kinetics of transferrin acidification in BALB/c 3T3 cells: Exposure to pH 6 followed by temperature-sensitive alkalinization during recycling. *Proc. Natl. Acad. Sci. USA 84*:7119–7123.
209. Schwartz, A. L., Strous, G. J. A. M., Slot, J. W., and Geuze, H. J. (1985). Immunoelectron microscopic localization of acidic intracellular compartments in hepatoma cells. *EMBO J. 4*:899–904.
210. Tycko, B., Keith, C. H., and Maxfield, F. R. (1983). Rapid acidification of endocytic vesicles containing asialoglycoprotein in cells of a human hepatoma line. *J. Cell Biol. 97*:1762–1776.
211. Davoust, J., Gruenberg, J., and Howell, K. E. (1987). Two threshold values of low pH block endocytosis at different stages. *EMBO J. 6*:3601–3609.
212. Herman, B., and Albertini, D. F. (1984). A time-lapse video image intensification analysis of cytoplasmic organelle movements during endosome translocation. *J. Cell Biol. 98*:565–576.
213. Geuze, H. J., Slot, J. W., Strous, G. J. A. M., Lodish, H. F., and Schwartz, A. L. (1983). Intracellular site of asialoglycoprotein receptor-ligand uncoupling: Double-label immunoelectron microscopy during receptor-mediated endocytosis. *Cell 32*:277–287.
214. Belcher, J. D., Hamilton, R. L., Brady, S. E., Hornick, C. A., Jaeckle, S., Schneider, W. J., and Havel, R. J. (1987). Isolation and characterization of three endosomal fractions from the liver of estradiol-treated rats. *Proc. Natl. Acad. Sci. USA 84*:6785–6789.
215. van Deurs, B., Pedersen, L. R., Sundan, A., Olsnes, S., and Sandvig, K. (1985). Receptor-mediated endocytosis of a ricin-colloidal gold conjugate in Vero cells: Intracellular routing to vacuolar and tubulo-vesicular portions of the endosomal system. *Exp. Cell Res. 159*:287–304.
216. Harding, C., Levy, M. A., and Stahl, P. (1985). Morphological analysis of ligand uptake and processing: The role of multivesicular endosomes and CURL in receptor–ligand processing. *Eur. J. Cell Biol. 36*:230–238.
217. Kessell, I., Holst, B. D., and Roth, T. F. (1989). Membranous intermediates in endocytosis are labile, as shown in a temperature-sensitive mutant. *Proc. Natl. Acad. Sci. USA 86*:4968–4972.
218. Oka, J. A., and Weigel, P. H. (1983). Microtubule-depolymerizing agents inhibit asialo-orosomucoid delivery to lysosomes but not its endocytosis or degradation in isolated rat hepatocytes. *Biochim. Biophys. Acta 763*:368–376.
219. Quintart, J., Courtoy, P. J., and Baudhuin, P. (1984). Receptor-mediated endocytosis in rat liver: Purification and enzymic characterization of low density organelles involved in uptake of galactose-exposing proteins. *J. Cell Biol. 98*:877–884.
220. Wall, D. A., and Hubbard, A. L. (1985). Receptor-mediated endocytosis of asialoglycoproteins by rat liver hepatocytes: Biochemical characterization of the endosomal compartments. *J. Cell Biol. 101*:2104–2112.
221. Berg, T., Kindberg, G. M., Ford, T., and Blomhoff, R. (1985). Intracellular transport of asialoglycoproteins in rat hepatocytes: Evidence for two subpopulations of lysosomes. *Exp. Cell Res. 161*:285–296.
222. Gruenberg, J., and Howell, K. E. (1989). Membrane traffic in endocytosis: Insights from cell-free assays. *Annu. Rev. Cell Biol. 5*:453–481.
223. Davey, J., Hurtley, S. M., and Warren, G. (1985). Reconstitution of an endocytic fusion event in a cell-free system. *Cell 43*:643–652.
224. Diaz, R., Mayorga, L., and Stahl, P. (1988). In vitro fusion of endosomes following receptor-mediated endocytosis. *J. Biol. Chem. 263*:6093–6100.
225. Gruenberg, J. E., and Howell, K. E. (1986). Reconstitution of vesicle fusions occurring in endocytosis with a cell-free system. *EMBO J. 5*:3091–3101.

226. Diment, S., and Stahl, P. (1985). Macrophage endosomes contain proteases which degrade endocytosed protein ligands. *J. Biol. Chem. 260*:15311–15317.
227. Schmid, S. L., and Rothman, J. E. (1985). Enzymatic dissociation of clathrin cages in a two-stage process. *J. Biol. Chem. 260*:10044–10049.
228. Chappell, T. G., Konforti, B. B., Schmid, S. L., and Rothman, J. E. (1987). The ATPase core of a clathrin uncoating protein. *J. Biol. Chem. 262*:746–751.
229. Aronson, N. N., Jr., and Kuranda, M. J. (1989). Lysosomal degradation of Asn-linked glycoproteins. *FASEB J. 3*:2615–2622.
230. Medh, J. D., and Weigel, P. H. (1991). Reconstitution of galactosyl receptor inactivation in permeabilized rat hepatocytes is ATP dependent. *J. Biol. Chem. 266*:8771–8778.
231. Fishman, J. B., Dickey, B. F., McCrory, M. F., and Fine, R. E. (1986). Reversible inactivation of vasopressin and angiotensin II binding to hepatocyte membranes by a calcium-dependent cytosolic protein. *J. Biol. Chem. 261*:5810–5816.
232. Gudas, J. M., and Hankinson, O. (1986). Reversible inactivation of the Ah receptor associated with changes in intracellular ATP levels. *J. Cell. Physiol. 128*:449–456.
233. Driscoll, W. J., Lee, Y. C., and Strott, C. A. (1990). Regulation of adrenocortical pregnenolone-binding protein activity by phosphorylation/dephosphorylation. Phosphatase mediated inactivation is reversed by cytosolic kinase. *J. Biol. Chem. 265*:12306–12311.
234. Auricchio, F., Migliaccio, A., Castoria, G., Rotondi, A., Di-Domenico, M., and Pagano, M. (1986). Activation–inactivation of hormone binding sites of the oestradiol-17 β receptor is a multiregulated process. *J. Steroid Biochem. 24*:39–43.
235. Boldt, D. H., Phillips, J. L., and Alcantara, O. (1987). Disparity between expression of transferrin receptor ligand binding and non-ligand binding domains on human lymphocytes. *J. Cell. Physiol. 132*:331–336.
236. Ventura, M. A., Louache, F., Rouis, M., Erlich, D., Goldstein, S., Testa, U., and Thomopoulos, P. (1987). Specific modulation of surface receptors in J.774 macrophages by anchorage. *J. Exp. Cell Res. 170*:290–299.
237. Shepherd, V. L., Abdolrasulnia, R., Garrett, M., and Cowan, H. B. (1990). Down-regulation of mannose receptor activity in macrophages after treatment with lipopolysaccharides and phorbol esters. *J. Immunol. 145*:1530–1536.
238. McAbee, D. D., Lear, M. C., and Weigel, P. H. (1991). Total cellular activity and distribution of a subpopulation of galactosyl receptors in isolated rat hepatocytes are differentially affected by microtubule drugs, monensin, low temperature, and chloroquine. *J. Cell. Biochem. 45*:59–68.
239. Blackshear, P. J., Nairn, A. C., and Kuo, J. F. (1988). Protein kinases 1988: A current perspective. *FASEB J. 2*:2957–2969.
240. Medh, J. D., Haynes, P. A., Weigel, P. H., and LaBelle, E. F. (1989). Ligand binding and internalization by the rat hepatic asialoglycoprotein receptor does not generate polyphosphoinositide derived second messengers. *Life Sci. 45*:2285–2294.
241. Schwartz, A. L. (1984). Phosphorylation of the human asialoglycoprotein receptor. *Biochem. J. 223*:481–486.
242. Fallon, R. J., and Schwartz, A. L. (1987). Mechanism of the phorbol ester-mediated redistribution of asialoglycoprotein receptor: Selective effects on receptor recycling pathways in HepG2 cells. *Mol. Pharmacol. 32*:348–355.
243. Clarneburg, R. (1983). Asialoglycoprotein receptor is uninvolved in clearing intact glycoproteins from rat blood. *Am. J. Physiol. 244*:G247–G253.
244. Wong, K.-L., Charlwood, P. A., Hatton, M. W. C., and Regoeczi, E. (1974). Studies of the metabolism of asialotransferrins: Evidence that transferrin does not undergo desialylation in vivo. *Clin. Sci. Mol. Med. 46*:763–774.
245. Springer, G. R., Cheingsong-Popov, R., Schirrmacher, V., Desai, P. R., and Tegtmeyer, H. (1983). Proposed molecular basis of murine tumor cell–hepatocyte interaction. *J. Biol. Chem. 258*:5702–5706.
246. Regoeczi, E., Chindemi, P. A., Debanne, M. T., and Charlwood, P. A. (1982). Partial resialylation of human asialotransferrin type 3 in the rat. *Proc. Natl. Acad. Sci. USA 79*:2226–2230.

247. Shih, L. Y., Kuerer, H. M., Chen, T. H., and Deposito, F. (1988). Strain difference in galactokinase level and susceptibility to the teratogenic effect of dietary galactose in mice: I. Teratogenic and embryopathic effect. *Teratology 38*:175–179.
248. Barondes, S. H. (1984). Soluble lectins: A new class of extracellular proteins. *Science 223*:1259–1264.
249. Runyan, R. B., Maxwell, G. D., and Shur, B. D. (1986). Evidence for a novel enzymatic mechanism of neural crest cell migration on extracellular glycoconjugate matrices. *J. Cell Biol. 102*:432–441.
250. Bayna, E. M., Shaper, J. H., and Shur, B. D. (1988). Temporally specific involvement of cell surface β-1,4galactosyltransferase during mouse embryo morula compaction. *Cell 53*:145–157.
251. Scully, N. F., Shaper, J. H., and Shur, B. D. (1987). Spatial and temporal expression of cell surface galactosyltransferase during mouse spermatogenesis and epididymal maturation. *Dev. Biol. 124*:111–124.
252. Hinek, A., Wrenn, D. S., Mecham, R. P., and Barondes, S. H. (1988). The elastin receptor: A galactosidase-binding protein. *Science 239*:1539–1541.
253. Krivan, H. C., Roberts, D. D., and Ginsburg, V. (1988). Many pulmonary pathogenic bacteria bind specifically to the carbohydrate sequence GalNAcβ1-4Gal found in some glycolipids. *Proc. Natl. Acad. Sci. USA 85*:6157–6161.
254. Samlowski, W. E., and Daynes, R. A. (1985). Bone marrow engraftment efficiency is enhanced by competitive inhibition of the hepatic asialoglycoprotein receptor. *Proc. Natl. Acad. Sci. USA 82*:2508–2512.
255. Wieser, R. J., and Oesch, F. (1988). Contact-dependent regulation of growth of diploid human fibroblasts is dependent upon the presence of terminal galactose residues on plasma membrane glycoproteins. *Exp. Cell Res. 176*:80–86.
256. Allen, H. J., Karakousis, C., Piver, M. S., Gamarra, M., Nava, H., Forsyth, B., Matecki, B., Jazayeri, A., Sucato, D., Kisailus, E., and DiCioccio, R. (1987). Galactoside-binding lectin in human tissues. *Tumor Biol. 8*:218–229.
257. Raz, A., and Lotan, R. (1987). Endogenous galactoside-binding lectins: A new class of functional tumor cell surface molecules related to metastasis. *Cancer Metastasis Rev. 6*:433–452.
258. Paietta, E., Hubbard, A. L., Wiernik, P. H., Diehl, V., and Stockert, R. J. (1987). Hodgkin's cell lectin: An ectosialyltransferase and lymphocyte agglutinant related to the hepatic asialoglycoprotein receptor. *Cancer Res. 47*:2461–2467.
259. Oka, J. A. and Weigel, P. H. (1991). Vanadate modulates the activity of a subpopulation of asialoglycoprotein receptors on isolated rat hepatocytes: Active surface receptors are internalized and replaced by inactive receptors. *Arch. Biochem. Biophys. 289*:362–370.

15

Molecular Biology of Glycosyltransferases and Glycosidases

Joseph T. Y. Lau and Terrance P. O'Hanlon *Roswell Park Cancer Institute, Buffalo, New York*

Although it has long been recognized that glycosylation is a common feature among extracellular and cell surface proteins and lipids, the biological roles of oligosaccharides on glycoconjugates remain elusive. It is quite clear that the overall physical properties of a glycoprotein, be it an enzyme, receptor, or ligand, comprise the sum of its parts, which include the polypeptide as well as the carbohydrate. Altered glycosylation may result in changes in stability, rate of proteolysis, and solubility, of the glycoprotein. It is also becoming increasingly evident that oligosaccharides on glycoconjugates are central to cellular recognition and are requisite to appropriate organismal function. For example, recent reports have implicated the role of complex oligosaccharides as well as particular glycosyltransferases in immune recognition [1–8], homing of lymphocytes to peripheral lymph nodes [9], and cellular migration during embryogenesis [10–12].

Glycosylation is not an absolute event. Microheterogeneity of oligosaccharide structures exist at any given glycosylation site on glycoproteins. A glycoprotein with multiple glycosylation sites, each site with its unique spectrum of oligosaccharides, exists in a pool of glycoforms; each glycoform differs from the other only by the oligosaccharide structures at a given glycosylation site. The biological contribution of complex carbohydrates, therefore, should be the sum of the glycoforms, so that the specific mixture of glycoforms found in a glycoprotein is species- and tissue-specific and that normal development and cellular differentiation is accompanied by alterations in glycosylation patterns [for a review on glycobiology, see [13].

Correct expression of oligosaccharide structures is dictated, at least in part, by activities of the *glycosyltransferases*, enzymes that mediate the synthesis of specific sugar linkages, and the *glycosidases*, enzymes that mediate the hydrolysis of specific linkages. Indeed, inappropriate expression of glycosyltransferase and glycosidase activities is reflected in the altered glycosylation patterns that accompany cellular transformation and oncogenesis [1,7]. Aberrant glycosylation is also manifested in other pathological states, such as rheumatoid arthritis [14,15], tuberculosis [16], Crohn's disease [17], and cystic fibrosis [18].

Further understanding of the biological functions of complex oligosaccharides and the regulatory pathways that govern their correct expression can be attained by detailed analysis of the glycosyltransferases and glycosidases. However, what is apparent is that the metabolism of complex oligosaccharides requires a bewildering array of enzymes. For example, ten or more sialyltransferases have been postulated to mediate the synthesis of an equal number of mammalian sialic acid linkages [19,20]. The same rationale applies to the galactosyltransferases, fucosyltransferases, and other glycosyltransferases. To date, in vitro enzyme activity has been demonstrated for only a few of these enzymes, and even fewer have been purified to homogeneity. The existence of the remaining enzymes is implied by the presence of oligosaccharides that could not be synthesized by any of the enzymes already characterized. Clearly, what is needed is a detailed knowledge of the structure and interrelationships of the enzymes that participate in the synthesis and degradation of oligosaccharides.

The recent application of molecular biological techniques to the study of glycosylation has enabled deduction, based upon nucleic acid sequence data, of the primary amino acid sequence structure of at least some glycosyltransferases. Specific antibody probes, generated against predicted peptides, in conjunction with nucleic acid probes, will facilitate a more complete assessment of the regulated expression of glycosyltransferases and glycosidases. High-level expression of cloned sequences in host bacterial or eukaryotic cells promises the availability of large quantities of enzymes suitable for biochemical analysis.

This chapter is not intended as an exhaustive summary of recent publications in the area. Rather, its purpose is to address the issues and current directions of the molecular biological approach in the analysis of glycosylation. In particular we will summarize current molecular approaches and insights in the isolation and characterization of glycosidase and glycosyltransferase gene sequences.

I. ISOLATION AND CHARACTERIZATION OF GLYCOSIDASE cDNA AND GENOMIC SEQUENCES

Glycosidases constitute a large heterogeneous array of enzymes that mediate the hydrolysis of glycosidic bonds with varying degrees of linkage specificity. In general, glycosidases can be classified into two broad categories. One type are the endoglycosidases, which catalyze the cleavage of specific internal glycosidic bonds. Endoglycosidase H, an enzyme that hydrolyzes the di-*N*-acetylchitobiose linkage in the core of *N*-glycosidically linked oligosaccharide, is a good example of an endoglycosidase. Although endoglycosidases have been extensively used as tools in the elucidation of complex carbohydrate structures, commercially available endoglycosidases are derived from lower eukaryotes. Mammalian endoglycosidases are not well studied. A second category is the exoglycosidases that catalyze the stepwise linkage-specific removal of sugars from the nonreducing termini of carbohydrate polymers. This group includes the neuraminidases, galactosidases, and fucosidases.

In mammalian cells, lysosomes are the primary sites for degradation of oligosaccharides by the acidic glycosidases. Because of the complexity of the linkages found within complex carbohydrates, many specific glycosidases are required for efficient catabolism. Several lysosomal storage diseases result from defective expression of specific glycosidases. In such congenital disorders, the intracellular accumulation of undegraded carbohydrate substrates leads to aberrant cellular structure and premature cell death [21,22]. Because of the considerable clinical implications, much attention has been focused on the glycosidases in question. One approach is the isolation and analysis of their cDNAs and genomic sequences. Probes generated by such studies can then be used to delineate the molecular basis of these inherited disorders. Immediate clinical benefits from these probes include more accurate and informative genetic screening and diagnosis.

B. β-Hexosaminidase

One enzyme that is the subject of intense research interest is β-hexosaminidase, a lysosomal exoglycosidase. A defect in this enzyme results in either Tay–Sachs or Sandhoff disease, clinically similar disorders, both manifesting considerable heterogeneity in severity and age of onset [23]. The α and β gene products encoded on human chromosome 15 [24] and chromosome 5 [25], respectively, are required for the normal expression of β-hexosaminidase. The α gene product is a 54-kd polypeptide. The β-subunit actually comprises two distinct chains, β-a (28 kd) and β-b (27 kd), that are produced by specific posttranslational cleavage of a 63-kd precursor molecule [26–28]. The catalytically active enzyme is a heterodimer that comprises one α- and one β-subunit (hexosaminidase A) or a homodimer of two β-subunits (hexosaminidase B). An additional minor isoform, hexosaminidase S (α-α), also exists. each hexosaminidase isoform differs in its range of substrates. Mutations in either the α-subunit (Tay–Sachs) or β-subunit (Sandhoff disease) results in accumulation of GM_2 ganglioside in brain and other tissues [29].

The initial cDNA clone isolated for the α-chain of β-hexosaminidase was only 240 base pairs (bp) in length, far short of an expected 2.1-kilobase (kb) full-length clone [30]. Isolation of this partial cDNA was successful, despite the low abundance of cellular hexosaminidase A mRNA. To overcome this problem, a cDNA library was subsequently generated from mRNA enriched for hexosaminidase sequences by polysome immunoselection, using specific antisera against β-hexosaminidase. The partial α-chain clone was then used as a probe to isolate another cDNA clone that contained the complete coding sequence of the α-chain [31]. The complete cDNA contained an open-reading frame of 1587 bp, corresponding to 529 amino acids. A putative signal sequence resides at the NH_2-terminus. An RNA blot analysis indicated that the α-chain is encoded by an approximately 2.1-kb mRNA. The α-chain gene is organized into 14 exons along 35 kb of genomic sequence [32]. A differential transcriptional termination event apparently generates two α-chain mRNAs that differ in the length of the 3′ untranslated region, resulting in a major mRNA of approximately 2.1 kb and a minor mRNA of approximately 2.6 kb [32].

Initial identification of a cDNA clone for the β-chain was achieved with synthetic oligonucleotide probes. These oligonucleotides were generated from partial peptide sequence information [33]. As with the α-chains, these initial β-chain cDNAs were only partial sequences. cDNAs containing the complete β-chain's coding regions were isolated using the initial clones [34,35]. The integrity of cDNAs containing a complete-coding region was demonstrated by the synthesis and assembly of a catalytically active hexosaminidase B in a cell-free system [36]. As with the α-chain, the β-chain–coding region is divided into 14 exons that are distributed over 40 kb of DNA. Amino acid sequences predicted by the cDNA clones reveal an identity of over 50% between the α- and β-chains [31,33]. This sequence similarity, together with the extensive sharing of intron/exon placement [34] suggests that duplication of a common ancestral gene generated the α- and β-chain genes.

Studies of Tay–Sachs disease among the Ashkenazi Jews indicated that β-hexosaminidase deficiency in some patients is due to greatly reduced or absent levels of α-chain mRNAs [30,37]. At least two distinct mutations underlie this form of Tay–Sachs disease. One of these is a G to C substitution in the 5′ donor site of intron 12. In addition to much reduced levels of α-chain mRNAs, the transcripts that remain are a heterogeneous population that variously contains introns 12 and 13. Among those mRNAs that contain intron 13, the transcripts are prematurely truncated and polyadenylated as a result of a polyadenylation signal within intron 13 [37]. The other mutation that likewise results in absence of α-chain transcripts is a 4-bp insertion in exon 11 [38]. However, in other Tay–Sachs cases, there are apparently normal levels of α-chain transcripts, despite the absence of β-hexosaminidase A activity. In one instance, a G to A substitution resulted in a change in amino acid residue 482 from glutamic acid to lysine. In this patient, the strong charge transversion at this position is thought to be responsible for the defective processing of the precursor [39]. Another point mutation results in a glycine to serine substitution at residue 269 that dramatically depresses catalytic activity of α-chains [40]. Less is known about the defects in the hexosaminidase β-chains that lead to Sandhoff disease. An early study documented the absence of the 2.2-kb β-chain mRNA in only one of three patients afflicted with Sandhoff disease [33].

B. α-L-Fucosidase

A molecular biological approach, similar to that for the β-hexosaminidase defect in G_{M2} gangliosidosis, has also been taken for other lysosomal storage diseases. A well-studied example is fucosidosis, a neurovisceral storage disease that is caused by defective α-L-fucosidase expression [41]. The mature native enzyme is a tetrameric homopolymer with a subunit relative molecular mass (M_r) of 50,000 Da. Molecular cloning of the cDNA for human α-L-fucosidase was initially achieved by immunoscreening a hepatoma cDNA library constructed in the expression vector phage, λgt11 [42,43]. The structural gene (*FUCA1*) for the fucosidase has been assigned to the short arm of human chromosome 1 [44] in the 1p36.1 to p34.1 regions [45,46]. A second locus (*FUCA2*) on chromosome 6 is apparently responsible for inherited quantitative variations in plasma enzyme activity [47,48]. However, *FUCA2* does not cross-hybridize to the liver α-fucosidase cDNA [49]. It is not known whether *FUCA2* encodes an α-fucosidase that is distinct from that specified by *FUCA1*.

Another genetic region that cross-hybridizes with the fucosidase cDNA has been identified on chromosome 2 [46,49]. Restriction-mapping analysis showed that this homologous region is colinear, with at least 70% of the α-fucosidase cDNA, suggesting that this is a processed pseudogene for the α-fucosidase [49]. It is not known whether this pseudogene is transcriptionally competent or participates in the expression of α-fucosidase.

C. α-Galactosidase A

Another clinically interesting lysosomal glycosidase is α-galactosidase A, an enzyme that hydrolyzes globotriosylceramide and related glycolipids [50]. α-Galactosidase deficiency results in Fabry's disease, an X-linked sphingolipidosis. The initial cDNA clone to α-galactosidase was isolated from a human liver λgt11 expression library using a monospecific antibody toward the purified enzyme [51]. The mature enzyme is encoded on a 1.45-kb mRNA, containing a coding region that specifies a 429-amino acid precursor protein [52,53]. The α-galactosidase mRNA is unusual in the apparent absence of a 3′ untranslated region [53,54]; two putative polyadenylation consensus signals, AATACA and ATTAAA, are located within the translated sequence [53]. The structural gene resides on 12 kb of DNA in the chromosomal region Xq22 [54].

D. Galactosialidosis

Defective glycosidases are not the sole cause of lysosomal disorders. Although mutations in the lysosomal β-galactosidase or neuraminidase structural genes result in Morquio B disease [23,55] and sialidosis [56], respectively, there is another genetically distinct locus that specifies correct expression of lysosomal β-galactosidase and neuraminidase. Galactosialidosis [57], an autosomal recessive disease, is characterized by concomitant deficiencies in neuraminidase and β-galactosidase. The defect in galactosialidosis is in the expression of a "protective protein." This protein appears to be essential for the full biological activities of both β-galactosidase and neuraminidase, although the mechanistic basis of this "protection" is not clear [58]. In normal cells, the 2-kb protective protein mRNA directs the synthesis of a 452-amino acid precursor that is processed in vivo to the mature heterodimer of a 32-kd and 20-kd polypeptides [59]. Quite paradoxical to its protective function, the protective protein bears extensive amino acid homology to yeast carboxypeptidase Y and the *KEX1* gene product, a carboxypeptidase B-like protease. In individuals exhibiting the various forms of galactosialidosis, the protective protein transcript is absent or present in much reduced levels in fibroblasts [59]. The precise molecular basis of galactosialidosis must await studies of the interactions of the protective protein with the β-galactosidase and neuraminidase, once the genes encoding the three components are cloned.

E. Other Glycosidases

In addition to the acidic glycosidases that participate in the normal lysosomal degradation of glycoconjugates, other glycosidases function in the neutral environment of the endoplasmic reticulum and Golgi networks. These glycosidases participate in the normal trimming and processing of the core *N*-glycosidically linked oligosaccharides [19,60]. In addition, cell surface and secreted glycosidases compose yet another group of enzymes that participate in normal tissue remodeling and in immunological response mechanisms. Studies of gene structures and the mechanisms regulating their expression should be the next step in elucidating this poorly understood group of enzymes that participate in the normal catabolism of oligosaccharide structures on complex glycoconjugates.

II. ISOLATION OF GLYCOSYLTRANSFERASE cDNA AND GENOMIC SEQUENCES

As with glycosidases, the classic approach to the molecular cloning of transferase sequences entails the screening of recombinant phage libraries with the appropriate immunological or nucleic acid probes. Although conceptually simplistic, the development of an unambiguous probe requires some prefatory biochemical knowledge of the target species. The development of monospecific antibody or the synthesis of oligonucleotides for target selection often required that investigators first obtain large quantities of purified enzyme. More often than not, this proved to be a formidable challenge; most tissues harbor very low levels of the transferases. A further complication is that enzyme activities are often membrane-associated and extremely difficult to solubilize. However, by selecting tissues or biological fluids known to be enriched for particular transferase activities, investigators were able to devise protocols that enabled the near homogeneous purification of a small number of transferase proteins. Efficient purification of solubilized protein activities often involved the use of affinity chromatography, in which native or fabricated substrate molecules were coupled to an immobilized support [20]. Partially purified enzyme preparations were further resolved and analyzed by classic methods. It is beyond the scope of this article to discuss all the transferases that have been purified to homogeneity, but rather, we will chronicle those species the isolation of which has prompted a successful search for the cognate gene sequence. For a comprehensive review on the isolation of glycosyltransferase activities see [20].

A. Sialyltransferases

Several glycosyltransferases involved in the terminal modification of *N*-linked oligosaccharide chains in animal cells have been purified to homogeneity and, subsequently, utilized to generate both immunological and nucleic acid probes for screening recombinant phage libraries. Weinstein et al. first reported the homogeneous purification of both the β-galactoside α2,6- and α2,3-sialyltransferases from rat liver, a tissue particularly enriched for the α2,6-sialyltransferase activity [61]. Purification of the α2,6-isoform involved isolation of the detergent-solubilized enzyme on a combination of CDP-hexanolamine-agarose and CDP-agarose affinity matrices. The 23,000-fold–purified protein was used to elicit monospecific antibody production in immunized rabbits. The affinity-purified anti-sialyltransferase antibody [62] was used to probe a λgt11 rat liver cDNA expression library. In this manner, a cDNA clone containing the complete-coding region for the α2,6-sialyltransferase was isolated [63]. The integrity of the clone was verified by comparing predicted amino acid sequence data with that derived from partial peptide sequencing of the purified enzyme preparation. The cDNA sequence predicts an open-reading frame of 403 amino acids. The sialyltransferase protein consists of a short NH_2-terminal region, followed by a single transmembrane domain. The COOH-terminal two-thirds of the polypeptide putatively contains the catalytic region.

B. Galactosyltransferases

Similarly, the β1,4-galactosyltransferase was purified from bovine milk by affinity chromatography on a combination of UDP-agarose and either *N*-acetylglucosamine or α-lactalbumin columns [64]. The purified protein was used to elicit specific antisera. Partial peptide sequence analysis of the pure enzyme generated sufficient information for the construction of oligonucleotide probes. The combined approach of specific antibody and oligonucleotide probes resulted in the recovery of a partial bovine β1,4-galactosyltransferase clone from a cDNA library [65]. Two groups have independently reported the isolation of the full-length murine isoform of β1,4-galactosyltransferase from an induced F9 λgt11 and mouse mammary λgt10 cDNA library, respectively [66,67]. Each group utilized partial bovine cDNA clones as hybridization probes in the isolation. Inferred amino acid sequence data denoted an 80% homology between the murine and bovine isoforms.

The human β1,4-galactosyltransferase was purified from human milk using a hexanolamine derivative–Sepharose 4B affinity matrix [68]. Partial peptide sequence information was used to design an oligonucleotide probe (60 nt) for the screening of a λgt10 cDNA library. The human galactosyltransferase clone contained a 1.7-kb insert fragment that spanned 783 bp of coding sequence: 3′ untranslated sequences constituted the remainder of the insert [69]. By using a 985-bp *Eco*RI fragment of the partial human clone, encompassing the 783-bp–coding domain, a full-length cDNA was later isolated using a combination of partial cDNA and oligonucleotide probes in screening a human placental cDNA library in λgt11 [70]. Together, the data suggest that human β-galactosyltransferase is specified by a 4.1-kb mRNA that contains a 1200-nt open-reading frame, followed by an extensive 3′untranslated region. In situ and somatic cell hybridization data have putatively localized the human β-galactosyltransferase gene to either the short arm of chromosome 9 [71] or chromosome 4 [72], respectively.

β-Galactosyltransferase protein sequences are closely conserved between species; 80% homology exists between murine, bovine, and human sequences [73]. Interestingly, no obvious sequence similarity exists between galactosyltransferase and other cloned glycosyltransferases (e.g., sialyltransferase). However, in an architecture that is shared with other glycosyltransferases, the galactosyltransferase protein consists of a short NH_2-terminal region, followed by a single membrane-spanning domain. The catalytic domain occupies the carboxyl two-thirds of the polypeptide. The implications of these findings will be subsequently discussed in greater detail.

C. Glycosyltransferases from Yeast

The ease with which yeast can be manipulated genetically and biochemically has afforded molecular glycobiologists an unique opportunity for the cloning of some glycosyltransferase genes. Yeast genetics is a prototypical system upon which complex regulatory events in higher eukaryotes may be practically modeled. At least in respect to the major endoplasmic reticulum (ER)-associated transferase events, the yeast system closely resembles the protein glycosylation machinery found in mammalian cells. It is the subsequent Golgi-related processing events that clearly distinguish the yeast apparatus from that of higher eukaryotes. For a detailed review of the yeast glycosylation system see [74].

The ALG, or asparagine-linked glycosylation, mutants of yeast are a group of recessive mutants that are impaired in dolicholpyrophosphate oligosaccharide biosynthesis (dol-P-P-$GlcNac_2Man_9Glc_3$). Several classes of ALG mutants, typified by the accumulation of core oligosaccharide synthetic intermediates, harbor a defective glycosyltransferase. The ALG series, with the noted exception of complementation group 7, was isolated by sugar suicide selection, whereby mutant cells cultured in the presence of radioactive mannose survive by virtue of their inability to incorporate mannose in *N*-linked glycan biosynthesis [75,76]. Enzyme and lectin probes were used to assess phenotypically the stage-specific block and to determine if the lesion could be attributed to a defective glycosyltransferase. To identify and recover the normal glycosyltransferase gene sequences, mutant cells were transformed with cloned genomic DNA from wild-type cells. Mutant cells that received complementing gene sequences exhibit revertant phenotypes. Wild-type DNA fragments that complement mutants can be characterized genotypically by two criteria. First, an episomally maintained complementing fragment must cosegregate with a selectable marker to eliminate the possibility that the revertant phenotype is generated by a spontaneous reversion of the mutant. Second, the isolated gene fragment, when altered in vitro and, subsequently, targeted for selective replacement of its normal endogenous homologue, should produce the original mutant phenotype. This verifies that the cloned fragment encodes the wild-type gene, rather than being merely a second-site suppressor of the original mutation.

Just such a set of mutants that are conditionally deficient in the capacity to synthesize various *N*-linked oligosaccharide structures was created by Robbins and co-workers [75,76]. At least three of these ALG mutants presumably encode mannosyltransferases responsible for the synthesis of core high mannose-type oligosaccharides associated with the ER compartment. Mutations in the *ALG1* gene, encoding a β1,4-mannosyltransferase, result in the accumulation of the dolichol pyrophosphoryl $(GlcNAc)_2$-processing intermediate at restrictive temperatures [75,76]. The gene was cloned by complementation and expressed in a bacterial system, whereby the lipid-linked $GlcNAc_2Man$ product could be detected in transformed cell lysates [77]. The β1,4-mannosyltransferase gene has been mapped to chromosome 4 of *Saccharomyces cerevisiae* and contains a 1.65-kb coding sequence [77]. The *ALG1* gene product is believed to be associated with the ER membrane. As with a number of the transferase genes, the *ALG1* transferase gene encodes at least two transcription products [74]. Another mutant, ALG2, accumulates the Man1-2GlcNAc2-PP-dolichol at restrictive temperatures. The lesion appears to be in another member of the mannosyltransferase family. Although *ALG2* mutation is less well characterized than *ALG1*, a DNA fragment from wild-type cells that complements the ALG2 mutant phenotype has been recovered [74].

Another group of mutants, ALG5, 6, and 8, are impaired for the terminal addition of glucose moieties to the growing core oligosaccharide chain [78,79]. The ALG5 mutant appears to be defective in the synthesis of Glc-P-dolichol, the donor for the terminal glucose residues. When microsomal membrane preparations from mutant strains were incubated in the presence of radiolabeled Glc-P-dolichol, only ALG5 efficiently catalyzed the addition of glucose to maturing core structures [74]. The ALG6 mutant appears to be deficient in the α1,3-glucosyltransferase activity and is a likely candidate for the eventual isolation of the transferase structural gene.

Unlike the majority of ALG mutants, ALG7 was not isolated by metabolite suicide selection. Rather, yeast cells, transformed with a high copy number plasmid bearing wild-type genomic fragments were selected by cultivating in the presence of lethal concentrations of tunicamycin. Tunicamycin, an antibiotic that inhibits the transfer of UDP-GlcNAc-1-P to dolichol phosphate acceptors, selects for transformed cells phenotypically resistant to the effects of the drug. A DNA fragment encoding the GlcNAc-1-P transferase, when present at high gene doses, is capable of surmounting the inhibitory effects of the antibiotic [80]. One such mutant, ALG7, was identified [81]. The yeast *ALG7* gene contains a 1.34-kb open-reading frame flanked by a seemingly versatile promoter region that harbors four transcriptional start sites as well as two putatively functional TATA elements [82]. The GlcNAc-1-P transferase is an essential enzyme, as documented by experiments in which constructed null mutations rendered cells nonviable [74].

D. Isolation of Mammalian Glycosyltransferase Genes by Complementation

A genetic approach similar to that for the yeast system can be used to isolate mammalian glycosyltransferase from cultured cells. A variety of stable animal cell lines exists that phenotypically represents mutations in nearly all steps of *N*-linked oligosaccharide biosynthesis [83]. Most notable among these is a battery of glycosylation mutants isolated from Chinese hamster ovary (CHO) cells [84]. Mutant selection takes advantage of lectins and their unique ability to bind to specific cell surface oligosaccharide structures, thereby obstructing cell growth. Mutant cells (Lec) that are defective in the expression of the specific oligosaccharide structures are resistant to growth retardation. As with yeast glycosylation mutants, obstruction of a glycosylation pathway may be the result of defective oligosaccharide modifying enzymes (i.e., glycosyltransferases), precursor biosynthesis/import (i.e., nucleotide sugars), or more indirect lesions, such as the inappropriate intracellular routing of cell surface glycoproteins. Obviously, this approach is limited to analysis of mutations that display phenotypically altered cell surface components. In addition, caution must be taken to standardize selective procedures; a variety of chemical and physical parameters (growth rate, temperature, serum composition, differentiation state, and others) can have profound effect on the nature of the glycans displayed on the cell surface.

The phenotypic classification of glycosylation mutations in CHO cells has been further refined by a number of means. A given Lec mutant can be screened for growth with a variety of lectins that vary in their specificity for sugar and linkage type. Mutant sensitivity utilizing a panel of lectins permits rapid categorization of newly isolated mutant variants into existing complementation groups [85]. Genotypically, complementation groups can be designated using somatic cell hybrids. In this approach, cell hybrids are constructed from mutants bearing selectable auxotrophic markers and screened for revertant phenotypes [84]. In this manner, fusion of mutants that share a common genotypic defect should not result in a revertant phenotype. In contrast, mutants that have dissimilar lesions (for example, a transport mutant and a glycosyltransferase mutant) should complement each other to regenerate the wild phenotype. Furthermore, classic biochemical analysis, such as enzymatic assays, can be used to delineate the molecular basis of the mutations. A detailed genotypic characterization of these Lec mutants is requisite for the molecular cloning of glycosyltransferase genes.

Identification of animal glycosyltransferase sequences by means of glycosylation mutants follows a strategy similar to that employed for the yeast system. cDNA or genomic restriction fragments are introduced into mutant CHO cells by high-efficiency transfection protocols [83]. Clones are selected based on the ability of the transfected sequences to restore wild-type properties to the mutant cells.

Following this strategy, the cloning of GlcNAc transferase I is based on a recessive CHO cell mutant line designated Lec1 that fails to modify *N*-linked oligomannose structures ($Man_5GlcNAc_2$) with β1,2-GlcNAc [86]. Biochemical evaluation of Lec1 suggests a defect in the *N*-acetylglucosaminyltransferase I gene. In another member of the Lec1 class, termed Lec1A, the same transferase activity is only partially impaired [87,88]. Presumably, assorted variations of a given mutant phenotype arise from different lesions residing within the same genetic locus. To isolate the GlcNAc transferase I gene, Lec1 cells were transfected with heterologous DNA sequences. Lec1 cells, because of their inability to express complex type *N*-glycans, are particularly sensitive for growth in concanavalin A (Con A), a lectin that recognizes high mannose structures. Primary transformants were selected for growth in the presence of Con A. These isolates exhibited parental CHO characteristics and actively expressed GlcNAc I transferase activity in vitro. Work is currently underway to recover and clone the gene responsible for the revertant phenotype [89].

The CHO cells exhibiting amplified expression of the tunicamycin-sensitive GlcNAc-1-P transferase activity [90] were isolated using a strategy similar to that utilized in isolation of the *ALG7* gene in yeast. Mutants capable of expressing 15 times the normal transferase activity were obtained by subjecting CHO cells to incremented concentrations of tunicamycin [91]. When genomic DNA from both mutant and wild-type cells was analyzed on Southern blots hybridized with a 1.8-kb yeast *ALG7* DNA probe, a three- to fourfold elevation in signal intensity was detected in tunicamycin-resistant clones. Continuous subculture of mutant cells in the presence of the antibiotic resulted in further elevation of the *ALG7* hybridization signal, suggesting that a gene amplification event was responsible. Amplification of the GlcNAc-1-P transferase gene sequence in tunicamycin resistant CHO cells was independently observed [92]. The amplified fragment was cloned using an *ALG7* gene probe from yeast, resulting in the isolation of a 6.6-kb fragment that partially encodes the hamster GlcNAc-1-P transferase. In analyzing Northern blots from CHO tunicamycin-resistant mutants, a family of RNA molecules ranging from 2.0 to 2.2 kb was detected [92]. Comparative sequence analysis of the animal cell and yeast genes reveal a 24-amino acid sequence that shares greater than 90% homology [92].

E. Other Approaches to Isolate Glycosyltransferase Genes

The use of animal cell mutants is a powerful means by which cDNA and gene sequences can be identified and isolated. Glycosyltransferase genes, because their mutations will likely result in phenotypic alterations at the cell surface, are especially amenable to lectin selection. In the CHO system, the use of lectins that selectively kill or retard the growth of cells bearing particular glycoform surface structures has proved an effective means of isolating glycosylation mutants.

A genetic approach that does not require the availability of mutant cell lines is being applied in the isolation of a β-galactoside (α-1,2)fucosyltransferase gene. This enzyme mediates the synthesis of blood group H Fucα1,2Gal-linkages that are normally not represented on the surface of mouse L cells. The L cells were cotransfected with a selectable marker and fragmented genomic DNA from a human cell line that is blood group H-positive. Transfected mouse L cells expressing blood group H surface antigen were screened by a procedure whereby positive transfectants were bound with a mouse anti-H IgM and adhered to culture dishes coated with goat antimouse sera. After extensive washing, adherent cells were subcultured in selective media and screened using a fluorescent-activated cell sorter (FACS). Blood group H-positive mouse cells were analyzed by conventional molecular and biochemical assay and were demonstrated to harbor a segment of foreign, human DNA presumed to direct the expression of an α1,2-fucosyltransferase activity [93,94].

Genetic approaches such as these promise to be efficient and powerful tools in the isolation of glycosyltransferase sequences. At least in theory, the principal and only requirement is a specific probe (i.e., lectin, antibody) that will permit unambiguous detection of a particular oligosaccharide phenotype.

A recently refined technique that will greatly facilitate the analysis of glycosyltransferase sequences is the polymerase chain reaction (PCR) [95–97]. The PCR is ideally suited for the rapid cloning of genomic and very low abundance mRNAs for which partial sequence information is known [98]. Briefly, the strategy involves the construction of oligonucleotide primers from partial sequence data or homologous cloned sequences and using them to amplify defined segments in target nucleic acid populations. Primer sequences are used to direct chain elongation along target sequences. The cyclic progression of anneal, elongate, melt, and anneal allows target sequences to be amplified several-millionfold. Amplified segments can be used directly for DNA sequence analysis and for the generation of hybridization probes for subsequent screening efforts [99,100]. A partial putatively mannosidase II genomic clone was obtained using the PCR technique [101].

From limited peptide sequence data, Robbins and co-workers engineered degenerate oligonucleotide primers and amplified the defined target sequence in rat liver DNA preparations. The integrity of the isolate was verified by DNA sequence analysis and is currently being used to screen a cDNA library in hopes of recovering a full-length clone. A novel use of PCR has been reported for the isolation of the human parvalbumin cDNA from a cerebellum mRNA population [102]. For this, a single specific oligonucleotide coupled with oligo(dT) was used for the PCR. Since the PCR amplification procedure requires only a single day for operation, one can readily appreciate the implications for gene cloning when only partial peptide sequence data is at hand. In light of the successful cloning of certain glycosyltransferase genes achieved with conventional technology, it seems but a matter of time before the full complement of cloned glycosylation genes are at hand.

III. STRUCTURAL AND REGULATORY IMPLICATIONS

The availability of cloned glycosyltransferase sequences has offered researchers an unique opportunity to study a host of biological processes for which the role of carbohydrate structure had long been implicated. The tools of molecular biology complement a more conventional arsenal of biochemical and serological approaches in probing the functional basis for glycoconjugate structure. Analysis of transferase gene sequences, a field that is still in its infancy, has already revealed a wealth of information on protein topology and structure. Available data suggest that the expression of glycosyltransferase genes is modulated by complex regulatory pathways, exerted at multiple points. In this section, we will outline recent advances made in the molecular analysis of glycosylation genes and prospects for future studies employing cloned genes as probes for biofunction.

A. Glycosyltransferase Protein and mRNA Structures

Currently, full-length cDNA clones for the β1,4-galactosyltransferase [66,67,70] and the α2,6-sialyltransferase [63] are available. Comparative analyses of these cloned sequences, as well as sequences to other glycosyltransferases, have yielded insights into the structure and interrelationship of the glycosyltransferases. An immediately striking observation is the close sequence conservation between homologous enzymes from different animal species. Murine and a partial bovine galactosyltransferase are nearly 80% identical on the amino acid level [66]. Comparison of human and murine galactosyltransferases predicts homologies of 91 and 88% on the protein and nucleic acid levels, respectively [70]. Sialyltransferase sequences from rat [63] and human [Lance, P., and Lau, J. Y. Y., unpublished observations] are likewise closely conserved. Together these data suggest functional requirements that mandate close evolutionary conservation. Interestingly, there is no sequence similarity between the different glycosyltransferases. What apparently is conserved, however, is the gross structural aspects of the glycosyltransferase proteins and mRNAs. Despite the absence of any nucleic acid sequence identity, both galactosyltransferase and sialyltransferase are encoded on large mRNAs that are approximately 4.0 kb and 4.7 kb in length, respectively. The protein-coding region of both mRNAs are flanked by an unusually long 3′ untranslated region (approximately 2.6 and 3.2 kb for galactosyltransferase and sialyltransferase, respectively). The open-reading frames predict a common polypeptide organization that specifies a short NH_2-terminal cytosolic extension, a 17- to 20-amino acid membrane-spanning region, followed by a large putatively catalytic COOH-domain. Similar topology has also been predicted for other glycosyltransferases, including the fucosyltransferase [93].

B. Hormonal Regulation

Little is known about the regulatory mechanisms that dictate the correct expression of glycosyltransferases. One well-studied example is the induction and secretion of the hepatic α2,6-sialyltransferase during the acute-phase reaction. In response to inflammation and systemic trauma, hepatocytes synthesize and secrete a number of sialylated glycoproteins including α_1-acid glycoprotein (AGP), haptoglobin, and fibrinogen [55]. As part of the inflammatory response, hepatic and serum levels of sialyltransferase activity are elevated [103,104]. The onset of the hepatic acute-phase response is controlled by a complex interplay between circulating glucocorticoids, cytokines, and a number of other yet undefined hepatocyte stimulatory factors (HSF) [105–114]. Several mechanisms, including transcriptional activation, stabilization of existing mRNA, enhanced translational efficiency, or allosteric modulation of enzyme activity, could singly or in combination account for the induction of transferase activities. Recently, the molecular basis for α2,6-sialyltransferase induction during the acute-phase reaction was addressed using probes constructed from cloned sialyltransferase sequences [115]. Hepatoma as well as primary hepatocytes responded to dexamethasone, a synthetic glucocorticoid, by a three- to fourfold induction of sialyltransferase activity. This induction was paralleled by an equivalent elevation of steady-state levels of sialyltransferase mRNA (Fig. 1). From similar experiments conducted in the presence of transcriptional and translational inhibitors, it appears that the hormone-mediated induction of sialyltransferase expression requires active transcription, but is not contingent upon de novo protein synthesis. Measurement of unprocessed transcripts in the nucleus, using a probe that recognizes only intron sequences, demonstrates that an increased level of primary transcripts accompanies sialyltransferase induction [Wang, X. C., and Lau, J. T. Y., unpublished observations]. Thus, one level of regulation for the hepatic α2,6-sialyltransferase gene is apparently the direct activation of transcription by steroid hormones. However, acute-phase induction of sialyltransferase causes not only an increased rate of enzyme synthesis, but also an increase in the rate of secretion of the soluble enzyme into the serum [103]. Current evidence favors the generation of a soluble enzyme from the membrane precursor by a proteolytic cleavage event. The mechanism that dictates the rate of soluble enzyme formation is not yet known. A number of factors, including glucocorticoids [116], are presumed to modulate the posttranslational targeting of secretory glycoproteins.

Likewise, β1,4-galactosyltransferase expression is responsive to hormonal stimulation. Rat parotid acinar cells cultured in the presence of isoprenaline, a β-adrenergic agonist, exhibit elevated enzyme activity as well as steady-state galactosyltransferase mRNA levels [117]. However, in contrast with hepatic sialyltransferase, parotid galactosyltransferase induction apparently requires both active transcription as well as on-going protein synthesis. Thus, the implication is that parotid galactosyltransferase gene activation requires the participation of short-lived, ancillary protein factors.

Current sequence data substantiated the notion that most transferase enzymes are intimately associated with cellular membranes. However, aside from the membrane-bound transferases commonly associated with the secretory apparatus, other glycosyltransferases exist on the surface of cellular membranes [12] or as soluble components of serum and mammary secretions [20,118,119]. A precursor-to-product relationship between the

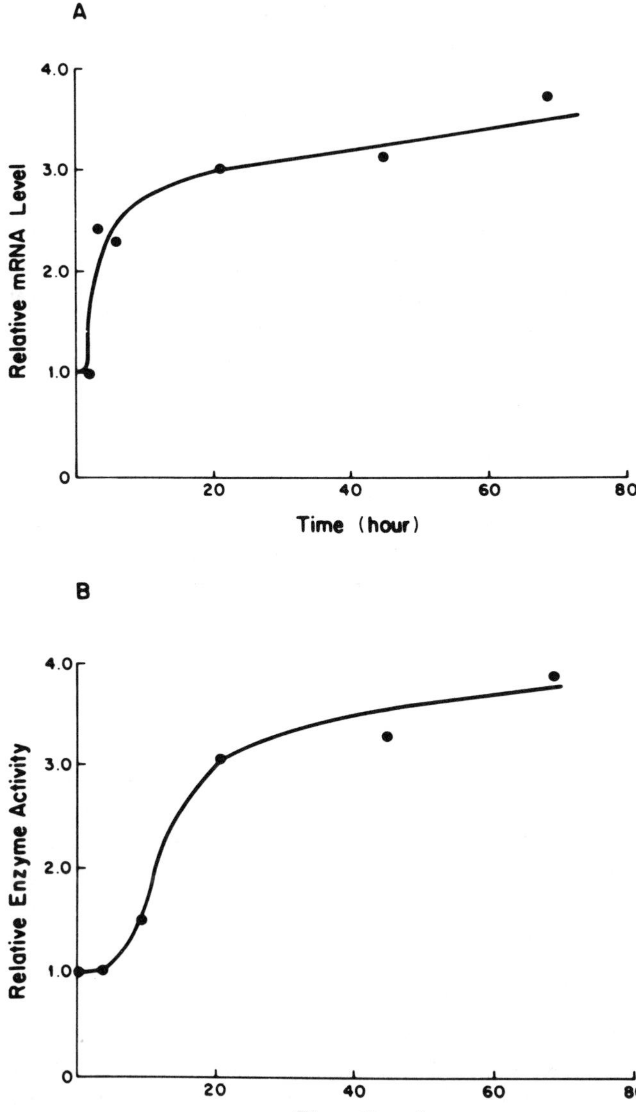

Figure 1 Effect of glucocorticoids on sialyltransferase expression. Rat hepatoma cells (Reuber H35) were cultured in the presence of the synthetic glucocorticoid, dexamethasone, for the length of time indicated. Time-dependent elevation of α2,6-sialyltransferase mRNA levels (A) was monitored by Northern blot hybridization of total cellular RNA. The signal intensity was quantitated by densitometry and normalised. Parallel cultures were used to assess α2,6-sialyltransferase enzymatic activity (B).

soluble and bound forms of the glycosyltransferases had long been suspected, but the molecular basis of this conversion event has remained elusive. For example, the mature membrane-bound α2,6-sialyltransferase is approximately 47 kd in mass [62]. A catalytically active, soluble form of the enzyme in serum has been estimated to range in size from 37 to 43 kd [120,121]. That the soluble sialyltransferase is derived from the membrane-associated form is substantiated by comparing NH_2-terminal amino acid sequence data of the soluble enzyme with the predicted polypeptide structure of the full-length clone. The data suggests a proteolytic cleavage event mapping 63 residues from the NH_2-terminus that liberates the soluble, catalytic domain from the hydrophobic anchor. Furthermore, the release of soluble sialyltransferase in vitro from an isolated Golgi preparation suggests that the protease responsible for the conversion is colocalized within the Golgi compartment [63]. Little is known about the regulatory mechanisms that might coordinate the secretion of the transferase. Interestingly, serum-bound levels of the α2,6-sialyltransferase activity accompany the hormonally induced hepatic inflammation response (acute phase) [103,105,106]. Primary sequence analysis of the cloned β1,4-galactosyltransferase has likewise revealed the existence of several potential proteolytic cleavage sites flanking the luminally bound catalytic domain [67]. The predicted topological arrangement of the cloned transferases and the proposed relationship between bound and secreted forms of the enzymes seem so far to have lent unity to the evolution of an otherwise diversified gene family.

C. Multiple Transcripts from Single Genes

Hormonal modulation of transcription is one way by which a steady-state level of glycosyltransferase expression is specified. Additional regulatory pathways exist that dictate tissue-specific and developmental-specific profiles of glycosyltransferase expression. Of considerable interest is the presence of multiple mRNAs representing sialyltransferase and galactosyltransferase gene sequences. Although the import of this transcript heterogeneity is unclear, the existence of these multiple transcripts has far-ranging regulatory as well as functional implications.

One intriguing observation originally arose from sequence analysis of the full-length β1,4-galactosyltransferase cDNA clone [67]. Nucleotide sequence information reveals the presence of two in-frame translational start codons (ATG). These predict the generation of two polypeptides, one shorter than the other by 13 amino acids. Primer extension and S1 protection analyses suggested that, in fact, cellular mRNA populations contained at least two forms of the β1,4-galactosyltransferase message. The longer transferase message includes both in-frame ATG codons, whereas the shorter transcript maps between the two ATG sites and, hence, contains only one ATG start codon. Probes designed selectively to recognize one or both forms of the message confirmed the existence of the divergent species. Together the data imply that the transferase gene specifies two populations of mRNA that may direct the synthesis of alternative forms of the enzyme. Although the actual implication of this event is now entirely unclear, the NH_2-terminal extension in the longer polypeptide is reminiscent of the cleavable signal sequence that directs nascent secretory or membrane-bound polypeptides to the ER. In a polypeptide-containing membrane-spanning domain, the additional presence of a signal peptide may influence topological orientation of the protein [122,123]. As mentioned earlier, galactosyltransferase polypeptide contains a single membrane-spanning domain. The presence or absence of a signal peptide may determine whether the catalytic domain of the transferase faces the cytosolic or luminal compartment. Therefore, one form of the enzyme could represent what we traditionally associate with luminally oriented *N*-linked glycoprotein biosynthesis, whereas the alternate form might participate in a yet obscure form of glycoconjugate biosynthesis within the cytosolic compartment. In light of recent findings on the synthesis of novel *O*-linked GlcNAc structures in the cytosol, the precedent exists for such a model [124–126]. It is not clear whether the alternate transcripts arise from an alternative splicing event or the differential utilization of upstream promoter elements. Nevertheless, the elucidation of the regulatory controls that govern the differential synthesis, as well as the tissue distribution of these related products, should prove interesting.

The molecular mechanisms that dictate the appropriate expression of glycosyltransferases in various tissues is still unknown. As a means of pioneering such an investigation, our laboratory sought to assess the expression of the α2,6-sialyltransferase mRNA in rodent tissues by RNA blot analysis (Fig. 2). It is quite apparent that tissue specificity extends beyond quantitative differences in expression. As expected, sialyltransferase transcripts are expressed to different levels in different tissues. Quite unexpectedly, different tissues appear to express qualitatively different sialyltransferase mRNAs. This difference is most profound in the kidney, in which the predominant sialyltransferase mRNA is substantially smaller than those present in other tissues. These alternatively sized mRNAs are apparently transcribed from the same sialyltransferase gene, since they all share the same 3′ untranslated domain; blot analysis of genomic DNA has demonstrated that this 3′ untranslated sequence resides in a unique genomic locus [127]. This hypothesis was further substantiated by isolation of a divergent cDNA clone from a rat kidney cDNA library. That this clone represents a product of alternate splicing is reaffirmed by genomic sequence data mapping the divergent boundary to an intron–exon junction of the α2,6-sialyltransferase gene. Moreover, mRNA protection analysis reveals the existence of a subpopulation of divergent sialyltransferase transcripts (Fig. 3). Comparative analyses of the divergent sialyltransferase mRNAs are currently in progress to understand their functional significance.

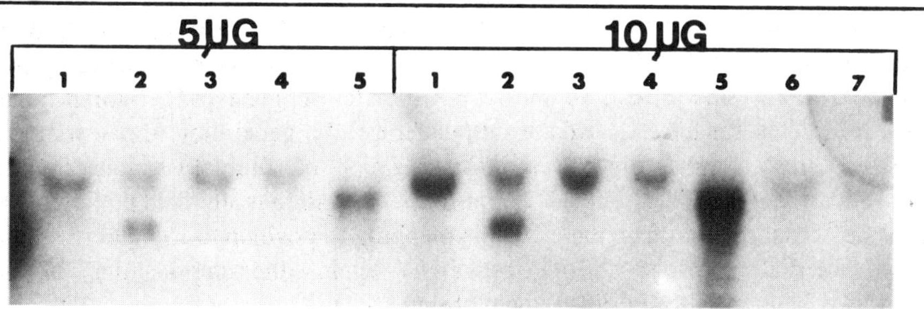

Figure 2 RNA blot analysis of rat tissues with hepatic α2,6-sialyltransferase gene sequences. Either 5 or 10 μg of total RNA isolated from rat submaxillary gland, kidney, lung, spleen, liver, heart, and brain (lanes 1–7, respectively) were fractionated in formaldehyde agarose gels and blotted on nylon membranes. Blots were hybridized with a hepatic α2,6-sialyltransferase cDNA probe. Notice that in addition to the observed quantitative differences of steady-state mRNA levels, their exists apparent tissue-specific, qualitative variations in sialyltransferase gene expression.

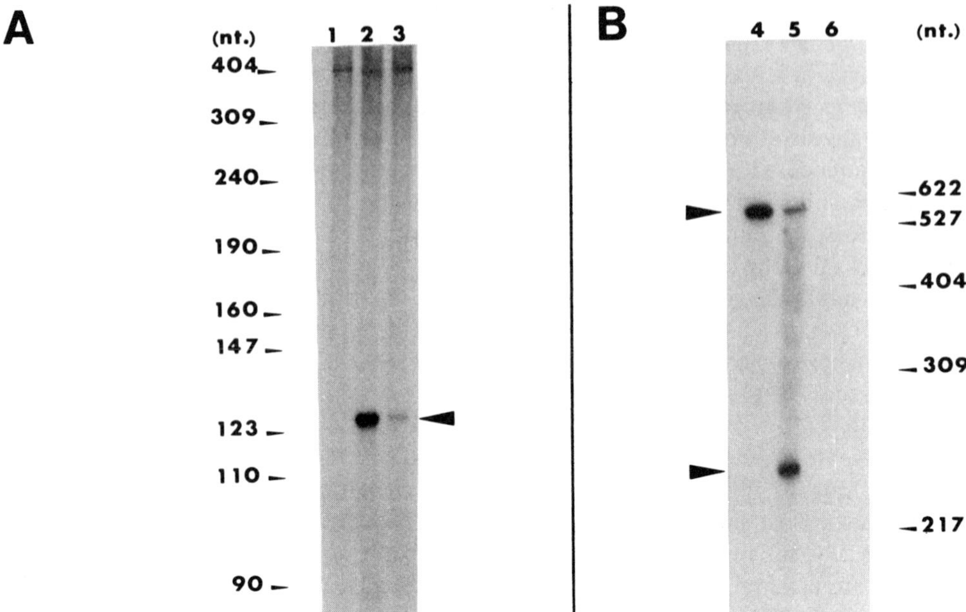

Figure 3 Nuclease S_1 protection analysis of rat tissue RNA. S_1 nuclease analysis was performed using varying lengths of cloned hepatic α2,6-sialyltransferase cDNA. Rat liver RNA (lanes 2, 4), kidney RNA (lanes 3, 5), and control tRNA (lanes 1, 6) were hybridized with sialyltransferase cDNA probes designed to protect either 125 nt (A) or 560 nt (B) and digested with S_1 nuclease. Note the apparent divergence in a subset of the kidney transcripts analyzed with the 560-nt probe. This divergent junction maps precisely to a known intron-exon boundary in the rat sialyltransferase gene and is consistent with the pattern of expression observed upon RNA blot analysis.

The generation of multiple, possibly alternatively spliced, messages from a single transferase gene is not unique to α2,6-sialyltransferase expression. The detection of several transcripts apparently derived from a single transferase gene has been previously documented for the yeast GlcNAc-1-P transferase [82], in which the source of the heterogeneity resides in both the 5′ and 3′ end of the *ALG7* transferase gene [82]. Similar to the mechanism postulated for the β1,4-galactosyltransferase, the generation of 5′ variants of the *ALG7* gene may result from differential utilization of alternative transcriptional promoter elements and start sites. The *ALG7* transcripts map to four alternate initiation sites within a 20-bp domain. Two upstream elements mapping to positions –157 and –139 may function as alternative promoter regions. Once again, the implications of this heterogeneous collection of transcripts remain obscure.

Heterogeneity exists in β1,4-galactosyltransferase mRNAs as well. In addition to the alternative transcriptional initiation variants described earlier, there are other variants that contain additional as well as deleted blocks of information. The boundaries between shared and divergent domains resemble splice junction consensus sequences, suggesting alternative usage of exon sequences [66]. It has been suggested that the expression of multiple galactosyltransferase transcripts may reflect a cassette-type model of gene expression [66]. One intriguing, but unsubstantiated, possibility is that catalytic or functional glycosyltransferase variants are specified by the different transcripts. One example is the β1,4-galactosyltransferase variant that resides on the surface of sperm heads and participates in the recognition event between egg and sperm [128]. The sperm head galactosyltransferase is catalytically similar to its Golgi galactosyltransferase homologue and is presumed to be derived from a common enzyme pool. It is of fundamental importance to determine if the Golgi-associated form and the sperm head galactosyltransferase isoforms are generated by differential utilization of the same gene sequence.

Generation of multiple mRNAs from single glycosyltransferase gene sequences may have a biological role, in addition to that of encoding functionally distinct enzyme variants. A consideration is the influence of cellular RNA-splicing pathways on the level of enzyme expression. For example, differential splicing may result in mRNAs that differ in their half-lives. In light of the apparently tissue-specific expression of divergent α2,6-sialyltransferase mRNAs, the functional implications for alternative processing of sialyltransferase sequences is intriguing. The functional and regulatory importance of these apparently posttranscriptional events in glycosyltransferase gene activity remains a topic for future investigation.

IV. PROSPECTS AND COMMENTS

Much of our current knowledge of glycosyltransferases and glycosidases has been acquired from classic biochemical approaches. From such studies we have gained a rudimentary working knowledge of the biochemical pathways involved in the synthesis and turnover of glycoconjugates. We lack, however, a precise and definitive understanding of the enzymes that participate in glycoconjugate metabolism, as well as the regulatory pathways that govern their expression. Without this understanding, an appreciation for the functionality of their catalytic end products cannot be achieved. Structural knowledge can be attained by the isolation and analysis of cDNA and genomic sequences. As with other enzymes, a more precise structure–function corollary can be inferred by studying the consequences of artificially induced mutations. Probes that are generated from cloned sequences provide indispensable tools in dissecting cellular regulatory and synthetic pathways. A detailed understanding of the glycosyltransferases and glycosidases and, hence, the expression of the glycoconjugates, can be achieved. The functional import of specific glycosidic linkages may likewise be addressed by the creation of mutants that are specifically devoid or enriched in a defined glycosyltransferase or glycosidase activity. Techniques for such an approach are becoming available in the form of mammalian homologous recombination systems [129–132] and germ-line transformation technology [133,134].

BIBLIOGRAPHY

1. Anderson, B., Davis, L. E., and Venegas, M. (1988). *Adv. Exp. Med. Biol. 228*:601–656 [review].
2. Mercurio, A. M. (1986). *Proc. Natl. Acad. Sci. USA 83*:2609–2613.
3. Dennis, J. W., and Laferte, S. (1985). *Cancer Res. 45*:6034–6040.
4. Dennis, J. W. (1985). *JNCI 74*:1111–1120.
5. Ahrens, P. B., and Ankel, H. (1987). *J. Biol. Chem. 262*:7575–7579.
6. Dennis, J. W., Laferte, S., Fukuda, M., Dell, A., and Carver, J. P. (1986). *Eur. J. Biochem. 161*:359–373.
7. Reading, C. L., and Hutchins, J. T. (1985). *Cancer Metastasis Rev. 4*:221–260 [review].

8. Dennis, J. W., Laferte, S., Waghorne, C., Breitman, M. L., and Kerbel, R. S. (1987). *Science 236*:582–585.
9. Weissman, I. L. (1986). *Bioessays 5*:112–116 [review].
10. Muramatsu, T. (1988). *J. Cell. Biochem. 36*:1–14 [review].
11. Bayna, E. M., Shaper, J. H., and Shur, B. D. (1988). *Cell 53*:145–157.
12. Shur, B. D. (1984). *Mol. Cell. Biochem. 61*:143–158 [review].
13. Rademacher, T. W., Parekh, R. B., and Dwek, R. A. (1988). *Annu. Rev. Biochem. 57*:785–838 [review].
14. Radoux, V., Menard, H. A., Begin, R., Decary, F., and Koopmann, W. J. (1987). *Arthritis Rheum. 30*:249–256.
15. Parekh, R. B., Dwek, R. A., Sutton, B., Fernandes, D. L., Leung, A., and Al, E. (1985). *Nature 316*:452–457.
16. Watkins, J., and Swannell, A. J. (1973). *Ann. Rheum. Dis. 32*:247–250.
17. Morgan, K. L. (1987). *Lancet 1*:1017–1019.
18. Morse, M. L. (1986). *Lancet 1*:625–626.
19. Schachter, H. (1986). *Biochem. Cell. Biol. 64*:163–181 [review].
20. Beyer, T. A., Sadler, J. E., Rearick, J. I., Paulson, J. C., and Hill, R. L. (1981). *Adv. Enzymol. 52*:23–175.
21. Kornfeld, S. (1986). *Clin. Invest. 77*:1–6.
22. Fukuda, M. N., Dell, A., and Scartezzin, P. (1987). *J. Biol. Chem. 262*:7195–7206.
23. O'Brien, J. S. (1983). In *The Metabolic Basis of Inherited Disease* (Stanbury, Wijngaarden, Fredrickson, Goldstein, and Brown, eds.). McGraw-Hill, New York, pp. 945–969.
24. Lalley, P. A., Rattazzi, M. C., and Shows, T. B. (1974). *Proc. Natl. Acad. Sci. USA 71*:1569–1573.
25. Gilberg, F., Kucherlapati, R., Cregan, R. P., Murnane, M. J., Darlington, G. J., and Ruddle, R. H. (1975). *Proc. Natl. Acad. Sci. USA 72*:263–267.
26. Tsui, F., Mahuran, D. J., Lowden, J. A., Mosmann, T., and Gravel, R. A. (1983). *J. Clin. Invest. 71*:965–973.
27. Hasilik, A., and Neufeld, E. F. (1980). *J. Biol. Chem. 255*:4937–4945.
28. Mahuran, D. J., Neote, K., Klavins, M. H., Leung, A., and Gravel, R. A. (1988). *J. Biol. Chem. 263*:4612–4618.
29. Dinarello, C. A. (1984). *Rev. Infect. Dis. 6*:51–95.
30. Myerowitz, R., and Proia, R. L. (1984). *Proc. Natl. Acad. Sci. USA 81*:5394–5398.
31. Myerowitz, R., Piekarz, R., Neufeld, E., Shows, T. B., and Suzuki, K. (1985). *Proc. Natl. Acad. Sci. USA 82*:;7830–7834.
32. Proia, R. L., and Soravia, E. (1987). *J. Biol. Chem. 262*:5677–5681.
33. O'Dowd, B. F., Quan, F., Willard, H. F., Lamhonwah, A., Korneluk, R. G., Lowden, J. A., Gravel, R. A., and Mahuran, D. J. (1985). *Proc. Natl. Acad. Sci. USA 82*:1184–1188.
34. Proia, R. L. (1988). *Proc. Natl. Acad. Sci. USA 85*:1883–1887.
35. Bapat, B., Ethier, M., Neote, K., Mahuran, D., and Gravel, R. A. (1988). *FEBS Lett. 237*:191–195.
36. Sonderfeld-Fresko, S., and Proia, R. L. (1988). *J. Biol. Chem. 263*:13463–13469.
37. Ohno, K., and Suzuki, K. (1988). *J. Biol. Chem. 263*:18563–18567.
38. Myerowitz, R., and Costigan, F. C. (1988). *J. Biol. Chem. 263*:18587–18589.
39. Nakano, T., Muscillo, M., Ohno, K., Hoffman, A. J., and Suzuki, K. (1988). *J. Neurochem. 51*:984–987.
40. Navon, R., and Proia, R. L. (1989). *Science 243*:1471–1474.
41. Durand, P., Borrone, C., and Della-Cella, G. (1966). *Lancet 1*:1313–1314.
42. Fukushima, H., Wet, D. R. J., and O'Brien, J. S. (1985). *Proc. Natl. Acad. Sci. USA 82*:1262–1265.
43. Wet, D. R. J., Fukushima, H., Dewju, N. N., Wilcox, E., O'Brien, J. S., and Helinski, D. R. (1984). *DNA 3*:437–447.
44. Turner, B. M., Smith, M., Turner, V. S., Kucherlapati, R. S., Ruddle, F. H., and Hirschborn, K. (1976). *Somat. Cell Genet. 4*:45–54.
45. Carritt, B., King, J., and Welch, J. M. (1982). *Ann. Hum. Genet. 46*:329–335.

46. Fowler, M. L., Nakai, H., Beyers, M. G., Fukushima, H., Eddy, R. L., Henry, W. M., Haley, L. L., O'Brien, J. S., and Shows, T. B. (1986). *Cytogenet. Cell Genet. 43*:103–108.
47. Eiberg, H., Mohr, J., and Nielsen, L. S. (1984). *Clin. Genet. 26*:23–29.
48. Murray, J. C., Sadler, E., Eddy, R. L., Shows, T. B., and Buetow, K. H. (1985). *Cytogenet. Cell Genet. 40*:707–715.
49. Carritt, B., and Welch, H. M. (1987). *Hum. Genet. 75*:248–250.
50. Desnick, R. J., and Sweeley, C. C. (1983). In *The Metabolic Basis of Inherited Diseases*, 5th ed. (Stanbury, Wyngaarden, Fredrickson, Goldstein, and Brown, eds.). MacGraw-Hill, New York, pp. 906–944.
51. Calhoun, D. H., Bishop, D. F., Bernstein, H. S., Quinn, M., Hantzopoulos, P., and Desnick, R. J. (1985). *Proc. Natl. Acad. Sci. USA 82*:7364–7368.
52. Bishop, D. F., Calhoun, D. H., Bernstein, H. S., Hantzopoulos, P., Quinn, M., and Desnick, R. J. (1986). *Proc. Natl. Acad. Sci. USA 83*:4859–4863.
53. Tsuji, S., Martin, B. M., Kaslow, D. C., Migeon, B. R., Choudary, P. V., Stubblefield, B. K., Mayor, J. A., Murray, G. J., Barranger, J. A., and Ginns, E. I. (1987). *Eur. J. Biochem. 165*: 275–280.
54. Bishop, D. F., Kornreich, R., and Desnick, R. J. (1988). *Proc. Natl. Acad. Sci. USA 85*:3903–3907.
55. Groebe, H., Krins, M., Schmidberger, H., von-Figura, K., Harer, K., Kresse, H., Paschke, E., Sewell, A., and Ullrich, K. (1980). *Am. J. Hum. Genet. 32*:258–272.
56. Cantz, M., Mendla, K., and Baumkotter, J. (1984). In *Recent Progress in Neurolipidoses and Allied Disorders* (M. T. Vanier, ed.). Fondation M. Merieux, Lyon, France, pp. 137–145.
57. Andria, G., Strisciuglio, P. G. P., Sly, W. S., and Dodson, W. E. (1981). In *Sialidases and Sialidoses* (G. Tettamanti, P. Durand, and S. DiDonato, eds.). Edi Ermes, Milan, pp. 379–395.
58. d'Azzo, A., Hoogeveen, A. T., Reuser, A. J. J., Robinson, D., and Galjaard, H. (1982). *Proc. Natl. Acad. Sci. USA 79*:4535–4539.
59. Galjart, N. J., Gillemans, N., Harris, A., Van der Horst, G. T. J., Verheijen, F. W., Galjaard, H., and d'Azzo, A. (1988). *Cell 54*:755–764.
60. Kornfeld, R., and Kornfeld, S. (1985). *Annu. Rev. Biochem. 54*:631–664.
61. Weinstein, J., de Souza e Silva, U., and Paulson, J. C. (1982). *J. Biol. Chem. 257*: 13835–13844.
62. Roth, J., Taatjes, D. J., Lucocq, J. M., Weinstein, J., and Paulson, J. C. (1985). *Cell 43*:287–295.
63. Weinstein, J., Lee, E. U., McEntee, K., Lai, P. H., and Paulson, J. C. (1987). *J. Biol. Chem. 262*:17735–17743.
64. Barker, R., Olsen, K. W., Shaper, J. H., and Hill, R. L. (1972). *J. Biol. Chem. 247*:7135–7147.
65. Shaper, N. L., Shaper, J. H., Neuth, J. L., Fox, J. L., Chang, H., Kirsh, I., and Hollis, G. F. (1986). *Proc. Natl. Acad. Sci. USA 83*:1573–1577.
66. Nakazawa, K., Ando, T., Kimura, T., and Narimatsu, H. (1988). *J. Biochem. 104*:165–168.
67. Shaper, N. L., Hollis, G. F., Douglas, J. G., Kirsch, I. R., and Shaper, J. H. (1988). *J. Biol. Chem. 263*:10420–10428.
68. Appert, H. E., Rutherford, T. J., Tarr, G. E., Thomford, N. R., and McCorquodale, D. J. (1986). *Biochem. Biophys. Res. Commun. 138*:224–229.
69. Appert, H. E., Rutherford, T. J., Tarr, G. E.. Wiest, J. S., Thomford, N. R., and McCorquodale, D. J. (1986). *Biochem. Biophys. Res. Commun. 139*:163–168.
70. Masri, K. A., Appert, H. E., and Fukuda, M. N. (1988). *Biochem. Biophys. Res. Commun. 157*:657–663.
71. Duncan, A. M. V., McCorquodale, M. M., Morgan, C., Rutherford, T. J., Appert, H. E., and McCorquodale, D. J. (1986). *Biochem. Biophys. Res. Commun. 141*:1185–1188.
72. Humphreys-Beher, M. G., Bunnell, B., Van Tuinen, P., Ledbetter, D. H., and Kidd, V. J. (1986). *Proc. Natl. Acad. Sci. USA 83*:8918–8922.
73. Connelly, S., and Manley, J. L. (1989). *Cell 57*:561–571.
74. Kukuruzinska, M. A., Bergh, M. L. E., and Jackson, B. J. (1987). *Annu. Rev. Biochem. 56*:915–944 [review].

75. Huffaker, T. C., and Robbins, P. W. (1982). *J. Biol. Chem. 257*:3203–3210.
76. Huffaker, T. C., and Robbins, P. W. (1983). *Proc. Natl. Acad. Sci. USA 80*:7466–7470.
77. Couto, J. R., Huffaker, T. C., and Robbins, P. W. (1983). *J. Biol. Chem. 259*:378–382.
78. Runge, K. W., Huffaker, T. C., and Robbins, P. W. (1984). *J. Biol. Chem. 259*:412–417.
79. Runge, K. W., and Robbins, P. W. (1986). *J. Biol. Chem. 261*:15582–15590.
80. Rine, J., Hansen, W., Hardeman, E., and Davis, R. W. (1983). *Proc. Natl. Acad. Sci. USA 80*:6750–6754.
81. Barnes, G., Hansen, W. J., Holcomb, C. L., and Rine, J. (1984). *Mol. Cell. Biol. 4*:2381–2388.
82. Kukuruzinska, M. A., and Robbins, P. W. (1987). *Proc. Natl. Acad. Sci. USA 84*:2145–2149.
83. Stanley, P. (1987). *Trends Genet. 3*:77–81.
84. Stanley, P. (1984). *Annu. Rev. Genet. 18*:525–552.
85. Stanley, P. (1985). *Mol. Cell. Biol. 5*:923–929.
86. Stanley, P., Narasimhan, S., Siminovitch, L., and Schachter, H. (1975). *Proc. Natl. Acad. Sci. USA 72*:3323–3327.
87. Chaney, W., and Stanley, P. (1986). *J. Biol. Chem. 261*:10551–10557.
88. Stanley, P., and Chaney, W. (1985). *Mol. Cell. Biol. 5*:1204–1211.
89. Ripka, J., Pierce, M., and Fregien, N. (1988). *J. Cell Biol. 107*:10a.
90. Criscuolo, B. A., and Krag, S. S. (1982). *J. Cell Biol. 92*:586–591.
91. Waldman, B. C., Oliver, C., and Krag, S. S. (1987). *J. Cell. Physiol. 131*:302–317.
92. Lehrman, M. A., Zhu, X., and Khounlo, S. (1988). *J. Biol. Chem. 263*:19796–19803.
93. Ernst, L. K., Rajan, V. P., Larsen, R. D., Ruff, M. M., and Lowe, J. B. (1989). *J. Biol. Chem. 264*:3436–3447.
94. Rajan, V. P., Larsen, R. D., Ajmera, S., Ernst, L. K., and Lowe, J. B. (1989). *J. Biol. Chem. 264*:11158–11167.
95. Saiki, R. K., Scharf, S., Faloona, F., Mullis, K. B., Horn, G. T., Erlich, H. A., and Arnheim, N. (1985). *Science 230*:1350–1354.
96. Mullis, K. B., and Faloona, F. A. (1987). *Methods Enzymol. 155*:335–350.
97. Scharf, S. J., Horn, G. T., and Erlich, H. A. (1986). *Science 233*:1076–1077.
98. Kawasaki, E. S., Clark, S. C., Coyne, M. Y., Smith, S. D., Champlin, R., Whitte, O. N., and McCormick, R. P. (1988). *Proc. Natl. Acad. Sci. USA 85*:5698–5702.
99. Wong, C., Dowling, C. E., Saiki, R. K., et al. (1987). *Nature 330*:384–386.
100. Engelke, D. R., Hoener, R. A., and Collins, F. S. (1988). *Proc. Natl. Acad. Sci. USA 85*: 544–548.
101. Moreman, K. W., and Robbins, P. W. (1988). *Annual Meeting Society for Complex Carbohydrates*, San Antonio, 30a.
102. Berchtold, M. W. (1989). *Nucleic Acids Res. 17*:453–454.
103. Kaplan, H. A., Woloski, B. M. R. N. J., Hellman, M., and Jamieson, J. C. (1983). *J. Biol. Chem. 258*:11505–11509.
104. Canonico, P. G., Little, J. S., Powanda, M. C., Bostian, K. A., and Beisel, W. R. (1980). *Infect. Immun. 29*:114–118.
105. Jamieson, J. C., Lammers, G., Janzen, R., and Woloski, B. M. R. N. J. (1987). *Comp. Biochem. Physiol. 87B*:11–15 [review].
106. Woloski, B. M. R. N. J., Gospodarek, E., and Jamieson, J. C. (1985). *Biochem. Biophys. Res. Commun. 130*:30–36.
107. van Dijk, W., Boers, W., Sala, M., Lasthuis, A., and Mookerjea, S. (1986). *Biochem. Cell. Biol. 64*:79–84.
108. Baumann, H., Won, K. A., and Jahreis, G. P. (1989). *J. Biol. Chem. 264*:8046–8051.
109. Ritchie, D. G., and Fuller, G. M. (1981). *Inflammation 5*:275–287.
110. Sanders, K. D., and Fuller, G. M. (1983). *Thromb. Res. 32*:133–145.
111. Sztein, M. B., Luger, T. A., and Oppenheim, J. J. (1982). *J. Immunol. 15*:87–90.
112. Baumann, H., and Mueller-Eberhard, U. (1987). *Biochem. Biophys. Res. Commun. 146*: 1218–1226.

113. Gulcher, J. R., Nies, D. E., Marton, L. S., and Stefansson, K. (1989). *Proc. Natl. Acad. Sci. USA* *86*:1588–1592.
114. Baumann, H., Hill, R. E., Sauder, D. N., and Jahreis, G. P. (1986). *J. Cell Biol.* *102*:370–383.
115. Wang, X. C., O'Hanlon, T. P., and Lau, J. T. Y. (1989). *J. Biol. Chem.* *264*:1854–1859.
116. Haffar, O. K., Aponte, G. W., Bravo, D. A., John, N. J., Hess, R. T., and Firestone, G. L. (1988). *J. Cell Biol.* *106*:1463–1474.
117. Humphreys-Beher, M. G. (1988). *Biochem. J.* *249*:357–362.
118. Hill, R. L., and Brew, K. (1975). *Adv. Enzymol.* *43*:411–490.
119. Smith, C. A., and Brew, K. (1977). *J. Biol. Chem.* *252*:7294–7299.
120. Weinstein, J., Souza, E. S. U., and Paulson, J. C. (1982). *J. Biol. Chem.* *257*:13845–13853.
121. Miagi, T., and Tsuiki, S. (1982). *Eur. J. Biochem.* *126*:253–361.
122. Wickner, W. T., and Lodish, H. F. (1985). *Science* *230*:400–407.
123. Mize, N. K., Andrews, D., and Lingappa, V. R. (1986). *Cell* *47*:711–719.
124. Hart, G. W., Holt, G. D., and Haltiwanger, R. S. (1988). *TIBS* *13*:380–384.
125. Holt, G. D., and Hart, G. W. (1986). *J. Biol. Chem.* *261*:8049–8057.
126. Capasso, J. M., Abeijon, C., and Hirschberg, C. B. (1988). *J. Biol. Chem.* *263*:19778–19782.
127. O'Hanlon, T. P., Lau, K. M., Wang, X. C., and Lau, J. T. Y. (1989). *J. Biol. Chem.* *264*:17389.
128. Lopez, L. C., and Shur, B. D. (1987). *J. Cell Biol.* *105*:1663–1670.
129. Thomas, K. R., and Capecchi, M. R. (1987). *Cell* *51*:503–512.
130. Dorin, J. R., Inglis, J. D., and Porteous, D. J. (1989). *Science* *243*:1357–1360.
131. Thompson, S., Clarke, A. R., Pow, A. M., Hooper, M. L., and Melton, D. W. (1989). *Cell* *56*:313–321.
132. Frohman, M. A., and Martin, G. R. (1989). *Cell* *46*:145–147.
133. Palmiter, R. D., and Brinster, R. L. (1986). *Annu. Rev. Genet.* *20*:465–499.
134. Lavitrano, M., Camaioni, A., Fazio, V. M., Dolci, S., Farace, M. G., and Spadafora, C. (1989). *Cell* *57*:717–723.

16

Plant Lectins: Molecular Biology, Synthesis, and Function

Marilynn E. Etzler *University of California, Davis, California*

Since their discovery in castor bean extracts in 1888, plant lectins have been found in a wide variety of species, ranging from the simplest algae to the higher plants. The abundance of many of these carbohydrate-binding proteins and the diversity of their carbohydrate specificities have made them valuable tools in the cellular localization, isolation, and characterization of complex carbohydrates [for review, see Lis and Sharon, 1986b]. Yet, despite our widespread use of plant lectins and a vast amount of information that has accumulated on the molecular properties and carbohydrate specificities of many of these proteins [for review see Goldstein and Poretz, 1986], we are only at the beginning stages of unraveling the answers to such basic questions as the relationship of lectin structure to function, the mode and regulation of lectin biosynthesis, and the physiological role of these molecules in the plant.

The last several years have witnessed the use of a variety of new approaches in addressing these questions. Many of these approaches have been made possible by the advent of recombinant DNA technology. In this chapter, I focus on these recent developments and relate them to current views and speculations on lectin structure, biosynthesis, and function.

I. DEFINITION

Information accumulated over the past several years makes it essential to reconsider the definition of lectins. This definition has undergone several transitions since the initial coining of the term "lectin" in 1954 [Boyd and Shapleigh, 1954]. The current definition [Nomenclature Committee of the IUB and IUB-IUPAC Joint Commission on Biochemical Nomenclature, 1981] emphasizes the ability of lectins to agglutinate cells and precipitate glycoconjugates. Molecular biological studies described in subsequent sections of this chapter have now made it clear that some carbohydrate-binding proteins that do not agglutinate cells nor precipitate glycoconjugates are definitely in this class of molecules. This review, therefore, will use the broader definition of lectins formulated by Kocourek and Horejsi (1983); *lectins* are defined as "proteins of non-immunoglobulin nature capable of specific recognition and reversible binding to carbohydrate moieties of complex carbohydrates without altering covalent structure of any of the recognized glycosyl ligands." It should be noted that by its emphasis on complex carbohydrates, this definition excludes some toxins and chemotactic proteins that bind only to simple sugars.

II. CONTRIBUTIONS OF RECOMBINANT DNA APPROACHES TO INFORMATION ON LECTIN STRUCTURE–FUNCTION RELATIONSHIPS

By the beginning of this decade, the abundance of lectins in the seeds of many plants, particularly legumes, had enabled investigators to determine the complete primary structures of a few lectins and the NH_2-terminal sequences of many others [for review see Goldstein and Poretz, 1986]. Within the past several years, our repertoire of lectin sequence information has been greatly expanded by the deduction of lectin primary structures from the nucleotide sequences of lectin complementary or genomic DNAs. Complete primary structures are now available for several plant lectins.

Comparisons of these amino acid sequences show the separation of lectins into categories that reflect the taxonomic classifications of the plants from which they are isolated. Although no homologies are yet evident in comparisons of sequences of lectins from different plant classes, sequence similarities among lectins within a particular plant class suggest that these molecules have a common evolutionary origin. Such taxonomic relationships are particularly clear among lectins from seeds of the Leguminoseae class of plants in which the greatest sequence similarities are found among lectins isolated from plants of the same subclass or tribe [for review, see Strosberg et al., 1986]. For example, approximately 73% of the lentil lectin sequence is identical with the corresponding sequences of pea and favin lectins, which are from plants of the same tribe, whereas only 53% of the lentil lectin sequence is identical with the sequence of the lectin from soybean, which is in a different tribe.

The three-dimensional structures of concanavalin A [Hardman and Ainsworth, 1972; Becker et al., 1975; Reeke et al., 1975], pea lectin [Meehan et al., 1982; Riskulov et al., 1984] and favin [Reeke and Becker, 1986] have been determined and found to be quite similar with one another, except for some differences consistent with differences in their carbohydrate specificities and quaternary structures [Reeke and Becker, 1986]. Comparisons of the primary structures of these lectins with the sequences of other legume seed lectins show a high degree of conservation at positions implicated in metal ion binding and subunit–subunit interactions of concanavalin A [Reeke and Becker, 1986; Strosberg et al., 1986; Schnell and Etzler, 1987]. Comparisons of predicted secondary [Foriers et al., 1981; Schnell and Etzler, 1988] and tertiary [Olsen, 1983] structures of several legume seed lectins suggest that the similarities in amino acid sequences among these lectins reflect a conservation in the secondary and tertiary structures of these proteins.

The foregoing structural information, crystallographic information available on other lectins [Wright et al., 1984; Montfort et al., 1987; Robertus et al., 1987], and the availability of lectin complementary and genomic DNAs now set the stage for site-directed mutagenesis studies that should help in the elucidation of the exact structural prerequisites for carbohydrate binding and specificity. Before such studies can be launched, however, it is essential to be able to express active lectin from the cloned

DNAs. As discussed in a later section of this chapter, many lectins are composed of more than one type of polypeptide chain and undergo various processing events during their biosynthesis. Such factors must be considered in designing systems for lectin expression.

Stubbs et al. (1986) expressed the pea lectin in *Escherichia coli* as a single polypeptide chain, containing both the α- and β-subunits. No detectable differences in carbohydrate-binding properties were found between the expressed lectin and the native pea lectin. The functional toxic A chain of ricin has also been expressed in *E. coli* [O'Hare et al., 1987; Piatek et al., 1988], and low levels of functional ricin B chain were expressed in COS-M6 cells [Chang et al., 1987]. With a different approach, Richardson et al. (1988) constructed a prericin B chain cDNA, used this cDNA to generate transcripts in vitro, and then microinjected the transcripts into *Xenopus* oocytes. The B chain was expressed in this system as a glycoprotein with its signal sequence removed and had carbohydrate-binding activity as well as the ability to associate with the A chain of the lectin to produce full cytotoxic activity.

The success in obtaining functional lectin in the foregoing studies paves the way for the site-directed mutagenesis studies mentioned earlier. It can be predicted that these molecular biological approaches may soon yield much new information on the relation of lectin structure to function.

III. LECTIN GENES

It has recently become apparent that many plants contain more than one lectin gene. The first clues to the presence of multiple lectin genes came from the finding that some plants contain more than one lectin and that in some cases these lectins have sequence differences most readily explained by the existence of separate genes [for review see Etzler, 1985, 1986]. The genetic basis for the isolectins of the wheat germ agglutinin was predicted from comparisons of isolectin contents of wheat species differing in their level of ploidy [Rice, 1976]. It has now been established that the various molecular forms of this dimeric lectin are due to the random combination of different subunits, each produced by a separate genome in the polyploid wheat [Peumans et al., 1982].

Hybridization studies of restriction digests of soybean genomic DNA with soybean seed lectin cDNA showed the presence of two lectin genes, designated *L1* and *L2* [Goldberg et al., 1983]. The *L1* gene is abundantly expressed during embryogenesis and encodes the seed lectin, whereas the function of the *L2* gene is not known; recent experiments suggest that *L2* may be an unexpressed pseudogene [Okamura et al., 1986]. Two proteins immunochemically related to the soybean lectin have been identified in the vegetative tissues of this plant [Vodkin and Raikhel, 1986]. It will be of interest to determine if one of these proteins, a 33-kd protein found in roots, embryo axis, and leaves, is the product of the *L2* gene.

A screen of 2284 soybean lines identified 14 lines defective in the expression of the soybean seed lectin [Pueppke, 1983]. Crosses between lectin-positive and lectin-negative cultivars established that the lectin was under the control of a dominant allele, *Le*, at a single locus; the lectin-defective lines are homozygous for the recessive allele, *le*, at this locus [Orf et al., 1978; Pull et al., 1978]. Although both lectin-positive and lectin-negative lines contained the *L1* and *L2* genes, the *L1* gene of lectin-negative lines is modified by the inclusion of a 3.4-kilobase (kb) insertion sequence in its coding region [Goldberg et al., 1983]. This insertion sequence has features resembling a transposable element [Vodkin et al., 1983], and it is thought to block the transcription of the *L1* gene in the lectin-negative lines, resulting in the reduction of lectin mRNA levels to about 0.01% of the levels found in the lectin-positive lines [Goldberg et al., 1983].

The *Ricinus communis* agglutinin and the toxin, ricin, are two closely related lectins present in the seeds of the castor bean plant. These lectins have similar, but distinct, carbohydrate specificities [Nicolson et al., 1974] and are each composed of two types of subunits, A and B, derived by posttranslational cleavage of a single precursor [Butterworth and Lord, 1983]. Comparisons of the amino acid sequences of the subunits of these two lectins and the cDNA sequences of their precursors have established that these lectins are encoded by separate, but related, genes [Ready et al., 1984; Roberts et al., 1985; Lamb et al., 1985]. Internal structural and sequence homologies within the ricin B chain have suggested that this portion of the gene arose by duplication before fusing with a gene encoding the A chain [Villafranca and Robertus, 1981; Lamb et al., 1985]. Homologies between the *R. communis* agglutinin and ricin suggest that the agglutinin may have arisen from the ricin gene by gene duplication [Ready et al., 1984; Roberts et al., 1985].

The ricin gene has been cloned, characterized, and found to contain no introns [Halling et al., 1985]. Southern blot analysis of restriction digests of castor bean DNA, using a ricin cDNA, identified at least five other ricinlike genes in the *R. communis* genome; this finding suggests the presence of a multigene lectin family [Halling et al., 1985]. Such a family of genes is consistent with findings of multiple forms of ricin and the agglutinin [Cawley et al., 1978] and minor variations found in comparisons of genomic, cDNA, and amino acid sequences of these lectins [Funatsu et al., 1978, 1979; Halling et al., 1985; Lamb et al., 1985; Roberts et al., 1985; Araki et al., 1986].

Two lectin genes have been isolated and characterized from the 'Tendergreen' cultivar of *Phaseolus vulgaris*. These two genes, designated *dlec1* and *dlec2*, were found to encode the E and L subunits of the seed lectin, phytohemagglutinin [Hoffman and Donaldson, 1985]. The E and L subunits differ in their affinities for erythrocyte and lymphocyte membrane receptors; tetramers composed of different combinations of these two subunits account for the five isolectins identified in this plant [Miller et al., 1973]. The *dlec1* and *dlec2* genes have 90% sequence identity between their mRNA and protein-coding regions and also show a high degree of conservation in their 5′ and 3′ flanking regions [Hoffman and Donaldson, 1985]. These two genes are separated by only about 4 kb and are in the same orientation on the chromosome; this close linkage, same orientation, and conservation in sequence suggest that they have arisen by gene duplication [Hoffman and Donaldson, 1985].

Two genes, *Pdlec1* and *Pdlec2*, similar to the *dlec1* and *dlec2* genes have also been characterized from *P. vulgaris* cv Pinto [Voelker et al., 1986]. The seeds of this cultivar are deficient in lectin production and produce only the L subunit [Pusztai et al., 1981]. Transcripts of both the *Pdlec1* and *Pdlec2* genes were found in developing cotyledons, but at levels substantially lower than amounts found in another cultivar. Sequence analysis of the *Pdlec1* gene showed an interruption in its coding frame caused by a single bp deletion, thereby accounting for the inability of the seeds to produce the E lectin subunit [Voelker et al., 1986]. Hybridization analyses of restriction digests of DNA indicate the presence of at least five lectin genes in the Pinto and Contender varieties of *P. vulgaris* [Horowitz, 1985] and four to six genes in the French bean variety [Hoffman, 1984]. A gene isolated from the French bean was found to encode a lectinlike protein with 43% sequence identity to the E subunit [Hoffman, 1984; Hoffman and Donaldson, 1985].

The lectins that have been identified from the foregoing multigene families are all expressed in the cotyledons of the plants during seed maturation. The *Dolichos biflorus* plant contains a family of lectins that are differentially expressed, both temporally and spatially [Etzler et al., 1986]. The amino acid sequences of two of these lectins, the seed lectin and DB58, have recently been deduced from their cDNA sequences and found to have 84% identity [Schnell and Etzler, 1987, 1988]. Yet, the seed lectin is present only in the cotyledons, where it appears during late maturation of the seeds, whereas the DB58 lectin appears during seedling development and is present only in the stems and leaves [Etzler et al., 1984; Roberts and Etzler, 1984]. These two lectins have recently been shown to be encoded by separate genes closely linked within the *D. biflorus* genome. Both the 5′ and 3′ flanking regions of these genes show a high degree of similarity with the exception of a 116-bp segment present only in the 5′ flanking region of the seed lectin gene [Harada et al., 1990]. A third lectin has been isolated from the roots of this plant; NH_2-terminal sequence analysis of this root lectin suggests that it may be encoded by a separate gene [Quinn and Etzler, 1987]. The different distributions of these lectins suggest that members of multigene lectin families may be adapted for different functions in different tissues. Such adaptation may invoke the need for separate regulatory controls of these genes.

All studies to date suggest that legume seed lectin synthesis is primarily regulated at the level of transcription [Hoffman et al., 1982; Goldberg et al., 1983; Higgins et al., 1983a; Walling et al., 1986; Okamura et al., 1986; Chappell and Chrispeels, 1986]. The soybean lectin gene was found to be correctly regulated, both temporally and spatially, upon introduction into tobacco plants, thereby suggesting that this lectin gene is under the control of a trans-acting factor(s) conserved during evolution of these two different plant classes [Okamuro et al., 1986].

A trans-acting DNA binding protein has recently been identified in soybean that interacts with a 60-bp region 125–185 nucleotides upstream of the transcription start site of the soybean lectin gene [Jofuku et al., 1987]. It is of interest that this region corresponds in location to a segment 99–212 nucleotides upstream of the transcription site of the *D. biflorus* seed lectin gene that is missing in the DB58 lectin gene that is not expressed in seeds [Harada et al., 1990].

The plant hormone, abscissic acid, has been found to enhance the expression of wheat germ agglutinin [Triplett and Quatrano, 1982; Raikhel and Quatrano, 1986] and the rice lectin [Peumans and Stinissen, 1983; Stinissen et al., 1984]. This hormone is important in promoting seed dormancy. Although work on the regulation of lectin genes is still in its infancy, it is clear from the foregoing studies that the future holds many promising developments in this area.

IV. LECTIN BIOSYNTHESIS

The biosynthetic pathways of several seed lectins have been established. In general, these lectins are synthesized on polysomes of the rough endoplasmic reticulum, undergo cotranslational glycosylation and removal of an NH_2-terminal signal sequence during translocation into the lumen of the endoplasmic reticulum, and then transit the Golgi en route to their final destination in the protein bodies [for review see Chrispeels, 1984]. Various posttranslational modifications occur during this biosynthetic process as described in the following sections.

A. Proteolytic Posttranslational Modifications

The two castor bean lectins, ricin and *R. communis* agglutinin, are each composed of two types of polypeptide chains, A and B [for review see Olsnes and Pihl, 1982]. In vitro translation studies have established that each of these lectins is synthesized as a single polypeptide preprolectin containing a signal sequence and both the A and B chains [Roberts and Lord, 1981; Butterworth and Lord, 1983]. Nucleotide sequencing of cDNAs for these precursors confirmed that each precursor is composed of a 24-amino acid NH_2-terminal signal, followed by a 266- 267-amino acid A chain, a 12-amino acid-linking peptide and a 262-amino acid B chain [Lamb et al., 1985; Roberts et al., 1985]. After cotranslational removal of the signal sequences, the prolectins must be proteolytically processed to obtain the native lectins. This processing occurs in the protein bodies [Lord, 1985]; an endoprotease activity that is capable of processing these precursors to the separate A and B chains has been demonstrated in these organelles [Harley and Lord, 1985].

The legume lectins can be grouped into several categories depending on the types of posttranslational proteolytic modifications they undergo during biosynthesis. The first group contains those lectins, such as favin and the pea lectin, that are composed of two small α-chains and two larger β-chains [for review see Goldstein and Poretz, 1986]. Each of these lectins is synthesized as a single polypeptide preprolectin containing an NH_2-terminal signal sequence, followed by the β-chain and then the α-chain [Hemperly et al., 1982; Higgins et al., 1983b]. As described earlier for the castor bean lectins, the pea and favin prolectins are posttranslationally proteolytically processed to yield the mature α- and β-chains of the lectins [Hemperly et al., 1982; Higgins et al., 1983a]. In the case of the pea lectin, this processing has been shown to occur in the protein bodies [Higgins et al., 1983a]. An additional proteolytic processing of the COOH-terminal ends of the α- and β-subunits of the pea lectin is inferred from comparisons of the cDNA sequence with the sequences of the mature subunits [Higgins et al., 1983b]. It has been suggested [Stubbs et al., 1986] that differential COOH-terminal cleavage of the α-chains could give rise to the two molecular forms of this lectin that are commonly found in peas [Entlicher et al., 1970].

The second category of legume lectins is currently represented by only a single lectin, concanavalin A. This lectin is a tetramer composed of identical polypeptide chains containing 237 amino acids [for review see Goldstein and Poretz, 1986]. Attempts to align the amino acid sequence of this lectin with the sequences of other legume lectins and prolectins revealed an interesting circular permutation in which the NH_2-terminal half (residues 1–118) of concanavalin A aligns with the COOH-terminal halves of other legume lectin sequences and the COOH-terminal half (residues 119–237) of concanavalin A aligns with the NH_2-terminal halves of the other lectins [Cunningham et al., 1979; Hemperly and Cunningham, 1983]. Initial attempts to explain this circular permutation focused on genomic rearrangements. However, recent evidence has established that this circular permutation is the result of a novel posttranslational modification, as described below.

The first clue to the origin of the permuted concanavalin A sequence came from the isolation and sequencing of a cDNA clone for this lectin [Carrington et al., 1985]. The amino acid sequence of the concanavalin A precursor, deduced from the nucleotide sequence of this clone, showed the presence of a 29-amino acid NH_2-terminal signal sequence, followed by residues 119–237 of the mature lectin sequence, a 15-amino acid segment not found in the mature lectin, residues 1–118 of the mature lectin sequence, and an additional nine amino acids as diagrammed in Figure 1. In an elegant biosynthetic study, Bowles et al. (1986) established that concanavalin A is synthesized as a glycosylated precursor that is subsequently modified by deglycosylation, a series of posttranslational proteolytic cleavages, and a peptide ligation, to form the mature lectin (see Fig. 1). This processing scheme is supported by pulse-chase experiments and the isolation and determination of the NH_2-terminal sequences of the various precursors.

Concanavalin A is the first example in eukaryotes in which there is evidence of a posttranslational ligation of two peptide precursors [Carrington et al., 1985]. It has been suggested that this process may be coupled with the cleavage of the nine-amino acid fragment from the COOH-terminal end of the precursor. Such a transpeptidation reaction would be energy-independent, and it has been found to occur during extracytoplasmic formation of peptides in prokaryotes [Bowles et al., 1986].

Computer-modeling studies support the possibility that the concanavalin A precursor might assume a three-dimensional structure, similar to that of the mature concanavalin A molecule, and that the various processing steps may thus occur without great perturbations of the conformation of the molecule [Bowles et al., 1986]. A similar prediction has been made for the pea prolectin, which has been found to have carbohydrate-binding activity [Higgins et al., 1983a]. The concanavalin A precursor does not bind carbohydrate; it has been proposed that the glycosylated 15-amino acid loop connecting residues 237–1 inhibits

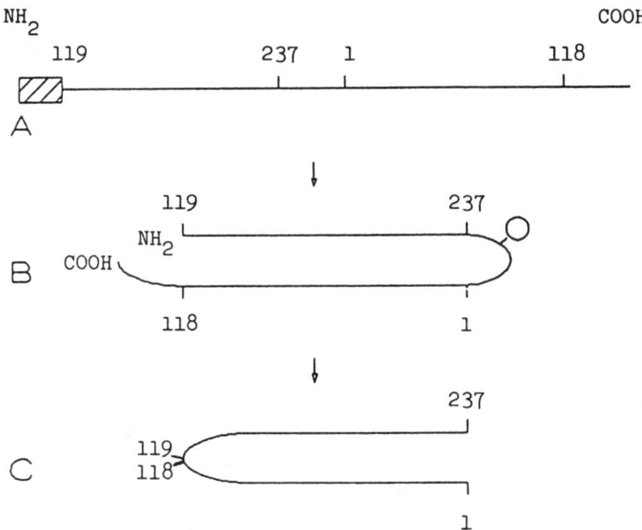

Figure 1 Relation of mature concanavalin A to its precursors as proposed by Carrington et al. (1985) and Bowles et al. (1986). The numbers refer to the corresponding residues in the mature lectin. (A) The initial mRNA translation product, which includes a 29-amino acid signal sequence (shaded box), a 15-amino acid segment between residues 237 and 1, and a COOH-terminal extension of nine amino acids. (B) The precursor formed upon cotranslational removal of the signal sequence and *N*-glycosylation (depicted by circle) of a consensus site within the 15-amino acid connecting segment. (C) Mature concanavalin A after posttranslational proteolytic cleavages that result in the removal of the connecting segment and COOH-terminal extension and a posttranslational ligation joining residues 118 and 119. It has been proposed that this ligation and the cleavage of the COOH-terminal extension occur as a transpeptidation reaction [Bowles et al., 1986].

the carbohydrate-binding activity, perhaps by obstructing access to the carbohydrate-binding site or altering the folding of this portion of the protein [Bowles et al., 1986].

The posttranslational proteolytic-processing events described in the foregoing are thought to occur in the protein bodies. It has been suggested that the synthesis of concanavalin A as an inactive precursor might be necessary to prevent it from interacting with glycosylated components during its transport in the cell [Bowles et al., 1986]; yet, the prolectin precursor of the pea lectin that has a carbohydrate specificity similar to that of concanavalin A has been found to be active [Higgins et al., 1983a].

A third category of legume lectin posttranslational processing is represented by the *P. vulgaris* phytohemagglutinin. This lectin is a tetramer composed of various proportions of two types of subunits, E and L. Each of these subunits is synthesized as a separate precursor that undergoes core *N*-glycosylation and removal of its signal sequence as it is translocated into the lumen of the endoplasmic reticulum [Vitale et al., 1984]. Although the exact point of subunit assembly has not been established, it is thought that this process occurs in the endoplasmic reticulum [Chrispeels, 1984]; the assembly of the subunits is random resulting in the production of five separate isolectins [Miller et al., 1973].

The biosynthesis of the *D. biflorus* seed lectin represents a modification of the foregoing biosynthetic scheme. This lectin is composed of equal amounts of two types of subunits, I and II, which differ from one another only by the absence of ten amino acids at the COOH-terminal end of subunit II [Carter and Etzler, 1975; Roberts et al., 1982; Schnell and Etzler, 1987]. Despite this structural similarity, carbohydrate-binding activity has been detected only with subunit I [Etzler et al., 1981]. Biosynthetic studies, both in vitro and in vivo, have established that the lectin is synthesized as a single glycosylated polypeptide precursor. The two types of lectin subunits arise from these precursors by differential COOH-terminal proteolytic processing [Schnell and Etzler, 1987; Schnell et al., 1987; Quinn and Etzler, 1989]. The mechanism that assures the exact 2:2 stoichiometry of these subunits in the lectin has not been established. One proposal is that subunit assembly precedes the processing step and leaves only two of the four subunits accessible to the proteolytic enzyme; however, other mechanisms cannot be ruled out at present.

B. Posttranslational Modifications of Carbohydrate Units

Most plant lectins are glycoproteins containing one or more *N*-linked oligosaccharides per monomer. Two main types of *N*-linked carbohydrate units have been identified: (1) a high mannose-type, identical with the $Man_9GlcNAc_2$ oligosaccharides found in other glycoproteins [Dorland et al., 1981]; and (2) a xylose–fucose type in which the pentasaccharide core has a D-xylose linked β1-2 to the β-mannosyl residue and an L-fucose linked α1-3 to the reducing terminal *N*-acetylglucosamine [Kitagaki-Ogawa et al., 1986; Ashford et al., 1987]. Many plant lectins contain either or both types of these carbohydrate units or various modifications of them [for review see Lis and Sharon, 1986c]. Structural analyses [Kimura et al., 1987] and in vivo biosynthetic studies [Vitale and Chrispeels, 1984; Faye et al., 1986] provide evidence that the xylose–fucose-type carbohydrate unit is derived from the high mannose-type by a series of posttranslational modifications that occur during the transit of the lectin through the Golgi. The type of carbohydrate unit at a particular position on the mature lectin is dependent upon its accessibility to the processing enzymes during transit [Faye et al., 1986].

Lectins from plants in the Solanaceae family contain carbohydrate units *O*-linked to hydroxyproline. In the case of the potato and *Datura stramonium* lectins, these carbohydrate units are tri- and tetraarabinofuranosides [for review see Lis and Sharon, 1986c]; a Galβ1-3Gal unit *O*-linked to serine has also been found in the *D. stramonium* lectin [Ashford et al., 1982]. The biosynthetic events that lead to the formation of these carbohydrate units have not yet been established.

Some lectins, such as the pea lectin, are not glycosylated. A comparison of the pea lectin amino acid sequence with that of favin shows that the favin glycosylation site, Asn-Ala-Thr, is altered to Asn-Ala-Ala in the pea lectin [Higgins et al., 1983b]. Concanavalin A and the wheat germ agglutinin are initially synthesized as glycosylated precursors [Herman et al., 1985; Bowles et al., 1986; Mansfield et al., 1988]. In concanavalin A, the glycosylation has been shown to occur on the 15-amino acid segment that is subsequently removed during posttranslational processing [Bowles et al., 1986]. A similar situation is thought to occur in the wheat germ agglutinin in which a consensus glycosylation site is present in the 15-amino acid COOH-terminal fragment that is posttranslationally removed [Raikhel and Wilkins, 1987; Mansfield et al., 1988].

C. Targeting

The signals that target the plant lectins to their proper destinations in the cell are not yet known. Many seed lectins have been found to be localized in protein bodies, which are the major storage protein organelles of dicotyledonous plants [for review see Etzler, 1986]. These protein bodies arise from vacuoles, which are functionally equivalent to the lysosomes of animal cells [van Der Wilden et al., 1980]. Unlike animal lysosomes, however, no evidence has been found for the involvement of mannose 6-phosphate signals in this targeting process. In fact, glycosylation has not been found to be essential for the correct transport of some lectins [Lord, 1985; Bollini et al., 1985].

Tague and Chrispeels (1987) have recently expressed the gene encoding the *P. vulgaris* L phytohemagglutinin in yeast and found that this lectin is correctly transported to the yeast vacuole. These transgenic experiments set the stage for the construction of various deletion and site directed mutants that may be used to investigate the prerequisites for targeting in this system. The near future will most probably witness a variety of such studies to establish the intracellular targeting signals of plant lectins.

Several of the foregoing biosynthetic schemes include posttranslational COOH-terminal proteolytic processing events. Perhaps these events are involved in the physiological function or targeting of these molecules to their destinations in the cell.

V. LECTIN FUNCTION

The biological role of plant lectins has been a question that has intrigued investigators for almost a century. The widespread distribution of these carbohydrate-binding proteins throughout the plant kingdom suggests that they have been conserved during evolution because of their importance to the survival of the plants. Yet, although many functions have been proposed for plant lectins over the years (summarized in Table 1), no biological role of plant lectins has yet been firmly established. The evidence for and against many of these roles has been reviewed [Etzler, 1985, 1986; van Driessche, 1988].

Until recently, attempts to determine the biological role of plant lectins have no doubt been hindered by the tendency to consider them as a single, discrete class of molecules that may have a common function. Recent findings of multigene lectin families and the differential expression of some of these lectins, both spatially and temporally, suggest that

Table 1 Some Biological Roles Proposed for Plant Lectins

Defense against bacterial [Jones, 1964], fungal [Mirelman et al., 1975], and viral [Partridge et al., 1976] pathogens or animal predators [Janzen et al., 1976]
Rhizobium–legume symbiosis [Krupe, 1956; Hamblin and Kent, 1973; Bohlool and Schmidt, 1974]
Packaging or mobilization of storage materials during seed maturation or germination [Weber and Neumann, 1980]
Maintenance of seed dormancy [Peumans et al., 1983]
Storage proteins [Boyd, 1963]
Carbohydrate transport [Krupe, 156; Boyd, 1963]
Carbohydrate-specific enzymes [Hankins and Shannon, 1978]
Mitogenic stimulators of plant embryonic cells [Howard et al., 1972]
Cell wall elongation [Kauss and Glaser, 1974]
Cell recognition between pollen and stigma or style [Golynskaya et al., 1976; Knox et al., 1976]

during evolution these different lectins may have been adapted for different functions. The multigene lectin family of *D. biflorus* is a case in point. Recent studies have shown that the genes for the DB58 lectin and the seed lectin from this plant are closely linked and exhibit about 90% similarity in nucleotide sequence [Harada et al., 1990]. Yet the *DB58* gene is under separate regulatory control and is expressed only in the stems and leaves, where it is noncovalently associated with the cell walls [Roberts and Etzler, 1984; Etzler et al., 1984; Schnell and Etzler, 1988], whereas the seed lectin is expressed during late seed maturation and is localized in the protein bodies of the cotyledons [Talbot and Etzler, 1978; Etzler et al., 1984]. A third lectin, DB46, has been isolated from the roots of this plant [Quinn and Etzler, 1987], and a lectinlike protein, DB57, has been isolated from the stems and leaves [Etzler et al., 1986]. The synthesis of DB57 in *D. biflorus* cell suspension cultures is enhanced by heat shock [Spadoro-Tank and Etzler, 1988]. The different distributions, cellular localizations, and modes of regulation of these lectins strongly support the possibility of their utilization for different roles within the plant.

In the foregoing example, it should be noted that other than the seed lectin, which accounts for about 10% of the soluble seed protein [Talbot and Etzler, 1978], the other lectins normally constitute only about 0.05% or less of the total protein of their respective tissues [Roberts and Etzler, 1984; Etzler et al., 1986]. Surveys of vegetative tissues of other plants have shown low levels of lectins in many of these tissues, some of which are encoded by genes other than the seed lectin genes [for review see Etzler, 1986]. It is essential, therefore, to consider a particular hypothesis for lectin function as it pertains to the appropriate lectins and not to discard it if all lectins do not fit into the framework. Keeping this consideration in mind, let us now turn to some of the roles that have been proposed for plant lectins and evaluate them in light of recent information.

An interesting hypothesis that has generated much controversy in recent years has been the possibility that lectins may be involved in rhizobia–legume symbiosis. This symbiosis is a multistep process that involves the adhesion of the bacteria to the roots, their internalization, and subsequent nodulation that results in the ability of the plant to fix nitrogen. There is a high degree of specificity of particular strains of rhizobia for certain legume species; the idea that lectins may play a role in determining the specificity of this process was suggested as early as 1956 [Krupe, 1956] and revived in the early 1970s [Hamblin and Kent, 1973; Bohlool and Schmidt, 1974].

The controversy about this hypothesis revolves around two main questions: (1) whether a correlation exists between the ability of rhizobial strain to nodulate a plant and the ability of the lectin from the host to recognize that rhizobial strain, and (2) whether lectin and the rhizobial receptor are present at the time and place of nodulation [for review, see Etzler, 1986]. Although much effort has been devoted to examining the abilities of various legume lectins to bind different strains of rhizobia, most of these studies have been conducted with the lectins from the seeds of the plants. These studies should now be reevaluated in light of the finding of lectins in legume roots that have different molecular properties than the seed lectins from the plants; these root lectins are most probably under the control of separate genes [Law and Strijdom, 1984; Quinn and Etzler, 1987]. Indeed, there is evidence that a soybean root lectin affects the nodulating ability of a mutant rhizobial strain and that this root lectin is present in the root exudates of a *lele* soybean line that does not express the seed lectin [Halverson and Stacey, 1985]. Recent studies also suggest that the binding of rhizobia to a soybean cell line originally derived from roots may be mediated by a lectin identified in these cells [Ho et al., 1986].

The possibility that lectins may play a role in the defense of plants against various pathogens and predators has also received widespread attention. Evidence in support of this role included the ability of chitin-binding lectins, such as the wheat germ agglutinin, to inhibit fungal growth and spore germination [Mirelman et al., 1975]. Although recent data suggest that this antifungal activity was due to the presence of contaminating chitinases in the lectin preparations [Schlumbaum et al., 1986], a defense role for legume seed lectins is strongly supported by recent studies of Osborn and co-workers [Osborn et al., 1986, 1988a,b]. These investigators have identified a protein, called arcelin, in the seeds of some wild lines of *P. vulgaris* that is genetically correlated with the resistance of these beans to insect predators [Osborn et al., 1986]. Arcelin and phytohemagglutinin have similar biochemical properties [Osborn et al., 1988b], and the cDNA sequence of arcelin has a high degree of similarity to the sequences of the two lectin genes [Osborn et al., 1988a]. These similarities, the frequencies at which these proteins occur, and the finding that the expression of these genes is tightly linked, suggest that arcelin is a lectin that arose by gene duplication and divergence from one of the lectin genes [Osborne et al., 1988a,b]. Both arcelin and phytohemagglutinin confer resistance to some insect predators when incorporated into artificial seeds; however, only arcelin is effective against the two most important bruchid pests of beans [Janzen et al., 1976; Osborn et al., 1988a].

The insecticidal activity of phytohemagglutinin and arcelin in artificial seeds was found only with high concentrations of these lectins [Janzen, 1976; Osborn et al., 1988a]. The high level of expression of legume seed lectin genes, therefore, may not be necessary for plant viability, but rather, may provide an evolutionary selective advantage to the plant. Such a case would explain the viability of legume strains with mutated seed lectin genes that are deficient in seed lectin expression [Pueppke, 1983]. Examination of such strains has shown the existence of other lectin genes and low levels of lectin transcripts in the seeds [Goldberg et al., 1983; Voelker et al., 1986]. Therefore, it is possible that the legume seed lectins may play some vital role for which low levels of lectin in the seeds are sufficient. The high levels of lectins found in legume seeds may simply reflect the adaptive advantage of defense against predators that was conferred by a chance acquisition of ability of an ancestral seed lectin gene to achieve high levels of expression. Regardless of the explanation for high levels of legume seed lectins, the abundance of these seed lectins has provided enormous benefits to carbohydrate chemists.

The oligomeric nature of most plant lectins endows them with a multivalence that enables them to cross-link glycoconjugate receptors on the surfaces of many animal cells [for review see Lis and Sharon, 1986a]. Such receptor clustering may be important in signal transduction, cell–cell interaction, or in a variety of events for which the formation of membrane microdomains may be essential. It is tempting to speculate that lectins may participate in such roles in plant cells. The soybean seed lectin has been found to modulate the membranes of soybean protoplasts [Metcalf et al., 1983], and there have been a few sporadic reports of weak mitogenic activity of plant seed lectins on plant cells [for review see Etzler, 1986]. However, at present, very little is known about natural endogenous receptors for plant lectins [for review see Etzler, 1986] or whether plant lectins interact with these receptors in vivo.

Currently, all of our hypotheses on plant lectin function have been centered on the carbohydrate-binding properties of these molecules. Although lectins were discovered fortuitously and are now defined on the basis of their carbohydrate-binding properties, carbohydrate binding may not be the only functional activity associated with these molecules. There are many examples in the biochemical literature of proteins that have ligand-binding sites and effector functions. Examples include such proteins as allosteric enzymes, immunoglobulins, and various cell surface receptors. In this context, it is of interest that a number of plant lectins have now been found to have hydrophobic ligand-binding sites, in addition to their carbohydrate-binding sites; these lectins include some legume lectins [Edelman and Wang, 1978; Roberts and Goldstein, 1982, 1982a,b; Yang et al., 1974], the wheat germ agglutinin, the potato lectin, and the *R. communis* lectin [Roberts and Goldstein, 1983a]. Although published evidence now indicates no interaction between the carbohydrate-binding and hydrophobic sites [Edelman and Wang, 1978; Roberts and Goldstein, 1982, 1983a], the possibility of such site–site interaction cannot yet be excluded.

Several naturally occurring ligands that have been found to bind to the hydrophobic sites of lectins include indoleacetic acid [Edelman and Wang, 1978], adenine, and adenine derivatives with cytokinin activity [Roberts and Goldstein, 1983a,b]. Both indoleacetic acid and cytokinin are plant hormones; the ability of lectins to bind to these hormones raises the possibility that they may be regulated by hormones or possibly function as plant hormone receptors.

For the toxic lectins, such as ricin, the carbohydrate-binding B chain is necessary for the uptake of the lectin into the cells and the eventual translocation of the toxic A chain into the cytoplasm [Youle et al., 1981; Olsnes and Pihl, 1982]. In eukaryotic cells, the A chain then inactivates the large ribosomal subunits by specifically cleaving a particular *N*-glycosidic bond in the 28S rRNA [Endo and Tsurugi, 1987]. Whether ricin plays some biological role in the castor bean seeds, other than to confer protection against predators, is not yet known. There is some evidence that the ricin gene, encoding both the A and B chains, was derived during evolution by a fusion of genes that initially had encoded separate carbohydrate-binding and toxic proteins [Ready et al., 1984]. The indirect influence of the B subunit on the separate activity of the A subunit supports the possibility that the carbohydrate binding properties of other lectins may modulate other functional activities of the same or different proteins.

VI. FUTURE DIRECTIONS

Many of the advances on plant lectins made within the last several years have been made possible by the advent of recombinant DNA technology. The proliferation of sequence information and the ability to alter these sequences by site-directed mutagenesis should rapidly lead to the acquisition of new insights into the relationship of lectin structure to function. One might even envision the possibility of eventually creating "designer lectins" to satisfy a wide variety of needs in glycoconjugate research.

Our growing awareness of the presence of multigene lectin families in plants is generating much interest in such areas as the regulation of lectin synthesis, targeting, and the biological function of these molecules. It is expected that the next several years will witness the increased utilization of transgenic studies and other molecular genetic techniques to address these questions.

We have come a long way since the discovery of lectins by Stillmark in 1888. During this first century, we have recognized the value of plant lectins to glycoconjugate biochemists and cell biologists. It is anticipated that as we begin this second century of exploration, we will rapidly build on the foundations we have laid and soon discover the biological value of the lectins to the plants.

VII. ADDENDUM

Since the completion of this chapter in November 1988, many new studies have been reported that contribute to our current status of information in this field. The author regrets that time restrictions did not allow the incorporation of this new information into the present chapter. However, it should be noted that as predicted in Section VI, many of these advances have occurred through the utilization of transgenic studies. One of the most exciting developments was the finding that the introduction of the pea lectin gene into clover roots enabled these roots to be nodulated by a *Rhizobium* strain that normally nodulates the roots of peas [Diaz et al., 1989]. Although the nodulation of the transformed roots was delayed and abnormal, the ability of the transferred gene to alter the host-specificity barrier has enhanced interest in the possible involvement of lectins in rhizobia–legume symbiosis.

REFERENCES

Araki, T., Yoshioka, Y., and Funatsu, G. (1986). The complete amino acid sequence of the B-chain of the *Ricinus communis* agglutinin isolated from large-grain castor bean seeds. *Biochim. Biophys. Acta 872*:277–285.

Ashford, D., Desai, N. N., Allen, A. K., Neuberger, A., O'Neill, M. A., and Selvendran, R. L. (1982). Structural studies of the carbohydrate moieties of lectins from potato (*Solanum tuberosum*) tubers and thorn-apple (*Datura stramonium*) seeds. *Biochem. J. 201*:199–208.

Ashford, D., Dwek, R. A., Welply, J. K., Amatayakul, S., Homans, S. W., Lis, H., Taylor, G. N., Sharon, N., and Rademacher, T. W. (1987). The β1-2-D-xylose and α1-3-L-fucose substituted *N*-linked oligosaccharides from *Erythrina cristagalli* lectin. Isolation, characterisation and comparison with other legume lectins. *Eur. J. Biochem. 166*:311–320.

Becker, J. W., Reeke, G. N., Wang, J. L., Cunningham, B. A., and Edelman, G. M. (1975). The covalent and three-dimensional structure of concanavalin A. III. Structure of the monomer and its interactions with metals and saccharides. *J. Biol. Chem. 250*:1513–1524.

Bohlool, B. B., and Schmidt, E. L. (1974). Lectins: A possible basis for specificity in the *Rhizobium*–legume root nodule symbiosis. *Science 185*:269–271.

Bollini, R., Ceriotti, A., Daminati, M. G., and Vitale, A. (1985). Glycosylation is not needed for the intracellular transport of phytohemagglutinin in developing *Phaseolus vulgaris* cotyledons and for the maintenance of its biological activities. *Physiol. Plant. 65*:15–22.

Bowles, D. J., Marcus, S. E., Pappin, D. J. C., Findlay, J. B. C., Eliopoulos, E., Maycox, P. R., and Burgess, J. (1986). Posttranslational processing of concanavalin A precursors in jackbean cotyledons. *J. Cell Biol. 102*:1284–1297.

Boyd, W. C. (1963). The lectins: Their present status. *Vox Sang. 8*:1–32.

Boyd, W. C., and Shapleigh, E. (1954). Antigenic relations of blood group antigens as suggested by tests with lectins. *J. Immunol. 73*:226–231.

Butterworth, A. G., and Lord, J. M. (1983). Ricin and *Ricinus communis* agglutinin subunits are all derived from a single-size polypeptide precursor. *Eur. J. Biochem. 137*:57–65.

Carrington, D. M., Auffret, A., and Hanke, D. E. (1985). Polypeptide ligation occurs during post-translational modification of concanavalin A. *Nature 313*:64–67.

Carter, W. G., and Etzler, M. E. (1975). Isolation and characterization of subunits from the predominant form of *Dolichos biflorus* lectin. *Biochemistry 14*:2685–2689.

Cawley, D. B., Hedblom, M. L., and Houston, L. L. (1978). Homology between ricin and *Ricinus communis* agglutinin: Amino-terminal sequence analysis and protein synthesis inhibition studies. *Arch. Biochem. Biophys. 190*:744–755.

Chang, M.-S., Russell, D. W., Uhr, J. W., and Vitetta, E. S. (1987). Cloning and expression of recombinant, functional ricin B chain. *Proc. Natl. Acad. Sci. USA 84*:5640–5644.

Chappell, J., and Chrispeels, M. J. (1986). Transcriptional and posttranscriptional control of phaseolin and phytohemagglutinin gene expression in developing cotyledons of *Phaseolus vulgaris. Plant Physiol. 81*:50–54.

Chrispeels, M. J. (1984). Biosynthesis, processing and transport of storage proteins and lectins in cotyledons of developing legume seeds. *Philos. Trans. R. Soc. Lond. B 304*:309–322.

Cunningham, B. A., Hemperly, J. J., Hopp, T. P., and Edelman, G. M. (1979). Favin versus concanavalin A: Circularly permuted amino acid sequences. *Proc. Natl. Acad. Sci. USA 76*: 3218–3222.

Diaz, C. L., Melchers, L. S., Hooykaas, P. J. J., Lugtenberg, B. J. J., and Kijne, J. W. (1989). Root lectin as a determinant of host–plant specificity in the *Rhizobium*–legume symbiosis. *Nature 338*:579–581.

Dorland, L., van Halbeek, H., Vliegenthart, J. F. G., Lis, H., and Sharon, N. (1981). Primary structure of the carbohydrate chain of soy-bean agglutinin. *J. Biol. Chem. 256*:7708–7711.

Edelman, G. M., and Wang, J. L. (1978). Binding and functional properties of concanavalin A and its derivatives. III. Interactions with indoleacetic acid and other hydrophobic ligands. *J. Biol. Chem. 253*:3016–3022.

Endo, Y., and Tsurugi, K. (1987). RNA *N*-glycosidase activity of ricin A-chain. Mechanism of action of the toxic lectin ricin on eukaryotic ribosomes. *J. Biol. Chem. 262*:8128–8130.

Entlicher, G., Kostir, J. V., and Kocourek, J. (1970). Studies on phytohemagglutinins. III. Isolation and characterization of hemagglutinins from the pea (*Pisum sativum* L.). *Biochim. Biophys. Acta 221*:272–281.

Etzler, M. E. (1985). Plant lectins: Molecular and biological aspects. *Annu. Rev. Plant Physiol. 36*:209–234.

Etzler, M. E. (1986). Distribution and function of plant lectins. In *The Lectins. Properties, Functions, and Applications in Biology and Medicine* (I. E. Liener, N. Sharon, and I. J. Goldstein, eds.). Academic Press, New York, pp. 371–435.

Etzler, M. E., Gupta, S., and Borrebaeck, C. (1981). Carbohydrate binding properties of the *Dolichos biflorus* lectin and its subunits. *J. Biol. Chem. 256*:2367–2370.

Etzler, M. E., MacMillan, S., Scates, S., Gibson, D. M., James, D. W., Jr., Cole, D., and Thayer, S. (1984). Subcellular localizations of two *Dolichos biflorus* lectins. *Plant Physiol. 76*:871–878.

Etzler, M. E., Quinn, J. M., Schnell, D. J., and Spadoro, J. P. (1986). Characterization of lectins from stems and leaves of *Dolichos biflorus*. In *Molecular Biology of Seed Storage Proteins and Lectins* (L. M. Shannon and M. J. Chrispeels, eds.). American Society of Plant Physiology, Rockville, Md., pp. 65–72.

Faye, L., Sturm, A., Bollini, R., Vitale, A., and Chrispeels, M. J. (1986). The position of the oligosaccharide side-chains of phytohemagglutinin and their accessibility to glycosidases determines their subsequent processing in the Golgi. *Eur. J. Biochem. 158*:655–661.

Foriers, A., Lebrun, E., van Rapenbusch, R., de Neve, R., and Strosberg, A. D. (1981). The structure of the lentil (*Lens culinaris*) lectin. Amino acid sequence determination and prediction of the secondary structure. *J. Biol. Chem. 256*:5550–5560.

Funatsu, G., Kimura, M., and Funatsu, M. (1979). Primary structure of Ala-chain of ricin D. *Agric. Biol. Chem. 43*:2221–2224.

Funatsu, G., Yoshitake, S., and Funatsu, M. (1978). Primary structure of Ile chain of ricin D. *Agric. Biol. Chem. 42*:501–503.

Goldberg, R. B., Hoschek, G., and Vodkin, L. O. (1983). An insertion sequence blocks the expression of a soybean lectin gene. *Cell 33*:465–475.

Goldstein, I. J., and Poretz, R. D. (1986). Isolation, physicochemical characterization, and carbohydrate-binding specificity of lectins. In *The Lectins. Properties, Functions, and Applications in Biology and Medicine* (I. E. Liener, N. Sharon, and I. J. Goldstein, eds.). Academic Press, New York, pp. 35–247.

Golynskaya, E. L., Bashikirova, N. V., and Tomchuk, N. N. (1976). Phytohemagglutinins of the pistil in *Primula* as possible proteins of generative incompatibility. *Sov. Plant Physiol. 23*:69–76.

Halling, K. C., Halling, A. C., Murray, E. E., Ladin, B. F., Houston, L. L., and Weaver, R. F. (1985). Genomic cloning and characterization of a ricin gene from *Ricinus communis*. *Nucleic Acids Res. 13*:8019–8033.

Halverson, L. J., and Stacey, G. (1985). Host recognition in the *Rhizobium*–soybean symbiosis. Evidence for the involvement of lectin in nodulation. *Plant Physiol. 77*:621–625.

Hamblin, J., and Kent, S. P. (1973). Possible role of phytohemagglutinin in *Phaseolus vulgaris* L. *Nature New Biol. 245*:28–30.

Hankins, C. N., and Shannon, L. M. (1978). The physical and enzymatic properties of a phytohemagglutinin from mung beans. *J. Biol. Chem. 253*:7791–7797.

Harada, J. J., Spadoro-Tank, J. P., Maxwell, J. C., Schnell, D. J., and Etzler, M. E. (1990). Comparison of the structures of two lectin genes differentially expressed in *Dolichos biflorus*. *J. Biol. Chem. 265*:4997–5001.

Hardman, K. D., and Ainsworth, C. D. (1972). Structure of concanavalin A at 2.4 A resolution. *Biochemistry 11*:4910–4919.

Harley, S. M., and Lord, J. M. (1985). In vitro endoproteolytic cleavage of castor bean lectin precursors. *Plant Sci. 41*:111–116.

Hemperly, J. J., and Cunningham, B. A. (1983). Circular permutation of amino acid sequences among legume lectins. *Trends Biochem. Sci. 8*:100–102.

Hemperly, J. J., Mostov, K. E., and Cunningham, B. A. (1982). In vitro translation and processing of a precursor form of favin, a lectin from *Vicia faba*. *J. Biol. Chem. 257*:7903–7909.

Herman, E. M., Shannon, L. M., and Chrispeels, M. J. (1985). Concanavalin A is synthesized as a glycoprotein precursor. *Planta 165*:23–29.

Higgins, T. J. V., Chandler, P. M., Zurawski, G., Button, S. C., and Spencer, D. (1983a). The biosynthesis and primary structure of pea seed lectin. *J. Biol. Chem. 258*:9544–9549.

Higgins, T. J. V., Chrispeels, M. J., Chandler, P. M., and Spencer, D. (1983b). Intracellular sites of synthesis and processing of lectin in developing pea cotyledons. *J. Biol. Chem. 258*: 9550–9552.

Ho, S.-C., Malek-Hedayat, S., Wang, J. L., and Schindler, M. (1986). Endogenous lectins from cultured soybean cells: Isolation of a protein immunologically cross-reactive with seed soybean agglutinin and analysis of its role in binding of *Rhizobium japonicum*. *J. Cell Biol. 103*:1043–1054.

Hoffman, L. M. (1984). Structure of a chromosomal *Phaseolus vulgaris* lectin gene and its transcript. *J. Mol. Appl. Genet. 2*:447–453.

Hoffman, L. M., and Donaldson, D. D. (1985). Characterization of two *Phaseolus vulgaris* phytohemagglutinin genes closely linked on the chromosome. *EMBO J. 4*:883–889.

Hoffman, L. M., Ma, Y., and Barker, R. F. (1982). Molecular cloning of *Phaseolus vulgaris* lectin mRNA and use of cDNA as a probe to estimate lectin transcript levels in various tissues. *Nucleic Acids Res. 10*:7819–7828.

Horowitz, J. (1985). A *Phaseolus mutation* results in a reduced level of lectin mRNA. *Mol. Gen. Genet. 198*:482–485.

Howard, I. K., Sage, H. J., and Horton, C. B. (1972). Studies on the appearance and location of hemagglutinins from a common lentil during the life cycle of the plant. *Arch. Biochem. Biophys. 149*:323–326.

Janzen, D. H., Juster, H. B., and Liener, I. E. (1976). Insecticidal action of the phytohemagglutinin in black beans on a bruchid beetle. *Science 192*:795–796.

Jofuku, K. D., Okamuro, J. K., and Goldberg, R. B. (1987). Interaction of an embryo DNA binding protein with a soybean lectin gene upstream region. *Nature 328*:734–737.

Jones, D. A. (1964). The lectin in the seeds of *Viccia cracca* L. II. A population study and a possible function for the lectin. *Heredity 19*:459–469.

Kauss, H., and Glaser, C. (1974). Carbohydrate-binding proteins from plant cell walls and their possible involvement in extension growth. *FEBS Lett. 45*:304–307.

Kimura, Y., Hase, S., Kobayashi, Y., Kyogoku, Y., Funatsu, G., and Ikenaka, T. (1987). Possible pathway for the processing of sugar chains containing xylose in plant glycoproteins deduced on structural analyses of sugar chains from *Ricinus communis* lectins. *J. Biochem. 101*: 1051–1054.

Kitagaki-Ogawa, H., Matsumoto, I., Seno, N., Takahashi, N., Endo, S., and Arata, Y. (1986). Characterization of the carbohydrate moiety of *Clerodendron trichotomum* lectins. Its structure and carbohydrate moieties. *Eur. J. Biochem. 161*:779–785.

Knox, R. B., Clarke, A., Harrison, S., Smith, P., and Marchalonis, J. J. 91976). Cell recognition in plants: Determinants of the stigma surface and their pollen interactions. *Proc. Natl. Acad. Sci. USA 73*:2788–2792.

Kocourek, J., and Horejsi, V. (1983). A note on the recent discussion on definition of the term "lectin." In *Lectins. Biology, Biochemistry, Clinical Biochemistry*, Vol. 3 (T. C. Bog-Hansen and G. A. Spengler, eds.). DeGruyter, Berlin/New York, pp. 3–6.

Krupe, M. (1956). *Blutgruppenspezifische Pflanzliche Eiweisskorper (Phytagglutine)*. Ferdinand Enke, Stuttgart, 131 pp.

Lamb, F. I., Roberts, L. M., and Lord, M. J. (1985). Nucleotide sequence of cloned cDNA coding for preproricin. *Eur. J. Biochem. 148*:265–270.

Law, I. J., and Strijdom, B. W. (1984). Properties of lectins in the root and seed of *Lotononis bainesii. Plant Physiol. 74*:773–778.

Lis, H., and Sharon, N. (1986a). Biological properties of lectins. In *The Lectins. Properties, Functions, and Applications in Biology and Medicine* (I. E. Liener, N. Sharon, and I. J. Goldstein, eds.). Academic Press, New York, pp. 265–291.

Lis, H., and Sharon, N. (1986b). Applications of lectins. In *The Lectins. Properties, Functions, and Applications in Biology and Medicine* (I. E. Liener, N. Sharon, and I. J. Goldstein, eds.). Academic Press, New York, pp. 294–370.

Lis, H., and Sharon, N. (1986c). Lectins as molecules and as tools. *Annu. Rev. Biochem. 55*:35–67.

Lord, M. J. (1985). Synthesis and intracellular transport of lectin and storage protein precursors in endosperm from castor bean. *Eur. J. Biochem. 146*:403–409.

Mansfield, M. A., Peumans, W. J., and Raikhel, N. V. (1988). Wheat-germ agglutinin is synthesized as a glycosylated precursor. *Planta 173*:482–489.

Meehan, E. J., Jr., McDuffie, J., Einspahr, H., Bugg, C. E., and Suddath, F. L. (1982). The crystal structure of pea lectin at 6 Å resolution. *J. Biol. Chem. 257*:13278–13282.

Metcalf, T. N., III, Wang, J. L., Schubert, K. R., and Schindler, M. (1983). Lectin receptors on the plasma membrane of soybean cells. Binding and lateral diffusion of lectins. *Biochemistry* *22*:3969–3975.

Miller, J. B., Noyes, C., Heinrikson, R., Kingdon, H. S., and Yachnin, S. (1973). Phytohemagglutinin mitogenic proteins. Structural evidence for a family of isomitogenic proteins. *J. Exp. Med.* *138*:939–951.

Mirelman, D., Galun, E., Sharon, N., and Lotan, R. (1975). Inhibition of fungal growth by wheat germ agglutinin. *Nature 256*:414–416.

Montfort, W., Villafranca, J. E., Monzingo, A. F., Ernst, S. R., Katzin, B., Rutenber, E., Xuong, N. H., Hamlin, R., and Robertus, J. D. (1987). The three-dimensional structure of ricin at 2.8 A. *J. Biol. Chem. 262*:5398–5403.

Nicolson, G. L., Blaustein, J., and Etzler, M. E. (1974). Characterization of two plant lectins from *Ricinus communis* and their quantitative interaction with a murine lymphoma. *Biochemistry* *13*:196–204.

Nomenclature Committee of IUB (NC-IUB) and IUB-IUPAC Joint Commission on Biochemical Nomenclature (JCBN) (1981). Newsletter. *Arch. Biochem. Biophys. 206*:458–462; *Eur. J. Biochem. 114*:1–4; *Hoppe-Seylers Z. Physiol. Chem. 362*:1–4; *J. Biol. Chem. 256*: 12–14.

O'Hare, M., Roberts, L. M., Thorpe, P. E., Watson, G. J., Prior, B., and Lord, J. M. (1987). Expression of ricin A chain in *Escherichia coli. FEBS Lett. 216*:73–78.

Okamuro, J. K., Jofuku, K. D., and Goldberg, R. B. (1986). Soybean seed lectin gene and flanking nonseed protein genes are developmentally regulated in transformed tobacco plants. *Proc. Natl. Acad. Sci. USA 83*:8420–8244.

Olsen, K. W. (1983). Prediction of three-dimensional structure of plant lectins from the domains of concanavalin A. *Biochem. Biophys. Acta 743*:212–218.

Olsnes, S., and Pihl, A. (1982). Toxic lectins and related proteins. In *Molecular Action of Toxins and Viruses*, Vol. 2 (P. Cohen and S. van Heyningen, eds.). Elsevier Biomedical Press, New York, pp. 51–105.

Orf, J. H., Hymowitz, T., Pull, S. P., and Pueppke, S. G. (1978). Inheritance of a soybean seed lectin. *Crop Sci. 18*:899–900.

Osborn, T. C., Alexander, D. C., Sun, S. S. M., Cardona, C., and Bliss, F. A. (1988a). Insecticidal activity and lectin homology of arcelin seed protein. *Science 240*:207–210.

Osborn, T. C., Blake, T., Gepts, P., and Bliss, F. A. (1986). Bean arcelin. 2. Genetic variation, inheritance and linkage relationships of a novel seed protein of *Phaseolus vulgaris* L. *Theor. Appl. Genet. 71*:847–855.

Osborn, T. C., Burow, M., and Bliss, F. A. (1988b). Purification and characterization of arcelin seed protein from common bean. *Plant Physiol. 86*:399–405.

Partridge, J., Shannon, L., and Gumpf, D. (1976). A barley lectin that binds free amino sugars. I. Purification and characterization. *Biochim. Biophys. Acta 451*:470–483.

Peumans, W. J., Stinissen, H. M., and Carlier, A. R. (1982). A genetic basis for the origin of six different isolectins in hexaploid wheat. *Planta 154*:562–567.

Peumans, W. J., and Stinisson, H. M. (1983). Gramineae lectins: Occurrence, molecular biology, and physiological function. In *Chemical Taxonomy, Molecular Biology, and Function of Plant Lectins* (I. J. Goldstein and M. E. Etzler, eds.). Alan R. Liss, New York, pp. 99–116.

Piatek, M., Lane, J. A., Laird, W., Bjorn, M. J., Wang, A., and Williams, M. (1988). Expression of soluble and fully functional ricin A chains in *Escherichia coli* is temperature-sensitive. *J. Biol. Chem. 263*:4837–4843.

Pueppke, S. G. (1983). Distribution and function of the 120,000 dalton soybean lectin in the genus *Glycine*. In *Lectins. Biology, Biochemistry, Clinical Biochemistry*, Vol. 3 (T. C Bog-Hansen and G. A. Spengler, eds.). DeGruyter, Berlin/New York, pp. 513–520.

Pull, S. P., Pueppke, S. G., Hymowitz, T., and Orf, J. H. (1978). Soybean lines lacking the 120,000-dalton seed lectin. *Science 200*:1277–1279.

Pusztai, A., Grant, G., and Stewart, J. C. (1981). A new type of *Phaseolus vulgaris* (cv. Pinto III) seed lectin: Isolation and characterization. *Biochem. Biophys. Acta 671*:146–154.

Quinn, J. M., and Etzler, M. E. (1987). Isolation and characterization of a lectin from the roots of *Dolichos biflorus. Arch. Biochem. Biophys. 258*:535–544.

Quinn, J. M., and Etzler, M. E. (1989). In vivo biosynthetic studies of the Dolichos biflorus seed lectin. *Plant Physiol. 91*:1382–1386.

Raikhel, N. V., and Wilkins, T. A. (1987). Isolation and characterization of a cDNA clone encoding wheat germ agglutinin. *Proc. Natl. Acad. Sci. USA 84*:6745–6749.

Ready, M., Wilson, K., Piatek, M., and Robertus, J. D. (1984). Ricin-like plant toxins are evolutionarily related to single-chain ribosome-inhibiting proteins from *Phytolacca. J. Biol. Chem. 259*:15252–15256.

Reeke, G. N., Jr., and Becker, J. W. (1986). Three-dimensional structure of favin: Saccharide binding—cyclic permutation in leguminous lectins. *Science 234*:1108–1111.

Reeke, G. N., Becker, J. W., and Edelman, G. M. (1975). The covalent and three-dimensional structure of concanavalin A. IV. Atomic coordinates, hydrogen bonding and quaternary structure. *J. Biol. Chem. 250*:1525–1547.

Rice, R. H. (1976). Wheat germ agglutinin. Evidence for a genetic basis for multiple forms. *Biochim. Biophys. Acta 444*:175–180.

Richardson, P. T., Gilmartin, P., Colman, A., Roberts, L. M., and Lord, J. M. (1988). Expression of functional ricin B chain in *Xenopus* oocytes. *Biotechnology 6*:565–570.

Riskulov, R. R., Dobrokhotova, Z. D., Kuzev, S. V., Lobsanov, Y. D., Lubnin, M. Y., Mokulskaya, T. D., Myshko, G. E., Proskudina, L. T., Rogacheva, M. M., Saprykina, L. F., Khrenov, A. A., and Mokulskii, M. A. (1984). Three-dimensional structure of lectin from pea (*Pisum satiyum*) at 5 A resolution. *FEBS Lett. 165*:97–101.

Roberts, D. M., and Etzler, M. E. (1984). Development and distribution of a lectin from the stems and leaves of *Dolichos biflorus. Plant Physiol. 76*:879–884.

Roberts, D. D., and Goldstein, I. J. (1982). Hydrophobic binding properties of the lectin from lima beans (*Phaseolus lunatus*). *J. Biol. Chem. 247*:11274–11277.

Roberts, D. D., and Goldstein, I. J. (1983a). Binding of hydrophobic ligands to plant lectins: Titration with arylamidonaphthalenesulfonates. *Arch. Biochem. Biophys. 224*:479–484.

Roberts, D. D., and Goldstein, I. J. (1983b). Binding of adenine and cytokinins to lima bean and other legume lectins. In *Chemical Taxonomy, Molecular Biology, and Function of Plant Lectins* (I. J. Goldstein and M. E. Etzler, eds.). Alan R. Liss, New York, pp. 131–141.

Roberts, L. M., Lamb, F. I., Pappin, D. J. C., and Lord, J. M. (1985). The primary sequence of *Ricinus communis* agglutinin. Comparison with ricin. *J. Biol. Chem. 260*:15682–15686.

Roberts, L. M., and Lord, M. J. (1981). The synthesis of *Ricinus communis* agglutinin. *Eur. J. Biochem. 119*:31–41.

Roberts, D. M., Walker, J., and Etzler, M. E. (1982). A structural comparison of the subunits of the *Dolichos biflorus* lectin by peptide mapping and carboxyl terminal amino acid analysis. *Arch. Biochem. Biophys. 218*:213–219.

Robertus, J. D., Piatek, M., Ferris, R., and Houston, L. (1987). Crystallization of ricin A chain obtained from a cloned gene expressed in *Escherichia coli. J. Biol. Chem. 262*:19–20.

Schlumbaum, A., Mauch, F., Vogeli, U., and Boller, T. (1986). Plant chitinases are potent inhibitors of fungal growth. *Nature 324*:365–367.

Schnell, D. J., Alexander, D. C., Williams, B. G., and Etzler, M. E. (1987). cDNA cloning and in vitro synthesis of the *Dolichos biflorus* seed lectin. *Eur. J. Biochem. 167*:227–231.

Schnell, D. J., and Etzler, M. E. (1987). Primary structure of the *Dolichos biflorus* seed lectin. *J. Biol. Chem. 262*:7220–7225.

Schnell, D. J., and Etzler, M. E. (1988). cDNA cloning, primary structure, and in vitro biosynthesis of the DB58 lectin from *Dolichos biflorus. J. Biol. Chem. 263*:14648–14653.

Spadoro-Tank, J. P., and Etzler, M. E. (1988). Heat shock enhances the synthesis of a lectin-related protein in *Dolichos biflorus* cell suspension cultures. *Plant Physiol. 88*:1131–1135.

Stinissen, H. M., Peumans, W. J., and De Langhe, E. (1984). Abscissic acid promotes lectin biosynthesis in developing and germinating rice embryos. *Plant Cell Rep. 3*:55–59.

Strosberg, A. D., Buffard, D., Lauwereys, M., and Foriers, A. (1986). Legume lectins: A large family of homologous proteins. In *The Lectins. Properties, Functions, and Applications in Biology and Medicine* (I. E. Liener, N. Sharon, and I. J. Goldstein, eds.). Academic Press, New York, pp. 251–264.

Stubbs, M. E., Carver, J. P., and Dunn, R. J. (1986). Production of pea lectin in *Escherichia coli. J. Biol. Chem. 261*:6141–6144.

Tague, B. W., and Chrispeels, M. J. (1987). The plant vacuolar protein, phytohemagglutinin, is transported to the vacuole of transgenic yeast. *J. Cell Biol. 105*:1971–1979.

Talbot, C. F., and Etzler, M. E. (1978). Development and distribution of *Dolichos biflorus* lectin as measured by radioimmunoassay. *Plant Physiol. 61*:847–850.

Triplett, B. A., and Quatrano, R. S. (1982). Timing, localization, and control of wheat germ agglutinin synthesis in developing wheat embryos. *Dev. Biol. 91*:491–496.

Van der Wilden, W., Herman, E. M., and Chrispeels, M. J. (1980). Protein bodies of mung bean cotyledons as autophagic organelles. *Proc. Natl. Acad. Sci. USA 77*:428–432.

Van Driessche, E. (1988). Structure and function of Leguminoseae lectins. In *Advances in Lectin Research*, Vol. 1 (H. Franz, ed.). VEB Verlag Volk und Gesundheit, pp. 73–134.

Villafranca, J. E., and Robertus, J. D. (1981). Ricin B chain is a product of gene duplication. *J. Biol. Chem. 256*:554–556.

Vitale, A., and Chrispeels, M. J. (1984). Transient *N*-acetylglucosamine in the biosynthesis of phytohemagglutinin: Attachment in the Golgi apparatus and removal in protein bodies. *J. Cell Biol. 99*:133–140.

Vitale, A., Ceriotti, A., Bollini, R., and Chrispeels, M. J. (1984). Biosynthesis and processing of phytohemagglutinin in developing bean cotyledons. *Eur. J. Biochem. 141*:97–104.

Vodkin, L. O., and Raikhel, N. V. (1986). Soybean lectin and related proteins in seeds and roots of Le^+ and Le^- soybean varieties. *Plant Physiol. 81*:558–565.

Vodkin, L. O., Rhodes, P. R., and Goldberg, R. B. (1983). A lectin gene insertion has the structural features of a transposable element. *Cell 34*:1023–1031.

Voelker, T. A., Staswick, P., and Chrispeels, M. (1986). Molecular analysis of two phytohemagglutinin genes and their expression in *Phaseolus vulgaris* cv. Pinto, a lectin-deficient cultiva of the bean. *EMBO J. 5*:2075–3082.

Walling, L., Drews, G. N., and Goldberg, R. B. (1986). Transcriptional and post-transcriptional regulation of soybean seed protein mRNA levels. *Proc. Natl. Acad. Sci. USA 83*:2123–2127.

Weber, E., and Neumann, D. (1980). Protein bodies, storage organelles in plant seeds. *Biochem. Physiol. Pflanz. 175*:279–306.

Wright, C. S., Gavilanes, F., and Peterson, D. L. (1984). Primary structure of wheat germ agglutinin isolectin 2. Peptide order deduced from x-ray structure. *Biochemistry 23*:280–287.

Yang, D. C. H., Gall, W. E., and Edelman, G. M. (1974). Rotational correlation time of concanavalin A after interactions with a fluorescent probe. *J. Biol. Chem. 249*:7018–7023.

Youle, R. J., Murray, G. J., and Neville, D. M., Jr. (1981). Studies on the galactose-binding site of ricin and the hybrid toxin Man6P-ricin. *Cell 23*:551–559.

17

Microbial Lectins

Nechama Gilboa–Garber and Nachman Garber *Bar Ilan University, Ramat-Gan, Israel*

I. HISTORICAL OVERVIEW

Stillmark's discovery of a hemagglutinin from seeds of *Ricinus communis* [1], in 1888, is considered as the beginning of lectinology. Microbial lectins were discovered several years later [2–6]. A very detailed and most accurate description of the history of microbial lectinology up to 1975 is included in the highly recommended book of Gold and Balding on receptor specific proteins [2]. In the present chapter only some milestones will be described.

The first report on a fungal hemagglutinin was that of Kobert in 1891 [2]. The first demonstrations of bacterial hemagglutinins, in 1902, are attributed to Kraus and Ludwig [7], who demonstrated hemagglutination of rabbit erythrocytes by culture filtrates of *Staphylococcus aureus* and two *Vibrio* strains, as well as to Flexner, who found weak hemagglutinins in filtrates of *Pseudomonas aeruginosa* ("*Bacillus pyocyaneus*"), *Salmonella typhi* ("*Bacillus typhosus*"), and *Staphylococcus aureus* [2,6]. A year later, Kayser reported hemagglutination by a filtrate of *Escherichia coli*, and Pearce and Winne added *Vibrio cholerae* ("*B. cholerae*") and *Shigella dysenteriae* ("*B. dysenteriae*") to the list [2]. Following these descriptions of bacterial extracellular hemagglutinins, Guyot found, in 1908, cell-bound bacterial hemagglutination in *E. coli* [8], and Ford found thermostable and thermolabile hemagglutinins in extracts of the edible mushroom *Amanita solitaria* and in several species of the family Agaricaceae [2].

During the next years, many additional hemagglutinins were shown in various enterobacteria, and in extracts of many fungi [2]. The hemagglutinins of both fungi and bacteria were reported to be selective for erythrocytes of different animals and other cell types [2]. Rosenthal, who found that *E. coli* hemagglutinin agglutinated leukocytes, thrombocytes, spermatozoa, yeasts, spores of molds, and pollen [9,10], drew attention to two very important aspects: the possibility that this sperm-agglutinating hemagglutinin might cause sterility, and the dependence of the hemagglutinating activity on the bacterial culture conditions [9,10].

In 1941, Hirst [11] and McClelland and Hare [12], independently, described hemagglutination by influenza A and B viruses. On the basis of these observations, in 1944, Salk [cited in 13] devised an important simple and rapid method for the detection and quantitation of influenza virus particles.

Hemagglutinins of *Haemophilus* were described in filtrates and intact cells by Keogh et al. in 1947 [14], and they were purified by Warburton and Fisher in 1951 [15]. This was probably the first report on a bacterial hemagglutinin purification [2].

In 1948, Lamanna [16] found that toxic culture filtrates of *Clostridium botulinum* contained several hemagglutinins (A–E), that agglutinated red blood cells of different animals and humans. Nineteen years later, Das Gupta and Boroff, separated the type A botulinum hemagglutinin from the neurotoxin [2].

In 1951, Elo et al. [17], being acquainted with the finding of the blood group-specific pytohemagglutinins, described hemagglutinating fungal extracts with ABH specificity: anti-(A+B)>O in *Hygrophorus hypothejus* and *Psilocybe spadicea*; anti-B in *Marasmius oreades*, and anti-H (A2 and O > A1 and B) in *Amanita muscaria* [2,17]. Many additional saline extracts of mushrooms were then examined by various investigators, and some of them were also shown to have a potential for blood groups A, B, H, and I detection [2].

In 1954, Gottschalk showed that hemagglutinating influenza viruses could be detached from the erythrocytes by α-neuraminidase, leaving them deprived of the receptors for the viruses [18]. This finding indicated that sialic acid might be involved in the erythrocyte receptors for the virus.

The numerous observations on the microbial hemagglutinins [19] were not correlated with those of the plants [20]. Even when, in 1954, Boyd and Shapleigh [21] coined the term *lectins* for the latter, the microbial hemagglutinins were not included in it. Nevertheless, Krüpe, in 1956 [22], and Mäkelä, in 1957 [23], reviewed the works on the microbial hemagglutinins together with those on the phytohemagglutinins.

In 1955, the first report on inhibition of bacterial hemagglutinin by a simple sugar was presented by Collier and de Miranda [24], who showed that D-mannose (Man) strongly inhibited the hemagglutinating activity of *E. coli*. The same year was also blessed by the very important observation of Duguid et al. [25], that in various enterobacteria, including *E. coli*, there are hemagglutinins that are intimately associated with filaments called *fimbriae*. Following the observation of Collier and de Miranda [24], Duguid and Gillies, in 1957 [26], showed that the fimbrial hemagglutinins in dysentery bacilli and *E. coli* strains were also inhibited by mannose and by α-methylmannose. They noted the correlation between the adhesive action of fimbrial "adhesins" to various substances and their hemagglutinating activity. Moreover, they stated that: "The occurrence of hemagglutinating activity in many pathogenic bacteria suggests that the pertinent adhesive properties may have their natural functions in fixing the bacteria, or their toxins, to the surfaces of host cells" [26]. Duguid also distinguished between mannose-sensitive (MS), and mannose-resistant (MR) adhesins [27]. In 1966, Duguid et al. found that most *Salmonella* strains possess type 1, MS fimbriae, resembling those of *E. coli*, and that these bacteria adhere to leukocytes, intestinal epithelial cells, and unicellular fungi [28]. Brinton, who used the term *pili for* fimbriae, added important information on their synthesis, expression, purification, and physicochemical properties and showed "phase variation" in their production, depending upon the cultural growth conditions [29,30]. Later on, Old and Duguid also expanded their investigations on the function and genetic aspects of the bacterial fimbriae [31].

In 1957, Gold and Ilian [cited in 2] found that sialidase-treated pan(poly)agglutinable erythrocytes were more sensitive to hemagglutination by the type A hemagglutinin of *C. botulinum* than untreated or papainized red blood cells. Collee [32] showed a similar phenomenon with *C. welchii*. He demonstrated that this hemagglutinin was present both in the supernatant fluid and the bacillary deposit and was active over wide pH and temperature ranges. The intracellular hemagglutinin could be liberated by ultrasonic disintegration. This liberated hemagglutinin was assumed to be multivalent, but was not inhibitable (as also *C. botulinum* C and D hemagglutinins [2]) by carbohydrates.

In 1961, Kathan et al. [33] showed that sialic acid-dependent viral hemagglutination was inhibited by glycophorin (the major sialic acid-containing protein of human erythrocytes, which was isolated in soluble form and exhibited M and N blood group activity). The identity of influenza A hemagglutinin was established in 1969 by Laver and Valentine [34]. They separated two types of solubilized viral glycoproteins by electrophoresis and showed that one of them is the hemagglutinin, and the other is the sialidase [34,35].

In 1964, Heumann and Marx showed inhibition by mannose of hemagglutination caused by intact cells of *Pseudomonas echinoides* [36]. Eleven years earlier, Drimmer-Herrenheiser [37] described a pseudomonal hemagglutinin, that was undissociable from the bacterial cell surface and was highly dependent on the cultural conditions, but exhibited no sugar or blood group specificity.

In 1969, Sage and Connett [38] described the first purification of a mushroom hemagglutinin of *Agaricus campestris*, using DEAE-cellulose, phosphocellulose, and hydroxyapatite columns. The purified hemagglutinin was shown to be a 64-kd glycoprotein containing cysteine and 4% carbohydrate, but no inhibition by sugars was observed [2,38]. A year later, Pardoe et al. described inhibition by α-D-galactose (Gal) of the fungal anti-B hemagglutinin of *Polyporus fomentarius* [2].

The years 1972–1975 were very fruitful in yielding additional strong indications for the very close connections between the microbial hemagglutinins [39] and the plant lectins, leading to their inclusion with, and recognition as lectins.

During these years a group of bacterial hemagglutinins, exhibiting specificity for galactose or *N*-acetyl-D-galactosamine (GalNAc) and their derivatives were described and isolated as pure proteins. In 1972, Gilboa-Garber [40,41] described the purification and characterization of a galactophilic lectin from extracts of the bacterium *P. aeruginosa* (later designated PA-I). This was the first report on inhibition of a bacterial hemagglutinin by galactose (only mannose-sensitive bacterial hemagglutinins were known up to that time). This lectin resembled plant lectins in exhibiting a selective hemagglutination, sugar specificity, relative resistance to heating and to proteolysis, and dependence on divalent cations [40,41]. A year later, galactose sensitivity was also shown in the hemagglutinin of a *Streptomyces* sp., isolated by Fujita et al. [42,43] and in the hemagglutinin of *C. botulinum* type A, by Balding et al. [44]. In mushrooms, Presant and Kornfeld [45] showed specificity for β-D-Gal-1-3-D-GalNAc-serine in *Agaricus bisporus* hemagglutinin, Coulet et al. [2] demonstrated galactose-binding in the hemagglutinin of *Inocybe fastigiata*, Anstee [46] confirmed the galactose-specificity of *Fomes fomentarius*, and Fujita et al. described arabinose–xylose specificity in the hemagglutinin of *Aspergillus niger* [47]. Concurrently, the galactose and *N*-acetylgalactosamine specificities of the slime mold, *Dictyostelium*, hemagglutinins were also reported by Rosen et al. and Barondes [48–50].

L-Fucose (Fuc)-specific hemagglutinins were described by Coulet et al. in the fungi *Laccaria laccata* and *Pholiota squarrosa* [2]. Later on, Fuc/Man-binding lectins were shown in bacteria: in extracts of *P. aeruginosa* (designated PA-II) by Gilboa-Garber et al. [52,53], and in filtrates of *Streptomyces* by Kameyama et al. [54,55].

Hence, the sugar variability of the microbial hemagglutinins was already approaching that of the plant lectins. They also fulfilled the anticipation of Collier and Jacoeb [56] that bacterial hemagglutinins, reacting with receptors of various cell types, might be used as a tool for the study of cell receptors, as was shown by Fujita et al. with anti-B blood group-specific *Streptomyces* hemagglutinin [57].

In 1972, Gilboa-Garber et al. described the first purification of a microbial lectin (PA-I of *P. aeruginosa*) by affinity chromatography [58–60], following the introduction of this method for plant lectin purification by Agrawal and Goldstein, in 1965 [61]. A year later, affinity chromatography was also used by Cuatrecasas (one of the inventors of this important method) for the purification of *Vibrio cholerae* lectinoid exoenterotoxin [62]. Later on, the anti-B–specific lectin of *Streptomyces* sp. [43] and the Fuc/Man-binding lectins of *P. aeruginosa* (PA-II) [52,53] and *Streptomyces* sp. [54,55] were also purified by affinity chromatography.

In 1973, Gilboa-Garber and Mizrahi showed the binding of horseradish peroxidase to the pseudomonal PA-II lectin [51]. PA-II–Sepharose was shown to be as useful as concanavalin A (Con A)-Sepharose, for purification of this and other glycosylated enzymes [60]. The two *P. aeruginosa* lectins were further shown to be active, as plant lectins, in inducing mitogenic stimulation of human peripheral lymphocytes [63,64] and murine splenocytes [64].

The state of art in prokaryote lectinology in 1975 was described by Gold and Balding:

> Although haemagglutination by bacteria was first reported at the beginning of the twentieth century, progress was slow and biochemical research in this field is still in its infancy. The known bacterial haemagglutinins possess all the properties characteristic of "Receptor specific proteins." So far only *Cl. botulinum, P. aeruginosa, E. coli*, and *Streptomyces* haemagglutinins have, to our knowledge, been purified, or partially purified [2].

A year later, Liener [3] included the microbial hemagglutinins in his review on plant lectins. However, until 1978, the microbial lectins wide prevalence and importance had not yet been fully realized. Lectins were still described as "proteins found primarily in plants" [65] or as "carbohydrate-binding proteins of plants and animals" [4]. However, in 1978, Goldstein and Hayes stated that

> The discovery of hemagglutinins in such diverse sources as bacteria, fungi, lichens, fish roe, snails, eels and even mammals, has resulted in a broadening of the term 'lectin' to include carbohydrate-binding proteins without regard to their origin [4].

The recognition of the microbial hemagglutinins as lectins had an important impact: plant lectinologists, microbiologists, biochemists, and investigators in related fields became aware of the wide distribution and major importance (scientific as well as medical and industrial) of this new source of lectins [39,66]. This was further extended by the division of microbial hemagglutinins into those that fit and those that contravene Goldstein's et al. [67] definition of *lectins*—sugar-binding proteins of nonimmune origin, which agglutinate cells and/or precipitate glycoconjugates—which has been adopted by the International Union of Biochemistry [68]. This definition implies that a lectin must be divalent or multivalent. It excludes sugar-binding proteins possessing only one active site, which are still considered lectins by Kocourek and Hořejši [69], as well as hemagglutinating proteins not shown to be inhibited by simple sugars. Therefore, the term *adhesins* [26–28,70–84] has been used side by side with other restricting terms such as *hemagglutinins* [70,85–89], *lectin-like agglutinins, bacterial (surface) lectins* [90–92], and *fimbrial lectins* [93]. In this chapter the term *lectinoids* will be used for the lectinlike molecules that do not fulfill the prerequisites of multivalence or sugar specificity, but exhibit specific and reversible binding sites for cellular receptors and macromolecular products, home and bind to them, and affect their destiny.

The great potential of the microbial lectins is only now beginning to be fully realized, and the gap of knowledge between them and the plant lectins is steadily diminishing. It is now clear that the spectrum of sugar detection by the microbial "true lectins" and "lectinoids" and their sensitivity to subtle structural differences in glycoconjugates are even broader than those of the plant lectins (see Sects. II and V). They may be more extensively studied genetically (see Sect. III) than plant and animal lectins, owing to the minimal volume of maximal populations, the short generation time, the easier control of culture conditions, the detailed knowledge of the defined bacterial chromosome and plasmids, and the easier procedures for mutagenesis and for mutant detection and selection. Moreover, it is possible to clone the genes that encode microbial lectin production into microbial strains and higher organisms that lack them, and to clone plant and animal genes that encode lectin production into bacteria. Thus "neo," "adopted microbial lectins" are also produced and may be investigated and used for various purposes (see Sect. V).

The investigation of the physiological roles of the microbial lectins and lectinoids opens new fascinating trends not hitherto known in animal and plant lectinology (see Sects. IV and V). Included in these are the phenomenon of sugar-dependent microbial aggregation in pure populations and heteropopulations, the microbial adhesion for settlement in the host and establishment of either a symbiotic or parasitic relationship, the homing and the conditioning of the microbial toxin function on the target cells, and the function of microbial lectins and lectinoids as a trap, exposing the microorganisms to the host defence mechanisms.

Finally, the microbial lectin applications are numerous and multidisciplinary: scientific, agricultural, industrial, and medical. The induction of production of specific antibodies against them may also protect the animals, as a vaccine, against infection (see Sect. V).

II. CHARACTERIZATION, PURIFICATION, AND CARBOHYDRATE-BINDING SPECIFICITY OF MICROBIAL LECTINS AND LECTINOIDS

Most of the microbial lectins and lectinoids were discovered by hemagglutination tests. Their sugar specificity was shown and identified by hemagglutination inhibition with a series of simple and variously linked sugar derivatives. Later on, more refined procedures were introduced, using synthetic glycoconjugates and inert carriers that bear glycoconjugates, equilibrium dialysis techniques, as well as more accurate physicochemical measurements of the interactions of the lectins with the sugars. The location of the microbial lectins and adhesins is highly variable: many are associated with fimbriae, flagella, and the cell surface, or present in the periplasmic space (on the inner membrane) or, even more internally, whereas others are found in the microbial culture filtrate. There are examples for which the same lectin may be distributed in several compartments of the microorganism [60]. Surface-located lectins may sometimes be released into the medium by a mild treatment [75,76], periplasmic located ones may be unmasked by removing the outer membrane (conversion to spheroplasts or protoplasts) [94,95], and intrinsic lectins may be liberated by the disruption of the cells [53]. Moreover, some microorganisms produce several lectin types [60] and additional adhesins (e.g., hydrophobic hemagglutinins) [60,96–100]. Many of the microbial lectins have been purified by use of ammonium sulfate precipitation, heating (for removal of nonlectin heat-sensitive proteins), molecular sieves, charged polymers, and affinity chromatography (using insoluble, natural or chemically linked sugar-bearing supports, such as Sepharose, agarose, Sephadex and latex beads) as an essential step [53]. The relative molecular mass (M_r) of microbial lectins ranges from 11 kd, for the monomeric blood group B-specific, galactose-binding lectin of *Streptomyces* sp. [43], which contains two sugar-binding sites [101], or the subunit of the Fuc/Man-binding PA-II lectin of *P. aeruginosa* [52], up to 800 kd for *Streptomyces* SFL 100-2 Fuc/Man-binding particle, which consists of 12 identical subunits (each of M_r 68 kd) [102,103] and possesses only two sugar-binding sites per particle [103]. The microbial lectins, like most of the plant lectins [4], are generally poor in sulfur-containing amino acids and are relatively rich in acidic and hydroxylic amino acids [52,60]. They react with distinct carbohydrate groups, some of which are detected and others not hitherto known to be detected by plant and animal lectins. There is a great variation in the specificities of the hemagglutinins obtained from different microbial species and strains of the same species [39,66]. This property contributes to their application for elucidation of previously unknown special spatial conformations of sugars and sugar interrelations. Like the plant lectins [4,68], the microbial lectins may be divided to those that can be inhibited by simple monosaccharides and to those specific for more complicated saccharides. Some of them exhibit preference for hydrophobic derivatives of the sugars (probably owing to hydrophobic clefts near the sugar-binding site). Among the microbial lectins, some react with sugars present on most animal and microbial cells and other are more selective. The latter react with only a few animal cells and detect human blood group antigens such as the anti-B of *Streptomyces* [42,43] and the anti-P [82] and anti-M(N) of *E. coli* strains [88,89]. In this chapter (text and Table 1), only a few examples will be presented. More information may be found in Mirelman's book [39] and Sharon's review [66] on microbial lectins and agglutinins. For up-to-date information the readers should follow the current literature in which the very rapid development in this subject is widely covered.

Duguid et al. [25] described fimbriae-associated hemagglutinins in 31 and somatic (surface-associated) hemagglutinins in 5 of 47 *E. coli* strains. They realized the relation between the adhesive property and the hemagglutinating activity and proved that type 1 fimbrial hemagglutinin was inhibited by mannose and methyl-α-mannoside [26,28,70], as a soluble analogue competing with the cell surface residues. The modification of the hydroxyl groups at C-2, C-3, C-4, and C-6 positions in the mannose molecule caused loss of inhibitory activity. Similar reversibility of *E. coli* fimbrial adhesion was later shown by Salit and Gotschlich, who used purified fimbriae with monkey kidney cells [104,105], and by Ofek et al., who used *E. coli* cells with epithelial cells [106–108]. The type 1 fimbriae are distributed on the bacterial cells (100–400/cell) peritrichously. Their length varies form 0.2 to 1 μm, and their width is 7 nm. Brinton [29,30] purified them and showed that they consist of protein subunits (*pilin*, M_r 16.6 kd) composed of 163 amino acids (about 50% of them hydrophilic), allied in a right-handed helix, 7 nm in diameter, with a hollow core of 2–2.5 nm. The subunits are held together by hydrogen and hydrophobic bonds. They tend to aggregate into bundles and crystals. Klemm presented the nucleotide sequence of the *fim A* gene [109] with data according those of Brinton's amino acid analysis [66]. Firon et al. [110–112] examined many branched mannosylated oligosaccharides and showed that

Table 1 Examples of Some Simple Sugar Specificities of Microbial Lectins and Lectinoids

Sugar specificity	Microbial lectins and lectinoids
D-Mannose (without L-fucose)	Type 1 fimbrial and non fimbrial adhesins of E. coli and many other enteric bacteria *Corynebacterium parvum* Japanese encephalitis virus
L-Fucose, frequently together with L-galactose and D-mannose	*Streptomyces* spp. *Pseudomonas aeruginosa* PA-II *Vibrio cholerae* *Aleuria aurantia* (fungus) *Laccaria amethystina* (fungus) *Rhizoctonia solani* (Fungus)
D-Galactose frequently together with related sugars (e.g., D-fucose, L-rhamnose, lactose and *N*-acetyl-D-galactosamine)	*Pseudomonas aeruginosa* PA-I *Streptomyces* sp. (anti-B) *Actinomyces viscosus* *Actinomyces naeslundii* *Fusobacterium nucleatum* *Streptococcus mutans* *Dictyostelium purpureum* (slime mold) *Fomes fomentarius* (fungus) *Agaricus bisporus* (fungus)
D-Galactose-α1-4-D-galactose *N*-Acetyl-D-galactosamine	*Escherichia coli* P fimbriae *Shigella dysenteriae* (Shiga) toxin *Leptotrichia buccalis* *Dictyostelium* strains (slime molds) *Clitocybe nebularis* (fungus) *Entamoeba histolytica* (protozoon)
Arabinose > D-xylose D-Glucose Dextran (*N*-Acetyl-D-glucosamine)$_n$	*Aspergillus niger* (fungus) *Bartonella bacilliformis* *Streptococcus mutans* *Bordetella bronchiseptica* *Chlamydia trachomatis* *Pasteurella multocida* *Conidiobolus lamprauges* (fungus) *Entamoeba histolytica* (protozoon)
Sialic acids and their derivatives	Influenza viruses Polyoma viruses Sendai virus *Mycoplasma pneumoniae* *Mycoplasma gallisepticum* *Escherichia coli* *Streptococcus sanguis* *Pseudomonas aeruginosa*

For references and detailed information the readers are referred to Ref. 98.

Manα6[Manα3]Manα6[Manα3]ManαOMe, and Manα6[Manα3]Manα6[Manα2 Manα3]ManαOMe, the trisaccharide Manα3Manβ4GlcNAc, and the aromatic glycoside p-nitrophenyl-α-Man are strong inhibitors of yeast agglutination by type 1 fimbriae of *E. coli* 346 [66]. The pattern of inhibition of this bacterial hemagglutinin is different from that of the mannose-binding plant lectins of the Con A–lentil–pea group [4]. It is not inhibited by glucosides and exhibits lower affinity to Manα2Man than to methylαMan. It differs also from the mannose-binding lectins of *P. aeruginosa* PA-II [52,53,60] and *Streptomyces* [54,55], which exhibit a higher specificity for fucose than for mannose. The binding to mannose on mammalian cell surface was suggested to be a binding to glycoproteins, since mannose has not yet been found in mammalian glycolipids [66]. Later reports also described type 1 fimbriae of *E. coli* and other enterobacteria as polymers of polypeptide subunits (M_r 17 kd in *E. coli* and somewhat larger in the others), in which the MS hemagglutination is affected by adhesive components, distinct from the major structural fimbrial subunit, which is generally associated with the tip, but may also be distributed on other parts of the fimbriae [113]. In addition to the MS fimbrial (F) and nonfimbrial (NF) *E. coli* adhesins, there are several other lectinoid adhesins of *E. coli* strains [85,86,114–121]. The most widely investigated are those that bind digalactoside [72,83,115–118] and sialylated residues [119–125]. Galα1-4Gal was shown to bind to the P or Pap fimbriae of uropathogenic *E. coli* strains [72,73,83,115–118]. Examination of one such strain, with a thin-layer chromatogram binding assay, against 50 glycolipids, indicated that it bound to all species containing Galα1-4Gal in terminal or internal position, but not to any other glycolipid [72,83]. Molecular models showed a bend ("knee") in the saccharide chain at Galα1-4Gal with substituents projecting toward the concave side. Therefore, the convex side of the knee was suggested as the binding epitope (the exact structural part of the receptor necessary for the binding), which is partially hydrophobic [83,126]. The Shiga toxin is also specific for Galα1-4Gal sequence, but is more selective in relation to substituents, indicating that uropathogenic *E. coli* adhesin and Shiga toxin recognize different epitopes of this disaccharide [73,83]. The Galα1-4Gal sequence is the dominant sugar in the antigenicity of human blood group P, and agglutination assays confirmed the blood group P specificity of the P or Pap fimbrial adhesin. Treatment of human erythrocytes with sialidase did not diminish the hemagglutination with uropathogenic *E. coli* adhesin (in contrast with the sialic acid-binding S–fimbrial adhesin), whereas treatment with α-galactosidase diminished it. The *E. coli* pili of types 1, Pap, and K99 share common structural characteristics [73] (K99 [121,124] reacts with sialic acid-containing ceramide). Most *E. coli* pilins are approximately 14–22 kd in size, with a hydrophobic NH_2-terminus, cysteine loop in the NH_2-terminal half of the protein, and a penultimate tyrosine residue at the COOH-terminus. Normark et al. [73] concluded that

> In no case is the structural basis for the interaction between a bacterial adhesin and its receptor known. It is likely that the binding site will turn out to be hydrophobic clefts in the molecules as has been found in other adhesins [127], although it is expected that specific hydrogen binding occurs between amino acids and hydroxyl groups in the carbohydrate.

Binding may be created by critical amino acid residues that are discontinuous, rather than continuous in sequence, but brought close together in space.

Sialic acid-binding hemagglutinins were described in influenza viruses [34,35] before their demonstration in bacteria, such as *E. coli* strains, *Mycoplasma* [82,128–132],

P. aeruginosa [133,134], and *Bordetella bronchiseptica* [135], or in parasitic protozoa [100]. There are several types of sialophilic adhesins among *E. coli* fimbrial (e.g., S-fimbriae and M-fimbriae) hemagglutinins (which were previously named X-fimbriae), and among the NF adhesins. Parkkinen et al. described sialophilic adhesins [123] that bind NeuAc-α-2-3Gal [136]. Some of the sialophilic adhesins are associated with human extra-intestinal infections. They also interact with *N*-acetylglucosamine and blood group M (M-fimbriae) and N antigens. Their hemagglutinating activity is abolished by sialidase treatment of the red blood cells and inhibited by compounds such as poly-NeuAc-α-2-8NeuAc or by *N*-acetylgalactosamine and sialic acid (NeuAc). Glycophorin, which is also inhibitory [123], was suggested as the erythrocyte receptor. It has been shown that a G_{M2} gangliosidelike glycoconjugate acts as a receptor for *E. coli* CFA 1 (colonization factor antigen) and K99 adhesins. The major glycolipid of horse erythrocytes, NeuGcα2-3Galβ1-4Glc ceramide was also shown to act as a receptor for *E. coli* K99 [124]. It inhibited hemagglutination and could induce hemagglutination by coating of otherwise unreactive guinea pig erythrocytes. The CFA/I pilin is 14 kd and lacks the other common features described for type 1, Pap, and K99 pili. However, in hydrophobicity, this protein resembles other *E. coli* pilins.

The sialophilic adhesins of *Mycoplasma* spp.: *M. pneumoniae, M. gallisepticum*, and *M. genitalium*, are membrane proteins [82,128–132]. The major adhesin of *M. pneumoniae* P1 (M_r 190 kd) is concentrated mostly at the organelle, at the tip of the microorganisms, that is involved in the attachment to and invasion to the host epithelial cells [81,128–132].

The influenza virus sialophilic hemagglutinin (HA) is a glycoprotein inserted in and extended to 135 Å off the base of the viral surface membrane [13,35,137]. It is composed of two types of polypeptide chains: HA1 (36.334 kd) and HA2 (25.750 kd). These two chains are covalently attached by a sulfide bond, and three such complex subunits coil to form a trimer of M_r 220 kd [13,35,137]. The distal region of the HA is formed by HA1 and contains the receptor-binding site and the antigenically important region of the molecule [35,137]. Bromelin treatment of the virus yields the ectodomain containing all of HA1 and most of HA2, but missing the HA2 hydrophobic part, which anchors the hemagglutinin to the membrane. The released ectodomain's three-dimensional structure was determined by x-ray diffraction [13,35,137].

Although the fimbrial and surface microbial adhesins are not always independent hemagglutinating units (the hemagglutination may be induced by cooperative function of several adhesins present on the microbial surface), the lectins of *P. aeruginosa* [40,41,60] and *Streptomyces* spp. and of several additional microorganisms are unequivocally "true lectins."

The galactophilic lectin, PA-I, of *P. aeruginosa* ATCC 33347 [40,41] was purified by Gilboa-Garber et al., in 1972, using Sepharose affinity chromatography [58]. Its hemagglutinating activity is much stronger after papain or sialidase treatment of the human and other animal erythrocytes [40,41]. It also agglutinates other cell types: leukocytes, thrombocytes [41], bacteria [138], protozoa, and spermatozoa, and interacts with epithelial cells [60]. Its sugar specificity is melibiose > methyl-α-Gal > Gal > methyl-β-Gal > GalNAc. Raffinose is also inhibitory [52,60] and rhamnose is a weak inhibitor (in contrast with the *Streptomyces* galactophilic lectin). This lectin (subunit M_r about 12 kd) resembles oligomeric plant lectins in its reactions with sugar-bearing complexes and in physicochemical properties. It is relatively resistant to heating (up to 70°C), to proteolysis, and to extreme pH values (ranging from 4 to 11), as shown in *E. coli* O_{86} agglutination test [138,139]. Its activity depends on divalent cations [41]. The purified lectin amino acid composition [52] resembles that of regular plant lectins, being richer in aspartic and glutamic acids, followed by glycine, alanine, valine, leucine, threonine, serine, and isoleucine. It contains one cysteine and methionine per subunit. A positive reaction for sugars was obtained with phenol–sulfuric acid or anthrone reagents, and glucosamine was found in the amino acid analyses [52]. The lectin is sensitive to strong oxidizing treatments and sulfhydryl reagents [59]. Galactose and dithiothreitol (DTT) protect it from inactivation by them [59,60], indicating that the sensitive amino acid (which may be the cysteine) plays an essential role in the lectin activity [59,60].

The second *P. aeruginosa* lectin, PA-II, which exhibits Fuc/Man specificity, was discovered in cultures of the same *P. aeruginosa* isolate, subcultured under different conditions. It was first shown to be mannose-sensitive [52,53,60], as is the MS hemagglutinins of *E. coli*. However, in contrast with the latter, methyl-α-Man was a weaker inhibitor than mannose [52]. Later on, it was found to exhibit Fuc >> Man-type specificity [53,60]. This lectin may be easily purified by Man-Sepharose affinity chromatography [53]. It is more resistant than PA-I to heating, proteolysis, and a wide pH range, as shown by *E. coli* O_{128} agglutination test [138,139]. The subunit M_r and amino acid composition (but not their sequence) of PA-II is similar, in essence, to that of PA-I [52]. Localization studies indicated that the pseudomonal lectins are present mainly intracellularly, but also in the periplasm (attached to the inner membrane) and on the outer membrane [95]. The myxobacterial hemagglutinin was also localized in the periplasmic space and on the cell surface [94]. Environmental factors affect the equilibrium of the pseudomonal lectin compartmental distribution and induce Pa-I–PA-II phase variation. The externally exposed lectins may be shown by the use of rabbit antibodies against the purified *P. aeruginosa* lectin preparations (Fig. 1). These antibodies agglutinate intact *P. aeruginosa* cells, which produce the lectins, as well as homologous and some heterologous serotypes [140,141]. The presence of surface mannose-sensitive hemagglutinin was also shown by Speert et al. [142] and that of the galactose-sensitive one, was shown by Gilboa-Garber et al. [143], who found that intact cells of *P. aeruginosa* which reacted with galactose residues of both B and P blood group antigens, were inhibited by galactose [143]. It must be stated as a warrant that the presence of a mixture of sugar (Gal, Fuc/Man, and sialic)-binding and hydrophobic hemagglutinins on the surface of *P. aeruginosa* cells may complicate the identification of individual activities [144]. However, carbohydrates could still inhibit *P. aeruginosa* adhesion [143]. Recently, Ahmed et al. have examined 19 defined *P. aeruginosa* strains [145–147] and found lectins of the types PA-I and PA-II, as described by Gilboa-Garber et al., exhibiting the following order or affinities for galactose derivatives: methyl-α-Gal > *O*-nitrophenyl-α-Gal, melibiose > Gal > *p*-nitrophenyl-α-Gal, methyl-β-Gal > GalNAc > *O*-nitrophenyl-β-Gal [147]. They reported that PA-I is a glycoprotein (13.8% carbohydrates) of M_r 57 kd, composed of four identical subunits of M_r 14.2–14.5 kd. According to the results of fluorescence quenching, using 4-methylumbelliferyl-α-D-Gal, it possesses four identical carbohydrate-binding sites [147]. Their PA-II-type lectin also exhibits Fuc/Man specificity. They found it to be a very small glycoprotein (12.2% carbohydrate) present as a monomer of M_r 10.7 kd [147]. Recently, Garber et al. [148], using equilibrium dialysis and hemagglutination inhibition tests, have shown that PA-II exhibits the highest affinity for *p*-nitrophenyl-α-Fuc, followed by fucose > L-fucosylamine, L-galactose >> Man >>> D-fructose. The association constant, K_a of fucose was 1.5×10^6 M^{-1} and there was one fucose-binding site per subunit. The results obtained using equilibrium dialysis correlated well with those of hemagglutination inhibition tests for the fucose derivatives of

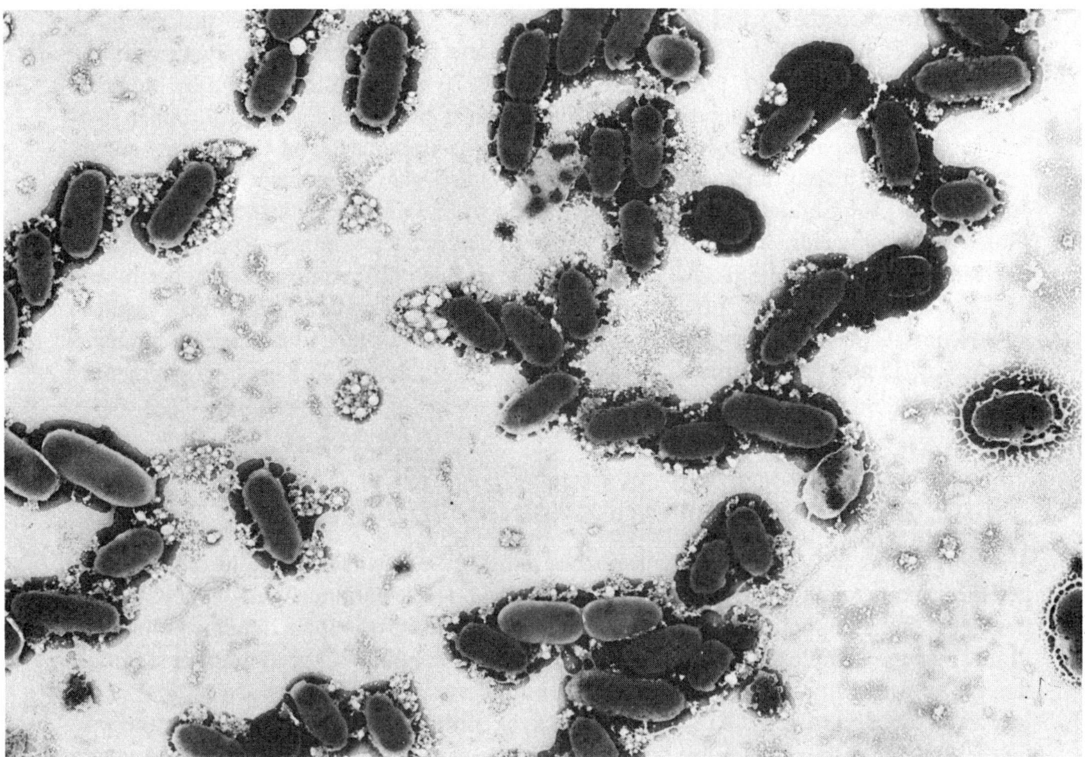

Figure 1 Agglutination of *P. aeruginosa* cells by rabbit antiserum produced against its purified lectin preparation (a mixture of PA-I and PA-II). (×8570 microscopic magnification.)

L-galactose, but not for mannose (which may indicate the possibility that fucose and mannose bind to different sites).

The galactophilic lectin of the *Streptomyces* sp., discovered by Fujita et al. [42,57], is distinguished by its antihuman blood group B specificity (agglutination titer for B erythrocytes 256 times higher than for A or O erythrocytes). Its blood group B specificity was ascribed to a higher affinity (60-fold) for galactose (the immunodominant sugar of B antigen) than for *N*-acetylgalactosamine (the immunodominant sugar of A antigen), indicating a particular specificity for the hydroxyl group at C-2 [42,43]. This lectin, which is insensitive to the sugar anomeric configuration, exhibits the following order of sugar affinity: L-rhamnose (6-deoxy-L-mannose) > methyl-β-Gal > methyl-α-Gal; Gal > L-arabinose > D-fucose. It was purified by Fujita et al. in 1975 [43], using insolubilized gum arabic and elution with 1 M NaCl, containing 1 M galactose. Its M_r is 11 kd. It contains large amounts of alanine, glycine, and valine (47% of the total amino acid residues), 1.8% neutral sugar, and no phenylalanine [43]. The lectin-dependent precipitation of gum arabic is observed at a pH range of 4–12 [43], and its binding to specific sugars induces a difference in the ultraviolet spectrum, indicating that tryptophan residues are clearly involved [101].

The *Streptomyces* sp. Fuc/Man-specific lectin (SFL), which was described by Kameyama et al., in 1979, was purified from a culture filtrate of a *Streptomyces* strain by affinity chromatography [54]. It is a monomer of M_r 70 kd, exhibiting the following order of sugar affinity: *p*-nitrophenyl-β-Fuc > Fuc > *p*-nitrophenyl-α-Fuc > *p*-nitrophenyl-α-Man > Man, D-tagatose > D-lyxose > D-fructose > D-arabinose. The O erythrocytes are agglutinated by it five times more strongly than B and nine times more strongly than A erythrocytes [54]. Similar lectins were found to be widely distributed among *Streptomyces* strains [55]. However, a special strain of *Streptomyces* was found to produce special Fuc/Man-specific lectin SFL no. 100-2, having a rather different sugar specificity and outstandingly high M_r (800 kd; composed of 12 identical subunits of M_r 68 kd). Electron microscopic observations showed that the lectin was associated with particles having a diameter of 25 nm with twofold rotational symmetry (four arms) [102] and possessing only two sugar-binding sites per particle [103].

III. BIOSYNTHESIS OF MICROBIAL LECTINS AND LECTINOIDS, ITS GENETIC DETERMINATION AND PHASE VARIATION

Microorganisms are of great advantage for studies on lectin and lectinoid biosynthesis, genetics, and regulation by environmental conditions (phase variation). Their short generation time, minimal population volume (unlimited number of replications), defined and limited genetic complex, and highly controlled growth conditions, are all of utmost importance. The possibility to induce mutations in one or several genes involved in coding for the lectins or adhesins, and the structures that bear or control them, contribute enormously to this field. Moreover, the possibilities of cloning these genes in defined recipient bacteria and using various genetic manipulations are a powerful new generation of tools for the exploration of lectin and lectinoid biosynthesis and its control, as well as for understanding their structure, properties, and physiological roles. The most detailed genetic studies were performed with the *E. coli* adhesins (described in the previous section).

The fact that the adhesive properties, associated with the MS fimbriae or pili, are determined genetically and are subjected to phase variation was reported soon after their discovery [30]. During the last several years, there has been an important breakthrough in the study of the genetic determinants coding for various *E. coli* fimbrial and nonfimbrial (afimbrial) adhesins [72–77,109,149–176]. They have been characterized on the molecular level with mutants and cloning techniques using plasmids carrying part of or the entire gene clusters for transformation to *E. coli* K-12 strain or to other strains and deficient mutants. DNA probes carrying the gene encoding the fimbrial subunit were used by Buchanan et al. [149] to detect homologous sequences of DNA from different enterobacteria, and it was found that the sequences encoding type 1 fimbriae within the family were not conserved [149].

DNA probes specific for different regions of the adhesive determinants (for hybridizations with DNA sequences coding for the other adhesive types) were constructed [172,173]. The pili–adhesin complex is usually encoded by a large chromosomal gene cluster or large conjugative plasmids. The genes that specify type 1 and P or Pap-type pili [72,73,150,152,155–169] have been cloned and mapped to distinct loci on the *E. coli* chromosome: type 1 pili at 98 min [30] and Pap pili at 63 min [72,73]. The genetic determination of adhesive function in type 1 and P fimbriae of *E. coli* is distinct from the gene encoding the fimbrial subunit [152,161,162]. Different regions in the gene clusters are involved in coding the following processes: (1) the production of the pilin or fimbrillin (the fimbriae or pili protein) subunit; (2) the biogenesis of the fimbriae; (3) the production of the associated adhesive unit of type 1 [109,150,152,161,162], type P [72,73,158–160, 165–167,169], type S [171–174], or C (MR type 1); (4) the control of the transcription; and (5) the production of the NH_2-terminal part of the fimbrillin subunit [168,170]. The operon coding for the major pilin gene and allied genes are separated from the gene encoding the hemagglutinating unit (adhesin). Bacteria deleted for the major pilin gene may still mediate MS- and P-specific hemagglutination [72,73,165,169], whereas inactivation of genes outside the major pilin gene and the allied genes, produce piliated bacteria that are negative for sugar binding [72,73,158,160]. There are transposon-insertional mutants and subclones that produce fimbriae without the adhesive unit, and others that lack fimbriae, but are adhesive. Klemm [109] has shown that *fim A* gene encodes the subunit of type 1 fimbriae and that *fim B* and *fim E* control its "on–off" phase variation in *E. coli* [154].

The *Pap* gene cluster was shown by Normark et al. to encode nine polypeptides [72,73,158–160,165–167,169,170] that combine to specify pilus formation with the digalactoside-specific adhesin (the genes are designated by *pap* and their products by *Pap*). The adhesin is coded by *papE* and *papF* (minor pilin), and *papG* (adhesin) genes, present in the distal part of the gene cluster, distinct from the operon, which encodes the piliation. The latter includes *papI* and *papB* (regulatory proteins), *papA* (major pilin), *papH* (pilus growth terminator), *papC* (assembly platform or base), and *papD*, which is the transport protein. Bacteria carrying lesion in the *papE, F*, and *G* genes (which are dispensable for pili formation) are piliated, but do not bear the adhesin. On the other hand, bacterial deleted for the gene *papA* (which is the major pilin subunit gene) are still able to mediate P or MS hemagglutination [72,73,165,169]. The genes *papC* and *papD* (transporter), which are present in the middle of the cluster are required for the localization of both the pili and the adhesin on the cell surface. In each case, the protein encoded by the *pap* gene is designated Pap polypeptide. The deduced structure of PapC shares features of *E. coli* outer membrane protein and is suggested to form a transmembrane channel or a pore, through which the polarization is initiated. The level of this protein correlates the number of pili per cell. The deduced structure of PapH indicates it to be a pilinlike protein, with a unique proline-rich NH_2-terminal extension, believed to mediate termination of the pilus growth and the anchoring of the pilus structure to the cell. The level of PapC and PapH (the polypeptide produced as the expression of *papC* and *papH* genes, respectively) determines the number of pili per cell and the length of the pili. In this model, PapH interacts with both papC and the pilus, using its unique NH_2-terminal for binding the pilus to the membrane and prevention of further PapA polymerization. The PapA/PapH ratio is thought to determine the pili length. It was shown that PapB, A, H, and C are transcribed as an operon, which contains a transcriptional attenuator, positioned between Pap A and Pap H, and governing the stoichiometry of PapA versus PapH and PapC in the cell. The reader is referred to the very interesting dissertations of Norgren, Lindberg, and Lund [167,169,170], and related publications [72–74,158–160,165,166], which describe the details of the genetic aspects and the assembly of pili and fimbrial adhesins. Ott et al. [172,173] described the comparison of the genetic determinant coding for the S-fimbrial adhesin (*sfa*) of *E. coli* to the other chromosomally encoded types 1 and P fimbrial determinants. They constructed DNA probes specific for different regions of the *sfa* and hybridized them with DNA sequences coding for P (F8 and F13), MS hemagglutinating type 1 (*F1A*), and *F1C* fimbriae. Their results indicated that the *sfa* and *F1C* DNA determinants exhibited homology along their entire lengths, but the P and type 1 fimbrial determinants exhibited homology only to the regions of the *sfa* cluster responsible for the control of transcription and, to a minor extent, to regions coding for proteins involved in biogenesis or adhesion of the fimbriae and for the NH_2-terminal part of the fimbrillin subunit.

In addition to the on–off phase variation, a variation from type 1 to S fimbriae was also observed in *E. coli* in vitro and in vivo [177–180], Saukkonen et al. [180] showed, that 6 hr after inoculation of type 1 fimbriae bearing *E. coli* cells, the entire population recovered from rats had shifted to type S (55%) and nonfimbriated (45%) forms.

The PA-I and PA-II lectins of *P. aeruginosa*, which are not linked to fimbriae, are characteristic, genetically governed, secondary metabolism products. They are produced in the late stationary stage, with highest activity on the third to fourth days of the culture growth. Although the detailed genetic basis of their production is not yet revealed, it embraces two very interesting aspects: (1) A strong correlation (possibly coregulation) between the production of these lectins and extracellular protease (elastase), pyocyanin, and hemolysin [64] (but not lipase, acetylcholinesterase, and exotoxin A); and (2) phase variations of the two types: the on–off type and PA-I–PA-II variation [52,53].

The correlation between the production of the pseudomonal lectins and the aforedescribed cytotoxic extracellular activities was shown in lectin-poor mutants and strains as well as in lectin-devoid strains [181]. Gilboa-Garber et al. have shown that (1) strains of *P. aeruginosa* that were devoid of any lectin activity were also devoid of protease (elastase), hemolysin, and pyocyanin activities in the culture filtrates, but could still exhibit full lipase or exotoxin A activities [181]. (2) Strains that exhibited lower lectin activity were also weaker producers of the foregoing extracellular factors [181]. (3) Mutants, obtained by UV or chemical mutagens (using *N*-methyl-*N′*-nitro-N-nitrosoguanidine), exhibiting lower protease activity (together with lower pyocyanin and hemolysin), were found to be weaker lectin producers [181]. Mutants entirely lacking protease or lectin activity were not observed. (4) Genetically defined mutants of *P. aeruginosa* wild strain PAO 1 (obtained from Dr. B. Wretlind), which produced low protease activity and less pyocyanin, exhibited reduced lectin activity. Revertants of these mutants and hyperproducers of protease, exhibited maximal lectin activity [182]. Hence, it may be stated that in *P. aeruginosa* there is a close genetic coregulation or a coherent physiological linkage between the biosynthesis of the lectins PA-I and PA-II and the production of the extracellular protease (elastase) and pyocyanin. The accurate basis of this phenomenon is very interesting, since it may be a key to understanding the lectins' general function—coexistence or cofunction of adhesins or hemagglutinins and lytic or toxic enzymes, such as protease (elastase, peptidases), chitinase, sialidase, mannosidase, and other glycosidases, various esterases, hemolysins and other toxins—which has also been reported in various other microorganisms [2,174,183–187], in plants [2,188], and in animals (see Sect. VI).

On–off phase variation is a very prominent trait of *P. aeruginosa* lectin production. Subculturing of bacterial isolates obtained from patients, culture medium composition, growth temperature, pH, and aeration, all are signals for it. In addition to this on–off variation, there is also the very interesting phenomenon of variation from one lectin type to another type. An example for this variation is "PA-I to PA-II" variation (simultaneous production of both lectins is also encountered) observed by Gilboa-Garber et al. [52,53] and used for the selective or preferential production of one or the other lectin [53]. Environmental conditions controlling the bidirectional phase variation of PA-I to PA-II are known and are reproducible, but they have not yet been characterized on the molecular level. Another interesting aspect is the localization control of the lectin molecules between the intracellular, periplasmic space, cell surface, and the extracellular medium: "in–out" variation. This phenomenon was observed in many microbial cultures, for which the presence of cell-bound or cell-free lectins were reported by the same or different investigators [2]. A cell-bound adhesin of *E. coli* was liberated by heating [75,76]. *Pseudomonas aeruginosa* lectins change location under varying growth conditions, and heating liberates a certain fraction of them (personal observations). A cell wall-associated, chitin-binding hemagglutinin (endo-CLA) was liberated from the fungus *Conidiobolus lamprauges* [189–192] by treatment with β-glucanase, but not with chitinase, cellulase, protease, or several other treatments.

The phase and location variations of the lectins or lectinoid adhesins may enable the microbe to evade unnecessary, or unfavorable, lectin-mediated interactions with extracellular targets (cells or macromolecules), with phagocytes, and with endogenous structures. Surface-bound lectins and adhesins are necessary when the microorganisms need nutrients, fixation to solid substrates, or autoaggregation for protection, function, or exposure to oxygen. Under different conditions, they may disturb the bacteria, by binding them to undesired support (agar), to phagocytes [193,194], and to other cells [195]. Then, losing the adhesins to the medium may enable the microorganisms to evade such unfavorable conditions. It may also enable the microorganisms to get rid of constrictions imposed by lectins on the structural rigid organization of cell envelopes and other cellular components (thus increasing cell flexibility).

IV. BIOLOGICAL FUNCTIONS OF THE MICROBIAL LECTINS AND LECTINOIDS (FINDINGS AND HYPOTHESES)

The policy of the editors "to mold the overview with the current findings in the field in such a way as to bring to light fresh ideas and new hypotheses for the specialist, while presenting structure–function correlate to the student" was adopted by the authors in this section. The subject of the biological functions of the microbial lectins and lectinoids is most fascinating and not yet fully understood. However, they are multifunctional, highly interesting, and important for the understanding of the function of many biochemical, biological, medical, and agricultural systems, and have great biotechnological merit. Therefore, proved and possible functions will be dealt together to communicate and stimulate ideas. For the sake of simplicity, lectins and lectinoids will be included together, since they exhibit similar functions and effects and may have similar applications.

A. Microbial Adhesion is a Multicomponential Phenomenon Involving Lectins and Lectinoids

Adhesion of microorganisms to various macromolecular substrates, to other microorganisms (homologous or heterologous), and to cells or tissues of higher organisms is crucial for their survival, settlement, invasion, and function. The microbial sugar-binding adhesins are important units, but not the only ones in this multicomponential phenomenon. Adhesion may involve several components functioning directly or indirectly together or apart (Figs. 2 and 3). They include the following direct interactions:

1. Hydrophobic interactions—not involving sugars
2. Hydrophilic interactions—not involving sugars or their derivatives (e.g., interactions with amino acids or charged groups)
3. Interactions of microbial lectins or adhesins with sugar groups of the target cellular or macromolecular substrate

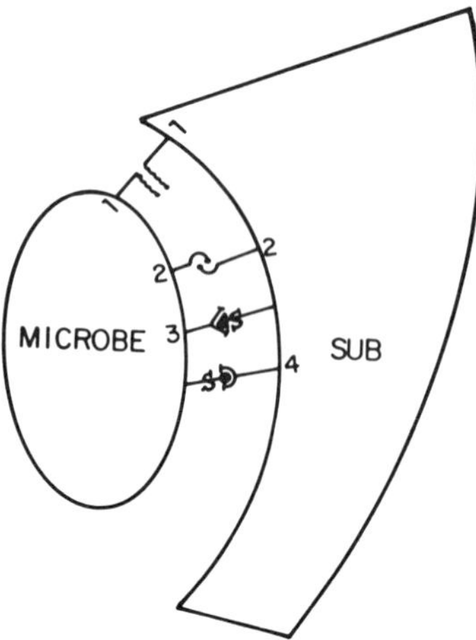

Figure 2 Types of direct interactions between adhering microbial cell (microbe) and a cellular or macromolecular substrate (sub). (1) Hydrophobic, (2) hydrophilic, (3) microbial lectin or lectinoid adhesin and sub sugar (solid triangles) and (4) sub lectin (or lectinoid) with the microbe sugar (solid circles).

4. Interactions of the target cellular or macromolecular substrate lectins or lectinoids with the microbial sugars

The indirect interactions are mediated by receptor-binding molecules (lectins and lectinoids) that function like antibodies, reacting with both the microbes and the target cellular or macromolecular substrate. Microbial adhesion is too important to rely on only one type of the described interactions. Generally, several types of interactions ensure the adhesion of microorganisms to different substrates under various environmental conditions. There are instances of monotypic adhesion (abolished by a single simple sugar, e.g., the MS adhesin of type 1 fimbriae of Enterobacteriaceae [28]). However, in most instances, the adhesion involves a mixture of interactions, functioning concomitantly, or alternatively (under different conditions). In these cases, the determination of the sugar specificity of the adhesive factors involved is more cumbersome and may lead to contradictory results and interpretations as well as to futile arguments.

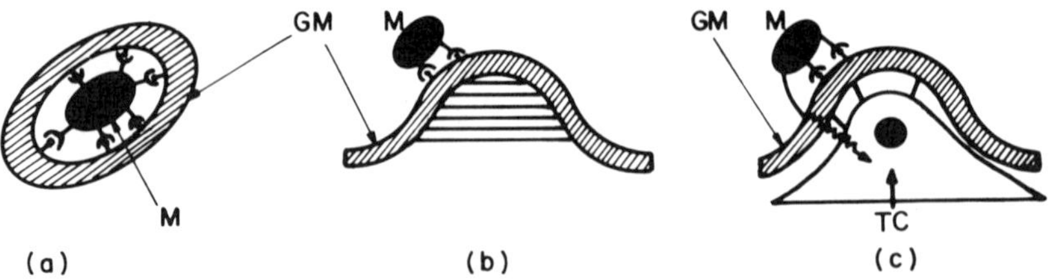

Figure 3 Direct interaction of microbial cell (M) lectins or lectinoid adhesins with extracellular glycosylated matrix (GM) of: (a) its own, (b) covering inorganic support, and (c) target cells (TC).

The most famous example for the mixed-type interactions is that of the nitrogen-fixing bacteria (the rhizobia) with the legum root hairs. This interaction is crucial for the establishment of their symbiosis, which is of utmost importance for the nitrogen cycle (enabling its utilization) in nature [196–201]. In this system, lectins of the plant root hairs may specifically bind sugars on the *Rhizobium* cell surface and the microbial lectins (or lectinoid adhesins) may react with sugars on the root hairs [202,203]. In addition, hydrophobic and amino acid-binding adhesive factors may also function [198]. Another important example is the phagocytosis of bacteria by phagocytic cells, during which the lectin of the bacterium [204–217], the phagocyte [194,214,218], or an external lectin (of different bacterium) [219], may function. The type 1 pili were reported to stimulate the oxidative metabolism of polymorphonuclear leukocytes [220] and to trap the bacteria to them and to other phagocytes [204–208]. Urinary Tamm–Horsfall glycoprotein, which interacts with them, protects the bacteria from phagocytosis [221]. The phagocytosis in vertebrates is further ensured by the most sophisticated, and highly efficient and specific immunoglobulins, which function like a highly specialized type of lectin. An example of a mixed-type parasitic adhesion is that of actinomycetes, which are harbored in the human oral cavity and are involved in gingivitis, periodontitis, and caries [93,222–237]. In addition to galactose and lactose binding (by their type 2 pili), hydrophobic interaction, by their type 1 pili and other hydrophobic components, has also been described [93,222–237]. These two pili types of the actinomycetes were shown to be involved in the interaction of the actinomycetes with the other oral-colonizing bacteria, such as streptococci [224–228], and with the mammalian epithelial cell surface [229–237]. The lectin or lectinoid role in the adhesion may be dominant and crucial in some instances, under certain circumstances [238–241], but absent in others. Hydrophobic interactions may replace it in some adhesion phenomena and at certain periods, with almost identical efficiency. Hence, the absence of microbial adhesins in certain bacteria is not a proof for their unimportance. Furthermore, there are reports of hydrophobic groups or segments next to the sugar-binding sites in many of the lectins and lectinoids. This feature may strengthen the interaction and ensure its efficiency, as it does in enzyme–substrate interactions. In certain instances, it has been shown [238] that injured epithelial cells (by viruses or medical treatment) facilitate bacterial adherence.

Lectin-dependent microbial adhesion may involve lectins or lectinoids that are associated with either the cell membrane or with supermolecular extracellular structures, such as the fimbriae [28] or pili, flagella, and capsules [127]. In some instances, the structures bearing the adhesins could be separated from the cells [104,105]. The detached structures could cause hemagglutination correlating the original activity of the intact bacteria, and the adhesins were further purified and analyzed. It must be emphasized that in all the foregoing adhesion phenomena, the initial interaction may be reversed by sugars or related components, which react reversibly and specifically with the binding sites. However, this initial step conditions the occurrence of a cascade of reactions that are already irreversible.

B. Microbial Lectin and Adhesin Interactions with Subcellular and Acellular Sugar-Containing Matrix

Adhesion of microorganisms to a sugar-bearing matrix may serve them in several ways (see Fig. 3):

1. Interaction with their own extracellular polysaccharides, mucoid secretions, spores, cysts, or other, and with cellular (structural and functional) glycosylated macromolecules: The microbial lectins and lectinoids may serve in the organization of the cell envelopes and

extracellular matrix, in holding it near the cell, and even in keeping active contact with it (see Fig. 3). At the same time, the same lectins may also function in intracellular organization and reorganization of the microbe's own membranal peptidoglycan, polysaccharide, and mucoid compounds. The slime mold *Dictyostelium discoideum* lectin, discoidin, interacts with its extracellular matrix, as well as with foreign microbial glycoconjugates [242,243]. *Amoeba histolytica* [100,244] and *Conidiobolus lamprauges* [189–192] $(GlcNAc)_n$-binding lectins interact with their own chitinous constituents of the cyst and the cell wall, respectively, as well as with those of other organisms.

2. Microbial adhesion to sugar-bearing macromolecules present on solid surfaces: This adhesion may serve as a recognition signal for microbial migration and settlement in ecosystems that harbor similar microorganisms, or that are rich in foreign organic materials, required for the microbial nutrition (see Fig. 3). Fucose-containing macromolecules of algal origin are recognized by marine *Vibrio* spp. [84], and alginic acid and related mucopolymers are recognized in aquatic ecosystems by *Pseudomonas* spp.

3. Microbial adhesion to acellular sugar-bearing molecules secreted by various cells, or present on their surface, including epithelial cells, which form the outer layers of mucous membranes covering the oral, respiratory [238], urinary, alimentary, and other cavities [235]: The acellular glycomatrix assures the nutritional requirement of the microorganisms and supplies them a mechanical support that mediates their contact with the desired target cells. It may also defend them from phagocytosis [221]: mucosal secretions, fibronectin, and Tamm–Horsfall urinary glycoprotein, as well as other extracellular matrix (ECM) constituents may function as the acellular molecules in this category. This mechanism is important in many natural macro-, marine, or terrestrial, ecosystems, as well as in animal mucosal cavities and in plants.

C. Microbial Autoaggregation in Homogeneous Populations

Microbial aggregation, which may involve lectinoid adhesins, was described in bacteria, such as myxobacteria [87,94], in slime molds [48–50,242,243], and in other microorganisms. Generally, this phenomenon is triggered by an environmental signal or effect, including a deficiency of a certain (or essential) nutrient, exposure to a certain substance, or changed conditions [48,87,181]. The aggregation may serve for the formation of new multicellular structures [50]. There are reports on correlation between the appearance of the lectins or lectinoids on the microbial cell surface and the initiation of aggregation [48,50,87]. There are reports describing mutants devoid of the adhesins and lacking the aggregating properties, for which the addition of the specific sugars, at the critical stage, prevents the aggregation of the wild types. However, sometimes, there is no such clear cut correlation, since additional mechanisms (e.g., other hydrophilic and hydrophobic interactions, not involving sugars), which ensure the aggregation (see Sect. IV.A), may also function in these systems. Autoaggregation may also serve for the construction of a pellicle on the nutritional medium, for gaining access to oxygen, or for avoidance of environmental disturbance or nutritional shortages. Kijne et al. [202] reported that mannose-specific lectin of *Rhizobium leguminosarum* contributes to autoagglutination of these bacterial cells shortly after nutritional limitation in the culture medium. This autoagglutination can be prevented by the addition of 0.2 M mannose solution to the culture medium before the autoagglutination begins. *Pseudomonas aeruginosa* hemagglutinins induced autoagglutination in their own old culture cells [Gilboa-Garber, unpublished results].

D. Microbial Coaggregation in Mixed Populations

Coaggregation is a specific bacterial adherence to microorganisms of different species and genera. Coaggregation caused by hemagglutinins was described by Collier and Jacoeb in 1955 [56]. They demonstrated specific adsorption of the hemagglutinin of *E. coli* by strains of *S. typhi, Proteus*, and *E. coli*. They distinguished different *E. coli* strains by their adsorption of the hemagglutinin from the culture medium in which another *E. coli* had been grown [2,56].

The galactose- and lactose-binding type 2 fimbriae, of *Actinomyces viscosus* and *A. naeslundii* were found to interact with receptors on *Streptococcus sanguis* and *S. mitis* [224,228]. Weiss et al. have shown that the gram-negative oral bacterium *Bacteriodes loeschei* PK 1295 forms multigeneric aggregates by bridging between otherwise noncoaggregating gram-positive microorganisms such as *S. sanguis* 34 (lactose-sensitive) and *A. israelii* PK14 (lactose-resistant). These interactions, which contribute to the establishment of the mixed microbial population in the human dental system, were assumed as contributing to the formation of dental plaques.

E. Microbial Adhesion in Symbiosis and Commensalism

Extensive literature has been published on the role of plant lectin contribution to the important symbiosis between rhizobia and the legumes [196–203]. The role of bacterial lectins or lectinoids in the interactions with the legume has also been described. Vesper and Bauer [cited in 203] have shown that attachment of *R. japonicum* to soybean root is inhibitable by D-galactose. Antiserum, or purified IgG, against "isolated" *R. japonicum* fimbriae inhibited the attachment by about 90% and nodulation by about 90 and 80%, respectively. Kijne et al. described mannose-specific lectin in the recognition of pea roots by *Rhizobium leguminosarum* [202] and attributed to them the autoagglutination of the bacterial cells after nutritional limitation in the culture medium. Similar symbiosis or commensalism can also be established in animals enjoying the products of the symbiotic and commensalic microorganisms, supplying them with vitamins, essential amino acids, and such, and with local defense against pathogenic microorganism invasion. Antibodylike lectin function of one microorganism against the other, for the benefit of the host, was described by Sudakevitz and Gilboa-Garber [219], who showed that in vitro exposure of *E. coli* O_{86} to *Pseudomonas* PA-I lectin and of *E. coli* O_{128} to the PA-II lectin led to their increased phagocytosis by human polymorphonuclear leukocytes [219]. In this system, the pseudomonal lectin function resembled that of opsonins. Pretreatment of the leukocytes (instead of the bacteria) with the same lectins depressed the phagocytosis of the untreated *E. coli* strains by them [219].

F. Microbial Lectins, Lectinoid Adhesins, and Lectinoid Toxins in Parasitism

Microbial adhesion to cells, as first step in disease, has been extensively investigated since the beginning of this century. The pathogenic adhesion is a multicomponential phenomenon in which the lectins and lectinoids may serve the pathogenic microorganisms in several ways:

1. Enable their binding to the host cells [234] or to ECM (mucins): This protects them from being washed off by the host excretions (urine, feces, saliva, tears) and from the host defense mechanisms, and provides their nutrients. Aronson et al. [239] reported that methyl-α-mannoside prevented *E. coli* adhesion to murine and rat bladder cells.

2. Enable their homing to the host-sensitive target cells: MS adhesins were found to prevail in entero- and uropathogenic *E. coli* strains [26–28,71,241], whereas most *E. coli* isolates of patients suffering from pyelonephritis were shown to contain P or Pap type fimbriae, exhibiting the digalactoside Gal-α1-4Gal specificity or X-type adhesins [77–79,83,115–118]. Vero cytotoxin-producing *E. coli* strains are associated with sporadic cases and outbreaks of hemorrhagic colitis and with hemolytic uremic syndrome in humans [185]. The S-fimbriae, which bind to sialic acid-containing glycoconjugates, are present on many *E. coli* strains that cause infant meningitis [180,245].
3. Enable their lytic factors and toxins to function on the host-sensitive cells and ECM: These factors may decompose components of the ECM and of the host cells and lead to either an extensive or to a limited lysis, which enables the parasite invasion. The linkage between the lectin (or lectinoid) molecule and the lytic or toxic factor may be on cell level as well as on the molecular level. There are many examples for microbial surface lectins or lectinoid adhesins that home to the sensitive target cells and condition the activity of the lytic factors:
   a. Bacteriophages adhere to glycosylated receptors on bacteria [246–251] and invade them by means of a lytic process. The adhesion may be prevented by plant [246,247] and by microbial [251] lectins, which compete with them on the glycosylated receptors, or by the specific sugar [248–250].
   b. Viruses, such as influenza A [13,35,137], and other microorganisms contain sialophilic lectins linked to sialidase, which enable them to affect the host cell, fuse with its membrane, and invade it.
   c. *Pseudomonas aeruginosa* lectins, present on the surface of the bacterium, facilitate the focusing of its proteolytic and other lytic and toxic activities on the sensitive cells [60,181].
   d. *Escherichia coli* adhesins were shown to be linked to hemolysin activity not only on the cell surface, but also at the genetic level [186,245,252–256].
   e. Fungi, such as *Conidiobolus lamprauges* [189–192] of the Entomophthorales, which contain many pathogens of insects and vertebrates, and *Candida albicans* [257], which is a human pathogen, were shown to exhibit a profound proteolytic, β-GlcNAcase and chitinase activities in the vicinity of their chitin-binding lectins. The chitin-binding lectin of *C. lamprauges* binds to the chitin of the insect cuticle for initiation of the parasite attack. Another example is the adhesion of the nematophagous fungi [258], which are important for regulation of nematode populations in the soil. These fungi attack living nematodes or their eggs and utilize them as a source of food. The fungus *Arthrobotrys oligospora* binds by means of a lectin, which detects GalNAc-containing carbohydrate on the nematode cuticle. After adherence, the process is irreversible and the parasite penetrates the nematode cuticle. The nematode dies and the fungus survives.

There are numerous additional examples of the "conditioning" service (homing, anchoring, positioning, and even enabling penetration into the cells for getting access to the sensitive components) given by the microbial surface lectins and lectinoids to several microbial toxic and lytic factors, enabling their action on the target cells. A similar, private service is given by soluble lectinoid units to soluble secreted microbial macromolecules, such as the exotoxins and exoenzymes, that affect the host cells [183,259–267]. Among the molecules enjoying this service are several highly potent exotoxins [266]. They are composed of a lectinoid subunit B, which binds glycosylated macromolecules, and a toxic subunit A. Examples are the exotoxins of *Corynebacterium diphtheriae*; *P. aeruginosa* (exotoxin A) and *Shigella*, which binds to cell glycoproteins; and the cholera, *E. coli* heat-labile (LT), tetanus, and botulinum toxins, which bind to gangliosides [260–267].

The subunits A of diphtherial and pseudomonal toxins inhibit protein syntheses of NAD-dependent ADP ribosylation of polypeptide elongation factor 2 (EF-2). The shigella toxin inhibits protein synthesis by irreversible ribosomal inactivation. Ribosome-inactivating toxins, acting as lectinoid-bound RNA *N*-glycosidase (see Sect. VI), are well known in plants (ricin, abrin, modeccin). The cholera and *E. coli* LT toxins induce NAD-dependent ADP ribosylation of a regulatory protein, leading to a prolonged activation of adenylate cyclase. The cAMP activates specific protein kinase and results in alteration of ion transport in the intestinal cells, exhibited in diarrhea [266].

In most of the aforementioned toxins, separation of B from A prevents the toxic action of the A subunit on the intact target cells, but does not affect its function on the sensitive cell-free intracellular target systems. The B (lectinoid) domain serves not only as a binding unit, but also exerts alterations in the target cell membrane. It induces redistribution of the cell surface receptors into aggregates (clusters) and caps. The diphtherial toxin (M_r 60 kd) subunits A (21 kd) and B (39 kd) are produced as a single polypeptide chain that is later cleaved by proteolytic nicking to produce a two-chain molecule linked together by a disulfide bridge. In cholera and *E. coli* (LT) toxins the two chains are synthesized separately in a precursor form. They assemble in a ratio of 1:5 A/B subunits and the complex is transported through the cell membrane. The cholera holotoxin (M_r 84 kd), consists of one A subunit (M_r 29 kd) and five B subunits (M_r 11 kd) arranged in a pentameric structure around the A chain. The two chains of *E. coli* LT are synthesized separately and then assemble together 1:5 A/B, which is transported through the cell membrane to be associated with it. The B subunits of both cholera and *E. coli* LT toxins bind to monosialoganglioside G_{M1} oligosaccharide NeuNAc-Gal-GalNAc-Gal. The tetanus and botulinum (types A, B, C1, C2, D, E, F, and G) neurotoxins are synthesized as a single polypeptide chain (M_r 150 kd) that is sensitive to proteolytic cleavage, resulting in a two-chain molecule: a heavy chain (M_r 95 kd in tetanus and 100 kd in botulinum) and a light chain (55 kd in tetanus and 50 kd in botulinum). The heavy chain contains the binding determinants that are most strongly inhibited by gangliosides of the type G_{T1b} with the trisialosyl oligosaccharide NeuNAc-NeuNAc-Gal-GalNAc-(NeuNAc)Gal (which inhibits both toxins) or the disialosyl G_{D1b} with the oligosaccharide NeuNAc-NeuNAc-Gal-GalNAc-Gal (which inhibits the tetanus toxin). The pseudomonal exotoxin A is a single chain of M_r 66kd containing disulfide bridges, and the shigella holotoxin M_r is 64 kd composed of one A (32 kd) and five B subunits (6.5 kd each), exhibiting β-1-4 linked GlcNAc specificity [266].

G. Lectin and Lectinoid-Mediated Phagocytosis of Microorganisms

Lectins may contribute to the protection of organisms against infection by trapping the pathogenic microorganisms. The microorganisms may be trapped by their own lectins or lectinoids [193,194,204–207,209–217], by those of unicellular organisms [244] and multicellular host phagocytic cells (macrophages, granulocytes, and such), and hepatocytes [194,214,218], or by those of other microorganisms [219]. The lectin-mediated adhesion leads to phagocytosis of the microorganisms. This phagocytosis is similar to the phagocytosis mediated by antibodies, opsonins, or complement components, and may culminate in digestion of the microorganisms by the phagocyte lytic enzymes. The process may be very important for organisms that are devoid of a developed active immune system or suffer from temporal or periodic immunological deficiency. The lectin-mediated phagocytosis was named *lectinophagocytosis* by Ofek and Sharon [194]. It relies on a multicomponential complex, involving cell-bound lectins and lectinoids of either the phagocytic cells or the microbe (*direct lectinophagocytosis*). Free lectins produced by the host (e.g., invertebrate hemolymph lectins and circulating mammalian lectins), or by other microorganisms may also contribute to it (*indirect lectin-mediated phagocytosis*) [219]. Bar-Shavit et al. [204] demonstrated that mannose residues on phagocytes are receptors for the attachment of *E. coli* and *S. typhi* strains. Silverblatt et al. [206] demonstrated that susceptibility of *E. coli* to phagocytosis was dependent on MS pili. Other investigators have also described similar mannose-dependence in *E. coli* phagocytosis by polymorphonuclear leukocytes [205,207] and macrophages [193,209]. Wileman et al. identified the macrophage mannose receptor as a 175-kd membrane protein [215]. A similar phenomenon of the contribution of a microbial fimbrial lectin to its phagocytosis has also been described in *Actinomyces*, the type 2 fimbriae of which are trapped by granulocytes. Sudakevitz and Gilboa-Garber [219] reported that PA-I and PA-II lectins of *P. aeruginosa* specifically increase the phagocytosis of enteropathogenic strains of *E. coli* by human granulocytes. Treatment of the leukocytes by the same lectins interfered with the phagocytosis of the untreated *E. coli* strains. Similar mechanisms may also be involved in the host protection by "normal microflora" against pathogenic microorganisms. Evasion of the lectin- or adhesin-bearing bacteria from the lectinophagocytosis trap may be achieved by the on–off phase variation [194,195] or by localization variation. The involvement of lectin–carbohydrate interaction in the recognition stage initiating the in vitro phagocytosis can be shown by its inhibition in the presence of simple sugars that are specific for the lectins. The recognition stage is followed by an oxygen burst [220] and degranulation of the phagocytes for release of their lytic enzymes. These phenomena are already irreversible.

H. Induction of Structural and Functional Alterations in Cell Membranes and Stimulation or Activation of Various Cellular Processes

Lectins of plants, animals, and microorganisms have been known to induce agglutination of various cells since their discovery [1–4]. However, the intensive study of their effects on cell membranes and cell metabolism is due to the very important observation of Nowell [268] that peripheral lymphocytes are transformed, in the presence of *Phaseolus vulgaris* lectin, to highly active and dividing cells. This stimulating observation led to numerous studies on lectins' effects on cell membranes and cell metabolism, and examination of their ability to stimulate mitogenesis in lymphocytes. Lectin binding to lymphocytes leads to redistribution of the membrane receptors to aggregates (clusters) and caps [269,270] and induces a cascade of reactions leading to increased synthesis of proteins and nucleic acids. The microbial lectins exhibit similar effects: the cholera toxin B subunit [84] and *P. aeruginosa* lectins [271] induce clusters and capping in lymphocytes (Fig. 4). *Bordetella pertussis* hemagglutinin [272] and pseudomonal lectins resemble the *P. vulgaris* hemagglutinin [268] in inducing mitogenic stimulation in human peripheral lymphocytes [63,64]. PA-II stimulates also murine splenocytes [64]. *Entamoeba histolytica* [100,273] and mycoplasmal cells [82] stimulate human lymphocytes. Similarly, microbial lectins stimulate the metabolism and phagocytic activity of free-living unicellular organisms [60,274,275]. The physiological implications of the microbial lectin stimulatory effects depend on the type of the stimulated cell. If helper-T lymphocytes or unicellular phagocytic

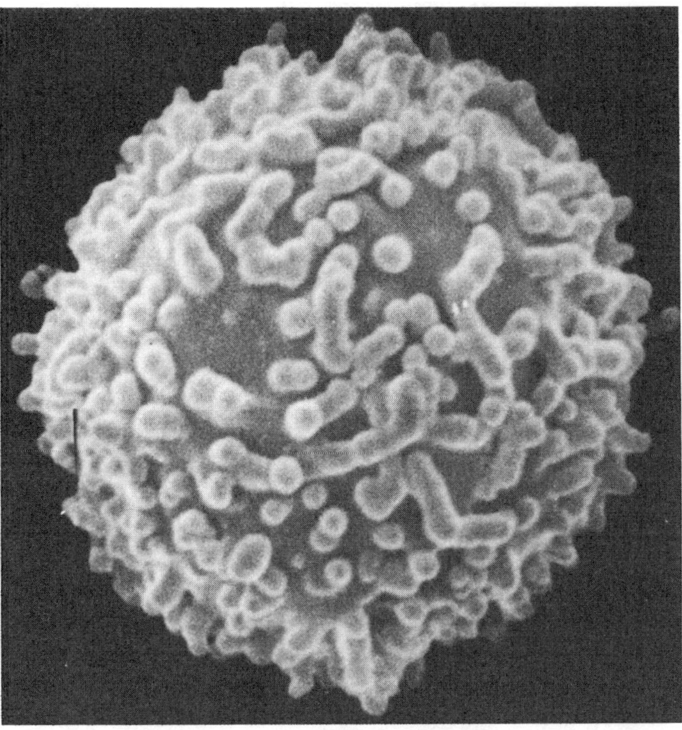

(A)

Figure 4 Scanning electron micrograph (×32,000) of a human peripheral lymphocyte: (A) without treatment, and (B, on the next page) after 60 min incubation with protoplasts prepared from *P. aeruginosa* bearing PA-I lectin. (From Ref. 271.)

cells are stimulated, it may strengthen the resistance of the host against the invading microorganisms, but if suppressor-T lymphocytes are stimulated, the host resistance is reduced. Although most of the studies examined lectins' effects on foreign cells, no less important are the effects of lectins in the cells that produce them.

V. APPLICATIONS OF MICROBIAL LECTINS AND LECTINOIDS

Since the first discovery of lectin-induced hemagglutination and toxic effects [1], the lectin applications grew steadily and are still developing very rapidly. As described in the introduction to this chapter, the finding of lectin-induced hemagglutination was followed by the demonstration that lectins could differentiate between erythrocytes of different animals and could be used in blood group determination and microbial identification. The discovery of the lectins' sugar specificity, led to the use of lectins as probes for sugars in solutions, on macromolecules, and on cells. Thus, lectins became an important tool for study of macromolecular and cellular structure, including the composition of blood group antigens [276]. They also became very useful for cytochemical and histochemical methods, based on labeled lectins, which enabled the study of cellular structure in normal and altered

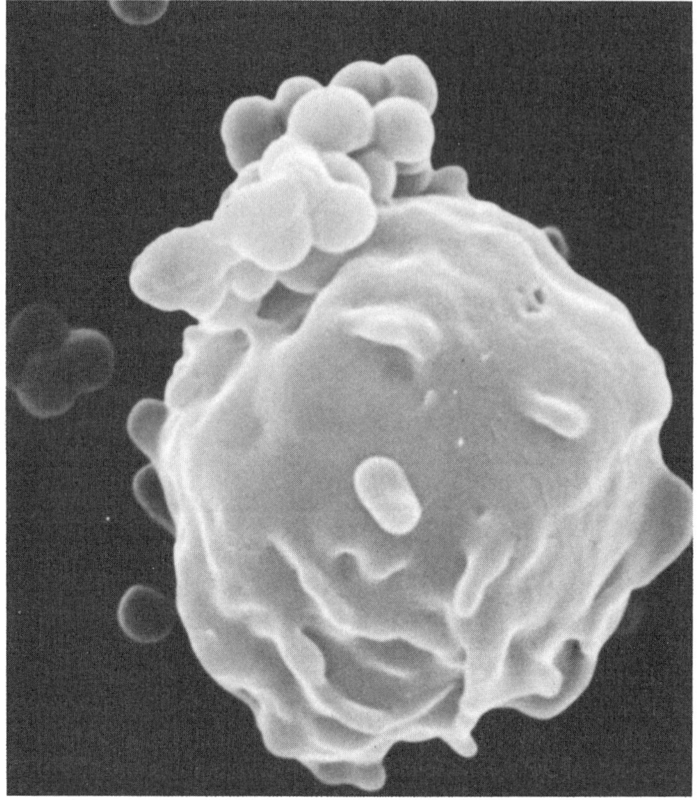

(B)

Figure 4 (Continued)

conditions and diagnosis of abnormal states. The differential reaction with cells enabled the use of lectins for macromolecule purification and cell separation, depending on the presence of sugars on them and the cell tendency to aggregate.

The finding by Nowell [268] of the dramatic, mitogenic lectin effect, led to the use of lectin-induced mitogenic stimulation of peripheral lymphocytes for diagnosis of genetic abnormalities and for gene studies and localization, by their hybridization with the chromosomes. Moreover, it opened a broad spectrum of investigations on cell membrane receptors and their interactions with lectins, hormones, and other activating factors, and on the changes, induced by these interactions, in cell membranes, metabolism, life cycle, and destiny. Various regulatory mechanisms of the cells were discovered by lectin applications [277], and various regulatory extracellular factors were produced (including interleukins, interferon, and other lymphokines). The latter are of crucial importance for the immune system and the resistance to disease. Cancer research and therapy are also nourished on lectins, which are very useful for detection of the alterations in the malignant cells, for reducing the cell tumorigenicity, and for prognosis of the immune status of the patient (exhibited in reduced lymphocyte stimulation). Addition of lectins to various other cellular systems (e.g., the fertilization process and the nervous system) have highly important consequences. Although most of the described investigations were based on the use of plant lectins, the microbial lectins may be as useful as the plant lectins and even more so. The microbial lectins may be useful in vaccines against microbial diseases, in induction of cell fusion, in the study of lectin genetics and production, and in studies on lectin function. The lectinoid toxins may be used for treatment of human, animal, and plant diseases (cancer and other), and may be implanted in plants (by cloning) for transfer of resistance to insects and various other parasites. In the present chapter, only few illustrating examples are discussed with the hope that the reader will be stimulated to look for additional ones in the literature and in his or her laboratory.

A. Detection of Sugars in Solutions, on Macromolecules, and on Cells

The variability of microbial lectin sugar specificity is already approaching that of plant and animal lectins. Examples were described in the introduction to this chapter and in Table 1. More examples may be found in Mirelman's book [39] and in the current literature, which is steadily growing. Detection of sugars by microbial lectins is based on the same principles as used with plant lectins. The microbial lectins also agglutinate cells that bear the specific sugars. Their agglutinating activity is inhibited only by the specific sugars and related derivatives in free or bound forms. Lectins labeled with fluorescent dyes, enzymes [51], radioactive compounds, or biotin–avidin [278], may be produced, and their interactions with cells and macromolecules may be followed by the conventional immunological–lectinological techniques (microscopic, ELISA, or cell sorting by FACS). An example is PA-II of *P. aeruginosa*, which exhibits an outstandingly high affinity for fucose [148] and, therefore, may serve as highly sensitive probe for this sugar in solutions or macromolecules and cells. Since it also reacts with mannose, a mannose-specific lectin (*Lens culinaris*, Con A, or *Pisum sativum*) has to be used as a control.

B. Detection of Defined Cell Receptors and Blood Group Antigens

Since the discovery of the microbial hemagglutinins, in the beginning of this century, they were shown to be able to distinguish between erythrocytes of different animals [2]. Fungal extracts with ABH blood group specificity were described a short time after the finding of the blood group-specific plant hemagglutinins (see Sect. I). The lectin of the fungus *Conidiobolus lamprauges* was shown to be highly specific for human erythrocyte [279]. Therefore, in addition to indicating the existence of highly specific human antigens, it may be useful for police investigations [279]. Viral hemagglutinins specifically interact with glycophorin, and the tetanus toxin binds to cell gangliosides [259]. In 1971, cholera toxin was also shown to be deactivated by a ganglioside (later shown to be G_{M1}) [260–265]. A year later, Fujita et al. demonstrated the first blood group-specific bacterial lectin, the anti-B of *Streptomyces* sp. [57]. The special orientation of part of the α-galactoside as the dominant sugar in blood group B receptor is recognized by this lectin, despite its insensitivity to α- or β-linkage of this sugar. The streptomycete lectin specificity differs from that of anti-B antibody in several additional features:

1. Its hemagglutinating activity is inhibited by galactose, whereas the antibody is inhibited only by the galactose-containing oligosaccharide hapten.
2. It reacts with erythrocytes of blood group B 256 times more strongly than with A and reacts weakly with O, whereas the anti-B antibody is entirely selective for B antigen-bearing cells.

3. It also reacts with receptors unrelated to the B blood group—such as the receptor of *Lactobacillus casei* for bacteriophage PL-1—because the high affinity of this lectin for L-rhamnose [251], not shared by the anti-B antibodies.

However, if a compatible dilution of the lectin solution is used, this lectin may serve as a good anti-B reagent for blood typing [42,57]. The galactophilic lectin PA-I of *P. aeruginosa* [60] also reacts with the α-galactoside of blood group B cells, but it exhibits a lower selectivity for these erythrocytes. It also differs from the streptomycete anti-B lectin in lower affinity for L-rhamnose (personal observations). Therefore, it cannot be used as a blood group B reagent in agglutination tests. Nevertheless, its adsorption on B cells and its reaction with saliva of secretors of AB and B types is significantly higher than that obtained with A and O cells or the respective saliva [280]. Moreover, Garber et al. [138] have shown that this lectin exhibits a high selective agglutinating activity towards *E. coli* O_{86} (Fig. 5), which bears blood group B-like antigen (cross-reacting with human anti-B serum). PA-I distinguishes between *E. coli* O_{86} and *E. coli* O_{128}, which bears blood group H(O)-like antigen (reacting with Ulex-I anti-H). The reason for the lack of higher specificity of PA-I for the erythrocyte blood group B antigen is due to its reaction with additional galactose-containing receptors, present on the erythrocyte, including those of T antigen (in desialylated red blood cells) and the P system (P^K) [143] and with *N*-acetylgalactosamine. Lectins that react with α-galactosyl of both B and P^K blood groups were also described in animals such as salmon. In addition to its enterotoxin, which binds G_{M1} in animal cell membranes [260–266], *V. cholerae* also produces lectins [84] that are inhibited by fucose, D-arabinose, and by a preparation of bullfrog gastric asialoglycoprotein with human blood group B+H antigenicity. It is not inhibited by bovine brain ganglioside, which inhibits the binding of the bacterial toxin [261–266]. Kameyama et al. [54,55] and Matsui et al. [102,103] showed that the fucose-specific lectin produced by *Streptomyces* sp. 16-3 (SFL 16-3), which is completely inhibitable by 5 mM fucose and 20 mM mannose, specifically reacts with poly(glycerol)-ceramide as the predominant receptor. Since this glycolipid does not contain mannose, the real receptor should be considered to be fucose. They showed that the fucosyl residue of this compound is essential for the binding, whereas the crude erythrocyte glycoprotein fraction exhibits only slight inhibition [54,55]. PA-II of *P. aeruginosa* exhibits a similar specificity, but its affinity for fucose is considerably higher than for mannose. In *E. coli* strains the mannose-sensitive adhesins (either fimbrial F1M or nonfimbrial) were reported to bind to the erythrocyte and leukocyte glycoproteins [91]. The P or Pap adhesins were reported to bind to digalactoside-containing erythrocyte glycosphingolipids and to be highly specific for the P^K blood group antigen of P1, P2, P^K, and p system [116–118]. The sialophilic adhesins of *E. coli* strains, *Bordetella bronchiseptica, P. aeruginosa*, and *Mycoplasma* were shown to react with both the major glycoprotein, glycophorin A [123] and its M and N blood groups [88,89], or with I and i antigen types [281], with G_{M2} ganglioside (CFA1 and K99 adhesins) [119–121,282,283], and with the major glycolipid (glycosphingolipid of horse erythrocytes): *N*-glycolylneuraminyl- containing ceramide. Microbial lectins and lectinoids with blood group reactivity may also contribute to elucidation of blood group antigenic composition. The Anton blood group antigen was shown in the erythrocyte receptor for *Haemophilus influenzae* [284], and a newly described blood group antigen, Dr [285], present in most human erythrocytes was detected by *E. coli* fimbrial adhesin, which was not inhibitable by simple sugars.

The contribution of the lectins to the study of cell receptors is not limited to only their detection and structural analysis, but they also serve for the study of their function.

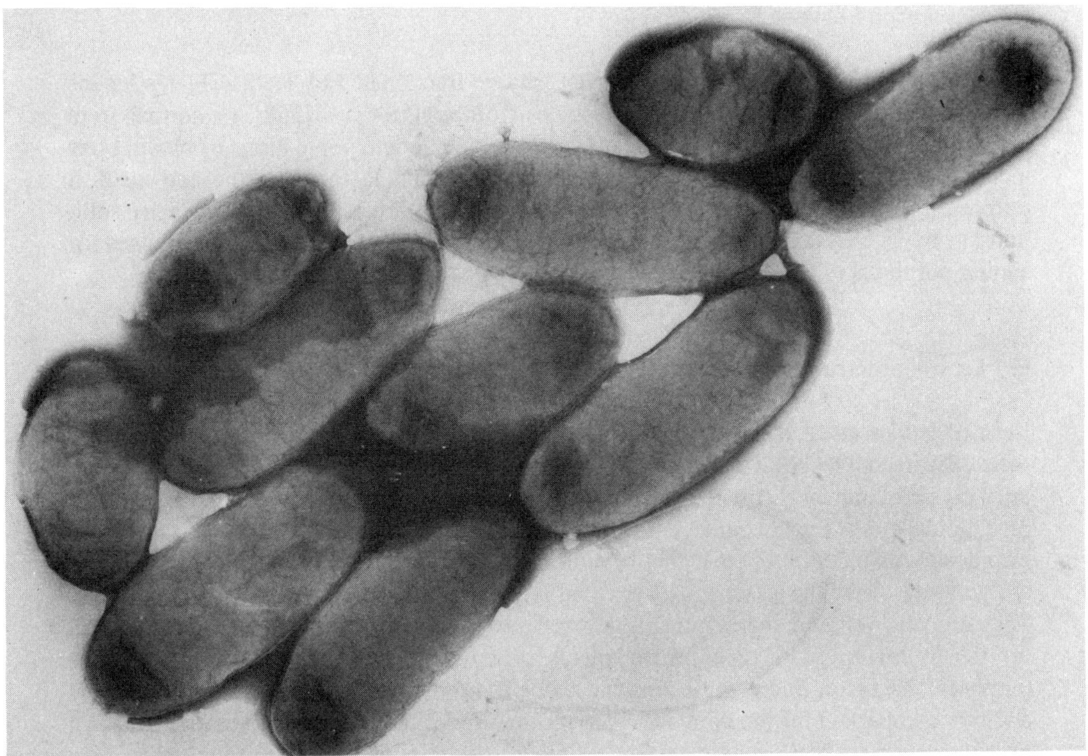

Figure 5 Agglutination of *E. coli* O_{86} cells by *P. aeruginosa* PA-I lectin (×10,860 electron microscopic magnification).

The L-rhamnose-containing polysaccharide receptor for bacteriophage PL-1 was identified by the streptomycete lectin, which blocked its adherence to *Lactobacillus casei* [251].

C. Identification of Bacteria

Microbial lectins may be used like the plant lectins as an aid for identification of bacteria. Agglutination of the bacteria, or fixation of labeled lectins by them, may be used for this aim. The *P. aeruginosa* lectins were shown to distinguish between the enteropathogenic strains of *E. coli* O_{86} and O_{128}. PA-I strongly agglutinates the first bacterium, whereas PA-II does so with the second one [138,139].

D. Separation Between Molecules and Purification of Glycosylated Macromolecules

One of the most useful applications of the plant lectin Con A is the purification of glycoproteins (including hormones and enzymes) on lectin-Sepharose columns. Microbial lectins, which bind mannose, may also be used for this purpose. The *P. aeruginosa* PA-II, linked to Sepharose [60], was used for purification of hyaluronidase, horseradish peroxidase, and additional glycosylated enzymes. The results obtained with it were similar to those attained with Con A.

E. Separation Between Cells

There are many mixed cell populations that may be fractionated by lectins. Plants lectins have been useful for separation of leukocytes from erythrocytes [268], or normal from transformed cells, and of the various leukocyte or lymphocyte populations of blood, bone marrow, and lymphatic organs. The separations are valuable for scientific and medical purposes. They may be performed by sedimentation techniques, using high-density solutions, or by labeled cell-sorting techniques (FACS). Microbial lectins may also be used for similar purposes.

F. Labeling of Special Glycosylated Cellular Components for Their Tracing

Labeled lectins are very useful for the tracing of glycosylated residues and receptors on cells. The lectin labeling may be a fluorescent dye that is detectable by fluorescence microscope. It may be peroxidase, which is stained with diaminobenzidine and H_2O_2 [51], or colloidal gold, ferritin, and radioactive biotin–avidin complex [278], which may be examined with electron microscope, or radioactive tracers, respectively. These techniques are very useful for the study of cell and tissue structures and for tracing sugars on them. They are very valuable for examination of various conditions and disease states as well as for the follow-up of therapeutic treatments. Microbial lectins may be used for these purposes. The use of microbial lectins and adhesins for cytochemical assays would provide additional important information about the mechanism of the parasite differential adhesion to its host tissues. Numerous pathological processes are associated with modification (in content or distribution) of cell surface glycoconjugates. The affinity of lectins toward specific sugars permits their use for histopathological assays. The diagnosis of cancer and storage diseases by the plant lectins, Con A and Ulex-I, may be also performed by microbial lectins.

Lectins are useful as neuronal tracers for mapping neuronal connective patterns [286–288]. The neuronal tracer is injected at a point of interest in the nervous system. It is taken up by the neurons and transported from the axonal terminations back to the cell body (retrograde transport) or can be taken by the cell soma and dendrites and transported forward to the axonal target areas (anterograde transport). Peroxidase-labeled plant and microbial lectins were used as neuroanatomical tracers [286,287] for evaluation of CNS pathways. They were found to be very useful for revealing even sparse projections, since their passive diffusion is very low [286]. The cholera lectinoid toxin and other microbial toxins were found to be sensitive retrograde transported markers [287,288].

G. Alterations of Intercellular Interactions

Intercellular interactions may be augmented or depressed by microbial lectins, as well as by plant lectins. For example, the adhesion of rhizobia to legum root hairs is increased by *P. aeruginosa* PA-II lectin [60,289], but decreased by sugars and competing plant lectins [203]. Another example is the increased phagocytosis of enteropathogenic *E. coli* strains by human polymorphonuclear cells, when the microbial cells are treated by *P. aeruginosa* lectins, as opposed to depressed phagocytosis when the granulocytes are treated by the same lectins [60,219].

H. Detection of Alterations in Glycosylated Cell Membrane Components

Lectins are very useful tools to detect quantitative and qualitative alterations of cell membrane components and changes in their distribution. Transformed cancer cells frequently exhibit higher agglutinability [290–293]. Altered cell agglutinability may supply information on the composition and distribution of the glycosylated compounds and their sialylation. Change in adsorption of lectins onto cells may also indicate alterations in these components. The adsorption may be assayed by inhibition of hemagglutinating activity of the examined lectin preparation or by the use of labeled lectins. The use of the labeled lectins enables cytochemical localization of changes in the glycosylated sugars involved in the lectin binding. Altered agglutinability of transformed cells by lectins also enables their separation from the normal cells.

I. Reduction of Cancer Cell Tumorigenicity

Plant lectins reduce tumorigenicity of cancer cells [291]. The PA-I of *P. aeruginosa* also significantly reduces the tumorigenicity of Lewis lung carcinoma (3LL) cells injected subcutaneously to C57Bl mice [292,293]. The lectin treatment of these cells reduced their tumorigenicity better than Con A and phytohemagglutinin, with preservation of their immunogenicity.

J. Stimulation of Cell Metabolism, Growth, and Functions, Enabling Studies of Their Biochemical Background

Microbial lectins affect the metabolism, growth rate, and function of various cells. An example of the microbial lectin effect is the stimulation of growth rate and phagocytic activity of the free-living, unicellular organisms: *Chlamydomonas, Euglena*, and *Tetrahymena*, by *P. aeruginosa* lectins [274,275]. The lectin-induced stimulation of phagocytic activity in *T. pyriformis* was correlated with the cAMP effect [275].

K. Activation of Macrophages, Polymorphonuclear Leukocytes, and Other Phagocytic Cells

Microbial lectins and adhesins were shown to induce activation of oxidative [220] and other metabolic pathways, leading to increased antibacterial and cytolytic efficiency of the macrophages, polymorphonuclear leukocytes, and other phagocytic cells. In these activities they resemble the effects of plant lectins.

L. Mitogenic Stimulation of Human Peripheral Lymphocytes

Mitogenic stimulation of human peripheral lymphocytes by microbial lectins [63,64] is very similar to that obtained with plant lectins [268] and has the same merits. PA-I and PA-II lectins were used for mitogenic stimulation of human lymphocytes (preferentially T cells), not only for chromosomal study, but also for the diagnosis and evaluation of the immune status of patients suffering from cancer [60,294] and other chronic diseases. The mitogenic response of lymphocytes of cancer-bearing patients, in the presence of PA-I and PA-II, was considerably lower than that of healthy subjects [60,294]. The pseudomonal lectins were as useful as the plant lectins for this medical application [294].

M. Mitogenic Stimulation of Murine Splenocytes

Some of the lectins that stimulate human peripheral lymphocytes were also stimulatory for murine splenocytes. The *P. aeruginosa* PA-II lectin behaves as a lectin of this group [64].

N. Detection and Quantitation of Viral Particles

The use of the hemagglutinating activity for viral quantitation [cited in 13] is an application of microbial lectins that is not shared by plant lectins. It is important because the viruses are invisible by the light microscope.

O. Use as a Vaccine Against Bacterial Infections

The use of lectinoid bacterial toxins as highly useful vaccines is traceable to the beginning of this century, when the diphtheria toxoid [295], a denatured lectinoid toxin, was introduced as a vaccine. Diphtheria and tetanus toxoids are used today, on a large scale, for active human immunization against diphtheria and tetanus. The antitoxins produced were shown to be useful for protection of animals against the toxin only when administered with it, or very shortly after it, but not later. In the aforementioned vaccines the lectinoid–toxoid complex is employed. Purified pili of *E. coli*, gonococci, and *Bordetella pertussis* were also found to be useful as a vaccine [296–298]. Mice injected with *P. aeruginosa* purified lectins became resistant against an otherwise lethal infection [140,141]. Humoral response against *Amoeba histolytica* lectin was reported to act against amebiasis [299]. *Bordetella pertussis* purified hemagglutinins were also shown to confer some protection against infection with the intact microorganisms [298]. It is anticipated that the knowledge of lectin or lectinoid structure would enable the production of synthetic peptide vaccines.

P. Use for Reduction of Cancer Cell Tumorigenicity With Antigenic Preservation as a Vaccine

The PA-I lectin of *P. aeruginosa* was shown to be very effective in reduction of 3LL cell tumorigenicity. Inoculation of lectin-treated 3LL cells to C57Bl mice led to significantly lower tumor production and mortality [292,293]. The surviving animals became considerably more resistant to inoculation with untreated 3LL cells, which were tumorigenic and lethal for untreated animals [293].

Q. Homing and Introduction of Drugs and Other Components to Target Cells

The homing service of microbial lectins and lectinoids may determine the distribution of microbial diseases, and the sensitivity of animals and specific tissues to them. This postulation may be true for intact microorganisms, which adhere to the specific host cells by means of the lectins and adhesins, as well as for their lectinoid toxins that bind to the host-sensitive cells. It has been known from the beginning of this century that diphtheria toxin is capable of causing damage to almost all types of cells in the susceptible animal, including skin, muscle, liver, adrenals, and nerve. Tetanus and botulinum toxins, on the other hand, appear to act only on the nervous system. It was supposed that tetanus toxin acts upon the anterior horn cells of the central nervous system, with resulting characteristic tetanic muscular spasm, whereas the botulinum toxin appears to act on the myoneural junctions and causes paralysis of nerve endings (death from botulism is due to respiratory paralysis). Each of these clostridial toxins causes injury to tissues when administered in very small doses.

When small doses are injected, there is a characteristic latent period of 24–72 hr, during which the toxin travels from the nerve endings to the CNS, through axis cylinders [300,301]. Related to this subject is the old-novel use of lectins in neurobiological research as "suicide transport agents." Some plant lectins, such as ricin, modeccin, and volkensin [302], as well as the lectinoid microbial toxins, are taken up by axonal projections and, upon transport back to the cell, kill the projection neurons. This property also enables selective lesioning of neuronal populations [302]. Like the clostridial toxic components, other toxic and cytostatic components may be linked to the lectins and lectinoid units and be transported by them. Similar linkage of enzymes and drugs to the microbial lectins and lectinoids may be valuable for medical treatment of genetic deficiencies and various malignant and chronic diseases.

R. Study of the Expression of "Adopted" Plant, Animal, and Heteromicrobial Lectin Genes in Bacteria

The vast development in gene cloning during the last decades, included the genes coding for lectin production in eukaryotic cells and in various microorganisms. These genes were cloned in host bacteria and expressed in them as "novel" microbial lectins. This technique enables the detailed study of lectin structure and of the genetic complex responsible for its subunit composition, production, regulation, and final form. It also enables the study of the lectin contribution to viability and virulence of the organism that produces it. Conjugal DNA transfer by endogenous genetic elements, such as broad-host–range plasmids and bacteriophages is very widely used. Moreover, mutated genes may also be cloned [303,304] and allow exhibition of the significance of their products to microbial pathogenesis. The type 1 pili [149,150] and the type P of uropathogenic *E. coli* strains [303], which were harbored in other *E. coli* strains, were investigated. Fimbrial [153] and afimbrial [79] adhesins of other *E. coli* strains, differing from the P and MR pili, were also cloned [155], as well as the S fimbriae [163,171–173]. The cloning enabled the study of the genes involved in the pili and adhesin synthesis and the influence of the cloned *E. coli* genes on the pathogenicity in different animal models [174,175].

S. Transfer of Resistance to Pathogens by the Cloning of Microbial Lectinoid Toxins to Higher Organisms

Another important aspect of "adopted" microbial lectins is the transfer of microbial genes, encoding microbial lectinoid toxins, to plants and animals that are not affected by them. Microbial toxins were shown to be useful for protecting plants against insects [305,306]. Therefore, microbial genes coding lectinoid toxins may also be useful for defending plants and other sensitive hosts against attacks by various parasites [305,306].

VI. CONCLUSIONS

The microbial lectins and lectinoids resemble plant and animal lectins in their sugar specificities, properties, and functions (see Sects. I and II). They are also similar to them in their in vitro biological effects and applicability for numerous important scientific and medical purposes (see Sects. IV and V.A–S). The lectins originating from microorganisms, and those adopted by them through cloning, have a great advantage for genetic studies of their production and its regulation. They are also suitable for studies of the factors that affect lectin localization (see Sect. III). The microbial lectins are of unique interest owing to their contribution to the establishment of infection and disease and their potential use as vaccines (see Sect. V). In this chapter we wish to present a new aspect of microbial lectinology not hitherto developed. This is the contribution of the microbial lectinology to the understanding of general lectin function, which is still considered as a central unsolved question in the up-to-date lectinological literature [5,39]. From our observations in *P. aeruginosa* [60,181,182,307–310] and overviewing the literature, we have recently suggested [307–309] that a general role of lectins might be: "the conditioning (enabling) and control (prevention), of the virtually irreversible activity of key lytic enzymes (or toxins), which trigger cascades of cellular events, leading to lysis or reorganization of endogenous or exogenous target macromolecules or cells, under special (or unfavorable) conditions." The facilitation is performed by a reversible interaction, analogous to that of immunoglobulins, hormones, and positioning sites of lytic enzymes (LEPS). This proposal is based on our and others' experimental evidence, and supported by theoretical considerations of the lectin analogy with immunoglobulins, hormones, and enzymes. The suggestion is forwarded with oversimplification to stimulate thought and research in the direction of *a lectin–lytic enzyme pair* and to invest efforts to discover lectin-associated (structurally or functionally) lytic enzyme activities in microbes, plants, and animals.

A. Genetic Linkage Between Lectins and Lytic Activities

As described in Section III, genetic studies with *P. aeruginosa* strains and mutants revealed that the lectin production correlated with that of several lytic and toxic activities. The levels of PA-I and PA-II were found to correlate with the proteolytic and hemolytic activities and the production of pyocyanin by the bacterium [60,181,182,309,310]. Lectin-devoid strains were shown to lack protease, hemolysin, and pyocyanin activities, and lectin-deficient strains and mutants (obtained from a high–lectin-producing wild type) were poor producers of these lytic and toxic factors. Moreover, mutants selected by Wretlind et al. as deficient protease producers or as hyperproducers of protease, were found by us to be deficient or high lectin producers, respectively [182]. The deficient strains and mutants grew nicely under optimal growth conditions and exhibited normal acetylcholinesterase, exotoxin A, and lipase activities. Genetic association between MR adhesins and hemolysin was also described in *E. coli* strains [174,186,256]. The most significant expression of the genetic linkage between them is their copresence on the same molecule of tectinoid enzymes or toxins [188,266] as described below (C).

B. Physiological Linkage Between Lectins and Lytic Activities

Physiological linkage between lectins or lectinoids and lytic activities is exhibited in several ways:

1. Coherent production of the lectins and the lytic activities: In *P. aeruginosa* both are products mainly of secondary metabolism [311,312] that are not produced during the early logarithmic growth phase, when growth conditions are optimal, but at the late stationary growth stage [40,53,60]. Time correlation in the appearance of lectins and glycosidases (e.g., concanavalin A and α-mannosidase or α-galactosidase) and other lytic enzymes has also been described in plants and animals.
2. Coherent, cataboliclike, repressing regulation of the production of lectins and the lytic activities: This phenomenon was reported in the presence of high sugar concentrations and in the dependence on ions and trace metals concentration in *P. aeruginosa* [313,314].

A similar phenomenon of pseudocatabolic repression of type 1 fimbriae production in *E. coli* was described by Eisenstein and Dodd [315]. The fact that fimbriae and lectin production is governed by cataboliclike repression may be also envisioned as a possible indirect indication to their involvement in catabolic mechanisms.

3. Coherence of hemagglutinins and lytic or toxic activities in numerous biological preparations from microorganisms, plants, and animals [2]: There are numerous reports on difficulties encountered during purification of lectins or hemagglutinins, owing to their "stubborn" contamination with hemolysin, protease, neurotoxins, and such [2]. Even the inhibition of fungal growth, ascribed to wheat germ agglutinin [316], was also ascribed to chitinase contamination of the lectin preparation.

C. Structural Linkage Between Lectins and Lytic Activities

Structural linkage between lectins and lytic activities may be of several types:

1. Presence of sugar-binding site and lytic domain on the same molecule: Examples for such bifunctional molecules are numerous, including the powerful microbial and plant lectinoid toxins [317–322], that are composed of a binding site B and an active site A [266]. The latter exhibits a RNA *N*-glycosidase activity [319–322] or other activities leading to cell lysis [266,317,319]. Additional examples are the lectinoid enzymes described in plants [188]. Actually, the presence of lectins or lectinoids with lytic activities on the same molecule is a most highly significant expression of the genetic linkage between them and a most profound evidence for the physiological importance of lectin–enzyme association.
2. Contiguous coexistence of the lectin (or lectinoid) domains with hydrolytic activities on bridging macromolecules or cell surfaces: Examples are numerous, including the influenza A virus and other viruses' sialophilic lectins, adjacent to sialidase or protease activities [13,34,35], and the microbial surface lectins with adjacent lytic enzymes (see Sect. III).
3. Existence of lectin or lectinoid components on and in cells containing endocellular lytic enzymes [323,324] (e.g., lysosomal enzymes [324]) that may be secreted from the cells or function in them on targets from without (following their internalization by the aid of lectins; e.g., microbial cells in macrophages, other phagocytes, and hepatocytes).
4. Sugar-independent interactions between lectins and enzymes leading to their association: Such interactions were reported using immobilized Con A and the jack bean glycosidases, α-mannosidases and α-galactosidase (the elution of the enzyme from the lectin, here, is not by addition of sugars, but by changing the ionic concentration [325]).

D. Experimental Data Showing that Inhibition of the Reversible Lectin Interaction with the Target Macromolecules Prevents the Associated Irreversible Lytic Reactions

There is a conspicuous amount of experimental data indicating the dependence of lytic processes, induced by cell-free systems or by intact cells, on an initial reversible lectin interaction. Prevention of this interaction, which abolishes the irreversible lytic processes, may be obtained by several means: (1) addition of the specific sugar, (2) addition of a competing lectin that exhibits a similar sugar specificity, (3) enzymatic removal of the lectin receptor from the target cells, or (4) separation of the lectin domain from the lytic component [266,317–322].

Among the numerous examples are the following:

1. Inhibition of bacteriophage-induced bacteriolysis by addition of competing plant lectins [246,247,251] or sugars [248], which prevent the reversible bacteriophage binding to the bacterial surface [249,250]
2. Inhibition of influenza virus fusion with and lysis of the host cells [13,137] following sialic acid removal from the cells by sialidase treatment
3. Inhibition of the toxic effects of microbial and plant lectinoid enzymes and toxins by sugars: for example, cholera toxin is deactivated by gangliosides [260], galactose inhibits most of the plant and bacterial ricinlike toxin effects [266,317–322], and separation of the B domain of most of these toxins abolishes their cytotoxic effect, although the activity of the separated A domain on the cell-free sensitive system is preserved [266,317–322]
4. Inhibition of the decomposition of peripheral desialylated glycoproteins in hepatocytes by sugars that compete with them for hepatic lectins, or by enzymatic removal of the galactose residue of these glycoproteins, which is involved in their recognition by the liver lectins [323]
5. Inhibition of sheep red blood cell lysis by invertebrate hemolymph lytic systems following addition of sugars that interact with the hemolymph lectins [326]
6. Inhibition of bacteriolysis in vertebrate phagocytes by inhibition of the microbial or the phagocyte lectins by the respective sugars [194]

E. Analogy Between Lectins and Positioning Sites of Lytic Enzymes (LEPS), Immunoglobulins, and Hormones

The proposed *lectin–lytic enzyme cofunction* emphasizes the analogy between lectins and LEPS [307–309] that home to the substrate "positioning groups" [327] in a reversible manner and provide the anchoring and optimal exposure of the substrate-sensitive residue to the enzyme lytic site [327]. Examples are the anionic positioning site of acetylcholinesterase and the hydrophobic positioning site of chymotrypsin, which allow the hydrolysis of acetylcholine and proteins, respectively [308,327; Fig. 6]. It also emphasizes the analogy between lectins and immunoglobulins, when the first are envisioned as proteins that condition the lysis of antigen-bearing targets by enabling the lytic activities of complement or phagocytic cells. The interaction of immunoglobulins with cells or macromolecules also permits the activity of lytic key enzymes. The key enzymes trigger a cascade of reactions in the cells that bear the immunoglobulins or the cells that interact with them (hormonelike action), leading to their reorganization and to their secretion of active components. The analogy of the action of the lectin, LEPS, and immunoglobulin is not only in the mode of their interactions, but also in that these reversible interactions (with the specific residues of cells or macromolecules: glycosylated macromolecules, positioning groups, and antigens, respectively) are competitively inhibited by simpler free molecules (sugar, positioning groups, and haptens, respectively) and that the reactions that follow them are already irreversible. Additional supportive points for the analogy is that the lectins, lytic enzymes, and antibodies (as well as hormones) are generally produced constitutively, at a low level, whereas their vigorous production is mainly inducible, only when they are needed (!).

Moreover, the analogy is strongly supported by the fact that lectins or lectinoids constitute LEPS of certain enzymes [188] and toxins [266,317–322] and substitute immunoglobulins in enabling phagocytosis [194] and lysis of bacteria and erythrocytes [307–309]. The α-galactosidase of *Vicia faba*, which contains a lectinoid site [188] comparable with the anionic site of acetylcholinesterase or the hydrophobic positioning site of chymotrypsin [327], is a good example. An example for immunoglobulinlike function is the

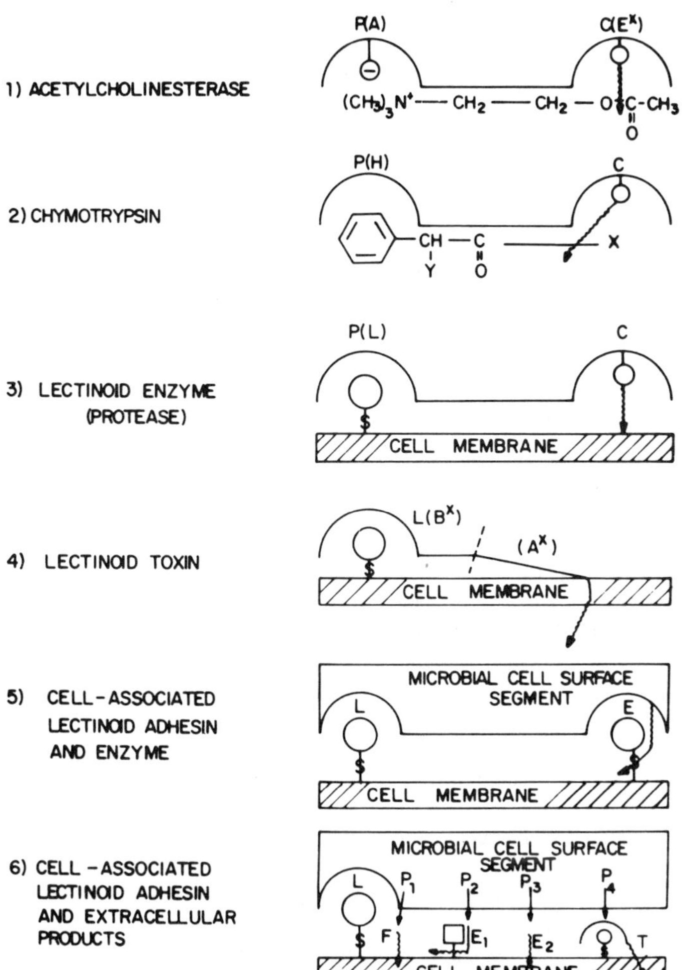

Figure 6 Analogy of lectin (or lectinoid) binding site (L in 4–6) to enzyme positioning site (P in 1–3) and its role in enabling maximal cytolytic and cytotoxic microbial function: (1) The positioning (anionic) site, P(A), and the catalytic (esteratic) site, C(E^x), of acetylcholinesterase. (2) The positioning (hydrophobic) site, P(H), and the catalytic (hydrolytic) site, C, of chymotrypsin. (3) The lectinoid positioning site, P(L), and the catalytic (hydrolytic) site, C, of hydrolytic enzymes: glycosidases, RNA N-glycosidase of the ricinlike plant and microbial toxins, protease, and others. (4) The lectinoid binding domain, L (B^x), and the active domain, A^x, of lectinoid toxins. (5) A lectin or lectinoid adhesin, L, adjacent to a lytic enzyme, E, which function on the same sugars (S). (6) Cell-associated lectin or lectinoid, L, which conditions the function of lytic and cytotoxic products of the microbe (P1–P4) on the target cell membrane. These products include factors (F), which act on the cell membranes; enzymes (E), which act on the surface of cell membrane (E1), or on the cell membrane internal components (E2); as well as toxins (T), which enter the cell.

lectin-mediated hemolysis in invertebrates comparable with antibody-mediated complement-induced hemolysis [307–309]. Another example of immunoglobulinlike, lectin-mediated function is the lectinophagocytosis (comparable with the opsonin-mediated phagocytosis observed in the same cells) [194]. The famous resistance of lectins to proteolysis [2–5], which may be the cofunctioning lytic activity, is highly compatible with the proposed theory.

ACKNOWLEDGMENT

The authors were supported by the Health Sciences Center fund of Bar-Ilan University, Ramat Gan, Israel, and by the USA-Israel Bi-National Science Foundation (BSF) grant no. 89-00454, Jerusalem.

REFERENCES

1. Stillmark, H. (1888). *Ueber Ricin, ein Giftiges Ferment aus dem Samen von* Ricinus *comm. L. und Einigen Anderen Euphorbiaceen*, Inaug. Diss., Dorpat University.
2. Gold, E. R., and Balding, P. (1975). *Receptor Specific Proteins. Plant and Animal Lectins*, Excerpta Medica, Amsterdam.
3. Liener, I. E. (1976). Phytohemagglutinins (phytolectins). *Annu. Rev. Plant Physiol. 27*:291–319.
4. Goldstein, I. J., and Hayes, C. E. (1978). The lectins: Carbohydrate-binding proteins of plants and animals. *Adv. Carbohydr. Chem. Biochem. 35*:127–340.
5. Liener, I. E., Sharon, N., and Goldstein, I. J., eds. (1986). *The Lectins. Properties, Functions and Applications in Biology and Medicine*. Academic Press, New York.
6. Kocourek, J. (1986). Historical background. In *The Lectins. Properties, Functions and Applications in Biology and Medicine* (I. E. Liener, N. Sharon, and I. J. Goldstein, eds.). Academic Press, New York, pp. 1–32.
7. Kraus, R., and Ludwig, S. (1902). Über bakterien Haemagglutinine und Antihaemagglutinine. *Wien. Klin. Wochenschr. 15*:120–121.
8. Guyot, G. (1908). Über die bacterielle Haemagglutination. *Zentralbl. Bakteriol. Abt. I Orig. 47*:640–653.
9. Rosenthal, L. (1931). Spermagglutination by bacteria. *Proc. Soc. Exp. Biol. 28*:827.
10. Rosenthal, L. (1943). Agglutinating properties of *Escherichia coli*. Agglutination of erythrocytes, leukocytes, thrombocytes, spermatozoa, spores of molds and pollen by strains of *E. coli*. *J. Bacteriol. 45*:545–550.
11. Hirst, G. K. (1941). The agglutination of red cells by allantoic fluid of chick embryos infected with influenza virus. *Science 94*:22–23.
12. McClelland, L., and Hare, R. (1941). The adsorption of influenza virus by red cells and a new in vitro method of measuring antibodies for influenza virus. *Can. J. Public Health 32*:530–538.
13. Markwell, M. A. K. (1986). Viruses as hemagglutinins and lectins. In *Microbial Lectins and Agglutinins: Properties and Biological Activity* (D. Mirelman, ed.). John Wiley & Sons, New York, pp. 21–53.
14. Keogh, E. V., North, E. A., and Warburton, M. F. (1947). Hemagglutinins of the *Haemophilus* group. *Nature 160*:63.
15. Warburton, M. F., and Fisher, S. (1951). The haemagglutinin of *Haemophilus pertussis* III. Extraction of the antigen from the bacteria and its stabilisation by adsorption. *Aust. J. Exp. Biol. Med. Sci. 29*:265–272.
16. Lamanna, C. (1948). Haemagglutination by botulinal toxin. *Proc. Soc. Exp. Biol. 69*:332.
17. Elo, J., Estola, E., and Malmstrom, N. (1951). On phytohemagglutinins present in mushrooms. *Ann. Med. Exp. Biol. Fenn. 29*:297–308.
18. Gottschalk, A. (1960). *The Chemistry and Biology of Sialic Acids and Related Substances*. Cambridge University Press, London.

19. Neter, E. (1956). Bacterial haemagglutination and haemolysis. *Bacteriol. Rev. 20*:166–188.
20. Uhlenbruck, G. (1987). Bacterial lectins: Mediators of adhesion. *Zentralbl. Bakteriol. Hyg. A 263*:497–508.
21. Boyd, W. C., and Shapleigh, E. (1954). Separation of individuals of any blood group into secretors and non-secretors by use of a plant agglutinin (lectin). *Blood 9*:1195–1198.
22. Krüpe, M. (1956). *Blutgruppenspezifische pflanzliche Eiweisskörper (Phytagglutinine)*. F. Enke, Stuttgart, pp. 1–131.
23. Mäkelä, O. (1957). Studies in haemagglutinins of Leguminosae seeds. *Ann. Med. Exp. Biol. Fenn. 35*(Suppl. 11):1—156.
24. Collier, W. A., and de Miranda, J. C. (1955). Bacterien-haemagglutination III. Die hemmung der coli-haemagglutination durch mannose. *Antonie Leeuwenhoek J. Microbiol. Serol. 21*: 133–140.
25. Duguid, J. P., Smith, I. W., Dempster, G., and Edmunds, P. N. (1955). Non-flagellar filamentous appendages ('fimbriae') and haemagglutinating activity in *Bacterium coli. J. Pathol. Bacteriol. 70*:335–348.
26. Duguid, J. P., and Gillies, R. R. (1957). Fimbriae and adhesive properties in dysentery bacilli. *J. Pathol. Bacteriol. 74*:397–411.
27. Duguid, J. P. (1959). Fimbriae and adhesive properties in *Klebsiella* strains. *J. Gen. Microbiol. 21*:271–286.
28. Duguid, J. P., Anderson, E. S., and Campbell, I. (1966). Fimbriae and adhesive properties in salmonellae. *J. Pathol. Bacteriol. 92*:107–138.
29. Brinton, C. C., Jr. (1959). Non-flagellar appendages of bacteria. *Nature 183*:782–786.
30. Brinton, C. C., Jr. (1965). The structure, function, synthesis and genetic control of bacterial pili and a molecular model for DNA and RNA transport in gram-negative bacteria. *Trans. N.Y. Acad. Sci. 27*:1003–1054.
31. Old, D. C., and Duguid, J. P. (1971). Selection of fimbriate transductants of *Salmonella typhimurium* dependent on motility. *J. Bacteriol. 107*:655–658.
32. Collee, J. G. (1961). The nature and properties of the hemagglutinin of *Clostridium welchii. J. Pathol. Bacteriol. 81*:297–312.
33. Kathan, R. H., Winzler, R. J., and Johnson, C. A. (1961). Preparation of an inhibitor of viral haemagglutination from human erythrocytes. *J. Exp. Med. 113*:37–45.
34. Laver, W. G., and Valentine, R. C. (1969). Morphology of the isolated hemagglutinin and neuraminidase subunits of influenza virus. *Virology 38*:105–119.
35. Wiley, D. C., and Skehel, J. J. (1987). The structure and function of the hemagglutinin membrane glycoprotein of influenza virus. *Annu. Rev. Biochem. 56*:365–394.
36. Heumann, W., and Marx, R. (1964). Feinstruktur und funktion der fimbrien bei dem sternbildenden bakterium *Psuedomonas echinoides. Arch. Mikrobiol. 47*:325–337.
37. Drimmer-Herrenheiser, H. (1953). Haemagglutination by a *Pseudomonas* of faecal origin. *Bull. Res. Counc. Isr. 2*:445.
38. Sage, H. J., and Connett, S. L. (1989). Studies on a hemagglutinin from the meadow mushroom. II. Purification, composition and structure of *Agaricus campestris* hemagglutinin. *J. Biol. Chem. 244*:4713–4719.
39. Mirelman, D., ed. (1986). *Microbial Lectins and Agglutinins: Properties and Biological Activity*. John Wiley & Sons, New York.
40. Gilboa-Garber, N. (1972). Inhibition of broad spectrum hemagglutinin from *Pseudomonas aeruginosa* by D-galactose and its derivatives. *FEBS Lett. 20*:242–244.
41. Gilboa-Garber, N. (1972). Purification and properties of hemagglutinin from *Pseudomonas aeruginosa* and its reaction with human blood cells. *Biochim. Biophys. Acta 273*:165–173.
42. Fujita, Y., Oishi, K., and Aida, K. (1973). Sugar specificity of anti-B hemagglutinin produced by *Streptomyces* sp. *Biochem. Biophys. Res. Commun. 53*:495–501.
43. Fujita, Y., Oishi, K., Suzuki, K., and Imahori, K. (1975). Purification and properties of an anti-B hemagglutinin produced by *Streptomyces* sp. *Biochemistry 14*:4465–4470.

44. Balding, P., Gold, E. R., Boroff, D. A., and Roberts, T. A. (1973). Observations on receptor specific proteins II. Haemagglutination and haemagglutination-inhibition reactions of *Clostridium botulinum* types A, C, D and E haemagglutinins. *Immunology 25*:773–782.
45. Presant, C. A., and Kornfeld, S. (1972). Characterization of the cell surface receptor for the *Agaricus bisporus* hemagglutinin. *J. Biol. Chem. 247*:6937–6945.
46. Anstee, D. J. (1972). The immunochemistry of cell surface galactosyl antigens. Ph.D. Thesis, Bristol, England.
47. Fujita, Y., Oishi, K., and Aida, K. (1974). A novel hemagglutinin produced by *Aspergillus niger. J. Biochem. 76*:1347–1349.
48. Rosen, S. D., Kafka, J. A., Simpson, D. L., and Barondes, S. H. (1973). Developmentally regulated, carbohydrate-binding protein in *Dictyostellium discoideum. Proc. Natl. Acad. Sci. USA 70*: 2554–2557.
49. Rosen, S. D., Reitherman, R. W., and Barondes, S. H. (1975). Distinct lectin activities from six species of cellular slime molds. *Exp. Cell Res. 95*:159–166.
50. Barondes, S. H. (1981). Lectins: Their multiple endogenous cellular functions. *Annu. Rev. Biochem. 50*:207–231.
51. Gilboa-Garber, N., and Mizrahi, L. (1973). Peroxidase attachment to Sepharose mediated by bacterial hemagglutinin of *Pseudomonas aeruginosa. Biochim. Biophys. Acta 317*: 106–113.
52. Gilboa-Garber, N., Mizrahi, L., and Garber, N. (1977). Mannose-binding hemagglutinins in extracts of *Pseudomonas aeruginosa. Can. J. Biochem. 55*:975–981.
53. Gilboa-Garber, N. (1982). *Pseudomonas aeruginosa* lectins. *Methods Enzymol. 83*:378–385.
54. Kameyama, T., Oishi, K., and Aida, K. (1979). Stereochemical structure recognized by the L-fucose-specific hemagglutinin produced by *Streptomyces* sp. *Biochim. Biophys. Acta 587*: 407–414.
55. Kameyama, T., Oishi, K., and Aida, K. (1981). Human erythrocyte receptor of L-fucose-specific lectin produced by *Streptomyces* sp. *Agric. Biol. Chem. 45*:975–980.
56. Collier, W. A., and Jacoeb, M. (1955). Bacterien–haemagglutination I. Versuche mit einem haemagglutinierenden stamm von *E. coli. Antonie Leeuwenhoek J. Microbiol. Serol. 21*:113.
57. Fujita, Y., Oishi, K., and Aida, K. (1972). Hemagglutination by culture broth of *Actinomycetes* and *Aspergillus. J. Gen. Appl. Microbiol. 18*:73–75.
58. Gilboa-Garber, N., Mizrahi, L., and Garber, N. (1972). Purification of the galactose-binding hemagglutinin of *Pseudomonas aeruginosa* by affinity column chromatography using Sepharose. *FEBS Lett. 28*:93–95.
59. Gilboa-Garber, N., and Garber, N. (1973). Effect of amino acid reagents on the galactose-binding hemagglutinin of *Pseudomonas aeruginosa*. In *Proceedings First International Congress Bacteriology*, Jerusalem, p. 188.
60. Gilboa-Garber, N. (1986). Lectins of *Pseudomonas aeruginosa*: Properties, biological effects and applications. In *Microbial Lectins and Agglutinins: Properties and Biological Activity* (D. Mirelman, ed.). John Wiley & Sons, New York, pp. 255–269.
61. Agrawal, B. B. L., and Goldstein, I. J. (1965). Specific binding of concanavalin A to cross-linked dextran gels. *Biochem. J. 96*:23c–25c.
62. Cuatrecasas, P., Parikh, I., and Hollenberg, M. D. (1973). Affinity chromatography and structural analysis of *Vibrio cholerae* enterotoxin–ganglioside agarose and the biological effects of ganglioside-containing soluble polymers. *Biochemistry 12*:4253–4264.
63. Sharabi, Y., and Gilboa-Garber, N. (1979). Mitogenic stimulation of human lymphocytes by *Pseudomonas aeruginosa* galactosephilic lectin. *FEMS Microbiol. Lett. 5*:273–276.
64. Avichezer, D., and Gilboa-Garber, N. (1987). PA-II, the L-fucose and D-mannose binding lectin of *Pseudomonas aeruginosa* stimulates human peripheral lymphocytes and murine splenocytes. *FEBS Lett. 216*:62–66.
65. Sharon, N. (1977). Lectins. *Sci. Am. 236*:108–119.

66. Sharon, N. (1986). Bacterial lectins. In *The Lectins, Properties, Functions, and Applications in Biology and Medicine* (I. E. Liener, N. Sharon, and I. J. Goldstein, eds.). Academic Press, New York, pp. 493–526.
67. Goldstein, I. J., Hughes, R. C., Monsigny, M., Osawa, T., and Sharon, N. (1980). What should be called a lectin? *Nature 285*:66.
68. Goldstein, I. J., and Poretz, R. D. (1986). Isolation, physicochemical characterization and carbohydrate-binding specificity of lectins. In *The Lectins, Properties, Functions, and Applications in Biology and Medicine* (I. E. Liener, N. Sharon, and I. J. Goldstein, eds.). Academic Press, New York, pp. 33–247.
69. Kocourek, J., and Hořejši, V. (1981). Defining a lectin. *Nature 290*:188.
70. Duguid, J. P., Clegg, S., and Wilson, M. I. (1979). The fimbrial and nonfimbrial haemagglutinins of *Escherichia coli. J. Med. Microbiol. 12*:213–227.
71. Duguid, J. P., and Old, D. C. (1980). Adhesive properties of Enterobacteriaceae. In *Bacterial Adherence. Receptors and Recognition*. Ser. B Vol. 6 (E. H. Beachey, ed.). Chapman & Hall, London, pp. 186–217.
72. Normark, S., Lark, D., Hull, R., Norgren, M., Båga, M., O'Hanley, P., Schoolnik, G., and Falkow, S. (1983). Genetics of digalactoside-binding adhesin from a uropathogenic *Escherichia coli* strain. *Infect. Immun. 41*:942–949.
73. Normark, S., Båga, M., Goransson, M., Lindberg, F. P., Lund, B., Norgren, M., and Uhlin, B.-E. (1986). Genetics and biogenesis of *Escherichia coli* adhesins. In *Microbial Lectins and Agglutinins: Properties and Biological Activity* (D. Mirelman, ed.). John Wiley & Sons, New York, pp. 113–143.
74. Lund, B., Lindberg, F. P., Båga, M., and Normark, S. (1985). Globoside-specific adhesins of uropathogenic *Escherichia coli* are encoded by similar trans-complementable gene clusters. *J. Bacteriol. 162*:1293–1301.
75. Stirm, S., Ørskov, F., Ørskov, I., and Mansa, B. (1967). Episome-carried surface antigen K88 of *Escherichia coli*: II. Isolation and chemical analysis. *J. Bacteriol. 93*:731–739.
76. Moch, T., Hoschutzky, H., Hacker, J., Kroncke, K. D., and Jann, K. (1987). Isolation and characterization of the α-sialyl-β-2,3-galactosyl-specific adhesin from fimbriated *Escherichia coli. Proc. Natl. Acad. Sci. USA 84*:3462–3466.
77. Walz, W., Schmidt, M. A., Labigne-Roussel, A. F., Falkow, S., and Schoolnik, G. (1985). AFA-I, a cloned afimbrial X-type adhesin from a human pyelonephritic *Escherichia coli* strain. Purification, and chemical, functional and serologic characterization. *Eur. J. Biochem. 152*:315–321.
78. Labigne-Roussel, A., and Falkow, S. (1988). Distribution and degree of heterogeneity of the afimbrial-adhesin-encoding operon (*afa*) among uropathogenic *Escherichia coli* isolates. *Infect. Immun. 56*:640–648.
79. Labigne-Roussel, A. F., Lark, D., Schoolnik, G., and Falkow, S. (1984). Cloning and expression of an afimbrial adhesin (AFA-1) responsible for P blood group-independent, mannose-resistant hemagglutination from a pyelonephritic *Escherichia coli* strain. *Infect. Immun. 46*: 251–259.
80. Goldhar, J., Perry, R., Golecki, J. R., Hoschutzky, H., Jann, B., and Jann, K. (1987). Nonfimbrial, mannose-resistant adhesins from uropathogenic *Escherichia coli* O83:K1:H4: and O14:K?:H11. *Infect. Immun. 55*:1837–1842.
81. Forestier, C., Welinder, K. G., Darfeuille-Michaud, A., and Klemm, P. (1987). Afimbrial adhesin from *Escherichia coli* strain 2230: Purification, characterization, and partial covalent structure. *FEMS Microbiol. Lett. 40*:47–50.
82. Razin, S. (1986). Mycoplasmal adhesins and lectins. In *Microbial Lectins and Agglutinins: Properties and Biological Activity* (D. Mirelman, ed.). John Wiley & Sons, New York, pp. 217–235.
83. Leffler, H., and Svanborg-Edén, C. (1986). Glycolipids as receptors for *Escherichia coli* lectins or adhesins. In *Microbial Lectins and Agglutinins: Properties and Biological Activity* (D. Mirelman, ed.). John Wiley & Sons, New York, pp. 83–111.

84. Booth, B. A., Sciortino, C. V., and Finkelstein, R. A. (1986). Adhesins of *Vibrio cholerae*. In *Microbial Lectins and Agglutinins: Properties and Biological Activity* (D. Mirelman, ed.). John Wiley & Sons, New York, pp. 169–182.
85. Sheladia, V. L., Chambers, J. P., Guevara, J., Jr., and Evans, D. J. (1982). Isolation, purification, and partial characterization of type V-A hemagglutinin from *Escherichia coli* GV-12, 01:H$^-$. *J. Bacteriol. 152*:757–761.
86. Williams, P. H., Knutton, S., Brown, M. G. M., Candy, D. C. A., and McNeish, A. S. (1984). Characterization of nonfimbrial mannose-resistant protein hemagglutinins of two *Escherichia coli* strains isolated from infants with enteritis. *Infect. Immun. 44*:592–598.
87. Zusman, D. R., Cumsky, M. G., Nelson, D. R., and Romeo, J. M. (1986). Myxobacterial hemagglutinin: A developmentally induced lectin from *Myxococcus xanthus*. In *Microbial Lectins and Agglutinins: Properties and Biological Activity* (D. Mirelman, ed.). John Wiley & Sons, New York, pp. 197–216.
88. Väisänen, V., Korhonen, T. K., Jokinen, M., Gahmberg, C. G., and Ehnholm, C. (1982). Blood group M specific haemagglutinin in pyelonephritogenic *Escherichia coli*. *Lancet 2*:1192.
89. Rhen, M., Klemm, P., and Korhonen, T. K. (1986). Identification of two new hemagglutinins of *Escherichia coli N*-acetyl-D-glucosamine-specific fimbriae and a blood group M-specific agglutinin by cloning the corresponding genes in *Escherichia coli* K-12. *J. Bacteriol. 168*:1234–1242.
90. Wadström, T., Trust, T. J., and Brooks, D. E. (1983). Bacterial surface lectins. In *Lectins Biology, Biochemistry, Clinical Biochemistry*, Vol. 3. (T. C. Bøg-Hansen and G. A. Spengler, eds.). Walter de Gruyter, Berlin, pp. 479–494.
91. Sharon, N., and Ofek, I. (1986). Mannose specific bacterial surface lectins. In *Microbial Lectins and Agglutinins: Properties and Biological Activity* (D. Mirelman, ed.). John Wiley & Sons, New York, pp. 55–81.
92. Sharon, N. (1987). Bacterial lectins, cell–cell recognition and infectious disease. *FEBS Lett. 217*:145–157.
93. Cisar, J. O. (1986). Fimbrial lectins of the oral actinomyces. In *Microbial Lectins and Agglutinins: Properties and Biological Activity* (D. Mirelman, ed.). John Wiley & Sons, New York, pp. 183–196.
94. Nelson, D. R., Cumsky, M. G., and Zusman, D. R. (1981). Localization of myxobacterial hemagglutinin in the periplasmic space and on the cell surface of *Myxococcus xanthus* during developmental aggregation. *J. Biol. Chem. 256*:12589–12595.
95. Glick, J., and Garber, N. (1983). The intracellular localization of *Pseudomonas aeruginosa* lectins. *J. Gen. Microbiol. 129*:3085–3090.
96. Glick, J., Garber, N., and Shohet, D. (1987). Surface haemagglutinating activity of *Pseudomonas aeruginosa*. *Microbios 50*:69–80.
97. Garber, N., Sharon, N., Shohet, D., Lam, J. S., and Doyle, R. J. (1985). Contribution of hydrophobicity to the hemagglutination reactions of *Pseudomonas aeruginosa*. *Infect. Immun. 50*:336–337.
98. Mirelman, D., and Ofek, I. (1986). Introduction to microbial lectins and agglutinins. In *Microbial Lectins and Agglutinins: Properties and Biological Activity* (D. Mirelman, ed.). John Wiley & Sons, New York, pp. 1–19.
99. Schoolnik, G. K., Rothbard, J. B., and Gotschlich, E. C. (1986). Structure–function analysis of gonococcal pili. In *Microbial Lectins and Agglutinins: Properties and Biological Activity* (D. Mirelman, ed.). John Wiley & Sons, New York, pp. 145–168.
100. Mirelman, D., and Ravdin, J. I. (1986). Lectins in *Entamoeba histolytica*. In *Microbial Lectins and Agglutinins: Properties and Biological Activity* (D. Mirelman, ed.). John Wiley & Sons, New York, pp. 319–334.
101. Fujita-Yamaguchi, Y., Oishi, K., Suzuki, K., and Imahori, K. (1982). Studies on carbohydrate binding to a lectin purified from *Streptomyces* sp. *Biochim. Biophys. Acta 701*: 86–92.

102. Matsui, I., Oishi, K., Kanaya, K., and Baba, N. (1985). The morphology of L-fucose, D-mannose specific lectin (SFL 100-2) produced by *Streptomyces* no. 100-2. *J. Biochem. 97*:399–408.
103. Matsui, I., and Oishi, K. (1986). Carbohydrate-binding properties of L-fucose, D-mannose-specific lectin (SFL 100-2) particles produced by *Streptomyces* sp. *J. Biochem. 100*:115–121.
104. Salit, I. E., and Gotschlich, E. C. (1977). Hemagglutination by purified type I *Escherichia coli* pili. *J. Exp. Med. 146*:1169–1181.
105. Salit, I. E., and Gotschlich, E. C. (1977). Type I *Escherichia coli* pili: Characterization of binding to monkey kidney cells. *J. Exp. Med. 146*:1182–1194.
106. Ofek, I., Mirelman, D., and Sharon, N. (1977). Adherence of *Escherichia coli* to human mucosal cells mediated by mannose receptors. *Nature 265*:623–625.
107. Ofek, I., and Beachey, E. H. (1978). Mannose binding and epithelial cell adherence of *Escherichia coli. Infect. Immun. 22*:247–254.
108. Ofek, I., Beachey, E. H., and Sharon, N. (1978). Surface sugars of animal cells as determinants of recognition in bacterial adherence. *Trends Biochem. Sci. 3*:159–160.
109. Klemm, P. (1984). The *fimA* gene encoding the type-1 fimbrial subunit of *Escherichia coli.* Nucleotide sequence and primary structure of the protein. *Eur. J. Biochem. 143*: 395–399.
110. Firon, N., Ofek, I., and Sharon, N. (1983). Carbohydrate specificity of the surface lectins of *Escherichia coli, Klebsiella pneumoniae,* and *Salmonella typhimurium. Carbohydr. Res. 120*:235–250.
111. Firon, N., Ofek, I., and Sharon, N. (1984). Carbohydrate-binding sites of the mannose-specific fimbrial lectins of enterobacteria. *Infect. Immun. 43*:1088–1090.
112. Firon, N., Ashkenazi, S., Mirelman, D., Ofek, I., and Sharon, N. (1987). Aromatic alpha-glycosides of mannose are powerful inhibitors of the adherence of type 1 fimbriated *Escherichia coli* to yeast and intestinal epithelial cells. *Infect. Immun. 55*:472–476.
113. Abraham, S. N., Goguen, J. D., and Beachey, E. H. (1988). Hyperadhesive mutant of type 1-fimbriated *Escherichia coli* associated with formation of fimH organelles (fimbriosomes). *Infect. Immun. 56*:1023–1029.
114. Evans, D. G,, and Evans, D. J., Jr. (1978). New surface-associated heat labile colonization factor antigen (CFA/II) produced by enterotoxigenic *Escherichia coli* of serogroups O6 and O8. *Infect. Immun. 21*:638–647.
115. Leffler, H., and Svanborg-Edén, C. (1980). Chemical identification of a glycosphingolipid receptor for *Escherichia coli* attaching to human urinary tract epithelial cells and agglutinating human erythrocytes. *FEMS Microbiol. Lett. 8*:127–134.
116. Källenius, G., Möllby, R., Svenson, S. B., Winberg, J., Lundblad, A., Svensson, S., and Cedergren, B. (1980). The P^K antigen as receptor for the haemagglutinin of pyelonephritic *Escherichia coli. FEMS Microbiol. Lett. 7*:297–302.
117. Bock, K., Breimer, M. E., Brignole, A., Hansson, G. C., Karlsson, K.-A., Larson, G., Leffler, H., Samuelsson, B. E., Strömberg, N., Svanborg-Edén, C., and Thurin, J. (1985). Specificity of binding of a strain of uropathogenic *Escherichia coli* to Gal α1-4Gal-containing glycosphingolipids. *J. Biol. Chem. 260*:8545–8551.
118. Väisänen, V., Elo, J., Tallgren, L. G., Siitonen, A., Makela, P. H., Svanborg-Edén, C., Källenius, G., Svenson, S. B., Hultberg, H., and Korhonen, T. (1981). Mannose-resistant hemagglutination and P antigen recognition are characteristic of *Escherichia coli* causing primary pyelonephritis. *Lancet 2*:1366–1369.
119. Faris, A., Lindahl, M., and Wadström, T. (1980). G_{M2}-like glycoconjugate as possible erythrocyte receptor for the CFA/I and K99 haemagglutinins of enterotoxigenic *Escherichia coli. FEMS Microbiol. Lett. 7*:265–269.
120. Lindahl, M., Faris, A., and Wadström, T. (1982). Colonization factor antigen on enterotoxigenic *Escherichia coli* is a sialic acid-specific lectin. *Lancet 2*:280.
121. Lindahl, M., and Wadström, T. (1983). Terminal *N*-acetylgalactosamine and sialic acid residues are recognized by K99 surface hemagglutinin of enterotoxigenic *E. coli. IRCS Med. Sci. 11*:790.

122. Korhonen, T. K., Väisänen-Rhen, V., Rhen, M., Pere, A., Parkkinen, J., and Finne, J. (1984). *Escherichia coli* fimbriae recognizing sialyl galactosides. *J. Bacteriol. 159*:762–766.
123. Parkkinen, J., Rogers, G. N., Korhonen, T., Dahr, W., and Finne, J. (1986). Identification of the *O*-linked sialyloligosaccharides of glycophorin A as the erythrocyte receptors for S-fimbriated *Escherichia coli. Infect. Immun. 54*:37–42.
124. Smit, H., Gaastra, W., Kamerling, J. P., Vliegenthart, J. F. G., and de Graaf, F. K. (1984). Isolation and structural characterization of the equine erythrocyte receptor for enterotoxigenic *Escherichia coli* K99 fimbrial adhesin. *Infect. Immun. 46*:578–584.
125. Klemm, P. (1982). Primary structure of the CFA1 fimbrial protein from human enterotoxigenic *Escherichia coli* strains. *Eur. J. Biochem. 124*:339–348.
126. Kabat, E. A. (1978). Dimensions and specificities of recognition sites on lectins and antibodies. *J. Supramol. Struct. 8*:79–88.
127. Ørskov, I., Birch-Andersen, A., Duguid, J. P., Stenderup, J., and Ørskov, F. (1985). An adhesive protein capsule of *Escherichia coli. Infect. Immun. 47*:191–200.
128. Sobelslavsky, O., Prescott, B., and Chanock, R. M. (1968). Adsorption of *Mycoplasma pneumoniae* to neuraminic acid receptors of various cells and possible role in virulence. *J. Bacteriol. 96*:695–705.
129. Powell, D. A., Hu, P. C., Wilson, M., Collier, A. M., and Baseman, J. B. (1976). Attachment of *Mycoplasma pneumoniae* to respiratory epithelium. *Infect. Immun. 13*:959–966.
130. Gabridge, M. G., and Taylor-Robinson, D. (1979). Interaction of *Mycoplasma pneumoniae* with human lung fibroblasts: Role of receptor sites. *Infect. Immun. 25*:455–459.
131. Baseman, J. B., Banai, M., and Kahane, I. (1982). Sialic acid residues mediated *Mycoplasma pneumoniae* attachment to human and sheep erythrocytes. *Infect. Immun. 38*: 384–391.
132. Glasgow, L. R., and Hill, R. L. (1980). Interaction of *Mycoplasma gallisepticum* with sialyl glycoproteins. *Infect. Immun. 30*:353–361.
133. Ramphal, R., and Pyle, M. (1983). Evidence for mucins and sialic acid as receptors for *Pseudomonas aeruginosa* in the lower respiratory tract. *Infect. Immun. 41*:339–344.
134. Hazlett, L. D., Moon, M., and Berk, R. S. (1986). In vivo identification of sialic acid as the ocular receptor for *Pseudomonas aeruginosa. Infect. Immun. 51*:687–689.
135. Ishikawa, H., and Isayama, Y. (1987). Evidence for sialyl glycoconjugates as receptors for *Bordetella bronchiseptica* on swine nasal mucosa. *Infect. Immun. 55*:1607–1609.
136. Parkkinen, J., Finne, J., Achtman, M., Väisänen, V., and Korhonen, T. K. (1983). *Escherichia coli* strains binding neuraminyl α-2-3 galactosides. *Biochem. Biophys. Res. Commun. 111*: 456–461.
137. Wharton, S. A. (1987). The role of influenza virus haemagglutinin in membrane fusion. *Microbiol. Sci. 4*:119–124.
138. Garber, N., Glick, J., Gilboa-Garber, N., and Heller, A. (1981). Interaction of *Pseudomonas aeruginosa* lectins with *Escherichia coli* strains bearing blood group determinants. *J. Gen. Microbiol. 123*:359–363.
139. Gilboa-Garber, N., Nir-Mizrahi, I., and Mizrahi, L. (1977). Specific agglutination of *Escherichia coli* $O_{128}B_{12}$ by the mannose-binding proteins of *Pseudomonas aeruginosa. Microbios 18*:99–109.
140. Gilboa-Garber, N., and Sudakevitz, D. (1982). The use of *Pseudomonas aeruginosa* lectin preparations as a vaccine. In *Advances in Pathology*, Vol. 1 (E. Levy, ed.). Pergamon Press, Oxford, pp. 31–33.
141. Sudakevitz, D., and Gilboa-Garber, N. (1987). Immunization of mice against various strains of *Pseudomonas aeruginosa* by using *Pseudomonas* lectin vaccine. *FEMS Microbiol. Lett. 43*:313–315.
142. Speert, D. P., Eftekhar, F., and Puterman, M. L. (1989). Nonopsonic phagocytosis of strains of *Pseudomonas aeruginosa* from cystic fibrosis patients. *Infect. Immun. 43*:1006–1011.
143. Gilboa-Garber, N., Sudakevitz, D., Sheffi, M., Levene, C., and Sela, R. (1991). Blood group specificity of the *Pseudomonas aeruginosa* lectins PA-I and PA-II. *J. Med. Sci. 27* (December issue in press).

144. Ko, H. L., Beuth, J., Soelter, J., Schroten, H., Uhlenbruck, G., and Pulverer, G. (1987). In vitro and in vivo inhibition of lectin mediated adhesion of *Pseudomonas aeruginosa* by receptor blocking carbohydrates. *Infection 15*:237–240.
145. Ahmed, H., Pal, R., Guha, A. K., and Chatterjee, B. P. (1986). Isolation, purification and specificity of the lectin from *Pseudomonas aeruginosa* bacteria (Habs strain H8). In *Lectins Biology, Biochemistry, Clinical Biochemistry*, Vol. 5 (T. C. Bøg-Hansen and E. van Driessche, eds.). Walter de Gruyter, Berlin, pp. 305–316.
146. Pal, R., Ahmed, H., and Chatterjee, B. P. (1987). Hemagglutination patterns of *Pseudomonas aeruginosa* strains of different serogroups, Habs and Fisher types. *Zentralbl. Bakteriol. Hyg. A 265*:299–304.
147. Pal, R., Ahmed, H., and Chatterjee, B. P. (1987). *Pseudomonas aeruginosa* bacteria Habs type 2a contains two lectins of different specificity. *Biochem. Arch. 3*:399–412.
148. Garber, N., Guempel, U., Gilboa-Garber, N., and Doyle, R. J. (1987). Specificity of the fucose-binding lectin of *Pseudomonas aeruginosa. FEMS Microbiol. Lett. 48*:331–334.
149. Buchanan, K., Falkow, S., Hull, R. A., and Hull, S. I. (1985). Frequency among Enterobacteriaceae of the DNA sequences encoding type 1 pili. *J. Bacteriol. 162*:799–803.
150. Clegg, S., Hull, S., Hull, R., and Pruckler, J. (1985). Construction and comparison of recombinant plasmids encoding type 1 fimbriae of members of the family Enterobacteriaceae. *Infect. Immun. 48*:275–279.
151. Korhonen, T. K., Väisänen, V., Saxén, H., Hultberg, H., and Svenson, S. B. (1982). P-antigen-recognizing fimbriae from human uropathogenic *Escherichia coli* strains. *Infect. Immun. 37*:286–291.
152. Minion, F. C., Abraham, S. N., Beachey, E. H., and Goguen, J. D. (1986). The genetic determinant of adhesive function in type 1 fimbriae of *Escherichia coli* is distinct from the gene encoding the fimbrial subunit. *J. Bacteriol. 165*:1033–1036.
153. Clegg, S. (1982). Cloning of genes determining the production of mannose-resistant fimbriae in a uropathogenic strain of *Escherichia coli* belonging to serotype O6. *Infect. Immun. 38*: 739–744.
154. Klemm, P. (1986). Two regulatory *fim* genes, *fimB* and *fimE*, control the phase variation of type 1 fimbriae in *Escherichia coli. EMBO J. 5*:1389–1393.
155. Hull, R. A., Gill, R. E., Hsu, P., Minshew, B. H., and Falkow, S. (1981). Construction and expression of recombinant plasmids encoding type 1 or D-mannose-resistant pili from a urinary tract infection *Escherichia coli* isolate. *Infect. Immun. 33*:933–938.
156. Orndorff, P. E., Spears, P. A., Schauer, D., and Falkow, S. (1985). Two modes of control of *pilA*, the gene encoding type 1 pilin in *Escherichia coli. J. Bacteriol. 164*:321–330.
157. O'Hanley, P., Lark, D., Normark, S., Falkow, S., and Schoolnik, G. K. (1983). Mannose-sensitive and Gal-Gal binding *Escherichia coli* pili from recombinant strains. Chemical, functional, and serological properties. *J. Exp. Med. 158*:1713–1719.
158. Norgren, M., Normark, S., Lark, D., O'Hanley, P., Schoolnik, G., Falkow, S., Svanborg-Edén, C., Båga, M., and Uhlin, B. E. (1984). Mutations in *Escherichia coli* cistrons affecting adhesion to human cells do not abolish Pap pili fiber formation. *EMBO J. 3*:1159–1166.
159. Båga, M., Normark, S., Hardy, J., O'Hanley, P., Lark, D., Olsson, O., Schoolnik, G., and Falkow, S. (1984). Nucleotide sequence of the *papA* gene encoding the Pap pilus subunit of human uropathogenic *Escherichia coli. J. Bacteriol. 157*:330–333.
160. Uhlin, B. E., Norgren, M., Båga, M., and Normark, S. (1985). Adhesion to human cells by *Escherichia coli* lacking the major subunit of a digalactoside-specific pilus-adhesin. *Proc. Natl. Acad. Sci. USA 82*:1800–1804.
161. Maurer, L., and Orndorff, P. E. (1985). A new locus, *pilE*, required for the binding of type 1 piliated *Escherichia coli* to erythrocytes. *FEMS Microbiol. Lett. 30*:59–66.
162. Maurer, L., and Orndorff, P. E. (1987). Identification and characterization of genes determining receptor binding and pilus length of *Escherichia coli* type 1 pili. *J. Bacteriol. 169*:640–645.
163. Hacker, J., Schmidt, G., Hughes, C., Knapp, S., Marget, M., and Goebel, W. (1985). Cloning and characterization of genes involved in production of mannose-resistant, neuraminidase-

susceptible (X) fimbriae from a uropathogenic O6:K15:H31 *Escherichia coli* strain. *Infect. Immun. 47*:434–440.

164. Van Die, I., Van Megen, I., Zuidweg, E., Hoekstra, W., De Ree, H., van den Bosch, H., and Bergmans, H. (1986). Functional relationship among the gene clusters encoding F7 1, F7 2, F9 and F11 fimbriae of human uropathogenic *Escherichia coli. J. Bacteriol. 167*:407–410.
165. Lindberg, F., Lund, B., and Normark, S. (1986). Gene products specifying adhesion of uropathogenic *Escherichia coli* are minor components of pili. *Proc. Natl. Acad. Sci. USA 83*:1891–1895.
166. Båga, M., Norgren, M., and Normark, S. (1987). Biogenesis of *E. coli* Pap pili: *PapH*, a minor pilin subunit involved in cell anchoring and length modulation. *Cell 49*:241–251.
167. Norgren, M. (1987). Genetic organization and biogenesis of Pap pili of uropathogenic *Escherichia coli. Umea Univ. Med. Diss. New Ser.* 183:1–36.
168. Klemm, P., Jørgensen, B. J., van Die, I., de Ree, H., and Bergmans, H. (1985). The *fim* genes responsible for synthesis of type 1 fimbriae in *Escherichia coli*, cloning and genetic organization. *Mol. Gen. Genet. 199*:410–414.
169. Lindberg, F. (1987). Specialized proteins at the pilus tip mediate Gal α1-4Gal-b specific binding of uropathogenic *Escherichia coli* to human cells. *Umea Univ. Med. Diss. New Ser.* 187.
170. Lund, B. (1987). The molecular mechanism of uroepithelial binding by uropathogenic *Escherichia coli. Umea Univ. Med. Diss. New Ser.* 194.
171. Schmoll, T., Hacker, J., and Goebel, W. (1987). Nucleotide sequence of the *sfaA* gene coding for the S-fimbrial protein subunit of *Escherichia coli. FEMS Microbiol. Lett. 41*:229–235.
172. Ott, M., Schmoll, T., Goebel, W., van Die, I., and Hacker, J. (1987). Comparison of the genetic determinant coding for the S-fimbrial adhesin (*sfa*) of *Escherichia coli* to other chromosomally encoded fimbrial determinants. *Infect. Immun. 55*:1940–1943.
173. Ott, M., Hacker, J., Schmoll, T., Jarchau, T., Korhonen, T. K., and Goebel, W. (1986). Analysis of the genetic determinants coding for the S-fimbrial adhesin (*sfa*) in different *Escherichia coli* strains causing meningitis or urinary tract infections. *Infect. Immun. 54*:646–653.
174. Hacker, J., Hof, H., Emody, L., and Goebel, W. (1986). Influence of cloned *Escherichia coli* hemolysin gene, S-fimbriae and serum resistance on pathogenicity in different animal models. *Microb. Pathogen. 1*:533–547.
175. Smith, H. W., and Linggood, M. A. (1971). Observation on the pathogenic properties of the K88, Hly and Ent plasmids of *Escherichia coli* with particular reference to porcin diarrhea. *J. Med. Microbiol. 4*:467–485.
176. Abraham, J. M., Freitag, C. S., Clements, J. R., and Eisenstein, B. I. (1985). An invertible element of DNA controls phase variation of type 1 fimbriae of *Escherichia coli. Proc. Natl. Acad. Sci. USA 82*:5724–5727.
177. Rhen, M., Mäkelä, P. H., and Korhonen, T. K. (1983). P-fimbriae of *Escherichia coli* are subject to phase variation. *FEMS Microbiol. Lett. 19*:267–271.
178. Nowicki, B., Vuopio-Varkila, J., Viljanen, P., Korhonen, T. K., and Mäkelä, P. H. (1986). Fimbrial phase variation and systemic *Escherichia coli* infection studied in the mouse peritonitis model. *Microb. Pathogen. 1*:335–347.
179. Nowicki, B., Rhen, M., Vaisanen-Rhen, V., Pere, A., and Korhonen, T. K. (1985). Kinetics of phase variation between S and type-1 fimbriae of *Escherichia coli. FEMS Microbiol. Lett. 28*:237–242.
180. Saukkonen, K. M. J., Nowicki, B., and Leinonen, M. (1988). Role of type 1 and S fimbriae in the pathogenesis of *Escherichia coli* O18:K1 bacteremia and meningitis in the infant rat. *Infect. Immun. 56*:892–897.
181. Gilboa-Garber, N. (1983). The biological functions of *Pseudomonas aeruginosa* lectins. In *Lectins Biology, Biochemistry, Clinical Biochemistry*, Vol. 3 (T. C. Bøg-Hansen and G. A. Spengler, eds.). Walter de Gruyter, Berlin, pp. 495–502.
182. Gilboa-Garber, N., and Wretlind, B. (1986). Proteolytic activity and pyocyanin correlate intracellular lectins in "hypo" and "hyper"-producing mutants of *Pseudomonas aeruginosa*. In *Proceedings 14th International Congress Microbiology*. Manchester, Engl., p. 63.

183. Booth, B. A., Boesman-Finkelstein, M., and Finkelstein, R. A. (1983). *Vibrio cholerae* soluble hemagglutinin/protease is a metalloenzyme. *Infect. Immun. 42*:639–644.
184. Gaastra, W., and de Graaf, F. K. (1982). Host-specific fimbrial adhesins of noninvasive enterotoxigenic *Escherichia coli* strains. *Microbiol. Rev. 46*:129–161.
185. Sherman, P., Soni, R., and Karmali, M. (1988). Attaching and effacing adherence of Vero cytotoxin-producing *Escherichia coli* to rabbit intestine in vivo. *Infect. Immun. 56*: 756–761.
186. Berger, H., Hacker, J., Juarez, A., Hughes, C., and Goebel, W. (1982). Cloning of the chromosomal determinants encoding hemolysin production and mannose-resistant hemagglutination in *Escherichia coli. J. Bacteriol. 152*:1241–1247.
187. O'Brien, A. D., and Holmes, R. K. (1987). Shiga and Shiga-like toxins. *Microbiol. Rev. 51*:206–220.
188. Dey, P. M., and Pridham, J. B. (1986). *Vicia faba* α-galactosidases with lectin activities: An overview. In *Lectins*, Vol. 5 (T. C. Bøg-Hansen and E. van Driessche, eds.). Walter de Gruyter, Berlin, pp. 161–170.
189. Ishikawa, F., Oishi, K., and Aida, K. (1979). Chitin-binding hemagglutinin produced by *Conidiobolus* strains. *Appl. Environ. Microbiol. 37*:1110–1112.
190. Ishikawa, F., Oishi, K., and Aida, K. (1981). Purification and some properties of chitin-binding hemagglutinin from *Conidiobolus lamprauges. Agric. Biol. Chem. 45*:557–564.
191. Ishikawa, F., Oishi, K., and Aida, K. (1983). Purification and immunological properties of chitin-binding hemagglutinin and β-*N*-acetylglucosaminidase produced by *Conidiobolus lamprauges. Agric. Biol. Chem. 47*:149–151.
192. Ishikawa, F., Oishi, K., and Aida, K. (1983). Chitin-binding hemagglutinin associated with cell wall of *Conidiobolus lamprauges. Agric. Biol. Chem. 47*:587–592.
193. Bar-Shavit, Z., Goldman, R., Ofek, I., Sharon, N., and Mirelman, D. (1980). Mannose-binding activity of *Escherichia coli*: A determinant of attachment and ingestion of the bacteria by macrophages. *Infect. Immun. 29*:417–424.
194. Ofek, I., and Sharon, N. (1988). Lectinophagocytosis: A molecular mechanism of recognition between cell surface sugars and lectins in the phagocytosis of bacteria. *Infect. Immun. 56*: 539–547.
195. Maayan, M. C., Ofek, I., Medalia, O., and Aronson, M. (1985). Population shift in mannose-specific fimbriated phase of *Klebsiella pneumoniae* during experimental urinary tract infection in mice. *Infect. Immun. 49*:785–789.
196. Hamblin, J., and Kent, S. P. (1973). Possible role of phytohaemagglutinin in *Phaseolus vulgaris L. Nature 245*:28–30.
197. Bohlool, B. B., and Schmidt, E. L. (1974). Lectins: A possible basis for specificity in the *Rhizobium*–legume root nodule symbiosis. *Science 185*:269–271.
198. Dazzo, F. B., and Hollingsworth, R. E. (1984). Trifoliin A and carbohydrate receptors as mediators of cellular recognition in the *Rhizobium trifolii*–clover symbiosis. *Biol. Cell 51*: 267–274.
199. Sequeira, L. (1978). Lectins and their role in host-pathogen specificity. Annu. Rev. Phytopathol. 16:453–481.
200. Schmidt, E. L. (1979). Initiation of plant root-microbe interactions. *Annu. Rev. Microbiol. 33*:355–376.
201. Bauer, W. D. (1981). Infection of legumes by rhizobia. *Annu. Rev. Plant Physiol. 32*: 407–449.
202. Kijne, J. W., van der Schaal, I. A. M., Diaz, C. L., and van Iren, F. (1983). Mannose-specific lectin and the recognition of pea roots by *Rhizobium leguminosarum*. In *Lectins Biology, Biochemistry, Clinical Biochemistry*, Vol. 3 (T. C. Bøg-Hansen and G. A. Spengler, eds.). Walter de Gruyter, Berlin, pp. 521–538.
203. Dazzo, F. B., Kijne, J. W., Haahtela, K., and Korhonen, T. K. (1986). Fimbriae, lectins, and agglutinins of nitrogen fixing bacteria. In *Microbial Lectins and Agglutinins: Properties and Biological Activity* (D. Mirelman, ed.). John Wiley & Sons, New York, pp. 237–254.

204. Bar-Shavit, Z., Ofek, I., Goldman, R., Mirelman, D., and Sharon, N. (1977). Mannose residues on phagocytes as receptors for the attachment of *Escherichia coli* and *Salmonella typhi*. *Biochem. Biophys. Res. Commun.* *78*:455–460.
205. Rottini, G., Cian, F., Soranzo, M. R., Albrigo, R., and Patriarca, P. (1979). Evidence for the involvement of human polymorphonuclear leucocyte mannose-like receptors in the phagocytosis of *Escherichia coli*. *FEBS Lett.* *105*:307–312.
206. Silverblatt, F. J ., Dreyer, J. S., and Schauer, S. (1979). Effect of pili on susceptibility of *Escherichia coli* to phagocytosis. *Infect. Immun.* *24*:218–223.
207. Mangan, D. F., and Snyder, I. S. (1979). Mannose-sensitive interaction of *Escherichia coli* with human peripheral leukocytes in vitro. *Infect. Immun.* *26*:520–527.
208. Mangan, D. F., and Snyder, I. S. (1979). Mannose-sensitive stimulation of human leukocyte chemiluminescence by *Escherichia coli*. *Infect. Immun.* *26*:1014–1019.
209. Warr, G. A. (1980). A macrophage receptor (mannose/glucosamine)–glycoproteins of potential importance in phagocytic activity. *Biochem. Biophys. Res. Commun.* *93*:737–745.
210. Vandenbroucke-Grauls, C. M. J. E., Thijssen, H. M. W. M., and Verhoef, J. (1984). Interaction between human polymorphonuclear leucocytes and *Staphylococcus aureus* in the presence and absence of opsonins. *Immunology* *52*:427–435.
211. Björkstén, B., and Wadström, T. (1982). Interaction of *Escherichia coli* with different fimbriae and polymorphonuclear leukocytes. *Infect. Immun.* *38*:298–305.
212. Blumenstock, E., and Jann, K. (1982). Adhesion of piliated *Escherichia coli* strains to phagocytes: Differences between bacteria with mannose-sensitive pili and those with mannose-resistant pili. *Infect. Immun.* *35*:264–269.
213. Svanborg-Edén, C., Bjursten, L.-M., Hull, R., Hull, S., Magnusson, K.-E., Moldovano, Z., and Leffler, H. (1984). Influence of adhesins on the interaction of *Escherichia coli* with human phagocytes. *Infect. Immun.* *44*:672–680.
214. Sharon, N. (1984). Surface carbohydrates and surface lectins are recognition determinants in phagocytosis. *Immunol. Today* *5*:143–147.
215. Wileman, T. E., Lennartz, M. R., and Stahl, P. D. (1986). Identification of the macrophage mannose receptor as a 175-kDa membrane protein. *Proc. Natl. Acad. Sci. USA* *83*:2501–2505.
216. Sandberg, A. L., Mudrick, L. L., Cisar, J. O., Brennan, M. J., Mergenhagen, S. E., and Vatter, A. E. (1986). Type 2 fimbrial lectin-mediated phagocytosis of oral *Actinomyces* spp. by polymorphonuclear leukocytes. *Infect. Immun.* *54*:472–476.
217. Rodriguez-Ortega, M., Ofek, I., and Sharon, N. (1987). Membrane glycoproteins of human polymorphonuclear leukocytes that act as receptors for mannose-specific *Escherichia coli*. *Infect. Immun.* *55*:968–973.
218. Perry, A., Keisari, Y., and Ofek, I. (1985). Liver cell and macrophage surface lectins as determinants of recognition in blood clearance and cellular attachment of bacteria. *FEMS Microbiol. Lett.* *27*:345–350.
219. Sudakevitz, D., and Gilboa-Garber, N. (1982). Effect of *Pseudomonas aeruginosa* lectins on phagocytosis of *Escherichia coli* strains by human polymorphonuclear leucocytes. *Microbios* *34*:159–166.
220. Goetz, M. B., and Silverblatt, F. J. (1987). Stimulation of human polymorphonuclear leukocyte oxidative metabolism by type 1 pili from *Escherichia coli*. *Infect. Immun.* *55*: 534–540.
221. Kuriyama, S. M., and Silverblatt, F. J. (1986). Effect of Tamm–Horsfall urinary glycoprotein on phagocytosis and killing of type I-fimbriated *Escherichia coli*. *Infect. Immun.* *51*:193–198.
222. Ellen, R. P. (1976). Establishment and distribution of *Actinomyces viscosus* and *Actinomyces naeslundii* in the human oral cavity. *Infect. Immun.* *14*:1119–1124.
223. Costello, A. H., Cisar, J. O., Kolenbrander, P. E., and Gabriel, O. (1979). Neuraminidase-dependent hemagglutination of human erythrocytes by human strains of *Actinomyces viscosus* and *Actinomyces naeslundii*. *Infect. Immun.* *26*:563–572.

224. McIntire, F. C., Vatter, E. A., Baros, J., and Arnold, J. (1978). Mechanism of coaggregation between *Actinomyces viscosus* T14V and *Streptococcus sanguis* 34. *Infect. Immun. 21*: 978–988.
225. McIntire, F. C., Crosby, L. K., Barlow, J. J., and Matta, K. L. (1983). Structural preferences of β-galactoside-reactive lectins on *Actinomyces viscosus* T14V and *Actinomyces naeslundii* WVU45. *Infect. Immun. 41*:848–850.
226. Cisar, J. O., Kolenbrander, P. E., and McIntire, F. C. (1979). Specificity of coaggregation reactions between human oral streptococci and strains of *Actinomyces viscosus* and *Actinomyces naeslundii. Infect. Immun. 24*:742–752.
227. Cisar, J. O., David, V. A., Curl, S. H., and Vatter, A. E. (1984). Exclusive presence of lactose-sensitive fimbriae on a typical strain (WVU45) of *Actinomyces naeslundii. Infect. Immun. 46*:453–458.
228. Kolenbrander, P. E., and Williams, B. L. (1981). Lactose-reversible coaggregation between oral actinomycetes and *Streptococcus sanguis. Infect. Immun. 33*:95–102.
229. Saunders, J. M., and Miller, C. H. (1980). Attachment of *Actinomyces naeslundii* to human buccal epithelial cells. *Infect. Immun. 29*:981–989.
230. Revis, G. J., Vatter, A. E., Crowle, A. J., and Cisar, J. O. (1982). Antibodies against the Ag2 fimbriae of *Actinomyces viscosus* T14V inhibit lactose-sensitive bacterial adherence. *Infect. Immun. 36*:1217–1222.
231. Heeb, M. J., Costello, A. H., and Gabriel, O. (1982). Characterization of a galactose-specific lectin from *Actinomyces viscosus* by a model aggregation system. *Infect. Immun. 38*: 993–1002.
232. Heeb, M. J., Marini, A. M., and Gabriel, O. (1985). Factors affecting binding of galacto ligands to *Actinomyces viscosus* lectin. *Infect. Immun. 47*:61–67.
233. Gabriel, O., Heeb, M. J., and Hinrichs, M. (1984). Interaction of the surface adhesins of the oral *Actinomyces* spp. with mammalian cells. In *Molecular Interactions in Oral Microbial Adherence and Aggregation* (S. E. Mergenhagen and B. Rosan, eds.). American Society for Microbiology, Washington, D.C., pp. 45–52.
234. Beachey, E. H. (1981). Bacterial adherence: Adhesin-receptor interactions mediating the attachment of bacteria to mucosal surfaces. *J. Infect. Dis. 143*:325–345.
235. Rosenberg, M., Rosenberg, E., Judes, H., and Weiss, E. (1983). Bacterial adherence to hydrocarbons and to surfaces in the oral cavity. *FEMS Microbiol. Lett. 20*:1–5.
236. Brennan, M. J., Cisar, J. O., Vatter, A. E., and Sandberg, A. L. (1984). Lectin-dependent attachment of *Actinomyces naeslundii* to receptors on epithelial cells. *Infect. Immun. 46*: 459–464.
237. Brennan, M. J., Cisar, J. O., and Sandberg, A. L. (1986). A 160-kilodalton epithelial cell surface glycoprotein recognized by plant lectins that inhibit the adherence of *Actinomyces naeslundii. Infect. Immun. 52*:840–845.
238. Ramphal, R., Small, P. A., Shands, J. W., Jr., Fischlsweiger, W., and Small, P. A., Jr. (1980). Adherence of *Pseudomonas aeruginosa* to tracheal cells injured by influenza infection or by endotracheal intubation. *Infect. Immun. 27*:614–619.
239. Aronson, M., Medalia, O., Schori, L., Mirelman, D., Sharon, N., and Ofek, I. (1979). Prevention of colonization of the urinary tract of mice with *Escherichia coli* by blocking of bacterial adherence with methyl α-D-mannopyranoside. *J. Infect. Dis. 139*:329–332.
240. Fader, R. C., and Davis, C. P. (1980). Effect of piliation on *Klebsiella pneumoniae* infection in rat bladders. *Infect. Immun. 30*:554–561.
241. Korhonen, T. K., Väisänen, V., Kallio, P., Nurmiaho-Lassila, E.-L., Ranta, H., Siitonen, A., Elo, J., Svenson, S. B., and Svanborg-Edén, C. (1982). The role of pili in adhesion of *Escherichia coli* to human urinary tract epithelial cells. *Scand. J. Infect. Dis. 33*(Suppl.):26–31.
242. Cooper, D. N. W., Haywood-Reid, P. L., Springer, W. R., and Barondes, S. H. (1986). Bacterial glycoconjugates are natural ligands for the carbohydrate binding site of discoidin I and influence its cellular compartmentalization. *Dev. Biol. 114*:416–425.

243. Springer, W. R., Cooper, D. N. W., and Barondes, S. H. (1984). Discoidin I is implicated in cell-substratum attachment and ordered cell migration of *Dictyostelium discoideum* and resembles fibronectin. *Cell 39*:557–564.
244. Mirelman, D. (1987). Ameba–bacterium relationship in amebiasis. *Microbiol. Rev. 51*: 272–284.
245. Korhonen, T. K., Valtonen, M. V., Parkkinen, J., Väisänen-Rhen, V., Finne, J., Ørskov, F., Ørskov, I., Svenson, S. B., and Mäkelä, P. H. (1985). Serotypes, hemolysin production, and receptor recognition of *Escherichia coli* strains associated with neonatal sepsis and meningitis. *Infect. Immun. 48*:486–491.
246. Archibald, A. R., and Coapes, H. E. (1972). Blocking of bacteriophage receptor sites by concanavalin A. *J. Gen. Microbiol. 73*:581–585.
247. Birdsell, D. C., and Doyle, R. J. (1973). Modification of bacteriophage 0_{25} adsorption to *Bacillus subtilis* by concanavalin A. *J. Bacteriol. 113*:198–202.
248. Watanabe, K., and Takesue, S. (1975). Use of L-rhamnose to study irreversible adsorption of bacteriophage PL-1 to a strain of *Lactobacillus casei*. *J. Gen. Virol. 28*:29–35.
249. Watanabe, K., Takesue, S., and Ishibashi, K. (1977). Reversibility of the adsorption of bacteriophage PL-1 to the cell walls isolated from *Lactobacillus casei*. *J. Gen. Virol. 34*: 189–194.
250. Watanabe, K., Takesue, S., and Ishibashi, K. (1980). Electron microscopic studies on the interaction between *Lactobacillus* phage PL-1 and cell walls isolated from its host cells. *J. Gen. Appl. Microbiol. 26*:413–420.
251. Ishibashi, K., Takesue, S., Watanabe, K., and Oishi, K. (1982). Use of lectins to characterize the receptor sites for bacteriophage PL-1 of *Lactobacillus casei*. *J. Gen. Microbiol. 128*: 2251–2259.
252. Hughes, C., Hacker, J., Roberts, A., and Goebel, W. (1983). Hemolysin production as a virulence marker in symptomatic and asymptomatic urinary tract infections caused by *Escherichia coli*. *Infect. Immun. 39*:546–551.
253. Mackman, N., Nicaud, J.-M., Gray, L., and Holland, I. B. (1986). Secretion of haemolysin by *Escherichia coli*. *Curr. Top. Microbiol. Immunol. 125*:159–181.
254. Waalwijk, C., MacLaren, D. M., and de Graaf, J. (1983). In vivo function of hemolysin in the nephropathogenicity of *Escherichia coli*. *Infect. Immun. 42*:245–249.
255. Welch, R. A., Dellinger, E. P., Minshew, B., and Falkow, S. (1981). Haemolysin contributes to virulence of extra-intestinal *E. coli* infections. *Nature 294*:665–667.
256. Low, D., David, V., Lark, D., Schoolnik, G., and Falkow, S. (1984). Gene clusters governing the production of hemolysin and mannose-resistant hemagglutination are closely linked in *Escherichia coli* serotype O4 and O6 isolates from urinary tract infections. *Infect. Immun. 43*:353–358.
257. Rüchel, R. (1981). Properties of a purified proteinase from the yeast *Candida albicans*. *Biochim. Biophys. Acta 659*:99–113.
258. Nordbring-Hertz, B. (1988). Nematophagous fungi: Strategies for nematode exploitation and for survival. *Microbiol. Sci. 5*:108–116.
259. van Heyningen, W. E., and Miller, P. A. (1961). The fixation of tetanus toxin by ganglioside. *J. Gen. Microbiol. 24*:107–119.
260. van Heyningen, W. E., Carpenter, C. C. J., Pierce, N. F., and Greenough, W. B. III. (1971). Deactivation of cholera toxin by ganglioside. *J. Infect. Dis. 124*:415–418.
261. Cuatrecasas, P. (1973). Interaction of *Vibrio cholerae* enterotoxin with cell membranes. *Biochemistry 12*:3547–3558.
262. Holmgren, J. (1973). Comparison of the tissue receptors for *Vibrio cholerae* and *Escherichia coli* enterotoxins by means of gangliosides and natural cholera toxoid. *Infect. Immun. 8*: 851–859.
263. Staerk, J., Ronneberger, H. J., Wiegandt, H., and Ziegler, W. (1974). Interaction of ganglioside $G_{Gtet}1$ and its derivatives with choleragen. *Eur. J. Biochem. 48*:103–110.

264. Fishman, P. H., Moss, J., and Osborne, J. C., Jr. (1978). Interaction of choleragen with the oligosaccharide of ganglioside G_{M1}: Evidence for multiple oligosaccharide binding sites. *Biochemistry 17*:711–716.
265. Eidels, L., Proia, R. L., and Hart, D. A. (1983). Membrane receptors for bacterial toxins. *Microbiol. Rev. 47*:596–620.
266. Keusch, G. T., Donohue-Rolfe, A., and Jacewicz, M. (1986). Sugar binding bacterial toxins. In *Microbial Lectins and Agglutinins: Properties and Biological Activity* (D. Mirelman, ed.). John Wiley & Sons, New York, pp. 271–295.
267. Richards, R. L., Moss, J., Alving, C. R., Fishman, P. H., and Brady, R. O. (1979). Choleragen (cholera toxin): A bacterial lectin. *Proc. Natl. Acad. Sci. USA 76*:1673–1676.
268. Nowell, P. C. (1960). Phytohemagglutinin: An initiator of mitosis in cultures of normal human leukocytes. *Cancer Res. 20*:462–466.
269. Yahara, I., and Edelman, G. M. (1973). The effects of concanavalin A on the mobility of lymphocyte surface receptors. *Exp. Cell Res. 81*:143–155.
270. Yahara, I., and Edelman, G. M. (1975). Modulation of lymphocyte receptor mobility by locally bound concanavalin A. *Proc. Natl. Acad. Sci. USA 72*:1579–1583.
271. Glick, J., Malik, Z., and Garber, N. (1981). Lectin-bearing protoplasts of *Pseudomonas aeruginosa* induce capping in human peripheral blood lymphocytes. *Microbios 32*:181–188.
272. Arai, H., and Sato, Y. (1976). Separation and characterization of two distinct hemagglutinins contained in purified leukocytosis-promoting factor from *Bordetella pertussis. Biochim. Biophys. Acta 444*:765–782.
273. Salata, R. A., and Ravdin, J. I. (1985). *N*-Acetyl-D-galactosamine inhibitable lectin of *Entamoeba histolytica* II. Mitogenic activity for human lymphocytes. *J. Infect. Dis. 151*:812–822.
274. Sharabi, Y., and Gilboa-Garber, N. (1980). Interactions of *Pseudomonas aeruginosa* hemagglutinins with *Euglena gracilis, Chlamydomonas reinhardi* and *Tetrahymena pyriformis. J. Protozool. 27*:80–83.
275. Gilboa-Garber, N., and Sharabi, Y. (1980). Increase of growth-rate and phagocytic activity of *Tetrahymena* induced by *Pseudomonas* lectins. *J. Protozool. 27*:209–211.
276. Morgan, W. T. J., and Watkins, W. M. (1953). The inhibition of the haemagglutinins in plant seeds by human blood group substances and simple sugars. *Br. J. Exp. Pathol. 34*:94–103.
277. Cuatrecasas, P., and Tell, G. P. E. (1973). Insulin-like activity of concanavalin A and wheat germ agglutinin-direct interactions with insulin receptors. *Proc. Natl. Acad. Sci. USA 70*: 485–489.
278. Bayer, E. A., Skutelsky, E., and Wilchek, M. (1983). The ultrastructural visualization of cell surface glycoconjugates. *Methods Enzymol. 83*:195–215.
279. Oishi, K., and Ishikawa, F. (1986). Recognition of human erythrocytes by a chitin-binding lectin of a fungus, *Conidiobolus lamprauges*. In *Chitin in Nature and Technology* (R. Muzzarelli, C. Jeuniaux, and G. W. Goodey, eds.). Plenum Publishing, New York, pp. 269–275.
280. Gilboa-Garber, N., and Mizrahi, L. (1985). *Pseudomonas* lectin PA-I detects hybrid product of blood group AB genes in saliva. *Experientia 41*:681–682.
281. Looms, L. M., Uemura, K.-I., and Feizi, T. (1985). Interaction of *Mycoplasma pneumoniae* with erythrocyte glycolipids of I and i antigen types. *Infect. Immun. 47*:15–20.
282. Lindahl, M., Brossmer, R., and Wadström, T. (1987). Carbohydrate receptor specificity of K99 fimbriae of enterotoxigenic *Escherichia coli. Glycoconjugate J. 4*:51–58.
283. Jacobs, A. A. C., Simons, B. H., and de Graaf, F. K. (1987). The role of lysine-132 and arginine-136 in the receptor-binding domain of the K99 fibrillar subunit. *EMBO J. 6*:1805–1808.
284. van Alphen, L., Poole, J., and Overbeeke, M. (1986). The Anton blood group antigen is the erythrocyte receptor for *Haemophilus influenzae. FEMS Microbiol. Lett. 37*:69–71.
285. Nowicki, B., Moulds, J., Hull, R., and Hull, S. (1988). A hemagglutinin of uropathogenic *Escherichia coli* recognized the Dr blood group antigen. *Infect. Immun. 56*:1057–1060.
286. Staines, W. A., Kimura, H., Fibiger, H. C., and McGeen, E. G. (1980). Peroxidase-labeled lectin as a neuroanatomical tracer: Evaluation in a CNS pathway. *Brain Res. 195*:485–490.

287. Trojanowski, J. Q., Gonatas, J. O., and Gonatas, N. K. (1982). Horseradish peroxidase (HRP) conjugates of cholera toxin and lectins are more sensitive retrogradely transported markers than free HRP. *Brain Res. 231*:33–50.
288. Sawchenko, P. E., and Gerfen, C. R. (1985). Plant lectins and bacterial toxins as tools for tracing neuronal connections. *TINS 8*:378–384.
289. Gilboa-Garber, N., and Mizrahi, L. (1981). Binding of *Pseudomonas aeruginosa* lectins to *Rhizobium* sp. *J. Appl. Bacteriol. 50*:21–28.
290. Inbar, M., and Sachs, L. (1969). Interaction of carbohydrate binding protein concanavalin A with normal and transformed cells. *Proc. Natl. Acad. Sci. USA 63*:1418–1425.
291. Shoham, J., Inbar, M., and Sachs, L. (1970). Differential toxicity on normal and transformed cells in vitro and inhibition of tumor development in vivo by concanavalin A. *Nature 227*:1244–1246.
292. Gilboa-Garber, N., Avichezer, D., and Leibovici, J. (1986). Application of *Pseudomonas aeruginosa* lectin (PA-I) for cancer research. In *Lectins Biology, Biochemistry, Clinical Biochemistry*, Vol. 5 (T. C. Bøg-Hansen and E. van Driessche, eds.). Walter de Gruyter, Berlin, pp. 329–338.
293. Avichezer, D., Leibovici, J., Gilboa-Garber, N., and Michowitz, M. (1987). New galactophilic lectins reduce tumorigenicity and preserve immunogenicity of Lewis lung carcinoma cells. In *Lectures and Symposia 14th International Cancer Congress*, Budapest 1986, Vol. 2 (K. Lapis and S. Eckhardt, eds.). S. Karger, Basel, pp. 79–86.
294. Gilboa-Garber, N., Avichezer, D., and Rozenszajn, L. A. (1986). Stimulation of peripheral lymphocytes from cancer patients and healthy subjects by *Pseudomonas aeruginosa* lectin. *FEMS Microbiol. Lett. 34*:237–240.
295. Ramon, G. (1923). Sur le pouvoir floculant et sur les proprietes immunisantes d'une toxine diphtherique rendue anatoxique (anatoxine). *C. R. Acad. Sci. 177*:1338–1340.
296. Rutter, J. W., and Jones, G. W. (1973). Protection against enteric disease caused by *E. coli*: A model for vaccination with a virulent determinant. *Nature 242*:531–532.
297. Munoz, J. J., Arai, H., and Cole, R. L. (1981). Mouse protecting and histamine-sensitizing activities of pertussigen and fimbrial hemagglutinin from *Bordetella pertussis*. *Infect. Immun. 32*:243–250.
298. Sato, Y., Izumiya, K., Sato, H., Cowell, J. L., and Manclark, C. R. (1981). Role of antibody to leukocytosis-promoting factor hemagglutinin and to filamentous hemagglutinin in immunity to pertussis. *Infect. Immun. 31*:1223–1231.
299. Petri, W. A., Jr., Joyce, M. P., Broman, J., Smith, R. D., Murphy, C. F., and Ravdin, J. I. (1987). Recognition of the galactose- or *N*-acetylgalactosamine-binding lectin of *Entamoeba histolytica* by human immune sera. *Infect. Immun. 55*:2327–2331.
300. Marie, A., and Morax, V. (1902). Recherches sur l'absorption de la toxine tetanique. *Ann. Inst. Pasteur 16*:818–832.
301. Meyer, H., and Ranson, F. (1903). Untersuchung uber den Tetanus. *Arch. Exp. Pathol. Pharmacol. 49*:369–416.
302. Wiley, R. G., Stripe, F., and Oeltman, T. N. (1986). Modeccin and volkensin are effective suicide transport agents in rat CNS. *Neurose Abst. 423*:2.
303. Rhen, M., Tenhunen, J., Väisänen-rhen, V., Pere, A., Båga, M., and Korhonen, T. K. (1986). Fimbriation and P-antigen recognition of *Escherichia coli* strains harbouring mutated recombinant plasmids encoding fimbrial adhesins of the uropathogenic *E. coli* strain KS71. *J. Gen. Microbiol. 132*:71–77.
304. Romeo, J. M., and Zusman, D. R. (1987). Cloning of the gene for myxobacterial hemagglutinin and isolation and analysis of structural gene mutations. *J. Bacteriol. 169*: 3801–3808.
305. Shields, R. (1987). Towards insect-resistant plants. *Nature 328*:12–18.
306. Vaeck, M., Reynaerts, A., Hofte, H., Jansens, S., De Beuckeleer, M., Dean, C., Zabeau, M., Van Montagu, M., and Leemans, J. (1987). Transgenic plants protected from insect attack. *Nature 328*:33–37.

307. Gilboa-Garber, N., and Garber, N. (1989). Lectin function is coupled to lytic enzymes. *Isr. J. Med. Sci. 25*:172.
308. Gilboa-Garber, N., and Garber, N. (1989). Microbial lectin cofunction with lytic activities as a model for a general basic lectin role. *FEMS Microbiol. Rev. 63*:211–222.
309. Gilboa-Garber, N. (1988). *Pseudomonas aeruginosa* lectins as a model for lectin production, properties, applications and functions. *Zentralbl. Bakteriol. Hyg. A 270*:3–15.
310. Gilboa-Garber, N., Buxenbaum, R., Mizrahi, L., and Avichezer, D. (1982). *Pseudomonas aeruginosa* strains and mutants which lack intracellular lectins are deficient in extracellular protease production. *Isr. J. Med. Sci. 18*:19.
311. Campbell, I. M. (1984). Secondary metabolism and microbial physiology. *Adv. Microb. Physiol. 25*:1–60.
312. Weinberg, E. D. (1970). Biosynthesis of secondary metabolites: Roles of trace metals. *Adv. Microb. Physiol. 4*:1–44.
313. Sudakevitz, D., Gilboa-Garber, N., and Mizrahi, L. (1979). Regulation of lectin production in *Pseudomonas aeruginosa* by culture medium composition. *Isr. J. Med. Sci. 15*:97.
314. Gilboa-Garber, N., Mizrahi, L., and Shainberg, D. (1982). Effect of glucose, mannose and galactose on the production of *Pseudomonas aeruginosa* protease, lectins and other exocellular and endocellular products. *Isr. J. Med. Sci. 18*:8.
315. Eisenstein, B. I., and Dodd, D. C. (1982). Pseudocatabolic repression of type 1 fimbriae of *Escherichia coli. J. Bacteriol. 151*:1560–1567.
316. Mirelman, D., Galun, E., Sharon, N., and Lotan, R. (1975). Inhibition of fungal growth by wheat germ agglutinin. *Nature 256*:414–416.
317. Middlebrook, J. L., and Dorland, R. B. (1984). Bacterial toxins: Cellular mechanisms of action. *Microbiol. Rev. 48*:199–221.
318. Stephen, J., and Pietrowski, R. A. (1986). Bacterial toxins. In *Aspects of Microbiology 2*, 2nd ed. American Society for Microbiology, Washington, D.C.
319. Lord, J. M., and Roberts, L. M. (1987). Plant toxins. *Microbiol. Sci. 4*:376–378.
320. Olsnes, S. (1987). Closing on ricin action. *Nature 328*:474–475.
321. Endo, Y., Mitsui, K., Motizuki, M., and Tsurugi, K. (1987). The mechanism of action of ricin and related toxic lectins on eukaryotic ribosomes. *J. Biol. Chem. 262*:5908–5912.
322. Endo, Y., and Tsurugi, K. (1987). RNA *N*-glycosidase activity of ricin A-chain. *J. Biol. Chem. 262*:8128–8130.
323. Ashwell, G., and Morell, A. G. (1974). The role of surface carbohydrates in the hepatic recognition and transport of circulating glycoproteins. *Adv. Enzymol. 41*:99–128.
324. Kornfeld, S. (1986). Trafficking of lysosomal enzymes in normal and disease states. *J. Clin. Invest. 77*:1–6.
325. Charvátóva, D., Tichá, and Kocourek, J. (1991). α-D-galactosidase and α-D-mannosidase from jack bean (*Canavalia ensiformis*) and their interaction with concanavalin A. In *Lectins Biology, Biochemistry, Clinical Biochemistry*, Vol. 7 (T. C. Bøg-Hansen and J. Kocourek, eds.). Walter de Gruyter, Berlin (in press).
326. Komano, H., and Natori, S. (1985). Participation of *Sarcophaga peregrina* humoral lectin in the lysis of sheep red blood cells injected into the abdominal cavity of larvae. *Dev. Comp. Immunol. 9*:31–40.
327. Lehninger, A. L. (1982). *Principles of Biochemistry*. Worth Publishers, New York.

18

Invertebrate Lectins: Distribution, Synthesis, Molecular Biology, and Function

Gerardo R. Vasta *University of Maryland, Baltimore, Maryland*

I. HISTORICAL BACKGROUND

The development of the field of invertebrate lectin research arises just over a decade after Stillmark's dissertation on ricin [1]: in 1899, Camus published what can be considered the first report on a tissue lectin detected in an invertebrate species, consisting in a brief description of the agglutination of erythrocytes and milk globules from different vertebrate species by saline extracts of albumin gland of the snail, *Helix pomatia* [2]. A few years later, Noguchi [3] reported the presence of humoral agglutinins in the hemolymph of the horseshoe crab *Limulus polyphemus* and the lobster *Homarus americanus*. Further studies [4] showed that *Limulus* agglutinins were heterogeneous in specificity and were heat-labile. In 1912, Cantacuzène [5] published an article showing that hemolymph of the crab *Eupagurus prideauxii* could not only agglutinate and lyse rabbit and sheep erythrocytes, but could also agglutinate bacteria. He later reported [6] that hemagglutinins in the hemolymph of the spider crab *Maia squinado* could be induced by immunization and, in a later review [7], suggested that those agglutinins had opsonic activity for bacteria. Tyler [8] hypothesized that the physiological function of agglutinins detected in the seminal fluids of various invertebrates would relate to fertilization mechanisms, in particular, preventing interspecies cross-fertilization. From the foregoing, it becomes clear that during this early period of research, the goals of a great majority of the studies consisted of attempting to elucidate the participation of humoral and tissue agglutinins from invertebrates in their "natural" or "adaptive" immune response and fertilization events. From the mid-1940s and until the early 1960s there was a relative paucity in the published information in this field, and renewed interest in the question of invertebrate agglutinins developed when Cheng and Sanders [9] reported the presence of specific agglutinins in the plasma of the snail *Viviparus malleatus*.

In 1964, Johnson [10] reported the first blood group-specific lectin in an invertebrate species, the butter clam *Saxidomus giganteus*, originating a whole new field of study that was to expand consistently through the following decade. The search for blood group substance P1 in invertebrates led Prokop et al. [11] to find agglutinins specific for blood group A- and B-like substances in the snail *Cepaea* (*Helix*) *hortensis*. The discovery of potent anti-A agglutinins in the snails *H. pomatia* and *Otala lactea* stimulated further work on mollusk agglutinins, in particular, those lectins from gastropods that were specific for human blood group substances [12].

The possibility that invertebrate agglutinins would constitute the precursors of vertebrate immunoglobulins generated interest in elucidating their molecular structure. From work with the American horseshoe crab *L. polyphemus*, Marchalonis and Edelman [13] provided the first description of the biochemical properties and model for the molecular structure of an invertebrate lectin, and Fernández-Morán et al. [14] provided its first electron micrographs. In the same year, Cohen [15] discovered its novel specificity for sialic acids. Later studies focused on the purification and characterization of the molecular properties and fine carbohydrate specificity of the snail lectins, with particular interest in their practical application as specific reagents [16].

In the last decade, a renewed concern in their biological roles and evolution has become evident [17]. The finding that invertebrates exhibit lectins on the surface of phagocytic hemocytes, which can selectively bind to potentially pathogenic bacteria and fungi, has stimulated studies on these lectins as possible recognition molecules involved in defense mechanisms [18]. Their roles in cell aggregation of sponges and modulation of symbiotic relationships in tunicates and tridacnid clams has been explored in the past few years [19–21]. With the development of recombinant DNA technology, research in their gene structure, expression, and evolution has yielded important information [22,23].

II. DEFINITIONS AND TERMINOLOGY

On the basis of their binding and agglutinating properties, source, presumed biological role, or apparent analogy to other molecules (i.e., antibodies), animal agglutinins have been reported, since their discovery, under a variety of names including *hemagglutinins, heterophile agglutinins, receptor-specific proteins, heteroagglutinins, natural* or *normal antibodies, antibody-like molecules, protectins*, and others. Although some of these terms have been kept in use in the current literature until the present time, the term *lectin*, proposed by Boyd and Slapeigh in 1954 for plant agglutinins [24], has become of widespread use for their animal counterparts as well.

Along with the considerable expansion of the field in the past few years, scientists have realized the need to determine the boundaries of the term *lectin*, placing particular emphasis on which molecules should not be included. Two definitions of lectins were proposed that differ in the minimum number of binding sites and their biological activities—toxic, enzyme, or hormonelike—on the cells to which they bind. Goldstein et al. [25] propose that "A lectin (from the Latin *legere*: to choose) is a sugar-binding protein or glycoprotein of non-immune origin which agglutinates cells and/or precipitates glycoconjugates," excluding carbohydrate-binding enzymes and toxins. Kocourek and Hořejší [26] define *lectins* as ". . . sugar-binding proteins or glycoproteins of non-immune origin which are devoid of enzymatic activity towards the sugars to which they bind and do not require free glycosidic hydroxyl groups on these sugars for their binding," thus, including lectins with toxic activity such as those from *Ricinus communis* and *Abrus precatorius*. It should be remembered, however, that these are operational definitions based on the sugar-binding properties and biological activities of lectins, and for practical reasons do not consider aspects related to their molecular organization, structural affinities, or evolutionary history. In recent years, the experimental evidence on protein primary structure and gene organization suggest that very distinct, and probably unrelated, groups of molecules are included in those definitions. On the other hand, the nonimmune origins of some animal lectins should be reconsidered, at least in a broad sense, since certain lectins, such as those from the larva of the flesh fly *Sarcophaga peregrina* can be induced by a challenge, such as injection of erythrocytes or injury of the body wall [23].

III. OCCURRENCE AND BIOSYNTHESIS

Since the early work, the presence of lectins in fluids and tissues from invertebrates has been investigated by hemagglutination assays, with untreated or enzyme-treated erythrocytes, or precipitation reactions with glycoproteins and polysaccharides. Although erythrocytes exhibit a very limited number of different sugars on the cell surface, the variety of linkages and anomeric configurations provides a diverse array of carbohydrate moieties. Their modification by glycohydrolases, glycooxidases, and the use of proteases to uncover hindered residues, have constituted useful methods for expanding the variety of indicator cells. Nevertheless, it is likely that additional lectins specific for carbohydrate structures that are found less frequently than those on the erythrocyte cell surface have been neglected, and have yet to be detected in tissues or fluids already screened for their presence by conventional methods.

Currently, lectins have been identified in species of virtually all taxa, from viruses and bacteria to vertebrates, and probably occur in all living organisms. Useful lists of their taxonomic distributions have been published elsewhere [28–30].

A. Organ, Tissue, and Cellular Distribution

Although the first invertebrate lectin reported was identified in the extracts of the albumin gland of the snail *H. pomatia*, the bulk of the early work was done in body fluids or secretions, such as hemolymph and coelomic and seminal fluids. In organisms of small size, whole-body extracts, obtained by the use of blenders, mortars, and presses, were assayed for lectin activity. Further interest in identifying the tissues or organ source of the lectin activity led to systematic studies in which, for each species, and occasionally individuals, the different organs, tissues, and fluids were separately screened for the presence of lectins. As a result, those in which the lectins of interest occurred at high concentrations were identified, such as the *protectins*, blood group-specific lectins from the albumin gland and eggs of gastropod snails [11]; the *N*-acylaminosugar-binding lectins in plasma of crabs [15,31], lobsters [32,33], and horseshoe crabs [13,15]; hemocyte lectins from crayfish [34], oyster [35,36], cockroach [37,38], and blue crab [31,39]; lectins found in mucous secretions, such as those from the protochordate *Branchiostoma* [40], the squid *Loligo vulgaris* [41], and on sponge [42] and anemone [43] cell surfaces. For both intra- and extracellular lectins, it is important to consider that, although they may occur in particular cells or body fluids, their biosynthesis and secretion may be carried out by different cells or tissues. The lectins in the hemolymph of the fly *Sarcophaga* larva are produced in the fat body [44]. Those that occur in the hemolymph of the oyster [35,36], lobster [32,33], crayfish [34], and blue crab [31,39] are synthesized and secreted by the hemocytes. However, from the two lectins isolated from the blue crab *Callinectes sapidus* plasma, only one could be isolated from the hemocyte cytosolic fraction [39]. It should be realized also that, although certain lectins may occur at relatively high concentrations in certain tissues or organs, other "minor" lectins of different distribution may be present in the same organisms. Furthermore, multiple lectin species with similar or different sugar specificities are usually found in the fluid or tissue under study.

B. Site of Synthesis

So far, the only example for which undisputable evidence on the site of synthesis of an invertebrate hemolymph lectin has been obtained, is that of the larva from the flesh fly *S. peregrina*. It was initially shown that the lectin is produced in the fat body upon injury of the larval body wall or pupation and is released to the circulation where it would be recruited by the hemocytes. However, further studies [45] failed to show newly synthesized lectin in the fat body upon injury of the larvae, thus allowing the possibility that the lectin would be produced elsewhere and accumulated and processed to the active form in the fat body. The definitive proof of the site of synthesis was obtained by Northern hybridization of mRNA from the fat body of third instar, normal, and injured larvae with a probe consisting in a fragment from the coding region of *S. peregrina* lectin cDNA: only the fat body mRNA from the injured larvae showed the transcript [23].

Evidence for the site of synthesis of hemolymph lectins in other invertebrate species has been obtained through different approaches, including hemocyte immunostaining and fractionation, and protein purification. In the oyster *Crassostrea virginica*, hemocyte lectins are probably integral membrane components, for they copurify with the microsomal fractions [35,36]. In the blue crab *Callinectes sapidus*, we were able to purify a lectin from the hemocyte cytosolic fraction, although the lectin is present with the highest specific activity in the microsomal fractions [31,39]. In the cockroach, the disappearance and reappearance of hemocyte lectins in the presence and absence, respectively, of protein synthesis inhibitors in cell cultures suggest that those cells produce the lectins [37].

C. Variability

Substantial variability of the developmental, individual, sexual, population, geographic, and seasonal levels has been observed in lectin content and, although less frequently, in carbohydrate specificity. In some invertebrate species, the fact that the lectin content of tissues and body fluids varies during certain developmental stages has been the basis for suggesting that their biological role may be related to cell–cell and cell–substrate recognition and tissue organization, and that they are developmentally regulated. Variability in the presence or absence of lectins or their relative titers during metamorphosis or after experimental wounds has been observed in insect species such as the lepidopterans *Pieris brassicae* [46,47] and *Bombix mori* [48], and the flesh fly *S. peregrina* larva [44,45]. Quantitative changes in *N*-acylaminosugar-binding lectins from other arthropod species, such as crabs and horseshoe crabs, have been observed during the molting cycle [13,15], suggesting that humoral lectins may be involved in the transport of the *N*-acetylglucosamine subunits that constitute chitin. The lectins detected in eggs of the snail *Helix* are not present in the juvenile stages and reappear only in the albumin gland of the sexually mature adult, suggesting a role in reproduction [2] and perivitelline protection of the embryo [11,12]. Although lectins are present in both the egg and the adult form of the Japanese horseshoe crab, *Tachypleus gigas*, the carbohydrate specificity of the egg lectins is different from those found in the adult's hemolymph [49].

Qualitative and quantitative differences have been observed in the lectin content of invertebrate specimens collected in different biotopes. One of the most remarkable examples is represented by lectins from the hemolymph and albumin gland in the planorbid snails of the genus *Biomphalaria*, in particular *B. glabrata*, the intermediary host for *Schistosoma mansoni*. Samples collected in Puerto Rico, Belo Horizonte (Minas Geraes, Brazil), and Venezuela were devoid of lectin activity; those collected in Surinam and Salvador (Bahia, Brazil) exhibited hemolymph lectins [50]. Mello and Paraense [51] published a study on albumin gland lectins from *B. glabrata* and *B. tenagophila* specimens collected in several locations from Brazil and Venezuela, and they describe differences not only in presence or absence of lectin activity, but also in titers and specificity against human blood groups A_1, A_2, and B. Geographic differences were also reported for lectin contents of hemolymph, albumin gland, and eggs from planorbid snails by Michelson and Dubois [52]. Remarkable serological differences were detected between specimens of the snail *Helix aspersa* collected in Spain and France [53]. Lectins from specimens collected in Mar del Plata (Argentina), where this species was introduced about a century ago, showed specificity profiles closer to those from Spanish origin [54]. In the same study, we found that *Ampullaria* snails exhibit variable lectin activity related to individual, sex, developmental stage, and locality. Isolectins from single *H. pomatia* snails from two different geographic origins, Sweden and Germany, showed qualitative and quantitative differences in isoelectric focusing gels [55].

In general, no correlation has been found between the specificity of the lectins and the taxonomy of the species studied, except for snails from the families Helicidae and Fruticolidae [12,28], in which there is a high frequency of *N*-acetyl-D-galactosamine-specific lectins, and in clams of the genus *Tridacna*, which show antigalactan specificities [19,56]. In the few examples of Crustacea species for which the binding properties of serum lectins were determined, the studies resulted in specificities for *N*-acylaminosugars, even when multiple distinct lectins were present [31–33,39,57]. Within the Chelicerata (horseshoe crabs, scorpions, whip scorpions, and spiders), all species yet studied in the class Merostomata and orders Scorpionida and Uropygi, exhibit serum lectins that bind sialic acids or sialoconjugates [29]. Thus, the Chelicerata constitute the only group of high taxonomic rank that includes species that exhibit certain common patterns in the specificity of their humoral lectins. This homogeneity in lectin specificity is probably apparent only when the major lectin is considered; usually other minor components with different specificities are also present in any particular species. However, the carbohydrate specificity of the major lectin may constitute an adaptation for interaction with other organisms, pathogens or symbionts, with which that particular invertebrate taxon associates. The eggs from snails from the family Helicidae, generally incubated in the humid soil, may require protection from the pathogenic *Escherichia coli* and *Streptococcus* strains, to which the egg lectins strongly bind. Most members of the genus *Tridacna* (giant clams) exhibit galactosyl-binding lectins that interact specifically with their aged algal symbionts through their exposed galactosyl residues [19]. The sialic acid-binding lectins from chelicerates, such as horseshoe crabs, spiders, and scorpions, also bind 2-keto-3-deoxyoctonate (KDO), *N*-acetylglucosamine, and *N*-acetylmuramic acid, all components of the bacterial capsules or cell walls, in particular from the gram-negative pathogens [58]. The opposite situation is found in most protochordate species, in which multiple distinct plasma lectins exhibit a variety of carbohydrate specificities [59].

IV. MOLECULAR PROPERTIES

Considering that invertebrates are currently accepted as polyphyletic in origin, one should assume that their carbohydrate-binding molecules arose at different times and are products of very diverse evolutionary histories. Hence, it should not be surprising that it is virtually impossible to find a general pattern in the molecular structure of invertebrate lectins. Moreover, it is a very common situation that, in a particular organism, multiple lectins with distinct subunit size and carbohydrate specificity are present, sometimes even in the same tissue or fluid. Furthermore, it has been shown that a certain degree of microheterogeneity occurs in the structure of the protein subunits, yielding molecular species with diverse isolectric points, known as *isolectins*. As a result of this vast diversity in structure and carbohydrate specificity it should be borne in mind that to associate a particular invertebrate species with one specific lectin is, in most instances, an oversimplification quite far from reality. However, across the diverse invertebrate and vertebrate phyla, it is eventually possible to find examples exhibiting restricted structural similarity, that suggest that certain lectin molecules, or at least regions of their amino acid sequence, probably relevant to their biological role, have been conserved in evolution along lineages that gave origin to the vertebrate phyla. These sequence similarities occur as relatively long stretches of sequence with a high percentage of homology, usually structural domains that exhibit carbohydrate-binding properties, or as isolated residues that are highly conserved and are probably important in providing a framework in which stretches of variable residue are intercalated while a particular conformation in the molecule is maintained.

A. Structural Aspects

The molecular properties of selected invertebrate lectins from several taxa are summarized in Table 1: It must be remembered, however, that they should not necessarily be considered representative of lectins found in other species from the same taxonomic group, for the reasons explained earlier.

Most invertebrate lectins are oligomers that consist of subunits of equal or different size, held together by disulfide bonds or noncovalent interactions. Those subunits may be constituted by one or more polypeptide chains bound by interchain disulfide or noncovalent interactions as well; each polypeptide chain may be folded and stabilized by intrachain disulfide bonds.

The relative molecular masses (M_r) of the native lectin oligomers from invertebrates vary between 37,000 in the coelenterate *Cerianthus membranaceus* [43] and several hundred thousand daltons, 600,000 in the tunicate *Halocynthia roretzi* [60], 470,000 in *Tridacna maxima* [56] and approximately 400,000 in *L. polyphemus* [13]. However, it has been shown for several invertebrate species that humoral lectins have a tendency to aggregate spontaneously, increasing their size heterogeneity in their native state. This property is concentration-dependent, as established for lectins from the slug *Limax flavus* [61] and tunicate *Didemnum candidum* [62]. At concentrations under 1 mg/mL, the sialic acid-binding lectin from the slug *L. flavus* is a dimer (M_r 44,000) of two identical subunits (M_r 22,000), linked by noncovalent interactions. Over that concentration, the lectin tends to aggregate in large molecular species [61]. A very similar phenomenon occurs with DCL-I, a lectin from the tunicate *D. candidum*, as measured by sedimentation equilibrium [62].

Helix pomatia lectin comprises three dimers of M_r 26,000–30,000, bound by noncovalent interactions, each consisting of subunits (M_r 12,500–16,700, partially reduced; 13,000, completely reduced) linked by single disulfide bonds [16]. Each monomer has one binding site and one intrachain disulfide bond. The native M_r was calculated as 100,000 by sedimentation velocity and viscosity measurements [16], but yielded values 20% lower by sedimentation equilibrium (M_r 79,000) [63], and the native lectin may aggregate into multimers, a "superagglutinating" species of M_r 280,000 [53].

Very limited information is available on the mechanisms and specificity of the aggregation, but the interaction is noncovalent, specific, or nonspecific.

The oyster *Crassostrea gigas* has two distinct lectin specificities, gigalin E and H, that in the native state are present as large aggregates with M_r ranging from 500,000 to 1,600,000) that, in turn, comprise three types of subunit: α_1 (M_r 21,000), α_2 (M_r 22,500), and β (M_r 33,000). The β-subunit comprises polypeptides of M_r 15,500 and 17,500. Gigalin H is predominantly $\alpha_1\alpha_2$, whereas gigalin E is $\alpha_1\beta$. The aggregates are probably organized structures, since free reassociation from dissociated subunits never yields active multimers [64]. The lectin I from the sponge *Geodia cydonium* has a tendency to form large aggregates by specific interaction of its galactosyl-binding site with the oligosaccharide chain of a second lectin I molecule [65]. Limulin, the lectin from the horseshoe crab *L. polyphemus*, has a tendency to aggregate through a region close to the COOH-terminal of the subunit [66].

Relative to the subunit architecture of the native lectin molecule, the single polypeptide chain of M_r 37,000 found in *Cerianthus* [43] and the lectin II (M_r 15,000) from the sponge *Axinella polypoides* [67] constitute probably the simplest and smallest invertebrate lectins yet identified. Similar monomeric lectins have been identified in the rock shrimp *Squilla mantis* [68] and the scorpion *Heterometrus bengalensis* [69]. Higher levels of molecular complexity can be found in the lectin I (M_r 21,000) from *Axinella*, comprising a subunit of M_r 15,000 plus an undetermined fragment [67], and the galactosyl-binding lectins DCL-I and DCL-II from the tunicate *D. candidum*, globular proteins composed of four equal-sized subunits of M_r 14,000 and M_r 15,000, respectively [62]. The blue crab, *C. sapidus*, exhibits at least two serum lectins, CSL-I and CSL-II: native CSL-I occurs as two major molecular species of M_r 520,000 and 240,000 both of which comprise subunits of M_r 37,000 and 38,000; CSL-II is present as multimers of M_r 500,000, 280,000, and 230,000, all comprising subunits of M_r 31,000 and M_r 33,000 [39]. At least three distinct lectins are p resent in the acorn barnacle *Megabalanus rosa*: BRA-I (M_r 140,000) and BRA-II (M_r 330,000) are composed of multiple subunits, each consisting in two identical basic units of M_r 22,000 cross-linked by disulfide bonds; in BRA-III (M_r 64,000) the basic units are M_r 18,000 [70]. The *L. polyphemus* lectin, limulin, considered now to be a primitive C-reactive protein [71], is a hexameric structure of M_r 335,000–500,000 [13,71,72], which comprises equimolar amounts of three different glycoproteins, of M_r between 18,000 and 24,000, arranged in a complex manner that is not yet clear [71]. Marchalonis and Edelman [13] have proposed a model in which a ring-shaped structure is composed of six subunits, each one, in turn, comprising three polypeptide chains. Electron micrographs of the native molecule [14] confirmed the proposed model. Similar images, ring-shaped molecules of five or six subunits, have been shown in *Crassosstrea gigas* lectins [64]. These observations are in agreement with the hydrodynamic properties of most invertebrate lectin molecules in solution, since frictional ratios, usually close to the unit, suggest that they behave as globular proteins. Such is the case with *D. candidum* lectins [62].

Nevertheless, electron micrographs of other invertebrate lectin molecules, such as those from the sea urchin *Anthocidaris crassispina* [23] and the cockroach *Periplaneta americana* [74], yielded structures very different from those described in the foregoing.

Table 1 Molecular Properties of Selected Lectins from Invertebrates

Phylum/species	Lectin	Tissue/organ[a]	Nominal[b] specificity	Subunit M_r	Native M_r[c] (or $S^o_{20,w}$)	Carbohydrate[d] (%)	Divalent cations required	Ref.
Porifera								
Axinella polypoides	Lectin-I	Extract	D-Gal	21,000	2.6	0.5	No	67
	Lectin-II		D-Gal	15,000	2.8	0.5		
Aaptos papillata	Lectin-I	Extract	GlcNAc	12,000; 21,000;	3.5	0.0	ND	84
	Lectin-II			16,000	6.0	0.0	ND	
	Lectin-III			16,000	5.5	0.0	ND	
Geodia cydonium	Lectin-I	Extract	D-Gal	12,200; 13,000; 13,800	36,500	10.0	Ca^{2+}[e]	65,168
Mollusca								
Limax flavus	LFA	Body wall	NeuAc	22,000	44,000	ND	ND	61
Helix pomatia	HPA	Albumin gland	αGalNAc	13,000	79,500	7.0	No	16,63,87,88
Achatina fulica	Achatinin$_H$ (ATN_H)	Hemolymph	9-*O*-Ac NeuNAc	15,000	242,000	21.0	Ca^{2+}	169–171
Tridacna maxima	Tridacnin	Plasma	Galβ-1,6-Gal	10,000; 40,000	470,000	7.0	Ca^{2+}	56
Crassostrea gigas	Gigalin H ($\alpha_1\alpha_2$)	Plasma	Sialic acids	21000 (α_1); 22,000 (α_2)	500,000–1,600,000	20.0	Ca^{2+}[f]	64,96,122
	Gigalin E ($\alpha_1\beta$)	Plasma		21,000 (α_1); 33,000 (β)	500,000–1,600,000	20.0	Ca^{2+}[g]	

Arthropoda								
Heterometrus bengalensis		Plasma	ND	146,000	146,000	0.0	No	69
Carcinoscorpius rotundicauda	Carcinoscorpin	Plasma	Sialic acids, phosphorylcholine	27,000 28,000	420,000	5.8	Ca^{2+}	83,93,97,172
Limulus polyphemus	Limulin; Limulus CRP	Amoebocytes, plasma	Sialic acids, phosphorylcholine	18,000 24,000	400,000	3.6–24.0	Ca^{2+}	13–15,58,66, 71,72,79,80, 90–92,95,100
Homarus americanus	LAg-1	Plasma	Sialic acids	70,000	19S	ND	Ca^{2+}	32,33
	LAg-2		GalNAc	25,000	11S	ND	Ca^{2+}	
Megabalanus rosa	BRA-1	Coelomic fluid	ND	22,000	330,000	Yes[h]	Ca^{2+i}	70
	BRA-2			22,000	140,000	Yes[h]	Ca^{2+i}	
	BRA-3			18,000	64,000	Yes[h]	Ca^{2+i}	
Callinectes sapidus	CSL-I	Hemocytes, plasma	*N*-acylaminosugars	37,000; 38,000	520,000 240,000	ND	Ca^{2+}	31,39
	CSL-II	Plasma	*N*-acylaminosugars	31,000; 33,000	500,000; 280,000; 230,000	ND	Ca^{2+}	
Cancer antennarius		Plasma	9-*O*- and 4-*O*-acetylsialic acids	36,000	70,000 or more	ND	Ca^{2+}	57
Sarcophaga peregrina	Fly larva lectin	Fat body, plasma	D-Gal,lactose	23,000	190,000	ND[j]	ND	23,27,44,45,74
Pieris brassicae		Epidermal cell membranes	GlcNAc, chitobiose, chitotriose	23,000	43,000	ND	ND	46,47
Allomyrina dichotoma	Allo A-I	Larva hemolymph	Galβ1-4GlcNAc	17,500; 20,000	65,000	0.0	No	94,173
	Allo-A-II		β-Gal	19,000; 20,000	66,500	0.0	No	

Table 1 (Continued)

Phylum/species	Lectin	Tissue/organ[a]	Nominal[b] specificity	Subunit M_r	Native M_r[c] (or $S^o_{20,w}$)	Carbohydrate[d] (%)	Divalent cations required	Ref.
Echinodermata								
Anthocidaris crassispina	Echinoidin	Coelomic fluid	Galβ-1,3-GalNAc	13,000	300,000	3.2	Ca^{2+}	73,81
Chordata								
Didemnum candidum	DCL-I	Plasma	βD-Gal	14,000	56,600	0.0	Ca^{2+}	62,89
	DCL-II	Plasma	D-Gal	15,000	57,500		Ca^{2+}	
Polyandrocarpa misakiensis		Homogenate	D-Gal	14,000	14,000	0.0	Ca^{2+}	82

[a]Tissue or organ from which the lectin was isolated.
[b]Carbohydrate(s) that behave(s) as the best inhibitor(s).
[c]Only sedimentation coefficients are available.
[d]Covalently bound oligosaccharide.
[e]Not an absolute requirement, but modifies the interaction between lectin, aggregation factor, and cells.
[f]Required for binding to erythrocytes, but not for mucins.
[g]Probably required for maintaining the protein conformation, but not for binding.
[h]Carbohydrate is present but percentage values have not been determined.
[i]Not required for binding to ligand.
[j]Consensus sequences for glycosylation sites are present at the gene level.
ND: Not determined.

Echinoidin, a lectin from the sea urchin, *A. crassispina,* is a butterflylike-shaped protein with four globular domains of M_r 300,000 assembled from 22 to 24 subunits, each consisting of dimers (M_r 26,000) of polypeptide chains (M_r 13,000) cross-linked by disulfide bonds [73]. Lectins from the cockroach, *P. americana,* were rod-shaped molecules of 10-nm width and 50.5-nm length in average, constituted by helical repeating units [74].

In most invertebrates, a certain degree of microheterogeneity has been observed in isoelectric focusing gels [39,55,62]. *Helix pomatia* purified lectin was fractionated by isoelectric focusing, into at least 12 components (isolectins), each showing the same molecular mass (M_r 79,000) and serological properties. However, they differ slightly from each other in amino acid composition. The native isolectin molecules would be composed of diverse combinations of subunits, with different isoelectric points [55].

In general, amino acid compositions of invertebrate lectins follow patterns similar to those from lectins of plant and animal origin [75]: high proportions of acidic amino acids (glutamic and aspartic acid) and low content of half cystine and methionine. As a result, low isoelectric points are common among invertebrate lectins. Nevertheless, in some examples, although the contents of aspartic and glutamic acids may be high, compared with the basic amino acids, the isoelectric points are relatively high, suggesting that the acidic amino acids are in the amidated form. The lectins DCL-II, from the tunicate *D. candidum* [62], and LFA, from the slug *L. flavus* [61], that well illustrate this observation, focus at pH 9.2–10.2 and 9–9.5, respectively.

In comparisons of amino acid compositions as an estimate of intra- and interspecific relatedness between lectins, the limitations of both the analytical experimental methods and statistical parameters SΔQ [76] and SΔn [77], should be carefully considered when drawing any conclusion. Isolectins, such as those from the snail *H. pomatia,* usually show a high degree of similarity in amino acid compositions [55], but when distinct lectins are compared, even those isolated from the same organism and exhibiting similar carbohydrate specificities, disparate profiles may be observed. In the Japanese horseshoe crab, *Tachypleus tridentatus,* comparisons of amino acid compositions of five lectins isolated show that the lectin G_1 is similar to G_2 and G_3 to G_4; G_{N3} differs from all others in low contents of tyrosine and high content of valine [78]. The galactosyl-binding lectin DCL-I from *D. candidum* appears more closely related in its amino acid composition to a lectin from the tunicate *Halocynthia pyriformis* and mammalian C-reactive proteins than to a second lectin DCL-II of similar specificity isolated from the same organism [62].

Relatively few partial or complete amino acid sequences of invertebrate lectins have as yet been determined. The first report providing this type of information consisted of the partial NH_2-terminal sequence (76 residues) of the lectin from *L. polyphemus* and provided undisputable evidence for the lack of overall homology between this molecule and vertebrate immunoglobulins [79]. It failed, however, to provide support to the apparent microheterogeneity observed in the subunit structure. Partial NH_2-terminal amino acid sequence has been obtained for the galactosyl-binding lectin DCL-I from *D. candidum,* that showed a certain degree of similarity with the mammalian pentraxins, C-reactive protein and serum amyloid P [62], and the oyster *Crassostrea virginica* [18].

In recent years, complete amino acid sequences for a few invertebrate lectins have been obtained, from direct protein sequencing or deduced from cDNA nucleotide sequences, and have been extremely useful in providing information about their carbohydrate- and divalent cation-binding domains, glycosylation sites, secondary structure, hydropathic profiles, and possible homology with other molecules and, thus, about their evolution. So far, complete amino acid sequences have been published for lectins from the horseshoe crab, *L. polyphemus* [80]; the sea urchin, *A. crassispina* [81]; the acorn barnacle, *Megabalanus rosa* [70]; the larva from the fresh fly, *S. peregrina* [23]; and the galactosyl-binding lectin from the tunicate *Polyandrocarpa misakiensis* [82]. The three subunits of the *Limulus* lectin that have now been sequenced (1.1, 1.4, and 3.3), share an NH_2-terminal sequence of 44 residues, COOH-terminal sequence from residues 206 to 218, the position of the six half-cystines that form the three disulfide bonds (Cys^{38}–Cys^{101}; Cys^{88}–Cys^{120}; Cys^{183}–Cys^{217}), and the glycosylation sites (Asn^{123}). There is a 10–11% microheterogeneity for the entire protein, and it is apparent that other distinct subunits exist [80]. The subunits of echinoidin, the lectin from *A. crassispina*, consist of a total of 147 amino acids, with a glycosylation site on Ser^{38} and three interchain disulfide bonds (Cys^{3}–Cys^{14}; Cys^{31}–Cys^{141}; Cys^{116}–Cys^{132}). A seventh half-cystine, Cys^{2}, is involved in the interpolypeptide disulfide bond [81]. The subunit of BRA-3, a lectin from *M. rosa*, also contains seven half-cystines on a total of 133 amino acids [70]. The amino acid sequence of the α-subunit from the *S. peregrine* lectin has been deduced from the nucleotide sequence and consists of 260 amino acid residues, with two potential glycosylation sites in positions 222–224 (Asn-Ala-Thr) and 265–267 (Asn-Ala-Ser). Two putative signal peptides of 23 and 19 amino acids could be identified [23]. A galactose-binding lectin from the tunicate *P. misakiensis* (M_r 15,000) is a monomer composed of 125 amino acid residues, with two intrachain disulfide bonds (Cys^{21}–Cys^{119}; Cys^{96}–Cys^{111}) and no covalently bound carbohydrate [82]. The sequence analyses of these proteins have yielded interesting information on their homology with other molecules, and this will be discussed in Section VI.

Circular dichroism analysis or information obtained from the amino acid or cDNA sequences suggests that most invertebrate lectins contain a high proportion of β-structure. *Tridacna maxima* lectin [56] and *D. candidum* lectin DCL-I [62] exhibit 40 and 37% of β-structure, respectively. In contrast, the amount of α-helix is usually low: *T. maxima* lectin exhibits a 10% α-helix structure. As an exception, DCL-I contains a relatively high proportion of α-helix (29%). High amounts of β-structure (β-sheets and β-turns) and low amounts of α-helix have been determined in the acorn barnacle *M. rosa* [70] and the sea urchin *A. crassispina* lectins [73]. Low amounts of any secondary structure were found in lectins from the merostome *Carcinoscorpius rotundicauda* [83]; *L. polyphemus* lectin appears to lack any α-helix or β-structure [72].

The presence of covalently bound carbohydrate chains is a variable feature in invertebrate lectins, ranging from relatively high amounts, such as in *L. polyphemus* lectin (24%) [13], to moderate or relatively low amounts, such as lectins from the sponge *G. cydonium* lectin (9.9%) [65], the clam *T. maxima* (7%) [56], the sea urchin *A. crassispina* (3.2%) [73], the sponge *Axinella polypoides* (0.%) [67], or none, as in the tunicate *D. candidum* [72]. In lectins for which the primary structure has been predicted from the cDNA sequence, such as in the fly *S. peregrina* larva lectin [23], potential glycosylation sites have been determined by the occurrence of the proper consensus sequences (Asn-X-Thr or Asn-X-Ser), but the presence of covalently bound oligosaccharide chains in the mature protein has not yet been established. The function of the covalently bound oligosaccharide chain is unknown, but it is noteworthy that the lectin I from the sponge *Geodia cydonium* contains, besides two carbohydrate recognition sites, at least one receptor site for the nonreducing terminal galactosyl residue of the oligosaccharide portion of a second lectin molecule [65]. This self-aggregating property can be specifically inhibited by lactose and may be of biological significance.

B. Binding Properties

For the carbohydrate specificity of invertebrate lectins, inhibition of hemagglutination and quantitative precipitation of glycoconjugates (glycoproteins and polysaccharides) with simple or complex carbohydrates have been the conventional methods for characterization. Equilibrium dialysis and Scatchard-type plots have provided information about the affinity and number of binding sites present in the active molecule. The specificity of invertebrate lectins has been serologically correlated with the size and shape of the binding sites: lectins such as those from the antozoan *Cerianthus membranaceus*, which preferentially bind to relatively large moieties (such as the trisaccharide Gal[1-3(6)]Gal-Gal [43]), or from the sponge *Aaptos papillata*, which binds to chitotriose [84], are considered "very specific." Some invertebrate lectins exhibit specificity for disaccharides such as *Tridacna maxima* [αGal(1-6)Gal] [56], *Botrylloides leachii* [Gal(1-6)Glc] [85], and *Geodia cydonium* [βGal(1-3)Gal] lectins and would be considered moderately specific. Other lectins with relatively small binding sites, such as those from *L. polyphemus* [15] or *Limax flavus* [61], that bind sialic acids could be considered of broad spectrum or nonspecific, since they will agglutinate most cells. However, this is more a matter of how ubiquitous the ligand is—in this example, sialic acids—rather than the intrinsic binding properties of the lectin. The nominal specificity of a lectin is defined by the carbohydrate for which it shows the highest affinity, and this means that most lectins that are considered specific for one monosaccharide may also bind other carbohydrates that are structurally related, but with lower affinity. It should be remembered that the definition of the specificity of a lectin arises from utilitarian criteria, which, in most cases, do not reflect the actual spectra of substances to which the lectin binds.

Certain invertebrate lectins recognize the nonreducing terminal sugar and its anomeric linkage, as occurs with certain ABH blood group-specific lectins such as *H. pomatia* and *Cepaea nemoralis* (αGalNAc) [12] and eel serum lectin (αL-Fuc) [86]. *Helix pomatia* lectin binds strongly to A_1, A_2, and A_1B erythrocytes, but weakly, or not at all, to A_2B and A_3 [87]. The lectin precipitates with A blood group substance (αGalNAc) and group C streptococcal polysaccharides (αGalNAc), but not group A or teichoic acid from *Staphylococcus aureus* (αGalNAc). The precipitation is efficiently inhibited by αGalNAc and methyl-α-D-GalNAc, but even better with the blood group A active pentasaccharide [16,88]. However, certain anomalous binding properties have been difficult to explain: this lectin can be purified on a dextran, such as Sephadex G-100, a polymer of 1-6-D-glucose, by elution with low concentrations (0.0067 M) of *N*-acetylgalactosamine [53]. In the sponge, *Axinella polypoides*, both lectins I and II precipitate with human blood group A_1, A_2, B, H, and Le^a and the precursor I substance. Glycosides in which D-galactose is linked α1-6 are better inhibitors than free D-galactose for lectin I, but equal for lectin II. Moreover, the β-anomeric linkage is preferred to the α; α-*N*-acetylgalactosamine is a poor inhibitor and so is L-fucose. Since α- and β-D-galactose are the immunodominant residues of B, Le^a, and precursor I substances, respectively, it is likely that lectins I and II bind substance A and H through side chains possessing nonreducing terminal β-D-galactose [67]. The lectin I from the sponge *A. papillata* reacted with blood group substances that contain nonreducing terminal *N*-acetylglucosamine, whereas those with *N*-acetylgalactosamine were inactive. In addition, it was inhibited by D-GlcNAc and N, N′, N″, N‴-tetraacetylchitobiose, but not by D-GalNAc. The lectin was precipitated by the monofunctional hapten *p*-nitrophenyl-α-D-GalNAc, but not by the β-anomer, although it was a good inhibitor, suggesting that hydrophobic bonds are involved in these interactions. *Aaptos* lectin II was completely precipitated by blood group substances containing terminal D-GalNAc, D-GlcNAc, and sialic acids, but inhibition by GlcNAc and sialic acids was four times more efficient than GalNAc, and inhibition by *N,N',N''*-chitotriose was 13 times more efficient than GlcNAc. *Aaptos* lectin III was similar to lectin II [84].

Most invertebrate lectins bind to the nonreducing terminal residues from oligosaccharides from glycoconjugates (glycoproteins and glycolipids) and polysaccharides. However, those of the coelenterate *Cerianthus membranaceus* bind to inner-chain sugar residues [43].

Some lectins exhibit specificity for only certain hydroxyl groups in the pyranoside ring of the nonreducing terminal sugar. The binding site of the lectin DCL-I from *D. candidum* is relatively small and has clefts for hydroxyl groups on carbons C-2, C-3, and C-4 of the nonreducing terminal D-galactose residues, since only the anomeric configuration of the hydroxyl groups on those positions are recognized, L-arabinose behaves as an effective an inhibitor as D-galactose [89].

In certain instances, not only the terminal sugar is recognized, but also so is the substituent on the amino or hydroxyl residues. Good examples of these are the sialic acid-binding lectins from horseshoe crabs *C. rotundicauda* and *L. polyphemus*, and the crab *Cancer antennarius*. Carcinoscorpin is similar to limulin in that it prefers glycolylneuraminic acid (NeuGc) to acetylneuraminic acid (NeuAc) and agglutinates horse erythrocytes more efficiently than any others. Horse erythrocytes contain hematoside, a derivative of the G_{M3} ganglioside containing NeuGc, as the major glycoconjugate [90]. This also constitutes an example of a lectin that prefers carbohydrate moieties from glycolipids, rather those from glycoproteins. By the use of natural and synthetic derivatives of sialic acids (SA) we showed that certain substitutions on the hydroxyls do not affect inhibition of *L. polyphemus* lectin [91]. However, the lack of *N*-acetylation renders the structure noninhibitory [72]. Limulin also prefers glycosides, compared with free sialic acids and, in particular, certain linkages: the best inhibitors for the lectin are SA-α(2-3)(6)GalNAc [72,92]. This suggests that a preference exists for the *O*-glycosidically linked oligosaccharides, rather than the *N*-linked. Carcinoscorpin binds to glycoproteins depending on the total content and topography of the sialic acids but does not discriminate between *O*- and *N*-linked oligosaccharides [93]. The lectin of the crab *Cancer antennarius* recognizes only NeuNAc that is *O*-acetylated in positions 4, 7, and 9 [57].

At the opposite extreme of the spectrum, certain lectins will recognize only the substituent in the amino group. The lectin CSL-I of the crab *Callinectes sapidus* appears to recognize only the *N*-acylamino group irrespectively of the core structure, since it is efficiently inhibited by NeuNAc, GlcNAc, or *N*-acetylglutamic acid [39].

In certain invertebrate lectins, the carbohydrate specificity is complex, and although the participation of sialic acids in the carbohydrate moieties recognized has been demonstrated, it does not constitute the ligand per se. The lectin of the sea urchin *A. crassispina* [73] is not inhibited by simple sugars, but only by tryptic fragments of human erythrocyte membranes, mainly those carrying the structure NeuAcα(2-3)Galβ(1–3) (NeuAcα2-6)GalNAc-Ser. However, the removal of the sialic acid from the fragments does not affect their inhibitory capabilities, suggesting that the inner disaccharide structure linked to serine or threonine is important for the binding. The lectin Allo A-II, from the insect *Allomyrna dichotoma* [94], interacts with the disaccharide Galβ(1-4)GlcNAc, but if the galactose is substituted by NeuAc2-6 it will enhance the affinity. If the substitution is at the C-2 or C-3 of the galactosyl residue, with sialic acid or any other sugar, the lectin will no longer bind. However, the interaction can be reversed by lactose, but not by sialic acid, indicating that the terminal sialic acid has only a secondary role, probably providing a better conformation to the carbohydrate moiety.

Certain lectins, on the other hand, exhibit binding specificity for apparently structurally unrelated ligands. Such is the case of the *L. polyphemus* lectin that binds NeuNAc and phosphorylcholine [71]. With use of anti-idiotypic monoclonal antibodies generated against a phosphorylcholine-binding myeloma protein, we showed that those binding properties correspond to two distinct independent binding sites in the same native molecule [95]. Since the lectin oligomer is composed of three different polypeptide chain subunits [80], it is not yet clear whether those binding sites are located in the same or different subunits. A similar situation occurs with the lectin carcinoscorpin from the merostome *Carcinoscorpius rotundicauda* that binds to sialic acids and phosphorylcholine [93].

At least two different bindings sites were shown for *H. pomatia* lectin by mixed agglutination with blood group A and neuraminidase-treated O erythrocytes [53]. From observations made on gigalins E and H, the lectins from the oyster *Crassostrea gigas*, Olafsen [96] has suggested that multiple heterogeneous binding sites in the native forms of those lectins would result from aggregation in various combinations of the three basic subunits α_1, α_2, and β, of different binding properties.

Lectins from the merostomes *L. polyphemus* and *C. rotundicauda*, in addition to their sialic acid- and phosphorylcholine-binding properties, also show antiglucuronyl activity [93,97,98]. Although some ligands are structurally unrelated, the negative charge shared by sialic and glucuronic acids may be a relevant factor. Both lectins, limulin and carcinoscorpin, as well as those from several other Chelicerata species [29], also bind to 2-keto-3-deoxyoctonate (KDO) [29,58,93]. However, this property may relate to a certain "flexibility" of the sialic acid-binding site, as both sialic acids and KDO are keto-deoxy acids and, thus, structurally related.

This specificity for KDO and glucuronic acid observed in limulin, carcinoscorpin, and lectins from other chelicerate species may be relevant to their biological role as recognition or defense molecules. Very few bacteria produce sialic acids, but KDO and glucuronic acids are ubiquitous components of lipopolysaccharides (LPS) from gram-negative bacteria. Dorai et al. [97] have shown that carcinoscorpin binds to and agglutinates bacteria only if they contain KDO in their LPS: *E. coli* K12, which lacks O-antigenic chains, and *Salmonella minnessota* R595, which exhibits terminal KDO in its LPS, are both strongly agglutinated, and this can be inhibited by free KDO, whereas *Vibrio cholerae*, which lacks KDO in the LPS, and the gram-positive *Staphylococcus aureus* are not agglutinated.

Although a few lectins have been characterized as monomeric, incomplete, or nonagglutinating, such as two minor lectins from *L. polyphemus* [92] and a minor lectin from *Callinectes sapidus* [39], the great majority, as described earlier, exhibit multiple-binding sites. By equilibrium dialysis measurements on the interaction of the *H. pomatia* lectin with the blood group A active pentasaccharide, it was determined that the lectin has six binding sites, with an intrinsic association constant of 5×10^3 L/M at 25°C. However, the association constants calculated for the interaction between the lectin and A erythrocytes are many orders of magnitude higher, 5×10^{10} L/M, than the constant for the interaction with the blood group determinant, and this would be due to multivalent interactions. The cell-binding experiments have allowed the estimation of the number of A sites available at the cell surface, 1.4×10^6 for blood group A erythrocytes [87].

For carcinoscorpin, association constants calculated for the binding to sialylated ligand are 1.82×10^3 L/M. This lectin exhibits two types of binding sites, two of high affinity and six of low affinity, that show positive cooperativity: the binding of ligand to one site increases the affinity for the ligand of neighboring unoccupied binding sites in the early phase of multivalent binding [93].

As happens with many plant lectins, the binding of certain invertebrate lectins to their carbohydrate moieties requires the presence of Ca^{2+} or other divalent cations. Nevertheless, numerous reports describe lectins that are active in the absence of divalent cations, such as those from the tunicate *Botrylloides leachii* [99] and the sponge *Axinella polypoides* [67].

The lectins from *L. polyphemus* [13] and *Anthocidaris crassispina* [73] require Ca^{2+} for binding. In the later, the depletion of Ca^{2+} induces structural changes, decrease in sedimentation coefficient and increase in Stokes radius that reduces the molecule to a more asymmetric conformation. A similar decrease in sedimentation coefficient was observed in limulin in the absence of Ca^{2+} [100]. The primary structure of *L. polyphemus* lectin revealed that residues 139–153 clusters acidic amino acids and is very similar to the consensus sequence required for Ca^{2+} binding in calmodulin and related molecules [22,80]. For carcinoscorpin, Ca^{2+} is an absolute requirement and (the optimal Ca^{2+} concentration is 0.02 M) cannot be substituted by other divalent cations such as Mg^{2+} or Ba^{2+}. Moreover, Mn^{2+} and Sr^{2+} have an inhibitory effect on the lectin activity [93]. Calcium is also required for the binding of gigalin E to horse erythrocytes, whereas Zn^{2+}, accumulated by the oyster hemocytes, acts as an inhibitor, even at physiological concentrations. Olafsen [96] has postulated that this lectin may constitute an opsonin that is regulated by the inter- and extracellular concentrations of divalent cations.

V. GENE STRUCTURE

A. Gene Organization and Expression

In the past few years genetic approaches using genomic DNA and cDNA-cloning techniques have brought new insight not only in the molecular structure of animal lectins, but also in their gene organization, expression, and evolution.

The structures of genes encoding for lectins from invertebrates have been reported for only two species of arthropods: the lectin (C-reactive protein) from the horseshoe crab *L. polyphemus* [22] and the inducible lectin from the larva of the flesh fly *S. peregrina* [23]. The former is a constitutive plasma protein, with nominal specificities for sialic acids and phosphorylcholine, secreted by amoebocytes and present at relatively constant concentrations between 1 and 5 mg/mL. Its primary structure [80], binding specificity [71], and immunological cross-reactivity [95] suggest that it may be a primitive member of the pentraxin family from vertebrates that include oligomeric inducible or constitutive proteins such as C-reactive protein (CRP), serum amyloid protein (SAP), and hamster female protein (HFP). On the other hand, the lectin from the *S. peregrina* larva is inducible, increasing its concentration in plasma after a challenge, such as injury of the body wall or injection of erythrocytes [27,44]. The primary structure of this lectin [23] has shown homology to another invertebrate lectin, echinoidin, from the sea urchin and the C-type membrane and soluble lectins from vertebrates [81].

The structural organization of the genes encoding for these two lectins also show substantial differences and provide evidence pertaining to the levels at which protein diversity is generated. Screening of genomic DNA from *L. polyphemus* amoebocytes [22] has shown that its lectin (or C-reactive protein), is a protein that exhibits subunit heterogeneity (more than three distinct subunits), is encoded by multiple closely related genes that use the same promoter and start and stop signals, and that exhibit extensive homology with each other throughout their nucleotide sequence. No intervening sequences are present, and these three genes consist of a leader sequence of 72 nucleotides encoding for a 24-amino acid signal peptide followed by a continuous-reading frame of 654 nucleotides. The three genes exhibit single nucleotide changes, with or without changes in the amino acids encoded, that correlate with differences observed at the protein level. There is evidence that additional homologous genes would encode for a series of related proteins, the whole constituting a family of lectins. Human C-reactive protein, which is an acute-phase protein, exhibits an intron between the signal peptide and the coding sequence, containing DNA segments that could potentially adopt the z form. In addition, between the TATA box and the coding region it contains a *Drosophila* heat-shock consensus sequence. Both regions, which have been thought to regulate synthesis of CRP in the mammalian acute-phase response, are missing from the genes of *Limulus*. Limulus lectin and human CRP show approximately 25% overall identity in their nucleotide sequences, with two regions of conserved residues: One region (residues 52–67) would constitute the binding site for phosphorylcholine and the other (residues 139–153) would bind Ca^{2+} [22].

B. Posttranslational Processing

Sarcophaga lectin genes differ from those for limulus lectin in that only one gene codes for the complete lectin: the α-subunit is posttranslationally processed to generate the β-subunit. Hence, contrary to the situation observed in *Limulus*, the heterogeneity of *Sarcophaga* lectin subunits is not a consequence of multiple genes, but rather, a later process, and only one copy of a single gene can be detected in genomic DNA [23].

An interesting situation has been described for *Giardia intestinalis* lectins, which become active only after proteolytic cleavage by the mammalian host's pancreatic trypsin, allowing the parasite to attach to the duodenal mucosa [101].

VI. EVOLUTION

A. Preliminary Evidence

Our knowledge about the evolutionary history of invertebrate lectins has been very limited because of the relative lack of information on their primary structures. Although the only rigorous criterion for assessing homology among molecules is the comparison of amino acid and nucleotide sequences, preliminary evidence has been obtained from the statistical comparison of amino acid compositions and the sharing of serological determinants.

Although less rigorous than the comparison of amino acid or gene sequences, statistical methods for the comparison of amino acid compositions such as SΔQ or Marchalonis and Weltman [76] and SΔn of Cornish-Bowden [77] have given results equivalent to those obtained from the comparison of amino acid sequences [102]. Since data on the primary structure of most invertebrate lectins are largely fragmentary or as yet unavailable, these methods have been useful in providing preliminary information. SΔQ comparisons of amino acid compositions of lectins isolated from invertebrate, protochordate, and cyclostome species, with a data base comprising amino acid compositions of 252 proteins, have shown that DCL-I, a lectin from the tunicate *D. candidum* may be structurally related to a mammalian C-reactive protein and to the sialic acid-binding plasma lectin of the tunicate *Halocynthia pyriformis*. It also shows a considerable degree of similarity with an egg lectin from the lamprey (*Petromyzon marinus*), with lamprey and carp μ-chains and with carcinoscorpin (the sialic acid-binding lectin from the Indian horseshoe crab *C. rotundicauda*). A marginal degree of relatedness occurs with DCL-II, a second galactosyl-binding lectin from *D. candidum* plasma [62].

Another approach that has been used for obtaining preliminary evidence on the structural relationship of molecules is serological cross-reactivity, using both monoclonal and polyclonal antibodies. Although it can be very useful, it has serious limitations and tentative conclusions on evolutionary relationships should be analyzed bearing in mind that even point mutations can modify the specificity of antibodies or antigenicity of proteins, masking true homology or giving rise to analogy. Data on the comparison of amino acid compositions and NH_2-terminal sequences of DCL-I with acute-phase proteins were supported by a serological approach. Enzyme-linked immunosorbent assay (ELISA) experiments showed that antibodies made against human CRP cross-reacted specifically with DCL-I [62]. It is possible that the identities in several positions of the amino acid sequence observed in the NH_2-terminal stretch, compared on the DCL-I and CRP molecules, may account for the cross-reactivities observed. We reported elsewhere [95] that anti-idiotypic monoclonal antibodies made against the myeloma protein TEPC 15, cross-react with limulin, the sialic acid-binding lectin of the horseshoe crab *L. polyphemus* and CRP. However, the anti-idiotypic monoclonal antibodies did not cross-react with DCL-I. This was not surprising, since DCL-I binds galactosyl residues, whereas the other three molecules, TEPC 15, CRP, and limulin, bind phosphorylcholine. Polyclonal antibodies made against CRP cross-react with TEPC-15 and limulin, but also with DCL-I. This suggests that, although TEPC-15, CRP, and limulin might share common determinants related to their binding sites for phosphorylcholine, determinants shared by DCL-I and CRP are probably located in other regions of the molecule.

A computer analysis was carried out to identify short stretches of significant sequence identity that might be expected to account for the observed serological cross-reaction between CRP and limulin and also the binding of monoclonal antibodies to the TEPC-15 idiotype of both molecules [95]. Stretches of overlapping significant identity were found from residue 47 to 54 in limulin, 83 to 61 in CRP, and 60 to 71 in TEPC-15. Although the shared residues are not identical in the comparisons between limulin and CPR or TEPC-15, these stretches overlap and the V_H region of TEPC-15 that was identified corresponds to the juncture of the second complementarity determining region and the third framework segments [95]. Although it might be concluded that the sharing of relatively short stretches of sequence between lectins, CRP, and classic immunoglobulins is most probably an illustration of convergent evolution in which different molecules reactive with a common ligand are forced to use similar residues to form the combining site, another interpretation that might also be proposed to account for the sharing of idiotypes is the insertion of short DNA segments or minigenes [103] into different framework sequences.

B. Amino Acid and Nucleotide Sequences

Since the early 1980s, complete amino acid and gene sequences of invertebrate and vertebrate lectins have gradually become available from direct sequencing of the proteins, cDNA clones, or both protein and genomic DNA. These studies on the primary structure of lectins have provided relevant information pertaining to their possible evolutionary history. The amino acid sequence of the subunit of echinoidin, a lectin from coelomic fluid of the sea urchin *A. crassispina* showed that the COOH-terminal half was homologous with the carbohydrate-binding domain of two mannose-binding lectins A and C from rat liver in 35 and 32% identity, respectively, the chicken hepatic lectin, and the rat asialoglycoprotein receptor. This would suggest that the COOH-terminal region of the molecule containing the carbohydrate-binding site is an early development in the evolution of animal lectins [81]. It also shows homology to the central portion of the lectin from the fly *S. peregrina* (28% identity) [23], which in turn, shows 18% identity with the rat mannose-binding proteins [104]. The complete amino acid and gene sequences of three subunits of *L. polypheus* lectin became available only fairly recently [80]. They show homology (~25%) to human and rabbit CRP, SAP, and HFP, all mammalian pentraxins that show extensive homology with each other. The amino acid sequence of the lectin from the tunicate *Polyandrocarpa misakiensis* [82] shows about 20–30% homology with several invertebrate lectins, such as those from *S. peregrina, A. crassispina*, and *M. rosa*, and with several vertebrate lectins, included in the C-type category. In addition, it is noteworthy that the 35 NH_2-terminal residues of this tunicate lectin show about 40% homology with the partial sequence of the immunoglobulin κ-chain variable region [82]. We have showed elsewhere [105,106] that certain plasma components from tunicate species exhibit serological cross-reactivity with elasmobranch immunoglobulins. Although the only quantitative criterion for assessing homology among molecules and the existence of families is the comparison of their complete amino acid or gene sequences, the comparison of partial NH_2-terminal sequence can also supply important information, provided that a considerable number of identities exists between the molecules in question. The comparison of 21 residues of DCL-I NH_2-terminal amino acid sequence with mammalian acute-phase proteins is interesting in that several identities are found between DCL-I and CRP and SAP in a region that overlaps with the sequence stretches that exhibit several identities with limulin [62].

Within the vertebrates, lectins are oligomers of equal of distinct subunits, and their molecular organizations vary, based on whether they are soluble or integral membrane proteins. The subunits of membrane lectins vary in size, but their molecular structures follow a pattern common to most integral membrane proteins: a COOH-terminal domain, variable in length and located in the extracellular side of the membrane that carries the carbohydrate-binding site and eventually the glycosylation sites, followed by a short transmembrane hydrophobic region that anchors the protein to the membrane, and finally an intracytoplasmic NH_2-terminal short segment.

Although some true integral membrane lectins have been identified and characterized in invertebrate species, none has been sequenced so far. Hence, it is not possible to assess if they follow a domain pattern similar to that established for their vertebrate counterparts. However, similarities in this respect have been observed between some soluble vertebrate and invertebrate lectins [107].

In general, the molecular structures of soluble lectin subunits differ from the integral membrane lectins in that hydrophobic and hydrophilic regions alternate randomly throughout the length of the sequence. The COOH-terminal region usually contains the carbohydrate-binding site and the NH_2-terminal region usually has variable structure and may exhibit particular properties: it may consist in a fibrillar collagenlike structure that, in one example, interacts with complement components, a structure similar to the core protein of the cartilage proteoglycan that exhibits regions that interact with glycosaminoglycans domains similar to epidermal growth factors or even structures similar to those proteins that bind RNA.

Drickamer [107] has classified membrane and soluble animal lectins into two major groups, based on the primary structure of their carbohydrate-binding sites. A first group, the C-type animal lectins, includes lectins that require Ca^{2+} for binding and exhibit a set of conserved residues, approximately 15% of the recognition domain, among which are cysteine residues probably involved in disulfide bonds that are required to remain intact for binding activity. This constitutes a polypeptide framework in which stretches of sequences with variable residues would confer the carbohydrate specificity. This group includes integral membrane lectins, such as mammalian asialoglycoprotein receptors and the chicken hepatic lectin; and soluble lectins, such as mammalian liver and serum mannose-binding proteins; invertebrate lectins, such as those from the fly *S. peregrina* and the sea urchin *A. crassispina*, and other proteins, such as the 28–32-kd apoprotein of the pulmonary surfactant, the core protein of cartilage and fibroblast proteoglycans, the lymphocyte receptor for the Fc portion of IgE, the mannose receptor from pulmonary macrophages, and a fucose-specific hepatic membrane lectin. Lectins of the C-type exhibit little or no homology in domains other than the carbohydrate-recognition site. Those domains confer specific functional properties to each particular lectin, such as hydrophobic domains that anchor the protein to the plasma membrane, complement-binding sites, and areas for covalent binding of glycosaminoglycans.

The second group, S-type animal lectins, comprises lectins that require the presence of thiols for binding activity, but not Ca^{2+} or other divalent cations. No detergents are required for their extraction and, hence, they are considered soluble lectins, although they may remain membrane-associated through their carbohydrate-binding sites. Their location is mostly intracellular, in the cytoplasmic compartment. Most of the S-type lectins have specificity for β-galactosides and have similar subunit M_rs (14–16 kd). These lectins constitute the family of related molecules, that exhibit considerable similarities in the primary structure and, as the C-type lectins, exhibit an invariant residue pattern in the carbohydrate recognition domains. The conserved residues are different from the C-type lectins, and no cysteines are among them. Included in this group are the 14- to 16-kDa soluble tissue lectins found in muscle, lung, and brain. In addition, the nuclear protein CBP 35 and elastin should be also included in this group [for pertinent references see 107].

From the limited information about their primary structure now available, two major groups of invertebrate lectins can be discriminated: one group would include lectins such as those from the flesh fly *S. peregrina* and the sea urchin *A. crassispina*, that show significant homology to membrane-integrated or soluble vertebrate lectins. The second group would consist of lectins, such as those from *L. polyphemus* and *D. candidum*, that show homology to vertebrate pentraxins such as CRP, SAP, and HFP. Although the pentraxins constitute a well-defined family of proteins, it has been pointed out that they exhibit several lectin properties, and this could well justify considering them as such [108,109].

VII. BIOLOGICAL ROLE

From the previous sections of this chapter, it becomes clear that a considerable number of invertebrate lectins have been thoroughly investigated in their distribution, carbohydrate specificity, and molecular structure. However, their biological functions are still far from being elucidated. Their potential ligands, simple or complex carbohydrates, occur in all living cells and in biological fluids as well, suggesting that protein–carbohydrate interactions constitute a basic phenomena common to all organisms. Evidence has accumulated in support of the current idea that different biological functions, exogenous and endogenous, are to be found for the different lectin–carbohydrate ligand systems.

The biological roles of invertebrate lectins have been experimentally analyzed, in vivo or in vitro, only in relatively few examples. In most cases, their functions in the organism have been hypothesized from circumstantial evidence such as the lectin-binding properties and the nature of the ligand, their organ or tissue distribution, their presence in particular developmental stages, selected physicochemical properties, and homology with other molecules already characterized in their biological roles. One of the main difficulties in establishing the physiological function of lectins from invertebrates is the enormous taxonomic diversity that we are confronted with in the invertebrate phyla, which added to the variety of methodology and experimental approaches employed, has not allowed us to draw unequivocal conclusions. Recently, the finding that invertebrates exhibit lectins on the surface of phagocytic hemocytes, which can selectively bind to potentially pathogenic bacteria and fungi, has stimulated studies on these lectins as possible recognition molecules involved in cellular defense mechanisms [18]. In the past few years, their roles in symbiotic associations in clams and ascidians with algae, and cell aggregation of sponges have been successfully explored [20–22].

Since the discovery of invertebrate lectins [2], it has been considered reasonable to assume that their physiological functions are related to their carbohydrate-binding properties. The fact that their carbohydrate ligands are present, with wide or narrow distributions, in organisms, cells, and complex biomolecules, has led to assigning most lectins self- or non–self-recognition functions with variable degrees of specificity. These recognition function include their involvement in interactions with cells or extracellular materials from the same organism (self-recognition or recognition of endogenous ligands) and interactions with foreign particles or cells (non–self-recognition or recognition of exogenous ligands). In addition, it has been shown that several animal lectins exhibit other nonspecific-binding properties, including the capability for hydrophobic interactions, complement fixation, formation of aggregates, binding to divalent cations, and others, that are not mediated by protein–carbohydrate binding. These multiple properties have found experimental support through determinations of protein primary structures and gene sequences that showed the presence of multiple domains, exhibiting distinct properties in a single polypeptide chain, in what we now consider hybrid or mosaic molecules [107,110,111].

A wide variety of approaches, experimental methods, and criteria have been followed for proposing the possible biological roles of invertebrate lectins. The earliest criterion employed to infer the possible biological role of a lectin from an invertebrate species was based on its tissue distribution [2,8,11]. The relatively higher concentrations of lectins in a particular cell type, tissue, organ, or body fluid led researchers to speculate that their role should be related to the physiological function carried out in that particular location. Lectins present in hemolymph of arthropods were proposed to act as carriers for sugars, glycoconjugates, or Ca^{2+} [15,32,100] or antibodylike molecules or opsonins involved in humoral responses against pathogens [5,6,15,17,18,29]. Lectins present on invertebrate hemocyte (amoebocyte, phagocyte) surfaces were proposed to participate in cellular recognition and to trigger phagocytosis, nodulation, and encapsulation responses. The blood group-specific lectins found in the albumin gland and eggs of the gastropod snails were thought to protect the embryo from bacterial and fungal invasion and were thus termed protectins [11,12]. For both intra- and extracellular lectins, it is important to consider that, although they may occur in a particular tissue or body fluid, their biosynthesis and secretion may be carried out by specific cells or tissues and later transported to the tissues and fluids, or adsorbed or pinocytosed by those cells in which they are finally detected. The lectins in the hemolymph of the fly *S. peregrina* larva are produced in the fat body and secreted to the hemolymph [27,44], whereas those that occur in the hemolymph of the lobster, crayfish, oyster, blue crab, and horseshoe crab are probably synthesized in the hemocytes

[22,32,34,36,39]. It should be realized that, although certain lectins occur at relatively high concentrations in certain tissues or organs, they may also be present in others, at lower concentrations, were they carry out similar or different functions. In addition, it is usually true that other minor lectins with different distributions are present in the same organism. Furthermore, in most cases, multiple lectin species with similar or different specificities are found in a particular fluid or tissue. These observations indicate that tissue and organ distribution patterns of lectins are rather complex, and all hypothesis about physiological roles based on this criterion must be subject to careful examination and scrutiny.

In some invertebrate species, the lectin content of tissues and body fluids vary during certain developmental stages, metamorphosis, and molting cycle. These findings have supported the idea that their biological role may be related to cell–cell and cell–substrate recognition and tissue organization, and that they are developmentally regulated. Variability in the presence or absence of lectins or their relative titers during metamorphosis has been observed in insect species: Lectins are present in the eggs of the lepidopteran *Pieris brassicae*, but not in the adult [46,47]. In the flesh fly *S. peregrina*, larva lectins can be induced by wounds or they will appear naturally on pupation, but they are lacking in the adult [27,44]. In the silk moth *Bombix mori*, lectins are normally present in the larva, but their titers will decline in the last instar, increase on pupation, and later decline again [48]. Quantitative changes in *N*-acylaminosugar-binding lectins from other arthropod species, such as crabs and horseshoe crabs, have been observed during the molting cycle [13,15], suggesting that humoral lectins may be involved in the transport of the *N*-acetylglucosamine subunits that constitute chitin, a structural polysaccharide from the exoskeleton.

A very important source of information that has contributed toward elucidating the biological role of invertebrate lectins and their evolution has consisted in the analysis of their primary structure and gene organization and the search of homology with other molecules, or their domains, of well-defined physiological functions or binding properties. Since complete primary structures are known for only a few invertebrate lectins [23,70,80,81,82], the sharing of serological determinants and the comparison of amino acid compositions have provided interesting preliminary evidence in certain examples [62,95]. The physicochemical characterization, the elucidation of the fine carbohydrate specificity, and the study of the conformational changes produced in the lectins as a consequence of their concentration in solution, binding to ligand, presence of divalent cations, or other variables have contributed to the understanding, still very limited of their biological roles [16,32,39,44,49,61,71,79,80,100,107].

Thorough characterization of the specificity and heterogeneity of the binding sites of invertebrate lectins, and the structural features and distribution in nature of the carbohydrates to which they bind, has provided important information relating to their physiological function. It should be kept in mind that the nominal specificity of a lectin is usually defined by the carbohydrate exhibiting the highest inhibitory capability. Thus, lectins that are considered specific for one monosaccharide may also bind other carbohydrates that are structurally related, but with lower affinity. In most cases, the definition of the specificity of a lectin arises from practical criteria that do not reflect the actual spectra of substances to which the lectin binds, some of which may be biologically relevant, whereas others may not. This multiplicity in binding can be accomplished through structural similarities of the molecules recognized or the presence of more than one specific combining site per molecule. As an example, despite their nominal specificity for sialic acids, chelicerata lectins (horseshoe crabs, scorpions, and spiders) also bind, although with lower affinity, other structurally related and unrelated molecules, such as 2-keto-3-deoxyoctonate, *N*-acetyl-D-glucosamine, *N*-acetyl-D-galactosamine, D-galactose, glucuronic acid, colominic acid, and phosphorylcholine [29,58,71,95], all present in the bacterial cell wall or products of bacterial metabolism. As we reported elsewhere [95] in limulin, unrelated structures such as sialic acids and phosphorylcholine are apparently recognized by different binding sites. This multiplicity in lectin specificity and the nature and distribution of the molecules recognized would support the likelihood of lectins constituting a carbohydrate-based recognition system for potentially pathogenic microorganisms. Hence, although invertebrate lectins have not yet been shown to express a recombinatorial diversity, like that of antibodies, a limited diversity in recognition capabilities would be accomplished in the invertebrate by the occurrence of multiple lectins, with distinct specificities; the presence of more than one binding site, specific for different carbohydrates, in a single molecule; and by certain "flexibility" of the binding sites that would allow the recognition of a range of structurally related carbohydrates. To this, we could add the presence of calmodulinlike calcium-binding sites and, in particular, hydrophobic regions that may confer nonspecific-binding properties to the lectins in question [22].

The binding of invertebrate lectins from crude extracts, body fluids, or purified preparations to what we could consider biologically relevant ligands (as opposed to "model" ligands, such as vertebrate erythrocytes and mucins) has provided more direct evidence about their biological role. Those relevant ligands include polysaccharides and glycoconjugates that the lectins may interact with in the natural environment, and microorganisms, such as bacteria, algae, fungi, and protozoa (or the carbohydrate products from their metabolism), that may constitute natural negative (pathogens, parasites, or predators) or positive (commensals or symbionts) interactions. Several examples from this type of approach, some of which will be discussed in detail later in this section, have yielded important information. The binding of *Plasmodium* lectins to isolated glycoproteins from erythrocyte membranes has demonstrated not only the nature of the interaction [112,113], but also the species specificity of *Plasmodium falciparum* and *P. knowlesi* [114]. The selective binding of lectins from a particular invertebrate species to its natural bacterial pathogens or parasites has been analyzed in several systems, among those, the interaction of lectins from *Helix* snails [115], the blue crab *Callinectes sapidus* [39], and the lobster *Homarus americanus* [116] with *Streptococcus* strains, *Vibrio parahaemolyticus* isolates, and the parasite *Gaffkia hommarii*, respectively. In a similar manner, binding properties of lectins from *Hydra viridis* [117], *Tridacna* clams [19], some tunicate species [21], and *Halichondria* sponges [20] have allowed the characterization of symbiotic relationships with *Chlorella* algae, *Symbiodinium* dynoflagelates, *Prochloron* algae, and *Pseudomonas insolita*, respectively.

The identification, isolation, and characterization of endogenous ligands is a critical aspect for the elucidation of the biological role(s) of invertebrate lectins from which self-recognition functions are presumed. However, this has been accomplished only for a few lectins. Clues about the participation of those endogenous ligands in defined physiological mechanisms should contribute in that aspect.

One of the experimental observations that suggest the presence of naturally occurring ligands [39,62,72] in the body fluids and tissue extracts from invertebrates is that purification recoveries are usually higher than what would be expected from testing lectin binding or agglutination in the starting material. This indicates that lectin–ligand interactions are disrupted during some step of the purification procedure, very likely during affinity or size-exclusion chromatography and, eventually, the components separated. Invertebrate lectins that interact with endogenous ligands have been demonstrated in several examples: Lectins from the sponge *Geodia cydonium* regulate cell aggregation phenomena and formation of colonies by binding to the antiaggregation factors that inhibit interactions between those aggregation factors and their receptors [42,65]. In the horseshoe crab *Tachypleus gigas* embryo, lectins and their natural inhibitors—glycoproteins rich in *N*-acylhexosamines and sialic acids—are found in the perivitelline fluid, and it has been suggested that they play a role in embryonic development [49].

Undoubtedly, a very valuable approach has been to examine directly, in vitro and in vivo, the possible participation of humoral and cell-associated lectins and their protein–carbohydrate interactions in various physiological processes. These include developmental events, such as metamorphosis, molt, nutrition, tissue organization, and the various stages of defense reactions, recognition, opsonization, agglutination, phagocytosis, encapsulation, and clearance of nonself substances. Invertebrate responses to nonself exhibit both humoral and cellular components; that is, nonself-substances are agglutinated, opsonized, and phagocytosed by individual hemocytes, or, if too large to be endocytosed, are surrounded by a multilayered capsule of hemocytes and melanized. Although the humoral and cellular aspects of the invertebrate internal defense system have been described, the molecular mechanisms by which this system recognizes nonself is not well understood. In addition to general surface phenomena, such as wettability or hydrophobicity, it is likely that specific cell surface receptors also mediate nonself-recognition mechanisms in invertebrates. The presence of cell surface carbohydrate-binding specific receptors has been described on hemocytes from invertebrate species [31,34–39]. Relevant information has been obtained in vivo by injecting biotic or abiotic particles, or tagged soluble substances into the invertebrate, tracing their processing, clearance course and fate, and monitoring the resulting cellular and humoral changes. In vitro studies have included specific attachment of particles, previously exposed or not to plasma or purified lectins, to phagocytes (hemocytes, coelomocytes) and their endocytosis. The nonself substances or particles used in these experiments have consisted of abiotic materials, such as carmine, charcoal, latex (nonspecific recognition); proteins, and glycoconjugates in solution or covalently bond to Sepharose (specific recognition); or biotic materials, including virus, live or killed bacteria or parasites, and fresh or fixed erythrocytes.

Although important progress has been achieved in many invertebrate systems, with the results obtained by this approach, it has been relatively difficult to construe unquestionable interpretations, and even more to generalize about the molecular mechanisms that are operative in the physiological processes in question. There are many reasons for this, including the diversity and complexity of the experimental organisms, the possibility that several mechanisms may be operating simultaneously or even synergistically for accomplishing one function and, most important of all, our lack of knowledge on many biochemical and functional aspects of the molecules involved. The conflicting results found throughout the literature on participation of lectins in agglutination and opsonization, phagocytosis, and encapsulation [118–121] probably show that the molecular mechanisms that mediate those responses vary within the invertebrate species and the size, chemical nature, and surface properties of the nonself-particle in question.

The capacity of invertebrates to mount adaptive responses to nonself substances has constituted one of the questions that have challenged invertebrate biologists for many years: Do invertebrates exhibit enhanced levels of humoral (recognition or effector) defense molecules or cellular responses after a challenge by nonself-substances? And furthermore, upon a second challenge by the foreign substance, is the response (in its cellular and humoral components) specific and increased in its intensity relative to the initial one? That is, are the humoral or cellular components of the defense response inducible and, if so, is there any true memory? Although many conflicting reports exist in the literature, inducible lectins have been demonstrated only in a few examples that include the flesh fly *S. peregrina* [44], the oyster *Crassostrea gigas* [122], and the earthworm *Lumbricus terrestris* [123]. That the response may not be specific does not invalidate the possibility that the induced humoral lectins may function in defense as "acute-phase reactants" against the potential pathogens. In most cases in which the lectins have been induced successfully, this state does not last for more than hours to a few days, and no true immunological memory has yet been demonstrated.

A number of invertebrate, vertebrate, and plant lectins are capable of inducing, usually in a dose–response fashion, activation of mammalian lymphocytes and macrophages [124,126]. This interesting biological property is exerted, at least in the initial steps, through the carbohydrate-binding site of the lectin that triggers activation signals through carbohydrate moieties on the surface of the mammalian cells. In most examples, the mitogenic properties are relatively specific for both the lectin and the cell in question [124,126]. The biological implication of this property for the invertebrate source of the lectin has not yet been clarified. Like the mammalian lymphocytes and macrophages, invertebrate hemocytes exhibit a variety of sugars on the cell surface, as was shown by several laboratories by the use of plant lectins [127,128]. It is conceivable that, under certain circumstances (i.e., binding to foreign ligands), humoral lectins would bind to the hemocyte and link the particle to the phagocyte (opsonic effect); simultaneously the binding could provide the cell with the activation signal that would induce degranulation, release of lysosomal enzymes, or initiate phagocytosis.

The next paragraphs will describe the experimental evidence that has supported the proposal of particular biological roles for invertebrate lectins.

A. Exogenous Functions

1. *External Defense*

The presence of lectins on the cell surfaces of external epithelia or in mucous secretions of aquatic organisms, have suggested the idea that these lectins may constitute a first barrier of defense with the role of monitoring contacts with nonself-materials or cells and triggering tissue rejection reactions or preventing the penetration of infectious microorganisms to the internal system by immobilizing them. Several examples of lectins in external mucous secretions are known. In the protochordate *Branchiostoma lanceolatum* [40], cephalopods such as the squid *Loligo vulgaris* [41] or the octopus, *Octopus vulgaris* [129], and coelenterates, such as *Cerianthus membranaceus* [43], high concentrations of lectins are found in the mucus that covers the external epithelia. The viscosity of the secretion added to the lectin's agglutination properties and its continuous turnover would constitute the basis of this defense mechanism. In the horseshoe crab *L. polyphemus*, the hypodermal glands secrete a viscous material that covers the external surfaces of the animal [130]. The lectins contained in this secretion would function in the same manner as the protective lectins from mucus described in the foregoing, but it is also possible that by binding to bacteria it would favor the formation of biofilms that, in turn, would induce the larval settlement of organisms, such as barnacles, that constitute regular epibionts for this species.

2. Internal Defense

Lectins are present in the body fluids of virtually all invertebrates [131,132] and, since these taxa lack immunoglobulins, much effort was applied to show that humoral lectins might be their functional analogs [13,100]. The main argument against the possible involvement of lectins in defense functions has been their apparent lack of diversity in their combining sites. Invertebrate lectins have not yet been shown to express a recombinatorial diversity like that of antibodies. However, as explained earlier in this chapter, a limited diversity in recognition capabilities would be accomplished in the invertebrate by the occurrence of multiple lectins with distinct specificities; the aggregation of different specific subunits in a single oligomer; the presence of more than one binding site, specific for different carbohydrates, in a single subunit polypeptide chain; and by a certain "flexibility" of the binding sites that would enable the recognition of a range of structurally related carbohydrates. To this, we could add the presence of calmodulinlike calcium-binding sites and, in particular, hydrophobic regions that may confer nonspecific-binding properties to the lectins in question [22].

Agglutination and Clearance. Many purified invertebrate lectins actually agglutinate in vitro potentially pathogenic bacteria. The lectins from the blue crab *Callinectes sapidus* agglutinated 10 of 11 isolates of *V. parahaemolyticus* from several sources. Interestingly enough, the nonrecognized *Vibrio* strain has been isolated from *Callinectes* hemolymph [39]. In addition to the phosphorylcholine- and sialic acid-binding lectin, *L. polyphemus* hemolymph exhibits galactosyl-binding activity that could be responsible for agglutinating gram-positive bacteria [133]. It has been shown that *Crassostrea gigas* lectins agglutinate bacteria such as *V. anguillarum*, which are natural pathogens for this species [64]. The in vitro results of bacterial agglutination have been reflected in in vivo studies of clearance of bacteria or model ligands, such as erythrocytes and glycoconjugates [134]. Usually, after the injection of particles, the hemolymph lectin titers are decreased. A second injection of ligand at that time results in a slower clearance course, which can be brought to the regular rates if the particles have been preincubated with hemolymph. Injection of carbohydrates for which the lectins have high affinity decreases the clearance rates. As an example, hemolymph from *Aplysia californica* agglutinates, in vitro, four species of bacteria, which, when injected into the animal, are rapidly cleared from circulation. However, *Serratia marcescens*, which is not agglutinated in vitro, can be found in circulation several weeks after injection [135]. McCumber et al. [119] have shown that the clearance of particles, such as xenogeneic proteins and viruses, by the blue crab *C. sapidus* does not depend on the presence of agglutinins in plasma and that the tissue from which the particle is cleared varies with the type of particle, suggesting that functionally distinct cell subpopulations exist.

The mechanism through which the plasma agglutinins would function as defense molecules has not been experimentally demonstrated, but it appears reasonable to think that, once the pathogenic bacteria have reached the internal cavities in the invertebrate, they would be immediately agglutinated by the humoral lectins. Agglutination would immobilize the pathogen and would prevent its dispersion and penetration of other tissues. It is interesting here that invertebrate lectins are particularly resistant to proteases [12]. In addition, protease inhibitors are frequently found in hemolymph and invertebrate tissues [136], and this would prevent the dispersion of the clumped bacteria by action of their proteases on the lectins. Agglutination probably induces the release of bacterial products that may act as chemotactic factors, increasing migration and degranulation of hemocytes, which, in turn, stimulates phagocytosis or killing of pathogens by the effector molecules, such as lysozyme, bacteriocidal peptides, bacteriotoxins, lipases, and others, released by the hemocytes. The effects of chemotactic factors on hemocyte activity has been examined by Cheng and Howland [137]. The fact that the pathogens are agglutinated and immobilized could make them more susceptible to the local action of these effector molecules. It is possible that their action, in particular proteolytic and glycolytic, could make the pathogens "more foreign" to the hemocyte and facilitate their phagocytosis [138]. Furthermore, it is clear that agglutination of the pathogens increases the size of the foreign particle, favoring its phagocytosis by the hemocytes (free phagocytes) and results in multiple pathogens being endocytosed and, thereby, cleared from circulation in a single phagocytosis event. It also increases the chances of the particle being randomly trapped by the sessile phagocytes lining the hemal sinuses.

Opsonization and Phagocytosis. The opsonic effect, first described for vertebrates, implies the binding of the opsonin (immunoglobulins, complement components, or CRP in vertebrates) to the foreign particle, followed by binding of the complex to the phagocyte surface, thereby linking the particles to the phagocyte and promoting their endocytosis. An analogous mechanism for the participation of humoral lectins in phagocytosis in the oyster *Crassostrea virginica* was proposed more than two decades ago by Tripp [139]. Although hemocytes from the oyster *C. gigas* can phagocytose in saline in the absence of lectin, suggesting that the hemocytes have the appropriate recognition molecules on the surface, it has been shown that previous exposure of the particle to the lectin enhances its phagocytosis and clearance rates [140]. Similar results have been obtained by Renwrantz [121,129] on the snail *Helix* and the mussel *Mytilus*, but some laboratories have shown that, in certain invertebrate species, hemolymph lectins are not required at all for phagocytosis and clearance [118,119], whereas others have shown that previous binding of the lectin is an absolute precondition for phagocytosis [141,142]. This indicates that the requirement of opsonization varies within a wide range, depending on the nature of the nonself substance (size, hydrophobicity, surface charge, presence of carbohydrates) and the invertebrate model studied. Concerning the molecular mechanisms by which lectins could act as opsonins, the binding of the lectin to the nonself substance would occur through its carbohydrate-binding sites, as discussed earlier in relation to their agglutination properties. It should be added that certain nonagglutinating lectins from invertebrates, such as those isolated from *Limulus* [92] and *Callinectes* [39] could still act as opsonins. On the other hand, how the lectin–ligand complex binds to the surface of the phagocyte has not yet been elucidated. Several mechanisms could be proposed, based in the molecular structure of lectins and properties of the hemocyte cell surface. One obvious possibility is that the lectin–ligand complex, through available carbohydrate-binding sites of the lectin component, binds to similar sugars on the hemocyte surface, the presence of which has been demonstrated by the binding of heterologous lectins [127,128]. However, why those carbohydrate moieties are not continuously saturated with the humoral lectin is difficult to explain, but the answer may be related to the reversible nature of the protein–carbohydrate interactions and the relatively low affinities measured in animal lectin–ligand systems. It is also possible that the expression of those carbohydrate moieties or lectin receptors is modulated by humoral factors released under stress or trauma, as was shown for vertebrate systems [110,111]. Another distinct possibility, with considerable support from the experimental evidence is that the binding to ligand (or divalent cations) produces conformational changes in the lectin molecule [61,83,100] that may expose cryptic carbohydrate-binding sites with affinity for sugars on the hemocyte surface, or certain regions of the lectin molecule that are detected as foreign by hemocyte receptors (this foreignness being a reflection of changes in surface charge or, more likely, by increased hydrophobicity).

The biological implications of mitogenic properties for mammalian lymphocytes and macrophages [23,124,126] observed on many invertebrate lectins may be related to the opsonic effect. The positive activation (mitogenic) signal transmitted to the hemocyte by the binding of the lectin–ligand complex would promote rapid endocytosis or degranulation responses.

Similarities of the primary protein structure and the presence of shared serological determinants, including idiotopes, have shown that several invertebrate lectins, namely those from *Limulus* and the tunicates *D. candidum* and *Clavelina picta* are phylogenetically related to mammalian CRP [143], an opsonin for polymorphonuclear leukocytes [144]. Moreover, limulin can substitute mammalian CRP in the CRP-mediated platelet killing of *Schistosoma* larvae with little decrease in killing rates [145]. It is also very interesting that *D. candidum*, a tunicate species in which we have detected a galactosyl-binding lectin, DCL-I [89], that cross-reacts with human CRP, exhibits similar amino acid composition and certain degree of homology in the primary structure [62]; it also exhibits interleukin-1-like activity in its plasma [146], a humoral immunoregulatory molecule that, together with IL-6, induces synthesis and secretion of CRP by liver tissues in mammals [147].

Inducibility of Humoral Lectins. The question about the existence of adaptive defense responses in invertebrates has drawn the attention and efforts of many investigators during the last two decades. Results obtained in the attempt to induce humoral lectins (agglutinins/opsonins) by specific challenges, such as injections of bacteria or erythrocytes, or nonspecific stimuli, such as trauma, have been controversial. In general, the responses have been weak and short-lived. Two models, however, have yielded relatively clear-cut results: the induction of lectins in the larvae of the flesh fly *S. peregrina* by injury [23,27,44] and in the oyster *C. gigas*, a filter feeder, by exposing the mollusks to an environment with high concentration of bacteria [122]. In the *S. peregrina* larvae the lectins are induced by nonspecific stimuli, such as wounds in the body wall or injection of saline or erythrocytes, but higher titers are obtained with the latter [30,43]. That the response may not be specific does not invalidate the possibility that the induced humoral lectins may function in defense against the potential pathogens that may penetrate through the wound. In *Crassostrea*, lectin titers were increased by exposing oysters to waters with high concentration of *V. anguillarum*, a facultative pathogen that is agglutinated by the oyster hemolymph lectins [122]. For the mechanisms of induction of lectin activity, several possibilities should be considered: the first should be the de novo synthesis of the lectin upon the challenge; second, the activation of the lectin from an inactive precursor by enzymatic cleavage, as shown for *Giardia lamblia*, for which the lectin that binds the parasite to the intestinal mucosa is processed to the active form through cleavage by duodenal trypsin [101]. This mechanism has been proposed also for the inducible lectin of *S. peregrina* larvae [23,27,44]. A third mechanism was proposed, based on their dependence on Ca^{2+} [132] and other divalent cations for activity: the activation of the lectin would be a consequence of conformational changes induced by variable concentrations of Ca^{2+} [132].

Hemocyte-Associated Lectins. In some mollusk, arthropod, and tunicate species, certain humoral lectins are present on the outer surface of the hemocytes, free blood phagocytes of well-known involvement in defense functions [18,34,39,148]. Different populations of invertebrate hemocytes participate in a series of defense-related reactions, including phagocytosis, encapsulation, hemolymph clotting, and release of phenoloxidase and melanization in response to challenge with nonself-components. However, the molecular basis of those cellular mechanisms of recognition are not yet understood. Considering the possibility that these membrane recognition molecules may be lectins, the presence of carbohydrate-binding proteins on the cell membrane of invertebrate hemocytes was examined by several authors. The presence of hemocyte-associated lectins was shown initially by Tyson and Jenkin [36], who demonstrated that hemocytes from the crayfish *Cherax destructor* could recognize bacteria through a membrane molecule, probably related to the hemolymph agglutinins. Amirante and Mazzalai [37] showed the presence of heteroagglutinins in cell membrane and cytoplasm of hemocytes from the cockroach (*Leucophaea maderae*), by using immunofluorescence techniques, and suggested that granular hemocytes and spherule cells are responsible for the active synthesis of two hemagglutinins. Our research on *Crassostrea virginica* hemocyte and serum lectins [35] showed that the hemocyte membrane exhibited a lectin that shared the carbohydrate specificity of a fraction of the oyster plasma lectins. In a further report [36], we showed directly, by immunocytofluorescence and flow cytometry, that the oyster hemocyte lectins are located on the outer face of the cell membrane and exhibit irregular distributions, suggesting that "patching"- and "cappinglike" phenomena occur. Experimental evidence suggested that the hemocyte lectins are serologically related to plasma lectins, and that the distribution of hemocyte lectins on the cells indicates a heterogeneous cellular expression on a population basis [36]. Lectins are also present on the plasma membrane of hemocytes from the blue crab *C. sapidus* [31,39]. Renwrantz and Stahmer [121], working on *Mytilus edulis*, showed that, although the hemolymph lectins exhibit opsonic activity, the hemocytes also bear a lectin on the cell membrane that is serologically related to the soluble lectins.

In vivo evidence for the participation of serum lectins and, possibly, hemocyte-associated lectins in invertebrate cellular responses to nonself was obtained in the cockroach *Periplaneta americana*: Glycoproteins that were the most effective inhibitors for the purified plasma lectins, when injected in solution or bound to Sepharose beads in the hemocoele of the insect, produced a cellular response, nodulation and encapsulation, significantly higher than for glycoproteins that behaved as poor inhibitors for the purified lectins [38].

3. Perivitelline Protection

The role of lectins in defense against pathogens would be operative, not only in the adult invertebrates or their larval stages, but also extend to the protection of the egg and developing embryo. The lectins for which this role was first proposed were first discovered in the albumen gland of the gastropod snails *H. pomatia* and *H. hortensis* [11,12]. This accessory gland from the reproductive system, connects to the hermaphroditic duct and covers the fertilized oocyte with its secretion; it contains very high concentrations of *N*-acetylgalactosamine-binding lectins, which in turn, comprise about 12 isolectins [55]. Because the *Helix* lectins strongly agglutinate a wide variety of bacteria, including pathogenic streptococci, they were named *protectins* to indicate that their role was to protect the embryo against the bacterial infection during incubation in the humid soil [11,12,115]. In eggs of the Japanese horseshoe crab *Tachypleus gigas*, lectins are present in the perivitelline fluid, as well as their endogenous ligand, a heterogeneous glycoprotein that contains *N*-acylsugars, including sialic acids [49]. Lectins have also been found in the perivitelline fluid of eggs from the sea urchin *A. crassispina* [73,81].

4. Pathogen Attachment

It has been clearly demonstrated in several models that lectins present on bacterial or parasite cell surfaces bind to oligosaccharides present in glycoproteins or glycolipids from cell membranes of vertebrates or invertebrates. Thus, protein–carbohydrate interactions would mediate not only defense reactions of invertebrates against pathogens, but would also mediate the attachment of pathogens to tissues of the host or vector. Lectins with specificity for *N*-acetylglucosamine, L-fucose, D-galactose, and sialic acids have been reported on the bacterial cell surface, extracts, and pili, and are involved in the selective attachment of bacteria to certain animal tissues [149]. A parasitic protozoan, ***Plasmodium***, bears cell membrane-associated lectins that are specific for glycopeptides present on erythrocyte plasma membranes, suggesting that they are responsible for the selection of the particular blood cells these parasites penetrate and infest. The major red cell sialoglycoproteins (glycophorins A, B, and, possibly, C) are recognized by lectinlike *P. falciparum* membrane proteins, which have been identified (M_r 140,000, 70,000, and 35,000) and bind specifically to *N*-acetyl-D-glucosamine [112,113]. Different erythrocyte glycoproteins would be involved in the attachment of *P. falciparum* and *P. knowlesi* [114]. Lectins specific for *N*-acetylmannosamine, *N*-acetylgalactosamine, and D-galactose have been found in the crop, midgut, and hemolymph, respectively, in the reduvid insect *Rhodnius prolixus* [150]. Receptors for these lectins were detectable in epimastogote, but not in trypomastigote forms of *Trypanosoma cruzi*, the parasite protozoan causative agent of Chagas disease in humans for which *R. prolixus* is the intermediary host, suggesting that each lectin is highly specific and interacts selectively with developmental stages of *T. cruzi* [150]. *Entamoeba histolytica* trophozoites adhere to human colonic mucus glycoproteins and epithelial cells through a M_r 170,000 Gal/GalNAc–binding lectin [151]. The attachment of *Giardia lamblia* to the mammalian intestinal mucosa is mediated by a lectin from the parasite that becomes activated by proteolytic activity in the duodenum, followed by the binding to the appropriate carbohydrate moieties on the enteric surface [101].

5. Nutrition and Feeding

D-galactose–specific lectins, present on the plasma membrane of *Acanthoamoeba*, a free-living protozoa, are considered to be involved in the recognition and binding of particles to be phagocytosed [152]. The ABH blood group antigens and specific agglutinins detected in the gut of hematophagous diptera have been postulated to be related to dietary requirements [153].

Within the prokaryotes, the role of lectins on the surface of marine bacteria, as a sensory apparatus for monitoring the nutrient levels in the surrounding environment, has been proposed [154].

6. Symbiosis

It has been shown, in several examples, that specific interactions between invertebrate hosts and symbiotic bacteria and algae are mediated by lectins produced by the host, and possibly by the symbionts as well. The lectin would function by physically binding the microorganism to the invertebrate tissues, regulating its proliferation, and aiding in its processing and disposal. It has been proposed that aged algal symbionts, the dynoflagelate *Symbiodinium microadriaticum*, from the *Tridacna* giant clams are recognized by lectins present in the hemal sinuses through the exposed galactosyl residues on the aged algal cell surface, phagocytosed, and digested as a final stage in their symbiotic relationship [19]. The clam hemolymph lectin would also recognize the algal species and strain, as the clams live in waters where several species of potential symbiotic algae are present. Since the interaction *Tridacna–Symbiodinium* is specific, it has been postulated that the lectin could mediate the symbiont selection process [155]. A similar interaction would occur between *Hydra viridis*, the freshwater hydra, and the *Chlorella*-like symbionts [117]. The sponge *Halichondria panicea* exhibits a lectin that would constitute the binding molecule that interacts with the bacterial symbiont *Pseudomonas insolita* [20]. The experimental evidence indicates that the interaction sponge–bacteria is species-specific, and the bacteria requires the presence of the lectin to grow in culture, suggesting that the lectin not only anchors the symbiont to the sponge tissues, but probably stimulates its proliferation as well [20]. A symbiotic system occurs between the tunicate *Didemnum molle* and the prokaryotic algae *Prochloron*: colonies of *D. molle* that are rich in *Prochloron* produce large amounts of lectin, whereas colonies lacking the algae contain no lectin, suggesting that the lectin stabilizes the *Didemnum–Prochloron* interaction [21].

The peritrophic membrane of the blowfly *Calliphora erythrocephala* expresses a lectin with highly specificity for mannose in the lumen side, as well as mannose and glucose residues [156]. Lectins of similar specificity, and mannose residues as well, are found in the pili of *Proteus morganii* and *P. vulgaris*, two species of bacteria that densely pack the peritrophic membrane and that are believed to constitute symbionts or commensals. This system is very interesting because it would illustrate a reciprocal lectin–ligand (mannose) interaction mediating the mutualistic relationship between the bacteria and the invertebrate.

7. Settlement and Induction of Metamorphosis in Invertebrate Larvae

A very interesting model, reviewed by Maki and Mitchell [157], which provides new insight into lectin recognition functions, is the settlement and subsequent metamorphosis of larvae of the polychaete *Janua (Dexiospira) brasiliensis*, stimulated by exopolysaccharides of mixed bacterial films that contain *Pseudomonas marina*. The larval settlement can be blocked by inhibiting the larval lectin with glucose or treating the bacterial film with periodate (which oxidizes the terminal carbohydrate residues, i.e., glucose) or preexposing it to Con A, a lectin that binds glucose and mannose. After the larval settlement, the glucose-containing carbohydrate moiety, probably from a polysaccharide or glycoprotein, induces the metamorphosis of the polychaete. The sugar in solution has no effect on the free-swimming larva.

8. Fertilization

In an early report, Tyler [8] hypothesized that agglutinins he detected in seminal fluids of several invertebrate species would be involved in fertilization (or preventing interspecies cross-fertilization). Later work on the sea urchin has shown that "bindin," a lectinlike protein, present on sperm cells, has affinity for structures present on the oocyte surface [158]. These lectinlike proteins isolated from spermatozoa in the sea urchin, as well as the identification of their complementary carbohydrate receptors on the oocyte cell surface [158] provide a model for fertilization at the molecular level, which further research will prove if it is extensive to other taxa.

B. Endogenous Functions

Lectins that interact with endogenous ligands have been demonstrated in several invertebrate models: Lectins from the sponge *G. cydonium* regulate cell aggregation phenomena and formation of colonies by binding to the antiaggregation factors that inhibit interactions between those aggregation factors and their receptors [42,65]. In the horseshoe crab *T. gigas* embryo, lectins and their natural inhibitors, glycoproteins rich in *N*-acylhexosamines and sialic acids, are found in the perivitelline fluid, and it has been suggested that they play a role in embryonic development [49]. The variability of the lectin contents during metamorphosis in insects has suggested their participation in developmental events, such as tissue reorganization and resorption [23,27,30,44,46–48].

Lectins are present in pupal epidermal cells of the lepidopteran *Pieris brassicae*, but not in the adult [46,47]. In the silk moth *B. mori*, lectins are normally present in the larva, but their titers will decline in the last instar, increase on pupation, and later decline again [48]. In the flesh fly larva *S. peregrina*, lectins can be induced by wounds or they will appear naturally on pupation, but they are lacking in the adult [23,27,44]. Since *S. peregrina* lectin stimulates an in vitro macrophagelike murine cell line and induces the release of cytotoxic factors that produce tumor necrosis, it was proposed that the physiological role of this lectin would be to induce cell death during the pupal tissue reorganization and, possibly, the recognition of the dead tissue as nonself by the hemocytes. A possible biological role as a cell adhesion molecule for the lectin echinoidin, from the sea urchin *A. crassispina*, has been predicted from its primary sequence. It contains the tripeptide Arg-Gly-Asp, considered to be the basic functional unit in cell adhesion proteins, such as fibronectin and discoidin, in a hydrophilic region located in a β-turn, suggesting that it is exposed on the surface of the molecule [81].

VIII. CONCLUSIONS

The past few years of invertebrate lectin research have provided us with a growing corpus of information about their molecular structure, gene organization, evolutionary relationships, and biological roles.

The recent availability of protein and gene sequences of both invertebrate and vertebrate lectins has revealed their complex molecular and gene structures and has helped to explain the multiplicity of biological properties observed. The analysis of the primary structure of several lectins has shown that various functional domains can be identified throughout the sequence, including those related to collagens, calmodulin, complement, regulatory proteins, cell adhesion proteins, such as fibronectin, ribonucleoproteins; and common to all, the carbohydrate-binding domains consisting of a set of highly conserved residues or a "framework" among which the variable residues would confer the sugar specificity [107]. The comparison of their primary structures suggests that two major groups of invertebrate lectins can be discriminated: One group would include lectins such as those from *L. polyphemus* and *D. candidum* that show homology to vertebrate pentraxins, such as C-reactive protein, serum amyloid protein, and hamster female protein. Although the vertebrate pentraxins constitute a well-defined family of proteins, they exhibit several lectin properties [108]. The second group would comprise lectins such as those from the flesh fly *S. peregrina*, the tunicate *Polyandrocarpa misakiensis*, and the sea urchin *A. crassispina*, that show homology to the C-type soluble and membrane-integrated vertebrate lectins. A third group, consisting of the only bona fide family of animal lectin molecules, the β-galactoside-binding or S-type lectins from vertebrates, would represent the most conserved members.

The gene organization of the mannose-binding proteins [159] suggest that several mechanisms would have contributed to the origin and evolution of hybrid molecules that exhibit lectin domains, including exon shuffling and fusion, gene duplication, and exon migration, resulting in molecules with multiple biological properties. This evolutionary process comprising multiple domain structures resembles the one proposed for the components of the mammalian immune system as originating from cell adhesion molecules (CAM) [160]; however, in these systems, the protein–carbohydrate interaction plays only a secondary role.

Limited information, however, is available on the evolutionary precursors of invertebrate lectins or their carbohydrate-binding domains, but the hypotheses proposed by Kolb [161] and Parish [162] for recognition molecules would apply. According to Kolb [161] the recognition molecules required in the evolution from unicellular to multicellular organisms originated from sugar-binding enzymes that lost their enzymatic properties, while retaining their carbohydrate-binding sites, through gene duplication and mutation events. Parish [162] has proposed that the recognition molecules would have origin in random associations of glycosyltransferases in oligomers that would exhibit a cytophylic site and nonself-recognizing carbohydrate-binding sites in the same molecules. On the other hand, Roegener et al. [129] have proposed that octopus hemolymph lectins may have evolved form hemocyanin or a tyrosinaselike molecules, a hypothesis supported by Pistole's observations on *L. polyphemus* agglutinins [163].

Currently, strong experimental evidence has been obtained in support of a wide variety of physiological roles for lectins present in prokaryotes, invertebrates, and vertebrates. It becomes clear, however, that within a considerable functional diversity self- and nonself-recognition is a biological property inherent in animal lectins. In the unicellular organisms, such as bacteria and protozoa, lectins would participate, in the free-living forms, in feeding and nutrient monitoring and, in the pathogenic examples, lectins would promote their attachment to the vector or host tissues. Lectins from the Parazoa (sponges) have been shown to participate in cell aggregation phenomena and symbiosis. Finally, in the metazoan phyla lectins would participate in all those aforementioned functions, to which we could add humoral and cellular defense, somatic cell adhesion, fertilization, tissue reorganization, larval settlement, and induction of the metamorphosis, suggesting a multiplicity of biological roles that correlate with the increasing complexity of the organizational levels.

There is ample evidence for participation of lectins as mediators of immune mechanisms in invertebrates, in their functions as agglutinins, opsonins, and cell-associated recognition molecules. However, three very important questions have remained open. The first relates to their possible homology with vertebrate defense molecules (i.e., immunoglobulins, major histocompatibility complex products, complement components, and acute-phase proteins). Except for a few examples of serological cross-reactivity [105] and moderate homology [82], there is no further evidence for structural similarity between invertebrate lectins and immunoglobulins, but from the data discussed here, it is clear that certain invertebrate and vertebrate lectins are homologous to C-reactive protein [22,62,80,95] and share carbohydrate-binding properties with related humoral components, such as SAP [164], and other molecules involved in inflammatory and immune mechanisms, including the mouse lymph node homing receptor [110] and the endothelial leukocyte adhesion molecule [111]. Furthermore, it is noteworthy that vertebrates exhibit humoral lectins that can act as opsonins, such as the major lung surfactant SP 28-36 [165], or that exhibit complement-dependent bactericidal activity, such as the lipopolysaccharide-binding serum protein RaRf [166], or that can modulate immune responses, such as SAP [109,167]. The second question refers to the lectin specificity and the generation of diversity in the carbohydrate recognition domains. As explained earlier, invertebrate lectins exhibit a relatively limited diversity in recognition capabilities, when compared with immunoglobulins. Nevertheless, the question to be asked should be: How much diversity is required in a carbohydrate-based nonself-recognition system? Although isomery and a wide variety of possible glycosidic linkages endow oligosaccharides with an enormous potential for structural diversity, the actual number of different carbohydrate moieties found in nature is relatively limited and, in large proportion, ubiquitous. Hence, the problem is the avoidance of self-recognition by the defense molecule. The key of the lectin-mediated self- or nonself-recognition mechanisms may reside in the low affinities and the reversible nature of the lectin–ligand interaction and, in particularly important, the conformational changes occurring in the lectin upon binding to ligand or the presence of divalent cations. Finally, the third question pertains to the existence of adaptive responses and immunological memory in invertebrates. Some lectins have been shown to be inducible and probably more examples will be found in the future by challenging the invertebrate with the adequate stimulus and route. True immunological memory has not yet been found, but, as with the question of generation of diversity in recognition, the issue of memory arises in a mammalian immunology context. The evolutionary success of the invertebrate phyla suggests that, if true, the lack of immunological memory should not imply an inefficient defense system, but rather that it is not a requirement for their survival. Future progress in the field of invertebrate lectins will require that research efforts be continued toward the elucidation of their protein primary structures and gene organization, regulation, and expression. This, added to much needed studies pursuant to the nature of the forces that mediate the protein–carbohydrate interactions, their kinetics, and the structural changes produced in the lectin upon binding to ligand, and most importantly, the identification and characterization of the biologically relevant ligands, will provide solid support for the definitive understanding of the invertebrate lectins' biological roles and evolution.

ACKNOWLEDGMENTS

The author is grateful to Rose Marie Davage and Wanda Medina Stewart for typing the manuscript. Part of the research described in the text was supported by Grant DCB-8896234 from the National Science Foundation.

REFERENCES

1. Stillmark, H. (1888). Ricin, ein giftiges Ferment aus den Samen von *Ricinus communis* L. und einigen anderen Euphorbiaceen. Doctoral Thesis, University of Dorpat, Dorpat (Tartu).
2. Camus, M. L. (1899). Recherches experimentales sur une agglutinine produite par la glande de l'albumen chez l'*Helix pomatia*. *C. R. Hebd. Seances Acad. Sci. 129*:233.
3. Noguchi, H. (1902). The interaction of the blood of cold-blooded animals, with reference to haemolysis, agglutination and precipitation. *Univ. Penn. Med. Bull. 15*:295.
4. Noguchi, H. (1903). On the multiplicity of the serum haemagglutinins of cold-blooded animals. *Zentralbl. Bakteriol. I Abt. Orig. 34*:286.
5. Cantacuzène, J. (1912). Sur certains anticorps naturels observés chez *Eupagurus prideauxii*. *C. R. Soc. Biol. 73*:663.
6. Cantacuzène, J. (1919). Anticorps normaux et expérimentaux chez quelques invertebrés marins. *C. R. Soc. Biol. 82*:1087.
7. Cantacuzène, J. (1923). In célébration du 75ème Anniversaire de la Foundation de la Société de Biologie. p. 48. Masson and Cie, Paris. [Quoted by Gold, E. R., and Balding, P. (1975). *Receptor–Specific Proteins, Plant and Animal Lectins*. Excerpta Medica, Amsterdam.]

8. Tyler, A. (1946). Natural heteroagglutinins in the body fluids and seminal fluids of various invertebrates. *Biol. Bull. 90*:213.
9. Cheng, T. C., and Sanders, B. G. (1962). Internal defense mechanisms in molluscs and electrophoretic analysis of a natural hemagglutinin in *Viviparus malleatus* Reeve. *Proc. Penn. Acad. Sci. 36*:72.
10. Johnson, H. M. (1964). Human blood group A-specific agglutinins of the butter clam *Saxidomus giganteus. Science 146*:548.
11. Prokop, O., Rackwitz, A., and Schlesinger, D. (1965). A "new" human blood group receptor A_{hel} tested with saline extracts from *Helix hortensis* (garden snail). *J. Forensic Med. 12*: 108.
12. Prokop, O., and Uhlenbruck, G. (1969). *Blood and Serum Groups*. Wiley Interscience, New York, p. 69.
13. Marchalonis, J. J., and Edelman, G. M. (1968). Isolation and characterization of a hemagglutinin from *Limulus polyphemus. J. Mol. Biol. 32*:453.
14. Fernández-Morán, H., Marchalonis, J. J., and Edelman, G. (1968). Electron microscopy of hemagglutinin from *Limulus polyphemus. J. Mol. Biol. 32*:467.
15. Cohen, E. (1968). Immunologic observations of the agglutinins on the hemolymph of *Limulus polyphemus* and *Birgus latro. Trans. N.Y. Acad. Sci. 30*:427.
16. Hammarström, S., and Kabat, E. A. (1964). Purification and characterization of a blood group reactive haemagglutinin from the snail *Helix pomatia* and a study of its combining site. *Biochemistry 8*:2696.
17. Ey, P. L., and Jenkin, C. R. (1982). Molecular basis of self and non-self recognition among the invertebrates. In *The Reticuloendothelial System*, Vol. 3 (N. Cohen and M. M. Sigel, eds.). Plenum, New York, p. 321.
18. Vasta, G. R. (1987). Serum and hemocyte-associated lectins of the oyster *Crassostrea virginica*. In *Lectins* (T. C. Bøg-Hansen, E. van Driessche, eds.). Walter de Gruyter, New York, p. 677.
19. Uhlenbruck, G., and Steinhausen, G. (1977). Tridacnins: Symbiosis-profit or defence-purpose? *Dev. Comp. Immunol. 1*:183.
20. Müller, W. E. G., Zahn, R. K., Kurelec, B., Lucu, C., Müller, I., and Uhlenbruck, G. (1981). Lectin, a possible basis for symbiosis between bacteria and sponges. *J. Bacteriol. 145*:548.
21. Müller, W. E. G., Maidhof, A., Zahn, R. K., Conrad, J., Rose, T., Stafanovich, P., Müller, I., Friese, U., and Uhlenbruck, G. (1984). Biochemical basis for the symbiotic relationship *Didemnum–Prochloron* (Prochlorophyta). *Biol. Cell. 5*:381.
22. Nguyen, N. Y., Suzuki, A., Cheng, S.-M., Zon, G., and Liu, T.-Y. (1986). Isolation and characterization of *Limulus* C-reactive protein genes. *J. Biol. Chem. 261*:10450.
23. Takahashi, H., Komano, H., Kawaguchi, N., Nakanishi, S., and Natori, S. (1985). Cloning and sequencing of cDNA of *Sarcophaga peregrina* humoral lectin induced on injury of the body wall. *J. Biol. Chem. 260*:12228.
24. Boyd, W. C., and Slapeigh, E. (1954). Specific precipitating activity of plant agglutinins (lectins). *Science 119*:419.
25. Goldstein, I. J., Hughes, R. C., Monsigny, M., Osawa, T., and Sharon, N. (1980). What should be called a lectin? *Nature 285*:66.
26. Kocourek, J., and Hořejší, V. (1981). Defining a lectin. *Nature 290*:188.
27. Komano, H., and Natori, S. (1985). Participation of *Sarcophaga peregrina* lectin in the lysis of sheep red blood cells injected into the abdominal cavity of larvae. *Dev. Comp. Immunol. 9*:31.
28. Gold, E. R., and Balding, P. (1975). *Receptor-Specific Proteins, Plant and Animal Lectins*. Excerpta Medica, Amsterdam.
29. Vasta, G. R., and Marchalonis, J. J. (1983). Humoral recognition factors in the Arthropoda. The specificity of Chelicerata serum lectins. *Am. Zool. 23*:157.
30. Yeaton, R. W. (1981). Invertebrate lectins: I. Occurrence. *Dev. Comp. Immunol. 5*:391.
31. Cassels, F. J., Marchalonis, J. J., and Vasta, G. R. (1986). Heterogeneous humoral and hemocyte-associated lectins with *N*-acylamino sugar specificities from the blue crab *Callinectes sapidus* Rathbun. *Comp. Biochem. Physiol. 85B*:23.

32. Hall, J. L., and Rowlands, D. T. (1974). Heterogeneity of lobster agglutinins. I. Purification and physicochemical characterization. *Biochemistry 13*:821.
33. Hall, J. L., and Rowlands, D. T. (1974). Heterogeneity of lobster agglutinins. II. Specificity of agglutinin–erythrocyte binding. *Biochemistry 13*:828.
34. Tyson, C. J., and Jenkin, C. R. (1974). Phagocytosis of bacteria in vitro by haemocytes from the crayfish (*Parachaeraps bicarinatus*). *Aust. J. Exp. Biol. Med. Sci. 52*:341.
35. Vasta, G. R., Sullivan, J. T., Cheng, T. C., Marchalonis, J. J., and Warr, G. W. (1982). A cell membrane associated lectin of the oyster hemocyte. *J. Invertebr. Pathol. 40*:367.
36. Vasta, G. R., Cheng, T. C., and Marchalonis, J. J. (1984). A lectin on the hemocyte membrane of the oyster (*Crassostrea virginica*). *Cell. Immunol. 88*:475.
37. Amirante, G. A., and Mazzalai, F. G. (1978). Synthesis and localization of hemoagglutinins in hemocytes of the cockroach *Leucophaea maderae*, L. *Dev. Comp. Immunol. 2*:73.
38. Lackie, A. M., and Vasta, G. R. (1988). The role of galactosyl-binding lectin in the cellular-immune response of the cockroach *Periplaneta americana* (Dictyoptera). *Immunology 64*:353.
39. Cassels, F. C., and Vasta, G. R. (1990). Serum and hemocyte lectins from the blue crab *Callinectes sapidus*. A biochemical characterization. (Submitted).
40. Möeck, A., and Renwrantz, L. (1987). Lectin activity and a protease inhibitor in mucus from the skin of *Branchiostoma lanceolatum*. *J. Invertebr. Pathol. 49*:221.
41. Marty, H. J. (1974). Evidence and significance of a haemagglutinin from the skin of cephalopods. *Z. Immunitaets. Forsch. 148*:225.
42. Müller, W. E. G., Kurelec, B., Zahn, R. K., Müller, I., Vaith, P., and Uhlenbruck, G. (1979). Aggregation of sponge cells. Function of a lectin in its homologous biological system. *J. Biol. Chem. 254*:7479.
43. Koch, O. M., Lee, C. K., and Uhlenbruck, G. (1982). Cerianthin lectins: A new group of agglutinins from *Cerianthus membranaceus* (Singapore). *Immunobiology 163*:53.
44. Komano, H., Mizuno, D., and Natori, S. (1981). A possible mechanism of induction of insect lectin. *J. Biol. Chem. 257*:7087.
45. Takahashi, H., Kawaguchi, N., Obinata, M., and Natori, S. (1984). Activation of the secretion of specific proteins from fat body following injury to the body wall of *Sarcophaga peregrina* larvae. *Insect Biochem. 14*:713.
46. Mauchamp, B. (1982). Purification of an *N*-acetyl-D-glucosamine specific lectin (P.B.A.) from epidermal cell membranes of *Pieris brassicae* L. *Biochimie 64*:1001.
47. Mauchamp, B., and Hubert, M. (1984). Internalization of plasma membrane glycoconjugates and plasma membrane lectin into epidermal cells during pharate adult wing development of *Pieris brassicae* L.: Correlation with resorption of molting fluid components. *Biol. Cell 50*:285.
48. Suzuki, T., and S. Natori. (1983). Identification of a protein having hemagglutinating activity in the hemolymph of the silk worm, *Bombyx mori*. *J. Biochem. 93*:583.
49. Shishikura, F., and Sekiguchi, K. (1984). Studies on the perivitelline fluid of horseshoe crab embryo. 2. Purification of agglutinin-binding substance from the perivitelline fluid of *Tachypleus gigas* embryo. *J. Biochem. 96*:629.
50. Gilbertson, D. E., and Etges, F. J. (1967). Haemagglutinins in the hemolymph of planorbid snails. *Ann. Trop. Med. Parasitol. 61*:144.
51. Mello, M. T., and Paraense, W. L. (1978). Geographical differences in lectinic activity of albumen gland extracts of the planorbid snails *Biomphalaria glabrata* and *B. tenagophila*. *Inst. Med. Trop. 20*:115.
52. Michelson, E. H., and Dubois, L. (1977). Agglutinins and lysins in the molluscan family Planorbidae: A survey of hemolymph, egg masses and albumin gland extracts. *Biol. Bull. 153*:219.
53. Uhlenbruck, G., and Weiss, A. (1973). Studies on broad spectrum agglutinins. XIV. Heterogeneity of *Helix aspersa* agglutinins. *Z. Immunitasts Forsch. 145*:356.
54. Palatnik, M., Vasta, G. R., Fink, N. E., and Chiesa, M. E. (1975). Protectins in Argentine molluscs: Immunological and immunochemical aspects. In *Immunologic Phylogeny* (W. H. Hildeman and A. A. Benedict, eds.). Plenum Press, New York, pp. 19–28.

55. Vretblad, P., Hjorth, R., and Låås, T. (1979). The isolectins of *Helix pomatia*. Separation by isoelectric focusing and preliminary characterization. *Biochim. Biophys. Acta 579*:52.
56. Baldo, B. A., Sawyer, W. H., Stick, R. V., and Uhlenbruck, G. (1978). Purification and characterization of a galactan-reactive agglutinin from the clam *Tridacna maxima* (Röding) and a study of its combining site. *Biochem. J. 17*:467.
57. Ravindranath, M., Higa, H., Cooper, E., and Paulson, J. (1985). Purification and characterization of an *O*-acetylsialic acid-specific lectin from a marine crab *Cancer antennarius. J. Biol. Chem. 260*:8850.
58. Rostam-Abadi, H., and Pistole, T. C. (1982). Lipopolysaccharide-binding lectin from the horseshoe crab, *Limulus polyphemus*, with specificity for 2-keto-3-deoxyoctonate (KDO). *Dev. Comp. Immunol. 6*:209
59. Vasta, G. R., Warr, G. W., and Marchalonis, J. J . (1982). Tunicate lectins: Distribution and specificity. *Comp. Biochem. Physiol. 73*:887.
60. Yokosawa, H., Sawada, H., Abe, Y., Numakunai, J., and Ishii, S. (1982). Galactose-specific lectin in the hemolymph of solitary ascidian *Halocynthia roretzi*: Isolation and characterization. *Biochem. Biophys. Res. Commun. 107*:451.
61. Miller, R. L., Collawn, J. R., Jr., and Fish, W. W. (1982). Purification and macromolecular properties of a sialic acid-specific lectin from the slug *Limax flavus. J. Biol. Chem. 257*:7574.
62. Vasta, G. R., Hunt, J., Marchalonis, J. J., and Fish, W. W. (1986). Galactosyl-binding lectins from the tunicate *Didemnum candidum*. Purification and physicochemical characterization. *J. Biol. Chem. 261*:9174.
63. Hammarström, S., Westöö, A., and Bjork, I. (1972). Subunit structure of *Helix pomatia* A hemagglutinin. *Scand. J. Immunol. 1*:4.
64. Olafsen, J. A. (1988). Role of lectins in invertebrate humoral defense. *Am. Fish. Soc. Spec. Publ. 18*:189.
65. Müller, W. E. G., Conrad, J., Schroeder, C., Zahn, R. K., Kurelec, B., Dreesbach, K., and Uhlenbruck, G. (1983). Characterization of trimeric, self-recognizing *Geodia cydonium* lectin I. *Eur. J. Biochem. 133*:263.
66. Kehoe, J. M., Kaplan, R., and Li, S. S.-L. (1979). Functional implications of the covalent structure of limulin: An overview. *Prog. Clin. Biol. Res. 29*:617–623.
67. Bretting, H., and Kabat, E. A. (1976). Purification and characterization of the agglutinins from the sponge *Axinella polypoides* and a study of their combining sites. *Biochemistry 15*:3228.
68. Amirante, G. A., and Basso, V. (1984). Analytical study of lectins in *Squilla mantis* L. (Crustacea: Stomatopoda) using monoclonal antibodies. *Dev. Comp. Immunol. 8*:721.
69. Basu, P. S., Datta, P. K., Agarwal, P., Ray, M. K., and Datta, T. K. (1984). Purification and partial characterization of an erythroagglutinin from the hemolymph of scorpion, *Heterometrus bengalensis. Biochemie 66*:487.
70. Muramoto, K., Ogata, K., and Kamiya, H. (1985). Comparison of the multiple agglutinins of the acorn barnacle *Megabalanus rosa. Agric. Biol. Chem. 49*:85.
71. Robey, F. A., and Liu, T.-Y. (1981). Limulin: A C-reactive protein from *Limulus polyphemus. J. Biol. Chem. 256*:969.
72. Roche, A.-C., and Monsigny, M. (1979). Limulin (*Limulus polyphemus* lectin). Isolation, physicochemical properties, sugar specificity and mitogenic activity. *Progr. Clin. Biol. Res. 29*:603.
73. Giga, Y., Sutoh, K., and Ikat, A. (1985). A new multimeric hemagglutinin from the coelomic fluid of the sea urchin *Anthocidaris crassispina. Biochemistry 24*:4461.
74. Kubo, T., and Natori, S. (1987). Purification and some properties of a lectin from the hemolymph of *Periplaneta americana* (American cockroach). *Eur. J. Biochem. 168*:75.
75. Sharon, N., and Lis, H. (1972). Lectins: Cell-agglutinating and sugar specific proteins. *Science 177*:949.
76. Marchalonis, J. J., and Weltman, J. K. (1971). Relatedness among proteins: A new method of estimation and its application to immunoglobulins. *Comp. Biochem. Physiol. 38B*:609.

77. Cornish-Bowden, A. (1981). Interpretation of amino acid compositions. *Trends Biochem. Sci. 6*:217.
78. Shimizu, S., Ito, M., and Niwa, M. (1977). Lectins in the hemolymph of Japanese horseshoe crab *Tachypleus tridentatus. Biochim. Biohys. Acta 500*:71.
79. Kaplan, R., Li, S. S.-L., and Kehoe, J. M. (1977). Molecular characterization of limulin, a sialic acid binding lectin from the hemolymph of the horseshoe crab, *Limulus polyphemus. Biochemistry 16*:4297.
80. Nguyen, N. Y., Suzuki, A., Boykins, K. A., and Liu, T.-Y. (1986). The amino acid sequence of *Limulus* C-reactive protein. Evidence of polymorphism. *J. Biol. Chem. 261*:10456.
81. Giga, Y., Ikai, A., and Takahashi, K. (1987). The complete amino acid sequence of echinoidin, a lectin from the coelomic fluid of the sea urchin *Anthocidaris crassispina. J. Biol. Chem. 262*:6197.
82. Suzuki, A., Takagi, T., Furukohri, T., Kawamura, K., and Nakauchi, M. (1990). A calcium-dependent galactose-binding lectin from the tunicate *Polyandrocarpa misakiensis*. Isolation, characterization, and amino acid sequence. *J. Biol. Chem. 265*:1274.
83. Mohan, S., Dorai, D. T., Srimal, S., Bachhawat, B. K., and Das, M. K. (1984). Circular dichroism studies on carcinoscorpin, the sialic acid binding lectin of horseshoe crab *Carcinoscorpius rotunda cauda, Indian. J. Biochem. Biophys. 21*:151.
84. Brettine, H., Kabat, E., Liao, J., and Pereira, M. E. A. (1976). Purification and characterization of the agglutinins from the *Aaptos papillata* and a study of their combining sites. *Biochemistry 15*:5029.
85. Schluter, S. F., and Ey, P. L. (1984). The interaction of lactose-specific lectins with acid-treated and lactose-substituted Sepharose. *J. Immunol. Methods 66*:89.
86. Watkins, W. M., and Morgan, W. T. J. (1952). Neutralization of the anti-H in cell serum by simple sugars. *Nature 169*:825.
87. Hammarström, S. (1973). Binding of *Helix pomatia* A hemagglutinin to human erythrocytes and other cells. Influence of multivalent interaction on affinity. *Scand. J. Immunol. 2*:1.
88. Hammarström, S., and Kabat, E. A. (1971). Studies on specificity and binding properties of the blood group A reactive hemagglutinin from *Helix pomatia. Biochemistry 10*:1684.
89. Vasta, G. R., and Marchalonis, J. J. (1986). Galactosyl-binding lectins from the tunicate *Didemnum candidum*. Carbohydrate specificity and characterization of the combining site. *J. Biol. Chem. 261*:9182.
90. Maget-Dana, R., Roche, A.-C., and Monsigny, M. (1979). Ganglioside-limulin interactions. *Prog. Clin. Biol. Res. 29*:567.
91. Cohen, E., Vasta, G. R., Korytnyk, W., Petrie, C. R. III, and Sharma, M. (1984). Lectins of the Limulidae and hemagglutination-inhibition by sialic acid analogs and derivatives. *Prog. Clin. Biol. Res. 157*:55.
92. Roche, A.-C., Schauer, R., and Monsigny, M. (1975). Protein–sugar interactions: Purification by affinity chromatography of limulin, an *N*-acylneuraminidyl-binding protein. *FEBS Lett. 57*:245.
93. Mohan, S., Dorai, D. T., Srimal, S., and Bachhawat, B. K. (1982). Binding studies of a sialic acid-specific lectin from the horseshoe crab *Carcinoscorpius rotunda-cauda* with various sialoglycoproteins. *Biochem. J. 203*:253.
94. Yamashita, K., Umetsu, K., Suzuki, T., Iwaki, Y., Endo, T., and Kobata, A. (1988). Carbohydrate binding specificity of immobilized *Allomyrina dichotoma* lectin II. *J. Biol. Chem. 263*:17482.
95. Vasta, G. R., Marchalonis, J. J., and Kohler, H. (1984). An invertebrate recognition protein cross-reacts with an immunoglobulin idiotype. *J. Exp. Med. 159*:1270.
96. Olafsen, J. A. (1986). Invertebrate lectins: Biochemical heterogeneity as a possible key to their biological function. In *Immunity in Invertebrates* (M. Brehelin, ed.). Springer-Verlag, Berlin, p. 94.
97. Dorai, D. T., Srimal, S., Mohan, S., Bachhawat, B. K., and Balganesh, T. S. (1982). Recognition of 2-keto-3-deoxyoctonate in bacterial cells and lipopolysaccharides by the sialic acid binding lectin from the horseshoe crab *Carcinoscorpius rotunda cauda. Biochem. Biophys. Res. Commun. 104*:141.

98. Vaith, P., Uhlenbruck, G., and Holtz, G. (1980). Anti-glucuronyl activity of *Limulus polyphemus* agglutinin. In *Protides of Biological Fluids* (H. Peeters, ed.). Pergamon Press, New York, pp. 455–458.
99. Schluter, S. F. (1982). The isolation and characterization of three lactose binding proteins in the haemolymph of the protochordate *Botrylloides leachii*. Doctoral Dissertation. University of Adelaide.
100. Finstad, C. L., Good, R. A., and Litman, G. W. (1974). The erythrocyte agglutinin from *Limulus polyphemus* hemolymph: Molecular structure and biological function. *Ann. N.Y. Acad. Sci. 234*:170.
101. Lev, B., Ward, H., Keusch, G. T., and Pereira, M. E. A. (1986). Lectin activation in *Giardia lamblia* by host protease: A novel host–parasite infection. *Science 232*:71.
102. Mansour, M. H., Delange, R., and Cooper, E. L. (1985). Isolation, purification and amino acid composition of the tunicate hemocyte Thy-1 homolog. *J. Biol. Chem. 260*:2681.
103. Wu, T. T., and Kabat, E. A. (1982). Fourteen nucleotides in the second complementarity-determining region of a human heavy-chain variable region gene are identical with a sequence in a human D minigene. *Proc. Natl. Acad. Sci. USA 79*:5031.
104. Drickamer, K., Dordal, M. S., and Reynolds, L. (1986). Mannose-binding proteins isolated from rat liver contain carbohydrate-recognition domains linked to collagenous tails. Complete primary structures and homology with pulmonary surfactant apoproteins. *J. Biol. Chem. 261*:6878.
105. Vasta, G. R., and Marchalonis, J. J. (1987). Lectins from protochordates as putative recognition molecules. In *Developmental and Comparative Immunobiology* (E. L. Cooper, C. Langlet, and J. Bierne, eds.). Alan R. Liss, New York, pp. 23–32.
106. Rosenshein, I. L., Schluter, S. F., Vasta, G. R., and Marchalonis, J. J. (1985). Phylogenetic conservation of heavy chain determinants of vertebrates and protochordates. *Dev. Comp. Immunol. 9*:783.
107. Drickamer, K. (1988). Two distinct classes of carbohydrate recognition domains in animal lectins. *J. Biol. Chem. 263*:9557.
108. Uhlenbruck, G., Solter, J., Janssen, E., and Haupt, H. (1982). Anti-galactan and anti-haemocyanin specificity of CRP. *Ann. N.Y. Acad. Sci. 389*:476.
109. Hins, C. R. K., Collins, P. M., Baltz, M. L., and Pepys, M. B. (1985). Human serum amyloid P component, a circulating lectin with specificity for the cyclic 4,6-pyruvate acetal of galactose. Interactions with various bacteria. *Biochem. J. 225*:107.
110. Siegelman, M. H., van de Rijn, M., and Weissman, I. L. (1989). Mouse lymph node homing receptor cDNA clone encodes a glycoprotein revealing tandem interaction domain. *Science 243*:1165.
111. Bevilacqua, M. P., Stengelin, S., Gimbrone, M. A., Jr., and Seed, B. (1989). Endothelial leukocyte adhesion molecule-1: An inducible receptor for neutrophils related to complement, regulatory proteins and lectins. *Science 243*:1160.
112. Jungery, M., Boyle, D., Patel, T., Pasvol, G., and Weatherall, D. J. (1983). Lectin-like polypeptides of *Plasmodium falciparum* bind to red cell sialoglycoproteins. *Nature 301*:704.
113. Jungery, M., Pasvol, G., Newbold, C. I., and Weatherall, D. J. (1983). A lectin-like receptor is involved in invasion of erythrocytes by *Plasmodium falciparum. Proc. Natl. Acad. Sci. USA 80*:1018.
114. Miller, L. H., Haynes, J. D., McAuliffe, F. M., Shiroishi, T., Durocher, J. R., and McGinniss, S. M. H. (1977). Evidence for differences in erythrocyte surface receptors for the malarial parasites *Plasmodium falciparum* and *Plasmodium knowlesi. J. Exp. Med. 146*:277.
115. Kohler, A., and Prokop, O. (1967). Agglutination von Streptokokken der Gruppe C durch ein Agglutinin aus *Helix pomatia. Z. Immunitaets. forsch. 133*:50.
116. Cornick, J. W., and Stewart, J. E. (1968). Interaction of the pathogen *Gaffkya homari* with natural defense mechanisms of *Homarus americanus. J. Fish. Res. Board Can. 25*:695.
117. Meints, R. H., and Pardy, R. L. (1980). Quantitative demonstration of cell surface involvement in a plant–animal symbiosis: Lectin inhibition of reassociation. *J. Cell Sci. 43*:239.

118. Bayne, C. J., Moore, M. N., Carefoot, T. H., and Thompson, R. J. (1979). Hemolymph functions in *Mytilus californianus*: The cytochemistry of hemocytes and their responses to foreign implants and hemolymph factors in phagocytosis. *J. Invert. Pathol. 34*:1.
119. McCumber, L. J., Hoffman, E. M., and Clem, L. W. (1979). Recognition of viruses and xenogenic proteins by the blue crab, *Callinectes sapidus*: A humoral receptor for T2 bacteriophage. *J. Invert. Pathol. 33*:1.
120. Renwrantz, L., and Mohr, W. (1978). Opsonizing effect of serum and album in gland extract on the elimination of human erythrocytes from the circulation of *Helix pomatia. J. Invert. Pathol. 31*:164.
121. Renwrantz, L., and Stahmer, A. (1983). Opsonizing properties of an isolated hemolymph agglutinin and demonstration of lectin-like recognition molecules at the surface of hemocytes from *Mytilus edulis. J. Comp. Physiol. 149*:535.
122. Hardy, S. W., Fletcher, T. C., and Olafsen, J. A. (1977). Aspects of cellular and humoral defence mechanisms in the Pacific oyster, *Crassostrea gigas*. In *Developmental Immunobiology* (J. B. Solomon, and J. D. Horton, eds.). Elsevier/North Holland Biomedical Press, Amsterdam, p. 59.
123. Stein, E. A., and Cooper, E. L. (1983). Carbohydrate and glycoprotein inhibitors of naturally occurring and induced agglutinins in the earthworm, *Lumbricus terrestris. Comp. Biochem. Physiol. 76B*:197.
124. Roche, A.-C., Perredon, Y., Halpern, B., and Monsigny, M. (1977). Limulin (*Limulus polyphemus* lectin): Mitogenic effect on human peripheral lymphocytes. *Eur. J. Immunol. 7*:263.
125. Vasta, G. R., Marchalonis, J. J., and Decker, J. (1986). Binding and mitogenic properties of a galactosyl-specific lectin from the tunicate *Didemnum candidum* for murine thymocytes and splenocytes. *J. Immunol. 137*:3216.
126. Campbell, P. A., Hartman, A. L., and Abel, C. A. (1982). Stimulation of B cells but not T cells by a sialic acid-specific lectin. *Immunology 45*:155.
127. Renwrantz, L. R., and Cheng, T. C. (1977). Identification of agglutinin receptors on hemocytes of *Helix pomatia. J. Invertebr. Pathol. 29*:88.
128. Warr, G. W., Decker, J. M., Mandel, T. E., DeLuca, D., Hudson, R., and Marchalonis, J. J. (1977). Lymphocyte-like cells of the tunicate, *Pyura stolonifera*: Binding of lectins, morphological and functional studies. *Aust. J. Exp. Biol. Med. Sci. 55*:151.
129. Rögener, W., Renwrantz, L., and Uhlenbruck, G. (1986). Comparison of a hemolymph lectin from *Octopus vulgaris* with hemocyanin. *Comp. Biochem. Physiol. 85B*:119.
130. Stagner, J. I., and Redmond, J. R. (1975). The immunological mechanisms of the horseshoe crab, *Limulus polyphemus. Mar. Fish. Rev. 37*:11.
131. Cohen, E., and Vasta, G. R. (1982). Immunohematological significance of ubiquitous lectins. In *Developmental Immunology: Clinical Problems and Aging* (E. Cooper and M. A. B. Brazier, eds.). Academic Press, New York, p. 99.
132. Rögener, W., and Uhlenbruck, G. (1984). Invertebrate lectins: The biological role of a biological rule. *Dev. Comp. Immunol. 8*:159.
133. Uhlenbruck, G., Steinhausen, G., and Baldo, B. A. (1976). Galactans and anti-galactans from invertebrates. *Z. Naturforsch. 31*:205.
134. Van der Knapp, W. P. W. (1982). The internal defense system of the pond snail *Lymnaea stagnalis*. Doctoral Thesis, Amsterdam.
135. Pauley, G. B., Krassner, S. M., and Chapman, F. A. (1971). Bacterial clearance in the California sea hare, *Aplysia californica. J. Invertebr. Pathol. 18*:227.
136. Quigley, J. P., and Armstrong, P. B. (1983). An endopeptidase inhibitor, similar to mammalian α2-macroglobulin, detected in the hemolymph of an invertebrate, *Limulus polyphemus. J. Biol. Chem. 258*:7903.
137. Cheng, T. C., and Howland, K. H. (1979). Chemotactic attraction between hemocytes of the oyster, *Crassostrea virginica* and bacteria. *J. Invertebr. Pathol. 33*:204.

138. Cooper-Willis, C. A. (1979). Changes in the acid phosphatase levels in the haemocytes and haemolymph of *Patella vulgata* after challenge with bacteria. *Comp. Biochem. Biophys. A63*:627.
139. Tripp, M. R. (1960). Mechanisms of removal of injected microorganisms from the American oyster, *Crassostrea virginica. Gmelin, Biol. Bull. 119*:273.
140. Hardy, S. W., Grant, P. T., and Fletcher, T. C. (1977). A haemagglutinin in the tissue fluid of the Pacific oyster, *Crassostrea gigas*, with specificity for sialic acid residues in glycoproteins. *Experientia 33*:767.
141. Prowse, R. H., and Tait, N. N. (1969). In vitro phagocytosis by amoebocytes from the hemolymph of *Helix aspersa*. Evidence for opsonic factor(s) in serum. *Immunobiology 17*:437.
142. Stuart, A. E. (1968). The reticulo-endothelial apparatus of the lesser octopus, *Eledone cirrosa. J. Pathol. Bacteriol. 96*:401.
143. Vasta, G. R. (1990). Invertebrate lectins, C-reactive proteins and serum amyloid. Structural relationships and evolution. In *Defense Molecules* (J. J. Marchalonis and C. Reinisch, eds.). Alan R. Liss, New York, pp. 188–199.
144. Kilpatrick, J. M., and Volanakis, J. E. (1985). Opsonic properties of C-reactive protein. Stimulation by phorbol myristate acetate enables human neutrophils to phagocytize C-reactive protein-coated cells. *J. Immunol. 134*:3364.
145. Bout, D., Joseph, M., Pontent, M., Vorng, H. D., and Capron, A. (1986). Rat resistance to schistosomiasis: Platelet-mediated cytotoxicity induced by C-reactive protein. *Science 231*:153.
146. Beck, G., Vasta, G. R., Marchalonis, J. J ., and Habicht, G. S. (1989). Characterization of interleukin-1 activity in tunicates. *Comp. Biochem. Physiol. 92B*:93.
147. Oppenheim, J., Kovacs, E., Matsushima, K., and Durum, S. (1986). There is more than one interleukin-1. *Immunol. Today 7*:45.
148. Parrinello, N., and Arizza, V. (1989). Sugar specific cellular lectins of *Phallusia mamillata* hemocytes: Purification, characterization and evidence for cell surface localization. *Dev. Comp. Immunol. 13*:113.
149. Mirelman, D., ed. (1986). *Microbial Lectins and Agglutinins: Properties and Biological Activity*. John Wiley & Sons, New York.
150. Pereira, M. E. A., Andrade, A. F. B., and Ribeiro, J. M. C. (1981). Lectins of distinct specificity in *Rhodnius prolixus* interact selectively with *Trypanosoma cruzi. Science 211*:597.
151. Petri, W. A., Jr., Broman, J., Healy, G., Quinn, T., and Ravdin, J. I. (1989). Antigenic stability and immunodominance of the Gal/GalNac adherence lectin of *Entamoeba histolytica. Am. J. Med. Sci. 297*:163.
152. Brown, R. C., Bass, H., and Coombs, J. P. (1975). Carbohydrate binding proteins involved in phagocytosis by *Acanthamoeba. Nature 254*:434.
153. Chiesa, M. E., and Palatnik, M. (1976). Antigenos ABH-similes y aglutininas en el tracto digestivo de mosquitos. *Cienc. Cult. 28*:547.
154. Yu, C., Lee, A. M., and Roseman, S. (1987). The sugar specific adhesion/deadhesion apparatus of the marine bacterium *Vibrio furnissii* is a sensorium that continuously monitors nutrient levels in the environment. *Biochem. Biophys. Res. Commun. 149*:86.
155. Fitt, W. K., and Trench, R. K. (1981). Spawning, development, and acquisition of zooxanthellae by *Tridacna squamosa* (Mollusca, Bivalvia). *Biol. Bull. 161*:213.
156. Peters, W., Kolb, H., and Kolb-Bachofen, V. (1983). Evidence for a sugar receptor (lectin) in the peritrophic membranes of the blowfly larva, *Calliphora erythrocephala* Mg (Diptera). *Insect Physiol. 29*:275.
157. Maki, J. S., and Mitchell, R. (1986). The function of lectins among marine bacteria, invertebrates and algae, In *Microbial Lectins and Agglutinins: Properties and Biological Activity* (D. Mirelman, ed.). John Wiley & Sons, New York, p. 409.
158. Vacquier, V. D., and Moy, G. W. (1977). Isolation of bindin: The protein responsible for adhesion of sperm to sea urchin eggs. *Proc. Natl. Acad. Sci. USA 74*:2456.

159. Drickamer, K., and McCreary, V. (1987). Exon structure of a mannose-binding protein gene reflects its evolutionary relationship to the asialoglycoprotein receptor and non-fibrillar collagens. *J. Biol. Chem. 262*:2582.
160. Edelman, G. M. (1987). Cell adhesion and the evolutionary origins of immunity. *Immunol. Rev. 100*:11.
161. Kolb, H. (1977). On the phylogenetic origin of the immune system: A hypothesis. *Dev. Comp. Immunol. 1*:193.
162. Parish, C. R. (1977). Simple model for self–non-self discrimination in invertebrates. *Nature 261*:711.
163. Gilbride, K. J. M., and Pistole, T. G. (1981). The presence of copper in a purified lectin from *Limulus polyphemus*: Possible new role for hemacyanin, *Dev. Comp. Immuno. 5*:347.
164. Ikeda, K., Sannoh, T., Kawaski, N., Kawasaki, T., and Yamashina, I. (1987). Serum lectin with known structure activates complement through the classical pathway. *J. Biol. Chem. 262*:7451.
165. Haagsman, H. P., Hagwood, S., Sargeant, T., Buckley, D., Tyler White, R., Drickamer, K., and Benson, B. J. (1987). The major lung surfactant protein SP 28-36, is a calcium dependent, carbohydrate-binding protein. *J. Biol. Chem. 262*:13877.
166. Kawakami, M., Ihara, I., Suzuki, A., and Harada, Y. (1982). Properties of a new complement-dependent bactericidal factor specific for Ra chemotype *Salmonella* in sera of conventional and germ-free mice. *J. Immunol. 129*:2198.
167. Sarlo, K. T., and Mortensen, R. F. (1985). Enhanced interleukin (IL-1) production mediated by mouse serum amyloid P component. *Cell Immunol. 93*:398.
168. Conrad, J., Uhlenbruck, G., Zahn, R. K., Kurelec, B., Jericevic, B., and Muller, E. G. (1984). The role of lectin I and of glycoconjugates in recognition of cells from the siliceous sponge *Geodia cydonium. Biol. Cell 51*:287.
169. Basu, S., Sarkar, M., and Mandal, C. (1986). A single step purification of a sialic acid binding lectin (achatinin$_H$) from *Achatina fulica* snail. *Mol. Cell. Biochem. 71*:149.
170. Mandal, C., and Basu, S. (1987). An unique specificity of a sialic acid binding lectin achatinin$_H$ from the hemolymph of *Achatina fulica* snail. *Biochem. Biophys. Res. Commun. 148*:795.
171. Mandal, C., Basu, S., and Mandal, C. (1989). Physicochemical studies on achatinin$_H$, a novel sialic acid-binding lectin. *Biochem. J. 257*:65.
172. Dorai, D. T., Bachhawat, B. K., Bishayee, S., Kannan, K., and Rao, D. R. (1981). Further characterization of the sialic acid-binding lectin from the horseshoe crab *Carcinoscorpius rotunda cauda. Arch. Biochem. Biophys. 209*:325.
173. Umetsu, K., Kosaka, S., and Suzuki, T. (1984). Purification and characterization of a lectin from the beetle *Allomyrina dichotoma. J. Biol. Chem. 95*:239.

19

β-Galactoside-Binding Vertebrate Lectins: Synthesis, Molecular Biology, Function

Reuben Lotan *The University of Texas M. D. Anderson Cancer Center, Houston, Texas*

One of the posttranslational modifications of proteins is glycosylation, the enzymatic covalent attachment of carbohydrates as monosaccharides or oligosaccharide side chains to form glycoconjugates. These are present predominantly in membranes; however, they are also found in the cytoplasm and in nuclei [Hart et al., 1989; Paulson, 1989; Sharon and Lis, 1989]. Normal development, cell growth, and differentiation involve various types of cellular interactions among cells and between cells and exogenous soluble or insoluble macromolecules. Cell surface-exposed glycoconjugates are thought to play a role in these processes by mediating intercellular recognition and adhesion and by serving as receptors for growth factors, enzymes, and hormones [Brandley and Schnaar, 1986; Paulson, 1989; Rutishauser and Jessell, 1988; Sharon and Liss, 1989]. The different sugar residues that are present in glycoconjugates can be linked in specific sequences and types of anomeric linkages, and it has been suggested that they form very specific recognition codes [Paulson, 1989]. Complementary molecules, capable of binding specific carbohydrates, are required for recognition and for mediating the functional consequences of such a recognition. Indeed, lectins, carbohydrate-binding proteins of nonimmune origin, which are devoid of enzymatic activity, have been found in various vertebrate tissues [Barondes, 1986, 1988]. The presence of a lectinlike protein in vertebrate liver was observed by Ashwell and Morell more than 20 years ago, and this lectin, which is an integral membrane protein, was found to have a physiological role in the binding and endocytosis of circulating desialated glycoproteins by liver hepatocytes. The properties of this asialoglycoprotein receptor and other members of this family have been reviewed extensively [Barondes, 1986; Drickamer, 1988; Harford and Ashwell, 1982] and will not be discussed in this chapter.

Subsequently, different galactoside-binding lectins were discovered by Teichberg and associates (1975). These lectins have been studied in different laboratories and found to constitute a family of proteins that are distributed widely in species ranging from the electric eel to humans [Barondes, 1986; Barondes et al., 1988; Monsigny et al., 1988].

Within each species, these lectins exhibit distinct tissue distribution (e.g., spleen, thymus, heart, lung, bone marrow, skin, intestine, pancreas, liver, and muscle), and their levels are regulated during development [Barondes, 1986]. These lectins are distinguished from the asialoglycoprotein receptor [Drickamer, 1988; Harford and Ashwell, 1982] in that they require thiol reagents, they are not dependent on Ca^{2+}, and they are not integral membrane proteins [Barondes, 1986; Drickamer, 1988]. Their physiological function(s) is not known; however, their ability to bind galactosides and the changes in their level during embryogenesis and development, processes that are also accompanied by changes in glycoconjugates, suggest that the lectins play a role in recognition and cellular interactions.

This chapter will review the properties of the thiol-dependent group of galactoside-binding proteins and will emphasize recent developments in the molecular biology of the lectins and in the understanding of their structure and functions.

I. GENERAL PROPERTIES OF GALACTOSIDE-BINDING LECTINS

A. Purification of Galactoside-Binding Lectins

Galactoside-specific lectins have been purified from various tissues by affinity chromatography using different immobilized galactosides (e.g., *p*-aminophenyl β-D-galactoside; lactose) or galactoglycoproteins (e.g., asialofetuin) [Powell, 1980; Levi and Teichberg, 1981; Barondes and Leffler, 1987]. Since these lectins are soluble, their effective extraction from tissues and cultured cells does not require the use of any detergent. However, an optimal extraction requires homogenization of tissue in an aqueous buffered solution containing β-mercaptoethanol (4–14 mM), ethylenediaminetetraacetic acid (EDTA 2 mM), and lactose (0.1–0.3 M). The inclusion of lactose increased the yield of lectin, presumably by preventing binding of the soluble lectins to endogenous membrane-associated glycoconjugates [Briles et al., 1979]. The yield of lectins from various normal tissues ranged from 0.1 to 2 μg/mg total soluble protein in chicken, mouse, rat, and calf tissues [Barondes and Leffler, 1987; Briles et al., 1979; Cerra et al., 1985; Ohara and Yamagata, 1986], whereas, in rabbit tissues, the lectin is about ten times more abundant [Catt et al., 1987]. Some cultured cell lines produce larger amounts of lectin than the tissue of origin. For example, five lines of human lung fibroblasts had much higher lectin levels than lung tissue [Whitney et al., 1985]. This could be either because in the original tissue only some types of cell produce the lectins and those gave rise to the cultured cells, or because the in vitro growth conditions stimulated lectin synthesis. The yield of a 29-kd lectin from human lung was 60 μg/100 g lung tissue [Sparrow et al., 1987].

Figure 1 shows an example of a single-step purification of both 14-kd and 34-kd lectins from murine melanoma cells on immobilized asialofetuin. These lectins have been localized by indirect immunofluorescence at the cell surface of intact viable cells, and in the cytoplasm of permeabilized cells (Fig. 2). The two lectins can be separated by gel filtration [Roff and Wang, 1983], or by ion-exchange chromatography [Cerra et al., 1985; Sparrow et al., 1987].

The relative molecular masses (M_r) reported for the lectins purified in different laboratories and from different tissues and species vary from 12 to 15 kd for the low M_r lectin—and in this chapter they will be referred to as the 14-kd lectins for the sake of

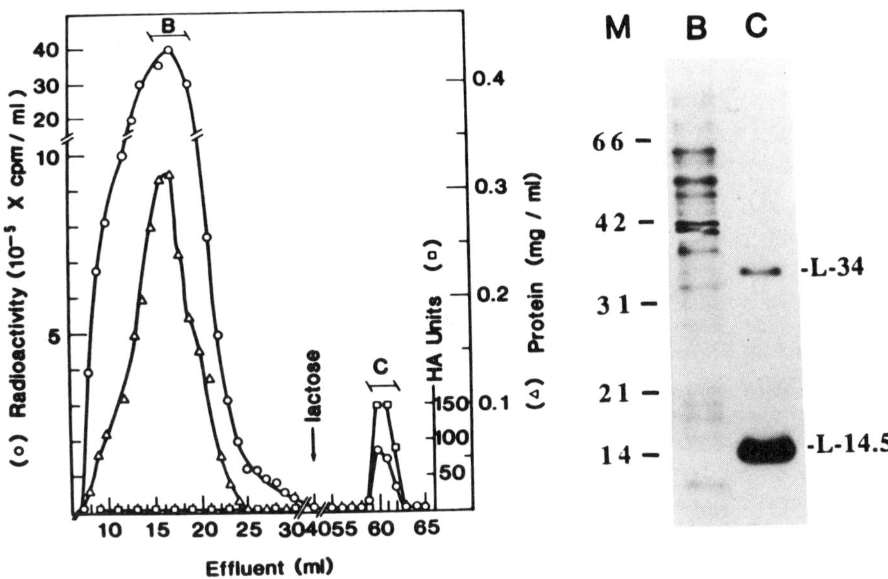

Figure 1 Purification of endogenous lectins from an extract of K-1735P melanoma cells by affinity chromatography. Cells were grown in four 10-cm–diameter tissue culture dishes and labeled for 12 hr by incubation in growth medium supplemented with L-[^{35}S]methionine (20 μCi/mL). The cells were detached, suspended (2.7 × 10^7 cells/3 mL) in MEPBS (calcium-free and magnesium-free phosphate-buffered saline containing 4 mM β-mercaptoethanol and 2 mM EDTA) supplemented with 0.3 M lactose, and homogenized on ice. The homogenate was certifuged at 12,000 *g* for 5 min at 4°C, the supernatant was collected and centrifuged further at 100,000 *g* for 60 min at 4°C. The supernatant fraction was dialyzed against MEPBS to remove the lactose and applied onto an asialofetuin-Affi-Gel 10 column (0.9 × 20 cm). The column was washed at 4°C with MEPBS until all the unbound radioactivity was eluted (after 40 mL of effluent). Then the bound material was eluted with 0.3 M lactose in MEPBS (arrow). Fractions (1 mL) were collected and analyzed for protein content, radioactivity, and hemagglutinating activity (HA). *One hemagglutinating activity unit* is defined as the highest dilution to effect agglutination of trypsinized rabbit erythrocytes when a 50 μL sample is diluted .1:1 (vol/vol) serially. The lactose-eluted fractions were dialyzed separately against MEPBS before analysis. Samples of the combined fractions designated B and C (first and second peaks, respectively) were analyzed by electrophoresis in a 12.5% polyacrylamide slab gel. The gel was processed for fluorography and the film was photographed. M, positions of molecular mass standard proteins (M_r × 10^{-3}).

uniformity. Likewise, the higher M_r lectins were reported to be of 29–35 kd, and they will be referred to as the 34-kd lectins.

Additional galactoside-binding lectins of 6.5 kd [Zalik et al., 1987], 16 kd [Barondes, 1988; Hirabayashi et al., 1987b; Oda and Kasai, 1983; Ohara and Yamagata, 1986; Roff et al., 1983], 17 kd [Leffler and Barondes, 1989], 18 kd [Cerra et al., 1985], 22 kd [Sparrow et al., 1987], 67 kd [Barondes, 1988, Hinek et al., 1988; Mecham et al., 1989], and 70 kd [Zalik et al., 1987] have been discovered. Although only limited information is available on these lectins, their properties will be described where appropriate.

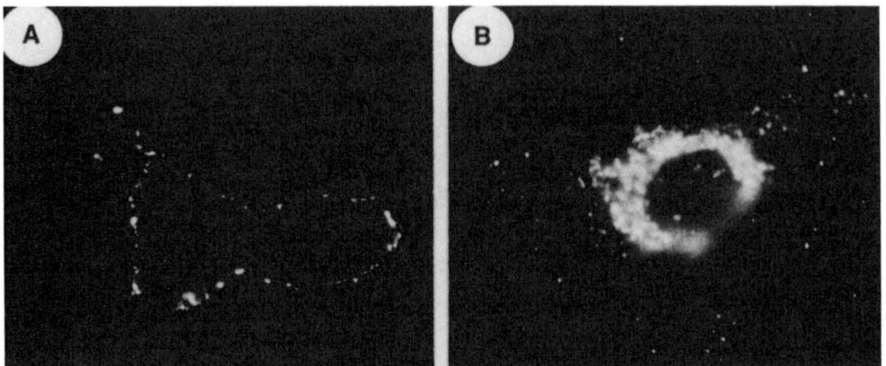

Figure 2 Localization of endogenous lectins on the surface of viable mouse UV-2237-IP3 fibrosarcoma cells (A) and in the intracellular compartment of fixed and permeabilized cells (B). The cells were grown on glass coverslips and incubated directly with monoclonal antilectin antibodies 5D7, either before (A) or after fixation with 3.5% formaldehyde, and permeabilization with 0.05% Nonidet P-40. Binding of the first antibody was localized by a further incubation with rhodamine-conjugated goat antimouse IgG and observation and photography under a fluorescence microscope. (Courtesy of Dr. A. Raz.)

B. Physicochemical Properties of the 14-kd Lectin

Studies of purified 14 kd lectins from different sources indicate that these proteins can be resolved into two or more molecular species by isoelectric focusing. Their pI values range from 4.6 to 5.8 [Allen et al., 1987a; Briles et al., 1979; Clerch et al., 1988; Hirabayashi et al., 1987b; Lotan et al., 1989; Sparrow et al., 1987]. The reason for the existence of molecules with different isoelectric points is unknown, since there is no evidence for posttranslational modifications, such as phosphorylation or glycosylation, of the lectins. Studies with rat lung lectin have demonstrated that lectin that is purified quickly shows a major component with a pI of 5.5, and upon storage, additional molecular species appear with a pI of 5.3 and lower [Clerch et al., 1988].

The majority of the 14-kd lectins appear to be dimeric under nondenaturing conditions [Briles et al., 1979; Caron et al., 1987; Clerch et al., 1988; Levi and Teichberg, 1981; Oda and Kasai, 1983; Powell, 1980; Whitney et al., 1986]. However, monomeric lectins have been found in chicken intestine [Beyer et al., 1980], rabbit bone marrow [Godsave et al., 1981], and 3T3 fibroblasts [Roff and Wang, 1983].

The requirement for thiol reagents for activity is characteristic of all the 14-kd lectins. Atmospheric oxygen can inactivate the electric eel lectin, which does not contain any cysteine residues, by oxidation of a tryptophan residue [Levi and Teichberg, 1981]. The oxidation of the tryptophan to oxindole could be prevented by lactose, suggesting that the tryptophan residue is located in or near the lactose-binding site. Other vertebrate lectins were found to contain several cysteine residues and their oxidation caused loss of lectin activity. The chicken skin lectin and the human placenta lectin have lost hemagglutinating activity or ability to bind to asialofetuin after treatment with *p*-chloromercuribenzoic acid [Oda and Kasai, 1983; Hirabayashi et al., 1987a,b]. The human fibroblast lectin was inactivated by incubation for 1 hr in cell growth medium without a reducing agent and lost its ability to bind lactosyl-Sepharose. This activity was regained after incubation with a reducing agent [Whitney et al., 1986]. Although activity was lost after treatment of the rat lung lectin with iodoacetate or ethyleneimine, alkylation with iodoacetamide yielded carboxyamidomethylated lectin, which was fully active in the absence of a reducing agent [Whitney et al., 1986]. The cysteine residues in the native human lectin do not form disulfide bonds because almost all of them could be titrated with 5,5′-dithio-bis(2-nitrobenzoic acid) in the presence of urea [Hirabayashi et al., 1987b]. Likewise, the cysteine residue of the rat lung lectin are in the reduced form. However, their oxidation may lead to the formation of intramolecular disulfide bonds in the absence of a reducing agent and to dramatic changes in secondary structure [Whitney et al., 1986]. In contrast with the electric eel lectin, the tryptophan residues of the rat lung lectin were not susceptible to oxidation [Whitney et al., 1986].

The sedimentation coefficient ($S_{20,w}$) of the dimeric rat lung lectin was 2.48 S, and the frictional ratio suggested that the lectin has an asymmetric elongated ellipsoid structure in solution [Clerch et al., 1988].

Additional properties of the 14-kd lectin were deduced from the amino acid sequence and will be discussed later.

C. Physicochemical Properties of the 34-kd Lectin

The first identification of the 34-kd lectin was the result of affinity chromatography of Triton X-100 extracts of 3T3 mouse fibroblasts on either immobilized lactosamine or immobilized asialofetuin. Three thiol-dependent and Ca^{2+}-independent lectins with polypeptide M_r of 13.5 kd, 16 kd, and 35 kd were purified by this procedure. The 35-kd lectin was also purified from mouse lungs and cultured human foreskin fibroblasts [Crittenden et al., 1984; Roff et al., 1983; Roff and Wang, 1983]. The 35-kd polypeptide could be separated from the smaller lectins by gel filtration and repurified by affinity chromatography, showing that it is a lectin. It was found to differ from the lower M_r lectins in antigenicity, suggesting that it was not a dimer of any of the smaller lectins [Roff et al., 1983; Roff and Wang, 1983]. The presence of a lectin with an M_r of 29 kd has been noticed when affinity-purified material from bovine heart and rat and bovine lung was analyzed by polyacrylamide gel electrophoresis, and it was assumed initially that this lectin was a dimeric form of the 14-kd lectin, which failed to dissociate completely [Cerra, 1984; DeWaard et al., 1976]. However, the peptide maps of the monomer and the hypothetical dimer were somewhat different [Cerra et al., 1984], and subsequent separation by ion-exchange chromatography, immunological studies, and analyses of carbohydrate binding have demonstrated that they were distinct proteins [Cerra et al., 1985; Leffler and Barondes, 1986]. More recently, a 29-kd lectin was characterized in human lung [Sparrow et al., 1987]. It seems that the 29-kd rat and human lectins and the mouse 35-kd lectins are homologous lectins because antibodies prepared against the 29-kd lectin from rat lung reacted with mouse melanoma cell 34-kd lectin [Lotan et al., 1989a]. Furthermore, coelectrophoresis of extracts of mouse, rat, and human cells on the same polyacrylamide slab gel followed by immunoblotting with antibodies against rat lung lectins indicated that the human and rat contained lectins that migrated as 31-kd polypeptides, whereas the mouse lectin migrated as a 34-kd polypeptide [Lotan, unpublished results]. As stated earlier, irrespective of their exact molecular mass these lectins will be referred to as the 34-kd lectins.

The mouse 3T3 lectin was resolved by isoelectric focusing into two components of pI 4.5 and 4.7, respectively [Roff and Wang, 1983]. Similar results were obtained with the

34-kd lectin from mouse melanoma cells [Lotan et al., 1989a]. A greater heterogeneity in pI was found with the human lung lectin, which was resolved into five components with acidic pI [Sparrow et al., 1987]. This heterogeneity does not seem to result from differential glycosylation of the components.

Additional properties of the 34-kd lectin were deduced from the amino acid sequence and will be discussed later.

II. MOLECULAR BIOLOGY OF GALACTOSIDE-BINDING LECTINS

The last few years have seen enormous advances in the molecular biology of vertebrate galactoside-binding lectins. The progress has been made through sequencing of purified proteins and cloning of the genes coding for the lectins, sequencing the cDNAs, and predicting the amino acid sequences and the protein structure. The pioneering and most detailed investigation has been carried out by Kasai and his colleagues on the chick embryo skin 14-kd lectin protein, cDNA, and gene [Oda and Kasai, 1983; Ohyama et al., 1986; Hirabayashi et al., 1987a; Ohyama and Kasai, 1988]. Partial and complete amino acid sequence, determined by protein sequencing or deduced from nucleotide sequences were elucidated subsequently for the 14-kd lectins from human lung [Gitt and Barondes, 1986; Hirabayashi et al., 1989], human hepatoma [Gitt and Barondes, 1986; Abbott and Feizi, 1989], human placenta [Paroutaud et al., 1987; Hirabayashi and Kasai, 1988; Couraud et al., 1989], human HL-60 promyelocytic leukemia [Couraud et al., 1989], bovine heart [Southan et al., 1987], bovine fibroblast cell line EBTr [Abbott et al., 1989], rat lung and uterus [Clerch et al., 1988], mouse UV2237-IP3 fibrosacroma [Raz et al., 1987a, 1988], mouse 3T3 fibroblasts [Wilson et al., 1989], and electric eel electric organ [Paroutaud et al., 1987]. The 34-kd lectin cDNA has been cloned and sequenced from mouse fibrosarcoma [Raz et al., 1987a, 1988, 1989], mouse 3T3 fibroblasts [Jia et al., 1987; Jia and Wang, 1988], and rat basophilic leukemia cells [Liu et al., 1985; Albrandt et al., 1987]. The amino acid sequences of the 14-kd and the 34-kd lectins are presented in Figures 3 and 4.

A. Molecular Biology of the 14-kd Lectin

1. The Chick Embryo 14-kd Lectin cDNA and Amino Acid Sequences

The first 14-kd lectin to be cloned was the one from chick embryo skin [Ohyama et al., 1986]. Tarsometatarsal skin from 18-day embryos was chosen as the source of mRNA for preparation of the cDNA clones because, at this stage, the embryo contains large amounts of the lectin [Oda and Kasai, 1983]. The strategy for selection of recombinant clones was based on the synthesis of oligonucleotide probes (17-mer, 32 oligonucleotides) complementary to the peptide sequence Phe-Asp-Tyr-Phe-Asp-Thr, which was determined by analysis of a proteolytic fragment of the purified lectin. Two clones to which the probe hybridized were sequenced and found to be partially overlapping. The amino acid sequence was deduced from the nucleotide sequences of these two clones [Ohyama et al., 1986]. The authentic amino acid sequence was determined using lectin purified from 15-day old chick embryo metatarsal skin (yield of 30 mg from 100 embryos), and it was identical with the deduced sequence, except that the NH_2-terminal residue in the protein acetylated serine, whereas the deduced NH_2-terminal amino acids were Met-Ser [Hirabayashi et al., 1987a]. It was suggested that methionine, the initiator of translation, is removed upon completion of the translation and then serine becomes the NH_2-terminal residue and is acetylated.

	1 10 20
HUMAN	M A C G L V A S N L N L K P G E C L R V R G E V A P D A K S
BOVINE	* A * * * *
RAT	* * * * * * * * * * * * * * * * * * K * * * * L * * * * * *
MOUSE	* * * * * * D Q Q A E S Q T * A M S Q S S * * * * S * D * *
CHICK	M S C Q * P * C T * * G * * * * Q R * T * K * I I * * N A * *
EL. EEL	S M N * V * D E R M S F * A * Q N * T * K * V P S I * S T N

	30 40 50
HUMAN	F V L N L G K D S N N L C L H F N P R F N A H G D A N T I V
BOVINE	* L * * * * * * D *
RAT	* *
MOUSE	* *
CHICK	* * M * * * * * * T H * G * * * * * * * D * * * * V * L * *
EL. EEL	* A I * V * N S A E D * A * * I * * * * D * * * * Q Q A V *

	60 70 80
HUMAN	C N S K D G G A W G T E Q R E A V F P F Q P G S V A E V C I
BOVINE	* * * * * A * * * * A * * * * S A * * * * * * * * V * * * *
RAT	* * * * * D * T * * * * * * * T A * * * * * * * I T * * * *
MOUSE	* * T * E D * T * * * * H * * P A * * * * * * * I T * * * *
CHICK	* * * * K M E E * * * * * * * T * * * * * K * A P I * I T F
EL. EEL	V * * F Q * * N * * * * * * * G G * * * K Q * E D F K I Q *

	90 100 110
HUMAN	T F D Q A N L T V K L P D G Y E F K F P N R L N L E A I N Y
BOVINE	S * N * T D * * I *
RAT	* * * * * D * * I * * * * * H * * * * * * * * * M * * * * *
MOUSE	* * * * * D * * I * * * * * H * * * * * * * * * M * * * * *
CHICK	S I N P S D * * * H * * - * H Q * S * * * * * G * S V F D *
EL. EEL	* * * S E E F R I I * * * * S * I H * * * N R Y M H F E G -

	120 130 134
HUMAN	M A A D G D F K I K C V A F D
BOVINE	L S * G * * * * * * * * * * E
RAT	* * * * * * * * * * * * * * E
MOUSE	* * * * * * * * * * * * * * E
CHICK	F D T H * * * T L R S * S W E
EL. EEL	E * R I Y S I E * *

Figure 3 Comparison of amino acid sequences of 14-kd lectins from several species. The numbering is based on the human lectin sequence. Residues that are identical with the human lectin are indicated by an asterisk (*) and those that are different are specified by the letter designation of amino acids. Residues that are underlined once (–) are identical in 5/6 of the 14-kd lectins and in the aligned 34-kd lectin, whereas residues that are underlined twice (=) are identical in all of the 14-kd lectins and in the 34-kd lectin (see Fig. 4). Dashes (-) indicate gaps introduced for optimal alignment (e.g., chicken lectin sequence position 102 and eel lectin sequence position 119). The data for the human [Hirabayashi and Kasai, 1988; Hirabayashi, et al., 1989; Couraud et al., 1989; Abbott and Feizi, 1989], bovine [Southan et al., 1987; Abbott et al., 1989], rat [Clerch, et al., 1988], and chick [Ohyama et al., 1986; Hirabayashi et al., 1987a], were obtained by both protein sequencing and cDNA sequencing. The data for the mouse lectin were deduced only from the cDNA sequence [Raz et al., 1988; Wilson et al., 1989], whereas the data for the electric eel were obtained from protein sequencing only [Paroutaud et al., 1987].

```
        1                 10                  20
CBP35 - R D S F S L N D A L A G S G N P N P Q G Y P G A W G N Q P
L-34  M A * T * * * * * * * * * * * * * * * * * * * * * * * * * *
IgEBP M A * G * * * * * * * * * * * * * * * R * W * * * * * * * *

      30                  40                  50
CBP35 G A G G Y P G A A Y P G A Y P G Q A P P G A Y P G Q A P P G
L-34  * * * * * * * * * * * * * * * * * * * * * * * * * * * * * *
IgEBP * * * * * * * * S * * * * * * * * * * * * G * * * * * * * S

      60                  70                  80
CBP35 A Y P G Q A P P S A Y P G P T A P G A Y P G P T A P G A Y P
L-34  * * * * * * * * * * * * * * * * * * * * * * * * * * * * * *
IgEBP * * * * P T G * * * * * * * * * * * * * * * * * * * * * F *

      90                  100                 110
CBP35 G Q P A P G A F P G Q P G A P G A Y P Q C S G G Y P A A G P
L-34  * S T * * * * * * * * * * * * * * * * S A P * * * * * * * *
IgEBP * Q P G - * P G A Y P S - * * * * * * S A P * A * * * T * *

      120                 130                 140
CBP35 Y G V P A G P L T V P Y D L P L P G G V M P R M L I T I M G
L-34  * * * * * * * * * * * * * * * * * * * * * * * * * * * * * *
IgEBP F * A * T * * * * * * * * M * * * * * * * * * * * * * * I *

      150                 160                 170
CBP35 T V K P N A N R I V L D F R R G N D V A F H F N P R F N E N
L-34  * * * * * * * * * * * * * * * * * * * * * * * * * * * * * *
IgEBP * * * * * * * S * T * N * K K * * * I * * * * * * * * * * *

      180                 190                 200
CBP35 N R R V L V C N T K Q D N N W G K E E R Q S A F P F E S G K
L-34  * * * * * * * * * * * * * * * * * * * * * * * * * * * * * *
IgEBP * * * * * * * * * * * * * * * * R * * * * * * * * * * * * *

      210                 220                 230
CBP35 P F K I Q V L V E A D H F K V A V N D A H L L Q Y N H R M K
L-34  * * * * * * * * A * E P * * * * * * * * * * * * * * * * * *
IgEBP * * * * * * * * E * D H * * * * * * * V * * * * * * * * * *

      240                 250                 260
CBP35 N L R E I S Q L G I S G D I T L T S A N H A M I
L-34  * * * * * * * * * * * * * * * * * * * * * * * *
IgEBP * * * * * * * * * * I * * * * * * * * S * * * *
```

Figure 4 Comparison of amino acid sequences of 34-kd lectins from mouse and rat. Residues that are identical in the three lectins are indicated by an asterisk (*), and those that are different are specified by the letter designation of amino acids. Residues that are underlined once (—) are identical in the three 34-kd lectins and in 5/6 of the 14-kd lectins, whereas residues that are underlined twice (=) are identical in all of the 34-kd lectins and all of the 14-kd lectins (see Fig. 3). Dashes (-) indicate gaps introduced for optimal alignment (e.g., IgEBP sequence position 94 and 102). The data were deduced from the cDNA sequence of CBP35 from mouse 3T3 cells [Jia and Wang, 1988], L-34 lectin from mouse fibrosarcoma cells [Raz et al., 1989], and IgEBP from the cDNA sequence of the IgE-binding protein from rat basophilic leukemia cells [Albrandt et al., 1987], which is the rat homologue of the 34-kd lectin [Laing et al., 1989].

There is no evidence for the presence of a cleavable NH_2-terminal leader (signal) peptide, characteristic of many secreted proteins, in the lectin cDNA.

The lectin is composed of 134-amino acid residues (see Fig. 3) with a calculated M_r of 14,970 D, which is close to the apparent M_r of 14,500, calculated from analysis of electrophoretic migration in polyacrylamide gels in the presence of sodium dodecyl sulfate (SDS) and a reducing agent. Internal sequence homologies (20–55%) were found among three regions of the lectin, spanning residues 3–30, 36–58, and 105–132, and it has been suggested that the lectin gene evolved by several gene duplications [Hirabayashi et al., 1987a]. This contention has been supported by the analysis of the gene structure as discussed later.

2. *The Sequence of 14-kd Lectins from Various Species and from Normal and Malignant Cells*

The complete amino acid sequence, determined either by deduction from cDNA sequences or by direct analysis of isolated protein, is now available for the 14-kd lectins from several species. The sequences of the lectins from human placenta [Hirabayashi and Kasai, 1988; Couraud et al., 1989], human lung [Hirabayashi et al., 1989], human HL-60 leukemia cells [Couraud et al., 1989], and human hepatoma [Abbott and Feizi, 1989] are very similar, if not identical. Likewise, the sequences of the lectin from bovine heart and bovine fibroblast cell line are identical [Abbott et al., 1989], those from rat lung and uterus are identical [Clerch et al., 1988], and those from 3T3 fibroblasts [Wilson et al., 1989] and mouse fibrosarcoma cells [Raz et al., 1988] are very similar. Thus, there seems to be no difference between the sequences of lectins derived from different tissues, or from normal and malignant cells in a given species.

The alignment of these sequences (see Fig. 3) demonstrates extensive homology, indicating that all these molecules constitute a distinct family. The degree of homology (percentage o identical sequence) between the sequence of the human lectin and those of the bovine, rat, mouse, chicken, and electric eel is 87.3, 90,3, 77.6, 54.5, and 43.3%, respectively. The differences between the lectin sequences of the different species probably reflect the phylogenic distance between them. Amino acid residues 3, 5, 14, 21, 30, 33, 35, 41, 43, 44, 46–49, 51–54, 59, 61, 68, 69, 73, 74, 77–79, 82, 100–103, 108–110, are identical in all of the 14-kd lectins (see Fig. 3). This conservation suggests that some or all of these residues are important for the lectins' function(s).

The internal sequence homology identified in the chicken 14-kd lectin was not detected in the rat lectin for which scores for sequence alignment of internal regions (residues 1–31, 31–63, 63–103, and 103–134) were not significantly higher than those for randomized sequences of the same amino acid composition [Clerch et al., 1988]. Likewise, the lectin from human placenta did not contain internal sequence homology [Hirabayashi and Kasai, 1988]. These differences between the chicken and mammalian lectins suggest a divergence during evolution.

A 40% homology between the 14-kd lectins (primarily residues 44–82) and the galactoside-binding 34-kd lectin (residues 171–208) has been detected as indicated in Figures 3 and 4 [Raz et al., 1988; Jia and Wang, 1988; Hirabayashi et al., 1989]. A similar homology between the 14-kd lectins and a low-affinity IgE-binding protein from rat basophilic leukemia, which has been identified as the rat galactoside-binding 34-kd lectin [Laing et al., 1989], has been detected by several investigators [Abbott et al., 1989; Raz et al., 1989; Wilson et al., 1989; Hirabayashi et al., 1989]. A tetrapeptide WGA(T)E [Tryp-Gly-Ala(Thr)-Glu (amino acid residues 68–71], identified as the epitope recognized by anti(bovine) lectin monoclonal antibodies, has been detected in all of the 14-kd lectins as well as in myelin basic protein, and antibodies prepared against myelin basic protein reacted with the lectin [Abbott et al., 1989]. It has been proposed that myelin basic protein may have a carbohydrate-binding site that includes the WGAE tetrapeptide [Abbott et al., 1989].

No homology was found between the sequence of the bovine lectin and those of several galactoside-binding proteins, including the B chain of the *Ricinus communis* plant agglutinin, discoidin I from *Dictiostelium discoideum*, a galactoside-binding protein from the tunicate *Didemnum candidum*, and the hepatic asialoglycoprotein receptor [Southan et al., 1987].

3. *Properties of the 14-kd Lectin Deduced from the Amino Acid Sequence*

The calculated molecular mass of the chicken lectin is 14,970 [Hirabayashi et al., 1987a] and for the human lectin 14,744 [Couraud et al., 1989].

A potential *N*-glycosylation site [Asn(92)-Pro(93)-Ser(94)] has been identified in the chicken lectin [Hirabayashi et al., 1987a] and in the human lectin [Asn(95)-Thr(96)-Leu(97)] [Hirabayashi and Kasai, 1988; Couraud et al., 1989]. However, no covalently-linked carbohydrates could be detected by colorimetric sugar analysis, no glycosylated amino acids were detected by sequence analysis, and the electrophoretic migration of the lectin was not altered by pretreatment with *N*-glycanase.

Computer analysis of the chicken lectin sequence data predicted that the secondary structure is 40% β-sheet, 30% reverse turn, and 10% α-helix, and the hydrophobicity profile indicates that the region between residues 62 and 75, where α-helical structure is predicted, is very hydrophilic [Hirabayashi et al., 1987a]. The chicken and the electric ell lectins exhibit only 39% homology; however, their predicted secondary structure is similar, as indicated by their propensity to form β-sheet structures and their hydropathy profiles [Paroutaud et al., 1987]. Although the rat and chicken lectins exhibit 57% sequence identity, a computer analysis of the rat lung lectin sequence data predicted a significantly different structure with a lower β-sheet (20%) and a higher α-helix (23%), but similar values were predicted for β-turn (33%), and random coil (24%). The predicted structure was in close agreement with that calculated from circular dichroism analysis of the purified rat lectin [Clerch et al., 1988]. The hydropathy score and profile calculated for the rat lectin indicate that there are no lengthy hydrophobic sequences that could intercalate in cell membranes. The localization of the chicken lectin in the extracellular compartment is compatible with the predicted high β-sheet content, which is one of the characteristics of extracellular proteins [Hirabayashi et al., 1987a].

All the 14-kd lectins require a reducing agent for activity. Interestingly, each of the human, bovine, and rat lectins contains six cysteine residues, the mouse five, the chick three, whereas the electric eel lectin does not contain any cysteine residues. Cysteine residues 2, 60, and 130 are conserved in the 14-kd lectin from most species and react readily with iodoacetic acid, indicating that they are present at the surface of the protein [Wilson et al., 1989]. Such a location for Cys-60 is compatible with the presence of this amino acid within the region assumed to constitute the galactoside binding site. However, the SH groups of the cysteine residues are not essential for lectin activity, as indicated by the maintenance of activity after modification of the SH groups with iodoacetamide [Whitney et al., 1986] and the conservation of galactoside-binding activity in the absence of reducing agents in lectin molecules bound covalently to Sepharose [Merkle and Cummings, 1988].

4. *The 14-kd Lectin Gene*

The 14-kd lectin gene has been cloned from a chick genomic library by screening with the cloned chick embryo skin lectin cDNA [Ohyama and Kasai, 1988]. The gene was found to contain four exons separated by three introns (Fig. 5). The introns inserted in the gene after nucleotide sequences coding for amino acid residues 2, 30, and 87. the introns span 2500 bp (1000 bp, 1200 bp, and 300 bp for introns 1, 2, and 3, respectively), and the exons' lengths are 67, 83, 172, and 215 (or 216) bp for exons I, II, III, and IV, respectively. Exon I encodes only two amino acids of the NH_2-terminus of the mature protein, in addition to the initiator methionine, whereas the three other exons each codes for one of the amino acid regions showing internal sequence homology [Hirabayashi et al., 1987a]. Exon II shows 42% sequence homology with exon III and 33.7% homology with exon IV, whereas exons II and IV share 45.8% homology. The internal homologies in nucleotide sequences and in the amino acid sequences of these regions of the gene and protein indicate that the lectin gene has evolved as a result of two duplications of an ancestral gene [Ohyama and Kasai, 1988]. The internal homology found in the chick lectin cDNA is not present in any of the mammalian lectins, suggesting that either this feature has not been conserved during evolution or that the chick lectin is not the ancestor of the lectin genes of higher species. Kasai suggested that the chicken 14-kd lectin gene and human, bovine, and rodent lectin genes have evolved from a common ancestral gene.

The transcription initiation site was determined to be an adenine residue located 58-bases upstream from the putative initiator ATG codon and a TATA boxlike sequence was identified 24- to 31-nucleotides upstream from the initiation codon by primer extension of the cDNA clone. The only ATG codon present in the upstream region was the one located before the NH_2-terminal amino acid (serine) of the mature protein residue, providing further evidence that the chicken lectin does not contain a cleavable signal peptide at its NH_2-terminus. Southern blot hybridization of the 14-kd lectin cDNA to genomic DNA, isolated from nuclei of 18-day chick embryo livers and digested with several restriction enzymes, labeled only DNA fragments of sizes that were in agreement with those expected from the restriction enzyme map of the cloned genomic DNA. These results indicate that the lectin gene is present in the chicken genome as a single-copy gene [Ohyama and Kasai, 1988]. A similar gene structure has been described for the human lectin gene [Gitt and Barondes, 1989].

Southern blot analyses of restricted genomic DNA isolated from C3H mouse lung and fibrosarcoma cells revealed similar hybridization patterns with the cDNA probe cloned from the fibrosarcoma cells [Raz et al., 1986a, 1988]. Similar results were obtained by Southern blot hybridization of a mouse cDNA clone from 3T3 cells to restricted genomic

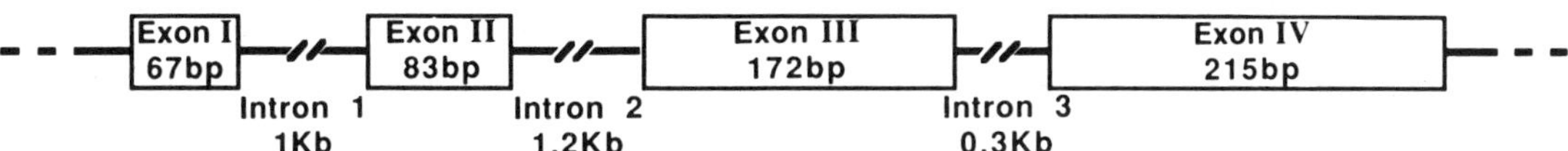

Figure 5 Structure of the chick 14-kd galactoside-binding lectin gene. The gene spans about 3.1 kb, which includes four exons and three introns. Exon I codes for three-amino acid residues, including the initiator methionine; exon II codes for residues 3–30; exon III codes for residues 31–87; and exon IV codes for residues 88–134 [Ohyama and Kasai, 1988]. Note that in Fig. 3 the residue numbers for the regions coded by the different exons are –2–1, 2–29, 30–86, and 87–134.

DNA from two strains of mice (DBA and Balb/c) and from the mouse EHS sarcoma [Wilson et al., 1989], by hybridization of bovine lectin cDNA with restricted human genomic DNA from hepatoma cells [Abbott and Feizi, 1989], and by hybridization of human HL-60 cDNA with endonuclease digested genomic DNA from HL-60 and human placenta [Couraud et al., 1989]. These independent studies indicate that only a single gene exists for the 14-kd lectin and that this gene is neither rearranged nor amplified in tumor cells.

It is interesting that the amino acid sequences suggested to constitute a part of the galactoside-binding site, including residues 68–74 [Paroutaud et al., 1987] and residues 44–82 [Raz et al., 1988] is present in exon II, which codes for residues 31–87 [Hirabayashi and Kasai, 1988].

5. Synthesis of the 14-kd Lectin

Various cells, including mouse 3T3 cells [Roff and Wang, 1983], lung fibroblasts, and endothelial cells [Whitney et al., 1985], human leukocytes [Allen et al., 1986], and tumor cells [Raz et al., 1987b; Lotan et al., 1989a] were found to produce lectins in vitro. The synthesis was monitored by including radioactive amino acids in the culture medium and electrophoretic–fluorographic analysis of the purified lectins (see Fig. 1). Some of the cell types, such as 3T3 fibroblasts, fibrosarcoma, and melanoma cells, produced both the 14-kd and the 34-kd lectins [Roff and Wang, 1983; Raz et al., 1987b; Lotan et al., 1989a], whereas other cells synthesized only the 14-kd lectin [Whitney et al., 1985; Allen et al., 1986].

The mouse fibrosarcoma 14-kd lectin cDNA clone hybridized in Northern blots of total fibrosarcoma RNA with a 750-bp mRNA [Raz et al., 1987a]. A transcript of the same size was detected by Northern blot analysis of lectin mRNA from human lymphoblastoid, myeloid, fibroblastic, monocytic, and hepatoma cell lines [Abbott and Feizi, 1989]. A lectin mRNA of about 700 bp was detected in RNA preparations from normal human placenta, human HL-60 promyelocytic leukemia cells, human SK hepatoma cells, human acute myelogenous leukemia KG1a cells, and human choriocarcinoma JEG-3 cells [Couraud et al., 1989]. A shorter mRNA (570 bp) was detected in neonatal mouse lung, in fibroblasts, in mouse erythroleukemia cell line, and in EHS tumor tissue [Wilson et al., 1989]. The cDNA contains a single open-reading frame of 405 bp to code for the 134-amino acid residues of a lectin with a predicted molecular mass of 14.6 kd [Couraud et al., 1989]. Thus, the mRNA contains short 5′ sequence and a 3′ polyadenylation tail.

In vitro translation of mRNA isolated from 18-day chick embryo skin in a rabbit reticulocyte lysate system, followed by immunoprecipitation with antilectin antibodies, demonstrated the synthesis of a 14-kd protein, which has the same size as the lectin isolated from the tissue. This observation suggests that lectin biosynthesis does not involve the production of a precursor protein, which contains a signal peptide and requires proteolytic processing to be become translocated into the endoplasmic reticulum [Ohyama et al., 1986]. An essentially similar result was obtained when lectin mRNA was selected from mouse fibrosarcoma poly(A)$^+$RNA by hybridization with cloned lectin cDNA, translated in vitro, and the protein immunoprecipitated with antilectin antibodies [Raz et al., 1987a]. The 14-kd lectin mRNA is translated on free ribosomes, like actin, in neonatal mouse and rat lung [Wilson et al., 1989]. This finding excludes the possibility that lectin is inserted cotranslationally into the endoplasmic reticulum membrane or lumen.

A recombinant lectin has been synthesized in *Escherichia coli* transformed with a vector containing the human HL-60 lectin cDNA and an inducible promoter [Couraud et al., 1989]. The lectin, which was isolated by affinity chromatography on lactose-Sepharose, had the same hemagglutinating activity, electrophoretic mobility, and immuno-reactivity as the lectin isolated from natural sources. The initiator methionine of the natural lectin molecules is removed and the NH_2-terminal amino acid (serine in chick and alanine in human, bovine, rat, and mouse lectins) is acetylated. The fully active recombinant lectin produced in *E. coli* contains a nonacetylated alanine at the NH_2-terminus, suggesting that the acetylation is not important for activity.

B. Molecular Biology of the 34-kd Lectin

1. The 34-kd Lectin cDNA and Amino Acid Sequences

The 34-kd lectin cDNA was cloned from a λgt11 expression library derived from mRNA prepared from mouse fibrosarcoma cells [Raz et al., 1987a]. The sequence of a partial cDNA [Raz et al., 1989]. The lectin cDNA was cloned independently from untransformed 3T3 cells [Jia et al., 1987], and fully sequenced [Jia and Wang, 1988]. The rat lectin cDNA was first cloned and partially sequenced by Liu and associates (1985), and then fully sequenced by Albrandt and co-workers (1987), who had not realized at the time that it is a galactoside-binding protein. They identified it as a low-affinity IgE-binding protein (IgEBP) of rat basophilic leukemia (RBL) cells. Subsequently, investigators who sequenced the 34-kd lectin cDNA and performed a computer-assisted search for proteins with sequence homologies, using a protein sequences data bank, found the homology between the mouse 34-kd lectin and the rat IgEBP [Laing et al., 1989; Raz et al., 1989]. That the IgEBP is indeed a galactoside-binding protein and that its binding to IgE is through IgE's carbohydrate chains has now been established [Laing et al., 1989].

Figure 4 shows the deduced amino acid sequences of the mouse 3T3 lectin (designated CBP35 for carbohydrate-binding protein 35), the mouse fibrosarcoma lectin (L-34), and the rat RBL leukemia cell IgEBP. The two mouse sequences differ in ten residue over 263 amino acids (95% homology). The rat lectin differs from both CPB35 and L-34 in 39/363 residues (85% homology). The rat 34-kd lectin sequence predicts a protein of 27,573 Da [Albrandt et al., 1987], and the mouse fibrosarcoma sequence predicts a protein of 30,255 Da [Raz et al., 1989]. The higher apparent size deduced from electrophoretic migration (31 kd for the IgEBP, and 34 kd for the fibrosarcoma lectin) may be the result of the unique two-domain structure of the lectin molecule [Raz et al., 1989].

The sequence of the 34-kd lectin and the predicted secondary structure have some characteristic features. The NH_2-terminal domain (residues 1–27) and the COOH-terminal domain (residues 128–263) have different structural properties, and each exhibits homologies to distinct proteins (Fig. 6). The NH_2-terminal domain contains preponderantly neutral residues organized in β-sheet structure and exhibits neutral hydropathy, being of neither a highly hydrophilic nor a hydrophobic nature. The NH_2-terminus does not contain a membrane-spanning hydrophobic signal sequence. In contrast, the COOH-terminal domain contains charged residues and features alternating reverse turns of both hydrophilic and hydrophobic regions, typical of globular proteins [Albrandt et al., 1987; Raz et al., 1989].

The NH_2-terminal domain includes a high proportion of proline (32/127), glycins (29/127), and alanine (24,127), most of which are found in a repetitive sequence YPGXXPGA between residues 34 and 107. The sequence between residues 12 and 125 shares 33.5% homology with the internal domain of the NH_2-terminus of collagen α-1(II) chain from bovine cartilage. The degree of homology increases to 65.5% if conserved

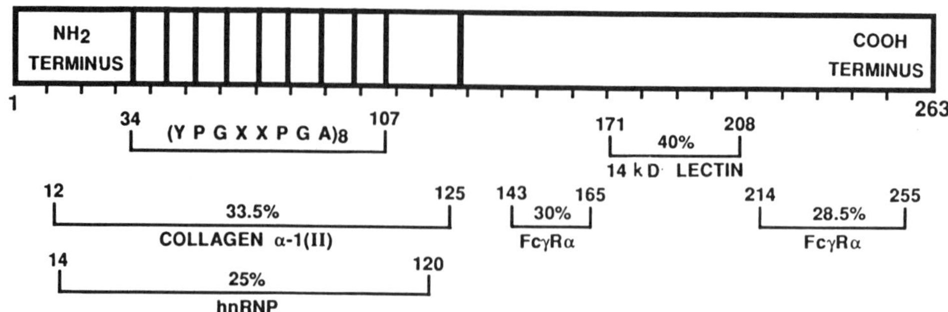

Figure 6 Structure of the 34-kd lectin and indication of homologies of defined domains to other proteins. The NH_2-terminal part of the lectin contains a repetitive sequence YPGXXPGA between residues 34 and 107 (residue 97 is F instead of Y) [Albrandt et al., 1987; Jia and Wang, 1988; Raz et al., 1989]. Residues 12–125 show 33.5% homology to bovine cartilage collagen α-1(II) [Raz et al., 1989], and residues 14–120 show 25% homology to human hnRNP [Jia and Wang, 1988]. Residues 171–208 in the COOH-terminal part of the lectin are 40% homologous to the 14-kd lectin [Raz et al., 1988; Jia and Wang, 1988]. Regions defined by residue 143–165 and 214–255 are homologous to FcγRα1 domain of the mouse macrophage–lymphocyte receptor for the Fc portion of IgG [Albrandt et al., 1987]. The scheme is modified from Raz et al. (1989).

amino acid substitutions are considered [Raz et al., 1989]. Since the repetitive sequence does not contain a glycine residue at every third position, this domain of the 34-kd lectin forms a β-sheet rather than the triple helical structure typical of collagen [Albrandt et al., 1987; Raz et al., 1989]. Nonetheless, the lectin is recognized by collagenase and is partially degraded by this enzyme [Raz et al., 1989]. In addition, the sequence of the NH_2-terminal domain shows homology to the sequences of a group of polypeptides of the heterogeneous nuclear ribonucleoprotein complex (hnRNP), which also exhibit distinct domains with nonuniform distribution of proline and glycine residues [Jia and Wang, 1988]. The detection of CBP35 in cell nuclei in association with hnRNP polypeptides lends further support to the suggestion that the 34-kd lectin is a component of the hnRNP [Laing and Wang, 1988].

Amino acid residues 171–208 in the COOH-terminal domain show 40% homology to residues 44–82 of the 14-kd lectins. Residues (designation according to the 14-kd lectin sequence in Fig. 3) His-44, Asn-46, Pro-47, Arg-48, Phe-49, Val-59, Asn-61, Trp-68, Gly-69, Glu-71, Arg-73, Phe-77, Pro-78, Phe79, and Gly-82, are identical in all the 14-kd lectins, as well as in the aligned sequence of the 34-kd lectin (compare Figs. 3 and 4). The conservation of this sequence in all the galactoside-binding lectins suggests that this region is part of the carbohydrate-binding site [Raz et al., 1988, 1989; Jia and Wang, 1988; Hirabayashi et al., 1989]. The putative carbohydrate-binding domain is flanked on both sides by regions (residues 143–165 and 214–255) that are homologous to FcγRα1, the mouse receptor for the Fc region of IgG [Albrandt et al., 1987; Raz et al., 1989]. Figure 6 shows the putative domain structure of the 34-kd lectin.

2. The 34-kd Lectin Gene

Because the 34-kd lectin gene has not been cloned, its structure is not known at this time. Southern blot analyses of genomic DNA from mouse fibrosarcoma UV2237-IP3 cells and from lungs of normal syngeneic C3H/HeJ mice, using the 34-kd lectin cDNA for hybridization, revealed the presence of a 6- and 3-kb *eco*RI fragment, a 5- and a 2.5-kd *Pst*I fragment, and three *Sac*I fragments of about 5, 6, and 8 kb, respectively [Raz et al., 1988]. There seemed to be no difference in the size and amount of the restricted DNA fragments between the malignant and the normal cells, suggesting that there is no rearrangement or amplification of the lectin gene after malignant transformation. The restriction pattern obtained for mouse and rat DNA was different, suggesting species differences in gene organization [Raz et al., 1988].

The homology of the COOH-terminal domain of IgEBP and FcγR has led to the suggestion that the two proteins are derived from a common ancestral gene [Albrandt et al., 1987]. The homology of the COOH-terminal domain of 34-kd lectin to the 14-kd lectin has led to the proposal that both lectins derived from the same ancestral gene, and the homology of the NH_2-terminal domain to collagen has led to the suggestions that the 34-kd lectin gene evolved as a result of the fusion of the 5′ end of the 14-kd lectin gene with the 3′ end of the NH_2-terminal domain of the collagen α-1 gene for form a chimeric gene and gene product [Raz et al., 1989].

3. *Synthesis of the 34-kd Lectin*

Mouse fibrosarcoma 34-kd lectin cDNA hybridized to a 1.65-kb mRNA. The predicted secondary structure of the mRNA suggests that there are two major stem-loop structures, which correspond to the two domains of the protein [Raz et al., 1989]. In vitro translation of selected lectin mRNA in a rabbit reticulocyte lysate system, followed by immunoprecipitation with antilectin antibodies and analysis by polyacrylamide gel electrophoresis and fluorography, resulted in the detection of a 34-kd protein product [Raz et al., 1987a]. Likewise, in vitro transcription of a full-length cDNA, followed by translation, yielded a 34-kd polypeptide that was immunoprecipitated by the antilectin antibodies [Raz et al., 1989].

Mouse 3T3 CBP35 lectin cDNA was inserted near the 3′ end of the β*Gal* gene in the λgt11 cloning vector, and a fusion protein was produced in *E. coli* infected with the vector and induced with isopropyl thiogalactoside (IPTG). The fusion protein, containing recombinant CBP35 lectin, accumulated intracellularly in an aggregated form, but could be solubilized with 0.1% SDS and 2 mM β-mercaptoethanol and, more interestingly, it could be recovered as a 30-kd protein after V8 protease digestion. This 30-kd protein was able to bind galactosamine-Sepharose and reacted with the anti-CBP35 antibodies, but not with anti βGal antibodies [Jia et al., 1987].

The full-length IgEBP cDNA [Albrandt et al., 1987] hybridized to a 1.1- to 1.3-kb mRNA in various normal rat tissues [Liu et al., 1985; Gritzmacher et al., 1988]. The frequence of IgEBP cDNA clones from RBL cell libraries indicated that the IgEBP mRNA is approximately 0.005% of the cell's total mRNA [Gritzmacher et al., 1988]. In vitro translation of sucrose gradient fractionated mRNA (13–14 S) from the RBL cells in the rabbit reticulocyte lysate system resulted in the synthesis of the IgEBP. This protein was adsorbed to immobilized IgE and could be immunoprecipitated by specific anti-IgEBP antibodies [Liu et al., 1985]. A similar protein was obtained by in vitro translation of mRNA selected by hybridization to partial IgEBP cDNA [Liu et al., 1985]. The mRNA transcribed in vitro from full-length IgEBP cDNA using the SP6 bacteriophage promoter, and subsequently used for in vitro translation, yielded a protein of 28.5 kd, which was adsorbed by IgE, as well as by asialofetuin-Sepharose and could be eluted with lactose [Albrandt et al., 1987; Laing et al., 1989].

Important information on the regulation of the synthesis of the 34-kd lectin in intact cells has been obtained recently. The level of CBP35 lectin in serum-starved, quiescent 3T3 cells increased within 8 hr after the cells were stimulated to proliferate by the addition of serum [Moutsatsos et al., 1987]. The increase in the protein was preceded by increases in the level of a 1.3 kb lectin mRNA, and in nuclear transcription rates of the gene [Agrwal et al., 1989]. The rapid induction by serum of CBP35 gene transcription 3T3 cells is similar to the regulation of several mitogen-activated genes (growth-related immediate early genes) and hnRNP polypeptides [Agrwal et al., 1989].

III. FUNCTIONS OF β-GALACTOSIDE-BINDING LECTINS

The conservation of galactoside-binding lectins during evolution from teleosts to birds and mammals, including humans, their presence in several tissues of diverse species, and their development regulation, strongly imply that they play a role in some physiological processes requiring protein–carbohydrate recognition [Barondes, 1986]. The function(s) of biologically active molecules can be deduced from their localization in tissues and body fluids at different stages of development, as well as from their distribution among subcellular compartments. The galactoside-binding lectins have been localized by immunofluorescence or immunohistochemical methods and immunoelectron microscopic methods using antilectin antibodies or by binding of fluorescently labeled glycoproteins or neoglycoproteins [Barondes, 1986; Gabius et al., 1986b; Monsigny et al., 1988; Moutstatsos et al., 1986; Raz et al., 1984]. Localization was also achieved by subcellular fractionation, followed by analysis of each fraction using quantitation of antibody binding [Sanford et al., 1982] or immunoblotting techniques [Moutstatsos et al., 1986]. By the use of one or more of these methods, galactoside-binding lectins have been identified in normal tissues, including heart, spleen, lung, liver, muscle, thymus, kidney, liver, skin, intestine, stomach, bone marrow, brain, and body fluids [Allen et al., 1987a,b; Briles et al., 1979; Caron et al., 1987; Crittenden et al., 1984; DeWaard et al., 1976; Gritzmacher et al., 1988; Harrison et al., 1984; Oda and Kasai, 1983; Powell, 1980; Teichberg et al., 1975]. The lectins have also been found in tumor biopsies, tumor homogenates, ascites fluids, and cultured tumor cells, including leukemia, melanoma, teratocarcinoma, colon carcinoma, neuroblastoma, renal carcinoma, hepatoma, cervical carcinoma, yolk sac tumor sarcomas, bladder carcinoma, breast carcinoma, lung carcinoma, testicular teratocarcinoma, and ovarian teratocarcinoma [Allen, 1987a,b; Gabius, 1987; Gabius et al., 1986b; Raz et al., 1984; Lotan and Raz, 1988a]. The ubiquitous nature of the lectins suggests that they play a fundamental function in different tissues and may also influence the behavior of malignant cells. Some of the ideas concerning lectin functions are discussed in the following sections.

A scheme showing the hypothetical and factual localization of lectins and their possible interactions with complementary proteins and glycoconjugates is shown in Figure 7.

A. Carbohydrate-Binding Specificity

The discovery that the electric eel electric organ contained lectin activity was accompanied by the establishment of the carbohydrate-binding specificity of this lectin, as well as that of similar lectins from various mammalian tissues for galactosides [Teichberg et al., 1975]. Several subsequent studies have addressed the carbohydrate-binding properties of these

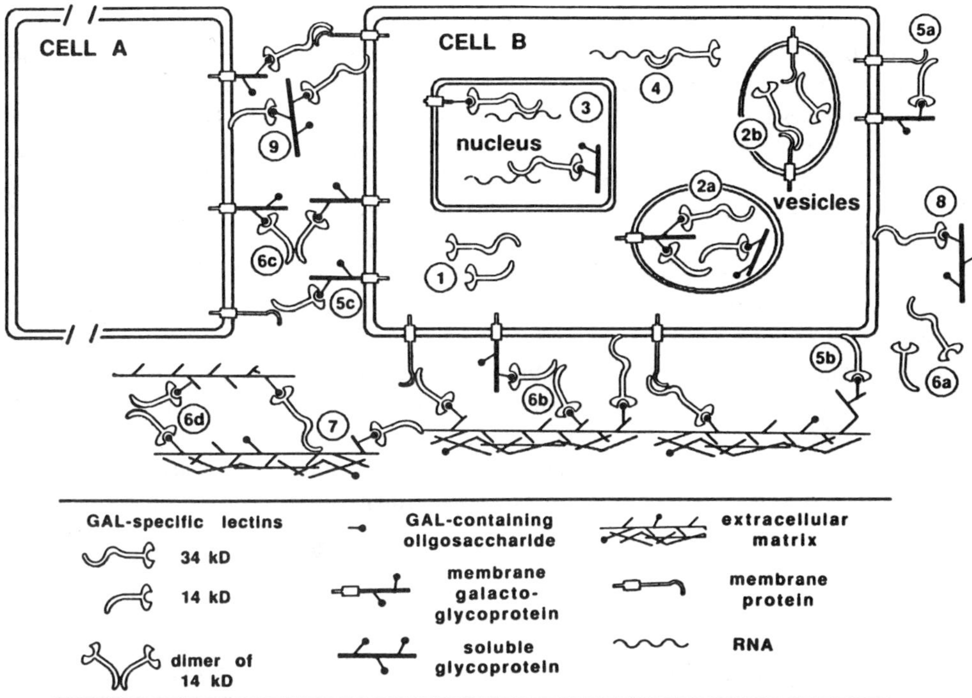

Figure 7 A scheme showing both hypothetical and factual cellular localization of the 14- and 34-kd galactoside-specific lectins and their interactions with complementary glycoconjugates (by protein–carbohydrate recognition), or with proteins (by protein–protein recognition). The lectins have been localized in the cytoplasmic compartment in the cytosol [1] [Barondes, 1986; Moutsatsos et al., 1986; Gritzmacher et al., 1988] or in vesicles [2] (see Fig. 2) [Barondes, 1986; Raz et al., 1984; Raz and Lotan, 1987]. In the vesicles the lectins may interact with soluble or membrane-associated glycoconjugates [2a] or with membrane-associated proteins [2b]. The 34-kd lectin has been localized in the nucleus [3] [Moutsatsos et al., 1986, 1987; Gritzmacher et al., 1988] in a carbohydrate-independent association with RNA and hnRNP [Laing and Wang, 1988]. It may also interact with nuclear glycoconjugates. Furthermore, the 34-kd lectin may be involved in transport of RNA from the nucleus to the cytoplasm [4] [Agrwal et al., 1989]. Both lectins have been detected at the cell surface (see Fig. 2) [Raz et al., 1984; Moutsatsos et al., 1986; Gritzmacher et al., 1988; Lotan et al., 1989a]. Their association with the cell surface may be by lectin–carbohydrate interaction with membrane galactoglycoproteins [5a, 5c, 6b, 6c], by lectin–protein interaction with membrane proteins [5a, 5c], or by some other interaction with the cell membrane [5b, 8]. Lectins have been detected in the extracellular compartment [6a, 6b, 6d, 7] [Barondes, 1986; Catt and Harrison, 1985; Harrison and Catt, 1986]. They may reach the cell exterior by being secreted [Barondes, 1986] as a result of cell breakage and leakage of cytoplasmic content [Briles et al., 1979], or by shedding of surface associated lectin. The cell surface-associated and externalized lectins may mediate cell–cell adhesion [5c, 6c, 9] and cell–ECM adhesion [5b, 6b]. Secreted or otherwise externalized soluble lectin molecules may interact with ECM components through lectin–carbohydrate [6d, 7] and lectin–protein [7] interactions and, thereby, play a role in organization of the ECM. Cell surface-associated lectins may bind soluble glycoproteins [8, 9], and such interactions may also mediate cell–cell adhesion [9]. All statements that are not accompanied by references are hypothetical and some of the referenced statements are not supported by definitive experimental evidence.

lectins in greater detail, using oligosaccharides with defined structure [Abbott et al., 1988; DeWaard et al., 1976; Childs and Feizi, 1979; Leffler and Barondes, 1986; Merkle and Cummings, 1989; Sparrow et al., 1987]. The specificity of the 14-kd lectin for different carbohydrates has been studied extensively for the bovine heart lectin [Abbott et al., 1988; De Waard et al., 1976; Childs and Feizi, 1979; Merkle and Cummings, 1988] and the rat and human lung lectins [Leffler and Barondes, 1986; Sparrow et al., 1987]. The studies by Barondes and his colleagues also compared the specificity of the 14-kd lectin with those of the 18- to 22-kd and the 29-kd (referred to in this chapter as 34-kd) lectins. Table 1 compares the relative activity of saccharides and glycopeptides selected from the aforementioned reports to demonstrate similarities and differences among lectins from different species (rat, bovine, and human) and different members of the galactoside-specific lectin family (the 14-kd and the 34-kd lectins). Overall the specificities of the different lectins are similar, with a few significant exceptions. The 14-kd and 29-kd (34-kd) lectins bind well to oligosaccharides containing terminal nonreducing β-galactose resides, but the penultimate sugar residue is also important, as exemplified by lactosamine (see compound 2 in Table 1), which is a better hapten than either lactose (compound 1) or compound 4. Substitution of lactose on C-2 of the galactose residue with fucose as in

Table 1 Lectin Inhibitory Activity of Saccharides and Glycopeptides Relative to Lactose

Compound	Structure	BL-14[a]	RL-14	HL-14	RL-29	HL-29
1	Galβ1-4Glc	1	1	1	1	1
2	Galβ1-4GlcNac	3.7	5	7.8	7	11.3
3	Galβ1-3GlcNac	4.4	1.2	3.1	2.7	6.3
4	Galβ1-3GalNac	1	0.04	<0.03	0.07	0.6
5	Galβ1-4Glc |1,2 Fucα	0.8	0.4	0.3	1.5	1.8
6	Galβ1-4Glc |1,3 Fucα	–[b]	0.02	<0.04	0.01	<0.02
7	GalNacα1-3Galβ1-4Glc |1,2 Fucα	–	0.3	0.3	25	32
8	NeuAcα2-3Galβ1-4Glc	0.4	0.3	0.6	0.3	1.2
9	NeuAcα2-6Galβ1-4Glc	–	<0.01	<0.03	<0.01	<0.05
10	Bovine erythrocyte gp[c]	8.2	NR[d]	NR	NR	NR
11	Adult erythrocyte LAG[e] gp	NR	16	11	90	146
12	Fetuin oligosaccharide	2	1.2	12	6	21
13	Asialofetuin oligosaccharide	NR	3	12	3	12

[a]BL-14, bovine heart lectin of 14 kd; RL-14, rat lung lectin of 14 kd; HL-14, human lung lectin of 14 kd; RL-29, rat lung lectin of 29 kd; HL-29, human lung lectin of 29 kd.
[b]No inhibition at the highest dose used (Abbott et al., 1988).
[c]gp, glycopeptide.
[d]NR, not reported.
[e]LAG, polylactosaminoglycan (Leffler and Barondes, 1986).
Source: The data on BL-14 are from Abbott et al. (1988) and DeWaard et al. (1976). The data on RL-14, and RL-29 are from Leffler and Barondes (1986), and those on HL-14 and HL-29 are from Sparrow et al. (1987).

compound 5 has only a small effect on binding by the lectins, whereas substitution with fucose on C-3 of the penultimate glucose (compound 6) is not compatible with the binding site requirements. The presence of a galactose residue at the terminal nonreducing position is not a prerequisite for binding to the lectins as substitution with sialic acid (NeuAc) residue in an α2-3 linkage to galactose (compound 8) has only a small effect on binding compared with lactose. In contrast, substitution with NeuAc in an α2-6 linkage to galactose (compound 9) is detrimental for binding to all the lectins. The accommodation of α2-3-linked NeuAc may be the reason for the small difference between the binding of fetuin oligosaccharides and the binding of asialofetuin oligosaccharides (see Table 1). These and other examples not included in Table 1 have led to the conclusion that the major determining factors in the binding of lactosamine to the lectins are unsubstituted an axial C-4 hydroxyl group and CH_2OH on C-6 of the galactose residue, and the equatorial C-3 or C-4 hydroxyl group on the penultimate *N*-acetylglucosamine or glucose residue [Abbott et al., 1988; Leffler and Barondes, 1986; Sparrow et al., 1987]. The lectin-binding site seems to accommodate compounds 2 and 3 with similar affinity. Because the spatial relation of the *N*-acetamido group to the C-4 hydroxyl group and the CH_2OH group at C-6 of the terminal galactose is different in compounds 2 and 3, it has been suggested that the acetamido group may serve to stabilize the ligand–receptor interaction, rather than be directly involved in the binding [Abbott et al., 1988]. The bovine lectin BL-14 exhibits a lower affinity for fucosyllactose substituted with GalNAcα1-3 (compound 7), which is a blood group A-active oligosaccharide, than the rat and human 14-kd lectins. More interesting is the finding that compound 7 is a much better hapten for the RL-29 and HL-29 than lactose or lactosamine. Thus, this compound shows differences in carbohydrate-binding specificity among 14-kd lectins from different species, as well as a difference between the 14-kd lectins and the 29-kd lectins within the same species. In addition, it was found that the 29-kd lectins bind β1-6-branched poly-*N*-acetyllactosaminoglycans consisting of repeating (-3Galβ1-4GlcNAcβ1-) units, such as those found in adult erythrocytes, with a much greater preference than they bind lactosamine (see Table 1). The 14-kd lectins also bind the lactosaminoglycans better than they bind lactosamine [Childs and Feizi, 1979; Leffler and Barondes, 1986], but less than the 29-kd lectins. It is puzzling that the lectins do not bind to human erythrocytes [DeWaard et al., 1976], although they do bind well to the erythrocyte LAG glycopeptides.

The results reported by Barondes and his colleagues were based on measuring the inhibition of the binding of ^{125}I-labeled lectins to immobilized asialofetuin [Leffler and Barondes, 1986; Sparrow et al., 1987]. The converse approach was taken by Merkle and Cummings (1988), who compared the binding of different glycopeptides to immobilized bovine heart lectin. Although the lectin was purified on immobilized asialofetuin, the immobilized lectin did not bind asialofetuin glycopeptides, which were bound well by immobilized *Ricinus communis* agglutinin. However, it did bind sialated or nonsialated poly-*N*-acetyllactosamine chains derived from Chinese hamster ovary (CHO) cells or from BW5147 mouse lymphoma cells. Furthermore, terminal β-galactosyl residues were not essential for binding the chains to the immobilized lectin, adn it was suggested that the length of the poly-*N*-acetyllactosamine chains may be important for the interaction with the immobilized lectin, rather than the terminal saccharide [Merkle and Cummings, 1988]. The finding that the major ligand for the chick embryonic skin 14-kd lectin is a polylactosaminoproteoglycan indicates that this is an important complementary structure for the function of the lectin in vivo [Oda and Kasai, 1984].

B. Function of Lectins at the Cell Surface

The carbohydrate moieties of cell surface glycoconjugates are modulated during differentiation and development [Feizi, 1985; Fukuda, 1985; Thorpe et al., 1988]. It has been suggested that some of these glycoconjugates serve as ligands for cell surface lectins in the process of cell recognition and adhesion in normal and malignant cells [Brandley and Schnaar, 1986; Lotan and Raz, 1988b; Raz and Lotan, 1987; Rutishauser and Jessell, 1988]. The 14-kd lectins have been detected on the surface of some cells including myoblasts [Barondes, 1986], erythroblasts [Harrison and Chesterton, 1980; Harrison and Catt, 1986]; lymphoid cells [Monsigny et al., 1988], mouse melanoma and fibrosarcoma cells (see Fig. 2), and human carcinoma, melanoma, and neuroblastoma cells [Raz et al., 1984]. Likewise, the 34-kd lectin has been detected on the surface of 3T3 fibroblasts [Moutstatsos et al., 1986], rat basophilic leukemia cells [Gritzmacher et al., 1988], and mouse melanoma cells [Lotan et al., 1989a]. The presence of these lectins at the cell surface has led to the proposal that they are involved in cellular interactions [Barondes, 1986; Harrison and Chesterton, 1980; Lotan and Raz, 1988b; Monsigny et al., 1988; Rutishauser and Jessell, 1988; Sharon and Lis, 1989].

1. Cell–Cell Recognition and Adhesion

Direct and indirect experimental evidence supports the contention that galactoside-binding lectins mediate intercellular recognition and adhesion. For example, a lectin purified form anemic rabbit bone marrow agglutinated rabbit erythroblasts, but not mature erythroid cells [Harrison and Chesterton, 1980]. Furthermore, the lectin agglutinated mouse erythroblasts, but not erythroleukemic cells. The erythroleukemia cells could be agglutinated by the lectin after they were induce to differentiate with dimethyl sulfoxide [Godsave et al., 1984]. It was suggested that the bone marrow lectin's function is to aggregate erythroblasts, which are clustered together around a macrophage nurse cell during erythropoiesis. Indeed, immunolocalization has demonstrated that the lectin is associated with the erythroblasts and with adjacent extracellular matrix (ECM) [Harrison and Chesterton, 1980; Godsave et al., 1984]. Since the bone marrow lectin is monomeric in physiological aqueous solutions, it is not clear how it can directly cross-link cell surface glycoproteins, as suggested [Harrison and Chesterton, 1980; Godsave et al., 1984]. It is possible that the monomeric lectin has a cell-binding domain that is distinct from the carbohydrate-binding domain, as observed in some other carbohydrate-binding proteins [Barondes, 1988] and, thus, it can serve as a bridge between cells, despite being monovalent at the galactoside-binding site (see Fig. 7; 5,8). A galactoside-specific 14-kd lectin from thymus was also implicated in cell differentiation. The lectin was found in the thymic epithelium, and it was suggested that it mediates the adhesion of immature thymocytes by binding to cell surface galacto-glycoconjugates on the T cells, thereby preventing them from leaving the thymic cortex until their maturation is completed. Since maturation of T cells is known to be accompanied by an increase in sialation of cell surface components, such a process may decrease the affinity of the binding to the lectin, or eliminate it altogether, allowing the mature cells to move to the thymic medulla or into the circulation [Sharon and Lis, 1989]. One of the first roles attributed to lectins was the mediation of myoblast fusion; however, the publication of conflicting reports supporting and opposing this hypothesis has diminished its appeal [Barondes, 1986].

A role for galactoside-binding lectins in lymphocyte homing has also been proposed. It has been shown that spleen cells bind to sections of mesenteric lymph node venules, probably to endothelial cells, and that the binding was inhibited specifically by β-galactosylated bovine serum albumin. A subpopulation of spleen cells that bears cell surface-associated β-galactoside–specific lectin homed to Peyer's patches after intravenous injection. This homing was inhibited when β-galactosylated bovine serum albumin was coinjected with the cells [Kieda and Monsigny, 1983; Monsigny et al., 1988]. Human peripheral blood leukocytes were agglutinated by human lectin isolated from pancreas and colonic mucosa. These cells are known to contain cell surface blood group Ii-active lactosaminoglycans, which could serve as ligands for the lectins [Allen et al., 1987b]. Human leukocytes also contain a galactoside-specific lectin, which may be secreted and act as an effector molecule by binding to the leukocytes themselves or to other cells [Allen et al., 1986]. Cultured lung alveolar fibroblasts and endothelial cells were found to synthesize the 14-kd lectin and to bind exogenous lectin, whereas alveolar macrophages did not synthesize such a lectin, but did possess binding site for the lectin. It has been suggested that if the fibroblasts or endothelial cells synthesize, secrete, and bind the lectin in vivo, it could mediate cell–cell or cell–ECM interactions. Such interactions could be important for lung remodeling, which occurs during postnatal development [Whitney et al., 1985]. A 14-kd lectin found in mammalian brain aggregated embryonal mouse brain cells in vitro, and this effect was inhibited by galactosides [Caron et al., 1987; Joubert et al., 1987]. That the endogenous brain cell surface-associated lectin mediates cell–cell adhesion was indicated by the inhibition of the self-aggregation of mechanically dissociated mouse brain cells with galactosides [Joubert et al., 1987]. The extent of self-aggregation of the brain cells, as well as the level of brain lectin, exhibited very similar changes during brain development, suggesting that the lectin was functioning in vivo as a mediator of intercellular recognition during embryogenesis and tissue differentiation [Joubert et al., 1987, 1988].

Lectin that was purified from bovine heart bound to cultured baby hamster kidney cells (BHK) and agglutinated them in solution [Stojanovic et al., 1983]. Lectin immobilized on inert glass promoted cell attachment and spreading. Although the lectin and cells are of different species, the findings support the suggested role of lectins in cell adhesion. Ricin-resistant BHK mutants, which exhibit a lower amount of cell surface galactose-containing components, bound less lectin and their agglutination was lower than that of wild-type cells. A difference between the BHK cell surface ligands for ricin and for the bovine 14-kd lectin was detected when it was found that ricin binding to neuraminidase-treated cells increased, whereas the binding of the mammalian lectin was not altered [Stojanovic et al., 1983]. This observation is compatible with the finding that the 14-kd lectin can bind sialyllactose (NeuAcα2-3Galβ1-4Glc) with an affinity that is similar for lactose (see Table 1) [Sparrow, 1987].

A role for lectins in intercellular adhesion requires that complementary, carbohydrate-containing molecules be present on the surface of adjacent cells. The identity of the physiologically relevant endogenous ligands for the cell surface lectins is known for only a few of the cell types for which lectins have been implicated in cell adhesion. For example, a polylactosaminoproteoglycan was identified as the endogenous ligand recognized in chick embryonic skin by externally added chicken skin lectin using photoaffinity-labeling techniques [Oda and Kasai, 1984]. Immobilized carboxamidomethylated rat lung lectin bound endogenous glycoproteins extracted with 1% Triton X-100 from lung homogenates. The glycoproteins included three components of 160–200 kd and a smaller component of 75 kd [Powell and Whitney, 1984]. Some lactosyl structures, including poly-*N*-acetyllactosamine-containing oligosaccharides, appear on subsets of developing neural cells, and their expression is concurrent with the expression of the 14-kd and 34-kd lectins in the neural cells [Reagan et al., 1986; Dodd et al., unpublished observation cited in Thorpe et al., 1988]. In sensory systems, the lectins are found in the cell bodies and terminals of the subset of sensory neurons that express lactosyl structures. This suggests that lectins and complementary lactose-containing glycoconjugates interact in or on the sensory neurons that synthesize both. Preliminary evidence suggests that the lectins are released from the sensory neurons and interact selectively with ligands on the surface of sensory and spinal neurons. The externalized lectins may mediate adhesive interactions and participate in the organization of sensory projections in the developing spinal cord [Reagan et al., 1986; Rutishauser and Jessell, 1988].

Poly-*N*-acetyllactosamine-containing oligosaccharides have been implicated in cell interactions in the preimplantation embryo following the finding that treatment of decompacted 8- to 16-cell mouse embryo cells with the enzyme endo-β-galactosidase, which cleaves such oligosaccharide chains, prolonged recompaction fivefold. It was suggested that endogenous carbohydrate-binding molecules, which bind poly-*N*-acetyllactosamine with high affinity, are present on the surface of the embryo cells and mediate cellular interactions [Rastan et al., 1985]. Indeed, the compaction process includes a late, Ca^{2+}-independent stage at which the putative lectin could function. The presence of galactoside-binding lectins in early mouse embryo has not been established; however, the 14-kd lectin was found in mouse embryonal carcinoma cells, and its level increased after retinoic acid-induced differentiation to primitive endoderm [Lotan et al., 1989b]. This lectin has been detected on the surface of the differentiated embryonal carcinoma cells [Amos et al., unpublished]. Poly-*N*-acetyllactosamine-containing oligosaccharides have been found to be prominent also in the early chick embryo, where they show remarkable changes in distribution in cells and ECM during development. It has been suggested that these structures can interact with endogenous lectins and play important role in morphogenetic events during embryogenesis [Thorpe et al., 1988]. The poly-*N*-acetyllactosaminoglycans were found in the embryo at apical aspects of epithelia and in the ECM [Thorpe et al., 1988], other studies have detected such carbohydrates also in integral membrane proteins [Feizi and Childs, 1987]. These findings suggest functions for cell surface lectins or secreted lectins in cell–cell and cell–ECM adhesion. The 14-kd lectin has been localized in situ in 13-day-old chick embryonic epidermis in the desmosomes, plasma membrane, and basement membrane, suggesting a role in cell–cell adhesion [Hirano et al., 1988].

The presence of galactoside-binding lectins between 25 and 45 kd has been observed in the earliest stages of chick embryo development, and the lectin inhibitor thiodigalactoside prevented aggregation of cells of the extraembryonic endoderm of the primitive chick embryo [Cook et al., 1979]. A further support for a role of the lectin in cell adhesion has come from the observations that decreased intercellular adhesion of aggregated extraembryonic endoderm cells from the early chick blastoderm during cavitation is accompanied by a release of lectin from the cells to the extracellular compartment [Milos and Zalik, 1986] and the finding that calcium-independent adhesion of extraembryonic endoderm cells is inhibited by the blastoderm β-D-galactoside-binding lectin and by β-galactosidase [Milos and Zalik, 1983]. It is unclear why the externalized lectin does not induce aggregation, but rather, inhibits it. One possible explanation is that the lectin is externalized as a complex with the endogenous carbohydrate-containing glycoconjugates, or that it is rapidly inactivated by oxidation in the extracellular compartment. Two galactoside-specific lectins of 6.5 kd and 70 kd have been purified from the chick blastoderm, and they are presumed to mediate the aforementioned interactions [Zalik et al., 1983, 1987]. Although they differ in molecular mass from the 14-kd and the 34-kd lectins, these chick lectins resemble the 14-kd–34-kd family of lectins in their extraction in aqueous buffers containing lactose, Ca^{2+}-independent action, requirements for thiols, and carbohydrate-binding specificity. It is not known, however, whether they share any antigenic determinants or sequence homology.

The formation of tumor metastases by circulating cancer cells correlates with an increased tendency of the cells to aggregate with other tumor cells or with host cells [Nicolson, 1984; Raz and Lotan, 1987]. Various tumors and tumor cell lines contain galactoside-specific lectins [Gabius et al., 1986b; Gabius, 1987; Raz and Lotan, 1981; Lotan and Raz, 1989a], and there is indirect evidence that these lectins may be involved in mediating intercellular adhesion of metastatic tumor cells [Raz and Lotan, 1987]. The 14-kd and 34-kd lectins were found to be present on the surface of tumor cells [Raz et al., 1984; Lotan et al., 1989a] and their levels were higher on the surface of tumor cells exhibiting a higher metastatic potential than on cells exhibiting a low metastatic potential [Raz and Lotan, 1987]. The binding of monoclonal antilectin antibodies to metastatic cells decreased homotypic aggregation in vitro and suppressed the ability of the cells to form lung metastases after intravenous injection in the tail vein of syngeneic mice. These results implicate the tumor cell surface lectins in cell adhesion and metastasis [Raz and Lotan, 1987].

2. Interactions of Cells with the Extracellular Matrix

The presence of lectins in the ECM has been demonstrated in several normal tissues [Barondes, 1986; Catt and Harrison, 1985; Harrison and Catt, 1986] and tumor biopsy specimens [Gabius et al., 1986a]. It is not clear how the lectins arrive at the extracellular site. Are they secreted, shed from the cell surface, or released as a result of cell breakage or turnover? Lectins may be produced at one site and be transported to another site in vivo, as indicated by the localization of the 14-kd lectin in the ECM of chick embryo developing kidney cells, which do not contain intracellular lectin and do not seem to be able to synthesize the lectin in culture [Didier et al., 1988]. The presence of lectin in human sera, ascites fluid, cyst fluid, and milk, detected by immunodiffusion, indicated that the lectin is re leased by cells in vivo and may be brought by the circulation to various organs. However, it is unknown whether the externalized lectin is active or inactive [Allen et al., 1987b].

In late fetal and early neonatal stages of rabbit development the 14-kd lectin was found to be abundant both within and around rounded fibroblasts that are synthesizing and secreting ECM components. This observation has led to the proposal that the lectin plays a role in the creation of an appropriate extracellular environment for differentiation [Catt et al., 1987]. In the adult rabbit bone marrow, the lectin was localized in the ECM, and it was proposed that the lectin mediates the association of erythroblasts with the hematopoietic microenvironment [Harrison and Catt, 1986]. Chicken muscle 14-kd lectin (CLL-I) is externalized late in muscle development, after myotube formation, and it was suggested that the extracellular lectin may participate in the organization of extracellular materials around myotubes by cross-linking them through their galactoside residues [Barondes, 1986].

The ECM contains several glycoprotein constituents that could serve as ligands for the lectins. A major component of the basement membrane ECM is laminin. This glycoprotein contains galactoside residues on *N*-linked poly-*N*-acetyllactosamine side chains [Arumugham et al., 1986] and is recognized by galactoside-specific plant lectins, such as that of *Griffonia simplicifolia* [Shibata et al., 1982]. It was found that immobilized bovine heart 14-kd lectin was able to bind laminin through laminin's poly-*N*-acetyllactosamine side chains [Cummings, 1989]. Both 14-kd and 31-kd lectins found in human colon carcinoma cells bound to laminin-Affigel and were eluted with lactose, but not with 0.5 M NaCl [Ohannesian et al., unpublished data], suggesting that these lectins bind laminin through galactosyl residues that are known to be present in the *N*-linked oligosaccharides of this matrix glycoprotein. Fetal fibronectin, another ECM glycoprotein component, also contains poly-*N*-acetyllactosamine side chains [Zhu and Lane, 1985] and could serve as a complementary partner for embryonal lectins in cell–matrix adhesion or in ECM organization.

Since most of the lectin is found in the cytoplasm, it is plausible to assume that wounding of tissues would release lectins from broken cells to the intercellular space at the wound site. This lectin might then be involved in the wound-healing process by participating in the organization of ECM components produced by connective tissue cells. It was argued that the sensitivity of the lectins to oxidation suggests that they act intracellularly [Wilson et al., 1989]. However, the 14-kd lectin from bovine heart remained active in the absence of reducing agents after it has been immobilized on agarose [Merkle and Cummings, 1988]. Thus, it is possible that by associating with some membrane proteins or components of the ECM the lectin can be protected from inactivation.

The 67-kd component of the elastin receptor on the surface of chondroblasts contains a galactoside-binding site as well as a site for protein–protein interaction with elastin. Lactose inhibited the formation of elastic fibers by chondroblasts and released the 67-kd lectin from the cell surface, without altering fibronectin organization. Therefore, it has been suggested that this lectin plays a role in the organization of elastic fibers in the ECM [Barondes, 1988; Hinek et al., 1988]. The mechanism by which the 67-kd elastin receptor organizes the matrix is unknown. It may use the galactoside-binding site for attachment to highly glycosylated microfibrillar proteins that are critical for elastic fiber assembly, and the elastin-binding site for presenting tropoelastin on the cell surface at an orientation that leads to fiber formation [Barondes, 1988; Hinek et al., 1988]. In addition, it has been reported that the 67-kd elastin-binding protein binds to laminin-Affigel and is recognized by antibodies prepared against the laminin receptor from human melanoma cells, suggesting that the 67-kd elastin receptor is capable of binding laminin and is related to the laminin receptor [Mecham et al., 1989]. Furthermore, antibodies to the 14-kd lectin from rat lung recognized the laminin receptor from human melanoma. As antibodies against the melanoma laminin receptor recognized the 14-kd lectin as well as the elastin receptor, it was suggested that the common epitope shared by the 14-kd lectin, the 67-kd elastin receptor, and the 67-kd laminin receptor is the conserved galactoside-binding site [Mecham et al., 1989]. Thus, the cell surface-associated 67-kd lectin can play a role in organization of the ECM by binding to cell surface galactosides and either elastin or laminin. The exact mechanism by which the 67-kd elastin/laminin receptor–lectin acts remains to be elucidated, as does the importance of the observation that the binding of galactoside decreases the affinity of the 67-kd lectin for elastin and of the laminin receptor to laminin, as evidenced by the ability of lactose to elute the two receptors from either immobilized laminin or immobilized elastin [Mecham et al., 1989]. It is possible that the 70-kd galactoside-binding lectin isolated from the early chick embryo is related to the 67-kd elastin and laminin receptors. It seems to be a peripheral protein in that it remained associated with the particular cell fraction after an initial extraction of the 6.5-kd lectin in the presence of lactose and required a prolonged extraction (60 hr) of the particulate material with lactose before it was solubilized [Zalik et al., 1983].

The ability to grow without anchorage to a solid substrate is one of the hallmarks of malignant transformation. It has been suggested that the ability of tumor cells to grow in semisolid agarose is facilitated by the expression of cell surface lectins, because a monoclonal antibody, which recognized both lectins, was capable of inhibiting the growth of several tumor cell lines in agarose, presumably by binding to cell surface-associated lectins [Lotan et al., 1985]. It has been proposed that the cell surface lectin can bind to the agarose, which is a polymer of galactose, and thus provide anchorage to the suspended tumor cells [Lotan et al., 1985].

3. *Cell Migration and Chemotaxis*

One of the roles suggested for carbohydrate structures of the poly-*N*-acetyllactosamine series is guiding cell migration or influencing cell orientation during morphogenetic events in the embryo by interacting with complementary structures (e.g., lectins) on cells or matrices [Thorpe et al., 1988]. Evidence for this hypothesis is not yet available. Results of immunolocalization of the 6.5-kd lectin from gastrulating chick blastoderm in the early chick embryo suggested that it is involved in cell migration. The lectin was found to be expressed as cells migrate from the ectoderm inward into the streak and emerge laterally to form the endoderm. Localized release of the lectin could modulate the transitory adhesions that occur as cells relocate. Furthermore, the lectin molecules tend to aggregate after they are released from cells, and the aggregated lectin could provide a substratum for cell spreading and migration [Zalik et al., 1987].

The elastin peptide VGVAPG and the laminin B1 chain peptide LGTIPG have been identified as the recognition sites for the 67-kd elastin and laminin receptors, respectively, and the two peptides induced a similar chemotactic response in ligament fibroblasts and human melanoma cells [Mecham et al., 1989]. Preincubation of the fibroblasts or tumor cells with lactose inhibited the movement of the cells to the peptides. It is possible that the cell surface receptors for elastin and laminin are mediating the chemotactic response, and that lactose released the receptors from the cell surface and thereby blocked chemotaxis.

4. *Mitogenic Activity and Immunological Modulation*

Galactoside-binding lectins of the 14-kd family bind to and stimulate or activate lymphoid cells and macrophages. The chicken lectins from liver and intestine were mitogenic to nonadherent mouse spleen cells. The responding cells were most likely of the B-cell lineage, since they were Thy-1–negative and because a similar effect was observed with spleen cells from nude mice (deficient in T cells), or from their heterozygote littermates. This effect was inhibited by asialofetuin and by lactose [Lipsick et al., 1980]. Likewise, a lectin from chick embryonic kidney was mitogenic to mouse lymph node lymphocytes [Pitts and Yang, 1980, 1981]. The ability to exert a mitogenic effect was also observed with the electric eel lectin. This lectin was found to bind to rabbit lymphocytes and enhance thymidine incorporation into DNA [Levi et al., 1983]. Furthermore, both the electric eel lectin and the mouse thymic lectin were found to bind and agglutinate selectively the immature subpopulation of mouse thymocytes, and it was suggested that the lectin, which is produced and secreted by thymic epithelial cells, can stimulate maturation of immature thymocytes into suppressor cells [Levi and Teichberg, 1983]. Since mammals contain galactoside-binding lectins in the spleen and thymus [Allen et al., 1987a; Briles et al., 1979; Levi and Teichberg, 1985; Monsigny et al., 1988], it has been suggested that the endogenous lectins may play a role in the stimulation of specific subpopulations of lymphoid cells in vivo [Lipsick et al., 1980]. This hypothesis needs to be substantiated in a syngeneic mammalian experimental model, because a chicken thymus lectin, which was found to bind to chicken thymocytes and agglutinate them, failed to induce mitogenic response or maturation into suppressor cells [Levi and Teichberg, 1985].

Purified placenta 14-kd lectin induced the production of cytotoxic factor(s) in cultured mouse macrophagelike cell line J774.1 and in human peripheral blood monocytes. This effect of the lectin was distinguished from that of bacterial lipopolysaccharide in that it was not inhibited by polymyxin B. Reports on the importance of cell surface *N*-linked oligosaccharides in the interactions of cytotoxic T cells and NK cells with target tumor cells and the demonstration of the presence of lectins on the surface of different lymphoid cells suggest that lectins may also mediate effector cell–target cell adhesion [Monsigny et al., 1988].

It was suggested that endogenous lectin may play a physiological role in activation of monocytes–macrophages to release cytotoxins [Kajigawa et al., 1986]. Because resident and thioglycolate-activated murine peritoneal macrophages contain both 14-kd and 34-kd galactoside-specific lectins [Lotan, unpublished], they may be involved in self-activation. This process might be triggered by a signal that would release the lectins into the cell exterior, followed by binding to cell surface glycoconjugates. Indeed, alveolar macrophages bind exogenous rat lung lectin produced by fibroblasts or endothelial cells. The release of such a lectin in the vicinity of lung macrophages during lung remodeling could stimulate them to phagocytose tissue remnants [Whitney et al., 1985]. Another way in which lectins may be involved in phagocytosis has been suggested for the process of erythropoiesis. When the maturing erythroblast undergoes enucleation, the extruded nuclei become surrounded by immunoreactive lectin, which is presumably membrane-associated, whereas the reticulocyte contains mainly an intracellular lectin. It is possible that the lectin coating the nucleus is involved in the subsequent phagocytosis of the nucleus by macrophages [Harrison and Catt, 1986] because it is well established that plant lectins stimulate phagocytosis [Sharon and Lis, 1989].

The mechanism by which the binding of the galactoside-specific lectins can modulate cell growth or cytotoxic activity is unknown. It is possible that the initial membrane events are similar for plant and vertebrate lectins. By extrapolation from studies on plant lectins, such as the galactoside-specific *R. communis* agglutinin, it is possible that the binding of vertebrate lectins to cell surface glycoconjugates causes a decrease in the mobility of galactose-containing oligosaccharides, followed by patching, capping, and signal transduction [Sharom and Ross, 1985].

Yet another effect on the immune response was attributed to the electric eel lectin. It inhibited the development of experimental autoimmune myasthenia gravis in rabbits injected with acetylcholine receptor, and its injection to myasthenic rabbits has led to a complete recovery. It has been suggested that the lectin stimulated suppressor cells that regulate immune response to self-antigens [Levi et al., 1983]. However, an analysis of 14 strains of mice failed to reveal a correlation between lectin level and mouse susceptibility to myasthenia gravis [Levi adn Teichberg, 1984].

It has been proposed that the stimulation of cell proliferation by lectins may not be restricted to lymphoid cells, but could also include cells in various embryonal tissues. This suggestion was based on the finding that the endogenous lectin is associated with and binds to the surface of embryonic cells and that lectin levels increase during development, when active cell proliferation occurs [Lipsick et al., 1980]. Indeed, the 14-kd galactoside-binding lectin, present in adult rat brain extracts, may be responsible for the stimulation of aggregation and growth of dissociated cultured mouse brain cells [Joubert et al., 1985]. Interestingly, the 14-kd lectin was prominent in adult rabbit tissues in areas surrounding proliferating and differentiating cells in intestinal crypts, hair follicles, and bone marrow [Catt and Harrison, 1985; Harrison and Catt, 1986].

Although not demonstrated directly for galactoside-binding lectins, the possibility should be considered that cell–cell and cell–ECM lectin-mediated adhesion might lead to growth inhibition in certain cell types, because it has been shown that the amount of cell surface galactosyl residues increases when murine fibroblasts reach confluence and become "contact-inhibited" [Nicolson and Lacorbiere, 1973], and immobilized *N*-linked plasma membrane glycopeptides, terminating with galactoside residues, inhibited the growth of human diploid fibroblasts [Wieser and Oesch, 1986].

C. Functions of Lectins in the Cytoplasm

Although most lectin localization studies have demonstrated that lectins are present in the greatest abundance in the cytoplasm (see Fig. 2) [Allen et al., 1987b; Barondes, 1986; Briles et al., 1979], there is no strong evidence for a function of the 14-kd lectins in this cellular compartment. In some cells, the lectins were localized in unidentified particles (see Fig. 2) [Raz et al., 1984], or in secretory vesicles [Barondes, 1986]. Whether they are associated with membrane glycoconjugates or are soluble in the lumen of the vesicles (see Fig. 7; 2a,2b) is not known. After homogenization of rat lung tissue the 14-kd lectin has been detected by antibodies in microsome, mitochondria, and plasma membrane fractions. Incubation with lactose decreased the amount of lectin associated with the membrane fractions, suggesting that the lectin is binding to membrane glycoconjugates, possibly during homogenization [Briles et al., 1979; Sanford et al., 1982]. Interestingly, the lungs of immature rats contained a large amount of soluble lectin that was extracted even without including lactose during homogenization, suggesting that the endogenous membrane glycoconjugates acquire lectin-binding determinants only after maturation. An analysis of the subcellular localization of IgEBP, the rat L-34 homologue, also demonstrated that it is prominent in the cytoplasm [Gritzmacher et al., 1988].

In the cytoplasmic compartment, the lectins may be involved in the transport of galactoside-containing glycoconjugates between different subcellular organelles. Alternatively, the lectin could present galactoglycoprotein acceptors in a multivalent form to sialyltransferases, as it has been shown that purified 14-kd lectin can stimulate β-D-galactoside α2-6-sialyltransferase of bovine colostrum in a cell-free system [Scudder et al., 1982]. The lectins of 3T3 fibroblasts (13.5, 16, and 35 kd) did not possess either a β-galactosidase or sialyltransferase activities [Roff and Wang, 1983]; however, the possibility that the lectins are enzymes for which assay conditions have not been established cannot be excluded.

It has been proposed that a lectin found in secretory granules of chicken intestinal goblet cells together with galactoside-containing mucin, may bind the mucin and organize it for secretion or after the complex has been secreted [Barondes, 1986]. It is also possible that the cytoplasmic lectin has no function inside the cell, and that the cytoplasm serves only as a reservoir for lectin destined for externalization at specific times during development, as was found in muscle cells [Barondes, 1986], or during wounding.

D. Functions of Lectins in the Nucleus

Galactoside-specific lectins have been detected in cell nuclei by indirect methods, including immunofluorescent staining with anti-14-kd lectin antibodies [Beyer and Barondes, 1980; Childs and Feizi, 1980; Carding et al., 1985], anti-CBP35 (L-34) lectin antibodies

[Moutsatsos et al., 1986, 1987], and binding of lactosylated fluorescent bovine serum albumin [Seve et al., 1985, 1986]. It has been proposed that the nuclear lectins may be involved in transport of glycoproteins or modulation of the activity of glycosylated nuclear enzymes, such as DNA polymerase [Seve et al., 1985]. The presence of neoglycoprotein-binding material in nuclear regions where RNA concentration was high suggested that the lectins may be involved in the synthesis, processing, and transport of RNA [Seve et al., 1986].

More direct evidence for a nuclear function has been reported for CBP35 the 34-kd lectin of 3T3 cells, which has been found in the nuclei of 3T3 mouse cells in association with the hnRNP complex [Laing and Wang, 1988]. This complex or particle includes several polypeptides, as well as hnRNA [Dreyfuss et al., 1988]. It was suggested that the site for interaction between CBP35 and the other hnRNP components is located in the NH_2-terminal domain of the molecule, which shows sequence homology to other hnRNP proteins [Laing and Wang, 1988; Jia and Wang, 1988]. CBP35 could be released from the nucleus by ribonuclease A treatment, suggesting that it is associated with RNA. It is not clear whether this is a direct binding or one mediated by other hnRNP polypeptides because CBP35 does not contain the eight-amino acid stretch K(R)GF(Y)G(A)F(Y)VxF(Y), which is highly conserved in RNA-binding proteins from various sources [Dreyfuss et al., 1988]. The identification of CBP35 as a component of the hnRNP complex implies that it plays a role in the functions attributed to hnRNP, including processing, packaging, and transport of mRNA from the nucleus to the cytoplasm [Agrwal et al., 1989]. Preliminary experiments in a cell-free system indicate that CBP35 is cotransported with mRNA from isolated nuclei [Laing et al., unpublished results cited in Agrwal et al., 1989]. The rapid induction of CBP35 transcription after mitogenic stimulation of quiescent 3T3 cells, which involves transition of the cell to the proliferative state, is compatible with its purported function in mRNA processing and transport.

The association of CBP35 with the other components of the hnRNP complex does not involve the carbohydrate-binding site, which is located in the COOH-terminal domain, as it was free to mediate the binding of the CBP35–hnRNP complex to immobilized galactoside [Laing and Wang, 1988]. It is unclear whether the ability to bind galactosides is contributing to the functions of CBP35 discussed earlier or to other functions. The presence of galactoside-containing glycoconjugates that could serve as endogenous ligands for the lectin in nuclei can be inferred from the reports on the presence in nuclei of ligands for the galactoside-specific plant lectin *R. communis* agglutinin [Burrus et al., 1988]. Some eukaryotic transcription factors [Jackson and Tjian, 1988] and nuclear cytoplasmic proteins [Hart et al., 1988] contain *O*-linked *N*-acetylglucosamine residues. Such proteins can serve as acceptors of galactose residues in a cell-free enzymatic glycosylation with galactosyltransferase. Although there is no evidence that they are galactosylated in the intact cell, this possibility cannot be excluded.

E. Lectin Structure–Function Relationships

Since the 14-kd lectin has been detected in the cell cytoplasm, in association with the cell surface membrane, and in the extracellular compartment, one would have expected to find some structural features that might explain the mechanism of translocation and secretion of this molecule. The high β-structure content is a characteristic of extracellular proteins, and this feature is compatible with the localization of some of the 14-kd lectin in the extracellular compartment. *N*-Acetylation is commonly found among intracellular proteins, and this is compatible with the cytoplasmic localization of the lectin in various cells. However, the absence of a cleavable signal peptide in the NH_2-terminus cannot explain the mechanism by which lectin molecules are secreted into the extracellular compartment. Seven of the first 15 amino acid residues at the NH_2-terminal are hydrophobic, and it has been suggested that this region plays a role in a unique secretion mechanism. Furthermore, the finding that the lectin mRNA is translated on free ribosomes and not on endoplasmic reticulum-bound ribosomes [Wilson et al., 1989], the site of synthesis of membrane-associated and secreted molecules, excludes the suggestion of the presence of an internal signal peptide for cotranslational insertion into the endoplasmic reticulum [Ohyama and Kasai, 1988; Wickner and Lodish, 1985]. These findings imply that lectin secretion must involve either a translocation of lectin into the endoplasmic reticulum after cytoplasmic translation, or an as yet undiscovered mechanism [Wilson et al., 1989].

Amino acid residues 44–82 in the 14-kd lectin (see Fig. 3) and the corresponding residues 171–208 in the 34-kd lectin are thought to be a part of the galactoside-binding site. Within this sequence is included the only tryptophan residue of the lectin (Trp-68). The presence of a tryptophan residue in the galactoside-binding site of the electric eel lectin has been proposed by Levi and Teichberg (1981) on the basis of their finding that lactose enhanced the fluorescence of the lectin and protected the lectin against inactivation and loss of fluorescence by oxidation. Furthermore, analysis of the pH dependence of lectin fluorescence in the absence and presence of lactose predicted the presence of a carboxyl residue with an abnormal p*K* in the binding site [Levi and Teichberg, 1981]. Indeed, there are two glutamic acid residues, Glu-71 and Glu-74, near Trp-68, and their proximity may affect their ionization, such that one of them will have an abnormal p*K* [Paroutaud et al., 1987]. However, in vertebrate lectins, other than the eel lectin, oxygen oxidizes the cysteine residues and causes lectin inactivation, without affecting the tryptophan residue [Whitney et al., 1986]. Thus, the involvement of Trp-68 in carbohydrate binding is not certain.

The 14-kd lectins from the electric eel [Levi and Teichberg, 1981], the bovine heart lectin [Briles et al., 1979], human lung [Powell, 1980; Allen et al., 1987b], and heart [Childs and Feizi, 1979] form dimeric molecules in nondenaturing buffers. A dimeric lectin should possess two carbohydrate-binding sites and could mediate intercellular (see Fig. 7; 6d), and cell–matrix adhesion (see Fig. 7; 6b), or cross-link ECM components (see Fig. 7; 6d). It is surprising that monomeric monovalent 14- and 34-kd lectins agglutinate red blood cells, apparently without undergoing dimerization [Roff and Wang, 1983; Harrison et al., 1984]. The monomeric rabbit bone marrow lectin exhibited at least a ten times lower specific hemagglutinating activity than dimeric lectins from several mammals [Catt et al., 1987]. Because the 34-kd lectin has a unique NH_2-terminal domain, which resembles collagen and is recognized as a substrate by collagenase [Raz et al., 1989], it is possible that it is also recognized by some collagen-binding proteins (collagen receptors) on the cell surface. A protein–protein interaction will leave the lectin's carbohydrate-binding site free for interaction with glycoconjugates on other cells or on extracellular glycoconjugates (see Fig. 7; 8). Such a possibility could also explain the ability of monomeric 14-kd rabbit bone marrow lectin to aggregate erythroblasts [Harrison et al., 1984]; however, there is no evidence for protein–protein recognition site on the 14-kd lectin molecule.

Only limited information was obtained on the function(s) of the 34-kd lectin. It is possible that this lectin shares all or a part of the activities described for the 14-kd lectin because both types of lectin have a similar affinity for some oligosaccharides (see Table 1). On the other hand, the 34-kd lectin exhibits a higher affinity for poly-*N*-acetyllactosaminoglycans and may have some unique functions not shared with the 14-kd lectin.

This may also be true for the 18- and the 22-kd lectins found in rat and human lung, respectively, which exhibit carbohydrate-binding specificities that are distinct from those of the 14-kd and the 34-kd lectins [Leffler and Barondes, 1986; Sparrow et al., 1987].

F. Development, Differentiation, Malignant Transformation, and Lectin Expression

The carbohydrate moieties of specific glycoconjugates are modulated during embryonal development, and these changes are thought to be important in the regulation of cell growth, differentiation, and intercellular interactions [Brandley and Schnaar, 1986; Feizi, 1985; Fukuda, 1985; Rutishauser and Jessell, 1988; Sharon and Lis, 1989; Thorpe et al., 1988]. Lectins may be involved in some of the carbohydrate-mediated processes as complementary partners in specific interactions. Thus, the regulation of such processes could be either by changes in carbohydrate structure, caused by alterations in glycosyltransferases [Schachter, 1986] or by modulation of the expression of carbohydrate-binding proteins. Indeed, changes in lectin activity during development of vertebrates have been documented in chicken, rabbit, and rat tissues [Barondes, 1986; Catt and Harrison, 1987; Levi and Teichberg, 1984]. That some of the changes in lectin expression correspond to changes in complementary carbohydrates is exemplified by the development of rat sensory neurons, in which 14- and 29-kd lectins were detected first on embryonic day 14 and 16, respectively, and continued to be expressed at subsequent stages, with the 14-kd lectin level declining somewhat after postnatal day 5. These changes were coincident with increases in the levels of lactoseries glycoconjugates [Regan et al., 1986]. There is almost no information on the regulation of the in vivo expression of the lectins at the mRNA level (e.g., in situ hybridization analysis). However, some information is available from analyses of tissue explants and cultured cells. Thus, a decrease in 14-kd lectin mRNA was observed in chick skin explants after treatment with retinol, which caused mucous metaplasia [Oda et al., 1989]. This was accompanied by a decrease in the level of the lectin in the tissue, as revealed by immunofluorescence labeling of skin sections. Interestingly, the levels of mRNA and immunofluorescent labeling in the dermis were not altered by retinol; therefore, it was suggested that the regulation of lectin expression by retinol was restricted to the keratinocytes [Oda et al., 1989].

Malignant transformation can result in changes in carbohydrate structure [Fukuda, 1985; Hakomori, 1989]. Likewise, changes have been observed in the level of lectins when transformed cells have been compared with their untransformed counterparts [Crittenden et al., 1984; Lotan and Raz, 1988a; Raz et al., 1987b]. More recently, the increased level of lectin mRNA was found to accompany malignant transformation [Raz et al., 1988; Agrwal et al., 1989]. The level of mRNA for the 14-kd lectin was about twofold higher in virally transformed 3T3-SvPy cells than in untransformed 3T3 cells. A more dramatic increase was found for the 34-kd lectin mRNA; it was higher (5- to 50-fold) in virally transformed 3T3-SvPy and NRK-RSV cells than in their untransformed counterparts, without evidence for gene amplification or rearrangement [Raz et al., 1988]. Likewise, the 34-kd lectin mRNA level in the virally transformed 3T3-KiMSV was higher than in 3T3 cells [Agrwal et al., 1989]. The latter study determined that the increased mRNA level was due to both an increased transcription rate and a higher stability of the mRNA in the transformed cells. Furthermore, among related fibrosarcoma sublines, the more metastatic one contained two- to fourfold higher levels of mRNA for the 14-kd and the 34-kd lectins [Raz et al., 1987a]. The amount of the 34-kd lectin increased in rat embryo cells after transfection with several oncogenes [Raz et al., 1987b]. Thus, the changes in mRNA correlated well with the changes in the level of the 34-kd lectin protein [Crittenden et al., 1984; Raz et al., 1987b].

Induction of differentiation of tumor cells in culture by various agents including $N^6,O^{2'}$-dibutyryl cyclic-AMP, retinoic acid, dimethyl sulfoxide, and butyrate was accompanied by changes in the level of either the 14- or the 34-kd lectins [Lotan et al., 1989a,b]. Differentiation was associated with decreased levels of a 34-kd lectin in the K-1735P and B16-F1 melanoma cells and decreased levels of a 14-kd lectin in S20 neuroblastoma, MDA-MB 175 breast carcinoma, and HL-60 and THP-1 leukemia cells. The level of a 14-kd lectin increased during differentiation of F-9 embryonal and KM12P colon carcinoma cells. Thus, tumor cell differentiation along specific pathways is accompanied by distinct modulation of lectin expression. These changes may represent a recapitulation of the regulation of lectin expression observed in normal embryonal development.

IV. CONCLUDING REMARKS

Major progress has been made over the last few years in the understanding of the structure of the galactoside-binding lectins. The sequence of both the 14-kd and the 34-kd lectins has been established in several species, and the gene structure is beginning to be unraveled. Detailed carbohydrate-binding specificity has also been determined. However, important questions remain to be answered. Foremost among these questions is the function(s) of the lectins in the nucleus, cytoplasm, cell surface, and the extracellular matrix. There are also questions concerning the mechanism by which lectins are secreted, the mechanism by which they are associated with the cell membrane, the reasons why many cells contain two or more galactoside-specific lectins, and the identity of the endogenous glycoconjugates with which the lectins interact during execution of their functions. The elucidation of the mechanisms by which the lectin genes are regulated during differentiation and development and the interplay between the carbohydrate-binding domain and other functional domain within the lectin molecule are also challenges for future research in this area.

ACKNOWLEDGMENT

The author was supported in part by a grant 2546 from the Council for Tobacco Research, USA, Inc.

REFERENCES

Abbott, W. M., and Feizi, T. (1989). Evidence that the 14 kDa soluble galactoside-binding lectin in man is encoded by a single gene. *Biochem. J. 259*:291–294.

Abbott, W. M., Hounsell, E. F., and Feizi, T. (1988). Further studies of oligosaccharide recognition by the soluble 13 kDa lectin of bovine heart muscle. *Biochem. J. 252*:283–287.

Abbott, W. M., Mellor, A., Edwards, Y., and Feizi, T. (1989). Soluble bovine galactose-binding lectin: cDNA cloning reveals the complete amino acid sequence and an antigenic relationship with the major encephalitogenic domain of myelin basic protein. *Biochem. J. 259*:283–290.

Agrwal, N., Wang, J. L., and Voss, P. G. (1989). Carbohydrate binding protein 35: Levels of transcription and mRNA accumulation in quiescent and proliferating cells. *J. Biol. Chem. 264*:17236–17242.

Albrandt, K., Orida, N. K., and Liu, F.-T. (1987). An IgE-binding protein with a distinctive repetitive sequence and homology with an IgG receptor. *Proc. Natl. Acad. Sci. USA 84*:6859–6863.

Allen, H. J. (1986). Binding specificity of a human leucocyte carbohydrate binding protein. *Immunol. Invest. 15*:379–392.

Allen, H. J., Cywinski, M., Palmberg, R., and DiCioccio, R. A. (1987a). Comparative analysis of galactoside-binding lectins isolated from mammalian spleens. *Arch. Biochem. Biophys. 256*: 523–533.

Allen, H. J., Kachurek, D., Bulera, S., Kisailus, E., and DiCioccio, R. (1986). In vitro synthesis of carbohydrate-binding proteins by human leucocytes. *Immunol. Invest. 15*:123–138.

Allen, H. J ., Karakousis, C., Piver, M. S., Gamarra, M., Nava, H., Forsyth, B., Matecki, B., Jazayeri, A., Sucato, D., Kisailus, E., and DiCioccio, R. (1987b). Galactoside-binding lectin in human tissues. *Tumor Biol. 8*:218–229.

Arumugham, R. G., Hsieh, T. C.-Y., Tanzer, M. L., and Laine, R. A. (1986). Structures of the asparagine-linked sugar chains of laminin. *Biochim. Biophys. Acta 883*:112–126.

Barondes, S. H. (1986). Vertebrate lectins: Properties and functions. In *The Lectins: Properties, Functions, and Applications in Biology and Medicine* (I. E. Liener, N. Sharon, and I. J. Goldstein, eds.). Academic Press, New York, pp. 437–466.

Barondes, S. H. (1988). Bifunctional properties of lectins: Lectins redefined. *Trends Biochem. Sci. 13*:480–482.

Barondes, S. H., Gitt, M. A., Leffler, H., and Cooper, D. N. W. (1988). Multiple soluble vertebrate galactoside-binding lectins. *Biochimie 70*:1627–1632.

Barondes, S. H., and Leffler, H. (1987). Soluble lactose-binding lectins from chicken and rat. *Methods Enzymol. 138*:510–515.

Beyer, E. C., and Barondes, S. H. (1980). Chicken tissue binding sites for purified chicken lectins. *J. Supramol. Struct. 13*:219–227.

Beyer, E. C., Zweig, S. E., and Barondes, S. H. (1980). Two lactose binding lectins from chicken tissues. Purified lectin from intestine is different from those in liver and muscle. *J. Biol. Chem. 255*:4236–4239.

Brandley, B. K., and Schnaar, R. L. (1986). Cell-surface carbohydrates in cell recognition and response. *J. Leukocyte Biol. 40*:97–111.

Briles, E. B., Gregory, W., Fletcher, P., and Kornfeld, S. (1979). Comparison of properties of β-galactoside-binding lectins from tissues of calf and chicken. *J. Cell Biol. 81*:528–537.

Burrus, G. R., Schmidt, W. N., Briggs, J. A., Hnilica, L. S., and Briggs, R. C. (1988). Lectin-binding proteins in nuclear preparations from rat liver and malignant tumors. *Cancer Res. 48*:551–555.

Caron, M., Joubert, R., and Bladier, D. (1987). Purification and characterization of a β-galactoside-binding soluble lectin from rat and bovine brain. *Biochim. Biophys. Acta 925*: 290–296.

Catt, J. W., and Harrison, F. L. (1985). Selective association of an endogenous lectin with connective tissues. *J. Cell Sci. 73*:347–359.

Cerra, R. F., Gitt, M. A., and Barondes, S. H. (1985). Three soluble rat β-galactoside-binding lectins. *J. Biol. Chem. 260*:10474–10477.

Cerra, R. F., Haywood-Reid, P. L., and Barondes, S. H. (1984). Endogenous mammalian lectin localized extracellularly in lung elastic fibers. *J. Cell Biol. 98*:1580–1589.

Childs, R. A., and Feizi, T. (1979). Calf heart lectin reacts with blood group Ii antigens and other precursor chains of the major blood group antigens. *FEBS Lett. 99*:175–179.

Childs, R. A., and Feizi, T. (1980). β-Galactoside-binding lectin of human and bovine tissues. *Cell Biol. Int. Rep. 4*:775.

Clerch, L. B., Whitney, P., Hass, M., Brew, K., Miller, T., Werner, R., and Massaro, D. (1988). Sequence of a full-length cDNA for rat lung β-galactoside-binding protein primary and secondary structure of the lectin. *Biochemistry 27*:692–699.

Cook, G. M. W., Zalik, S. E., Milos, N., and Scott, V. (1979). A lectin which binds specifically to β-D-galactoside groups is present at the earliest stages of chick embryo development. *J. Cell Sci. 38*:293–304.

Couraud, P. O., Casentini-Borocz, D., Bringman, T. S., Griffith, J., McGrogan, M., and Nedwin, G. E. (1989). Molecular cloning, characterization, and expression of a human 14-kDa lectin. *J. Biol. Chem. 264*:1310–1316.

Crittenden, S. L., Roff, C. F., and Wang, J. L. (1984). Carbohydrate-binding protein 35: Identification of the galactose-specific lectin in various tissues of mice. *Mol. Cell. Biol. 4*:1252–1259.

Cummings, R. D. (1989). Carbohydrate-binding specificities of plant and animal lectins. *Proc. Int. Symp. Lectins*, Abst. 14.

DeWaard, A., Hickman, S., and Kornfeld, S. (1976). Isolation and properties of β-galactoside binding lectins of calf heart and lung. *J. Biol. Chem. 251*:7581–7587.

Didier, E., Didier, P., Bayle, D., and Chevalier, M. (1988). Lectin activity and distribution of chicken lactose lectin I in the extracellular matrix of the chick developing kidney. *Cell Differ. 24*:83–96.

Dreyfuss, G., Swanson, M. S., and Pinol-Roma, S. (1988). Heterogeneous nuclear ribonucleoprotein particles and the pathway of mRNA formation. *Trends Biochem. Sci. 13*:86–91.

Drickamer, K. (1988). Two distinct classes of carbohydrate-recognition domains in animal lectins. *J. Biol. Chem. 263*:9557–9560.

Feizi, T. (1985). Demonstration by monoclonal antibodies that carbohydrate structures of glycoproteins and glycolipids are onco-developmental antigens. *Nature 314*:53–57.

Feizi, T., and Childs, R. A. (1987). Carbohydrates as antigenic determinants of glycoproteins. *Biochem. J. 245*:1–11.

Fukuda, M. (1985). Cell surface glycoconjugates as onco-differentiation markers in hematopoietic cells. *Biochim. Biophys. Acta. 780*:119–150.

Gabius, H.-J. (1987). Endogenous lectins in tumors and the immune system. *Cancer Invest. 5*:39–46.

Gabius, H.-J., Brehler, R., Schauer, A., and Cramer, F. (1986a). Localization of endogenous lectins in normal human breast, benign breast lesions and mammary carcinomas. *Virchows Arch. [B]. 52*:107–115.

Gabius, H.-J., Engelhardt, R., Rehm, S., Barondes, S. H., and Cramer, F. (1986b). Presence and relative distribution of the endogenous β-galactoside-specific lectins in different types of rat tumors. *Cancer J. 1*:19–21.

Gitt, M. A., and Barondes, S. H. (1986). Evidence that a human soluble β-galactoside-binding lectin is encoded by a family of genes. *Proc. Natl. Acad. Sci. USA 83*:7603–7607.

Gitt, M. A., and Barondes, S. H. (1989). Sequence and genomic structure of two members of the human lectin gene family. *Proc. Int. Symp. Lectins* Abst. 36.

Godsave, S. F., Harrison, F. L., and Chesterton, C. J. (1981). Molecular properties of a galaptin and its activity in normal and leukemic erythroid tissue. *J. Supramol. Struc. Cell. Biochem. 17*:223–230.

Gritzmacher, C. A., Robertson, M. W., and Liu, F. (1988). IgE-binding protein: Subcellular location and gene expression in many murine tissues and cells. *J. Immunol. 141*:2801–1806.

Hakomori, S. (1989). Aberrant glycosylation in tumors and tumor-associated carbohydrate antigens. *Adv. Cancer Res. 52*:257–331.

Harford, J., and Ashwell, G. (1982). The hepatic receptor for asialoglycoproteins. In *The Glycoconjugates*, Vol. 4, *Glycoproteins, Glycolipids, and Proteoglycans*, Part B (M. I. Horowitz, ed.). Academic Press, New York, pp. 27–55.

Harrison, F. L., and Chesterton, C. J. (1980). Factors mediating cell–cell recognition and adhesion. Galaptins, a recently discovered class of bridging molecules. *FEBS Lett. 122*:157–165.

Harrison, F. L., Fitzgerald, J. E., and Catt, J. W. (1984). Endogenous β-galactoside-specific lectins rabbit tissues. *J. Cell Sci. 72*:147–162.

Harrison, F. L., and Catt, J. W. (1986). Intra- and extracellular distribution of an endogenous lectin during erythropoiesis. *J. Cell Sci. 84*:201–212.

Hart, G. W., Holt, G. D., and Haltiwanger, R. S. (1988). Nuclear and cytoplasmic glycosylation: Novel saccharide linkages in unexpected places. *Trends Biochem. Sci. 13*:379–383.

Hinek, A., Wrenn, D. S., Mecham, R. P., and Barondes, S. H. (1988). The elastin receptor: A galactoside-binding protein. *Science 235*:1539–1541.

Hirabayashi, J., Ayaki, H., Soma, G., and Kasai, K. (1989). Cloning and nucleotide sequence of a full-length cDNA for human 14 kDa β-galactoside-binding lectin. *Biochim. Biophys. Acta 1008*:85–91.

Hirabayashi, J., and Kasai, K. (1988). Complete amino acid sequence of a β-galactoside-binding lectin from human placenta. *J. Biochem. 104*:1–4.

Hirabayashi, J., and Kasai, K. (1984). Human placenta β-galactoside-binding lectin. Purification and some properties. *Biochem. Biophys. Res. Commun. 122*:938–944.

Hirabayashi, J., Kawasaki, H., Suzuki, K., and Kasai, K. (1987a). Complete amino acid sequence of 14 kDa β-galactoside-binding lectin of chick embryo. *J. Biochem. 101*:775–787.

Hirabayashi, J., Kawasaki, H., Suzuki, K., and Kasai, K. (1987b). Further characterization and structural studies on human placenta lectin. *J. Biochem. 101*:987–995.

Hirano, H., Akimoto, Y., Kawakami, H., Oda, Y., and Kasai, K. (1988). Localization of endogenous β-galactoside-binding lectin in embryonic chick epidermis. In *Proceedings Kagoshima International Symposium Glycoconjugates in Medicine*, 8–13.

Jackson, S. P., and Tijan, R. (1988). O-glycosylation of eukaryotic transcription factors: Implications for mechanisms of transcriptional regulation. *Cell 55*:125–133.

Jia, S., Mee, R. P., Morford, G., Agrwal, N., Voss, P. G., Moutsatsos, I. K., and Wang, J. L. (1987). Carbohydrate-binding protein 35: Molecular cloning and expression of a recombinant polypeptide with lectin activity in *Escherichia coli. Gene 60*:197–204.

Jia, S., and Wang, J. L. (1988). Carbohydrate binding protein 35: Complementary DNA sequence reveals homology with proteins of the heterogeneous nuclear RNP. *J. Biol. Chem. 263*:6009–6011.

Joubert, R., Caron, M., and Bladier, D. (1987). Brain lectin-mediated agglutinability of dissociated cells from embryonic and postnatal mouse brain. *Dev. Brain Res. 36*:146–150.

Joubert, R., Caron, M., and Bladier, D. (1988). Distribution of β-galactoside specific lectin activities during pre- and postnatal mouse brain development. *Cell. Mol. Biol. 34*:79–87.

Joubert, R., Caron, M., Deugnier, M. A., Rioux, F., Sensenbrenner, M., and Bisconte, J. C. (1985). Effects of adult rat brain extracts on cultures of mouse brain cells: Consequence of the depletion in a carbohydrate-binding fraction. *Cell. Mol. Biol. 31*:131–138.

Kajikawa, T., Nakajima, Y., Hirabayashi, J., Kasai, K., and Yamazaki, M. (1986). Release of cytotoxin by macrophages on treatment with human placenta lectin. *Life Sci. 39*:1177–1181.

Kieda, C., and Monsigny, M. (1983). Adhesion of mouse spleen cells to lymph node veinules involves endogenous membrane lectins (sugar receptors) of the lymphocytes. In *Intercellular Communication in Leukocyte Function* (J. W. Parker, and R. L. O'Brien, eds.). John Wiley & Sons, New York, pp. 649–652.

Laing, J. G., Robertson, M. W., Gritzmacher, C. A., Wang, J. A., and Liu, F. (1989). Biochemical and immunological comparisons of carbohydrate-binding protein 35 and an IgE-binding protein. *J. Biol. Chem. 264*:1907–1910.

Laing, J. G., Robertson, M. W., Gritzmacher, C. A., Wang, J. A., and Liu, F. (1989). Biochemical and immunological comparisons of carbohydrate-binding protein 35 and an IgE-binding protein. *J. Biol. Chem. 264*:1907–1910.

Leffler, H., and Barondes, S. H. (1986). Specificity of binding of three soluble rat lung lectins to substituted and unsubstituted mammalian β-galactosides. *J. Biol. Chem. 261*:10119–10126.

Leffler, H., and Barondes, S. H. (1989). Newly discovered soluble galactoside binding lectins in rat and mouse intestine and species and tissue related variations in lectin expression. *Proc. Int. Symp. Lectins* Abst. 32.

Levi, G., Tarrab-Hazdai, R., and Teichberg, V. I. (1983). Prevention and therapy with electrolectin of experimental autoimmune myasthenia gravis in rabbits. *Eur. J. Immunol. 13*:500–507.

Levi, G., and Teichberg, V. I. (1981). Isolation and physicochemical characterization of electrolectin, a β-D-galactoside binding lectin from the electric organ of *Electrophorus electicus. J. Biol. Chem. 256*:5735–5740.

Levi, G., and Teichberg, V. I. (1983). Selective interactions of electrolectins from eel electric organ and mouse thymus with immature thymocytes. *Immunol. Lett. 7*:35–39.

Levi, G., and Teichberg, V. I. (1984). The distribution of electrolectin in mouse: Genetic and ontogenic variations. *Biochem. Biophys. Res. Commun. 119*:801–806.

Levi, G., and Teichberg, V. I. (1985). Isolation and characterization of chicken thymic electrolectin. *Biochem. J. 226*:379–384.

Lipsick, J. S., Beyer, E. C., Barondes, S. H., and Kaplan, N. O. (1980). Lectins from chicken tissues are mitogenic for Thy-1 negative murine spleen cells. *Biochem. Biophys. Res. Commun. 97*:56–61.

Liu, F-T, Albrandt, K., Mendel, E., Kulczycki, A., Jr., and Orida, N. K. (1985). Identification of an IgE-binding protein by molecular cloning. *Proc. Natl. Acad. Sci. USA 82*:4100–4104.

Lotan, R., Carralero, D., Lotan, D., and Raz, A. (1989a). Biochemical and immunological characterization of K-1735 melanoma galactoside-binding lectins and their modulation by differentiation inducers. *Cancer Res. 49*:1261–1268.

Lotan, R., Lotan, D., and Carralero, D. (1989). Modulation of galactoside-binding lectins in tumor cells by differentiation-inducing agents. *Cancer Lett. 48*:115–122.

Lotan, R., Lotan, D., and Raz, A. (1985). Inhibition of tumor cell colony formation in culture by a monoclonal antibody to endogenous lectins. *Cancer Res. 45*:4349–4353.

Lotan, R., and Raz, A. (1988a). Lectins in cancer cells. *Ann. N.Y. Acad. Sci. 551*:385–397.

Lotan, R., and Raz, A. (1988b). Endogenous lectins as mediators of tumor cell adhesion. *J. Cell. Biochem. 37*:107–117.

Mecham, R. P., Hinek, A., Griffin, G. L., Senior, R. M., and Liotta, L. A. (1989). The elastin receptor shows structural and functional similarities to the 67-kDa tumor cell laminin receptor. *J. Biol. Chem. 264*:16652–16657.

Merkle, R. K., and Cummings, R. D. (1988). Asparagine-linked oligosaccharides containing poly-*N*-acetyllactosamine chains are preferentially bound by immobilized calf heart agglutinin. *J. Biol. Chem. 263*:16143–16149.

Milos, N., and Zalik, S. E. (1983). Calcium-independent adhesion of extra-embryonic endoderm cells from the early chick blastoderm is inhibited by the blastoderm β-D-galactoside-binding lectin and by β-galactosidase. *Cell Differ. 12*:341–347.

Milos, N., and Zalik, S. E. (1986). Release of β-D-galactoside-binding lectins into the cavities of aggregates of chick extraembryonic endoderm cells. *Cell Differ. 18*:1–7.

Monsigny, M., Roche, A.-C., Kieda, C., Midoux, P., and Obernovich, A. (1988). Characterization and biological implications of membrane lectins in tumor, lymphoid and myeloid cells. *Biochimie 70*:1633–1649.

Moutsatsos, I. K., Davis, J. M., and Wang, J. L. (1986). Endogenous lectins from cultured cells: Subcellular localization of carbohydrate-binding protein 35 in 3T3 fibroblasts. *J. Cell Biol. 102*:477–483.

Moutsatsos, I. K., Wade, M., Schindler, M., and Wang, J. L. (1987). Endogenous lectins from cultured cells: Nuclear localization of carbohydrate-binding protein 35 in proliferating 3T3 fibroblasts. *Proc. Natl. Acad. Sci USA 84*:6452–6456.

Nicolson, G. L. (1984). Cell surface molecules and tumor metastasis. Regulation of metastatic diversity. *Exp. Cell Res. 150*:3–22.

Nicolson, G. L., and Lacorbiere, M. (1973). Cell contact-dependent increase in membrane D-galactopyranosyl-like residues on normal, but not virus- or spontaneously transformed, murine fibroblasts. *Proc. Natl. Acad. Sci. USA 70*:1672–1676.

Oda, Y., and Kasai, K. (1983). Purification and characterization of β-galactoside-binding lectin from chick embryonic skin. *Biochim. Biophys. Acta 761*:237–245.

Oda, Y., and Kasai, K. (1984). Photochemical cross-linking of β-galactoside-binding lectin to polylactosamino-proteoglycan of chick embryonic skin. *Biochem. Biophys. Res. Commun. 123*:1215–1220.

Oda, Y., Ohyama, Y., Obinata, A., Endo, H., and Kasai, K.-I. (1989). Endogenous β-galactoside-binding lectin expression is suppressed in retinol-induced mucous metaplasia of chick embryonic epidermis. *Exp. Cell Res. 182*:33–43.

Ohara, T., and Yamagata, T. (1986). Isolation and characterization of galactose-binding proteins from new-born mice. *Biochim. Biophys. Acta 884*:344–354.

Ohyama, Y., Hirabayashi, J., Oda, Y., Ohno, S., Kawasaki, H., Suzuki, K., and Kasai, K. (1986). Nucleotide sequence of chick 14K β-galactoside-binding lectin mRNA. *Biochem. Biophys. Res. Commun. 134*:51–56.

Ohyama, Y., and Kasai, K. (1988). Isolation and characterization of the chick 14K β-galactoside-binding lectin gene. *J. Biochem. 104*:173–177.

Paroutaud, P., Levi, G., Teichberg, V. I., and Strosberg, A. D. (1987). Extensive amino acid sequence homologies between animal lectins. *Proc. Natl. Acad. Sci. USA 84*:6345–6348.

Paulson, J. C. (1989). Glycoproteins: What are the sugar chains for? *Trends Biochem. Sci. 14*:272–276.

Pitts, M. J ., and Yang, D. C. H. (1980). Isolation of a developmentally regulated lectin from chick embryo. *Biochem. Biophys. Res. Commun. 95*:750–757.

Pitts, M. J., and Yang, D. C. H. (1981). Mitogenicity and binding properties of β-galactoside-binding lectin from chick-embryo kidney. *Biochem. J. 195*:435–439,

Powell, J. T. (1980). Purification and properties of lung lectin: Rat lung and human lung β-galactoside-binding proteins. *Biochem. J. 187*:123–129.

Powell, J. T., and Whitney, P. L. (1980). Postnatal development of rat lung. *Biochem. J. 188*:1–8.

Powell, J. T., and Whitney, P. L. (1984). Endogenous ligands of rat lung β-galactoside-binding protein (galaptin) isolated by affinity chromatography on carboxyamidomethylated–galaptin–Sepharose. *Biochem. J. 223*:769–774.

Rastan, S., Thorpe, S. J., Scudder, P., Brown, S., Gooi, H. C., and Feizi, T. (1985). Cell interactions in preimplantation embryos: Evidence for involvement of saccharides of the poly-*N*-acetyllactosamine series. *J. Embryol. Exp. Morphol. 87*:115–128.

Raz, A., and Lotan, R. (1981). Lectin-like activities associated with human and murine neoplastic cells. *Cancer Res. 41*:3642–3647.

Raz, A., Avivi, A., Pazerini, G., and Carmi, P. (1987a). Cloning and expression of cDNA for two endogenous UV-2237 fibrosarcoma lectin genes. *Exp. Cell Res. 173*:109–116.

Raz, A., Carmi, P., and Pazerini, G. (1988). Expression of two different endogenous galactoside-binding lectins sharing sequence homology. *Cancer Res. 48*:645–649.

Raz, A., and Lotan, R. (1987). Endogenous galactoside-binding lectins: A new class of functional tumor cell surface molecules related to metastasis. *Cancer Metastasis Rev. 6*:433–452.

Raz, A., Meromsky, L., Carmi, P., Karkash, R., Lotan, D., and Lotan, R. (1984). Monoclonal antibodies to endogenous galactose-specific tumor cell lectins. *EMBO J. 3*:2979–2984.

Raz, A., Meromsky, L., Zvibel, I., and Lotan, R. (1987b). Transformation-related changes in the expression of endogenous cell lectins. *Int. J. Cancer. 39*:353–360.

Raz, A., Pazerini, G., and Carmi, P. (1989). Identification of the metastasis-associated, galactoside-binding lectin as a chimeric gene product with homology to an IgE-binding protein. *Cancer Res. 49*:3489–3493.

Regan, L. J., Dodd, J., Barondes, S. H., and Jessell, T. M. (1986). Selective expression of endogenous lactose-binding lectins and lactoseries glycoconjugates in subsets of rat sensory neurons. *Proc. Natl. Acad. Sci. USA 83*:2248–2252.

Roff, C. F., and Wang, J. L. (1983). Endogenous lectins from cultured cells: Isolation and characterization of carbohydrate-binding proteins from 3T3 fibroblasts. *J. Biol. Chem. 258*:10657–10663.

Roff, C. F., Rosevear, P. R., Wang, J. L., and Barker, R. (1983). Identification of carbohydrate-binding proteins from mouse and human fibroblasts. *Biochem. J. 211*:625–629.

Rutishauser, U., and Jessell, T. M. (1988). Cell adhesion molecules in vertebrate neural development. *Physiol. Rev. 68*:819–857.

Sanford, G. L., Davis, L. D., and Powell, J. T. (1982). The subcellular localization of the β-galactoside-binding protein of rat lung. *Biochem. J. 204*:97–102.

Schachter, H. (1986). Biosynthetic controls that determine the branching and microheterogeneity of protein-bound oligosaccharides. *Biochem. Cell Biol. 64*:163–181.

Scudder, P., Childs, R. A., Feizi, T., Joziasse, D. H., Schiporat, W. E. C. M., and Van den Eijnden, D. H. (1982). Bovine heart lectin stimulates β-D-galactoside α-2,6 sialyltransferase of bovine colostrum. *Biochem. Biophys. Res. Commun. 104*:272–279.

Seve, A.-P., Hubert, J., Bouvier, D., Bourgeois, C., Midoux, P., Roche, A.-C., and Monsigny, M. (1986). Analysis of sugar-binding sites in mammalian cell nuclei by quantitative flow microfluorometry. *Proc. Natl. Acad. Sci. USA 83*:5997–6001.

Seve, A. P., Hubert, J., Bouvier, D., Boutielle, M., Maintier, C., and Monsigny, M. (1985). Detection of sugar-binding proteins in membrane-depleted nuclei. *Exp. Cell Res. 157*:533–538.

Sharon, N., and Lis, H. (1989). Lectins as cell recognition molecules. *Science* (in press).

Sharom, F. J., and Ross, T. E. (1985). Spin labeling of sialic acid and galactose residues on lymphocyte plasma membrane: Effects of lectins on oligosaccharide dynamics. *Mol. Immunol. 22*:521–530.

Shibata, S., Peters, B. P., Roberts, D. D., Goldstein, J., and Liotta, L. A. (1982). Isolation of laminin by affinity chromatography on immobilized *Griffonia simplicifolia* I lectin. *FEBS Lett. 142*: 194–198.

Southan, C., Aitken, A., Childs, R. A., Abbott, W. M., and Feizi, T. (1987). Amino acid sequence of β-galactoside-binding bovine heart lectin: Member of a novel class of vertebrate proteins. *FEBS Lett. 214*:301–304.

Sparrow, C. P., Leffler, H., and Barondes, S. H. (1987). Multiple soluble β-galactoside-binding lectins from human lung. *J. Biol. Chem. 262*:7383–7390.

Stojanovic, D., Hughes, R. C., Feizi, T., and Childs, R. A. (1983). Interactions of a mammalian β-galactoside-binding lectin with hamster fibroblasts. *J. Cell. Biochem. 21*:119–127.

Teichberg, V. I., Silman, I., Beitsch, D. D., and Resheff, G. (1975). A β-D-galactoside binding protein from electric organ tissue of *Electrophorus electricus. Proc. Natl. Acad. Sci. USA 72*:1383–1387.

Thorpe, S. J., Bellairs, R., and Feizi, T. (1988). Developmental patterning of carbohydrate antigens during early embryogenesis of the chick: Expression of antigens of the poly-*N*-acetyllactosamine series. *Development 102*:193–210.

Whitney, P., Maxwell, S., Ryan, U., and Massaro, D. (1985). Synthesis and binding of lactose-specific lectin by isolated lung cells. *Am. J. Physiol. 248*:258–264.

Whitney, P. L., Powell, J. T., and Sanford, G. L. (1986). Oxidation and chemical modification of lung β-galactoside-specific lectin. *Biochem. J. 238*:683–689.

Wickner, W. T., and Lodish, H. F. (1985). Multiple mechanisms of protein insertion into and across membranes. *Science 230*:400–406.

Wieser, R., and Oesch, F. (1986). Contact inhibition of growth of human diploid fibroblasts by immobilized plasma membrane glycoproteins. *J. Cell Biol. 103*:361–367.

Wilson, T. J. G., Firth, M. N., Powell, J. T., and Harrison, F. L. (1989). The sequence of the mouse 14 kDa β-galactoside-binding lectin and evidence for its synthesis on free cytoplasmic ribosomes. *Biochem. J. 261*:847–852.

Zalik, S. E., Milos, N., and Ledsham, I. (1983). Distribution of two β-galactoside-binding lectins in the gastrulating chick embryo. *Cell Differ. 12*:121–127.

Zalik, S. E., Thomson, L. W., and Ledsham, I. M. (1987). Expression of an endogenous galactose-binding lectin in the early chick embryo. *J. Cell Sci. 88*:483–493.

Zhu, B. C. R., and Laine, R. A. (1985). Polylactosamine glycosylation on human fetal placental fibronectin weakens the binding of fibronectin to gelatin. *J. Biol. Chem. 260*:4041–4045.

Index